Sylvia S. Mader
Michael Windelspecht
Appalachian State University

With contributions by
Lynn Preston
Tarrant County College

Twelfth Edition

HUMAN BIOLOGY

The McGraw·Hill Companies

HUMAN BIOLOGY, TWELFTH EDITION

Published by McGraw-Hill, a business unit of The McGraw-Hill Companies, Inc., 1221 Avenue of the Americas, New York, NY 10020.

This book is printed on acid-free paper.

1 2 3 4 5 6 7 8 9 0 QDB/QDB 1 0 9 8 7 6 5 4 3 2 1

ISBN 978-0-07-352546-4
MHID 0-07-352546-4

Vice President, Editor-in-Chief: *Marty Lange*
Vice President, EDP: *Kimberly Meriwether David*
Senior Director of Development: *Kristine Tibbetts*
Publisher: *Michael S. Hackett*
Senior Developmental Editor: *Rose M. Koos*
Senior Marketing Manager: *Tamara Maury*
Senior Project Manager: *April R. Southwood*
Senior Buyer: *Sandy Ludovissy*
Senior Media Project Manager: *Jodi K. Banowetz*
Senior Designer: *Laurie B. Janssen*
Cover Image: © *Dawn Kish/Gettyimages*
Senior Photo Research Coordinator: *Lori Hancock*
Photo Research: *Evelyn Jo Johnson*
Compositor: *Electronic Publishing Services Inc., NYC*
Typeface: *10/12 Palatino LT Std*
Printer: *Quad/Graphics*

Library of Congress Cataloging-in-Publication Data

Mader, Sylvia S.
Human biology / Sylvia S. Mader, Michael Windelspecht ; with contributions by Lynn Preston.—12th ed.
p. cm.
Includes index.
ISBN 978-0-07-352546-4—ISBN 0-07-352546-4 (hard copy : alk. paper) 1. Human biology.
I. Windelspecht, Michael, 1963- II. Title.
QP36.M2 2012
612—dc22

2010038108

www.mhhe.com

Brief Contents

CHAPTER 1
Exploring Life and Science 1

PART I
Human Organization 19

CHAPTER 2
Chemistry of Life 19

CHAPTER 3
Cell Structure and Function 43

CHAPTER 4
Organization and Regulation of Body Systems 65

PART II
Maintenance of the Human Body 91

CHAPTER 5
Cardiovascular System: Heart and Blood Vessels 91

CHAPTER 6
Cardiovascular System: Blood 115

CHAPTER 7
Lymphatic System and Immunity 133

Infectious Diseases Supplement 155

CHAPTER 8
Digestive System and Nutrition 169

CHAPTER 9
Respiratory System 196

CHAPTER 10
Urinary System 217

PART III
Movement and Support in Humans 239

CHAPTER 11
Skeletal System 239

CHAPTER 12
Muscular System 262

PART IV
Integration and Coordination in Humans 285

CHAPTER 13
Nervous System 285

CHAPTER 14
Senses 315

CHAPTER 15
Endocrine System 339

PART V
Reproduction in Humans 365

CHAPTER 16
Reproductive System 365

CHAPTER 17
Development and Aging 393

PART VI
Human Genetics 419

CHAPTER 18
Patterns of Chromosome Inheritance 419

CHAPTER 19
Cancer 447

CHAPTER 20
Patterns of Genetic Inheritance 467

CHAPTER 21
DNA Biology and Technology 491

PART VII
Human Evolution and Ecology 517

CHAPTER 22
Human Evolution 517

CHAPTER 23
Global Ecology and Human Interferences 545

CHAPTER 24
Human Population, Planetary Resources, and Conservation 567

About the Authors

Dr. Sylvia S. Mader has authored several nationally recognized biology texts published by McGraw-Hill. Educated at Bryn Mawr College, Harvard University, Tufts University, and Nova Southeastern University, she holds degrees in both Biology and Education. Over the years, she has taught at University of Massachusetts, Lowell, Massachusetts Bay Community College, Suffolk University, and Nathan Mathew Seminars. Her ability to reach out to science-shy students led to the writing of her first text, *Inquiry into Life,* that is now in its thirteenth edition. Highly acclaimed for her crisp and entertaining writing style, her books have become models for others who write in the field of biology.

Although her writing schedule is always quite demanding, Dr. Mader enjoys taking time to visit and explore the various ecosystems of the biosphere. Her several trips to the Florida Everglades and Caribbean coral reefs resulted in talks she has given to various groups around the country. She has visited the tundra in Alaska, the taiga in the Canadian Rockies, the Sonoran Desert in Arizona, and tropical rain forests in South America and Australia. A photo safari to the Serengeti in Kenya resulted in a number of photographs for her texts. She was thrilled to think of walking in Darwin's steps when she journeyed to the Galápagos Islands with a group of biology educators. Dr. Mader was also a member of a group of biology educators who traveled to China to meet with their Chinese counterparts and exchange ideas about the teaching of modern-day biology.

Dr. Michael Windelspecht serves as the Introductory Biology Coordinator at Appalachian State University in Boone, North Carolina, where he directs a program that enrolls over 4,500 nonscience majors annually. He was educated at the University of Maryland, Michigan State University, and the University of South Florida. As an educator, Dr. Windelspecht teaches not only introductory biology for nonmajors but also biology for science majors, genetics, and human genetics. In addition to his teaching assignments, Dr. Windelspecht is active in promoting the scientific literacy of secondary school educators. He has led multiple workshops on integrating water quality research into the science curriculum and has spent several summers teaching Pakistani middle school teachers.

As an author, Dr. Windelspecht has published five reference textbooks and multiple print and online lab manuals. He served as the series editor for a ten-volume work on the human body. For years, Dr. Windelspecht has been active in the development of multimedia resources for the online and hybrid science classrooms. Along with his wife Sandra, he owns a multimedia production company that actively develops and assesses the use of new technologies in the classroom.

Preface

One of the easiest ways to engage today's students in the sciences is to make the content relevant to their lives. From the latest developments in health and medicine, to environmental issues, students have a fundamental interest in the world around them. *Human Biology* was designed to integrate the topics of health, wellness, and the environment in a way that perfectly suits the nonmajors' course.

With this purpose in mind, the authors identified several goals that guided them through the revision of *Human Biology*, Twelfth Edition.

- **Homeostasis and Evolution** coverage increased
- **Genetics of human disease** coverage expanded
- **Case Studies** revised to include discussions of medical procedures and the genetics of human disease
- **Applications** added to enhance the relevancy of content for students
- **Media** assets integrated in textbook and in Connect™ Biology

Increased Coverage of Homeostasis and Evolution

Many of the Connecting the Concepts features at the end of each section address homeostasis and direct students to other sections of the textbook to emphasize how the body systems interact to make homeostasis possible. Throughout the textbook, new evolutionary diagrams have been added to indicate the importance of evolution in the study of human biology. In addition, many of the Connections and Misconceptions boxes now relate to evolutionary themes.

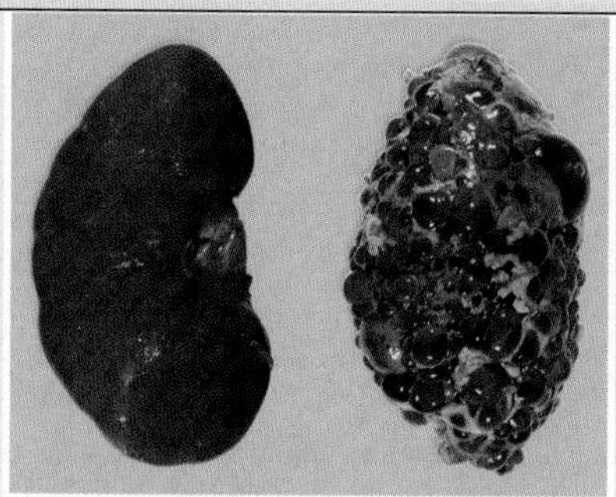

CHAPTER

10

Urinary System

CASE STUDY POLYCYSTIC KIDNEY DISEASE

Michael and Jada were excited about the birth of their child. Married for three years, they already had a healthy daughter, and they were happy that she would have a younger sister to play with. After Aiesha was born, however, it soon became clear that something was wrong. She weighed only 5 lb 4 oz at birth. The first time she urinated, there was an obvious tinge of blood in her urine. In addition, she also seemed to urinate much more frequently than was to be expected for an infant, and her blood pressure was higher than was normal. When her doctors performed ultrasound and magnetic resonance imaging (MRI) scans of her abdominal organs, they found that Aiesha had signs of polycystic kidney disease (PKD). Aiesha's doctors explained that in PKD, cysts (small, fluid-filled sacs) form within the collecting ducts of the nephrons in the interior of the kidneys. The ultrasound results indicated that both of Aiesha's kidneys were covered in cysts (see the kidney above and right) and that this usually meant that the cysts were present inside the kidneys as well. Michael and Jada were informed that the presence of these cysts explained Aiesha's symptoms. The doctors also told Michael and Jada that PKD would most likely cause Aiesha's kidneys to fail and that they should immediately prepare her for a kidney transplant. Because PKD is a genetic disorder, the physicians suggested that both parents undergo genetic tests to see if they were carriers for PKD.

As you read through the chapter, think about the following questions.

1. What is the role of the kidneys in the body?
2. How would problems in the collecting ducts of the nephrons cause kidney failure?
3. Why would problems with the kidneys result in blood in the urine and high blood pressure?

CHAPTER CONCEPTS

10.1 **The Urinary System**
In the urinary system, kidneys produce urine, which is stored in the bladder before being discharged from the body. The kidneys are major organs of homeostasis.

10.2 **Kidney Structure**
Microscopically, the kidneys are composed of kidney tubules (nephrons). These tubules have a blood supply that interacts with parts of the tubule as they produce urine.

10.3 **Urine Formation**
Urine is composed primarily of nitrogenous waste products and salts in water. Urine formation is a stepwise process.

10.4 **Kidneys and Homeostasis**
The kidneys are involved in the salt–water balance and the acid–base balance of the blood, in addition to excreting nitrogenous wastes.

10.5 **Kidney Function Disorders**
Various types of illnesses, including diabetes, kidney stones, and infections, can lead to renal failure. Hemodialysis is needed for the survival of patients with renal failure.

BEFORE YOU BEGIN

Before beginning this chapter, take a few moments to review the following discussions.

Section 2.2 What determines whether a solution is acidic or basic?

Section 3.3 How does water move across a plasma membrane?

Section 4.8 How do feedback mechanisms contribute to the maintenance of homeostasis?

Connecting the Concepts

The respiratory and circulatory systems cooperate extensively to maintain homeostasis in the body. For more on the interactions of these two systems, refer to the following discussions.

Section 5.5 outlines the circulatory pathways that move gases to and from the lungs.

Section 6.2 describes the role of the red blood cells in the transport of gases.

Improved Coverage of the Genetics of Human Disease

Throughout the book, the genetic basis of human diseases (such as Down syndrome, cystic fibrosis, and Huntington disease) is identified and discussed. Many of these conditions are now the focus of the chapter openers, in which the relationship between genetics and some of the more common human diseases is presented.

Revised Chapter-Opening Case Studies

One of the major features of this new edition of *Human Biology* is a complete reworking of the chapter-opening case studies. As with the chapter-opening material in previous editions, all case studies have a human focus and lead the students into the chapter in an engaging way. For the twelfth edition, many of the case studies include discussions of medical procedures and the genetic basis of human disease. Also, several questions have been added to each case study to integrate the topic of the case study with the material in the chapter. These questions may be assigned by the instructor to assess student understanding of the topic or to facilitate classroom discussions. Each case study is concluded at the end of the chapter to further integrate the chapter concepts, and many of the Thinking Critically About the Concepts questions at the end of the chapter combine case study concepts with chapter content. Students are challenged to thoughtfully integrate these ideas, and the answers to the questions are given in Appendix B.

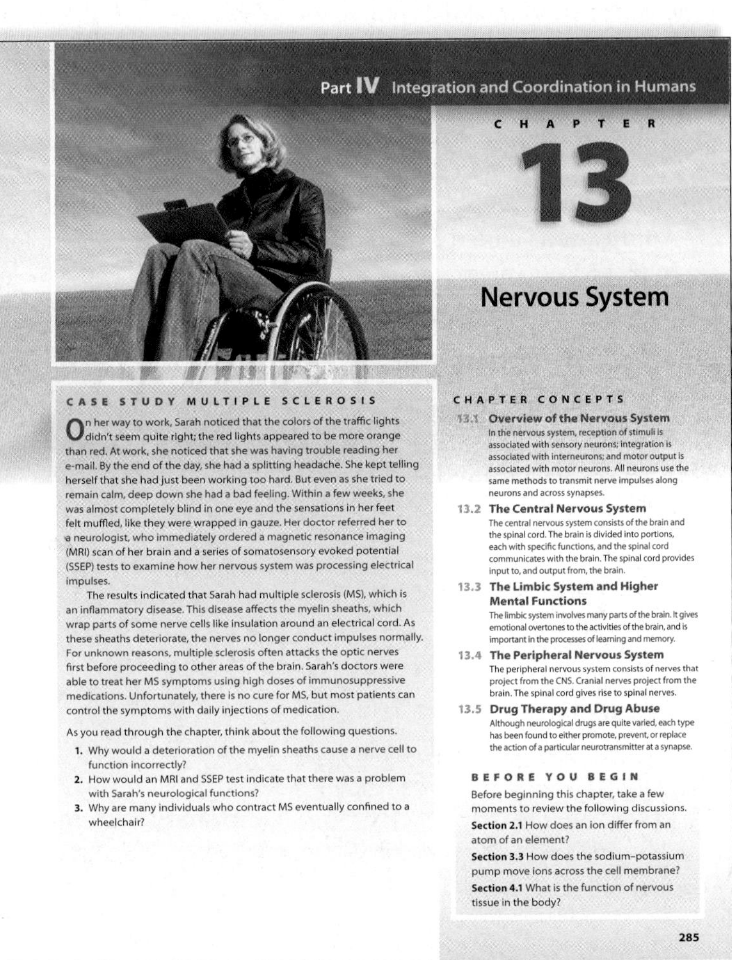

Part IV Integration and Coordination in Humans

CHAPTER 13

Nervous System

CASE STUDY MULTIPLE SCLEROSIS

On her way to work, Sarah noticed that the colors of the traffic lights didn't seem quite right; the red lights appeared to be more orange than red. At work, she noticed that she was having trouble reading her e-mail. By the end of the day, she had a splitting headache. She kept telling herself that she had just been working too hard. But even as she tried to remain calm, deep down she had a bad feeling. Within a few weeks, she was almost completely blind in one eye and the sensations in her feet felt muffled, like they were wrapped in gauze. Her doctor referred her to a neurologist, who immediately ordered a magnetic resonance imaging (MRI) scan of her brain and a series of somatosensory evoked potential (SSEP) tests to examine how her nervous system was processing electrical impulses.

The results indicated that Sarah had multiple sclerosis (MS), which is an inflammatory disease. This disease affects the myelin sheaths, which wrap parts of some nerve cells like insulation around an electrical cord. As these sheaths deteriorate, the nerves no longer conduct impulses normally. For unknown reasons, multiple sclerosis often attacks the optic nerves first before proceeding to other areas of the brain. Sarah's doctors were able to treat her MS symptoms using high doses of immunosuppressive medications. Unfortunately, there is no cure for MS, but most patients can control the symptoms with daily injections of medication.

As you read through the chapter, think about the following questions.

1. Why would a deterioration of the myelin sheaths cause a nerve cell to function incorrectly?
2. How would an MRI and SSEP test indicate that there was a problem with Sarah's neurological functions?
3. Why are many individuals who contract MS eventually confined to a wheelchair?

CHAPTER CONCEPTS

13.1 **Overview of the Nervous System**
In the nervous system, reception of stimuli is associated with sensory neurons; integration is associated with interneurons; and motor output is associated with motor neurons. All neurons use the same methods to transmit nerve impulses along neurons and across synapses.

13.2 **The Central Nervous System**
The central nervous system consists of the brain and the spinal cord. The brain is divided into portions, each with specific functions, and the spinal cord communicates with the brain. The spinal cord provides input to, and output from, the brain.

13.3 **The Limbic System and Higher Mental Functions**
The limbic system involves many parts of the brain. It gives emotional overtones to the activities of the brain, and is important in the processes of learning and memory.

13.4 **The Peripheral Nervous System**
The peripheral nervous system consists of nerves that project from the CNS. Cranial nerves project from the brain. The spinal cord gives rise to spinal nerves.

13.5 **Drug Therapy and Drug Abuse**
Although neurological drugs are quite varied, each type has been found to either promote, prevent, or replace the action of a particular neurotransmitter at a synapse.

BEFORE YOU BEGIN

Before beginning this chapter, take a few moments to review the following discussions.

Section 2.1 How does an ion differ from an atom of an element?

Section 3.3 How does the sodium–potassium pump move ions across the cell membrane?

Section 4.1 What is the function of nervous tissue in the body?

285

Applications to Enhance the Relevancy of Content

Connections and Misconceptions

In addition to the Focus readings, a feature of *Human Biology*, Twelfth Edition, is the Connections and Misconceptions applications. Throughout the text, some common questions that are brought up in human biology classrooms are explored, including:

Are tanning beds safe?
Are stem cells only found in embryos?
What is methylmercury and why is it dangerous?

These pieces will help students relate the content of the text to their everyday lives and to many of the topics encountered in the media.

Connections and Misconceptions

What causes cystic fibrosis?

In 1989, scientists determined that defects in a gene on chromosome 7 were the cause of cystic fibrosis (CF). This gene, called *CFTR* (cystic fibrosis conductance regulator), codes for a protein that is responsible for the movement of chloride ions across the membranes of cells that produce mucus, sweat, and saliva. Defects in this gene cause an improper water–salt balance in the excretions of these cells, which in turn leads to the symptoms of CF. To date, there are over 1,400 known mutations in the CF gene. This tremendous amount of variation in this gene accounts for the differences in the severity of the disease in CF patients.

By knowing the precise gene that causes the disease, scientists have been able to develop new treatment options for people with CF. At one time, an individual with CF rarely saw his or her twentieth birthday; now it is routine for people to live into their 30s and 40s. New treatments, such as gene therapy, are being explored for sufferers of CF.

Connections and Misconceptions

Does cranberry juice really prevent or cure a urinary tract infection?

Research has supported the use of cranberry juice to prevent urinary tract infections. It appears to prevent bacteria that would cause infection from adhering to the surfaces of the urinary tracts. However, cranberry juice has not been shown to be an effective treatment for an already existing urinary tract infection.

Video Cranberries vs. Bacteria

Media Integration

A significant new feature of this edition is the integration of content and animation, video, and audio assets. Virtually every section of the textbook is now linked to MP3 files, animations of biological processes, National Geographic or ScienCentral videos.

MP3 Files. These 3- to 5-minute audio files not only serve as a review of the material in the chapter but also assist the student in the pronunciation of scientific terms.

Animations. Drawing on McGraw-Hill's vast library of animations, the authors have selected animations that will enhance the student's understanding of complex biological processes.

Videos. Two different types of movies are integrated into this edition of the text. The ScienCentral videos are short news clips on recent advances in the sciences. The National Geographic videos provide the student with a glimpse of the complexity of life that normally would not be possible in the classroom.

Virtual Labs. Simulated experiments allow students to explore the topics covered in the chapter.

Biology Matters Bioethical Focus

Stem-Cell Research

In the human body, stem cells are analogous to immortal "parents." Their "offspring," called *daughter cells*, can remain as stem cells and potentially divide indefinitely. However, most daughter cells differentiate further, forming mature cells called *end cells*. Research using stem cells has remained a source of controversy since 1998, when scientists discovered how to isolate and grow human stem cells in the laboratory.

There are primarily two different types of stem cells: embryonic and adult. Advantages and disadvantages exist for each type. *Embryonic stem cells* are derived from fertilized embryos at various stages of development. Fertilized human ova stored in infertility clinics are often used as the source of embryonic stem cells. The use of these cells for research has sparked tremendous controversy, because many people believe these cells have the potential to become a human being. *Adult stem cells* are undifferentiated cells found in various body tissues, whose purpose is to repair or replace

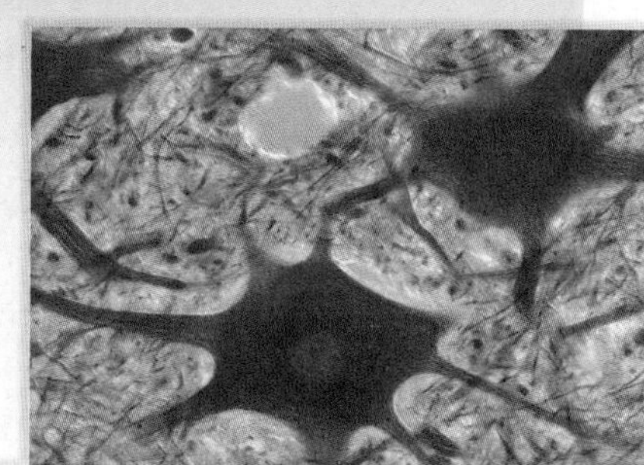

...tial cure for ...eurons such as these. Stem ...oducing neurons could be ...patients.

...bypass some of the con-...embryonic stem cells and, ...rapid ...g stem ...or this ...agazine ...ugh of

Video Heart Stem Cells

Video Making Brain Cells

...ould be spent on a therapy ...rs of intensive research"? ...ent on therapies that have ...? ...the ban on certain types of ...can proceed faster? ...iderations should be used ...r stem-cell therapy?

...tent/full/322/5909/1766

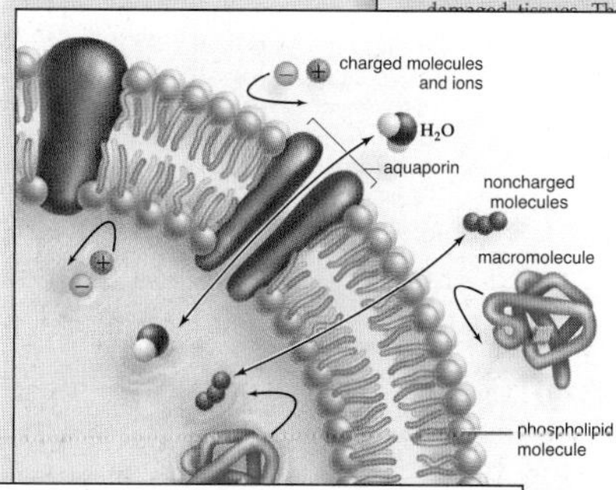

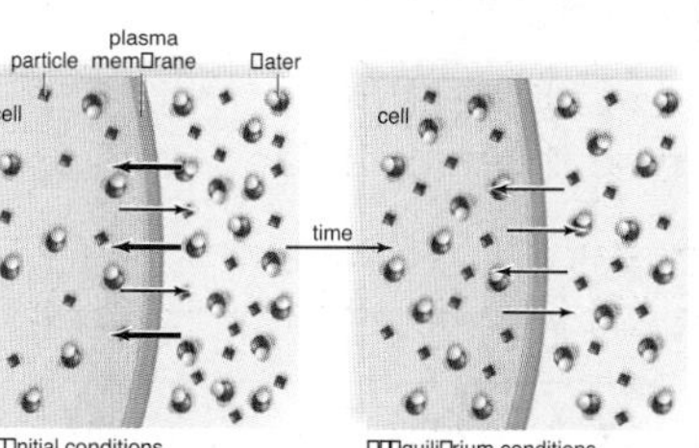

Figure 3.8 Diffusion across the plasma membrane.
a. When a substance can diffuse across the plasma membrane, it will move back and forth across the membrane, but the net movement will be toward the region of lower concentration. **b.** At equilibrium, equal numbers of particles and water have crossed in both directions, and there is no net movement.

...na membrane. ...ane whereas ...cross

...ws only certain ...n freely. There-...ely permeable ...as oxygen and ...ne easily. The ...reely cross the ...quaporins. Ions ...without more

...cules from an ...er concentra-...on is a passive ...ellular energy

Certain molecules can freely cross the plasma membrane by diffusion. When molecules can cross a plasma membrane, which way will they go? The molecules will move in both directions. But the *net movement* will be from the region of higher concentration to the region of lower concentration, until equilibrium is achieved. At equilibrium, as many molecules of the substance will be entering as leaving the cell (Fig. 3.8). Oxygen diffuses across the plasma membrane, and the net movement is toward the inside of the cell. This is because a cell uses oxygen when it produces ATP molecules for energy purposes.

Animation Diffusion Through Cell Membranes

MP3 Simple Diffusion

Osmosis

Osmosis is the net movement of water across a semipermeable membrane, from an area of higher concentration to an area of lower concentration. The membrane separates the two areas, and solute is unable to pass through the membrane. Water will tend to flow from the area that has less solute (and therefore more water) to the area with more solute (and therefore less water). **Tonicity** refers to the osmotic characteristics of a solution across a particular membrane, such as a red blood cell membrane.

Normally, body fluids are *isotonic* to cells (Fig. 3.9*a*). There is the same concentration of nondiffusible solutes and water on both sides of the plasma membrane. Therefore, cells maintain their normal size and shape. Intravenous solutions given in medical situations are usually isotonic.

Solutions that cause cells to swell or even to burst due to an intake of water are said to be *hypotonic*. A hypotonic solution has a lower concentration of solute and a higher concentration of water than the cells. If red blood cells are placed in a hypotonic solution, water enters the cells. They swell to bursting (Fig. 3.9*b*). *Lysis* is used to refer to the process of bursting cells. Bursting of red blood cells is termed *hemolysis*.

Solutions that cause cells to shrink or shrivel due to loss of water are said to be *hypertonic*. A hypertonic solution has a higher concentration of solute and a lower concentration of water than do the cells. If red blood cells are placed in a hypertonic solution, water leaves the cells; they shrink (Fig. 3.9*c*). The term *crenation* refers to red blood cells in this condition. These changes have occurred due to osmotic pressure. **Osmotic pressure** controls water movement in our bodies. For example, in the small and large intestines, osmotic pressure allows us to absorb the water in food and drink. In the kidneys, osmotic pressure controls water absorption as well.

Animation How Osmosis Works

MP3 Osmosis

Media Study Tools

www.mhhe.com/maderhuman12e

Enhance your study of this chapter with study tools and practice tests. Also ask your instructor about the resources available through ConnectPlus, including the media-rich eBook, interactive learning tools, and animations.

Virtual Lab

The virtual lab "Enzyme-Controlled Reactions" provides an interactive investigation of how environmental conditions regulate enzyme activity.

A Student's Guide to Using This Textbook: The Learning System

Pedagogical Features Facilitate Your Understanding of Biology

Case Study

Puts the content of the chapter in the context of a human event—often a medical condition—and leads you into the chapter in an interesting way. Each case study is accompanied by questions that assist you in integrating the topics into the chapter content. Thinking Critically About the Concepts questions at the end of the chapter also connect the case study to the chapter concepts.

CASE STUDY KNEE REPLACEMENT

Jackie was an outstanding athlete in high school, and even now, in her early 50s, she tried to stay in shape. But during her customary 3-mile jogs, she was having an increasingly hard time ignoring the pain in her left knee. She had torn some ligaments in her knee playing intramural and intermural volleyball in college, and it had never quite felt the same. In her 40s, she was able to control the pain by taking over-the-counter medications; but two years ago she had had arthroscopic surgery to remove some torn cartilage and calcium deposits. Now that the pain was getting worse than before, she knew her best option might be a total knee replacement.

Although it sounds drastic, replacing old, arthritic joints with new artificial ones is becoming increasingly routine. About 500,000 artificial knees were installed in U.S. patients in 2006, which represents a 65% increase from 2000. During the procedure, a surgeon removes bone from the bottom of the femur and the top of the tibia and replaces each with caps made of metal or ceramic, held in place with bone cement. A plastic plate is installed to allow the femur and tibia to move smoothly against each other, and a smaller plate is attached to the kneecap (patella) so that it can function properly.

As you read through the chapter, think about the following questions.

1. What is the role of cartilage in the knee joint?
2. What specific portions of these long bones are being removed during knee replacement?
3. Why does Jackie's physical condition make her an ideal candidate for knee replacement?

CHAPTER CONCEPTS

11.1 **Overview of the Skeletal System**
Bones are the organs of the skeletal system. The tissues of the system are compact and spongy bone, various types of cartilage, and fibrous connective tissue in the ligaments that hold bones together.

11.2 **Bone Growth, Remodeling, and Repair**
Bone is a living tissue that grows, remodels, and repairs itself. In all of these processes, some bone cells break down bone and some repair bone.

11.3 **Bones of the Axial Skeleton**
The axial skeleton lies in the midline of the body and consists of the skull, the hyoid bone, the vertebral column, and the rib cage.

11.4 **Bones of the Appendicular Skeleton**
The appendicular skeleton consists of the bones of the pectoral girdle, upper limbs, pelvic girdle, and lower limbs.

11.5 **Articulations**
Joints are classified according to their degree of movement. Synovial joints are freely movable.

BEFORE YOU BEGIN

Before beginning this chapter, take a few moments to review the following discussions.

Section 4.1 What is the role of connective tissue in the body?

Section 4.9 How does the skeletal system contribute to homeostasis?

Chapter Concepts

Provides a concise preview of the topics covered in each section.

NEW Before You Begin

Links the content of the chapter with material from earlier in the text. The questions designate important topics that you should understand before proceeding into the chapter.

NEW Learning Outcomes

Provide you with an overview of what you are to know. Your instructor can assign activities through Connect™ Biology to help you achieve these outcomes.

11.2 Bone Growth, Remodeling, and Repair

Learning Outcomes

Upon completion of this section, you should be able to

1. Summarize the process of ossification and list the types of cells involved.
2. Describe the process of bone remodeling.
3. Explain the steps in the repair of bone.

NEW Connecting the Concepts

Directs you to areas of the textbook that provide additional information on a topic. Many of the Connecting the Concepts features have been designed to enhance your understanding of homeostasis.

broken ends are wedged into each other. A spiral fracture occurs when the break is ragged due to twisting of a bone.

Connecting the Concepts

For more on bone development and the hormones that influence bone growth, refer to the following discussions.

Section 8.6 provides additional information on inputs of vitamin D and calcium in the diet.

Section 15.2 examines the role of growth hormones in the body.

Section 15.3 describes the action of the hormones calcitonin and PTH.

NEW Media Integration

Enhance your study of biology with media. Go to www.mhhe.com/maderhuman12e to access the animations, videos, and MP3 files referenced throughout this book. Ask your instructor about related quizzes that are available through Connect™ Biology.

Check Your Progress 11.2

1. Classify cells of the skeletal system into ones involved in bone growth, remodeling, and repair.
2. Describe how bone growth occurs during development.
3. Summarize the stages in the repair of bone.

Check Your Progress

Questions at the end of each section help you assess and/or apply your understanding of the material in the section. The questions progress in difficulty (red, yellow, green) to ensure you are going beyond memorization of content.

NEW Case Study Conclusion

The conclusion summarizes the opening case study in the context of the topics just covered in the chapter.

Media Study Tools

Provides a link to the *Human Biology* website, which contains practice tests, animations, and videos organized and integrated by chapter to help you succeed in your study of biology. The ConnectPlus™ platform provides a media-rich eBook, interactive learning tools, and access to the LearnSmart™ system for enhanced student performance.

NEW Virtual Labs

Referenced in some chapters, these virtual labs allow you to investigate topics associated with the content of the chapter from a scientific perspective.

Summarizing the Concepts

Provides an excellent overview of the chapter concepts using concise, bulleted summaries, summary tables, and key illustrations.

CASE STUDY CONCLUSION

Over the next few months, both Kevin and Mary dedicated hours to understanding the causes and treatments of Tay-Sachs disease. They learned that the disease is caused by a recessive mutation that limits the production of an enzyme called beta-hexosaminidase A. This enzyme is loaded into a newly formed lysosome by the Golgi apparatus. The enzyme's function is to break down a specific type of fatty acid chain called *gangliosides.* Gangliosides play an important role in the early formation of the neurons in the brain. Tay-Sachs disease occurs when the gangliosides overaccumulate in the neurons.

Though the prognosis for their child was intially poor—very few children with Tay-Sachs live beyond the age of four, the parents were encouraged to find out what advances in a form of medicine called *gene therapy* might be able to prolong the life of their child. In gene therapy, a correct version of the gene is introduced into specific cells in an attempt to regain lost function. Some initial studies using mice as a model had demonstrated an ability to reduce ganglioside concentrations by providing a working version of the gene that produced beta-hexosaminidase A to the neurons of the brain. Though research was still ongoing, it was a promising piece of information for both Kevin and Mary.

Media Study Tools

www.mhhe.com/maderhuman12e

Enhance your study of this chapter with study tools and practice tests. Also ask your instructor about the resources available through ConnectPlus, including the media-rich eBook, interactive learning tools, and animations.

Virtual Lab

The virtual lab "Enzyme-Controlled Reactions" provides an interactive investigation of how environmental conditions regulate enzyme activity.

Summarizing the Concepts

3.1 What Is a Cell?

- Cells, the basic units of life, come from pre-existing cells.
- Microscopes are used to view cells, which must remain small to have a favorable surface area-to-volume ratio.

3.2 How Cells Are Organized

The human cell is surrounded by a plasma membrane and has a central nucleus. Between the plasma membrane and the nucleus is the cytoplasm, which contains various organelles. Organelles in the cytoplasm have specific functions.

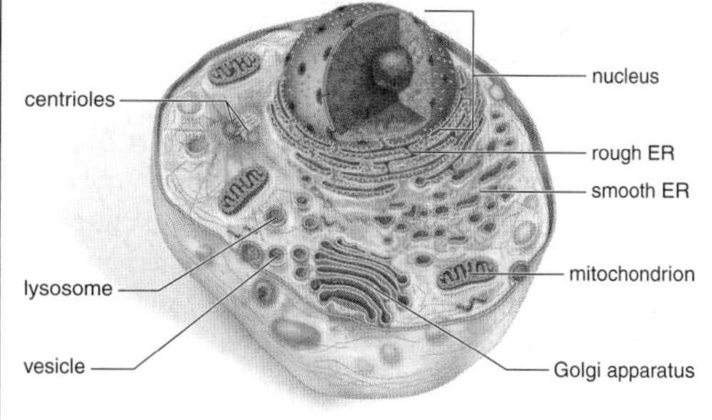

3.3 The Plasma Membrane and How Substances Cross It

The plasma membrane is a phospholipid bilayer that

- selectively regulates the passage of molecules and ions into and out of the cell.
- contains embedded proteins, which allow certain substances to cross the plasma membrane.

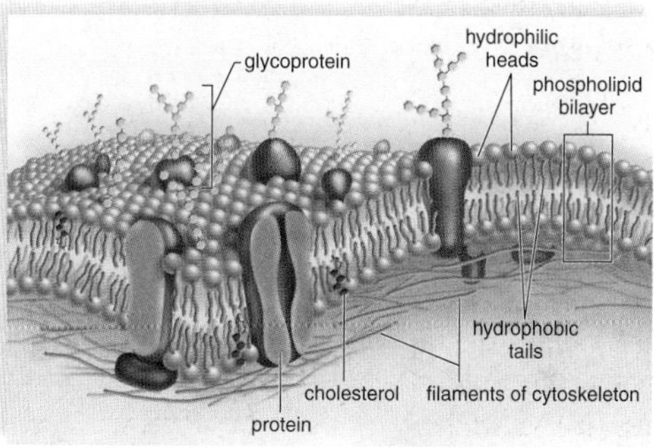

Passage of molecules into or out of cells can be passive or active.

- Passive mechanisms (no energy required) are diffusion (osmosis) and facilitated transport.
- Active mechanisms (energy required) are active transport and endocytosis and exocytosis.

3.4 The Nucleus and the Endomembrane System

- The nucleus houses DNA, which specifies the order of amino acids in proteins.
- Chromatin is a combination of DNA molecules and proteins that make up chromosomes.
- The nucleolus produces ribosomal RNA (rRNA).
- Protein synthesis occurs in ribosomes, small organelles composed of proteins and rRNA.

The Endomembrane System

The endomembrane system consists of the nuclear envelope, endoplasmic reticulum (ER), Golgi apparatus, lysosomes, and vesicles.

- Rough ER has ribosomes, where protein synthesis occurs.
- Smooth ER has no ribosomes and has various functions, including lipid synthesis.
- The Golgi apparatus processes and packages proteins and lipids into vesicles for secretion or movement into other parts of the cell.
- Lysosomes are specialized vesicles produced by the Golgi apparatus. They fuse with incoming vesicles to digest enclosed material, and they autodigest old cell parts.

3.5 The Cytoskeleton, Cell Movement, and Cell Junctions

The cytoskeleton consists of microtubules, actin filaments, and intermediate filaments that give cells their shape; and it allows organelles to move about the cell. Cilia and flagella, which contain microtubules, allow a cell to move.

Cell junctions connect cells to form tissues and to faciliate communication between cells.

3.6 Mitochondria and Cellular Metabolism

- Mitochondria have an inner membrane that forms cristae, which project into the matrix.
- Mitochondria are involved in cellular respiration, which uses oxygen and releases carbon dioxide.
- During cellular respiration, mitochondria convert the energy of glucose into the energy of ATP molecules.

Cellular Respiration and Metabolism

- A metabolic pathway is a series of reactions, each of which has its own enzyme.
- Enzymes bind their substrates in the active site.
- Sometimes enzymes require coenzymes (such as NAD^+), nonprotein molecules that participate in the reaction.

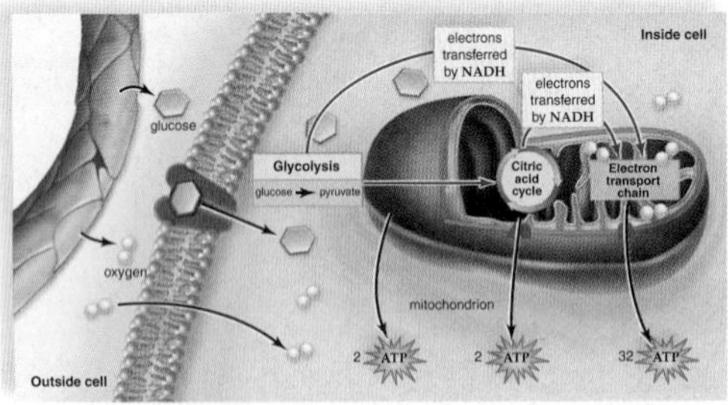

- Cellular respiration is the enzymatic breakdown of glucose to carbon dioxide and water.
- Cellular respiration includes three pathways: glycolysis (occurs in the cytoplasm and is anaerobic), the citric acid cycle (releases carbon dioxide), and the electron transport chain (passes electrons to oxygen).

Fermentation

- If oxygen is not available in cells, the electron transport chain is inoperative, and fermentation (which does not require oxygen) occurs.
- Fermentation produces very little ATP.

Understanding Key Terms

actin filament 55
active site 57
active transport 51
aerobic 59
anaerobic 58
cell theory 44
cellular respiration 57
centrosome 55
chromatin 53
chromosome 53
cilium 55
citric acid cycle 58
coenzyme 58
cytoplasm 46
cytoskeleton 55
diffusion 50
electron transport chain 59
endomembrane system 54
endoplasmic reticulum (ER) 53
eukaryotic cell 46
facilitated transport 51
fermentation 60
flagellum 55
fluid-mosaic model 49
glycolysis 58
Golgi apparatus 54
intermediate filament 55
lysosome 54
metabolism 57
microtubule 55
mitochondrion 57
NAD^+ (nicotinamide adenine dinucleotide) 58
nuclear envelope 53
nuclear pore 53
nucleolus 53
nucleoplasm 53
nucleus 53
organelle 46
osmosis 50
osmotic pressure 50
phagocytosis 52
plasma membrane 46
polyribosome 54
product 57
prokaryotic cell 46
reactant 57
ribosome 53
selectively permeable 46
substrate 57
tonicity 50
vesicle 54

Match the key terms to these definitions.

a. ________ Protein molecules form a shifting pattern within the fluid phospholipid bilayer.

b. ________ Diffusion of water through a selectively permeable membrane.

c. ________ The cell will allow some substances to pass through while not permitting others.

d. ________ Anaerobic breakdown of glucose that results in a gain of two ATP and end products, such as alcohol and lactate.

e. ________ Metabolic pathways that use energy from carbohydrate, fatty acid, and protein break down to produce ATP molecules.

Testing Your Knowledge of the Concepts

Complete the following questions.

1. Explain the three key concepts of the cell theory. (page 44)
2. Which type of microscope would you use to observe the swimming behavior of a flagellated protozoan? Explain. (page 45)
3. Describe how the eukaryotic cell gained mitochondria and chloroplasts. (page 46)
4. Invagination of plasma membrane produced what structures in eukaryotic cells not present in prokaryotic cells? (page 46)

Understanding Key Terms

Lists the boldface terms in the chapter and their page references. A matching exercise allows you to test your knowledge of the terms.

Testing Your Knowledge of the Concepts

Questions help you review material and prepare for tests. (See Appendix B for answers.)

Thinking Critically About the Concepts

In the case study at the beginning of the chapter, the child had malfunctioning lysosomes, which caused an accumulation of fatty acid in the system. Each part of a cell plays an important role in the homeostasis of the entire body.

1. What might occur if the cells of the body contained a malfunctioning mitochondria?
2. What would happen to homeostasis if enzymes were no longer produced in the body?
3. Knowing what you know about the function of a lysosome, what might occur if the cells' lysosomes were overproductive instead of malfunctioning?

Thinking Critically About the Concepts

Questions encourage you to apply what you have learned to the opening case study. (See Appendix B for answers.)

Biology Matters Readings

The Biology Matters readings in the twelfth edition collectively put the chapter concepts in the context of modern-day issues:

Health Focus readings review procedures and technology that can contribute to your well being.

Science Focus readings describe how experimentation and observations have contributed to our knowledge about the living world.

Bioethical Focus readings describe modern situations that call for value judgments and challenge you to develop a point of view.

Historical Focus readings help you better understand how the study of biology has evolved over time.

Biology Matters Bioethical Focus

DNA Fingerprinting and the Criminal Justice System

Traditional fingerprinting has been used for years to identify criminals and to exonerate those wrongly accused of crimes. The opportunity now arises to use DNA fingerprinting in the same way. DNA fingerprinting requires only a small DNA sample. This sample can come from blood left at the scene of the crime, semen from a rape case, or even a single hair root!

There have also been reported problems with sloppy laboratory procedures and the credibility of forensic experts. In one particular case, Curtis McCarty had been placed on death row three times by the same team of prosecutor and police lab analyst. After 21 years in prison, he was exonerated. The prosecutor has been accused of misconduct, and the police

Biology Matters Science Focus

Coloring Organisms Green; Green Fluorescent Proteins and Cells

Most cells lack any significant pigmentation. Thus, cell biologists frequently rely on dyes to produce enough contrast to resolve organelles and other cellular structures. The first of these dyes were developed in the nineteenth century from chemicals used to stain clothes in the textile industry. Since then, significant advances have occurred in the development of cellular stains.

In 2008, three scientists—Martin Chalfie, Roger Y. Tsien,

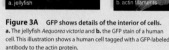

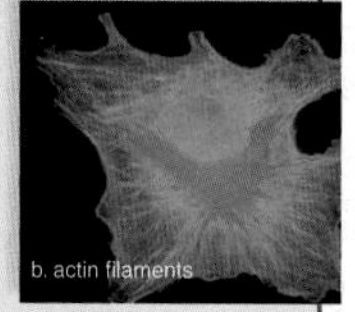

Figure 3A **GFP shows details of the interior of cells.** **a.** The jellyfish *Aequorea victoria* and **b.** the GFP stain of a human cell. This illustration shows a human cell tagged with a GFP-labeled antibody to the actin protein.

Biology Matters Health Focus

Pursuing Youthful Skin

More and more members of the "baby-boomer" generation are willing to spend lavishly for a youthful appearance. Over 30 million Americans have turned to Botox, laser treatments, and/or tanning to help obtain that vigorous, "healthy" look. But how safe and effective are these treatments?

Botox

Botox is a drug used to reduce the appearance of facial wrinkles and lines. Botox is the registered trade name for a derivative of

Biology Matters Historical Focus

The Syphilis Research Scandal of Tuskegee University

Several sections of this chapter have covered the process of science, the way that legitimate research should be conducted, and the importance of informed consent when using human research subjects. As professionals, scientists have a responsibility to design moral and ethical research. Unfortunately, as with all professionals, not all scientists are ethical. Documented cases of risky, life-threatening, and, in some cases, inhumane research on humans (often without the subject's consent or knowledge) blot scientific history. One of the most extreme examples of such "research" was that done by Dr. Josef Mengele, the handsome Nazi doctor called the "Angel of Death." Mengele tortured concentration camp prisoners in multiple horrible ways. Some were slowly frozen to death, others poisoned, and still others bled to death—all to fulfill Mengele's obscene notion of scientific inquiry.

Regrettably, the history of research in the United States is also stained by misconduct. One notorious example of unethical research involving human subjects began in the United States in 1932 and continued until 1972. This research was carried out by the Public Health Service (PHS). Investigators wished to study were poor, mostly illiterate, sharecropper farmers. None of the men were informed of their participation in a research study nor about available treatment options. The men were told that investigators were testing for and treating "bad blood." The phrase described a number of common illnesses, including anemia, that were widespread at the time. While they participated in the study, the men were offered medical exams, transportation to and from clinics, treatments for other ailments, food, and money for their burial expenses if necessary.

When the study first began, there were few available treat-

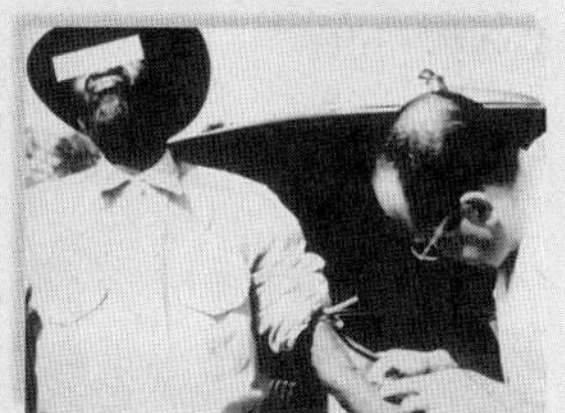

Figure 1A Poorly educated African Americans were recruited for the Tuskegee project with promises of free medical care.

Connections and Misconceptions

This unique feature presents the types of spontaneous inquiries that you may have as you study the workings of the human body. Questions and answers can be serious or funny, but each will capture your attention.

Connections and Misconceptions

What causes meningitis?

Meningitis is caused by an infection of the meninges by either a virus or a bacterium. Viral meningitis is less severe than bacterial meningitis, which in some cases can result in brain damage and death. Bacterial meningitis is usually caused by one of three species of bacteria: *Haemophilus influenzae* type b (Hib), *Streptococcus pneumoniae,* and *Neisseria meningitidis.* Vaccines are available for Hib bacteria and some forms of *S. pneumoniae* and *N. meningitidis.* The Centers for Disease Control (CDC) recommend that individuals between the ages of 11 and 18 be vaccinated against bacterial meningitis.

A Student's Guide to Using This Textbook: The Art Program

Vivid and Engaging Illustrations Bring the Study of Biology to Life

Combination Art

Drawings of structures are paired with micrographs to provide you with two perspectives: the explanatory clarity of line drawings and the realism of photos.

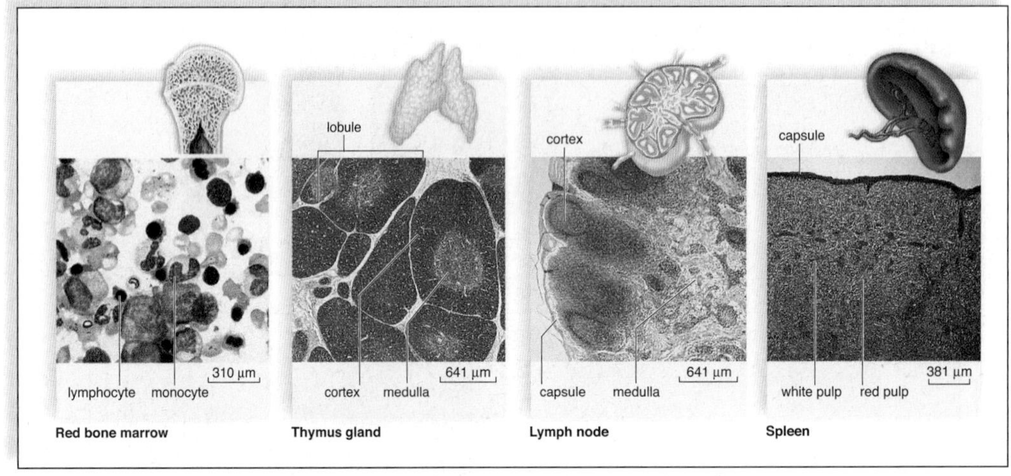

Multilevel Perspective

Such illustrations guide you from the more intuitive macroscopic level of learning to the functional foundations revealed through microscopic images.

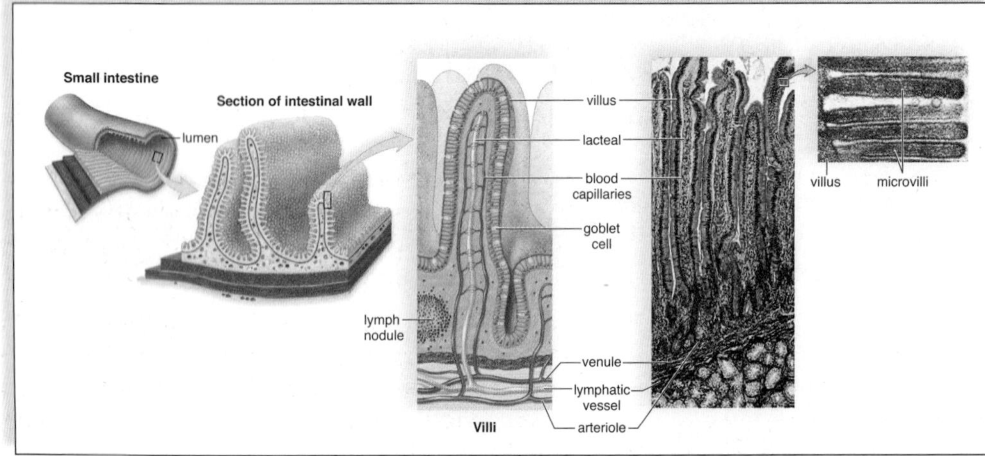

Icons

Icons orient you to the whole structure or process by providing small drawings that help you visualize how a particular structure is part of a larger one.

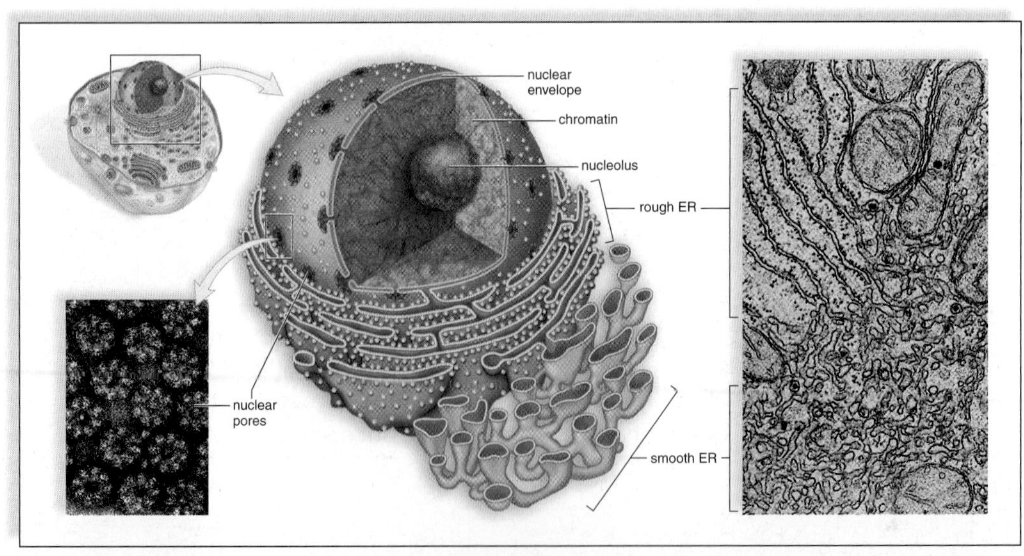

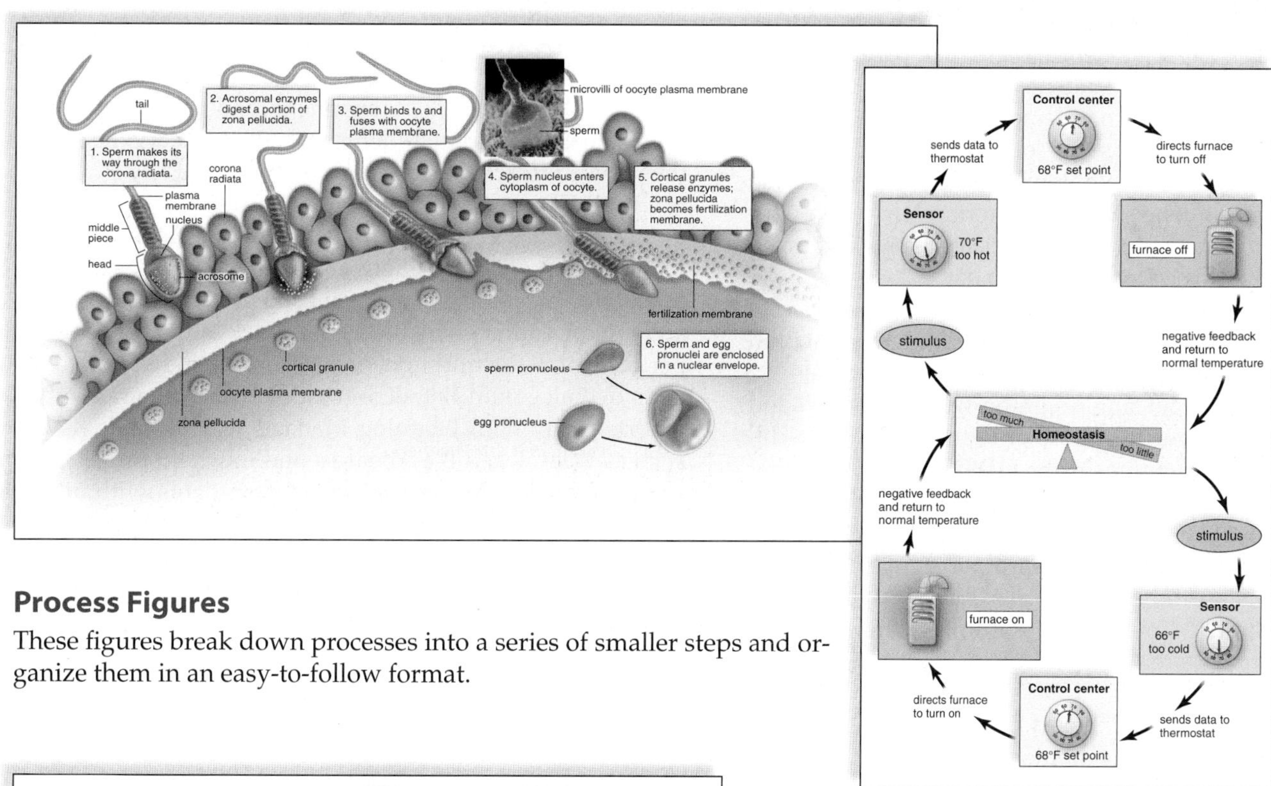

Process Figures

These figures break down processes into a series of smaller steps and organize them in an easy-to-follow format.

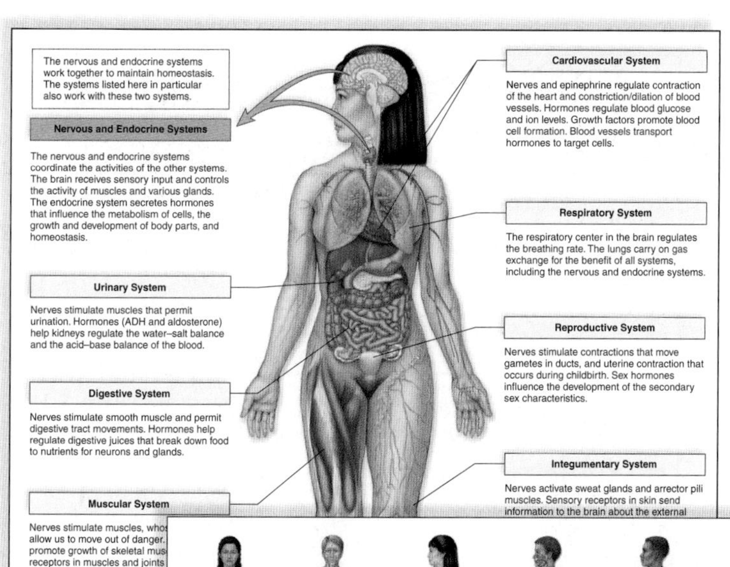

Nerves stimulate smooth muscle and permit digestive tract movements. Hormones help regulate digestive juices that break down food to nutrients for neurons and glands.

Nerves activate sweat glands and arrector pili muscles. Sensory receptors in skin send information to the brain about the external

Human Systems Work Together

Working-together illustrations use brief concise statements to tell you how various other systems help a featured system achieve homeostasis.

The working-together illustrations have been integrated into homeostasis sections making a united whole. The homeostasis sections show how the systems achieve homeostasis despite real-life experiences that could alter the internal environment. For example, see page 85.

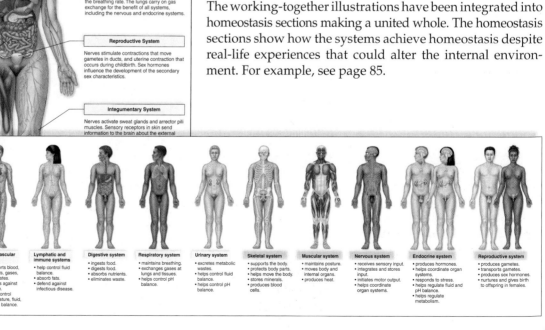

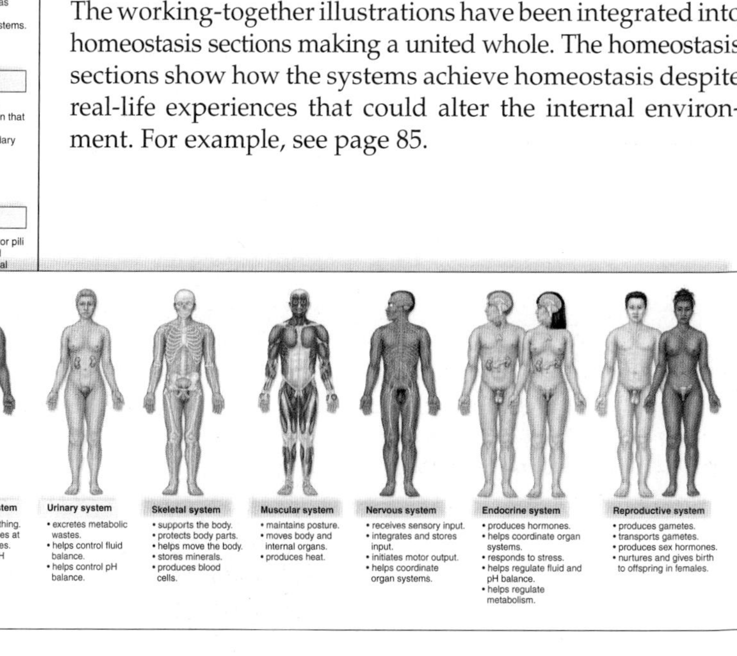

Teaching and Learning Tools

Do More

McGraw-Hill Higher Education and Blackboard Have Teamed Up

Blackboard, the Web-based course-management system, has partnered with McGraw-Hill to better allow students and faculty to use online materials and activities to complement face-to-face teaching. Blackboard features exciting social learning and teaching tools that foster more logical, visually impactful, and active learning opportunities for students. You'll transform your closed-door classrooms into communities where students remain connected to their educational experience 24 hours a day.

This partnership allows you and your students access to McGraw-Hill's Connect™ and Create™ right from within your Blackboard course—all with one single sign-on.

Not only do you get single sign-on with Connect™ and Create™, you also get deep integration of McGraw-Hill content and content engines right in Blackboard. Whether you're choosing a book for your course or building Connect™ assignments, all the tools you need are right where you want them—inside of Blackboard.

Gradebooks are now seamless. When a student completes an integrated Connect™ assignment, the grade for that assignment automatically (and instantly) feeds your Blackboard grade center.

McGraw-Hill and Blackboard can now offer you easy access to industry leading technology and content, whether your campus hosts it or we do. Be sure to ask your local McGraw-Hill representative for details.

LearnSmart™

LearnSmart™ is available as an integrated feature of McGraw-Hill Connect™ Biology and provides students with a GPS (Guided Path to Success) for your course. Using artificial intelligence, LearnSmart™ intelligently assesses a student's knowledge of course content through a series of adaptive questions. It pinpoints concepts the student does not understand and maps out a personalized study plan for success. This innovative study tool also has features that allow instructors to see exactly what students have accomplished and a built in assessment tool for graded assignments. Visit the following site for a demonstration. **www.mhlearnsmart.com**

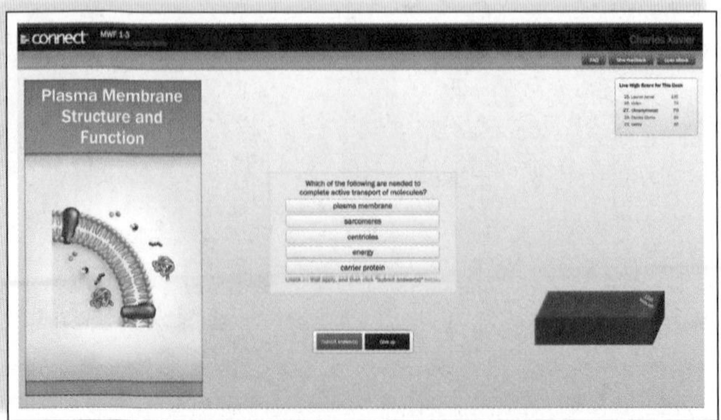

McGraw-Hill Connect™ Biology

www.mhhe.com/maderhuman12e

McGraw-Hill Connect™ Biology provides online presentation, assignment, and assessment solutions. It connects your students with the tools and resources they'll need to achieve success.

With Connect™ Biology, you can deliver assignments, quizzes, and tests online. A robust set of questions and activities are presented and aligned with the textbook's learning outcomes. As an instructor, you can edit existing questions and author entirely new problems. Track individual student performance—by question, assignment, or in relation to the class overall—with detailed grade reports. Integrate grade reports easily with Learning Management Systems (LMS), such as WebCT and Blackboard—and much more.

ConnectPlus™ Biology provides students with all the advantages of Connect™ Biology, plus 24/7 online access to an eBook. This media-rich version of the book is available through the McGraw-Hill Connect™ platform and allows seamless integration of text, media, and assessments.

To learn more, visit
www.mcgrawhillconnect.com

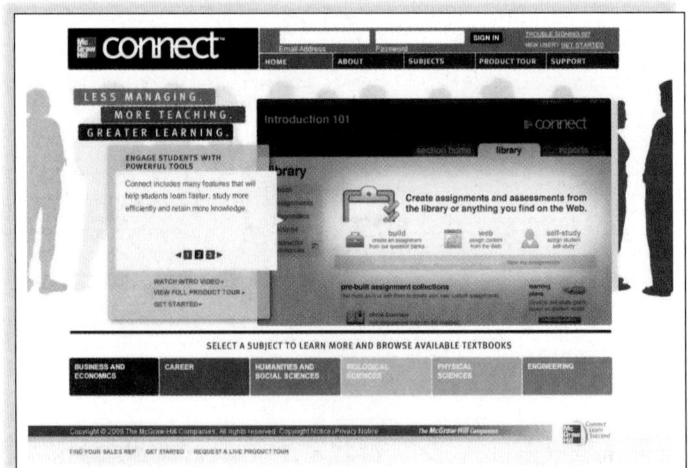

My Lectures—Tegrity

Tegrity Campus™ records and distributes your class lecture with just a click of a button. Students can view anytime/anywhere via computer, iPod, or mobile device. It indexes as it records your PowerPoint presentations and anything shown on your computer so students can use keywords to find exactly what they want to study. Tegrity is available as an integrated feature of McGraw-Hill Connect™ Biology and as a standalone.

Animations for a New Generation

Dynamic, 3D animations of key biological processes bring an unprecedented level of control to the classroom. Innovative features keep the emphasis on teaching rather than entertaining.

- An options menu lets you control the animation's level of detail, speed, length, and appearance, so you can create the experience you want.
- Draw on the animation using the whiteboard pen to highlight important areas.
- The scroll bar lets you fast forward and rewind while seeing what happens in the animation, so you can start at the exact moment you want.
- A scene menu lets you instantly jump to a specific point in the animation.
- Pop-ups add detail at important points and help students relate the animation back to concepts from lecture and the textbook.
- A complete visual summary at the end of the animation reminds students of the big picture.
- Animation topics include: Cellular Respiration, Photosynthesis, Molecular Biology of the Gene, DNA Replication, Cell Cycle and Mitosis, Membrane Transport, and Plant Transport.

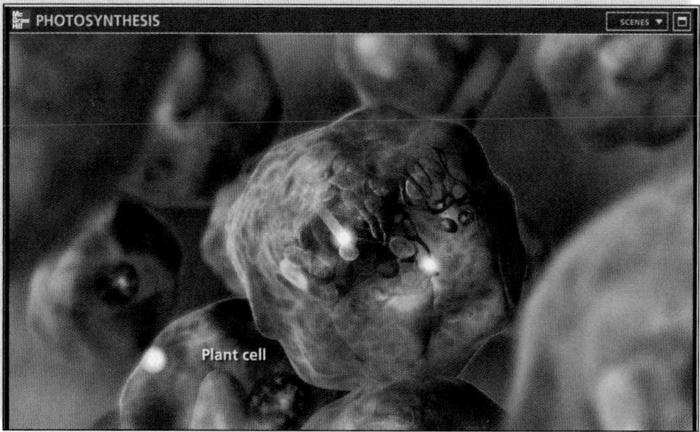

Create

With **McGraw-Hill Create™**, www.mcgrawhillcreate.com, you can easily rearrange chapters, combine material from other content sources, and quickly upload content you have written like your course syllabus or teaching notes. Find the content you need in Create by searching through thousands of leading McGraw-Hill textbooks. Arrange your book to fit your teaching style. Create even allows you to personalize your book's appearance by selecting the cover and adding your name, school, and course information. Order a Create book and you'll receive a complimentary print review copy in 3–5 business days or a complimentary electronic review copy (eComp) via e-mail in minutes. Go to www.mcgrawhillcreate.com today and register to experience how McGraw-Hill Create™ empowers you to teach *your* students *your* way.

Presentation Tools

Everything you need for outstanding presentations in one place.

www.mhhe.com/maderhuman12e

- *FlexArt Image Powerpoints*—including every piece of art that has been sized and cropped specifically for superior presentations, as well as labels that can be edited and flexible art that can be picked up and moved on key figures. Also included are tables, photographs, and unlabeled art pieces.
- *Lecture PowerPoints with Animations*—animations illustrating important processes are embedded in the lecture material.
- *Animation PowerPoints*—animations only are provided in PowerPoint.
- *Labeled JPEG Images*—Full-color digital files of all illustrations that can be readily incorporated into presentations, exams, or custom-made classroom materials.
- *Base Art Image Files*—unlabeled digital files of all illustrations.

Presentation Center

In addition to the images from your book, this online digital library contains photos, artwork, animations, and other media from an array of McGraw-Hill textbooks.

Computerized Test Bank

A comprehensive bank of test questions is provided within a computerized test bank powered by McGraw-Hill's flexible electronic testing program, **EZ Test Online.** A new tagging scheme allows you to sort questions by Bloom's difficulty level, learning outcome, topic, and section. With EZ Test Online, instructors can select questions from multiple McGraw-Hill test banks or author their own, and then either print the test for paper distribution or give it online.

Instructor's Manual

The instructor's manual contains chapter outlines, lecture enrichment ideas, and discussion questions.

Laboratory Manual

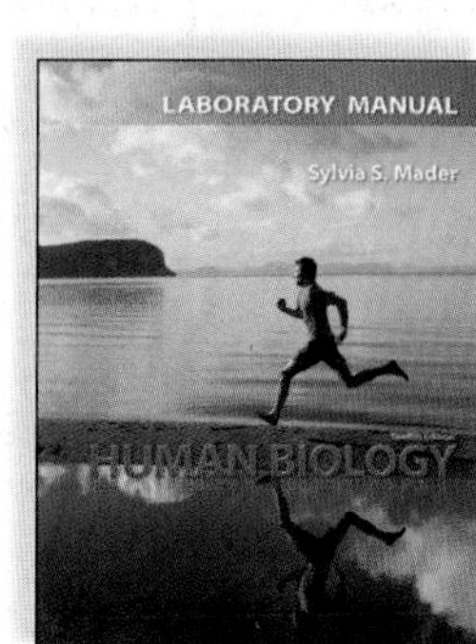

The *Human Biology Laboratory Manual* is written by Dr. Sylvia Mader. Every laboratory has been written to help students learn the fundamental concepts of biology and the specific content of the chapter to which the lab relates, as well as gain a better understanding of the scientific method.

Companion Website

www.mhhe.com/maderhuman12e

The Mader *Human Biology* companion website allows students to access a variety of free digital learning tools that include

- Chapter-level quizzing
- Animations and videos
- Vocabulary flashcards
- Virtual labs

Content Changes

Overview of Content Changes to *Human Biology,* Twelfth Edition

Chapter 1: Exploring Life and Science

This chapter previews the text by discussing the characteristics of life, principles of evolution, organization of the biosphere, and the scientific process. A new chapter opener on the characteristics of life has been added. Evolutionary trees are now included to indicate the three domains of life.

Part I: Human Organization

Chapter 2: Chemistry of Life includes a new illustration of relative pH values to improve student understanding of acid–base relationships and an illustration of the structure of fiber. The chapter also contains several new applications: a new case study on blood chemistry, a new application reading on the origin of elements, and two new Biology Matters: Health Focus readings on the importance of fiber and omega-3 fatty acids in the diet. **Chapter 3: Cell Structure and Function** contains a new case study on Tay–Sachs disease, a Biology Matters: Science Focus reading on the use of green fluorescent proteins in cell biology, and new application readings on cystic fibrosis and induced pluripotent stem cells. A virtual lab on enzymes is provided at the end of the chapter. **Chapter 4: Organization and Regulation of Body Systems** includes a new case study on artificial skin and new applications on the safety of tanning beds and the causes of meningitis.

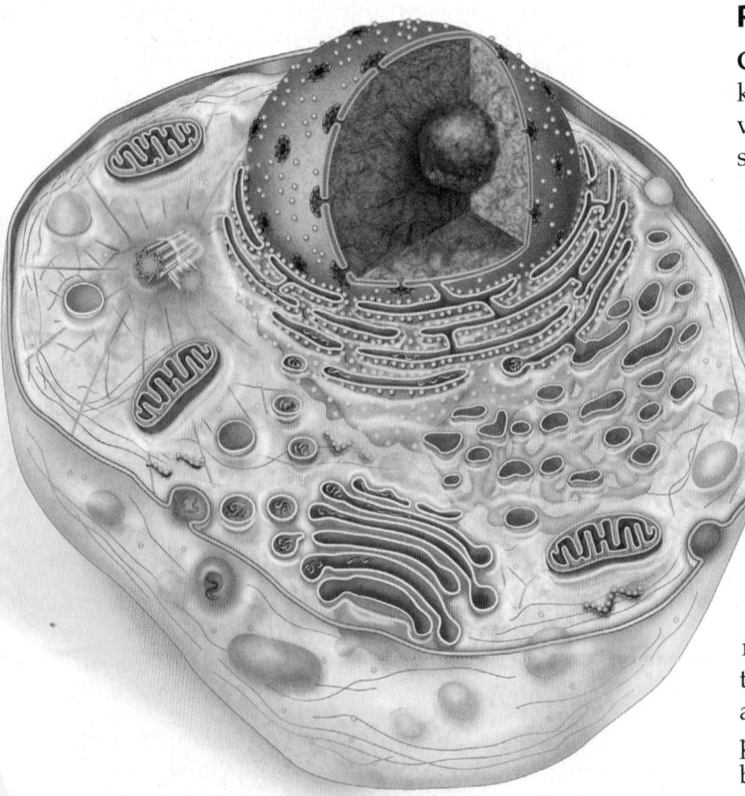

Part II: Maintenance of the Human Body

Chapter 5: Cardiovascular System: Heart and Blood Vessels includes a case study on peripheral artery disease and a new application reading on the development of the heart in a fetus. A virtual lab on blood pressure is included at the end of the chapter. **Chapter 6: Cardiovascular System: Blood** was revised to include a new case study on leukemia and application readings on stem cells, blood doping, Christmas disease, and Bombay syndrome. **Chapter 7: Lymphatic System and Immunity** now opens with a case study on lupus. In addition, application readings on the effects of refrigeration on the growth of bacteria and the mode of action of aspirin are now included. The **Infectious Diseases supplement** has been moved to follow Chapter 7 for a more logical progression and includes updated information on the extent of the HIV/AIDS epidemic, as well as information on the trials to develop an HIV/AIDS vaccine. **Chapter 8: Digestive System and Nutrition** opens with a new case study on gastroesophageal reflux disease (GERD), and a virtual lab on nutrition has been placed in the end-of-chapter material. In **Chapter 9: Respiratory System,** the opening material has been enhanced to indicate more of the tests for sleep apnea, and a new application reading on cystic fibrosis has been included. **Chapter 10: Urinary System** has been reorganized so that the discussions of homeostasis coincide with the coverage of the regulatory functions of the kidneys. A new case study on polycystic kidney disease (PKD) starts the chapter, and a new application reading on the causes of a floating kidney is provided.

Part III: Movement and Support in Humans

Chapter 11: Skeletal System opens with a case study on knee replacement surgery. A new application reading investigates the evolutionary reasons why human toes are shorter than fingers. The Biology Matters: Health Focus on osteoporosis has been updated with new recommendations on calcium and vitamin D intake. **Chapter 12: Muscular System** examines the tests used to detect muscular dystrophy in the opening case study. The application readings examine the number of muscles in the body, the interaction of hemoglobin and muscle tissue, the use of Botox to remove wrinkles, and the causes of muscle soreness following exercise. A virtual lab on muscle stimulation is provided at the end of the chapter.

Part IV: Integration and Coordination in Humans

Chapter 13: Nervous System begins with a case study on multiple sclerosis (MS) and includes a discussion of how the disease is diagnosed. A new application on the mode of action of aspirin is included, as well as new material on the pharmacology of methamphetamines. **Chapter 14: Senses** begins with a case study on the use of cochlear implants, and a new application reading focuses on the causes of pinkeye. **Chapter 15: Endocrine System** opens with a case study on the tests for diabetes mellitus.

Part V: Reproduction in Humans

Chapter 16: Reproductive System includes a case study that examines the causes and diagnosis of cervical cancer. New application readings for this chapter examine polyploidy in liver cells, hormone replacement therapy, and emergency contraceptive pills. **Chapter 17: Development and Aging** opens with a case study on pregnancy testing. The concept of females as the "default sex" is highlighted in a new application reading, and another new reading examines the genetic basis of longevity.

Part VI: Human Genetics

Chapter 18: Patterns of Chromosome Inheritance includes an increased focus on control of the cell cycle (Figure 18.4), and the role of cell cycle checkpoints. The discussion of Barr bodies has been moved from Chapter 3 to Chapter 18. The chapter opens with a new case study on cell cycle control and breast cancer, and an additional application reading examines the relationship between the age of a woman and the risk of Down syndrome in her children. A virtual lab on the cell cycle and cancer is included at the end of the chapter. **Chapter 19: Cancer** includes new illustrations on the roles of tumor suppressor and proto-oncogenes in the cell. Data on cancer types (Figures 19.5 and 19.7) have been updated. Chapter 19 opens with a case study on nephroblastoma, and a new application reading explores the potential link between transposons and cancer. **Chapter 20: Patterns of Genetic Inheritance** begins with a case study on phenylketonuria. The new application readings in the chapter examine the history of the Punnett square, the relationship between skin color and race, and fragile X syndrome. Virtual labs on the use of Punnett squares and sex-linked traits are included at the end of the chapter. **Chapter 21: DNA Biology and Technology** includes a new figure (21.13) that diagrams the stages in the regulation of gene expression. A case study that examines the development of insulin using recombinant DNA technology opens the chapter, and the topics of microRNA, and the universal nature of the genetic code, are included in new application readings. Two virtual labs, classifying using biotechnology and knocking out genes, are included at the end of the chapter.

Part VII: Human Evolution and Ecology

Chapter 22: Human Evolution presents the new classification of humans within the Hominins in Section 22.4 and opens with a revised case study to include new information on the Neandertals. The chapter also includes new application readings on vestigial organs in humans, intelligent design, and artificial selection. **Chapter 23: Global Ecology and Human Interferences** begins with a scenario of the future consequences of human threats to the environment. The chapter also includes information on greenhouse gases, and ecosystems on the floor of the ocean. A virtual lab on modeling ecosystems is provided at the end of the chapter. **Chapter 24: Human Population, Planetary Resources, and Conservation** includes an updated case study with additional information on overfishing, and application readings on methylmercury and the topic of birthrates and death rates in developed and developing countries.

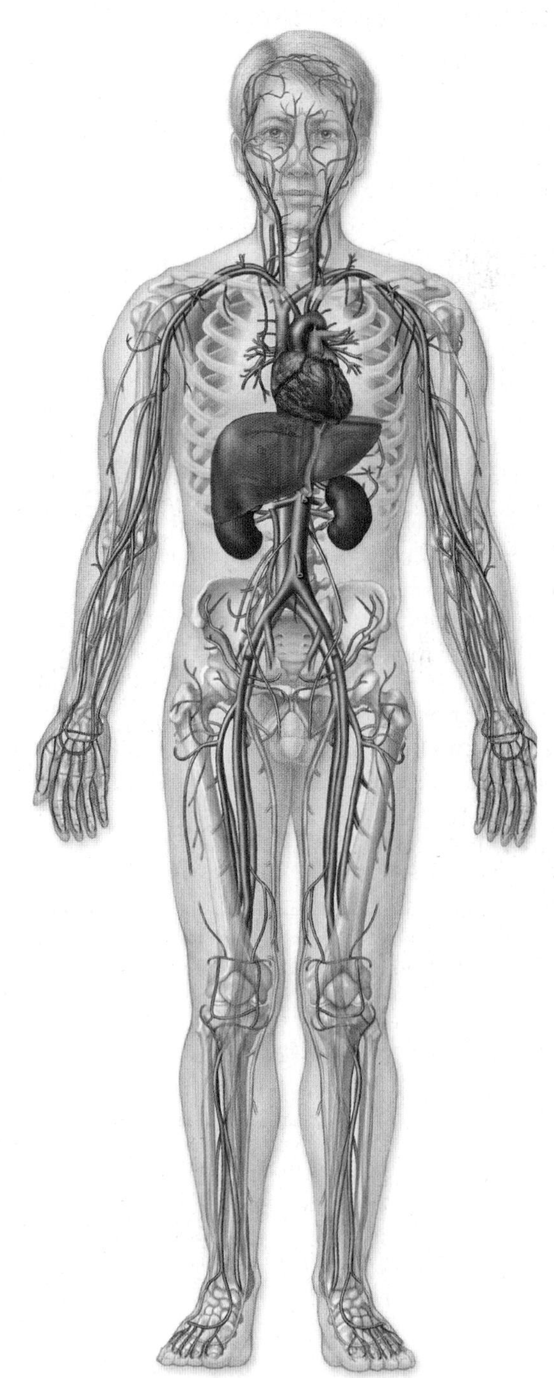

Acknowledgments

Dr. Sylvia Mader represents one of the icons of science education. Her dedication to her students, coupled to her clear, concise, writing style, has benefited the education of thousands of students over the past three decades. It is an honor to continue her legacy, and to bring her message to the next generation of students. Throughout each chapter, I have striven to ensure that the material was written and illustrated in the familiar Mader style.

A project such as this could never be completed without the work of a coordinated group. As always, the McGraw-Hill professionals guided this revision, assisting in all aspects. From beginning brainstorming sessions to completed text, this team supplied creativity, advice, and support whenever it was needed. The Developmental Editor, Rose Koos, and Project Manager, April Southwood, provided invaluable assistance throughout the entire process. Their dedication to providing a quality educational product is evident throughout this text. I would also like to thank the Publisher, Michael Hackett, for the opportunity to contribute to the scientific literacy of our students.

Fresh, appealing new photos are a feature of this book, which students and professors alike will enjoy. Jo Johnson and Lori Hancock did a superb job of finding just the right photographs and micrographs. The design of the book is the result of the creative talents of Laurie Janssen and many others who assisted in deciding the appearance of each element in the text. Marketing Manager Tamara Maury directed the marketing team whose work is second to none. Eric Weber, the Digital Product Manager, played an integral role in developing the ConnectPlus resources for this text.

I am extremely grateful to my contributing author for this edition, Lynn Preston of Tarrant County College. Lynn assisted me with this project from beginning to end. Together with the editorial team, she supplied ideas and content for the many updates, new features, and new illustrations that enrich this twelfth edition of *Human Biology*. Finally, this edition of *Human Biology* would not be of the same excellent quality without the suggested changes from the many reviewers listed in the following sections.

Who I am, as an educator and an author, is a direct reflection of what I have learned from my students. Education is a two-way street, and it is my honest opinion that both my professional life and my personal life have been enriched by interactions with my students. They have encouraged me to learn more, teach better, and never stop questioning the world around me. As is the case with educators, we strive to make the world a better place, and I thank McGraw-Hill Higher Education, Dr. Sylvia Mader, and especially my wife Sandra for the opportunity to make this dream a reality.

Michael Windelspecht, Ph.D.
Blowing Rock, North Carolina

360° Development

McGraw-Hill's **360° Development Process** is an ongoing, never-ending, market-oriented approach to building accurate and innovative print and digital products. It is dedicated to continual large-scale and incremental improvement driven by multiple customer feedback loops and checkpoints. This is initiated during the early planning stages of our new products, and intensifies during the development and production stages, then begins again upon publication in anticipation of the next edition.

This process is designed to provide a broad, comprehensive spectrum of feedback for refinement and innovation of our learning tools, for both student and instructor. The 360° Development Process includes market research, content reviews, course- and product-specific symposia, accuracy checks, and art reviews. We appreciate the expertise of the many individuals involved in this process.

Ancillary Authors

Test Bank and Instructor's Manual: Kimberly Lyle-Ippolito, *Anderson University*
Lecture Outlines/Image PowerPoints: Lynn Preston, *Tarrant County College*
FlexArt Manuscript: Sharon Thoma, *University of Wisconsin—Madison*
eBook Quizzes: Jennifer Burtwistle, *Northeast Community College*
Connect Question Bank:
Krissy Johnson, *Appalachian State University*
Alex James, *Appalachian State University*

Twelfth Edition Reviewers

Michael Adams, *Pasco-Hernando Community College*
Bill Bassman, *Touro College*
Erwin Bautista, *University of California, Davis*
Mario Ciani, *Mercy College*
Frank Conrad, *Metropolitan State College of Denver*
Christina Costa, *Mercy College*
Angela Crocker, *Erie Community College*
Maria Dell, *Santa Monica College*
Elizabeth Desy, *Southwest Minnesota State University*
Miriam Flaum, *Touro College*
Ferdinand Gomez, *Florida International University*
Cole Hawkins, *Solano Community College*
Michael Kalafatis, *Cleveland State University*
Pushkar Kaul, *Clark Atlanta University*
Mary Jane Keith, *Wichita State University*
Edwin Klibaner, *Touro College*
Carol Mack, *Erie Community College, North Campus*
Jennifer McCoy, *Wichita State University*
Nicole Okazaki, *Weber State University*
Robert Okazaki, *Weber State University*
Chuma Okere, *Clark Atlanta University*
Jill O'Malley, *Erie Community College, City Campus*
Jacqueline Pal, *California State University*
Daniel Peña, *Dutchess Community College*
Linda Peters, *Holyoke Community College*
E. Sarahi Ramirez, *Harrisburg Area Community College*
Susan Rohde, *Triton College*
Mary Ann Sadler, *St. Charles Community College*
Tobili Sam-Yellowe, *Cleveland State University*

Twelfth Edition Reviewers

Roy Silcox, *Brigham Young University*
Joshua Smith, *Missouri State University*
Robert Smith, *McHenry County College*
Mike Squires, *Columbus State Community College*
Francis Sullivan, *Metropolitan State College of Denver*
Kent Thomas, *Wichita State University*
Michael Troyan, *Pennsylvania State University*
Wendy Vermillion, *Columbus State Community College*
David Waddell, *California State University*
Miryam Wahrman, *William Paterson University*
Robert Wiggers, *Stephen F. Austin State University*
Jessica Wooten, *Franklin University*
Amber Wyman, *Finger Lakes Community College*
Dianne York, *Lincoln University of Pennsylvania*
Marlena Yost, *Mississippi State University*

Eleventh Edition Reviewers

Tamatha R. Barbeau, *Francis Marion University*
Bill Radley Bassman, *Touro College, Stern College*
Frank J. Conrad, *Metropolitan State College of Denver*
Valentina David, *Bethune-Cookman University*
Maria M. Dell, *Santa Monica College*
Charles J. Dick, *Pasco-Hernando Community College*
Thomas J. Franco, *Erie Community College, North Campus*
Judith E. Goedert, *City College of San Francisco*
Melodye Gold, *Bellevue Community College*
Mary Louise Greeley, *Salve Regina University*
Virginia Gutierrez-Osborne, *Fresno City College*
Martin Hahn, *William Paterson University*
Rebecca J. Heick, *St. Ambrose University*
Jonathan P. Hubbard, *Hartnell College*
Edwin Klibaner, *Touro College*
Robert A. Krebs, *Cleveland State University*
Nicole Okazaki, *Weber State University*
Phillip A. Ortiz, *Empire State College, State University of New York*
Polly K. Phillips, *Florida International University*
Nancy K. Prentiss, *University of Maine at Farmington*
Nicholas Roster, *Northwestern Michigan College*
Megan E. Thomas, *University of Nevada, Las Vegas*
Wendy Vermillion, *Columbus State Community College*
Jagan Valluri, *Marshall University*

Previous Edition Reviewers and Contributors

Rita Alisauskas, *County College of Morris*
Deborah Allen, *Jefferson College*
Elizabeth Balko, *SUNY-Oswego*
Tamatha R. Barbeau, *Francis Marion University*
Marilynn R. Bartels, *Black Hawk College*
Erwin A. Bautista, *University of California, Davis*
Robert D. Bergad, *Metropolitan State University*
Hessel Bouma III, *Calvin College*
Frank J. Conrad, *Metropolitan State College of Denver*
William Cushwa, *Clark College*
Debbie A. Zetts Dalrymple, *Thomas Nelson Community College*
Diane Dembicki, *Dutchess Community College*
Charles J. Dick, *Pasco-Hernando Community College*
Kristiann M. Dougherty, *Valencia Community College*
David A. Dunbar, *Cabrini College*
William E. Dunscombe, *Union County College*
David Foster, *North Idaho College*
David E. Fulford, *Edinboro University of Pennsylvania*
Sandra Grauer, *Limestone College*
Mary Louise Greeley, *Salve Regina University*
Esta Grossman, *Washtenaw Community College*
Gretel M. Guest, *Alamance Community College*
Martin E. Hahn, *William Paterson University*
Rosalind C. Haselbeck, *University of San Diego*
Timothy P. Hayes, *Marshall University*
Mark F. Hoover, *Penn State Altoona*
Anna K. Hull, *Lincoln University*
Laurie A. Johnson, *Bay College*
Mary King Kananen, *Penn State Altoona*
Patricia Klopfenstein, *Edison Community College*
J. Kevin Langford, *Stephen F. Austin State University*
Lee H. Lee, *Montclair State University*
Edwin Lephart, *Brigham Young University*
Martin A. Levin, *Eastern Connecticut State University*
Nardos Lijam, *Columbus State Community College*
William J. Mackay, *Edinboro University of Pennsylvania*
Terry R. Martin, *Kishwaukee College*
Deborah J. McCool, *Penn State Altoona*
V. Christine Minor, *Clemson University*
Nick Nagle, *Metropolitan State College of Denver*
Roger C. Nealeigh, *Central Community College-Hastings*
Polly K. Phillips, *Florida International University*
Shawn G. Phippen, *Valdosta State University*
Mason Posner, *Ashland University*
Donna R. Potacco, *William Paterson University*
Mary Celeste Reese, *Mississippi State University*
Jill D. Reid, *Virginia Commonwealth University*
Kay Rezanka, *Central Lakes College*
April L. Rottman, *Rock Valley College*
Deborah B. Schulman, *Cleveland State University*
Lois Sealy, *Valencia Community College*
Jia Shi, *Skyline College*
Mark Smith, *Chaffey College*
Alicia Steinhardt, *Hartnell Community College West Valley Community College*
Lei Lani Stelle, *Rochester Institute of Technology*
Kenneth Thomas, *Northern Essex Community College*
Chad Thompson, *SUNY-Westchester Community College*
Jamey Thompson, *Hudson Valley Community College*
Doris J. Ward, *Bethune-Cookman College*
Susan Weinstein, *Marshall University*

Dave Cox, *Lincoln Land Community College*
Patrick Galliart, *North Iowa Area Community College*
Sandra Grauer, *Limestone College*
Sharron Jenkins, *Purdue University North Central*
Jill Kolodsick, *Washtenaw Community College*
Edwin Lephart, *Brigham Young University*
Susannah Nelson Longenbaker, *Columbus State Community College*
Debbie J. McCool, *Penn State Altoona*
Jodi Rymer, *Christine Wildsoet Laboratory University of California, Berkeley*
Linda D. Smith-Staton, *Pellissippi State Technical Community College*
Linda Strause, *University of California–San Diego*
Michael Thompson, *Middle Tennessee State University*

Contents

Preface v
Student's Guide viii
Teaching and Learning Tools xiv
Content Changes xvi
Acknowledgments xviii
Readings xxiii

CHAPTER 1

Exploring Life and Science 1

1.1 The Characteristics of Life 2
1.2 Humans Are Related to Other Animals 6
1.3 Science as a Process 9
1.4 Making Sense of a Scientific Study 13
1.5 Science and Social Responsibility 14

PART I

Human Organization 19

CHAPTER 2

Chemistry of Life 19

2.1 From Atoms to Molecules 20
2.2 Water and Living Things 24
2.3 Molecules of Life 28
2.4 Carbohydrates 29
2.5 Lipids 31
2.6 Proteins 34
2.7 Nucleic Acids 37

CHAPTER 3

Cell Structure and Function 43

3.1 What Is a Cell? 44
3.2 How Cells Are Organized 46
3.3 The Plasma Membrane and How Substances Cross It 49
3.4 The Nucleus and Endomembrane System 53
3.5 The Cytoskeleton, Cell Movement, and Cell Junctions 55
3.6 Mitochondria and Cellular Metabolism 57

CHAPTER 4

Organization and Regulation of Body Systems 65

4.1 Types of Tissues 66
4.2 Connective Tissue Connects and Supports 66
4.3 Muscular Tissue Moves the Body 69
4.4 Nervous Tissue Communicates 70
4.5 Epithelial Tissue Protects 72
4.6 Integumentary System 75
4.7 Organ Systems, Body Cavities, and Body Membranes 79
4.8 Homeostasis 84

PART II

Maintenance of the Human Body 91

CHAPTER 5

Cardiovascular System: Heart and Blood Vessels 91

5.1 Overview of the Cardiovascular System 92
5.2 The Types of Blood Vessels 93
5.3 The Heart Is a Double Pump 94
5.4 Features of the Cardiovascular System 100
5.5 Two Cardiovascular Pathways 103
5.6 Exchange at the Capillaries 104
5.7 Cardiovascular Disorders 105

CHAPTER 6

Cardiovascular System: Blood 115

6.1 Blood: An Overview 116
6.2 Red Blood Cells and Transport of Oxygen 118
6.3 White Blood Cells and Defense Against Disease 121
6.4 Platelets and Blood Clotting 123
6.5 Blood Typing and Transfusions 125
6.6 Homeostasis 127

CHAPTER 7

Lymphatic System and Immunity 133

7.1 Microbes, Pathogens, and You 134
7.2 The Lymphatic System 137
7.3 Innate Defenses 140
7.4 Acquired Defenses 143
7.5 Acquired Immunity 147
7.6 Hypersensitivity Reactions 150

INFECTIOUS DISEASES SUPPLEMENT 155

S.1 AIDS and Other Pandemics 156
S.2 Emerging Diseases 166
S.3 Antibiotic Resistance 167

CHAPTER 8

Digestive System and Nutrition 169

8.1 Overview of Digestion 170
8.2 First Part of the Digestive Tract 172
8.3 The Stomach and Small Intestine 175
8.4 Three Accessory Organs and Regulation of Secretions 178
8.5 The Large Intestine and Defecation 180
8.6 Nutrition and Weight Control 183

CHAPTER 9

Respiratory System 196

9.1 The Respiratory System 197
9.2 The Upper Respiratory Tract 198
9.3 The Lower Respiratory Tract 200
9.4 Mechanism of Breathing 201
9.5 Control of Ventilation 205
9.6 Gas Exchanges in the Body 206
9.7 Respiration and Health 208

CHAPTER 10

Urinary System 217

10.1 The Urinary System 218
10.2 Kidney Structure 220
10.3 Urine Formation 225
10.4 Kidneys and Homeostasis 228
10.5 Kidney Function Disorders 233

PART III

Movement and Support in Humans 239

CHAPTER 11

Skeletal System 239

11.1 Overview of the Skeletal System 240
11.2 Bone Growth, Remodeling, and Repair 242
11.3 Bones of the Axial Skeleton 248
11.4 Bones of the Appendicular Skeleton 253
11.5 Articulations 255

CHAPTER 12

Muscular System 262

12.1 Overview of the Muscular System 263
12.2 Skeletal Muscle Fiber Contraction 267
12.3 Whole Muscle Contraction 271
12.4 Muscular Disorders 276
12.5 Homeostasis 278

PART IV

Integration and Coordination in Humans 285

CHAPTER 13

Nervous System 285

13.1 Overview of the Nervous System 286
13.2 The Central Nervous System 293
13.3 The Limbic System and Higher Mental Functions 299
13.4 The Peripheral Nervous System 302
13.5 Drug Therapy and Drug Abuse 306

CHAPTER 14

Senses 315

14.1 Overview of Sensory Receptors and Sensations 316
14.2 Proprioreceptors, Cutaneous Receptors, and Pain Receptors 318
14.3 Senses of Taste and Smell 320
14.4 Sense of Vision 322
14.5 Sense of Hearing 328
14.6 Sense of Equilibrium 332

CHAPTER 15

Endocrine System 339

15.1 Endocrine Glands 340
15.2 Hypothalamus and Pituitary Gland 345
15.3 Thyroid and Parathyroid Glands 349
15.4 Adrenal Glands 351
15.5 Pancreas 354
15.6 Other Endocrine Glands 358
15.7 Homeostasis 360

PART V

Reproduction in Humans 365

CHAPTER 16

Reproductive System 365

16.1 Human Life Cycle 366
16.2 Male Reproductive System 367
16.3 Female Reproductive System 371
16.4 The Ovarian Cycle 374
16.5 Control of Reproduction 379
16.6 Sexually Transmitted Diseases 385

CHAPTER 17

Development and Aging 393

17.1 Fertilization 394
17.2 Pre-Embryonic and Embryonic Development 395
17.3 Fetal Development 401
17.4 Pregnancy and Birth 407
17.5 Development After Birth 410

PART VI

Human Genetics 419

CHAPTER 18

Patterns of Chromosome Inheritance 419

18.1 Chromosomes 420
18.2 The Cell Cycle 422
18.3 Mitosis 424
18.4 Meiosis 428
18.5 Comparison of Meiosis and Mitosis 433
18.6 Chromosome Inheritance 437

CHAPTER 19

Cancer 447

19.1 Cancer Cells 448
19.2 Causes and Prevention of Cancer 452
19.3 Diagnosis of Cancer 455
19.4 Treatment of Cancer 459

CHAPTER 20

Patterns of Genetic Inheritance 467

20.1 Genotype and Phenotype 468
20.2 One- and Two-Trait Inheritance 469
20.3 Inheritance of Genetic Disorders 476
20.4 Beyond Simple Inheritance Patterns 479
20.5 Sex-Linked Inheritance 484

CHAPTER 21

DNA Biology and Technology 491

21.1 DNA and RNA Structure and Function 492
21.2 Gene Expression 496
21.3 DNA Technology 503
21.4 Genomics 511

PART VII

Human Evolution and Ecology 517

CHAPTER **22**

Human Evolution 517

22.1 Origin of Life 518
22.2 Biological Evolution 520
22.3 Classification of Humans 526
22.4 Evolution of Hominins 530
22.5 Evolution of Humans 533

CHAPTER **23**

Global Ecology and Human Interferences 545

23.1 The Nature of Ecosystems 547
23.2 Energy Flow 550
23.3 Global Biogeochemical Cycles 552

CHAPTER **24**

Human Population, Planetary Resources, and Conservation 567

24.1 Human Population Growth 568
24.2 Human Use of Resources and Pollution 570
24.3 Biodiversity 579
24.4 Working Toward a Sustainable Society 586

Appendix A:
Periodic Table of the Elements A-1

Appendix B:
Answer Key A-2

Glossary G-1
Credits C-1
Index I-1

Readings

Biology Matters **Bioethical Focus**

Stem-Cell Research 61
Cardiovascular Disease Prevention: Who Pays for an Unhealthy Lifestyle? 110
Bans on Smoking 209
Anabolic Steroid Use 280
Medical Marijuana Use 309
Noise Pollution 331
Using Growth Hormones to Treat Pituitary Dwarfism 348
Male and Female Circumcision 373
Should Infertility Be Treated? 383
The Differences Between Reproductive and Therapeutic Cloning 404
Selecting Children 443
Preimplantation Genetic Diagnosis 475
DNA Fingerprinting and the Criminal Justice System 505
Effects of Biocultural Evolution on Population Growth 538
Guaranteeing Access to Safe Drinking Water 556

Biology Matters **Health Focus**

Fiber in the Diet 31
The Omega-3 Fatty Acids 33
Good and Bad Cholesterol 34
Lactate and the Athlete 60
Pursuing Youthful Skin 83
What to Know When Giving Blood 124
Heartburn (GERD) 174
Swallowing a Camera 182
Searching for a Magic Weight-Loss Bullet 184
When Zero Is More Than Nothing 188
Questions About Smoking, Tobacco, and Health 212
Urinary Difficulties Due to an Enlarged Prostate 222
Urinalysis 226
You Can Avoid Osteoporosis 246
Exercise, Exercise, Exercise 275
Correcting Vision Problems 327
Preventing Transmission of STDs 388
Alzheimer Disease 412
Prevention of Cancer 454
Shower Check for Cancer 457
Are Genetically Engineered Foods Safe? 510

Biology Matters **Science Focus**

Robert Koch 10
Coloring Organisms Green; Green Fluorescent Proteins and Cells 46
Nerve Regeneration and Stem Cells 72
Face Transplantation 78
The Challenges of Developing an AIDS Vaccine 163
Lab-Grown Bladders 235
Identifying Skeletal Remains 252
Rigor Mortis 276
Mosaics, Barr Bodies, and Breast Cancer 438
Homo floresiensis 534
Ozone Shield Depletion 562
Mystery of the Vanishing Bees 584

Biology Matters **Historical Focus**

The Syphilis Research Scandal of Tuskegee University 16
Surgeon Without a Degree: Vivien Theodore Thomas (1910–1985) 107
Making Blood Transfusion Possible: Karl Landsteiner (1868–1943) 128
Mary Mallon: The Most Dangerous Woman in America 136
Osteoarthritis and Joint-Replacement Surgery 257
Iron Horse: Lou Gehrig (1903–1941) 277
Artist and Scientist: Santiago Ramon y Cajal (1852–1934) 292
Surviving Diabetes Mellitus 357
An End to "Laudable Pus" 409
The Immortal Henrietta Lacks 462
Hemophilia: The Royal Disease 482
Overlooked Genius: Rosalind Franklin 493

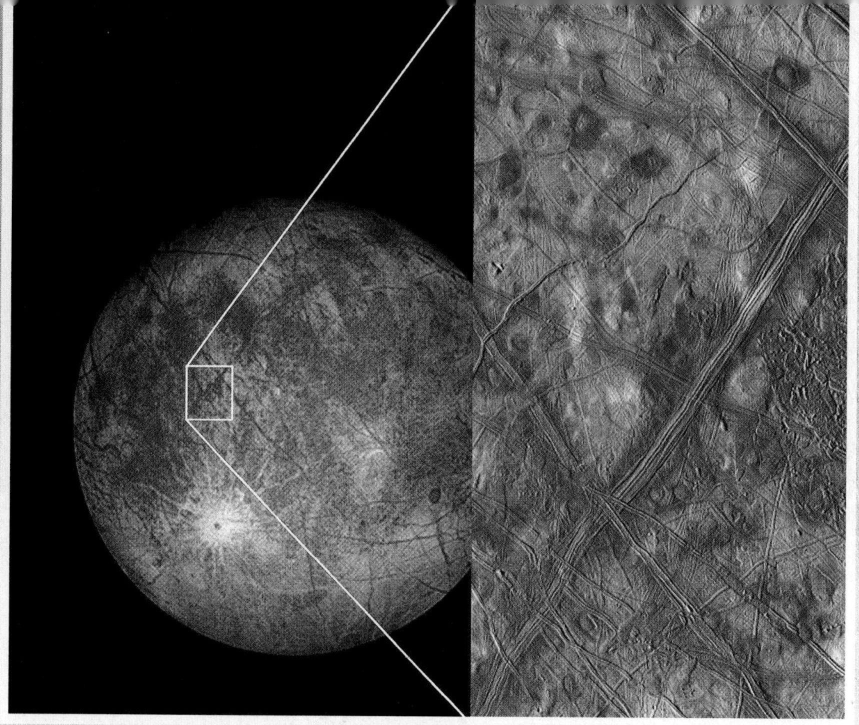

CHAPTER

1

Exploring Life and Science

CHAPTER CONCEPTS

1.1 The Characteristics of Life
The process of evolution accounts for the diversity of living things and why living things share the same basic characteristics of life.

1.2 Humans Are Related to Other Animals
Humans are eukaryotes and are further classified as mammals in the animal kingdom. We differ from other mammals, including apes, by our highly developed brain, upright stance, creative language, and the ability to use a wide variety of tools.

1.3 Science as a Process
Biologists use the scientific process when they study the natural world. A hypothesis is formulated and tested to arrive at a conclusion. Theories explain how the natural world is organized.

1.4 Making Sense of a Scientific Study
Data are more easily understood if results are presented in the form of a graph and are accompanied by a statistical analysis.

1.5 Science and Social Responsibility
Scientific investigations and technology have always been influenced by human values. Everyone has a responsibility to ensure that science and technology are used for the good of all.

CASE STUDY THE SEARCH FOR LIFE

What do Europa, Titan, and Earth all have in common? Besides being part of our solar system, they are all at the frontline of our species' effort to understand the nature of life.

Europa is one of the larger moons of Jupiter, and it has had held a special fascination for astronomers since Galileo first described it in 1610. Today, Europa is one of the prime candidates to harbor life outside of Earth. Geologists believe that under Europa's icy exterior lies a vast ocean of water. Having analyzed past comet impact sites on the surface of Europa, geologists also believe that this ocean contains the basic ingredients for life, including carbon and possibly even free oxygen. Europa's ocean is warmed by a constant tug of war between it and Jupiter, and the ocean is protected by an ice casing almost 19 kilometers thick.

Titan is the second-largest satellite in the solar system, larger than even our moon. Although it is in orbit around Saturn, and thus located some distance from the influence of the sun, Titan has become a focal point for the study of extraterrestrial life since the NASA space probe *Cassini–Huygens* first arrived at Saturn in 2004. *Cassini* has detected the presence of the building blocks of life on Titan, including lakes of methane and ammonia, and vast deposits of hydrogen and carbon compounds called hydrocarbons.

On Earth, scientists are exploring the extreme environments near volcanoes and deep-sea thermal vents to get a better picture of what life may have looked like under the inhospitable conditions that dominated at the time we now know life first began on our planet. Already, in the past few years, marine biologists have discovered new forms of life that cannot only live off of hydrogen sulfide, a deadly gas to most life, but also thrive under extreme pressure and temperatures.

As you read through the chapter, think about the following questions.

1. What are the basic characteristics that define life?
2. What evidence would you look for on Europa or Titan that would tell you that life may have existed on these moons in the past?
3. What does it tell us if we discover life on Europa or Titan and it has similar characteristics to life on Earth? What if it is very different?

1.1 The Characteristics of Life

Learning Outcomes

Upon completion of this section, you should be able to

1. Explain the basic characteristics that are common to all living things.
2. Describe the levels of organization of life.
3. Summarize how the terms *homeostasis, metabolism, development,* and *adaptation* all relate to living organisms.
4. Recognize the special relationship between life and evolution.

The science of **biology** is the study of living organisms and their environments. All living things (Fig. 1.1) share seven basic characteristics. Living things (1) are organized, (2) acquire materials and energy, (3) reproduce, (4) grow and develop, (5) are homeostatic, (6) respond to stimuli, and (7) have an evolutionary history.

Living Things Are Organized

Figure 1.2 illustrates that **atoms** join together to form the **molecules** that make up a cell. A **cell** is the smallest structural and functional unit of an organism. Some organisms are single cells. Humans are **multicellular** because they are composed of many different types of cells. A nerve cell is one of the types of cells in the human body. It has a structure suitable to conducting a nerve impulse.

A **tissue** is a group of similar cells that perform a particular function. Nervous tissue is composed of millions of nerve cells that transmit signals to all parts of the body. Several types of tissues make up an **organ,** and each organ belongs to an **organ system.** The organs of an organ system work together to accomplish a common purpose. The brain works with the spinal cord to send commands to body parts by way of nerves. **Organisms,** such as trees and humans, are a collection of organ systems.

The levels of biological organization extend beyond the individual. All the members of one **species** (group of interbreeding organisms) in a particular area belong to a **population.** A tropical grassland may have a population of zebras, acacia trees, and humans, for example. The interacting populations of the grasslands make up a **community.** The community of populations interacts with the physical environment to form an **ecosystem.** Finally, all the Earth's ecosystems make up the **biosphere.**

Figure 1.1 All life shares common characteristics.
From the simplest one-celled organisms to complex plants and animals, all life shares seven basic characteristics.

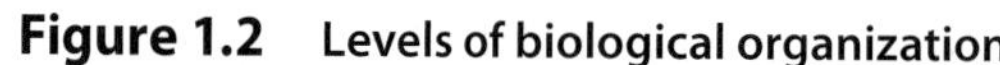

Figure 1.2 Levels of biological organization.
Living organisms are organized. The smallest unit of living organisms is the cell. The sum of all living things—and the locations that they inhabit—is called the biosphere.

Biosphere
Regions of the Earth's crust, waters, and atmosphere inhabited by living things

↑

Ecosystem
A community plus the physical environment

↑

Community
Interacting populations in a particular area

↑

Population
Organisms of the same species in a particular area

↑

Organism
An individual; complex individuals contain organ systems

↑

Organ System
Composed of several organs working together

↑

Organ
Composed of tissues functioning together for a specific task

↑

Tissue
A group of cells with a common structure and function

↑

Cell
The structural and functional unit of all living things

Molecule
Union of two or more atoms of the same or different elements

Atom
Smallest unit of an element composed of electrons, protons, and neutrons

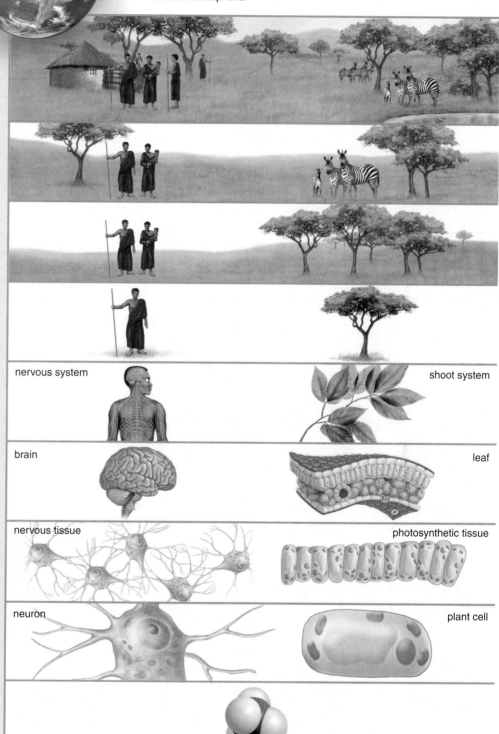

Living Things Acquire Materials and Energy

Humans, like all living organisms, cannot maintain their organization or carry on life's activities without an outside source of materials and energy. Humans and other animals acquire materials and energy when they eat food (Fig. 1.3).

Food provides nutrient molecules, which are used as building blocks or for energy. It takes energy (work) to maintain the organization of the cell and of the organism. Some nutrient molecules are broken down completely to provide the necessary energy to convert other nutrient molecules into the parts and products of cells. The term **metabolism** describes all of the chemical reactions that occur within a cell.

> **Connections and Misconceptions**
>
> **How many cells are in your body?**
>
> The number of cells in a human body varies depending on the size of the person and whether cells have been damaged or lost. However, most estimates suggest that there are well over 100 trillion cells in a human body.

a.

b.

c.

Figure 1.3 Humans and other animals must acquire energy. **a.** Humans eat plants and animals they raise for food. **b.** A red-tailed hawk captures prey to feed its young. **c.** Some animals feed only on plants, such as these grazing sheep.

The ultimate source of energy for the majority of life on Earth is the sun. Plants, algae, and some bacteria are able to harvest the energy of the sun and convert it to chemical energy by a process called **photosynthesis.** Photosynthesis produces organic molecules, such as sugars, that serve as the basis of the food chain for many other organisms, including humans and all other animals.

Life Is Homeostatic

For the metabolic pathways within a cell to function correctly, the environmental conditions of the cell must be kept within strict operating limits. The ability of a cell or an organism to maintain an internal environment that operates under specific conditions is called **homeostasis.** In humans, many of our organ systems work to maintain homeostasis. For example, human body temperature normally fluctuates slightly between 36.5 and 37.5°C (97.7 and 99.5°F) during the day. In general, the lowest temperature usually occurs between 2 A.M. and 4 A.M., and the highest usually occurs between 6 P.M. and 10 P.M. However, activity can cause the body temperature to rise, and inactivity can cause it to decline. The cardiovascular system and the nervous system work together to maintain a constant temperature. However, the body's ability to maintain a normal temperature is somewhat dependent on the external temperature. Even though we can shiver when we are cold and perspire when we are hot, we will die if the external temperature becomes overly cold or hot.

This text emphasizes how all the systems of the human body help maintain homeostasis. The digestive system takes in nutrients, and the respiratory system exchanges gases with the environment. The cardiovascular system distributes nutrients and oxygen to the cells and picks up their wastes. The metabolic waste products of cells are excreted by the urinary system. The work of the nervous and endocrine systems is critical because these systems coordinate the functions of the other systems. Throughout the text, the Connecting the Concepts sections will provide you with links to more information on homeostasis.

Living Things Respond to Stimuli

Homeostasis would be impossible without the ability of the body to respond to stimuli. Response to external stimuli is more apparent to us because it does involve movement, as when we quickly remove a hand from a hot stove. Certain sensory receptors also detect a change in the internal environment, and then the central nervous system brings about an appropriate response. When you are startled by a loud noise, your heartbeat increases, which causes your blood pressure to increase. If blood pressure rises too high, the brain directs blood vessels to dilate, helping to restore normal blood pressure.

Living things respond to external stimuli, often by moving toward or away from a stimulus, such as the sight of food. Living things use a variety of mechanisms to move, but movement in humans and other animals is dependent upon their nervous and musculoskeletal systems. The leaves of plants track the passage of the sun during the day; when a houseplant is placed near a window, its stems bend to face the sun. The movement of an animal, whether self-directed or in response to a stimulus, constitutes a large part of its behavior. Some behaviors help us acquire food and reproduce.

Living Things Reproduce and Develop

Reproduction is a fundamental characteristic of life. Cells come into being only from pre-existing cells, and all living things have parents. When living things **reproduce,** they create a copy of themselves and ensure the continuance of their own kind. Following the fertilization of the egg by a sperm cell, the resulting zygote undergoes a rapid period of growth and development. This is common in almost all living organisms. Figure 1.4*a* illustrates that an acorn progresses to a seedling before it becomes an adult oak tree. In humans, growth occurs as the fertilized egg develops into a fetus (Fig. 1.4*b*). **Growth,** recognized by an increase in size and often the number of cells, is a part of development. In humans, **development** includes all the changes that occur from the time the egg is fertilized until death; therefore, it includes all the changes that occur during childhood, adolescence, and adulthood. Development also includes the repair that takes place following an injury.

The purpose of reproduction is to pass on a copy of the genetic information to the offspring. DNA contains the hereditary information that directs not only the structure of each cell but also its function. The information in the DNA is contained within **genes,** short sequences of hereditary material that specify the instructions for a specific trait. Before reproduction occurs, DNA is replicated so that an exact copy of each gene may be passed on to the offspring. When humans reproduce, a sperm carries genes contributed by a male into the egg, which contains genes contributed by a female. The genes direct both growth and development so that the organism will eventually resemble the parents. Sometimes, minor variations in these genes, called **mutations,** may result in an organism making it better suited for its environment. These mutations are the basis of evolutionary change.

Living Things Adapt and Evolve

Evolution is the process by which a species changes through time. When a new variation arises that allows certain members of the species to capture more resources, these members tend to survive and have more offspring than the other, unchanged members. Therefore, each successive generation will include

a.

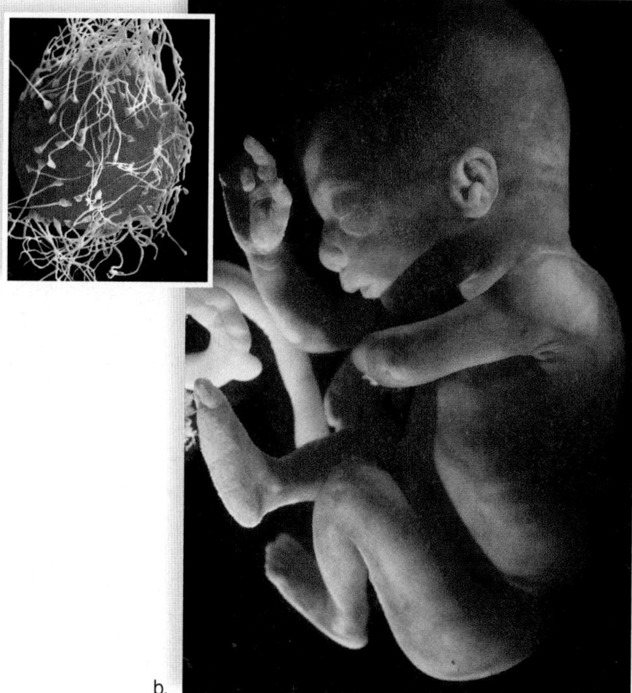

b.

Figure 1.4 Growth and development define life.
a. A small acorn becomes a tree, and **(b)** following fertilization, an embryo becomes a fetus by the process of growth and development.

more members with the new variation, which represents an **adaptation** to the environment. Consider, for example, a red-tailed hawk, which catches and eats rabbits. A hawk can fly, in part, because it has hollow bones to reduce its weight and flight muscles to depress and elevate its wings. When a hawk dives, its strong feet take the first shock of the landing and its long, sharp claws reach out and hold onto the prey. All these characteristics are a hawk's adaptations to its way of life.

Evolution, which has been going on since the origin of life and which will continue as long as life exists, explains both the unity and the diversity of life. All organisms share the same characteristics of life because their ancestry can be traced to the first cell or cells. Organisms are diverse because they are adapted to different ways of life.

Connecting the Concepts

Both homeostasis and evolution are central themes in the study of biology. For more examples of homeostasis and evolution, refer to the following discussions.

Section 4.8 explains how body temperature is regulated.

Section 9.5 investigates how the nervous system controls the rate of respiration.

Section 10.4 explores the role of the kidneys in fluid and salt homeostasis.

Check Your Progress 1.1

1. Describe the basic characteristics of life.
2. Summarize how each characteristic of life contributes to homeostasis.
3. Explain why living things are organized.

1.2 Humans Are Related to Other Animals

Learning Outcomes

Upon completion of this section, you should be able to

1. Summarize the place of humans in the overall classification of living organisms.
2. Describe the relationship between humans and the biosphere, and the role of culture in shaping that relationship.

Biologists classify living things as belonging to one of three **domains**. The evolutionary relationships of these domains are presented in Figure 1.5. Two of these domains, domain Bacteria and domain Archaea, contain prokaryotes,

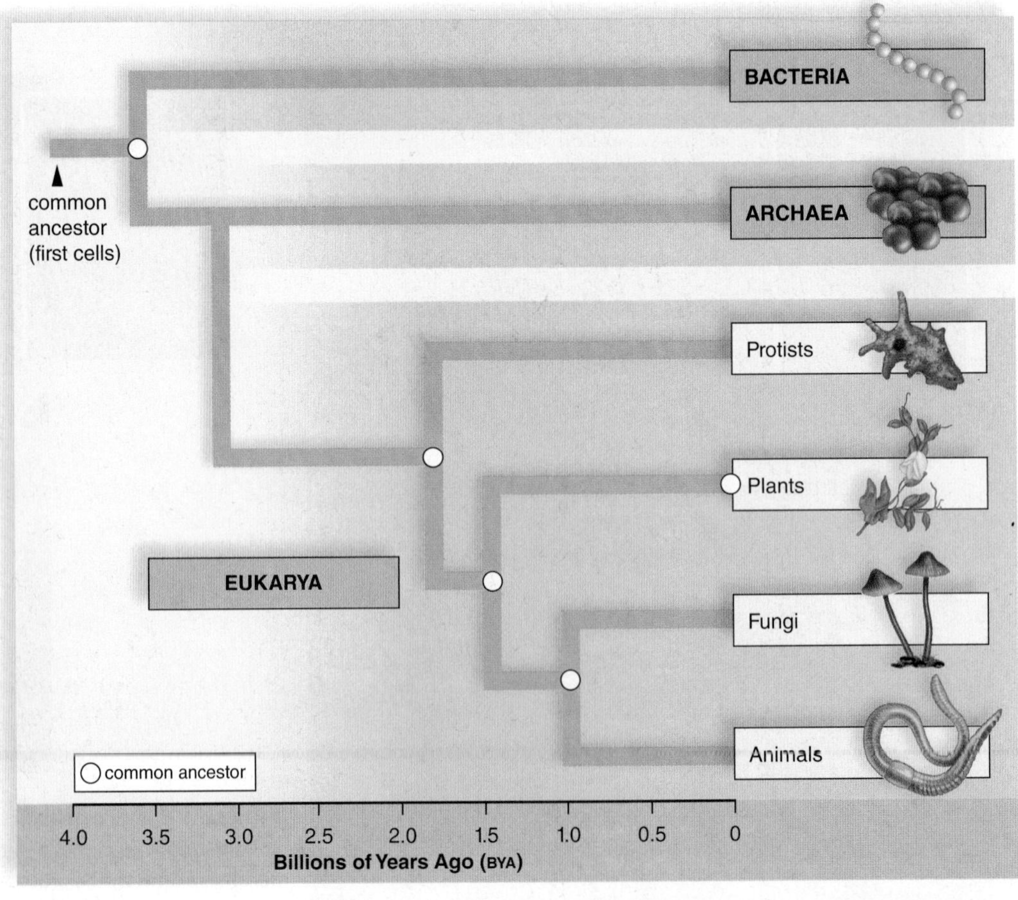

Figure 1.5 The evolutionary relationships of the three domains of life.
Living organisms are classified into three domains: Bacteria, Archaea, and Eukarya. A geologic time scale is provided on the bottom for reference.

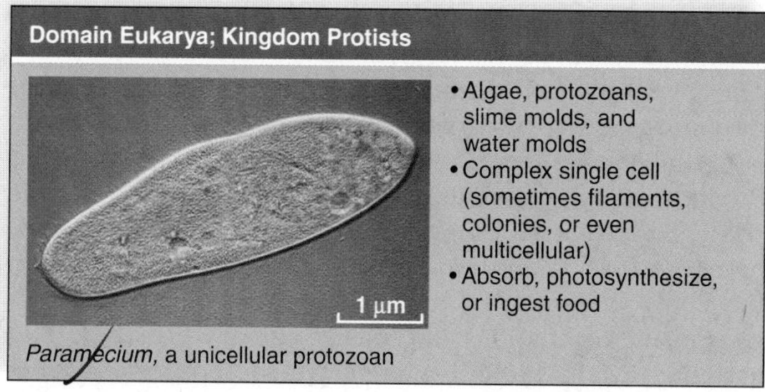

Domain Eukarya; Kingdom Protists
1 µm
• Algae, protozoans, slime molds, and water molds
• Complex single cell (sometimes filaments, colonies, or even multicellular)
• Absorb, photosynthesize, or ingest food

Paramecium, a unicellular protozoan

Domain Eukarya; Kingdom Animals
• Sponges, worms, insects, fishes, frogs, turtles, birds, and mammals
• Multicellular with specialized tissues containing complex cells
• Ingest food

Vulpes, a red fox

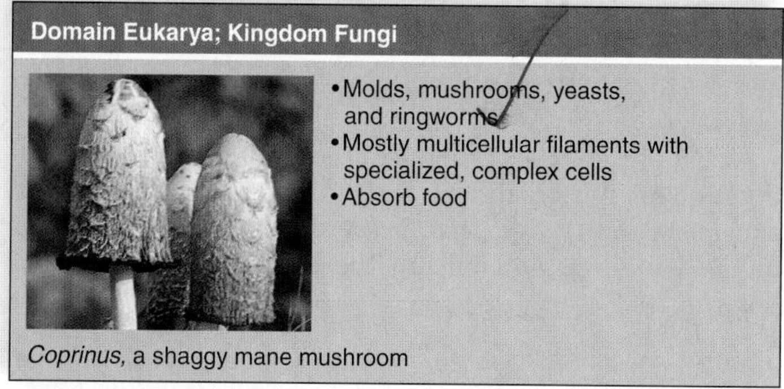

Domain Eukarya; Kingdom Fungi
• Molds, mushrooms, yeasts, and ringworms
• Mostly multicellular filaments with specialized, complex cells
• Absorb food

Coprinus, a shaggy mane mushroom

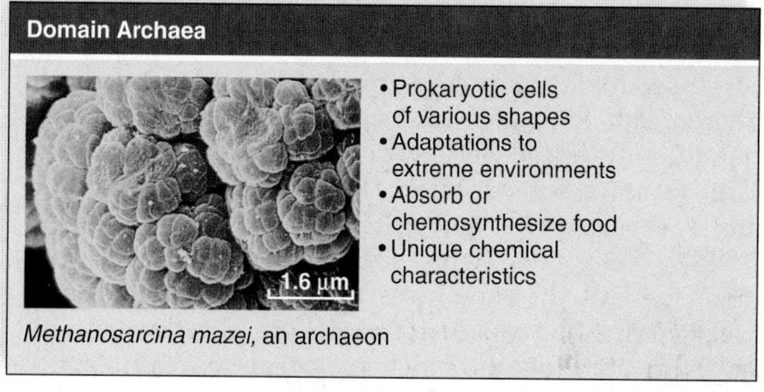

Methanosarcina mazei, an archaeon

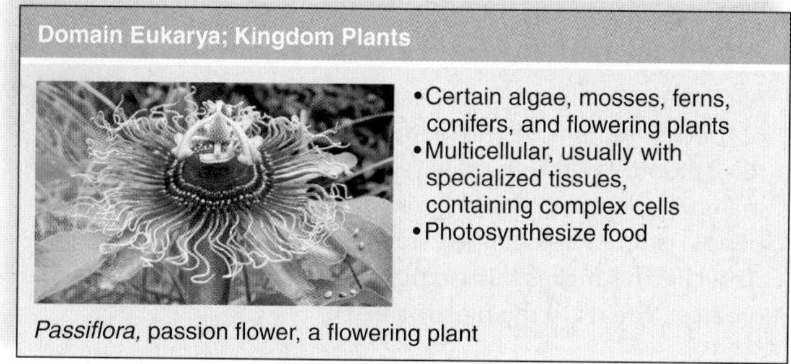

Domain Eukarya; Kingdom Plants
• Certain algae, mosses, ferns, conifers, and flowering plants
• Multicellular, usually with specialized tissues, containing complex cells
• Photosynthesize food

Passiflora, passion flower, a flowering plant

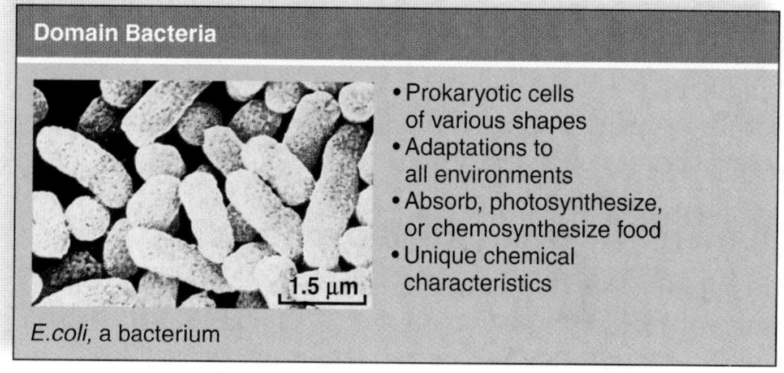

Domain Bacteria
1.5 µm
• Prokaryotic cells of various shapes
• Adaptations to all environments
• Absorb, photosynthesize, or chemosynthesize food
• Unique chemical characteristics

E.coli, a bacterium

Figure 1.6 The classification of life.
This figure provides some of the characteristics of the organisms of each of the major domains and kingdoms of life. Humans belong to the domain Eukarya and kingdom Animalia.

one-celled organisms that lack a nucleus. Organisms in the third domain, Eukarya, are classified as being members of one of four **kingdoms** (Fig. 1.6)—plants, fungi, animals, and protists. Most organisms in kingdom Animalia are **invertebrates,** such as the earthworm, insects, and mollusks. **Vertebrates** are animals that have a nerve cord protected by a vertebral column, which gives them their name. Fish, reptiles, amphibians, and birds are all vertebrates. Vertebrates with hair or fur and mammary glands are classified as mammals. Humans, raccoons, seals, and meerkats are examples of mammals.

Human beings are most closely related to apes. We are distinguished from apes by our (1) highly developed brains, (2) completely upright stance, (3) creative language, and (4) ability to use a wide variety of tools. Humans did not evolve from apes; apes and humans share a common, apelike ancestor.

Today's apes are our evolutionary cousins. Our relationship to apes is analogous to you and your first cousin being descended from your grandparents. We could not have evolved from our cousins because we are contemporaries—living on Earth at the same time.

Humans Have a Cultural Heritage

Human beings have a cultural heritage in addition to a biological heritage. **Culture** encompasses human activities and products passed on from one generation to the next outside of direct biological inheritance. Among animals, only humans have a language that allows us to communicate information and experiences symbolically. We are born without knowledge of an accepted way to behave, but we gradually acquire this knowledge by adult instruction and imitation of role models. Members of the previous generation pass on their beliefs, values, and skills to the next generation. Many of the skills involve tool use, which can vary from how to hunt in the wild to how to use a computer. Human skills have also produced a rich heritage in the arts and sciences. However, a society highly dependent on science and technology has its drawbacks as well. Unfortunately, this cultural development may mislead us into believing that humans are somehow not part of the natural world surrounding us.

Humans Are Members of the Biosphere

All living things on Earth are part of the biosphere, a living network that spans the surface of the Earth into the atmosphere and down into the soil and seas. Although humans can raise animals and crops for food, we depend on the environment for many services. Without microorganisms that decompose, the waste we create would soon cover the Earth's surface. Some species of bacteria can clean up pollutants like heavy metals and pesticides.

Freshwater ecosystems, such as rivers and lakes, provide fish to eat, drinking water, and water to irrigate crops. The water-holding capacity of forests prevents flooding, and the ability of forests and other ecosystems to retain soil prevents soil erosion. Many of our crops and prescription drugs were originally derived from plants that grew naturally in an ecosystem. Some human populations around the globe still depend on wild animals as a food source. And we must not forget that almost everyone prefers to vacation in the natural beauty of an ecosystem.

Humans Threaten the Biosphere

The human population tends to modify existing ecosystems for its own purposes (Fig. 1.7). Humans clear forests and grasslands to grow crops. Later, houses are built on what was once farmland. Clusters of houses become small towns that often grow into cities. The overuse of water supplies by large human populations can result in desertification, or the expansion of desert regions (Fig. 1.7*b*). Human activities have altered almost all ecosystems and reduced biodiversity (the number of different species present). The present **biodiversity** of our planet has been estimated to be as high as 15 million species. So far, under 2 million have been identified and named. It is estimated that we are now losing as many as 400 species per day due to human activities. Many biologists are alarmed about the present rate of **extinction** (death of a species). They believe it may eventually rival the rates of the five mass extinctions that occurred earlier in our planet's history. The dinosaurs became extinct during the last mass extinction 65 million years ago.

One of the major bioethical issues of our time is preservation of the biosphere and biodiversity. If we adopt a conservation ethic that preserves the biosphere and biodiversity, we will ensure the continued existence of our species.

a.

b.

Figure 1.7 Humans negatively influence many ecosystems. **a.** When humans build cities, diversity is lost. Notice the absence of a variety of plants/trees. **b.** An overuse of water resources can lead to desertification.

Connections and Misconceptions

How many humans are there?

As of the end of 2008, it was estimated that there were over 6.7 billion humans on the planet. Each of those humans needs food, shelter, clean water and air, and materials to maintain a healthy lifestyle. We add an additional 75 million people per year—that is like adding ten New York Cities per year! This makes human population growth one of the greatest threats to the biosphere.

Connecting the Concepts

To learn more about the preceding material, refer to the following discussions.

Chapter 22 examines recent developments in the study of human evoution.

Chapter 23 provides a more detailed look at ecosystems.

Chapter 24 details some of the emerging threats that humans pose to the biosphere.

Check Your Progress 1.2

1. Define the term *biosphere*.
2. Explain why it is important to know the evolutionary relationships between organisms.
3. Summarize how the increase in the human population affects our biosphere.

1.3 Science as a Process

Learning Outcomes

Upon completion of this section, you should be able to

1. Describe the general process of the scientific method.
2. Distinguish between a control group and an experimental group in a scientific test.
3. Recognize the importance of scientific journals in the reporting of scientific information.

Science is a way of knowing about the natural world. When scientists study the natural world, they aim to be objective, rather than subjective. Objective observations are supported by factual information, whereas subjective observations involve personal judgment. For example, the fat content of a particular food would be an objective observation of a nutritional study. Reporting about the good or bad taste of the food would be a subjective observation. It is difficult to make objective observations and conclusions because we are often influenced by our prejudices. Scientists must keep in mind that scientific conclusions can change because of new findings. New findings are often made because of recent advances in techniques or equipment.

Importance of Scientific Theories in Biology

Science is not just a pile of facts. The ultimate goal of science is to understand the natural world in terms of scientific theories. **Scientific theories** are concepts that tell us about the order and the patterns within the natural world—in other words, how the natural world is organized. For example, following are some of the basic theories of biology.

Theory	Concept
Cell	All organisms are composed of cells, and new cells only come from pre-existing cells.
Homeostasis	The internal environment of an organism stays relatively constant.
Genes	Organisms contain coded information that dictates their form, function, and behavior.
Ecosystem	Populations of organisms interact with each other and the physical environment.
Evolution	All organisms have a common ancestor, but each is adapted to a particular way of life.

Evolution is the unifying concept of biology because it makes sense of what we know about living things. For example, the theory of evolution enables scientists to understand the variety of living things and their relationships. It explains common structural features, physiology, patterns of development, and behaviors. The theory of evolution has been supported by so many observations and experiments for over a hundred years that some biologists refer to it as the **principle** of evolution. This term is the preferred terminology for theories generally accepted as valid by an overwhelming number of scientists.

The Scientific Method Has Steps

Unlike other types of information available to us, scientific information is acquired by a process known as the **scientific method.** The approach of individual scientists to their work is as varied as the scientists. For the sake of discussion, it is possible to speak of the scientific method as consisting of certain steps (Fig. 1.8).

After making initial observations, a scientist will, most likely, study any previous **data,** results and conclusions reported by previous research. Imagination and creative thinking also help a scientist formulate a **hypothesis.** The hypothesis becomes the basis for more observation and/or experimentation. The new data help a scientist come

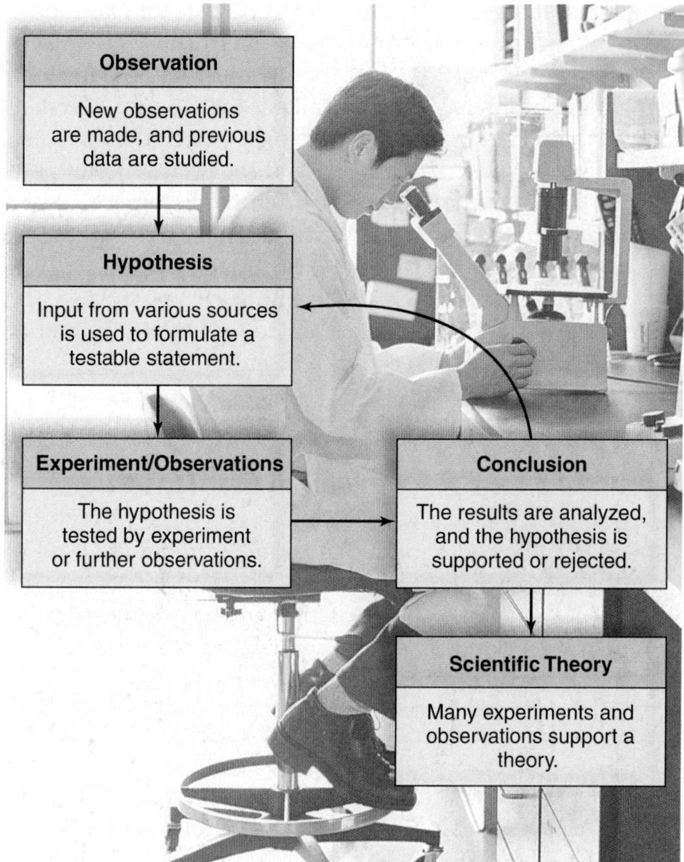

Figure 1.8 The scientific method.
On the basis of new and/or previous observations, a scientist formulates a hypothesis. The hypothesis is tested by further observations and/or experiments, and new data either support or do not support the hypothesis. The return arrow indicates that a scientist often chooses to retest the same hypothesis or to test a related hypothesis. Conclusions from many different but related experiments may lead to the developme of a scientific theory. For example, studies pertaining to developme anatomy, and fossil remains all support the theory of evolution

to a **conclusion** that either supports or does not support the hypothesis. Hypotheses are always subject to modification, so they can never be proven true; however, they can be proven untrue. When the hypothesis is not supported by the data, it must be rejected; therefore, some think of the body of science as what is left after alternative hypotheses have been rejected.

Science is different from other ways of knowing by its use of the scientific method to examine a phenomenon. Any suggestions about the natural world not based on data gathered by employing the scientific method cannot be accepted as within the realm of science. Scientific theories are concepts based on a wide range of observations and experiments.

How the Cause of Ulcers Was Discovered

Let's take a look at how the cause of ulcers was discovered so we can get a better idea of how the scientific method works. In 1974, Barry James Marshall was a young resident physician at Queen Elizabeth II Medical Center in Perth, Australia. There he saw many patients who had bleeding stomach ulcers. A pathologist at the hospital, Dr. J. Robin Warren, told him about finding a particular bacterium, now called *Helicobacter pylori,* near the site of peptic ulcers (open sores in the stomach). Using the computer networks available at that time, Marshall compiled much data showing a possible correlation between the presence of *Helicobacter pylori* and the occurrence of both gastritis (inflammation of the stomach) and stomach ulcers. On the basis of these data, Marshall formulated a hypothesis: *Helicobacter pylori* is the cause of gastritis and ulcers.

Marshall decided to make use of Koch's postulates, the standard criteria that must be fulfilled to show that a pathogen (bacterium or virus) causes a disease.

Koch's Postulates

- The suspected pathogen (virus or bacterium) must be present in every case of the disease;
- the pathogen must be isolated from the host and grown in a lab dish;
- the disease must be reproduced when a pure culture of the pathogen is inoculated into a healthy susceptible host; and
- the same pathogen must be recovered again from the experimentally infected host.

The First Two Criteria

By 1983, Marshall had fulfilled the first and second of Koch's criteria. He was able to isolate *Helicobacter pylori* from ulcer patients and grow it in the laboratory. (Success was achieved only after a petri dish was inadvertently left in the incubator for six, instead of two, days.) Further, he had determined that bismuth, the active ingredient in Pepto-Bismol, could destroy the bacteria in a petri dish.

Despite presentation of these findings to the scientific community, most physicians continued to believe that stomach acidity and stress were the causes of stomach ulcers. In those days, patients were usually advised to make drastic changes in their lifestyle or seek psychiatric counseling to "cure" their ulcers. Many scientists believed that no bacterium would be able to survive the normal acidity of the stomach.

The Last Two Criteria

Marshall had a problem in fulfilling the third and fourth of Koch's criteria. He had been unable to infect guinea pigs and rats with the bacteria because the bacteria just did not flourish in the intestinal tract of those animals. Marshall was not able to use human subjects because our society does not condone the use of humans as experimental subjects in dangerous or life-threatening research. Marshall was so determined to support his hypothesis that, in 1985, he decided to perform the experiment on himself! To the disbelief of those in the lab that day, he and another volunteer swallowed a foul-smelling and -tasting solution of *Helicobacter pylori.* Within the week, they felt lousy and were vomiting up their stomach contents. Examination by endoscopy showed that their stomachs were now inflamed, and biopsies of the stomach lining contained the suspected bacterium (Fig. 1.9). Their symptoms abated without need of medication, and they never developed an ulcer. At Marshall's next talk, he challenged his audience to refute his hypothesis. Many tried, but ultimately the investigators supported his findings.

The Conclusion

In science, many experiments that involve a considerable number of subjects are required before a conclusion can be reached. By the early 1990s, at least three independent studies involving hundreds of patients had been published showing that antibiotic therapy could eliminate *Helicobacter*

Biology Matters **Science Focus**

Robert Koch

Robert Koch (1843–1910) was a German microbiologist who helped verify the germ theory of disease and established the standard as to whether an organism causes a particular disease. Koch was a scientist who studied anthrax, a type of soil-living bacterium that causes disease in animals and humans. Anthrax was used as a biological weapon in the United States by unknown persons following the terrorist attacks in 2001. Koch expanded on Louis Pasteur's work by establishing a series of steps, now known as Koch's postulates, to identify whether a specific bacterium is responsible for a disease. His procedures are still widely used by scientists and medical professionals.

Koch is credited not only with his work on the germ theory but also with the development of several important scientific instruments. He was one of the first to use an incubator, an instrument that maintains a steady temperature for growing bacteria outside of an organism. He also was the first to use gelatin as a medium for growing bacteria in dishes.

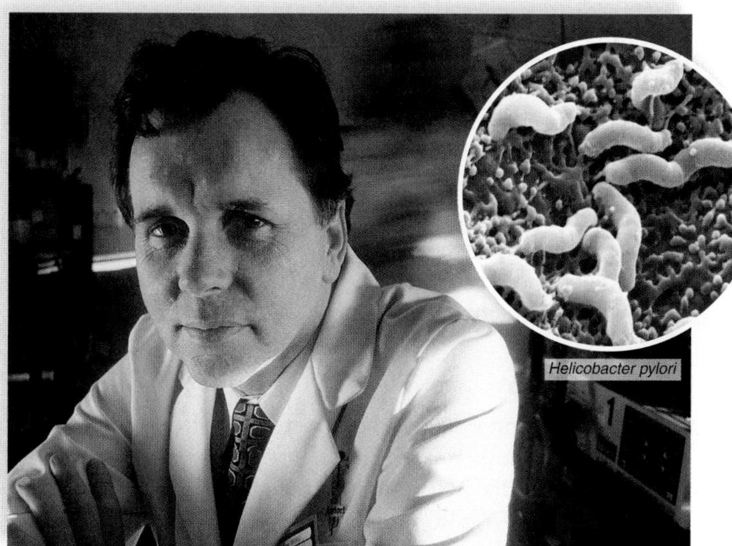

Figure 1.9 Dr. Barry Marshall and the cause of stomach ulcers.
Dr. Barry Marshall, pictured here, fulfilled Koch's postulates to show that *Helicobacter pylori* is the cause of peptic ulcers. The inset shows the presence of the bacterium in the stomach.

pylori from the intestinal tract and cure patients of ulcers wherever they occurred in the tract.

Dr. Marshall received all sorts of prizes and awards, but he and Dr. Warren were especially gratified to receive a Nobel Prize in Medicine in 2005. The Nobel committee reportedly thanked Marshall and Warren for their "pioneering discovery," stating that peptic ulcer disease now could be cured with antibiotics and acid-secretion inhibitors rather than becoming a "chronic, frequently disabling condition."

How to Do a Controlled Study

The work that Marshall and Warren did was largely observational. Often, scientists perform an **experiment,** a series of procedures to test a hypothesis. As an example, let's say investigators want to determine which of two antibiotics best treats an ulcer. When scientists do an experiment, they try to vary just the **experimental variables,** in this case, the medications being tested. A **control group** is not given the medications, but one or more **test groups** are given the medications. If by chance, the control group shows the same results as a test group, the investigators immediately know the results of their study are invalid because it would mean the medications may have nothing to do with the results.

The study depicted in Figure 1.10 shows how investigators may study this hypothesis.

Hypothesis: Newly discovered antibiotic B is a better treatment for ulcers than antibiotic A, which is in current use.

Investigators who perform clinical research must obtain informed consent from their subjects before proceeding with the research. The informed consent ensures that subjects know details about the research and that their participation is voluntary. The risks and benefits involved in participating in the research are all outlined. It is important to reduce the number of possible variables (differences) such as sex, weight, other illnesses, and so forth between the groups. Therefore, the investigators *randomly* divide a very large group of volunteers (Fig. 1.10*a*) equally into the three groups. The hope is that any differences will be distributed evenly among the three groups. This is more likely to occur if the investigators have a large number of volunteers.

The three groups will be treated (Fig. 1.10*b*) as follows:

Control group: Subjects with ulcers are not treated with either antibiotic.

Test group 1: Subjects with ulcers are treated with antibiotic A.

Test group 2: Subjects with ulcers are treated with antibiotic B.

After the investigators have determined that all volunteers do suffer from ulcers, they will want the subjects to think they are all receiving the *same* treatment. This is an additional way to protect the results from any influence other than the medication. To achieve this end, the subjects in the control group can receive a **placebo,** a treatment that appears to be the same as that administered to the other two groups but contains no medication. In this study, the use of a *placebo* would help ensure the same dedication by all subjects to the study.

The Results

After two weeks of administering the same amount of medication (or placebo) in the same way, the intestinal tract of each subject is examined to determine if ulcers are still present. Endoscopy, depicted in the photograph in Figure 1.10*c*, is one possible way to examine a patient for the presence of ulcers. This procedure is performed under sedation and involves inserting an endoscope—a small, flexible tube with a tiny camera on the end—down the throat and into the stomach. It allows the doctor to see the lining of the stomach and check for possible ulcers. Tests performed during an endoscopy can also determine if *Helicobacter pylori* is present.

Endoscopy is somewhat subjective so it is probably best if the examiner is not aware of which group the subject is in. Otherwise, the prejudice of the examiner may influence the examination. When neither the patient nor the examiner is aware of the specific treatment, it is called a *double-blind* study.

In this study, the investigators may decide to determine effectiveness of the medication by the percentage of people who no longer have ulcers. So, if 20 people out of 100 still have ulcers, the medication is 80% effective. The difference in effectiveness is easily read in the graph portion of Figure 1.10*d*.

Conclusion: On the basis of their data, the investigators conclude that their hypothesis has been supported.

Publication of Scientific Studies

Scientific studies are customarily published in a scientific journal, so that all aspects of a study are available to the scientific community. Before information is published in scientific journals, it is typically reviewed by experts. These people ensure that the research is credible, accurate, unbiased, and well executed.

Another scientist should be able to read about an experiment in a scientific journal, repeat the experiment in a different location, and get the same (or very similar) results. Each article begins with a short synopsis of the study so scientists can quickly find the articles of greatest interest to them. The materials and methods of performing each study are clearly outlined so that researchers can more easily repeat the work. Some articles are rejected for publication by reviewers when they believe there is something questionable about the design or manner in which an experiment was conducted.

Further Study

As mentioned previously, the conclusion of one experiment often leads to another experiment. Scientists reading the study described in Figure 1.10 may decide that it would be important to test the difference in the ability of antibiotic A and B to kill *Helicobacter pylori* in a petri dish. Or, they may want to test which medication is more effective in women than men, and so forth. The need for scientists to expand on findings explains why science changes and the findings of yesterday may be improved upon tomorrow.

Scientific Journals Versus Other Sources of Information

The information in many scientific journals is highly regarded by scientists because of the review process and because it is "straight from the horse's mouth," so to speak. The investigator who did the research is generally the primary author of a published study. Reading the actual results of the experiment tends to prevent the possibility of misinformation and/or bias. Do you remember playing "pass the message" when you were young? When someone started a message and it was passed around a circle of people, the last person to hear the message rarely received the original message. As each person passed along the message, information was added to or deleted from the original. That same thing may happen to scientific information when it is published in magazines or books or reported by someone other than the original investigator.

Unfortunately, the studies in scientific journals may be technical and difficult for a layperson to read and understand. The general public typically relies on secondary sources of information for its science news. Sometimes, the information may be out of context or misunderstood by the reporter, and the result is transmission of misinformation. Ideally, a reference to the original source (scientific journal article) will be provided so that information can be verified. Remember also that it often

a.

State Hypothesis:
Antibiotic B is a better treatment for ulcers than antibiotic A.

Large number of subjects were selected.

Subjects were divided into three groups.

b.

Perform Experiment:
Groups were treated the same except as noted.

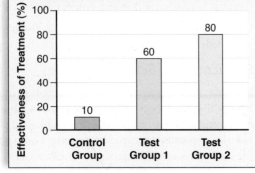

Control group: received placebo

Test group 1: received antibiotic A

Test group 2: received antibiotic B

c.

Collect Data:
Each subject was examined for the presence of ulcers.

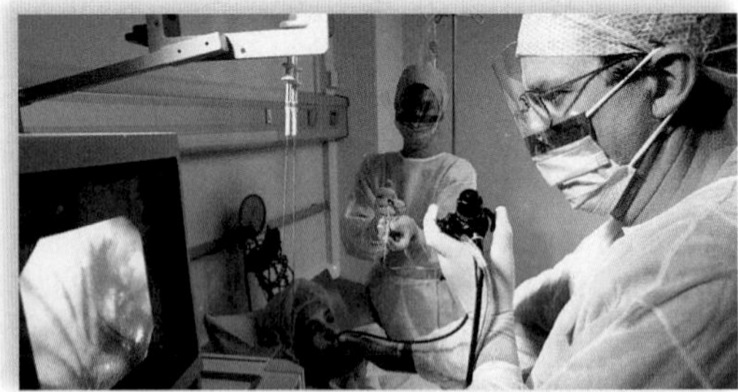

d.

Effectiveness of Treatment (%)

100, 80, 60, 40, 20, 0

Control Group: 10

Test Group 1: 60

Test Group 2: 80

Conclusion: Hypothesis is supported: Antibiotic B is a better treatment for ulcers than antibiotic A.

Figure 1.10 Design of a scientific study.
In this controlled laboratory experiment to test the effectiveness of a medication in humans, subjects were divided into three groups. The control group received a placebo and no medication. Test group 1 received antibiotic A and test group 2 received antibiotic B. The results are depicted in a graph, and it shows that antibiotic B was found to be a more effective treatment than antibiotic A for the treatment of ulcers.

takes years to do enough experiments for the scientific community to accept findings as well founded. Be wary of claims that have only limited data to support them and any information that may not be supported by repeated experimentation.

People should be especially careful about scientific information available on the Internet, which is not well regulated. Reliable, credible scientific information can often be found at websites with URLs (uniform resource locators or Web addresses) containing .edu (for educational institution), .gov (for government sites such as the National Institutes for Health or Centers for Disease Control), and .org (for nonprofit organizations such as the American Lung Association or the National Multiple Sclerosis Society). Unfortunately, quite a bit of scientific information on the Internet is intended to entice people into purchasing some sort of product for weight loss, prevention of hair loss, or similar maladies. These websites usually have URLs ending with .com or .net. It pays to question and verify the information from these websites with another source (primary, if possible).

Connecting the Concepts

For more information on the topics presented in this section, refer to the following discussions.

Section S.3 discusses how resistance to antibiotics occurs.

Section 8.3 provides more information on ulcers.

Check Your Progress 1.3

1. Describe each step of the scientific method.
2. Explain why a controlled study is an important part of the experimental design.
3. List a few pros and cons of using a scientific journal versus other sources of information.

1.4 Making Sense of a Scientific Study

Learning Outcomes

Upon completion of this section you should be able to

1. Explain the difference between anecdotal and testimonial data.
2. Interpret information that is presented in a scientific graph.
3. Recognize the importance of statistical analysis to the study of science.

When evaluating scientific information, it is important to consider the type of data given to support it. Anecdotal data, which consist of testimonials by individuals rather than results from a controlled, clinical study, are never considered reliable data. An example of anecdotal data would be claims from people that a particular diet helped them lose weight. Obviously, this doesn't mean the diet will work for everyone. Testimonial data are suspect because the effect of whatever is under discussion may not have been studied with a large number of subjects or a control group.

We must also keep in mind that just because two events occur at the same time, one factor may not be the cause of the other. Dr. Marshall had this problem when his data largely depended on finding *Helicobacter pylori* at the site of ulcers. More data were needed before the scientific community could conclude that *Helicobacter pylori* was the cause of an ulcer. Similarly, that a human papillomavirus (HPV) infection usually precedes cervical cancer could be viewed only as limited evidence that HPV causes cervical cancer. In this instance, HPV has turned out to be a cause of cervical cancer, but not all correlations (relationships) turn out to be causations. For example, scientific studies do not support the well-entrenched belief that exposure to cold temperatures results in colds. Instead, we now know that viruses cause colds.

What to Look For

Although most everyone who examines a scientific paper is tempted to first read the abstract (synopsis) at the beginning and then skip to the conclusion at the end of the study, it is a good idea to also examine the investigators' methodology and results. The methodology tells us how the scientists conducted their study, and the results tell us what facts (data) were discovered. Always keep in mind that the conclusion is not the same as the data. The conclusion is an interpretation of the data by the individuals who authored the paper. Although scientific papers are reviewed by other scientists, ultimataly it is up to us to decide if the conclusion is justified by the data.

Graphs

Data are often depicted in the form of a bar graph (see Fig. 1.10*d*) or a line graph (Fig. 1.11). A graph shows the relationship between two quantities, such as the taking of an antibiotic and

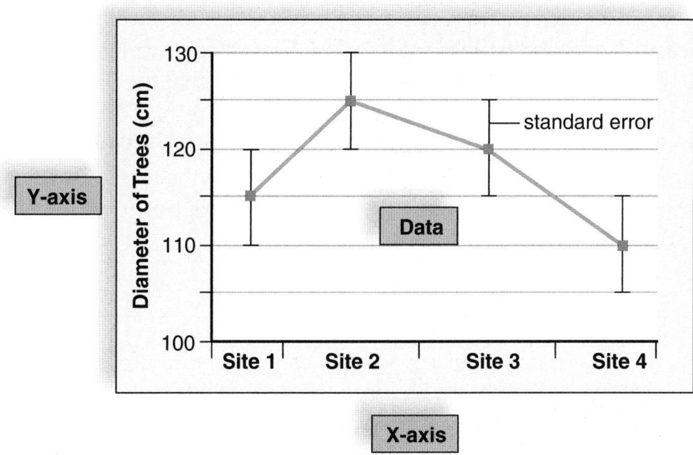

Figure 1.11 The presentation of scientific data. This line graph shows that the diameter of tree trunks varied at four different places. The bars above each data point represent the variation, or standard error, in the results.

the disappearance of an ulcer. As in Figure 1.10, the experimental variable (study groups) is plotted on the x-axis (horizontal), and the result (effectiveness) is plotted along the y-axis (vertical). Graphs are useful tools to summarize data in a clear and simplified manner. For example, Figure 1.10*d* immediately shows that antibiotic B produced the best results.

The title and labels can assist you in reading a graph; therefore, when looking at a graph, first check the two axes to determine what the graph pertains to. For example, in Figure 1.11, we can see that the investigators were studying tree trunk diameters at four sites. By looking at this graph, we know that trees with the greatest diameter are found at site 2, and we can also see to what degree the tree trunk diameters differed between the sites.

Statistical Data

Most authors who publish research articles use statistics to help them evaluate their experimental data. In statistics, the **standard error** tells us how uncertain a particular value is. Suppose you predict how many hurricanes Florida will have next year by calculating the average number during the past ten years. If the number of hurricanes per year varies widely, your standard error will be larger than if the number per year is usually about the same. In other words, the standard error tells you how far off the average could be. If the average number of hurricanes is four and the standard error is ± 2, then your prediction of four hurricanes is between two and six hurricanes. In Figure 1.11, the standard error is represented by the bars above and below each data point. This provides a visual indication of the statistical analysis of the data.

Statistical Significance

When scientists conduct an experiment, there is always the possibility that the results are due to chance or due to some factor other than the experimental variable. Investigators take into account several factors when they calculate the probability value (p) that their results were due to chance alone. If the probability value is low, researchers describe the results as statistically significant. A probability value of less than 5% (usually written as $p < 0.05$) is acceptable, but, even so, keep in mind that the lower the p value, the *less* likely that results are due to chance. Therefore, the lower the p value, the *greater* the confidence the investigators and you can have in the results. Depending on the type of study, most scientists like to have a p value of <0.05, but p values of <0.001 are common in many studies.

Connecting the Concepts

For more examples of how graphs can be used to present scientific data, refer to the following discussions.

Figure 7.14 demonstrates how immunizations increase antibody concentration in the blood.

Figure 13.3 shows the relationship between an action potential and voltage across a cell membrane.

Figure 24.1 illustrates human population growth.

Check Your Progress 1.4

1. Explain why anecdotal data are not used in controlled, clinical studies.
2. Explain why it is important to study an investigator's methodology and results.
3. Summarize how the use of graphs and statistics aids in data analysis.

1.5 Science and Social Responsibility

Learning Outcomes

Upon completion of this section you should be able to

1. Recognize the importance of ethics in scientific studies.
2. Dicuss the need for the general public to have a general understanding of science and its relationship to society.

As we have learned in this chapter, science is a systematic way of acquiring knowledge about the natural world. Biologists are scientists that study living things, from the tiniest microbes to the tallest trees (Fig. 1.12). Religion, aesthetics, and ethics are other ways in which human beings seek order in the natural world. Science differs from these other ways of knowing and learning by its process, based on the scientific method. Science considers hypotheses that can be tested only by experimentation and observation. Only after an immense amount of data has been gathered do scientists arrive at a scientific theory, a well-found concept about the natural world. Knowing this should help you realize that all scientific theories have merit.

Science is a slightly different endeavor from technology, but science is the driving force behind technology. **Technology** is the application of scientific knowledge to the interests of humans. As members of Western civilization, we have always believed that science and technology offer us ways to improve our lives. Building houses, paving roads, growing crops, and curing illnesses all depend on technology. Just think of how many things you use each day made of plastic and you will know how much technology means to your daily life. Even the field of nuclear physics is of direct benefit to human beings. The findings of nuclear physics play a role in cancer therapy, medical imaging, and homeland security. It has also given us nuclear power and the atomic bomb. In other words, science and technology are not risk free; uncontrolled technology can result in unanticipated side effects.

Science and Technology: Benefits Versus Risks

Investigations into cell structure and genes led to the biotechnology revolution of current times. We now know how to manipulate genes. If you have diabetes and are using insulin, it was produced by genetically modified (GM) bacteria. Despite its benefits, there are risks associated with biotechnology. Ecologists are concerned that GM crops could

a.

b.

Figure 1.12 Biologists work in many environments.
Data collection can be done **(a)** in the laboratory or **(b)** in the field. Biologists discover basic information about the natural world, including the effects of technology on human health and the environment.

endanger the biosphere. For example, many farmers are now planting GM cotton that produces an insect-killing toxin. Though insect pests are their intended target, toxins may also kill the natural predators that feed on insect pests. It may not be wise to kill off friendly predatory insects. People are now eating GM foods and/or foods that have been manufactured by using GM products. Some people are concerned about the possible effects of GM foods on human health.

In medicine, gene technology raises even more difficult ethical issues. These include whether humans should be cloned or whether gene therapy should be used to modify the inheritance of people, even before they are born. A current debate centers around the use of stem cells to cure human illnesses, such as spinal cord injuries or Alzheimer disease. You may not appreciate this debate until you know that early human embryos are composed only of cells called embryonic stem cells. Embryonic stem cells are genetically capable of becoming any type of tissue needed to cure a human illness. Should human embryos be dismantled and used for this purpose? It means they will never have the opportunity to become human beings.

Everyone Is Responsible

Science, technology, and society have interacted throughout human history, and scientific investigation and technology have always been affected by human values. Studying science, such as human biology, can give citizens the background they need to fully participate in ethical debates. All citizens should assume this responsibility because everyone, not just scientists, needs to be involved in making value judgments about the proper use of technology.

Should windmills be placed in Nantucket Sound or other bodies of water to reduce our use of oil, which causes global warming? To participate in this debate, you have to know what windmills might do to the ecology of Nantucket Sound, what global warming is, and what the evidence for global warming is. In other words, you need to review the data of scientists and only then participate in the debate about how technology should be used in this particular instance.

Should the rise in human population be curtailed to help preserve biodiversity? You might hear on the news that we are in a biodiversity crisis—the number of extinctions expected to occur in the near future is unparalleled in the history of the Earth. To help answer this question, you will want to know how large the human population is, what the benefits of biodiversity are, how it is threatened, and what the threat is to humankind if biodiversity is not preserved.

Scientists can inform and educate us, but they need not bear the burden of making these decisions alone because science does not make value judgments. This is the job of all of us.

Connecting the Concepts

The topic of bioethics is covered in several of the boxed readings in this text. For some examples of bioethical discussions that involve biotechnology, refer to the following discussions.

Chapter 17 contains a bioethical focus on whether humans should be cloned.

Chapter 21 examines the increased role of DNA fingerprinting in the justice system.

Check Your Progress 1.5

1. List examples that demonstrate both the risks and benefits of technology.
2. Discuss how human values affect science and technology.
3. Discuss the following: In your opinion, which members of society should be responsible for deciding how technology should be used?

Biology Matters Historical Focus

The Syphilis Research Scandal of Tuskegee University

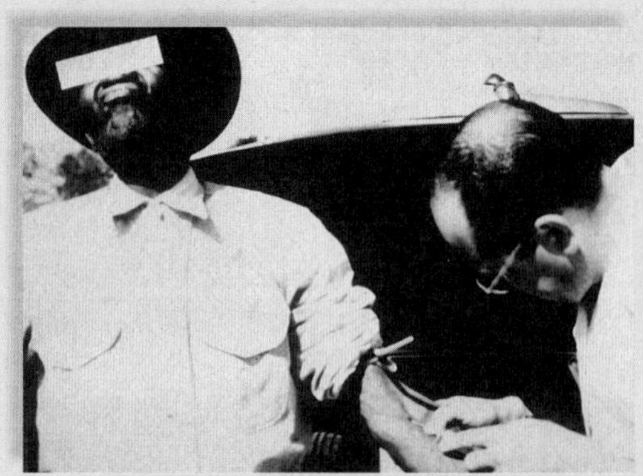

Figure 1A Poorly educated African Americans were recruited for the Tuskegee project with promises of free medical care.

Several sections of this chapter have covered the process of science, the way that legitimate research should be conducted, and the importance of informed consent when using human research subjects. As professionals, scientists have a responsibility to design moral and ethical research. Unfortunately, as with all professionals, not all scientists are ethical. Documented cases of risky, life-threatening, and, in some cases, inhumane research on humans (often without the subject's consent or knowledge) blot scientific history. One of the most extreme examples of such "research" was that done by Dr. Josef Mengele, the handsome Nazi doctor called the "Angel of Death." Mengele tortured concentration camp prisoners in multiple horrible ways. Some were slowly frozen to death, others poisoned, and still others bled to death—all to fulfill Mengele's obscene notion of scientific inquiry.

Regrettably, the history of research in the United States is also stained by misconduct. One notorious example of unethical research involving human subjects began in the United States in 1932 and continued until 1972. This research was carried out by the Public Health Service (PHS). Investigators wished to study the progression of syphilis, a sexually transmitted bacterial disease, in African-American males. African-American males were of interest because of diseases associated with African Americans, such as sickle-cell anemia. There was interest in determining if race had an impact on the progression of syphilis.

An untreated syphilis infection has three distinct stages. During the first stage, sores or ulcers appear on the genitals of the infected individual. The bacteria associated with the sores can infect individuals who come into contact with the sores. Stage-two syphilis typically develops several weeks later. Its symptoms include those associated with the flu: fever, headaches, and joint pain. If the disease shows up in a third stage, severe complications associated with the nervous system occur. Paralysis and insanity are common, and organ failure will eventually kill. Untreated pregnant women can pass syphilis to a fetus. Congenital syphilis in an infant causes physical deformity and mental retardation.

The syphilis study began in 1932 at the Tuskegee Institute. Initially, 600 African-American males were enrolled, with 399 males infected with syphilis and 201 uninfected males. The men were poor, mostly illiterate, sharecropper farmers. None of the men were informed of their participation in a research study nor about available treatment options. The men were told that investigators were testing for and treating "bad blood." The phrase described a number of common illnesses, including anemia, that were widespread at the time. While they participated in the study, the men were offered medical exams, transportation to and from clinics, treatments for other ailments, food, and money for their burial expenses if necessary.

When the study first began, there were few available treatments for syphilis. Compounds containing mercury and arsenic were used, but all were toxic to the patient.

Originally, the study sought to determine if these toxic compounds helped, or if untreated individuals fared the same as those treated. The study was intended to last six months but persisted for decades. Research continued on these men, even after penicillin became the accepted treatment for syphilis in 1947. Treatment with penicillin was never offered to the study's participants, even though it would have cured them of the disease.

Objections to the Tuskegee study began in the mid-1960s. One critic deemed the project "bad science" because some men had been partially treated during the study. In July 1972, the Associated Press (AP) publicized the story of the syphilis project. The study was finally discontinued. A class-action lawsuit settlement covering medical and burial expenses for all living participants was reached in 1974. President Clinton offered a national apology to survivors and their families in 1997. Even now, there are some family members of the study's participants receiving benefits from that lawsuit.

CASE STUDY CONCLUSION

In this chapter, you have explored some of the basic characteristics of life as we know it. The question is, How can we apply our knowledge of life on Earth to detect life on other planets? Most likely, the life on these moons is not highly organized. Most scientists believe that simple multicellular organisms may be the only life forms that can survive this distance from the sun. Thus, future missions to Europa and Titan will likely look for evidence of life acquiring materials and using energy. When you eat food, you produce carbon dioxide and other waste products. Living organisms on other planets should do the same. By studying the extreme environments of Europa, Titan, and our own planet, we may be better able to define the basic properties of life and what it really means when we say that something "is alive."

Media Study Tools

www.mhhe.com/maderhuman12e

Enhance your study of this chapter with study tools and practice tests. Also ask your instructor about the resources available through ConnectPlus, including the media-rich eBook, interactive learning tools, and animations.

Summarizing the Concepts

1.1 The Characteristics of Life

All living organisms share common characteristics; they

- have levels of organization—atoms, molecules, cells, tissues, organs, organ systems, organisms, populations, community, ecosystem, and biosphere;
- acquire materials and energy from the environment;
- reproduce and develop;
- maintain homeostasis;
- respond to stimuli; and
- have an evolutionary history and are adapted to a way of life.

1.2 Humans Are Related to Other Animals

The classification of living things mirrors their evolutionary relationships. Humans are mammals, a type of vertebrate in domain Eukarya. Humans differ from other mammals, including apes, by their

- highly developed brains;
- completely upright stance;
- creative language; and
- ability to use a wide variety of tools.

Humans Have a Cultural Heritage

Language, tool use, values, and information are passed on from one generation to the next.

Humans Are Members of the Biosphere

Humans depend on the biosphere for its many services, such as absorption of pollutants, sources of water and food, prevention of soil erosion, and natural beauty. Human population growth and use of resources threaten the biodiversity of the biosphere.

1.3 Science as a Process

The scientific method consists of

- making an observation;
- formulating a hypothesis;
- carrying out experiments and observations;
- coming to a conclusion; and
- presenting results of the study for peer review.

Over time, widely accepted concepts and ideas that explain patterns in the natural world may become theories.

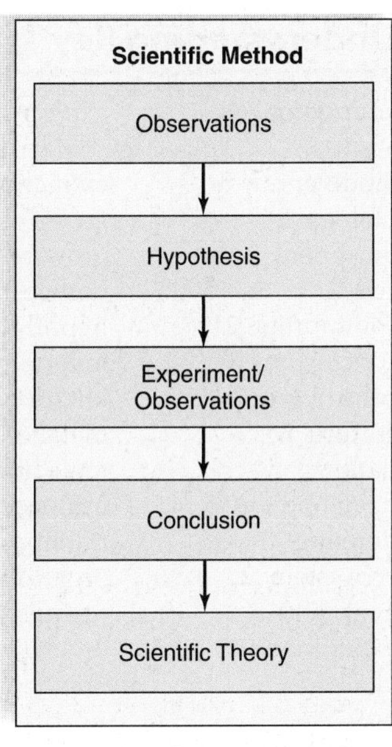

How the Cause of Ulcers Was Discovered

Dr. Marshall followed Koch's postulates to show that *Helicobacter pylori* causes ulcers. When he was unable to infect an animal with the bacterium, he and another volunteer infected themselves. His persistence led to clinical studies that showed antibiotics can cure ulcers.

How to Do a Controlled Study

- A large number of subjects are divided randomly into groups.
- The test group/s is/are exposed to an experimental variable.
- A control group is not exposed to an experimental variable and is given a placebo.
- All groups are otherwise treated the same, and it is best if the subjects and the technicians do not know what group they are in.
- The results and conclusion are published in a scientific journal.

Scientific Journals Versus Other Sources of Information

Primary sources of information are best; if secondhand sources are used, the reader needs to carefully evaluate the source. The Internet is not regulated, although URLs that end in .edu, .gov, and .org most likely provide reliable scientific information.

1.4 Making Sense of a Scientific Study

- Beware of anecdotal and correlation data because they need further study to be substantiated.
- Both bar and line graphs clearly show the relationship between quantities.

Statistical Data

The standard error tells us how uncertain a particular value is. The statistical significance tells us how trustworthy the results are. The lower the statistical significance, the less likely the results are due to chance and the more likely they are due to the experimental variable.

1.5 Science and Social Responsibility

- Scientific information is based on observation and experimentation. Therefore, scientists need not make value judgments for us.
- The use of modern technology has its risks, and all citizens need to be able to make informed decisions regarding how and when technology should be used.

Understanding Key Terms

adaptation 6	experimental variable 11	photosynthesis 4
atom 2	extinction 8	placebo 11
biodiversity 8	gene 5	population 2
biology 2	growth 5	principle 9
biosphere 2	homeostasis 4	reproduce 5
cell 2	hypothesis 9	science 9
community 2	invertebrate 7	scientific method 9
conclusion 10	kingdom 7	scientific theory 9
control group 11	metabolism 4	species 2
culture 7	molecule 2	standard error 14
data 9	multicellular 2	technology 14
development 5	mutation 5	test group 11
domains 6	organ 2	tissue 2
ecosystem 2	organism 2	vertebrate 7
evolution 5	organ system 2	
experiment 11		

Match the key terms to these definitions.

a. ________ Zone of air, land, and water at the surface of the Earth in which living organisms are found.

b. ________ Smallest unit of a human being and all living things.

c. ________ Concept supported by a broad range of observations, experiments, and conclusions.

d. ________ An internal environment that normally varies within only certain limits.

e. ________ An artificial situation devised to test a hypothesis.

Testing Your Knowledge of the Concepts

1. Name and describe the basic characteristics of life. (pages 2–6)
2. What is homeostasis, and why is it important? Give some examples that show how systems work together to maintain homeostasis. (pages 4–5)
3. How do human activities threaten the biosphere? (page 8)
4. Discuss the importance of a scientific theory, and describe several theories basic to understanding biological principles. (page 9)
5. With reference to the steps of the scientific method, explain how scientists arrive at a theory. (page 9)
6. What are Koch's postulates, and what are they used for? (page 10)
7. What is a control group, and what is the importance of a control group in a controlled study? (page 11)
8. How do science and technology improve our lives? (pages 14–15)

In questions 9–12, match each description with the correct characteristic of life from the key.

Key:

a. Life is organized.
b. Living things reproduce and grow.
c. Living things respond to stimuli.
d. Living things have an evolutionary history.
e. Living things acquire materials and energy.

9. Human heart rate increases when the person is scared.
10. Humans produce only humans.
11. Humans need to eat for building blocks and energy.
12. Similar cells form tissues in the human body.
13. The level of organization that includes two or more tissues that work together is a/an
    a. organ.
    b. tissue.
    c. organ system.
    d. organism.
14. The level of organization most responsible for the maintenance of homeostasis is the ____________ level.
    a. cellular
    b. organ system
    c. organ
    d. tissue
15. The level of organization that includes all the populations in a given area along with the physical environment would be a/an
    a. community.
    b. ecosystem.
    c. biosphere.
    d. tribe.
16. Which best describes the evolutionary relationship between humans and apes?
    a. Humans evolved from apes.
    b. Humans and chimpanzees evolved from apes.
    c. Humans and apes evolved from a common apelike ancestor.
    d. Chimpanzees evolved from humans.

In questions 17–19, match the explanation with a theory in the key.

Key:

a. homeostasis
b. cell
c. evolution
d. gene

17. All living things share a common ancestor.
18. The internal environment remains relatively stable.
19. Organisms contain coded information that dictates form and function.

Thinking Critically About the Concepts

1. Viruses are generally lumped into a "germs" grouping with bacteria. Viruses are composed of a small amount of genetic material (DNA or RNA) wrapped in a protein coat. Can something so simple be considered a living organism? Why aren't viruses mentioned in the system of classification covered in Section 1.2? (Hint: Consider the shared characteristics of all living organisms.)
2. Can anecdotal data ever be a considered a reliable data source for scientific research? Why or why not? If not, what purpose can anecdotal data serve?
3. In the case of Europa and Titan, if life were found to exist there, would that change our definition of the basic characteristics of life? Would that change our definition of a biosphere?

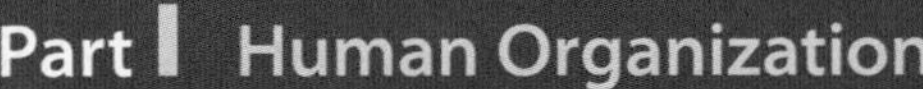

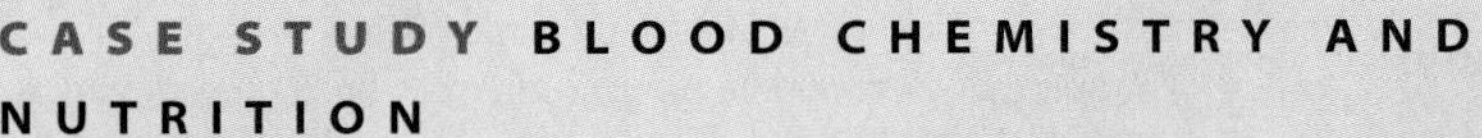

CHAPTER 2

Chemistry of Life

CASE STUDY BLOOD CHEMISTRY AND NUTRITION

As a 26-year-old male, David knew that he was slightly overweight and definitely not in the physical condition that he had been five years earlier. Still, he considered himself to be healthy; therefore, he only reluctantly agreed to the blood test at his annual physical exam. A few weeks later, David's doctor called and asked David to come in for a review of the results.

In the office, David's doctor explained that he had some real concerns about the blood test results. The doctor explained that David had a total cholesterol value of 218 and a blood triglyceride value of 150 mg/dL. Furthermore, his good cholesterol (HDL) was low (45 mg/dL) and his bad cholesterol (LDL) was high (130 mg/dL). If these values remained uncorrected, there would be an increased risk of heart disease and atherosclerosis and, potentially, diseases such as diabetes and cancer in David's future. David's doctor recommended that he reduce his dietary fat intake, increase his exercise, and come back in three months for a follow-up visit. If his blood lipids did not come into acceptable ranges, David's doctor was going to put him on Lipitor, a cholesterol-reducing medication.

David realized that he had received a wake-up call, and he was curious about some of the terms the doctor mentioned during the visit. For example, what is the difference between good and bad cholesterol? And what is a triglyceride? In doing some online research, he discovered that actually there is no such thing as "good" and "bad" cholesterol. The LDL and HDL actually refer to a type of protein that is involved in lipid transport in the body. David was confused, because his doctor had referred to these as being cholesterol. David realized that he not only knew very little about the content of his food but also was in the dark as to what these nutrients did in his body.

As you read through the chapter, think about the following questions.

1. What is it about cholesterol that would make it increase the risk for the diseases the doctor mentioned?
2. Is all cholesterol bad for the body?
3. What are the roles of proteins in the body, and how do they relate to David's cholesterol values?

CHAPTER CONCEPTS

2.1 From Atoms to Molecules
All matter is composed of atoms, which react with one another to form molecules.

2.2 Water and Living Things
The properties of water make life, as we know it, possible. Living things are affected adversely by water that is too acidic or too basic.

2.3 Molecules of Life
Carbohydrates, lipids, proteins, and nucleic acids are macromolecules with specific functions in cells.

2.4 Carbohydrates
Glucose is blood sugar, and humans store glucose as glycogen. Cellulose is plant material that is a source of fiber in the diet.

2.5 Lipids
Fats and oils, which provide long-term energy, differ by consistency and can have a profound effect on our health. Other lipids, such as the sterols, function differently in the body.

2.6 Proteins
Proteins have numerous and varied functions in cells. Their shape suits their function.

2.7 Nucleic Acids
DNA is our genetics material. RNA serves as a helper to DNA. ATP is an energy molecule that is used by the cell to do metabolic work.

BEFORE YOU BEGIN

Before beginning this chapter, take a few moments to review the following discussions.

Section 1.1 What are the basic characteristics of all living organisms?

Figure 1.2 What is the difference between an atom and a molecule?

2.1 From Atoms to Molecules

Learning Outcomes

Upon completion of this section, you should be able to

1. Distinguish between atoms and elements.
2. Illustrate the structure of an atom.
3. Define an isotope and summarize its application in both medicine and biology.
4. Distinguish between ionic and covalent bonds.

Matter refers to anything that takes up space and has mass. It is helpful to remember that matter can exist as a solid, a liquid, or a gas. Then we can realize that not only are we humans composed of matter, but so are the water we drink and the air we breathe.

Elements

An **element** is one of the basic building blocks of matter; an element cannot be broken down by chemical means. Considering the variety of living and nonliving things in the world, it's remarkable that there are only 92 naturally occurring elements. It is even more surprising that over 90% of the human body is composed of just four elements: carbon, nitrogen, oxygen, and hydrogen. Even so, other elements, such as iron, are important to our health. Iron-deficiency anemia results when the diet doesn't contain enough iron for the making of hemoglobin. Hemoglobin serves an important function in the body because it transports oxygen to our cells.

Each element has a name and a symbol. For example, carbon has been assigned the atomic symbol C, and iron has been assigned the symbol Fe. Some of the symbols we use for elements are derived from Latin. For example, the symbol for sodium is Na because *natrium,* in Latin, means "sodium." Likewise, the symbol for iron is Fe because *ferrum* means "iron." Chemists arrange the elements in a *periodic* table, so named because all the elements in a column show *periodicity.* To show periodicity means that each column of elements behaves similarly during chemical reactions. For example, all the elements in column VII (7) undergo the same type of chemical reactions, for reasons we will soon explore. Figure 2.1 shows only some of the first 36 elements in the periodic table, but a complete table is available in Appendix A. As you study each box in the table, note that the atom's symbol is in the center. The number above the symbol is the **atomic number.** The elements in each row are arranged in order, according to increasing atomic number. Finally, the number below each symbol is that element's **atomic mass.** Atomic mass is the average mass of all atoms of a particular element.

Atoms

An **atom** is the smallest unit of an element that still retains the chemical and physical properties of the element. The same name is given to the element and the atoms of the element.

I	II	III	IV	V	VI	VII	VIII
1 (atomic number) **H** (atomic symbol) 1.008						atomic mass	2 **He** 4.003
3 **Li** 6.941	4 **Be** 9.012	5 **B** 10.81	6 **C** 12.01	7 **N** 14.01	8 **O** 16.00	9 **F** 19.00	10 **Ne** 20.18
11 **Na** 22.99	12 **Mg** 24.31	13 **Al** 26.98	14 **Si** 28.09	15 **P** 30.97	16 **S** 32.07	17 **Cl** 35.45	18 **Ar** 39.95
19 **K** 39.10	20 **Ca** 40.08	31 **Ga** 69.72	32 **Ge** 72.59	33 **As** 74.92	34 **Se** 78.96	35 **Br** 79.90	36 **Kr** 83.60

Figure 2.1 A Portion of the Periodic Table of Elements. The number on the top of each square is the atomic number, which increases from left to right. The letter symbols represent each element and are sometimes abbreviations of Greek or Latin names. Below the symbol is the value for atomic mass. A complete periodic table can be found in Appendix A.

Connections and Misconceptions

Where do elements come from?

We are all familiar with elements. Iron, sodium, oxygen, and carbon are all common terms in our lives, but where do they originate from?

Normal chemical reactions do not produce elements. The majority of the heavier elements, such as iron, are produced only by the intense chemical reactions within stars. When these stars reach the end of their life, they explode, producing a supernova. Supernovas scatter the heavier elements into space, where they eventually are involved in the formation of planets. The late astronomer and philosopher Carl Sagan (1934–1996) frequently referred to humans as "star stuff." In many ways, we—all living organisms—are formed from elements that originated within the stars.

Though it is possible to split an atom, an atom is the smallest unit to enter into chemical reactions. For our purposes, it is satisfactory to think of each atom as having a central **nucleus** and pathways about the nucleus called **orbitals.** Even though an atom is extremely small, it contains even smaller subatomic particles. The subatomic particles, called **protons** and **neutrons,** are located in the nucleus, and **electrons** orbit about the nucleus in the orbitals (Fig. 2.2). Most of an atom is empty space. If we could draw an atom the size of a football stadium, the nucleus would be like a gumball in the center of the field, and the electrons would be tiny specks whirling about in the upper stands.

Animation Atomic Structure

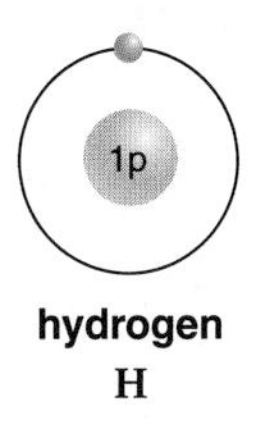

Subatomic Particles		
Particle	**Charge**	**Atomic Mass**
Proton	+1	1
Neutron	0	1
Electron	−1	0.00055

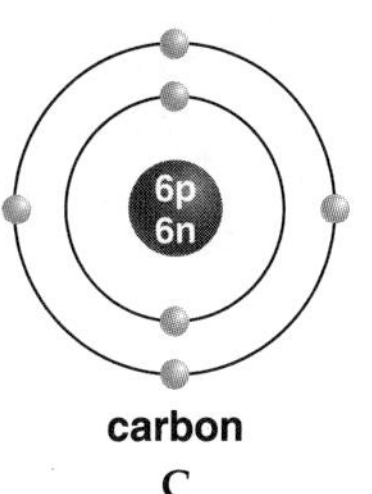

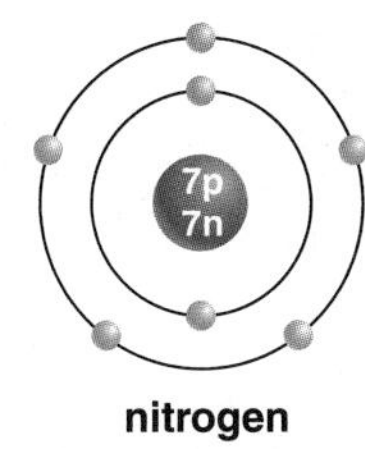

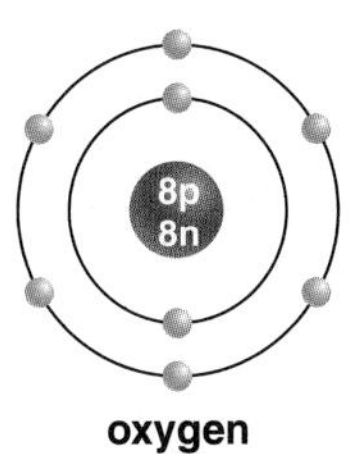

Figure 2.2 **The Atomic Structure of Select Elements.**
Notice how the protons (p) and neutrons (n) are located in the nucleus and the electrons (blue dots) are found in energy oribitals around the nucleus.

Protons carry a positive (+) charge, and electrons have a negative (−) charge. To be electrically neutral, an atom of any given element must have the same number of protons and electrons, so that the positive and negative charges are balanced. The atomic number (the number written *above* each chemical symbol in Fig. 2.1) tells how many protons—and therefore how many electrons—an atom has when it is electrically neutral. For example, the atomic number of carbon is six; therefore, a neutral atom of carbon has six protons and six electrons.

How many electrons are there in each orbital of an atom? Any given orbital can hold only two electrons. Each orbital is filled with electrons according to its energy level, with the lowest energy orbitals filled first (just like the lowest seats in the football stadium fill first, so spectators can get a good look at the team!). The innermost orbital has the lowest energy level and holds the first two electrons. After that, orbitals are arranged into groups called *shells.* The first shell has only one orbital. After that, each shell for the first 18 atoms illustrated in Figure 2.1 can hold up to eight electrons. Using this information, we can determine how many electrons are in the outer shells of the atoms shown in Figure 2.2. Hydrogen (H) has only one orbital that contains one electron. Helium fills the first orbital—the first shell—with its two electrons. Carbon (atomic number of six) has two shells, and the outer shell has four electrons. Nitrogen (atomic number of seven) has two shells, and the outer shell has five electrons. Oxygen (atomic number of eight) has two shells, and the outer shell has six electrons. Elements with greater than 18 electrons possess additional shells and thus can accommodate additional electrons.

The **mass** of an atom represents its quantity of matter. The subatomic particles are so light that their mass is indicated by special designations called *atomic mass units* (abbreviated amu). Protons and neutrons are each assigned one atomic mass unit, whereas electrons have an exceedingly small mass unit (0.00055 amu or 1/1,836th the mass of a proton). The sum of protons and neutrons in the atom's nucleus is called the **mass number.** (Don't confuse an element's *mass number* with its *atomic mass,* the number below each symbol in Fig. 2.1.) For example, the mass number for one form of carbon is 6 protons + 6 neutrons = 12. The mass number is written as a superscript before the atomic symbol, so the symbol for carbon with a mass number of 12 is ^{12}C.

Isotopes

Isotopes of the same type of atom have the same number of protons (thus, the same atomic number) but different numbers of neutrons. Therefore, their mass numbers are different. For example, the element carbon 12 (^{12}C) has six neutrons, carbon 13 (^{13}C) has seven neutrons, and carbon 14 (^{14}C) has eight neutrons. You can determine the number of neutrons for an isotope by subtracting the atomic number (carbon's atomic number is 6; see Fig. 2.1) from the mass number.

Unlike the other two isotopes of carbon, carbon 14 is unstable and breaks down over time. As carbon 14 decays, it releases various types of energy in the form of rays and subatomic particles; therefore, it is a **radioisotope.** The radiation given off by radioisotopes can be detected in various ways. You may be familiar with the use of a Geiger counter to detect radiation.

Animation
Half-Life

Low Levels of Radiation

The importance of chemistry to biology and medicine is nowhere more evident than in the many uses of radioisotopes. A radioisotope behaves the same chemically as the stable isotopes of an element. This means that you can put a small amount of radioisotope in a sample and it becomes a **tracer** by which to detect molecular changes.

Specific tracers are used in imaging the body's organs and tissues. For example, after a patient drinks a solution containing a minute amount of iodine 131, it becomes concentrated in the thyroid—the only organ to take up iodine to make the hormone thyroxine. A subsequent image of the thyroid indicates whether it is healthy in structure and function (Fig. 2.3*a*). The use of positron-emission tomography (PET) is a way to determine the comparative activity of tissues. Radioactively labeled glucose, which emits a subatomic particle known as a *positron,* can be injected into the body. The radiation given off is detected by sensors and analyzed by a computer. The result is a color image that shows which tissues took up glucose and are metabolically active (Fig. 2.3*b*). A PET scan of the brain can help diagnose a brain tumor, Alzheimer disease, epilepsy, and can determine whether a stroke has occurred.

Video
Nuclear Medicine

High Levels of Radiation

Radioactive substances in the environment can harm cells, damage DNA, and cause cancer. The release of radioactive particles following a nuclear power plant accident can

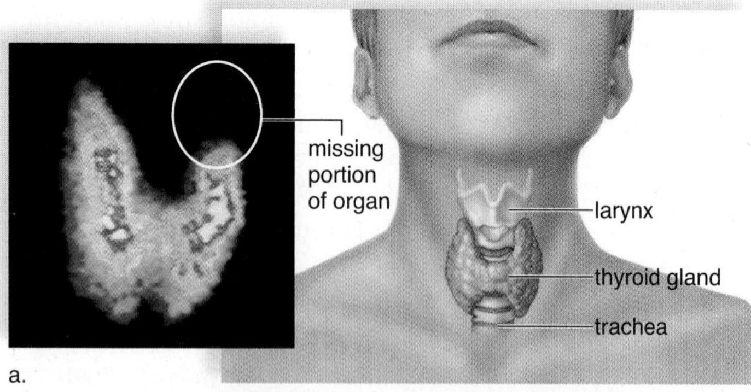

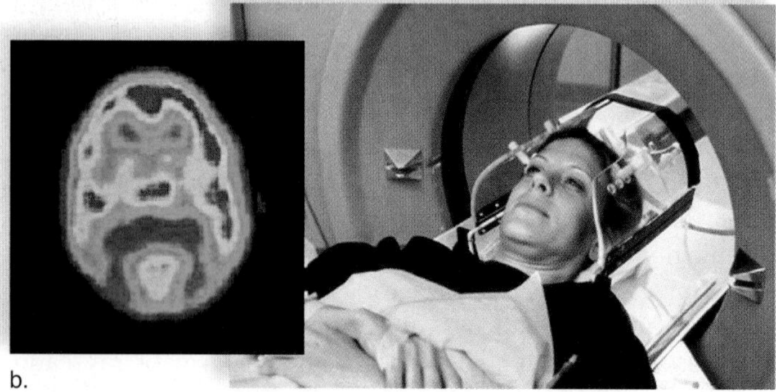

Figure 2.3 Medical uses for low-level radiation.
a. The missing area (indicated by the arrow) in this thyroid scan indicates the presence of a tumor that does not take up radioactive iodine. **b.** A PET (positron-emission tomography) scan reveals which portions of the brain are most active (red surrounded by light green).

have far-reaching and long-lasting effects on human health. However, the effects of radiation can also be put to good use. Radiation from radioisotopes has been used for many years to kill bacteria and viruses and to sterilize medical and dental products. Increasingly, the technology is being used to increase the safety of our food supply. By using specific types of radiation, food can be sterilized without irradiating or damaging the food (Fig 2.4*a*).

Radiation can be used to ensure public safety against bacterial infection. In the wake of the terrorist attacks of 9/11, radiation was used for a time to sterilize the U.S. mail because of fear of contamination by possible pathogens such as anthrax bacteria.

The ability of radiation to kill cells is often applied to cancer cells. Radioisotopes can be introduced into the body in a way that allows radiation to destroy only cancer cells, with little risk to the rest of the body (Fig. 2.4*b*). Another form of high-energy radiation, X-rays, can be used for medical diagnosis and cancer therapy.

Molecules and Compounds

Atoms often bond with one another to form a chemical unit called a **molecule.** A molecule can contain atoms of the same type, as when an oxygen atom joins with another oxygen atom to form oxygen gas. Or, the atoms can be different, as

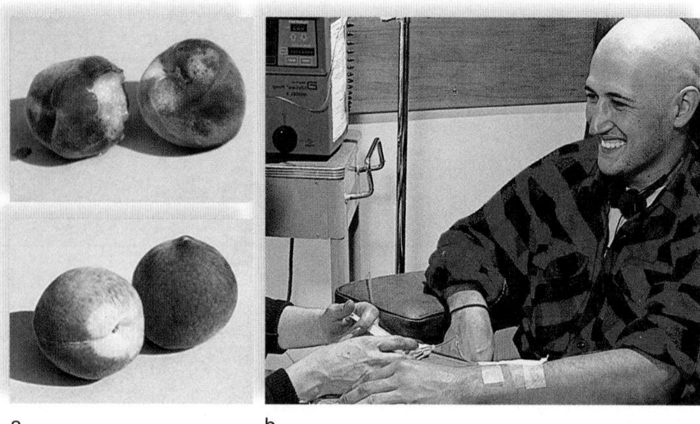

Figure 2.4 Uses of high-level radiation.
a. Radiation can be used to sterilize food by killing bacteria and fungi.
b. Physicians can use radiation therapy to kill cancer cells.

when an oxygen atom joins with two hydrogen atoms to form water. When the atoms are different, a **compound** is formed. Two types of bonds join atoms: the ionic bond and the covalent bond.

Ionic Bonding

Atoms with more than one shell are most stable when the outer shell contains eight electrons. During an ionic reaction, atoms give up or take on an electron or electrons to achieve a stable outer shell.

Figure 2.5 depicts a reaction between a sodium (Na) atom and a chlorine (Cl) atom. Sodium, with one electron in the outer shell, reacts with a single chlorine atom. Why? Once the reaction is finished and sodium loses one electron to chlorine, sodium's outer shell will have eight electrons. Similarly, a chlorine atom, which has seven electrons already, needs to acquire only one more electron to have a stable outer shell.

Animation
Ionic Bonds

Ions are particles that carry either a positive (+) or negative (−) charge. When the reaction between sodium and chlorine is finished, the sodium ion carries a positive charge because it now has one less electron than protons, and the chloride ion carries a negative charge because it now has one more electron than protons:

Sodium Ion (Na^+)	Chloride Ion (Cl^-)
11 protons (+)	17 protons (+)
10 electrons (−)	18 electrons (−)
One (+) charge	One (−) charge

The attraction between oppositely charged sodium ions and chloride ions forms an **ionic bond.** The resulting compound, sodium chloride, is table salt, which we use to enliven the taste of foods.

In contrast to sodium, why would calcium, with two electrons in the outer shell, react with two chlorine atoms?

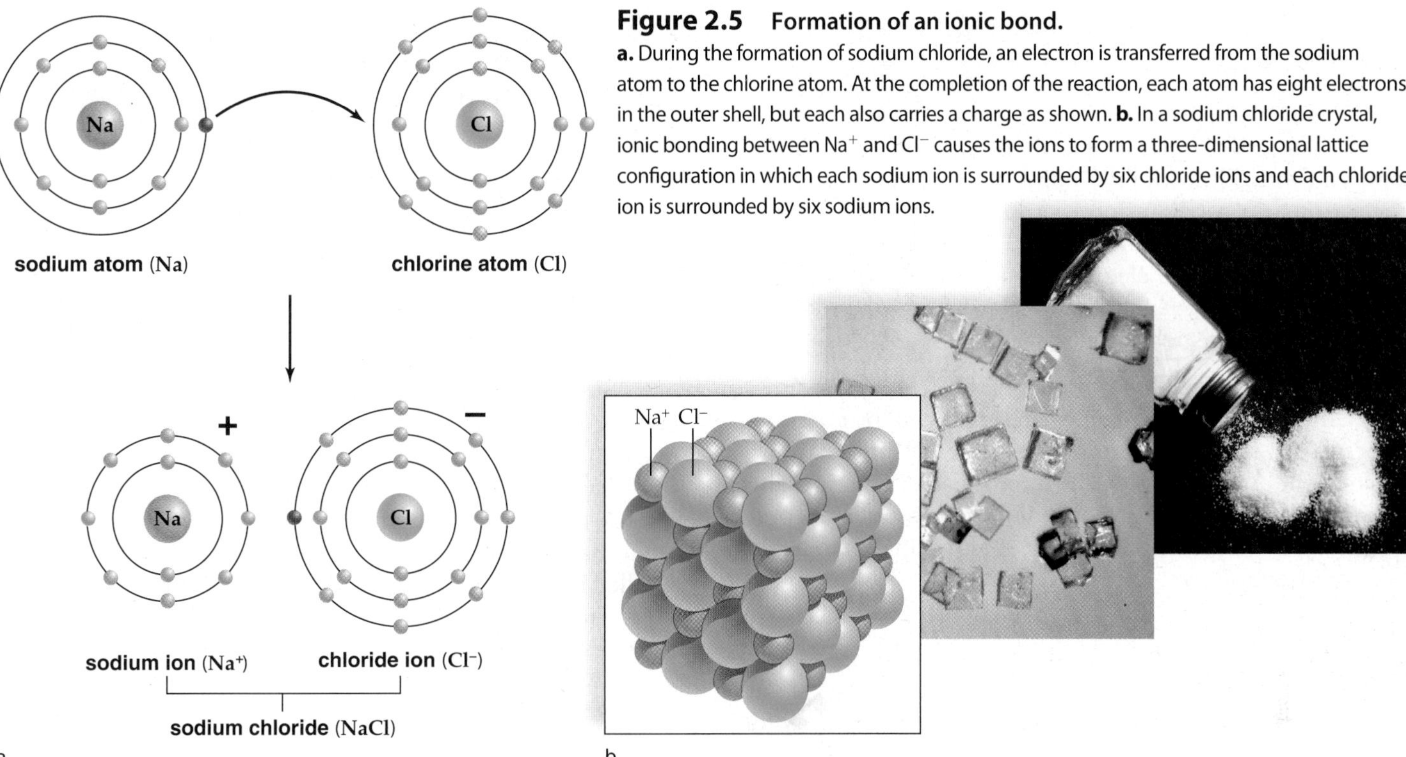

Figure 2.5 Formation of an ionic bond.
a. During the formation of sodium chloride, an electron is transferred from the sodium atom to the chlorine atom. At the completion of the reaction, each atom has eight electrons in the outer shell, but each also carries a charge as shown. **b.** In a sodium chloride crystal, ionic bonding between Na^+ and Cl^- causes the ions to form a three-dimensional lattice configuration in which each sodium ion is surrounded by six chloride ions and each chloride ion is surrounded by six sodium ions.

Whereas calcium needs to lose two electrons, each chlorine, with seven electrons already, requires only one more electron to have a stable outer shell. The resulting salt ($CaCl_2$) is called calcium chloride. Calcium chloride is used as a deicer in northern climates.

The balance of various ions in the body is important to our health. Too much sodium in the blood can contribute to high blood pressure. Calcium deficiency leads to rickets (a bowing of the legs) in children. Too much or too little potassium results in heartbeat irregularities and can be fatal. Bicarbonate, hydrogen, and hydroxide ions are all involved in maintaining the acid–base balance of the body (see pages 26–27).

Covalent Bonding

Atoms share electrons in **covalent bonds.** The overlapping, outermost shells in Figure 2.6 indicate that the atoms are sharing electrons. Just as two hands participate in a handshake, each atom contributes one electron to the shared pair. These electrons spend part of their time in the outer shell of

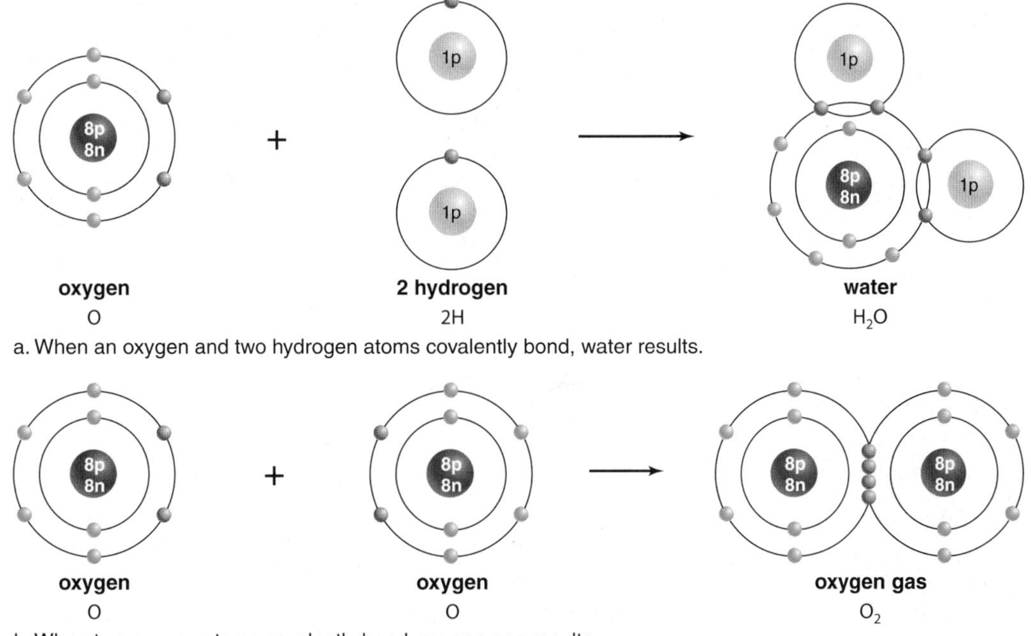

Figure 2.6 Covalent bonds.
Covalent bonds allow atoms to fill their outer shells by sharing electrons. Because the electrons are being shared, it is necessary to count the electrons in the outer shell as belonging to both bonded atoms. Hydrogen is most stable with two electrons in the outer shell; oxygen is most stable with eight electrons in the outer shell. Therefore, the molecular formula for water is **(a)** H_2O, and for oxygen gas it is **(b)** O_2.

each atom; therefore, they are counted as belonging to both bonded atoms.

Double and Triple Bonds Besides a single bond, in which atoms share only a pair of electrons, a double or a triple bond can form. In a double bond, atoms share two pairs of electrons; in a triple bond, atoms share three pairs of electrons. For example, in Figure 2.6*b*, each oxygen atom (O) requires two more electrons to achieve a total of eight electrons in the outer shell. Four electrons are placed in the outer, overlapping shells in the diagram.

Structural and Molecular Formulas Covalent bonds can be represented in a number of ways. In contrast to the diagrams in Figure 2.6, structural formulas use straight lines to show the covalent bonds between the atoms. Each line represents a pair of shared electrons. Molecular formulas indicate only the number of each type of atom making up a molecule. A comparison follows:

Structural formula: H—O—H, O=O

Molecular formula: H_2O, O_2

What would be the structural and molecular formulas for carbon dioxide? Carbon, with four electrons in the outer shell, requires four more electrons to complete its outer shell. Each oxygen, with six electrons in the outer shell, needs only two electrons to complete its outer shell. Therefore, carbon shares two pairs of electrons with each oxygen atom, and the formulas are as follows:

Structural formula: O=C=O

Molecular formula: CO_2

Connecting the Concepts

The study of biology has a firm foundation in chemistry. To see this relationship in more detail, refer to the following discussions.

Section 3.3 describes how cells move ions across cell membranes.

Section 10.3 explains how our urinary system uses ions to maintain homeostasis.

Check Your Progress 2.1

1. How is an atom organized?
2. One isotope of calcium is ^{40}Ca, whereas a second is ^{48}Ca. How many neutrons are found in each isotope? (Hint: The atomic number of calcium is 20.)
3. What are radioisotopes, and how can they benefit humans?
4. What are two basic types of bonds formed between atoms to form molecules?

Animation Ionic and Covalent Bonds

MP3 Chemical Bonding

2.2 Water and Living Things

Learning Outcomes

Upon completion of this section, you should be able to

1. Describe how hydrogen bonds are formed.
2. List the properties of water.
3. Summarize the structure of the pH scale and the importance of buffers to biological systems.

Water is the most abundant molecule in living organisms, usually making up about 60–70% of the total body weight. Furthermore, the physical and chemical properties of water make life as we know it possible.

In water, the electrons spend more time circling the oxygen (O) atom than the hydrogens because oxygen, the larger atom, has a greater ability to attract electrons than do the smaller hydrogen (H) atoms. The negatively charged electrons are closer to the oxygen atom, so the oxygen atom becomes slightly negative. In turn, the hydrogens are slightly positive. Therefore, water is a **polar** molecule; the oxygen end of the molecule has a slight negative charge (δ^-), and the hydrogen end has a slight positive charge (δ^+).

In Figure 2.7*a*, the diagram on the left shows a structural model of water, and the one on the right is called a space-filling model.

Hydrogen Bonds

A **hydrogen bond** is the attraction of a slightly positive, covalently bonded hydrogen to a slightly negative atom in the vicinity. These usually occur between a hydrogen and either an oxygen or nitrogen atom. A hydrogen bond is represented by a dotted line because it is relatively weak and can be broken rather easily.

In Figure 2.7*b*, you can see that each hydrogen atom, being slightly positive, bonds to the slightly negative oxygen atom of another water molecule.

Properties of Water

Water molecules are cohesive, meaning that they cling together, because of their polarity and hydrogen bonding. Polarity and hydrogen bonding cause water to have many characteristics beneficial to life.

Animation Water Properties

1. *Water is a liquid at room temperature. Therefore, we are able to drink it, cook with it, and bathe in it.* Compounds with a low molecular weight are usually gases at room temperature. For example, oxygen (O_2), with a molecular weight of 32, is a gas; but water, with a molecular weight of 18, is a liquid. The hydrogen bonding between water molecules keeps water a liquid and not a gas at room temperature. Water does not boil and become a gas until 100°C, one of the reference points for the Celsius temperature scale (see the chart inside the back cover). Without

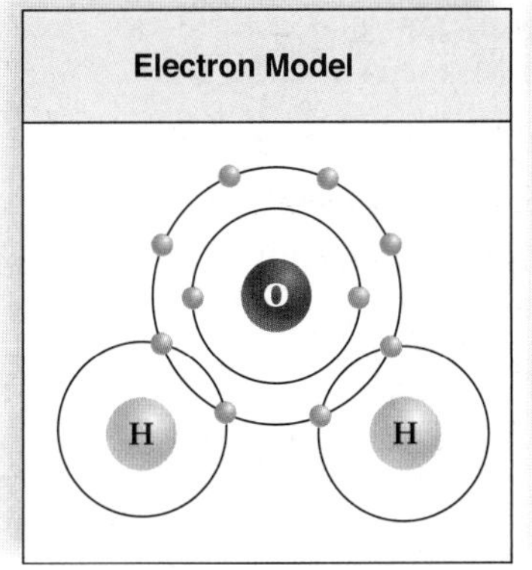

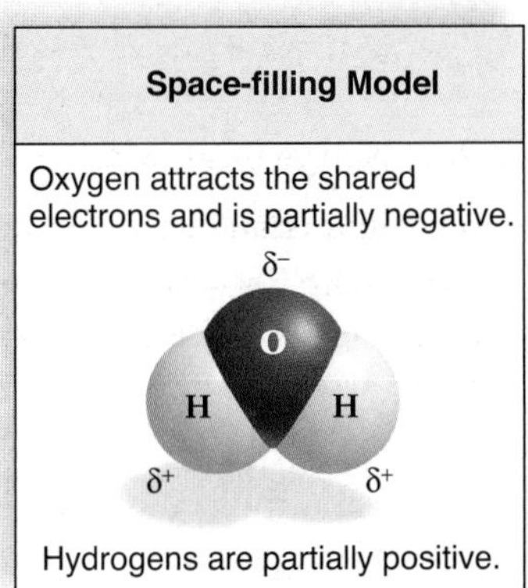

Space-filling Model
Oxygen attracts the shared electrons and is partially negative.
δ−
O
H
H
δ+
δ+
Hydrogens are partially positive.

a. Water (H_2O)

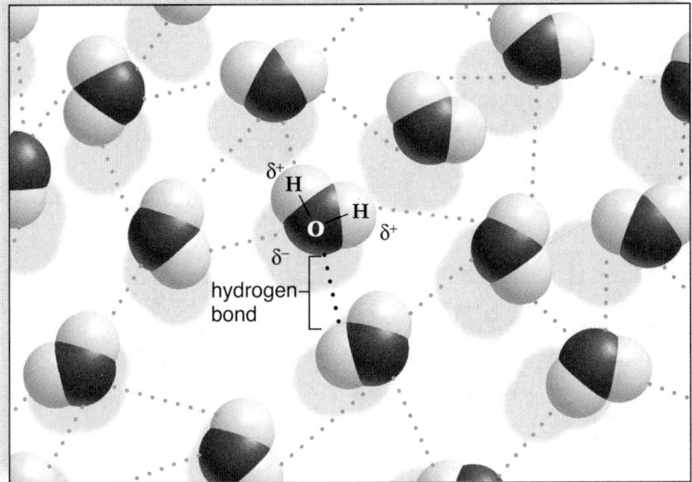

b. Hydrogen bonding between water molecules

Figure 2.7 Hydrogen bonds and water molecules.
a. Two models for the structure of water. The diagram on the left shows the sharing of electrons between the oxygen and hydrogen atoms. The diagram on the right illustrates that water is a polar molecule because electrons are not equally shared. Electrons move closer to oxygen, creating a partial negative charge, whereas hydrogen has a partial positive charge. **b.** The partial charges allow hydrogen bonds (dotted lines) to form temporarily between water molecules.

hydrogen bonding between water molecules, our body fluids—and indeed our bodies—would be gaseous!

2. *The temperature of liquid water rises and falls slowly, preventing sudden or drastic changes.* The many hydrogen bonds that link water molecules cause water to absorb a great deal of heat before it boils (Fig. 2.8*a*). A **calorie** of heat energy raises the temperature of 1 g of water 1°C.[1] This is about twice the amount of heat required for other covalently bonded liquids. On the other hand, water holds heat, and its temperature falls slowly. Therefore, water protects us and other organisms from rapid temperature changes and helps us maintain our normal internal temperature. This property also allows great bodies of water, such as oceans, to maintain a relatively constant temperature. Water is a good temperature buffer.
3. *Water has a high heat of vaporization, keeping the body from overheating.* It takes a large amount of heat to vaporize

[1]Be aware that a calorie—spelled with a lowercase *c*—is not the same as a Calorie (capital C), which is used to measure the heat energy of food. A Calorie is 1,000 calories, or one kilocalorie.

a.

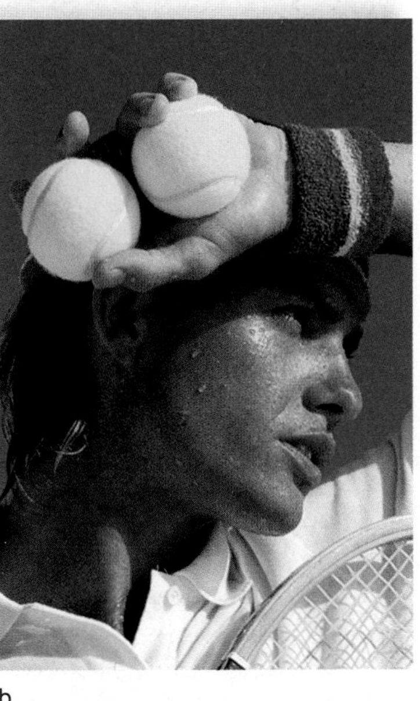

b.

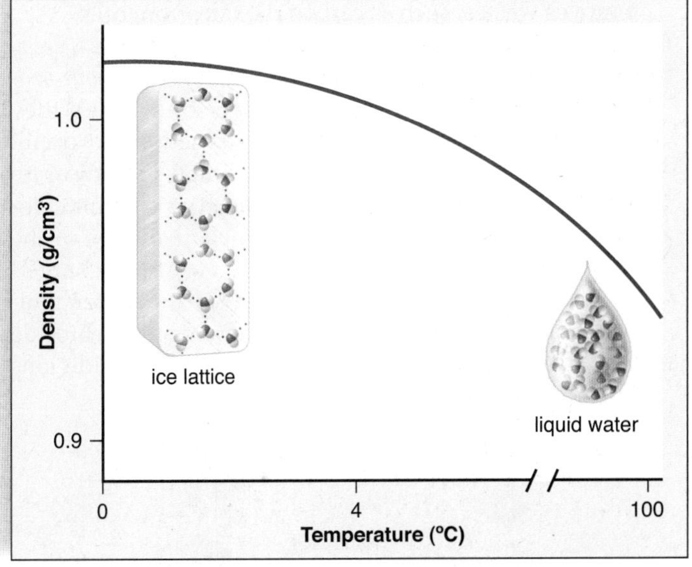

c.

Figure 2.8 A few of the properties of water.
a. Water becomes gaseous at 100°C. If it was a gas at a lower temperature, life could not exist. **b.** Body heat vaporizes sweat. In this way, bodies cool when the temperature rises. **c.** Ice is less dense than water, so it forms a protective layer on top of ponds during the winter.

water, that is, change water to steam (Fig. 2.8*a*). (Converting 1 g of the hottest water to steam requires an input of 540 calories of heat energy.) This property of water helps moderate the Earth's temperature so that life can continue to exist. Also, in a hot environment, most mammals sweat, and the body cools as body heat is used to evaporate sweat, mostly liquid water (Fig. 2.8*b*).

4. *Frozen water is less dense than liquid water, so ice floats on water.* Remarkably, water is more dense at 4°C than at 0°C. Most substances contract when they solidify. By contrast, water expands when it freezes because in ice, water molecules form a lattice in which the hydrogen bonds are farther apart than in liquid water (Fig. 2.8*c*). This is why cans of soda burst when placed in a freezer or why frost heaves make northern roads bumpy in the winter. Also, because ice is lighter than cold water, bodies of water freeze from the top down. The ice acts as an insulator to prevent the water below it from freezing. Thus, aquatic organisms are protected and have a better chance of surviving the winter.
5. *Water molecules are cohesive, yet still flow freely. Therefore, liquids fill vessels, such as blood vessels.* Water molecules cling together because of hydrogen bonding. You've seen that cohesive property of water in your everday life if you've ever washed drinking glasses and then stacked them while wet. Big mistake! The glasses will stick together as if glued because of hydrogen bonding in the film of water on the glass. Yet, though its molecules cling to each other, water still is able to flow freely. Water's cohesive property allows dissolved and suspended molecules to be evenly distributed throughout a system. Therefore, water is an excellent transport medium. Within our bodies, blood fills our arteries and veins because it is 92% water. After blood transports oxygen and nutrients to cells, these molecules are used to produce cellular energy. Blood also removes wastes, such as carbon dioxide, from cells.
6. *Water is a solvent for polar (charged) molecules and thereby facilitates chemical reactions both outside and inside our bodies.* When ions and molecules disperse in water, they move about and collide, allowing reactions to occur. Therefore, water is a solvent that facilitates chemical reactions. For example, when a salt such as sodium chloride (NaCl) is put into water, the negative ends of the water molecules are attracted to the positively charged sodium ions, and the positive ends of the water molecules are attracted to the negatively charged chloride ions. This causes the sodium ions and the chloride ions to separate and to dissolve in water:

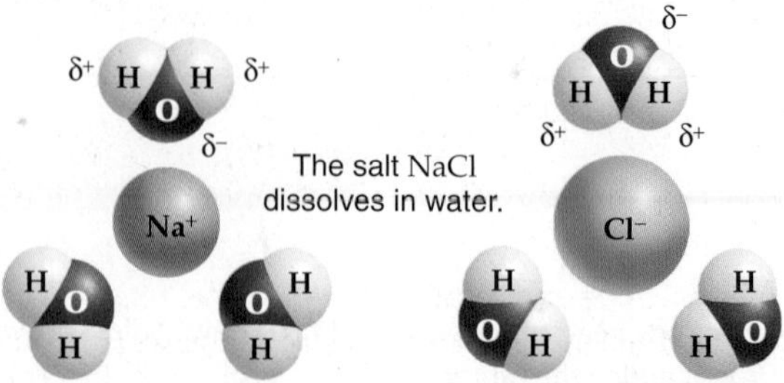

Connections and Misconceptions

How do lungs stay open and keep from collapsing?

Our lives depend on water's cohesive property. A thin film of water coats the surface of the lungs and the inner chest wall. This film allows the lungs to stick to the chest wall, keeping the lungs open so we can breathe.

Ions and molecules that interact with water are said to be **hydrophilic,** that is, "water-loving." Nonionized and nonpolar molecules that do not interact with water are said to be **hydrophobic,** that is, "water-fearing."

Acids and Bases

When water molecules dissociate (break up), they release an equal number of hydrogen ions (H^+) and hydroxide ions (OH^-):

$$\underset{\text{water}}{H{-}O{-}H} \rightleftharpoons \underset{\text{hydrogen ion}}{H^+} + \underset{\text{hydroxide ion}}{OH^-}$$

Only a few water molecules at a time dissociate, and the actual number of H^+ or OH^- is 10^{-7} moles/liter. A **mole** is a unit of scientific measurement for atoms, ions, and molecules.

MP3 Water and pH

Acidic Solutions (High H^+ Concentrations)

Lemon juice, vinegar, tomatoes, and coffee are all acidic solutions. What do they have in common? **Acids** are substances that dissociate in water, releasing hydrogen ions (H^+). The acidity of a substance depends on how fully it dissociates in water. For example, an important inorganic acid is hydrochloric acid (HCl), which dissociates in this manner:

$$HCl \longrightarrow H^+ + Cl^-$$

Dissociation of HCl is almost complete. If hydrochloric acid is added to a beaker of water, the number of hydrogen ions (H^+) increases greatly. Therefore, HCl is called a *strong acid.* Hydrochloric acid is produced by the stomach and aids in food digestion.

Basic Solutions (Low H^+ Concentrations)

Milk of magnesia and ammonia are commonly known basic substances. **Bases** are substances that either take up hydrogen ions (H^+) or release hydroxide ions (OH^-). For example, an important base is sodium hydroxide (NaOH), which dissociates almost completely in this manner:

$$NaOH \longrightarrow Na^+ + OH^-$$

If sodium hydroxide is added to a beaker of water, the number of hydroxide ions increases. Thus, sodium hydroxide is called a *strong base.* Sodium hydroxide is also called *lye* and is contained in many drain-cleaning products.

You should not taste strong acids or bases because they are destructive to cells. Many household cleansers, such as ammonia or bleach, have poison symbols and carry a strong warning not to ingest the product.

pH Scale

The **pH scale** is used to indicate the acidity and basicity (alkalinity) of a solution. Pure water with an equal number of hydrogen ions (H^+) and hydroxide ions (OH^-) has a pH of exactly 7.

The pH scale was devised to simplify discussion of the hydrogen ion concentration [H^+] and consequently of the hydroxide ion concentration [OH^-]. It eliminates the use of cumbersome numbers. To understand the relationship between hydrogen ion concentration and pH, consider the following:

[H^+] (moles per liter)			pH
0.000001	=	1×10^{-6}	6
0.0000001	=	1×10^{-7}	7
0.00000001	=	1×10^{-8}	8

You will notice that to determine pH, each number is first expressed in scientific notation. The negative value of the exponent (remember, the exponent is the superscript number) equals pH. So for a solution of 0.000001 moles per liter of hydrogen ions, the scientific notation is 1×10^{-6}, and the pH is $-[-6]$ or pH 6.

Of the two values above and below pH 7, which one indicates a higher hydrogen ion concentration than pH 7 and, therefore, refers to an acidic solution? A number with a smaller negative exponent indicates a greater quantity of hydrogen ions (H^+) than one with a larger negative exponent. Therefore, the pH 6 solution is an acidic solution.

Basic solutions have fewer hydrogen ions (H^+) than hydroxide ions. Of the three values, the pH 8 solution is a basic solution because it indicates a lower hydrogen ion concentration [H^+] (greater hydroxide ion concentration) than the pH 7 solution.

The pH scale (Fig. 2.9) ranges from 0 to 14. As we move toward a higher pH, each unit has 10 times the basicity of the previous unit; and as we move toward a lower pH, each unit has 10 times the acidity of the previous unit.[2]

MP3 **Acid-Base Balances**

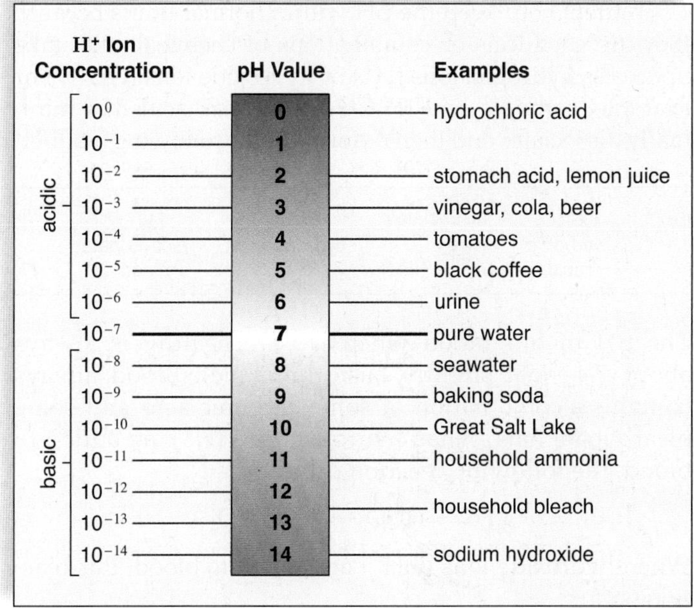

Figure 2.9 **The pH scale.**
The pH scale ranges from 0 to 14, with 0 being the most acidic and 14 being the most basic. A solution at pH 7 (neutral pH) has equal amounts of hydrogen ions (H^+) and hydroxide ions (OH^-). An acidic pH has more H^+ than OH^- and a basic pH has more OH^- than H^+. The pH of familiar solutions can be seen on the scale.

Buffers

In living things, the pH of body fluids needs to be maintained within a narrow range or else health suffers. Normally, pH stability is possible because the body and the environment have **buffers** to prevent pH changes. A problem arises when precipitation in the form of rain or snow becomes so acidic that the environment runs out of natural buffers in the soil or water. Rain normally has a pH of about 5.7, but very acidic rain—sometimes with a pH as low as 2.1—is recorded every year across the Appalachian Mountains. Why is the pH of the rain so low? Rain becomes acidic because the burning of fossil fuels emits sulfur dioxides (SO_2) and nitrogen oxides (NO_2) into the atmosphere. They combine with water to produce sulfuric acid (H_2SO_4) and nitric acid (HNO_3). Acid deposition destroys statues and kills forests (Fig. 2.10). It also leads to fish kills in lakes and streams.

a.

b.

Figure 2.10 **The effects of acid rain.**
Acid rain formed during the burning of fossil fuels causes **(a)** statues to deteriorate and **(b)** trees to die.

[2] pH is defined as the negative log of the hydrogen ion concentration [H^+]. In the logarithmic scale, a change from one whole number to another represents a change by a factor of 10. Therefore, a pH of 5.0 is 10 times more acidic than a pH of 6.0.

Buffers help keep the pH within normal limits because they are chemicals or combinations of chemicals that take up excess hydrogen ions (H^+) or hydroxide ions (OH^-). For example, carbonic acid (H_2CO_3) is a weak acid that minimally dissociates and then re-forms in the following manner:

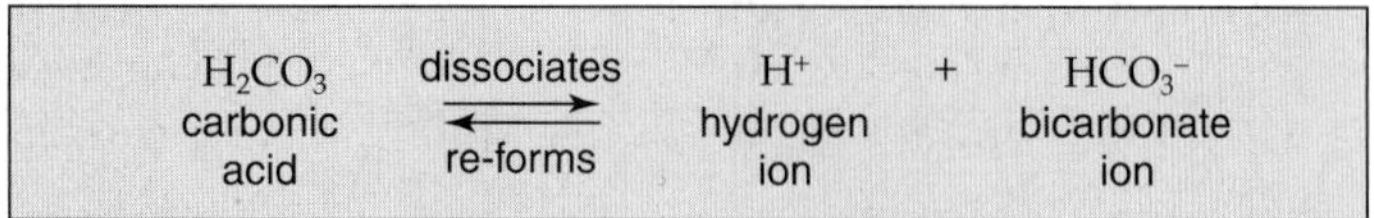

The pH of our blood when we are healthy is always about 7.4—just slightly basic (alkaline). Blood always contains a combination of some carbonic acid and some bicarbonate ions. When hydrogen ions (H^+) are added to blood, the following reaction occurs:

$$H^+ + HCO_3^- \longrightarrow H_2CO_3$$

When hydroxide ions (OH^-) are added to blood, this reaction occurs:

$$OH^- + H_2CO_3 \longrightarrow HCO_3^- + H_2O$$

These reactions prevent any significant change in blood pH.

Connecting the Concepts

The properties of water play an important role in human physiology and maintaining homeostasis. For a few examples, refer to the following discussions.

Section 5.6 provides a description of how the body moves nutrients using the circulatory system.

Section 10.4 examines how the kidneys maintain the water balance of the body.

Section 10.4 explores the relationship between the water balance of the body and homeostasis.

Check Your Progress 2.2

1. List the characteristics of water that help support life.
2. Contrast the hydrogen ion concentration of acids and bases.
3. Summarize why the pH of the environment (soil and water) and the body rarely changes.

2.3 Molecules of Life

Learning Outcomes

Upon completion of this section, you should be able to

1. List the four classes of organic molecules that are found in cells.
2. Describe the processes by which the organic molecules are assembled and disassembled.

Four categories of **organic molecules**—carbohydrates, lipids, proteins, and nucleic acids—are unique to cells. In biology, **organic** doesn't refer to how food is grown; it refers to a molecule that contains carbon (C) and hydrogen (H) and is usually associated with living things.

Each type of organic molecule in cells is composed of subunits. When a cell constructs a **macromolecule**, a molecule that contains many subunits, it uses a **dehydration reaction,** a type of synthesis reaction. During a dehydration reaction, an —OH (hydroxyl group) and an —H (hydrogen atom), the equivalent of a water molecule, are removed as the molecule forms (Fig. 2.11*a*). The reaction is reminiscent of a train whose length is determined by how many boxcars it has hitched together. To break down macromolecules, the cell uses a **hydrolysis reaction** in which the components of water are added during the breaking of the bond between the molecules (Fig. 2.11*b*).

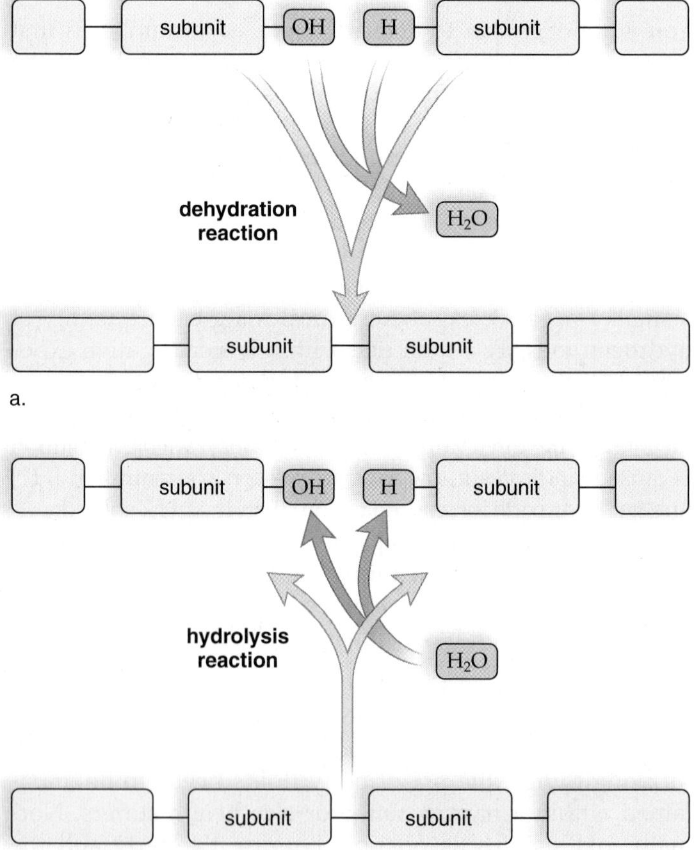

Figure 2.11 **The breakdown and synthesis of macromolecules.** **a.** Subunits bond in a dehydration reaction. Water is given off and a macromolecule forms. **b.** In the reverse reaction, macromolecules are divided into subunits by hydrolysis.

Connecting the Concepts

The organic nutrients not only play an important role in the building of cells but also serve as sources of energy. For more information, refer to the following discussions.

Section 3.6 explains how the body uses organic nutrients as energy sources.

Sections 8.1–8.4 explore how the digestive system processes the organic nutrients and prepares them for transport.

Check Your Progress 2.3

1. Explain the difference between an organic and an inorganic molecule.
2. List the four classes of molecules that are unique to cells.
3. Describe the type of reaction that occurs during the synthesis of a macromolecule.

2.4 Carbohydrates

Learning Outcomes

Upon completion of this section, you should be able to

1. Summarize the basic chemical properties of a carbohydrate.
2. State the roles of carbohydrates in human physiology.
3. Compare the structure of simple and complex carbohydrates.
4. Explain the importance of fiber in the diet.

Carbohydrate molecules are characterized by the presence of the atomic grouping H—C—OH, in which the ratio of hydrogen atoms (H) to oxygen atoms (O) is approximately 2:1. This ratio is the same as the ratio in water (*hydros* in Greek means "water," so the name "hydrates of carbon" seems appropriate). **Carbohydrates,** first and foremost, function for quick and short-term energy storage in all organisms, including humans.

Simple Carbohydrates

If a carbohydrate is made up of just one ring and its number of carbon atoms is low (from five to seven), it is called a simple sugar, or **monosaccharide.** The designation **pentose** means a 5-carbon sugar, and the designation **hexose** means a 6-carbon sugar. **Glucose,** the hexose our bodies use as an immediate source of energy, can be written in any one of these ways:

$C_6H_{12}O_6$

Other common hexoses are fructose, found in fruits, and galactose, a constituent of milk.

A **disaccharide** (*di,* "two"; *saccharide,* "sugar") is made by joining only two monosaccharides together by a dehydration reaction. Maltose is a disaccharide that is formed by a dehydration reaction between two glucose molecules (Fig. 2.12). When our hydrolytic digestive juices break down maltose, the result is two glucose molecules. When glucose and fructose join, the disaccharide sucrose forms. Sucrose, ordinarily derived from sugarcane and sugar beets, is commonly known as table sugar.

Complex Carbohydrates (Polysaccharides)

Macromolecules such as starch, glycogen, and cellulose are **polysaccharides** that contain many glucose units.

Starch and **glycogen** are readily stored forms of glucose in plants and animals, respectively. Some of the

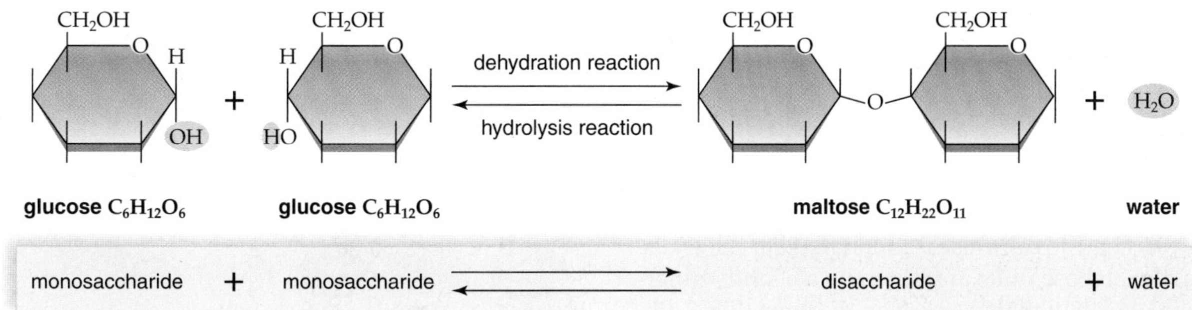

Figure 2.12 The synthesis and breakdown of a dissacharide.
Maltose, a dissacharide, is formed by a dehydration reaction between two glucose molecules. The breakdown of maltose occurs following a hydrolysis reaction and the addition of water.

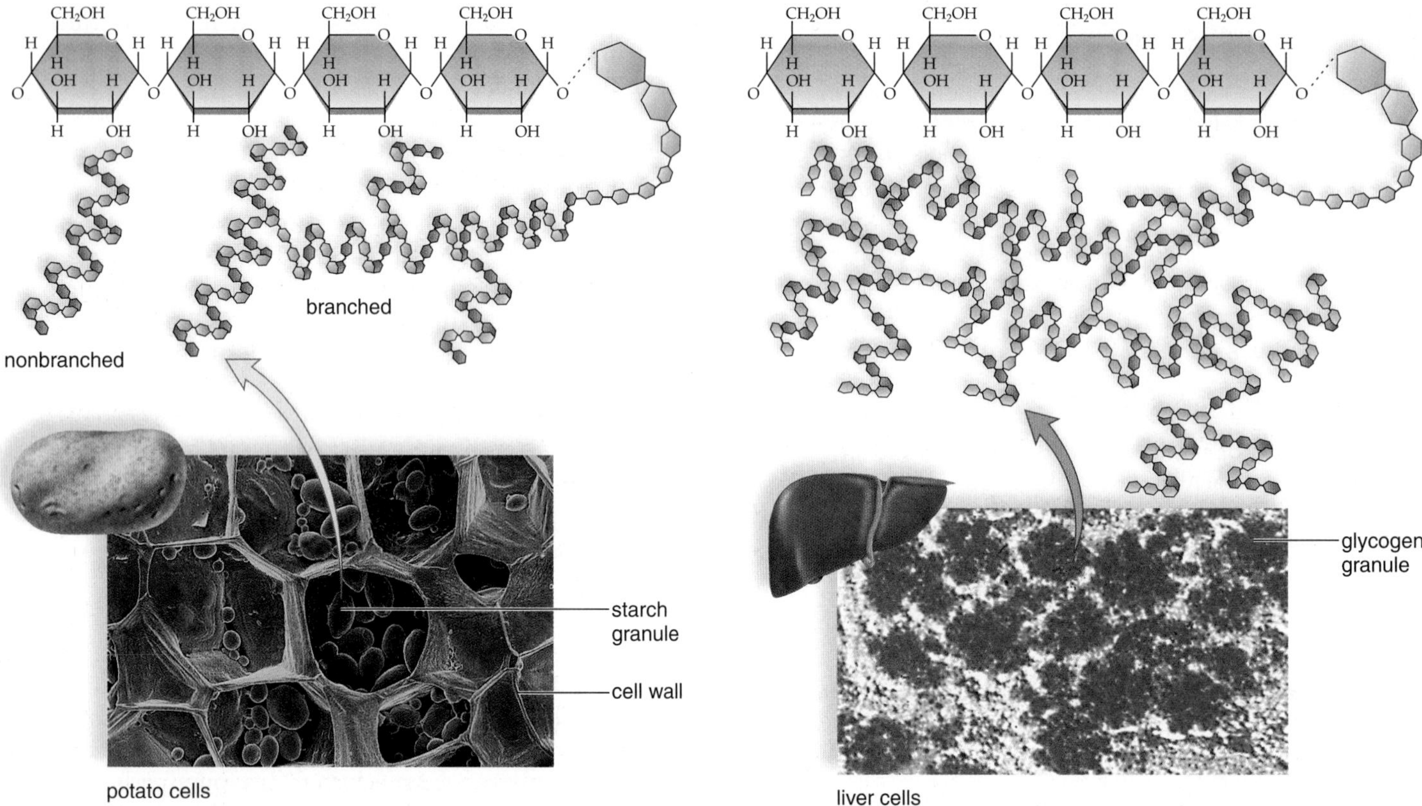

Figure 2.13 Starch is a plant complex carbohydrate. Starch has straight chains of glucose molecules. Some chains are also branched, as indicated. The electron micrograph shows starch granules in potato cells.

Figure 2.14 Glycogen is an animal complex carbohydrate. Glycogen is more branched than starch. The electron micrograph shows glycogen granules in liver cells.

macromolecules in starch are long chains of up to 4,000 glucose units. Starch and glycogen have slightly different structures. Starch has fewer side branches, or chains, than does glycogen. A side chain of glucose branches off from the main chain, as shown in Figures 2.13 and 2.14. Flour, used for baking and typically obtained from grinding wheat, is high in starch. Pasta and potatoes are also high-starch foods.

After we eat starchy foods, such as potatoes, bread, and cake, glucose enters the bloodstream and, normally, the liver stores glucose as glycogen. The release of the hormone insulin from the pancreas promotes the storage of glucose as glycogen. In between eating, the liver releases glucose so that, normally, the blood glucose concentration is always about 0.1%.

The polysaccharide **cellulose** is found in plant cell walls. In cellulose, the glucose units are joined by a slightly different type of linkage than that in starch or glycogen (Fig. 2.15). Though this might seem to be a technicality, it is important because we are unable to digest foods containing this type of linkage; therefore, cellulose largely passes through our digestive tract as fiber, or roughage.

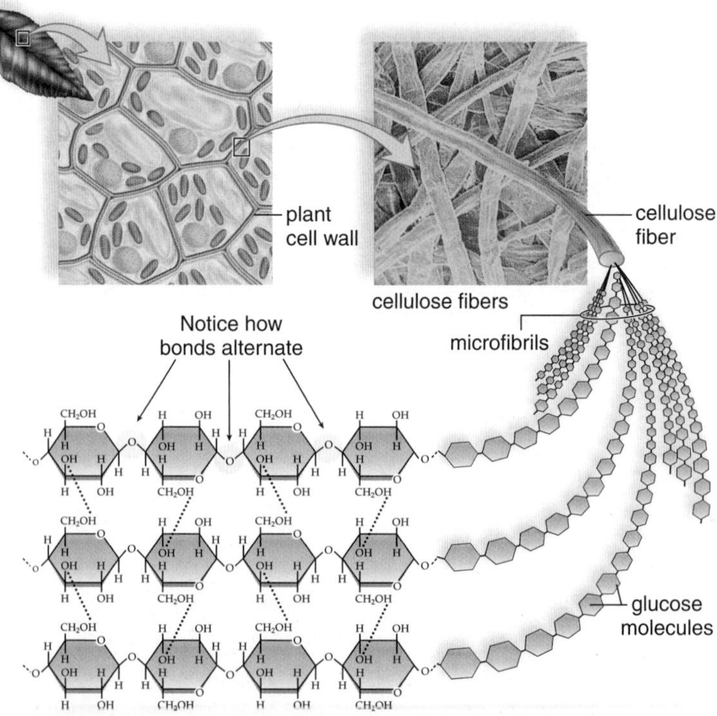

Figure 2.15 Fiber is a plant complex carbohydrate. Unlike starch, we don't get energy directly from fiber. This is because of the arrangements of the chemical bonds in the fiber chains. However, fiber is an important component of our diet and helps keep our digestive system healthy.

Biology Matters **Health Focus**

Fiber in the Diet

Fiber, also called *roughage,* is mainly composed of the undigested carbohydrates that pass through the digestive system. Most fiber is derived from the structural carbohydrates of plants. This includes such material as cellulose, pectins, and lignin. Fiber is not truly a nutrient because we do not use it directly for energy or building cells, but it is an extremely important component of our diet. Fiber adds bulk to material in the intestines, keeping the colon functioning normally, and it binds many types of harmful chemicals in the diet, including cholesterol, and prevents them from being absorbed.

There are two basic types of fiber—insoluble and soluble. Soluble fiber dissolves in water and acts in the binding of cholesterol. Soluble fiber is found in many fruits, as well as oat grains. Insoluble fiber provides the bulk to the fecal material and is found in bran, nuts, seeds, and whole wheat foods.

An adult male should bring in around 38 grams per day of fiber, whereas the average adult female should consume approximately 25 grams per day. One serving of whole grain bread (one slice) provides about 3 grams of fiber, and a single serving of beans (1/2 cup) contains 4 to 5 grams of fiber. A diet high in fiber has been shown to reduce the risk of cardiovascular disease, diabetes, colon cancer, and diverticulosis.

Connecting the Concepts

In humans, carbohydrates primarily serve as energy molecules, although fiber does act to ensure the health of the digestive system. For more information on the interaction of the human body with carbohydrates, refer to the following discussions.

Section 3.6 explores the use of carbohydrates for energy at the cellular level.

Section 8.3 examines how the digestive system processes carbohydrates.

Section 8.6 explains how fiber promotes the health of the digestive system.

Check Your Progress 2.4

1. Explain the function of simple carbohydrates and polysaccharides in humans.
2. Describe the difference in structure between a simple carbohydrate and the various complex carbohydrates.
3. Summarize the benefits of fiber in the diet.

MP3 Carbohydrates

2.5 Lipids

Learning Outcomes

Upon completion of this section, you should be able to

1. Compare the structure of fats, phospholipids, and steroids.
2. State the function of each class of lipids.

Lipids are diverse in structure and function, but they have a common characteristic: They do not dissolve in water. Their low solubility in water is due to an absence of polar groups. They contain little oxygen and consist mostly of carbon and hydrogen atoms.

Lipids contain more energy per gram than other biological molecules; therefore, fats in animals and oils in plants function well as energy-storage molecules. Others (phospholipids) form a membrane so that the cell is separated from its environment and has inner compartments as well. Steroids are a large class of lipids that includes, among other molecules, the sex hormones.

Fats and Oils

The most familiar lipids are those found in fats and oils. **Fats,** usually of animal origin (e.g., lard and butter), are solid at room temperature. **Oils,** usually of plant origin (e.g., corn oil and soybean oil), are liquid at room temperature. Fat has several functions in the body. It is used for long-term energy storage, insulates against heat loss, and forms a protective cushion around major organs. Steroids are formed from smaller lipid molecules and function as chemical messengers.

Emulsifiers can cause fats to mix with water. They contain molecules with a nonpolar end and a polar end. The molecules position themselves about an oil droplet so that their polar ends project outward. The droplet *disperses* in water, which means that **emulsification** has occurred. Emulsification takes place when dirty clothes are washed with soaps or detergents. Also, prior to the digestion of fatty foods, fats are emulsified by bile. The liver manufactures bile and the gallbladder stores it.

Fats and oils form when one glycerol molecule reacts with three fatty acid molecules (Fig. 2.16). A fat is sometimes called a **triglyceride** because of its three-part structure, or the term *neutral fat* can be used because the molecule is nonpolar and carries no charges.

Waxes are molecules made up of one fatty acid combined with another single organic molecule, usually an alcohol (chemists refer to "alcohols" as an entire group of molecules that includes drinking alcohol and rubbing alcohol). Waxes

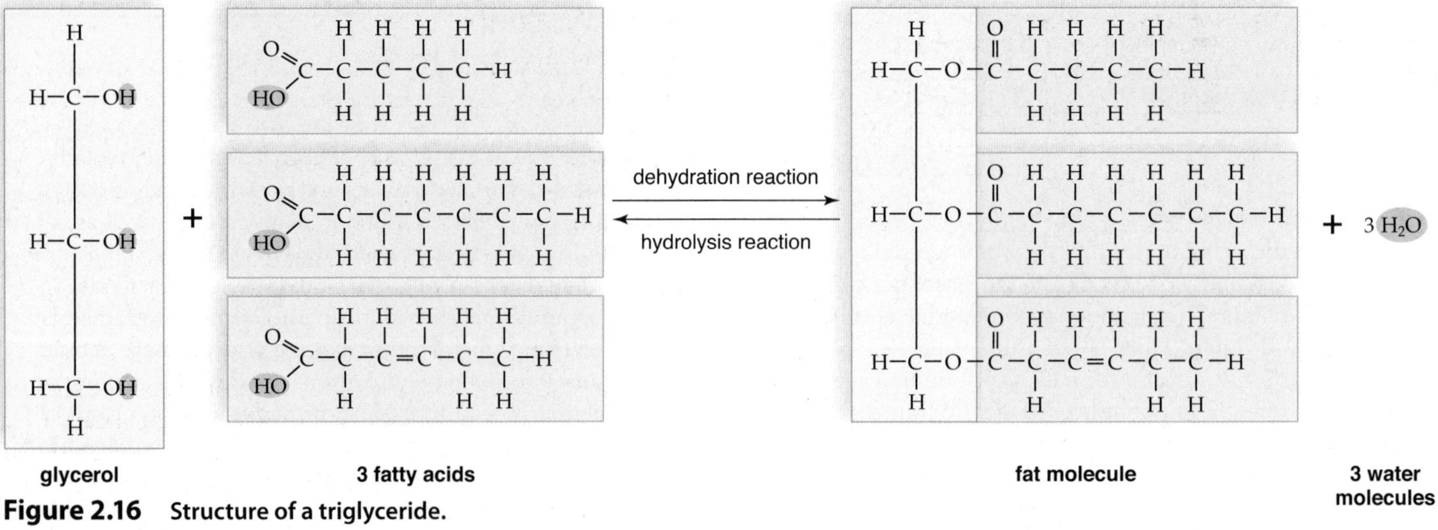

Figure 2.16 **Structure of a triglyceride.**
Triglycerides are formed when three fatty acids combine with glycerol by dehydration synthesis reactions. The reverse reaction starts digestion of fat; hydrolysis introduces water, and fatty acid–glycerol bonds are broken.

prevent loss of moisture from body surfaces. Cerumen, or ear wax, is a very thick wax produced by glands lining the outer ear canal (see Chap. 14). It protects the ear canal from irritation and infection by trapping particles, bacteria, and viruses. When ear wax is completely washed away by swimming or diving, the result is a painful "swimmer's ear."

Saturated, Unsaturated, and Trans-Fatty Acids

A **fatty acid** is a carbon–hydrogen chain that ends with the acidic group —COOH (Fig. 2.16, *left*). Most of the fatty acids in cells contain 16 or 18 carbon atoms per molecule, although smaller ones with fewer carbons are also known. Fatty acids are either saturated or unsaturated. **Saturated fatty acids** have no double bonds between the carbon atoms. The chain is saturated, so to speak, with all the hydrogens it can hold. **Unsaturated fatty acids** have double bonds in the carbon chain wherever the number of hydrogens is less than two per carbon (Fig. 2.17).

In general, oils, present in cooking oils and bottle margarines, are liquids at room temperature because the presence of a double bond creates a bend in the fatty acid chain. Such kinks prevent close packing between the hydrocarbon chains and account for the fluidity of oils. On the other hand, butter, which contains saturated fatty acids and no double bonds, is a solid at room temperature.

Saturated fats, in particular, contribute to the disease *atherosclerosis.* Atherosclerosis is caused by formation of lesions, or *atherosclerotic plaques,* on the inside of blood vessels. The plaques narrow blood vessel diameter, choking off the blood and oxygen supply to tissues. Atherosclerosis is the primary cause of cardiovascular disease (heart attack and stroke) in the United States. Even more harmful than naturally occurring saturated fats are the so-called **trans fats** (*trans* in Latin means "across"; Fig. 2.17), which are created artificially using vegetable oils. Trans fats may be partially hydrogenated to make them semisolid. Complete hydrogenation of oils causes all double bonds to become saturated. Partial hydrogenation does not saturate all bonds. It reconfigures some double bonds, but some of the hydrogen atoms end up on different sides of the chain. Trans fats are found in shortenings and solid margarines. They also occur in processed foods (snack foods, baked goods, and fried foods).

Current dietary guidelines from the American Heart Association (AHA) advise replacing trans fats with unsaturated oils. In particular, monounsaturated oils (like olive oil, with one double bond in the carbon chain) are recommended. Polyunsaturated oils (many double bonds in the carbon chain) such as corn oil, canola oil, and safflower oil also fit in the AHA guidelines.

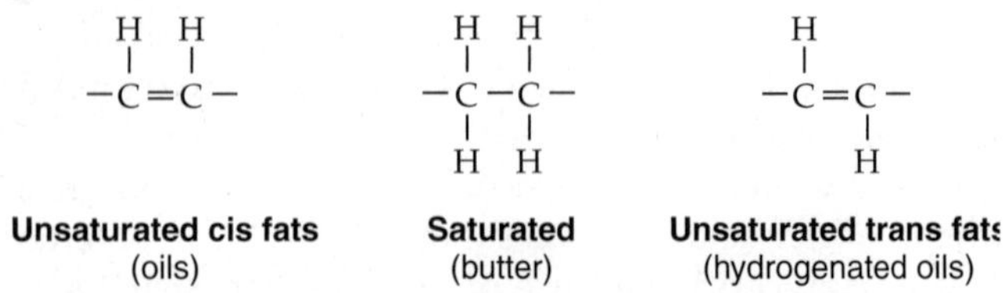

Figure 2.17 **Comparison of saturated, unsaturated, and trans fats.**
Saturated fats have no double bonds between carbons in the fatty acid. Unsaturated fats have one or more double bonds in the fatty acid. For a fat to be a trans fat, the hydrogens need to be on opposite sides of the carbon–carbon double bond.

Dietary Fat

For good health, the diet should include some fat; but, for the reasons stated, the first thing to do when looking at a nutrition label is to check the total amount of fat per serving. The total recommended amount of fat in a 2,000-calorie diet is 65 grams. That information results in the % Daily Value (DV) given in the sample nutrition label for macaroni and cheese in Figure 2.18. As of January 2006, food manufacturers are required to list the amount of trans fats greater than 0.5 g in the nutrition label for a food.

Animation
Food Label

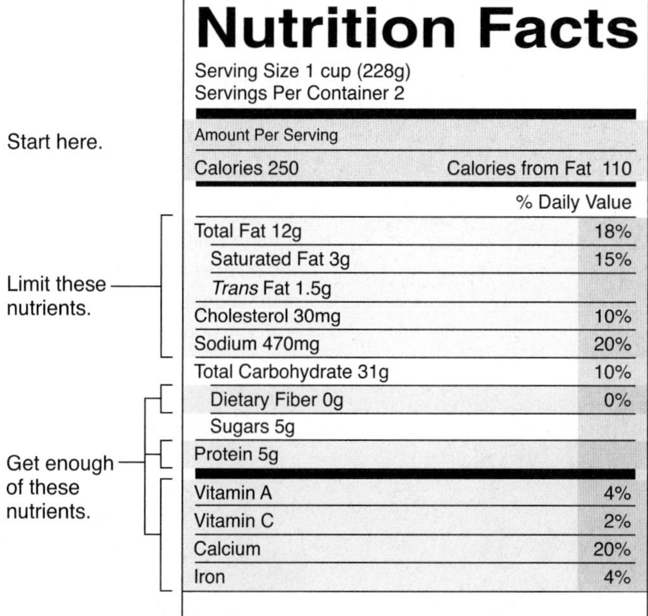

Nutrition Facts

Serving Size 1 cup (228g)
Servings Per Container 2

Amount Per Serving	
Calories 250	Calories from Fat 110
	% Daily Value
Total Fat 12g	18%
Saturated Fat 3g	15%
Trans Fat 1.5g	
Cholesterol 30mg	10%
Sodium 470mg	20%
Total Carbohydrate 31g	10%
Dietary Fiber 0g	0%
Sugars 5g	
Protein 5g	
Vitamin A	4%
Vitamin C	2%
Calcium	20%
Iron	4%

Figure 2.18 Understanding a food label.
Food labels provide some important information about the product. Each of the items listed on the label is referenced to the % Daily Value, which is based on a 2,000-calorie diet. In general, total fat, cholesterol, and sodium should be limited in the diet.

Biology Matters Health Focus

The Omega-3 Fatty Acids

Not all fats are bad. In fact, some of them are essential to our health. A special class of unsaturated fatty acids, the omega-3 fatty acids, is considered both an essential and developmentally important nutrient. The name *omega-3* (also called *n-3 fatty acids*) is derived from the location of the double bond in the carbon chain.

The three important omega-3 fatty acids are linolenic acid (ALA), docosahexaenoic acid (DHA), and eicosapentaenoic acid (EPA). Omega-3 fatty acids are a major component of the fatty acids in the brain, and adequate amounts of them appear to be important in children and young adults. A diet that is rich in these fatty acids also offers protection against cardiovascular disease, and research is ongoing with regard to other health benefits. DHA may reduce the risk of Alzheimer disease. DHA and EPA may be manufactured from APA in small amounts within our bodies. Some of the best sources of omega-3 fatty acids are cold-water fish such as salmon and sardines. Flax oil, also called *linseed oil,* is an excellent plant-based source of omega-3 fatty acids.

Although the fatty acids are an important component of the diet, nutritionists warn not to overdo the diet with excessive supplements as the omega-3s may cause health issues when taken in large doses.

Phospholipids

Phospholipids have a phosphate group (Fig. 2.19). They are constructed like fats, except that in place of the third fatty acid, there is a phosphate group or a grouping that contains both phosphate and nitrogen. These molecules are not electrically neutral, as are fats, because the phosphate and nitrogen-containing groups are ionized. They form the polar (hydrophilic) head of the molecule, and the rest of the molecule becomes the nonpolar (hydrophobic) tails. (Remember that *hydrophilic* means "water-loving" and *hydrophobic* is "water-fearing.")

Phospholipids are the primary components of cellular membranes. They spontaneously form a *bilayer* (a sort of molecular "sandwich") in which the hydrophilic heads (the sandwich "bread") face outward toward watery solutions, and the tails (the sandwich "filling") form the hydrophobic interior (Fig. 2.19*b*).

Steroids

Steroids are lipids that have an entirely different structure from those of fats. Steroid molecules have a backbone of four fused carbon rings. Each one differs primarily by the attached molecules, called *functional groups,* attached to the rings. Cholesterol is a component of an animal cell's plasma membrane and is the precursor of several other steroids, such as the sex hormones estrogen and testosterone. The liver usually makes all the cholesterol the body needs.

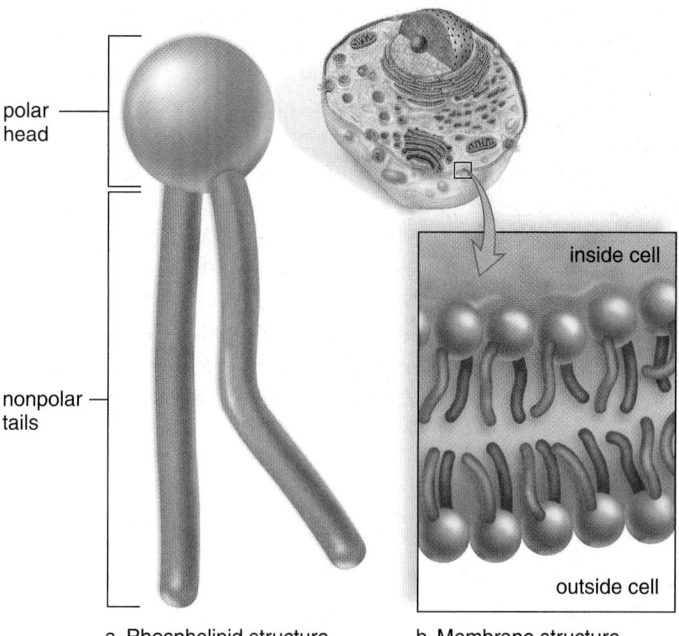

Figure 2.19 Structure of a phospholipid.
a. Phospholipids are structured like fats with one fatty acid is replaced by a polar phosphate group. Therefore, the head is polar, whereas the tails are nonpolar. **b.** This causes the molecules to arrange themselves in a "sandwich" arrangement when exposed to water—polar phosphate groups on the outside of the layer, nonpolar lipid tails on the inside of the layer.

Dietary sources should be restricted because elevated levels of cholesterol, saturated fats, and trans fats are linked to atherosclerosis, a disease of the blood vessels in which fatty plaques accumulate inside blood vessel linings and reduce blood flow (see page 206).

The male sex hormone, testosterone, is formed primarily in the testes; the female sex hormone, estrogen, is formed primarily in the ovaries. Testosterone and estrogen differ only by the functional groups attached to the same carbon backbone. However, they have a profound effect on the body and the sexuality of humans and other animals (Fig. 2.20). The taking of anabolic steroids, usually to build muscle strength, is illegal because the side effects are harmful to the body (see Chap. 12, Bioethical Focus).

Connecting the Concepts

Fats and lipids have a variety of uses in the human body. To see how they interact with the various systems of the body, refer to the following discussions.

Section 5.7 provides more information on atherosclerosis.
Section 8.3 explores the digestion and absorption of fats.
Section 15.1 examines how some lipids act as hormones in the body.

Biology Matters Health Focus

Good and Bad Cholesterol

Blood tests to analyze your lipid profile are part of many annual medical exams. After your annual exam, your doctor tells you that your total cholesterol is 210, your HDL value is low (34), and your LDL value is high (84). You know that you need to get your total cholesterol below 200, the threshold of a healthy diet; but like most people, you have no idea what the other two numbers mean. Then you remember that LDL is commonly referred to as "bad" cholesterol, and HDL is called the "good" cholesterol. In actuality, these molecules are not forms of cholesterol; they are types of proteins. The lipoproteins in the body serve as a form of fat and cholesterol carrier, moving these nutrients around as needed. An LDL is a lipoprotein that is full of triglycerides and cholesterol, whereas an HDL is basically empty. Thus, a high LDL value indicates that your carriers were usually full, meaning that your diet must be providing too many of these nutrients. After additional research, you find out that other factors, such as the amount of dietary fiber, daily exercise, and even genetics, can play a role in regulating "good" and "bad" levels of these lipoproteins. Furthermore, the numbers that the doctor gave you were actually concentrations of these molecules in your blood (in milligrams per deciliter, or mg/dL). In today's world, it is important that we all understand the terminology associated with our own medical history.

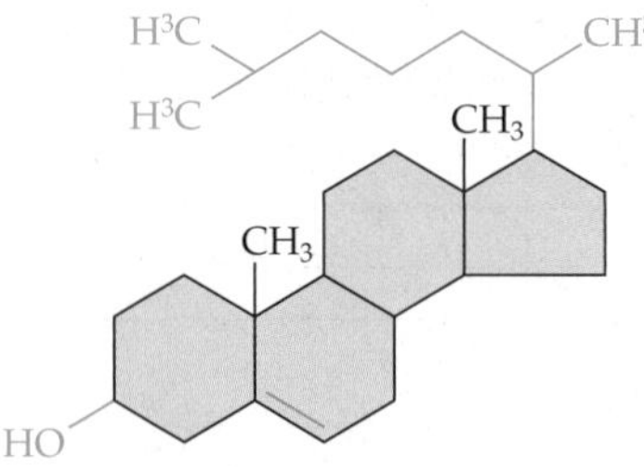

a. Cholesterol

b. Testosterone

OH
CH3
HO

c. Estrogen

Figure 2.20 **Examples of steroids.** **a.** All steroids are made from cholesterol and have four carbon rings. Compare the structure of **(b)** testosterone and **(c)** estrogen, and notice the slight changes in their attached groups (shown in blue).

Check Your Progress 2.5

1. State the function of fats and oils in the human body.
2. List the uses of phospholipids and steroids in the body.
3. Hypothesize what would happen if all fats were removed from the body.

2.6 Proteins

Learning Outcomes

Upon completion of this section, you should be able to

1. Describe the structure of an amino acid.
2. Explain how amino acids are combined to form proteins.
3. Summarize the four levels of protein structure.

Proteins are of primary importance in the structure and function of cells. Some of their many functions in humans follow.

Hair is a protein.

Support: Some proteins are structural proteins. Keratin, for example, makes up hair and nails. Collagen lends support to ligaments, tendons, and skin.

Enzymes: Enzymes bring reactants together and thereby speed chemical reactions in cells. They are specific for one particular type of reaction and only function at body temperature.

Transport: Channel and carrier proteins in the plasma membrane allow substances to enter and exit cells. Some other proteins transport molecules in the blood of animals; **hemoglobin** in red blood cells is a complex protein that transports oxygen.

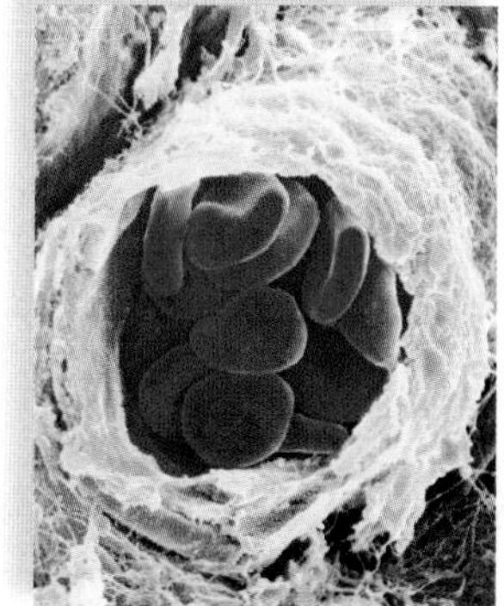
Hemoglobin is a protein.

Defense: Antibodies are proteins. They combine with foreign substances, called antigens. In this way, they prevent antigens from destroying cells and upsetting homeostasis.

Hormones: Hormones are regulatory proteins. They serve as intercellular messengers that influence the metabolism of cells. The hormone insulin regulates the content of glucose in the blood and in cells. The presence of growth hormone determines the height of an individual.

Motion: The contractile proteins actin and myosin allow parts of cells to move and cause muscles to contract. Muscle contraction facilitates the movement of animals from place to place.

Muscle contains protein.

The structures and functions of vertebrate cells and tissues differ according to the type of proteins they contain. For example, muscle cells contain actin and myosin, red blood cells contain hemoglobin, and support tissues contain collagen.

Amino Acids: Subunits of Proteins

Proteins are macromolecules with **amino acid** subunits. The central carbon atom in an amino acid bonds to a hydrogen atom and also to three other groups of atoms. The name *amino acid* is appropriate because one of these groups is an —NH_2 (amino group) and another is a —COOH (carboxyl group, an acid). The third group is the *R* group for an amino acid:

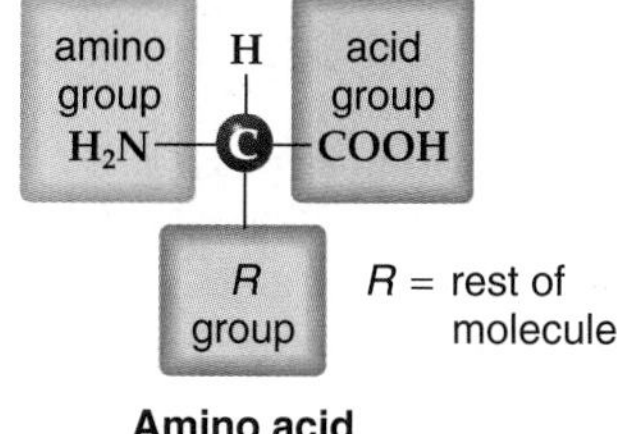

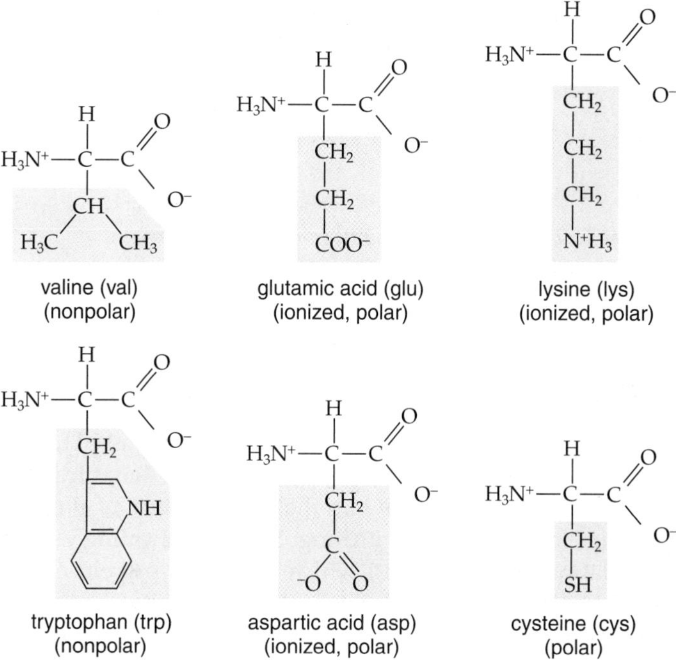

Figure 2.21 **The structure of a few amino acids.**
Amino acids all have an amine group (H_3N^+), an acid group (COO^-), and an *R* group, all attached to the central carbon atom. The *R* groups (screened in blue) are all different. Some *R* groups are nonpolar and hydrophobic; others are polar and hydrophilic. Still others are polar and ionized.

Amino acids differ according to their particular *R* group. The *R* groups range in complexity from a single hydrogen atom to a complicated ring compound. Some *R* groups are polar and some are not. Also, the amino acid cysteine ends with an —SH group, which often serves to connect one chain of amino acids to another by a disulfide bond, —S—S—. Several amino acids commonly found in cells are shown in Figure 2.21.

Peptides

Figure 2.22 shows how two amino acids join by a dehydration reaction between the carboxyl group of one and the amino group of another. The covalent bond between two amino acids is called a **peptide bond.** When three or more amino acids are linked by peptide bonds, the chain that results is called a **polypeptide.** The atoms associated with the peptide bond share the electrons unevenly because oxygen attracts electrons more than nitrogen. Therefore, the hydrogen attached to the nitrogen has a slightly positive charge (δ^+), whereas the oxygen has a slightly negative charge (δ^-):

O^{δ^-} — peptide bond
—C—N—
H^{δ^+}

δ^- = slightly negative
δ^+ = slightly positive

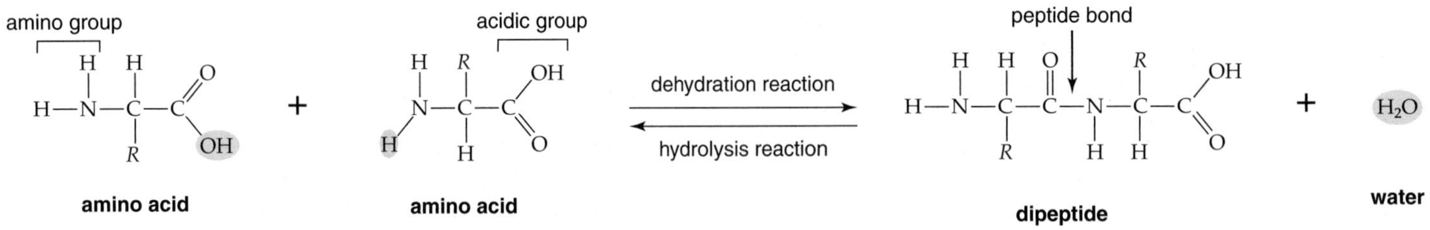

amino group
acidic group
peptide bond
dehydration reaction
hydrolysis reaction
amino acid
amino acid
dipeptide
H_2O
water

Figure 2.22 Synthesis and breakdown of a protein.
Amino acids join by peptide bonds using a dehydration reaction, and a water molecule is given off. In the reverse reaction, peptide bonds are broken by hydrolysis, and a water molecule is introduced.

Shape of Proteins

Proteins cannot function unless they have a specific shape. When proteins are exposed to extremes in heat and pH, they undergo an irreversible change in shape called **denaturation.** For example, we are all aware that the addition of vinegar (an acid) to milk causes curdling. Similarly, heating causes coagulation of egg whites, which contain a protein called albumin. Denaturation occurs because the normal bonding between the *R* groups has been disturbed. Once a protein loses its normal shape, it is no longer able to perform its usual function. Researchers recognize a change in protein shape is responsible for both Alzheimer disease or Creutzfeldt–Jakob disease (the human form of mad cow disease).

Animation Protein Denaturation

Levels of Protein Organization

The structure of a protein has at least three levels of organization and can have four levels (Fig. 2.23). The first level, called the *primary structure,* is the linear sequence of the amino acids joined by peptide bonds. Each particular polypeptide has its own sequence of amino acids.

The *secondary structure* of a protein comes about when the polypeptide takes on a certain orientation in space. Once amino acids are assembled into a polypeptide, the resulting C=O section between amino acids in the chain is polar, having a partially negative charge. (Remember that oxygen holds on to electrons longer than carbon, and that's what causes the partially negative charge.) Hydrogen bonding is possible between the C=O of one amino acid and the N—H of another amino acid in a polypeptide. Coiling of the chain results in an α (alpha) helix, or a right-handed spiral , and a folding of the chain results in a pleated sheet. Hydrogen bonding between peptide bonds holds the shape in place.

The *tertiary structure* of a protein is its final, three-dimensional shape. In enzymes, the polypeptide bends and twists in different ways. In most enzymes, the hydrophobic portions are packed on the inside and the hydrophilic portions are on the outside where they can make contact with water. The tertiary structure of the enzymes determines what types of molecules with which they will interact. The tertiary shape of a polypeptide is maintained by various types of bonding between the *R* groups; covalent, ionic, and hydrogen bonding all occur.

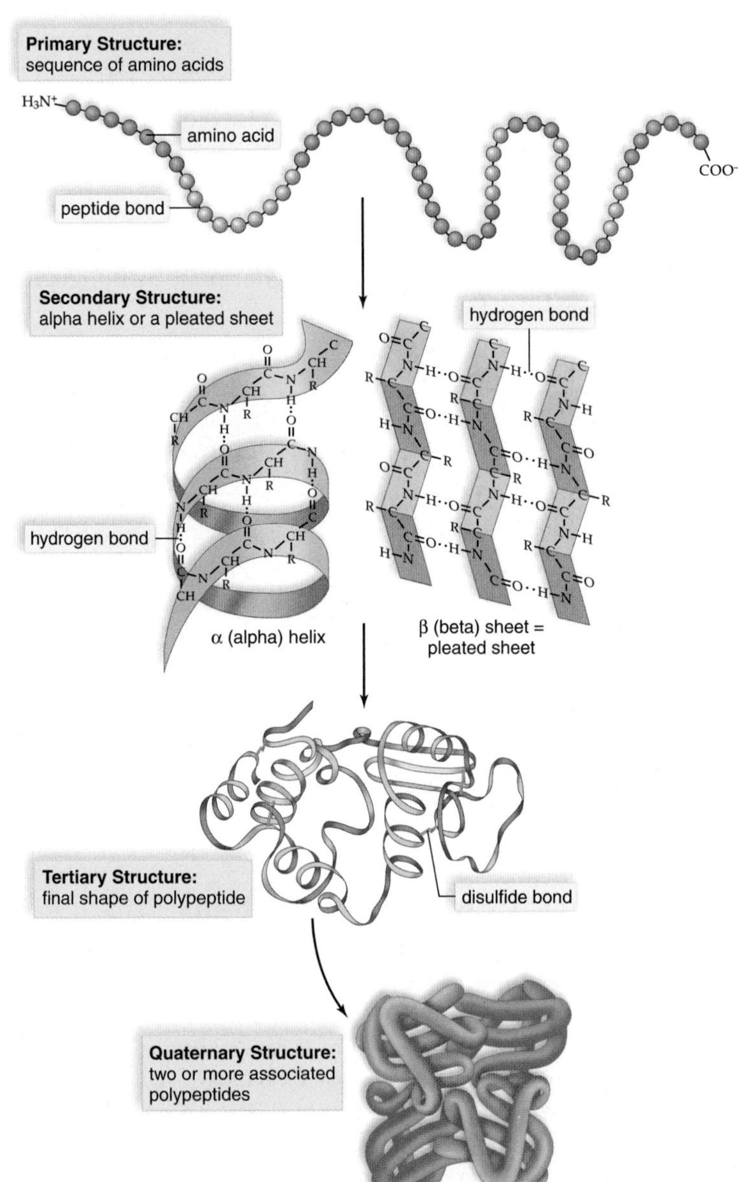

Figure 2.23 Levels of protein structure.
The structure of proteins can differ significantly. Primary structure, the sequence of amino acids, determines secondary and tertiary structure. Quaternary structure is created by assembling smaller proteins into a large structure.

Some proteins have only one polypeptide, and others have more than one polypeptide, each with its own primary, secondary, and tertiary structures. These separate polypeptides are arranged to give these proteins a fourth level of structure, termed the *quaternary structure.* **Hemoglobin** is a complex protein having a quaternary structure; many enzymes also have a quaternary structure. Each of four polypeptides in hemoglobin are tightly associated with a nonprotein *heme* group. A heme group contains an iron (Fe) atom that binds to oxygen; in that way, hemoglobin transports O_2 to the tissues.

Connecting the Concepts

Almost every function of the body is somehow connected to the activity of a protein. For more information on these processes, refer to the following discussions.

Section 7.1 gives more information on how misfolded proteins may cause disease.

Section 8.3 explains how the digestive system processes proteins.

Section 13.1 explores how some proteins are used as neurotransmitters in the nervous system.

Section 21.2 examines the process of protein synthesis in a cell.

Check Your Progress 2.6

1. Describe the major functions of proteins in organisms.
2. Explain the structure of an amino acid.
3. Describe how the shape of a protein relates to its function.

2.7 Nucleic Acids

Learning Outcomes

Upon completion of this section, you should be able to

1. Explain the differences between RNA and DNA.
2. Summarize the role of ATP in cellular reactions.

The two types of nucleic acids are **DNA (deoxyribonucleic acid)** and **RNA (ribonucleic acid)** (Fig. 2.24). Early investigators called them *nucleic acids* because they were first detected in the nucleus of cells. The discovery of the structure of DNA has had an enormous influence on biology and on society in general. DNA stores genetic information in the cell and in the organism. DNA replicates and transmits this information when each cell—and each organism—reproduces. Researchers are beginning to understand how genes function and are working on ways to manipulate them. The science of biotechnology is largely devoted to altering the genes in living organisms.

Function of DNA and RNA

Each DNA molecule contains many genes, and genes specify the sequence of the amino acids in proteins. RNA is an intermediary that conveys DNA's instructions regarding the amino acid sequence in a protein. If DNA's information is faulty, illness can result. The relationship between a gene, a protein, and an illness is illustrated by sickle-cell disease. In sickle-cell disease,

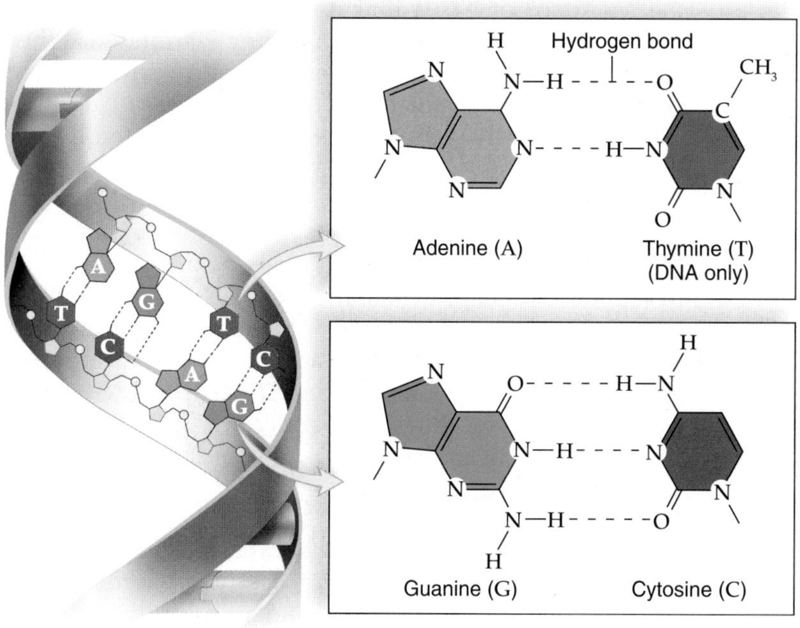

Figure 2.24 **The structure of DNA and RNA.**
a. In DNA, adenine and thymine are a complementary base pair. Note the hydrogen bonds that join them (like the "steps" in a spiral staircase). Likewise, guanine and cytosine can pair. **b.** RNA has uracil instead of thymine, so complementary base pairing isn't possible.

the individual's red blood cells are sickle-shaped. This occurs because in one particular spot in the hemoglobin molecule, an amino acid called valine substitutes for an amino acid called glutamine. Exchanging one amino acid for another—a seemingly small change—makes red blood cells lose their normal round, flexible shape and become weak and easily torn. Profound effects on the person's health result. When these abnormal red blood cells go through small blood vessels, they clog the flow of blood and break apart. Sickle-cell disease is another cause of anemia, and it also results in pain and organ damage.

How the Structure of DNA and RNA Differs

Though both DNA and RNA are polymers of nucleotides, there are some small differences in the types of subunits each contains and in their final structure. These differences give DNA and RNA their unique functions in the body.

Nucleotide Structure

Each **nucleotide** is a molecular complex of three types of subunit molecules—phosphate (phosphoric acid), a pentose (5-carbon) sugar, and a nitrogen-containing base:

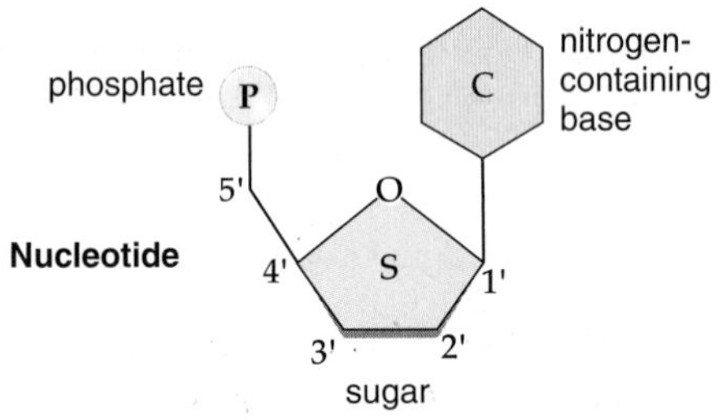

The nucleotides in DNA contain the sugar deoxyribose, and the nucleotides in RNA contain the sugar ribose; this difference accounts for their respective names (Table 2.1). There are four different types of bases in DNA: **adenine (A), thymine (T), guanine (G),** and **cytosine (C).** The base can have two rings (adenine or guanine) or one ring (thymine or cytosine). In RNA, the base **uracil (U)** replaces the base thymine. These structures are called *bases* because their presence raises the pH of a solution.

Table 2.1 DNA Structure Compared to RNA Structure

	DNA	RNA
Sugar	Deoxyribose	Ribose
Bases	Adenine, guanine, thymine, cytosine	Adenine, guanine, uracil, cytosine
Strands	Double-stranded with base pairing	Single-stranded
Helix	Yes	No

Polynucleotide Structure

The nucleotides link to make a polynucleotide called a *strand,* which has a backbone made up of phosphate–sugar–phosphate–sugar. The bases project to one side of the backbone. The nucleotides of a gene occur in a definite order, and so do the bases. After many years of work, researchers now know the sequence of the bases in human DNA—the human genome. This breakthrough is expected to lead to improved genetic counseling, gene therapy, and medicines to treat the causes of many human illnesses.

DNA is double-stranded, with the two strands twisted about each other in the form of a *double helix* (see Fig. 2.24*a*). In DNA, the two strands are held together by hydrogen bonds between the bases. When coiled, DNA resembles a spiral staircase. When unwound, it resembles a stepladder. The uprights (sides) of the ladder are made entirely of phosphate and sugar molecules, and the rungs of the ladder are made only of **complementary paired bases.** Thymine (T) always pairs with adenine (A), and guanine (G) always pairs with cytosine (C). Complementary bases have shapes that fit together.

Complementary base pairing allows DNA to replicate in a way that ensures that the sequence of bases will remain the same. This is important because it is the sequence of bases that determines the sequence of amino acids in a protein. RNA is single-stranded. When RNA forms, complementary base pairing with one DNA strand passes the correct sequence of bases to RNA (Fig. 2.24*b*). RNA is the nucleic acid directly involved in protein synthesis.

ATP: An Energy Carrier

In addition to being the subunits of nucleic acids, nucleotides have metabolic functions. When adenosine (adenine plus ribose) is modified by the addition of three phosphate groups instead of one, it becomes **ATP (adenosine triphosphate),** which is an energy carrier in cells.

Structure of ATP Suits Its Function

ATP is a high-energy molecule because the last two phosphate bonds are unstable and easily broken. Usually in cells, the last phosphate bond is hydrolyzed, leaving the molecule **ADP (adenosine diphosphate)** and

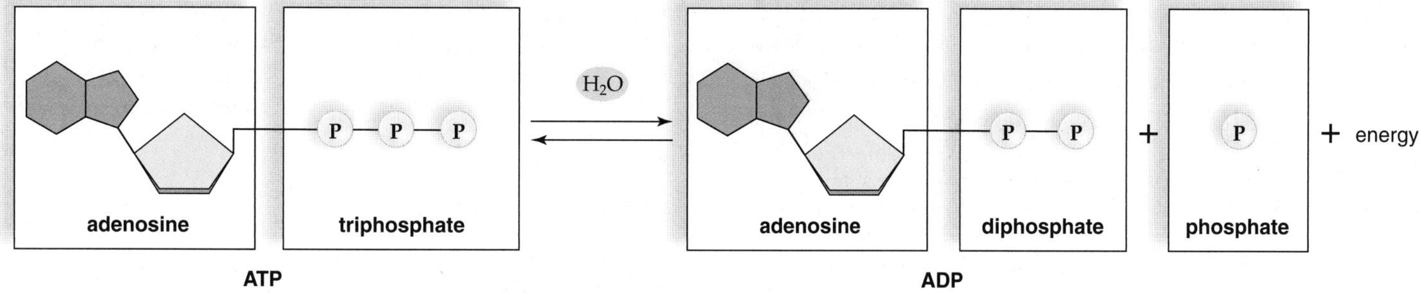

Figure 2.25 ATP is the universal energy currency of cells.
ATP is composed of the base adenosine and three phosphate groups (called a triphosphate). When cells need energy, ATP is hydrolyzed (water is added) forming ADP and Ⓟ. Energy is released. To recycle ATP, energy from food is required and the reverse reaction occurs: ADP and Ⓟ join to form ATP, and water is given off.

a molecule of inorganic phosphate Ⓟ (Fig. 2.25). The energy released by ATP breakdown is used by the cell to synthesize macromolecules, such as carbohydrates and proteins. In muscle cells, the energy is used for muscle contraction; and in nerve cells, it is used for the conduction of nerve impulses. After ATP breaks down, it can be recycled by adding Ⓟ to ADP. Notice in Figure 2.25 that an input of energy is required to re-form ATP.

MP3 ATP

Glucose Breakdown Leads to ATP Buildup

A glucose molecule contains too much energy to be used as a direct energy source in cellular reactions. Instead, the energy of glucose is converted to that of ATP molecules. ATP contains an amount of energy that makes it usable to supply energy for chemical reactions in cells. Muscles use ATP energy and produce heat when they contract. This is the heat that warms the body.

As we shall see in Chapter 3, oxygen is involved in the breakdown of glucose. Insufficient oxygen limits glucose breakdown and limits ATP buildup.

MP3 Nucleic Acids

Connecting the Concepts

As the information-carrying and energy molecules of the body, the nucleic acids play an important role in how our cells, tissues, and organs function. For more information on this class of molecules, refer to the following discussions.

Section 21.1 provides a more detailed look at the structure of DNA and RNA.

Sections 18.1–18.5 explain the relationship between cell division and patterns of inheritance.

Section 21.2 explores how DNA contains the information to make proteins.

Section 21.4 examines how advances in biotechnology are giving scientists the ability to manipulate DNA in the laboratory.

Check Your Progress 2.7

1. Describe the structure of ATP.
2. State the type of bond that joins the bases within a DNA double helix. Why are these bonds used?
3. Compare the structure of DNA and RNA. What impact do these differences have on their function?

CASE STUDY CONCLUSION

After three months of work, David felt more prepared for his visit with his physician. Not only had he made some important adjustments to his diet by limiting the amount of dietary fat and watching the cholesterol content of food but he had also increased his weekly exercise regime. More importantly, he now had a better understanding of the terms the doctor had used in his initial visit. David now recognized that *cholesterol* was an important molecule in his body; however, because it was also hydrophobic, it could cause problems with his circulatory system. Furthermore, he also understood what his doctor meant by the terms *good* and *bad cholesterol*. David's doctor was actually referring to lipoproteins, a form of protein that transports lipids and cholesterol in the blood. A high level of LDL—the "bad cholesterol"—meant that his body had an excess of fat to be transported, and HDL represented empty transport proteins. Ideally, low LDL and high HDL values signified a healthy cardiovascular system and a reduced risk for a number of diet-related diseases.

Media Study Tools

www.mhhe.com/maderhuman12e

Enhance your study of this chapter with study tools and practice tests. Also ask your instructor about the resources available through ConnectPlus, including the media-rich eBook, interactive learning tools, and animations.

Virtual Lab

The virtual lab "Nutrition" provides a more detailed look at how your diet provides you with daily amounts of nutrients such as protein, carbohydrates, calories, and fats. The lab allows you to select a variety of foods for breakfast, lunch, dinner, and snacks; and it charts how these selections influence your % Daily Value (DV) of these nutrients.

Summarizing the Concepts

2.1 From Atoms to Molecules

- Matter is composed of elements; each element is made up of just one type of atom.
- An atom's mass is based on the number of protons and neutrons in the nucleus, as well as the electrons orbiting the nucleus.
- An atom's chemical properties depend on the number of electrons in its orbitals and outer shell.
- Atoms react by forming ionic bonds or covalent bonds.

2.2 Water and Living Things

- Water is a liquid, instead of a gas, at room temperature.
- Water heats and freezes slowly, moderating temperatures and allowing bodies to cool by vaporizing water.
- Frozen water is less dense than liquid water, so ice floats on water.
- Water is cohesive and fills tubular vessels, such as blood vessels. A thin film of water allows the lungs to adhere to the chest wall.
- Water is the universal solvent because of its polarity.
- pH is determined by the hydrogen ion concentration [H^+]. Acids increase H^+ but decrease the pH of water, and bases decrease H^+ but increase the pH of water.

2.3 Molecules of Life

Carbohydrates, lipids, proteins, and nucleic acids are macromolecules with specific functions in cells.

2.4 Carbohydrates

- Simple carbohydrates are monosaccharides or disaccharides.
- Glucose is a 6-carbon sugar used by cells for quick energy.
- Complex carbohydrates are polysaccharides. Starch, glycogen, and cellulose are polysaccharides containing many glucose units.
- Plants store glucose as starch, whereas animals store glucose as glycogen.
- Cellulose forms plant cell walls. Cellulose is dietary fiber. Fiber plays an important role in digestive system health.

Organic molecules	Examples	Monomers	Functions
Carbohydrates	Monosaccharides, disaccharides, polysaccharides	Glucose	Immediate energy and stored energy; structural molecules
Lipids	Fats, oils, phospholipids, steroids	Glycerol; Fatty acid	Long-term energy storage; membrane components
Proteins	Structural, enzymatic, carrier, hormonal, contractile	amino group H_2N; acid group COOH; *R* group; Amino acid	Support, metabolic, transport, regulation, motion
Nucleic acids	DNA, RNA	phosphate; base; Nucleotide	Storage of genetic information

2.5 Lipids

- Fats and oils, which function in long-term energy storage, contain glycerol and three fatty acids.
- Fatty acids can be saturated or unsaturated.
- Plasma membranes contain phospholipids.
- Steroids are complex lipids composed of three interlocking rings. Testosterone and estrogen are steroids.
- Cholesterol is tranported by proteins called lipoproteins (LDLs and HDLs).

2.6 Proteins

- Proteins are structural proteins (keratin, collagen), hormones, or enzymes that speed chemical reactions.
- Proteins account for cell movement (actin, myosin), enable muscle contraction (actin, myosin), or transport molecules in blood (hemoglobin).
- Proteins are macromolecules with amino acid subunits.
- A peptide is composed of two amino acids, and a polypeptide contains many amino acids.
- A protein has levels of structure:
  - A primary structure is determined by the sequence of amino acids that forms a polypeptide.
  - A secondary structure is an α (alpha) helix or pleated sheet.
  - A tertiary structure occurs when the secondary structure forms a three-dimensional, globular shape.
  - A quaternary structure occurs when two or more polypeptides join to form a single protein.

2.7 Nucleic Acids

- Nucleic acids are macromolecules composed of nucleotides. Nucleotides are composed of a sugar, a base, and a phosphate. DNA and RNA are polymers of nucleotides.
- DNA contains the sugar deoxyribose; contains the bases adenine, guanine, thymine, and cytosine; is double-stranded; and forms a helix.
- RNA contains the sugar ribose; contains the bases adenine, guanine, uracil, and cytosine; and does not form a helix.
- ATP is a high-energy molecule because its bonds are unstable.
- ATP undergoes hydrolysis to ADP + Ⓟ, which releases energy used by cells to do metabolic work.

Understanding Key Terms

acid 26
adenine (A) 38
ADP (adenosine diphosphate) 38
amino acid 35
atom 20
atomic mass 20
atomic number 38
ATP (adenosine triphosphate) 38
base 26
buffer 27
calorie 25
carbohydrate 29
cellulose 30
complementary paired bases 38
compound 22
covalent bond 24
cytosine (C) 38
dehydration reaction 28
denaturation 36
disaccharide 29
DNA (deoxyribonucleic acid) 37
electron 20
element 20
emulsification 31
fat 31
fatty acid 32
glucose 29
glycogen 29
guanine (G) 38
hemoglobin 37
hexose 29
hydrogen bond 24
hydrolysis reaction 28
hydrophilic 26
hydrophobic 26
ion 22
ionic bond 22
isotope 21
lipid 31
macromolecule 28
mass 21
mass number 21
matter 20
mole 26
molecule 22
monosaccharide 29
neutron 20
nucleotide 38
nucleus 20
oil 31
orbital 20
organic 28
organic molecule 28
pentose 29
peptide bond 35
phospholipid 33
pH scale 27
polar 24
polypeptide 35
polysaccharide 29
protein 35
proton 20
radioisotope 21
RNA (ribonucleic acid) 37
saturated fatty acid 32
starch 29
steroid 33
thymine (T) 38
tracer 21
trans fat 32
triglyceride 31
unsaturated fatty acid 32
uracil (U) 38

Match the key terms to these definitions.

a. ________ Breaking up of fat globules into smaller droplets by the action of bile salts or any other emulsifier.

b. ________ Charged particle that carries a negative or positive charge.

c. ________ Chemical bond in which atoms share one or more pairs of electrons.

d. ________ Type of molecule that interacts with water by dissolving in water and/or forming hydrogen bonds with water molecules.

e. ________ Weak bond that arises between a slightly positive hydrogen atom of one molecule and a slightly negative atom of another molecule or between parts of the same molecule.

Testing Your Knowledge of the Concepts

1. Name the subatomic particles of the atom. Describe their charge, atomic mass, and location in the atom. (pages 20–21)
2. Why can a radioisotope be used as a tracer in the human body? Give an example. (pages 21–22)
3. Explain the difference between an ionic bond and a covalent bond. (pages 22–24)
4. Relate the properties of water to its polarity and hydrogen bonding between water molecules. (pages 24–25)
5. On the pH scale, which numbers indicate a basic solution? An acidic solution? A neutral solution? What makes a solution basic, acidic, or neutral? (pages 26–27)
6. What are buffers, and why are they important to life? (pages 27–28)
7. Name, describe, and give an example of each class of carbohydrate. What is the main function of each of the three classes? (pages 29–30)
8. What are the subunits of a triglyceride? What is the difference between a saturated and an unsaturated fatty acid? What are the functions of fats in the body? (pages 31–32)
9. How does the structure of a phospholipid differ from a triglyceride? Describe the arrangement of phospholipids in the plasma membrane. (page 33)
10. What are the building blocks of proteins? How do these components bond together to make a protein? (pages 34–35)
11. Discuss the primary, secondary, and tertiary structures of proteins. Why are these structures so important? (pages 36–37)
12. What structural differences are there between DNA and RNA? (pages 37–38)
13. What type of reaction releases the energy of an ATP molecule? Explain. (page 39)
14. The atomic number gives the
    a. number of neutrons in the nucleus.
    b. number of protons in the nucleus.
    c. weight of the atom.
    d. number of protons in the outer shell.
15. Isotopes differ in their
    a. number of protons.
    b. atomic number.
    c. number of neutrons.
    d. number of electrons.

16. Which type of bond results from the complete transfer of electrons from one atom to another?
   a. covalent
   b. ionic
   c. hydrogen
   d. neutral
17. Which of the following properties of water is not due to hydrogen bonding between water molecules?
   a. Water prevents large temperature changes.
   b. Ice floats on water.
   c. Water is a solvent for polar molecules and ionic compounds.
   d. Water flows and fills spaces because it is cohesive.
18. If a chemical accepted H^+ from the surrounding solution, the chemical would be
   a. a base.
   b. an acid.
   c. a buffer
   d. both a and c are correct
19. What is true of a solution that goes from pH 5 to pH 8?
   a. The H^+ concentration decreases as the solution becomes more basic.
   b. The H^+ concentration increases as the solution becomes more acidic.
   c. The H^+ concentration decreases as the solution becomes more acidic.
20. An example of a polysaccharide used for energy storage in humans is
   a. cellulose.
   b. glycogen.
   c. cholesterol.
   d. starch.
21. Saturated and unsaturated fatty acids differ in the
   a. number of carbon-to-carbon double bonds.
   b. consistency at room temperature.
   c. number of hydrogen atoms present.
   d. All of these are correct.
22. The difference between one amino acid and another is found in the ________ group.
   a. amino
   b. carboxyl
   c. *R*
   d. All of these are correct.
23. An example of a hydrolysis reaction is
   a. amino acid + amino acid $\longrightarrow$ dipeptide + H_2O.
   b. dipeptide + H_2O $\longrightarrow$ amino acid + amino acid.
   c. denaturation of a polypeptide.
   d. Both b and c are correct.
24. The helix and pleated sheet form of a protein is its ________ structure.
   a. secondary
   b. tertiary
   c. primary
   d. quaternary
25. An RNA nucleotide differs from a DNA molecule in that RNA has
   a. ribose sugar.
   b. a phosphate molecule.
   c. a uracil base.
   d. Both a and c are correct.

In questions 26–29, match each subunit with the molecule in the key.

Key:

a. fat
b. polysaccharide
c. polypeptide
d. DNA, RNA

26. Glucose
27. Nucleotide
28. Glycerol and fatty acids
29. Amino acid

In questions 30–33, match the molecular category with a molecule in the preceding key.

30. Lipids
31. Proteins
32. Nucleic acids
33. Carbohydrates
34. Label this diagram using these terms: dehydration reaction, hydrolysis reaction, subunits, macromolecule.

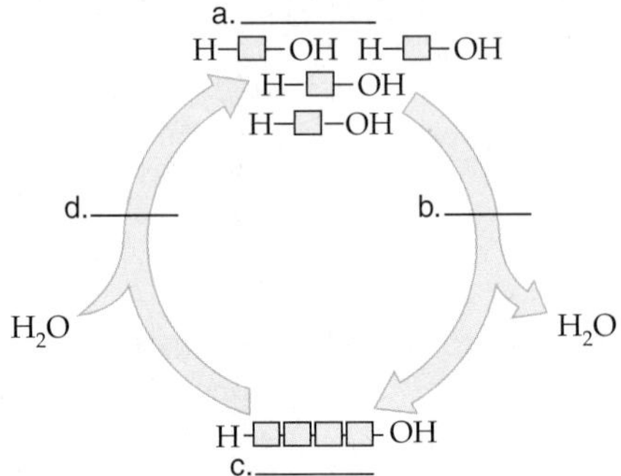

Thinking Critically About the Concepts

The case study in this chapter included information about healthy cholesterol levels in the blood. There are several biological molecules essential to life, and the body needs them in certain amounts in order to function properly. Cholesterol is one of these molecules. Sometimes, the body can make these molecules for us; other times, we need to get them through our diet. Either way, the balancing of "just enough" versus "too much" is the role of homeostasis.

1. If the doctor prescribed cholesterol-lowering medicine for David, why did he also have to change his diet and exercise program? Why would the medication alone not be enough?
2. Cholesterol is a needed substance for the human body; we use it as a base for certain hormones and as a support structure in our plasma membranes. If it is needed in the body, how can it also be harmful?
3. Name another substance that is needed in the body but can be detrimental to homeostasis in large quantities.
4. The presence and absence of certain molecules in the blood can alter the pH of the blood. Why would it be important to maintain a steady blood pH?
5. Substances that the body makes or we ingest through the diet, like carbohydrates and lipids, can be stored in different areas for future use. Why is storage an important part of homeostasis?

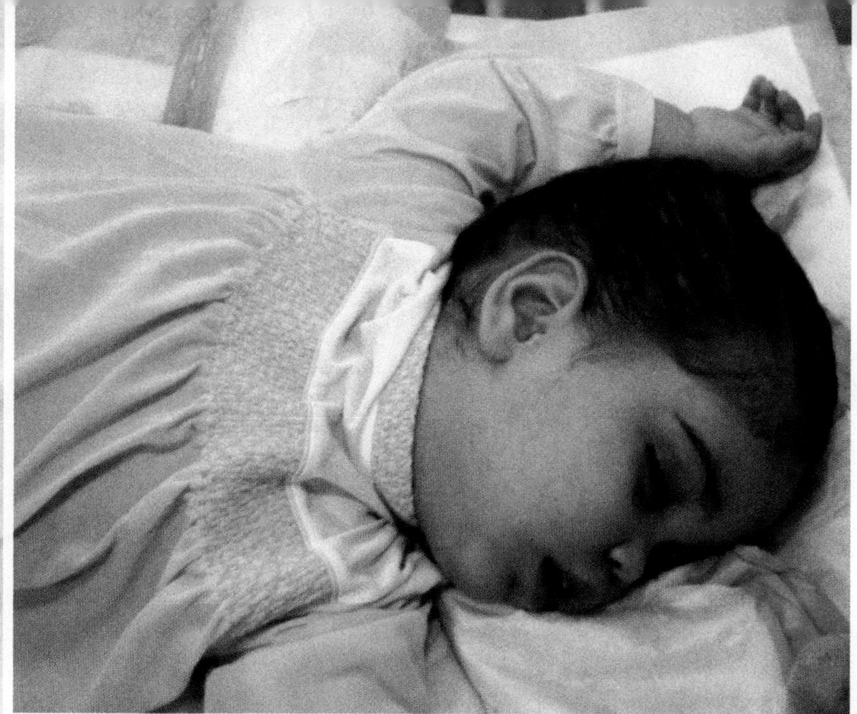

CHAPTER

3

Cell Structure and Function

CASE STUDY WHEN CELLS MALFUNCTION

Mary and Kevin first noticed that something was wrong with their newborn about four months after birth. Whereas most newborns rapidly strengthen and are developing the ability to hold their head up and are demonstrating hand–eye coordination, their baby seemed to be weakening. In addition, Mary began to sense that something was wrong when their baby started having trouble swallowing his formula. After consulting with their pediatrician, Mary and Kevin decided to bring their child to a local pediatric research hospital to talk with physicians trained in newborn developmental disorders.

After a series of tests that included blood work and a complete physical examation, the specialists at the research center informed Kevin and Mary that the symptoms their newborn was exhibiting were characteristic of a condition called Tay–Sachs disease. This condition is a rare metabolic disorder that causes one of the internal components of the cell, the lysosome, to malfunction. Because of this malfunction, fatty acids were accumulating in the cells of their child. These accumulations were causing the neurons to degrade, producing the symptoms noted by the parents.

What puzzled the research team was the fact that neither Kevin nor Mary were of Eastern European descent. Populations from this area are known to have a higher rate of the mutation that causes Tay–Sachs disease. However, genetic testing of both Kevin and Mary indicated that they were carriers for the trait, meaning that though they each had one normal copy of the gene associated with Tay–Sachs disease, each carried a defective copy as well. Only one good copy of the gene is needed for the lysosome to function correctly. Unfortunately, each had passed on a copy of the defective gene to their child.

Despite the poor prognosis for their child, both Kevin and Mary were determined to learn more about how this defect caused the lysosome to malfunction and about what treatments were being developed to prolong the life span of a child with Tay–Sachs disease.

As you read through the chapter, think about the following questions.

1. What organelle produces the lysosomes?
2. What is the role of the lysosome in a normally functioning cell?
3. Why would a malfunction in the lysosome cause an accumulation of fatty acids in the cell?

CHAPTER CONCEPTS

3.1 What Is a Cell?
Cells are the basic units of life. Cell size is limited by the surface area-to-volume ratio.

3.2 How Cells Are Organized
Human cells have a plasma membrane, cytoplasm, and a nucleus. The cytoplasm contains several types of organelles that carry out specific functions.

3.3 The Plasma Membrane and How Substances Cross It
The structure of the plasma membrane influences its permeability. Passive and active transport mechanisms, diffusion, transport by carriers, and the use of vesicles allow substances to cross the plasma membrane.

3.4 The Nucleus and Endomembrane System
The nucleus stores the genetic material. Ribosomes act as sites for protein synthesis. The endomembrane system acts as a series of interchangeable organelles that manufacture and modify proteins—and other organic molecules—for use by the cell.

3.5 The Cytoskeleton, Cell Movement, and Cell Junctions
The cytoskeleton is composed of fibers that maintain the shape of the cell and assist the movement of organelles. Cilia and flagella contain microtubules and can move using ATP energy. In tissues, cells are connected by junctions that allow for coordinated activities.

3.6 Mitochondria and Cellular Metabolism
Mitochondria are the sites of cellular respiration, an aerobic process that produces the majority of ATP for a cell. Fermentation, an aerobic process, produces only two ATP per glucose molecule.

BEFORE YOU BEGIN

Before beginning this chapter, take a few moments to review the following discussions.

Section 2.2 What properties of water make it a crucial molecule for life as we know it?

Sections 2.3 to 2.7 What are the basic roles of carbohydrates, fats, proteins, and nucleic acids in the cell?

Section 2.7 What is the role of ATP in a cell?

3.1 What Is a Cell?

Learning Outcomes

Upon completion of this section, you should be able to

1. State the basic principles of the cell theory.
2. Explain how the surface area-to-volume ratio limits cell size.
3. Summarize the role of microscopy in the study of cells.

All organisms, including humans, are composed of cells. This is not apparent until you compare unicellular organisms with the tissues of multicellular ones under the microscope. The cell theory, one of the fundamental principles of modern biology, was not formulated until after the invention of the microscope in the seventeenth century.

Most cells are small and can be seen only under a microscope. The small size of cells means that they are measured using the smaller units of the metric system, such as the *micrometer* (μm). A micrometer is 1/1,000 millimeter. The micrometer is the common unit of measurement for people who use microscopes professionally (see the inside back cover of this text for a complete list of metric units). Most human cells are about 100 μm in diameter, about the width of a human hair. The internal contents of a cell are even smaller and, in most cases, may only be viewed using powerful microscopes.

The Cell Theory

As stated by the **cell theory,** *a cell is the basic unit of life.* Nothing smaller than a cell is alive. A unicellular organism exhibits the seven characteristics of life we discussed in Chapter 1. There is no smaller unit of life able to reproduce, respond to stimuli, remain homeostatic, grow and develop, take in and use materials from the environment, and become adapted to the environment. In short, life has a cellular nature.

All living things are made up of cells. Although it may be apparent that a unicellular organism is necessarily a cell, what about multicellular ones? Humans are multicellular. Is there any tissue in the human body not composed of cells? At first, you might be inclined to say that bone is not composed of cells. However, if you were to examine bone tissue under the microscope, you would be able to see that it, too, is composed of cells surrounded by material they have deposited. Cells look different—a blood cell looks different than a nerve cell. They both look different than a cartilage cell (Fig. 3.1). Despite having certain parts in common, cells in a multicellular organism are specialized in structure and function.

New cells arise only from pre-existing cells. Until the nineteenth century, most people believed in spontaneous generation, that is, that nonliving objects could give rise to living organisms. For example, maggots were thought to arise from meat hung in the butcher shop. Maggots often did arise from meat to which flies had access. However, people did not realize that the maggots (living) did not arise from the meat (nonliving). A series of experiments by Francesco Redi in the seventeenth century demonstrated that meat that was placed within sealed containers did not generate maggots. In other words, life did not generate spontaneously. In 1864, the French scientist Louis Pasteur conducted a now-classic set of experiments using bacterial cells. His experiments proved conclusively that spontaneous generation of life from nonlife was not possible.

When mice or humans reproduce, a sperm cell joins with an egg cell to form a zygote. This is the first cell of a new multicellular organism. By reproducing, parents pass a copy of their genes onto their offspring. The genes contain the instructions that allow the zygote to grow and develop into the complete organism.

Figure 3.1 **Cells vary in structure and function.** A cell's structure is related to its function. Despite differences in appearance, all exchange substances with their environment.

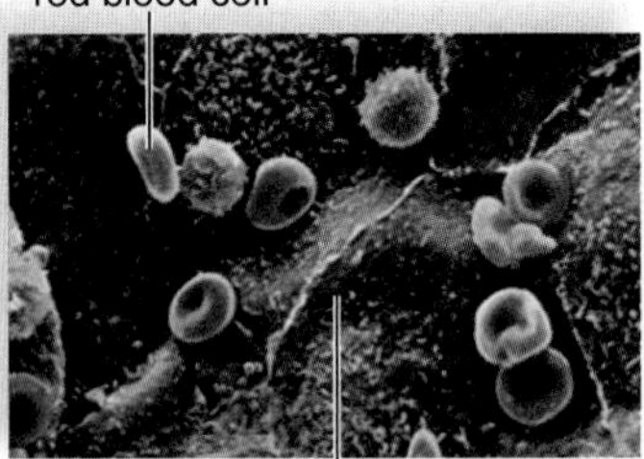

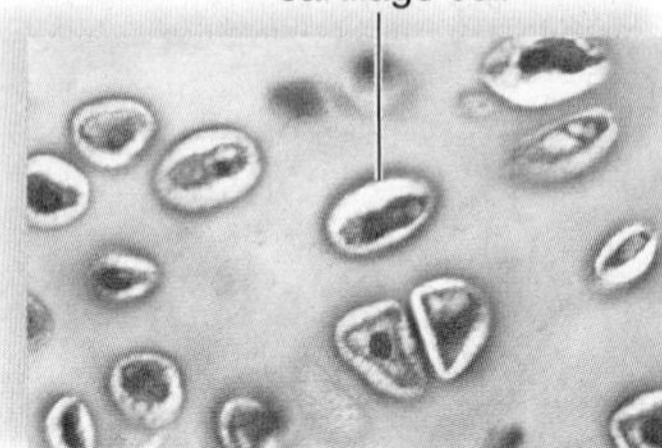

Cell Size

A few cells, such as a hen's egg or a frog's egg, are large enough to be seen by the naked eye. In comparison, a human egg cell is around 100 μm in size, placing it right at the limit of what can be viewed by our eyes. However, most cells are much smaller. The small size of cells is explained by considering the surface area-to-volume ratio of cells. Nutrients enter a cell—and waste exits a cell—at its surface. Therefore, the greater the amount of surface, the greater the ability to get material in and out of the cell. A large cell requires more nutrients and produces more waste than a small cell. Yet, as cells become larger in volume, the proportionate amount of surface area actually decreases. You can see this by comparing the two cubes in Figure 3.2.

We would expect, then, that there would be a limit to how large an actively metabolizing cell can become. Once a hen's egg is fertilized and starts metabolizing, it divides repeatedly without increasing in size. Cell division restores the amount of surface area needed for adequate exchange of materials.

Microscopy

Micrographs are photographs of objects most often obtained by using the compound light microscope, the transmission electron microscope, or the scanning electron microscope

	1 mm	2 mm
Surface area (square mm)	$6 \times 1\ mm^2 = 6\ mm^2$	$6 \times 4\ mm^2 = 24\ mm^2$
Volume (cubic mm)	$(1\ mm)^3 = 1\ mm^3$	$(2\ mm)^3 = 8\ mm^3$
$\frac{\text{Surface area}}{\text{Volume}}$	$\frac{6}{1}$	$\frac{24}{8} = \frac{3}{1}$

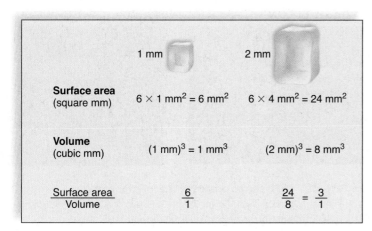

Figure 3.2 Surface area-to-volume ratio limits cell size. Large cells are unable to efficiently exchange nutrients and waste with their environment because of a decreased surface area-to-volume ratio. As the cell increases in size, the surface area increases by the square of the width and the volume increases by the width cubed (expressed as w^3).

(Fig. 3.3). A *compound light microscope* uses a set of glass lenses and light rays passing through the object to magnify objects. The image can be viewed directly by the human eye.

The *transmission electron microscope* makes use of a stream of electrons to produce magnified images. The human eye cannot see the image. Therefore, it is projected onto a fluorescent screen or photographic film to produce an image that can be viewed. The magnification produced by a transmission electron microscope is much higher than that of a light microscope. Also, this microscope has the ability to produce enlarged images with greater detail. In other words, the transmission electron microscope has a higher resolving power—the ability to distinguish between two adjacent points. Table 3.1 provides a comparison of the resolving power of the eye, the light microscope, and the transmission electron microscope.

A *scanning electron microscope* provides a three-dimensional view of the surface of an object. A narrow beam of electrons is scanned over the surface of the specimen, which is coated with a thin layer of metal. The metal gives off secondary electrons, which are collected to produce a television-type picture of the specimen's surface on a screen.

Table 3.1 Resolving Power of The Eye and Common Microscopes

	Magnification	Resolving Power
Eye	N/A	0.1 mm (100 μm)
Light microscope	1,000×	0.0001 mm (0.1 μm)
Transmission electron microscope	50,000×	0.000001 mm (0.01 μm)

As you no doubt will discover in the laboratory, the light microscope has the ability to view living specimens—this is not true of the electron microscope. Because electrons cannot travel very far in air, a strong vacuum must be maintained along the entire path of the electron beam. Cells are often treated before being viewed under a microscope. In electron microscopy, cells are treated so that they do not decompose and are then embedded into a matrix. The matrix allows the researcher to slice the cell into very thin pieces. Because most cells are transparent, they are often stained with colored dyes before before being viewed under a light microscope. Certain cellular components take up the dye more than other components; therefore, contrast is enhanced.

A similar approach is used in electron microscopy, except in this case the sample is treated with electron-dense metals to provide contrast. The metals do not provide color, so electron micrographs may be colored after the micrograph is obtained. The expression "falsely colored" means that the original micrograph was colored after it was produced.

Connecting the Concepts

For more on the human egg cells and red blood cells mentioned in this section, refer to the following discussions.

Section 6.2 discusses how red blood cells transport gases within the circulatory system.

Section 6.6 provides an overview of how red blood cells help maintain homeostasis in the body.

Section 17.1 examines the complex structure of a human egg cell.

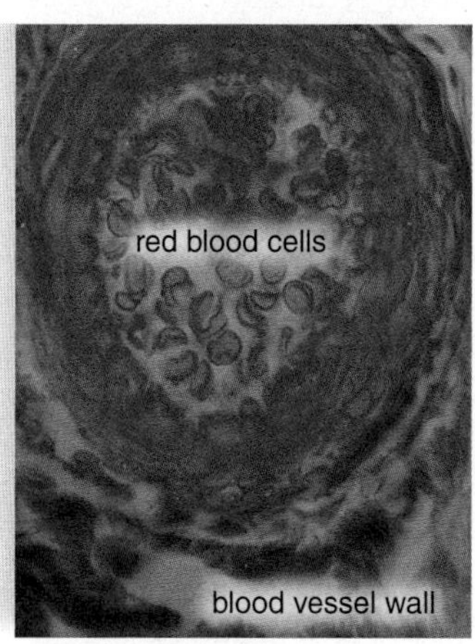

a. Light micrograph

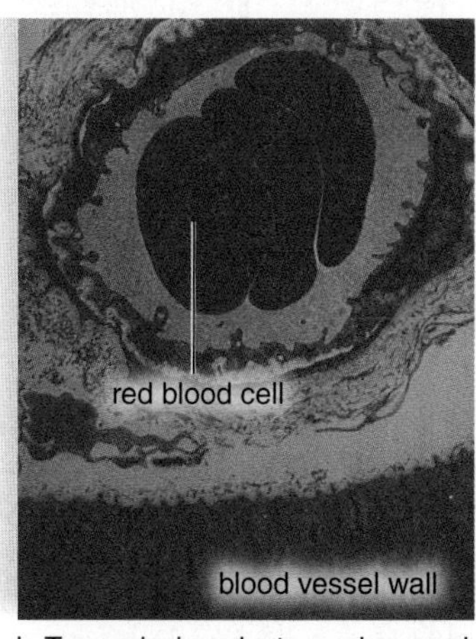

b. Transmission electron micrograph

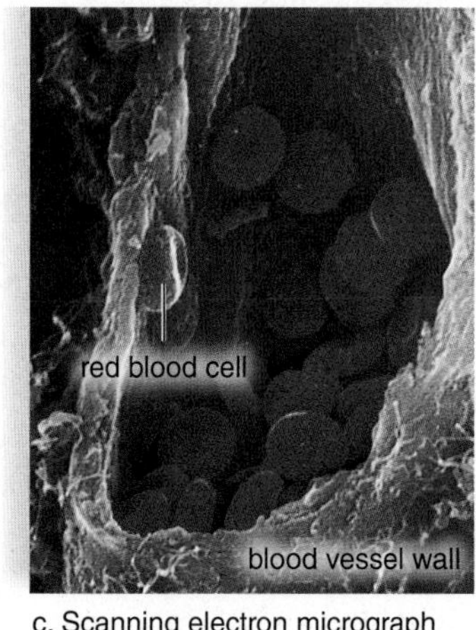

c. Scanning electron micrograph

Figure 3.3 Micrographs of human red blood cells. Each of these micrographs is of a human red blood cell.
a. Light micrograph (LM) of many cells in a large vessel (stained). **b.** Transmission electron micrograph (TEM) of just three cells in a small vessel (colored). **c.** Scanning electron micrograph (SEM) gives a three-dimensional view of cells and vessels (colored).

Biology Matters Science Focus

Coloring Organisms Green; Green Fluorescent Proteins and Cells

Most cells lack any significant pigmentation. Thus, cell biologists frequently rely on dyes to produce enough contrast to resolve organelles and other cellular structures. The first of these dyes were developed in the nineteenth century from chemicals used to stain clothes in the textile industry. Since then, significant advances have occurred in the development of cellular stains.

In 2008, three scientists—Martin Chalfie, Roger Y. Tsien, and Osamu Shimomura—earned the Nobel Prize in Chemistry or Medicine for their work with a protein called *green fluorescent protein,* or GFP. GFP is a bioluminescent protein found in the jellyfish *Aequorea victoria,* commonly called the crystal jelly. The crystal jelly is a native of the West Coast of the United States. Normally, this jellyfish is transparent. However, when disturbed, special cells in the jellyfish release a fluorescent protein called aequorin. Aequorin fluoresces with a green color. The research teams of Chalfie, Tsien, and Shimomura were able to isolate the fluorescent protein from the jellyfish and develop it as a molecular tag. These tags can be generated for almost any protein within the cell, revealing not only its cellular location but also how its distribution within the cell may change as a result of a response to its environment. Figure 3A shows how a GFP-labeled antibody can be used to identify the cellular location of the actin proteins in a human cell. Actin is one of the prime components of the cell's microfilaments, which in turn are part of the cytoskeleton of the cell. This image shows the distribution of actin in a human cell.

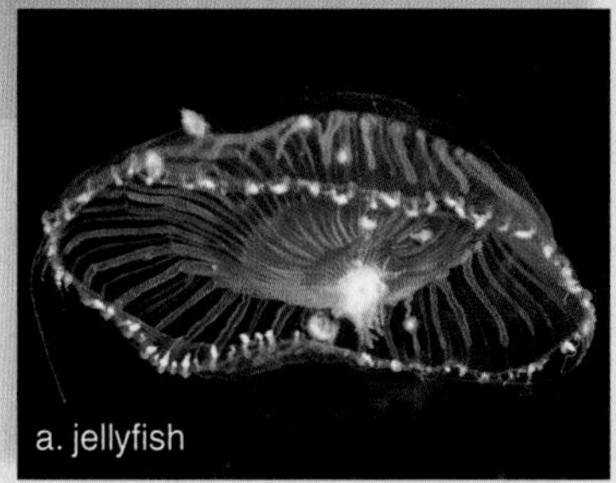

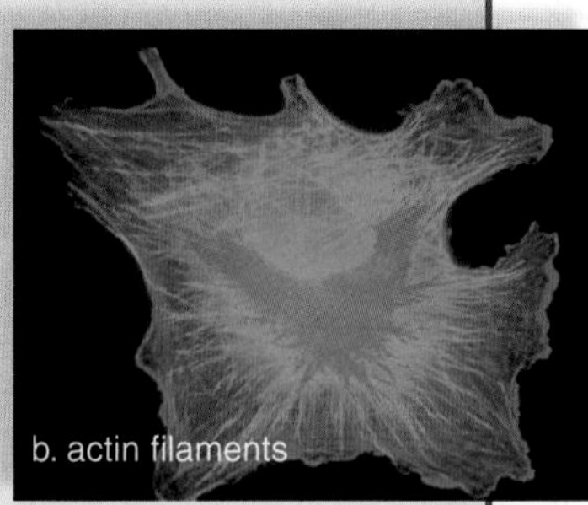

Figure 3A **GFP shows details of the interior of cells.** **a.** The jellyfish *Aequorea victoria* and **b.** the GFP stain of a human cell. This illustration shows a human cell tagged with a GFP-labeled antibody to the actin protein.

Check Your Progress 3.1

1. Describe the cell theory and state its importance to the study of biology.
2. Explain how a cell's size relates to its function.
3. Compare and contrast the information that may be obtained from a light microscope and an electron microscope.

3.2 How Cells Are Organized

Learning Outcomes

Upon completion of this section, you should be able to

1. Identify the components of a human cell and state its function.
2. Distinguish between the structure of a prokaryotic cell and that of a eukaryotic cell.
3. Summarize how eukaryotic cells evolved from prokaryotic cells.

Biologists classify cells into two broad categories—the prokaryotes and eukaryotes. The prokaryotic group includes the bacteria; the eukaryotic group consists of animals, plants, fungi, and some single-celled organisms. Despite their differences, both types of cells have a **plasma membrane,** an outer membrane that regulates what enters and exits a cell. The plasma membrane is a phospholipid bilayer. This bilayer is a "sandwich" made of two layers of phospholipids. Their polar phosphate molecules form the top and bottom surfaces of the bilayer, and the nonpolar lipid lies in between. The phospholipid bilayer is **selectively permeable,** which means it allows certain molecules—but not others—to enter the cell. Proteins scattered throughout the plasma membrane play important roles in allowing substances to enter the cell. All types of cells also contain **cytoplasm,** which is a semifluid medium that contains water and various types of molecules suspended or dissolved in the medium. The presence of proteins accounts for the semifluid nature of the cytoplasm. The cytoplasm contains **organelles.** Originally, the term *organelle* referred to only membranous structures, but we will use it to include any well-defined subcellular structure. Eukaryotic cells have many different types of organelles.

Internal Structure of Eukaryotic Cells

The most prominent organelle within the **eukaryotic cell** is a **nucleus,** a membrane-enclosed structure in which DNA is found. **Prokaryotic cells** (such as bacterial cells) lack a nucleus. Although the DNA of prokaryotic cells is centrally placed within the cell, it is not surrounded by a membrane.

Eukaryotic cells also possess organelles, each type of which has a specific function (see Figure 3.4). Many organelles are surrounded by a membrane, which allows compartmentalization of the cell. This keeps the various cellular activities separated from one another.

MP3 Cellular Organelles

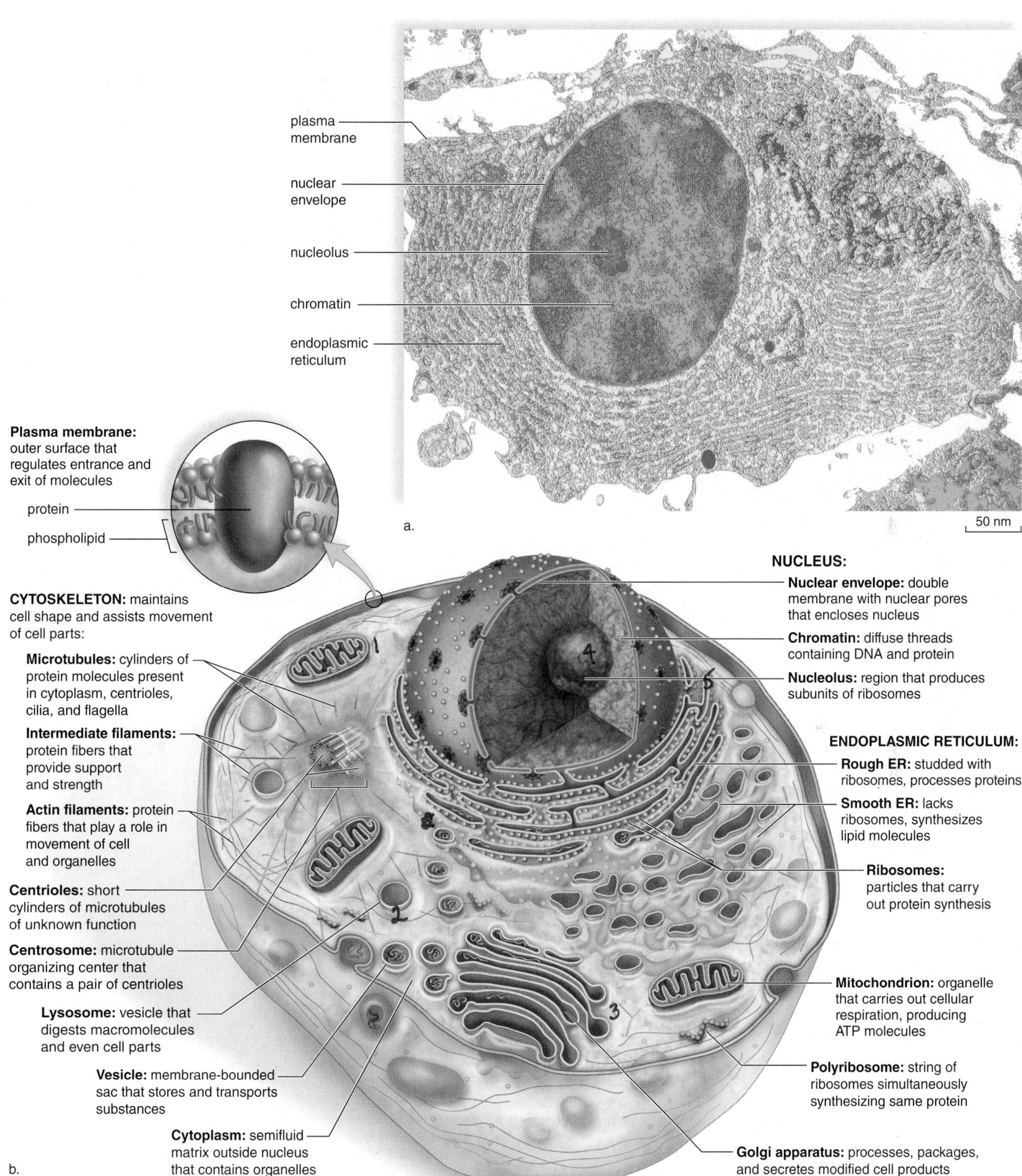

plasma membrane
nuclear envelope
nucleolus
chromatin
endoplasmic reticulum
a.
50 nm
Plasma membrane: outer surface that regulates entrance and exit of molecules
protein
phospholipid
CYTOSKELETON: maintains cell shape and assists movement of cell parts:
Microtubules: cylinders of protein molecules present in cytoplasm, centrioles, cilia, and flagella
Intermediate filaments: protein fibers that provide support and strength
Actin filaments: protein fibers that play a role in movement of cell and organelles
Centrioles: short cylinders of microtubules of unknown function
Centrosome: microtubule organizing center that contains a pair of centrioles
Lysosome: vesicle that digests macromolecules and even cell parts
Vesicle: membrane-bounded sac that stores and transports substances
Cytoplasm: semifluid matrix outside nucleus that contains organelles
b.
NUCLEUS:
Nuclear envelope: double membrane with nuclear pores that encloses nucleus
Chromatin: diffuse threads containing DNA and protein
Nucleolus: region that produces subunits of ribosomes
ENDOPLASMIC RETICULUM:
Rough ER: studded with ribosomes, processes proteins
Smooth ER: lacks ribosomes, synthesizes lipid molecules
Ribosomes: particles that carry out protein synthesis
Mitochondrion: organelle that carries out cellular respiration, producing ATP molecules
Polyribosome: string of ribosomes simultaneously synthesizing same protein
Golgi apparatus: processes, packages, and secretes modified cell products

Figure 3.4 The structure of a typical eukaryotic cell.

a. A transmission electron micrograph of the interior structures of a cell. **b.** The structure and function of the components of a eukaryotic cell.

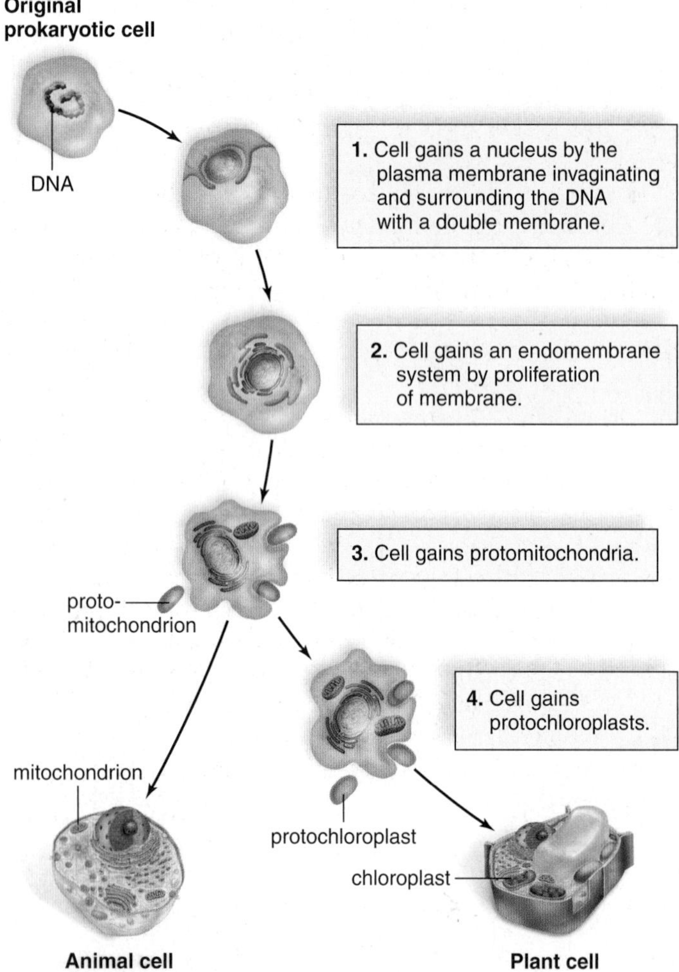
Original prokaryotic cell
DNA
1. Cell gains a nucleus by the plasma membrane invaginating and surrounding the DNA with a double membrane.
2. Cell gains an endomembrane system by proliferation of membrane.
3. Cell gains protomitochondria.
proto-mitochondrion
4. Cell gains protochloroplasts.
mitochondrion
protochloroplast
chloroplast
Animal cell
Plant cell

Figure 3.5 The evolution of eukaryotic cells.
Invagination of the plasma membrane of a prokaryotic cell could have created the nucleus. Later, the cell gained organelles, some of which may have been independent prokaryotes.

Evolutionary History of the Eukaryotic Cell

Figure 3.5 shows that the first cells to arise were prokaryotic cells. Prokaryotic cells today are represented by the bacteria and archaea, which differ mainly by their chemistry. Bacteria are well known for causing diseases in humans, but they also have great environmental and commercial importance. The archaea are known for living in extreme environments that may mirror the first environments on Earth. These environments are too hot, too salty, and/or too acidic for the survival of most cells. The eukaryotic cell is believed to have evolved from the archaea.

The internal structure of eukaryotic cells is believed to have evolved, as shown in Figure 3.5. The nucleus could have formed by *invagination* of the plasma membrane, a process whereby a pocket is formed in the plasma membrane. The pocket would have enclosed the DNA of the cell, thus forming its nucleus. Surprisingly, some of the organelles in eukaryotic cells may have arisen by engulfing prokaryotic cells. The engulfed prokaryotic cells were not digested; rather, they then evolved into different organelles. One of these events would have given the eukaryotic cell a mitochondrion. Mitochondria are organelles that carry on cellular respiration. Another such event may have produced the chloroplast. Chloroplasts are found in cells that carry out photosynthesis. This process is often called *endosymbiosis*.

Animation Endosymbiosis

Early prokaryotic organisms, such as the archaeans, were well adapted to life on the early Earth. The environment that they evolved in contained conditions that would be instantly lethal to life today. The atmosphere contained no oxygen; instead, it was filled with carbon monoxide and other poisonous gases; the temperature of the planet was greater than 200°F; and there was no ozone layer to protect organisms from damaging radiation from the sun.

Despite these conditions, prokaryotic life survived and in doing so gradually adapted to Earth's environment. In the process, most of the archaea bacteria went extinct. However, we now know that some are still around and can be found in some of the most inhospitable places on the planet, such as thermal vents and salty seas. The study of these ancient bacteria is still shedding light on the early origins of life.

Connections and Misconceptions

How old are the bacteria?

Scientists now recognize that the first cells on Earth were the prokaryotes. This ancient group of organisms first appeared on the planet over 3.5 billion years ago. Sometimes that amount of time can be hard to visualize, so a geologic timescale has been provided in the animation "Geologic History of Earth."

Animation Geologic History of Earth

Connecting the Concepts

The material in this section summarizes some previous concepts of eukaryotic and prokaryotic cells and the role of phospholipids in the cell membrane. For more information, refer to the following discussions.

Section 1.2 illustrates the difference in the classification of eukaryotic and prokaryotic cells.

Section 2.5 provides a review of lipids, including the phospholipids.

Check Your Progress 3.2

1. List the three main components of a eukaryotic cell.
2. Describe the main differences between a eukaryotic and a prokaryotic cell.
3. Describe the possible evolution of the nucleus, mitochondria, and chloroplast.

3.3 The Plasma Membrane and How Substances Cross It

Learning Outcomes

Upon completion of this section, you should be able to

1. Describe the structure of the cell membrane and list the type of molecules found in the membrane.
2. Distinguish between diffusion, osmosis, and facilitated diffusion, and state the role of each in the cell.
3. Explain how tonicity relates to the direction of water movement across a membrane.
4. Compare passive- and active-transport mechanisms.
5. State how eukaryotic cells move large molecules across membranes.

A human cell, like all cells, is surrounded by an outer plasma membrane (Fig. 3.6). The plasma membrane marks the boundary between the outside and the inside of the cell. The integrity and function of the plasma membrane are necessary to the life of the cell.

The plasma membrane is a phospholipid bilayer with attached or embedded proteins. A phospholipid molecule has a polar head and nonpolar tails (see Fig. 2.19). When phospholipids are placed in water, they naturally form a spherical bilayer. The polar heads, being charged, are hydrophilic (attracted to water). They position themselves to face toward the watery environment outside and inside the cell. The nonpolar tails are hydrophobic (not attracted to water). They turn inward toward one another, where there is no water.

At body temperature, the phospholipid bilayer is a liquid. It has the consistency of olive oil. The proteins are able to change their position by moving laterally. The **fluid-mosaic model** is a working description of membrane structure. It states that the protein molecules form a shifting pattern within the fluid phospholipid bilayer. Cholesterol lends support to the membrane.

Short chains of sugars are attached to the outer surface of some protein and lipid molecules. These are called *glycoproteins* and *glycolipids,* respectively. These carbohydrate chains, specific to each cell, help mark the cell as belonging to a particular individual. They account for why people have different blood types, for example. Other glycoproteins have a special configuration that allows them to act as a receptor for a chemical messenger, such as a hormone. Some plasma membrane proteins form channels through which certain substances can enter cells. Others are either enzymes that catalyze reactions or carriers involved in the passage of molecules through the membrane.

MP3 Membrane Structure

Figure 3.6 Organization of the plasma membrane. A plasma membrane is composed of a phospholipid bilayer in which proteins are embedded. The hydrophilic heads of phospholipids are a part of the outside surface and the inside surface of the membrane. The hydrophobic tails make up the interior of the membrane. Note the plasma membrane's asymmetry—carbohydrate chains are attached to the outside surface, and cytoskeleton filaments are attached to the inside surface. Cholesterol lends support to the membrane.

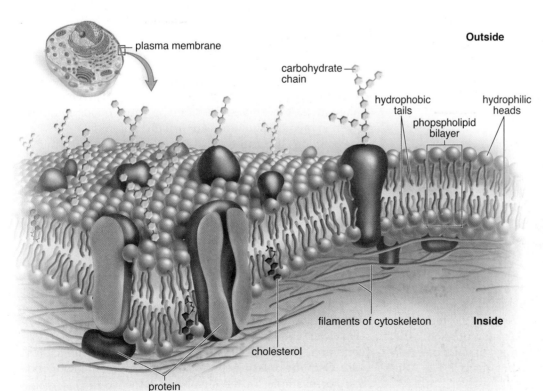

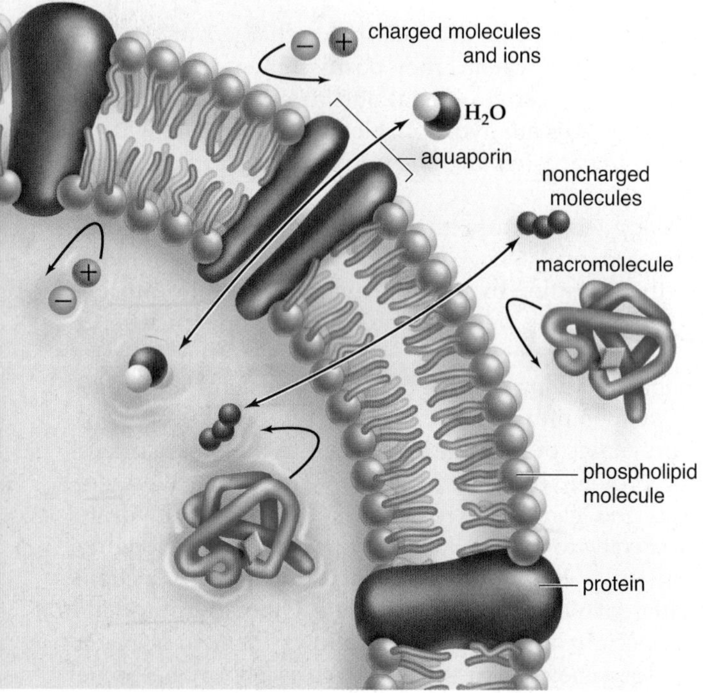

Figure 3.7 Selective permeability of the plasma membrane. Small, uncharged molecules are able to cross the membrane whereas large or charged molecules cannot. Water travels freely across membranes through aquaporins.

Plasma Membrane Functions

The plasma membrane keeps a cell intact. It allows only certain molecules and ions to enter and exit the cytoplasm freely. Therefore, the plasma membrane is said to be selectively permeable (Fig. 3.7). Small, lipid-soluble molecules, such as oxygen and carbon dioxide, can pass through the membrane easily. The small size of water molecules allows them to freely cross the membrane by using protein channels called *aquaporins*. Ions and large molecules cannot cross the membrane without more direct assistance, which will be discussed later.

Diffusion

Diffusion is the random movement of molecules from an area of higher concentration to an area of lower concentration, until they are equally distributed. Diffusion is a passive way for molecules to enter or exit a cell. No cellular energy is needed to bring it about.

Certain molecules can freely cross the plasma membrane by diffusion. When molecules can cross a plasma membrane, which way will they go? The molecules will move in both directions. But the *net movement* will be from the region of higher concentration to the region of lower concentration, until equilibrium is achieved. At equilibrium, as many molecules of the substance will be entering as leaving the cell (Fig. 3.8). Oxygen diffuses across the plasma membrane, and the net movement is toward the inside of the cell. This is because a cell uses oxygen when it produces ATP molecules for energy purposes.

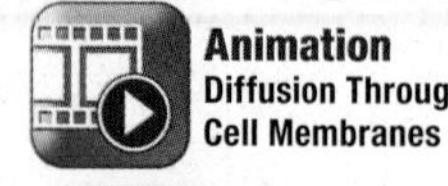

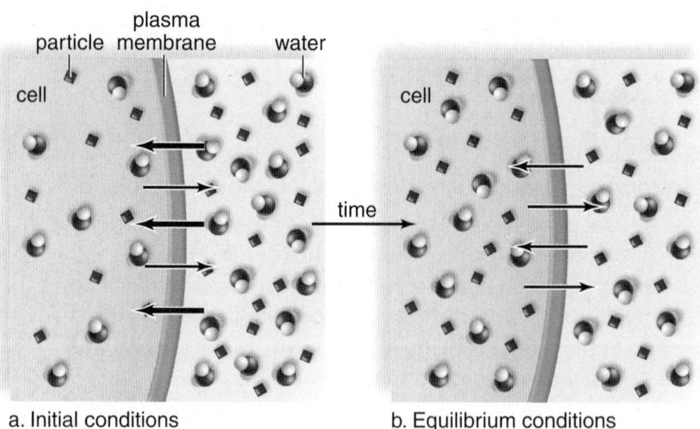

Figure 3.8 Diffusion across the plasma membrane.
a. When a substance can diffuse across the plasma membrane, it will move back and forth across the membrane, but the net movement will be toward the region of lower concentration. **b.** At equilibrium, equal numbers of particles and water have crossed in both directions, and there is no net movement.

Osmosis

Osmosis is the net movement of water across a semipermeable membrane, from an area of higher concentration to an area of lower concentration. The membrane separates the two areas, and solute is unable to pass through the membrane. Water will tend to flow from the area that has less solute (and therefore more water) to the area with more solute (and therefore less water). **Tonicity** refers to the osmotic characteristics of a solution across a particular membrane, such as a red blood cell membrane.

Normally, body fluids are *isotonic* to cells (Fig. 3.9*a*). There is the same concentration of nondiffusible solutes and water on both sides of the plasma membrane. Therefore, cells maintain their normal size and shape. Intravenous solutions given in medical situations are usually isotonic.

Solutions that cause cells to swell or even to burst due to an intake of water are said to be *hypotonic*. A hypotonic solution has a lower concentration of solute and a higher concentration of water than the cells. If red blood cells are placed in a hypotonic solution, water enters the cells. They swell to bursting (Fig. 3.9*b*). *Lysis* is used to refer to the process of bursting cells. Bursting of red blood cells is termed *hemolysis*.

Solutions that cause cells to shrink or shrivel due to loss of water are said to be *hypertonic*. A hypertonic solution has a higher concentration of solute and a lower concentration of water than do the cells. If red blood cells are placed in a hypertonic solution, water leaves the cells; they shrink (Fig. 3.9*c*). The term *crenation* refers to red blood cells in this condition. These changes have occurred due to osmotic pressure. **Osmotic pressure** controls water movement in our bodies. For example, in the small and large intestines, osmotic pressure allows us to absorb the water in food and drink. In the kidneys, osmotic pressure controls water absorption as well.

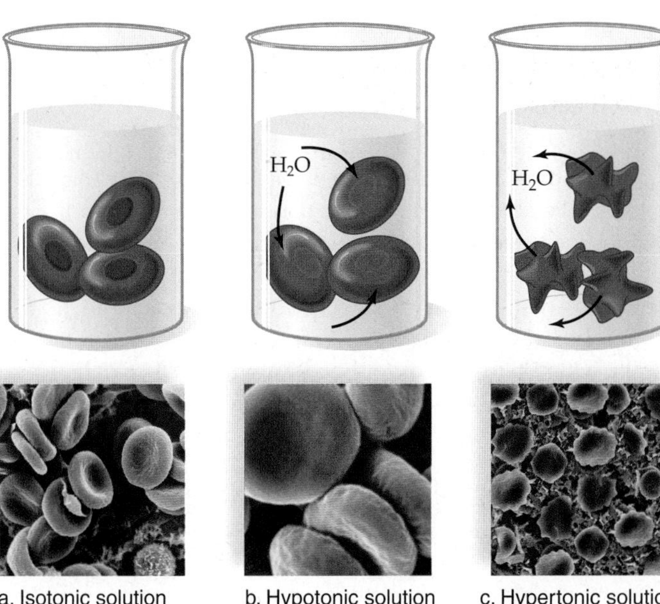

a. Isotonic solution (same solute concentration as in cell)
b. Hypotonic solution (lower solute concentration than in cell)
c. Hypertonic solution (higher solute concentration than in cell)

Figure 3.9 **Effects of changes in tonicity on red blood cells.** **a.** In an isotonic solution, cells remain the same. **b.** In a hypotonic solution, cells gain water and may burst (lysis). **c.** In a hypertonic solution, cells lose water and shrink (crenation).

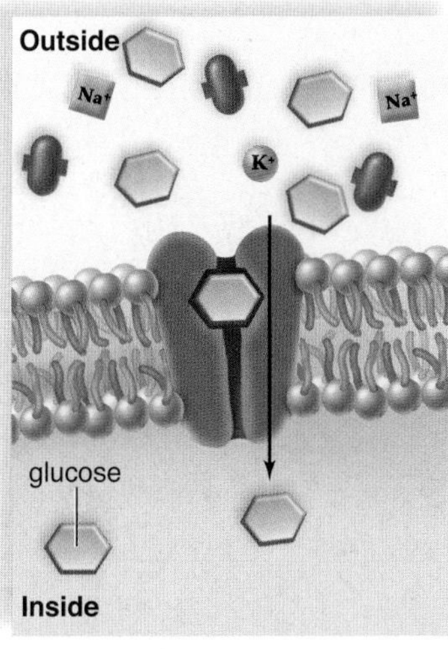

Figure 3.10 **Faciliated transport across a cell membrane.**
This is a passive form of transport in which substances move down their concentration gradient through a protein carrier. In this example, glucose (green) moves into the cell by facilitated transport. The end result will be an equal distribution of glucose on both sides of the membrane.

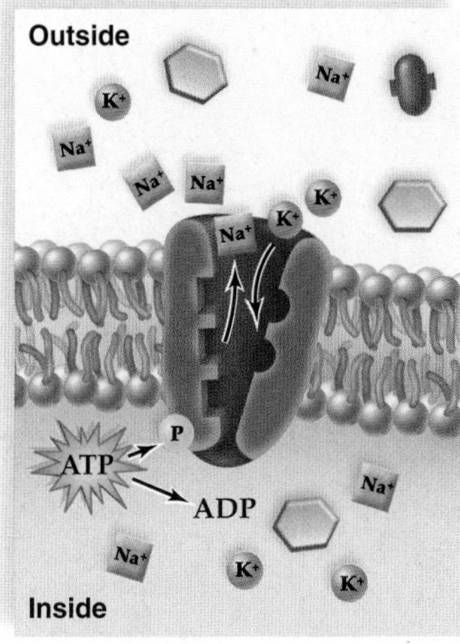

Figure 3.11 **Active transport and the sodium–potassium pump.**
This is a form of transport in which a molecule moves from low concentration to high concentration. It requires a protein carrier and energy. Na^+ exits and K^+ enters the cell by active transport, so Na^+ will be concentrated outside and K^+ will be concentrated inside the cell.

Facilitated Transport

Many solutes do not simply diffuse across a plasma membrane. They are transported by means of protein carriers within the membrane. During **facilitated transport,** a molecule is transported across the plasma membrane from the side of higher concentration to the side of lower concentration (Fig. 3.10). This is a passive means of transport because the cell does not need to expend energy to move a substance down its concentration gradient. Each protein carrier, sometimes called a *transporter,* binds only to a particular molecule, such as glucose. Type 2 diabetes results when cells lack a sufficient number of glucose transporters.

Animation How Facilitated Diffusion Works

Active Transport

During **active transport,** a molecule is moving from a *lower* to *higher* concentration. One example is the concentration of iodine ions in the cells of the thyroid gland. In the digestive tract, sugar is completely absorbed from the gut by cells that line the intestines. In another example, water homeostasis is maintained by the kidneys by the active transport of sodium ions (Na^+) by cells lining kidney tubules.

Active transport requires a protein carrier and the use of cellular energy obtained from the breakdown of ATP. When ATP is broken down, energy is released. In this case, the energy is used to carry out active transport. Proteins involved in active transport often are called *pumps.* Just as a water pump uses energy to move water against the force of gravity, energy is used to move substances against their concentration gradients. One type of pump active in all cells moves sodium ions (Na^+) to the outside and potassium ions (K^+) to the inside of the cell (Fig. 3.11). This type of pump is associated especially with nerve and muscle cells.

The passage of salt (NaCl) across a plasma membrane is of primary importance in cells. First, sodium ions are pumped across a membrane. Then, chloride ions diffuse through channels that allow their passage. In cystic fibrosis, a mutation in these chloride ion channels causes them to malfunction. This leads to the symptoms of this inherited (genetic) disorder.

Connections and Misconceptions

Can you drink seawater?

Seawater is hypertonic to our cells. Seawater contains approximately 3.5% salt, whereas our cells contain 0.9%. Once the salt entered the blood, your cells would shrivel up and die as they lost water trying to dilute the excess salt. Your kidneys can only produce urine that is slightly less salty than seawater, so you would dehydrate providing the amount of water necessary to rid your body of the salt. In addition, salt water contains high levels of magnesium ions, which cause diarrhea and further dehydration.

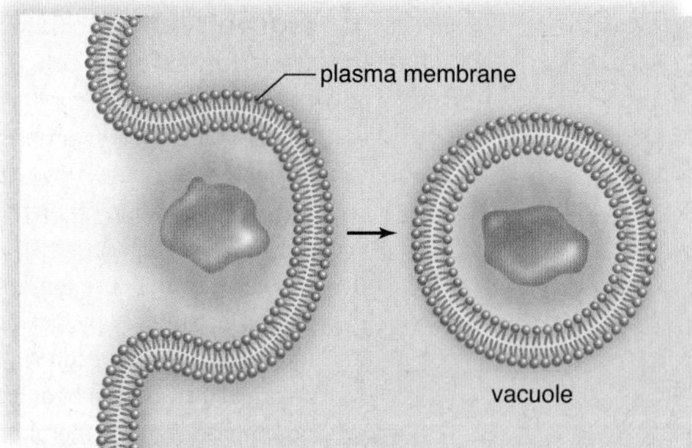

a. Phagocytosis

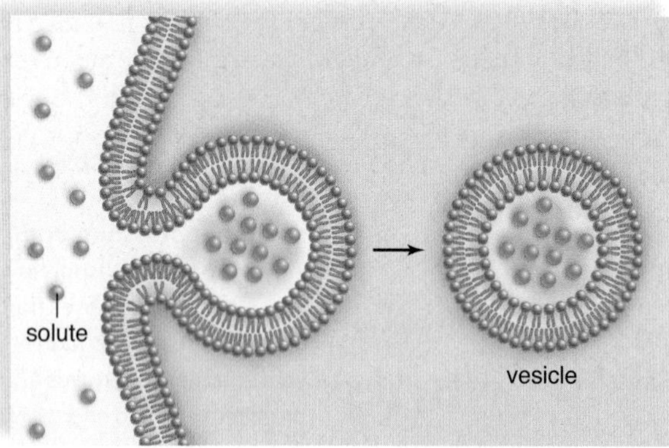

b. Pinocytosis

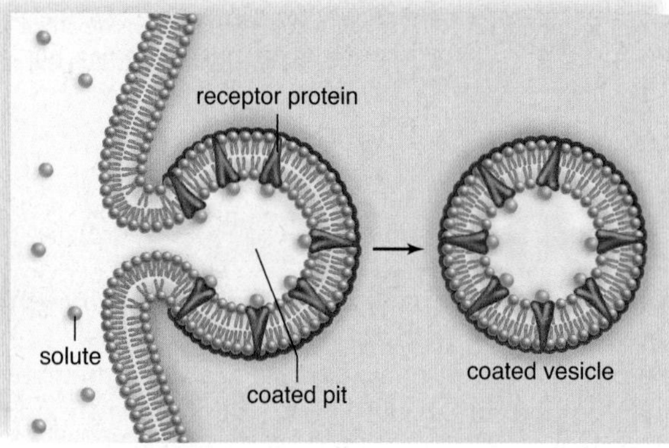

c. Receptor-mediated endocytosis

Figure 3.12 Movement of large molecules across the membrane.

a. Large substances enter a cell by endocytosis. **b.** Large substances exit a cell by exocytosis. **c.** In receptor-mediated endocytosis, molecules first bind to specific receptors and are then brought into the cell by endocytosis.

Endocytosis and Exocytosis

During *endocytosis,* a portion of the plasma membrane invaginates, or forms a pouch, to envelop a substance and fluid. Then the membrane pinches off to form an endocytic vesicle inside the cell (Fig. 3.12*a*). Some white blood cells are able to take up pathogens (disease-causing agents) by endocytosis. Here the process is given a special name, **phagocytosis.** Usually, cells take up molecules and fluid, and then the process is called *pinocytosis*. An inherited form of cardiovascular disease occurs when cells fail to take up a combined lipoprotein and cholesterol molecule from the blood by pinocytosis.

Connections and Misconceptions

What causes cystic fibrosis?

In 1989, scientists determined that defects in a gene on chromosome 7 were the cause of cystic fibrosis (CF). This gene, called *CFTR* (cystic fibrosis conductance regulator), codes for a protein that is responsible for the movement of chloride ions across the membranes of cells that produce mucus, sweat, and saliva. Defects in this gene cause an improper water–salt balance in the excretions of these cells, which in turn leads to the symptoms of CF. To date, there are over 1,400 known mutations in the CF gene. This tremendous amount of variation in this gene accounts for the differences in the severity of the disease in CF patients.

By knowing the precise gene that causes the disease, scientists have been able to develop new treatment options for people with CF. At one time, an individual with CF rarely saw his or her twentieth birthday; now it is routine for people to live into their 30s and 40s. New treatments, such as gene therapy, are being explored for sufferers of CF.

During *exocytosis,* a vesicle fuses with the plasma membrane as secretion occurs (Fig. 3.12*b*). Later in the chapter, we will see that a steady stream of vesicles move between certain organelles, before finally fusing with the plasma membrane. This is the way that signaling molecules, called *neurotransmitters,* leave one nerve cell to excite the next nerve cell or a muscle cell.

A special form of endocytosis uses a receptor, a special form of membrane protein, on the surface of the cell to concentrate specific molecules of interest for endocytosis. This process is called *receptor-mediated endocytosis* (Fig. 3.12*c*). An inherited form of cardiovascular disease occurs when cells fail to take up a combined lipoprotein and cholesterol molecule from the blood by receptor-mediated endocytosis.

Animation Endocytosis and Exocytosis

Connecting the Concepts

The movement of materials across a cell membrane is crucial to the maintenance of homeostasis for many organ systems in humans. For some examples, refer to the following discussions.

Section 8.3 examines how nutrients, including glucose, are moved into the cells of the digestive system.

Section 10.4 investigates how the movement of salts by the urinary system maintains blood homeostasis.

Section 20.2 explains the patterns of inheritance associated with cystic fibrosis.

Check Your Progress 3.3

1. Describe the structure and overall function of the plasma membrane.
2. Compare and contrast diffusion and osmosis.
3. Discuss the various ways materials can enter and leave cells.

3.4 The Nucleus and Endomembrane System

Learning Outcomes

Upon completion of this section, you should be able to

1. Recognize the structure of the nucleus and its role as the storage place of the genetic information.
2. Summarize the function of the organelles of the endomembrane system.
3. Explain the role and location of the ribosomes.

The nucleus and several organelles are involved in the production and processing of proteins. The endomembrane system is a series of membrane organelles that function in the processing of materials for the cell.

The Nucleus

The **nucleus,** a prominent structure in cells, stores genetic information (Fig. 3.13). Every cell in the body contains the same genes. Genes are segments of DNA that contain information for the production of specific proteins. Each type of cell has certain genes turned on and others turned off. DNA, with RNA acting as an intermediary, specifies the proteins in a cell. Proteins have many functions in cells, and they help determine a cell's specificity.

Chromatin is the combination of DNA molecules and proteins that make up the **chromosomes.** Chromatin can coil tightly to form visible chromosomes during *meiosis* (cell division that forms reproductive cells in humans) and *mitosis* (cell division that duplicates cells). Most of the time, however, the chromatin is uncoiled. Individual chromosomes cannot be distinguished and the chromatin appears grainy in electron micrographs of the nucleus. Chromatin is immersed in a semifluid medium called the **nucleoplasm.** A difference in pH suggests that nucleoplasm has a different composition from cytoplasm.

Micrographs of a nucleus do show one or more dark regions of the chromatin. These are nucleoli (sing., **nucleolus**), where ribosomal RNA (rRNA) is produced. This is also where rRNA joins with proteins to form the subunits of ribosomes.

The nucleus is separated from the cytoplasm by a double membrane known as the **nuclear envelope.** This is continuous with the **endoplasmic reticulum (ER),** a membranous system of saccules and channels discussed in the next section. The nuclear envelope has **nuclear pores** of sufficient size to permit the passage of ribosomal subunits out of the nucleus and proteins into the nucleus.

Ribosomes

Ribosomes are organelles composed of proteins and rRNA. Protein synthesis occurs at the ribosomes. Ribosomes are often attached to the endoplasmic reticulum; but they also may occur

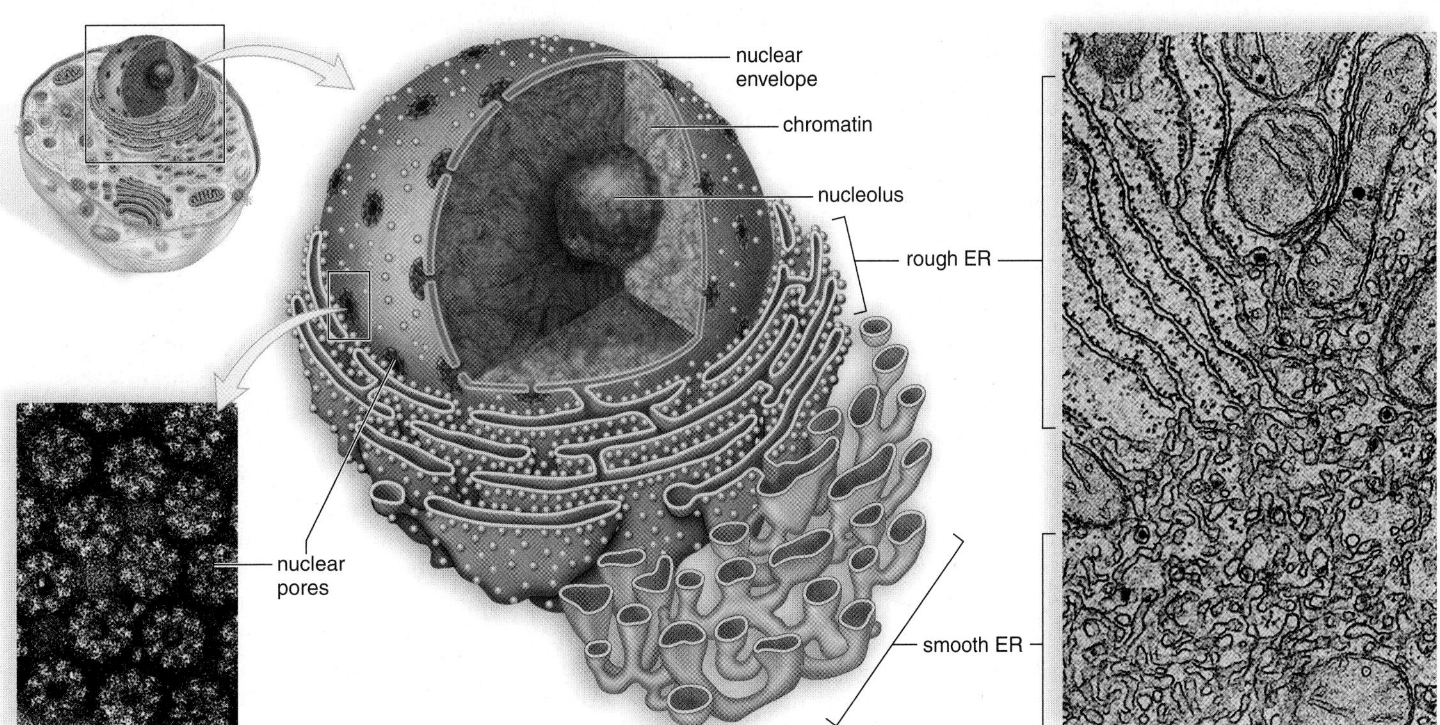

Figure 3.13 The nucleus and endoplasmic reticulum.
The nucleus contains chromatin. Chromatin has a special region called the nucleolus, where rRNA is produced and ribosome subunits are assembled. The nuclear envelope contains pores (TEM, *left*) that allow substances to enter and exit the nucleus to and from the cytoplasm. The nuclear envelope is attached to the endoplasmic reticulum (TEM, *right*), which often has attached ribosomes, where protein synthesis occurs.

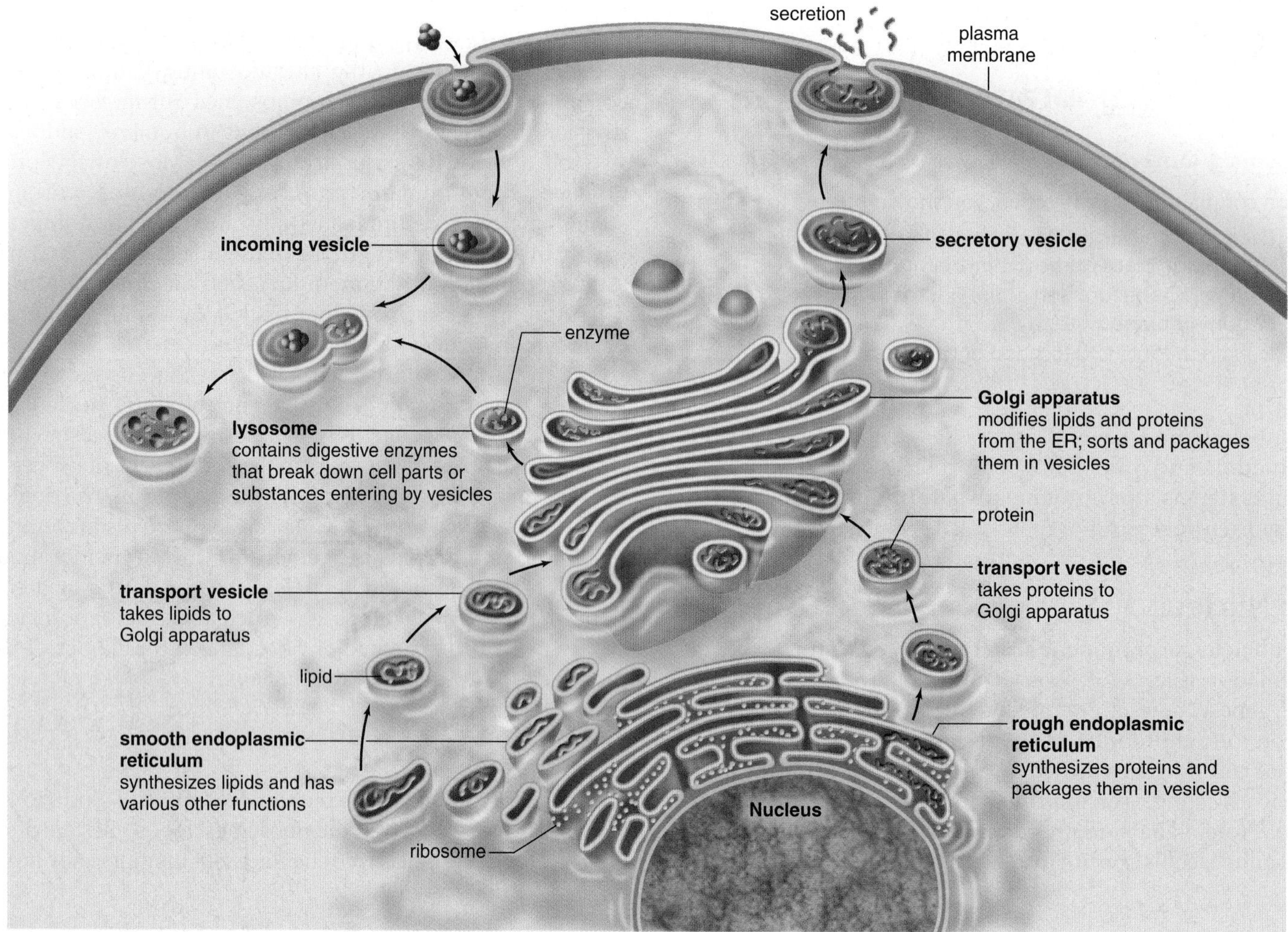

Figure 3.14 **The endomembrane system.**
The organelles in the endomembrane system work together to produce, modify, secrete, and digest proteins and lipids.

free within the cytoplasm, either singly or in groups called **polyribosomes.** Proteins synthesized at ribosomes attached to the endoplasmic reticulum have a different destination from that of proteins manufactured at ribosomes free in the cytoplasm.

The Endomembrane System

The **endomembrane system** consists of the nuclear envelope, the endoplasmic reticulum, the Golgi apparatus, lysosomes, and **vesicles** (tiny membranous sacs) (Fig. 3.14). This system compartmentalizes the cell so that chemical reactions are restricted to specific regions. The vesicles transport molecules from one part of the system to another.

The Endoplasmic Reticulum

The endoplasmic reticulum has two portions. Rough ER is studded with ribosomes on the side of the membrane that faces the cytoplasm. Here, proteins are synthesized and enter the ER interior, where processing and modification begin. Some of these proteins are incorporated into membrane, and some are for export. Smooth ER, continuous with rough ER, does not have attached ribosomes. Smooth ER synthesizes the phospholipids that occur in membranes and has various other functions, depending on the particular cell. In the testes, it produces testosterone. In the liver, it helps detoxify drugs.

The ER forms transport vesicles in which large molecules are transported to other parts of the cell. Often, these vesicles are on their way to the plasma membrane or the Golgi apparatus.

The Golgi Apparatus

The **Golgi apparatus** is named for Camillo Golgi, who discovered its presence in cells in 1898. The Golgi apparatus consists of a stack of slightly curved saccules, whose appearance can be compared to a stack of pancakes. Here, proteins and lipids received from the ER are modified. For example, a chain of sugars may be added to them. This makes them glycoproteins and glycolipids, molecules often found in the plasma membrane.

The vesicles that leave the Golgi apparatus move to other parts of the cell. Some vesicles proceed to the plasma membrane, where they discharge their contents. In all, the Golgi apparatus is involved in processing, packaging, and secretion.

Lysosomes

Lysosomes, membranous sacs produced by the Golgi apparatus, contain *hydrolytic enzymes*. Lysosomes are found in all cells of the body but are particularly numerous in white blood cells that engulf disease-causing microbes. When a lysosome fuses with such an endocytic vesicle, its contents are digested by lysosomal enzymes into simpler subunits that then enter the cytoplasm. In a process called autodigestion, parts of a cell may be broken down by the lysosomes. Some human diseases are caused by the lack of a particular lysosome enzyme. Tay–Sachs disease occurs when an undigested substance collects in nerve cells, leading to developmental problems and death in early childhood.

Animation Lysosomes

Connecting the Concepts

For a more detailed look at how the organelles of the endomembrane system function, refer to the following discussions.

Section 17.5 contains information on how aging is related to the breakdown of cellular organelles.

Section 20.2 explores the patterns of inheritance associated with Tay–Sachs disease.

Section 21.2 provides a more detailed look at how ribosomes produce proteins.

Check Your Progress 3.4

1. Describe the functions of the following organelles: endoplasmic reticulum, Golgi apparatus, lysosomes.
2. Explain how the nucleus, ribosomes, and rough endoplasmic reticulum contribute to protein synthesis.
3. Describe the function of the endomembrane system including the formation and actions of transport vesicles.

3.5 The Cytoskeleton, Cell Movement, and Cell Junctions

Learning Outcomes

Upon completion of this section, you should be able to

1. Explain the role of the cytoskeleton in the cell.
2. List the major protein fibers in the cytoskeleton.
3. State the role of flagella and cilia in human cells.
4. Compare the function of adhesion junctions, gap junctions, and tight junctions in human cells.

It took a high-powered electron microscope to discover that the cytoplasm of the cell is crisscrossed by several types of protein fibers collectively called the **cytoskeleton** (see Fig. 3.4). The cytoskeleton helps maintain a cell's shape and either anchors the organelles or assists their movement, as appropriate.

In the cytoskeleton, **microtubules** are much larger than actin filaments. Each is a cylinder that contains rows of a protein called tubulin. The regulation of microtubule assembly is under the control of a microtubule organizing center called the **centrosome** (see Fig. 3.4). Microtubules help maintain the shape of the cell and act as tracks along which organelles move. During cell division, microtubules form spindle fibers, which assist the movement of chromosomes. **Actin filaments,** made of a protein called *actin*, are long, extremely thin fibers that usually occur in bundles or other groupings. Actin filaments are involved in movement. Microvilli, which project from certain cells and can shorten and extend, contain actin filaments. **Intermediate filaments,** as their name implies, are intermediate in size between microtubules and actin filaments. Their structure and function differ according to the type of cell.

Cilia and Flagella

Cilia (sing., **cilium**) and **flagella** (sing., **flagellum**) are involved in movement. The ciliated cells that line our respiratory tract sweep debris trapped within mucus back up the throat. This helps keep the lungs clean. Similarly, ciliated cells move an egg along the oviduct, where it will be fertilized by a flagellated sperm cell (Fig. 3.15). Motor molecules, powered by ATP, allow the microtubules in cilia and flagella to interact and bend and, thereby, move.

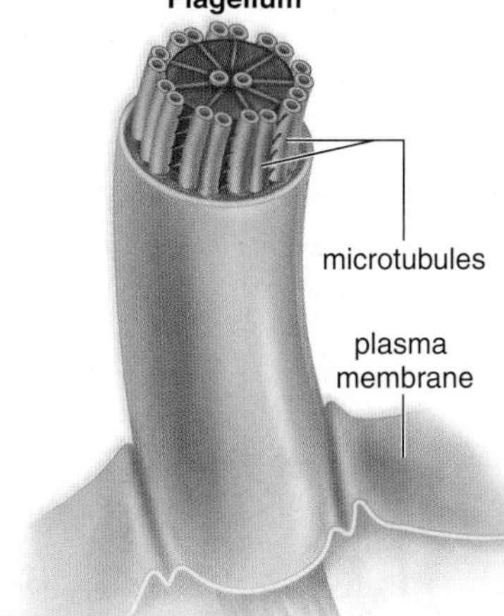

b.

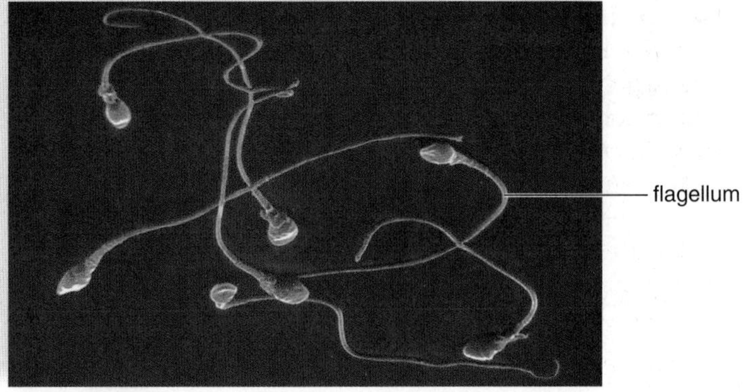

c.

Figure 3.15 **Structure and function of the flagella and cilia.** Human reproduction is dependent on the normal activity of cilia and flagella. **a.** Both cilia and flagella have an inner core of microtubules within a covering of plasma membrane. **b.** Cilia within the oviduct move the egg to where it is fertilized by a flagellated sperm. **c.** Sperm have very long flagella.

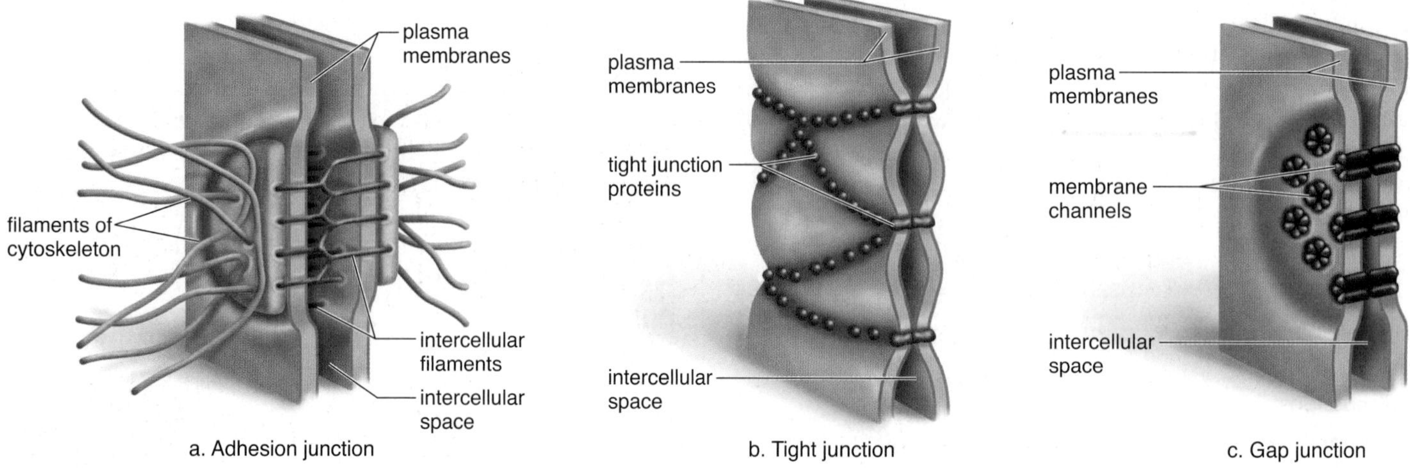

Figure 3.16 Junctions between cells.
a. Adhesion junctions mechanically connect cells. **b.** Tight junctions form barriers with the external environment. **c.** Gap junctions allow for communication between cells.

The importance of normal cilia and flagella is illustrated by the occurrence of a genetic disorder. Some individuals have an inherited genetic defect that leads to malformed microtubules in cilia and flagella. Not surprisingly, these individuals suffer from recurrent and severe respiratory infections. The ciliated cells lining respiratory passages fail to keep their lungs clean. They are also unable to reproduce naturally due to the lack of ciliary action to move the egg in a female or the lack of flagella action by sperm in a male.

Connections and Misconceptions

How fast does a human sperm swim?

Individual sperm speeds vary considerably and are greatly influenced by environmental conditions. However, in recent studies, researchers found that some human sperm could travel at top speeds of approximately 20 cm/hour. This means that these sperm could reach the female ovum in less than an hour. Scientists are interested in sperm speed so that they can design new contraceptive methods.

Junctions Between Cells

As we will see in the next chapter, human tissues are known to have junctions between their cells that allow them to function in a coordinated manner. Figure 3.16 illustates the three main types of cell junctions in human cells.

Adhesion junctions serve to mechanically attach adjacent cells. In these junctions, the cytoskeletons of two adjacent cells are interconnected. They are a common type of junction between skin cells. In *tight junctions,* connections between the plasma membrane proteins of neighboring cells produce a zipperlike barrier. These types of junctions are common in the digestive system and the kidney, where it is necessary to contain fluids (digestive juices and urine) within a specific area. *Gap junctions* serve as communication portals between cells. In these junctions, channel proteins of the plasma membrane fuse, allowing easy movement between adjacent cells.

Connecting the Concepts

The cytoskeleton of the cell plays an important role in many aspects of our physiology. To explore this further, refer to the following discussions.

Section 9.1 investigates how the ciliated cells of the respiratory system function.

Section 16.2 explains the role of the flagellated sperm cell in reproduction.

Section 18.1 explores how the cytoskeleton is involved in cell division.

Check Your Progress 3.5

1. List the three types of fibers found in the cytoskeleton.
2. Describe the structure of cilia and flagella and state the function of each.
3. List the types of junctions found in animal cells and provide a function for each.

3.6 Mitochondria and Cellular Metabolism

Learning Outcomes

Upon completion of this section, you should be able to

1. Identify the key structures of a mitochondrion.
2. Summarize the relationship between the mitochondria and energy-generating pathways of the cell.
3. Summarize the roles of glycolysis, citric acid cycle, electron transport chain, and fermentation in energy generation.
4. Illustrate the stages of the ATP cycle.

Mitochondria (sing., **mitochondrion**) are often called the powerhouses of the cell. Just as a powerhouse burns fuel to produce electricity, the mitochondria convert the chemical energy of glucose products into the chemical energy of ATP molecules. In the process, mitochondria use up oxygen and give off carbon dioxide. Therefore, the process of producing ATP is called **cellular respiration.** The structure of mitochondria is appropriate to the task. The inner membrane is folded to form little shelves called *cristae*. These project into the *matrix*, an inner space filled with a gel-like fluid (Fig. 3.17). The matrix of a mitochondrion contains enzymes for breaking down glucose products. ATP production then occurs at the cristae. Protein complexes that aid in the conversion of energy are located in an assembly-line fashion on these membranous shelves.

The structure of a mitochondrion supports the hypothesis that they were originally prokaryotes engulfed by a cell. Mitochondria are bounded by a double membrane, as a prokaryote would be if taken into a cell by endocytosis. Even more interesting is the observation that mitochondria have their own genes—and they reproduce themselves!

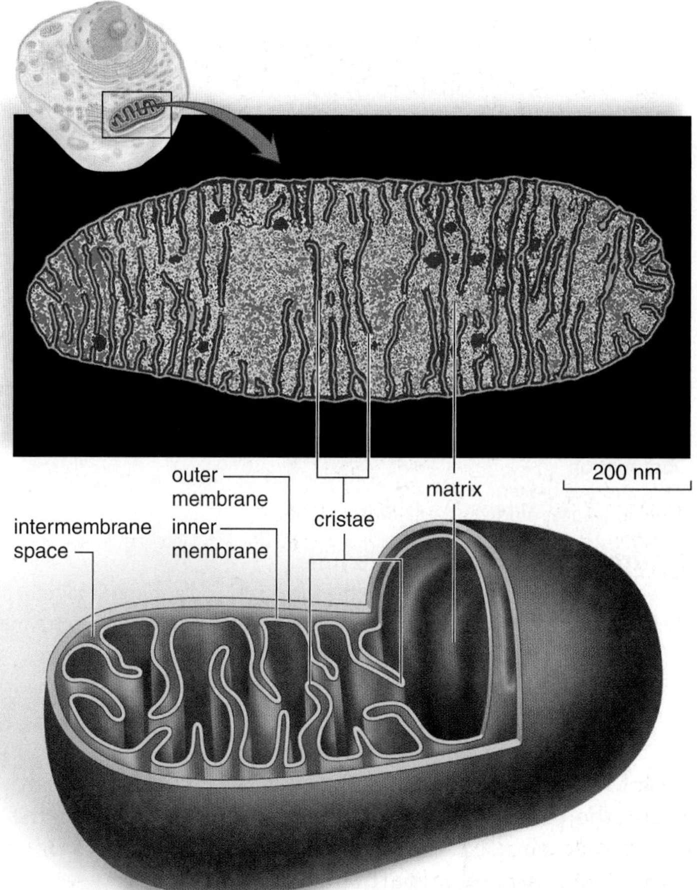

Figure 3.17 The structure of a mitochondrion.
A mitochondrion (TEM, *above*) is bounded by a double membrane, and the inner membrane folds into projections called cristae. The cristae project into a semifluid matrix that contains many enzymes.

Cellular Respiration and Metabolism

Cellular respiration is an important component of **metabolism,** which includes all the chemical reactions that occur in a cell. Often, metabolism requires metabolic pathways and is carried out by enzymes sequentially arranged in cells:

$$A \xrightarrow{1} B \xrightarrow{2} C \xrightarrow{3} D \xrightarrow{4} E \xrightarrow{5} F \xrightarrow{6} G$$

The letters, except *A* and *G*, are **products** of the previous reaction and the **reactants** for the next reaction. *A* represents the beginning reactant, and *G* represents the final product. The numbers in the pathway refer to different enzymes. *Each reaction in a metabolic pathway requires a specific enzyme.* The mechanism of action of enzymes has been studied extensively because enzymes are so necessary in cells.

Animation Biochemical Pathways

Metabolic pathways are highly regulated by the cell. One type of regulation is *feedback inhibition.* In feedback inhibition, one of the end products of the metabolic pathway interacts with an enzyme early in the pathway. In most cases, this feedback slows down the pathway so that the cell does not produce more product than it needs.

Animation Feedback Inhibition of Biochemical Pathways

Enzymes

The reactant/s that participate/s in the reaction is/are called the enzyme's **substrate/s.** Enzymes are often named for their substrates. For example, lipids are broken down by lipase, maltose by maltase, and lactose by lactase.

Enzymes have a specific region, called an **active site,** where the substrates are brought together so they can react. An enzyme's specificity is caused by the shape of the active site. Here the enzyme and its substrate/s fit together in a specific way, much as the pieces of a jigsaw

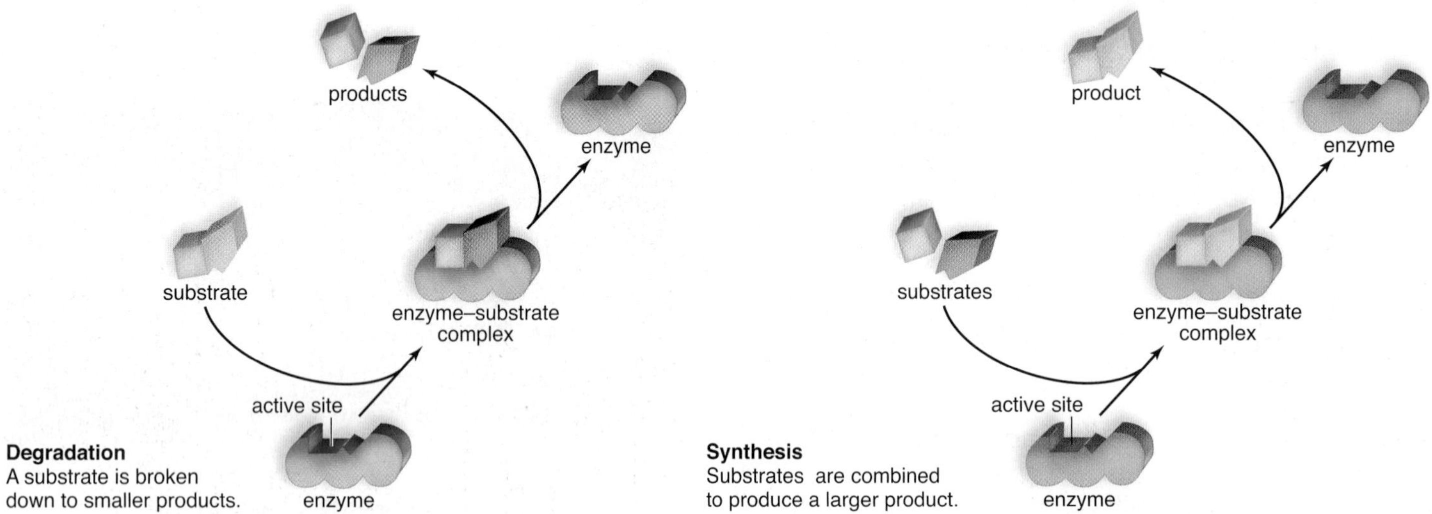

Figure 3.18 **Action of an enzyme.**
An enzyme has an active site, where the substrates and enzyme fit together in such a way that the substrates are oriented to react. Following the reaction, the products are released and the enzyme is free to act again. Some enzymes carry out degradation, in which the substrate is broken down to smaller products. Other enzymes carry out synthesis, in which the substrates are combined to produce a larger product.

puzzle fit together (Fig. 3.18). After one reaction is complete, the product or products are released. The enzyme is ready to be used again. Therefore, a cell requires only a small amount of a particular enzyme to carry out a reaction. A chemical reaction can be summarized in the following manner:

$$E + S \longrightarrow ES \longrightarrow E + P$$

where E = enzyme, S = substrate, ES = enzyme–substrate complex, and P = product. An enzyme can be used over and over again.

- **Coenzymes** are nonprotein molecules that assist the activity of an enzyme and may even accept or contribute atoms to the reaction. It is interesting that vitamins are often components of coenzymes. The vitamin niacin is a part of the coenzyme **NAD^+ (nicotinamide adenine dinucleotide),** which carries hydrogen (H) and electrons.

Animation
How the NAD^+ Works

Cellular Respiration

After blood transports glucose and oxygen to cells, cellular respiration begins. Cellular respiration breaks down glucose to carbon dioxide and water. Three pathways are involved in the breakdown of glucose—glycolysis, the citric acid cycle, and the electron transport chain (Fig. 3.19). These metabolic pathways allow the energy within a glucose molecule to be slowly released so that ATP can be gradually produced. Cells would lose a tremendous amount of energy if glucose breakdown occurred all at once. Much energy would be lost as heat. When humans burn wood or coal, the energy escapes all at once as heat. But a cell gradually "burns" glucose, and energy is captured as ATP.

MP3
Cellular Respiration

- **Glycolysis** means "sugar splitting." During glycolysis, glucose, a six-carbon (C_6) molecule, is split so that the result is two three-carbon (C_3) molecules of *pyruvate.* Glycolysis, which occurs in the cytoplasm, is found in most every type of cell. Therefore, this pathway is believed to have evolved early in the history of life.

Glycolysis is termed **anaerobic,** because it requires no oxygen. This pathway can occur in microbes that live in bogs or swamps or our intestinal tract, where there is no oxygen. During glycolysis, hydrogens and electrons are removed from glucose, and NADH results. The breaking of bonds releases enough energy for a net yield of two ATP molecules.

Pyruvate is a pivotal molecule in cellular respiration. When oxygen is available, the molecule enters mitochondria and is completely broken down. When oxygen is not available, fermentation occurs (discussion follows).

Animation
Glycolysis

- The **citric acid cycle,** also called the *Krebs cycle,* completes the breakdown of glucose. As this cyclical series of enzymatic reactions occurs in the matrix of mitochondria, carbon dioxide is released. Hydrogen and electrons are carried away by NADH. In addition, the citric acid cycle also produces two ATP per glucose molecule.

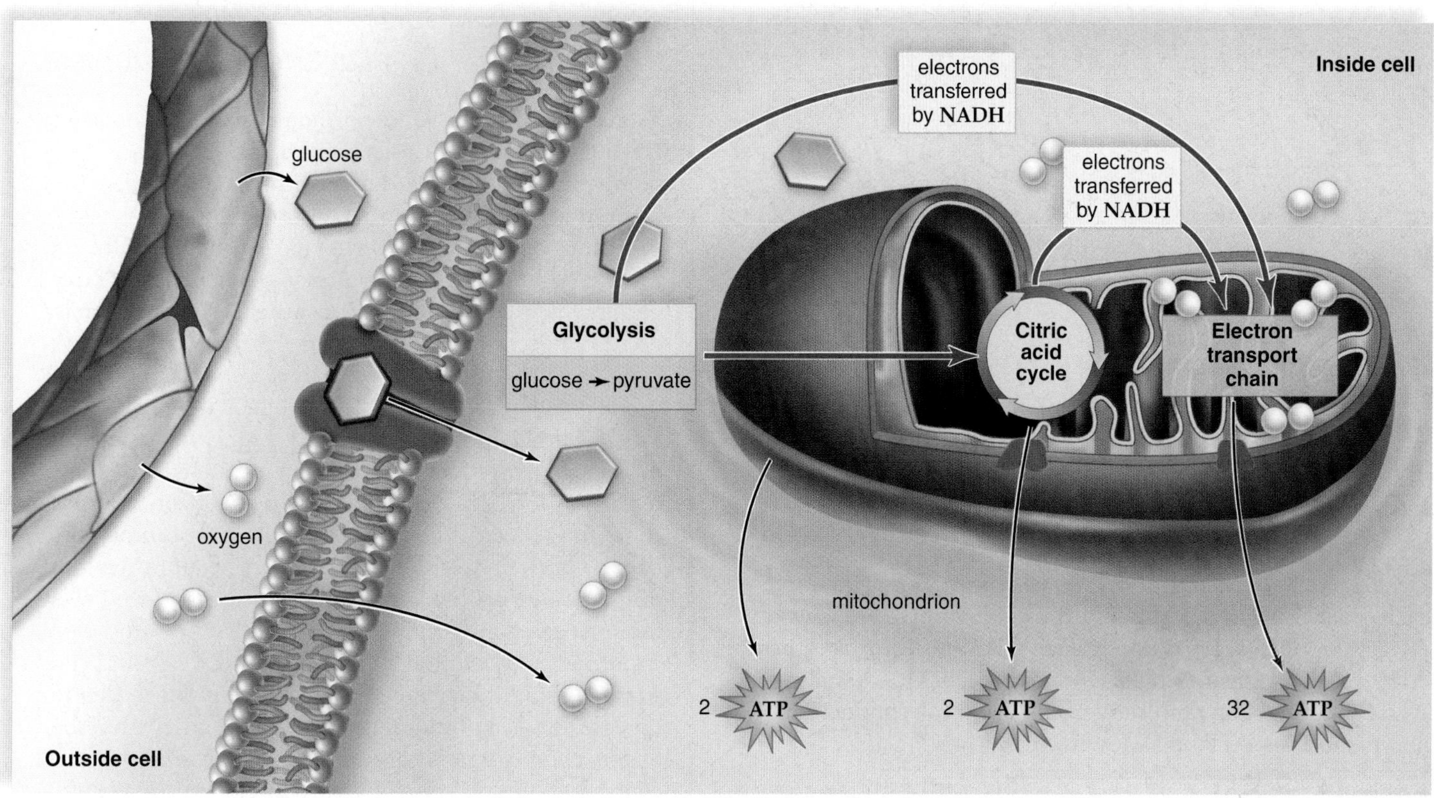

Figure 3.19 **Production of ATP.**
Glucose enters a cell from the bloodstream by facilitated transport. The three main pathways of cellular respiration (glycolysis, citric acid cycle, and electron transport chain) all produce ATP, but most is produced by the electron transport chain. NADH carries electrons to the electron transport chain from glycolysis and the citric acid cycle. ATP exits a mitochondrion by facilitated transport.

So far, we have considered only carbohydrates as a possible fuel for cellular respiration. But what about fats and proteins? Fats are digested to glycerol and fatty acids. When our cells run out of glucose, they primarily substitute fatty acids for glucose as an energy source. Fatty acids yield C–C molecules that can enter the citric acid cycle.

Proteins are digested to amino acids whose carbon chain can easily enter the citric acid cycle. However, first the amino group has to be removed from the carbon chain. This step is primarily carried out in the liver because the liver has the appropriate enzymes to process amino groups. The liver converts the amino groups to urea, which is excreted by the kidneys. When you eat a lot of protein, your body begins to use amino acids as an energy source. Your kidneys work overtime excreting all that nitrogen.

Animation How the Krebs Cycle Works

- NADH molecules from glycolysis and the citric acid cycle deliver electrons to the **electron transport chain.** The members of the electron transport chain are carrier proteins grouped into complexes. These complexes are embedded in the cristae of a mitochondrion. Each carrier of the electron transport chain accepts two electrons and passes them on to the next carrier. The hydrogens carried by NADH molecules will be used later.

High-energy electrons enter the chain and, as they are passed from carrier to carrier, the electrons lose energy. Low-energy electrons emerge from the chain. Oxygen serves as the final acceptor of the electrons at the end of the chain. After oxygen receives the electrons, it combines with hydrogens and becomes water.

The presence of oxygen makes the citric acid cycle and the electron transport chain **aerobic.** Oxygen does not combine with any substrates during cellular respiration. Breathing is necessary to our existence, and the sole purpose of oxygen is to receive electrons at the end of the electron transport chain.

The energy, released as electrons pass from carrier to carrier, is used for ATP production. It took many years for investigators to determine exactly how this occurs, and the details are beyond the scope of this text. Suffice it to say that the inner mitochondrial membrane contains an ATP–synthase complex that combines ADP + Ⓟ to produce ATP. The ATP–synthase complex produces about 32 ATP per glucose molecule.

Animation Electron Transport System and ATP Synthesis

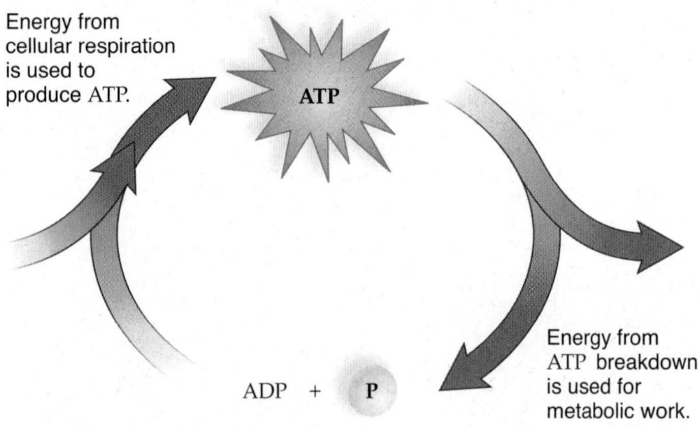

Figure 3.20 The ATP cycle.
The breakdown of glucose by cellular respiration transfers energy to form ATP. ATP is used for energy-requiring reactions, such as muscle contraction. ATP breakdown also gives off heat. Additional food energy rejoins ADP and P to form ATP again.

Each cell produces ATP within its mitochondria; therefore, each cell uses ATP for its own purposes. Figure 3.20 shows the ATP cycle. Glucose breakdown leads to ATP buildup, and then ATP is used for the metabolic work of the cell. Muscle cells use ATP for contraction, and nerve cells use it for conduction of nerve impulses. ATP breakdown releases heat.

Fermentation

Fermentation is an anaerobic process, meaning that it does not require oxygen. When oxygen is not available to cells, the electron transport chain soon becomes inoperative. This is because oxygen is not present to accept electrons. In this case, most cells have a safety valve so that some ATP can still be produced. Glycolysis operates as long as it is supplied with "free" NAD^+—NAD^+ that can pick up hydrogens and electrons. Normally, NADH takes electrons to the electron transport chain and, thereby, is recyled to become NAD+. However, if the system is not working due to a lack of oxygen, NADH passes its hydrogens and electrons to pyruvate molecules, as shown in the following reaction:

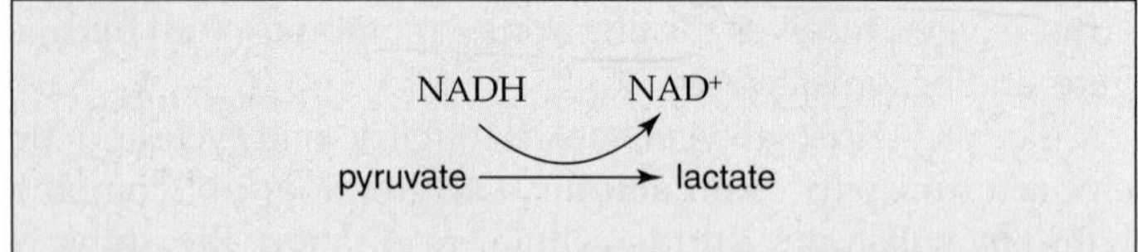

This means that the citric acid cycle and the electron transport chain do not function as part of fermentation. When oxygen is available again, lactate can be converted back to pyruvate and metabolism can proceed as usual.

Fermentation can give us a burst of energy for a short time, but it produces only two ATP per glucose molecule. Also, fermentation results in the buildup of lactate. Lactate is toxic to cells and causes muscles to cramp and fatigue. If fermentation continues for any length of time, death follows.

Fermentation takes its name from yeast fermentation. Yeast fermentation produces alcohol and carbon dioxide (instead of lactate). When yeast is used to leaven bread, carbon dioxide production makes the bread rise. When yeast is used to produce alcoholic beverages, it is the alcohol that humans make use of.

Biology Matters Health Focus

Lactate and the Athlete

Exercise is a dramatic test of homeostatic mechanisms. During exercise, the mitochondria of our muscle cells require much oxygen. They also produce an increased amount of carbon dioxide. No doubt, if you run as fast as you can, even for a short time, you are out of breath. You are in oxygen deficit—your muscles have run out of oxygen and have started fermenting instead. Aerobic exercise occurs when you can manage to get a steady supply of oxygen to your muscle cells so that oxygen deficit does not occur. Athletes are better at this than nonathletes. Why?

The number of mitochondria is higher in the muscles of persons who train. Therefore, an athlete is more likely to rely on the citric acid cycle and the electron transport chain to generate ATP. The citric acid cycle can be powered by fatty acids instead of glucose. Therefore, the level of glucose in the blood remains at a normal level, even though exercise is occurring.

Muscle cells with few mitochondria do not start consuming O_2 until they are out of ATP and ADP concentration is high. After endurance training, the large number of mitochondria start consuming O_2 as soon as the ADP concentration starts rising. This is due to muscle contraction and breakdown of ATP. This faster rise in O_2 uptake at the onset of exercise means that the oxygen deficit is less, and the formation of lactate due to fermentation is less.

As mentioned, the body is able to process lactate and change it back to pyruvate. Again, this is the oxygen deficit—the amount of oxygen it takes to rid the body of lactate. Athletes incur less of an oxygen deficit than nonathletes.

Connecting the Concepts

For additional information on the processing of nutrients for energy, refer to the following discussions.

Sections 2.3 to 2.5 provide a more detailed look at the carbohydrates and other energy nutrients.

Section 8.3 explores how the small intestine processes nutrients for absorption.

Section 8.6 examines how the body uses nutrients other than carbohydrates as fuel.

Check Your Progress 3.6

1. Summarize the roles of enzymes in chemical reactions.
2. Describe the three basic steps of cellular respiration. Include the starting and ending molecules for each step.
3. Hypothesize what would happen to homeostasis if each of three major steps of cellular respiration were missing.

Biology Matters Bioethical Focus

Stem-Cell Research

In the human body, stem cells are analogous to immortal "parents." Their "offspring," called *daughter cells*, can remain as stem cells and potentially divide indefinitely. However, most daughter cells differentiate further, forming mature cells called *end cells*. Research using stem cells has remained a source of controversy since 1998, when scientists discovered how to isolate and grow human stem cells in the laboratory.

There are primarily two different types of stem cells: embryonic and adult. Advantages and disadvantages exist for each type. *Embryonic stem cells* are derived from fertilized embryos at various stages of development. Fertilized human ova stored in infertility clinics are often used as the source of embryonic stem cells. The use of these cells for research has sparked tremendous controversy, because many people believe these cells have the potential to become a human being. *Adult stem cells* are undifferentiated cells found in various body tissues, whose purpose is to repair or replace damaged tissues. The use of adult stem cells is generally accepted. However, adult stem cells lack the flexibility of embryonic stem cells, because adult stem cells form far fewer types of end cells.

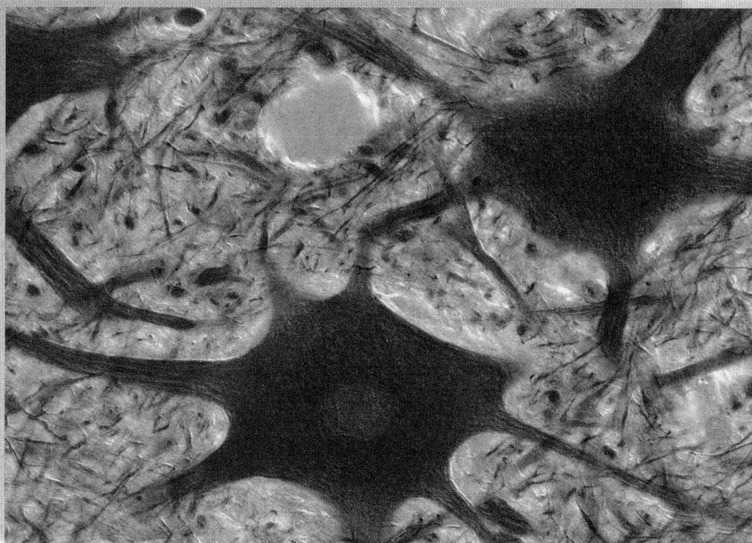

Figure 3B Stem cells as a potential cure for Parkinson disease.
Parkinson disease results in the loss of neurons such as these. Stem cells that differentiate into dopamine-producing neurons could be transplanted into the brains of Parkinson patients.

With all the time and money spent on stem-cell research, how close are we to using stem cells for the cure of disease? Let's use Parkinson disease as an example. Parkinson disease is a progressive motor control disorder, triggered by the death of certain neurons in the brain (Fig. 3B). These neurons are responsible for releasing the neurotransmitter dopamine onto specific brain cells that control movement. (This is why Parkinson patients are often treated with L-dopa, which is converted into dopamine.) It is now possible to cause stem cells in the laboratory to differentiate into neurons that produce dopamine. To be used for transplant purposes, however, the stem cells must produce enough end cells for transplant. Further, the cells must survive after the transplant and function correctly for the remainder of the patient's life. Finally, transplanted cells must not harm the patient. The usual risks of surgery would still exist for the transplant recipient: damage to healthy tissue, bleeding, infection.

A possible solution was introduced in 2008 when researchers first developed the use of *induced pluripotent stem cells*, or *iPS cells*. These cells are normal cells of the body that have been chemically "convinced" to return to an undifferentiated state. In other words, it is now possible to induce adult cells of the body to form stem cells. By doing so, researchers hope to be able to bypass some of the controversies surrounding the use of embryonic stem cells and, in the process, develop a more rapid and effective method of obtaining stem cells to fight specific diseases. For this groundbreaking work, *Science Magazine* was awarded its 2008 Breakthrough of the Year Award.*

Video Heart Stem Cells

Video Making Brain Cells

Decide Your Opinion

1. How much time and money should be spent on a therapy that may only work after "years of intensive research"? Would this money be better spent on therapies that have a higher likelihood of success?
2. Should the president remove the ban on certain types of stem cells so that this research can proceed faster?
3. What criteria and ethical considerations should be used to select Parkinson patients for stem-cell therapy?

* "Breakthrough of the Year: Reprogramming Cells," *Science* 322, no. 5909 (2008), http://www.sciencemag.org/cgi/content/full/322/5909/1766 (accessed November 8, 2009).

CASE STUDY CONCLUSION

Over the next few months, both Kevin and Mary dedicated hours to understanding the causes and treatments of Tay–Sachs disease. They learned that the disease is caused by a recessive mutation that limits the production of an enzyme called beta-hexosaminidase A. This enzyme is loaded into a newly formed lysosome by the Golgi apparatus. The enzyme's function is to break down a specific type of fatty acid chain called *gangliosides.* Gangliosides play an important role in the early formation of the neurons in the brain. Tay–Sachs disease occurs when the gangliosides overaccumulate in the neurons.

Though the prognosis for their child was intially poor—very few children with Tay–Sachs live beyond the age of four, the parents were encouraged to find out what advances in a form of medicine called *gene therapy* might be able to prolong the life of their child. In gene therapy, a correct version of the gene is introduced into specific cells in an attempt to regain lost function. Some initial studies using mice as a model had demonstrated an ability to reduce ganglioside concentrations by providing a working version of the gene that produced beta-hexosaminidase A to the neurons of the brain. Though research was still ongoing, it was a promising piece of information for both Kevin and Mary.

Media Study Tools

www.mhhe.com/maderhuman12e

Enhance your study of this chapter with study tools and practice tests. Also ask your instructor about the resources available through ConnectPlus, including the media-rich eBook, interactive learning tools, and animations.

Virtual Lab

The virtual lab "Enzyme-Controlled Reactions" provides an interactive investigation of how environmental conditions regulate enzyme activity.

Summarizing the Concepts

3.1 What Is a Cell?

- Cells, the basic units of life, come from pre-existing cells.
- Microscopes are used to view cells, which must remain small to have a favorable surface area-to-volume ratio.

3.2 How Cells Are Organized

The human cell is surrounded by a plasma membrane and has a central nucleus. Between the plasma membrane and the nucleus is the cytoplasm, which contains various organelles. Organelles in the cytoplasm have specific functions.

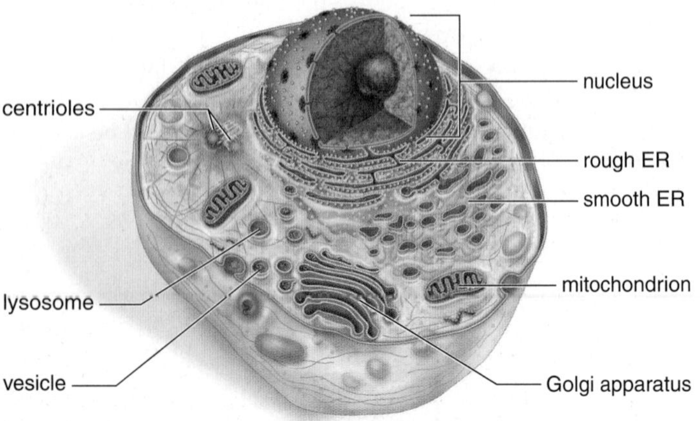

3.3 The Plasma Membrane and How Substances Cross It

The plasma membrane is a phospholipid bilayer that

- selectively regulates the passage of molecules and ions into and out of the cell.
- contains embedded proteins, which allow certain substances to cross the plasma membrane.

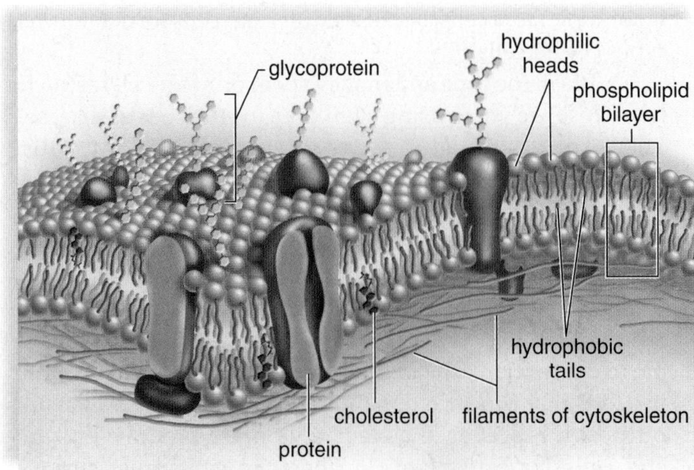

Passage of molecules into or out of cells can be passive or active.

- Passive mechanisms (no energy required) are diffusion (osmosis) and facilitated transport.
- Active mechanisms (energy required) are active transport and endocytosis and exocytosis.

3.4 The Nucleus and the Endomembrane System

- The nucleus houses DNA, which specifies the order of amino acids in proteins.
- Chromatin is a combination of DNA molecules and proteins that make up chromosomes.
- The nucleolus produces ribosomal RNA (rRNA).
- Protein synthesis occurs in ribosomes, small organelles composed of proteins and rRNA.

The Endomembrane System

The endomembrane system consists of the nuclear envelope, endoplasmic reticulum (ER), Golgi apparatus, lysosomes, and vesicles.

- Rough ER has ribosomes, where protein synthesis occurs.
- Smooth ER has no ribosomes and has various functions, including lipid synthesis.
- The Golgi apparatus processes and packages proteins and lipids into vesicles for secretion or movement into other parts of the cell.
- Lysosomes are specialized vesicles produced by the Golgi apparatus. They fuse with incoming vesicles to digest enclosed material, and they autodigest old cell parts.

3.5 The Cytoskeleton, Cell Movement, and Cell Junctions

The cytoskeleton consists of microtubules, actin filaments, and intermediate filaments that give cells their shape; and it allows organelles to move about the cell. Cilia and flagella, which contain microtubules, allow a cell to move.

Cell junctions connect cells to form tissues and to faciliate communication between cells.

3.6 Mitochondria and Cellular Metabolism

- Mitochondria have an inner membrane that forms cristae, which project into the matrix.
- Mitochondria are involved in cellular respiration, which uses oxygen and releases carbon dioxide.
- During cellular respiration, mitochondria convert the energy of glucose into the energy of ATP molecules.

Cellular Respiration and Metabolism

- A metabolic pathway is a series of reactions, each of which has its own enzyme.
- Enzymes bind their substrates in the active site.
- Sometimes enzymes require coenzymes (such as NAD^+), nonprotein molecules that participate in the reaction.

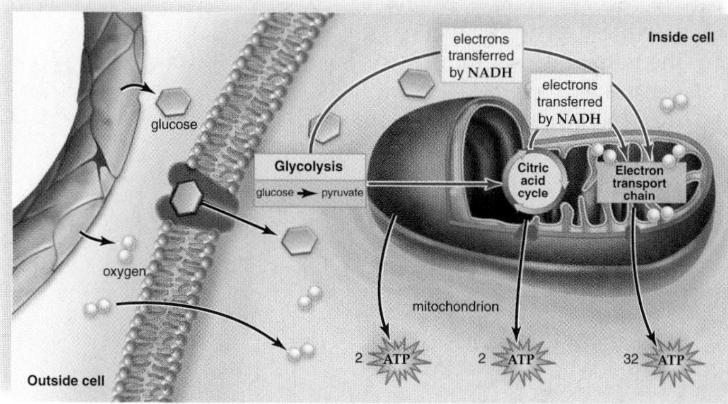

- Cellular respiration is the enzymatic breakdown of glucose to carbon dioxide and water.
- Cellular respiration includes three pathways: glycolysis (occurs in the cytoplasm and is anaerobic), the citric acid cycle (releases carbon dioxide), and the electron transport chain (passes electrons to oxygen).

Fermentation

- If oxygen is not available in cells, the electron transport chain is inoperative, and fermentation (which does not require oxygen) occurs.
- Fermentation produces very little ATP.

Understanding Key Terms

actin filament 55
active site 57
active transport 51
aerobic 59
anaerobic 58
cell theory 44
cellular respiration 57
centrosome 55
chromatin 53
chromosome 53
cilium 55
citric acid cycle 58
coenzyme 58
cytoplasm 46
cytoskeleton 55
diffusion 50
electron transport chain 59
endomembrane system 54
endoplasmic reticulum (ER) 53
eukaryotic cell 46
facilitated transport 51
fermentation 60
flagellum 55
fluid-mosaic model 49
glycolysis 58
Golgi apparatus 54
intermediate filament 55
lysosome 54
metabolism 57
microtubule 55
mitochondrion 57
NAD^+ (nicotinamide adenine dinucleotide) 58
nuclear envelope 53
nuclear pore 53
nucleolus 53
nucleoplasm 53
nucleus 53
organelle 46
osmosis 50
osmotic pressure 50
phagocytosis 52
plasma membrane 46
polyribosome 54
product 57
prokaryotic cell 46
reactant 57
ribosome 53
selectively permeable 46
substrate 57
tonicity 50
vesicle 54

Match the key terms to these definitions.

a. ________ Protein molecules form a shifting pattern within the fluid phospholipid bilayer.

b. ________ Diffusion of water through a selectively permeable membrane.

c. ________ The cell will allow some substances to pass through while not permitting others.

d. ________ Anaerobic breakdown of glucose that results in a gain of two ATP and end products, such as alcohol and lactate.

e. ________ Metabolic pathways that use energy from carbohydrate, fatty acid, and protein break down to produce ATP molecules.

Testing Your Knowledge of the Concepts

Complete the following questions.

1. Explain the three key concepts of the cell theory. (page 44)
2. Which type of microscope would you use to observe the swimming behavior of a flagellated protozoan? Explain. (page 45)
3. Describe how the eukaryotic cell gained mitochondria and chloroplasts. (page 46)
4. Invagination of plasma membrane produced what structures in eukaryotic cells not present in prokaryotic cells? (page 46)
5. How does the organization of the plasma membrane relate to its function? (pages 50–52)

6. Describe four different ways materials can enter and or leave a cell. (pages 50–51)
7. For the following cell organelles, describe the structure and function of each: nucleus, nucleolus, ribosomes, endoplasmic reticulum (rough and smooth), Golgi apparatus, lysosomes, centrioles, and mitochondria. (pages 53–55)
8. What would happen to a eukaryotic cell without a cytoskeleton? (pages 55–56)
9. Describe an enzyme and coenzyme. Explain the mechanism of enzyme function, particularly the relationship of shape to its activity. (pages 57–58)
10. Which stage of cellular respiration produces the most ATP? Explain. (pages 57–59)
11. The small size of cells is best correlated with
    a. the fact that they are self-reproducing.
    b. an adequate surface area for exchange of materials.
    c. their vast versatility.
    d. All of these are correct.
12. A phospholipid has a head and two tails. The tails are found
    a. at the surfaces of the membrane.
    b. in the interior of the membrane.
    c. spanning the membrane.
    d. where the environment is hydrophobic.
    e. Both b and d are correct.
13. Facilitated diffusion differs from diffusion in that facilitated diffusion
    a. involves the passive use of a carrier protein.
    b. involves the active use of a carrier protein.
    c. moves a molecule from a low to a high concentration.
    d. involves the use of ATP molecules.
14. When a cell is placed in a hypotonic solution,
    a. solute exits the cell to equalize the concentration on both sides of the membrane.
    b. water exits the cell toward the area of lower solute concentration.
    c. water enters the cell toward the area of higher solute concentration.
    d. solute exits and water enters the cell.

In questions 15–18, match each function to the proper organelle in the key.

Key:

a. mitochondrion
b. nucleus
c. Golgi apparatus
d. rough ER

15. Packaging and secretion
16. ATP production (powerhouse of cell)
17. Protein synthesis
18. Control center for the cell
19. Vesicles carrying proteins for secretion move from the ER to the
    a. smooth ER.
    b. lysosomes.
    c. Golgi apparatus.
    d. nucleolus.
20. The active site of an enzyme
    a. is identical to that of any other enzyme.
    b. is the part of the enzyme where the substrate can fit.
    c. can be used over and over.
    d. is where the coenzyme binds.
    e. Both b and c are correct.
21. The oxygen required by cellular respiration becomes part of which molecule?
    a. ATP
    b. H_2O
    c. pyruvate
    d. CO_2
22. Fermentation results in a buildup of
    a. NADH.
    b. ATP.
    c. lactic acid.
    d. NAD^+.
    e. carbon dioxide.
23. Use these terms to label the following diagram of the plasma membrane: carbohydrate chain, filaments of the cytoskeleton, hydrophilic heads, hydrophobic tails, protein (used twice), phospholipid bilayer.

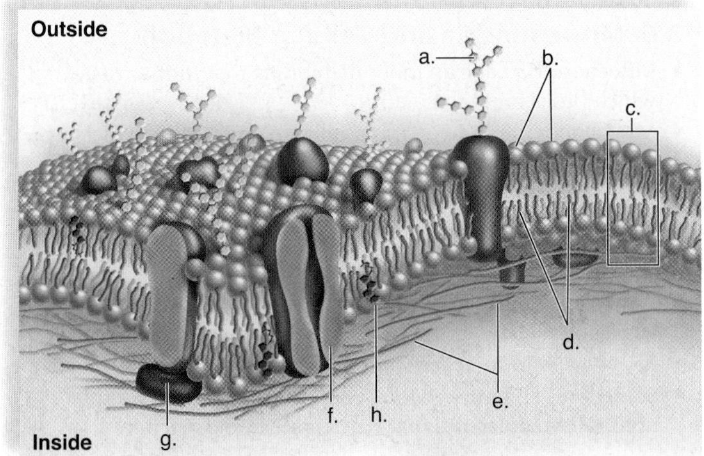

24. Use these terms to label the following diagram: substrates, enzyme (used twice), active site, product, and enzyme–substrate complex.

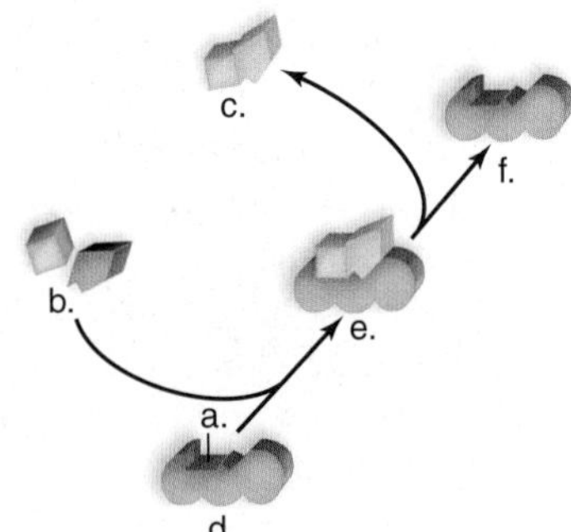

Thinking Critically About the Concepts

In the case study at the beginning of the chapter, the child had malfunctioning lysosomes, which caused an accumulation of fatty acid in the system. Each part of a cell plays an important role in the homeostasis of the entire body.

1. What might occur if the cells of the body contained a malfunctioning mitochondria?
2. What would happen to homeostasis if enzymes were no longer produced in the body?
3. Knowing what you know about the function of a lysosome, what might occur if the cells' lysosomes were overproductive instead of malfunctioning?

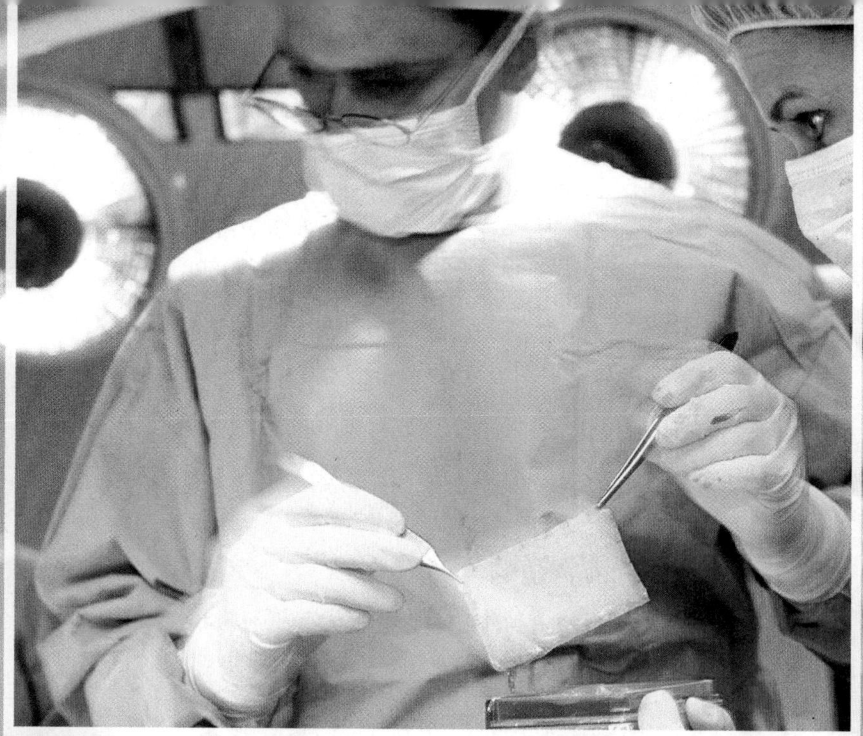

CHAPTER

4

Organization and Regulation of Body Systems

CASE STUDY ARTIFICIAL SKIN

When Kristen awoke in the hospital, she discovered that the fire in her home had produced third-degree burns over a large portion of her legs. The doctors informed Kristen that the burns on her legs covered too much of an area for autografting, the traditional grafting techniques that remove skin from other parts of her body to cover the burn areas. Another available option was allografting, where skin is removed from another person, or a cadaver, and used to cover the burn areas. However, Kristen's specialists were not eager to take that route because complications often develop due to rejection of the foreign tissue or infections.

Instead, the doctors recommended a relatively new technique—artificial skin. Just a decade ago, the concept of artificial skin may have been found only in a science-fiction movie, but recent developments in medicine have made the use of artificial skin a reality. The purpose of using artificial skin is not to permanently replace the damaged tissue; rather, the procedure is designed to protect the damaged tissue and allow time for the patient's skin to heal itself.

The first step of the procedure, after the removal of the burned tissue, is to cover the wound with the artificial skin. This skin contains collagen, a connective tissue, and an adhesive-like carbohydrate that allows the artificial skin to bind to the underlying tissue. Initially, the artificial skin contains a plastic wrapping that simulates the epidermis and protects the tissue from water loss and infection. The next step is to remove a small sample of epidermal cells from an unburned area of skin on the patient's body. These are taken to a laboratory and placed in incubators to grow sheets of skin. Once ready, the plastic covering on the patient is replaced by the sheets of epidermal cells. Over time, the laboratory-grown artificial skin is integrated into the newly growing skin.

As you read through the chapter, think about the following questions.

1. What types of tissue are normally found in skin?
2. Why would burn damage to the skin represent such a serious challenge to Kristen's health?
3. Why would it be more difficult to produce new *dermis* in the laboratory than new *epidermis*?

CHAPTER CONCEPTS

4.1 Types of Tissues
The body contains four types of tissues: connective, muscular, nervous, and epithelial.

4.2 Connective Tissue Connects and Supports
Connective tissues bind and support body parts.

4.3 Muscular Tissue Moves the Body
Muscle tissue moves the body and its parts.

4.4 Nervous Tissue Communicates
Nervous tissue transmits information throughout the body.

4.5 Epithelial Tissue Protects
Epithelial tissues line cavities and cover surfaces.

4.6 Integumentary System
The skin is the largest organ and plays an important role in maintaining homeostasis.

4.7 Organ Systems, Body Cavities, and Body Membranes
Organ systems contain multiple organs that interact to carry out a process.

4.8 Homeostasis
Homeostasis maintains the internal environment and is made possible by feedback mechanisms.

BEFORE YOU BEGIN

Before beginning this chapter, take a few moments to review the following discussions.

Section 1.1 How do tissues and organs fit into the levels of biological organization?

Section 3.2 How are cells structured?

Section 3.3 How are cells linked together to form tissues?

4.1 Types of Tissues

Learning Outcome

Upon completion of this section, you should be able to

1. List the four types of tissues and provide a general function for each.

Recall from the material on the levels of biological organization (see Fig. 1.2) that cells are composed of molecules; a tissue is a group of similar cells; an organ contains several types of tissues; and several organs are found in an organ system. In this chapter, we will further explore the tissue, organ, and organ system levels of organization.

A **tissue** is composed of specialized cells of the same type that perform a common function in the body. The tissues of the human body can be categorized into four major types:

Connective tissue binds and supports body parts.
Muscular tissue moves the body and its parts.
Nervous tissue receives stimuli and conducts nerve impulses.
Epithelial tissue covers body surfaces and lines body cavities.

Cancers are classified according to the type of tissue from which they arise. Sarcomas are cancers arising in muscular or connective tissue (especially bone or cartilage). Leukemias are cancers of the blood. Lymphomas are cancers of lymphoid tissue. Carcinomas, the most common type, are cancers of epithelial tissue. The chance of developing cancer in a particular tissue is related to the rate of cell division. Both epithelial cells and blood cells reproduce at a high rate. Thus, carcinomas and leukemias are common types of cancers.

Connecting the Concepts

For more information on cancer, refer to the following discussions.

Section 6.3 examines leukemia as a form of blood cancer.

Section 19.1 provides additional information on the most common types of cancer.

Check Your Progress 4.1

1. List the four major tissue types found in the human body.
2. Describe how cancers are classified.
3. Explain why carcinomas and leukemias are the most common cancer types.

4.2 Connective Tissue Connects and Supports

Learning Outcomes

Upon completion of this section, you should be able to

1. Describe the primary types of connective tissue and provide a function for each.
2. Compare the structure and function of bone and cartilage.
3. Differentiate between blood and lymph.

Connective tissue is diverse in structure and function. Even so, all types have three components: specialized cells, ground substance, and protein fibers. These components are shown in Figure 4.1, a diagrammatic representation of loose fibrous connective tissue. The ground substance is a noncellular material that separates the cells. It varies in consistency from solid (bone) to semifluid (cartilage) to fluid (blood).

The fibers are of three possible types. White **collagen fibers** contain collagen, a protein that gives them flexibility and strength. **Reticular fibers** are very thin collagen fibers, highly branched proteins that form delicate supporting networks. Yellow **elastic fibers** contain elastin, a protein that is not as strong as collagen but is more elastic (meaning that it can return to its original shape; elastic fibers may stretch over 100 times their relaxed size without damaging the proteins). Inherited connective tissue disorders arise when people inherit genes that lead to malformed fibers. For example, in Marfan syndrome, there are mutations in the fibrillin gene. Fibrillin is a component of elastic fibers. The mutation

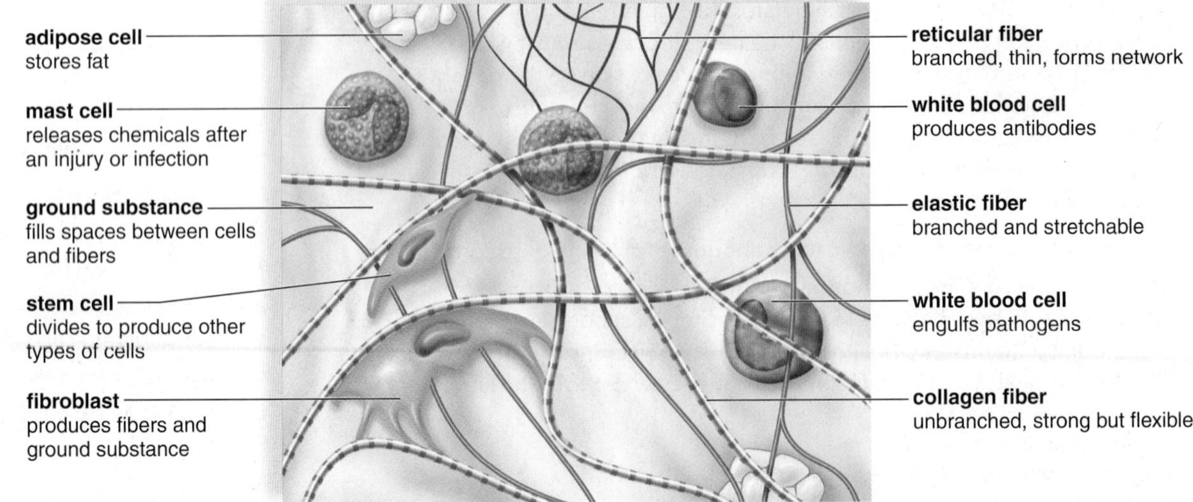

Figure 4.1 **Components of connective tissue.** All connective tissues have three components: specialized cells, ground substance, and protein fibers. Loose fibrous connective tissue is shown here.

results in decreased elasticity in connective tissues normally rich in elastic fibers, such as the aorta. Individuals with this disease often die from aortic rupture, which occurs when the aorta cannot expand in response to increased blood pressure.

Fibrous Connective Tissue

Fibrous tissue exists in two forms: loose fibrous tissue and dense fibrous tissue. Both loose fibrous and dense fibrous connective tissues have cells called **fibroblasts** located some distance from one another and separated by a jelly-like ground substance containing white collagen fibers and yellow elastic fibers (Fig. 4.2). **Matrix** is a term that includes ground substance and fibers.

Loose fibrous connective tissue, also called areolar tissue, supports epithelium and many internal organs. Its presence in lungs, arteries, and the urinary bladder allows these organs to expand. It forms a protective covering enclosing many internal organs, such as muscles, blood vessels, and nerves.

Adipose tissue is a special type of loose connective tissue in which the cells enlarge and store fat. Adipose tissue has little extracellular matrix. Its cells are crowded, and each is filled with liquid fat. The body uses this stored fat for energy, insulation, and organ protection. Adipose tissue also releases a hormone called *leptin,* which regulates appetite-control centers in the brain. Adipose tissue is primarily found beneath the skin, around the kidneys, and on the surface of the heart.

Dense fibrous connective tissue contains many collagen fibers packed together. This type of tissue has more specific functions than does loose connective tissue. For example, dense fibrous connective tissue is found in **tendons,** which connect muscles to bones, and in **ligaments,** which connect bones to other bones at joints.

Specialized Connective Tissue

There are several types of specialized connective tissue. Cartilage and bone are supportive connective tissues. In both tissues, the extracellular matrix is solid. In fluid connective tissues, such as blood and lymph, the matrix surrounding the cells is a liquid.

Cartilage

In **cartilage,** the cells lie in small chambers called *lacunae* (sing., **lacuna**), separated by a solid, yet flexible, matrix. This matrix is formed by cells called *chondroblasts* and *chondrocytes.* Unfortunately, because this tissue lacks a direct blood supply, it heals slowly. The three types of cartilage are distinguished by the type of fiber found in the matrix.

Hyaline cartilage (Fig. 4.2), the most common type of cartilage, contains only fine collagen fibers. The matrix has a glassy, translucent appearance. Hyaline cartilage is found in the nose and at the ends of the long bones and the ribs, and it forms rings in the walls of respiratory passages. The fetal skeleton also is made of this type of cartilage. Later, the cartilaginous fetal skeleton is replaced by bone.

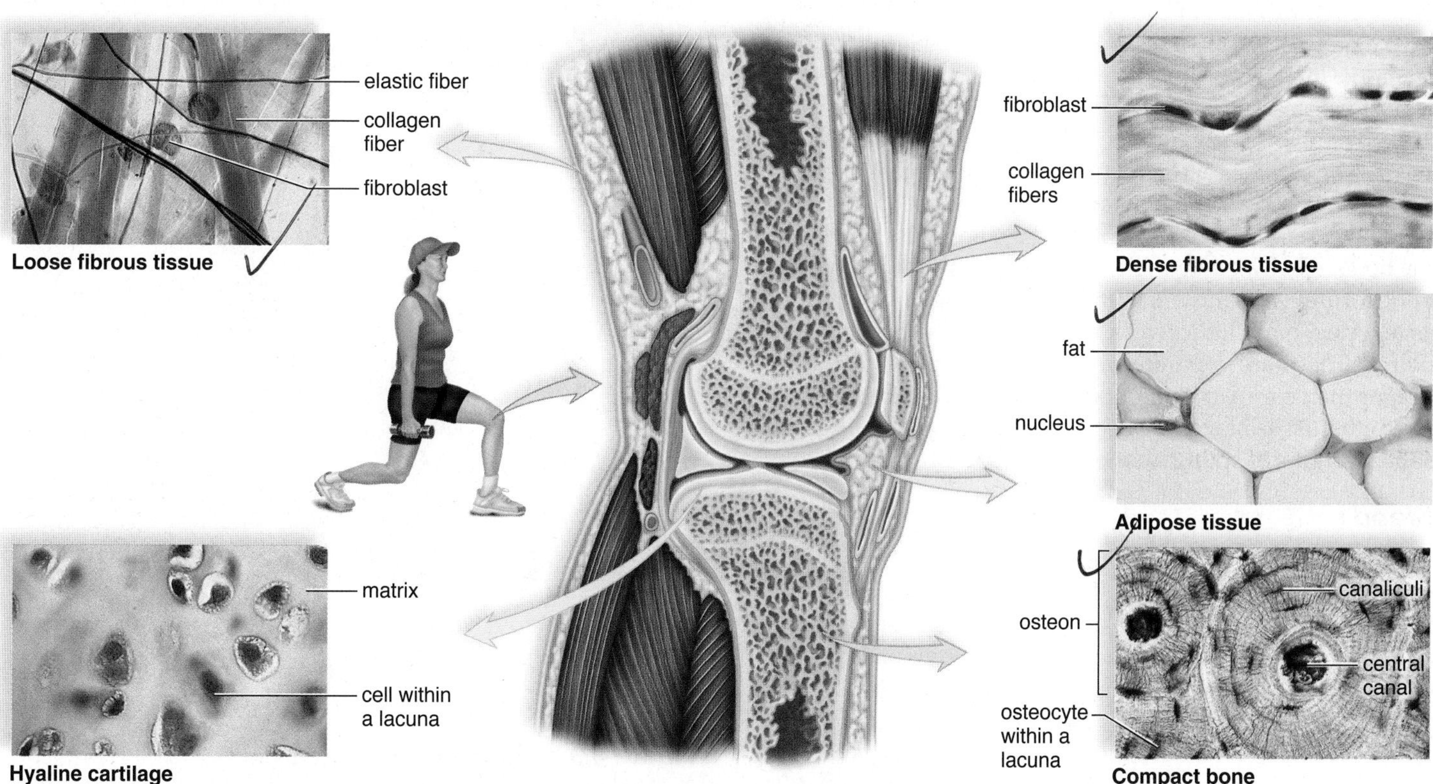

Figure 4.2 Connective tissues in the knee.
Most types of connective tissue may be found in the knee.

Connections and Misconceptions

Does dieting get rid of fat cells?

Unfortunately, no. Dieting can shrink the size of the fat cells (adipocytes), but their number stays the same. This means that it is easier to regain the weight if diet and exercise are not maintained. Liposuction is the only way to remove fat cells from the body.

Elastic cartilage has more elastic fibers than hyaline cartilage. For this reason, it is more flexible and is found, for example, in the framework of the outer ear.

Fibrocartilage has a matrix containing strong collagen fibers. Fibrocartilage is found in structures that withstand tension and pressure, such as the disks between the vertebrae in the backbone and the cushions in the knee joint.

Bone

Bone is the most rigid connective tissue. It consists of an extremely hard matrix of inorganic salts, notably calcium salts. These salts are deposited around protein fibers, especially collagen fibers. The inorganic salts give bone rigidity. The protein fibers provide elasticity and strength, much as steel rods do in reinforced concrete. Cells called *osteoblasts* and *osteoclasts* are responsible for forming the matrix in bone tissue.

Compact bone makes up the shaft of a long bone (Fig. 4.2). It consists of cylindrical structural units called *osteons* (see Chap. 11). The central canal of each osteon is surrounded by rings of hard matrix. Bone cells are located in lacunae between the rings of matrix. In the central canal, nerve fibers carry nerve impulses, and blood vessels carry nutrients that allow bone to renew itself. Thin extensions of bone cells within canaliculi (minute canals) connect the cells to each other and to the central canal.

The ends of the long bones are composed of spongy bone covered by compact bone. Spongy bone also surrounds the bone marrow cavity. This, in turn, is covered by compact bone forming a "sandwich" structure. **Spongy bone** appears as an open, bony latticework with numerous bony bars and plates. These are separated by irregular spaces. Although lighter than compact bone, spongy bone is still designed for strength. Just as braces are used for support in buildings, the solid portions of spongy bone follow lines of stress.

Blood

Blood represents a fluid connective tissue. **Blood,** which consists of formed elements (Fig. 4.3) and plasma, is located in blood vessels. Blood transports nutrients and oxygen to **tissue fluid.** Tissue fluid bathes the body's cells and removes carbon dioxide and other wastes. Blood helps distribute heat and also plays a role in fluid, ion, and pH balance. The systems of the body help keep blood composition and chemistry within normal limits.

The formed elements of blood each have specific functions. The **red blood cells (erythrocytes)** are small, biconcave, disk-shaped cells without nuclei. The presence of the red pigment hemoglobin makes the cells red, which, in turn, makes the blood red. Hemoglobin is composed of four units. Each unit is composed of the protein globin and a complex iron-containing structure called *heme.* The iron forms a loose association with oxygen; therefore, red blood cells transport oxygen.

White blood cells (leukocytes) may be distinguished from red blood cells because they have a nucleus. Without staining, leukocytes would be translucent. There are many different types of white blood cells, but all are involved in protecting the body from infection. Some white blood cells are *generalists,* meaning that they will respond to any foreign invader in the body. These are phagocytic cells, because they engulf infectious agents such as bacteria. Others are more specific and may either produce antibodies (molecules that combine with foreign substances to inactivate them) or may directly attack specific invading agents or infected cells in the body.

Platelets (thrombocytes) are not complete cells. Rather, they are fragments of giant cells present only in bone marrow. When a blood vessel is damaged, platelets form a plug that seals the vessel, and injured tissues release molecules that help the clotting process.

Lymph

Lymph is also a fluid connective tissue. Lymph is a clear (sometimes faintly yellow) fluid derived from the fluids surrounding the tissues. It contains white blood cells. Lymphatic vessels absorb excess tissue fluid and various dissolved solutes in the tissues. They transport lymph to particular vessels of the cardiovascular system. Lymphatic vessels absorb fat molecules from the small intestine. Lymph nodes, composed of fibrous connective tissue, occur along the length of lymphatic vessels. Lymph is cleansed as it passes through lymph nodes, in particular, because white blood cells congregate there. Lymph nodes enlarge when you have an infection.

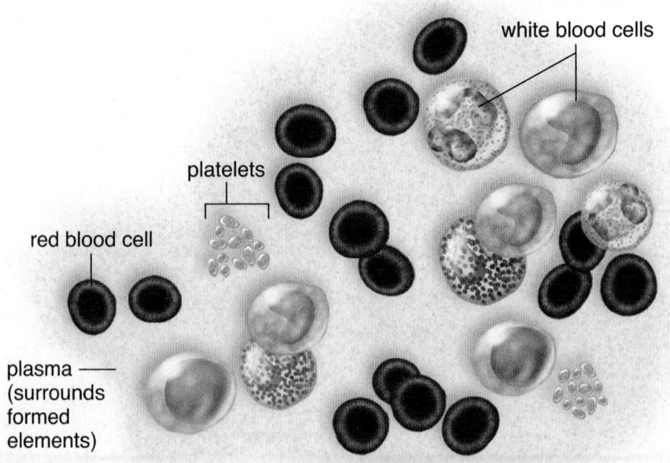

Figure 4.3 The formed elements of blood.
Red blood cells, which lack a nucleus, transport oxygen. Each type of white blood cell has a particular way to fight infections. Platelets, fragments of a particular cell, function in helping to seal injured blood vessels.

Types of Connective Tissue

- Fibrous connective tissue
  - **Loose** Fibers create loose, open framework
  - **Dense** Fibers are densely packed
- Specialized connective tissue
  - Supportive connective tissue
    - **Cartilage** Solid yet flexible matrix
    - **Bone** Solid and rigid matrix
  - Fluid connective tissue
    - **Blood** Contained in blood vessels
    - **Lymph** Contained in lymphatic vessels

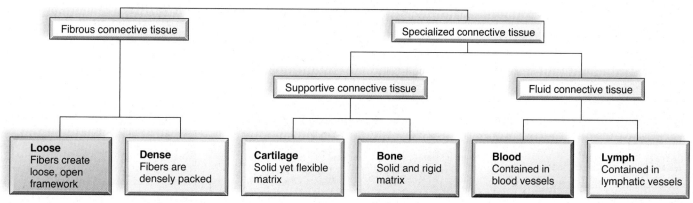

Figure 4.4 Types of connective tissue.
Connective tissue is divided into two main categories—fibrous and specialized.

Connecting the Concepts

The tissue types discussed in this section are examined in greater detail later in the book. For more information, refer to the following discussions.

Section 6.1 details the types of formed elements found in blood.

Section 7.2 explains the role of the lymphatic system in moving lymph through the body.

Section 11.1 provides a more detailed examination of cartilage and bone.

Check Your Progress 4.2

1. List the two main categories of connective tissue, and provide some examples of each type.
2. List the features of the different types of fibrous and supportive connective tissue and where each can be found in the human body.
3. Describe how each of the two fluid connective tissues are important to homeostasis.

4.3 Muscular Tissue Moves the Body

Learning Outcome

Upon completion of this section you should be able to

1. Distinguish between the three types of muscles with regard to location and function in the body.

Muscular (contractile) tissue is composed of cells called *muscle fibers.* Muscle fibers contain filaments made of proteins called *actin* and *myosin.* The interaction of actin and myosin accounts for movement. The three types of vertebrate muscular tissue are skeletal, smooth, and cardiac.

Skeletal muscle is also called *voluntary muscle* (Fig. 4.5*a*). It is attached by tendons to the bones of the skeleton. When it contracts, body parts move. Contraction of skeletal muscle is under voluntary control and occurs faster than in the other muscle types. Skeletal muscle fibers are cylindrical and long—sometimes they run the length of the muscle. They arise during development when several cells fuse, resulting in one fiber with multiple nuclei. The nuclei are located at the periphery of the cell, just inside the plasma membrane. The fibers have alternating light and dark bands that give them a **striated,** or striped, appearance. These bands are due to the placement of actin filaments and myosin filaments in the cell.

Smooth (visceral) muscle is so named because the cells lack striations (Fig. 4.5*b*). The spindle-shaped cells each have a single nucleus. These cells form layers in which the thick middle portion of one cell is opposite the thin ends of adjacent cells. Consequently, the nuclei form an irregular pattern in the tissue. Smooth muscle is not under conscious or voluntary control. Therefore, it is said to be involuntary. Smooth muscle is found in the walls of viscera (intestine, bladder, and other internal organs) and blood vessels. It contracts more slowly than skeletal muscle but can remain contracted for a longer time. When the smooth muscle of the bladder contracts, urine is sent into a tube called the *urethra,* which takes it to the outside. When the smooth muscle of the blood vessels contracts, blood vessels constrict, helping to raise blood pressure.

Cardiac muscle (Fig. 4.5*c*) is found only in the walls of the heart. Its contraction pumps blood and accounts for the heartbeat. Cardiac muscle combines features of both smooth and skeletal muscle. Like skeletal muscle, it has striations, but the contraction of the heart is involuntary for the most part. Cardiac muscle cells also differ from skeletal muscle cells in that they usually have a single, centrally placed nucleus. The cells are branched and seemingly fused, one with another. The heart appears to be composed of one large interconnecting mass of muscle cells. Cardiac muscle cells are separate and individual, but they are bound end to end at **intercalated disks.** These are areas where folded plasma membranes between two cells contain adhesion junctions and gap junctions (see Section 3.5).

Skeletal muscle
- has striated cells with multiple nuclei.
- occurs in muscles attached to skeleton.
- functions in voluntary movement of body.

muscle fiber
striation
nucleus
250×

a.

Smooth muscle
- has spindle-shaped cells, each with a single nucleus.
- cells have no striations.
- functions in movement of substances in lumens of body.
- is involuntary.
- is found in blood vessel walls and walls of the digestive tract.

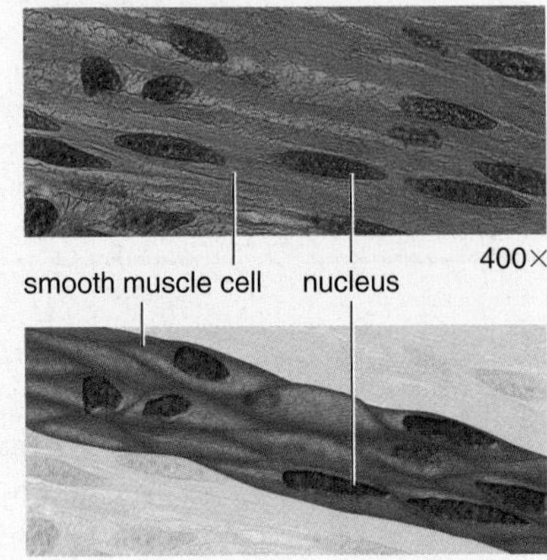

b.

Cardiac muscle
- has branching, striated cells, each with a single nucleus.
- occurs in the wall of the heart.
- functions in the pumping of blood.
- is involuntary.

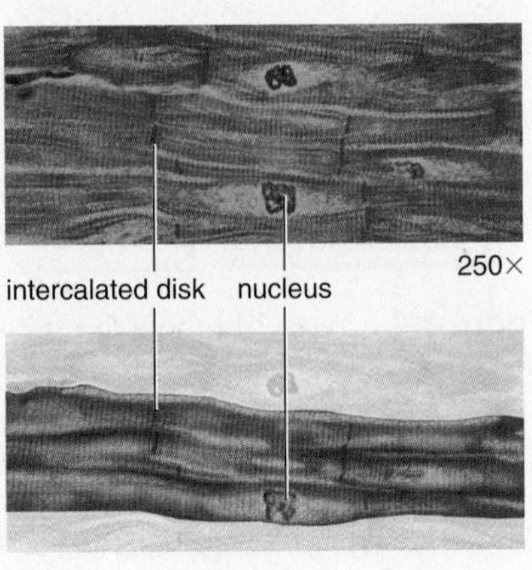

c.

Figure 4.5 **The three types of muscle tissue.**
a. Skeletal muscle is voluntary and striated. **b.** Smooth muscle is involuntary and nonstriated. **c.** Cardiac muscle is involuntary and striated.

Connecting the Concepts

Muscle tissue plays an important role in our physiology. For more information on each of the three types of muscles, refer to the following discussions.

Section 5.3 provides a more detailed look at how the heartbeat is generated.

Section 8.1 illustrates how smooth muscle lines the digestive tract.

Section 12.2 examines the strucure and function of skeletal muscle.

Check Your Progress 4.3

1. Explain the difference in the structure and function of skeletal, smooth, and cardiac muscle.
2. Describe where each type of muscle fiber is found in the body.
3. Explain why smooth muscle and cardiac muscle are involuntary and summarize what advantage this provides to homeostasis.

4.4 Nervous Tissue Communicates

Learning Outcomes

Upon completion of this section you should be able to

1. Distinguish between neurons and neuroglia cells.
2. Describe the structure of a neuron.

Nervous tissue consists of nerve cells, called neurons, and neuroglia, the cells that support and nourish the neurons.

Neurons

A **neuron** is a specialized cell that has three parts: dendrites, a cell body, and an axon (Fig. 4.6). A *dendrite* is an extension that receives signals from sensory receptors or other neurons. The *cell body* contains most of the cell's cytoplasm and the nucleus. An *axon* is an extension that conducts nerve impulses. Long axons are covered by myelin, a white fatty substance. The term *fiber*[1] is used here to refer to an axon along with its myelin sheath, if it has one. Outside the brain and spinal cord, fibers bound by connective tissue form **nerves.**

The nervous system has three functions: sensory input, integration of data, and motor output. Nerves conduct signals from sensory receptors to the spinal cord and the brain, where integration, or processing, occurs. However, the phenomenon called sensation occurs only in the brain. Nerves also conduct signals from the spinal cord and brain to muscles and glands and other organs. This triggers a characteristic response from each tissue. For example, muscles contract and glands secrete. In this way, a coordinated response to the original sensory input is achieved.

Neuroglia

In addition to neurons, nervous tissue contains neuroglia. **Neuroglia** are cells that outnumber neurons nine to one and take up more than half the volume of the brain. Although the

[1] In connective tissue, a fiber is a component of the matrix; in muscle tissue, a fiber is a muscle cell; in nervous tissue, a fiber is an axon.

Connections and Misconceptions

How fast is a reflex?

A reflex is a built-in pathway that allows the body to react quickly to a response. One example, the knee jerk, or patellar reflex, is tested by tapping just below the kneecap. The lower leg will then involuntarily kick forward. The reaction is designed to protect the thigh muscle from excessive stretch. The knee-jerk reflex is an example of a simple stretch reflex. There is only one pathway required: the stretch sensation (caused by tapping the knee), to the spinal cord, to the leg muscle. The whole circuit is complete within a millisecond—or 1/1,000 second!

primary function of neuroglia is to support and nourish neurons, research is being conducted to determine how much they directly contribute to brain function. Neuroglia do not have long extensions (axons or dendrites). However, researchers are gathering evidence that neuroglia do communicate among themselves and with neurons, even without these extensions. Examples of neuroglia found in the brain are microglia, astrocytes, and oligodendrocytes (Fig. 4.6). *Microglia,* in addition to supporting neurons, engulf bacterial and cellular debris. *Astrocytes* provide nutrients to neurons and produce a hormone known as glial-derived neutrophic factor (GDNF). This growth factor is currently undergoing clinical trials as a therapy for Parkinson disease and other diseases caused by neuron degeneration. *Oligodendrocytes* form the myelin sheaths around fibers in the brain and spinal cord. Outside the brain, Schwann cells are the type of neuroglia that encircle long nerve fibers and form a myelin sheath.

Connecting the Concepts

Nervous tissue plays an important role in transmitting the signals needed to maintain homeostasis. For more information on how neurons work, refer to the following discussions.

Section 13.1 provides a more detailed examination of how neurons function.

Section 14.1 discusses how neurons are involved in sensation.

Check Your Progress 4.4

1. Describe the structure and function of a neuron.
2. Discuss the different types of neuroglial cells and the function of each.
3. Explain how the neurons and neuroglial cells work together to make nervous tissue function.

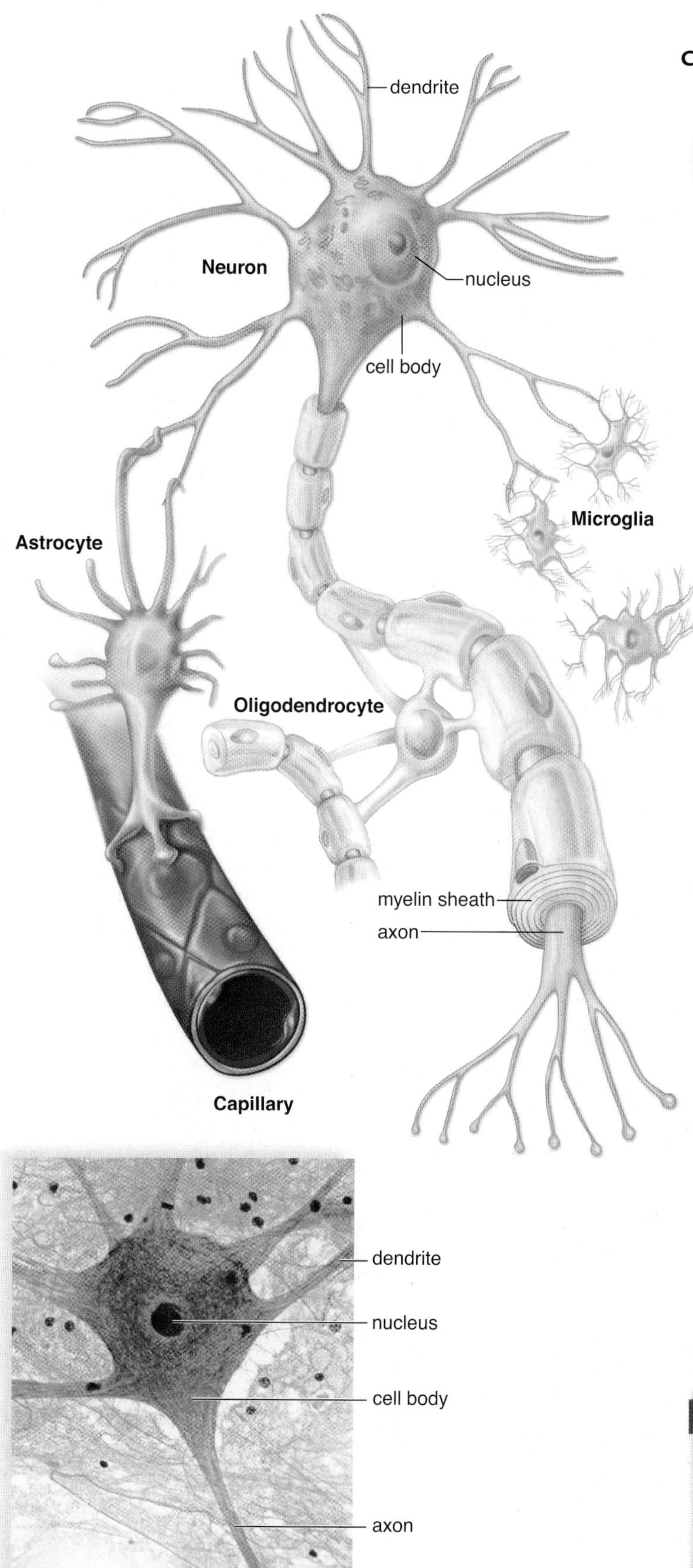

Figure 4.6 A neuron and examples of supporting neuroglia cells.
Neurons conduct nerve impulses. Neuroglia support and service neurons. Microglia are a type of neuroglia that become mobile in response to inflammation and phagocytize debris. Astrocytes lie between neurons and a capillary. Therefore, substances entering neurons from the blood must first pass through astrocytes. Oligodendrocytes form the myelin sheaths around fibers in the brain and spinal cord.

Biology Matters Science Focus

Nerve Regeneration and Stem Cells

In humans, axons outside the brain and spinal cord can regenerate—but not those inside these organs (Fig. 4A). After injury, axons in the human central nervous system (CNS) degenerate, resulting in permanent loss of nervous function. Interestingly, about 90% of the cells in the brain and the spinal cord are not even neurons. They are glial cells. In nerves outside the brain and spinal cord, the glial cells are Schwann cells that help axons regenerate. The glial cells in the CNS are oligodendrocytes and astrocytes, and they inhibit axon regeneration.

The spinal cord does contain its own stem cells. When the spinal cord is injured in experimental animals, these stem cells do proliferate. But instead of becoming functional neurons, they become glial cells. Researchers are trying to understand the process that triggers the stem cells to become glial cells. In the future, this understanding would allow manipulation of stem cells into neurons.

In early experiments with neural stem cells in the laboratory, scientists at Johns Hopkins University caused embryonic stem (ES) cells to differentiate into spinal cord motor neurons, the type of nerve cell that causes muscles to contract. The motor neurons then produced axons. When grown in the same dish with muscle cells, the motor neurons formed neuromuscular junctions and even caused muscle contractions. The cells were then transplanted into the spinal cords of rats with spinal cord injuries. Some of the transplanted cells survived for longer than a month within the spinal cord. However, no improvement in symptoms was seen and no functional neuron connections were made.

4.5 Epithelial Tissue Protects

Learning Outcomes

Upon completion of this section you should be able to

1. State the role of epithelial cells in the body.
2. Distinguish between the different forms of epithelial tissue with regard to location and function.

Epithelial tissue, also called epithelium (pl., epithelia), consists of tightly packed cells that form a continuous layer. Epithelial tissue covers surfaces and lines body cavities. Usually, it has a protective function. It can also be modified to carry out secretion, absorption, excretion, and filtration.

Epithelial cells are exposed to the environment on one side. On the other side, they are bounded by a **basement membrane.** The basement membrane should not be confused with

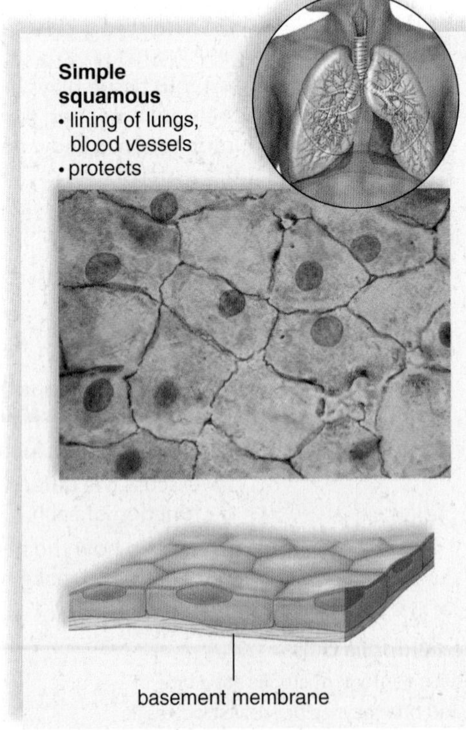

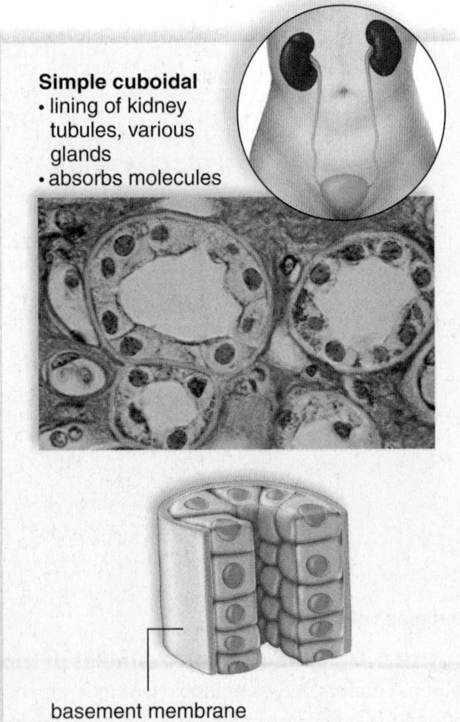

Figure 4.7 The basic types of epithelial cells.
Basic epithelial tissues found in humans are shown, along with locations of the tissue and the primary function of the tissue at these locations.

In later experiments, this same research group tried the experiment again. This time, the paralyzed rats were first treated with drugs to overcome inhibition from the central nervous system. A nerve growth factor was added as well. These techniques significantly increased the success of the transplanted neurons. Amazingly, axons of transplanted neurons reached the muscles, formed neuromuscular junctions, and provided partial relief from the paralysis.

Research is being done on the use of the body's own stem cells within the CNS to repair damaged neurons. Studies on the transplantation of laboratory-grown stem cells continue as well. Many questions remain, but the current results are promising.

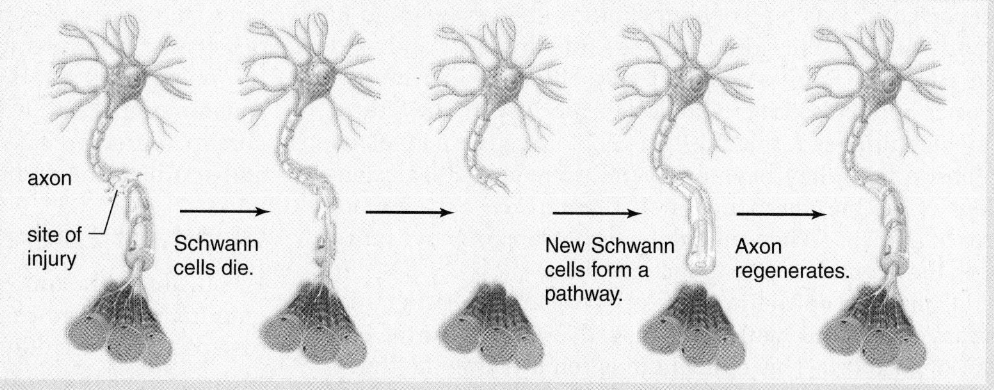

Figure 4A Regeneration of nerve cells. Outside the CNS, nerves regenerate because new neuroglia called Schwann cells form a pathway for axons to reach a muscle. In the CNS, comparable neuroglia called oligodendrocytes do not have this function.

the plasma membrane (in the cell) or the body membranes that line the cavities of the body. Instead, the basement membrane is a thin layer of various types of carbohydrates and proteins that anchors the epithelium to underlying connective tissue.

Simple Epithelia

Epithelial tissue is either simple or stratified. Simple epithelia have only a single layer of cells (Fig. 4.7) and are classified according to cell type. **Squamous epithelium,** composed of flattened cells, is found lining the air sacs of lungs and walls of blood vessels. Its shape and arrangement permit exchanges of substances in these locations. Oxygen–carbon-dioxide exchange occurs in the lungs, and nutrient–waste exchange occurs across blood vessels in the tissues.

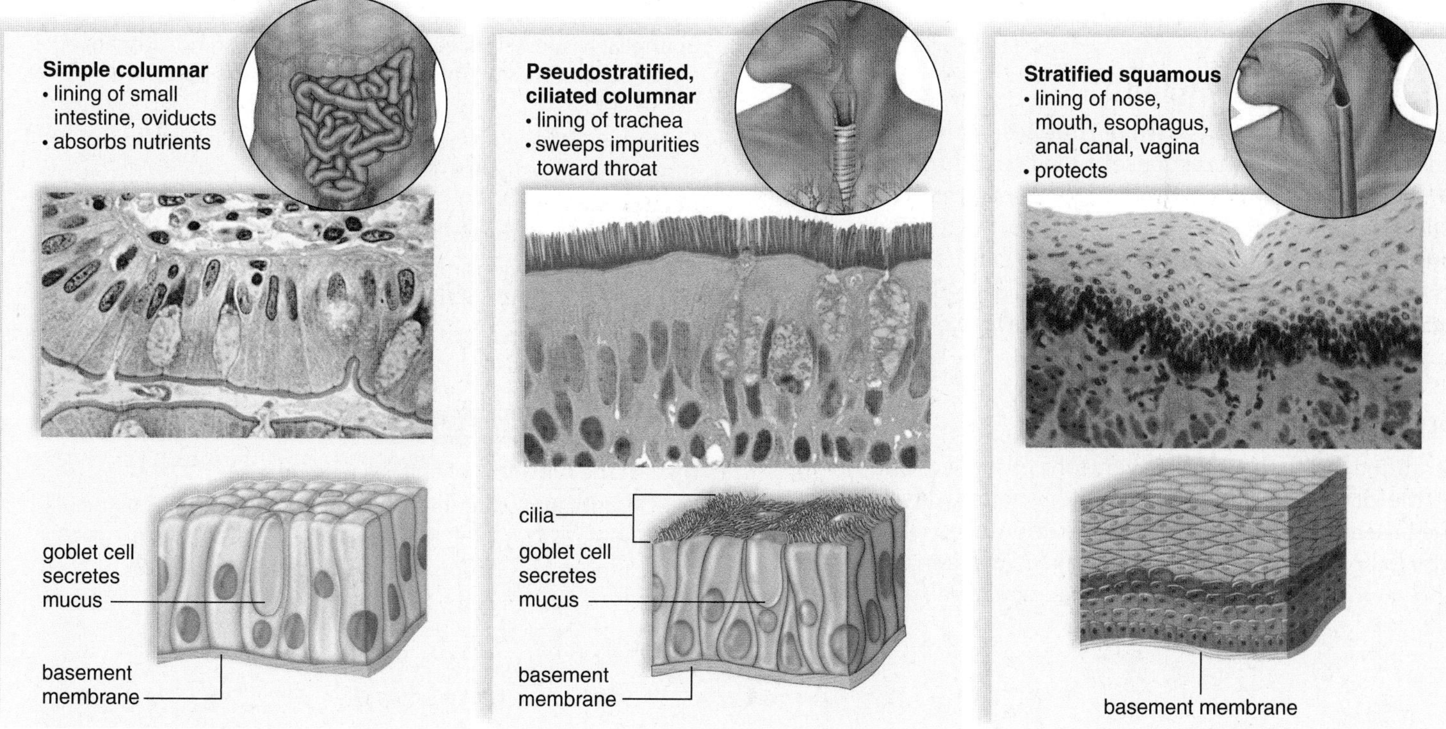

Figure 4.7—*continued.*

Cuboidal epithelium consists of a single layer of cube-shaped cells. This type of epithelium is frequently found in glands, such as the salivary glands, the thyroid gland, and the pancreas. Simple cuboidal epithelium also covers the ovaries and lines kidney tubules, the portions of the kidney in which urine is formed. When cuboidal cells are involved in absorption, they have microvilli (minute cellular extensions of the plasma membrane). These increase the surface area of the cells. When cuboidal cells function in active transport, they contain many mitochondria.

Columnar epithelium has cells resembling rectangular pillars or columns, with nuclei usually located near the bottom of each cell. This epithelium is found lining the digestive tract, where microvilli expand the surface area and aid in absorbing the products of digestion. Ciliated columnar epithelium is found lining the oviducts, where it propels the egg toward the uterus, or womb.

Pseudostratified Columnar Epithelium

Pseudostratified columnar epithelium (Fig. 4.7) is so named because it appears to be layered (*pseudo*, "false"; *stratified*, "layers"). However, true layers do not exist because each cell touches the basement membrane. In particular, the irregular placement of the nuclei creates the appearance of several layers, whereas only one exists. The lining of the windpipe, or trachea, is pseudostratified ciliated columnar epithelium. A secreted covering of mucus traps foreign particles. The upward motion of the cilia carries the mucus to the back of the throat, where it may either be swallowed or expectorated (spit out). Smoking can cause a change in the secretion of mucus and can inhibit ciliary action, resulting in a chronic inflammatory condition called *bronchitis*.

Transitional Epithelium

The term *transitional epithelium* implies changeability, and this tissue changes in response to tension. It forms the lining of the urinary bladder, the ureters (tubes that carry urine from the kidneys to the bladder), and part of the urethra (the single tube that carries urine to the outside). All are organs that may need to stretch. When the bladder is distended, this epithelium stretches, and the outer cells take on a squamous appearance.

Stratified Epithelia

Stratified epithelia have layers of cells piled one on top of the other (see Fig. 4.7). Only the bottom layer touches the basement membrane. The nose, mouth, esophagus, anal canal, the outer portion of the cervix (adjacent to the vagina), and vagina are lined with stratified squamous epithelium. Cancer of the cervix is detectable by doing a *pap smear*. Cells lining the cervix are smeared onto a slide later examined to detect any abnormalities.

As we shall see, the outer layer of skin is also stratified squamous epithelium, but the cells have been reinforced by keratin, a protein that provides strength. Stratified cuboidal and stratified columnar epithelia also are found in the body.

Glandular Epithelium

When an epithelium secretes a product, it is said to be glandular. A **gland** can be a single epithelial cell, as in the case of mucus-secreting goblet cells, or a gland can contain many cells. Glands with ducts that secrete their product onto the outer surface (e.g., sweat glands and mammary glands) or into a cavity (e.g., salivary glands) are called **exocrine glands.** Ducts can be simple or compound:

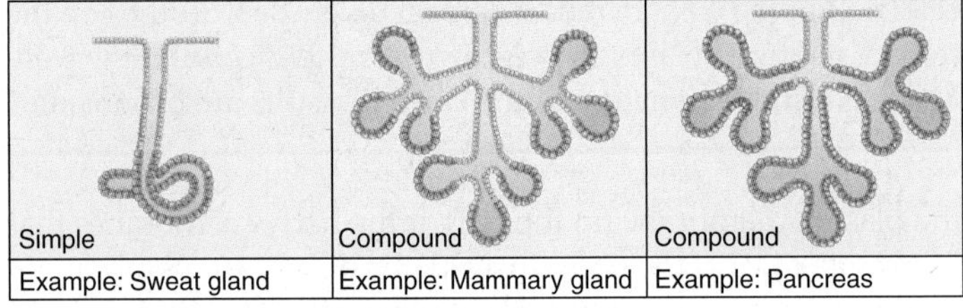

Glands that have no ducts are appropriately known as the ductless glands, or endocrine glands. **Endocrine glands** (e.g., pituitary gland and thyroid) secrete hormones directly into the bloodstream.

Connecting the Concepts

Epithelial tissue is involved in the operation of most organs of the body. For more information, refer to the following discussions.

Section 8.3 describes how specilaized epethlial cells in the stomach secrete hydrochloric acid.

Section 9.6 examines how gas exchange occurs across the epithelial cells of the lungs.

Section 15.1 provides more information on the endocrine and exocrine glands of the body.

Check Your Progress 4.5

1. List the functions of epithelial tissue.
2. Describe the structures of each type of epithelial tissue.
3. Summarize how the structure of some epithelial tissue relates to its function. Give some specific examples.

4.6 Integumentary System

Learning Outcomes

Upon completion of this section, you should be able to

1. Describe the structure and function of human skin.
2. Compare the function of the epidermis and dermis.
3. Identify the function of the acccessory organs associated with the skin.

In some cases, a tissue is associated with a particular organ. For example, nervous tissue is typically associated with the brain. But an **organ** is composed of two or more types of tissues working together to perform particular functions. The skin is an organ comprised of all four tissue types: epithelial, connective, muscular, and nervous tissue. An **organ system** contains many different organs that cooperate to carry out a process, such as the digestion of food. The skin has several accessory organs (hair, nails, sweat glands, and sebaceous glands) and, therefore, is sometimes referred to as the **integumentary system.**

Skin is the most conspicuous system in the body because it covers the body. In an adult, the skin has a surface area of about 1.8 square meters (over 19.5 square feet). It accounts for nearly 15% of the weight of an average human. The skin has numerous functions. It protects underlying tissues from physical trauma, pathogen invasion, and water loss. It also helps regulate body temperature. Therefore, skin plays a significant role in homeostasis**,** the relative constancy of the internal environment. The skin even synthesizes certain chemicals that affect the rest of the body. Skin contains sensory receptors such as touch and temperature receptors. Thus it helps us to be aware of our surroundings and communicate with others.

Regions of the Skin

The skin has two regions: the epidermis and the dermis (Fig. 4.8). A **subcutaneous layer**, sometimes called the *hypodermis,* is found between the skin and any underlying structures, such as muscle or bone.

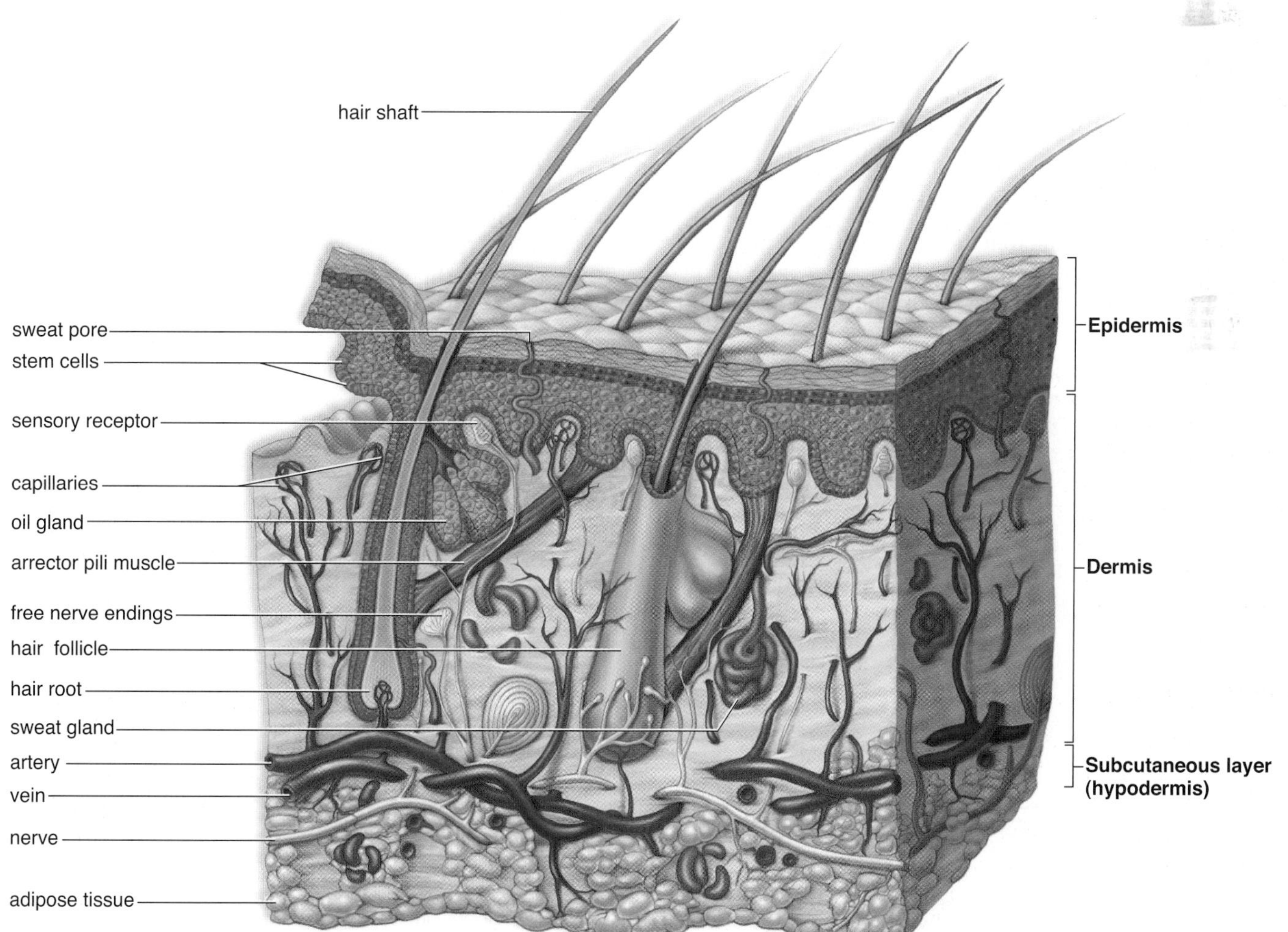

Figure 4.8 **Anatomy of human skin.**
Skin consists of two regions: the epidermis and the dermis. A subcutaneous layer (the hypodermis) lies below the dermis.

The Epidermis

The **epidermis** is made up of stratified squamous epithelium. New epidermal cells for the renewal of skin are derived from stem (basal) cells. The importance of these stem cells is observed when there is an injury to the skin. If an injury, such as a burn, is deep enough to destroy stem cells, then the skin can no longer replace itself. As soon as possible, the damaged tissue is removed and skin grafting begins. The skin needed for grafting is usually taken from other parts of the patient's body. This is called *autografting,* as opposed to allografting. In *allografting,* the graft is received from another person and is sometimes obtained from cadavers. Autografting is preferred because rejection rates are low. If the damaged area is extensive, it may be difficult to acquire enough skin for autografting. In that case, small amounts of epidermis are removed and cultured in the laboratory. This produces thin sheets of skin that can be transplanted back to the patient (see the chapter opener).

Newly generated skin cells become flattened and hardened as they push to the surface (Fig. 4.9). Hardening takes place because the cells produce *keratin,* a waterproof protein. These cells are also called keratinocytes. Outer skin cells are dead and keratinized, so the skin is waterproof. This prevents water loss and helps maintain water homeostasis. The skin's waterproofing also prevents water from entering the body when the skin is immersed. Dandruff occurs when the rate of keratinization in the skin of the scalp is two or three times the normal rate. A thick layer of dead keratinized cells, arranged in spiral and concentric patterns, forms fingerprints and footprints that are genetically unique.

Two types of specialized cells are located deep in the epidermis. **Langerhans cells** are macrophages, white blood cells that phagocytize infectious agents and then travel to lymphatic organs. There they stimulate the immune system to react to the pathogen. **Melanocytes,** lying deep in the epidermis, produce melanin. This is the main pigment responsible for skin color. The number of melanocytes is about the same in all individuals, so variation in skin color is due to the amount of melanin produced and its distribution. When skin is exposed to the sun, melanocytes produce more melanin. This protects the skin from the damaging effects of the ultraviolet (UV) radiation in sunlight. The melanin is passed to other epidermal cells, and the result is tanning. In some people, this results in the formation of patches of melanin called freckles. Another pigment, called carotene, is present in epidermal cells and in the dermis. It gives the skin of some Asian populations its yellowish hue. The pinkish color of fair-skinned people is due to the pigment hemoglobin in the red blood cells in the blood vessels of the dermis.

Some ultraviolet radiation does serve a purpose, however. Certain cells in the epidermis convert a steroid related to cholesterol into **vitamin D** with the aid of UV radiation. However, only a small amount of UV radiation is needed. Vitamin D leaves the skin and helps regulate both calcium and phosphorus metabolism in the body. Calcium and phosphorus have a variety of roles and are important in the proper development and mineralization of the bones.

Skin Cancer Whereas we tend to associate a tan with good health, it signifies that the body is trying to protect itself from the dangerous rays of the sun. Too much ultraviolet radiation is dangerous and can lead to skin cancer. Basal cell carcinoma (Fig. 4.10*a*), derived from stem cells gone awry, is the more common type of skin cancer and is the most curable. Melanoma (Fig. 4.10*b*), the type of skin cancer derived from melanocytes, is extremely serious. To prevent skin cancer, you should stay out of the

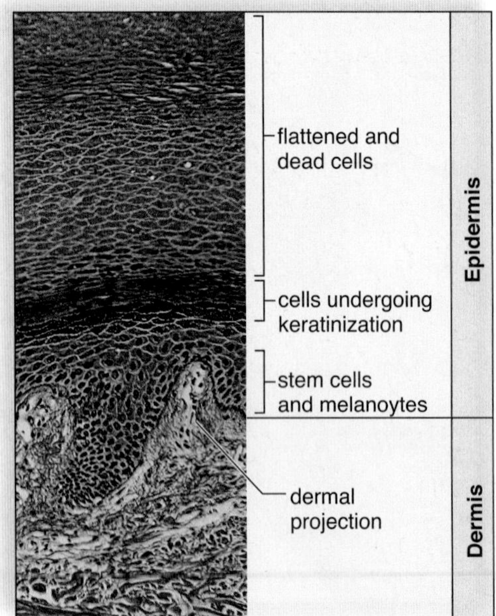

Figure 4.9 A light micrograph of human skin.
The keratinization of cells is shown in this image.

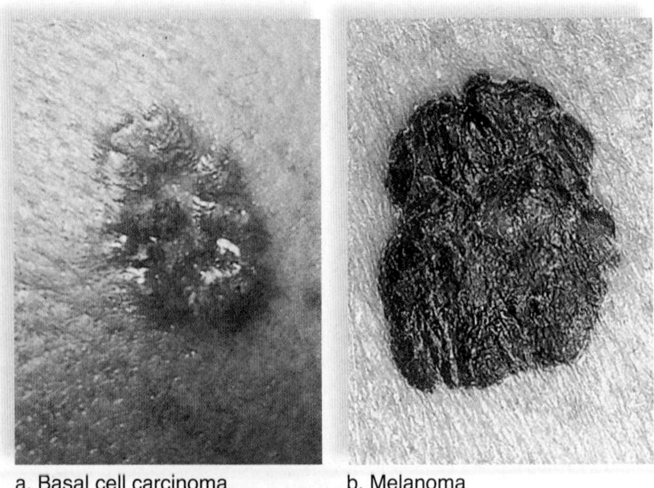

Figure 4.10 Cancers of the skin.
a. Basal cell carcinoma derived from stem cells and **(b)** melanoma derived from melanocytes are types of skin cancer.

sun between the hours of 10 A.M. and 3 P.M. When you are in the sun, follow these guidelines.

- Use a broad-spectrum sunscreen that protects from both UVA (long-wave) and UVB (short-wave) radiation and has a sun protection factor (SPF) of at least 15. (This means that if you usually burn, for example, after a 20-minute exposure, it will take 15 times longer, or 5 hours, before you will burn.)
- Wear protective clothing. Choose fabrics with a tight weave and wear a wide-brimmed hat.
- Wear sunglasses that have been treated to absorb UVA and UVB radiation.

Also, avoid tanning machines; even if they use only high levels of UVA radiation, the deep layers of the skin will become more vulnerable to UVB radiation.

Video Melanoma Marker

The Dermis

The **dermis** is a region of dense fibrous connective tissue beneath the epidermis. (*Dermatology* is a branch of medicine that specializes in diagnosing and treating skin disorders.) The dermis contains collagen and elastic fibers. The collagen fibers are flexible but offer great resistance to overstretching. They prevent the skin from being torn. The elastic fibers maintain normal skin tension but also stretch to allow movement of underlying muscles and joints. The number of collagen and elastic fibers decreases with age and with exposure to the sun. This causes the skin to become less supple and more prone to wrinkling. The dermis also contains blood vessels that nourish the skin. When blood rushes into these vessels, a person blushes. When blood is minimal in them, a person turns "blue." Blood vessels in the dermis play a role in temperature regulation. If body temperature starts to rise, the blood vessels in the skin will dilate. As a result, more blood is brought to the surface of the skin for cooling. If the outer temperature cools, the blood vessels constrict, so less blood is brought to the skin's surface.

The sensory receptors—primarily in the dermis—are specialized for touch, pressure, pain, hot, and cold. These receptors supply the central nervous system with information about the external environment. The sensory receptors also account for the use of the skin as a means of communication between people. For example, the touch receptors play a major role in sexual arousal.

The Subcutaneous Layer

Technically speaking, the subcutaneous layer beneath the dermis is not a part of skin. It is a common site for injections. This layer is composed of loose connective tissue and adipose tissue, which stores fat. Fat is a stored source of energy in the body. Adipose tissue helps to thermally insulate the body from either gaining heat from the outside or losing heat from the inside. A well-developed subcutaneous layer gives the body a rounded appearance and provides protective padding against external assaults. Excessive development of the subcutaneous layer accompanies obesity.

Connections and Misconceptions

Are tanning beds safe?

According to a study sponsored by the World Health Organization (WHO),* individuals who use tanning beds prior to the age of 30 increase their chances of developing skin cancer by 75%. Though exposure to UV radiation does increase vitamin D levels slightly, it also results in an increased formation of cataracts (clouding of the cornea of the eyes) and a suppressed immune system. The UV radiation from tanning beds is known to damage the DNA of cells deep in the skin and can reduce the amount of collagen in the connective tissues, resulting in premature aging of the skin. For these reasons, the WHO has issued a request for increased government monitoring of tanning beds and has classified them as "known to be carcinogenic to humans."

*"Sunbeds, Tanning, and UV Exposure," World Health Organization, http://www.who.int/mediacentre/factsheets/fs287/en (accessed June 15, 2010).

Accessory Organs of the Skin

Nails, hair, and glands are structures of epidermal origin, even though some parts of hair and glands are largely found in the dermis. **Nails** are a protective covering of the distal part of fingers and toes, collectively called *digits* (Fig. 4.11). Nails grow from special epithelial cells at the base of the nail in the portion called the *nail root*. The *cuticle* is a fold of skin that hides the nail root. The whitish color of the half-moon shaped base, or *lunula*, results from the thick layer of cells in this area. The cells of a nail become keratinized as they grow out over the nail bed.

Hair follicles begin at a bulb in the dermis and continue through the epidermis where the hair shaft extends beyond

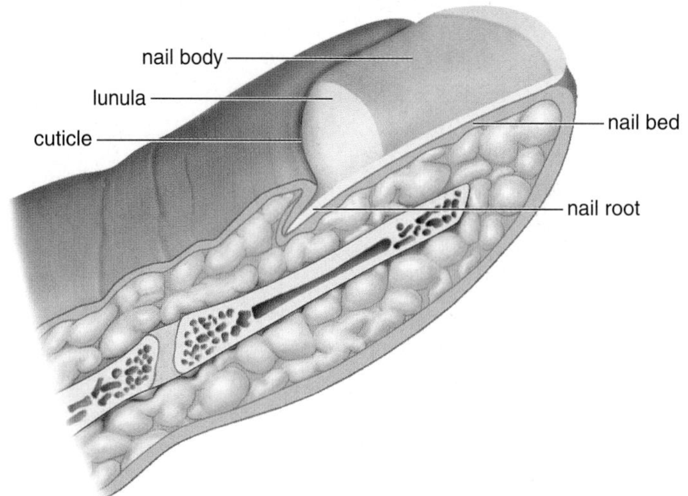

Figure 4.11 Anatomy of a human nail.
Cells produced by the nail root become keratinized, forming the nail body.

Biology Matters **Science Focus**

Face Transplantation

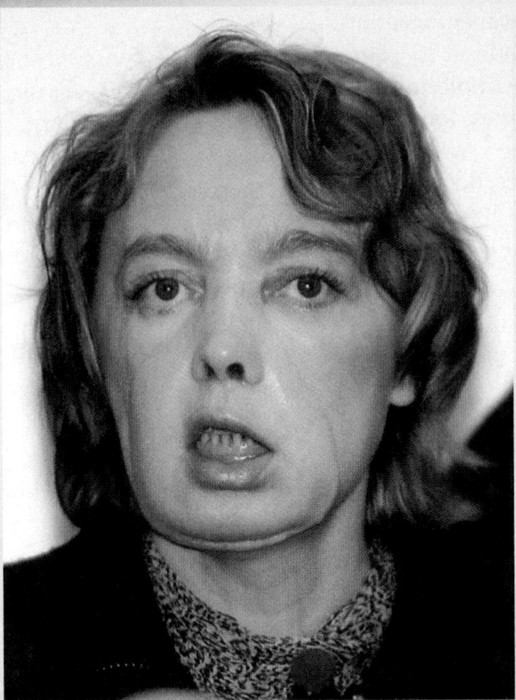

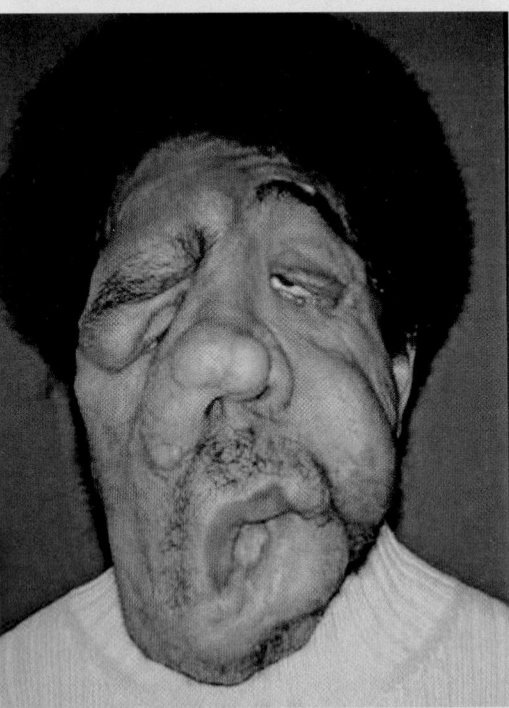

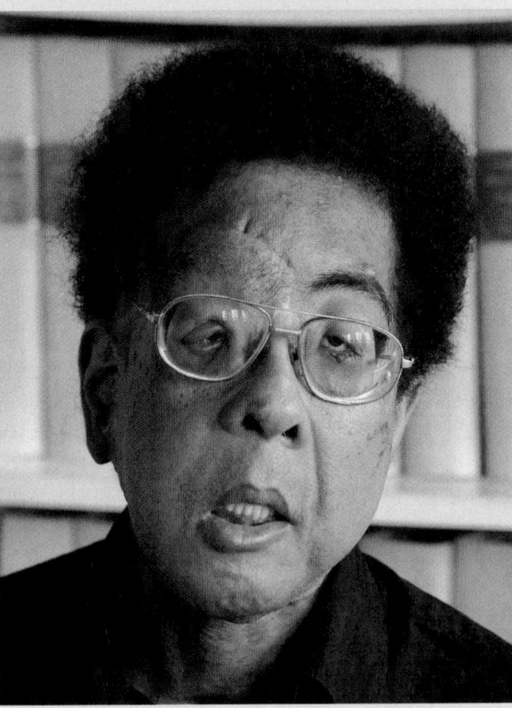

Figure 4B Face transplant recipients.
Isabelle Dinoire (*top, left* and *right*) and Pascal Coler (*bottom, left* and *right*) are both examples of successful face transplants.

In 2005, a French surgical team led by Professors Bernard Devauchelle and Jean Michel Dubernard was able to perform the world's first partial face transplant. The recipient was a woman, Isabelle Dinoire, severely disfigured by a dog mauling. Muscles, veins, arteries, nerves, and skin were transplanted onto the lower half of Isabelle's face (Fig. 4B, *top left*). The donor's lips, chin, and nose were transplanted. The donor was a brain-dead patient whose family had agreed to donate all their loved one's organs and tissues. The donor shared Isabelle's blood type and was a good tissue match. Eighteen months after the surgery (Fig. 4B, *top right*), Isabelle was able to eat, drink, and smile.

In 2008, a surgical team at Henri-Mondor Hospital in France was able to perform the first full face transplant. The patient Pascal Coler suffered from a condition called neurofibromatosis, which caused tumors to grow on his face, producing severe disfiguration (Fig. 4B, *bottom left*). As was the case with Isabelle, tissue from a donor's face was used to reconstruct Pascal's appearance (Fig. 4B, *lower right*).

Although the ability to do these types of transplant has existed for some time, doctors remain concerned about the ethical aspects of the procedure. Organ transplantation has always involved some moral concerns, because the donor must still be alive when the organs are harvested. Historically, face transplants have been a "quality-of-life" issue and not a "life-or-death" surgery. However, this attitude is changing as injured soldiers returning from wars in Afghanistan and Iraq are being considered for face transplants in the United States. In addition, recipients of face transplants must frequently undergo extensive counseling to prepare themselves emotionally for the "new face," and all recipients must spend the remainder of their lives on immunosuppressive drugs.

the skin (see Fig. 4.8). A dark hair color is largely due to the production of true melanin by melanocytes present in the bulb. If the melanin contains iron and sulfur, hair is blond or red. Graying occurs when melanin cannot be produced, but white hair is due to air trapped in the hair shaft.

Contraction of the *arrector pili muscles* attached to hair follicles causes the hairs to "stand on end" and goosebumps to develop. Epidermal cells form the root of a hair, and their division causes a hair to grow. The cells become keratinized and die as they are pushed farther from the root.

Each hair follicle has one or more **oil glands** (see Fig. 4.8), also called *sebaceous glands,* which secrete sebum. *Sebum* is an oily substance that lubricates the hair within the follicle and the skin. The oil secretions from sebaceous glands are acidic and retard the growth of bacteria. If the sebaceous glands fail to discharge (usually because they are blocked with keratinocytes), the secretions collect and form "whiteheads." Over time, the sebum in a whitehead oxidizes to form a"blackhead." Acne is an inflammation of the sebaceous glands that most often occurs during adolescence due to hormonal changes.

Sweat glands (see Fig. 4.8), also called *sudoriferous glands,* are numerous and present in all regions of skin. A sweat gland is a tubule that begins in the dermis and either opens into a hair follicle or, more often, opens onto the surface of the skin. Sweat glands play a role in modifying body temperature. When body temperature starts to rise, sweat glands become active. Sweat absorbs body heat as it evaporates. Once the body temperature lowers, sweat glands are no longer active.

MP3
Skin and Its Tissues

Connecting the Concepts

For more information on the role of the skin in human physiology, refer to the following discussions.

Section 8.6 provides more information on the relationship between vitamin D and bone calcium homeostasis.

Sections 19.1 and 19.2 explain how cells develop into cancer cells and present some of the different forms of cancer.

Section 22.5 explores some of evolutionary reasons for variations in skin color.

Check Your Progress 4.6

1. Compare the structure and function of the epidermis and dermis.
2. Explain how each accessory organ of the skin aids in homeostasis.
3. Hypothesize why structures like sensory receptors and substances for skin color are found in the dermis and not the epidermis.

4.7 Organ Systems, Body Cavities, and Body Membranes

Learning Outcomes

Upon completion of this section, you should be able to

1. State the function of each organ system in the human body.
2. Identify the major cavities of the human body.
3. Name the body membranes and provide a function for each.

This section examines how the organs of the body form organ systems. Some of these organ systems, such as the respiratory system, occupy specific cavities of the body, and others, such as the muscular and circulatory systems, are found throughout the body. The organs and cavities of the body are lined with membranes that often secrete fluid to aid in the physiology of the organ or organ system.

Organ Systems

Figure 4.12 illustrates the organ systems of the human body. It should be emphasized that just as organs work together in an organ system, so do organ systems work together in the body. In some cases, it is arbitrary to assign a particular organ to one system when it also assists the functioning of many other systems.

Integumentary System

The major organ of the integumentary system is the skin. It also includes nails, hairs, muscles that move hairs, the oil and sweat glands, blood vessels, and nerves leading to sensory receptors. As discussed in Section 4.6, this system has many homeostatic functions.

Cardiovascular System

In the **cardiovascular system,** the heart pumps blood and sends it out under pressure into the blood vessels. In humans, blood is always contained in blood vessels, never leaving these vessels unless the body suffers an injury.

While blood is moving throughout the body, it distributes heat produced by the muscles. Blood transports nutrients and oxygen to the cells and removes their waste molecules, including carbon dioxide. Despite the movement of molecules into and out of the blood, it has a fairly constant volume and pH. This is particularly due to exchanges in the lungs, the digestive tract, and the kidneys. The red blood cells in blood transport oxygen, and the white blood cells fight infections. Platelets are involved in blood clotting.

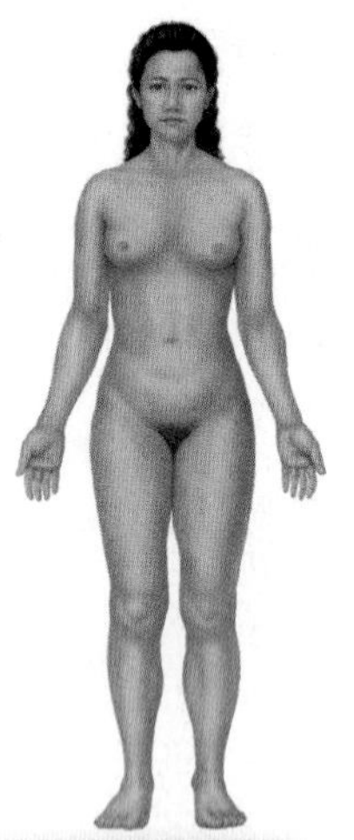

Integumentary system

- protects body.
- receives sensory input.
- helps control temperature.
- synthesizes vitamin D.

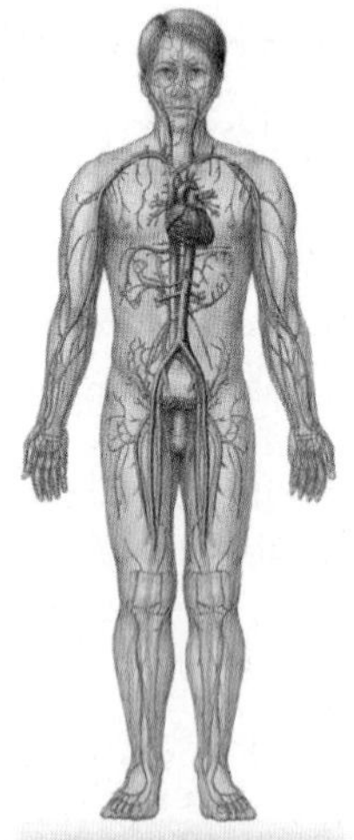

Cardiovascular system

- transports blood, nutrients, gases, and wastes.
- defends against disease.
- helps control temperature, fluid, and pH balance.

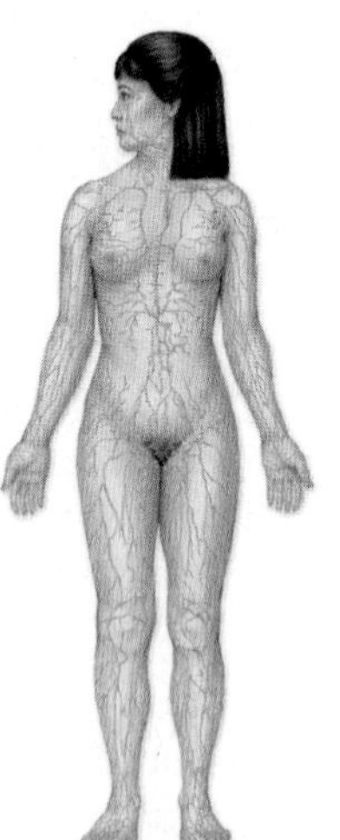

Lymphatic and immune systems

- help control fluid balance.
- absorb fats.
- defend against infectious disease.

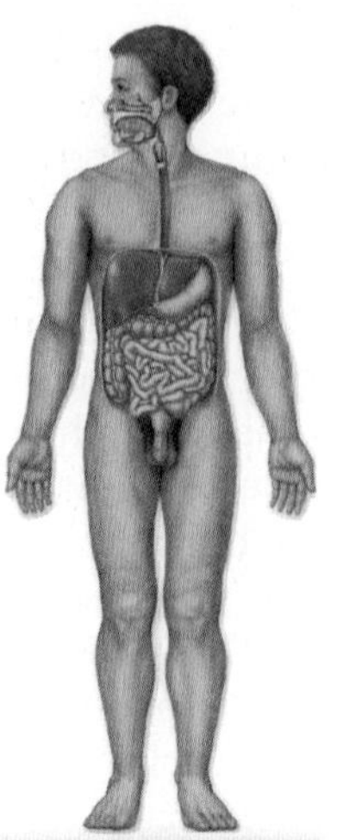

Digestive system

- ingests food.
- digests food.
- absorbs nutrients.
- eliminates waste.

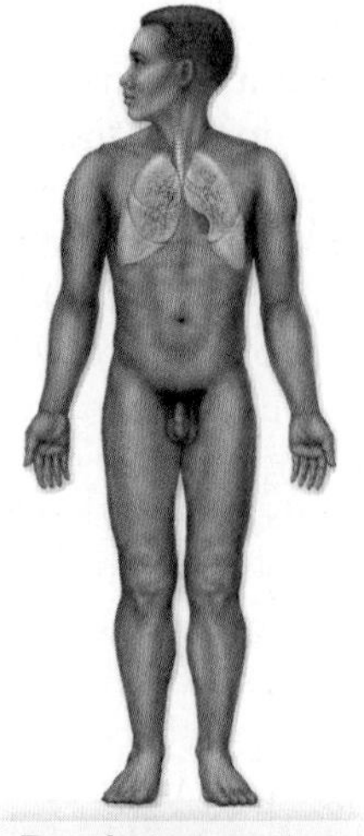

Respiratory system

- maintains breathing.
- exchanges gases at lungs and tissues.
- helps control pH balance.

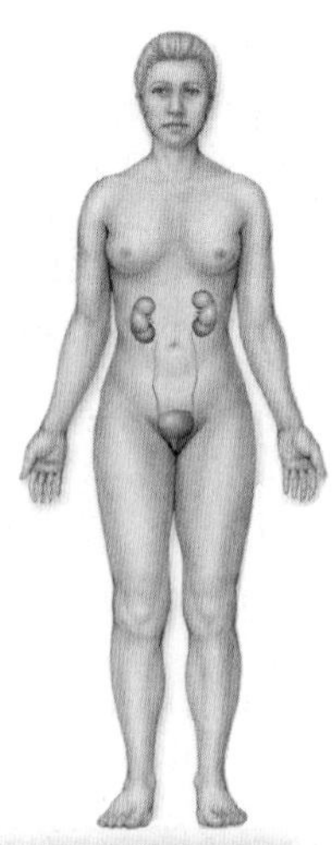

Urinary system

- excretes metabolic wastes.
- helps control fluid balance.
- helps control pH balance.

Figure 4.12 Organ systems of the body.

Lymphatic and Immune Systems

The **lymphatic system** consists of lymphatic vessels, lymph nodes, the spleen, and other lymphatic organs. This system collects excess tissue fluid and plays a role in absorbing fats and transporting lymph to cardiovascular veins. It also purifies lymph and stores lymphocytes, the white blood cells that produce antibodies.

The **immune system** consists of all the cells in the body that protect us from disease. The lymphocytes, in particular, belong to this system.

Digestive System

The **digestive system** consists of the mouth, esophagus, stomach, small intestine, and large intestine (colon). It also includes these associated organs: teeth, tongue, salivary glands, liver, gallbladder, and pancreas. This system receives food and digests it into nutrient molecules, which can enter the cells of the body. The nondigested remains are eventually eliminated.

Respiratory System

The **respiratory system** consists of the lungs and the tubes that take air to and from them. The respiratory system brings oxygen into the body and removes carbon dioxide from the body at the lungs. The removal of carbon dioxide helps adjust the acid–base balance of the blood.

Urinary System

The **urinary system** contains the kidneys, the urinary bladder, and the tubes that carry urine. The kidneys rid the body of metabolic wastes, particularly nitrogenous wastes. They also help regulate the salt–water balance and acid–base balance of the blood.

Skeletal System

The bones of the **skeletal system** protect body parts. For example, the skull forms a protective encasement for the brain, as does the rib cage for the heart and lungs. The skeleton helps move the body because it serves as a place of attachment for the skeletal muscles.

The skeletal system also stores minerals, notably calcium, and produces blood cells within red bone marrow.

Muscular System

In the **muscular system,** skeletal muscle contraction maintains posture and accounts for the movement of the body and its parts. Cardiac muscle contraction results in the heartbeat. The walls of internal organs, such as the bladder, contract due to the presence of smooth muscle. Muscle contraction releases heat, which helps warm the body.

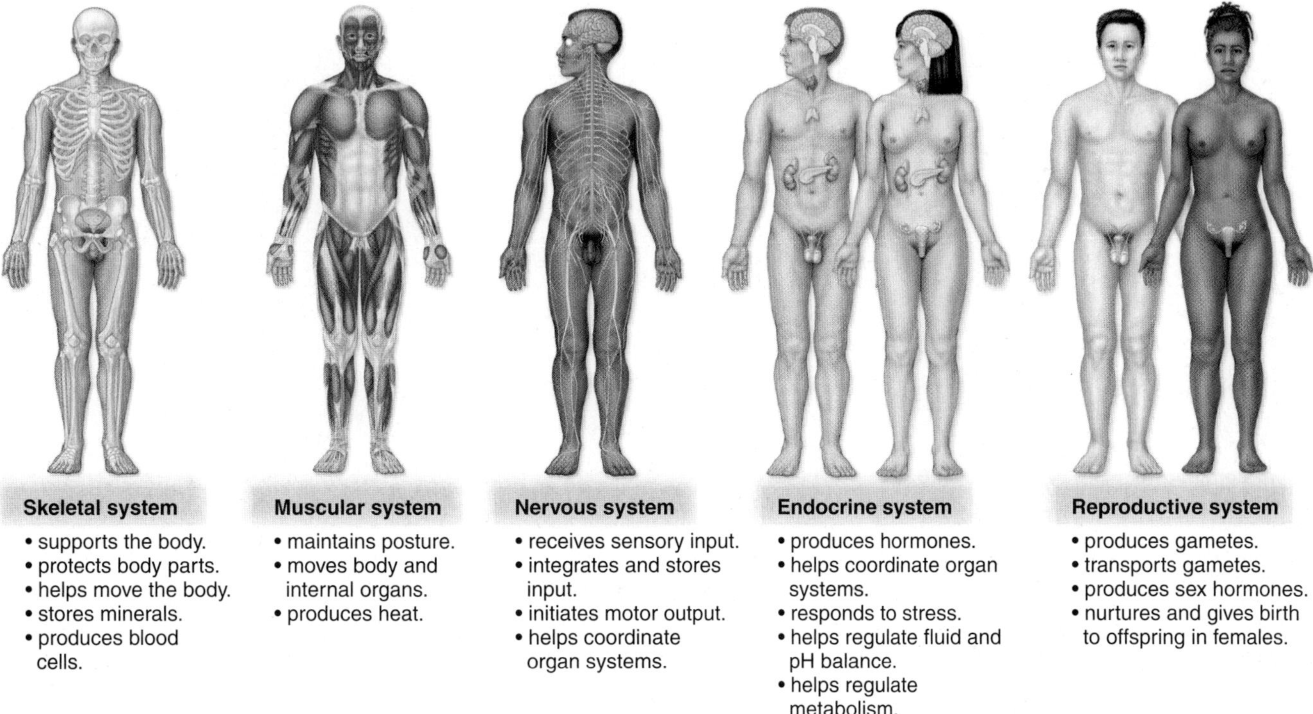

Skeletal system
• supports the body.
• protects body parts.
• helps move the body.
• stores minerals.
• produces blood cells.
Muscular system
• maintains posture.
• moves body and internal organs.
• produces heat.
Nervous system
• receives sensory input.
• integrates and stores input.
• initiates motor output.
• helps coordinate organ systems.
Endocrine system
• produces hormones.
• helps coordinate organ systems.
• responds to stress.
• helps regulate fluid and pH balance.
• helps regulate metabolism.
Reproductive system
• produces gametes.
• transports gametes.
• produces sex hormones.
• nurtures and gives birth to offspring in females.

Figure 4.12—*continued.*

Nervous System

The **nervous system** consists of the brain, spinal cord, and associated nerves. The nerves conduct nerve impulses from sensory receptors to the brain and spinal cord, where integration occurs. Nerves also conduct nerve impulses from the brain and spinal cord to the muscles and glands, allowing us to respond to both external and internal stimuli.

Endocrine System

The **endocrine system** consists of the hormonal glands, which secrete chemical messengers called *hormones* into the bloodstream. Hormones have a wide range of effects, including regulation of cellular metabolism, regulation of fluid and pH balance, and helping us respond to stress. Both the nervous and endocrine systems coordinate and regulate the functioning of the body's other systems. The endocrine system also helps maintain the functioning of the male and female reproductive organs.

Reproductive System

The **reproductive system** has different organs in the male and female. The male reproductive system consists of the testes, other glands (such as the prostrate), and various ducts that conduct semen to and through the penis. The testes produce sex cells called *sperm.* The female reproductive system consists of the ovaries, oviducts, uterus, vagina, and external genitals. The ovaries produce sex cells called *eggs.* When a sperm fertilizes an egg, an offspring begins development.

Body Cavities

The human body is divided into two main cavities: the ventral cavity and the dorsal cavity (Fig. 4.13*a*). Called the *coelom* in early development, the *ventral cavity* later becomes the thoracic, abdominal, and pelvic cavities. The thoracic cavity contains the lungs and the heart. The thoracic cavity is separated from the abdominal cavity by a horizontal muscle called the **diaphragm.** The stomach, liver, spleen, pancreas, gallbladder, and most of the small and large intestines are in the abdominal cavity. The pelvic cavity contains the rectum, the urinary bladder, the internal reproductive organs, and the rest of the small and large intestine. Males have an external extension of the abdominal wall called the *scrotum,* which contains the testes.

The *dorsal cavity* has two parts. (1) The cranial cavity within the skull contains the brain. (2) The vertebral canal, formed by the vertebrae, contains the spinal cord.

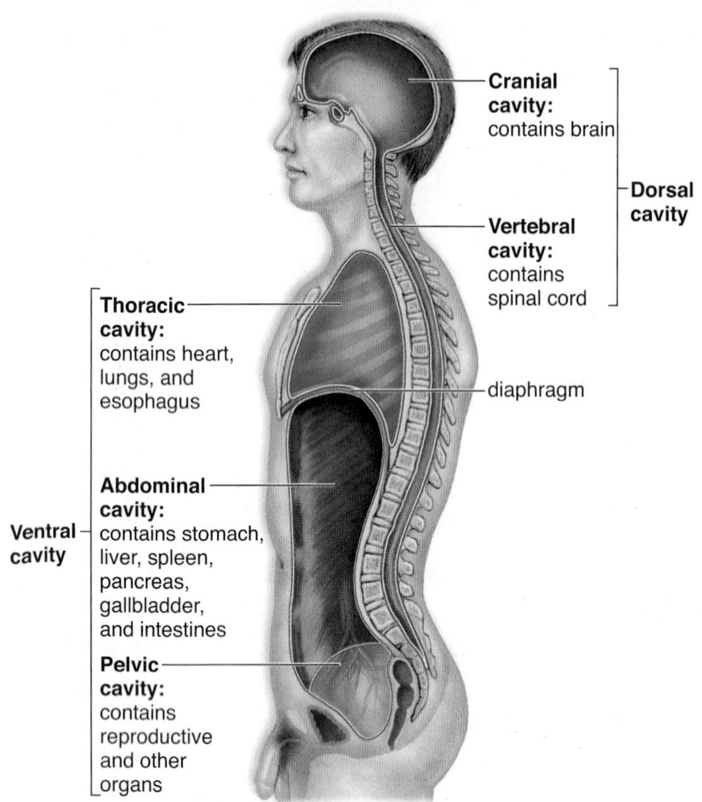

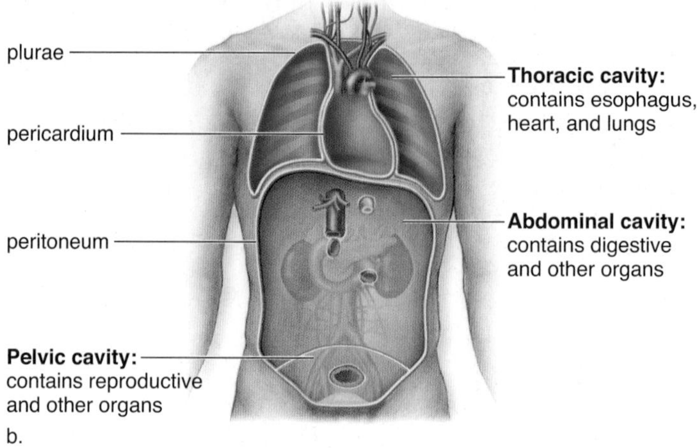

Figure 4.13 Body cavities of humans.
a. Side view. The posterior or dorsal (toward the back) cavity contains the cranial cavity and the vertebral canal. The brain is in the cranial cavity, and the spinal cord is in the vertebral canal. In the anterior or ventral (toward the front) cavity, the diaphragm separates the thoracic cavity from the abdominal cavity. The heart and lungs are in the thoracic cavity; the other internal organs are in either the abdominal cavity or the pelvic cavity. **b.** Frontal view of the thoracic cavity, showing serous membranes.

Body Membranes

Body membranes line cavities and the internal spaces of organs and tubes that open to the outside. The body membranes are of four types: mucous, serous, and synovial membranes and the meninges.

Mucous membranes line the tubes of the digestive, respiratory, urinary, and reproductive systems. They are composed of an epithelium overlying a loose fibrous connective tissue layer. The epithelium contains specialized cells that secrete mucus. This mucus ordinarily protects the body from invasion by bacteria and viruses. Hence, more mucus is secreted and expelled when a person has a cold and has to blow her/his nose. In addition, mucus usually protects the walls of the stomach and small intestine from digestive juices. This protection breaks down when a person develops an ulcer.

Serous membranes line and support the lungs, the heart, and the abdominal cavity and its internal organs (Fig. 4.13*b*). They secrete a watery fluid that keeps the membranes lubricated. Serous membranes support the internal organs and compartmentalize the large thoracic and abdominal cavities.

Serous membranes have specific names according to their location. The pleurae (sing., **pleura**) line the thoracic cavity and cover the lungs. The pericardium forms the pericardial sac and covers the heart. The peritoneum lines the abdominal cavity and covers its organs. A double layer of peritoneum, called mesentery, supports the abdominal organs and attaches them to the abdominal wall. **Peritonitis** is a life-threatening infection of the peritoneum.

Synovial membranes composed only of loose connective tissue line the cavities of freely movable joints. They secrete synovial fluid into the joint cavity. This fluid lubricates the ends of the bones so that they can move freely. In rheumatoid arthritis, the synovial membrane becomes inflamed and grows thicker, restricting movement.

The **meninges** (sing., meninx) are membranes found within the dorsal cavity. They are composed only of connective tissue and serve as a protective covering for the brain and spinal cord. Meningitis is a life-threatening infection of the meninges.

Connections and Misconceptions

What causes meningitis?

Meningitis is caused by an infection of the meninges by either a virus or a bacterium. Viral meningitis is less severe than bacterial meningitis, which in some cases can result in brain damage and death. Bacterial meningitis is usually caused by one of three species of bacteria: *Haemophilus influenzae* type b (Hib), *Streptococcus pneumoniae,* and *Neisseria meningitidis.* Vaccines are available for Hib bacteria and some forms of *S. pneumoniae* and *N. meningitidis.* The Centers for Disease Control (CDC) recommend that individuals between the ages of 11 and 18 be vaccinated against bacterial meningitis.

Connecting the Concepts

Each of the organ systems listed in this section is covered in greater detail in later chapters of the text. For more information on body cavities and body membranes, refer to the following discussions.

Section 7.3 describes how mucous membranes assist the immune response of the body.

Section 9.4 illustrates how the thoracic cavity is involved in breathing.

Section 11.5 explains how synovial joints (and their associated membranes) allow movement.

Check Your Progress 4.7

1. Summarize the overall function of each of the body systems and explain how each aids in homeostasis.
2. Describe the location of the two major body cavities.
3. List the four types of body membranes and describe the structure and function of each.

Biology Matters Health Focus

Pursuing Youthful Skin

More and more members of the "baby-boomer" generation are willing to spend lavishly for a youthful appearance. Over 30 million Americans have turned to Botox, laser treatments, and/or tanning to help obtain that vigorous, "healthy" look. But how safe and effective are these treatments?

Botox

Botox is a drug used to reduce the appearance of facial wrinkles and lines. Botox is the registered trade name for a derivative of botulinum toxin A, a protein toxin produced by the bacterium *Clostridium botulinum.* Botox stops communication between motor nerves and muscles, causing muscle paralysis. Botox treatments were approved by the U.S. Food and Drug Administration (FDA) for use as cosmetic treatments in 2002. Treatments are direct injections under the skin, where the toxin causes facial muscle paralysis. The injections reduce the appearance of wrinkles and lines that appear as a result of normal facial muscle movement. However, Botox treatment is not without side effects. Excessive drooling or a slight rash around the injection site are among the milder side effects of treatment. Spreading of Botox from the injection site may also paralyze facial muscles unintended for treatment. In a few cases, muscle pain and weakness have resulted. Though rare, more serious side effects, including allergic reactions, may also occur. When performed in a medical facility by a licensed physician, Botox treatment is generally considered safe and effective.

Laser Treatments

Laser treatments can be used for treatment of acne, wrinkles, and spider veins, as well as for removal of birthmarks, scars, tattoos, and hair. Lasers use a concentrated beam of light energy directed at the tissue. When the light beam hits the tissue, it is absorbed and converted to heat, and the tissue is destroyed. Laser treatments vary as to the energy of the laser and the depth of the skin removed. The beam of light is so precise that it can be used to treat small areas without affecting adjacent areas of the skin. Each treatment is short, usually between 15 and 60 minutes, but several treatments are necessary. New, more youthful appearing skin grows over the treated area. Deeper treatments, such as for wrinkle treatments or scar removal, may require extensive healing times (a month or longer). Redness, blistering, and the possibility of infection are all side effects. Other problems may include lightening or darkening of the skin, but these are usually temporary. The effectiveness of the procedure depends on the type of laser chosen and the experience and skill of the person operating it. As with Botox treatment, it is important for the procedure to be performed in a medical facility by a licensed physician.

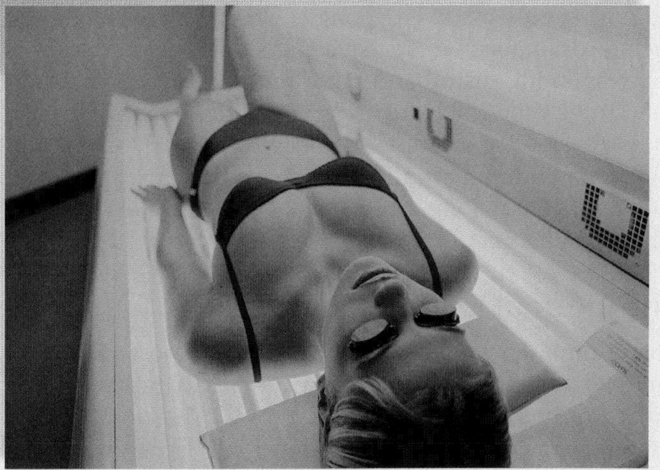

Figure 4C Tanning can damage skin permanently. In addition to aging of the skin, tanning can cause cancer. Tanning booths are now recognized as causing some forms of cancer.

Tanning

Whether you get that dark, glowing look from a tanning salon (Fig. 4C) or from the beach, there is no such thing as a safe tan. Darker, tanned skin is due to the extra production of the pigment melanin, a sign of skin damage. Instead of leading to a youthful appearance, tanning can lead to premature wrinkling as well as skin cancer. Damage to the skin accumulates over time—the longer the exposure, the greater the damage. Almost 90% of the changes people associate with aging, including leathery skin, dark spots, and wrinkles, are commonly attributed to sun damage—hardly the youthful complexion originally hoped for! More than 90% of skin cancers are due to sun exposure. Excessive sun exposure is especially dangerous for young people—those in their teens and twenties. The use of tanning beds among young people increases the risk of melanoma by 75%. Getting a dark tan or a burn while young may result in skin cancer in later life.

4.8 Homeostasis

Learning Outcomes

Upon completion of this section, you should be able to

1. Define *homeostasis* and provide an example.
2. Distinguish between positive and negative feedback mechanisms.

Homeostasis is the body's ability to maintain a relative constancy of its internal environment by adjusting its physiological processes. Even though external conditions may change dramatically, we have physiological mechanisms that respond to disturbances and limit the amount of internal change. Conditions usually stay within a narrow range of normalcy. For example, blood glucose, pH levels, and body temperature typically fluctuate during the day, but not greatly. If internal conditions change to any great degree, illness results.

The Internal Environment

The internal environment has two parts: blood and tissue fluid. Blood delivers oxygen and nutrients to the tissues and carries carbon dioxide and wastes away. Tissue fluid, not blood, bathes the body's cells. Therefore, tissue fluid is the medium through which substances are exchanged between cells and blood. Oxygen and nutrients pass through tissue fluid on their way to tissue cells from the blood. Then carbon dioxide and wastes are carried away from the tissue cells by the tissue fluid and are brought back into the blood. The cooperation of body systems is required to keep these substances within the range of normalcy in blood and tissue fluid.

The Body Systems and Homeostasis

The nervous and endocrine systems are particularly important in coordinating the activities of all the other organ systems as they function to maintain homeostasis (Fig. 4.14). The nervous system is able to bring about rapid responses to any changes in the internal environment. The nervous system issues commands by electrochemical signals rapidly transmitted to effector organs, which can be muscles, such as skeletal muscles, or glands, such as sweat and salivary glands. The endocrine system brings about slower responses, but they generally have more lasting effects. Glands of the endocrine system, such as the pancreas or the thyroid, release hormones. *Hormones,* such as insulin from the pancreas, are chemical messengers that must travel through the blood and tissue fluid to reach their targets.

The nervous and endocrine systems together direct numerous activities that maintain homeostasis, but all the organ systems must do their part to keep us alive and healthy. Picture what would happen if any component of the cardiovascular, respiratory, digestive, or urinary system failed (Fig. 4.14). If someone is having a heart attack, the heart is unable to pump the blood to supply cells with oxygen. Or think of a person who is choking. The trachea (or windpipe) is blocked, so no air can reach the lungs for uptake by the blood. Unless the obstruction is removed quickly, cells will begin to die as the blood's supply of oxygen is depleted. When the lining of the digestive tract is damaged, as in a severe bacterial infection, nutrient absorption is impaired and cells face an energy crisis. It is important not only to maintain adequate nutrient levels in the blood but also to eliminate wastes and toxins. The liver makes urea, a nitrogenous end product of protein metabolism. Urea and other metabolic wastes are excreted by the kidneys, the urine-producing organs of the body. The kidneys rid the body of nitrogenous wastes and also help to adjust the blood's water–salt and acid–base balances.

A closer examination of how the blood glucose level is maintained helps us understand homeostatic mechanisms. When a healthy person consumes a meal and glucose enters the blood, the pancreas secretes the hormone insulin. Then glucose is removed from the blood as cells take it up. In the liver, glucose is stored in the form of glycogen. This storage is beneficial because later, if blood glucose levels drop, glycogen can be broken down to ensure that the blood level remains constant. Homeostatic mechanisms can fail. In diabetes mellitus, the pancreas cannot produce enough insulin or the body cells cannot respond appropriately to it. Therefore, glucose does not enter the cells and they must turn to other molecules, such as fats and proteins, to survive. This, along with too much glucose in the blood, leads to the numerous complications of diabetes mellitus.

Another example of homeostasis is the ability of the body to regulate the acid–base balance of the body. When carbon dioxide enters the blood, it combines with water to form carbonic acid. However, the blood is buffered, and pH stays within normal range as long as the lungs are busy excreting carbon dioxide. These two mechanisms are backed up by the kidneys, which can rid the body of a wide range of acidic and basic substances and, therefore, adjust the pH.

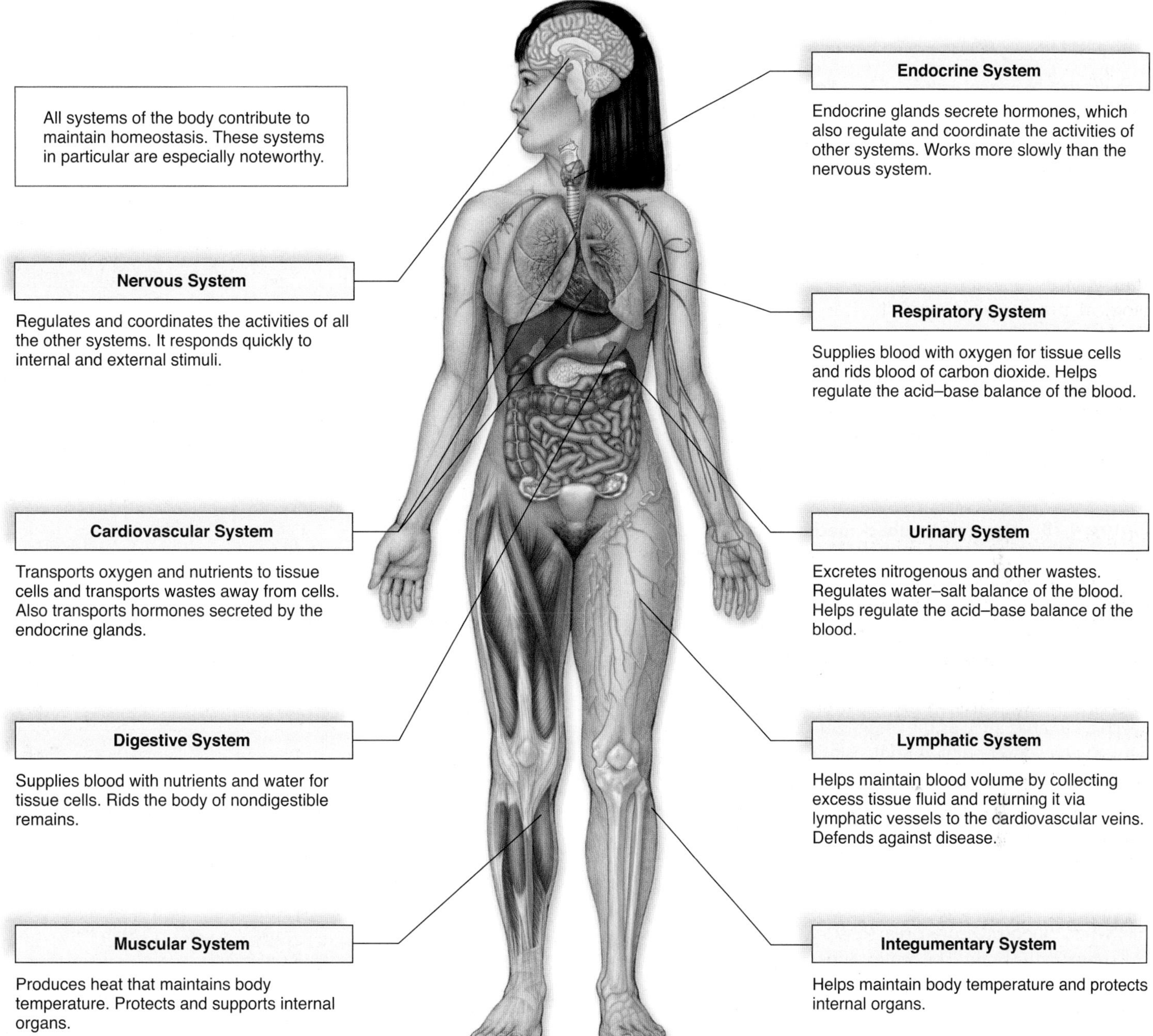

All systems of the body contribute to maintain homeostasis. These systems in particular are especially noteworthy.
Nervous System
Regulates and coordinates the activities of all the other systems. It responds quickly to internal and external stimuli.
Cardiovascular System
Transports oxygen and nutrients to tissue cells and transports wastes away from cells. Also transports hormones secreted by the endocrine glands.
Digestive System
Supplies blood with nutrients and water for tissue cells. Rids the body of nondigestible remains.
Muscular System
Produces heat that maintains body temperature. Protects and supports internal organs.
Endocrine System
Endocrine glands secrete hormones, which also regulate and coordinate the activities of other systems. Works more slowly than the nervous system.
Respiratory System
Supplies blood with oxygen for tissue cells and rids blood of carbon dioxide. Helps regulate the acid–base balance of the blood.
Urinary System
Excretes nitrogenous and other wastes. Regulates water–salt balance of the blood. Helps regulate the acid–base balance of the blood.
Lymphatic System
Helps maintain blood volume by collecting excess tissue fluid and returning it via lymphatic vessels to the cardiovascular veins. Defends against disease.
Integumentary System
Helps maintain body temperature and protects internal organs.

Figure 4.14 **Homeostasis by the organ systems of the human body.**
All the organ systems contribute to homeostasis in many ways. Some of the main contributions of each system are given in this illustration.

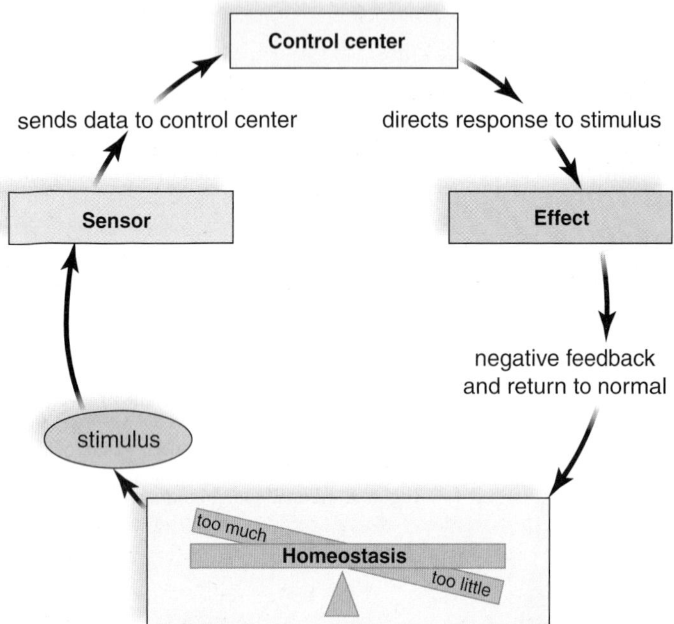

Figure 4.15 **Negative feedback mechanisms.**
This diagram shows how the basic elements of a feedback mechanism work. A sensor detects the stimulus, and a control center brings about an effect that resolves, or corrects, the stimulus.

Negative Feedback

Negative feedback is the primary homeostatic mechanism that keeps a variable, such as the blood glucose level, close to a particular value, or set point. A homeostatic mechanism has at least two components: a sensor and a control center (Fig. 4.15). The sensor detects a change in the internal environment. The control center then brings about an effect to bring conditions back to normal. Then the sensor is no longer activated. In other words, a **negative feedback** mechanism is present when the output of the system resolves or corrects the original stimulus. For example, when blood pressure rises, sensory receptors signal a control center in the brain. The center stops sending nerve signals to muscle in the arterial walls. The arteries can then relax. Once the blood pressure drops, signals no longer go to the control center.

Mechanical Example

A home heating system is often used to illustrate how a more complicated negative feedback mechanism works (Fig. 4.16). You set the thermostat at 68°F. This is the *set point.* The thermostat contains a thermometer, a sensor that detects when the room temperature is above or below the set point. The thermostat also contains a control center. It turns the furnace off when the room is warm and turns it on when the room is cool. When the furnace is off, the room cools a bit. When the furnace is on, the room warms a bit. In other words, typical of negative feedback mechanisms, there is a fluctuation above and below normal.

Human Example: Regulation of Body Temperature

The sensor and control center for body temperature are located part of the brain called the *hypothalamus.* A negative feedback mechanism prevents change in the same direction. Body temperature does not get warmer and warmer because warmth brings about a change toward a lower body temperature. Likewise, body temperature does not get continuously colder. A body temperature below normal brings about a change toward a warmer body temperature.

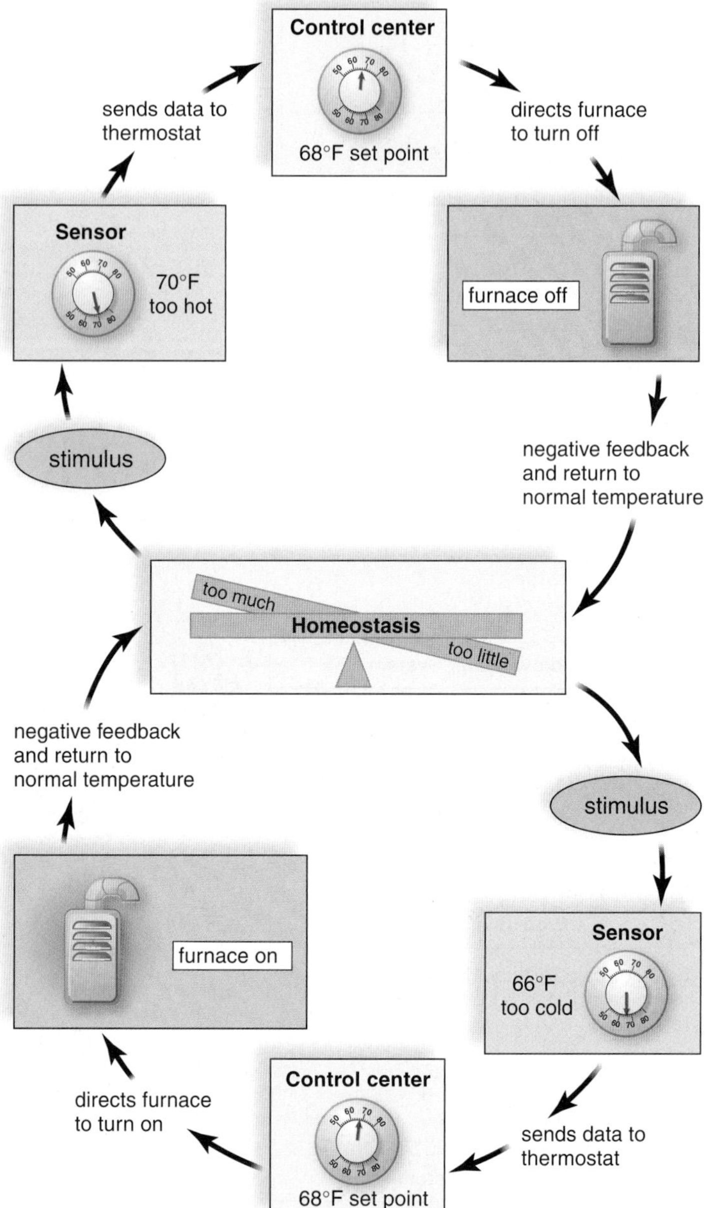

Figure 4.16 **Action of a complex negative feedback mechanism.**
This diagram shows how room temperature is returned to normal when the room becomes too hot (*above*) or too cold (*below*). The thermostat contains both the sensor and the control center. *Above:* The sensor detects that the room is too hot, and the control center turns the furnace off. The stimulus is resolved, or corrected, when the temperature returns to normal. *Below:* The sensor detects that the room is too cold, and the control center turns the furnace on. Once again, the stimulus is resolved, or corrected, when the temperature returns to normal.

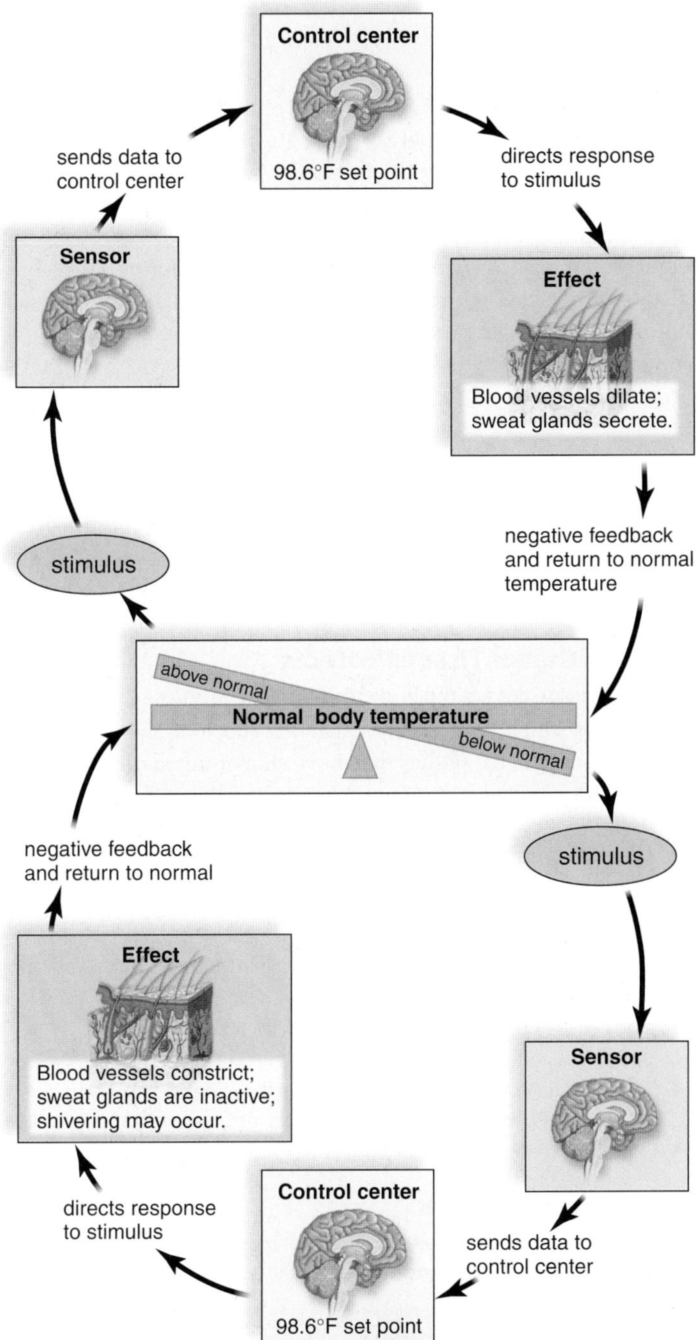

Figure 4.17 **Body temperature homeostasis.**
Above: When body temperature rises above normal, the hypothalamus senses the change and causes blood vessels to dilate and sweat glands to secrete so that temperature returns to normal. *Below:* When body temperature falls below normal, the hypothalamus senses the change and causes blood vessels to constrict. In addition, shivering may occur to bring temperature back to normal. In this way, the original stimulus is resolved, or corrected.

Above-Normal Temperature When the body temperature is above normal, the control center directs the blood vessels of the skin to dilate (Fig. 4.17, *above*). This allows more blood to flow near the surface of the body, where heat can be lost to the environment. In addition, the nervous system activates the sweat glands, and the evaporation of sweat helps lower body temperature. Gradually, body temperature decreases to 98.6°F.

Below-Normal Temperature When the body temperature falls below normal, the control center directs (via nerve impulses) the blood vessels of the skin to constrict (Fig. 4.17, *below*). This conserves heat. If body temperature falls even lower, the control center sends nerve impulses to the skeletal muscles and shivering occurs. Shivering generates heat, and gradually body temperature rises to 98.6°F. When the temperature rises to normal, the control center is inactivated.

Positive Feedback

Positive feedback is a mechanism that brings about an increasing change in the same direction. When a woman is giving birth, the head of the baby begins to press against the cervix (entrance to the womb), stimulating sensory receptors there. When nerve signals reach the brain, the brain causes the pituitary gland to secrete the hormone oxytocin. Oxytocin travels in the blood and causes the uterus to contract. As labor continues, the cervix is increasingly stimulated, and uterine contractions become stronger until birth occurs.

A positive feedback mechanism can be harmful, as when a fever causes metabolic changes that push the fever higher. Death occurs at a body temperature of 113°F because cellular proteins denature at this temperature and metabolism stops. However, positive feedback loops such as those involved in childbirth, blood clotting, and the stomach's digestion of protein assist the body in completing a process that has a definite cutoff point.

Connecting the Concepts

The maintenance of homeostasis is an important function of all organ systems. For more information on the organ systems, refer to the following discussions.

Section 6.6 describes how the cardiovascular system helps maintain homeostasis.

Section 10.4 explores the role of the urinary system in maintaining homeostasis.

Section 13.2 describes the role of the hypothalamus as part of the central nervous sytem.

Check Your Progress 4.8

1. Define *homeostasis* and explain why it is important to body function.
2. Summarize how each body system contributes to homeostasis.
3. Distinguish between negative feedback and positive feedback mechanisms in regard to homeostasis.

CASE STUDY CONCLUSION

Skin is a very complex organ, consisting of all of the four types of tissues. By developing an understanding of how tissues interact to form organs, it is now possible to develop synthetic tissues such as artificial skin. Though scientists are now expanding these techniques to include the development of artificial dermal tissue, the complexity of the interactions of the various tissues (nervous, muscle, connective) in this layer has presented some obstacles. However, there have been some important advances. Recently, scientists have developed newer forms of artificial skin that release antibiotics directly onto the healing tissue, further protecting the patient against life-threatening infections. Research on artificial tissues is not confined to epithelial tissue. Scientists are actively exploring the development of artificial cardiac tissue and artificial replacement organs such as the kidney, liver, and lungs. Like Kristen's artificial skin, these will not be plastic substitutes but rather, organic material that integrates with the patient's living cells to replace damaged tissue.

Media Study Tools

www.mhhe.com/maderhuman12e

Enhance your study of this chapter with study tools and practice tests. Also ask your instructor about the resources available through ConnectPlus, including the media-rich eBook, interactive learning tools, and animations.

Summarizing the Concepts

4.1 Types of Tissues

Human tissues are categorized into four groups:

- Connective
- Muscular
- Nervous
- Epithelial

4.2 Connective Tissue Connects and Supports

Connective tissues have cells separated by a matrix that contains ground substance and fibers (e.g., collagen fibers). Examples include

- loose fibrous connective tissue, including adipose tissue, and dense fibrous connective tissue (tendons and ligaments);
- cartilage and bone (matrix for cartilage is solid, yet flexible; matrix for bone is solid and rigid); and
- blood (matrix is a liquid called plasma) and lymph.

4.3 Muscular Tissue Moves the Body

Muscular tissue is of three types: skeletal, smooth, and cardiac.

- Skeletal and cardiac muscle are striated.
- Cardiac and smooth muscle are involuntary.
- Skeletal muscle is found in muscles attached to bones.
- Smooth muscle is found in internal organs.
- Cardiac muscle makes up the heart.

4.4 Nervous Tissue Communicates

- Nervous tissue is composed of neurons and several types of neuroglia.
- Each neuron has dendrites, a cell body, and an axon. Axons conduct nerve impulses.

4.5 Epithelial Tissue Protects

Epithelial tissue covers the body and lines its cavities.

- Types of simple epithelia are squamous, cuboidal, and columnar.
- Certain epithelial tissues may have cilia or microvilli.
- Stratified epithelia have many layers of cells, with only the bottom layer touching the basement membrane.
- Glandular epithelia secretes a product either into ducts or into the blood.

4.6 Integumentary System

Skin and its accessory organs comprise the integumentary system. Skin has two regions:

- The epidermis contains stem cells, which produce new epithelial cells.
- The dermis contains epidermally derived glands and hair follicles, nerve endings, blood vessels, and sensory receptors.
- A subcutaneous layer (hypodermis) lies beneath the skin.

4.7 Organ Systems

Organs make up organ systems, summarized in the table on this page. Some organs are found in particular body cavities.

Table 4.1 Organ Systems

Transport	**Integumentary**
Cardiovascular (heart and blood vessels)	Skin and accessory organs
Lymphatic and Immune (lymphatic vessels)	
Maintenance	**Motor**
Digestive (e.g., stomach, intestines)	Skeletal (bones and cartilage)
Respiratory (tubes and lungs)	Muscular (muscles)
Urinary (tubes and kidneys)	
Control	**Reproduction**
Nervous (brain, spinal cord, and nerves)	Reproductive (tubes and testes in males; tubes and ovaries in females)
Endocrine (glands)	

4.8 Homeostasis

Homeostasis is the relative constancy of the internal environment, tissue fluid, and blood. All organ systems contribute to homeostasis.

- The cardiovascular, respiratory, digestive, and urinary systems directly regulate the amount of gases, nutrients, and wastes in the blood, keeping tissue fluid constant.
- The lymphatic system absorbs excess tissue fluid and functions in immunity.
- The nervous system and endocrine system regulate the other systems.

Negative Feedback

Negative feedback mechanisms keep the environment relatively stable. When a sensor detects a change above or below a set point, a control center brings about an effect that reverses the change and returns conditions to normal. Examples include

- regulation of blood glucose level by insulin,
- regulation of room temperature by a thermostat and furnace, and
- regulation of body temperature by the brain and sweat glands.

Positive Feedback

In contrast to negative feedback, a positive feedback mechanism brings about rapid change in the same direction as the stimulus and does not achieve relative stability. These mechanisms are useful under certain conditions, such as during birth.

Understanding Key Terms

adipose tissue 67
basement membrane 72
blood 68
cardiac muscle 69
cardiovascular system 79
cartilage 67
collagen fiber 66
columnar epithelium 74
compact bone 68
connective tissue 66
cuboidal epithelium 74
dense fibrous connective tissue 67
dermis 77
diaphragm 81
digestive system 80
elastic cartilage 68
elastic fiber 66
endocrine gland 74
endocrine system 81
epidermis 76
epithelial tissue 72
exocrine gland 74
fibroblast 67
fibrocartilage 68
gland 74
hair follicle 77
homeostasis 84
hyaline cartilage 67
immune system 80
integumentary system 75
intercalated disk 69
lacuna 67
Langerhans cell 76
ligament 67
loose fibrous connective tissue 67
lymph 68
lymphatic system 80
matrix 67
melanocyte 76
meninges 82
mucous membrane 82
muscular system 80
muscular (contractile) tissue 69
nail 77
negative feedback 86
nerve 70
nervous system 81
nervous tissue 70
neuroglia 70
neuron 70
oil gland 79
organ 75
organ system 75
peritonitis 82
platelet (thrombocyte) 68
pleura 82
positive feedback 87
pseudostratified columnar epithelium 74
red blood cell (erythrocyte) 68
reproductive system 81
respiratory system 80
reticular fiber 66
serous membrane 82
skeletal muscle 69
skeletal system 80
skin 75
smooth (visceral) muscle 69
spongy bone 68
squamous epithelium 73
striated 69
subcutaneous layer 75
sweat gland 79
synovial membrane 82
tendon 67
tissue 66
tissue fluid 68
urinary system 80
vitamin D 76
white blood cell (leukocyte) 68

Match the key terms to these definitions.

a. ________ Dense fibrous connective tissue that joins bone to bone at a joint.

b. ________ Outer region of the skin composed of keratinized stratified squamous epithelium.

c. ________ Cancer of epithelial tissue.

d. ________ Relative constancy of the body's internal environment.

e. ________ Porous bone found at the ends of long bones where blood cells are formed.

Testing Your Knowledge of the Concepts

1. What are the functions of the four major tissue types? (page 67)
2. What features do all connective tissues have in common? Why? (page 66)
3. Explain why skeletal muscle is voluntary and cardiac muscle and smooth muscle are involuntary. (pages 69–70)
4. What are the two types of cells found in nervous tissue? Briefly describe the structure and function of each. (pages 70–71)
5. How are epithelial tissues classified? Describe each major type, and give at least one location for each type and the reason why it is found in that location. (pages 72–73)
6. Explain why the skin is sometimes referred to as the integumentary system. (page 75)
7. Referring to Figure 4.12, list each organ system, the major organs, major functions of each, and how each aids in homeostasis. (pages 80–85)
8. What organs of the body are found in the thoracic cavity? The abdominal cavity? (pages 81–82)
9. List the types of membranes found in the body, their functions, and their locations. (page 82)
10. Why is homeostasis defined as the "*relative* constancy of the internal environment"? Does negative feedback or positive feedback tend to promote homeostasis? Explain. (pages 84–86)
11. Which of the following is *not* a type of fibrous connective tissue?
    a. hyaline cartilage
    b. areolar tissue
    c. tendons and ligaments
    d. adipose tissue

12. What type of cartilage is found in the rib cage and walls of the respiratory passages?
    a. fibrocartilage
    b. elastic cartilage
    c. ligamentous cartilage
    d. hyaline cartilage
13. Blood is a/an ________ tissue because it has a ________.
    a. connective; gap junction
    b. muscular; ground substance
    c. epithelial; gap junction
    d. connective; ground substance
14. This type of muscle contains striations.
    a. smooth muscle
    b. skeletal muscle
    c. cardiac muscle
    d. both b and c
    e. all of the above
15. Which of the following forms the myelin sheath around nerve fibers outside the brain and spinal cord?
    a. microglia
    b. neurons
    c. Schwann cells
    d. astrocytes
16. Which of these is not a type of epithelial tissue?
    a. simple cuboidal and stratified columnar
    b. bone and cartilage
    c. stratified squamous and simple squamous
    d. pseudostratified and transitional
    e. All of these are epithelial tissues.
17. What type of epithelial tissue is found in the walls of the urinary bladder to provide it with the ability to distend?
    a. simple cuboidal epithelium
    b. transitional epithelium
    c. pseudostratified columnar epithelium
    d. stratified squamous epithelium
18. Without melanocytes, skin would
    a. be too thin.
    b. lack nerves.
    c. not tan.
    d. not be waterproof.
19. Which of the following is a function of skin?
    a. temperature regulation
    b. manufacture of vitamin D
    c. protection from invading pathogens
    d. All of these are correct.
20. Fluid balance is a primary goal of which system?
    a. endocrine
    b. lymphatic
    c. digestive
    d. integumentary
21. The skeletal system functions in
    a. blood cell production.
    b. mineral storage.
    c. movement.
    d. All of these are correct.
22. Which system helps control pH balance?
    a. digestive
    b. urinary
    c. respiratory
    d. Both b and c are correct.
23. Which type of membrane is found lining systems open to the outside environment, such as the respiratory system?
    a. serous
    b. mucous
    c. synovial
    d. meningeal
24. Which allows rapid change in one direction and does not achieve stability?
    a. homeostasis
    b. positive feedback
    c. negative feedback
    d. All of these are correct.
25. Which of the following is an example of negative feedback?
    a. Uterine contractions increase as labor progresses.
    b. Insulin decreases blood sugar levels after a meal is eaten.
    c. Sweating increases as body temperature drops.
    d. Platelets continue to plug an opening in a blood vessel until blood flow stops.

Thinking Critically About the Concepts

In the hierarchy of biological organization, you have learned that groups of cells make tissues and two or more tissue types compose an organ. In this chapter, the four types of tissues (connective, muscular, nervous, and epithelial) have been discussed in detail. The skin is an organ system referred to as the integumentary system that contains all four tissue types. In addition to the epidermis and dermis, the integumentary system also includes accessory structures such as nails, sweat glands, sebaceous glands, and hair follicles. Each of these components of the skin aids in the various functions of the integumentary system. In the case study, Kristen has burns severe enough to need artificial skin treatment. This treatment will help Kristen's skin repair itself while mimicking some of the functions the integumentary system does for homeostasis.

1. The doctor diagnosed Kristen's burn as severe. Which of the following best describes a severe burn versus a superficial burn?
   a. Superficial burns include the epidermis, dermis, and hypodermis layers.
   b. Severe burns only include the layers of the epidermis.
   c. Severe burns include anything including and below the dermis.
   d. Superficial burns occur on the limbs only.
2. What accessory structures and tissues are damaged in a severe burn? Why?
3. What types of functions will the artificial skin perform while Kristen's own skin is repairing itself?
4. What effects can a severe burn have on overall homeostasis of the body? Give a few examples.
5. Which structures in the dermis will have the slowest repair time compared to others? Which might never repair themselves fully? Why?
6. Without the integumentary system, what might happen to the functions of the cardiovascular system? The nervous system?

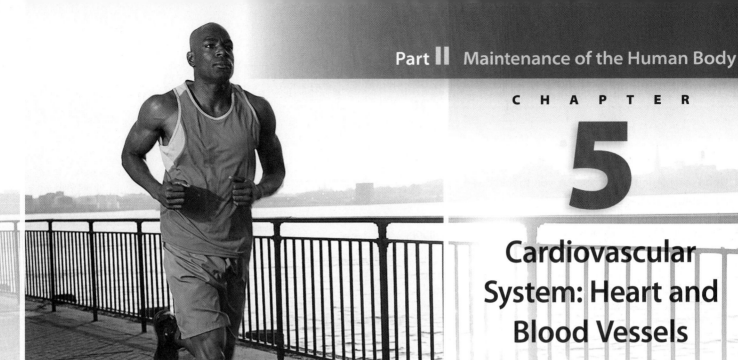

CHAPTER 5

Cardiovascular System: Heart and Blood Vessels

CHAPTER CONCEPTS

5.1 Overview of the Cardiovascular System
In the cardiovascular system, the heart pumps blood, and blood vessels transport blood about the body. Exchanges between the cardiovascular system and tissues occur at the capillaries.

5.2 The Types of Blood Vessels
Arteries, which branch into arterioles, move blood from the heart to capillaries. Capillaries empty into venules. Venules join to form veins that return blood to the heart.

5.3 The Heart Is a Double Pump
The heart's right side pumps blood to the lungs, and the left side pumps blood to the rest of the body.

5.4 Features of the Cardiovascular System
Blood pressure decreases as blood moves from the arteries to the capillaries. Skeletal muscle contraction largely accounts for movement of blood in the veins.

5.5 Two Cardiovascular Pathways
The pulmonary circuit takes blood from the heart to the lungs and back to the heart. The systemic circuit carries blood from the heart to all other organs, and then returns blood to the heart.

5.6 Exchange at the Capillaries
At tissue capillaries, nutrients and oxygen are exchanged for carbon dioxide and other wastes, permitting body cells to remain alive and healthy.

5.7 Cardiovascular Disorders
Hypertension and atherosclerosis are disorders that can lead to stroke, heart attack, and aneurysm. Ways to prevent and treat cardiovascular disease are constantly being improved.

CASE STUDY PERIPHERAL ARTERIAL DISEASE

Ted is a 44-year-old who enjoys running in half-marathons. He trains two to three times per week, running 5 miles a session and lifting weights in his neighborhood gym. He follows a healthy diet and does not smoke or drink alcohol. Ted was diagnosed with diabetes mellitus at age 12 and is insulin-dependent but through exercise and diet has controlled his diabetes well all his life. Ted noticed for several days that his legs were numb, and he had intermittent periods of pain from both hips down to his toes. He decided it was a result of his earlier overzealous workout. When the symptoms persisted, he went to see his doctor. After his exam, the doctor ordered a CT scan and an MRA (magnetic resonance angiography), which through the use of powerful magnets, radio waves, and a contrast dye injected into his blood vessels would give a clear picture of Ted's vessels. When the results returned, Ted's doctor informed him his symptoms were caused by peripheral arterial disease (PAD). PAD is a type of atherosclerosis in which plaque builds up in artery walls. Plaque is commonly composed of fatty substances like cholesterol, calcium, and fibrous tissues, which are naturally found in blood. Whereas atherosclerosis can occur in any artery of the body, PAD normally strikes the limbs. In Ted's case, the major arteries supplying oxygenated blood to the legs and pelvis were blocked with plaque causing poor circulation leading to the pain and numbness he was experiencing.

As you read through the chapter, think about the following questions.

1. How does a blood vessel become blocked and what does that do to circulation?
2. What will happen to tissues that do not receive proper circulation?
3. How can a healthy person like Ted develop atherosclerosis?

BEFORE YOU BEGIN

Before beginning this chapter, take a few moments to review the following discussions.

Section 3.3 How do substances cross plasma membranes?

Section 4.1 What are the types of tissues?

Section 4.7 What are the roles of the various organ systems?

5.1 Overview of the Cardiovascular System

Learning Outcomes

Upon completion of this section, you should be able to

1. Identify the two components of the cardiovascular system.
2. Summarize the functions of the cardiovascular system.
3. Explain the purpose of the lymphatic system in circulation.

The cardiovascular system consists of (1) the heart, which pumps blood, and (2) the blood vessels, through which the blood flows. The beating of the heart sends blood into the blood vessels. In humans, blood is always contained within blood vessels.

Circulation Performs Exchanges

Even though circulation of blood depends on the beating of the heart, the purpose of circulation is to serve cells. Remember that cells are surrounded by tissue fluid that is used to exchange substances between the blood and cells. Blood removes waste products from tissue fluid. Blood also brings tissue fluid the oxygen and nutrients cells require to continue their existence. Blood would not be able to continue to perform this function if it did not become oxygenated in the lungs (exchanging carbon dioxide waste for needed oxygen) and cleansed by the kidneys (Fig. 5.1). At the lungs, blood drops off carbon dioxide and picks up oxygen, as indicated by the two arrows in and out of the lungs in Figure 5.1.

Gas exchange is not the only function of blood. Nutrients enter the bloodstream at the intestines and transport much-needed substances to the body's cells. Blood is purified of its wastes at the kidneys, and water and salts are retained as needed. The liver is important, because it takes up amino acids from the blood and returns needed proteins. Liver proteins transport substances such as fats in the blood. The liver also removes **toxins** and chemicals that may have entered the blood at the intestines, and its colonies of white blood cells destroy bacteria and other pathogens. Thousands of miles of blood vessels, which form an intricate circuit reaching almost every cell of the body, move the blood and its contents through the body to and from all the body's organs.

Functions of the Cardiovascular System

In this chapter, we will see that

1. contractions of the heart generate blood pressure, which moves blood through blood vessels;
2. blood vessels transport the blood from the heart into arteries, capillaries, and veins; and blood then returns to the heart so the circuit can be completed;
3. gas exchange (pickup of carbon dioxide waste and drop-off of oxygen for the cells) occurs at the smallest-diameter vessels, the capillaries; and
4. the heart and blood vessels regulate blood flow, according to the needs of the body.

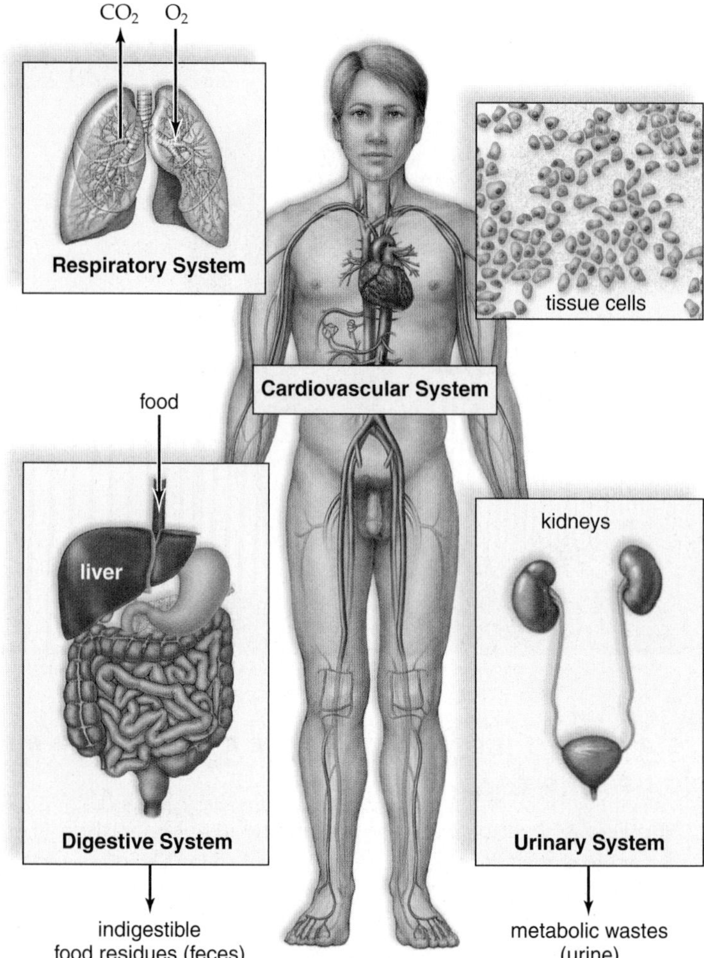

Figure 5.1 The cardiovascular system and homeostasis. The cardiovascular system transports blood throughout the body and, with the help of other systems in the body, it maintains favorable conditions for the cells of the body.

The **lymphatic system** (see chapter 7) assists the cardiovascular system because lymphatic vessels collect excess tissue fluid and return it to the cardiovascular system. When exchanges occur between blood and tissue fluid, fluid collects in the tissues. This excess fluid enters lymphatic vessels, which start in the tissues and end at cardiovascular veins in the shoulders. As soon as fluid enters lymphatic vessels, it is called *lymph.* Lymph, you will recall, is a fluid tissue, as is blood (see Chapter 4).

Connections and Misconceptions

If you must stand all day, why do shoes that are loose in the morning feel so tight by evening?

Capillaries are the small, thin vessels used for tissue gas exchange, but they can also leak due to these characteristics. When you stand for long periods, gravity increases the pressure inside the capillaries. Tissue fluid leaks into tissue spaces—and your shoes get too snug. When you lie down at night, your feet are level with your heart. Excess tissue fluid that caused swelling now filters back into lymphatic vessels, which empty into shoulder veins, and fluid returns to the cardiovascular system. By morning, your feet have lost their excess water, and those shoes are loose again.

Check Your Progress 5.1

1. Detail the two parts of the cardiovascular system.
2. Describe the functions of the cardiovascular system.
3. Explain how the lymphatic system interacts with the cardiovascular system.

Connecting the Concepts

As the primary transport system in the human body, the cardiovascular system interacts with all of the major organ systems. In doing so, it plays an important role in homeostasis. For more information on these interactions, refer to the following discussions.

Section 4.8 investigates the role of the cardiovascular system in homeostasis.

Section 7.2 takes a greater look at how the lymphatic system interacts with the cardiovascular system.

Section 8.3 examines how the cardiovascular system moves nutrients that are absorbed by the small intestine.

Section 10.2 explores the interaction of the cardiovascular system with the urinary system.

5.2 The Types of Blood Vessels

Learning Outcomes

Upon completion of this section, you should be able to

1. Describe the structure and function of the three types of blood vessels.
2. Explain how blood flow is regulated in each of the three types of blood vessels.

The structure of the three types of blood vessels (arteries, capillaries, and veins) is appropriate to their function (Fig. 5.2).

The Arteries: From the Heart

The arterial wall has three layers. The innermost layer is a thin layer of cells called *endothelium.* Endothelium is surrounded by a relatively thick middle layer of smooth muscle and elastic tissue. The artery's outer layer is connective tissue. The strong walls of an artery give it support when blood enters under pressure; the elastic tissue allows an artery to expand to absorb the pressure. **Arterioles** are small arteries barely visible to the naked eye. Whereas the middle layer of arterioles has some elastic tissue, it is composed mostly of smooth muscle. These muscle fibers encircle the arteriole. When the fibers contract, the vessel constricts; when these muscle fibers relax, the vessel dilates. The constriction or dilation of arterioles controls

Figure 5.2 **Structure of a capillary bed.**
A capillary bed is a maze of tiny vessels that lies between an arteriole and a venule. When precapillary sphincter muscles are relaxed, the capillary bed opens and blood flows through the capillaries. Otherwise, blood flows through a shunt directly from the arteriole to the venule. As blood passes through a capillary in the tissues, it gives up its oxygen (O_2). Therefore, blood goes from being O_2-rich in the arteriole to being O_2-poor in the vein. (Note: Blood vessels are colored red if they carry O_2-rich blood and blue if they carry O_2-poor blood. Gas exchange is reversed in the pulmonary circuit, as described in Section 5.5.)

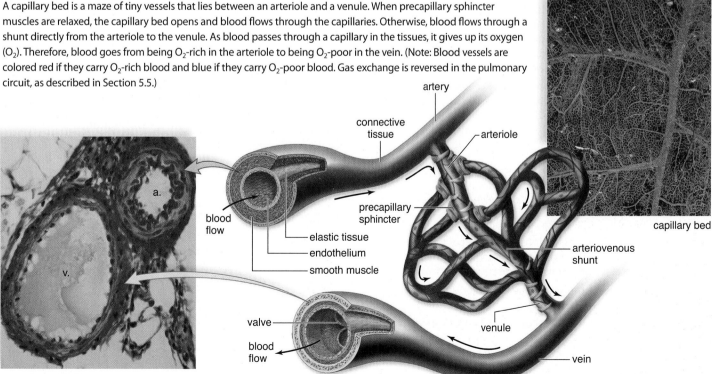

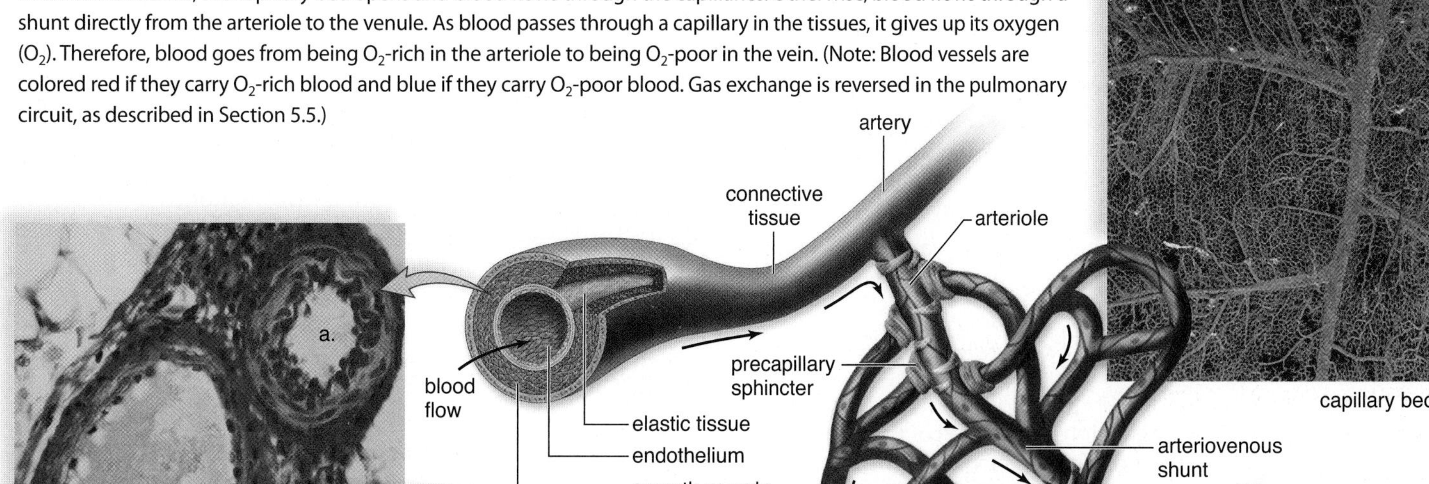

blood pressure. When arterioles constrict, blood pressure rises. Dilation of arterioles causes blood pressure to fall.

The Capillaries: Exchange

Arterioles branch into capillaries. Each capillary is an extremely narrow, microscopic tube with a wall composed only of endothelium. Capillary endothelium is formed by a single layer of epithelial cells with a basement membrane. Although capillaries are small, their total surface area in humans is about 6,300 square meters. Capillary beds (networks of many capillaries) are present in all regions of the body, so no cell is far from a capillary and thus not far from gas exchange with blood. In the tissues, only certain capillaries are open at any given time. For example, after eating, the capillaries supplying the digestive system are open, whereas most serving the muscles are closed. Rings of muscle called **precapillary sphincters** control the blood flow through a capillary bed (see Fig. 5.2). Constriction of the sphincters closes the capillary bed. When a capillary bed is closed, the blood moves to an area where gas exchange is needed, going directly from arteriole to venule through a pathway called an **arteriovenous shunt.**

Animation Fluid Exchange Across the Walls of Capillaries

The Veins: To the Heart

Venules are small veins that drain blood from the capillaries and then join to form a vein. The walls of venules (and veins) have the same three layers as arteries. However, there is less smooth muscle in the middle layer of a vein and less connective tissue in the outer layer. Therefore, the wall of a vein is thinner than that of an artery.

Veins often have **valves,** which allow blood to flow only toward the heart when open and prevent backward flow of blood when closed. Valves are extensions of the inner-wall layer and are found in the veins that carry blood against the force of gravity, especially the veins of the lower extremities. The walls of veins are thinner, so they can expand to a greater extent. At any one time, about 70% of the blood is in the veins. In this way, the veins act as a blood reservoir. If blood is lost due to hemorrhaging, nervous stimulation causes the veins to constrict, providing more blood to the rest of the body.

Connecting the Concepts

Blood vessels act as conduits for the movement of blood, and in doing so, blood vessels interact with all of the tissues and cells of the body. For more information on blood and blood vessels, refer to the following discussions.

Section 6.1 examines the composition of blood and the function of its various components.

Figure 9.11 illustrates the major arteries and veins involved in moving gases within the body.

Figure 10.6 demonstrates how the capillaries in the kidney assist in the filtration of the blood.

Check Your Progress 5.2

1. List and detail the different types of blood vessels.
2. Describe the differences in function of all the blood vessels.
3. Hypothesize what areas of the body would have more precapillary sphincters than others and why.

5.3 The Heart Is a Double Pump

Learning Outcomes

Upon completion of this section, you should be able to

1. Identify the structures and chambers of the human heart.
2. Diagram the flow of blood through the human heart.
3. Explain the internal and external controls of the heartbeat.

The **heart** is a cone-shaped, muscular organ located between the lungs, directly behind the sternum (breastbone). The heart is tilted so that the apex (the pointed end) is oriented to the left (Fig. 5.3). To approximate the size of your heart, make a fist, then clasp the fist with your opposite hand. The major portion of the heart is the interior wall of tissue called the **myocardium,** consisting largely of cardiac muscle tissue. The muscle fibers of myocardium are branched. Each fiber is tightly joined to neighboring fibers by structures called *intercalated disks* (Fig. 5.4*b*). The intercalated disks also include cell junctions like gap junctions and desmosomes. **Gap junctions** are used to aid in simultaneous contractions of the cardiac fibers. **Desmosomes** include arrangements of protein fibers that tightly hold the membranes of adjacent cells together and prevent overstretching. The heart is surrounded by the **pericardium,** a thick, membranous sac that supports and protects the heart. The inside of the pericardium secretes pericardial fluid (a lubrication fluid), and the pericardium slides smoothly over the heart's surface as it pumps the blood.

Internally, a wall called the **septum** separates the heart into a right side and a left side (Fig. 5.4*a*). The heart has four chambers. The two upper, thin-walled atria (sing., **atrium**) are called the right atrium and the left atrium. Each atrium has a wrinkled, earlike flap on the outer surface called an *auricle.* The two lower chambers are the thick-walled **ventricles,** called the right ventricle and the left ventricle (see Fig. 5.4*a*).

Heart valves keep blood flowing in the right direction and prevent its backward movement. The valves that lie between the atria and the ventricles are called the **atrioventricular (AV) valves.** These valves are supported by strong fibrous strings called **chordae tendineae.** The chordae tendineae are attached to papillary muscles that project from the ventricular walls. Chordae anchor the valves, preventing them from inverting when the heart contracts.

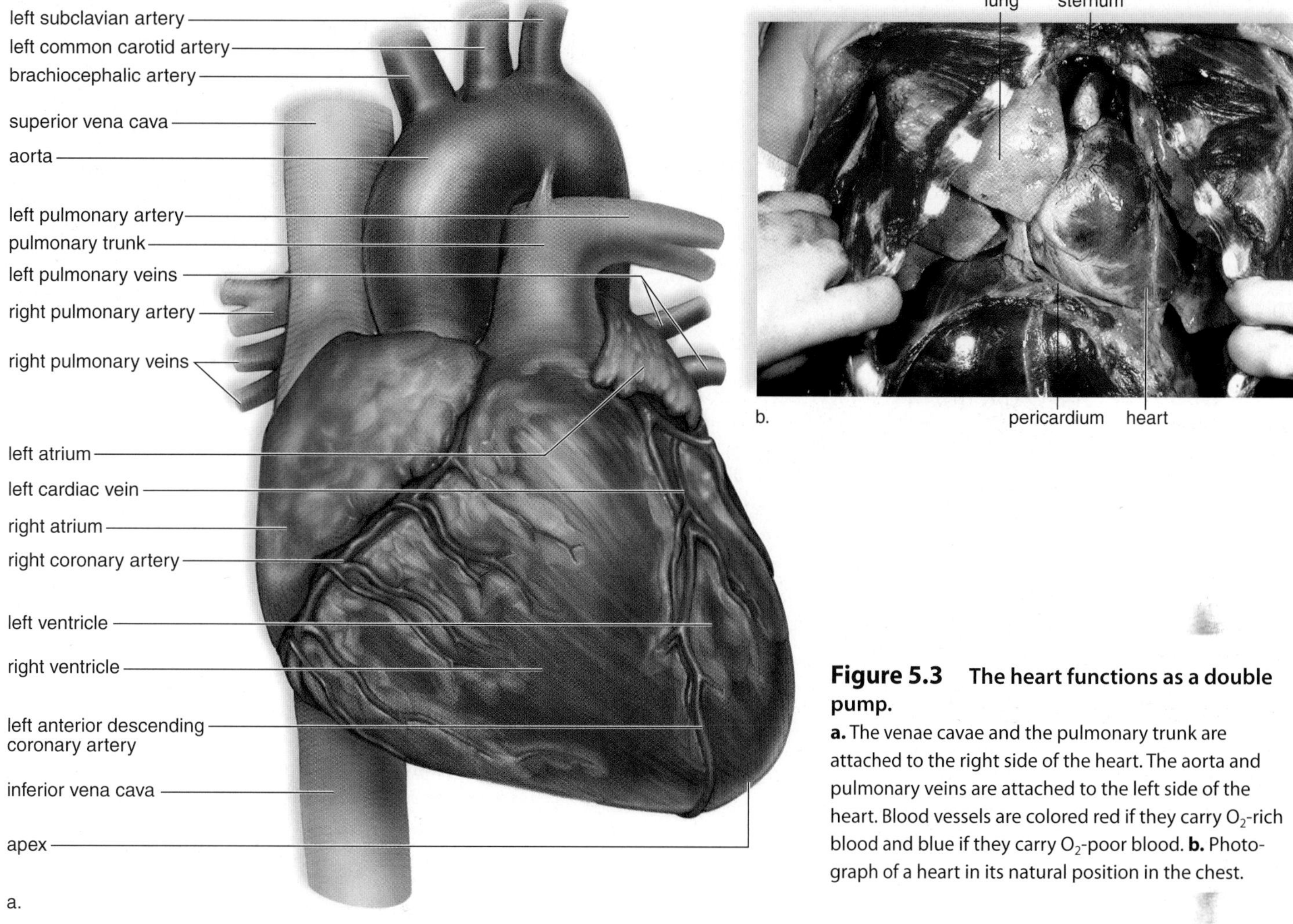

Figure 5.3 The heart functions as a double pump.
a. The venae cavae and the pulmonary trunk are attached to the right side of the heart. The aorta and pulmonary veins are attached to the left side of the heart. Blood vessels are colored red if they carry O_2-rich blood and blue if they carry O_2-poor blood. **b.** Photograph of a heart in its natural position in the chest.

The AV valve on the right side is called the *tricuspid* valve because it has three flaps, or cusps. The AV valve on the left side is called the *bicuspid* valve because it has two flaps. The bicuspid valve is commonly referred to as the *mitral* valve, because it has a shape like a bishop's hat, or miter. The remaining two valves are the **semilunar valves,** with flaps shaped like half-moons. These valves lie between the ventricles and their attached vessels. The semilunar valves are named for their attached vessels: The pulmonary semilunar valve lies between the right ventricle and the pulmonary trunk. The aortic semilunar valve lies between the left ventricle and the aorta.

Coronary Circulation: The Heart's Blood Supply

The myocardium—the middle, muscular layer of the three layers of the walls of the heart—receives oxygen and nutrients from the coronary arteries. Likewise, wastes are removed by the cardiac veins. The blood that flows through the heart contributes little to either nutrient supply or waste removal.

The **coronary arteries** (see Fig. 5.3) serve the heart muscle itself. These arteries are the first branches off the aorta. They originate just above the aortic semilunar valve. They lie on the exterior surface of the heart, where they divide into diverse arterioles. The coronary capillary beds join to form venules which converge to form the cardiac veins, which empty into the right atrium. Because they have a very small diameter, they can become easily clogged leading to coronary artery disease (see Case Study).

If these vessels become completely clogged, oxygen and nutrients such as glucose, will not reach the muscles of the heart. This may result in a *myocardial infarction*, or heart attack. Coronary artery disease may be treated with medication, coronary bypass surgery (see page 108), or angioplasty (page 108).

Animation
Heart Stem Cells

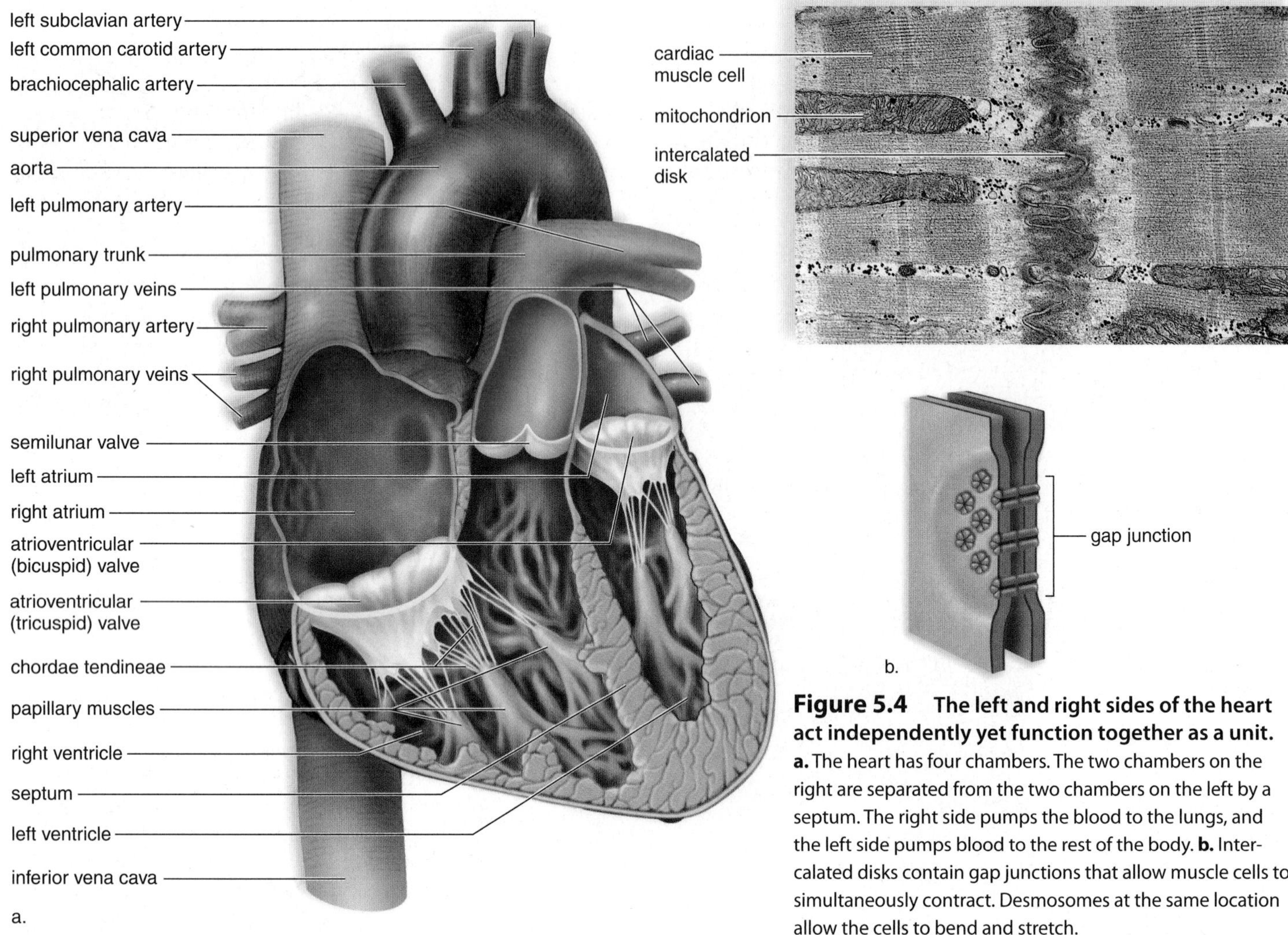

Figure 5.4 The left and right sides of the heart act independently yet function together as a unit. **a.** The heart has four chambers. The two chambers on the right are separated from the two chambers on the left by a septum. The right side pumps the blood to the lungs, and the left side pumps blood to the rest of the body. **b.** Intercalated disks contain gap junctions that allow muscle cells to simultaneously contract. Desmosomes at the same location allow the cells to bend and stretch.

Passage of Blood Through the Heart

Recall that intercalated disks (see Fig. 5.4*b*) join fibers of cardiac muscle cells, allowing them to communicate with each other. By sending electrical signals between cells, both atria and then both ventricles contract simultaneously. We can trace the path of blood through the heart and body in the following manner.

- The superior vena cava and the inferior vena cava (see Fig. 5.4*a*) carry oxygen-poor blood from body veins to the right atrium.
- The right atrium contracts, sending blood through an atrioventricular valve (the tricuspid valve) to the right ventricle. (Remember, the left atrium contracts at the same time.)
- The right ventricle contracts, pumping blood through the pulmonary semilunar valve into the pulmonary trunk. The pulmonary trunk, which carries oxygen-poor blood, divides into two **pulmonary arteries,** which go to the lungs.
- Pulmonary capillaries within the lungs allow gas exchange. Oxygen enters the blood; carbon dioxide waste is excreted from the blood.
- Four **pulmonary veins,** which carry oxygen-rich blood, enter the left atrium.
- The left atrium pumps blood through an atrioventricular valve (the bicuspid [mitral] valve) to the left ventricle.
- The left ventricle contracts, sending blood through the aortic semilunar valve into the aorta. (The right ventricle is contracting at the same time.)
- Large arteries, smaller arteries, and arterioles supply tissue capillaries. Tissue capillaries drain into increasingly larger veins. Veins drain into the superior and inferior vena cava, and the cycle starts again.

Animation Cardiac Cycle

From this description, you can see that oxygen-poor blood never mixes with oxygen-rich blood and that blood must go through the lungs to pass from the right side to the left side of the heart. The heart is a double pump because the right ventricle of the heart sends blood through the lungs, and the left ventricle sends blood throughout the body. Thus, the left ventricle has the harder job of pumping blood.

The atria have thin walls, and each pumps blood into the ventricle right below it. The ventricles are thicker, and they pump blood into arteries (pulmonary artery and aorta) that travel to other parts of the body. The thinner myocardium of the right ventricle pumps blood to the lungs nearby in the thoracic cavity. The left ventricle has a thicker wall with more cardiac muscle cells than the right ventricle, and this enables it to pump blood out of the heart with enough force to send it through the body.

The pumping of the heart sends blood under pressure out into the arteries. The left side of the heart pumps with greater force, so blood pressure is greatest in the aorta. Blood pressure then decreases as the total cross-sectional area of arteries and then arterioles increases (see Fig. 5.8). A different mechanism, aside from blood pressure, is used to move blood in the veins.

The Heartbeat Is Controlled

Each heartbeat is called a **cardiac cycle** (Fig. 5.5). Recall that when the heart beats, first the two atria contract at the same time. Next, the two ventricles contract at the same time. Then, all chambers relax. **Systole,** the working phase, refers to contraction of the chambers, and **diastole,** the resting phase, refers to relaxation of the chambers (Fig. 5.5). The heart contracts, or beats, about 70 times a minute on average in a healthy adult, with each heartbeat lasting about 0.85 second with a normal resting rate varying from 60 to 80 beats per minute.

There are two audible heartbeat sounds referred to as "lub-dup." The first sound, "lub," occurs when increasing pressure of blood inside a ventricle forces the cusps of the AV valves to slam shut (Fig. 5.5*b*). In contrast, the pressure of blood inside a ventricle causes the semilunar valves (pulmonary and aortic) to open. The "dup" occurs when the ventricles relax, and blood in the arteries flows backward momentarily, causing the semilunar valves to close (Fig. 5.5*c*). A heart murmur, or a slight swishing sound after the "lub," is often due to leaky valves, which allow blood to pass back into the atria after the AV valves have closed. Faulty valves can be surgically corrected.

MP3
Cardiac Cycle

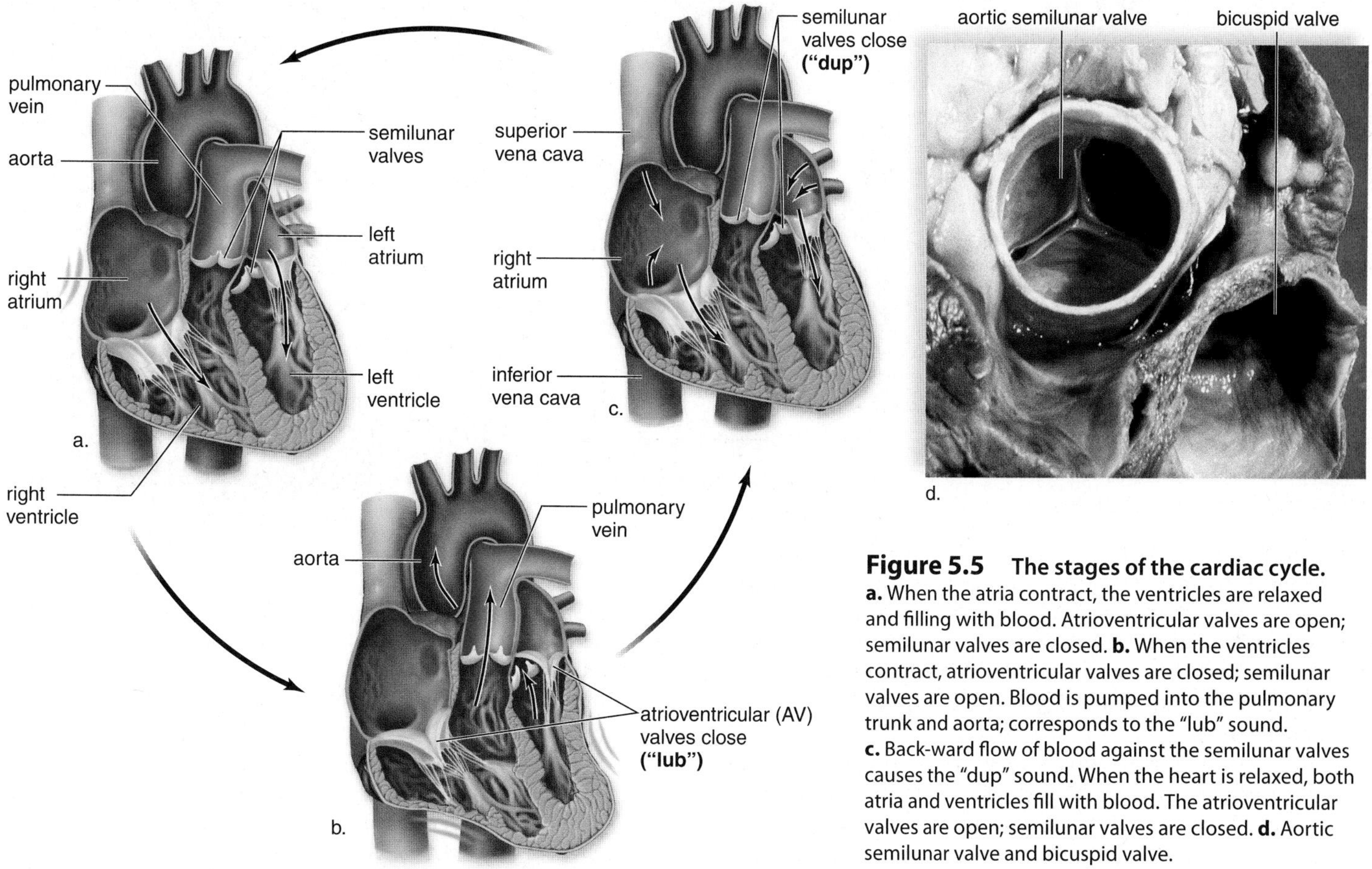

pulmonary vein
aorta
right atrium
right ventricle
semilunar valves
left atrium
left ventricle
a.
superior vena cava
right atrium
inferior vena cava
c.
semilunar valves close **("dup")**
aortic semilunar valve
bicuspid valve
d.
aorta
pulmonary vein
atrioventricular (AV) valves close **("lub")**
b.

Figure 5.5 The stages of the cardiac cycle.
a. When the atria contract, the ventricles are relaxed and filling with blood. Atrioventricular valves are open; semilunar valves are closed. **b.** When the ventricles contract, atrioventricular valves are closed; semilunar valves are open. Blood is pumped into the pulmonary trunk and aorta; corresponds to the "lub" sound. **c.** Back-ward flow of blood against the semilunar valves causes the "dup" sound. When the heart is relaxed, both atria and ventricles fill with blood. The atrioventricular valves are open; semilunar valves are closed. **d.** Aortic semilunar valve and bicuspid valve.

Internal Control of Heartbeat

The rhythmic contraction of the atria and ventricles is due to the internal (intrinsic) conduction system of the heart. Nodal tissue is a unique type of cardiac muscle located in two regions of the heart. Nodal tissue has both muscular and nervous characteristics. The **SA (sinoatrial) node** is located in the upper dorsal wall of the right atrium. The **AV (atrioventricular) node** is located in the base of the right atrium very near the septum (Fig. 5.6*a*). The SA node initiates the heartbeat and automatically sends out an excitation signal every 0.85 second. This causes the atria to contract. When signal impulses reach the AV node, there is a slight delay that allows the atria to finish their contraction before the ventricles begin their contraction.

The signal for the ventricles to contract travels from the AV node through the two branches of the **atrioventricular (AV) bundle** before reaching the numerous and smaller **Purkinje fibers.** The AV bundle, its branches, and the Purkinje fibers work efficiently because gap junctions (tiny channels built into intercalated disks) allow electrical current to flow from cell to cell (see Fig. 5.4*b*).The SA node is called the **pacemaker** because it regulates heartbeat. If the SA node fails to work properly, the heart still beats due to signals generated by the AV node. But the beat is slower (40 to 60 beats per minute). To correct this condition, it is possible to implant an artificial pacemaker, which automatically gives an electrical stimulus to the heart every 0.85 second.

Animation
Conducting System of Heart

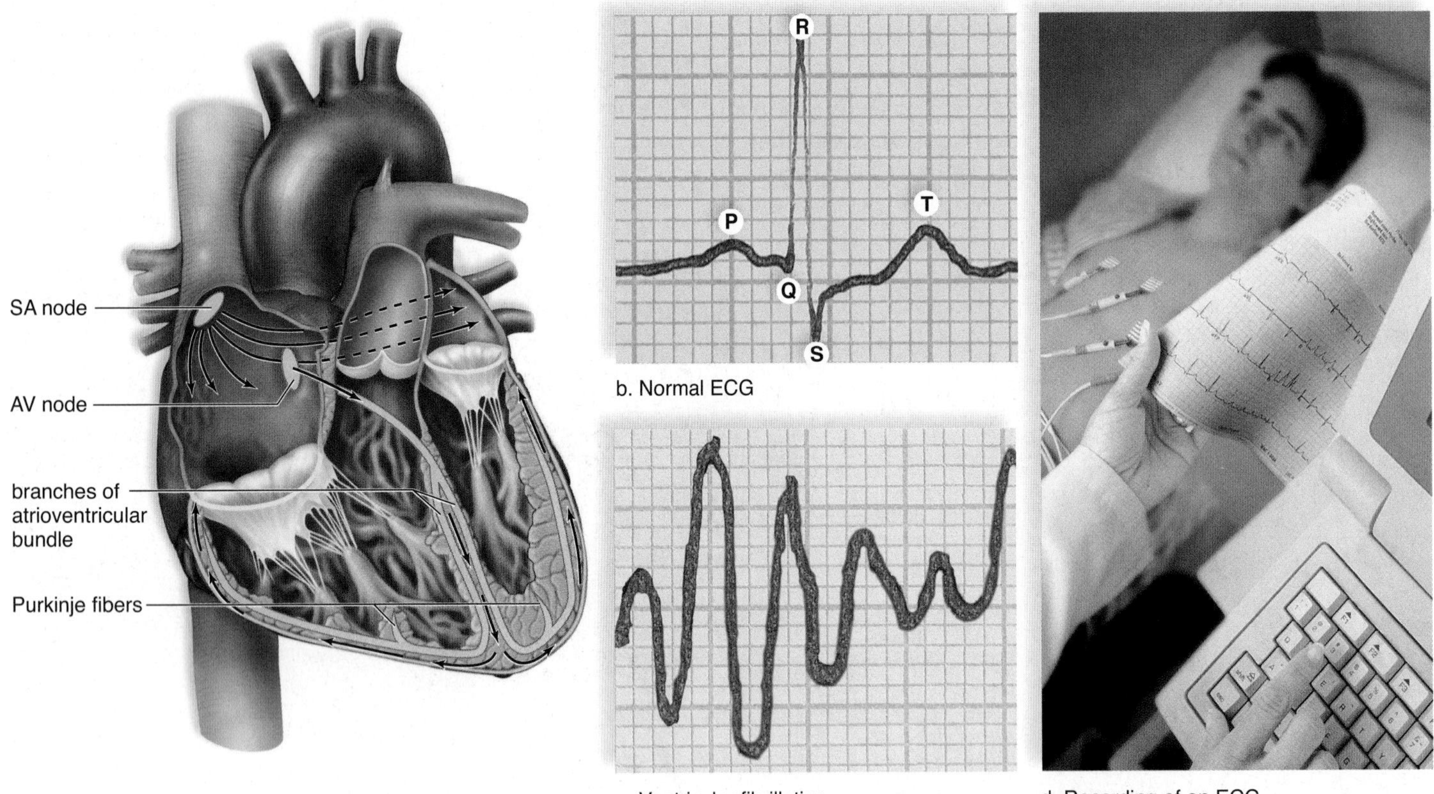

Figure 5.6 An electrical signal pathway through the heart.
a. The SA node sends out a stimulus (black arrows), which causes the atria to contract. When this stimulus reaches the AV node, it signals the ventricles to contract. The electrical signal passes down the two branches of the atrioventricular bundle to the Purkinje fibers. Thereafter, the ventricles contract. **b.** A normal ECG indicates that the heart is functioning properly. The P wave occurs just prior to atrial contraction; the QRS complex occurs just prior to ventricular contraction; and the T wave occurs when the ventricles are recovering from contraction. **c.** Ventricular fibrillation produces an irregular ECG due to irregular stimulation of the ventricles. **d.** The recording of an ECG.

Connections and Misconceptions

Automatic external defibrillators (AEDs) can be found in airports and other public places. These are for heart emergencies, such as heart attacks. Should they be used only by a paramedic?

When an emergency happens in a public place like an airport or shopping mall, anyone can use an AED. If a person collapses, perhaps suffering from ventricular fibrillation, the computerized device will make the decisions. It will explain, step-by-step, how to first check for breathing and pulse. Next, it will explain how to apply pads to the chest to deliver the shock, if necessary.

Once the chest pads are attached, the computer analyzes heart activity to determine if a shock is needed. Strong electrical current is applied for a short time to defibrillate the heart. The rescuer moves back, pushing a button when prompted. Voice instructions also explain how to do cardiopulmonary resuscitation (CPR) until paramedics arrive.

However, it's a good idea to first be familiar with CPR and AED use. The Red Cross and many hospitals regularly offer introductory and refresher classes. With training, you might be able to save a person's life!

External Control of Heartbeat

The body has an external (extrinsic) way to regulate the heartbeat. A cardiac control center in the medulla oblongata, a portion of the brain that controls internal organs, can alter the beat of the heart by way of the parasympathetic and sympathetic portions of the nervous system. As studied in Chapter 1, the parasympathetic division promotes those functions associated with a resting state. The sympathetic division brings about those responses associated with fight or flight. It makes sense that the parasympathetic division decreases SA and AV nodal activity when we are inactive. By contrast, the sympathetic division increases SA and AV nodal activity when we are active or excited.

The hormones epinephrine and norepinephrine, released by the adrenal medulla, also stimulate the heart. During exercise, for example, the heart pumps faster and stronger due to sympathetic stimulation and the release of epinephrine and norepinephrine.

Animation
Baroceptor Control of Heartbeat

The Electrocardiogram Is a Record of the Heartbeat

An **electrocardiogram (ECG)** is a recording of the electrical changes that occur in myocardium during a cardiac cycle. Body fluids contain ions that conduct electrical currents. Therefore, the electrical changes in myocardium can be detected on the skin's surface. When an ECG is administered, electrodes placed on the skin are connected by wires to an instrument that detects the myocardium's electrical changes. Thereafter, a pen rises or falls on a moving strip of paper. Figure 5.6*b* depicts the pen's movements during a normal cardiac cycle.

When the SA node triggers an impulse, the atrial fibers produce an electrical change called the P wave. The P wave indicates the atria are about to contract. After that, the QRS complex signals that the ventricles are about to contract. The electrical changes that occur as the ventricular muscle fibers recover produce the T wave.

Various types of abnormalities can by detected by an ECG. One of these, called *ventricular fibrillation,* is caused by uncoordinated, irregular electrical activity in the ventricles. Compare Figure 5.6*b* with Figure 5.6*c* and note the irregular line illustrated in 5.6*c*. Ventricular fibrillation is of special interest because it can be caused by an injury, heart attack, or drug overdose. It is the most common cause of sudden cardiac death in a seemingly healthy person over age 35. Once the ventricles are fibrillating, coordinated pumping of the heart ceases and body tissues quickly become oxygen starved. Normal electrical conduction must be reestablished as quickly as possible or the person will die. A strong electrical current is applied to the chest for a short time in a process called *defibrillation.* In response, all heart cells discharge their electricity at once. Then, the SA node may be able to reestablish a coordinated beat.

MP3
Cardiac Conduction System

Connecting the Concepts

For more information on the heart as a muscle and to gain a greater understanding of how our heart develops and ages, refer to the following discussions.

Figure 12.1 compares the structure of cardiac muscle to other muscle tissues.

Figure 17.7 illustrates the heart of a developing fetus.

Section 17.5 explores some of the changes to the cardiovascular system as we age.

Check Your Progress 5.3

1. Illustrate the flow of blood through the heart.
2. Explain what causes the "lub" and the "dup" sounds of a heartbeat.
3. Summarize the internal and external controls of the heartbeat.

5.4 Features of the Cardiovascular System

Learning Outcomes

Upon completion of this section, you should be able to

1. Explain how blood pressure differs in veins, arteries, and capillaries.
2. Distinguish between systolic and diastolic pressure.
3. Analyze systolic and diastolic pressures to assess the cardiovascular health of an individual.

When the left ventricle contracts, blood is sent out into the aorta under pressure. A progressive decrease in pressure occurs as blood moves through the arteries, arterioles, capillaries, venules, and finally the veins. Blood pressure is highest in the aorta. By contrast, pressure is lowest in the superior and inferior venae cavae, which enter the right atrium (Fig. 5.4*a*).

Pulse Rate Equals Heart Rate

The surge of blood entering the arteries causes their elastic walls to stretch, but then they almost immediately recoil. This rhythmic expansion and recoil of an arterial wall can be felt as a **pulse** in any artery that runs close to the body's surface. It is customary to feel the pulse by placing several fingers on either the radial artery (near the outer border of the palm side of the wrist) or the carotid artery (located on either side of the trachea in the neck).

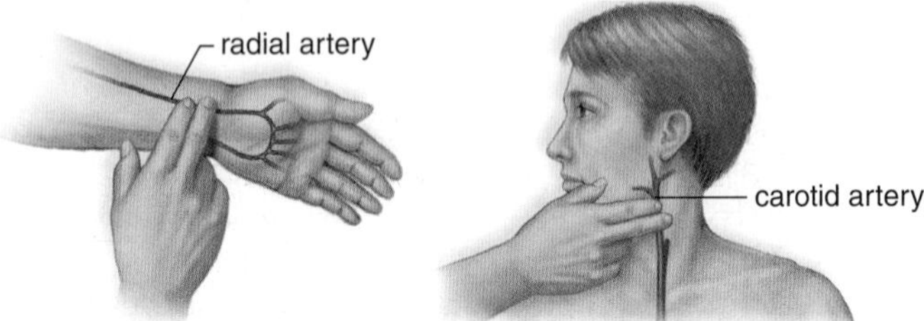

Normally, the pulse rate indicates the heart rate because the arterial walls pulse whenever the left ventricle contracts. The pulse rate is usually 70 beats per minute in a healthy adult but can vary between 60 and 80 beats per minute.

Blood Flow Is Regulated

The beating of the heart is necessary to homeostasis because it creates the pressure that propels blood in the arteries and the arterioles. Arterioles lead to the capillaries where exchange with tissue fluid takes place.

Blood Pressure Moves Blood in Arteries

Blood pressure is the pressure of blood against the wall of a blood vessel. A sphygmomanometer (blood pressure instrument) can be used to measure blood pressure, usually in the brachial artery of the arm (Fig. 5.7). The highest arterial pressure, called the **systolic pressure,** is reached during ejection of blood from the heart. The lowest arterial pressure, called the **diastolic pressure,** occurs while the heart ventricles are relaxing. Blood pressure is measured in millimeters mercury (mm Hg). Normal resting blood pressure for a young adult should be slightly lower than 120 mm Hg over 80 mm Hg, or 120/80, but these values can vary somewhat and still be within the range of normal blood pressure (Table 5.1). The number 120 represents the systolic pressure, and 80 represents the diastolic pressure. High blood pressure is called *hypertension,* and low blood pressure is called *hypotension.*

Both systolic and diastolic blood pressure decrease with distance from the left ventricle because the total cross-sectional area of the blood vessels increases—there are more arterioles than arteries. The decrease in blood pressure causes the blood velocity to gradually decrease as it flows toward the capillaries.

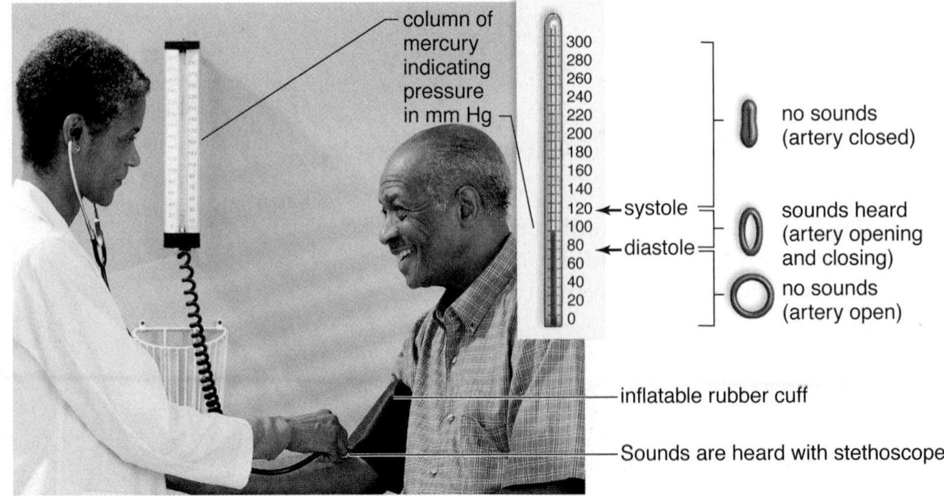

Figure 5.7 Using a sphygmomanometer to determine blood pressure. The sphygmomanometer cuff is first inflated until the artery is completely closed by its pressure. Next, the pressure is gradually reduced. The clinician listens with a stethoscope for the first sound, indicating that blood is moving past the cuff in an artery. This is systolic blood pressure. The pressure in the cuff is further reduced until no sound is heard, indicating that blood is flowing freely through the artery. This is diastolic pressure. The procedure can be done manually or by computerized sphygmomanometers.

Table 5.1 Normal Values for Adult Blood Pressure

	Top Number (Systolic)	Bottom Number (Diastolic)
Hypotension	Less than 95	Less than 50
Normal	Below 120	Below 80
Prehypertension	120–139	80–89
Stage 1 hypertension*	140–159	90–99
Stage 2 hypertension*	160 or more	100 or more

*Patient will need medication and must maintain a healthy lifestyle.

Blood Flow Is Slow in the Capillaries

There are many more capillaries than arterioles, and blood moves slowly through the capillaries (Fig. 5.8). This is important because the slow progress allows time for the exchange of substances between the blood in the capillaries and the surrounding tissues. Any needed changes in flow rate are adjusted by the opening and closing of the precapillary sphincters.

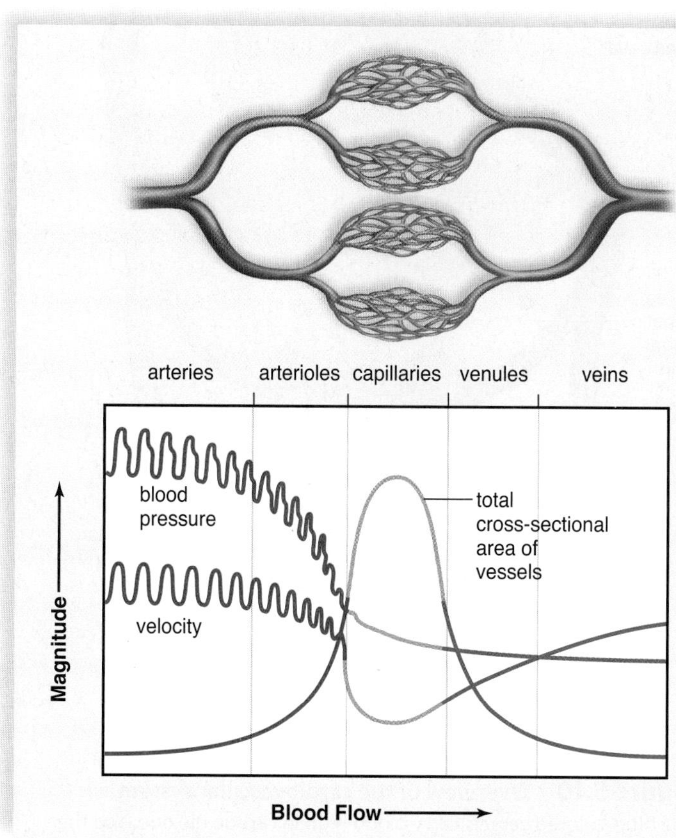

Figure 5.8 **Cross-sectional areas of blood vessels.**
In capillaries, blood is under minimal pressure and has the least velocity. Blood pressure and blood velocity drop off because capillaries have a greater total cross-sectional area than arterioles.

Blood Flow in Veins Returns Blood to Heart

By studying Figure 5.8, it can be seen that the velocity of blood flow increases from capillaries to veins. As an analogy, imagine a single narrow street emptying its cars into a fast multilane highway. Likewise, as capillaries empty into veins, blood can travel faster. However, it's also apparent from Figure 5.8 that blood pressure is minimal in venules and veins. Blood pressure thus plays only a small role in returning venous blood to the heart. Venous return is dependent upon three additional factors:

1. the **skeletal muscle pump,** dependent on skeletal muscle contraction;
2. the **respiratory pump,** dependent on breathing; and
3. valves in veins.

The skeletal muscle pump functions every time a muscle contracts. When skeletal muscles contract, they compress the weak walls of the veins. This causes the blood to move past a valve (Fig. 5.9). Once past the valve, blood cannot flow backward. The importance of the skeletal muscle pump in moving blood in the veins can be demonstrated by forcing a person to stand rigidly still for an hour or so. Fainting may occur because lack of muscle contraction causes blood to collect in the limbs. Poor venous return deprives the brain of needed oxygen. In this case, fainting is beneficial because the resulting horizontal position aids in getting blood to the head.

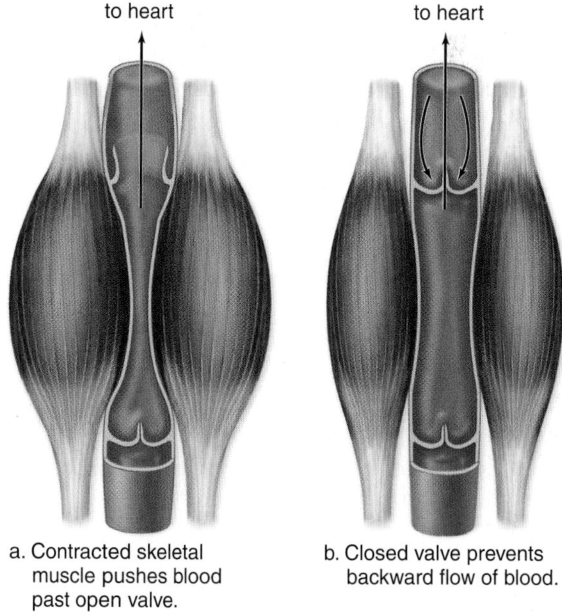

Figure 5.9 **The skeletal muscle pump.**
a. Pressure on the walls of a vein, exerted by skeletal muscles, increases blood pressure within the vein and forces a valve open. **b.** When external pressure is no longer applied to the vein, blood pressure decreases, and back pressure forces the valve closed. Closure of the valves prevents the blood from flowing backward. Therefore, veins return blood to the heart.

Connections and Misconceptions

What causes varicose veins to develop? Why do some pregnant women develop them and others do not?

Veins are thin-walled tubes divided into many separate chambers by vein valves. Excessive stretching occurs if veins are overfilled with blood. For example, if a person stands in one place for a long time, leg veins can't drain properly and blood pools in them. As the vein expands, vein valves become distended and fail to function. These two mechanisms cause the veins to bulge and be visible on the skin's surface. Obesity, sedentary lifestyle, female gender, genetic predisposition, and increasing age are risk factors for varicose veins. Hemorrhoids are varicose veins in the rectum.

Pregnancy in particular can cause leg veins to become distended, because the fetus in the abdomen compresses the large abdominal veins. Once again, leg veins can't drain properly and the valves malfunction. Women who don't develop varicose veins during pregnancy may just be genetically "lucky."

Most varicose veins are only a cosmetic problem. Occasionally they will cause pain, muscle cramps, or itching and become a medical problem. There are safe and effective ways to get rid of them, so see your doctor.

The respiratory pump works much like an eyedropper. When the dropper's suction bulb is released, the bulb expands. Fluid travels from higher pressure in the bottom of glass tube, moving up the tube toward the lower pressure at the top. When we inhale, the chest expands, and this reduces pressure in the thoracic cavity. Blood will flow from higher pressure (in the abdominal cavity) to lower pressure (in the thoracic cavity). When we exhale, the pressure reverses, but again the valves in the veins prevent backward flow.

MP3 Blood Flow & Blood Pressure

Connecting the Concepts

The maintenance of blood pressure is an important aspect of homeostasis. Blood pressure is influenced by a variety of factors, including diet and internal controls. For examples, refer to the following discussions.

Section 8.6 explores how lipids and minerals—namely, sodium—in the diet influence blood pressure.

Section 10.4 explains how the water–salt balance and an enzyme called *renin* influence blood pressure.

Check Your Progress 5.4

1. Explain what the pulse rate of a person indicates.
2. Compare and contrast the pressure of blood flow in the veins, arteries, and capillaries.
3. Explain why valves are needed in the veins.

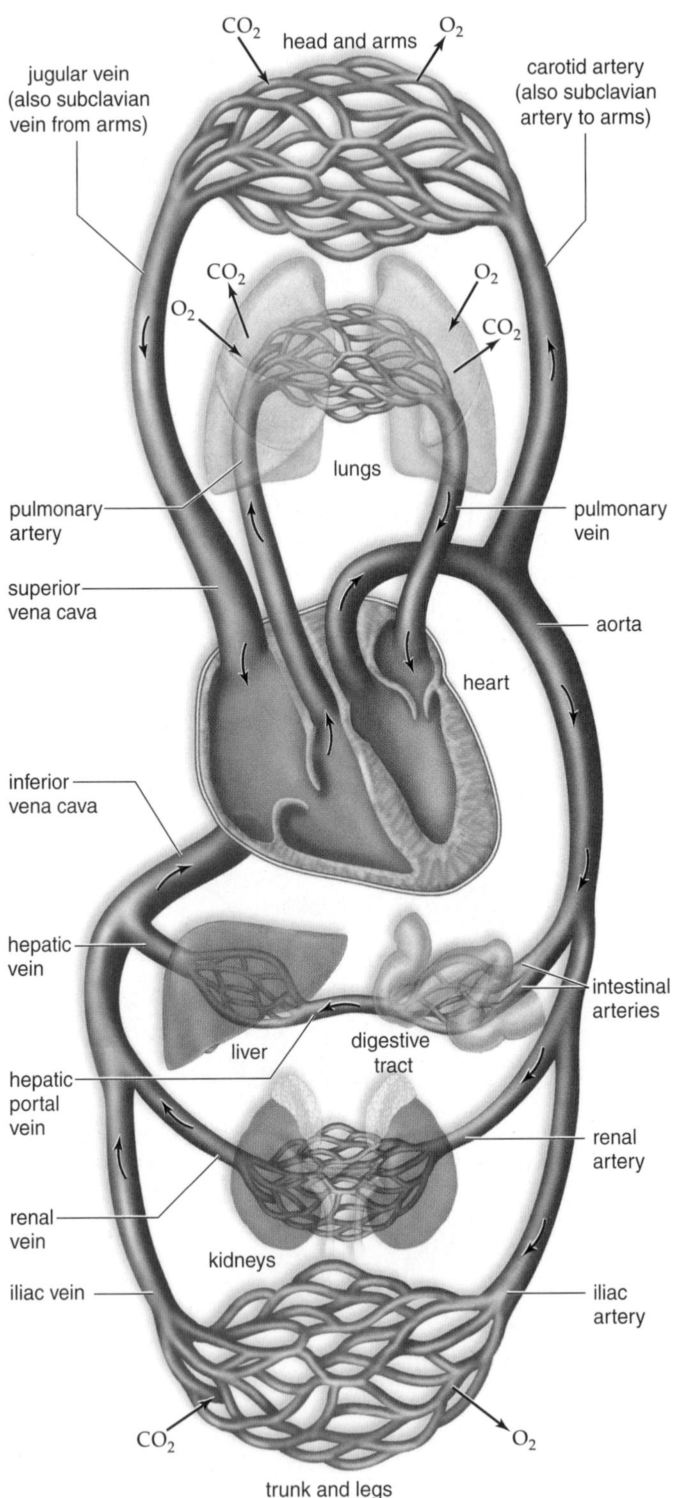

Figure 5.10 Overview of the cardiovascular system.
The blue-colored vessels carry blood high in carbon dioxide, and the red-colored vessels carry blood high in oxygen; the arrows indicate the flow of blood. Compare this diagram, useful for learning to trace the path of blood, with Figure 5.11 to realize that arteries and veins go to all parts of the body. Also, there are capillaries in all parts of the body. No cell is located far from a capillary.

5.5 Two Cardiovascular Pathways

Learning Outcomes

Upon completion of this section, you should be able to

1. Compare blood flow in the pulmonary and systemic circuits.
2. Identify the major arteries and veins of both the pulmonary and systemic circuits.
3. Compare the oxygen content of the blood in the arteries and veins of the pulmonary and systemic circuits.
4. Explain the location and purpose of the hepatic portal system.

The blood flows in two circuits: the **pulmonary circuit,** which circulates blood through the lungs, and the **systemic circuit,** which serves the needs of body tissues (Fig. 5.10). As we shall see, both circuits are necessary to homeostasis.

The Pulmonary Circuit: Exchange of Gases

The path of blood through the lungs can be traced as follows: Blood from all regions of the body first collects in the right atrium and then passes into the right ventricle, which pumps it into the pulmonary trunk. The pulmonary trunk divides into the right and left pulmonary arteries, which branch as they approach the lungs. The arterioles take blood to the pulmonary capillaries, where carbon dioxide is given off and oxygen is picked up. Blood then passes through the pulmonary venules, which lead to the four pulmonary veins that enter the left atrium. Blood in the pulmonary arteries is oxygen-poor but blood in the pulmonary veins is oxygen-rich, so it is not correct to say that all arteries carry blood high in oxygen and all veins carry blood low in oxygen (as people tend to believe). Just the reverse is true in the pulmonary circuit.

The Systemic Circuit: Exchanges with Tissue Fluid

The systemic circuit includes all of the arteries and veins shown in Figure 5.10. (For simplicity, some blood vessels are not shown.) The heart pumps blood through 60,000 miles of blood vessels to deliver nutrients and oxygen and remove wastes from all body cells.

The largest artery in the systemic circuit, the **aorta,** receives blood from the heart; the largest veins, the **superior** and **inferior venae cavae,** return blood to the heart. The superior vena cava collects blood from the head, the chest, and the arms, and the inferior vena cava collects blood from the lower body regions. Both enter the right atrium.

Tracing the Path of Blood

It's easy to trace the path of blood in the systemic circuit by beginning with the left ventricle, which pumps blood into the aorta. Branches from the aorta go to the organs and major body regions. For example, this is the path of blood to and from the lower legs:

left ventricle—aorta—common iliac artery—
femoral artery—lower leg capillaries—
lower leg veins--femoral vein—common iliac vein—
inferior vena cava—right atrium

When tracing blood, mention the aorta, the proper branch of the aorta, the region, and the vein returning blood to the vena cava. In many instances, the artery and the vein that serve the same region are given the same name (Fig. 5.11). What happens in between the artery and the vein? Arterioles from the artery branch into capillaries, where exchange takes place, and then venules join into the vein that enters a vena cava.

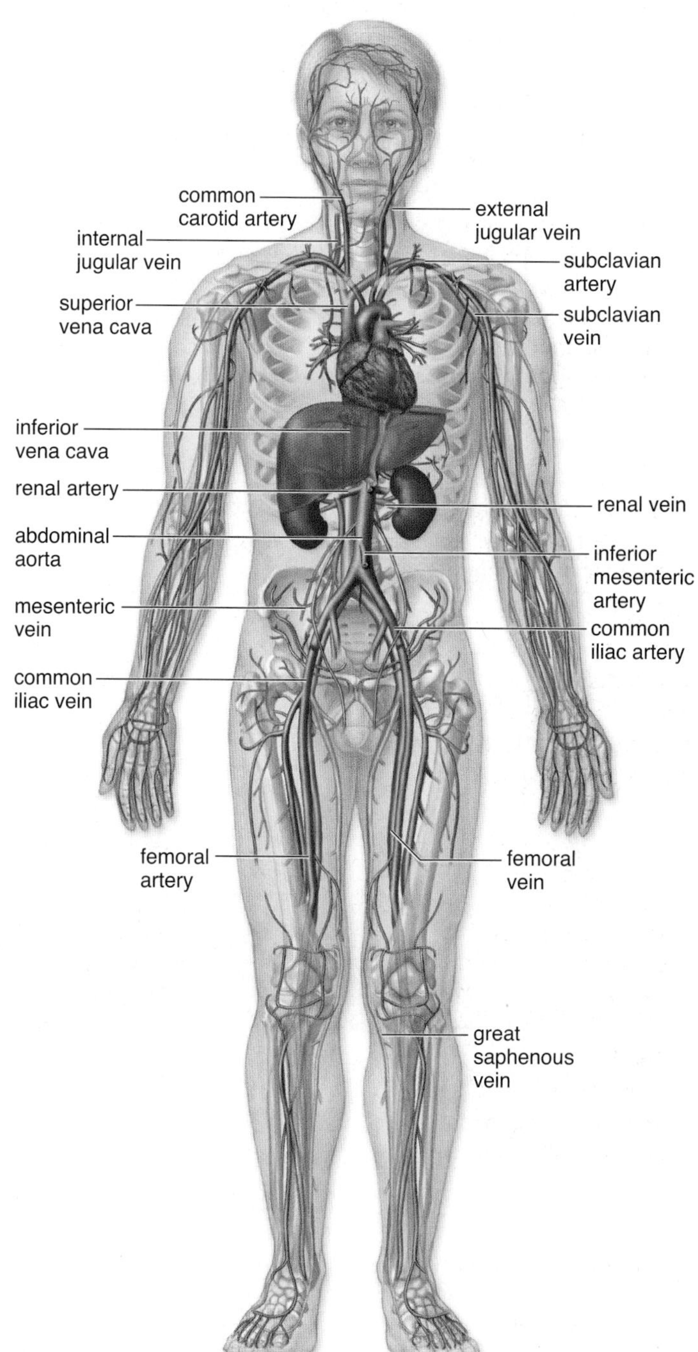

Figure 5.11 The major arteries and veins of the systemic circuit. This illustration offers a realistic representation of major blood vessels (arteries and veins) of the systemic circuit.

Hepatic Portal System: Specialized for Blood Filtration

The **hepatic portal vein** (see Fig. 5.10) drains blood from the capillary beds of the digestive tract to a capillary bed in the liver. (A so-called *portal* system always lies between two capillary beds.) The blood in the hepatic portal vein is oxygen-poor, but rich in glucose, amino acids, and other nutrients absorbed by the small intestine. The liver stores glucose as glycogen, and it either stores amino acids or uses them immediately to manufacture blood proteins. The liver also purifies the blood of toxins and pathogens that have entered the body by way of the intestinal capillaries. After blood has filtered slowly through the liver, it is collected by the **hepatic vein** and returned to the inferior vena cava.

Connecting the Concepts

During fetal development, minor changes in the flow of blood occur due to a bypassing of the lungs. For more on this and additional information on the hepatic portal system and pulmonary circuits, refer to the following discussions.

Figure 8.8 illustrates how the hepatic portal vein brings nutrients to the liver.

Figure 9.11 diagrams the transport of oxygen and carbon dioxide in the pulmonary and systemic circuits.

Figure 17.7 shows how the systemic circuit interacts with the placenta during development.

Check Your Progress 5.5

1. Illustrate the flow of blood in the pulmonary circuit.
2. Describe the path of blood from the heart to the digestive tract and back to the heart by way of the hepatic portal vein.
3. Compare the relative oxygen content of the blood flowing in the pulmonary artery with that in the pulmonary vein.

5.6 Exchange at the Capillaries

Learning Outcomes

Upon completion of this section, you should be able to

1. Describe the processes that move materials across the walls of a capillary.
2. Explain what happens to the excess fluid that leaves the capillaries.

Two forces control movement of fluid through the capillary wall: blood pressure, which tends to cause fluids in the blood to move from capillary to tissue spaces, and osmotic pressure, which tends to cause water to move in the opposite direction. At the arterial end of a capillary, blood pressure (30 mm Hg) is higher than the osmotic pressure of blood (21 mm Hg) (Fig. 5.12). Osmotic pressure is created by the

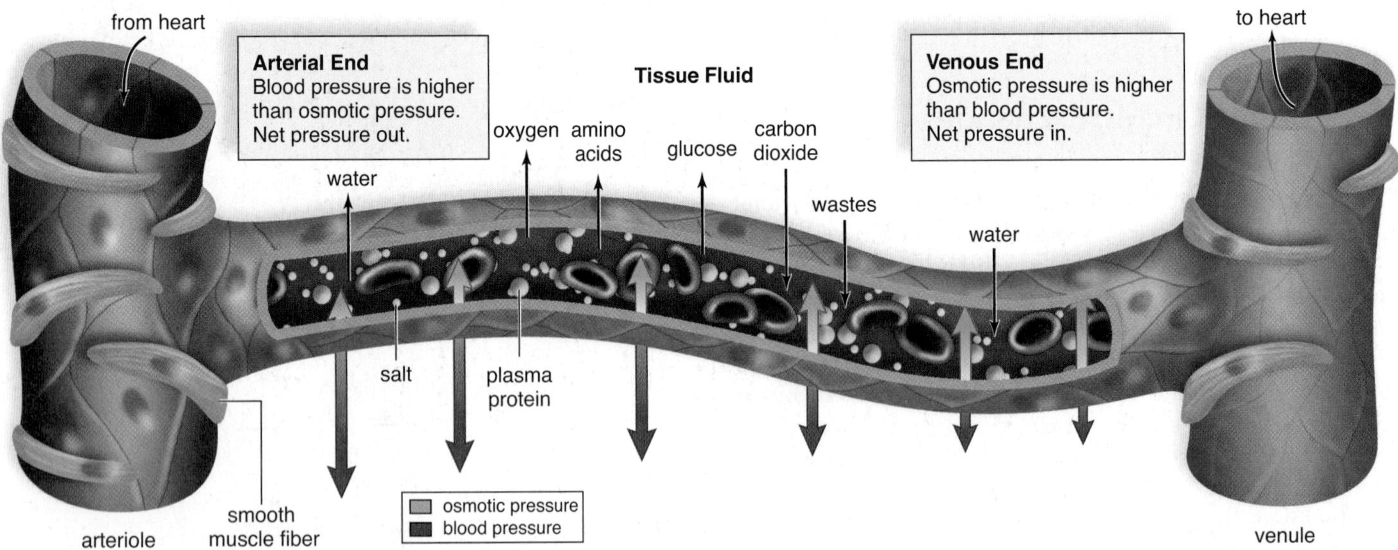

Figure 5.12 The movement of solutes in a capillary bed.
The capillary shows the exchanges that take place and the forces that aid the process. At the arterial end of a capillary, the blood pressure is higher than the osmotic pressure. Tissue fluid tends to leave the bloodstream. In the midsection, solutes, including oxygen (O_2) and carbon dioxide (CO_2), diffuse from high to low concentration. Carbon dioxide and wastes diffuse into the capillary while nutrients and oxygen enter the tissues. At the venous end of a capillary, the osmotic pressure is higher than the blood pressure. Tissue fluid tends to reenter the bloodstream. The red blood cells and the plasma proteins are too large to exit a capillary.

presence of solutes dissolved in plasma, the liquid fraction of the blood. Dissolved plasma proteins are of particular importance in maintaining the osmotic pressure. Most plasma proteins are manufactured by the liver. Blood pressure is higher than osmotic pressure at the arterial end of a capillary, so water exits a capillary at the arterial end.

Midway along the capillary, where blood pressure is lower, the two forces essentially cancel each other, and there is no net movement of fluid. Solutes now diffuse according to their concentration gradient: Oxygen and nutrients (glucose and amino acids) diffuse out of the capillary; carbon dioxide and wastes diffuse into the capillary. Red blood cells and almost all plasma proteins remain in the capillaries. The substances that leave a capillary contribute to tissue fluid, the fluid between the body's cells. Plasma proteins are too large to readily pass out of the capillary. Thus, tissue fluid tends to contain all components of plasma, except much lower amounts of protein.

At the venule end of a capillary, blood pressure has fallen even more. Osmotic pressure is greater than blood pressure, and fluid tends to move back. Almost the same amount of fluid that left the capillary returns to it, although some excess tissue fluid is always collected by the lymphatic capillaries (Fig. 5.13). Tissue fluid contained within lymphatic vessels is called **lymph.** Lymph is returned to the systemic venous blood when the major lymphatic vessels enter the subclavian veins in the shoulder region.

MP3
Capillary Exchange and Bulk Flow

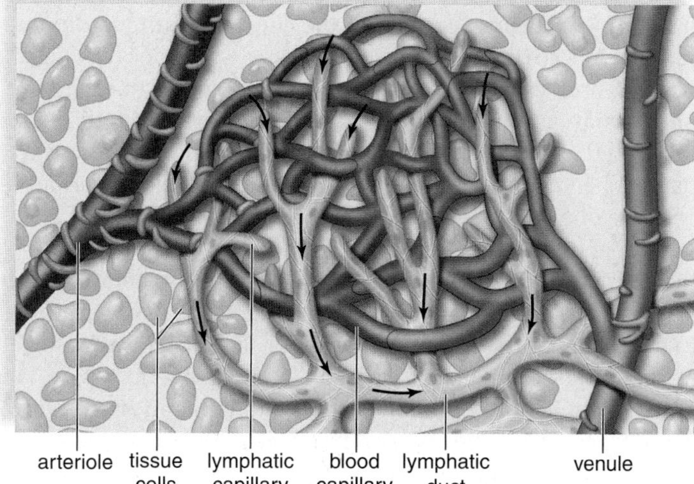

Figure 5.13 A lymphatic bed within a capillary bed. Lymphatic capillary beds lie alongside blood capillary beds. When lymphatic capillaries take up excess tissue fluid, it becomes lymph. Lymph returns to the cardiovascular system through cardiovascular veins in the chest. Precapillary sphincters can shut down a blood capillary, and blood then flows through the shunt.

Connecting the Concepts

The movement of materials in and out of the capillaries plays an important role in the function of a number of other body systems. For examples of the importance of capillaries, refer to the following discussions.

Section 7.2 provides additional information on the role of the lymphatic system.

Section 9.6 explains how gas exchange occurs across the walls of the capillaries in the lungs.

Section 10.3 explores how the urinary system establishes gradients to move waste materials out of the capillaries for excretion.

Check Your Progress 5.6

1. Explain what happens to the excess fluid created during capillary exchange.
2. Describe the exchange of materials across the walls of a capillary.
3. Summarize what occurs when blood and osmotic pressure change at the venous end of a capillary.

5.7 Cardiovascular Disorders

Learning Outcomes

Upon completion of this chapter, you should be able to

1. Explain the underlying causes of cardiovascular disease in humans.
2. Summarize how advances in medicine can treat cardiovascular disorders.

Cardiovascular disease (CVD) is the leading cause of untimely death in the Western countries. Modern research efforts have resulted in improved diagnosis, treatment, and prevention. This section discusses the range of advances that have been made, first in correcting vascular disorders and then in correcting heart disorders. The Bioethical Focus, *Cardiovascular Disease Prevention,* describes risky behaviors that one should avoid to possibly prevent CVD from developing in the first place.

Disorders of the Blood Vessels

Hypertension and atherosclerosis often lead to stroke or heart attack, due to an artery blocked by a blood clot or clogged by plaque. Treatment involves removing the blood clot or prying open the affected artery. Another possible outcome is an **aneurysm,** a burst blood vessel. An aneurysm can be prevented by replacing a blood vessel that is about to rupture with an artificial one.

High Blood Pressure

Hypertension occurs when blood moves through the arteries at a higher pressure than normal. Also known as high blood pressure, hypertension is sometimes called a silent killer. It may not be detected until it has caused a heart attack, stroke, or even kidney failure. Hypertension is present when the systolic blood pressure is 140 or greater or the diastolic blood pressure is 90 or greater. Though systolic and diastolic pressures are both important, diastolic pressure is emphasized when medical treatment is being considered.

The best safeguard against developing hypertension is regular blood pressure checks and a lifestyle that lowers the risk of CVD. If hypertension is present, prescription drugs can help lower blood pressure. Diuretics cause the kidneys to excrete more urine, ridding the body of excess fluid. In addition, hormones (the body's chemical messengers) that raise blood pressure can be inactivated. Drugs called beta-blockers and ACE (angiotensin-converting enzyme) inhibitors help to control hypertension caused by hormones.

Hypertension is often seen in individuals who have **atherosclerosis.** Atherosclerosis is caused by formation of lesions, or *atherosclerotic plaques,* on the inside of blood vessels. The **plaques** narrow blood vessel diameter, choking off blood and oxygen supply to the tissues (Fig. 5.14). In most instances, atherosclerosis begins in early adulthood and develops progressively through middle age, but symptoms may not appear until an individual is 50 or older. To prevent the onset and development of atherosclerosis, the American Heart Association and other organizations recommend a diet low in saturated fat and cholesterol but rich in omega-3 polyunsaturated fatty acids.

Atherosclerotic plaques can cause a clot to form on the irregular, roughened arterial wall. As long as the clot remains stationary, it is called a **thrombus.** If the thrombus dislodges and moves along with the blood, it is called an **embolus.** A **thromboembolism** consists of a clot first carried in the bloodstream that then becomes completely stationary when it lodges in a small blood vessel. If a thromboembolism is not treated, the life-threatening complications described in the next section can result.

Research has suggested several possible causes for atherosclerosis aside from hypertension. Chief among these, as discussed in the aforementioned Bioethical Focus, *Cardiovascular Disease Prevention,* are smoking and a diet rich in lipids and cholesterol. Research also indicates that a low-level bacterial or viral infection that spreads to the blood may cause an injury that starts the process of atherosclerosis. Surprisingly, such an infection may originate with gum diseases or be due to *Helicobacter pylori* (the bacterium that causes ulcers). People who have high levels of C-reactive protein, which occur in the blood following a cold or injury, are more likely to have a heart attack.

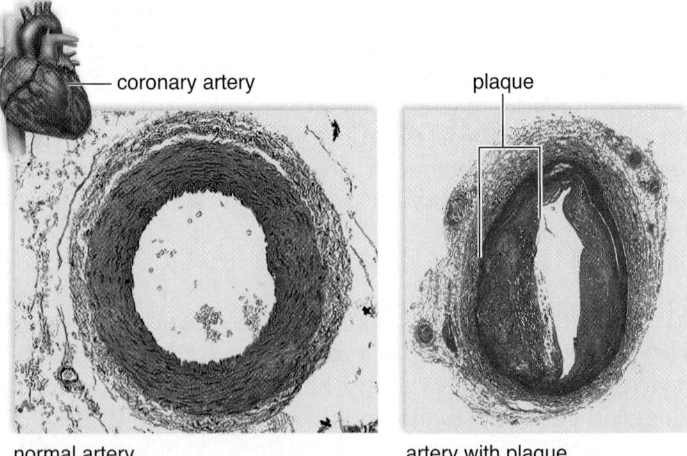

Figure 5.14 Atherosclerotic plaque and its effect on vessel diameter.
When an atherosclerotic plaque forms in a coronary artery, a heart attack is more apt to occur because of restricted blood flow.

Stroke, Heart Attack, and Aneurysm

Stroke, heart attack, and aneurysm are associated with hypertension and atherosclerosis. A cerebrovascular accident (CVA), also called a **stroke,** often results when a small cranial arteriole bursts or is blocked by an embolus. Lack of oxygen causes a portion of the brain to die, and paralysis or death can result. A person is sometimes forewarned of a stroke by a feeling of numbness in the hands or the face, difficulty in speaking, or temporary blindness in one eye.

A myocardial infarction (MI), also called a **heart attack,** occurs when a portion of the heart muscle dies due to lack of oxygen. If a coronary artery becomes partially blocked, the individual may then suffer from **angina pectoris.** Characteristic symptoms of angina pectoris include a feeling of pressure, squeezing, or pain in the chest. Pressure and pain can extend to the left arm, neck, jaw, shoulder, or back. Nausea and vomiting, anxiety, dizziness, and shortness of breath may accompany the chest discomfort. Nitroglycerin or related drugs dilate blood vessels and help relieve the pain. When a coronary artery is completely blocked, perhaps because of a thromboembolism, a heart attack occurs.

An aneurysm is a ballooning of a blood vessel, most often the abdominal artery or the arteries leading to the brain. Atherosclerosis and hypertension can weaken the wall of an artery to the point that an aneurysm develops. If a major vessel such as the aorta bursts, death is likely. It is possible to replace a damaged or diseased portion of a vessel, such as an artery, with a plastic tube. Cardiovascular function is preserved because exchange with tissue cells can still take place at the capillaries. In the future, it may be possible to use vessels made in the laboratory by injecting a patient's cells inside an inert mold.

Dissolving Blood Clots

Medical treatment for a thromboembolism includes the use of tissue plasminogen activator (t-PA), a biotechnology drug that converts plasminogen, a protein found in blood, into plasmin, an enzyme that dissolves blood clots. Tissue plas-

Biology Matters Historical Focus

Surgeon Without a Degree: Vivien Theodore Thomas (1910–1985)

On a warm summer day in 1941, professors, students, and visitors at Johns Hopkins University Hospital stopped to stare at an unusual sight. Dressed in a laboratory jacket, a young African-American man crossed the campus. In that era of strict racial segregation and Jim Crow laws, people of African-American descent on a college campus were maids or janitors. Bathrooms, drinking fountains, hotels, and even baseball teams were segregated by race. African-American patients were treated at the hospital but had to sneak in by a rear entrance. To see a black man outfitted as a researcher was an extraordinary sight—but Vivien Thomas was an extraordinary man.

Thomas had determined in high school that he would pursue a medical career. After a year in college as a premedical student, the stock market crash and bank failures of 1929 shattered his dream. Thomas's college education and medical school were never completed. Instead, Thomas worked at Vanderbilt University in Nashville serving as a laboratory assistant to surgeon Alfred Blalock.

After putting in 16-hour days in the laboratory, Thomas tutored himself in anatomy (the study of body structures) and physiology (the study of bodily function). Next, he taught himself surgical techniques using experimental animals, including how to administer anesthesia and sew blood vessels. Thomas became a brilliant researcher, devising a number of surgical instruments. His respirator, which took over breathing when the chest was cut open, allowed surgeries that had not been possible previously.

When Blalock was offered a position at the prestigious Johns Hopkins University Hospital in Baltimore, he accepted upon one condition—that Vivien Thomas be allowed to accompany him. There, Vivien and Blalock continued their collaboration in medical research and surgical methods. Vivien helped to devise the Blalock–Taussing surgery, named after Blalock and the heart physician Helen Taussig.* This operation helped to save the lives of many babies with severe heart defects. During many complicated surgeries, Viven stood at Blalock's right hand. There, he patiently advised his boss on proper technique. Blalock's contemporary referred to Vivien as "the surgical glove on Blalock's experimental hand."

Though they could enjoy each other's company in the hospital lab, the two men's close relationship ended at the laboratory door. Vivien was paid and classified as a janitor, earning extra money by serving drinks at Blalock's parties. His salary was so low that at one point he threatened to seek work as a carpenter to make more money. Only then did Blalock see to it that his assistant received a raise.

In 1976, Johns Hopkins University awarded Vivien Thomas an honorary degree. Today, his portrait hangs where it should, right next to that of Alfred Blalock in the Blalock Clinical Sciences Building at Johns Hopkins. To the end of his life, Thomas remained a very humble man. Of his life and work, Thomas wrote: "As for me, I just work here ... I've thoroughly enjoyed the role I have played and only tried to be me."

* Helen Taussig was also a pioneer. A specialist in the care of children with heart defects, she was also the first female president of the American Heart Association.

minogen activator is also being used for stroke patients, but with less success. Some patients experience life-threatening bleeding in the brain. A better treatment might be new biotechnology drugs that act on the plasma membrane to prevent brain cells from releasing and/or receiving toxic chemicals caused by the stroke.

If a person has symptoms of angina or a stroke, aspirin may be prescribed. Aspirin lowers the probability of clot formation. There is evidence that aspirin protects against first heart attacks, but there is no clear support for taking aspirin every day to prevent strokes in symptom-free people. Physicians warn that long-term use of aspirin might have harmful effects, including bleeding in the brain.

Treating Clogged Arteries

Cardiovascular disease used to require open-heart surgery and, therefore, a long recuperation time and a long unsightly scar that could occasionally ache. Now, bypass surgery can be accomplished by using robotic technology

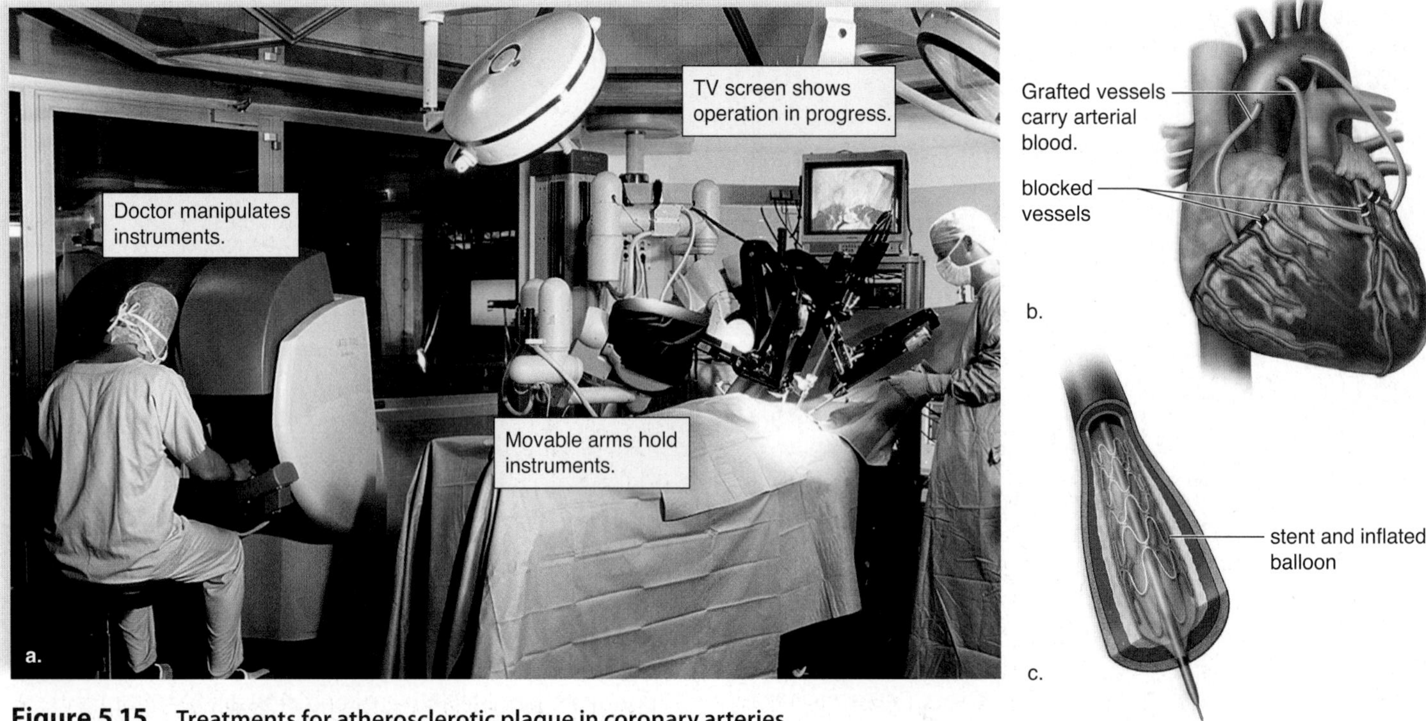
TV screen shows
operation in progress.
Doctor manipulates
instruments.
Movable arms hold
instruments.
a.
Grafted vessels
carry arterial
blood.
blocked
vessels
b.
stent and inflated
balloon
c.

Figure 5.15 **Treatments for atherosclerotic plaque in coronary arteries.**
a. Robotic surgery techniques can assist in coronary bypass operations. **b.** In this procedure, blood vessels (leg veins of arteries placed in the chest) are stitched to coronary arteries, bypassing the obstruction. **c.** In a stent procedure, a balloon-tipped catheter guides the stent (a cylinder of expandable metal mesh) to the obstructed area. When the balloon is inflated, the stent expands and opens the artery.

(Fig. 5.15*a*). A video camera and instruments are inserted through small cuts, while the surgeon sits at a console and manipulates interchangeable grippers, cutters, and other tools attached to movable arms above the operating table. Looking through two eyepieces, the surgeon gets a three-dimensional view of the operating field. Robotic surgery is also used in valve repairs and other heart procedures.

A **coronary bypass operation** is one way to treat an artery clogged with plaque. A surgeon takes a blood vessel—usually a vein from the leg—and stitches one end to the aorta and the other end to a coronary artery located past the point of obstruction. Figure 5.15*b* shows a triple bypass in which three blood vessel segments have been used to allow blood to flow freely from the aorta to cardiac muscle by way of the coronary artery. Since 1997, gene therapy has been used instead of a coronary bypass to grow new blood vessels that will carry blood to cardiac muscle. The surgeon need only make a small incision and inject many copies of the gene that codes for vascular endothelial growth factor (VEGF) between the ribs directly into the area of the heart that most needs improved blood flow. The VEGF encourages new blood vessels to sprout out of an artery. If collateral blood vessels do form, they transport blood past clogged arteries, making bypass surgery unnecessary. About 60% of patients who undergo the procedure do show signs of vessel growth within two to four weeks.

Another alternative to bypass surgery is available, namely, the stent (Fig. 5.15*c*). A stent is a small metal mesh cylinder that holds a coronary artery open after a blockage has been cleared. Until a couple of years ago, stenting was a second step following angioplasty. During **angioplasty,** a plastic tube was inserted into an artery of an arm or leg and then guided through a major blood vessel toward the heart. When the tube reached the region of plaque in an artery, a balloon attached to the end of the tube was inflated, forcing the vessel open. Now, instead, a stent with an inner balloon is pushed into the blocked area. When the balloon inside the stent is inflated, it expands, locking itself in place. Some patients go home the same day or, at most, after an overnight stay. The stent is more successful when it is coated with a drug that seeps into the artery lining and discourages cell growth. Uncoated stents can close back up in a few months, but the drug used in the coated ones successfully discourages closure. Blood clotting might occur after stent placement, so recipients have to take anticlotting medications.

Heart Failure

When a person has **heart failure,** the heart no longer pumps as it should. Heart failure is a growing problem because people who used to die from heart attacks now survive but are left with damaged hearts. Often the heart is oversized, not because the cardiac wall is stronger but because it is sagging and swollen. One idea is to wrap the heart in a fabric sheath to prevent it from getting too big. This might allow better pumping, similar to the way a weight lifter's belt restricts and reinforces stomach muscles. But a failing heart can have other problems, such as an abnormal heart rhythm.

To counter that condition, it's possible to place an implantable cardioverter-defibrillator (ICD) just beneath the skin of the chest. This device can sense both an abnormally slow and an abnormally fast heartbeat. If the former, the ICD generates the missing beat like a pacemaker does. If the latter, it sends the heart a sharp jolt of electricity to slow it down. If the heart rhythm becomes erratic, the ICD sends an even stronger shock—like a defibrillator does.

Heart Transplants

Although heart transplants are now generally successful, many more people are waiting for new hearts than there are organs available. Today, only about 2,200 heart transplants are done annually, though many thousands of people could use them. Genetically altered pigs may one day be used as a source of hearts because of the shortage of human hearts. Also, bone marrow stem cells have been injected into the heart, and researchers report that they apparently become cardiac muscle!

Today, a left ventricular assist device (LVAD), implanted in the abdomen, is an alternative to a heart transplant. A tube passes blood from the left ventricle to the device, which pumps it on to the aorta. A cable passes from the device through the skin to an external battery, which the patient must tote around. Soon, as many as 20,000 people a year may be outfitted with an LVAD. Instead of an LVAD, a few pioneers have volunteered to receive a Jarvik 2000, which is a pump inserted inside the left ventricle. The Jarvik is powered by an external battery no larger than a C battery.

Even fewer patients have received a so-called **total artificial heart (TAH),** such as that shown in Figure 5.16. An internal battery and controller regulate the pumping speed, and an external battery powers the device by passing electricity through the skin via external and internal coils. A rotating centrifugal pump moves silicon hydraulic fluid between left and right sacs to force blood out of the heart into the pulmonary trunk and the aorta. All recipients, thus far, have been near death, and most have lived for only a short time. It is possible that once healthier patients receive total artificial hearts, the survival rate will improve. Different types of TAHs are being investigated in animals.

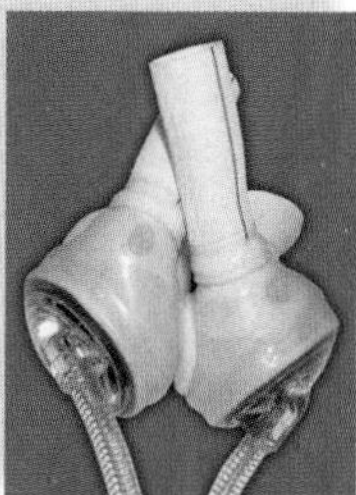

Figure 5.16 **An artifical heart.**
The CardioWest temporary total artificial heart moves blood in the same manner as a natural heart. The controller is implanted into the patient's abdomen. The heart is powered by an internal rechargeable battery, with an external battery as a backup system.
Courtesy of SynCardia Systems, Inc.

Connecting the Concepts

Many cardiovascular disorders are caused by incorrect diet or a sedentary lifestyle. Others are the result of genetic abnormalities. For more on the relationships between diet and cardiovascular help and how scientists identify genes associated with cardiovascular disease, refer to the following discussions.

Section 8.6 explores how each class of nutrient relates to a healthy lifestyle.

Table 8.4 lists methods of lowering saturated fats and cholesterol in the diet.

Sections 21.3 and 21.4 explain how biotechnology and genomics are identifying new genes associated with disease.

Check Your Progress 5.7

1. Summarize the cardiovascular disorders that are common in humans.
2. Summarize the treatments that are available for cardiovascular disorders.
3. Discuss why CVD is the leading source of death in Western countries.

Connections and Misconceptions

How does a heart develop in a fetus?

A fetal heart will begin forming from tissues in the chest of the fetus first by forming a tube structure that will later develop into the chambers of the heart. By the fifth week of fetal development, the heart is formed; it is too small to hear but it can be seen on an ultrasound. By week 10, the heart is fully developed and has a heartbeat rate of 150–195 beats per minute. The sounds of the heart at this point are often referred to as *fetal heart tones* (*FHTs*) and are very rapid due to the heart's small size; but eventually they will settle into a rate of 120–160 beats per minute (usually after the twelfth week) and may even beat in-synch with the mother's rate. The fetal heart can be heard through the end of the first trimester and into the begining of the second using an amplification machine called a *Doppler instrument* that bounces harmless sound waves off the heart. After that point, depending on the position of the fetus, a stethoscope can detect the heartbeat as the fetus and its heart get larger.

Biology Matters Bioethical Focus

Cardiovascular Disease Prevention: Who Pays for an Unhealthy Lifestyle?

Cardiovascular disease (CVD) is not only the number one killer in the United States today; it is also one of the most expensive. More than $431 billion is spent annually for health care and lost productivity, according to the American Heart Association. Family history of heart attack under age 55, male gender, and ethnicity (people of African-American descent are at great risk) are unalterable risk factors for CVD. However, most cases of CVD are preventable (Fig. 5A). Preventable risk factors include

- Use of tobacco: Nicotine constricts arterioles, increasing blood pressure. As a result, the heart must pump harder to propel blood. Carbon monoxide in smoke decreases the blood's oxygen-carrying ability. (Tobacco use also causes many different cancers, as you'll discover in Chapter 19.)
- Drug and alcohol abuse: Stimulants, like cocaine and amphetamines, can cause irregular heartbeat and lead to heart attacks. Intravenous (IV) drug use may cause cerebral blood clots and stroke. Alcohol abuse can destroy body organs, including the heart. (However, low to moderate alcohol consumption actually decreases CVD risk.)
- Obesity and a sedentary lifestyle: Extra body tissue requires an additional blood supply, increasing the heart's workload. Hypertension develops as the heart works harder to pump blood. Obesity also increases the risk of type 2 diabetes. Diabetes causes blood vessel damage and atherosclerosis. Lack of exercise contributes to obesity.
- Poor diet: A diet high in saturated fats and cholesterol is a risk for CVD. Cholesterol is ferried in the blood by two proteins: LDL (low-density lipoprotein, the "bad" lipoprotein) and HDL (high-density lipoprotein, the "good" lipoprotein). LDL carries its cholesterol to deposit in tissues, but HDL carries cholesterol to the liver where it can be metabolized. Elevated blood LDL and/or low blood HDL levels can contribute to cardiovascular disease. A diet low in saturated fat and cholesterol can help to restore LDL and HDL to recommended levels. Drugs called statins can further lower LDL level if necessary.
- Stress: A stress-filled lifestyle contributes directly to CVD and may also cause a person to overeat, avoid exercise, start or increase smoking, or abuse drugs and alcohol.
- Poor dental hygiene: Avoiding the dentist or dental clinic results in gum inflammation. Microbes from infected gums can enter the bloodstream and trigger formation of atherosclerotic plaques.

With few exceptions, these risk factors are personal choices.

Figure 5A **Risk factors for CVD that are under an individual's control.**
Unhealthy lifestyle choices—smoking, drinking alcohol to excess, obesity, and a sedentary lifestyle—are factors that a person has the power to change.

Who Pays for Treatment?

People who practice risky behaviors already pay more: for example, they pay extra for health or life insurance. Tobacco and alcohol taxes also help to defray health-care costs. However, insurance companies may not cover pre-existing conditions, including CVD. And what about the uninsured? In these cases, treatment costs are borne by taxpayers. Ultimately, we all pay for CVD treatment—and we all have an incentive to ensure that individuals adopt healthy lifestyles—or pay more for treatment if they refuse.

Is Legislation Needed?

Several organizations, such as the World Health Organization, have advocated legislation to help pay for CVD and its consequences. One proposed example is the so-called "fat tax" on foods with high fat and/or poor nutritional value. The tax would pay for treatment of obesity-related disease, including CVD. Other initiatives stress education of both adults and children regarding healthy lifestyle choices. However, critics oppose this type of legislation, insisting that rewarding healthy behaviors and penalizing risky behaviors are beyond the scope of government.

Decide Your Opinion

1. Would you support charging higher insurance premiums or taxes to people who don't practice a healthy lifestyle?
2. Should public funding be used for prevention programs if money could be saved in the future?
3. The public does subsidize health care that disproportionately benefits the less healthy, so do you believe that financial interest trumps personal freedom in this matter? Why or why not?

CASE STUDY CONCLUSION

Ted wondered why he was affected by heart disease. He was relatively young, he followed a healthy diet, and he was an athlete. His doctor explained that the plaque buildup consisted of fat and other substances, including cholesterol, that formed on the inner walls of the arteries, narrowing them and restricting blood flow. Clots can be composed of many substances, but an elevated level of cholesterol in the blood can be a common condition for clotting. Being diabetic is a risk factor for PAD as well, even though Ted's diabetes was under control and had been for years. Ted still did not understand why he was diagnosed with PAD, given that his diabetes was in check, he exercised, and he ate a healthy diet. The doctor explained that our bodies make cholesterol naturally and we also obtain cholesterol from the foods we eat. Ted had a very low cholesterol diet, and after some testing, it was determined that even though he had a healthy lifestyle, Ted had an elevated level of cholesterol in his blood that could predispose him to inappropriate clot formations. Ted needs to keep treating his PAD because it can double his risk for a heart attack. Now Ted is on a statin medication to help lower his blood cholesterol level, he maintains his diet and exercise program, he continues to monitor his diabetes daily, and he is aware of any PAD symptoms that may arise. With proper lifestyle, regular doctor visits, medications, and an increased knowledge regarding heart disease, Ted can lead a healthy life for many years.

Media Study Tools

www.mhhe.com/maderhuman12e

Enhance your study of this chapter with study tools and practice tests. Also ask your instructor about the resources available through ConnectPlus, including the media-rich eBook, interactive learning tools, and animations.

Virtual Lab

The virtual lab "Blood Pressure" provides an interactive look at how factors such as age and gender influence the risk of hypertension.

Summarizing the Concepts

5.1 Overview of the Cardiovascular System

The cardiovascular system consists of the heart and blood vessels.

The heart pumps blood, and blood vessels take blood to and from capillaries, where exchanges of nutrients for wastes occur with tissue cells. Blood is refreshed at the lungs, where gas exchange occurs; at the digestive tract, where nutrients enter the blood; and the kidney, where wastes are removed from blood.

5.2 The Types of Blood Vessels

The Arteries Arteries (and arterioles) take blood away from the heart. Arteries have the thickest walls, which allows them to withstand blood pressure.

The Capillaries Exchange of substances occurs in the capillaries.

capillary bed

The Veins Veins (and venules) take blood to the heart. Veins have relatively weak walls with valves that keep the blood flowing in one direction.

5.3 The Heart Is a Double Pump

The heart has a right and left side. Each side has an atrium and a ventricle. Valves keep the blood moving in the correct direction.

Passage of Blood Through the Heart

- The right atrium receives O_2-poor blood from the body, and the right ventricle pumps it into the pulmonary circuit (to the lungs).
- The left atrium receives O_2-rich blood from the lungs, and the left ventricle pumps it into the systemic circuit.

The Heartbeat Is Controlled

During the cardiac cycle, the SA node (pacemaker) initiates the heartbeat by causing the atria to contract. The AV node conveys the stimulus to the ventricles, causing them to contract. The heart sounds, "lub-dup," are due to the closing of the atrioventricular valves, followed by the closing of the semilunar valves.

5.4 Features of the Cardiovascular System

Pulse The pulse indicates the heartbeat rate.

Blood Pressure Moves Blood in Arteries Blood pressure caused by the beating of the heart accounts for the flow of blood in the arteries.

Blood Flow Is Slow in the Capillaries The reduced velocity of blood flow in capillaries facilitates exchange of nutrients and wastes in the tissues.

Blood Flow in Veins Returns Blood to the Heart Blood flow in veins is caused by skeletal muscle contraction, the presence of valves, and respiratory movements.

5.5 Two Cardiovascular Pathways

The cardiovascular system is divided into the pulmonary circuit and the systemic circuit.

The Pulmonary Circuit: Exchange of Gases

In the pulmonary circuit, blood travels to and from the lungs.

The Systemic Circuit: Exchanges with Tissue Fluid

In the systemic circuit, the aorta divides into blood vessels that serve the body's organs and cells. Venae cavae return O_2-poor blood to the heart.

5.6 Exchange at the Capillaries

This diagram illustrates capillary exchange in tissues of the body—not including the gas-exchanging surfaces of the lungs.

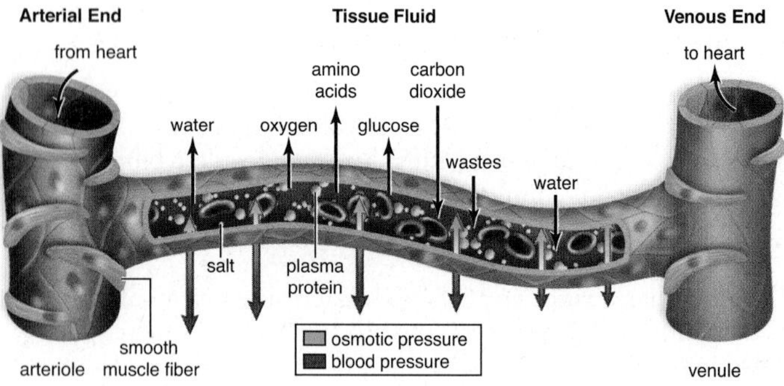

- At the arterial end of a cardiovascular capillary, blood pressure is greater than osmotic pressure; therefore, fluid leaves the capillary.
- In the midsection, oxygen and nutrients diffuse out of the capillary, and carbon dioxide and other wastes diffuse into the capillary.
- At the venous end, osmotic pressure created by the presence of proteins exceeds blood pressure, causing most of the fluid to reenter the capillary. Some fluid remains as interstitial (tissue) fluid.

Excess fluid not picked up at the venous end of the cardiovascular capillary enters the lymphatic capillaries.

- Lymph is tissue fluid contained within lymphatic vessels.
- The lymphatic system is a one-way system. Its fluid is returned to blood by way of a cardiovascular vein.

5.7 Cardiovascular Disorders

Cardiovascular disease is the leading cause of death in the Western countries.

- Hypertension and atherosclerosis can lead to stroke, heart attack, or an aneurysm.
- Following a heart-healthy diet, getting regular exercise, maintaining a proper weight, and not smoking reduce cardiovascular disease risk.

Understanding Key Terms

aneurysm 105
angina pectoris 106
angioplasty 108
aorta 103
arteriole 93
arteriovenous shunt 94
atherosclerosis 106
atrioventricular (AV) bundle 98
atrioventricular (AV) valve 94
atrium 94
AV (atrioventricular) node 98
blood pressure 100
cardiac cycle 97
chordae tendineae 94
coronary artery 95
coronary bypass operation 108
desmosome 94
diastole 97
diastolic pressure 100
electrocardiogram (ECG) 99
embolus 106
gap junction 94
heart 94
heart attack 106
heart failure 108
hepatic portal vein 104
hepatic vein 104
hypertension 106
inferior vena cava 103
lymph 105
lymphatic system 92
myocardium 94
pacemaker 98
pericardium 94
plaque 106
precapillary sphincter 94
pulmonary artery 96
pulmonary circuit 103
pulmonary vein 96
pulse 100
Purkinje fibers 98
respiratory pump 101
SA (sinoatrial) node 98
semilunar valve 95
septum 94
skeletal muscle pump 101
stroke 106
superior vena cava 103
systemic circuit 103
systole 97
systolic pressure 100
thromboembolism 106
thrombus 106
tissue fluid 105
total artificial heart (TAH) 109
toxin 92
valve 94
ventricle 94
venule 94

Match the key terms to these definitions.

a. ________ Relaxation of a heart chamber.

b. ________ Large systemic vein that returns blood from body areas below the diaphragm.

c. ________ Rhythmic expansion and recoil of arteries resulting from heart contraction; can be felt from outside the body.

d. ________ Vessel that takes blood from capillaries to a vein.

e. ________ That part of the cardiovascular system that serves body parts and does not include the gas-exchanging surfaces in the lungs.

Testing Your Knowledge of the Concepts

1. What are the two parts of the cardiovascular system, and what are the functions of each part? (page 92)
2. Explain where exchanges occur in the body and the importance of those exchanges. (page 92)
3. Which of the three types of blood vessels are most numerous? Explain. (page 93)

4. Describe the structure of the heart, including the chambers and valves. (pages 94–97)
5. What is the function of the septum in the heart? What would happen if the heart had no septum? (page 94)
6. Trace the path of blood through the heart, including chambers, valves, and vessels the blood travels through. (pages 96–97)
7. Describe the cardiac cycle, using the terms *systole* and *diastole.* What are the roles of the SA node and the AV node in the cardiac cycle? (pages 97–98)
8. Distinguish between the internal and external controls of the heartbeat. Explain how an ECG relates to the cardiac cycle. (pages 98–99)
9. In what vessel is the blood pressure the highest? The lowest? In what vessel is blood flow rate the fastest? The slowest? Why are the pressure and rate in the capillaries important to capillary function? (pages 100–101)
10. Explain why skeletal muscle contraction has an effect on venous flow but not arterial flow. (pages 101–102)
11. Distinguish between the two cardiovascular pathways. (page 103)
12. Trace the pathway of blood to and from the brain in the systemic circuit. (page 103)
13. Describe the process by which nutrients are exchanged for wastes across a capillary, using glucose and carbon dioxide as examples. (pages 104–105)
14. What is the most probable association between high blood pressure and a heart attack? With this association in mind, what type of diet might help prevent a heart attack? (page 106)

In questions 15–20, match the descriptions to the blood vessel in the key. Answers may be used more than once.

Key:

a. venules	d. arteries
b. veins	e. arterioles
c. capillaries	

15. Drain blood from capillaries
16. Empty into capillaries
17. May contain valves
18. Take blood away from the heart
19. Sites for exchange of substances between blood and tissue fluid
20. Rate of blood flow the slowest
21. During ventricular diastole,
    a. blood flows into the aorta.
    b. the ventricles contract.
    c. the semilunar valves are closed.
    d. Both a and b are correct.
22. When the atria contract, the blood flows
    a. into the attached blood vessels.
    b. into the ventricles.
    c. through the atrioventricular valves.
    d. to the lungs.
    e. Both b and c are correct.
23. Which of these associations is mismatched?
    a. left ventricle—aorta
    b. right ventricle—pulmonary trunk
    c. right atrium—vena cava
    d. left atrium—pulmonary artery
    e. Both b and c are incorrectly matched.
24. Which statement is not correct concerning the heartbeat?
    a. The atria contract at the same time.
    b. The ventricles relax at the same time.
    c. The AV valves open at the same time.
    d. The semilunar valves open at the same time.
    e. First the right side contracts; then the left side contracts.
25. Accumulation of plaque in an artery wall is
    a. an aneurysm.
    b. angina pectoris.
    c. atherosclerosis.
    d. hypertension.
    e. a thromboembolism.
26. Label the following diagram of the cardiovascular system using this alphabetized list:

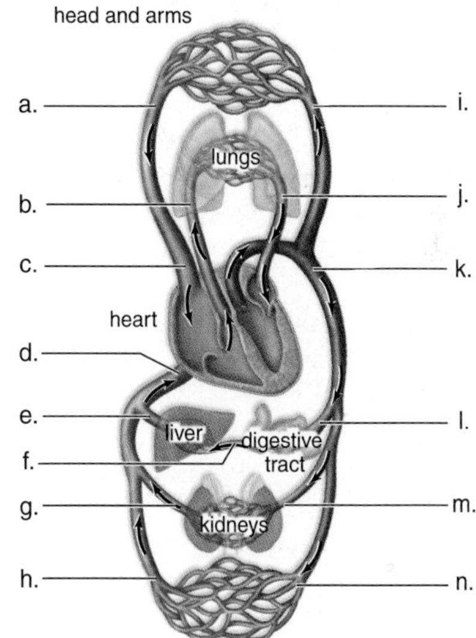

aorta	hepatic portal vein
carotid artery	hepatic vein
iliac artery	pulmonary artery
iliac vein	pulmonary vein
inferior vena cava	renal artery
jugular vein	renal vein
mesenteric arteries	superior vena cava

27. Label the following diagram showing the forces involved with capillary exchange. Use either blood pressure or osmotic pressure to label arrows a–d.

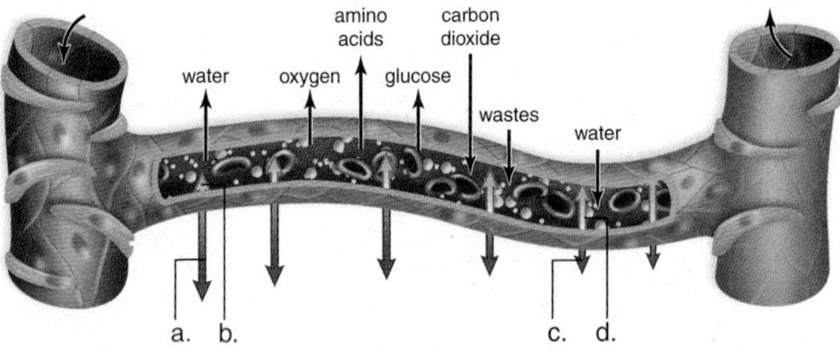

Thinking Critically About the Concepts

The cardiovascular system is an elegant example of the concept that structure supports function. Each type of blood vessel has a specific job. Each vessel's physical characteristics enable it to do that job. The muscle walls of the right and left ventricle vary in thickness depending on where they pump the blood. When organ structure is damaged or changed (as arteries are in atherosclerosis), the organ's ability to perform its function may be compromised. Homeostatic conditions, such as blood pressure, may be affected as well. Dietary and lifestyle choices can either prevent damage or harm the cardiovascular system.

1. In the Case Study, what symptoms did Ted misdiagnose before going to the doctor?
2. What were Ted's risk factors for cardiovascular disease?
3. Why is a blood test important when diagnosing a heart attack?
4. a. Why don't aortic aneurysms occur more frequently?
   b. Why do aortic aneurysms typically result in death?
5. a. What happens to the ventricular walls of someone with hypertrophic cardiomyopathy? (Hint: Consider the prefix *hyper*—what does it mean to describe a person?)
   b. How would this disease affect the heart's ability to pump blood efficiently?
6. Lymphatic vessels transport tissue fluid back to the cardiovascular system. The pressure in the lymphatic vessels is low. Do you expect lymphatic vessels to have valves? Why or why not?

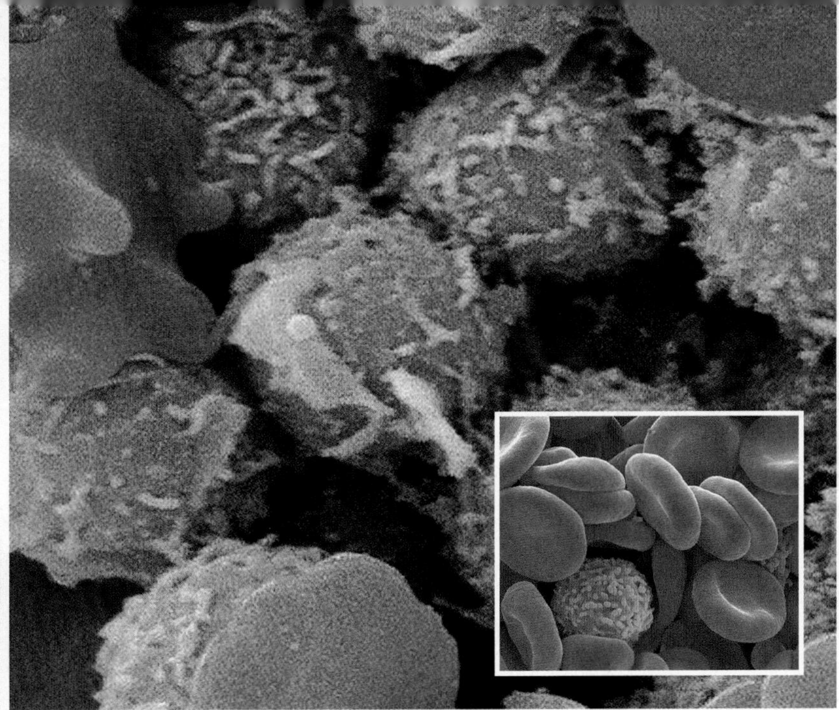

CHAPTER

6

Cardiovascular System: Blood

CASE STUDY CANCER OF THE BLOOD—LEUKEMIA

Ben was a 20-year-old placekicker on the football team at the university. He had been active all of his life, from playing baseball and football as a teenager to participating in intense intramural activities in college. He considered himself to be in fine health, and he worked hard to maintain his position on the team.

At the start of his second season, Ben noticed that he did not have as much energy as he was accustomed to. Practices seemed to take a lot more out of him than they normally did. He was frequently tired and often went to bed right after his practice. During the football drills, he often experienced a severe shortness of breath. Rather than risk his place on the team, Ben tried to cover up these symptoms; but over the next few weeks, his health worsened and he noticed swelling in his neck, armpits, and groin. In addition, bruises that he received in practice seemed to be taking a lot longer to heal. Ben became concerned, and when he approached his trainers, they immediately sent him to the team physician.

After completing a physical exam, Ben's physician ordered a complete blood count, or CBC test. The CBC test determines the total quantity of red and white blood cells, the amount of hemoglobin in the blood, and the number of platelets. The results of these tests indicated that Ben had a very high white blood cell count, but that these cells were not functioning correctly. The team physician immediately referred Ben to a specialist. A bone marrow biopsy confirmed what the doctors suspected; Ben had a type of cancer called acute lymphocytic leukemia (ALL).

As you read through the chapter, think about the following questions.

1. What is the role of white blood cells in the body?
2. Why would the doctors order a bone marrow biopsy if they suspected a blood disorder?
3. Why is leukemia considered to be a form of cancer?

CHAPTER CONCEPTS

6.1 Blood: An Overview
Blood, a liquid connective tissue, is a transport medium with a wide variety of functions in the body.

6.2 Red Blood Cells and Transport of Oxygen
Red blood cells contain hemoglobin, which transports oxygen and helps transport carbon dioxide.

6.3 White Blood Cells and Defense Against Disease
A variety of white blood cells, or leukocytes, exists. Collectively, these cells help the body fight infection.

6.4 Platelets and Blood Clotting
Platelets are cell fragments that maintain homeostasis by repairing breaks in blood vessels and preventing loss of blood.

6.5 Blood Typing and Transfusions
Blood typing is based on specific antigens found on the surface of red blood cells.

6.6 Homeostasis
Blood and the cardiovascular system play an important role in the maintenance of homeostasis in the body.

BEFORE YOU BEGIN

Before beginning this chapter, take a few moments to review the following discussions.

Section 4.2 Why is blood considered to be a connective tissue?

Section 4.8 What is homeostasis?

Section 5.6 How does the blood exchange nutrients and wastes with the tissues of the body?

6.1 Blood: An Overview

Learning Outcomes

Upon completion of this section you should be able to

1. List the functions of blood in the human body.
2. Compare the composition of formed elements and plasma in the blood.

In Chapter 5, we learned that the cardiovascular system consists of the heart, which pumps the blood, and the blood vessels, which conduct blood around the body. In this chapter, we will learn about the functions and composition of blood.

Functions of Blood

The human heart is an amazing muscular pump. With each beat, the human heart pumps approximately 75 mL of blood. On average, the heart beats 70 times per minute. Thus, the heart pumps roughly 5,250 mL per minute—75 mL/beat × 70 beats/minute—circulating the body's entire blood supply once each minute! If needed (when exercising, for example), the heart can cycle blood throughout the body even faster. The functions of blood fall into three categories: transport, defense, and regulation.

Blood is the primary transport medium. Blood acquires oxygen in the lungs and distributes it to tissue cells. Similarly, blood picks up nutrients from the digestive tract for delivery to the tissues. In its return trip to the lungs, blood transports carbon dioxide. Every time a person exhales, the carbon dioxide waste is eliminated. Blood also transports other wastes, such as excess nitrogen from the breakdown of proteins, to the kidneys for elimination. Blood exchanges nutrients and wastes with tissues by capillary exchange (see Fig. 5.12). In doing so, it maintains homeostasis by keeping the composition of tissue fluid within normal limits.

In addition to the transport of nutrients and wastes, various organs and tissues secrete hormones into the blood. Blood transports these to other organs and tissues, where they serve as signals that influence cellular metabolism. Blood is well suited for its role in transporting substances. Proteins in the blood help transfer hormones to the tissues. Special proteins called lipoproteins (also known as HDL and LDL) carry lipids, or fats, throughout the body. Most important, hemoglobin (found in red blood cells) is specialized to combine with oxygen and deliver it to cells. Hemoglobin also assists in transferring waste carbon dioxide back to the lungs.

Carbon monoxide (chemical formula CO) is a colorless, odorless gas, which can also bind to hemoglobin. Hundreds of people die accidentally from CO poisoning, caused by malfunction of a fuel-burning appliance or by improper venting of CO fumes. When CO is present, it takes the place of oxygen (O_2) in hemoglobin. As a result, cells are starved of oxygen. Tragically, treatment may be delayed, because symptoms of poisoning—headache, body ache, nausea, dizziness, drowsiness—can be mistaken for the "flu." Government guidelines recommend that *any* equipment producing CO be checked regularly to ensure proper ventilation. Finally, CO detectors, like smoke detectors, should be installed and used properly.

Blood defends the body against pathogen invasion and blood loss. Certain blood cells are capable of engulfing and destroying pathogens by a process called *phagocytosis* (see Chap. 3).

Other white blood cells produce and secrete antibodies into the blood. An **antibody** is a protein that combines with and disables specific pathogens. Disabled pathogens can then be destroyed by the phagocytic white blood cells.

When an injury occurs, blood clots and defends against blood loss. Blood clotting involves platelets (described in Section 6.4) and proteins. For example, prothrombin and fibrinogen are two inactive blood proteins. They circulate constantly, ready to be activated to form a clot if needed. Without blood clotting, we could bleed to death even from a small cut.

Blood has regulatory functions. Blood plays an important role in homeostasis. Blood helps regulate body temperature by picking up heat, mostly from active muscles, and transporting it about the body. If the body becomes too warm, blood is transported to dilated blood vessels in the skin. Heat disperses to the environment, and the body cools to a normal temperature.

The liquid portion of blood, its plasma, contains dissolved salts and proteins. These solutes create blood's *osmotic pressure,* which keeps the liquid content of the blood high (see Chap. 3, page 50, for a review of osmosis). In this way, blood plays a role in helping to maintain its own water–salt balance.

Blood buffers, body chemicals that stabilize blood pH, regulate the body's acid-base balance and keep it at a relatively constant pH of 7.4.

Composition of Blood

Blood is a tissue, and, like any tissue, it contains cells and cell fragments (Fig. 6.1). Collectively, the cells and cell fragments are called the **formed elements.** The cell and cell fragments are suspended in a liquid called **plasma.** Therefore, blood is classified as a liquid connective tissue.

The Formed Elements

The formed elements are red blood cells, white blood cells, and platelets. These are produced in red bone marrow, which can be found in most bones of a child but only in certain bones of an adult. Red bone marrow contains pluripotent stem cells, the parent cells that divide and give rise to all of the various types of blood cells (see Fig. 6.1). Stem cells continue to be the focus of a tremendous amount of research. Scientists are discovering that under the right conditions in the laboratory, stem cells can be "coaxed" into becoming a greater variety of cell types—cells that might have the potential to treat human diseases such as diabetes and Alzheimer disease, among many others (see Chap. 21 for a complete discussion of stem-cell research).

Red blood cells are two to three times smaller than white blood cells, but there are many more of them. There are millions of red blood cells and only thousands of white blood cells in a mm^3 of blood, about this size: ■

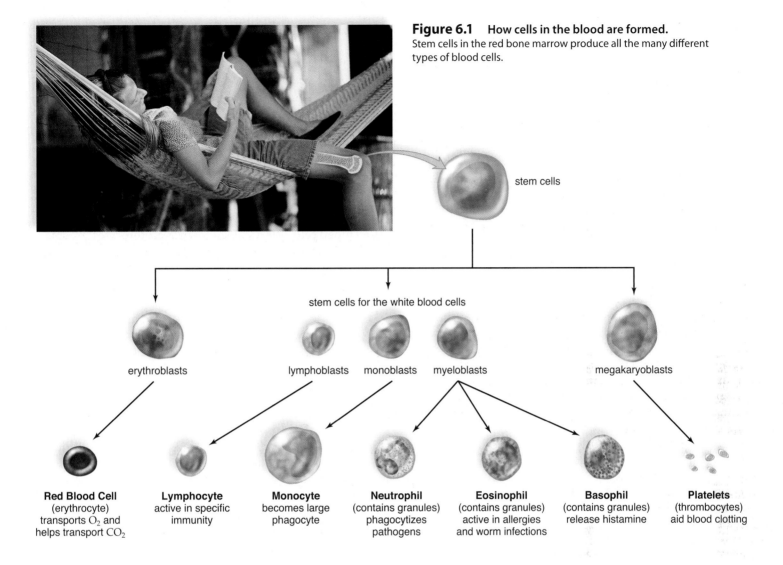

Figure 6.1 How cells in the blood are formed.
Stem cells in the red bone marrow produce all the many different types of blood cells.

stem cells

stem cells for the white blood cells

erythroblasts

lymphoblasts

monoblasts

myeloblasts

megakaryoblasts

Red Blood Cell (erythrocyte) transports O_2 and helps transport CO_2

Lymphocyte active in specific immunity

Monocyte becomes large phagocyte

Neutrophil (contains granules) phagocytizes pathogens

Eosinophil (contains granules) active in allergies and worm infections

Basophil (contains granules) release histamine

Platelets (thrombocytes) aid blood clotting

Connections and Misconceptions

Are stem cells only present in embryos?

Actually, this is a common misconception brought on by a media focus on embryonic stem (ES) cells. In reality, any actively dividing tissue requires a form of stem cell to act as the original source of cells. The difference is that many ES cells are totipotent—meaning that they retain the ability to form almost any type of cell—though most adult stem cells are pluripotent. Pluripotent cells have undergone an additional stage of specialization and, therefore, can only produce a limited variety of new cell types.

Plasma

Plasma is the liquid medium for carrying various substances in the blood. It also distributes the heat generated as a by-product of metabolism, particularly muscle contraction. About 91% of plasma is water (Fig. 6.2). The remaining 9% of plasma consists of various salts (ions) and organic molecules. Salts are dissolved in plasma.

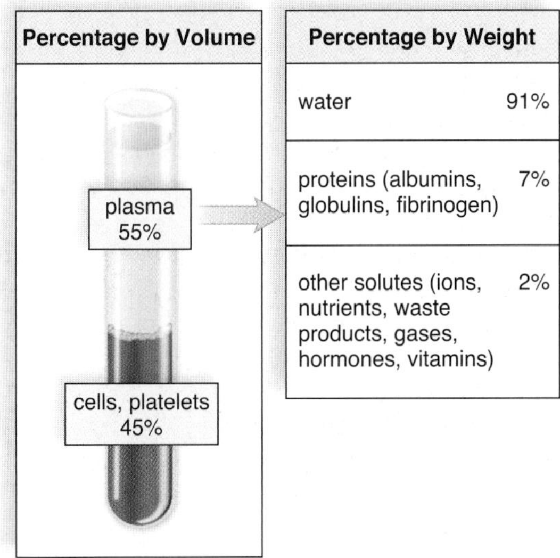

Figure 6.2 The composition of blood plasma.
Plasma, the liquid portion of blood, is mainly water and proteins. However, many solutes such as nutrients, vitamins, and hormones are transported in plasma.

As mentioned previously, salts and plasma proteins maintain the osmotic pressure of blood. Salts also function as buffers that help maintain blood pH. Small organic molecules such as glucose and amino acids are nutrients for cells; urea is a nitrogenous waste product on its way to the kidneys for excretion.

The most abundant organic molecules in blood are called the **plasma proteins.** The liver produces the majority of the plasma proteins. The plasma proteins have many functions that help maintain homeostasis. Like salts, they are able to take up and release hydrogen ions. Therefore, they help keep blood pH around 7.4. Plasma proteins are too large to pass through capillary walls. They remain in the blood, establishing an osmotic gradient between blood and tissue fluid. This **osmotic pressure** is a force that prevents excessive loss of plasma from the capillaries into tissue fluid.

Three major types of plasma proteins are the **albumins, globulins,** and **fibrinogen.** Albumins are the most abundant plasma proteins and contribute most to plasma's osmotic pressure. They also combine with and help transport other organic molecules. The globulins are of three types called alpha, beta, and gamma globulins. Alpha and beta globulins also combine with and help transport substances in the blood such as hormones, cholesterol, and iron. Gamma globulins are also known as antibodies and are produced by white blood cells called lymphocytes, not by the liver. Gamma globulins are important in fighting disease-causing pathogens (see Section 6.3). Fibrinogen is an inactive plasma protein. Once activated, fibrinogen forms a blood clot (see Section 6.4).

Connecting the Concepts

For more information on the topics presented in this section, refer to the following discussions.

Section 2.5 contains a Biology Matters: Health Focus box on the differences between HDLs and LDLs in the blood.

Section 4.2 examines why blood is classified as a connective tissue.

Section 7.4 explores how white blood cells are involved in the immune response.

Check Your Progress 6.1

1. Describe each function of blood as it relates to homeostasis.
2. List the different types of plasma proteins and explain why each is important.
3. Describe why blood is a connective tissue.

6.2 Red Blood Cells and Transport of Oxygen

Learning Outcomes

Upon completion of this section you should be able to

1. Explain the role of hemoglobin in gas transport.
2. Compare the transport of oxygen and carbon dioxide by red blood cells.
3. Summarize the role of erythropoietin in red blood cell production.

Red blood cells (RBCs), also known as **erythrocytes,** are small (usually between 6 to 8 micrometers), biconcave disks that lack a nucleus when mature. They occur in great quantity; there are 4–6 million red blood cells per mm^3 of whole blood.

How Red Blood Cells Carry Oxygen

Red blood cells are highly specialized for oxygen (O_2) transport. RBCs contain **hemoglobin (Hb),** a pigment that makes red blood cells and blood red. The globin portion of hemoglobin is a protein that contains four highly folded polypeptide chains. The heme part of hemoglobin is an iron-containing group in the center of each polypeptide chain (Fig. 6.3). The iron combines reversibly with oxygen. This means that heme accepts O_2 in the lungs and then lets go of it in the tissues. By contrast, carbon monoxide (CO) combines with the iron of heme and then will not easily let go.

Each hemoglobin molecule can transport four molecules of O_2, and each RBC contains about 280 million hemoglobin molecules. This means that each red blood cell can carry over a billion molecules of oxygen.

Red blood cells are an excellent example of structure suiting function. They have no nucleus; their biconcave shape comes about because they lose their nucleus during maturation (see Fig. 6.1). The biconcave shape of RBCs gives them a greater surface area for the diffusion of gases into and out of the cell. All of the internal space of RBCs is used for transport of oxygen. Aside from having no nucleus, RBCs also lack most organelles, including mitochondria. RBCs anaerobically produce ATP, and they do not consume any of the oxygen they transport.

When oxygen binds to heme in the lungs, hemoglobin assumes a slightly different shape and is called **oxyhemoglobin.** In the tissues, heme gives up this oxygen, and hemoglobin resumes its former shape, called **deoxyhemoglobin.** The released oxygen diffuses out of the blood into tissue fluid and then into cells.

How Red Blood Cells Help Transport Carbon Dioxide

After blood picks up carbon dioxide (CO_2) in the tissues, about 7% is dissolved in plasma. If the percentage

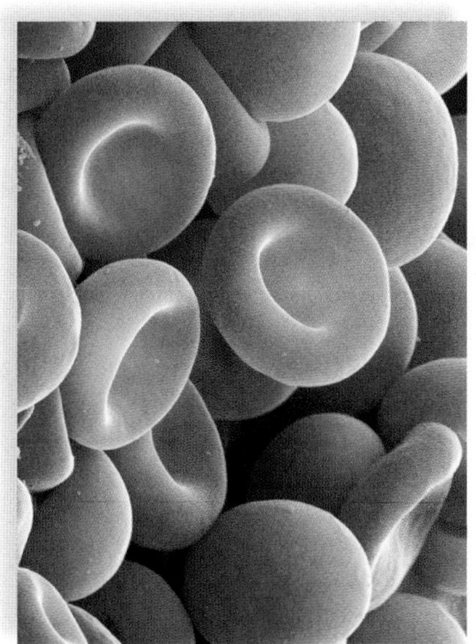
a. Red blood cells

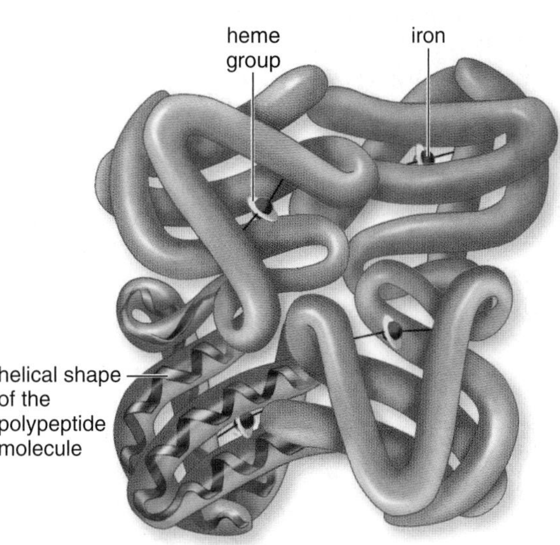

b. Hemoglobin molecule

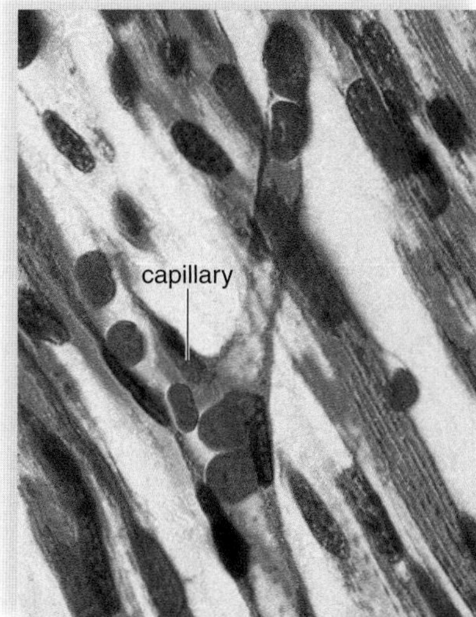

c. Blood capillary

Figure 6.3 Red blood cells and the structure of hemoglobin.
a. Red blood cells are biconcave disks containing many molecules of hemoglobin. **b.** Hemoglobin contains two types of polypeptide chains (blue, purple), forming the molecule's globin portion. An iron-containing heme group is in the center of each chain. Oxygen combines loosely with iron when hemoglobin is oxygenated. **c.** Red blood cells move single file through the capillaries.

of plasma CO_2 were higher than 7%, plasma would be carbonated and bubble like a soda. Instead, hemoglobin directly transports about 25% of CO_2, combining it with the globin protein. Hemoglobin-carrying CO_2 is termed *carbaminohemoglobin*.

The remaining CO_2 (about 68%) is transported as the bicarbonate ion (HCO_3^-) in the plasma. Consider this equation:

$$\underset{\text{carbon dixide}}{CO_2} + \underset{\text{water}}{H_2O} \rightleftharpoons \underset{\text{carbonic acid}}{H_2CO_3} \rightleftharpoons \underset{\text{hydrogen ion}}{H^+} + \underset{\text{bicarbonate ion}}{HCO_3^-}$$

Carbon dioxide moves into RBCs, combining with cellular water to form carbonic acid (arrows pointing left to right illustrate this part of the reaction). An enzyme found inside RBCs, called carbonic anhydrase, speeds the reaction. Carbonic acid quickly separates, or *dissociates,* to form hydrogen ions (H^+) and bicarbonate ions (HCO_3^-). The bicarbonate ions diffuse out of the RBCs to be carried in the plasma. The H^+ from this equation binds to globin, the protein portion of hemoglobin. Thus, hemoglobin assists plasma proteins and salts in keeping the blood pH constant. When blood reaches the lungs, the reaction is reversed (arrows pointing right to left). Hydrogen ions and bicarbonate ions reunite to re-form carbonic acid. The carbonic anhydrase enzyme also speeds this reverse reaction. Carbon dioxide diffuses out of the blood and into the airways of the lungs, to be exhaled from the body.

Red Blood Cells Are Produced in Bone Marrow

The RBC stem cells in the bone marrow divide and produce new cells that differentiate into mature RBCs (see Fig. 6.1). As red blood cells mature, they lose their nucleus and acquire hemoglobin. Because they lack a nucleus, RBCs are unable to replenish important proteins and repair cellular damage. Therefore, red blood cells only live about 120 days. When they age, red blood cells are phagocytized by white blood cells (macrophages) in the liver and spleen.

It is estimated that about 2 million RBCs are destroyed per second, and therefore, an equal number must be produced to keep the red blood cell count in balance. When red blood cells are broken down, hemoglobin is released. The globin portion of hemoglobin is broken down into its component amino acids, which are recycled by the body. The majority of the iron is recovered and returned to the bone marrow for reuse, although some is lost and must be replaced in the diet. The rest of the heme portion of the molecule undergoes chemical degradation and is excreted by the liver and kidneys. If the liver fails to excrete heme, it accumulates in tissues, causing a condition called *jaundice.* In jaundice, the skin and whites of the eyes turn yellow. Likewise, when skin is bruised, the chemical breakdown of heme causes the skin to change color from red/purple to blue to green to yellow. The body has a way to boost the number of RBCs when insufficient oxygen is being delivered to the cells. The kidneys release a hormone called **erythropoietin**

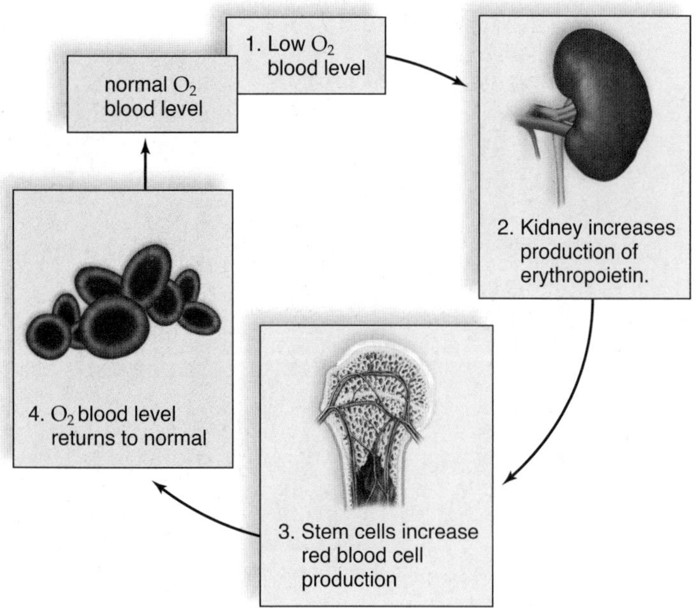
1. Low O_2 blood level
normal O_2 blood level
2. Kidney increases production of erythropoietin.
3. Stem cells increase red blood cell production
4. O_2 blood level returns to normal

Figure 6.4 Response of the kidneys to a decrease in blood oxygen concentration.
The kidneys release increased amounts of erythropoietin whenever the oxygen capacity of the blood is reduced. Erythropoietin stimulates the red bone marrow to speed up its production of red blood cells, which carry oxygen.

Connections and Misconceptions

What is blood doping?

Blood doping is any method of increasing the normal supply of RBCs for the purpose of delivering oxygen more efficiently, reducing fatigue, and giving athletes a competitive edge. To accomplish blood doping, athletes can inject themselves with EPO some months before the competition. These injections will increase the number of RBCs in their blood. Several weeks later, four units of their blood are removed and centrifuged to concentrate the RBCs. The concentrated RBCs are reinfused shortly before the athletic event. Blood doping is a dangerous, illegal practice. Several cyclists died in the 1990s from heart failure, probably due to blood that was too thick with cells for the heart to pump.

(EPO), which stimulates the stem cells in bone marrow to produce more red blood cells (Fig. 6.4). The liver and other tissues also produce EPO for the same purpose.

Animation Hemoglobin Breakdown

Disorders Involving Red Blood Cells

When there is an insufficient number of red blood cells or the cells do not have enough hemoglobin, the individual suffers from **anemia** and has a tired, run-down feeling. Iron, vitamin B_{12}, and the B vitamin folic acid are necessary for the production of red blood cells. *Iron-deficiency anemia* is the most common form. It results from inadequate intake of dietary iron, which causes insufficient hemoglobin synthesis. A lack of vitamin B_{12} causes *pernicious anemia,* in which stem-cell activity is reduced due to inadequate DNA production. As a consequence, fewer red blood cells are produced. *Folic-acid-deficiency anemia* also leads to a reduced number of RBCs, particularly during pregnancy. Pregnant women should consult with their health-care provider about the need to increase their intake of folic acid, because a deficiency can lead to birth defects in the newborn.

Hemolysis is the rupturing of red blood cells. In hemolytic anemia, the rate of red blood cell destruction increases. **Sickle-cell disease** is a hereditary condition in which the individual has sickle-shaped red blood cells that tend to rupture as they pass through the narrow capillaries. The RBCs look like those below. The problem arises because the protein in two of the four chains making up hemoglobin is abnormal. The life expectancy of sickle-shaped red blood cells is about 90 days instead of 120 days.

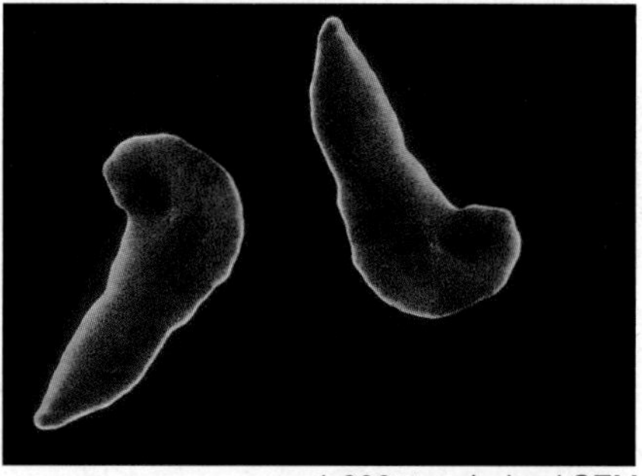
1,600×, colorized SEM

Connecting the Concepts

For additional information on the topics presented in this section, refer to the following discussions.

Section 9.6 details the process of gas exchange in the lungs.

Section 10.4 explains how the kidneys assist in maintaining red blood cell homeostasis.

Section 20.2 describes the patterns of inheritance associated with sickle-cell disease.

Check Your Progress 6.2

1. Describe the anatomy of an RBC and how that anatomy leads to the cell's ability to transport oxygen.
2. Describe the pros and cons of a mature RBC not containing a nucleus.
3. Detail a few blood disorders, including their causes and symptoms.

6.3 White Blood Cells and Defense Against Disease

Learning Outcomes

Upon completion of this section you should be able to

1. Explain the function of white blood cells in the body.
2. Distinguish between granular and agranular leukocytes.
3. Describe some of the disorders associated with white blood cells.

White blood cells (leukocytes) differ from red blood cells in that they are usually larger, have a nucleus, lack hemoglobin, and are translucent unless stained. White blood cells are not as numerous as red blood cells. There are only 5,000–11,000 per mm^3 of blood. White blood cells are derived from stem cells in the red bone marrow, where most types mature. There are several types of white blood cells (Fig. 6.5), and the production of each type is regulated by a protein called a **colony-stimulating factor (CSF).** In a person with normally functioning bone marrow, the numbers of white blood cells can double within hours, if needed. White blood cells are able to squeeze through pores in the capillary wall; therefore, they are also found in tissue fluid and lymph (Fig. 6.6).

White blood cells fight infection; thus, they are an important part of the immune system. The **immune system,** discussed in Chapter 7, consists of a variety of cells, tissues, and organs that defend the body against pathogens, cancer cells, and foreign proteins. Many white blood cells live only a few days—they probably die while fighting pathogens. Others live for months or even years.

White blood cells have various ways to fight infection. Certain ones are very good at phagocytosis. During phagocytosis, a projection from the cell surrounds a pathogen and literally engulfs it. A vesicle containing the pathogen is formed inside the cell. Lysosomes attach and empty their digestive enzymes into the vesicle. The enzymes breakdown the pathogen. Other white blood cells produce antibodies, proteins that combine with antigens. An **antigen** is a cell or other substance foreign to the individual. The antibody–antigen pair is then marked for destruction, again by phagocytosis. We will be describing antigens and antibodies involved in blood typing and coagulation in Section 6.5.

White Blood Cells	Function
Granular leukocytes	
• Neutrophils 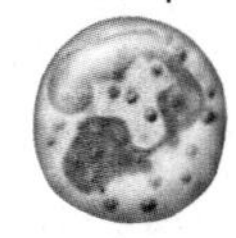	Phagocytize pathogens and cellular debris.
• Eosinophils 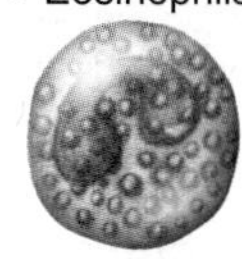	Use granule contents to digest large pathogens, such as worms, and reduce inflammation.
• Basophils 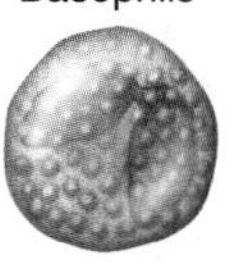	Promote blood flow to injured tissues and the inflammatory response.
Agranular leukocytes	
• Lymphocytes 	Responsible for specific immunity; B cells produce antibodies; T cells destroy cancer and virus-infected cells.
• Monocytes 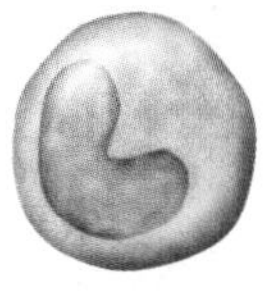	Become macrophages that phagocytize pathogens and cellular debris.

Figure 6.5 **Some examples of white blood cells.** Neutrophils, eosinophils, and basophils are granular leukocytes. Lymphocytes and monocytes have few, if any, granules (agranular).

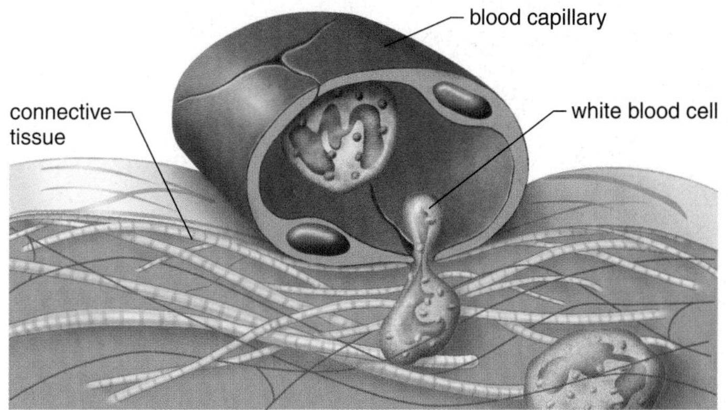

Figure 6.6 **Movement of white blood cells into the tissue.** White blood cells can squeeze between the cells of a capillary wall and enter body tissues.

Types of White Blood Cells

White blood cells are classified into the **granular leukocytes** and the **agranular leukocytes** (Fig. 6.5). Granulocytes have noticeable cytoplasmic granules, which can be easily seen when the cells are stained and examined with a microscope. Granules, like lysosomes, contain various enzymes and proteins. Agranulocytes contain only sparse, fine granules, which are not easily viewed under a microscope.

Granular Leukocytes

The granular leukocytes include neutrophils, eosinophils, and basophils.

Neutrophils account for 50–70% of all white blood cells. Therefore, they are the most abundant of the white blood cells. They have a multilobed nucleus, so they are called *polymorphonuclear leukocytes,* or "polys." The granules of neutrophils are not easily stained with acidic red dye, nor with basic purple dye. (This accounts for their name, neutrophil.) Neutrophils can be recognized by their numerous light-pink granules. Neutrophils are usually first responders to bacterial infection, and their intense phagocytic activity is essential to overcoming an invasion by a pathogen.

Eosinophils have a bilobed nucleus. Their large, abundant granules take up eosin and become a red color. (This accounts for their name, eosinophil.) Not much is known specifically about the function of eosinophils, but they increase in number in the event of a parasitic worm infection or an allergic reaction.

Basophils have a U-shaped or lobed nucleus. Their granules take up the basic stain and become a dark-blue color. (This accounts for their name, basophil.) In the connective tissues, basophils (and similar cells called **mast cells**) release histamine associated with allergic reactions. Histamine dilates blood vessels but constricts the air tubes that lead to the lungs, which is what happens during an asthma attack when someone has difficulty breathing.

Agranular Leukocytes

The agranular leukocytes include the lymphocytes and the monocytes. Lymphocytes and monocytes do not have granules and have nonlobular nuclei. They are sometimes called the mononuclear leukocytes.

Lymphocytes account for 25–35% of all white blood cells. Therefore, they are the second most abundant type of white blood cell. Lymphocytes are responsible for specific immunity to particular pathogens and toxins (poisonous substances). The lymphocytes are of two types: B cells and T cells. Mature B cells called plasma cells produce antibodies, the proteins that combine with target pathogens and mark them for destruction. Some T cells (cytotoxic T cells) directly destroy pathogens. The AIDS virus attacks one of several types of T cells. In this way, the virus causes immune deficiency, an inability to defend the body against pathogens. B lymphocytes and T lymphocytes are discussed more fully in Chapter 7.

Monocytes are the largest of the white blood cells. After taking up residence in the tissues, they differentiate into even larger macrophages. In the skin, they become dendritic cells. Like the neutrophils, macrophages and dendritic cells are active phagocytes, destroying pathogens, old cells, and cellular debris. Macrophages and dendritic cells also stimulate other white blood cells, including lymphocytes, to defend the body.

Disorders Involving White Blood Cells

Immune deficiencies are sometimes inherited. Children have **severe combined immunodeficiency disease (SCID)** when the stem cells of white blood cells lack an enzyme called adenosine deaminase. Without this enzyme, B and T lymphocytes do not develop and the body cannot fight infections. About 100 children are born with the disease each year. Injections of the missing enzyme can be given twice weekly, but a bone marrow transplant from a compatible donor is the best way to cure the disease.

Cancer is due to uncontrolled cell growth. **Leukemia,** which means "white blood," refers to a group of cancers that involve uncontrolled white blood cell proliferation. Most of these white blood cells are abnormal or immature. Therefore, they are incapable of performing their normal defense functions. Each type of leukemia is named for the type of cell dividing out of control. For example, lymphocytic leukemia involves abnormal lymphocyte proliferation.

An Epstein–Barr virus (EBV) infection of lymphocytes is the cause of **infectious mononucleosis,** so named because the lymphocytes are mononuclear. EBV, a member of the herpes virus family, is one of the most common human viruses. Symptoms of infectious mononucleosis are fever, sore throat, and swollen lymph glands. Although symptoms usually disappear in one or two months without medication, EBV remains dormant and hidden in a few cells in the throat and blood for the rest of a person's life. Stress can reactivate the virus. Reactivation means that a person's saliva can pass on the infection to someone else, as with intimate kissing. This is why mononucleosis is called the "kissing disease."

Connecting the Concepts

White blood cells play an important role in the defense against disease. For more information on these cells, refer to the following discussions.

Section 7.3 describes the role of the monocytes in the innate immune response.

Section 7.4 details how the lymphocytes are involved in the acquired defenses.

Section S.1 explores how the HIV virus infects white blood cells, resulting in the disease called AIDS.

Check Your Progress 6.3

1. Explain why there are different types of white blood cells.
2. Describe the structure and function of each type of white blood cell.
3. Summarize three disorders of white blood cells.

6.4 Platelets and Blood Clotting

Learning Outcomes

Upon completion of this section you should be able to

1. Explain how blood clotting relates to homeostasis.
2. List the steps in the formation of a blood clot.
3. Describe disorders associated with blood clotting.

Platelets (thrombocytes) result from fragmentation of large cells, called **megakaryocytes,** in the red bone marrow. Platelets are produced at a rate of 200 billion a day, and the blood contains 150,000–300,000 per mm^3. These formed elements are involved in the process of blood **clotting,** or coagulation. Also involved are the plasma proteins prothrombin and fibrinogen, manufactured in the liver and deposited in the blood. Vitamin K is necessary to the production of prothrombin.

Blood Clotting

The blood-clotting process helps the body maintain homeostasis in the cardiovascular system by ensuring that the plasma and formed elements remain within the blood vessels. At least 12 clotting factors and calcium ions (Ca^{2+}) participate in the formation of a blood clot.

Clot formation is initiated when a blood vessel is damaged (Fig. 6.7). At the site of the break in the blood vessel, platelets and damaged tissue release **prothrombin activator,** which, with the assistance of calcium ions, converts the plasma protein **prothrombin** to thrombin. **Thrombin,** in turn, acts as enzyme "scissors" to cut two short amino acid chains from each fibrinogen molecule. The fibrin fragments then join end to end, forming long threads of **fibrin.** Fibrin threads wind around the platelet plug in the damaged area of the blood vessel and provide the framework for the clot. Red blood cells trapped within the fibrin threads make the clot appear red. A fibrin clot is temporary. Once blood vessel repair starts, an enzyme called plasmin destroys the fibrin network so tissue cells can grow.

After blood clots, a yellowish fluid called **serum** escapes from the clot. It contains all the components of plasma except fibrinogen and prothrombin.

Disorders Related to Blood Clotting

An insufficient number of platelets is called **thrombocytopenia.** Thrombocytopenia is either due to low platelet production in bone marrow or increased breakdown of platelets outside the marrow. A number of conditions, including leukemia, can lead to thrombocytopenia. It can also be drug-induced. Symptoms include bruising, rash, and nosebleeds or bleeding in the mouth. Gastrointestinal bleeding or bleeding in the brain are possible complications.

If the lining of a blood vessel becomes roughened, a clot can form spontaneously inside an unbroken blood vessel. Most often, roughening occurs because an atherosclerotic plaque has formed (see Chapter 5). Rarely, the vessel lining is damaged during placement of an intravenous tube. The spontaneous clot is called a *thrombus* (pl., thrombi) if it remains stationary inside the blood vessel. Sitting for long periods, as when traveling, can also cause thrombus formation. If the clot dislodges and travels in the blood, it is called an embolus. If a **thromboembolism** is not treated, blood flow to the tissues can stop completely. Heart attack or stroke can result, as discussed in Chapter 5.

Hemophilia is an inherited clotting disorder that causes a deficiency in a clotting factor. There are many forms of the disorder. Hemophilia A, caused by deficiency of clotting factor VIII, is more likely to occur in boys than in girls. Hemophilia A is caused by an abnormal copy of the factor VIII production gene, found on the X chromosome. This hemophilia arises when a boy has an abnormal gene on his single X chromosome. Girls only need one normal gene between their two X chromosomes to make the normal amounts of clotting factor VIII (see Chapter 20). In hemophilia, the slightest bump can

1. Blood vessel is punctured.

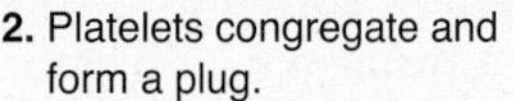

2. Platelets congregate and form a plug.

3. Platelets and damaged tissue cells release prothrombin activator, which initiates a cascade of enzymatic reactions.

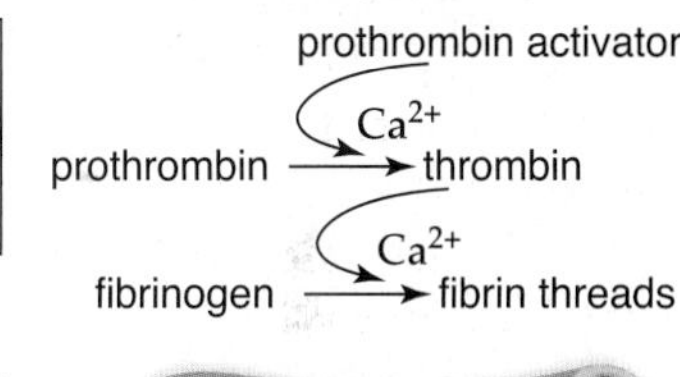

4. Fibrin threads form and trap red blood cells.

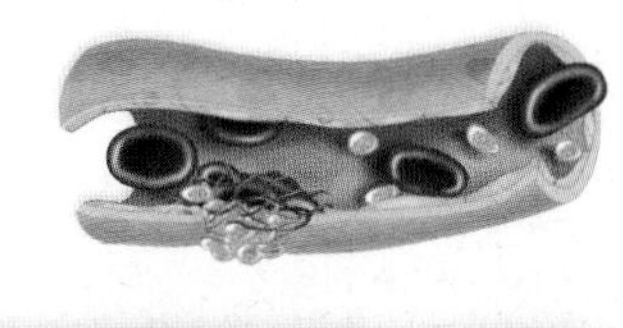

a. Blood-clotting process

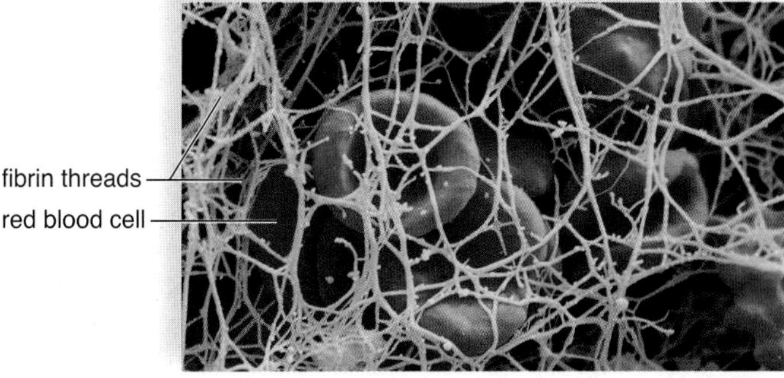

b. Blood clot 4,400×

Figure 6.7 **The steps in the formation of a blood clot.**
a. Platelets and damaged tissue cells release prothrombin activator, which acts on prothrombin in the presence of Ca^{2+} (calcium ions) to produce thrombin. Thrombin acts on fibrinogen in the presence of Ca^{2+} to form fibrin threads.
b. A scanning electron micrograph of a blood clot shows red blood cells caught in the fibrin threads.

Biology Matters Health Focus

What to Know When Giving Blood

Congratulations! You've decided to help another person—perhaps an accident victim, a newborn infant, someone with sickle-cell anemia, or a leukemia patient. Donating blood is fairly easy and will take approximately 1 hour of your time.

The Procedure

After you register, an attendant will ask private and confidential questions about your health and lifestyle and will answer your questions. Your temperature, blood pressure, and pulse will be recorded. A drop of your blood is tested to ensure that you're not anemic.

The supplies used for your donation are sterile and are used only for you. You can't be infected with a disease when donating blood. When the actual donation is started, you may feel a brief "sting." The procedure takes about 10 minutes, and you will have given about a pint of blood (Fig. 6A). Your body replaces the liquid part (plasma) in hours and the cells in a few weeks.

You will have several opportunities both prior to and after giving blood to let Red Cross officials know whether you consider your blood to be safe. Immediately after you donate, you are given a number to call if you decide that your blood may not be safe to give to another person. Donated blood is tested for syphilis bacteria and AIDS antibodies, as well as hepatitis and other viruses. You are notified if tests are positive, and your blood won't be used if it could make someone ill. However, you should *never* use the process of a blood donation to get tested for any medical condition, especially AIDS. It is possible to have a negative result for AIDS antibodies and yet still spread the virus, because forming antibodies takes several weeks after exposure.

The Cautions

Some medications and medical conditions have waiting periods before you can donate blood. Before giving blood, you should inform the medical staff if you meet any of the conditions in the following list:

- You have recently had an infection or a fever.
- You have taken or are taking drugs that slow blood clotting. You should wait 48 hours after taking aspirin or aspirin-related drugs.
- You have had malaria, have taken drugs for malaria prevention, or have traveled to malaria-prone countries.
- You have a medical history of hepatitis or tuberculosis.
- You have been treated for syphilis or gonorrhea in the last 12 months.

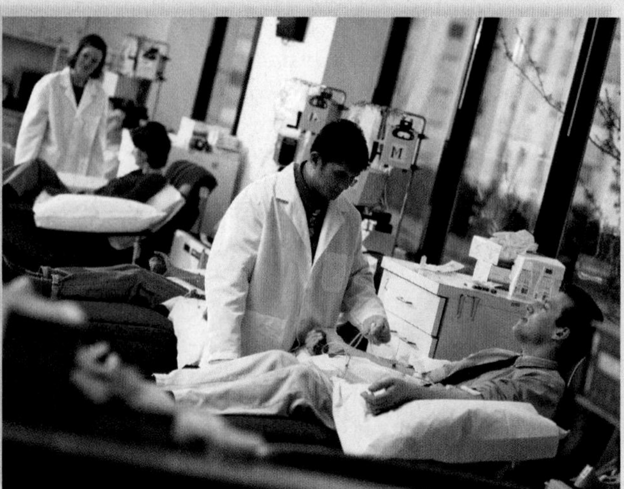

Figure 6A Donating blood can help save a life.

- You have AIDS, have had a positive HIV test, or are at risk for getting an HIV infection due to one of the following:
  - You have ever injected illegal drugs.
  - You have taken clotting factor concentrates for hemophilia.
  - You have been given money or drugs for sex since 1977.
  - You had a sexual partner within the last year who did any of the above things.

For men:

- You had sex *even once* with another man since 1977, *or* had sex with a female prostitute within the last year.

For women:

- You had sex with a male or female prostitute within the last year, *or* had a male sexual partner who had sex with another man *even once* since 1977.

When You're Finished

An area is provided to relax after donation, with drinks and snacks to enjoy. Plan to rest for a few minutes.

Most people feel fine while they give blood and afterward, but a few donors have an upset stomach or feel faint or dizzy after donation. Resting, drinking fluid, and eating a snack usually help. Occasionally, bruising, redness, and pain occur at your donation site, so avoid strenuous exercise and lifting for a day or so. Very rarely, a person may have muscle spasms and/or suffer nerve damage. You should contact your physician if you have any concerns about potential side effects of donating blood.

For more information about blood donation, including eligibility requirements, visit the Red Cross website at www.redcrossblood.org.

cause bleeding into the joints. Cartilage degeneration in the joints and absorption of underlying bone can follow. The most frequent cause of death is bleeding into the brain with accompanying neurological damage. Regular injections of factor VIII can successfully treat the disease.

Connections and Misconceptions

What are some other forms of clotting disorders?

In addition to hemophilia A, there are several other forms of clotting disorders. Hemophilia B, also called *Christmas disease* after the first person to be diagnosed with the disease (Stephen Christmas in 1952), is caused by a deficiency in clotting factor IX. Von Willebrand disease is caused by a deficiency in the gluelike protein that holds the platelets together at the site of the damaged blood vessel. All of these disorders may be partially treated with injections of the missing proteins.

Connecting the Concepts

For more information on the disorders discussed in this section, refer to the following discussions.

Section 5.7 describes how blood clots relate to disorders of the cardiovascular system.

Section 20.4 explores the inheritance of hemophilia in the royal families of Europe.

Check Your Progress 6.4

1. List the components of the blood that are involved in the formation of a blood clot.
2. Briefly describe the stages of blood clotting.
3. Briefly describe a few blood-clotting disorders.

6.5 Blood Typing and Transfusions

Learning Outcomes

Upon completion of this section you should be able to

1. Explain what determines blood types in humans.
2. Predict the compatibility of blood types for a transfusion.
3. Summarize the role of Rh factor in hemolytic disease of the newborn.

A **blood transfusion** is the transfer of blood from one individual to another. For transfusions to be done safely, blood must be typed so that **agglutination** (clumping of red blood cells) does not occur when blood from different people is mixed. Blood typing usually involves determining the ABO blood group and whether the individual is Rh-negative (Rh–) or Rh-positive (Rh+).

ABO Blood Groups

Only certain types of blood transfusions are safe because the plasma membranes of red blood cells carry glycoproteins that can be antigens to other individuals. An antigen is any substance that is foreign to a person's body. ABO blood typing is based on the presence or absence of two possible antigens, called type A antigen and type B antigen. Whether these antigens are present or not depends on the particular inheritance of the individual.

In Figure 6.8, type A antigen and type B antigen are given different shapes and colors. Study each figure to see

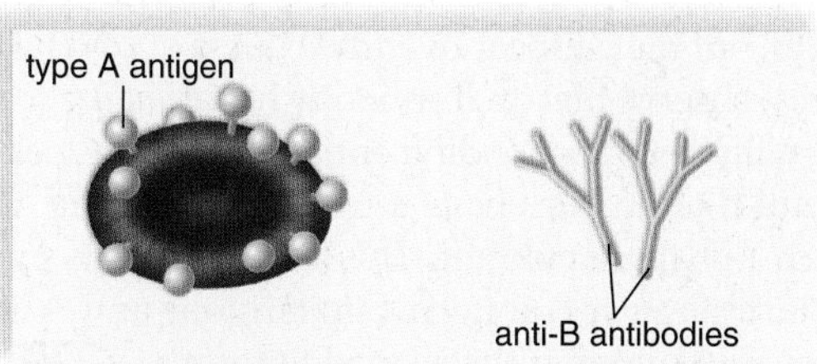

Type A blood. Red blood cells have type A surface antigens. Plasma has anti-B antibodies.

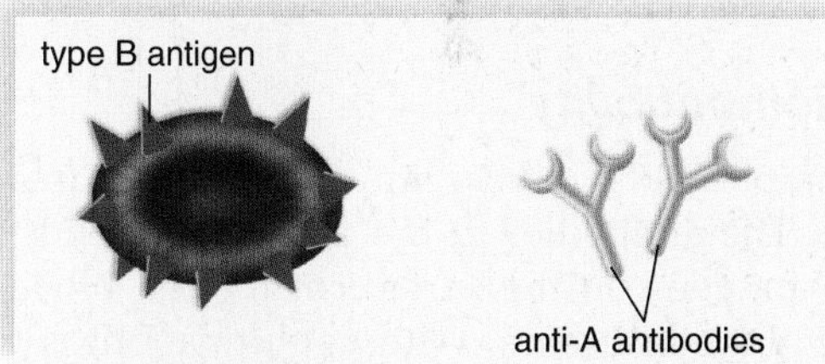

Type B blood. Red blood cells have type B surface antigens. Plasma has anti-A antibodies.

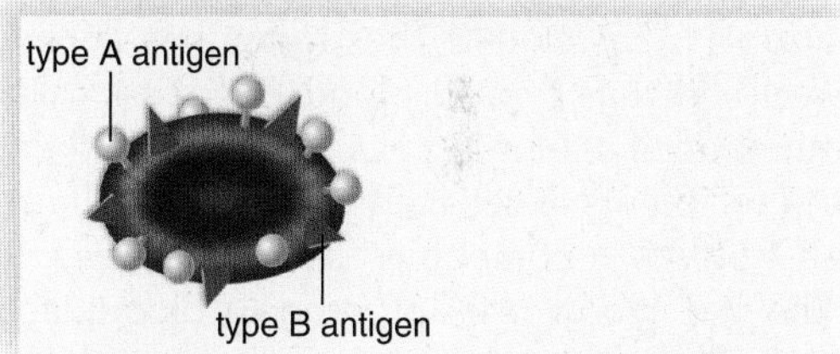

Type AB blood. Red blood cells have type A and type B surface antigens. Plasma has neither anti-A nor anti-B antibodies.

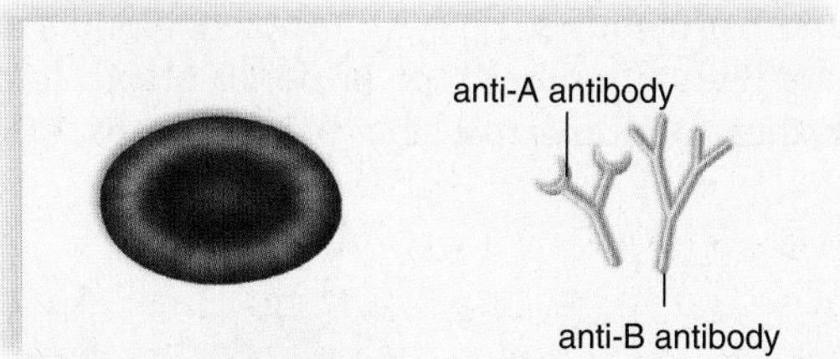

Type O blood. Red blood cells have neither type A nor type B surface antigens. Plasma has both anti-A and anti-B antibodies.

Figure 6.8 **The ABO blood type system.**
In the ABO system, blood type depends on the presence or absence of antigens A and B on the surface of red blood cells. In these drawings, A and B antigens are represented by different shapes on the red blood cells. The possible anti-A and anti-B antibodies in the plasma are shown for each blood type. An anti-B antibody cannot bind to an A antigen, and vice versa.

the type(s) of antigens found on the red blood cell plasma membrane. As you might expect, type A blood has the A antigen (blue spheres) and type B blood has the B antigen (violet triangles). If you guessed that type AB blood has both A and B antigens, you're correct. Finally, notice that type O blood has neither antigen on its red blood cells.

Each blood type has antibodies (yellow Y-shaped molecules) that correspond to the *opposite* blood type. Thus, an individual with type A blood has anti-B antibodies in the plasma, and a person with type B blood has anti-A antibodies in the plasma. Further, a person with type O blood has both antibodies in the plasma and someone with type AB blood lacks both antibodies (Fig. 6.8). Anti-A and/or anti-B antibodies are not present at birth, but they appear over the course of several months.

Finally, observe that each antibody has a *binding site* that will combine with its corresponding antigen in a tight lock-and-key fit. The anti-B antibodies have a triangular binding site on the top of each Y-shaped molecule. This binding site fits snugly with the purple, triangular B antigen. Similarly, the anti-A antibodies have a spherical binding site shaped to form a perfect fit with the A antigen. The presence of these antibodies can cause agglutination.

MP3 Blood Groupings

Blood Compatibility

Blood compatibility is very important when transfusions are done. The antibodies in the plasma must not combine with the antigens on the surface of the red blood cells or else agglutination occurs. With agglutination, anti-A antibodies have combined with type A antigens, and anti-B antibodies have combined with type B antigens. Therefore, agglutination is expected if the donor has type A blood and the recipient has type B blood (Fig. 6.9). What about other combinations of blood types? Try out all other possible donors and recipients to see if agglutination will occur.

Type O blood is sometimes called the *universal donor* because the red blood cells of type O blood lack A and B antigens. Type O donor blood should not agglutinate with any other type of recipient blood. Likewise, type AB blood is sometimes called *universal recipient* blood because the plasma lacks A and B antibodies. Type AB recipient blood should not agglutinate with any other type of donor blood. In practice, however, there are other possible blood groups, aside from ABO blood groups. Before blood can be safely transfused from one person to another, it is necessary to physically combine donor blood with recipient blood on a glass slide, then observe whether agglutination occurs. This procedure, called blood-type crossmatching, takes only a few minutes. It is done before every blood transfusion is performed. Type O blood donation without crossmatching is performed only in an emergency, when blood loss is severe and the patient's survival is at stake.

Connections and Misconceptions

What is the most common ABO blood type? What is the least common ABO blood type?

It depends on your genetic inheritance. In the United States and other countries with a large immigrant population, the most common ABO type is type O. However, type A blood is most common in several European countries, including Portugal, Armenia, Switzerland, and Norway. Worldwide, type B blood is common in central Asia, India, and the Far East. Worldwide, the least common ABO type is type AB.

Rh Blood Groups

The designation of blood type usually also includes whether the person has or does not have the Rh factor on the red blood cell. Rh– individuals normally do not have antibodies to the Rh factor, but they make them when exposed to the Rh factor.

If a mother is Rh– and the father is Rh+, a child can be Rh+. During a pregnancy, Rh+ antigens can leak across the placenta into the mother's bloodstream. The presence of these Rh+ antigens causes the mother to produce anti-Rh antibodies (Fig. 6.10). Usually, in a subsequent pregnancy with another Rh+ baby, the anti-Rh antibodies may cross the placenta and destroy the unborn child's red blood cells. This is called *hemolytic disease of the newborn (HDN)* because hemolysis starts in the womb and continues after the baby is born. Due to red blood cell destruction, the baby will be severely anemic. Excess hemoglobin breakdown products in the blood can lead to brain damage and mental retardation, or even death.

The Rh problem is prevented by giving Rh– women an Rh immunoglobulin injection no later than 72 hours after giving

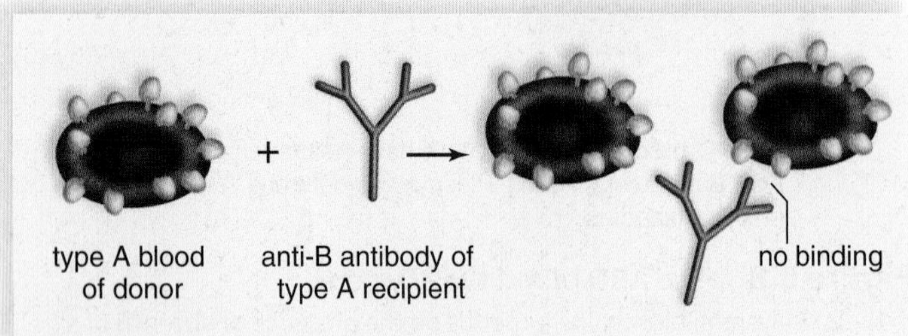

a. No agglutination

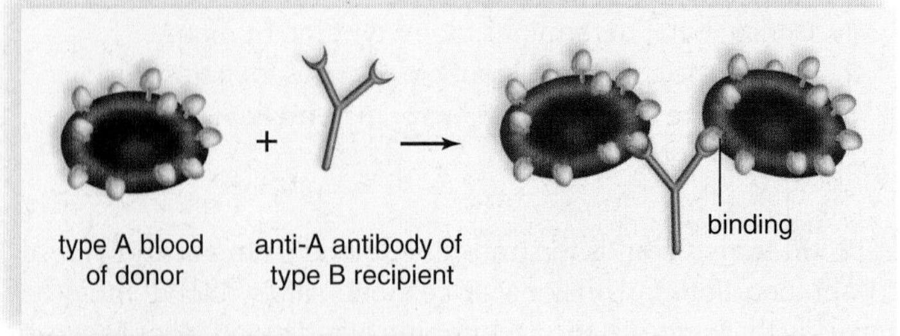

b. Agglutination

Figure 6.9 Blood compatibility and agglutination.
No agglutination occurs in (**a**) because anti-B antibodies cannot combine with the A antigen. Agglutination occurs in (**b**) when anti-A antibodies in the recipient combine with A antigen on donor red blood cells.

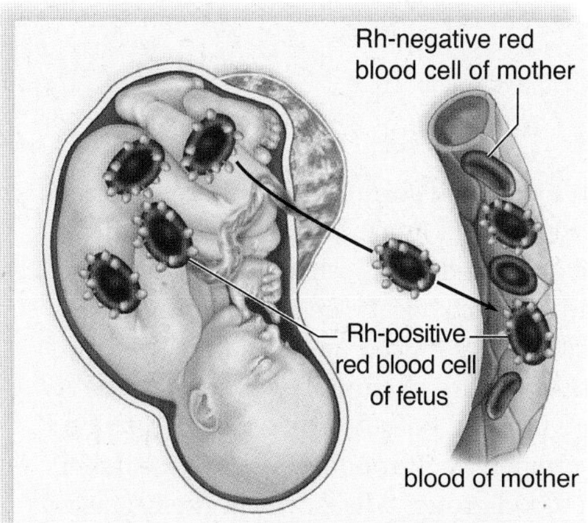

a. Fetal Rh-positive red blood cells leak across placenta into mother's bloodstream.

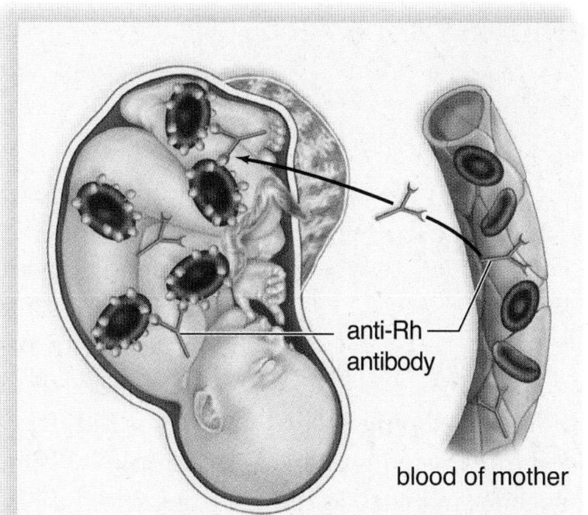

b. Mother forms anti-Rh antibodies that cross the placenta and attack fetal Rh-positive red blood cells.

Figure 6.10 **Rh factor disease (hemolytic disease of the newborn).** **a.** Due to a pregnancy in which the child is Rh-positive, an Rh-negative mother can begin to produce antibodies against Rh-positive red blood cells. **b.** Usually in a subsequent pregnancy, these antibodies can cross the placenta and cause hemolysis of an Rh-positive child's red blood cells.

birth to an Rh+ child. This injection contains anti-Rh antibodies that attack any of the baby's red blood cells in the mother's blood before these cells can stimulate her immune system to produce her own antibodies. This treatment is termed RhoGAM, and does not harm the newborn's red blood cells. Treatment is not beneficial if the woman has already begun to produce antibodies. Therefore, the timing of the injection is most important.

Connections and Misconceptions

Is ABO blood typing accurate?

Usually, the answer is yes. However, there are some interesting genetic disorders that may complicate traditional ABO blood typing. For example, individuals with Bombay syndrome lack the enzyme to correctly attach A and B antigens to the surface of the red blood cells. These individuals may carry the genes to produce the A and B antigens on the surface of the red blood cells, but since they are not attached, they may appear to have type O blood.

Connecting the Concepts

For more information on the chemistry and genetics of blood types, refer to the following discussions.

Section 3.3 describes the structure and function of glycoproteins.

Section 20.3 examines patterns of inheritance associated with human ABO blood types.

Check Your Progress 6.5

1. Explain what determines blood type and list the four different types of blood.
2. Solve the following: If a person has type A blood, to whom may he or she donate blood?
3. Explain what causes hemolytic disease of the newborn.

6.6 Homeostasis

Learning Outcomes

Upon completion of this section you should be able to

1. Summarize how the cardiovascular system interacts with other body systems to maintain homeostasis.

Think of how important it is to homeostasis for all the human systems to work together. For example, imagine one of your worst nightmares as a student. After studying all night for a final exam, you oversleep the next morning. You wake up just a few minutes before class begins. Immediately, your heart starts beating faster and your blood pressure rises, due to sympathetic *nervous system* stimulation (Chap. 13). Hormones, such as epinephrine (adrenaline) from the adrenal glands, pour into the bloodstream to prolong your body's physical response to stress. The *endocrine system* is at work (Chap. 15).

Frantic, you throw on your clothes and sprint to the bus. Your *muscular* and *respiratory systems* meet the demand (Chaps. 9 and 12). As you run, your muscles require more oxygen. Breathing and heart rate increase to meet the demand. Carbon dioxide generated by hard-working muscles is carried in the blood to the lungs so it can be expelled. To eliminate the heat produced by your muscles, you sweat and blood flow to your skin increases.

The nightmare continues. In your mad dash up the steps to class, you fall, skinning your knee. At first, blood flows from the wound, but a clot soon seals off the injured area. As a scab forms, healing has already begun. Any injury that breaks the skin can allow harmful bacteria into the body, but the immune system usually deals quickly with invaders. Later, you might notice that your knee is not only skinned but also black and blue. Bruises form when blood leaks out of damaged vessels underneath the skin and then clots.

Biology Matters Historical Focus

Making Blood Transfusion Possible: Karl Landsteiner (1868–1943)

"Time is so short and there is so much to do. We must hurry."
Karl Landsteiner

It's hard to imagine, in this age of rapidly developing medical technology, that not too long ago blood transfusion was impossible. Injuries with severe bleeding, such as an injury to the chest or abdomen, were usually fatal. Likewise, most perished after limb amputations (though amazingly enough, historical accounts and primitive prosthetic limbs show that a lucky few survived). Uncontrolled bleeding afer childbirth claimed the lives of many women. Early transfusion experiments involved the blood transfer from animals to humans and were a dismal failure. Similarly, human-to-human transfusion was likely to kill the blood recipient. Microscopic examinations of transfused red blood cells showed that the cells clumped together in a process called agglutination. Subsequently, the cells broke open and released their hemoglobin, causing shock and death. After a few such incidents, transfusions were abandoned.

The work of Dr. Karl Landsteiner made blood transfusion possible. Using human blood samples, Landsteiner first separated the cells from the liquid (called serum). Mixing the cells from one donor with serum from a second donor sometimes caused blood cells to agglutinate and then break. However, at other times, when blood and serum from different donors were mixed, the cells remained intact. Landsteiner concluded that molecules found on the red blood cell membranes divided blood into three groups, which he labeled A, B, and O. Mixing type A with type B always produced agglutination, but mixing bloods of the same type (for example, type A with type A) never did. Blood type O could be safely mixed with all blood samples without producing agglutination. Later researchers described the fourth blood group, AB, as well as the Rh factor and other molecules now used for blood typing.

Landsteiner's work enabled the first successful human transfusion in 1907. Adding anticoagulants (chemicals that prevent blood from clotting) made blood storage and blood banking possible. Transfusions saved many lives during both world wars. Victims of traumatic injury or hemorrhage following childbirth could now survive, even after a tremendous blood loss. Chest and abdominal surgeries, previously impossible because of potential blood loss, became routine.

The use of blood typing to identify criminals and their victims, called forensic serology, also evolved from Dr. Landsteiner's work. Paternity testing based on blood typing was now possible as well (although it has since been largely replaced by much more definitive DNA testing). Landsteiner, called the Father of Modern Serology (the science of blood identification), was awarded the Nobel Prize in Medicine in 1930.

You make it to class just in time. As you settle down to take the exam, your heartbeat and breathing gradually slow, because the muscle need is not as great. As you concentrate, blood flow increases in certain highly active parts of your brain. The brain is a very demanding organ in terms of both oxygen and glucose.

You start to feel hungry. There is little blood glucose to provide cellular fuel, so glycogen stored in the liver is broken down to make glucose available to blood and body cells. Finally, you return the exam and eat breakfast afterward. When most nutrients from your meal are absorbed by the *digestive system,* they enter the bloodstream (Chap. 8). However, fats enter lacteals, vessels of the *lymphatic system,* for transport to the blood (Chap. 7).

Notice in this scenario how many organ systems interacted with the cardiovascular system. The nervous, endocrine, muscular, respiratory, digestive, and lymphatic systems helped to serve cellular needs. While all of this was occurring, the urinary system (Chap. 10) assisted in maintaining homeostasis by regulating ion concentrations in the blood and removing waste material. In other words, to promote homeostasis, all of the organ systems must work together.

How Body Systems Work Together

Figure 6.11 summarizes how these and other systems cooperate with the cardiovascular system. You previously learned that the body's internal environment contains blood and tissue fluid. Tissue fluid originates from blood plasma but contains very little plasma protein. Tissue fluid is absorbed by lymphatic capillaries, after which it is referred to as lymph. The lymph courses through lymphatic vessels, eventually returning to the venous system. Thus, the cardiovascular and lymphatic systems are intimately linked.

Homeostasis is possible only if the cardiovascular system delivers oxygen from the lungs, as well as nutrients from the digestive system, to the tissue fluid surrounding cells. Simultaneously, the cardiovascular system also removes metabolic wastes, delivering waste to excretory organs.

The three components of the muscular system make essential contributions to blood movement. Cardiac muscle contractions circulate blood throughout the body. Contraction or relaxation of the smooth muscle in blood vessel walls changes vessel diameter and helps to maintain the correct blood pressure. Further, skeletal muscle contraction compresses both cardiovascular and lymphatic veins. Lymph

All systems of the body work with the cardiovascular system to maintain homeostasis. These systems in particular are especially noteworthy.
Cardiovascular System
Heart pumps the blood. Blood vessels transport oxygen and nutrients to the cells of all the organs and transport wastes away from them. The blood clots to prevent blood loss. The cardiovascular system also specifically helps the other systems as mentioned below.
Digestive System
Blood vessels deliver nutrients from the digestive tract to the cells. The digestive tract provides the molecules needed for plasma protein formation and blood cell formation. The digestive system absorbs the water needed to maintain blood pressure and the Ca^{2+} needed for blood clotting.
Urinary System
Blood vessels transport wastes to be excreted. Kidneys excrete wastes and help regulate the water-salt balance necessary to maintain blood volume and pressure and help regulate the acid-base balance of the blood.
Muscular System
Muscle contraction keeps blood moving through the heart and in the blood vessels, particularly the veins.
Nervous System
Nerves help regulate the contraction of the heart and the constriction/dilation of blood vessels.
Endocrine System
Blood vessels transport hormones from glands to their target organs. The hormone epinephrine increases blood pressure; other hormones help regulate blood volume and blood cell formation.
Respiratory System
Blood vessels transport gases to and from lungs. Gas exchange in lungs supplies oxygen and rids the body of carbon dioxide, helping to regulate the acid-base balance of blood. Breathing aids venous return.
Lymphatic System
Capillaries are the source of tissue fluid, which becomes lymph. The lymphatic system helps maintain blood volume by collecting excess tissue fluid (i.e., lymph), and returning it via lymphatic vessels to the cardiovascular veins.
Skeletal System
The rib cage protects the heart, red bone marrow produces blood cells, and bones store Ca^{2+} for blood clotting.

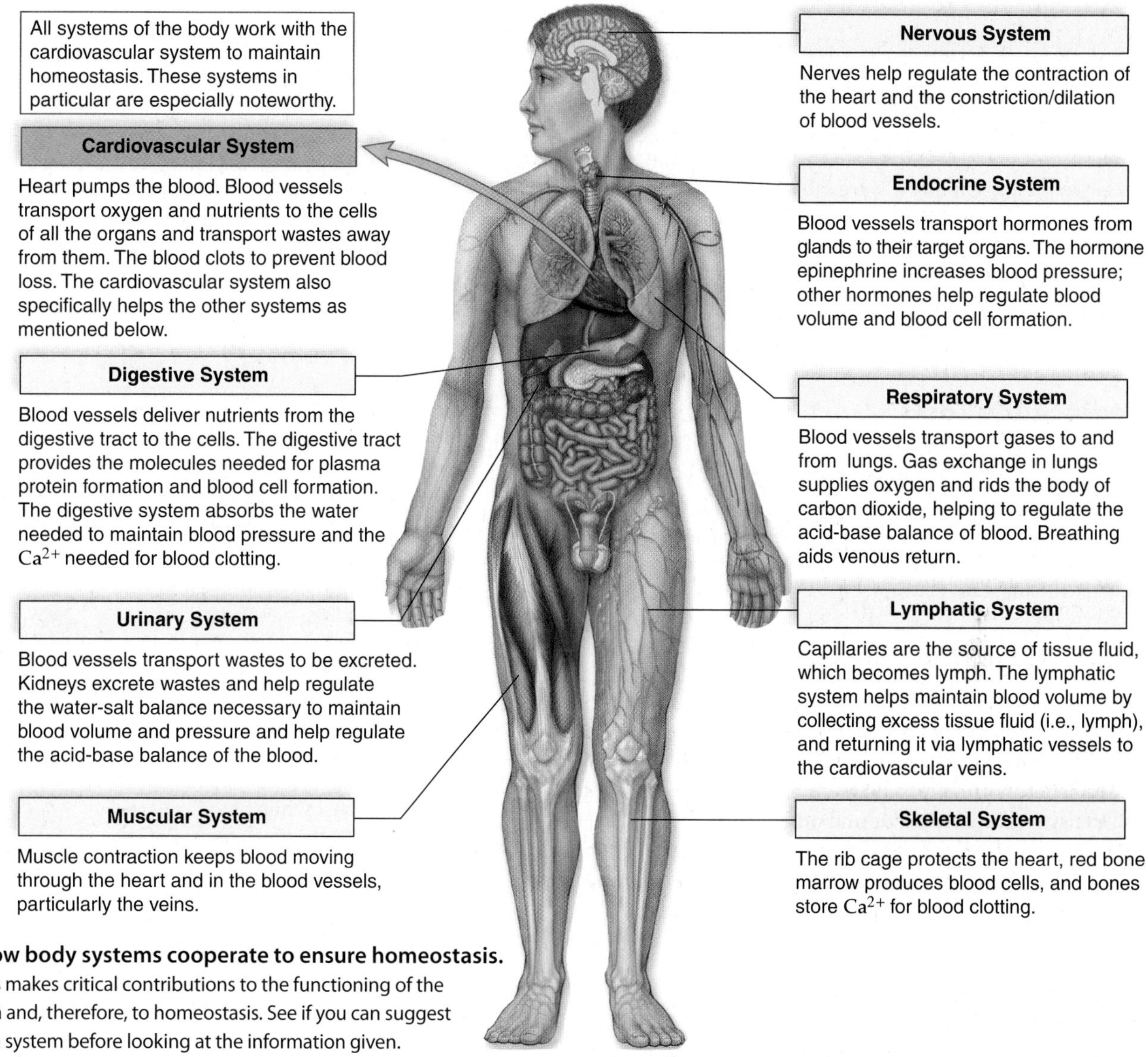

Figure 6.11 How body systems cooperate to ensure homeostasis. Each of these systems makes critical contributions to the functioning of the cardiovascular system and, therefore, to homeostasis. See if you can suggest contributions by each system before looking at the information given.

returns to cardiovascular veins, and blood in the cardiovascular veins drains back to the heart. The circulation of tissue fluid and blood is then complete.

The skeletal and endocrine systems are vital to cardiovascular homeostasis. Red bone marrow produces blood cells. Without the needed blood cells, the person becomes anemic and lacks an immune response. In addition, bones contribute calcium ions (Ca^{2+}) to the process of blood clotting. Without blood clotting, bleeding from an injury (even something as simple as skinning a knee) could be fatal. Both blood cell production and bone calcium release are regulated by hormones. Once again, the endocrine system cooperates with the cardiovascular system.

Finally, we must not forget that the urinary system has functions besides producing and excreting urine. The kidneys help regulate the acid–base and salt–water balances of blood and tissue fluid. Erythropoietin, a hormone produced by the kidneys, stimulates red blood cell production. The urinary system joins the muscular, skeletal, and endocrine systems to maintain the internal environment.

Connecting the Concepts

For more information on how the cardiovascular system interacts with other body systems, refer to the following discussions.

Section 7.2 examines the interaction of the cardiovascular system and lymphatic system.

Section 10.3 explores how the urinary system removes wastes from the blood.

Check Your Progress 6.6

1. Explain how the functions of the cardiovascular system contribute to homeostasis.
2. State how each system listed in Figure 6.11 interacts with the cardiovascular system.
3. List a few examples of disorders that affect homeostasis and describe why.

CASE STUDY CONCLUSION

Leukemia is a cancer of the blood. Like any cancer, leukemia is caused by uncontrolled cell growth. In Ben's case, his symptoms could be explained by the fact that the white blood cells in his bone marrow were multiplying uncontrollably and blocking out the ability of the red blood cells and platelets to complete their normal tasks. Therefore, he felt tired and was unable to heal correctly. The treatment for leukemia almost always involves chemotherapy—the use of chemicals to kill the cells that are growing uncontrollably and restore the normal balance of formed elements in the blood. Increasingly, people with leukemia are using bone marrow stem-cell transplants. In this case, either chemotherapy or radiation treatment (sometimes both) is used to kill all of the bone marrow cells. Stem cells from a compatible donor are then inserted into the bone marrow, where they will hopefully establish a new population of healthy blood cells and platelets.

Media Study Tools

www.mhhe.com/maderhuman12e

Enhance your study of this chapter with study tools and practice tests. Also ask your instructor about the resources available through ConnectPlus, including the media-rich eBook, interactive learning tools, and animations.

Summarizing the Concepts

6.1 Blood: An Overview

Blood functions to

- transport hormones, oxygen, and nutrients to cells;
- transport carbon dioxide and other wastes from cells;
- fight infections and perform various regulatory functions;
- maintain blood pressure and regulate body temperature; and
- keep the pH of body fluids within normal limits.

These functions help maintain homeostasis.

Blood has two main components: plasma and formed elements (red blood cells, white blood cells, and platelets).

Plasma

- 91% of blood plasma is water.
- Plasma proteins (albumins, globulins, and fibrinogen) are mostly produced by the liver.
- Plasma proteins maintain osmotic pressure and help regulate pH. Albumins transport other molecules, globulins function in immunity, and prothrombin and fibrinogen enable blood clotting.

6.2 Red Blood Cells and Transport of Oxygen

Red blood cells lack a nucleus. Instead, they contain hemoglobin, which combines with oxygen and transports it to the tissues. Hemoglobin assists in carbon dioxide transport, as well.

red blood cells

Red blood cell production is controlled by the blood oxygen concentration. When oxygen concentration decreases, the kidneys increase production of the hormone erythropoietin. In response, more red blood cells are produced by the bone marrow.

6.3 White Blood Cells and Defense Against Disease

White blood cells are larger than red blood cells. They have a nucleus and are translucent unless stained. White blood cells are either granular leukocytes or agranular leukocytes.

- The granular leukocytes are eosinophils, basophils, and neutrophils. Neutrophils are abundant, respond first to infections, and phagocytize pathogens.

eosinophil

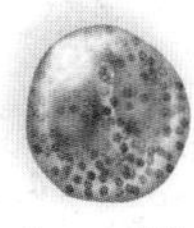
basophil

neutrophil

- The agranular leukocytes include monocytes and lymphocytes. Monocytes are the largest white blood cells. They can become macrophages that phagocytize pathogens and cellular debris. Lymphocytes (B cells and T cells) are responsible for specific immunity.

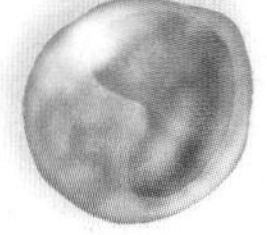
monocyte

lymphocyte

All blood cells are produced within red bone marrow from stem cells. Red blood cells live about 120 days and are eventually destroyed in the liver and spleen. Most white blood cells survive for only a few days, but some persist for months to years.

6.4 Platelets and Blood Clotting

Platelets result from fragmentation of megakaryocytes in the red bone marrow and function in blood clotting.

Blood Clotting

Platelets and two plasma proteins, prothrombin and fibrinogen, function in blood clotting, an enzymatic process. Fibrin threads that trap red blood cells result from the enzymatic reaction.

6.5 Blood Typing and Transfusions

Blood typing usually involves determining the ABO blood group and whether the person is Rh– or Rh+. Determining blood type is necessary for transfusions so that agglutination of red blood cells does not occur.

ABO Blood Groups

ABO blood typing determines the presence or absence of type A antigen and type B antigen on the surface of red blood cells.

- **Type A Blood** Type A surface antigens; plasma has anti-B antibodies.

- **Type B Blood** Type B surface antigens; plasma has anti-A antibodies.
- **Type AB Blood** Both type A and type B surface antigens; plasma has neither anti-A nor anti-B antibodies (universal recipient).
- **Type O Blood** Neither type A nor type B surface antigens. Plasma has both anti-A and anti-B antigens (universal donor).
- **Agglutination** Agglutination occurs if the corresponding antigen and antibody are mixed (i.e., if the donor has type A blood and the recipient has type B blood).

Rh Blood Groups

The Rh antigen must also be considered when transfusing blood. It is very important during pregnancy because an Rh– mother may form antibodies to the Rh antigen while carrying or after the birth of an Rh+ child. These antibodies can cross the placenta to destroy the red blood cells of an Rh+ child.

6.6 Homeostasis

Homeostasis depends upon the cardiovascular system because it serves the needs of the cells. Other body systems are also critical to cardiovascular system function:

- The digestive system supplies nutrients.
- The respiratory system supplies oxygen and removes carbon dioxide from the blood.
- The nervous and endocrine systems help maintain blood pressure. Endocrine hormones regulate red blood cell formation and calcium balance.
- The lymphatic system returns tissue fluid to the veins.
- Skeletal muscle contraction (skeletal system) and breathing movements (respiratory system) propel blood in the veins.

Understanding Key Terms

agglutination 125
agranular leukocyte 122
albumin 118
anemia 120
antibody 116
antigen 121
basophil 122
blood doping 120
blood transfusion 125
clotting 123
colony-stimulating factor (CSF) 121
deoxyhemoglobin 118
eosinophil 122
erythropoietin (EPO) 119
fibrin 123
fibrinogen 118
formed element 116
globulin 118
granular leukocyte 122
hemoglobin (Hb) 118
hemolysis 120
hemophilia 123
immune system 121
infectious mononucleosis 122
leukemia 122
lymphocyte 123
mast cell 122
megakaryocyte 123
monocyte 122
neutrophil 122
osmotic pressure 118
oxyhemoglobin 118
plasma 116
plasma protein 118
platelet (thrombocyte) 123
prothrombin 123
prothrombin activator 123
red blood cell (erythrocyte) 118
serum 123
severe combined immunodeficiency disease (SCID) 122
sickle-cell disease 120
thrombin 123
thrombocytopenia 123
thromboembolism 123
white blood cell (leukocyte) 121

Match the key terms to these definitions.

a. ________ Iron-containing protein in red blood cells that combines with and transports oxygen.

b. ________ Component of blood that is either cellular or derived from a cell.

c. ________ Liquid portion of blood.

d. ________ A group of cancerous conditions that involve uncontrolled white blood cell proliferation.

e. ________ Plasma protein converted to thrombin during the steps of blood clotting.

Testing Your Knowledge of the Concepts

1. The transport function of blood is dependent on what components? The defense function of blood is dependent on what components? The regulatory functions of blood are dependent on what components? (pages 116–17)
2. Why is the osmotic pressure of blood important to overall homeostasis? (page 118)
3. Describe the structure and functions of a red blood cell, including the molecule hemoglobin. What is the role of red blood cells in the blood? How is the production of red blood cells regulated? (pages 118–20)
4. Briefly describe the different types of white blood cells by structure and function. (pages 121–122)
5. What formed element is crucial to blood clotting? What other substances are necessary for clotting? Explain the steps that take place when blood clots. (page 123)
6. List and describe three disorders involving RBCs, three disorders involving WBCs, and two disorders involving platelets. (pages 120, 122, and 123)
7. For each type of ABO blood, give the antigen or antigens and antibody or antibodies present. List which blood type each can receive and to which type each can be given. (pages 125–126)
8. Explain what occurs in hemolytic disease of the newborn. (page 127)
9. Choose five body systems, and briefly explain how they are critical to cardiovascular system function. (pages 128–129)

In questions 10–14, match each description with a component of blood in the key. Answers can be used more than once.

Key:

a. red blood cells
b. white blood cells
c. red and white blood cells and platelets
d. plasma

10. Antigens in plasma membrane determine blood type
11. Transport oxygen
12. Includes monocytes
13. Contains plasma proteins
14. Formed elements

In questions 15–19, match each description with a white blood cell in the key.

Key:

a. lymphocytes
b. neutrophils
c. basophils
d. monocytes
e. eosinophils

15. U-shaped nucleus; blue-stained granules that release histamine
16. Includes B cells and T cells that provide specific immunity
17. Bilobed nucleus, red-stained granules, allergic reactions, and parasitic worms
18. Largest; no granules; become macrophages
19. Most abundant; multilobed nucleus; first responders to invasion
20. Which of the plasma proteins contribute/s most to osmotic pressure?
    a. albumin
    b. globulins
    c. erythrocytes
    d. fibrinogen
21. Stem cells are responsible for
    a. red blood cell production.
    b. white blood cell production.
    c. platelet production.
    d. the production of all formed elements.
22. Which hemoglobin component is recovered for reuse following red blood cell destruction?
    a. heme
    b. globin
    c. iron
    d. Both b and c are correct.
23. When the oxygen capacity of the blood is reduced,
    a. the liver produces more bile.
    b. the kidneys release erythropoietin.
    c. the bone marrow produces more red blood cells.
    d. sickle-cell disease occurs.
    e. Both b and c are correct.
24. Which of the following is not true of white blood cells?
    a. formed in red bone marrow
    b. carry oxygen and carbon dioxide
    c. can leave the bloodstream and enter tissues
    d. can fight disease and infection
25. Which of the following is in the correct sequence for blood clotting?
    a. prothrombin activator, prothrombin, thrombin
    b. fibrin threads, prothrombin activator, thrombin
    c. thrombin, fibrinogen, fibrin threads
    d. prothrombin, clotting factors, fibrinogen
    e. Both a and c are correct.
26. Theoretically, a person with type AB blood should be able to receive
    a. type B and type AB blood.
    b. type O and type B blood.
    c. type A and type O blood.
    d. All of these are correct.
27. If a person has type B^- blood, it means there are
    a. anti-A antibodies in the plasma.
    b. no B antigens on the red blood cells.
    c. Rh antigens on the red blood cells.
    d. no Rh antigens on the red blood cells.
    e. Both a and d are correct.

Thinking Critically About the Concepts

After reading the homeostasis section of this chapter, you should better understand the interactions between different organ systems that are required for homeostasis. No system ever operates alone. Keep in mind what you've learned about the cardiovascular system as you continue your studies. You may be surprised by how often you hear about concepts from Chapters 5 and 6 in upcoming chapters.

Ben, from the case study, suffers from leukemia. By causing huge numbers of abnormal white blood cells to be produced, the disease disrupts homeostasis in all organ systems. Thrombocytopenia, or platelet deficiency, may result in a fatal hemorrhage. Diseased white blood cells can't battle infections. Most important, when bone marrow produces leukemia cells instead of red blood cells, tissues cannot receive the oxygen they require. Without aggressive treatment, Ben's disease would be fatal.

1. Carbon monoxide (CO) is a deadly, odorless, colorless gas. Hemoglobin in red blood cells binds much more closely to CO than it does to oxygen. Hemoglobin's ability to transport oxygen is severely compromised in the presence of CO.
    a. A malfunctioning furnace is one potential cause for CO poisoning. What are some other situations in which you could be exposed to carbon monoxide? What safeguards should be in place when around these situations?
    b. If cells are deprived of oxygen, what is the effect on the production of cellular energy? Be specific.
2. There are three specific nutrients mentioned in this chapter that are necessary for the body to form red blood cells.
    a. Can you name all three?
    b. What are good food sources for these nutrients?
3. What type of organic molecule is hemoglobin? What nutrient is required to form hemoglobin? (Review Chap. 2 if necessary.)
4. The hormone erythropoietin is produced by the kidneys whenever more red blood cells are necessary. Think of several reasons for the body needing more red blood cells.
5. Athletes who abuse erythropoietin have many more red blood cells than usual.
    a. After examining Figure 6.3 and imagining many more red blood cells than usual in this capillary, explain why an athlete might die from having too many red blood cells.
    b. Why is death more likely at night when an athlete is sleeping, rather than during the day when he or she is active? Base your explanation on the heart rate.
6. After a trip to a country close to the equator, a student returns covered with mosquito bites. She has a low fever for a week, and blood tests show a very high level of eosinophils. Can you think of a disease that might cause these symptoms? Base your answer on what you know about tropical diseases.
7. Considering what you have learned from the "Homeostasis" section of this chapter, you should be able to think of at least two other organ systems that interact with the cardiovascular system (and the blood in particular) to regulate blood oxygen. List two organ systems and describe how they work with the cardiovascular system to make oxygen available to body cells.

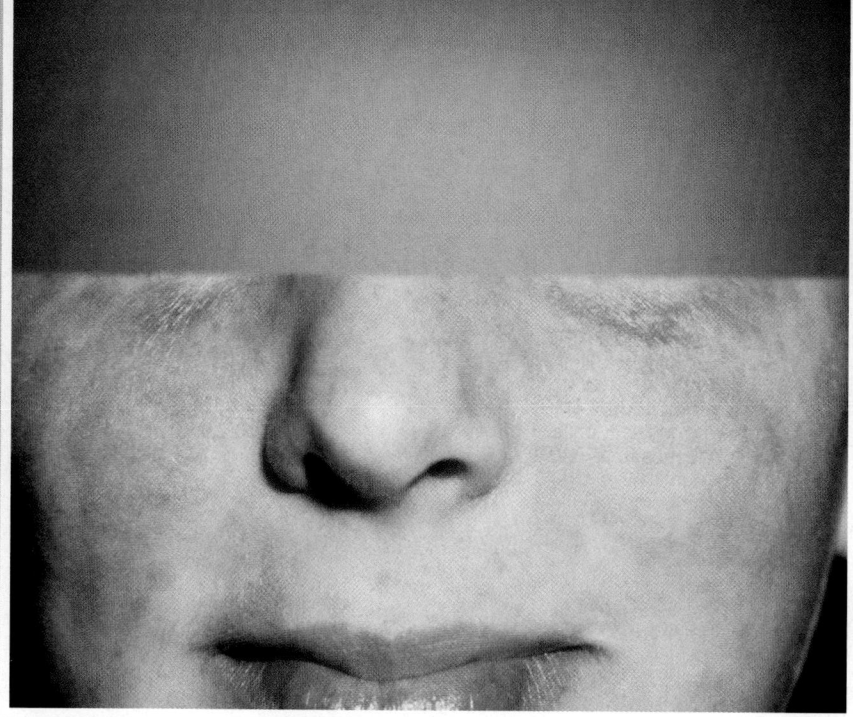

CHAPTER

7

Lymphatic System and Immunity

CASE STUDY LUPUS

Abigail is a healthy and active 12-year-old girl. She has a pretty unremarkable patient history that included nothing but the normal childhood diseases—croup when she was an infant, chicken pox a few years ago, several ear infections, and a couple of bad bouts of the flu. Over the last few months, she has been complaining to her mother that she was tired, her arms and legs ached, and her knees and elbows were bothering her. Her mother immediately thought it was "growing pains" and dismissed the symptoms. Within a few weeks, Abigail developed a rash on her cheeks and across the bridge of her nose, which resembled a butterfly shape. Her mother believed it was due to the new face soap she had switched to. Thinking Abigail might be sensitive to an ingredient in the soap, her mother quickly switched back to the old brand. The rash did not subside. Abigail suddenly began developing ulcers in her mouth, which interfered with eating and drinking. Within a few more weeks, Abigail also began experiencing some digestive issues—stomachaches after eating, periodic bouts of diarrhea, and a noticeable weight loss. Within another few weeks, Abigail began losing handfuls of hair. Her mother took her to the pediatrician.

Dr. Koos did a full exam on Abigail and ran a battery of tests over the next few days. The tests included a complete blood count (CBC), urinalysis, various protein assays, and an ANA test (antinuclear antibody), which is a test commonly used to aid in the diagnosis of many different autoimmune disorders. Once the results were in and a diagnosis was finally obtained, Dr. Koos explained that Abigail had lupus, an autoimmune disease.

As you read through the chapter, think about the following questions.

1. What is an autoimmune disease?
2. Are the symptoms Abigail experienced common to autoimmune diseases?
3. How is an autoimmune disease acquired?

CHAPTER CONCEPTS

7.1 Microbes, Pathogens, and You
Bacteria and viruses are pathogens that cause human diseases.

7.2 The Lymphatic System
The lymphatic vessels return excess tissue fluid to cardiovascular veins. The lymphatic organs (red bone marrow, thymus, lymph nodes, and spleen) are important to immunity.

7.3 Innate Defenses
Innate defenses are barriers that prevent pathogens from entering the body and mechanisms able to deal with minor invasions.

7.4 Acquired Defenses
Acquired defenses specifically counteract an invasion in two ways: by producing antibodies and by outright killing of abnormal cells.

7.5 Acquired Immunity
The two main types of acquired immunity are immunization by vaccines and the administration of prepared antibodies.

7.6 Hypersensitivity Reactions
The immune system is associated with allergies, tissue reaction, and autoimmune disorders. Treatment is available for these, but research goes forward into finding new and better cures.

BEFORE YOU BEGIN

Before beginning this chapter, take a few moments to review the following discussions.

Section 3.2 What are some differences between prokaryotic and eukaryotic cells?

Section 6.3 What is the role of white blood cells in defense against pathogens?

7.1 Microbes, Pathogens, and You

Learning Outcomes

Upon completion of this section you should be able to

1. Distinguish between a prokaryotic and eukaryotic cell.
2. Identify the structures of a prokaryotic cell.
3. Describe the structure of a general virus.

Microbes (microscopic organisms, such as bacteria, viruses, and protists) are widely distributed in the environment. They cover both inanimate objects and the surfaces of plants and animals. They are plentiful on and within our bodies. Many of the activities of microbes are useful to humans. We eat foods produced by bacteria every day. They contribute to the production of yogurt, cheese, bread, beer, wine, and many pickled foods. Today, drugs available through biotechnology are produced by bacteria. Microbes help us in still another way. Without the activity of decomposers, the biosphere (including ourselves) would cease to exist. When a tree falls to the forest floor, it eventually rots because decomposers, including bacteria and fungi, break down the remains of dead organisms to inorganic nutrients. Plants need these inorganic nutrients to make the many molecules that become food for us.

Despite all these benefits, there are certain bacteria and viruses known as **pathogens,** or disease-causing agents. The body has three lines of defense against invasion:

1. Barriers to entry, such as the skin and mucous membranes of body cavities, act to prevent pathogens from gaining entrance into the body.
2. First responders, such as the phagocytic white blood cells, act to prevent an infection after an invasion has occurred due to a pathogen getting past a barrier and into the body.
3. Acquired defenses overcome an infection by killing the particular disease-causing agent that has entered the body. Acquired defenses also protect us against cancer.

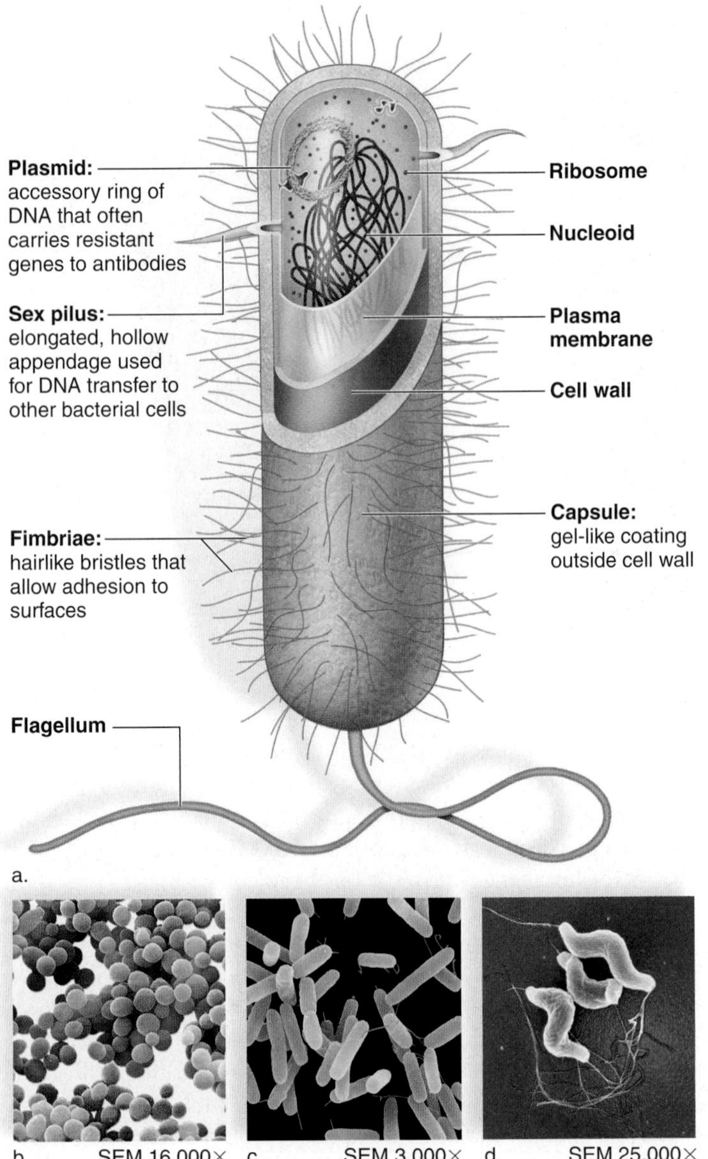

Figure 7.1 **Typical shapes of bacteria.**
a. Bacterial features, especially the four with definitions, contribute to the ability of bacteria to cause disease. Bacteria occur in three shapes. **b.** *Staphylococcus aureus* is a sphere-shaped bacterium that causes toxic shock syndrome. **c.** *Pseudomonas aeruginosa* is a rod-shaped bacterium that causes urinary tract infections. **d.** *Campylobacter jejuni* is a curve-shaped bacterium that causes food poisoning.

Bacteria

Bacteria are single-celled prokaryotes that do not have a nucleus. Figure 7.1 illustrates the main features of bacterial anatomy and shows the three common shapes: **coccus** (spherical-shaped), **bacillus** (rod-shaped), and **spirillum** (curved, sometimes spiral-shaped). Bacteria have a cell wall that contains a unique amino disaccharide. The "cillin" antibiotics, such as penicillin, interfere with the production of the cell wall. The cell wall of some bacteria is surrounded by a **capsule** that has a thick, gummy consistency. Capsules often allow bacteria to stick to surfaces such as teeth. They also prevent phagocytic white blood cells from taking them up and destroying them.

Motile bacteria usually have long, very thin appendages called flagella (sing., **flagellum**). The flagella rotate 360° and cause the bacterium to move backward. Some bacteria have **fimbriae,** stiff fibers that allow the bacteria to adhere to surfaces such as host cells. Fimbriae allow a bacterium to cling to and gain access to the body. In contrast, a **pilus** is an elongated hollow appendage used to transfer DNA from one cell to another. Genes that allow bacteria to be resistant to antibiotics can be passed from one to the other through a pilus by a process called *conjugation.*

Animation Antibiotic Resistance

Bacteria are independent cells that are capable of performing many diverse functions. Their DNA is packaged in a chromosome that occupies the center of the cell. Many bacteria also have small circular pieces of DNA called **plasmids.** Genes that allow bacteria to be resistant to antibiotics are often located in a plasmid. Abuse of antibiotic therapy increases the number of resistant bacterial strains that are difficult to kill, even with antibiotics.

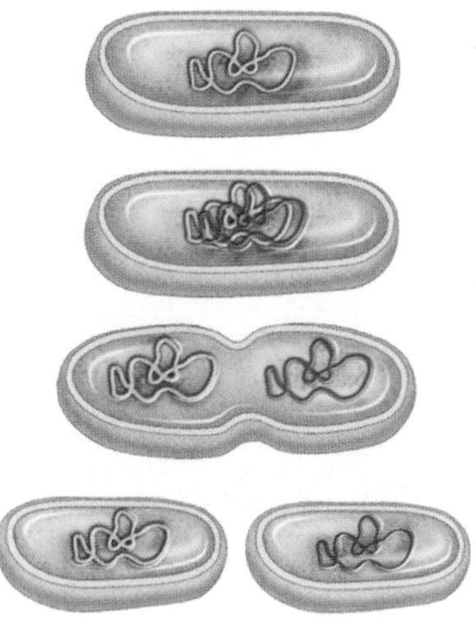

Figure 7.2 Binary fission.
When bacteria reproduce, division (fission) produces two (binary) cells identical to the original cell.

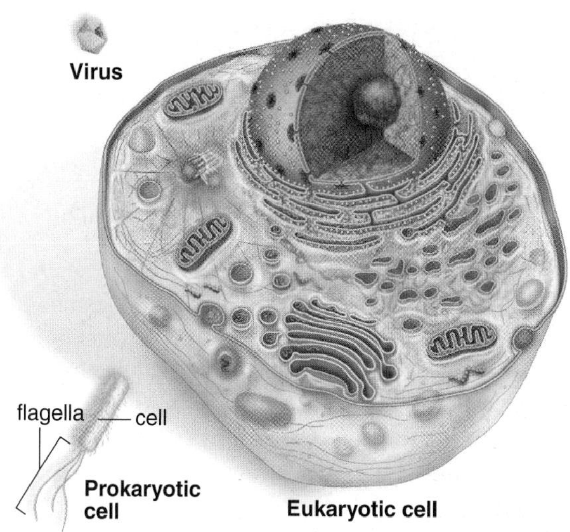

Figure 7.3 Comparative sizes of viruses, bacteria, and eukaryotic cells.
Viruses are tiny noncellular particles, whereas bacteria are small independent cells. Eukaryotic cells are more complex and larger because they contain a nucleus and many organelles.

Bacteria reproduce by a process called *binary fission* (Fig. 7.2). The single, circular chromosome attached to the plasma membrane is copied. Then the chromosomes are separated as the cell enlarges. The newly formed plasma membrane and cell wall separate the cell into two cells. Bacteria can reproduce rapidly under favorable conditions, with some species doubling their numbers every 12 minutes.

Animation
Binary Fission

Strep throat, tuberculosis, gangrene, gonorrhea, and syphilis are well-known bacterial diseases. Not only does growth of bacteria cause disease but some bacteria release molecules called **toxins** that inhibit cellular metabolism. For example, it is important to have a tetanus shot because the bacteria that causes this disease, *Clostridium tetani,* produces a toxin that prevents relaxation of muscles. In time, the body contorts because all the muscles have contracted; if no medical treatment is available, suffocation can occur.

Viruses

Viruses bridge the gap between the living and the nonliving. Outside a host, viruses are essentially chemicals that can be stored on a shelf. But when the opportunity arises, viruses replicate inside cells, and during this period, they clearly appear to be alive.

Viruses are acellular—not composed of cells. They are parasites and do not live independently. Viruses cause diseases such as colds, flu, measles, chicken pox, polio, rabies, AIDS, genital warts, and genital herpes.

Virus particles are about one-quarter the size of a bacterium, about one-hundredth the size of a eukaryotic cell (Fig. 7.3). A virus always has two parts: an outer capsid composed of protein units and an inner core of nucleic acid (Fig. 7.4). A virus carries the genetic information needed to reproduce itself. In contrast to cellular organisms, the viral genetic material need not be double-stranded DNA, nor even DNA. Indeed, some viruses, such as HIV and

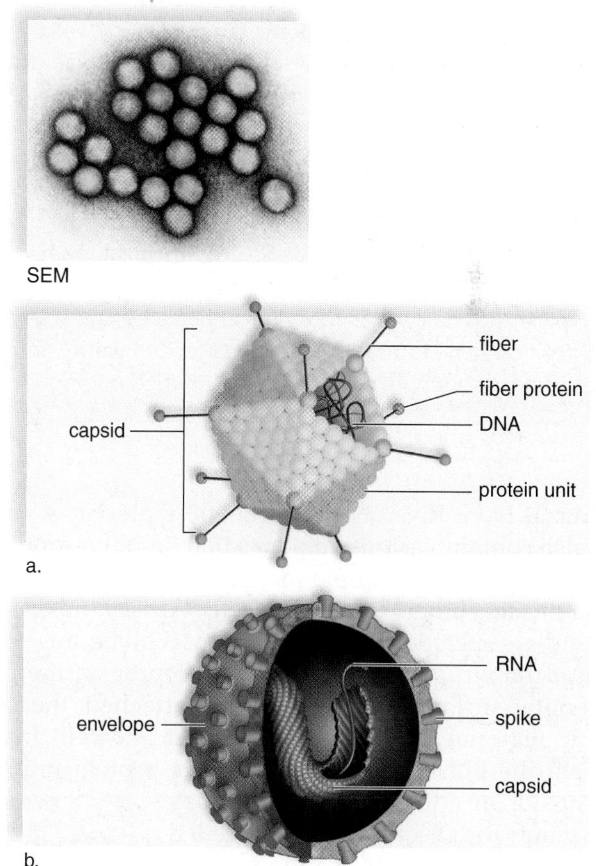

Figure 7.4 Typical virus structures.
Despite their diversity, all viruses have an outer capsid, composed of protein subunits, and a nucleic acid core, composed of either DNA or RNA but not both. **a.** Adenoviruses cause colds, and **(b)** influenza viruses cause the flu. Influenza viruses are surrounded by an envelope with spikes.

Biology Matters Historical Focus

Mary Mallon: The Most Dangerous Woman in America

Mary Mallon, also known as Typhoid Mary, was a cook in New York from 1900 to 1915. But she was also a carrier of the bacteria that cause typhoid fever (often referred to as typhoid), even though she was healthy. She probably had a mild case of typhoid fever earlier in her life and recovered. However, in this age before antibiotics, Mary still carried the bacteria in her body. As a result, she infected those who ate her cooking. From 1900 to 1907, Mary worked in seven different households. There, she caused 22 cases of typhoid fever and was responsible for one child's death. When Mary left the households, the typhoid fever cases stopped.

Typhoid is a bacterial disease spread via food or water contaminated by sewage. Infected patients shed the bacteria in their feces or urine. The disease is spread if an infected person prepares food or beverages with unwashed hands. High fever, headache, weakness, and severe diarrhea are symptoms of typhoid. These symptoms can persist for weeks. Further, the patient can tear the weakened intestinal wall during recovery and then die from excessive bleeding. The fatality rate in the early 1900s was approximately 10% (Fig. 7A).

When public officials first suspected Mary was the cause of the outbreaks, they approached her to ask for urine and fecal samples. Enraged, her reply was to attack them with a kitchen knife. In 1907, public health officials took her into custody by force. She was sent to a quarantine island located in New York's East River. Mary never believed that she was a typhoid carrier—after all, she was perfectly healthy. Imprisoned against her will, she always maintained her innocence.

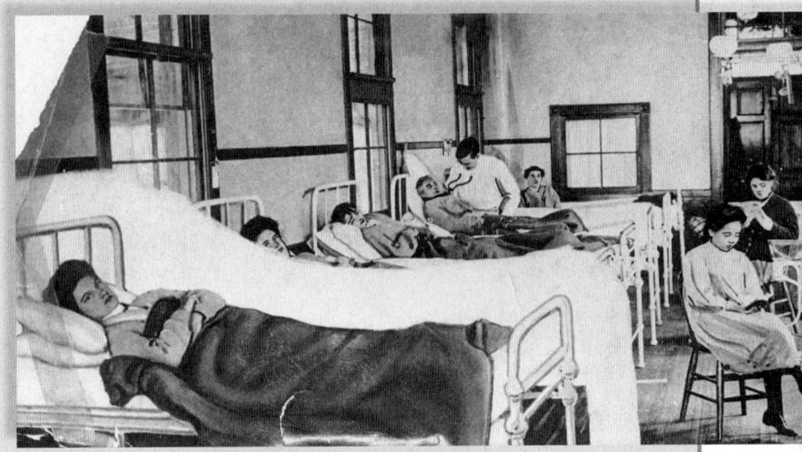

Figure 7A **Mary Mallon, the "most dangerous woman in America."**
As a typhoid carrier, Mary spread typhoid to others. In an age before antibiotics, typhoid had a high fatality rate.

In 1910, she was released, with orders that she never again work in food preparation. She was trained to work as a laundress instead and directed to check in regularly with health officials. However, after her release, Mary disappeared. In 1915, there was an outbreak of typhoid at the Sloane Maternity Hospital in Manhattan. Twenty-two people sickened and two died. The hospital had recently hired a cook named Mrs. Brown—Mary, using an alias. She was sent back to the same cottage on North Brother Island, where she was permanently imprisoned. Mary lived for another 23 years, dying in 1938.

influenza, have RNA as their genetic material. A virus may also contain various enzymes that help it reproduce.

Viruses are microscopic pirates, commandeering the metabolic machinery of a host cell. Viruses gain entry into and are specific to a particular host cell because portions of the virus are specific for a receptor on the host cell's outer surface. Once the virus is attached, the viral genetic material (DNA or RNA) enters the cell. Inside the cell, the nucleic acid codes for the protein units in the capsid. In addition, the virus may have genes for special enzymes needed for the virus to reproduce and exit from the host cell. In large measure, however, a virus relies on the host's enzymes and ribosomes for its own reproduction.

Animation Viral Entry into Host Cells

Video How Viruses Attack

Connections and Misconceptions

Does refrigeration kill bacteria?

The answer is no. The speed at which bacteria reproduce depends on a number of factors, including moisture in the environment and temperature. At the temperatures found in most refrigerators and freezers, bacterial growth is slowed but the bacteria are not killed. Once the temperature returns to a favorable level, the bacteria resume normal cell division. The only way to kill most food-related bacteria is by using high temperatures, such as those used in boiling or thorough cooking.

Animation Food Pathogens & Temperature

Connections and Misconceptions

Are antibacterial cleansers really effective?

Studies show that using antibacterial soap when washing your hands is no more effective in preventing the spread of bacteria than using pure soap and water. In some cases, the main ingredient in most antibacterial soaps, triclosan, may actually cause some bacteria to become resistant to certain drugs like amoxicillin, which is used to kill bacteria. Waterless hand sanitizers (which are alcohol-based) are recommended to be used in conjunction with, not as a replacement for, regular hand washing. Proper hand washing and sanitizing can prevent the spread of colds, flus, and certain diarrhea disorders when done often and properly.

Prions

Prions, proteinaceous infectious particles, cause a group of degenerative diseases of the nervous system, also called wasting diseases. Originally thought to be viral diseases, prions cause Creutzfeldt–Jakob disease (CJD) in humans; scrapie in sheep; and bovine spongiform encephalopathy (BSE), commonly called **mad cow disease,** in cattle. These infections are apparently transmitted by ingestion of brain and nerve tissues from infected animals. Prion proteins are believed to play the role of a "housekeeper" in the brains of healthy individuals. However, in some people, a "rogue" form of the prion protein folds into a new shape and in the process loses its original function. The "rogue" protein is able to refold normal prion proteins into the new shape and thus cause disease. Two molecular models of a protein are shown in Figure 7.5.

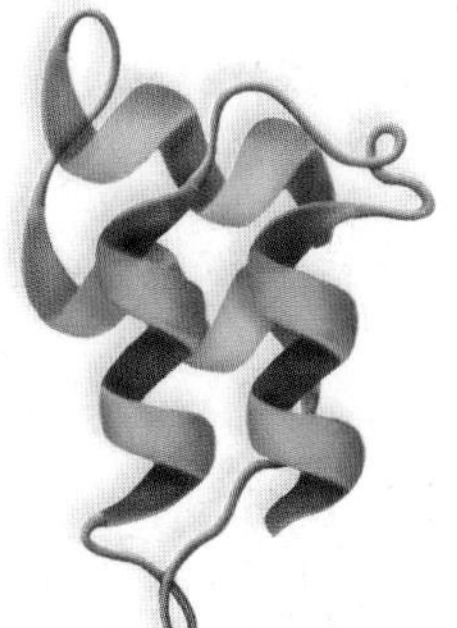

Figure 7.5 Comparison of the normal prion protein and "rogue" prion protein.
The model on the left shows the normal prion protein composed of alpha helices (coils). The model on the right shows the "rogue" prion protein with β-pleated sheets (flat arrows).

We can imagine that the one on the left is the normal protein and the one on the right is the prion. Nervous tissue is lost, and calcified plaques show up in the brain due to activity of prions. Thankfully, the incidence of prion diseases in humans is very low.

Animation How Prions Arise

Connecting the Concepts

For more information on the topics presented in this section, refer to the following discussions.

Section S.3 explores the problem of antibiotic resistance in medicine.

Section 16.6 explains the link between bacteria and some sexually transmitted diseases.

Section 19.2 examines how some viruses may cause cancer.

Check Your Progress 7.1

1. Describe what makes a virus or bacteria a pathogen.
2. Detail the structures in bacteria that can be associated with virulence, the ability to cause disease.
3. Summarize why viruses are considered parasites that always cause disease.

7.2 The Lymphatic System

Learning Outcomes

Upon completion of this section, you should be able to

1. Describe the structure of the lymphatic system.
2. Summarize how the lymphatic system contributes to homeostasis.
3. Explain how the lymphatic system interacts with the circulatory system.

The lymphatic system consists of lymphatic vessels and the lymphatic organs. This system, closely associated with the cardiovascular system, has four main functions that contribute to homeostasis: (1) Lymphatic capillaries absorb excess tissue fluid and return it to the bloodstream; (2) in the small intestines, lymphatic capillaries called lacteals absorb fats in the form of lipoproteins and transport them to the bloodstream; (3) the lymphatic system is responsible for the production, maintenance, and distribution of lymphocytes; and (4) the lymphatic system helps defend the body against pathogens.

Lymphatic Vessels

Lymphatic vessels form a one-way system of capillaries to vessels and, finally, to ducts. These vessels take lymph to cardiovascular veins in the shoulders (Fig. 7.6). As you recall from Chapter 6, lymphatic capillaries take up excess tissue fluid. Tissue fluid is mostly water, but it also contains solutes (i.e., nutrients, electrolytes, and oxygen) derived from plasma. Tissue fluid also contains cellular products (i.e., hormones, enzymes, and wastes) secreted by cells. The fluid inside lymphatic vessels is called **lymph.** Lymph is usually a colorless liquid, but after a meal, it appears creamy because of its lipid content.

The lymphatic system has two ducts (highways): the thoracic duct and the right lymphatic duct. The larger thoracic duct returns lymph collected from the body below the thorax, the left arm, and left side of the head and neck. The thoracic duct drains fluid into the left subclavian vein. The right lymphatic duct returns lymph from the right arm and right side of the head and neck into the right subclavian vein.

The construction of the larger lymphatic vessels is similar to that of cardiovascular veins, including the presence of valves. The movement of lymph within lymphatic capillaries is largely dependent upon skeletal muscle contraction. Lymph forced through lymphatic vessels as a result of muscular compression is prevented from flowing backward by one-way valves.

Lymphatic Organs

Lymphatic organs are divided into primary: red bone marrow and the thymus gland; and secondary: lymph nodes and spleen (Fig. 7.7).

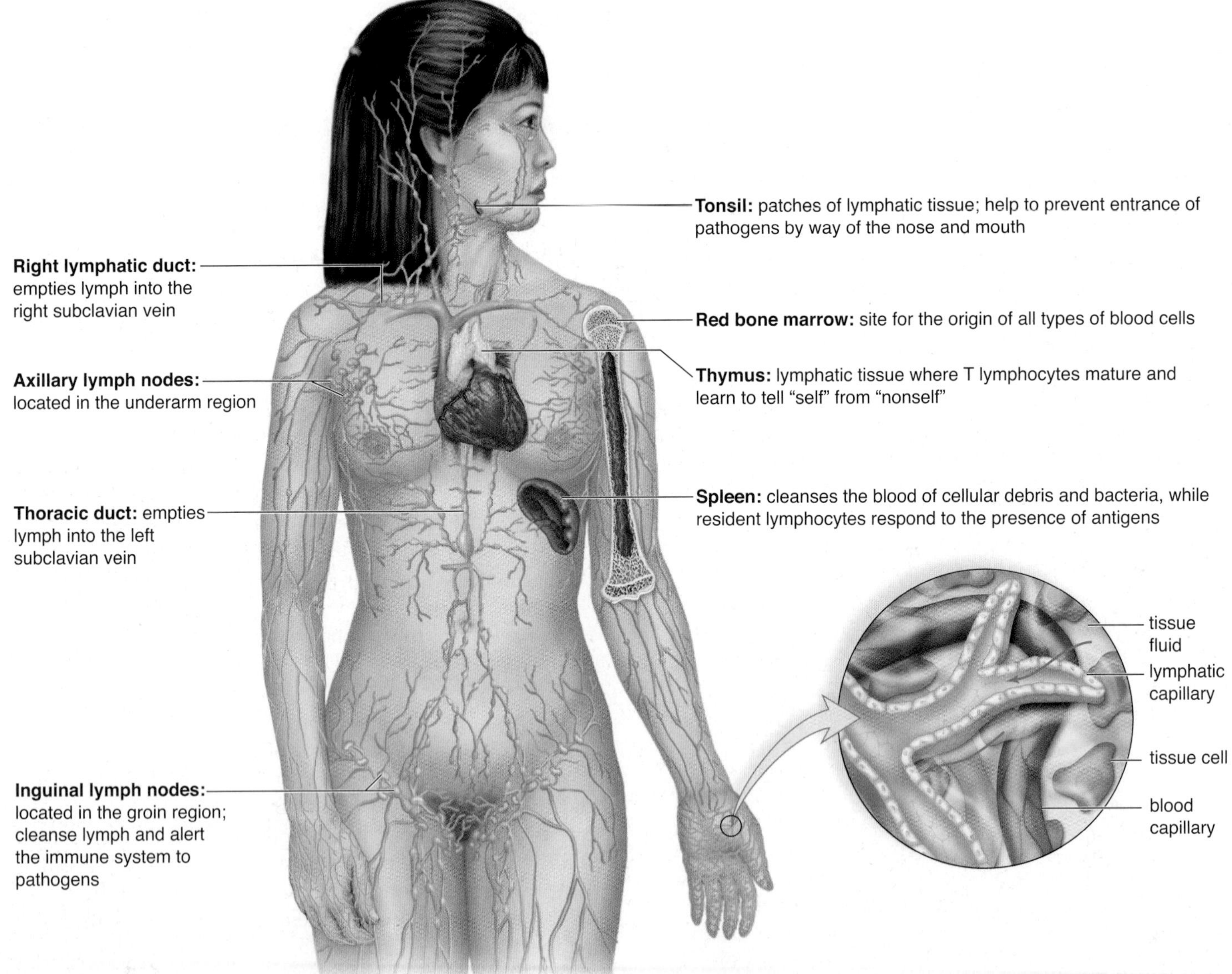

Figure 7.6 **Functions of the lymphatic system components.**
Lymphatic vessels drain excess fluid from the tissues and return it to the cardiovascular system. The enlargement shows that lymphatic vessels, like cardiovascular veins, have valves to prevent backward flow. The lymph nodes, spleen, thymus gland, and red bone marrow are the main lymphatic organs that assist immunity.

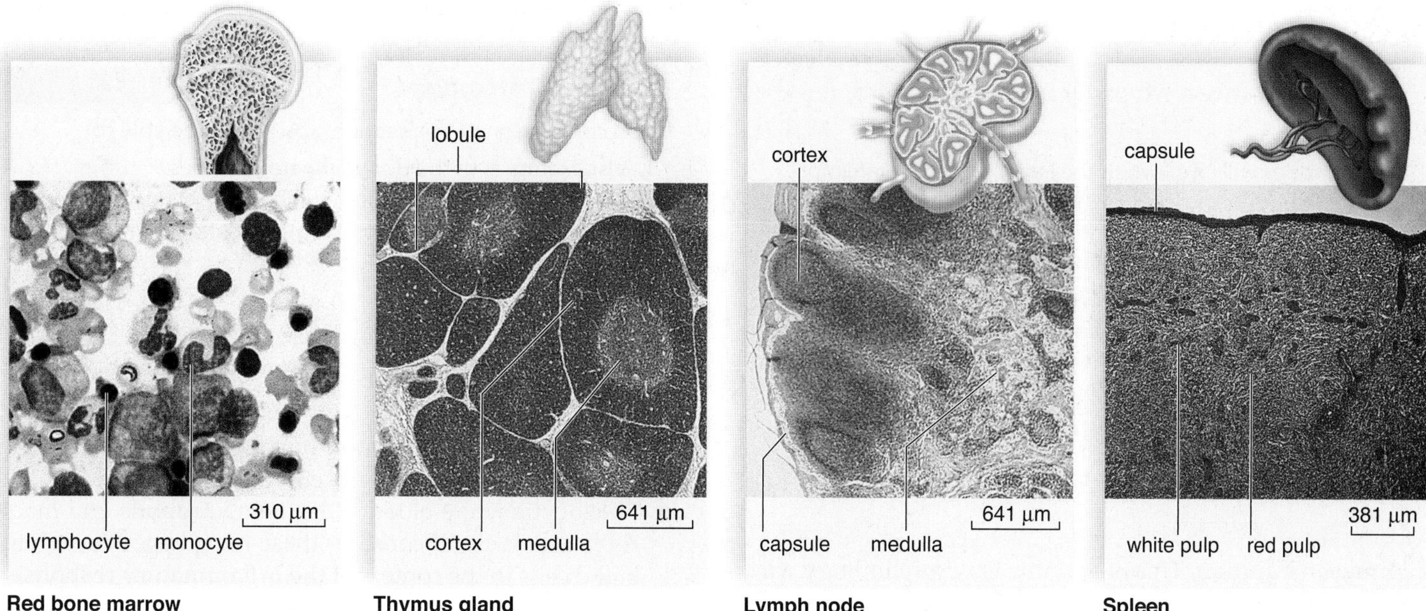

Figure 7.7 **Tissue samples from primary and secondary lymphatic organs.**
Left: Red bone marrow and the thymus gland are the primary lymphatic organs. Blood cells, including lymphocytes, are produced in red bone marrow. B cells mature in the bone marrow, but T cells mature in the thymus. *Right:* Lymph nodes and the spleen are secondary lymphatic organs. Lymph is cleansed in lymph nodes, and blood is cleansed in the spleen.

The Primary Lymphatic Organs

Red bone marrow produces all types of blood cells. In a child, most bones have red bone marrow; in an adult, it is limited to the sternum, vertebrae, ribs, part of the pelvic girdle, and the upper ends of the humerus and femur. In addition to the red blood cells, bone marrow produces the various types of white blood cells: neutrophils, eosinophils, basophils, lymphocytes, and monocytes. Lymphocytes are either **B cells (B lymphocytes)** or **T cells (T lymphocytes).** B cells mature in the bone marrow, but the T cells mature in the thymus. Any B cell that reacts with cells of the body is removed in the bone marrow and does not enter the circulation. This ensures that the B cells do not harm normal cells of the body.

The soft, bilobed **thymus gland** is located in the thoracic cavity between the trachea and the sternum, superior to the heart. The thymus will begin shrinking in size before puberty and is noticeably smaller in an adult than in a child.

The thymus has two functions: (1) The thymus gland produces thymic hormones, such as thymosin, thought to aid in the maturation of T lymphocytes. Thymosin may also have other functions in immunity. (2) Immature T lymphocytes migrate from the bone marrow through the bloodstream to the thymus, where they mature. Only about 5% of these cells ever leave the thymus. These T lymphocytes have survived a critical test: If any show the ability to react with the individual's cells, they die in the thymus. If they have potential to attack a pathogen, they can leave the thymus. The thymus is absolutely critical to immunity because without mature properly functioning T cells, the body's response to specific pathogens is poor or absent.

Secondary Lymphatic Organs

The secondary lymphatic organs are the spleen, the lymph nodes, and other organs containing lymphoid tissue. These include the tonsils, Peyer's patches, and the appendix.

The **spleen** filters blood. The spleen, the largest lymphatic organ, is located in the upper left region of the abdominal cavity posterior to the stomach. Connective tissue divides the spleen into regions known as white pulp and red pulp (Fig. 7.7). The red pulp, which surrounds venous sinuses (cavities), is involved in filtering the blood. Blood entering the spleen must pass through the sinuses before exiting. Here, macrophages that are like powerful vacuum cleaners engulf pathogens and debris, such as worn-out red blood cells.

The spleen's outer capsule is relatively thin, and an infection or a blow can cause the spleen to burst. Although the spleen's functions are replaced by other organs, a person without a spleen is often slightly more susceptible to infections and may have to receive antibiotic therapy indefinitely.

Lymph nodes, which occur along lymphatic vessels, filter lymph. Connective tissue forms a capsule and also divides a lymph node into compartments. Each compartment contains a sinus that increases in size toward the center of the node. As lymph courses through the sinuses, it is exposed to macrophages, which engulf pathogens and debris. Lymphocytes, also present in sinuses, fight infections and attack cancer cells.

Lymph nodes are named for their location. For example, inguinal nodes are in the groin, and axillary nodes are in the armpits. Physicians often feel for the presence of swollen, tender lymph nodes in the neck as evidence that the body is

Connections and Misconceptions

What does it mean when my lymph nodes are swollen?

Lymph nodes will swell when they are fighting an infection. When the body is invaded by a bacteria or virus, individual nodes can swell from 1/2 to 2 inches in diameter. Cuts, burns, bites, rashes, and any break in the skin can cause an infection and thus the swelling of your lymph nodes. Swollen nodes near the groin mean infection on a leg or lower abdomen; in your armpits mean arms or chest; and swollen nodes on the front of your neck mean an infection in the face, ears, nose, or throat. Some diseases, like chicken pox, can cause all your nodes to swell.

fighting an infection. This is a noninvasive, preliminary way to help make such a diagnosis.

Lymphatic nodules are concentrations of lymphatic tissue not surrounded by a capsule. The **tonsils** are patches of lymphatic tissue located in a ring about the pharynx. The tonsils perform the same functions as lymph nodes; but because of their location, they are the first to encounter pathogens and antigens that enter the body by way of the nose and mouth.

Peyer's patches are located in the intestinal wall and tissues within the appendix, a small extension of the large intestine, and encounter pathogens that enter the body by way of the intestinal tract.

Animation Lymphatic System

Connecting the Concepts

As noted, the lymphatic system plays a role in the movement of fats and the return of excess fluid to the circulatory system. For more information on these functions, refer to the following discussions.

Section 5.1 examines how the lymphatic system interacts with the circulatory system.

Section 8.3 explores how the lymphatic system is involved in the processing of fat in the diet.

Figure 8.6 diagrams the location of the lymphatic vessels in the small intestine.

Check Your Progress 7.2

1. List the four main functions of the lymphatic system that contribute to homeostasis.
2. Detail the differences between a primary and a secondary lymphatic organ and give an example of each.
3. Predict what might happen to the body if the lymphatic ducts did not allow lymph to drain.

7.3 Innate Defenses

Learning Outcomes

Upon completion of this section, you should be able to

1. List examples of the body's innate defenses.
2. Summarize the events in the inflammatory response.
3. Explain the role of the complement system.

Immunity, the ability to combat diseases and cancer, includes two innate lines of defense. These act indiscriminately against all pathogens:

- Barriers to entry. The body puts up barriers that aid in the prevention of pathogen entry.
- Phagocytic white blood cells, the neutrophils and macrophages. We will consider these two phagocytic white blood cells in the context of the **inflammatory response,** a special reaction of the body when first invaded. Protective proteins are also part of this line of defense.

Barriers to Entry

The body has built-in barriers, both physical and chemical, that serve as the first line of defense against an infection by pathogens.

Skin and Mucous Membranes

The intact skin is generally an effective physical barrier that prevents infection. Mucous membranes lining the respiratory, digestive, reproductive, and urinary tracts are also physical barriers to entry by pathogens. For example, the ciliated cells that line the upper respiratory tract sweep mucus and trapped particles up into the throat, where they can be swallowed, spit, or coughed out.

Chemical Barriers

The chemical barriers to infection include the secretions of sebaceous (oil) glands of the skin. These secretions contain chemicals that weaken or kill certain bacteria on the skin.

Perspiration, saliva, and tears contain an antibacterial enzyme called **lysozyme.** Saliva also helps to wash microbes off the teeth and tongue, and tears wash the eyes. Similarly, as urine is voided from the body, it flushes bacteria from the urinary tract.

The acid pH of the stomach inhibits growth or kills many types of bacteria. At one time, it was thought that no bacterium could survive the acidity of the stomach. But now we know that ulcers are caused by the bacterium *Helicobacter pylori* (see Chapter 1). Similarly, the acidity of the vagina and its thick walls discourages the presence of pathogens.

Connections and Misconceptions

How do antihistamines work?

Once histamine is released from mast cells, it binds to receptors on other body cells. There, the histamine causes the familiar symptoms associated with infections and allergies: sneezing, itching, runny nose, and watery eyes. Antihistamines work by blocking the receptors on the cells so that histamine can no longer bind. For allergy relief, antihistamines are most effective when taken before exposure to the allergen.

Resident Bacteria

Finally, a significant chemical barrier to infection is created by the normal flora, microbes that usually reside in the mouth, intestine, and other areas. By using available nutrients and releasing their own waste, these resident bacteria prevent potential pathogens from taking up residence. For this reason, chronic use of antibiotics can make a person susceptible to pathogenic infection by killing off the normal flora.

Inflammatory Response

The inflammatory response exemplifies the second line of defense against invasion by a pathogen. Inflammation employs mainly neutrophils and macrophages to surround and kill (engulf by phagocytosis) pathogens trying to get a foothold inside the body. Protective proteins are also involved. Inflammation is usually recognized by its four hallmark symptoms: redness, heat, swelling, and pain (Fig. 7.8).

The four signs of the inflammatory response are due to capillary changes in the damaged area, and all serve to protect the body. Chemical mediators, such as **histamine,** released by damaged tissue cells and **mast cells,** cause the capillaries to dilate and become more permeable. Excess blood flow due to enlarged capillaries causes the skin to redden and become warm. Increased temperature in an inflamed area tends to inhibit growth of some pathogens. Increased blood flow brings white blood cells to the area. Increased permeability of capillaries allows fluids and proteins, including blood-clotting factors, to escape into the tissues. Clot formation in the injured area prevents blood loss. The excess fluid in the area presses on nerve endings, causing the familiar pain associated with swelling. Together, these events summon white blood cells to the area.

As soon as the white blood cells arrive, they move out of the bloodstream into the surrounding tissue. The neutrophils are first and actively phagocytize debris, dead cells, and bacteria they encounter. The many neutrophils attracted to the area can usually localize any infection and keep it from spreading. If neutrophils die off in great quantity, they become a yellow-white substance called **pus.**

When an injury is not serious, the inflammatory response is short-lived and the healing process will quickly return the affected area to a normal state. Nearby cells secrete chemical factors to ensure the growth (and repair) of blood vessels and new cells to fill in the damaged area.

If, on the other hand, the neutrophils are overwhelmed, they call for reinforcements by secreting chemical mediators called **cytokines.** Cytokines attract more white blood cells to the area, including monocytes. Monocytes are

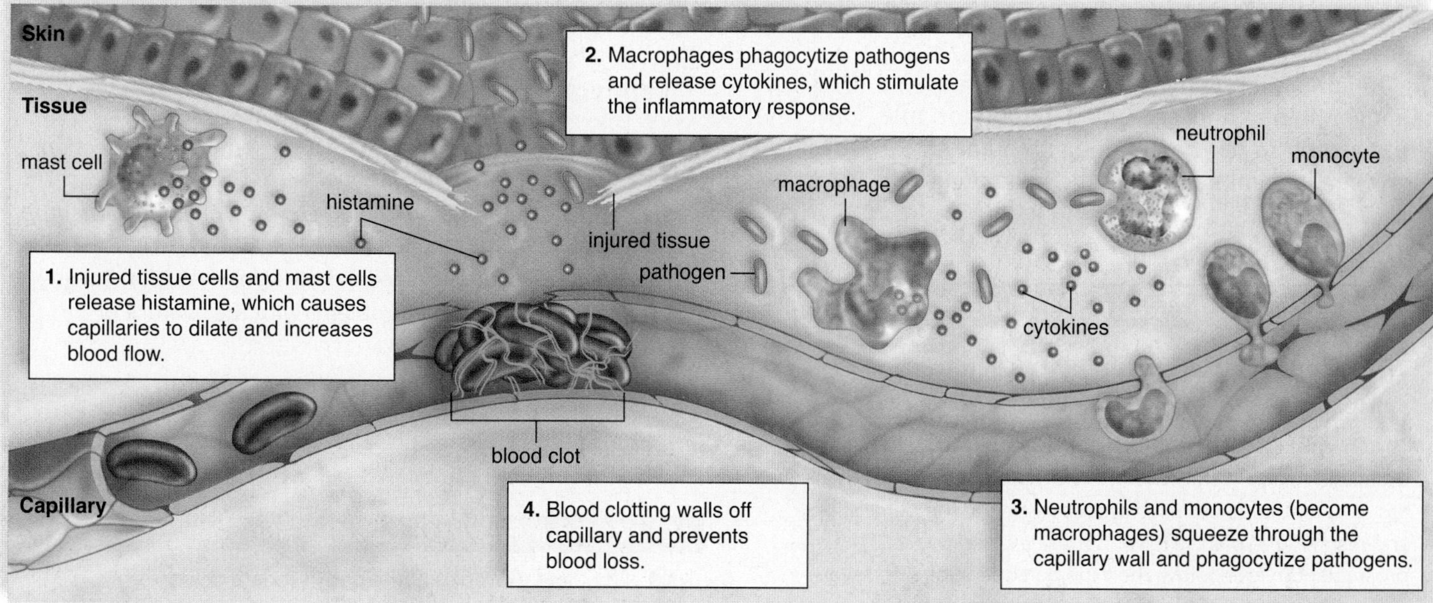

Figure 7.8 **Steps of the inflammatory response.**
1. Due to capillary changes in a damaged area and the release of chemical mediators, such as histamine by mast cells, an inflamed area exhibits redness, heat, swelling, and pain. **2.** Macrophages release cytokines, which stimulate the inflammatory and other immune responses. **3.** Monocytes and neutrophils squeeze through capillary walls from the blood and phagocytize pathogens. **4.** A blood clot can form a seal in a break in a blood vessel.

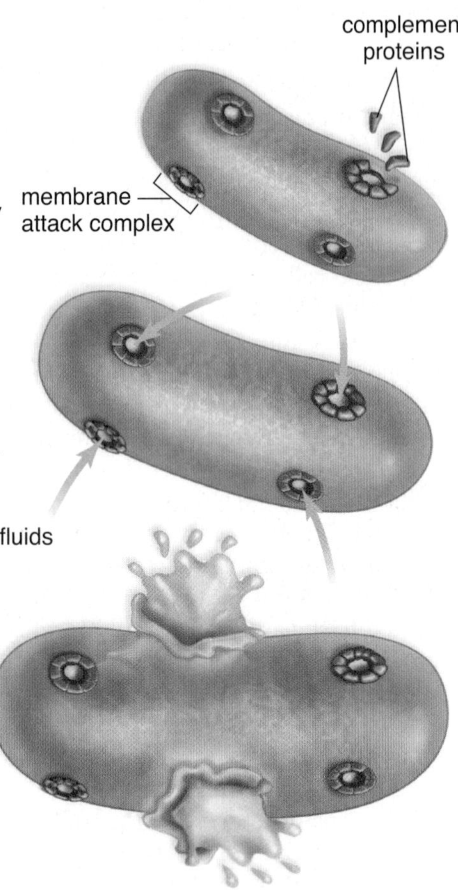

Figure 7.9 The killing of a bacteria by the complement system. When complement proteins in the blood plasma are activated by an immune response, they form a membrane attack complex that makes holes in bacterial cell walls and plasma membranes, allowing fluids to enter until the cell eventually bursts.

longer-lived cells that become **macrophages,** even more powerful phagocytes than neutrophils. Macrophages can enlist the help of lymphocytes to carry out specific defense mechanisms.

Inflammation is the body's natural response to an irritation or injury and serves an important role. Once the healing process has begun, inflammation rapidly subsides. However, in some cases, chronic inflammation may last for weeks, months, or even years if an irritation or infection cannot be overcome. Inflammatory chemicals may cause collateral damage to the body, in addition to killing the invaders. Should an inflammation persist, anti-inflammatory medications, such as aspirin, ibuprofen, or cortisone, can minimize the effects of various chemical mediators.

Protective Proteins

The **complement system,** often simply called complement, is composed of a number of blood plasma proteins designated by the letter C and a number. The complement proteins "complement" certain immune responses, which accounts for their name. For example, they are involved in and amplify the inflammatory response because certain complement proteins can bind to mast cells and trigger histamine release. Others can attract phagocytes to the scene. Some complement proteins bind to the surface of pathogens already coated with antibodies, which ensures that the pathogens will be phagocytized by a neutrophil or macrophage.

Certain other complement proteins join to form a **membrane attack complex** that produces holes in the surface of bacteria. Fluids then enter the bacterial cell to the point that they burst (Fig. 7.9).

Interferons are proteins produced by virus-infected cells as a warning to noninfected cells in the area. Interferon binds to receptors of noninfected cells, causing them to prepare for possible attack by producing substances that interfere with viral replication. Interferons are used as treatment in certain viral infections, such as hepatitis C.

Animation Antiviral Activity of Interferon

Connections and Misconceptions

Why is aspirin used to alleviate so many symptoms?

The chemicals in aspirin decrease the body's ability to make prostaglandins. Prostaglandins, made by most of the body's tissues, are substances used as messengers in the perception and response to pain, fever, and muscle contractions. If the prostaglandin level is low, then the perception of pain, fever, and muscle contractions will be lowered. Aspirin also decreases the production of certain substances needed in the beginning stages of blood clotting, which is why it is prescribed to patients with certain cardiovascular clotting disorders.

Connecting the Concepts

For more information on the topics presented in this section, refer to the following discussions.

Section 4.7 examines how the skin forms a physical barrier to the exterior environment.

Section 6.3 provides additional details on white blood cells.

Section S.3 discusses the problems with the overuse of antibiotics.

Check Your Progress 7.3

1. List some examples of the body's innate defenses.
2. Describe the blood cells associated with innate defenses, and detail how they function.
3. Discuss how the complement proteins got their name.

7.4 Acquired Defenses

Learning Outcomes

Upon completion of this section, you should be able to

1. Explain the role of an antigen in the acquired defenses.
2. Summarize the process of antibody-mediated immunity and list the cells involved in the process.
3. Summarize the process of cell-mediated immunity and list the cells involved in the process.

When innate defenses have failed to prevent an infection, acquired defenses come into play. Acquired defenses overcome an infection by doing away with the particular disease-causing agent that has entered the body. Acquired defenses also protect against cancer.

How Acquired Defenses Work

Acquired defenses respond to large molecules, normally protein structures, called **antigens** that the immune system recognizes as foreign to the body. Fragments of bacteria, viruses, molds, or parasitic worms can all be antigenic. Further, abnormal plasma membrane proteins produced by cancer cells may also be antigens. We do not ordinarily become immune to our own normal cells, so it is said that the immune system is able to distinguish self from nonself.

Acquired defenses primarily depend on the action of lymphocytes, which differentiate as either B cells (B lymphocytes) or T cells (T lymphocytes). B cells and T cells are capable of recognizing antigens because they have specific antigen receptors. These antigen receptors are plasma membrane proteins whose shape allows them to combine with particular antigens. Each lymphocyte has only one type of receptor. It is often said that the receptor and the antigen fit together like a lock and key. We encounter millions of different antigens during our lifetime, so we need a diversity of B cells and T cells to protect us against them. Remarkably, this diversification occurs during the maturation process. Millions of specific B cells and/or T cells are formed, increasing the likelihood that at least one will recognize any possible antigen.

B cells differentiate into other types of cells, and so do T cells. The several types of cells we will be discussing are listed in Table 7.1 for easy reference.

B Cells and Antibody-Mediated Immunity

The receptor on a B cell is called a **B-cell receptor (BCR).** The **clonal selection model** (Fig. 7.10) states that an antigen

Table 7.1 Immune Cell Type and Function

Cell	Function
B cells	Produce plasma cells and memory cells
Plasma cells	Produce specific antibodies
Memory cells	Ready to produce antibodies in the future
T cells	Regulate immune response; produce cytotoxic T cells and helper T cells
Cytotoxic T cells	Kill virus-infected cells and cancer cells
Helper T cells	Regulate immunity
Memory T cells	Ready to kill in the future

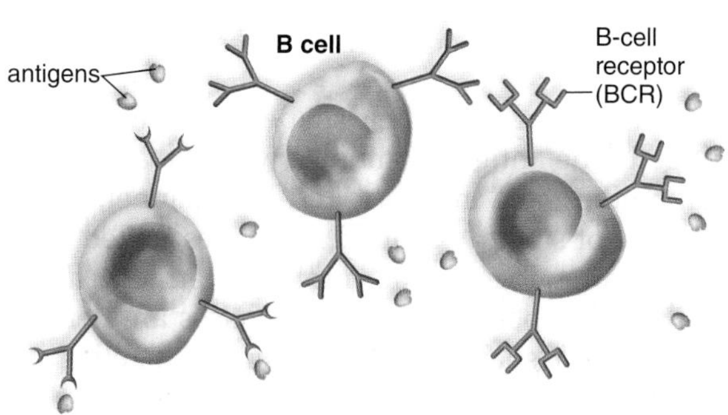

a. **Activation:** When a B cell receptor binds to an antigen activation occurs.

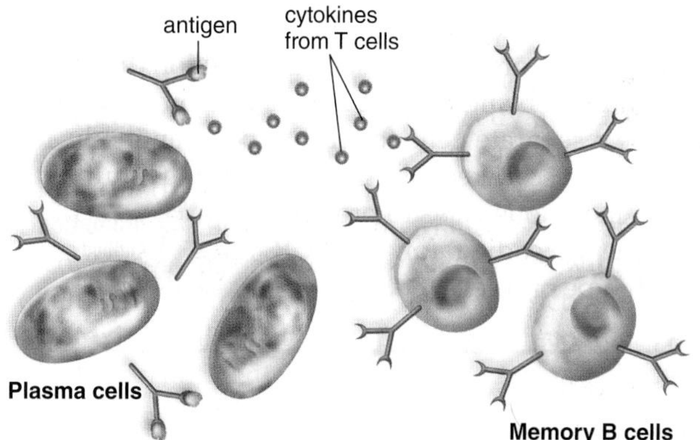

b. **Clonal expansion**–During clonal expansion, cytokines secreted by helper T cells stimulate B cells to clone mostly into plasma cells or memory cells.

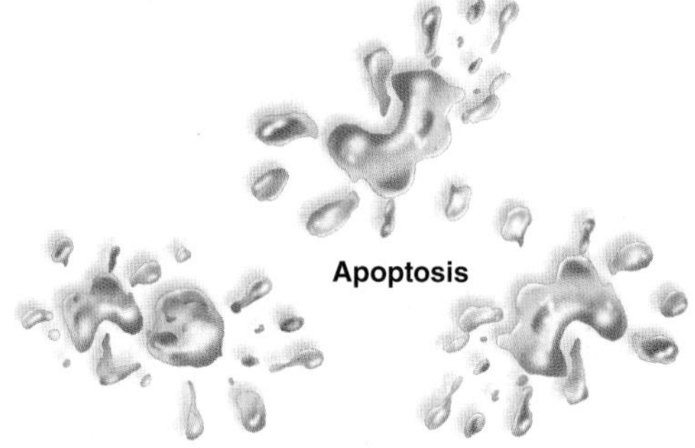

c. **Apoptosis**–Apoptosis, or programmed cell death, occurs to plasma cells left in the system after the infection has passed.

Figure 7.10 Antibody-mediated immunity or the clonal selection model for B cells.
a. Activation of a B cell occurs when its B-cell receptor (BCR) combines with an antigen (colored green). In the presence of cytokines produced by T cells, the B cell undergoes clonal expansion **(b.)**, producing many plasma cells that secrete antibodies specific to the antigen. **c.** After the infection, plasma cells undergo apoptosis.

Media Study Tools

www.mhhe.com/maderhuman12e

Enhance your study of this chapter with study tools and practice tests. Also ask your instructor about the resources available through ConnectPlus, including the media-rich eBook, interactive learning tools, and animations.

Summarizing the Concepts

7.1 Microbes, Pathogens, and You

Microbes perform valuable services, but they also cause disease.

Bacteria

- Bacteria are prokaryotic cells that cause disease by multiplying in hosts and also by producing toxins.

Viruses

- Viruses are noncellular particles consisting of a protein coat and a nucleic acid core.
- Viruses take over the machinery of the host to reproduce.

7.2 The Lymphatic System

The lymphatic system consists of lymphatic vessels that return lymph to cardiovascular veins.

The primary lymphatic organs are

- the red bone marrow, where all blood cells are made and the B lymphocytes mature; and
- the thymus gland, where T lymphocytes mature.

The secondary lymphatic organs are

- the spleen; lymph nodes; and other organs containing lymphoid tissue, such as the tonsils, Peyer's patches, and the appendix. Blood is cleansed of pathogens and debris in the spleen. Lymph is cleansed of pathogens and debris in the nodes.

7.3 Innate Defenses

Immunity involves innate and acquired defenses. The innate defenses include

- barriers to entry; and
- the inflammatory reaction, which involves the phagocytic neutrophils and macrophages and protective proteins.

7.4 Acquired Defenses

Acquired defenses require B cells and T cells, also called B lymphocytes and T lymphocytes.

B Cells and Antibody-Mediated Immunity

- Activated B cells undergo clonal selection with production of plasma cells and memory B cells, after their B-cell receptor combines with a specific antigen.
- Plasma cells secrete antibodies and eventually undergo apoptosis. Plasma cells are responsible for antibody-mediated immunity.
- An antibody is usually a Y-shaped molecule that has two binding sites for a specific antigen.
- Memory B cells remain in the body and produce antibodies if the same antigen enters the body at a later date.

T Cells and Cell-Mediated Immunity

- For a T cell to recognize an antigen, the antigen must be presented by an antigen-presenting macrophage, along with an HLA (human leukocyte antigen).
- Activated T cells undergo clonal expansion until the illness has been stemmed. Then, most of the activated T cells undergo apoptosis. A few cells remain, however, as memory T cells.
- The two main types of T cells are cytotoxic T cells and helper T cells.
    - Cytotoxic T cells kill virus-infected cells or cancer cells on contact because they bear a nonself protein.
    - Helper T cells produce cytokines and stimulate other immune cells.

7.5 Acquired Immunity

- Active (long-lived) immunity can be induced by vaccines when a person is well and in no immediate danger of contracting an infectious disease. Active immunity depends upon the presence of memory cells in the body.
- Passive immunity is needed when an individual is in immediate danger of succumbing to an infectious disease. Passive immunity is short-lived because the antibodies are administered to—and not made by—the individual.
- Monoclonal antibodies, produced by the same plasma cell, have various functions, from detecting infections to treating cancer.
- Cytokines, including interferon, are a form of passive immunity used to treat AIDS and to promote the body's ability to recover from cancer.

7.6 Hypersensitivity Reactions

Allergic responses occur when the immune system reacts vigorously to substances not normally recognized as foreign.

- Immediate allergic responses, usually consisting of coldlike symptoms, are due to the activity of antibodies.
- Delayed allergic responses, such as contact dermatitis, are due to the activity of T cells.
- Tissue rejection occurs when the immune system recognizes a tissue as foreign.
- Immune deficiencies can be inherited or can be caused by infection, chemical exposure, or radiation.
- Autoimmune disorders occur when the immune system reacts to tissues/organs of the individual as if they were foreign.

Understanding Key Terms

active immunity 147
allergen 150
allergy 150
anaphylactic shock 150
antibody-mediated immunity 144
antibody titer 148
antigen 143
antigen-presenting cell (APC) 145
apoptosis 144
autoimmune disease 151
bacillus 134
bacteria 134
B cell (B lymphocyte) 139
B-cell receptor (BCR) 143
capsule 134
cell-mediated immunity 146
clonal selection model 143
coccus 134
complement system 142
cytokine 141
cytotoxic T cell 145
delayed allergic response 150
fimbriae 134
flagellum 134
gamma globulin 148
hay fever 150
helper T cell 146
histamine 141
human leukocyte antigen (HLA) 145
immediate allergic response 150
immunity 140
immunization 148
immunosuppressive 150
inflammatory response 140
interferon 142
interleukin 149
lymph 138
lymphatic nodule 140
lymphatic organ 138
lymph node 139
lysozyme 140
macrophage 142
mad cow disease 134
major histocompatibility complex (MHC) 145
mast cell 141
membrane attack complex 142
memory T cell 146
monoclonal antibody 148
multiple sclerosis (MS) 151
myasthenia gravis 151
passive immunity 147
pathogen 134
Peyer's patches 140
pilus 134
plasma cell 144
plasmid 134
prion 137
pus 141
red bone marrow 139
rheumatic fever 151
rheumatoid arthritis 151
severe combined immunodeficiency disease (SCID) 150
spirillum 134
spleen 139
systemic lupus erythematosus (SLE) 151
T cell (T lymphocyte) 139
T-cell receptor (TCR) 145
thymus gland 139
tonsils 140
toxin 135
vaccine 148
virus 135
xenotransplantation 150

Match the key terms to these definitions.

a. ________ Antigen prepared in such a way that it can promote active immunity without causing disease.

b. ________ Series of proteins in plasma that form an innate defense mechanism against pathogen invasion.

c. ________ Foreign substance, usually a protein, that stimulates the immune system to react, such as by producing antibodies.

d. ________ Process of programmed cell death involving a cascade of specific cellular events leading to the death and destruction of the cell.

e. ________ Type of lymphocyte that matures in the thymus gland and exists in three varieties, one of which kills antigen-bearing cells outright.

Testing Your Knowledge of the Concepts

1. Describe the basic characteristics of bacteria. Explain how five particular features contribute to the ability of bacteria to cause disease. (pages 134–135)
2. What is the structure of a virus? Is a virus living? Explain how a virus is able to reproduce. (pages 135–36)
3. What are prions, and how do they cause disease? (page 137)
4. Why are the red bone marrow and the thymus gland termed primary lymphatic organs? (page 139)
5. In what ways are the spleen and lymph nodes similar, and in what ways are they different? (pages 139–140)
6. How do innate defenses differ from acquired defenses? (pages 140, 143)
7. What is the first line of defense? Describe several methods of protection. (page 140)
8. What is the second line of defense? Describe the various cells and chemicals and their roles in protecting the body. (pages 140–141)
9. What are the four signs of inflammation, and what causes them? (pages 141–142)
10. What two types of cells are involved in providing acquired defenses against pathogens? What type of immunity does each provide? (pages 143–144)
11. What is the clonal selection model as it applies to B cells? What becomes of the clones that are produced? (page 145–146)
12. Describe the structure of an antibody. What are the five main classes, where are they found, and what are their functions? (pages 144–145)
13. Explain how T cells recognize an antigen. What are the types of T cells, and how do they function in immunity? (pages 145–146)
14. How is active immunity achieved? How is passive immunity achieved? (pages 147–148)
15. Discuss the production of monoclonal antibodies and their applications. (pages 148–149)
16. Discuss allergies, tissue rejection, and autoimmune diseases as they relate to the immune system. (pages 150–151)
17. Which of the following is a function of the spleen?
    a. produces T cells
    b. removes worn-out red blood cells
    c. produces immunoglobulins
    d. produces macrophages
    e. regulates the immune system
18. Which of the following is a function of the thymus gland?
    a. production of red blood cells
    b. secretion of antibodies
    c. production and maintenance of stem cells
    d. site for the maturation of T lymphocytes

For questions 19–23, match the lymphatic organs in the key to the description of its location.

Key:

a. lymph nodes
b. Peyer's patches
c. spleen
d. thymus gland
e. tonsils

19. Arranged in a ring around the pharynx
20. In the intestinal wall
21. Occur along lymphatic vessels
22. Upper left abdominal cavity, posterior to the stomach
23. Superior to the heart, posterior to the sternum

24. Which of the following is a function of the secondary lymphatic organs?
    a. transport of lymph
    b. clonal selection of B cells
    c. located where lymphocytes encounter antigens
    d. All of these are correct.
25. Defense mechanisms that function to protect the body against many infectious agents are called
    a. acquired.
    b. innate.
    c. barriers to entry.
    d. immunity.
26. Which of the following is most directly responsible for the increase in capillary permeability during the inflammatory reaction?
    a. pain
    b. white blood cells
    c. histamine
    d. tissue damage
27. Which of the following is *not* a goal of the inflammatory reaction?
    a. brings more oxygen to damaged tissues
    b. decreases blood loss from a wound
    c. decreases the number of white blood cells in the damaged tissues
    d. prevents entry of pathogens into damaged tissues
28. Which of the following is *not* correct concerning interferon?
    a. Interferon is a protective protein.
    b. Virus-infected cells produce interferon.
    c. Interferon has no effect on viruses.
    d. Interferon can be used to treat certain viral infections.
29. Which one of these does *not* pertain to B cells?
    a. have passed through the thymus
    b. have specific receptors
    c. are responsible for antibody-mediated immunity
    d. synthesize and liberate antibodies
30. Which of these pertain/s to T cells?
    a. have specific receptors
    b. are of more than one type
    c. are responsible for cell-mediated immunity
    d. stimulate antibody production by B cells
    e. All of these are correct.
31. During a secondary immune response,
    a. antibodies are made quickly and in great amounts.
    b. antibody production lasts longer than in a primary response.
    c. B cells become plasma cells.
    d. All of these are correct.
32. Active immunity may be produced by
    a. having a disease.
    b. receiving a vaccine.
    c. receiving gamma globulin injections.
    d. Both a and b are correct.
    e. Both b and c are correct.
33. Which of the following is *not* an innate body defense?
    a. complement
    b. the skin and mucous membranes
    c. antibodies
    d. lysozyme

Thinking Critically About the Concepts

An allergic response is an overreaction of the immune system in response to an antigen. Such responses always require a prior exposure to the antigen. The reaction can be immediate (within seconds to minutes) or delayed (within hours). Insect venom reactions often involve the development of hives and itching. Asthmalike symptoms include shortness of breath and wheezing. Decreased blood pressure will eventually cause loss of consciousness. This immediate, severe allergic effect, called anaphylaxis, can be fatal if untreated. In addition to insect venom, food allergens such as those in milk, peanut butter, and shellfish may also elicit life-threatening symptoms. In these cases, IgE antibodies, histamine, and other inflammatory chemicals are the culprits involved in making the reaction so severe.

1. How are B cells involved in immediate allergic responses?
2. An allergist is a doctor that treats people with known allergies. What type of treatments do you think are used? Why?
3. Why do you think certain allergens affect some people more than others? Why are some people asymptomatic to a particular antigen whereas others are affected?
4. Think of an analogy (something you're already familiar with) for the barrier defenses like your skin and mucous membranes.
5. Someone bitten by a poisonous snake should be given some antivenom (antibodies) to prevent them from dying. If they are bitten by the same type of snake three years after the initial bite, will they have immunity to the venom, or should they get another shot of antivenom? Justify your response with an explanation of the type of immunity someone gains from a shot of antibodies.

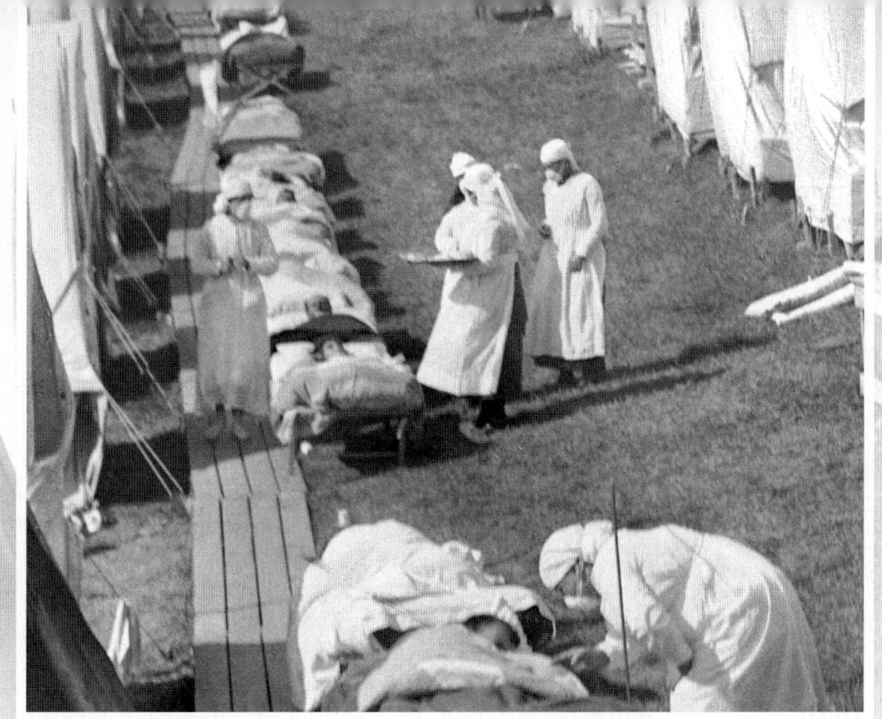

Infectious Diseases Supplement

THE SCIENCE OF EPIDEMIOLOGY

Epidemiology is the study of diseases in populations, including the causes, distribution, and control of these diseases. Humans have a long history of struggle with infectious diseases—diseases caused by pathogens, which include certain bacteria, viruses, fungi, parasites, protozoans, and prions. The term pathogen comes from the Greek term meaning "suffering."

Infectious disease outbreaks are described in the earliest written documents. Smallpox outbreaks were described 10,000 years ago in the Nile Valley. Malaria was described 4,000 years ago in China. The Chinese treated it with the Qinghao plant—the source of the drug currently used to treat malaria. Outbreaks of bubonic plague—the so-called "Black Death"—in Europe decimated the continent's population. The spread of the disease, as well as the public response to it, literally determined European history.

In more modern times, the battle between infectious agents and humans includes the development of vaccinations. The Chinese first developed vaccines around 200 BCE. In 1796, the smallpox vaccine was introduced to the Western world. Death rates from this virus dropped dramatically. The disease was eradicated from the human population in 1977. A rabies vaccine was developed in 1885, and vaccines for diphtheria, typhus, tetanus, and polio soon followed. Child mortality rates plummeted and remain low in countries where vaccination is routine.

The death rates from infectious diseases dropped still further in the 1940s with the development of antibiotics. With the developments of vaccines and antibiotics, scientists hoped they could eradicate other pathogens or, at least, significantly limit the suffering caused by infectious diseases. However, infectious diseases still kill large numbers of people. According to the World Health Organization (WHO), infectious diseases kill over 13 million people annually. In developing countries, infectious diseases account for 54% of deaths. Lower respiratory infections, followed by HIV/AIDS, diarrheal diseases, malaria, and tuberculosis, are currently the leading causes of infectious disease deaths.

CHAPTER CONCEPTS

S.1 AIDS and Other Pandemics
Epidemiology is the study of diseases in populations. Fields of study include the characteristics of the pathogens, how they cause disease, how the disease spreads, and treatment and prevention strategies for the disease. HIV/AIDS, tuberculosis, and malaria are currently pandemics.

S.2 Emerging Diseases
Emerging diseases include diseases that have never before been seen, as well as those previously recognized in a small number of people in isolated settings. Diseases that have been present throughout history, but not known to be caused by a pathogen, are also considered to be emerging diseases. Reemerging diseases are previously known diseases undergoing resurgence, often due to human carelessness.

S.3 Antibiotic Resistance
Misuse of antibiotics has led to the selection of antibiotic resistant organisms. Some organisms have developed multidrug resistances, and these organisms are very difficult to treat.

BEFORE YOU BEGIN

Before beginning this chapter, take a few moments to review the following discussions.

Section 7.1 What is the difference in the structure of a bacteria and a virus?

Section 7.4 What is the role of a T cell in the immune response?

Section 7.5 How do immunizations protect an individual against disease?

S.1 AIDS and Other Pandemics

Learning Outcomes

Upon completion of this section, you should be able to

1. Distinguish between an outbreak, epidemic, and pandemic.
2. Describe the phases of an HIV infection.
3. List the stages in the HIV life cycle.
4. Summarize the barriers in the development of an effective HIV vaccination.
5. Describe the causes of tuberculosis and malaria.

A disease is classified as an **epidemic** if there are more cases of the disease than expected in a certain area for a certain period. The number of cases that constitute an epidemic depends on what is expected. For example, a few cases of a very rare disease may constitute an epidemic, whereas a larger number of a very common disease may not be. If the epidemic is confined to a local area, it is usually called an **outbreak.** Global epidemics are called **pandemics** (Fig. S.1) HIV/AIDS, tuberculosis, malaria, and influenza are all examples of current pandemics. Organizations such as the Centers for Disease Control and Prevention (CDC) and the WHO monitor and respond to the threats of **infectious diseases**. These organizations are primarily responsible for determining whether an outbreak has reached epidemic or pandemic levels.

HIV/AIDS

Acquired immunodeficiency syndrome (AIDS) is caused by a virus known as the **human immunodeficiency virus (HIV).** There are two main types of HIV: HIV-1 and HIV-2. HIV-1 is the more widespread, virulent form of HIV. Of the two types of HIV, HIV-2 corresponds to a type of immunodeficiency virus found in the green monkey, which lives in western Africa. In addition, researchers have found a virus identical to HIV-1 in a subgroup of chimpanzees once common in west-central Africa. Perhaps HIV viruses were originally found only in nonhuman primates. They could have mutated to HIV after humans ate nonhuman primates for meat.

HIV can infect cells with particular surface receptors. Most importantly, HIV infects and destroys cells of the immune system, particularly helper T cells and macrophages. As the number of helper T cells declines, the body's ability to fight an infection also declines. As a result, the person becomes ill with various diseases. AIDS is the advanced stage of HIV infection, in which a person develops one or more of a number of opportunistic infections. An **opportunistic infection** is one that has the *opportunity* to occur only because the immune system is severely weakened.

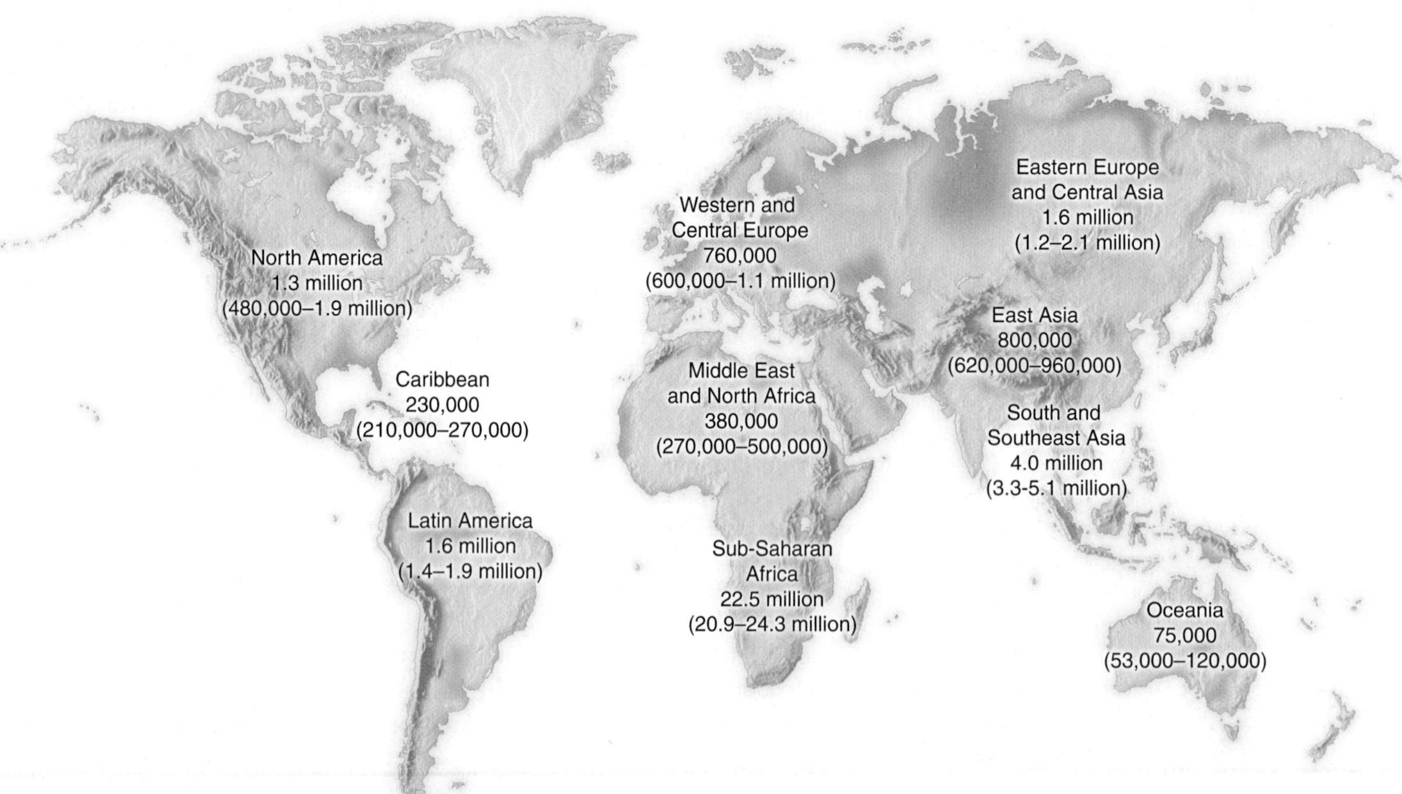

Figure S.1 Worldwide cases of HIV/AIDS.
HIV/AIDS occurs in all continents of the globe. According to 2007 statistics, approximately 33.2 million people are infected.

Origin of and Prevalence of HIV

It is generally accepted that HIV originated in Africa and then spread to the United States and Europe by way of the Caribbean. However, the exact dates of the first human cases of HIV are still being investigated. Recent molecular analyses of the HIV virus suggest that the virus may have first infected humans sometime between 1884 and 1924. Direct evidence of HIV in humans has been obtained from tissue and blood samples taken in the 1950s and 1960s. HIV has been found in a preserved 1959 blood sample taken from a man who lived in an African country now called the Democratic Republic of the Congo. British scientists have been able to show that AIDS came to their country perhaps as early as 1959. They examined the preserved tissues of a Manchester seaman who died that year and concluded that he most likely died of AIDS. Similarly, it is thought that HIV entered the United States on numerous occasions as early as the 1950s. But the first documented case is a 15-year-old male who died in Missouri in 1969, with skin lesions now known to be characteristic of an AIDS-related cancer. Doctors froze some of his tissues because they could not identify the cause of death. Researchers also want to test the preserved tissue samples of a 49-year-old Haitian who died in New York in 1959 of the type of pneumonia now known to be AIDS-related.

Throughout the 1960s, it was customary in the United States to list leukemia as the cause of death in immunodeficient patients. Most likely, some of these people actually died of AIDS. HIV is not extremely infectious. Thus, it took several decades for the number of AIDS cases to increase to the point that AIDS became recognizable as a specific and separate disease. The name *AIDS* was coined in 1982, and HIV was found to be the cause of AIDS in 1983–84.

Worldwide, estimates of HIV/AIDS infection rates and deaths are updated every two to three years. As of 2007, (Table S.1), an estimated 33.2 million people were living with HIV infection. Among the 2.5 million new HIV infections, nearly 20% are in people under the age of 15. Table S.1 also tells us that 2.1 million people died during 2007; but since the beginning of the epidemic, over 25 million people have died of AIDS. Today, at least 0.8% of the adults in the world have an HIV infection.

As we can deduce from studying Figure S.1 and Table S.1, most people infected with HIV live in the developing (poor, low- to middle-income) countries. The hardest-hit regions are shown in Figure S.2.

Phases of an HIV Infection

HIV occurs as several subtypes. HIV-1C is prominent in Africa, and HIV-1B causes most infections in the United States. The following description of the phases of HIV infection pertains to an HIV-1B infection. The helper T cells and macrophages infected by HIV are called *CD4 cells* because they display a molecule called CD4 on their surface. With the destruction of CD4 cells, the immune system is significantly impaired. After all, macrophages present the antigen to helper T cells. In turn, helper T lymphocytes coordinate the immune response. B lymphocytes are stimulated to produce antibodies, and cytotoxic T cells destroy cells infected with a virus. In the United States, one of the most common causes of AIDS deaths is from *Pneumocystis jiroveci* pneumonia (PCP); in Africa, tuberculosis kills more HIV-infected people than any other AIDS-related illness.

In 1993, the U.S. Centers for Disease Control and Prevention (CDC) issued clinical guidelines for the classification of HIV to help clinicians track the status, progression, and phases of HIV infection. The classification of HIV infection will be discussed in three categories (or phases); the system is based on two aspects of a person's health—the CD4 T-cell count and the history of AIDS-defining illnesses.

Category A: Acute Phase A person in category A typically has no apparent symptoms (asymptomatic), is highly

Table S.1 HIV Global Statistics, 2007

	People Living with HIV	New Infection	AIDS Deaths	Adult Prevalence (Percentage)
Sub-Saharan Africa	22.5 million	1.7 million	1.6 million	5.0
Asia	4.8 million	432,000	302,000	0.4
Latin America	1.6 million	100,000	58,000	0.5
Caribbean	230,000	17,000	11,000	1.0
North American, Western and Central Europe	2 million	77,000	33,000	0.9
Eastern Europe and Central Asia	1.6 million	150,000	55,000	0.9
North Africa and the Middle East	380,000	35,000	25,000	0.3
Oceania	75,000	14,000	1,200	0.4
Total	**33.2 million**	**2.5 million**	**2.1 million**	**0.8%**

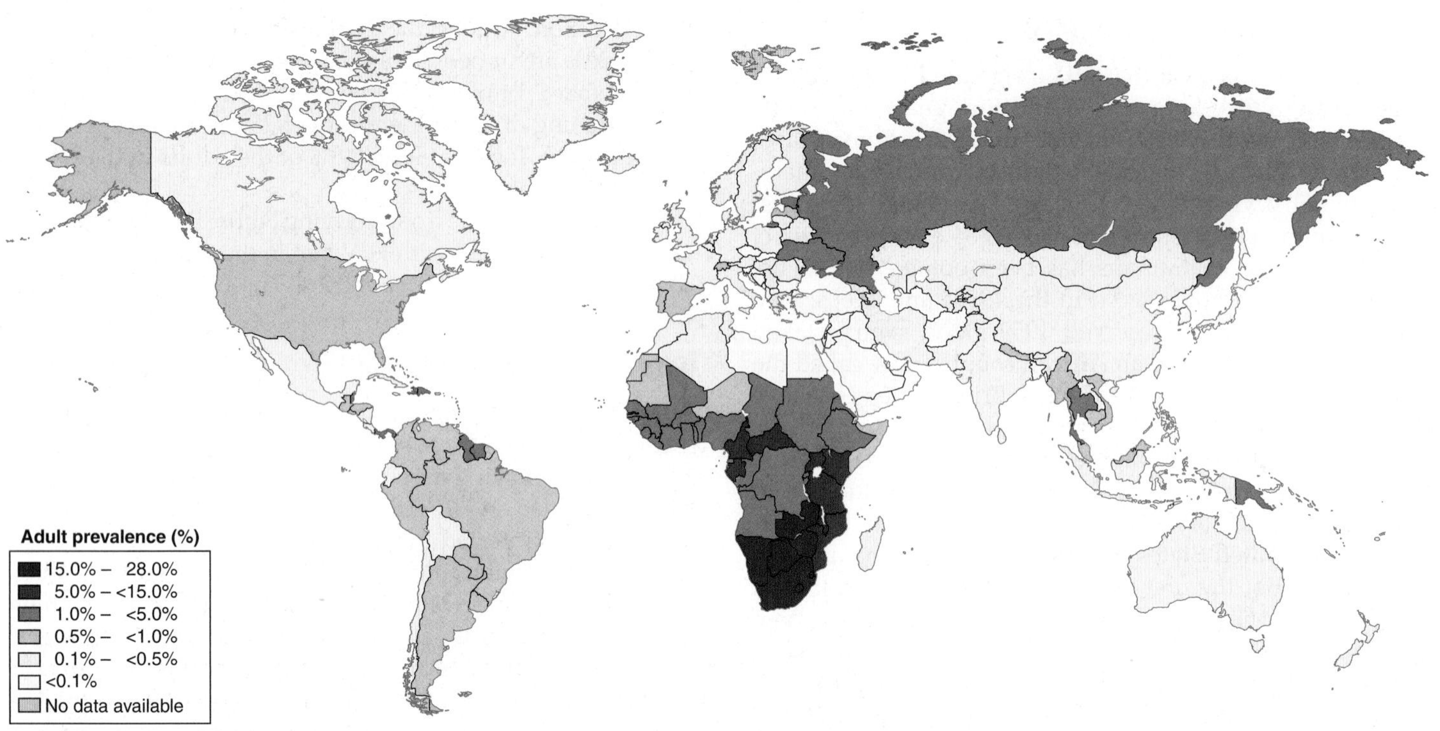

Figure S.2 **A Global View of HIV Infection.**
This map show the prevalence of HIV infections in adults. Some of the most infected areas are in Sub-Saharan Africa and Southeast Asia.

infectious, and has a CD4 T-cell count that has never fallen below 500 cells per cubic millimeter (cells/mm^3) of blood, which is sufficient for the immune system to function normally (Fig. S.3). A normal CD4 T-cell count is at least 800 cells/mm^3.

Today, investigators are able to track not only the blood level of CD4 T cells but also the viral load. The viral load is the number of HIV particles in the blood. At the start of an HIV-1B infection, the virus replicates ferociously, and the killing of CD4 T cells is evident because the blood level of these cells drops dramatically. During the first few weeks of infection, some people (1–2%) develop flulike symptoms (fever, chills, aches, swollen lymph nodes) that may last an average of two weeks. After this, a person may remain "symptom free" for years. At the beginning of this acute phase of infection, an HIV antibody test is usually negative because it generally takes an average of 25 days before there are detectable levels of HIV antibodies in body fluids.

After time, the body responds to the infection by increased activity of immune cells, and the HIV blood test becomes positive. During this phase, the number of CD4 T cells is greater than the viral load (Fig. S.3); but some investigators believe that a great unseen battle is going on. The body is staying ahead of the hordes of viruses entering the blood by producing as many as 1 to 2 billion new helper T lymphocytes each day. This is called the "kitchen-sink model" for CD4 T-cell loss. The sink's faucet (production of new CD4 T cells) and the sink's drain (destruction of CD4 T cells) are wide open. As long as the body can produce enough new CD4 T cells to keep pace with the destruction of these cells by HIV and by cytotoxic T cells, the person has a healthy immune system that can deal with the infection. In other words, a person in category A would have no history of conditions listed in categories B and C.

Category B: Chronic Phase A person in category B would have a CD4 T-cell count between 499 and 200 cells/mm^3 and one or more of a variety of symptoms related to an impaired immune system. The symptoms might include yeast infections of the mouth or vagina, cervical dysplasia (precancerous abnormal growth), prolonged diarrhea, thick sores on the tongue (hairy leukoplakia), or shingles (to list a few). Swollen lymph nodes, unexplained persistent or recurrent fevers, fatigue, coughs, or diarrhea are often seen as well. During this chronic stage of infection, the number of HIV particles is on the rise (Fig. S.3). However, they do not as yet have any of the conditions listed for category C.

Category C: AIDS A person in category C is diagnosed with AIDS. When a person has AIDS, the CD4 T-cell count has fallen below 200 cells/mm^3 or the person has developed one or more of the 25 AIDS-defining illnesses (or opportunistic infections) described by the CDC's list of conditions in the 1993 AIDS surveillance case definition. Persons with AIDS die from one or more opportunistic diseases rather than from the HIV infection. Recall that an opportunistic illness

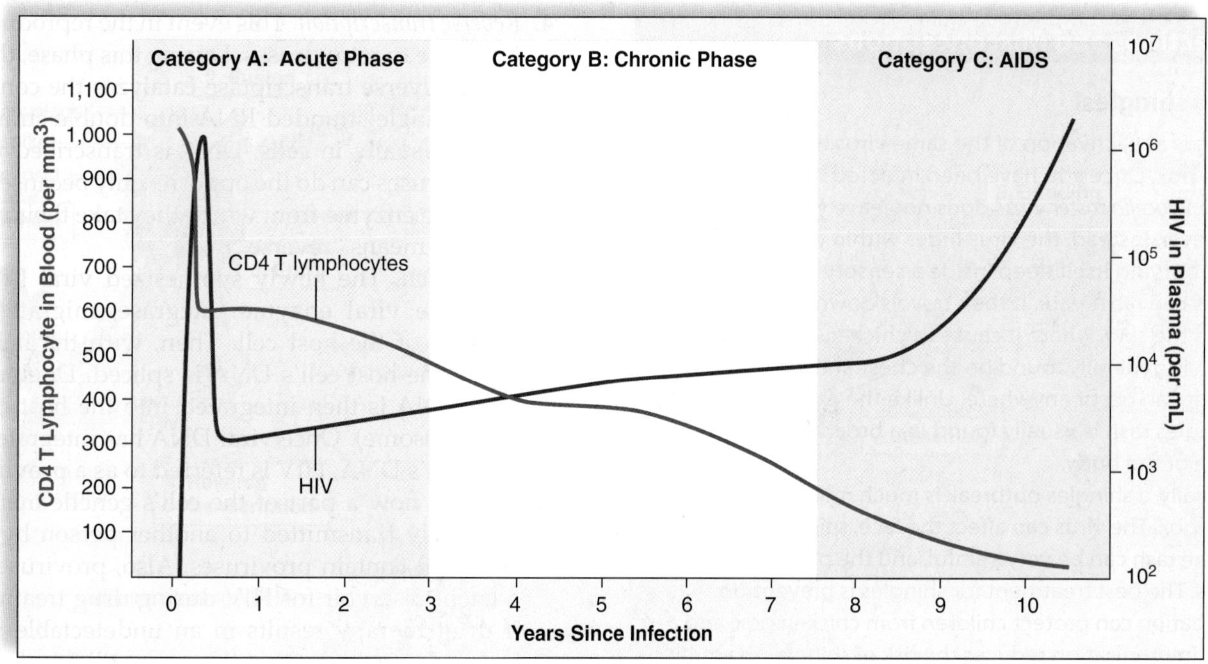

Figure S.3 Stages of an HIV infection.

In category A individuals, the number of HIV particles in plasma rises upon infection and then falls. The number of CD4 T lymphocytes falls but stays above 400/mm^3. In category B individuals, the number of HIV particles in plasma is slowly rising and the number of T lymphocytes is decreasing. In category C individuals, the number of HIV particles in plasma rises dramatically as the number of T lymphocytes falls below 200/mm^3.

occurs only when the immune system is weakened. These diseases include the following.

- *Pneumocystis jiroveci* pneumonia—a fungal infection of the lungs
- *Mycobacterium tuberculosis*—a bacterial infection usually of lymph nodes or lungs but may be spread to other organs
- Toxoplasmic encephalitis—a protozoan parasitic infection, often seen in the brain of AIDS patients
- Kaposi's sarcoma—an unusual cancer of the blood vessels, which gives rise to reddish purple, coin-sized spots and lesions on the skin
- Invasive cervical cancer—a cancer of the cervix, which spreads to nearby tissues

Once one or more of these opportunistic infections have occurred, the person will remain in category C. Newly developed drugs can treat opportunistic diseases. Still, most AIDS patients are repeatedly hospitalized due to weight loss, constant fatigue, and multiple infections. Death usually follows in two to four years. Although there is still no cure for AIDS, many people with HIV infection are living longer, healthier lives due to the expanding use of antiretroviral therapy.

HIV Structure

HIV consists of two single strands of RNA (its nucleic acid genome); various proteins; and an envelope, which it acquires from its host cell (Fig. S.4). The virus's genetic material is protected by a series of three protein coats: the nucleocapsid, capsid, and matrix. Within the matrix are three

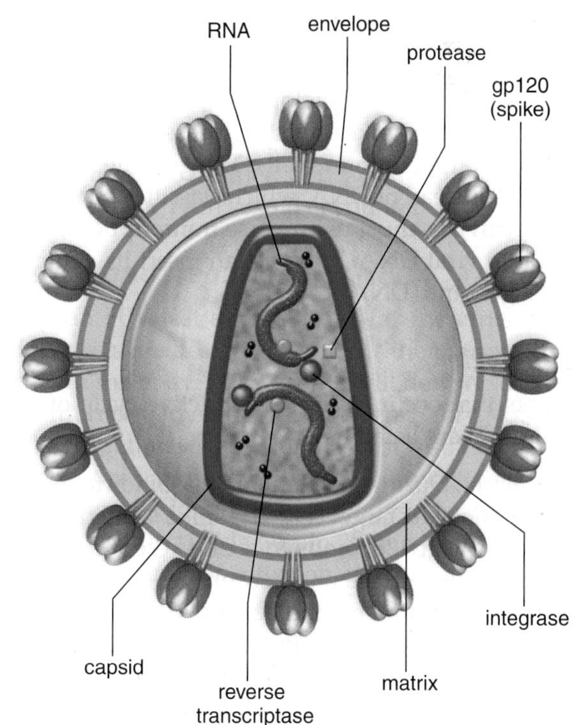

Figure S.4 The structure of the human immunodeficiency virus. HIV is composed of two strands of RNA, a capsid, and an envelope containing spike proteins.

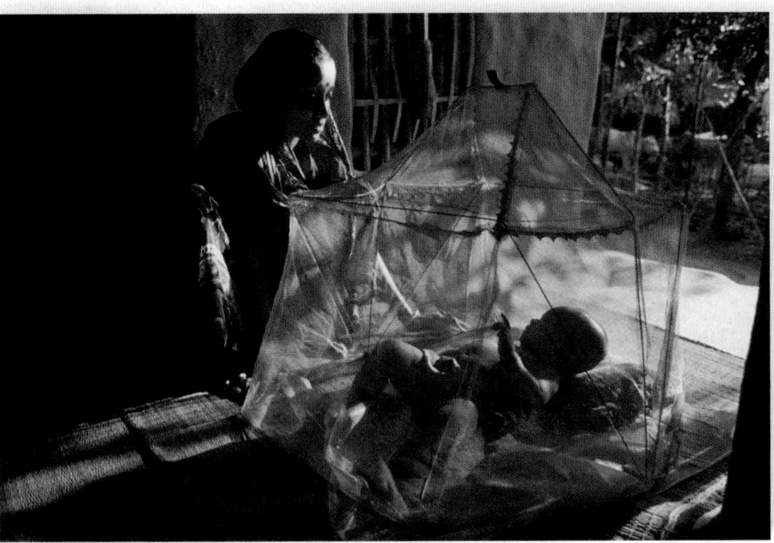

Figure S.9 Preventing the transmission of malaria. Studies show that the use of insecticide-treated mosquito nets reduces the risk of malaria transmission.

Health organizations are working on the prevention of infection through vector control. Strategies include eliminating the mosquito by removing its breeding sites and by insecticide fogging of large areas. Additional efforts are aimed at preventing humans from being bitten by the mosquito, using simple mosquito nets. The use of insecticide-treated mosquito nets for children has reduced the incidence of malaria (Fig. S.9).

Drug-resistant *Plasmodium* and insecticide-resistant *Anopheles* are becoming significant problems. *P. falciparum* and *P. vivax* have developed strains that are resistant to the antimalarial drugs. Efforts to develop a malaria vaccine are ongoing.

Connecting the Concepts

For more information on the topics presented in this section, refer to the following discussions.

Section 6.2 examines the role of erythrocytes in the body.

Section 7.1 describes the structure of a typical virus.

Section 9.3 explores the structure of the lungs.

Check Your Progress S.1

1. Describe the differences between an outbreak, epidemic, and pandemic and give an example of each.
2. Summarize the HIV replication cycle and list the types of cells this virus infects.
3. Explain the role of the mosquito in the malarial life cycle.
4. Discuss how viruses and bacteria may develop resistance to drugs.

S.2 Emerging Diseases

Learning Outcomes

Upon completion of this section, you should be able to

1. Define the term *emerging disease.*
2. List some examples of emerging diseases.

In the past several years, avian influenza (H5N1), swine flu (H1N1), and severe acute respiratory syndrome (SARS) have generated a lot of press. These are considered new or **emerging diseases.** The National Institute of Allergy and Infectious Diseases (NIAID) lists 18 **pathogens** that are newly recognized in the last two decades and five additional pathogens that are considered to be reemerging. Reemerging diseases are ones that have reappeared after a significant decline in incidence. *Streptococcus*, the bacteria that causes strep throat and other infections, is considered to be a reemerging pathogen due to increasing resistance to antibiotics. Finally, there are diseases that have been known throughout human history but had not been known to be caused by an infectious agent or the pathogen had never been identified. Ulcers caused by *Helicobacter pylori* (recognized in 1983) are an example.

Where do emerging diseases come from? Some of these diseases may result from new and/or increased exposure to animals or insect populations that act as vectors for disease. Changes in human behavior and use of technology can result in new diseases. SARS is thought to have arisen in Guandong, China, due to consumption of civets, a type of exotic cat considered a delicacy (Fig. S.10). The civets were possibly infected by exposure to horseshoe bats sold in open markets. Legionnaires' disease emerged in 1976 due to contamination of a large air-conditioning system in a hotel. The bacteria thrived in the cooling tower used as the water

Figure S.10 A potential source of the SARS epidemic. It appears that civets may have been the source of the SARS epidemic. Civet meat is considered a delicacy in some parts of China.

source for the air-conditioning system. In addition, globalization results in the transport of diseases all over the world that were previously restricted to isolated communities. The first SARS cases were reported in southern China the week of November 16, 2002. By the end of February 2003, SARS had reached nine countries/provinces, mostly through airline travel. Some pathogens mutate and change hosts, jumping from birds to humans, for example. Before 1997, avian flu was thought to affect only birds. A mutated strain jumped to humans in the 1997 outbreak. To control that epidemic, officials killed 1.5 million chickens to remove the source of the virus.

NIAID also monitors reemerging diseases. Reemerging diseases have been known in the past but were thought to have been controlled. Diseases in this category include known diseases that are spreading from their original geographic location or diseases that have suddenly increased in incidence. A change in geographic location could be due to global warming allowing expansion of habitats for insect vectors. Reemerging diseases can also be due to human carelessness, as in the abuse of antibiotics or poorly implemented vaccination programs. This allows previously controlled diseases to resurge.

Connecting the Concepts

For more information on the topics in this section, refer to the following discussions.

Section 1.3 examines the research that identified *H. pylori* as a cause of ulcers.

Section 7.1 provides additional information on the H1N1 virus.

Check Your Progress S.2

1. Define *emerging disease,* and give an example.
2. Explain how emerging diseases arise.
3. Explain what may be done to reduce the threat of emerging and reemerging diseases.

S.3 Antibiotic Resistance

Learning Outcomes

Upon completion of this section, you should be able to

1. Summarize how a pathogen becomes resistant to an antibiotic.
2. Explain the significance of XDR TB and MRSA.

Some well-known pathogens are becoming more difficult to fight due to the advent of **antibiotic resistance.** Just four years after penicillin was introduced in 1943, bacteria began developing resistance to it. The use of antibiotics does not cause humans to become resistant to the drugs. Instead, pathogens become resistant. There are some organisms in a population naturally resistant to the drug (Fig. S.11). They have acquired this resistance through mutations or interactions with other organisms. The drug regimen kills the susceptible ones while leaving naturally resistant ones to multiply and repopulate the patient's body. The new population is then resistant to the drug. Tuberculosis, malaria, gonorrhea, *Staphylococcus aureus,* and enterococci (or group D *Streptococcus*) are a few of the diseases and organisms connected to antibiotic resistance.

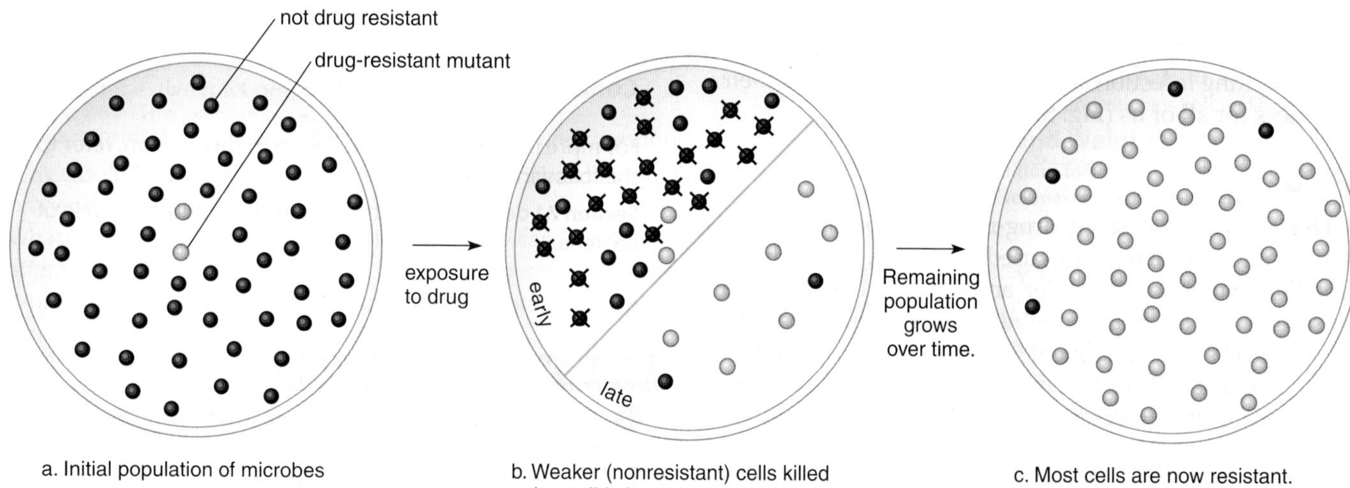

Figure S.11 Development of antibiotic resistance.
a. In a microbe population, random mutation can result in cells with drug resistance. **b.** Drug use kills all nonresistant microbes, leaving only the stronger, resistant cells to survive. **c.** The new microbe population is now mostly resistant to the antibiotic.

Biology Matters Health Focus

Swallowing a Camera

During a traditional endoscopy procedure, the doctor uses an endoscope (a retractable, tubelike instrument with an embedded camera) to examine the patient's GI tract. The PillCam™ has become a viable alternative to traditional endoscopy. With a gulp of water, PillCam is swallowed, and it travels through the digestive system. Instead of spending an uncomfortable half day or more at the doctor's office, a patient visits the doctor in the morning, swallows the camera, puts on the recording device, and goes about his or her daily routine.

Propelled by the normal muscular movement of the digestive system, PillCam embarks on a 4- to 8-hour journey through the digestive system. As it travels through the stomach, the twists and turns of the small intestine, and the large intestine, PillCam continuously captures high-quality, wide-angle film footage of its journey and beams this information to the recording device worn by the patient.

At the end of the day, PillCam reaches the end of its journey, and it is defecated. Later, the recording device is returned to the doctor's office so that the data may be retrieved. The doctor can view PillCam's journey as a 90-minute movie.

Like the food that we eat, PillCam traverses the numerous twists and turns of the intestine with ease. This provides a more accurate diagnosis, because a larger portion of the GI tract can be examined. PillCam does not create discomfort in the patient, so the considerable risks involved with using anaesthetics and painkillers are eliminated. And finally, the doctor does not need to be present for the entire procedure, saving valuable time and money for both doctor and patient.

Colonoscopy is a routine procedure used to examine and diagnose colon cancer. It employs a colonoscope, an instrument similar to an endoscope, inserted through the anus into the large intestine. It is capable of removing precancerous tissue before it becomes invasive. PillCam cannot be used to remove tissue samples for analysis, so its use as a replacement for colonoscopy is somewhat limited.

Figure 8B An endoscopy procedure using a PillCam. PillCam traverses through the entire GI tract, taking pictures after it is swallowed.

Polyps and Cancer

The colon is subject to the development of **polyps,** small growths arising from the epithelial lining. Polyps, whether benign or cancerous, can be removed surgically. If colon cancer is detected while still confined to a polyp, the expected outcome is a complete cure. The National Cancer Institute estimates over 146,000 new cases of colon cases are diagnosed per year in the United States. Some investigators believe that dietary fat increases the likelihood of colon cancer, because dietary fat causes an increase in bile secretion. It could be that intestinal bacteria convert bile salts to substances that promote the development of cancer. On the other hand, fiber in the diet seems to inhibit the development of colon cancer and regular elimination reduces the time that the colon wall is exposed to any cancer-promoting agents in feces.

Connections and Misconceptions

Why do you get nauseaus?

Nausea is the queasy, uneasy feeling you get in your stomach in response to some medications, injuries, headaches, motion sickness, psychological stresses, and hormone changes. Nausea is normally felt before vomiting. When we are nauseaus, we lose muscle tone in our stomach and our intestines begin to reflux, moving its contents into our stomach, causing that queasy feeling, and sometimes resulting in vomiting.

One diagnostic tool for all of these disorders is an endoscopy. As discussed in the Health Focus, *Swallowing a Camera,* endoscopy by means of a flexible tube (and camera) inserted into the GI tract, usually from the anus, is gradually being replaced by the PillCam, a camera you swallow.

Connecting the Concepts

For more information on the large intestine, refer to the following discussions.

Section 2.4 describes the role of fiber as a complex carbohydrate.

Section 7.3 examines how the bacteria in the large intestine act as an innate defense against disease.

Section 19.2 outlines some of the more common causes of cancer.

Check Your Progress 8.5

1. Describe the different parts of the large intestine and provide the function for each.
2. Detail how the functions of the large intestine contribute to homeostasis. Give some examples.
3. Discuss how a disorder of the large intestine can affect homeostasis overall. Give a few examples.

8.6 Nutrition and Weight Control

Learning Outcomes

Upon completion of this section, you should be able to

1. Calculate a BMI value and interpret its relationship to your overall health.
2. Identify the role of each class of nutrient in the human body.

Obesity—being grossly overweight—has doubled in the United States in only 20 years. According to the Centers for Disease Control (CDC), there are 34 states with a 25% or higher population of obese adults and 30 states with 30% or higher population of obese children. These statistics are of great concern because excess body fat is associated with a higher risk for premature death, type 2 diabetes, hypertension, cardiovascular disease, stroke, gallbladder disease, respiratory dysfunction, osteoarthritis, and certain types of cancer.

The Health Focus, *Searching for the Magic Weight-Loss Bullet,* tells us about the various ways people have tried to keep their weight under control. The conclusion is that achieving or maintaining a healthy weight requires not only eating a variety of healthy foods but also exercising. In other words, to reverse a trend toward obesity, eat fewer calories by making wiser food choices and be more active.

How Obesity Is Defined

Today, obesity is often defined as having a **body mass index (BMI)** of 30 or greater. If you know your height and weight, you can determine your BMI using the table in Figure 8.11. The BMI can also be calculated by dividing weight in pounds (lb) by height in inches (in.) squared and multiplying by a conversion factor of 703:

$$(\text{weight in pounds}/\text{height in inches})^2 \times 703$$

Most people find that using the table is a lot easier. As a general rule,

a healthy BMI = 18.5 to 24.9;
an overweight BMI = 25 to 29.9;
an obese BMI = 30 or higher; and
a morbidly obese BMI = 40 or higher.

Your BMI gives you an idea of how much of your weight is due to adipose tissue, commonly known as fat. In general, the taller you are, the more you could weigh without it being due to fat. Using BMI in this way works for most people, especially if they tend to be sedentary. But your BMI number should be used only as a general guide. It does not take into account fitness, bone structure, or gender. For example, a weight lifter might have an obese BMI, not because of the amount of body fat, but because of increased bone and muscle weight.

Animation BMI

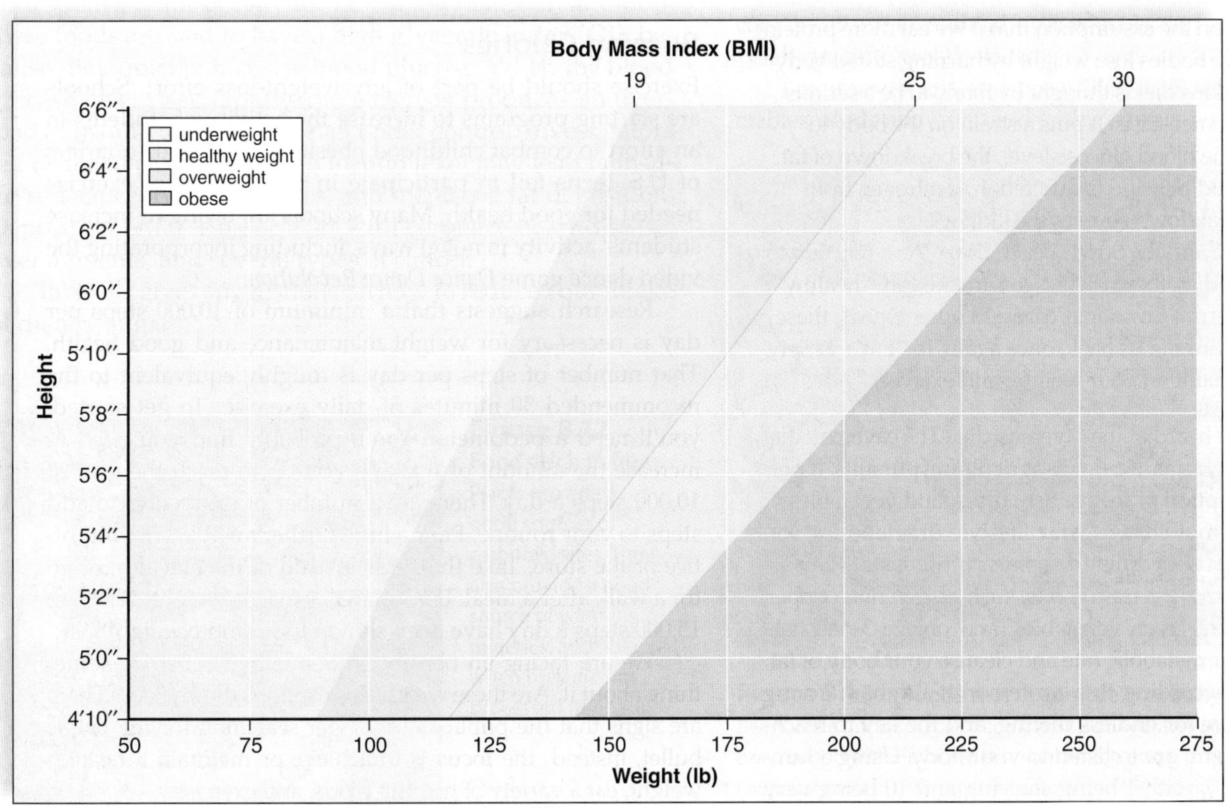

Source: U.S. Department of Agriculture: *Dietary Guidelines for Americans, 2005*

Figure 8.11 The body mass index chart. Match your weight with your height, then determine your body mass index (BMI). Healthy BMI = 18.5 to 24.9; overweight BMI = 25 to 29.9; obese BMI = 30 to 39.9; morbidly obese BMI = 40 or more.

Other abnormal eating practices include binge-eating disorder and muscle dysmorphia. Many obese people suffer from **binge-eating disorder,** a condition characterized by episodes of overeating without purging. Stress, anxiety, anger, and depression can trigger food binges. A person suffering from **muscle dysmorphia** (Fig. 8.16*c*) thinks his or her body is underdeveloped. Body-building activities and a preoccupation with diet and body form accompany this condition. Each day, the person may spend hours in the gym working out on muscle-strengthening equipment. Unlike anorexia nervosa and bulimia, muscle dysmorphia affects more men than women.

Connections and Misconceptions

How can you easily determine the serving size of a certain food?

The food pyramid (www.mypyramid.gov) provides good information about the serving size for most foods you might eat. For example, about 12 baby carrots is a serving (cup) of carrots. Some good analogies for serving sizes will help. A serving of meat (protein) should be the size of a deck of cards. A potato should approximate the size of a lightbulb or a computer mouse, whereas an apple might be the size of a baseball. A single serving of cheese should resemble the size of a pair of dice.

Connecting the Concepts

For more information on how the body utilizes nutrients, refer to the following discussions.

Section 11.1 explains how vitamin D acts as a hormone that influences bone growth.

Section 14.4 examines the role of vitamin A in vision.

Section 19.2 examines how a diet with adequate levels of vitamins A and C may help prevent cancer.

Check Your Progress 8.6

1. Briefly describe and give an example of each class of nutrients.
2. Discuss why carbohydrates and fats might be the cause of the obesity epidemic today.
3. Detail why it is important to overall homeostasis to have a balanced diet.

CASE STUDY CONCLUSION

Nicole's test results indicated that she had gastroesophageal reflux disease, or GERD, as well as a duodenal ulcer. GERD is a condition in which food and liquid travel backward up the esophagus from the stomach [see the Health Focus, *Heartburn (GERD),* early in this chapter]. After you swallow food, it mixes with very acidic gastric juices to chemically digest the food so it can enter the intestine where nutrients can be removed. When this mixture travels up the esophagus, the acidity of the digestive juices irritates the lining of the esophagus causing heartburn and lesions. If not treated, the esophagus can begin to deteriorate. The duodenal ulcer Nicole had was a raw area on the duodenum. The ulcer is caused by the same acidic digestive juices affecting this area that affected her esophagus. Additionally, Nicole was tested for levels of *Helicobacter pylori* (*H. pylori*). This is a type of bacteria that commonly lives in the digestive tracts of humans. Some people have elevated levels of the bacteria that can cause heartburn and reflux as well as ulcerations of the stomach and duodenum. Her levels of *H. pylori* were significant. Dr. Winch gave Nicole a few medications: an antibiotic to decrease the *H. pylori* level; a sucralfate, a medication that coats the area of the ulceration so it can heal; and protonix, a medication to reduce the amount of acid the stomach makes. The doctor also suggested a few lifestyle changes that can help. Drinking and smoking have been shown to worsen GERD and ulcers. Dr. Winch also showed Nicole studies that suggested that a healthy diet and moderate exercise aided in managing the levels of not only acidic gastric juices produced but also *H. pylori.* Dr. Winch also cautioned Nicole against taking OTC medications that can cause stomach inflammations, such as aspirin, ibuprofen, and naproxen. By managing her diet, lifestyle, and medications, Nicole can get her GI tract back in working order.

Media Study Tools

www.mhhe.com/maderhuman12e

Enhance your study of this chapter with study tools and practice tests. Also ask your instructor about the resources available through ConnectPlus, including the media-rich eBook, interactive learning tools, and animations.

Virtual Lab

The virtual lab "Nutrition" provides an interactive investigation of how your food choices relate to your daily intake of select nutrients.

Summarizing the Concepts

8.1 Overview of Digestion

The organs of the digestive system are located within the GI tract. The processes of digestion require ingestion, digestion, movement, absorption, and elimination. All parts of the tract have four layers, called the mucosa, submucosa, muscularis, and serosa.

8.2 First Part of the Digestive Tract

- In the mouth, teeth chew the food, saliva contains salivary amylase for digesting starch, and the tongue forms a bolus for swallowing.
- Both the mouth and the nose lead into the pharynx. The pharynx opens into both the food passage (esophagus) and air passage (trachea, or windpipe). During swallowing, the opening into the nose is blocked by the soft palate, and the epiglottis covers the trachea. Food enters the esophagus, and peristalsis begins. The esophagus moves food to the stomach.

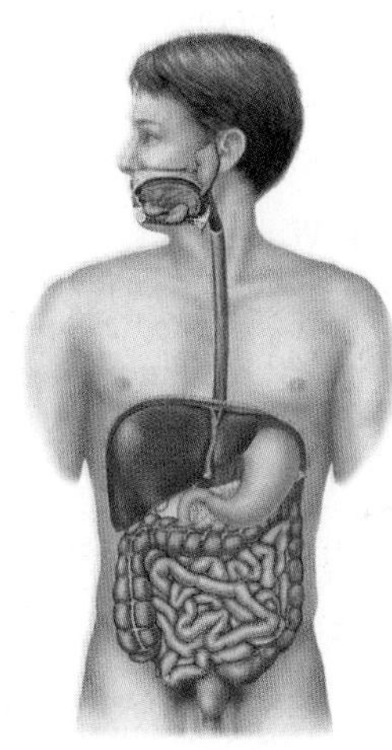

8.3 The Stomach and Small Intestine

- The stomach expands and stores food and also churns, mixing food with the acidic gastric juices. This juice contains pepsin, an enzyme that digests protein.
- The duodenum of the small intestine receives bile from the liver and pancreatic juice from the pancreas. Bile emulsifies fat and readies it for digestion by lipase.
- The pancreas produces enzymes that digest starch (pancreatic amylase), protein (trypsin), and fat (lipase). The intestinal enzymes finish the process of chemical digestion.
- Small nutrient molecules are absorbed at the villi in the walls of the small intestine.

8.4 The Accessory Organs and Regulation of Secretions

Three accessory organs of digestion send secretions to the duodenum via ducts. These organs are the pancreas, liver, and gallbladder.

- The pancreas produces pancreatic juice, which contains digestive enzymes for carbohydrate, protein, and fat.
- The liver produces bile, destroys old blood cells, detoxifies blood, stores iron, makes plasma proteins, stores glucose as glycogen, breaks down glycogen to glucose, produces urea, and helps regulate blood cholesterol levels.
- The gallbladder stores bile, produced by the liver. The secretions of digestive juices are controlled by the nervous system and by hormones.

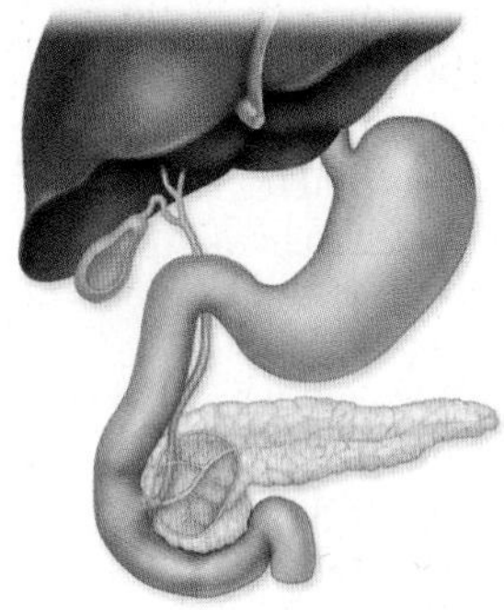

- Gastrin produced by the lower part of the stomach stimulates the upper part of the stomach to secrete pepsin.
- Secretin and CCK produced by the duodenal wall stimulate the pancreas to secrete its juices and the gallbladder to release bile.

8.5 The Large Intestine and Defecation

- The large intestine consists of the cecum; the colon (including the ascending, transverse, and descending colon); and the rectum, which ends at the anus.
- The large intestine absorbs water, salts, and some vitamins; forms the feces; and carries out defecation.
- Disorders of the large intestine include diverticulosis, irritable bowel syndrome, inflammatory bowel disease, polyps, and cancer.

8.6 Nutrition and Weight Control

The nutrients released by the digestive process should provide us with adequate energy, essential amino acids and fatty acids, and all necessary vitamins and minerals. Today, obesity is on the increase, possibly because people eat too much food and make improper food choices. Obesity is associated with many illnesses, including type 2 diabetes and cardiovascular disease. The food guide pyramid shows foods to emphasize and foods to minimize for good health.

- Carbohydrates are necessary in the diet, but simple sugars and refined starches cause a rapid release of insulin that can lead to type 2 diabetes.
- Proteins supply essential amino acids.
- Unsaturated fatty acids, particularly the omega-3 fatty acids, are protective against cardiovascular disease.

- Saturated fatty acids and trans fats contribute to heart disease.
- Vitamins and minerals are also required by the body in certain amounts.

Understanding Key Terms

absorption 171
anorexia nervosa 191
anus 180
appendix 171
bile 176, 178
bilirubin 180
binge-eating disorder 192
body mass index (BMI) 183
bolus 173
bulimia nervosa 191
cecum 180
cholesterol 178
chyme 176
cirrhosis 179
colon 180
constipation 181
defecation 180
dental caries 173
diaphragm 175
diarrhea 181
digestion 170
diverticulosis 171
duodenum 176
elimination 171
epiglottis 173
esophagus 173
essential amino acids 185
essential fatty acids 186
fiber 180
gallbladder 178
gallstone 178
gastric gland 176
glottis 173
glycemic index (GI) 185
hard palate 172
heartburn 174
hemorrhoid 181
hepatitis 179
hormone 178
hydrolyze 170
ingestion 170
jaundice 179
lacteal 177
lactose intolerance 178
large intestine 180
lipase 176
liver 178
lumen 171
mineral 187
movement 171
mucosa 171
muscle dysmorphia 192
muscularis 171
nutrient 185
obesity 183
oblique 176
osteoporosis 188
pancreas 178
pancreatic amylase 178
pepsin 176
periodontitis 173
peristalsis 170
peritonitis 171
pharynx 173
polyp 182
rectum 180
rugae 176
salivary amylase 172
salivary gland 172
serosa 171
small intestine 176
soft palate 172
sphincter 174
stomach 175
submuscosa 171
trypsin 178
urea 178
vermiform appendix 180
villus 177
vitamin 189

Match the key terms to these definitions.

a. ________ Essential requirement in the diet, needed in small amounts. Often a part of a coenzyme.

b. ________ Fat-digesting enzyme secreted by the pancreas.

c. ________ Lymphatic vessel in an intestinal villus; it aids in the absorption of fats.

d. ________ Muscular tube for moving swallowed food from the pharynx to the stomach.

e. ________ Organ attached to the liver that serves to store and concentrate bile.

Testing Your Knowledge of the Concepts

1. Argue that absorption is the most important of the five processes of digestion over the other four processes. (pages 170–171)
2. List the main organs of the digestive tract, and state the contribution of each to the digestive process. (pages 170–72)
3. Discuss the absorption of the products of digestion into the lymphatic and cardiovascular systems. (page 177)
4. Name the enzymes involved in the digestion of starch, protein, and fat, and tell where these enzymes are active and what they do. (page 177)
5. Why are the pancreas, liver, and gallbladder considered *accessory* organs of digestion and not organs of digestion? (pages 178–179)
6. Name and state the functions of the hormones that assist the nervous system in regulating digestive secretions. (page 180)
7. What is the chief contribution of each of these in the body: carbohydrates, proteins, fats, fruits, and vegetables? (pages 185–186)
8. Which three eating disorders involve binge eating? How are these three disorders different from one another? (pages 191–192)
9. Tracing the path of food in the following list (a–f), which step is out of order first?
   a. mouth
   b. pharynx
   c. esophagus
   d. small intestine
   e. stomach
   f. large intestine
10. Which association is incorrect?
    a. mouth—starch digestion
    b. esophagus—protein digestion
    c. small intestine—starch, lipid, protein digestion
    d. stomach—food storage
    e. liver—production of bile
11. Why can a person not swallow food and talk at the same time?
    a. To swallow, the epiglottis must close off the trachea.
    b. The brain cannot control two activities at once.
    c. To speak, air must come through the larynx to form sounds.
    d. A swallowing reflex is only initiated when the mouth is closed.
    e. Both a and c are correct.
12. Which association is incorrect?
    a. pancreas—produces alkaline secretions and enzymes
    b. salivary glands—produce saliva and amylase
    c. gallbladder—produces digestive enzymes
    d. liver—produces bile
13. Peristalsis occurs
    a. from the mouth to the small intestine.
    b. from the beginning of the esophagus to the anus.
    c. only in the stomach.
    d. only in the small and large intestine.
    e. only in the esophagus and stomach.

14. Bile
    a. is an important enzyme for the digestion of fats.
    b. cannot be stored.
    c. is made by the gallbladder.
    d. emulsifies fat.
    e. All of these are correct.

15. Which of the following is not a function of the liver in adults?
    a. produces bile
    b. detoxifies alcohol
    c. stores glucose
    d. produces urea
    e. makes red blood cells

16. The large intestine
    a. digests all types of food.
    b. is the longest part of the intestinal tract.
    c. absorbs water.
    d. is connected to the stomach.
    e. is subject to hepatitis.

In questions 17–21, match each function to an organ in the key.

Key:

a. mouth
b. esophagus
c. stomach
d. small intestine
e. large intestine

17. Removes nondigestible remains
18. Serves as a passageway
19. Stores food
20. Absorbs nutrients
21. Receives food
22. The amino acids that must be consumed in the diet are called essential. Nonessential amino acids
    a. can be produced by the body.
    b. are only needed occasionally.
    c. are stored in the body until needed.
    d. can only be found in the diet; the body cannot synthesize these amino acids.

In questions 23–28, match each statement to an answer in the key. Answers may be used more than once. Some may have more than one answer.

Key:

a. gastrin
b. secretin
c. CCK
d. All of these are correct.
e. None of these is correct.

23. Stimulates gallbladder to release bile
24. Hormone carried in bloodstream
25. Stimulates the stomach to digest protein
26. Enzyme that digests food
27. Secreted by duodenum
28. Secreted by the stomach

In questions 29–33, match each statement to a vitamin or mineral in the key.

Key:

a. calcium
b. vitamin K
c. sodium
d. iodine
e. vitamin A

29. Needed to make thyroid hormone
30. Needed for night vision
31. Needed for bones, teeth, and muscle contraction
32. Needed for nerve conduction, pH, and water balance
33. Needed for making clotting proteins

Thinking Critically About the Concepts

Bariatric surgery is a type of medical procedure that reduces the size of the stomach and enables food to bypass a section of the small intestine. The surgery is generally done when obese individuals have unsuccessfully tried numerous ways to lose weight and their health is compromised by their weight. There are many risks associated with the surgery, but it helps a number of people lose a considerable amount of weight and ultimately improve their overall health. After people undergo the surgery, there are several lifestyle changes they must make to avoid nutritional deficiencies and to compensate for the small size of their stomach.

1. a. Why do some people who have had bariatric surgery process their food in a blender (or have to chew thoroughly) before swallowing their food?
   b. Why should people who have had bariatric surgery drink liquids between meals rather than with meals?
2. What risk is there to the esophagus after bariatric surgery?

CHAPTER

9

Respiratory System

CHAPTER CONCEPTS

9.1 The Respiratory System
The respiratory system is divided into two regions, the upper and lower respiratory tracts.

9.2 The Upper Respiratory Tract
Air is warmed and filtered in the nose, then moves across the pharynx to pass through the glottis into the larynx.

9.3 The Lower Respiratory Tract
The trachea leads to the bronchial tree, which ends in the lungs. Gas exchange occurs in the air sacs called alveoli.

9.4 Mechanism of Breathing
During inspiration, expansion of the chest moves air into the lungs; during expiration, air leaves the lungs.

9.5 Control of Ventilation
The respiratory center in the brain automatically regulates breathing by relyingon special receptors to detect changes in the pH of the blood.

9.6 Gas Exchanges in the Body
In the lungs, carbon dioxide leaves blood and oxygen enters the blood to be carried by hemoglobin. In the tissues, oxygen leaves the blood and carbon dioxide, to be carried largely as bicarbonate ions, enters the blood.

9.7 Respiration and Health
Several illnesses are associated with the upper and lower respiratory tracts.

BEFORE YOU BEGIN

Before beginning this chapter, take a few moments to review the following discussions.

Section 2.2 What do the values of the pH scale indicate?

Section 5.5 What is the path of the blood entering and leaving the lungs?

Section 6.2 How are red blood cells involved in transporting oxygen and carbon dioxide?

CASE STUDY SLEEP APNEA

For weeks, Justin had felt rundown and tired. He had just moved into the city and started a new job. Even though the past few weeks had been hectic, he had been trying to get at least 7 hours of sleep each night. However, when he woke up each morning, he still felt tired. He was finding it more difficult each day to concentrate on his job, and he found himself forgetting important details of meetings. Concerned that his lack of sleep might cause problems at work, Justin scheduled an appointment with his doctor.

Justin's doctor decided to do a complete physical exam. He checked Justin's weight and blood pressure and asked Justin a series of questions about his diet; medicines he was taking; and, finally, his sleep habits. Though Justin was in general a healthy individual, he reported that he had put on a few pounds since leaving college and starting his job and that his sleep problems seemed to be even worse if he had a few drinks before going to bed. Justin also mentioned that he rarely slept through the night and often awoke suddenly with a feeling of being out of breath.

Given these symptoms, the doctor suggested that Justin may be suffering from sleep apnea. He explained to Justin that there are two types of sleep apnea—obstructive sleep apnea and central sleep apnea. Central sleep apnea is usually caused as a result of an illness or injury to the central nervous system and is associated with neurological problems in the brain. Because Justin did not have a history of any of these conditions, his doctor focused on obstructive sleep apnea. In this condition, the airways in the upper respiratory system become blocked. To test for a sleep disorder, the doctor referred Justin to the local sleep center for a polysomnogram. This test examines blood oxygen levels during sleep, electrical patterns in the brain, eye movement, heart and breathing rate, and the sleeping position of the individual.

As you read through the chapter, think about the following questions.

1. What structures of the upper respiratory system might contribute to obstructive sleep apnea?
2. Why does the polysomnogram monitor brain function as well as breathing and heart rates?
3. Why might Justin's weight gain and occasional alcohol use contribute to the condition?

9.1 The Respiratory System

Learning Outcomes

Upon completion of this section, you should be able to

1. Summarize the role of the respiratory system.
2. Distinguish between inspiration and expiration.
3. Identify the structures of the human respiratory system.

The organs of the respiratory system ensure that oxygen enters the body and carbon dioxide leaves the body (Fig. 9.1). During **inspiration,** or inhalation (breathing in), air is conducted from the atmosphere to the lungs by a series of cavities, tubes, and openings, illustrated in Figure 9.1. During **expiration,** or exhalation (breathing out), air is conducted from the lungs to the atmosphere by way of the same structures.

Ventilation is another term for breathing that includes both inspiration and expiration. Once ventilation has occurred, the respiratory system depends on the cardiovascular system to transport oxygen (O_2) from the lungs to the tissues and carbon dioxide (CO_2) from the tissues to the lungs.

Gas exchange is necessary because the cells of the body carry out cellular respiration to make energy in the form of ATP. During cellular respiration, cells use up O_2 and produce CO_2. The respiratory system provides these cells with O_2 and removes CO_2.

MP3
Respiratory Structure and Function

Connecting the Concepts

The respiratory and circulatory systems cooperate extensively to maintain homeostasis in the body. For more on the interactions of these two systems, refer to the following discussions.

Section 5.5 outlines the circulatory pathways that move gases to and from the lungs.

Section 6.2 describes the role of the red blood cells in the transport of gases.

Check Your Progress 9.1

1. Trace the path of air from the nasal cavities to the lungs.
2. Distinguish between *inspiration* and *expiration* and what is occurring in the respiratory system during each.
3. Describe the functions of the respiratory system.

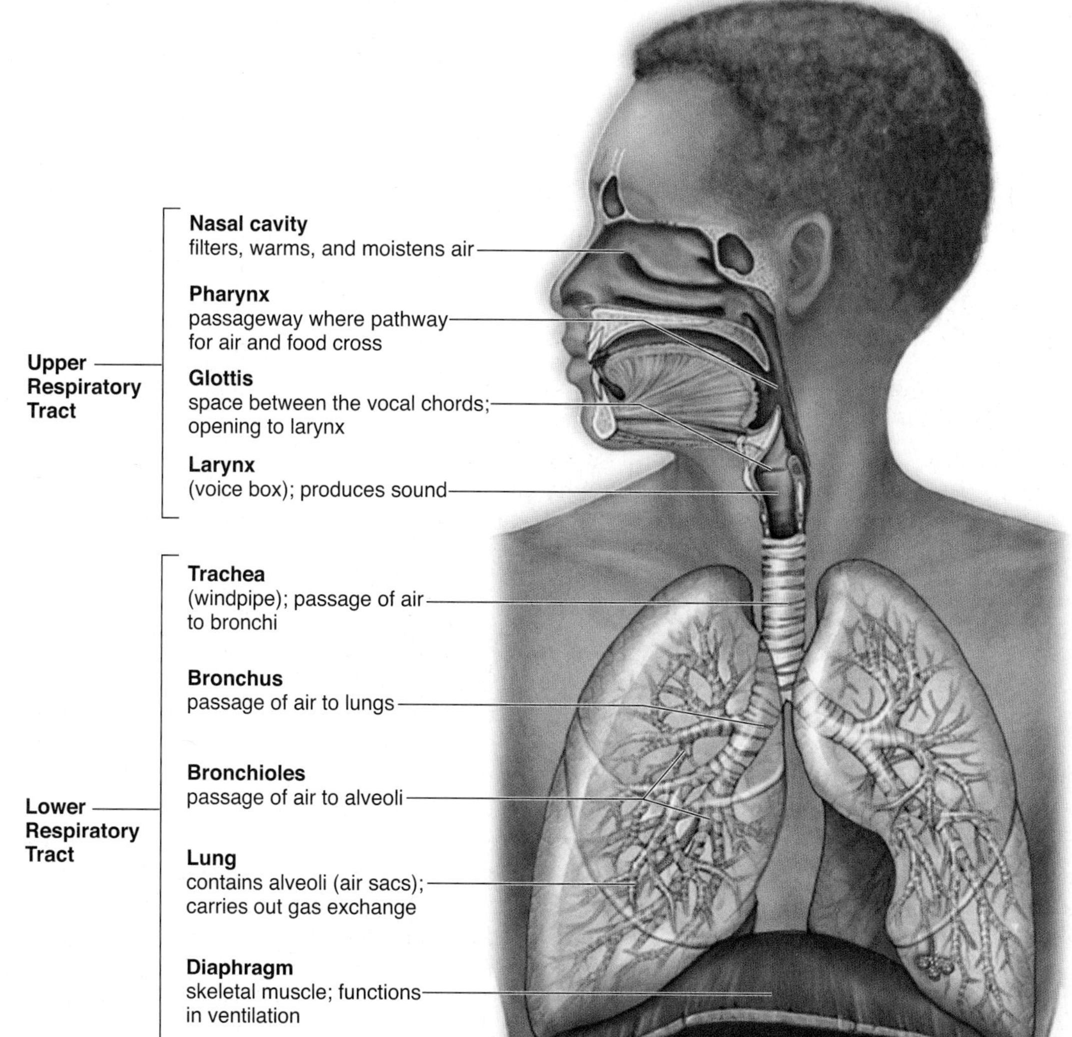

Figure 9.1 **The human respiratory tract.**
The respiratory tract extends from the nose to the lungs. Note the organs in the upper respiratory tract and the ones in the lower respiratory tract.

9.2 The Upper Respiratory Tract

Learning Outcomes

Upon completion of this section, you should be able to

1. Summarize the role of the nose, pharynx, and larynx in respiration.
2. Identify the structures of the upper respiratory system and provide their function.
3. Explain how sound is produced by the larynx.

The nasal cavities, pharynx, and larynx are the organs of the upper respiratory tract (Fig. 9.2).

The Nose

The nose opens at the nares (nostrils) that lead to the **nasal cavities.** The nasal cavities are narrow canals separated from each other by a septum composed of bone and cartilage (Fig. 9.2).

Air entering the nasal cavities is met by large stiff hairs that act as a screening device. The hairs filter the air and trap small particles (dust, mold spores, pollen, etc.) so they don't enter air passages. The rest of the nasal cavities are lined by mucous membrane. The mucus secreted by this membrane helps trap dust and move it to the pharynx, where it can be swallowed or expectorated by coughing or spitting. Under the mucous layer is the submucosa. The submucosa contains a large number of capillaries that help warm and moisten the incoming air, but they also make us more susceptible to nose bleeds if the nose suffers an injury. When we breathe out on a cold day, the moisture in the air condenses so that we can see our breath.

In the narrow upper recesses of the nasal cavities are special ciliated cells that act as odor receptors. Nerves lead from these cells to the brain, where the impulses generated by the odor receptors are interpreted as smell.

Connections and Misconceptions

Is stifling a sneeze harmful?

Scientists have found that air travels at 100 miles an hour during a sneeze. This is enough force to propel sneeze droplets up to 12 ft away from the person sneezing. If a sneeze is stifled, the air is forced into the eustachian tube and middle ear, potentially causing damage to the middle ear.

The tear (lacrimal) glands drain into the nasal cavities by way of tear ducts. When you cry, your nose runs as tears drain from the eye surface into the nose. The nasal cavities also connect with the sinuses (cavities) of the skull. At times, fluid may accumulate in these sinuses, causing an excess of pressure, resulting in a sinus headache.

Air in the nasal cavities passes into the nasopharynx, the upper portion of the pharynx. Connected to the nasopharynx are tubes called **auditory (eustachian) tubes** that connect to the middle ear. When air pressure inside the middle ears equalizes with the air pressure in the nasopharynx, the auditory tube openings may create a "popping" sensation. When in a plane taking off or landing, some people chew gum or yawn in an effort to move air from the ears and prevent the popping.

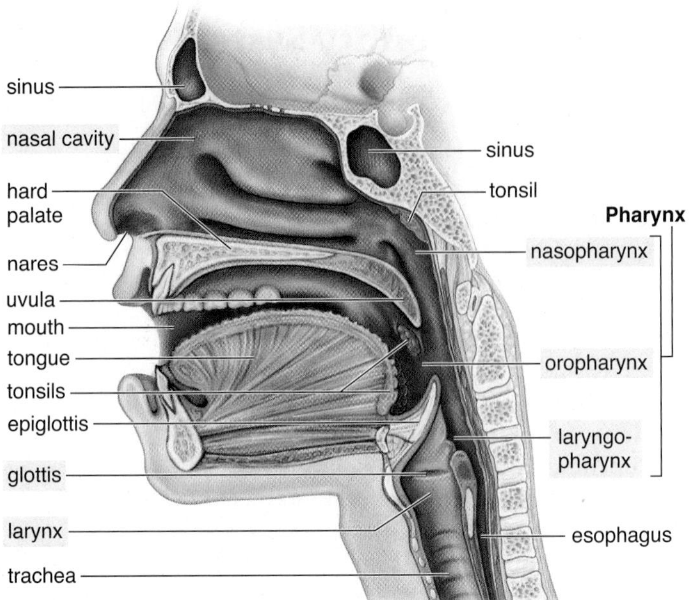

Figure 9.2 **The upper respiratory tract.**
This drawing shows the path of air from the nasal cavities to the trachea, in the lower respiratory tract. The designated structures are in the upper respiratory tract.

The Pharynx

The **pharynx** is a funnel-shaped passageway that connects the nasal and oral cavities to the larynx. Therefore, the pharynx, commonly referred to as the "throat," has three parts: the nasopharynx, where the nasal cavities open above the soft palate; the oropharynx, where the oral cavity opens; and the laryngopharynx, which opens into the larynx.

The **tonsils** form a protective ring at the junction of the oral cavity and the pharynx. The tonsils contain lymphocytes, which protect against invasion of inhaled foreign antigens. The tonsils are the primary defense during breathing, because inhaled air passes directly over the tonsils. In the tonsils, B cells and T cells are prepared to respond to antigens that may subsequently invade internal tissues and fluids. Therefore, the respiratory tract assists the immune system in maintaining homeostasis.

In the pharynx, the air passage and the food passage lie parallel to each other and share a common opening in the laryngopharynx. The larynx is normally open, allowing air to pass, but the esophagus is normally closed and opens only when a person swallows. If someone swallows and some of the food enters the larynx, coughing occurs in an effort to dislodge the flood. If the passageway remains blocked, the Heimlich maneuver (Fig. 9.3) may be used to dislodge food blocking the **epiglottis** and airway.

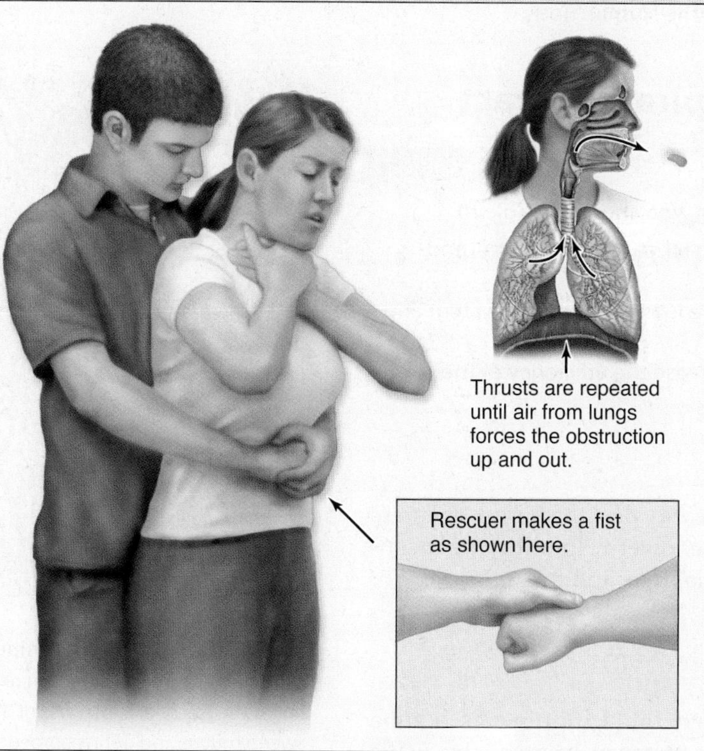

Rescuer stands behind the victim and wraps arms around the victim below the rib cage.
Victim should be positioned so that the head, arms, and chest are over the rescuer's arms.
Fist is used to make upward subdiaphragm thrusts that go upward rather than in.
Thrusts are repeated until air from lungs forces the obstruction up and out.
Rescuer makes a fist as shown here.
Note: Recommended for use in conscious adults and children over one year.

Figure 9.3 **The Heimlich maneuver.** The Heimlich maneuver is used when someone's ability to breathe is prevented by an obstruction blocking the airway. The steps associated with the Heimlich maneuver are shown in the illustration.

The Larynx

The **larynx** is a cartilaginous structure that serves as a passageway for air between the pharynx and the trachea. The larynx can be pictured as a triangular box whose apex, the Adam's apple (or laryngeal prominence), is located at the front of the neck. The larynx is called the *voice box* because it houses the vocal cords. The **vocal cords** are mucosal folds supported by elastic ligaments, and the slit between the vocal cords is called the **glottis** (Fig. 9.4). When air is expelled through the glottis, the vocal cords vibrate, producing sound. At the time of puberty, the growth of the larynx and the vocal cords is much more rapid and accentuated in the male than in the female, causing the male to have a more prominent Adam's apple and a deeper voice. The voice "breaks" in the young male due to his inability to control the longer vocal cords.

The high or low pitch of the voice is regulated when speaking and singing by changing the tension on the vocal cords. The greater the tension, as when the glottis becomes narrower, the higher the pitch. When the glottis is wider, the pitch is lower (Fig. 9.4, *right*). The loudness or intensity of the voice depends upon the amplitude of the vibrations—the degree to which the vocal cords vibrate.

Ordinarily, when food is swallowed, the larynx moves upward against the epiglottis, a flap of tissue that prevents food from passing into the larynx. You can detect the movement of the larynx by placing your hand gently on your larynx and swallowing.

Connecting the Concepts

Several organ systems interact within the region of the upper respiratory tract. For more information on these systems, refer to the following discussions.

Section 7.2 describes the organization of the lymphatic system and the role of lymphatic tissue.

Section 8.2 illustrates the connection between the upper digestive and upper respiratory systems.

Section 14.3 examines how the respiratory system contributes to the sense of smell.

Check Your Progress 9.2

1. Describe the function and location of each of the structures of the upper respiratory tract.
2. Name and briefly describe the body systems that have connections with the pharynx.
3. Describe the two pathways that cross in the pharynx.

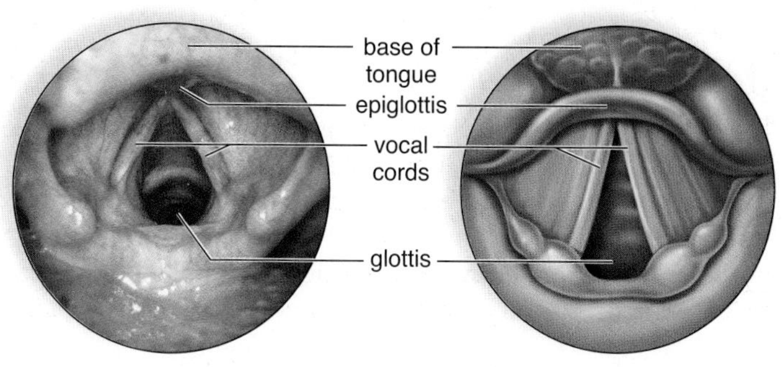

Figure 9.4 **The vocal cords.** Viewed from above, the vocal cords can be seen to stretch across the glottis, the opening to the trachea. When air is expelled through the glottis, the vocal cords vibrate, producing sound. The glottis is narrow when we produce a high-pitched sound, and it widens as the pitch deepens.

9.3 The Lower Respiratory Tract

Learning Outcomes

Upon completion of this section, you should be able to

1. Summarize the role of the trachea, bronchial tree, and lungs in respiration.
2. Identify the structures of the lower respiratory system and provide their function.
3. Explain how the alveoli increase the efficiency of the respiratory system.

Once the incoming air makes it way past the larynx, it enters the lower respiratory tract. The lower respiratory tract consists of the trachea, the bronchial tree, and the lungs.

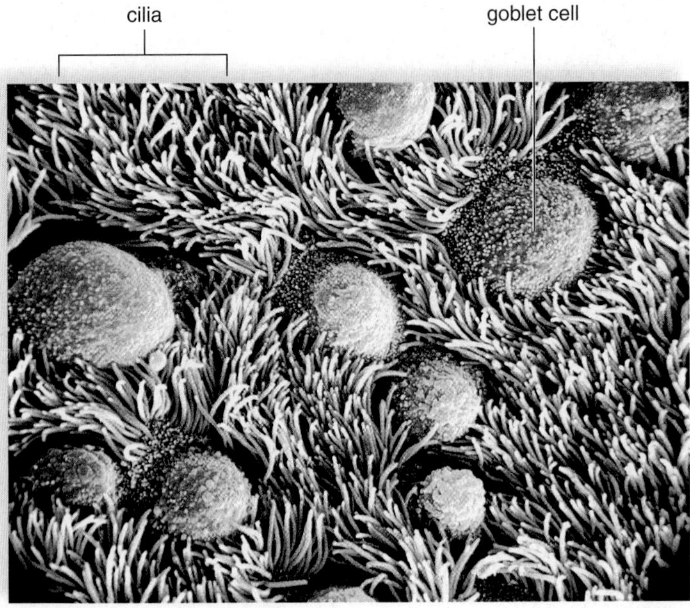

Figure 9.5 **The cells lining the trachea.**
Scanning electron micrograph of the surface of the mucous membrane lining the trachea, consisting of goblet cells and ciliated cells. The cilia sweep mucus and debris embedded in it toward the pharynx, where it is swallowed or expectorated.

The Trachea

The **trachea,** commonly called the windpipe, is a tube connecting the larynx to the primary bronchi. Its walls consist of connective tissue and smooth muscle reinforced by C-shaped cartilaginous rings. The rings prevent the trachea from collapsing.

The trachea lies anterior to the esophagus. It is separated from the esophagus by a flexible muscular wall. This orientation allows the esophagus to expand when swallowing. The mucous membrane that lines the trachea has an outer layer of pseudostratified ciliated columnar epithelium (see Fig. 4.7) and goblet cells (Fig. 9.5). The goblet cells produce mucus, which traps debris in the air as it passes through the trachea. The mucus is then swept toward the pharynx and away from the lungs by the cilia that project from the epithelium.

When one coughs, the tracheal wall contracts, narrowing its diameter. Therefore, coughing causes air to move more rapidly through the trachea, helping to expel mucus and foreign objects. Smoking is known to destroy the cilia; consequently, the soot in cigarette smoke collects in the lungs. Smokers often develop heavy coughs as a result. Smoking is discussed more fully in the Health Focus, *Questions About Smoking, Tobacco, and Health,* in Section 9.7.

If the trachea is blocked because of illness or the accidental swallowing of a foreign object, a breathing tube can be inserted by way of an incision made in the trachea. This tube acts as an artificial air intake and exhaust duct. The operation is called a **tracheostomy.**

The Bronchial Tree

The trachea divides into right and left primary bronchi (sing., **bronchus**), which lead into the right and left lungs (see Fig. 9.1). The bronchi branch into a few secondary bronchi that also branch, until the branches become about 1 mm in diameter and are called **bronchioles.** The bronchi resemble the trachea in structure. As the bronchial tubes divide and subdivide, their walls become thinner, and the small rings of cartilage are no longer present. During an asthma attack, the smooth muscle of the bronchioles contracts, causing bronchiolar constriction and characteristic wheezing. Each bronchiole leads to an elongated space enclosed by a multitude of air pockets or sacs called alveoli (sing., **alveolus**) (Fig. 9.6).

The Lungs

The **lungs** are paired, cone-shaped organs in the thoracic cavity. In the center of the thoracic cavity are the trachea, heart, thymus, and esophagus. The lungs are on either side of the trachea. The right lung has three lobes, and the left lung has two lobes, allowing room for the heart, which points left. Each lobe is further divided into lobules, and each lobule has a bronchiole serving many alveoli.

The lungs follow the contours of the thoracic cavity, including the diaphragm, the muscle that separates the thoracic cavity from the abdominal cavity. Each lung is enclosed by pleurae (sing., **pleura**), two layers of serous membrane that produces serous fluid. The parietal pleura adheres to the thoracic cavity wall, and the visceral pleura adheres to the surface of the lung. *Surface tension* is the tendency for water molecules to cling to one another due to hydrogen bonding between molecules. Surface tension holds the two pleural layers together. Therefore, the lungs must follow the movement of the thorax when breathing occurs. When someone has pleurisy, these layers are inflamed. Breathing, sneezing, and coughing are quite painful because the layers rub against each other.

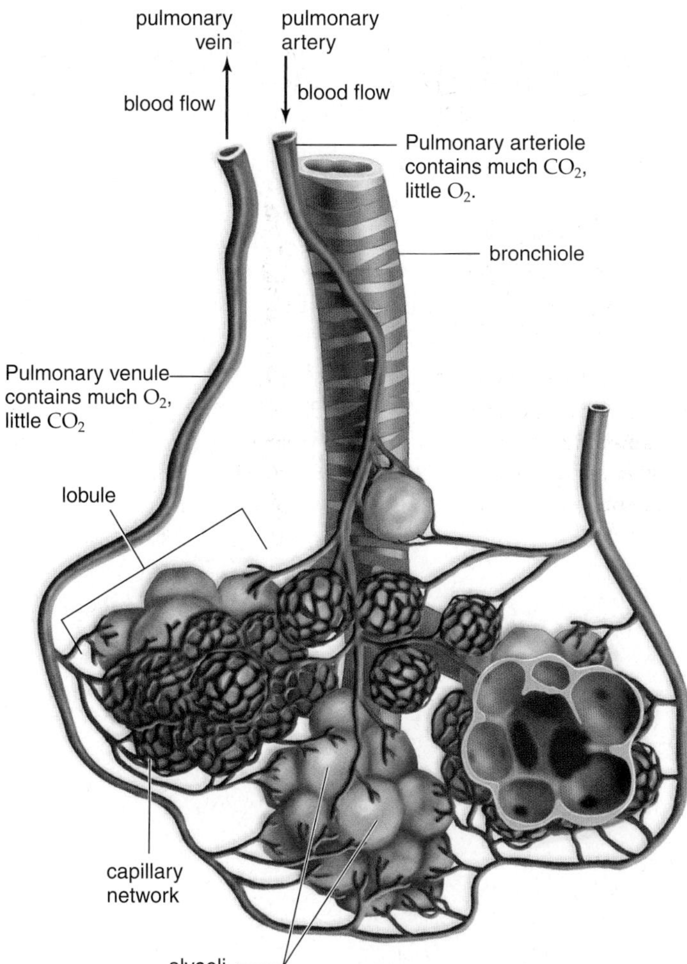

Figure 9.6 **Pulmonary circulation to and from the lungs.** The lungs consist of alveoli surrounded by an extensive capillary network. The pulmonary artery carries O_2-poor (CO_2-rich) blood (colored blue), and the pulmonary vein carries O_2-rich blood (colored red).

The Alveoli

The lungs have about 300 million alveoli, with a total cross-sectional area of 50–70 m^2. That's about the size of a tennis court. Each alveolar sac is surrounded by blood capillaries. The walls of the sac and the capillaries are largely simple squamous epithelium (see Fig. 4.7). Gas exchange occurs between air in the alveoli and blood in the capillaries. Oxygen diffuses across the alveolar wall and enters the bloodstream, and carbon dioxide diffuses from the blood across the alveolar wall to enter the alveoli (see Fig. 9.6).

The alveoli of human lungs are lined with a **surfactant,** a film of lipoprotein that lowers the surface tension of water and prevents the alveoli from closing. The lungs collapse in some newborn babies—especially premature infants—who lack this film. The condition, called **infant respiratory distress syndrome,** is now treatable by surfactant replacement therapy.

Connecting the Concepts

For more information on the structures of the thoracic cavity, refer to the following discussions.

Section 5.3 examines the structure of the heart.

Section 7.2 describes the function of the thymus gland as an organ of the lymphatic system.

Section 8.2 summarizes the role of the esophagus in the digestive system.

Check Your Progress 9.3

1. Briefly describe the functions of the organs of the lower respiratory system.
2. Detail the structures of the lower respiratory system that participate in gas exchange.
3. Discuss what might occur to overall homeostasis if the alveoli did not function properly.

9.4 Mechanism of Breathing

Learning Outcomes

Upon completion of this section, you should be able to

1. Contrast the processes of inspiration and expiration during ventilation.
2. Define the terms *tidal volume, vital capacity,* and *residual volume* in relation to ventilation.
3. Summarize the purpose of the inspiratory and expiratory reserve volumes.

Ventilation, or breathing, has two phases. The process of **inspiration,** also called inhalation, moves air into the lungs; the process of **expiration,** also called exhalation, moves air out of the lungs. To understand ventilation, the manner in which air enters and exits the lungs, it is necessary to remember the following facts:

1. Normally, there is a continuous column of air from the pharynx to the alveoli of the lungs.
2. The lungs lie within the sealed thoracic cavity. The rib cage, consisting of the ribs joined to the vertebral column posteriorly and to the sternum anteriorly, forms the top and sides of the thoracic cavity. The intercostal muscles lie between the ribs. The diaphragm and connective tissue form the floor of the thoracic cavity.
3. The lungs adhere to the thoracic wall by way of the pleura. Any space between the two pleurae is minimal due to the surface tension of the fluid between them.

Inspiration

Inspiration is the active phase of ventilation because this is the phase in which the diaphragm and the external intercostal muscles contract (Fig. 9.7*a*). In its relaxed state, the diaphragm is dome-shaped. During inspiration, it contracts and becomes a flattened sheet of muscle. Also, the external intercostal muscles contract, causing the rib cage to move upward and outward.

Following contraction of the diaphragm and the external intercostal muscles, the volume of the thoracic cavity is larger than it was before. As the thoracic volume increases, the lungs increase in volume as well because the lung adheres to the wall of the thoracic cavity. As the lung volume increases, the air pressure within the alveoli decreases, creating a partial vacuum. In other words, alveolar pressure is now less than atmospheric pressure (air pressure outside the lungs). Air will naturally flow from outside the body into the respiratory passages and into the alveoli, because a continuous column of air reaches into the lungs.

Air comes into the lungs because they have already opened up; air does not force the lungs open. This is why it is sometimes said that *humans inhale by negative pressure.* The creation of a partial vacuum in the alveoli causes air to enter the lungs. Whereas inspiration is the active phase of breathing, the actual flow of air into the alveoli is passive. Just as with bellows used to fan a fire, when the handles of the bellows are pulled apart, air automatically flows into the bellows. When the pressure to pull apart the bellows is released, air flows out as the bellows close (Fig. 9.8).

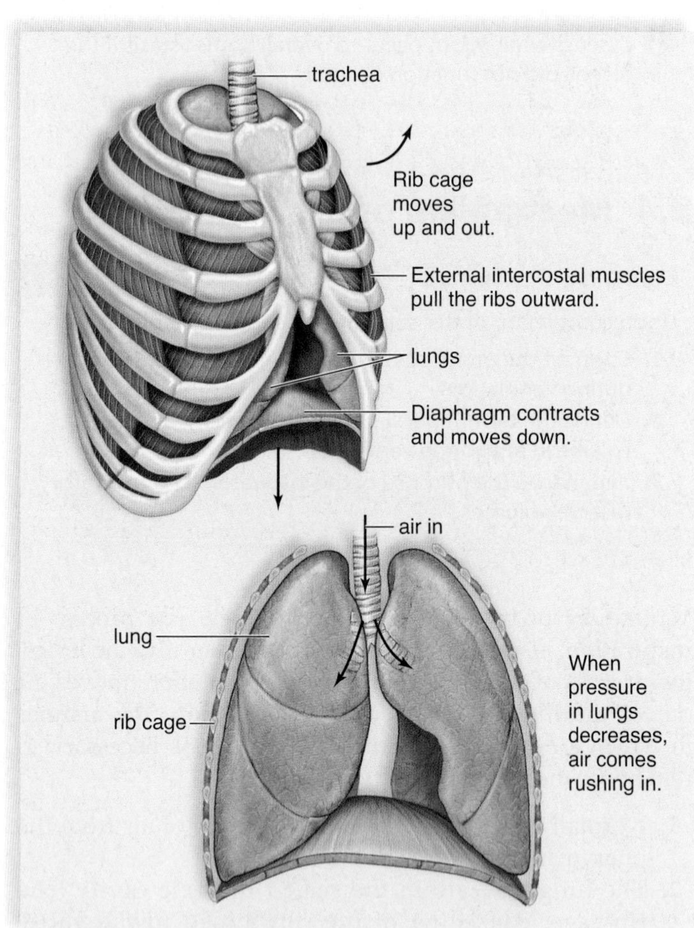

a. Inspiration

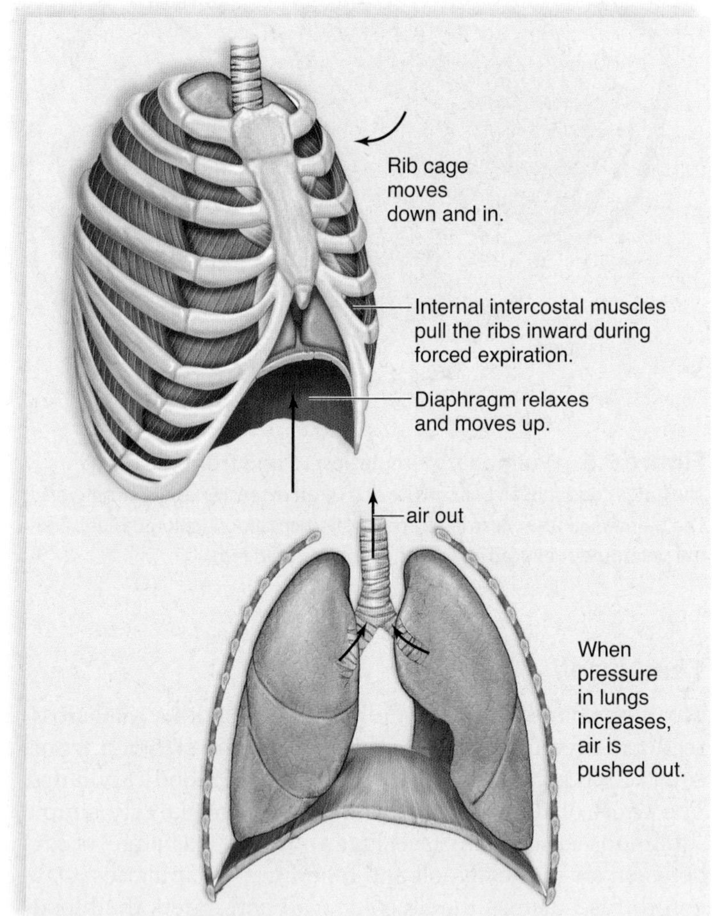

b. Expiration

Figure 9.7 The thoracic cavity during inspiration and expiration.
a. During inspiration, the thoracic cavity and lungs expand so that air is drawn in. **b.** During expiration, the thoracic cavity and lungs resume their original positions and pressures. Now air is forced out.

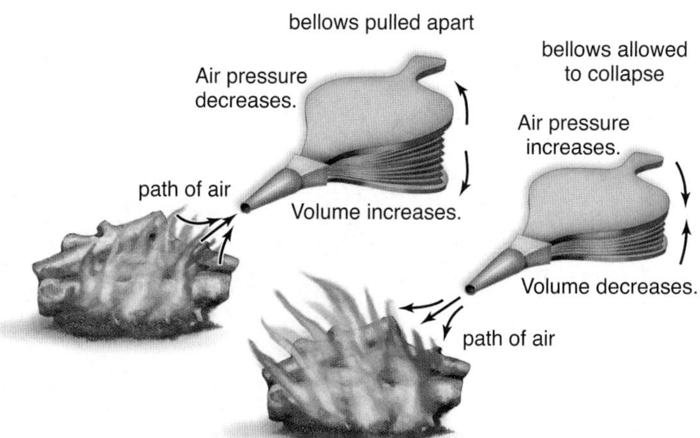

Figure 9.8 The relationship between air pressure and volume.
When the bellows are pulled apart, the volume in the bellows increases and the air pressure decreases. When the bellows collapse, the volume decreases and air pressure increases.

Expiration

Usually, expiration is the passive phase of breathing, and no effort is required to bring it about. During expiration, the diaphragm and external intercostal muscles relax. The rib cage returns to its resting position, moving down and inward (Fig. 9.7*b*). The elastic properties of the thoracic wall and lung tissue help them to recoil. In addition, the lungs recoil because the surface tension of the fluid lining the alveoli tends to draw them closed (see Chap. 2). If we continue the analogy of the bellows, imagine simply allowing the open bellows to fall shut. As the volume of the bellows decreases, the air pressure inside increases. Now air flows out.

What keeps the alveoli from collapsing as a part of expiration? Recall that the presence of surfactant lowers the surface tension within the alveoli. Also, as the lungs recoil, the pressure between the pleura decreases, and this tends to make the alveoli stay open. The importance of the reduced intrapleural pressure is demonstrated when in an accident the thoracic cavity is punctured (a "punctured lung"). Air now enters the intrapleural space, causing the lung to collapse.

Maximum Inspiratory Effort and Forced Expiration

If you recall the last time you exercised vigorously—perhaps running in a race, or even just climbing all those stairs to your classroom—you probably remember that you were breathing a lot harder than normal during and immediately after that heavy exercise. Maximum inspiratory effort involves muscles of the back, chest, and neck. This increases the size of the thoracic cavity to larger than normal, thus allowing maximum expansion of the lungs.

Expiration can also be forced. The maximum inspiratory efforts of heavy exercise are accompanied by forced expiration. Forced expiration is also necessary to sing, blow air into a trumpet, or blow out birthday candles. Contraction of the internal intercostal muscles can force the rib cage to move downward and inward. Also, when the abdominal wall muscles contract, they push on the abdominal organs. In turn, the organs push upward against the diaphragm and the increased pressure in the thoracic cavity helps expel air.

Volumes of Air Exchanged During Ventilation

As ventilation occurs, air moves into the lungs from the nose or mouth during inspiration and then moves out of the lungs during expiration. A free flow of air to and from the lungs is vitally important. Therefore, a technique has been developed that allows physicians to determine if there is a medical problem that prevents the lungs from filling with air upon inspiration and releasing it from the body upon expiration. This technique is illustrated in Figure 9.9, which shows the measurements recorded by a spirometer when a person breathes as directed by a technician. The actual numbers mentioned in the following discussion about lung volumes are averages. These numbers are affected by gender, height, and age, so your lung volumes may be different from the values stated here.

Tidal Volume Normally, when we are relaxed, only a small amount of air moves in and out with each breath, similar, perhaps, to the tide at the beach. This amount of air, called the **tidal volume,** is only about 500 mL.

Vital Capacity It is possible to increase the amount of air inhaled and, therefore, the amount exhaled by deep breathing. The maximum volume of air that can be moved in plus the maximum amount that can be moved out during a single breath is called the **vital capacity.** It is called vital capacity because your life depends on breathing, and the more air you can move, the better off you are. A number of different illnesses discussed at the end of this chapter can decrease vital capacity.

Inspiratory and Expiratory Reserve Volume As noted previously, we can increase inspiration by expanding the chest and also by lowering the diaphragm to the maximum extent possible. Forced inspiration (**inspiratory reserve volume**) usually adds another 2,900 mL of inhaled air, and that's quite a bit more than a tidal volume of only 500 mL!

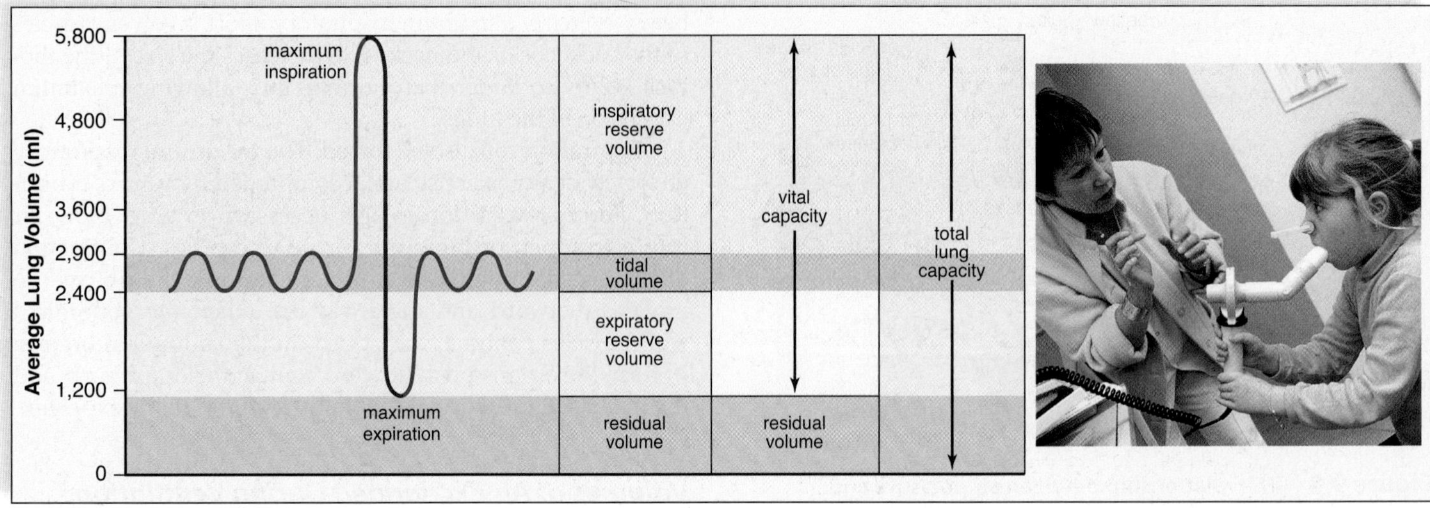

Figure 9.9 **Measuring the vital capacity of the lungs.**
A spirometer is an instrument that measures the amount of air inhaled and exhaled with each breath. During inspiration, there is an upswing, and during expiration, there is a downswing. Vital capacity (red) is measured by taking the deepest breath and then exhaling as much as possible.

We can increase expiration by contracting the abdominal and thoracic muscles. This so-called **expiratory reserve volume** is usually about 1,400 mL of air. You can see from Figure 9.9 that vital capacity is the sum of tidal, inspiratory reserve, and expiratory reserve volumes.

Residual Volume It is a curious fact that some of the inhaled air never reaches the lungs; instead, it fills the nasal cavities, trachea, bronchi, and bronchioles (see Fig. 9.1). These passages are not used for gas exchange; therefore, they are said to contain **dead air space.** To ensure that newly inhaled air reaches the lungs, it is better to breathe slowly and deeply.

Also, note in Figure 9.9 that even after a very deep exhalation, some air (about 1,000 mL) remains in the lungs. This is called the **residual volume.** The residual volume is the amount of air that can't be exhaled from the lungs. In some lung diseases to be discussed later, the residual volume gradually increases because the individual has difficulty emptying the lungs. Increased residual volume will cause the expiratory reserve volume to be reduced. As a result, vital capacity is decreased as well.

MP3 Ventilation

Connections and Misconceptions

What happens when "the wind gets knocked out of you"?

The "wind may get knocked out of you" following a blow to the upper abdomen in the area of the stomach. There is a network of nerves in that region called the solar plexus. Trauma to this area can cause the diaphragm to experience a sudden, involuntary, and painful contraction called a spasm. It's not possible to breathe while the diaphragm experiences this spasm—hence, the feeling of being breathless. The pain and inability to breathe stop once the diaphragm relaxes.

Connecting the Concepts

To understand how the lungs develop as we age, refer to the following discussions.

Section 17.3 describes the role of the lungs in fetal development.

Section 17.5 examines the effects of aging on the ability of the lungs to exchange gases.

Check Your Progress 9.4

1. Explain how the volume (size) of the thoracic cavity affects the pressure in the lungs.
2. Distinguish between the different volumes of air exchanged during ventilation and describe when each is used.
3. Discuss what effect insufficient expiration might have on overall homeostasis.

9.5 Control of Ventilation

Learning Outcomes

Upon completion of this section, you should be able to

1. Explain how the nervous system controls the process of breathing.
2. Explain the role of chemoreceptors and pH levels in regulating breathing rate.

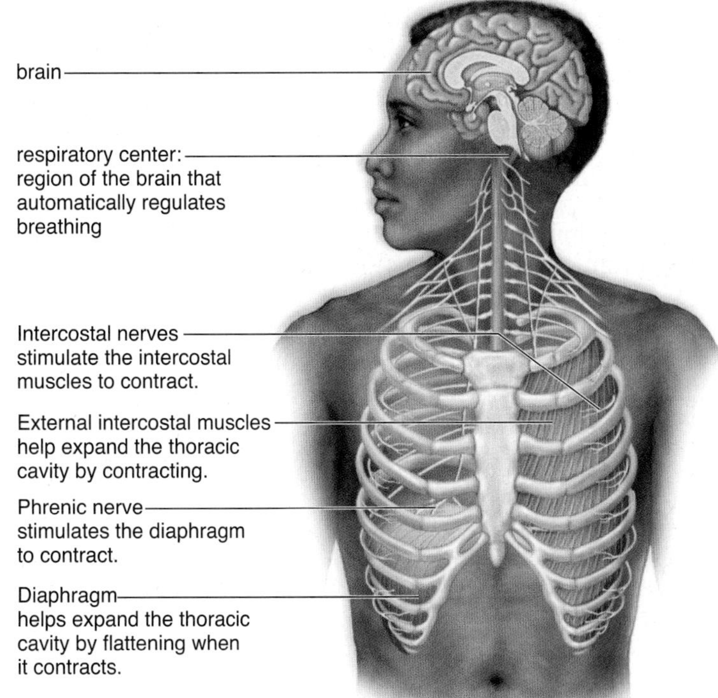

brain
respiratory center: region of the brain that automatically regulates breathing
Intercostal nerves stimulate the intercostal muscles to contract.
External intercostal muscles help expand the thoracic cavity by contracting.
Phrenic nerve stimulates the diaphragm to contract.
Diaphragm helps expand the thoracic cavity by flattening when it contracts.

Figure 9.10 **The control of breathing by the respiratory center.** During inspiration, the respiratory center, located in the medulla oblongata, stimulates the external intercostal (rib) muscles to contract via the intercostal nerves and stimulates the diaphragm to contract via the phrenic nerve. The thoracic cavity and then the lungs expand and air comes rushing in. Expiration occurs due to a lack of stimulation from the respiratory center to the diaphragm and intercostal muscles. As the thoracic cavity and then the lungs resume their original size, air is pushed out.

Breathing is controlled in two ways. It is under both nervous and chemical control.

Nervous Control of Breathing

Normally, adults have a breathing rate of 12 to 20 ventilations per minute. The rhythm of ventilation is controlled by a **respiratory control center** located in the medulla oblongata of the brain. The respiratory control center automatically sends out nerve signals to the diaphragm and the external intercostal muscles of the rib cage, causing inspiration to occur (Fig. 9.10). When the respiratory center stops sending nerve signals to the diaphragm and the rib cage, the muscles relax and expiration occurs.

Sudden infant death syndrome (**SIDS**), or crib death, claims the life of about 2,000 infants a year in the United States. An infant under one year of age is put to bed seemingly healthy, and sometime while sleeping, the child stops breathing. Though the precise cause of SIDS is not known, scientists have ruled out vaccinations, vomiting, and infections as factors. Most research is focusing on miscommunication between the respiratory center of the brain and the lungs and, possibly, problems with heart function.

Although the respiratory center automatically controls the rate and depth of breathing, its activity can be influenced by nervous input. We can voluntarily change our breathing pattern to accommodate activities such as speaking, singing, eating, swimming under water, and so forth. Following forced inspiration, stretch receptors in the airway walls respond to increased pressure. These receptors initiate inhibitory nerve impulses. The impulses travel from the inflated lungs to the respiratory center. This temporarily stops the respiratory center from sending out nerve signals. In this manner, excessive stretching of the elastic tissue of the lungs is prevented.

Connections and Misconceptions

How long can people hold their breath?

Most individuals can hold their breath for 1–2 minutes. With practice, many are able to hold their breath for up to 3 minutes. Free divers, individuals who compete to see how long they can hold their breath while diving in water, can hold their breath for 5 minutes or more.

Researchers hope to more closely investigate divers' use of hyperventilation and lung "packing" to make free diving possible. Lung packing begins with very deep breathing. The diver then "packs," adding more air by additional breathing through the mouth. The use of hyperventilation or lung packing to prolong breath holding should never be tried by untrained individuals. People have drowned by attempting to do so.

Chemical Control of Breathing

As you know, working cells produce carbon dioxide, which enters the blood. There, carbon dioxide combines with water, forming an acid, which breaks down and gives off hydrogen ions (H^+). These hydrogen ions can change the pH of the blood. **Chemoreceptors** are sensory receptors in the body that are sensitive to chemical composition of body fluids. Two sets of chemoreceptors sensitive to pH can cause breathing to speed up. A centrally placed set is located in the medulla oblongata of the brain stem, and a peripherally placed set is in the circulatory system. Carotid bodies, located in the carotid arteries, and aortic bodies, located in the aorta, are sensitive to blood pH. These chemoreceptors are not strongly affected by low oxygen (O_2) levels. Instead, they are stimulated when the carbon dioxide entering the blood is sufficient to change blood pH.

When the pH of the blood becomes more acidic (decreases), the respiratory center increases the rate and depth of breathing. With an increased breathing rate, more carbon dioxide is removed from the blood. The hydrogen ion concentration returns to normal, and the breathing rate returns to normal.

Most people are unable to hold their breath for more than 1 minute. When you hold your breath, metabolically produced carbon dioxide begins accumulating in the blood. As a result, H^+ accumulates and the blood becomes more acidic. The respiratory center, stimulated by the chemoreceptors, is able to override a person's voluntary inhibition of respiration. Breathing resumes, despite attempts to prevent it.

MP3
Control of Respiration

Connecting the Concepts

For more on the structures that regulate breathing, refer to the following discussions.

Section 13.2 examines the location and function of the medulla oblongata.

Section 14.1 describes how chemoreceptors help maintain homeostasis.

Check Your Progress 9.5

1. Explain why we automatically breath 12 to 20 times a minute. Describe what might happen to homeostasis if those numbers were only 2 to 5 times a minute.
2. Describe the nervous system's control of the respiratory system.
3. Discuss why it's not possible to hold your breath for more than a minute or so.

9.6 Gas Exchanges in the Body

Learning Outcomes

Upon completion of this section, you should be able to

1. Distinguish between external and internal respiration.
2. Summarize the chemical processes that are involved in external and internal respiration.
3. Identify the role of carbonic anhydrase and carbaminohemoglobin in respiration.

Gas exchange is critical to homeostasis. Oxygen needed to produce energy must be supplied to all the cells, and carbon dioxide must be removed from the body during gas exchange. As mentioned previously, respiration includes the exchange of gases not only in the lungs but also in the tissues (Fig. 9.11).

The principles of diffusion govern whether O_2 or CO_2 enters or leaves the blood in the lungs and in the tissues. Gases exert pressure, and the amount of pressure each gas exerts is called its partial pressure, symbolized as P_{O_2} and P_{CO_2}. If the partial pressure of oxygen differs across a membrane, oxygen will diffuse from the higher to lower partial pressure.

Animation
Changes in the Partial Pressure of Oxygen and Carbon Dioxide

External Respiration

External respiration refers to the exchange of gases between air in the alveoli and blood in the pulmonary capillaries (see Fig. 9.6 and Fig. 9.11*a*). Blood in the pulmonary capillaries has a higher P_{CO_2} than atmospheric air. Therefore, *CO_2 diffuses out of the plasma into the lungs.* Most of the CO_2 is carried in plasma as **bicarbonate ions** (HCO_3^-). In the low-P_{CO_2} environment of the lungs, the reaction proceeds to the right.

$$\underset{\text{hydrogen ion}}{H^+} + \underset{\text{bicarbonate ion}}{HCO_3^-} \longrightarrow \underset{\text{carbonic acid}}{H_2CO_3} \xrightarrow{\text{carbonic anhydrase}} \underset{\text{water}}{H_2O} + \underset{\text{carbon dioxide}}{CO_2}$$

The enzyme **carbonic anhydrase** speeds the breakdown of carbonic acid (H_2CO_3) in red blood cells.

What happens if you hyperventilate (breathe at a high rate) and therefore push this reaction far to the right? The blood will have fewer hydrogen ions, and alkalosis, a high blood pH, results. In that case, breathing is inhibited, and you may suffer from various symptoms ranging from dizziness to continuous contractions of the skeletal muscles. You may have heard that you should inhale and exhale from a paper bag after hyperventilating. Doing so increases the CO_2 in your blood, because you're inhaling the CO_2 you just exhaled into the bag. This restores a normal blood pH. What happens if you hypoventilate (breathe at a low rate) and this reaction does not occur? Hydrogen ions build up in the blood and acidosis occurs. Buffers may compensate for the low pH, and breathing most likely increases. Extreme changes in blood pH affect enzyme function, which may lead to coma and death.

The pressure pattern for O_2 during external respiration is the reverse of that for CO_2. Blood in the pulmonary capillaries is low in oxygen, and alveolar air contains a higher partial pressure of oxygen. Therefore, *O_2 diffuses into plasma and then into red blood cells in the lungs.* Hemoglobin takes up this oxygen and becomes **oxyhemoglobin** (HbO_2).

Animation
Gas Exchange During Respiration

$$\underset{\text{deoxyhemoglobin}}{Hb} + \underset{\text{oxygen}}{O_2} \longrightarrow \underset{\text{oxyhemoglobin}}{HbO_2}$$

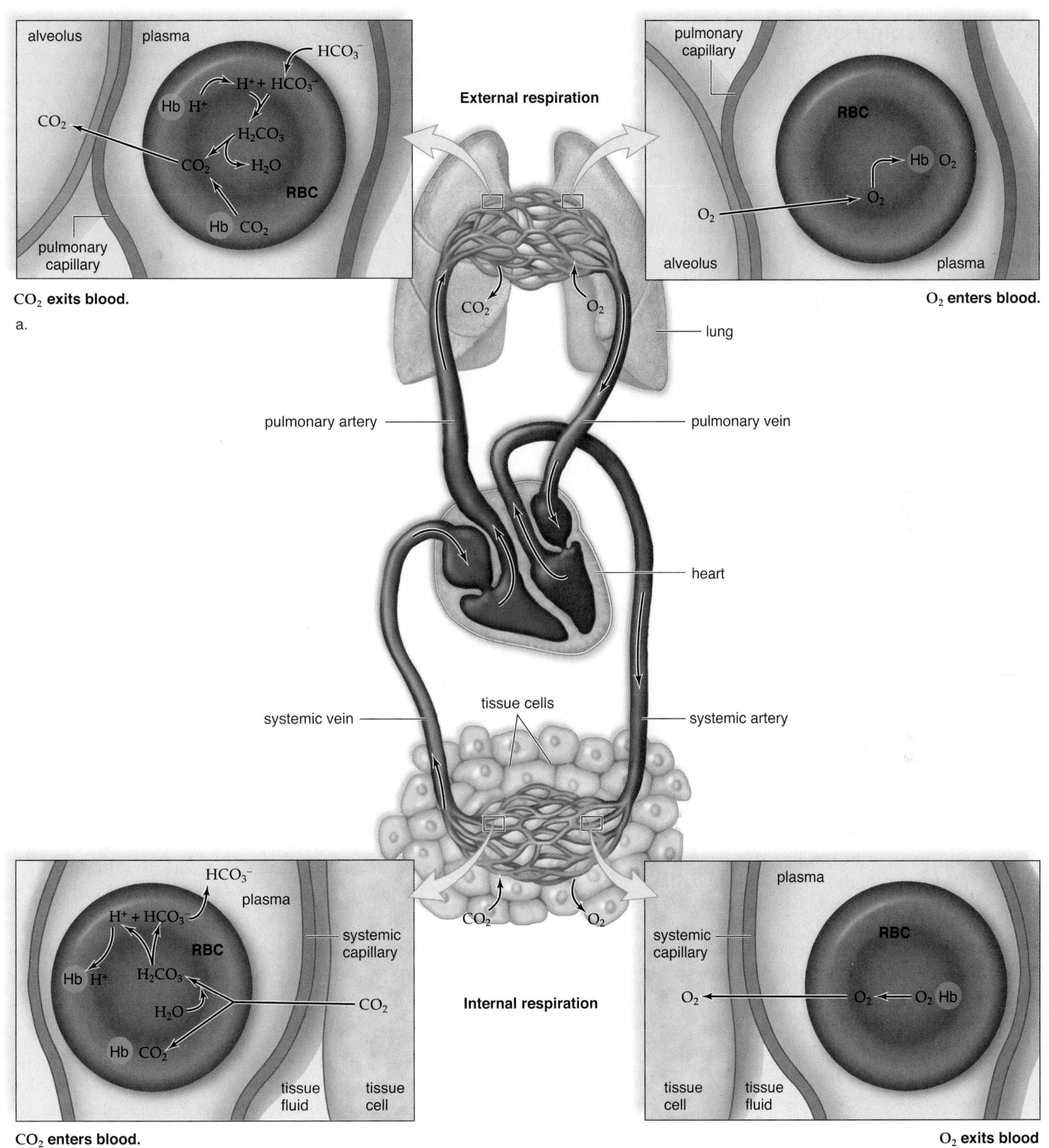

CO_2 exits blood.

a.

O_2 enters blood.

CO_2 enters blood.

b.

O_2 exits blood

Figure 9.11 Movement of gases during external and internal respiration.
a. During external respiration in the lungs, HCO_3^- is converted to CO_2, which exits the blood. O_2 enters the blood and hemoglobin (Hb) carries O_2 to the tissues. **b.** During internal respiration in the tissues, O_2 exits the blood, and CO_2 enters the blood. Most of the CO_2 enters red blood cells, where it becomes the bicarbonate ion, carried in the plasma. Some hemoglobin combines with CO_2 and some combines with H^+.

Internal Respiration

Internal respiration refers to the exchange of gases between the blood in systemic capillaries and the tissue cells. In Figure 9.11*b*, internal respiration is shown at the bottom. Blood entering systemic capillaries is a bright red color because red blood cells contain oxyhemoglobin. The temperature in the tissues is higher and the pH is slightly lower (more acidic), so oxyhemoglobin naturally gives up oxygen. After oxyhemoglobin gives up O_2, it diffuses out of the blood into the tissues.

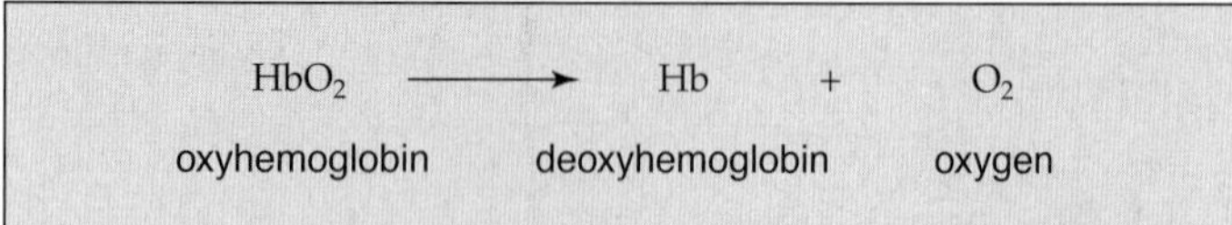

Oxygen diffuses out of the blood into the tissues because the P_{O_2} of tissue fluid is lower than that of blood. The lower P_{O_2} is due to cells continuously using up oxygen in cellular respiration (see Fig. 3.16). *Carbon dioxide diffuses into the blood from the tissues* because the P_{CO_2} of tissue fluid is higher than that of blood. Carbon dioxide is produced during cellular respiration and collects in tissue fluid.

After CO_2 diffuses into the blood, most enters the red blood cells, where a small amount is taken up by hemoglobin, forming **carbaminohemoglobin** ($HbCO_2$). In plasma, CO_2 combines with water, forming carbonic acid (H_2CO_3), which dissociates to hydrogen ions (H^+) and bicarbonate ions (HCO_3^-).

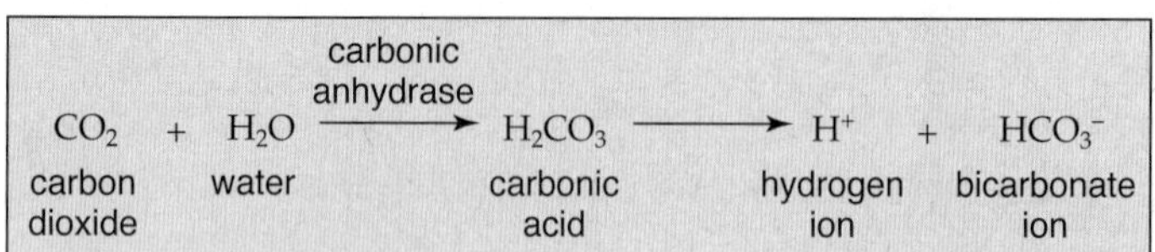

The enzyme carbonic anhydrase, mentioned previously, speeds the reaction in red blood cells. Bicarbonate ions (HCO_3^-) diffuse out of red blood cells and are carried in the plasma. The globin portion of hemoglobin combines with excess hydrogen ions produced by the overall reaction, and Hb becomes HHb, called **reduced hemoglobin.** In this way, the pH of blood remains fairly constant. Blood that leaves the systemic capillaries is a dark maroon color because red blood cells contain reduced hemoglobin.

MP3 Gas Exchange

Connecting the Concepts

For more on the mechanisms by which respiration maintains homeostasis, refer to the following discussions.

Section 2.2 explores the relationship between H^+ concentration and pH.

Section 6.1 examines how components of the blood plasma help buffer the pH of the blood.

Section 6.2 provides additional information on hemoglobin and red blood cells.

Check Your Progress 9.6

1. Describe the differences between external respiration and internal respiration.
2. Describe how hemoglobin functions in the transport of both oxygen and carbon dioxide.
3. Detail the influence of P_{O_2} on both external and internal respiration.

9.7 Respiration and Health

Learning Outcomes

Upon completion of this section, you should be able to

1. Identify the symptoms and causes of selected upper respiratory tract infections.
2. Identify the symptoms and causes of selected lower respiratory tract disorders.
3. Summarize the relationship between smoking, cancer, and emphysema.

The respiratory tract is constantly exposed to environmental air. The quality of this air and whether it contains infectious pathogens such as bacteria and viruses can affect our health.

Upper Respiratory Tract Infections

Upper respiratory infections (URIs) can spread from the nasal cavities to the sinuses, middle ears, and larynx. What we call "strep throat" is a primary bacterial infection caused by *Streptococcus pyogenes* that can lead to a generalized URI and even a systemic (affecting the body as a whole) infection. The symptoms of strep throat are severe sore throat, high fever, and white patches on a dark red throat. Strep throat is bacterial so it can be treated successfully with antibiotics.

Sinusitis

Sinusitis develops when nasal congestion blocks the tiny openings leading to the sinuses (see Fig. 9.2). Symptoms include postnasal discharge and facial pain that worsens when the patient bends forward. Pain and tenderness usually occur over the lower forehead or over the cheeks. In the latter, toothache is also a complaint. Successful treatment depends on restoring proper drainage of the sinuses. Even a hot shower and sleeping upright can be helpful. Nasal spray decongestants and oral antihistamines also alleviate the symptoms of sinusitis. Sprays may be preferred because they treat the symptoms without the side effects, such as drowsiness, of oral medicine. However, nasal sprays can become habit forming if used for long periods. Persistent sinusitis should be evaluated by a health-care professional.

Biology Matters Bioethical Focus

Bans on Smoking

In 1964, the surgeon general of the United States made it known to the general public that smoking was hazardous to our health; thereafter, a health warning was placed on packs of cigarettes. At that time, 40.4% of adults smoked, but by 1990, only about 26% of adults smoked. In the meantime, however, the public became aware that passive smoking—just being in the vicinity of someone who is smoking—can also lead to cancer and other health problems. By now, many state and local governments have passed legislation that bans smoking in public places such as restaurants, elevators, public meeting rooms, and in the workplace. The United States Navy is considering a ban on smoking on submarines and ships, and several towns are contemplating legislation banning smoking completely within their borders.

Is legislation that restricts the freedom to smoke ethical? Or is such legislation akin to racism and creating a population of second-class citizens segregated from the majority on the basis of a habit? Are the desires of nonsmokers being allowed to infringe on the rights of smokers? Or is this legislation one way to help smokers become nonsmokers? One study showed that workplace bans on smoking reduce the daily consumption of cigarettes among smokers by 10%.

Bans on smoking have the potential for a significant economic impact. Legislation might ban smoking in family-style restaurants, but allow bars and restaurants associated with casinos to allow smoking. In certain states, a smoking ban in businesses has cost some underage employees their jobs. A bar can allow smoking if the age limit to enter is 21, but that means employees under 21 have to be let go. Is it fair to treat some businesses differently or to deny people their jobs? The selling of tobacco and even the increased need for health care it generates help the economy. One smoker writes, "Smoking causes people to drink more, eat more, and leave larger tips. Smoking also powers the economy of Wall Street." Is this a reason to allow smoking to continue? Or should we require all places of business to put in improved air-filtration systems? Would that do away with the dangers of passive smoking?

Does legislation that bans smoking in certain areas represent government invasion of our privacy? If yes, is reducing the chance of cancer a good enough reason to allow the government to invade our privacy? Some people are prone to cancer more than others. Should we all be regulated by the same legislation?

Decide Your Opinion

1. Is legislation that bans smoking in public places creating a group of second-class citizens whose rights are being denied? Should the rights of the nonsmoker be more important than the smoker in this instance?
2. Should we be concerned about passing and following legislation that possibly puts a damper on the economy, even if it does improve the health of people?
3. Are bans on smoking an invasion of our privacy? If so, is prevention of cancer in certain persons a good enough reason to risk a possible invasion of our privacy?

Otitis Media

Otitis media is an infection of the middle ear. This infection is considered here because it is a complication often seen in children who have a nasal infection. Infection can spread by way of the auditory tube from the nasopharynx to the middle ear. Pain is the primary symptom of a middle-ear infection. A sense of fullness, hearing loss, vertigo (dizziness), and fever may also be present. Antibiotics are prescribed if necessary, but physicians are aware today that overuse of antibiotics can lead to resistance of bacteria to antibiotics. Tubes (called tympanostomy tubes) are sometimes placed in the eardrums of children with multiple recurrences to help prevent the buildup of pressure in the middle ear and the possibility of hearing loss. Normally, the tubes fall out with time.

Tonsillitis

Tonsillitis occurs when the tonsils become inflamed and enlarged. The tonsil in the posterior wall of the nasopharynx is often called the adenoid. If tonsillitis occurs frequently and enlargement makes breathing difficult, the tonsils can be removed surgically in a **tonsillectomy.** Fewer tonsillectomies are performed today than in the past because we now know that the tonsils trap many of the pathogens that enter the pharynx. Therefore, they are a first line of defense against invasion of the body.

Laryngitis

Laryngitis is an infection of the larynx with accompanying hoarseness, leading to the inability to talk in an audible voice. Usually, laryngitis disappears with treatment of the URI. Persistent hoarseness without the presence of a URI is one of the warning signs of cancer and should be looked into by a physician.

Lower Respiratory Tract Disorders

Lower respiratory tract disorders include infections, restrictive pulmonary disorders, obstructive pulmonary disorders, and lung cancer.

Lower Respiratory Infections

Acute bronchitis is an infection of the primary and secondary bronchi. Usually, it is preceded by a viral URI that has led to a secondary bacterial infection. Most likely, a nonproductive cough has become a deep cough that expectorates mucus and perhaps pus.

Pneumonia is a viral or bacterial infection of the lungs in which the bronchi and alveoli fill with thick fluid (Fig. 9.12). Most often, it is preceded by influenza. High fever and chills with headache and chest pain are symptoms of pneumonia. Rather than being a generalized lung infection, pneumonia may be localized in specific lobules of the lungs. Obviously, the more lobules involved, the more serious is the infection. Pneumonia can be caused by a bacterium that is usually held in check but has gained the upper hand due to stress and/or reduced immunity. AIDS patients are subject to a particularly rare form of pneumonia caused by a fungus named *Pneumocystis jiroveci* (formerly *Pneumocystis carinii*). Pneumonia of this type is almost never seen in individuals with a healthy immune system.

Pulmonary tuberculosis is a bacterial disease that was called *consumption* at one time. When the bacteria (*Mycobacterium tuberculosis*) invade the lung tissue, the cells build a protective capsule around the foreigners, isolating them from the rest of the body. This tiny capsule is called a tubercle. If the resistance of the body is high, the imprisoned organisms die, but if the resistance is low, the organisms eventually can be liberated. If a chest X-ray detects active tubercles, the individual is put on appropriate drug therapy to ensure the localization of the disease and the eventual destruction of any live bacteria. It is possible to tell if a person has ever been exposed to tuberculosis with a *tuberculin* test. This procedure uses a

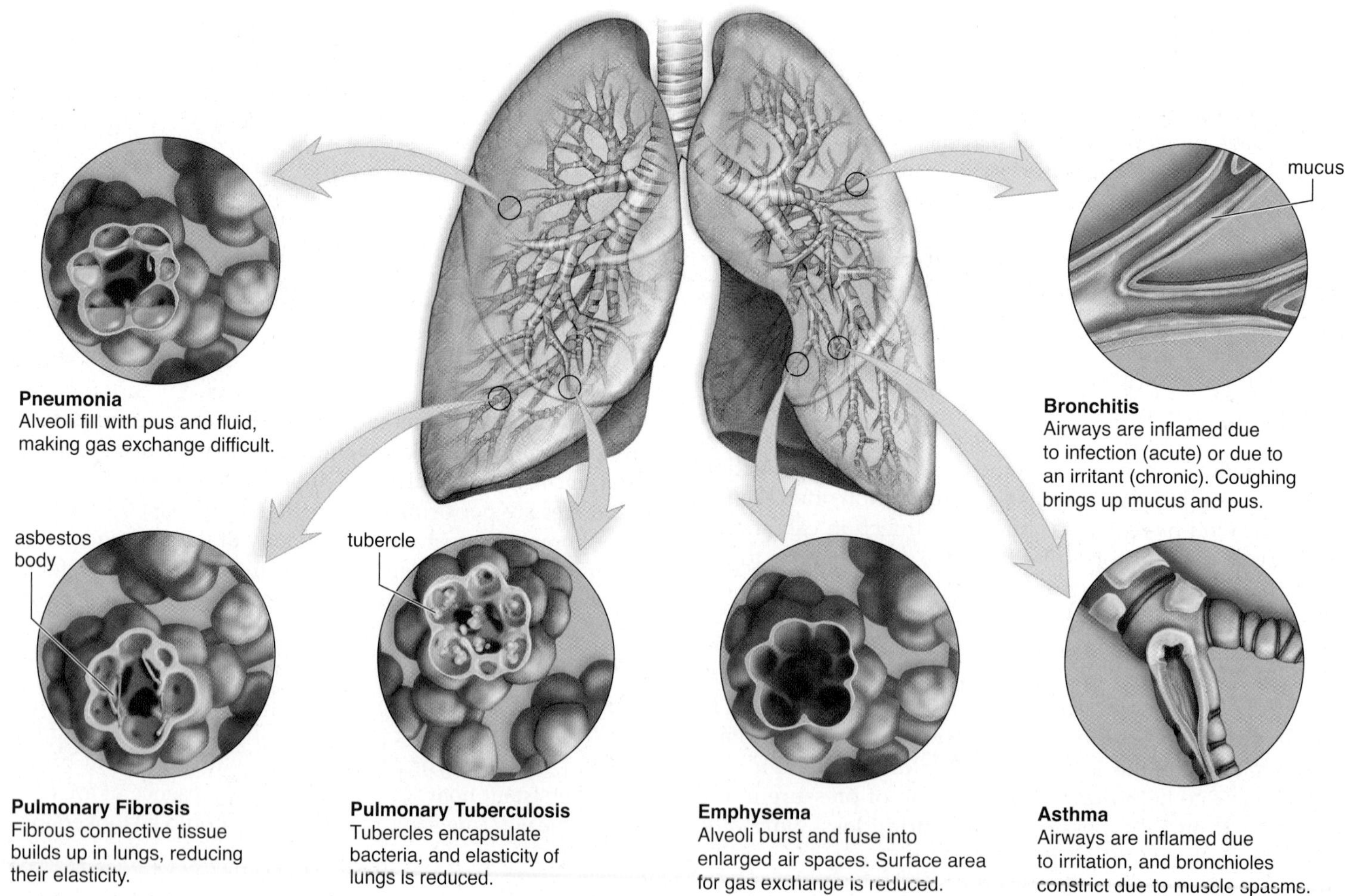

mucus
Pneumonia
Alveoli fill with pus and fluid, making gas exchange difficult.
Bronchitis
Airways are inflamed due to infection (acute) or due to an irritant (chronic). Coughing brings up mucus and pus.
asbestos body
tubercle
Pulmonary Fibrosis
Fibrous connective tissue builds up in lungs, reducing their elasticity.
Pulmonary Tuberculosis
Tubercles encapsulate bacteria, and elasticity of lungs is reduced.
Emphysema
Alveoli burst and fuse into enlarged air spaces. Surface area for gas exchange is reduced.
Asthma
Airways are inflamed due to irritation, and bronchioles constrict due to muscle spasms.

Figure 9.12 Some diseases and disorders of the respiratory system.
Exposure to infectious pathogens and/or polluted air, including tobacco smoke, causes the diseases and disorders shown here.

Connections and Misconceptions

What is cystic fibrosis?

Cystic fibrosis is a genetic disease in which chloride ion transporters in mucus-producing epithelial cells do not function correctly. Because of this, the mucous secretions are often very thick and easily clog small structures such as the alveoli. Though cystic fibrosis is usually considered to be just a disease of the respiratory system, because this is where the symptoms usually first appear, it can also cause problems in organs such as the pancreas. Just a few decades ago, people with cystic fibrosis usually died before the age of 20. New advances in medicine and treatments now enable individuals to live into their 30s and 40s.

Video
Good Poison

highly diluted bacterial extract injected into the patient's skin. If a tuberculin test is positive, an X-ray will be done to confirm an active disease. A person may have a positive test but no active disease.

Restrictive Pulmonary Disorders

In restrictive pulmonary disorders, vital capacity is reduced because the lungs have lost their elasticity. Inhaling particles such as silica (sand), coal dust, asbestos, and fiberglass can lead to **pulmonary fibrosis,** a condition in which fibrous connective tissue builds up in the lungs. The lungs cannot inflate properly and are always tending toward deflation. Breathing asbestos is also associated with the development of cancer. Asbestos was formerly used widely as a fireproofing and insulating agent, so unwarranted exposure has occurred. It has been projected that 2 million deaths caused by asbestos exposure—mostly in the workplace—will occur in the United States between 1990 and 2020.

Obstructive Pulmonary Disorders

In obstructive pulmonary disorders, air does not flow freely in the airways and the time it takes to inhale or exhale maximally is greatly increased. Several disorders, including chronic bronchitis, emphysema, and asthma, are collectively referred to as chronic obstructive pulmonary disease (COPD) because they tend to recur.

In **chronic bronchitis,** the airways are inflamed and filled with mucus. A cough that brings up mucus is common. The bronchi have undergone degenerative changes, including the loss of cilia and their normal cleansing action. Under these conditions, an infection is more likely to occur. Smoking is the most frequent cause of chronic bronchitis. Exposure to other pollutants can also cause chronic bronchitis.

Emphysema is a chronic and incurable disorder in which the alveoli are distended and their walls damaged. As a result, the surface area available for gas exchange is reduced. Emphysema, most often caused by smoking, is often preceded by chronic bronchitis. Air trapped in the lungs leads to alveolar damage and a noticeable ballooning of the chest. The elastic recoil of the lungs is reduced, so not only are the airways narrowed but the driving force behind expiration is also reduced. The victim is breathless and may have a cough. The surface area for gas exchange is reduced, so less oxygen reaches the heart and the brain. Even so, the heart works furiously to force more blood through the lungs, and an increased workload on the heart can result. Lack of oxygen to the brain can make the person feel depressed, sluggish, and irritable. Exercise, drug therapy, supplemental oxygen, and giving up smoking may relieve the symptoms and possibly slow the progression of emphysema. Severe emphysema may be treated by lung transplantation or lung volume reduction surgery (LVRS). During LVRS, a third of the most diseased lung tissue is removed. The removal enables the remaining tissue to function better. The result is an increase in patients' breathing ability and lung capacity.

Asthma is a disease of the bronchi and bronchioles that is marked by wheezing, breathlessness, and sometimes a cough and expectoration of mucus. The airways are unusually sensitive to specific irritants, which can include a wide range of allergens such as pollen, animal dander, dust, tobacco smoke, and industrial fumes. Even cold air can be an irritant. When exposed to the irritant, the smooth muscle in the bronchioles undergoes spasms. It now appears that chemical mediators given off by immune cells in the bronchioles cause the spasms. Most asthma patients have some degree of bronchial inflammation that further reduces the diameter of the airways and contributes to the seriousness of an attack. Asthma is not curable, but it is treatable. Special inhalers can control the inflammation and possibly prevent an attack, and other types of inhalers can stop the muscle spasms should an attack occur.

Lung Cancer

Lung cancer is more prevalent in men than in women but has surpassed breast cancer as a cause of death in women. The increase in the incidence of lung cancer in women is directly related to increased numbers of women who smoke. Autopsies on smokers have revealed the progressive steps by which the most common form of lung cancer develops. The first event appears to be thickening and callusing of the cells lining the bronchi. (Callusing occurs whenever cells are exposed to irritants.) Then cilia are lost, making it impossible to prevent dust and dirt from settling in the lungs. Following this, cells with atypical nuclei appear in the callused lining. A tumor consisting of disordered cells with atypical nuclei is considered cancer

Biology Matters Health Focus

Questions About Smoking, Tobacco, and Health

Is cigarette smoking really addictive?

Yes. The nicotine in cigarette smoke causes addiction to smoking. Nicotine is an addictive drug (just like heroin and cocaine). Small amounts make the smoker want to smoke more. Smokers usually suffer withdrawal symptoms when they stop. Also, nicotine can affect the mood and nature of the smoker. The younger a person is when he or she begins to smoke, the more likely he or she is to develop an addiction to nicotine.

What are some of the short-term and long-term effects of smoking cigarettes?

Short-term effects include shortness of breath and nagging coughs, diminished ability to smell and taste, premature aging of the skin, and increased risk of sexual impotence in men. Smokers tend to tire easily during physical activity. Long-term effects include many types of cancer, heart disease, aneurysms, bronchitis, emphysema, and stroke. Smoking contributes to the severity of pneumonia and asthma.

Does smoking cause cancer?

Yes. Tobacco use accounts for about one-third of cancer deaths in the United States. Smoking causes almost 90% of lung cancers. Smoking also causes cancers of the oral cavity, pharynx, larynx (voice box), and esophagus. It contributes to the development of cancers of the bladder, pancreas, cervix, kidney, and stomach. Smoking is also linked to the development of some leukemias.

Why do smokers have "smoker's cough"?

Cigarette smoke contains chemicals that irritate the air passages and lungs. When a smoker inhales these substances, the body tries to protect itself by producing mucus and coughing. The nicotine in smoke decreases the sweeping action of cilia (hairlike formations lining the airways), so some of the poisons in the smoke remain in the lungs.

If you smoke but do not inhale, is there any danger?

Yes. Wherever smoke touches living cells, it does harm. Even if smokers don't inhale, they are breathing the smoke as secondhand smoke and are still at risk for lung cancer. Pipe and cigar smokers, who often do not inhale, are at an increased risk for lip, mouth, tongue, and several other cancers.

Does cigarette smoking affect the heart?

Yes. Smoking increases the risk of heart disease, the number one cause of death in the United States. Cigarette smoking is the biggest risk factor for sudden heart death. Smokers who have a heart attack are more likely to die within an hour of the heart attack than nonsmokers. Cigarette smoke at very low levels (much lower than the levels that cause lung disease) can cause damage to the heart.

How does smoking affect pregnant women and their babies?

Smoking during pregnancy is linked with a greater chance of miscarriage, premature delivery, stillbirth, infant death, low birth weight, and sudden infant death syndrome (SIDS). Up to 10% of infant deaths would be prevented if pregnant women did not smoke. When a pregnant woman smokes, she really is smoking for two because the nicotine, carbon monoxide, and other dangerous chemicals in smoke enter her bloodstream and then pass into the baby's body. This prevents the baby from getting essential nutrients and oxygen for growth.

What are the dangers of environmental tobacco smoke?

Environmental tobacco smoke (ETS) causes about 3,000 lung cancer deaths and about 35,000 to 40,000 deaths from heart disease each year in healthy nonsmokers. Children whose parents smoke are more likely to suffer from asthma, pneumonia or bronchitis, ear infections, coughing, wheezing, and increased mucus production in the first two years of life.

Are chewing tobacco and snuff safe alternatives to cigarette smoking?

No. The juice from smokeless tobacco is absorbed directly through the lining of the mouth. This creates sores and white patches that often lead to cancer of the mouth. Smokeless tobacco users greatly increase their risk of other cancers, including those of the pharynx (throat). Other effects of smokeless tobacco include harm to teeth and gums.

How can people stop smoking?

There are a number of organizations, including the American Cancer Society and the American Lung Association, that offer suggestions for how to quit smoking. Both organizations also offer support groups to people interested in quitting. There is even advice about how people can help their friends quit smoking. Nicotine Anonymous offers a 12-step program for kicking a nicotine addiction, modeled after the 12-step Alcoholics Anonymous program. Recovering smokers have reported that a support system is very important.

Biology Matters Health Focus *(continued)*

What products are available to replace nicotine while attempting to quit?

Several over-the-counter products are available to replace nicotine. These include nicotine patches, gum, and lozenges. These products are designed to supply enough nicotine to alleviate the withdrawal symptoms people experience during attempts to stop smoking. A nicotine nasal spray and nicotine inhaler are also available by prescription. The delivery of nicotine to the brain by these prescription drugs is comparable to that of a cigarette. These drugs can gradually wean heavy smokers away from nicotine. However, nicotine is still a drug, with the same danger for overdose.

Are there new methods to help people stop smoking?

More than ten smoking-cessation drugs have recently been approved by the FDA, and others are in development phases. One such drug, Chantix, affects areas in the brain stimulated by nicotine. Chantix may lessen withdrawal symptoms by mimicking the effects of nicotine. Additional aids to smoking cessation include a vaccine that stimulates the synthesis of antibodies to nicotine. The antibodies bind to nicotine, preventing the stimulatory effects that result in addiction. This vaccine is being tested in clinical trials.

in situ (at one location). A normal lung versus a lung with cancerous tumors is shown in Figure 9.13. A final step occurs when some of these cells break loose and penetrate other tissues, a process called metastasis. Now the cancer has spread. The original tumor may grow until a bronchus is blocked, cutting off the supply of air to that lung. The entire lung then collapses, the secretions trapped in the lung spaces become infected, and pneumonia or a lung abscess (localized area of pus) results. The only treatment that offers a possibility of cure is to remove a lobe or the whole lung before metastasis has had time to occur. This operation is called **pneumonectomy.** If the cancer has spread, chemotherapy and radiation are also required.

The Bioethical Focus, *Bans on Smoking,* lists the various illnesses, including cancer, apt to occur when a person smokes. Research indicates that passive smoking—exposure to smoke produced by others who are smoking (secondhand smoke)—can also cause lung cancer and other illnesses associated with smoking. If a person stops voluntary smoking and avoids passive smoking, and if the body tissues are not already cancerous, the lungs may return to normal over time.

Connecting the Concepts

For more on the diseases that influence the respiratory system, refer to the following discussions.

Sections 19.1 and 19.2 examine the biology of cancer cells and the major causes of cancer.

Section S.1 provides a more detailed look at tuberculosis and the H1N1 flu epidemic.

Section S.3 explains the causes and consequences of antibiotic resistance.

Check Your Progress 9.7

1. Name and describe the symptoms of some common respiratory infections and disorders of the upper respiratory tract and of the lower respiratory tract.
2. Detail how each of the common respiratory infections in the preceding question can be treated.
3. Describe the three respiratory disorders commonly associated with smoking tobacco.

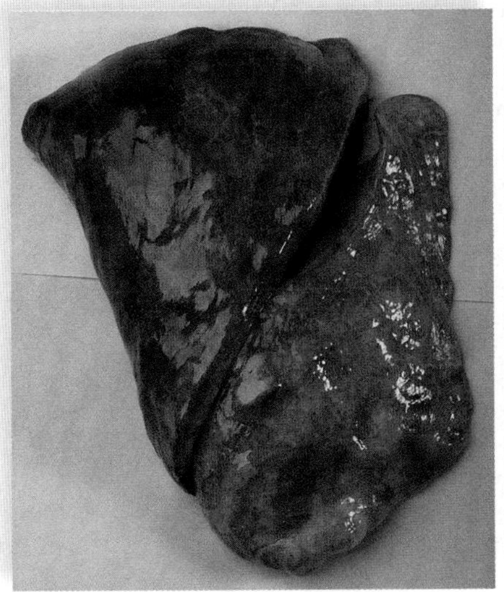

a. Normal lung

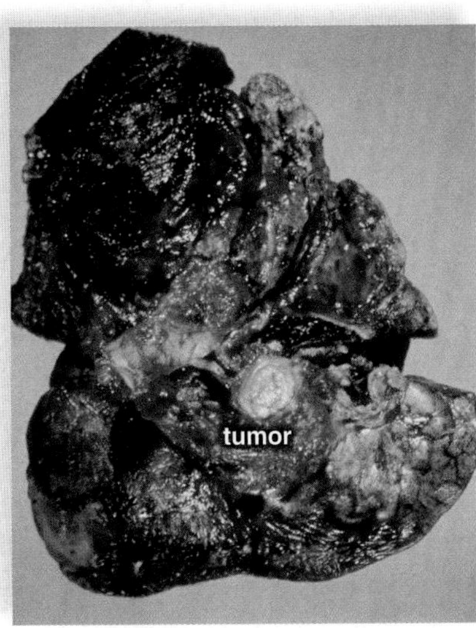

b. Lung cancer

Figure 9.13 **Effect of smoking on a human lung.** **a.** Normal lung, note the healthy red color. **b.** Lungs of a heavy smoker. Notice how black the lungs are except where cancerous tumors have formed.

CASE STUDY CONCLUSION

The results of the polysomnogram treatment indicated that Justin's condition was most likely due to periods of sleep apnea during the evening. Over the 7-hour test period, Justin experienced an average of five apnea events per hour. His blood oxygen concentration was also low during these periods. The brain wave tests did not indicate anything abnormal, suggesting that the problem was associated with obstructive sleep apnea, not central sleep apnea.

As a first level of treatment, Justin's doctor suggested that he increase his exercise, shed the extra pounds, and reduce his use of alcohol. In addition, he scheduled Justin to be fitted for a device called a CPAP (continuous positive airway pressure) mask. The device delivers a constant flow of air into the upper respiratory tract. This serves to keep the airways open, reducing the frequency of apnea events. Justin would have to use the device nightly, because the CPAP mask was not a cure for obstructive sleep apnea. If diet and exercise did not restore Justin's normal sleep patterns, the other option would be surgery to remove some of the soft tissues in the pharynx region, a procedure called an uvulopalatopharyngoplasty (UPPP). However, Justin's doctor was confident that his changes in lifestyle and use of the CPAP mask would help reduce Justin's occurrences of sleep apnea.

Media Study Tools

www.mhhe.com/maderhuman12e

Enhance your study of this chapter with study tools and practice tests. Also ask your instructor about the resources available through ConnectPlus, including the media-rich eBook, interactive learning tools, and animations.

Summarizing the Concepts

9.1 The Respiratory System

The respiratory tract consists of the nose, the pharynx, the larynx, the trachea, the bronchi, the bronchioles, and the lungs.

9.2 The Upper Respiratory Tract

Air from the nose enters the pharynx and passes through the glottis into the larynx:

- nose: filters and warms the air
- pharynx: opening into parallel air and food passageways
- larynx: the voice box that houses the vocal cords

9.3 The Lower Respiratory Tract

- The trachea (windpipe) is lined with goblet cells and ciliated cells.
- The bronchi, along with the pulmonary arteries and veins, enter the lungs.
- The lungs consist of the alveoli, air sacs surrounded by a capillary network.

9.4 Mechanism of Breathing

Breathing involves inspiration and expiration of air.

Inspiration

The diaphragm lowers, and the rib cage moves upward and outward; the lungs expand, and air rushes in.

Expiration

The diaphragm relaxes and moves up. The rib cage moves down and in; pressure in the lungs increases; air is pushed out of the lungs.

Respiratory volumes can be measured:

- **Tidal Volume** The amount of air that normally enters and exits with each breath.
- **Vital Capacity** The amount of air that moves in plus the amount that moves out with maximum effort.
- **Inspiratory and Expiratory Reserve Volume** The difference between normal amounts and the maximum effort amounts of air moved.
- **Residual Volume** The amount of air that stays in the lungs when we breathe.

9.5 Control of Ventilation

The respiratory center in the brain automatically causes us to breathe 12–20 times a minute. Extra carbon dioxide in the blood can decrease the pH; if so, chemoreceptors alert the respiratory center, which increases the rate of breathing.

9.6 Gas Exchanges in the Body

Both external and internal respiration depend on diffusion. Hemoglobin activity is essential to the transport of gases and, therefore, to external and internal respiration.

External Respiration

- CO_2 diffuses out of plasma into lungs.
- O_2 diffuses into the plasma and then into red blood cells in the capillaries. O_2 is carried by hemoglobin.

Internal Respiration

- O_2 diffuses out of the blood into the tissues.
- CO_2 diffuses into the blood from the tissues. CO_2 is carried in the plasma as the bicarbonate ion (HCO_3^-).

9.7 Respiration and Health

A number of illnesses are associated with the respiratory tract.

Upper Respiratory Tract Infections

- Infections of the nasal cavities, sinuses, throat, tonsils, and larynx are all upper respiratory tract infections.
- These include sinusitis, otitis media, tonsillitis, and laryngitis.

Lower Respiratory Tract Disorders

- Lower respiratory infections include acute bronchitis, pneumonia, and pulmonary tuberculosis.
- Restrictive pulmonary disorders are exemplified by pulmonary fibrosis.
- Obstructive pulmonary disorders are exemplified by chronic bronchitis, emphysema, and asthma.
- Smoking can eventually lead to lung and other cancers.

Understanding Key Terms

acute bronchitis 210
alveolus 200
asthma 211
auditory (eustachian) tube 198
bicarbonate ion 206
bronchiole 200
bronchus 200
carbaminohemoglobin 208
carbonic anhydrase 206
chemoreceptor 205
chronic bronchitis 211
dead air space 204
emphysema 211
epiglottis 198
expiration 197, 201
expiratory reserve volume 204
external respiration 206
glottis 199
infant respiratory distress syndrome 201
inspiration 197, 201
inspiratory reserve volume 203
internal respiration 208
laryngitis 209
larynx 199
lung cancer 211
lungs 200
nasal cavity 198
otitis media 209
oxyhemoglobin 206
pharynx 198
pleura 200
pneumonectomy 213
pneumonia 210
pulmonary fibrosis 211
pulmonary tuberculosis 210
reduced hemoglobin 208
residual volume 204
respiratory control center 205
sinusitis 208
sudden infant death syndrome (SIDS) 205
surfactant 201
tidal volume 203
tonsillectomy 209
tonsillitis 209
tonsils 198
trachea 200
tracheostomy 200
ventilation 197
vital capacity 203
vocal cord 199

Match the key terms to these definitions.

a. ________ Common passageway for both food intake and air movement, located between the mouth and the esophagus.

b. ________ Chemical in the lungs that reduces the surface tension of water to keep the alveoli from collapsing.

c. ________ Fold of tissue across the glottis within the larynx; creates vocal sounds when it vibrates.

d. ________ Form in which most of the carbon dioxide is transported in the bloodstream.

e. ________ Stage during breathing when air is pushed out of the lungs.

Testing Your Knowledge of the Concepts

1. Name the three parts of the pharynx. (page 198)
2. How is the structure of the trachea important for respiration, as well as digestion? (page 200)
3. Describe the structure of an alveolus, and explain how it is suited for gas exchange. (pages 210–11)
4. What are the steps of inspiration and expiration? How is breathing controlled? (pages 202–203, 205–206)
5. Using Figure 9.9, list and define the volumes and capacities of air movement. (pages 203–204)
6. Describe what occurs during external respiration, using two important equations. What is the driving force for the gas exchange? (pages 206–207)
7. Describe what occurs during internal respiration, using two important equations. What is the driving force for the gas exchange? (page 208)
8. Describe several upper and lower respiratory tract disorders (other than cancer). If appropriate, explain why breathing is difficult with these conditions. (pages 208–211)
9. List the steps by which lung cancer develops. (page 211)
10. Which of these is anatomically incorrect?
    a. The nose has two nasal cavities.
    b. The pharynx connects the nasal and oral cavities to the larynx.
    c. The larynx contains the vocal cords.
    d. The trachea enters the lungs.
    e. The lungs contain many alveoli.
11. How is inhaled air modified before it reaches the lungs?
    a. It must be humidified.
    b. It must be warmed.
    c. It must be filtered.
    d. All of these are correct.
12. What is the name of the structure that prevents food from entering the trachea?
    a. glottis
    b. septum
    c. epiglottis
    d. Adam's apple

In questions 13–17, match each description with a structure in the key.

Key:

a. pharynx
b. glottis
c. larynx
d. trachea
e. bronchi
f. bronchioles

13. Branched tubes that lead from bronchi to the alveoli
14. Reinforced tube that connects larynx with bronchi
15. Chamber behind oral cavity and between nasal cavity and larynx
16. Opening into larynx
17. Divisions of the trachea that enter lungs
18. Which of these is incorrect concerning inspiration?
    a. Rib cage moves up and out.
    b. Diaphragm contracts and moves down.
    c. Pressure in lungs decreases, and air comes rushing in.
    d. The lungs expand because air comes rushing in.

19. Air enters the human lungs because
    a. atmospheric pressure is lower than the pressure inside the lungs.
    b. atmospheric pressure is greater than the pressure inside the lungs.
    c. although the pressures are the same inside and outside, the partial pressure of oxygen is lower within the lungs.
    d. the residual air in the lungs causes the partial pressure of oxygen to be lower than it is outside.
20. The maximum volume of air that can be moved in and out during a single breath is called the
    a. expiratory and inspiratory reserve volume.
    b. residual volume.
    c. tidal volume.
    d. vital capacity.
    e. functional residual capacity.
21. If air enters the intrapleural space (the space between the pleura),
    a. a lobe of the lung can collapse.
    b. the lungs could swell and burst.
    c. the diaphragm will contract.
    d. nothing will happen because air is needed in the intrapleural space.
22. The enzyme carbonic anhydrase
    a. causes the blood to be more basic in the tissues.
    b. speeds up the conversion of carbonic acid to carbon dioxide and water, and the reverse.
    c. actively transports carbon dioxide out of capillaries.
    d. is active only at high altitudes.
    e. All of these are correct.
23. Hemoglobin assists transport of gases by
    a. combining with oxygen.
    b. combining with CO_2.
    c. combining with H^+.
    d. being present in red blood cells.
    e. All of these are correct.
24. In humans, the respiratory center
    a. is stimulated by carbon dioxide.
    b. is located in the medulla oblongata.
    c. controls the rate of breathing.
    d. All of these are correct.
25. Which of the following is not true of obstructive pulmonary disorders?
    a. Air does not flow freely in the airways.
    b. Vital capacity is reduced due to loss of lung elasticity.
    c. Disorders may include chronic bronchitis, emphysema, and asthma.
    d. Ventilation takes longer to occur.
26. Label this diagram of the human respiratory tract.

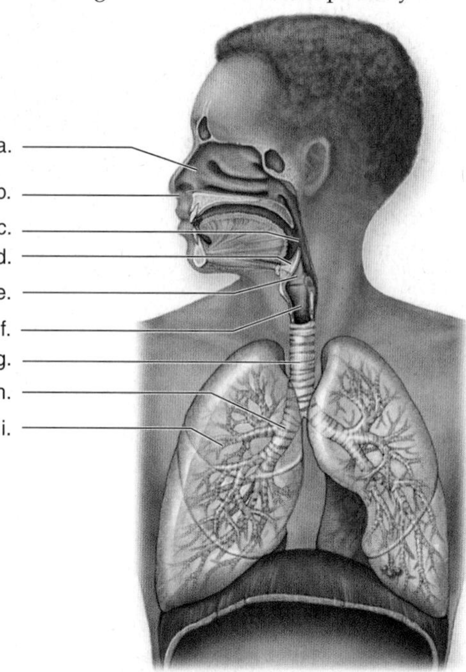

Thinking Critically About the Concepts

Children may also suffer from obstructive sleep apnea (OSA). Symptoms experienced during the daytime include breathing through the mouth and difficulty focusing. Children don't often have the excessive sleepiness during the day that adults like Justin have. Children with OSA often have enlarged tonsils or adenoids. Removal of the tonsils and adenoids often alleviates the problem. CPAP might be necessary if the OSA continues after surgery.

1. Why would enlarged tonsils or adenoids cause OSA?
2. Explain the expression, "The food went down the wrong tube," by referring to structures along the path of air.
3. Long-term smokers often develop a chronic cough.
    a. What purpose does the smokers' cough serve?
    b. Why are nonsmokers less likely to develop a chronic cough?
4. Professional singers who need to hold a long note while singing must exert a great deal of control over their breathing. What muscle(s) needs/need conditioning to acquire better control over their breathing?
5. Why would someone who nearly drowned have a blue tint to his or her skin?

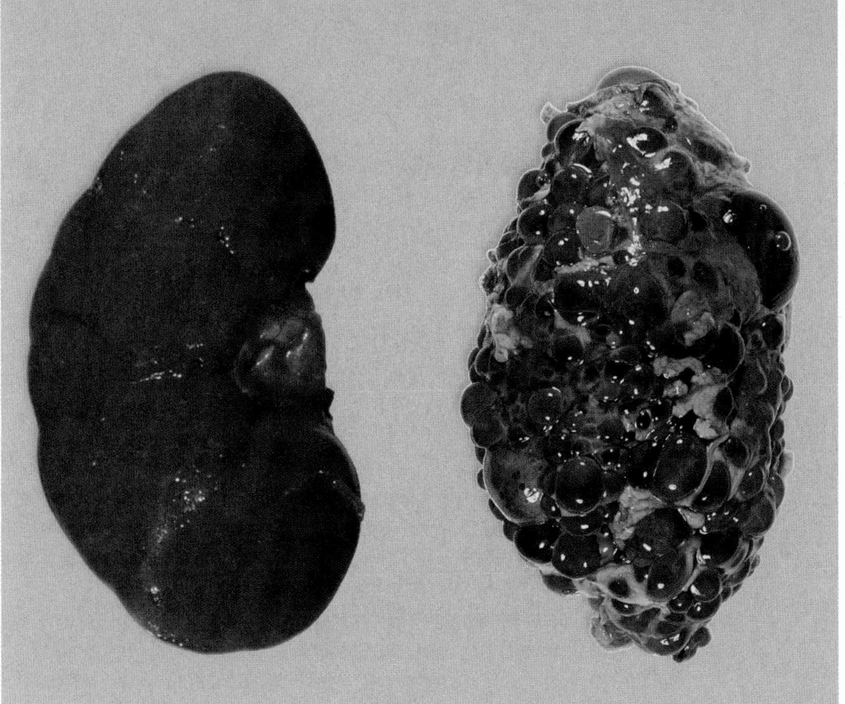

CHAPTER

10

Urinary System

CASE STUDY POLYCYSTIC KIDNEY DISEASE

Michael and Jada were excited about the birth of their child. Married for three years, they already had a healthy daughter, and they were happy that she would have a younger sister to play with. After Aiesha was born, however, it soon became clear that something was wrong. She weighed only 5 lb 4 oz at birth. The first time she urinated, there was an obvious tinge of blood in her urine. In addition, she also seemed to urinate much more frequently than was to be expected for an infant, and her blood pressure was higher than was normal. When her doctors performed ultrasound and magnetic resonance imaging (MRI) scans of her abdominal organs, they found that Aiesha had signs of polycystic kidney disease (PKD). Aiesha's doctors explained that in PKD, cysts (small, fluid-filled sacs) form within the collecting ducts of the nephrons in the interior of the kidneys. The ultrasound results indicated that both of Aiesha's kidneys were covered in cysts (see the kidney above and right) and that this usually meant that the cysts were present inside the kidneys as well. Michael and Jada were informed that the presence of these cysts explained Aiesha's symptoms. The doctors also told Michael and Jada that PKD would most likely cause Aiesha's kidneys to fail and that they should immediately prepare her for a kidney transplant. Because PKD is a genetic disorder, the physicians suggested that both parents undergo genetic tests to see if they were carriers for PKD.

As you read through the chapter, think about the following questions.

1. What is the role of the kidneys in the body?
2. How would problems in the collecting ducts of the nephrons cause kidney failure?
3. Why would problems with the kidneys result in blood in the urine and high blood pressure?

CHAPTER CONCEPTS

10.1 The Urinary System
In the urinary system, kidneys produce urine, which is stored in the bladder before being discharged from the body. The kidneys are major organs of homeostasis.

10.2 Kidney Structure
Microscopically, the kidneys are composed of kidney tubules (nephrons). These tubules have a blood supply that interacts with parts of the tubule as they produce urine.

10.3 Urine Formation
Urine is composed primarily of nitrogenous waste products and salts in water. Urine formation is a stepwise process.

10.4 Kidneys and Homeostasis
The kidneys are involved in the salt–water balance and the acid–base balance of the blood, in addition to excreting nitrogenous wastes.

10.5 Kidney Function Disorders
Various types of illnesses, including diabetes, kidney stones, and infections, can lead to renal failure. Hemodialysis is needed for the survival of patients with renal failure.

BEFORE YOU BEGIN

Before beginning this chapter, take a few moments to review the following discussions.

Section 2.2 What determines whether a solution is acidic or basic?

Section 3.3 How does water move across a plasma membrane?

Section 4.8 How do feedback mechanisms contribute to the maintenance of homeostasis?

10.1 The Urinary System

Learning Outcomes

Upon completion of this section, you should be able to

1. Identify the organs of the urinary system and state their function.
2. Summarize the functions of the urinary system.

The urinary system is the organ system of the body that plays a major role in maintaining the salt, water, and pH homeostasis of the blood. Collectively, these organs carry out the process of **excretion,** or the removal of metabolic waste from the body. These metabolic waste materials are the by-products of the normal activities of the cells and tissues. In comparison to excretion, defecation—a process of the digestive system (Chapter 8)—eliminates undigested food and bacteria in the form of feces. Excretion in humans is performed by the formation and discharge of urine from the body.

Organs of the Urinary System

The urinary system consists of the kidneys, ureters, urinary bladder, and urethra (Fig. 10.1).

Kidneys

The **kidneys** are paired organs located near the small of the back on either side of the vertebral column. They lie in depressions beneath the peritoneum, where they receive some protection from the lower rib cage. The kidneys are bean-shaped and reddish-brown in color. The fist-sized organs are covered by a tough capsule of fibrous connective tissue, called a renal capsule. Masses of adipose tissue adhere to each kidney. The concave side of a kidney has a depression where a **renal artery** enters and a **renal vein** and a ureter exit the kidney. The renal artery transports blood to be filtered to the kidneys, and the renal vein carries filtered blood away from the kidneys.

Connections and Misconceptions

What is a "floating kidney"?

A floating kidney, a condition also known as nephroptosis, occurs when the kidney becomes detached from its position and moves freely beneath the peritoneum. A floating kidney may develop in people who are very thin, or in someone who has recently received a sharp blow to the back. When the kidney becomes dislodged, it may form a kink in the ureter, causing urine to back up into the kidney. This can potentially result in damage to the structures inside the kidney. Surgery can correct a floating kidney by reattaching it to the abdominal wall.

Ureters

The **ureters** conduct urine from the kidneys to the bladder. They are small, muscular tubes about 25 cm long and 5 mm in diameter. The wall of a ureter has three layers: an inner mucosa (mucous membrane), a smooth muscle layer, and

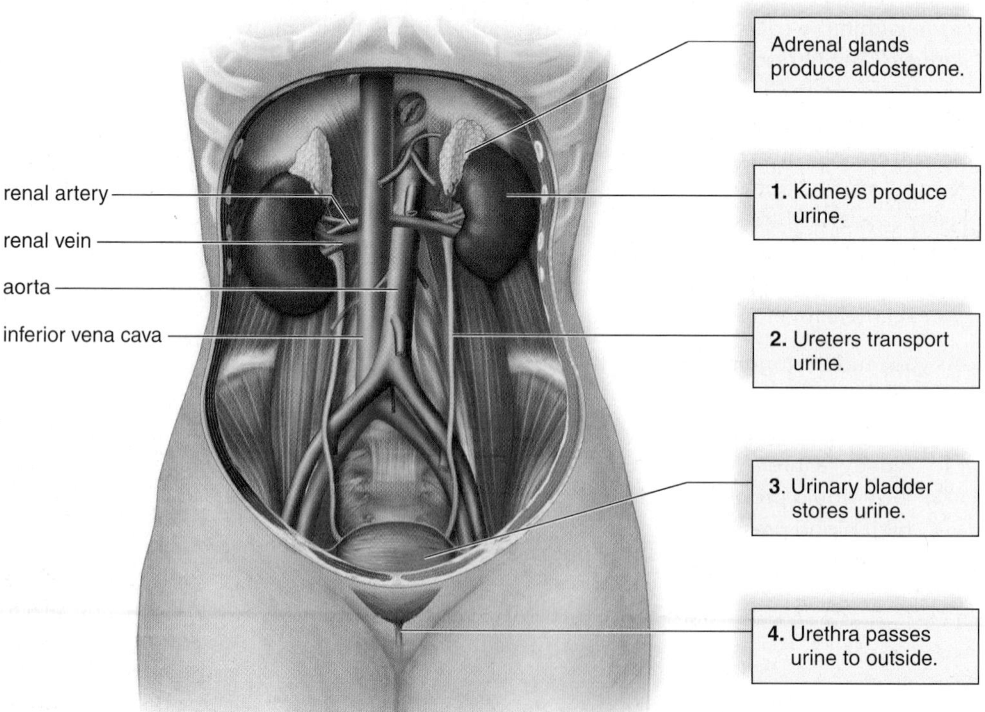

Figure 10.1 **The urinary system.** The kidneys, ureters, urinary bladder, and urethra. The adrenal glands are part of the endocrine system.

an outer fibrous coat of connective tissue. Peristaltic contractions cause urine to enter the bladder even if a person is lying down. Urine enters the bladder in spurts that occur at the rate of one to five per minute.

Urinary Bladder

The **urinary bladder** stores urine until it is expelled from the body. The bladder has three openings: two for the ureters and one for the urethra, which drains the bladder (Fig. 10.2).

The bladder wall is expandable because it contains a middle layer of circular fibers of smooth muscle and two layers of longitudinal smooth muscle. The epithelium of the mucosa becomes thinner, and folds in the mucosa called *rugae* disappear as the bladder enlarges. The bladder's rugae are similar to those of the stomach. A layer of transitional epithelium enables the bladder to stretch and contain an increased volume of urine.

The bladder has other features that allow it to retain urine. After urine enters the bladder from a ureter, small folds of bladder mucosa act like a valve to prevent backward flow. Two sphincters in close proximity are found where the urethra exits the bladder. The internal sphincter occurs around the opening to the urethra. It is composed of smooth muscle and is involuntarily controlled. An external sphincter is composed of skeletal muscle that can be voluntarily controlled.

When the urinary bladder fills to about 250 mL with urine, stretch receptors are activated by the enlargement of the bladder. These receptors send sensory nerve signals to the spinal cord. Subsequently, motor nerve impulses from the spinal cord cause the urinary bladder to contract and the sphincters to relax so that urination, also called **micturition,** is possible (Fig. 10.2).

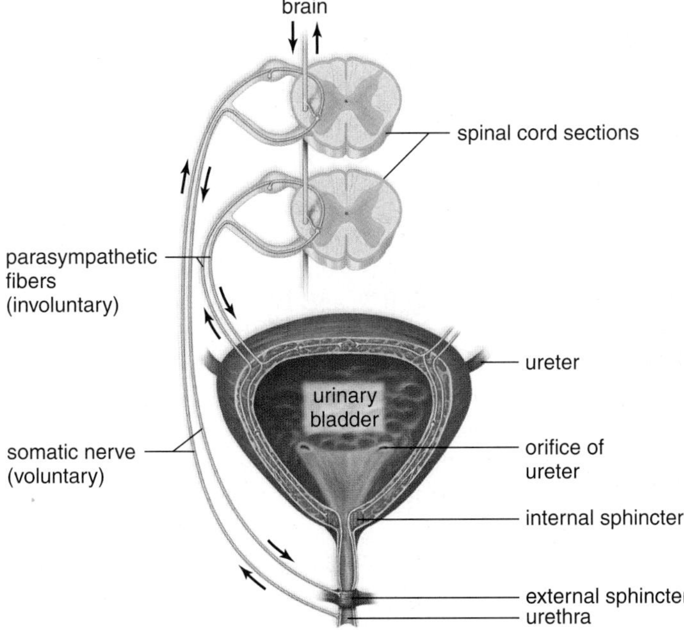

Figure 10.2 Sensory impulses trigger a desire to urinate. As the bladder fills with urine, sensory impulses go to the spinal cord and then to the brain. The brain can override the urge to urinate. When urination occurs, motor nerve impulses cause the bladder to contract and the sphincters to relax.

Connections and Misconceptions

What is an overactive bladder?

In an overactive bladder the muscles of the bladder contract, even though the bladder may not be full. Muscle contraction causes strong feelings of urgency to go to the bathroom. Use of medications, such as Detrol LA and Ditropan XL, control the symptoms by blocking nerve signals to the bladder and calming the bladder's muscle contractions. An overactive bladder can also be treated without medication by urinating at set times of the day or by doing exercises to strengthen the muscles that control urination.

Urethra

The **urethra** is a small tube that extends from the urinary bladder to an external opening. Therefore, its function is to remove urine from the body. The urethra has a different length in females than in males. In females, the urethra is only about 4 cm long. The short length of the female urethra makes bacterial invasion of the urinary tract easier. In males, the urethra averages 20 cm when the penis is flaccid (limp, nonerect). As the urethra leaves the male urinary bladder, it is encircled by the prostate gland. The prostate sometimes enlarges, restricting the flow of urine in the urethra. The Health Focus, *Urinary Difficulties Due to an Enlarged Prostate,* discusses this problem in men.

In females, the reproductive and urinary systems are not connected. However, in males, the urethra carries urine during urination and sperm during ejaculation.

Functions of the Urinary System

As the kidneys produce urine, they carry out the following functions that contribute to homeostasis.

Excretion of Metabolic Wastes

The kidneys excrete metabolic wastes, notably nitrogenous wastes. Urea is the primary nitrogenous end product of metabolism in human beings, but humans also excrete some ammonium, creatinine, and uric acid.

Urea is a by-product of amino acid metabolism. The breakdown of amino acids in the liver releases ammonia, a compound that is very toxic to cells. The liver rapidly combines the ammonia with carbon dioxide to produce urea, which is much less harmful.

Creatine phosphate is a high-energy phosphate reserve molecule in muscles. The metabolic breakdown of creatine phosphate results in **creatinine.**

The breakdown of nucleotides, such as those containing adenine and thymine, produces **uric acid.** Uric acid is rather insoluble. If too much uric acid is present in blood, crystals form and precipitate out. Crystals of uric acid sometimes collect in the joints, producing a painful ailment called **gout.**

Maintenance of Water–Salt Balance

A principal function of the kidneys is to maintain the appropriate water–salt balance of the blood. As we shall see, blood volume is intimately associated with the salt balance of the body. As you know, salts, such as NaCl, have the ability to cause osmosis—the diffusion of water, in this case, into the blood. The more salts there are in the blood, the greater the blood volume and the greater the blood pressure. In this way, the kidneys are involved in regulating blood pressure.

The kidneys also maintain the appropriate level of other ions, such as potassium ions (K^+), bicarbonate ions (HCO_3^-), and calcium ions (Ca^{2+}), in the blood.

Maintenance of Acid–Base Balance

The kidneys regulate the acid–base balance of the blood. For a person to remain healthy, the blood pH should be just about 7.4. The kidneys monitor and help control blood pH, mainly by excreting hydrogen ions (H^+) and reabsorbing the bicarbonate ions (HCO_3^-) as needed to keep blood pH at 7.4. Urine usually has a pH of 6 or lower because our diet often contains acidic foods.

Secretion of Hormones

The kidneys assist the endocrine system in hormone secretion. The kidneys release renin, an enzyme that leads to aldosterone secretion. Aldosterone is a hormone produced by the adrenal glands (see Fig. 10.1), which lie atop the kidneys. As described in Section 10.4, aldosterone is involved in regulating the water–salt balance of the blood.

Erythropoietin (EPO) is a hormone secreted by the kidneys. When blood oxygen decreases, EPO increases red blood cell synthesis by stem cells in the bone marrow (see Chap. 6). When the concentration of red blood cells increases, blood oxygen increases also. Genetically engineered EPO is sometimes prescribed for people in kidney failure (see Chap. 21 for a description of the genetic engineering process). Failing kidneys produce less EPO, resulting in fewer red blood cells and symptoms of fatigue. Supplemental EPO will increase red blood cell synthesis and energy levels.

Additional Functions of the Kidneys

The kidneys also reabsorb filtered nutrients and synthesize vitamin D. Vitamin D is a molecule that promotes calcium ion (Ca^{2+}) absorption from the digestive tract.

Connecting the Concepts

For more on the interaction of the urinary system with other body systems, refer to the following discussions.

Section 6.1 describes the composition of blood.

Section 12.3 describes how muscles use creatine phosphate as an energy molecule.

Section 15.4 examines the structure and function of the adrenal glands.

Check Your Progress 10.1

1. List and briefly describe the functions of the organs of the urinary system.
2. Summarize the processes the kidneys perform to maintain homeostasis.
3. Predict what might occur to overall homeostasis if the kidneys could not excrete metabolic waste products from the body.

10.2 Kidney Structure

Learning Outcomes

Upon completion of this section, you should be able to

1. Identify the structures of a human kidney.
2. Identify the structures of a nephron and state the function of each.

When a kidney is sliced lengthwise, it is possible to see that many branches of the renal artery and vein reach inside the kidney (Figs. 10.3*a* and 10.4). If the blood vessels are removed, it is easier to identify the three regions of a kidney. (1) The **renal cortex** is an outer, granulated layer that dips down in between a radially striated inner layer called the renal medulla. (2) The **renal medulla** consists of cone-shaped tissue masses called renal pyramids. (3) The **renal pelvis** is a central space, or cavity, continuous with the ureter (Fig. 10.3*b*, *c*).

Microscopically, the kidney is composed of over 1 million **nephrons,** sometimes called renal, or kidney, tubules (Fig. 10.3*d*). The nephrons filter the blood and produce urine. Each nephron is positioned so that the urine flows into a collecting duct. Several nephrons enter the same collecting duct. The collecting ducts eventually enter the renal pelvis.

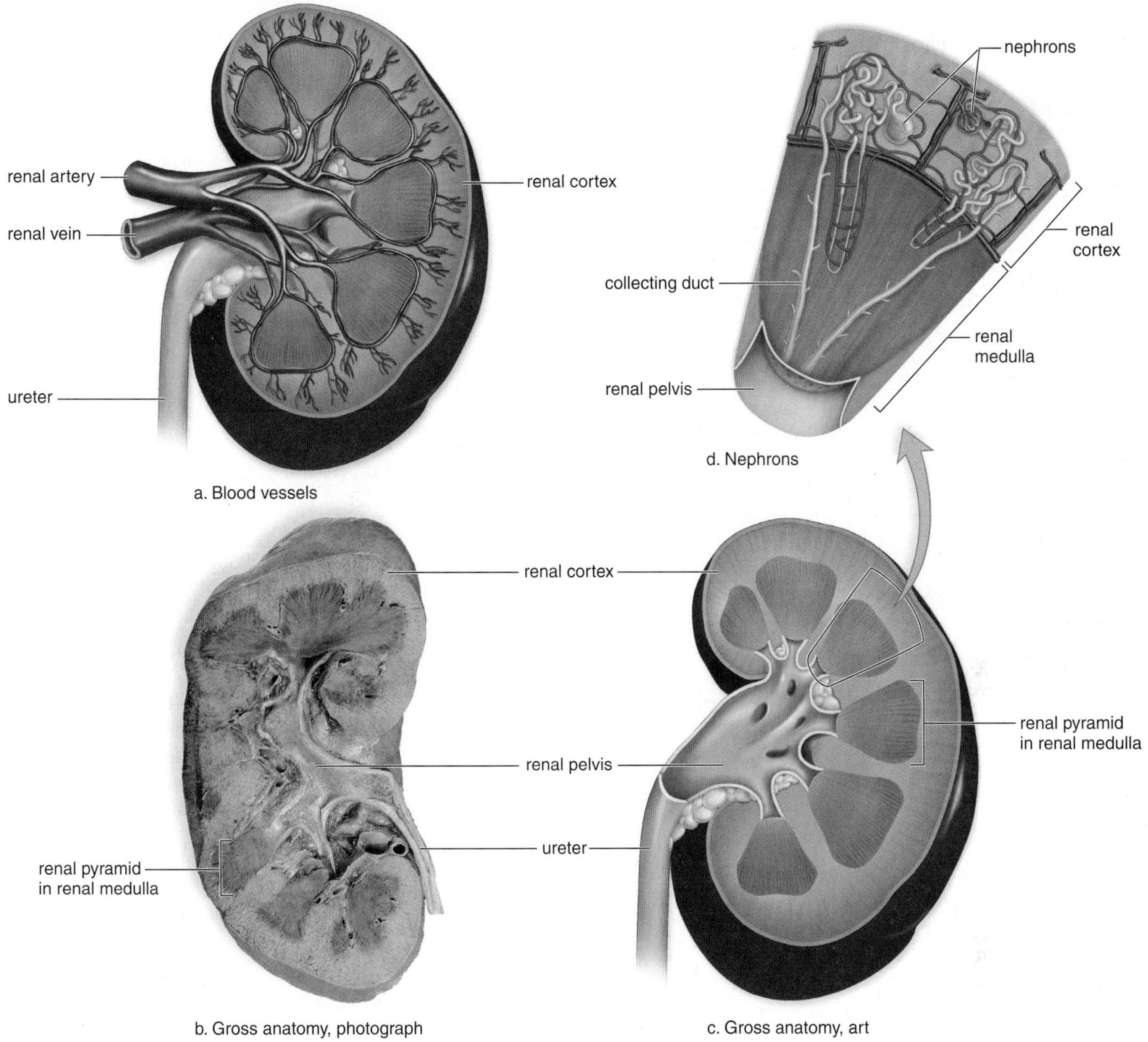

renal artery
renal vein
ureter
renal cortex

a. Blood vessels

nephrons
renal cortex
collecting duct
renal medulla
renal pelvis

d. Nephrons

renal cortex
renal pelvis
ureter
renal pyramid in renal medulla
renal pyramid in renal medulla

b. Gross anatomy, photograph

c. Gross anatomy, art

Figure 10.3 The anatomy of a human kidney.
a. A longitudinal section of the kidney showing the blood supply. The renal artery divides into smaller arteries, and these divide into arterioles. Venules join to form small veins, which join to form the renal vein. **b.** and **c.** The same section without the blood supply. Now it is easier to distinguish the renal cortex; the renal medulla; and the renal pelvis, which connects with the ureter. The renal medulla consists of the renal pyramids. **d.** An enlargement showing the placement of nephrons.

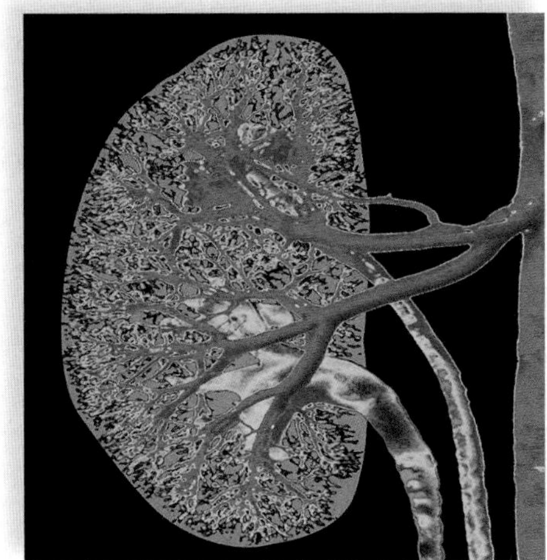

Figure 10.4
The blood vessels of the kidney.
A procedure called an angiogram injects a contrast medium that is opaque to X-rays into the blood vessels, which highlights them when an X-ray of the area is taken.

Biology Matters Health Focus

Urinary Difficulties Due to an Enlarged Prostate

The prostate gland, part of the male reproductive system, surrounds the urethra at the point where the urethra leaves the urinary bladder (Fig. 10A). The prostate gland produces and adds a fluid to semen as semen passes through the urethra within the penis. At about age 50, the prostate gland often begins to enlarge, growing from the size of a walnut to that of a lime or even a lemon. This condition is called benign prostatic hyperplasia (BPH). As it enlarges, the prostate squeezes the urethra, causing urine to back up—first into the bladder, then into the ureters, and finally, perhaps, into the kidneys.

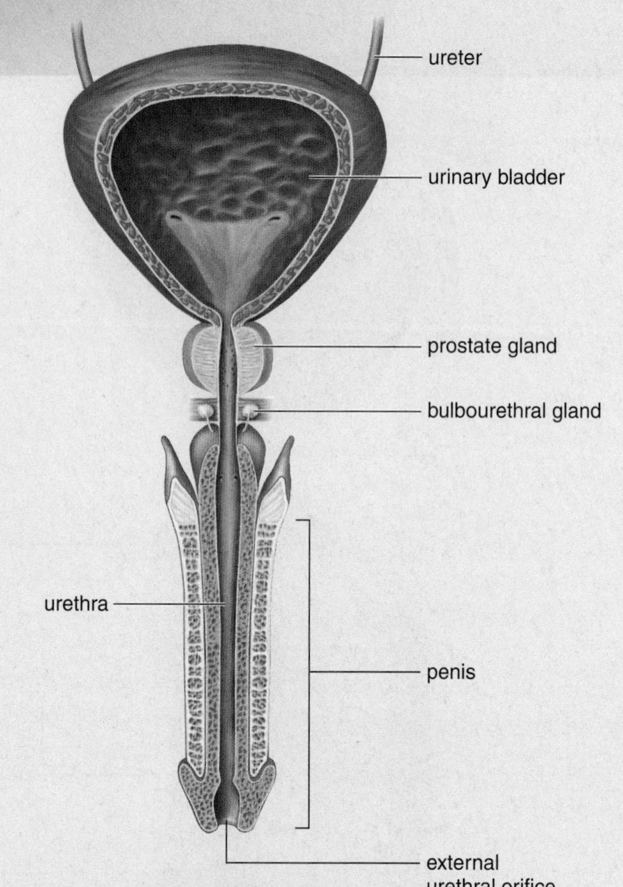

Figure 10A **Location of the prostate gland.**
Note the position of the prostate gland, which can enlarge to obstruct urine flow.

Treatment Emphasis Is on Early Detection

The treatment for BPH can involve (1) a more invasive procedure to reduce the size of the prostate or (2) taking a drug that is expected to shrink the prostate and/or improve urine flow. Prostate tissue can be destroyed by applying microwaves to a specific portion of the prostate. In many cases, however, a physician may decide that prostate tissue should be removed surgically. In some cases, rather than performing abdominal surgery, which requires an incision of the abdomen, the physician gains access to the prostate via the urethra. This operation, called transurethral resection of the prostate (TURP), requires careful consideration because one study found that the death rate during the five years following TURP is much higher than that following abdominal surgery.

Some drug treatments recognize that prostate enlargement is due to a prostate enzyme (5-alpha-reductase) that acts on the male sex hormone testosterone, converting it into a substance that promotes prostate growth. Growth is fine during puberty, but continued growth in an adult is undesirable. Three substances, one a nutrient supplement and two prescription drugs, interfere with the action of the enzyme that promotes growth. Saw palmetto, sold in tablet form as an over-the-counter nutrient supplement, is derived from a plant of the same name. This drug should not be taken unless the need for it is confirmed by a physician, but it is particularly effective during the early stages of prostate enlargement. The prescription drugs finasteride and dutasteride are more powerful inhibitors of the enzyme, but patients complain of erectile dysfunction and loss of libido while on the drugs.

Two other medications have a different mode of action. Nafarelin prevents the release of LH, which when present leads to testosterone production. When it is administered, approximately half of the patients report relief of urinary symptoms even after drug treatment is halted. However, again the patients experience erectile dysfunction and other side effects, such as hot flashes. The drug terazosin, on the market for hypertension because it relaxes arterial walls, also relaxes muscle tissue in the prostate. Improved urine flow was experienced by 70% of the patients taking this drug. However, the drug has no effect on the prostate's overall size.

Many men are concerned that BPH may be associated with prostate cancer, but the two conditions are not necessarily related. BPH occurs in the inner zone of the prostate, whereas cancer tends to develop in the outer area. If prostate cancer is suspected, blood tests and a biopsy, in which a tiny sample of prostate tissue is surgically removed, will confirm the diagnosis.

Enlarged Prostate and Cancer

Although prostate cancer is the second most common cancer in men, it is not a major killer. Typically, prostate cancer is so slow growing that the survival rate is about 98% if the condition is detected early.

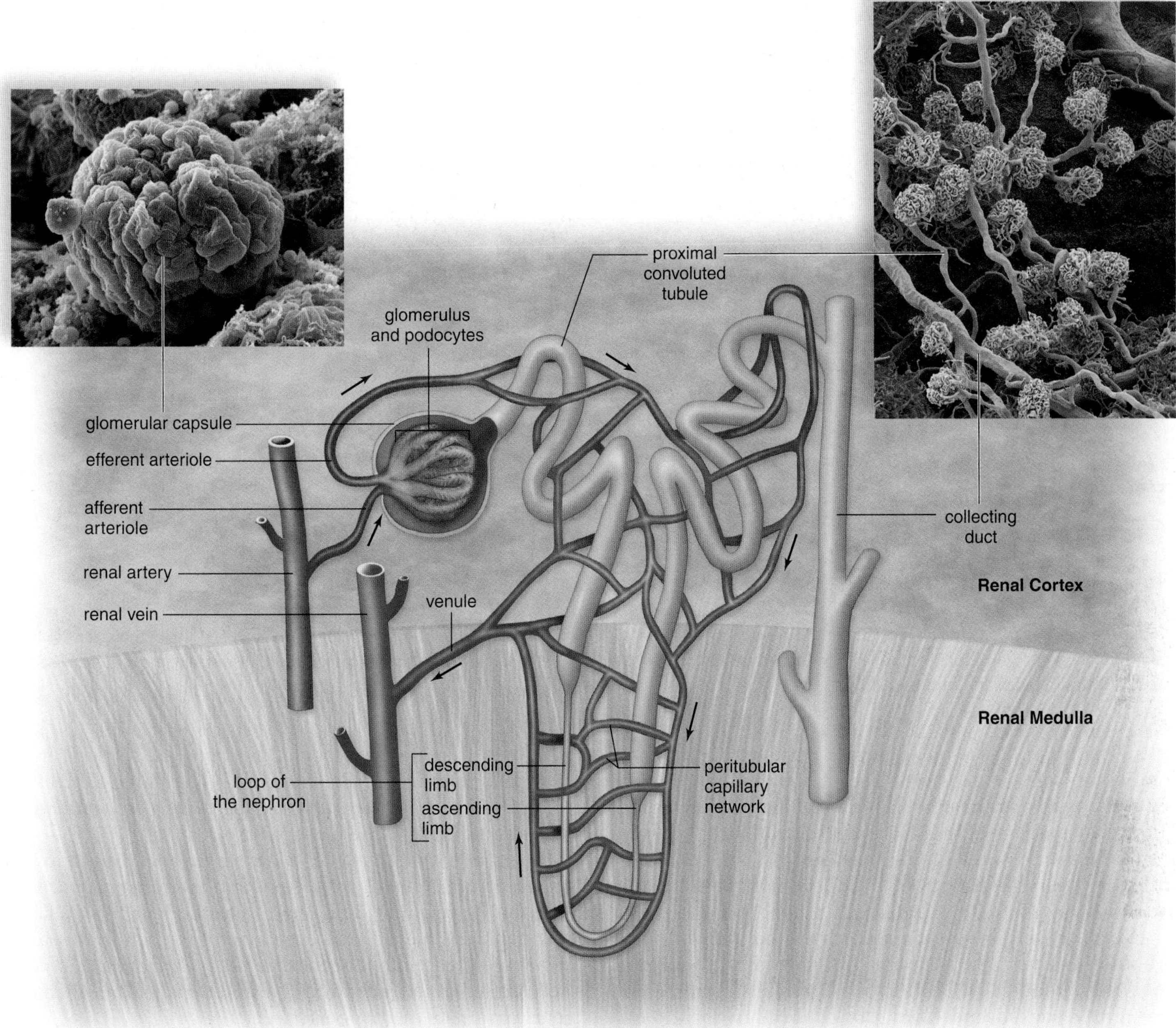

Figure 10.5 The structure of a nephron.
A nephron is made up of a glomerular capsule, the proximal convoluted tubule, the loop of the nephron, the distal convoluted tubule, and the collecting duct. The photomicrographs show the microscopic anatomy of these structures. You can trace the path of blood about the nephron by following the arrows.

Anatomy of a Nephron

Each nephron has its own blood supply including two capillary regions (Fig. 10.5). From the renal artery, an afferent arteriole transports blood to the **glomerulus,** a knot of capillaries inside the glomerular capsule. Blood leaving the glomerulus is carried away by the efferent arteriole. Blood pressure is higher in the glomerulus because the efferent arteriole is narrower than the afferent arteriole. The efferent arteriole divides and forms the **peritubular capillary network,** which surrounds the rest of the nephron. Blood from the efferent arteriole travels through the peritubular capillary network. Then the blood goes into a venule that carries blood into the renal vein.

Parts of a Nephron

Each nephron is made up of several parts (see Fig. 10.5). Some functions are shared by all parts of the nephron. However, the specific structure of each part is especially suited to a particular function.

First, the closed end of the nephron is pushed in on itself to form a cuplike structure called the **glomerular capsule** (Bowman's capsule). The outer layer of the glomerular capsule is composed of squamous epithelial

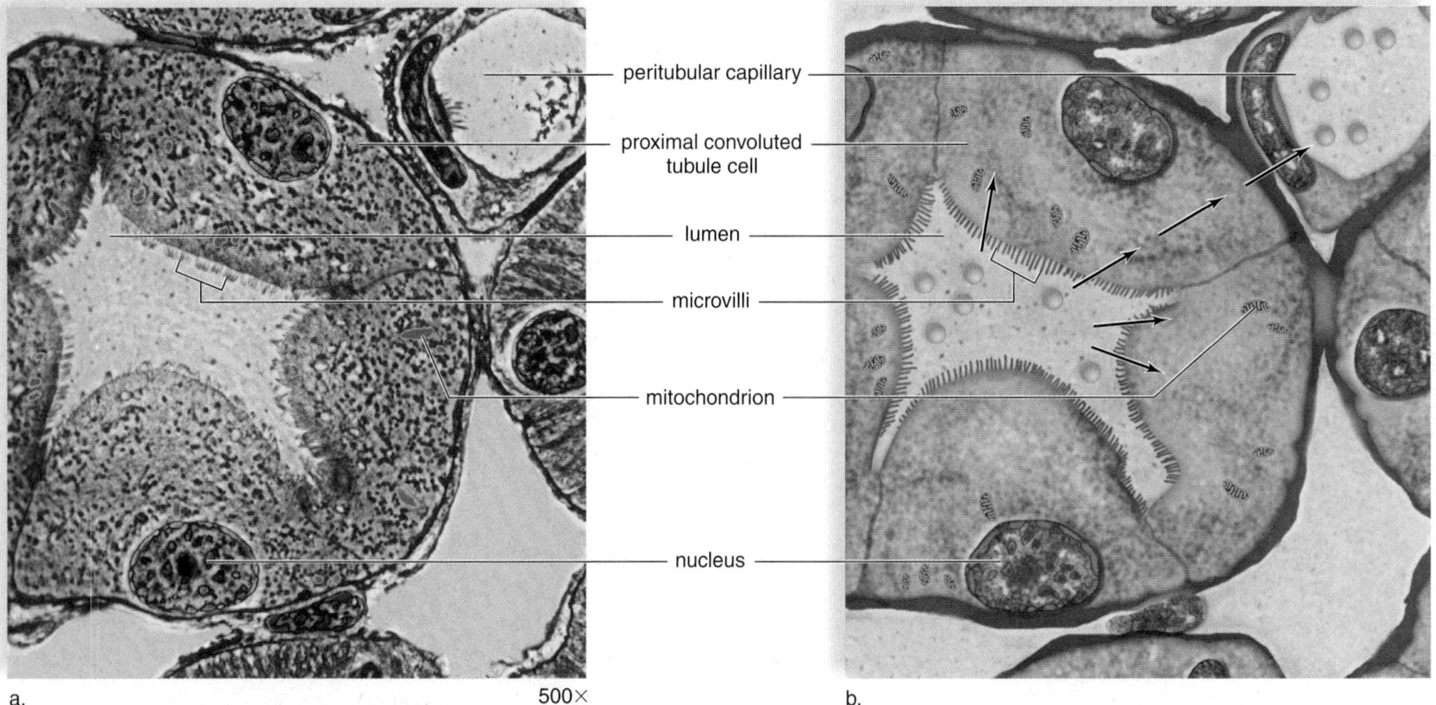

Figure 10.6 The specialized cells of the proximal convoluted tubule.
a. This photomicrograph shows that the cells lining the proximal convoluted tubule have a brushlike border composed of microvilli, which greatly increase the surface area exposed to the lumen. The peritubular capillary network surrounds the cells. **b.** Diagrammatic representation of (a) shows that each cell has many mitochondria, which supply the energy needed for active transport, the process that moves molecules (green) from the lumen of the tubule to the capillary, as indicated by the arrows.

cells. The inner layer is made up of podocytes that have long cytoplasmic extensions. The podocytes cling to the capillary walls of the glomerulus and leave pores that allow easy passage of small molecules from the glomerulus to the inside of the glomerular capsule. This process, called glomerular filtration, produces a filtrate of the blood.

Next, there is a **proximal convoluted tubule.** The cuboidal epithelial cells lining this part of the nephron have numerous microvilli, about 1 μm in length, that are tightly packed and form a brush border (Fig. 10.6). A brush border greatly increases the surface area for the tubular reabsorption of filtrate components. Each cell also has many mitochondria, which can supply energy for active transport of molecules from the lumen to the peritubular capillary network.

Simple squamous epithelium appears as the tube narrows and makes a U-turn called the **loop of the nephron** (loop of Henle). Each loop consists of a descending limb and an ascending limb. The descending limb of the loop allows water to diffuse into tissue surrounding the nephron. The ascending limb actively transports salt from its lumen to interstitial tissue. As we shall see, this activity facilitates the reabsorption of water by the nephron and collecting duct.

The cuboidal epithelial cells of the **distal convoluted tubule** have numerous mitochondria, but they lack microvilli. This means that the distal convoluted tubule is not specialized for reabsorption. Instead, its primary function is ion exchange. During ion exchange, cells reabsorb certain ions, returning them to the blood. Other ions are secreted from the blood into the tubule. The distal convoluted tubules of several nephrons enter one collecting duct. Many **collecting ducts** carry urine to the renal pelvis.

As shown in Figure 10.5, the glomerular capsule and the convoluted tubules always lie within the renal cortex. The loop of the nephron dips down into the renal medulla. A few nephrons have a very long loop of the nephron, which penetrates deep into the renal medulla. Collecting ducts are also located in the renal medulla, and together they give the renal pyramids their appearance.

Connecting the Concepts

For more information on the topics presented in this section, refer to the following discussions.

Section 4.5 illustrates the various forms of epithelial tissue and provides a function for each.

Section 5.2 summarizes the function of arterioles, capillaries, and venules in the circulatory system.

Check Your Progress 10.2

1. Name the three major areas of a kidney.
2. Describe the different parts of a nephron with regard to structure and function.
3. Discuss why the nephrons have such an intricate capillary system.

10.3 Urine Formation

Learning Outcomes

Upon completion of this section, you should be able to

1. Summarize the three processes involved in the formation of urine.
2. List the components of the glomerular filtrate.
3. Describe how tubular reabsorption processes nutrient and salt molecules.
4. Explain the substances that are removed from the blood by tubular secretion.

Figure 10.7 gives an overview of urine formation, which is divided into three processes.

Glomerular Filtration

Glomerular filtration occurs when whole blood enters the glomerulus by way of the afferent arteriole. Due to glomerular blood pressure, water and small molecules move from the glomerulus to the inside of the glomerular capsule. This is a filtration process because large molecules and formed elements are unable to pass through

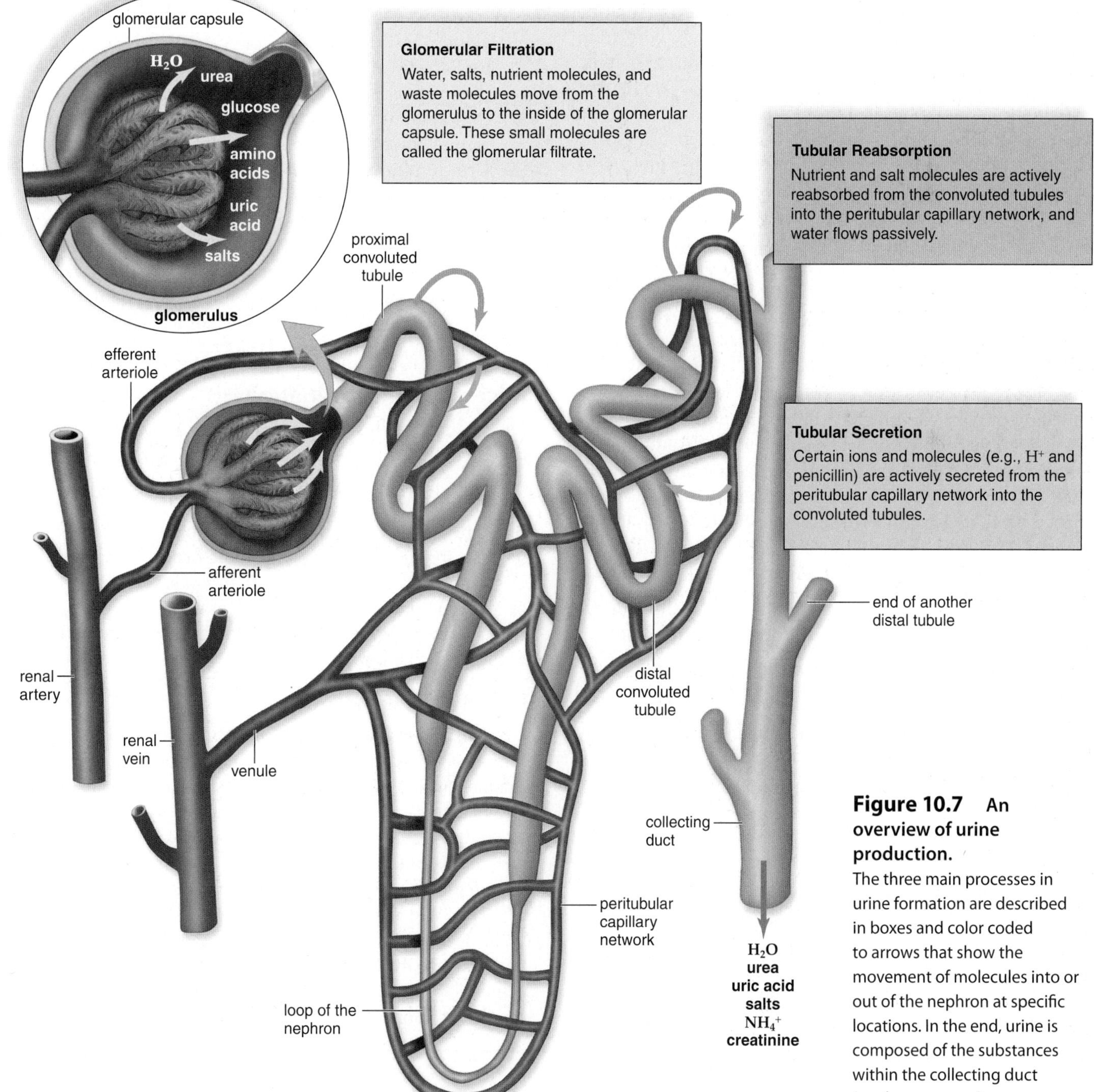

glomerular capsule
H_2O
urea
glucose
amino acids
uric acid
salts
glomerulus
Glomerular Filtration
Water, salts, nutrient molecules, and waste molecules move from the glomerulus to the inside of the glomerular capsule. These small molecules are called the glomerular filtrate.
Tubular Reabsorption
Nutrient and salt molecules are actively reabsorbed from the convoluted tubules into the peritubular capillary network, and water flows passively.
proximal convoluted tubule
efferent arteriole
Tubular Secretion
Certain ions and molecules (e.g., H^+ and penicillin) are actively secreted from the peritubular capillary network into the convoluted tubules.
afferent arteriole
end of another distal tubule
renal artery
distal convoluted tubule
renal vein
venule
collecting duct
peritubular capillary network
H_2O
urea
uric acid
salts
NH_4^+
creatinine
loop of the nephron

Figure 10.7 An overview of urine production. The three main processes in urine formation are described in boxes and color coded to arrows that show the movement of molecules into or out of the nephron at specific locations. In the end, urine is composed of the substances within the collecting duct (see brown arrow).

Biology Matters Health Focus

Urinalysis

A routine urinalysis can detect abnormalities in urine color, concentration, and content. Significant information about the general state of one's health, particularly in the diagnosis of renal and metabolic diseases is provided by this analysis also. A complete urinalysis consists of three phases of examination: physical, chemical, and microscopic.

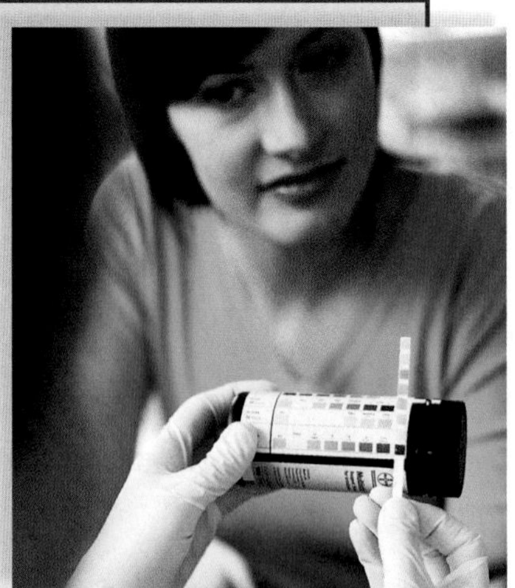

Figure 10B **A urinalysis provides an indication of a person's overall health.** This patient's urine has been tested and the multiple-test stick is being compared to a reference chart. The dark brown pad shows that the patient has glucose in her urine; this indicates that she has diabetes mellitus.

Physical Examination

During a physical examination, a clinician may examine urine color, clarity, and odor. The color of normal, fresh urine is usually pale yellow. However, the color may vary from almost colorless (dilute) to dark yellow (concentrated). Pink, red, or smoky brown urine is usually a sign of bleeding that may be due to a kidney, bladder, or urinary tract infection. Liver disorders can also produce dark brown urine. The clarity of normal urine may be clear or cloudy. However, a cloudy urine sample is also characteristic of abnormal levels of bacteria. Finally, urine odor is usually "nutty" or aromatic; but a foul-smelling odor is characteristic of a urinary tract infection. A sweet, fruity odor is characteristic of glucose in the urine or diabetes mellitus. Urine's odor is also affected by the consumption of garlic, curry, asparagus, and vitamin C.

Chemical Examination

The chemical examination is often done with a dipstick, a thin strip of plastic impregnated with chemicals that change color upon reaction with certain substances present in urine (Fig. 10B). The color change on each segment of the dipstick is compared with a standardized color chart. Dipsticks can be used to determine urine's specific gravity; pH; and content of glucose, bilirubin, urobilinogen, ketone, protein, nitrite, blood, and white blood cells (WBCs) in the urine.

1. *Specific gravity* is an indicator of how well the kidneys are able to adjust tonicity in urine. Normal values for urine specific gravity range from 1.002 to 1.035. A high value (concentrated urine) may be a result of dehydration or diabetes mellitus.
2. Normal *urine pH* can be as low as 4.5 and as high as 8.0. In patients with kidney stone disease, urine pH has a direct effect on the type of stones formed.
3. *Glucose* is normally not present in urine. If present, diabetes mellitus is suspected.
4. *Bilirubin* (a by-product of hemoglobin degradation) is not normally present in the urine. *Urobilinogen* (a by-product of bilirubin degradation) is normally present in very small amounts. High levels of bilirubin or urobilinogen may indicate liver disease.
5. *Ketones* are not normally found in urine. Urine ketones are a by-product of fat metabolism. The presence of ketones in the urine may indicate diabetes mellitus or use of a low-carbohydrate diet, such as the Atkins diet.
6. Plasma *proteins* should not be present in urine. A significant amount of urine protein (proteinuria) is usually a sign of kidney damage.
7. Urine typically does not contain *nitrates*. The presence of nitrites is a sign of urinary tract infection.
8. It is not abnormal for blood to show up in the urine when a woman is menstruating. Blood present in the urine at other times may indicate a bacterial infection or kidney damage.
9. The chemical test for *WBCs* is normally negative. A high urine WBC count usually indicates a bacterial infection somewhere in the urinary tract.

Microscopic Examination

Finally, in a microscopic examination, the urine is centrifuged, and the sediment (solid material) is examined under a microscope. When renal disease is present, the urine often contains an abnormal amount of cellular material. Urinary casts are sediments formed by the coagulation of protein material in the distal convoluted tubule or the collecting duct. Casts may be a sign of many different disorders, depending on the type present. The presence of crystals in the urine is characteristic of kidney stones, kidney damage, or problems with metabolism.

Forensic Analysis

Urinalysis is also used by the federal government to screen potential employees for the use of numerous illegal drugs. Business establishments conduct similar random testing. Drugs associated with date rape (such as Rohypnol, known as "roofies," and GHB, or gamma hydroxybutyrate) can be detected in the urine of victims to determine if they were drugged.

There are more than 100 different tests that can be done on urine. Many different factors such as diet, kidney function, and other health disorders can affect what ends up in one's urine. Consequently, a urinalysis can provide a clinician with important information about a patient's overall health.

Table 10.1 Reabsorption from Nephrons

Substance	Amount Filtered (per Day)	Amount Excreted (per Day)	Reabsorption (%)
Water (L)	180	1.8	99.0
Sodium (g)	630	3.2	99.5
Glucose (g)	180	0.0	100.0
Urea (g)	54	30.0	44.0

L = liters, g = grams

the capillary wall. In effect, then, blood in the glomerulus has two portions: the filterable components and the nonfilterable components.

Filterable Blood Components	Nonfilterable Blood Components
Water	Formed elements (blood cells and platelets)
Nitrogenous wastes	Plasma proteins
Nutrients	
Salts (ions)	

The nonfilterable components leave the glomerulus by way of the efferent arteriole. The **glomerular filtrate** inside the glomerular capsule now contains the filterable blood components in approximately the same concentration as plasma.

As indicated in Table 10.1, nephrons in the kidneys filter 180 liters of water per day, along with a considerable amount of small molecules (such as glucose) and ions (such as sodium). If the composition of urine were the same as that of the glomerular filtrate, the body would continually lose water, salts, and nutrients. Therefore, we can conclude that the composition of the filtrate must be altered as this fluid passes through the remainder of the tubule.

Tubular Reabsorption

Tubular reabsorption occurs as molecules and ions are passively and actively reabsorbed from the nephron into the blood of the peritubular capillary network. The osmolarity of the blood is maintained by the presence of plasma proteins and salt. When sodium ions (Na^+) are actively reabsorbed, chloride ions (Cl^-) follow passively. The reabsorption of salt (NaCl) increases the osmolarity of the blood compared with the filtrate. Therefore, water moves passively from the tubule into the blood. About 65% of Na^+ is reabsorbed at the proximal convoluted tubule.

Nutrients such as glucose and amino acids return to the peritubular capillaries almost exclusively at the proximal convoluted tubule. This is a selective process because only molecules recognized by carrier proteins are actively reabsorbed. Glucose is an example of a molecule that ordinarily is completely reabsorbed because there is a plentiful supply of carrier proteins for it. However, every substance has a maximum rate of transport. After all its carriers are in use, any excess in the filtrate will appear in the urine. In **diabetes mellitus,** because the liver and muscles fail to store glucose as glycogen, the blood glucose level is above normal and glucose appears in the urine. The presence of excess glucose in the filtrate raises its osmolarity. Therefore, less water is reabsorbed into the peritubular capillary network. The frequent urination and increased thirst experienced by people with untreated diabetes are due to less water being reabsorbed from the filtrate into the blood.

We have seen that the filtrate that enters the proximal convoluted tubule is divided into two portions: components reabsorbed from the tubule into blood, and components not reabsorbed that continue to pass through the nephron to be further processed into urine.

Reabsorbed Filtrate Components	Nonreabsorbed Filtrate Components
Most water	Some water
Nutrients	Much nitrogenous waste
Required salts (ions)	Excess salts (ions)

The substances not reabsorbed become the tubular fluid, which enters the loop of the nephron.

Connections and Misconceptions

Can you drink urine if there is no other source of water available?

You're probably familiar with news reports of trapped individuals who survived for days without water by drinking their urine. If you're a fan of *Man vs. Wild* on the Discovery Channel, you may remember an episode in the Australian Outback when Bear Grylls drank his urine.

In some cultures, people drink urine as an alternative medicine or for cosmetic reasons. This practice is known as urine therapy. Urine from someone who is healthy should be sterile and, therefore, drinkable. However, urine contains any number of other substances besides water, like salts and metabolites of various drugs. If these substances are highly concentrated, you may become dehydrated more quickly. Many survival guides advise against drinking your urine.

Tubular Secretion

Tubular secretion is a second way by which substances are removed from blood and added to the tubular fluid. Hydrogen ions (H^+), creatinine, and drugs such as penicillin are some of the substances moved by active transport from blood into the kidney tubule. In the end, urine contains substances that have undergone glomerular filtration but have not been reabsorbed and substances that have undergone tubular secretion. Tubular secretion is now known to occur along the length of the kidney tubule.

Connecting the Concepts

For more information on the topics presented in this section, refer to the following discussions.

Section 3.3 describes the process of osmosis and the movement of ions across a cell membrane.

Section 6.1 lists the substances found in the plasma component of blood.

Check Your Progress 10.3

1. List the three major processes in urine formation and state the location in the nephron where each occurs.
2. Detail the functions of glomerular filtration, tubular reabsorption, and tubular secretion.
3. Predict what might happen to the overall function of a nephron if glomerular filtration could not occur.

10.4 Kidneys and Homeostasis

Learning Outcomes

Upon completion of this section, you should be able to

1. Summarize how the kidney maintains the water–salt balance of the body.
2. State the purpose of ADH, ANH, and aldosterone in homeostasis.
3. Explain how the kidneys assist in the maintenance of the pH levels of the blood.

The kidneys play a major role in homeostasis, from the maintenance of the water–salt balance in the body to regulating the pH of the blood. In doing so, the kidneys interact with every other organ system of the human body (Fig.10.8).

Kidneys Excrete Waste Molecules

In Chapter 8, we compared the liver to a sewage treatment plant because it removes poisonous substances from the blood and prepares them for excretion. Similarly, the liver produces urea, the primary nitrogenous end product of humans, which is excreted by the kidneys. If the liver is a sewage treatment plant, the tubules of the kidney are like the trucks that take the sludge, prepared waste, away from the town (the body).

Metabolic waste removal is absolutely necessary for maintaining homeostasis. The blood must constantly be cleansed of the nitrogenous wastes, end products of metabolism. The liver produces urea, and muscles make creatinine. These wastes, and also uric acid from the cells, are carried by the cardiovascular system to the kidneys. The urine-producing kidneys are responsible for the excretion of nitrogenous wastes. They are assisted to a limited degree by the sweat glands in the skin, which excrete perspiration, a mixture of water, salt, and some urea. In times of kidney failure, urea is excreted by the sweat glands and forms a so-called uremic frost on the skin.

Water–Salt Balance

Most of the water found in the filtrate is reabsorbed into the blood before urine leaves the body. All parts of a nephron and the collecting duct participate in the reabsorption of water. The reabsorption of salt always precedes the reabsorption of water. In other words, water is returned to the blood by the process of osmosis. During the process of reabsorption, water passes through water channels, called **aquaporins**, within a plasma membrane protein.

Sodium ions (Na^+) are important in plasma. Usually more than 99% of the Na^+ filtered at the glomerulus is returned to the blood. The kidneys also excrete or reabsorb other ions, such as potassium ions (K^+), bicarbonate ions (HCO_3^-), and magnesium ions (Mg^{2+}) as needed.

Reabsorption of Salt and Water from Cortical Portions of the Nephron

The proximal convoluted tubule, the distal convoluted tubule, and the cortical portion of the collecting ducts are present in the renal cortex. Most of the water (65%) that enters the glomerular capsule is reabsorbed from the nephron into the blood at the proximal convoluted tubule. Na^+ is actively reabsorbed, and Cl^- follows passively. Aquaporins are always open and water is reabsorbed osmotically into the blood.

All systems of the body work with the urinary system to maintain homeostasis. These systems are especially noteworthy.

Urinary System

As an aid to all the systems, the kidneys excrete nitrogenous wastes and maintain the water–salt balance and the acid–base balance of the blood. The urinary system also specifically helps the other systems.

Cardiovascular System

Production of renin by the kidneys helps maintain blood pressure. Blood vessels transport nitrogenous wastes to the kidneys and carbon dioxide to the lungs. The buffering system of the blood helps the kidneys maintain the acid–base balance.

Digestive System

The liver produces urea excreted by the kidneys. The yellow pigment found in urine, called urochrome (breakdown product of hemoglobin), is produced by the liver. The digestive system absorbs nutrients, ions, and water. These help the kidneys maintain the proper level of ions and water in the blood.

Muscular System

The kidneys regulate the amount of ions in the blood. These ions are necessary to the contraction of muscles, including those that propel fluids in the ureters and urethra.

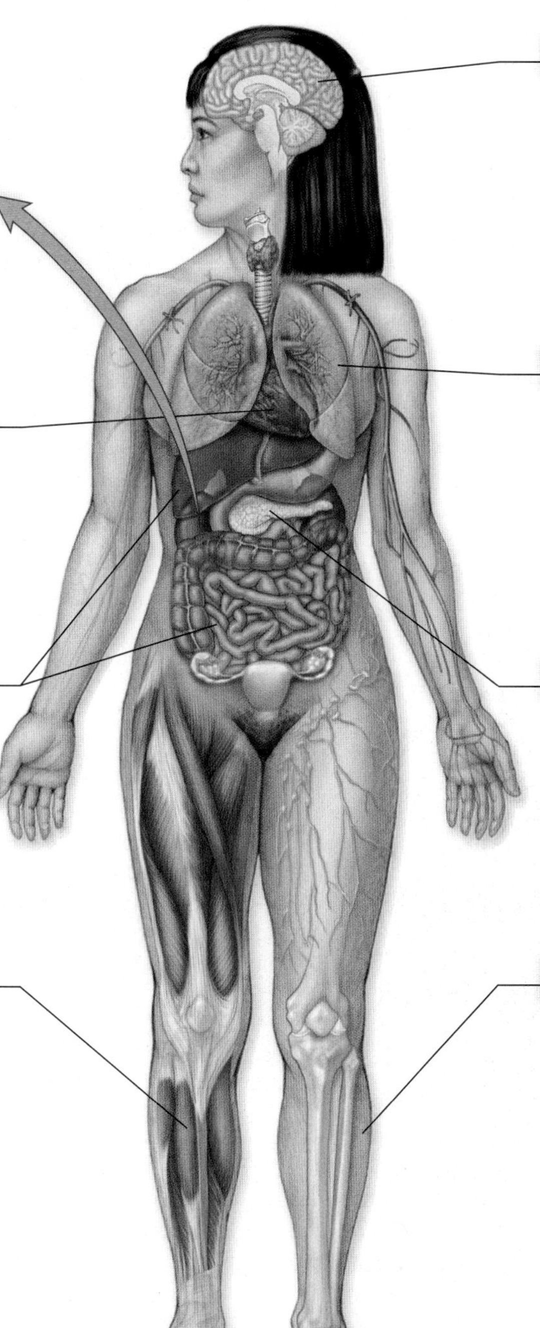

Nervous System

The kidneys regulate the amount of ions (e.g., K^+, Na^+, Ca^{2+}) in the blood. These ions are necessary for nerve impulse conduction. The nervous system controls urination.

Respiratory System

The kidneys help the lungs by exhaling carbon dioxide as bicarbonate ions, and the lungs help the kidneys maintain the acid–base balance of the blood by exhaling carbon dioxide.

Endocrine System

The kidneys produce renin, leading to the production of aldosterone, a hormone that helps the kidneys maintain the water–salt balance. The kidneys produce the hormone erythropoietin, and they change vitamin D to a hormone. The posterior pituitary secretes ADH, which regulates water retention by the kidneys.

Integumentary System

Sweat glands excrete perspiration, a solution of water, salt, and some urea.

Figure 10.8 The urinary system and homeostasis.
The urinary system works primarily with these systems to bring about homeostasis.

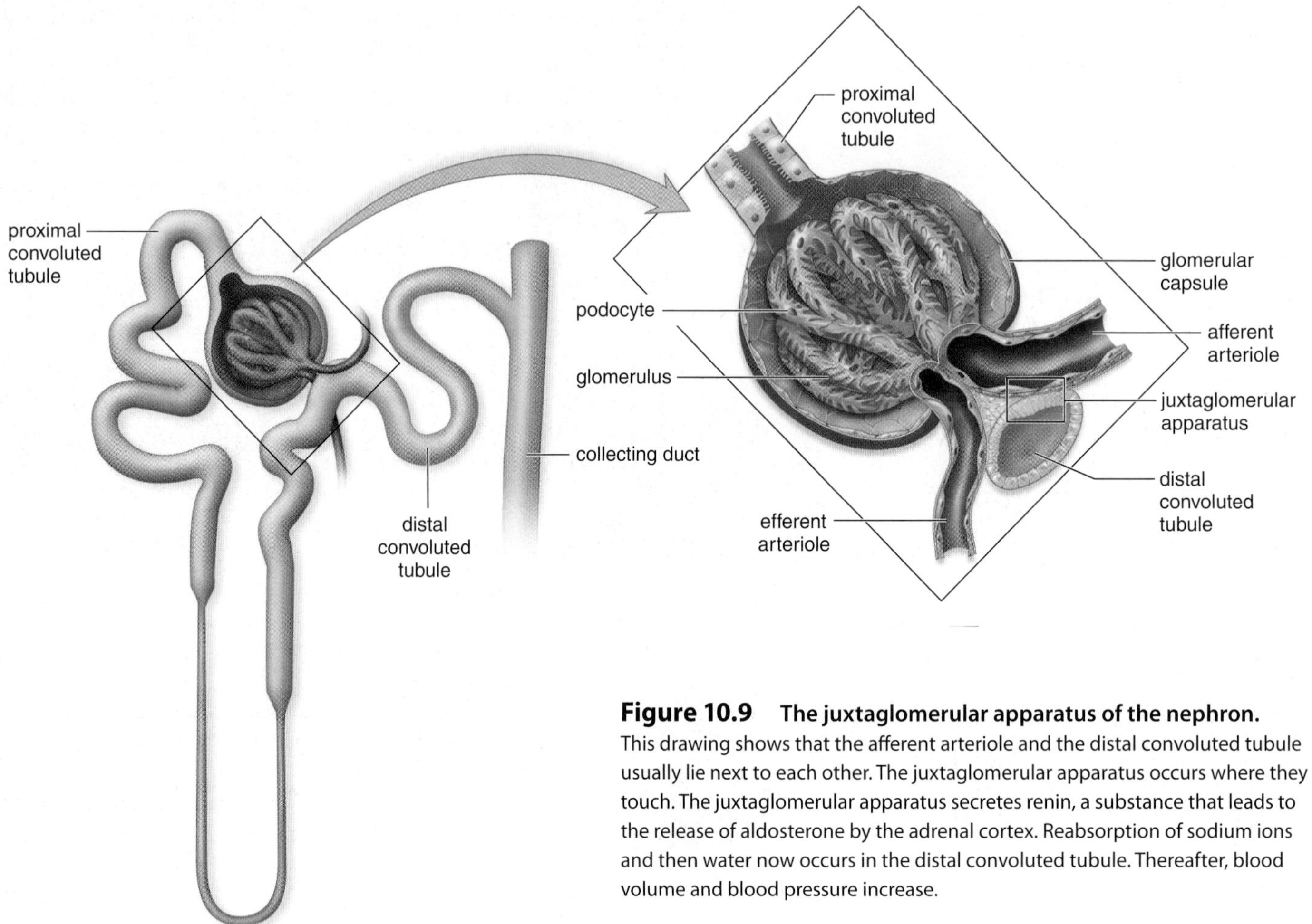

Figure 10.9 The juxtaglomerular apparatus of the nephron. This drawing shows that the afferent arteriole and the distal convoluted tubule usually lie next to each other. The juxtaglomerular apparatus occurs where they touch. The juxtaglomerular apparatus secretes renin, a substance that leads to the release of aldosterone by the adrenal cortex. Reabsorption of sodium ions and then water now occurs in the distal convoluted tubule. Thereafter, blood volume and blood pressure increase.

Hormones regulate the reabsorption of sodium and water in the distal convoluted tubule. **Aldosterone** is a hormone secreted by the adrenal glands, which sit atop the kidneys. This hormone promotes ion exchange at the distal convoluted tubule. Potassium ions (K^+) are excreted, and sodium ions (Na^+) are reabsorbed into the blood. The release of aldosterone is set into motion by the kidneys. The **juxtaglomerular apparatus** is a region of contact between the afferent arteriole and the distal convoluted tubule (Fig. 10.9). When blood volume (and, therefore, blood pressure) falls too low for filtration to occur, the juxtaglomerular apparatus can respond to the decrease by secreting **renin.** Renin is an enzyme that ultimately leads to secretion of aldosterone by the adrenal glands. Research scientists speculate that excessive renin secretion—and thus, reabsorption of excess salt and water—might contribute to high blood pressure.

Aquaporins are not always open in the distal convoluted tubule. Another hormone called **antidiuretic hormone (ADH)** must be present. ADH is produced by the hypothalamus and secreted by the posterior pituitary according to the osmolarity of the blood. If our intake of water has been low, ADH is secreted by the posterior pituitary. Water moves from the distal convoluted tubule and the collecting duct into the blood.

Atrial natriuretic hormone (ANH) is a hormone secreted by the atria of the heart when cardiac cells are stretched due to increased blood volume. ANH inhibits the secretion of renin by the juxtaglomerular apparatus and the secretion of aldosterone by the adrenal glands. Its effect, therefore, is to promote the excretion of sodium ions (Na^+), called *natriuresis.* Normally, salt reabsorption creates an osmotic gradient that causes water to be reabsorbed. Thus, by causing salt excretion, ANH causes water excretion, too. If ANH is present, less water will be reabsorbed, even if ADH is also present.

Reabsorption of Salt and Water from Medullary Portions of the Nephron

The ability of humans to regulate the tonicity of their urine is dependent on the work of the medullary portions of the nephron (loop of the nephron) and the collecting duct.

The Loop of the Nephron A long loop of the nephron, which typically penetrates deep into the renal medulla, is made up of a descending limb and an ascending limb. Salt (NaCl)

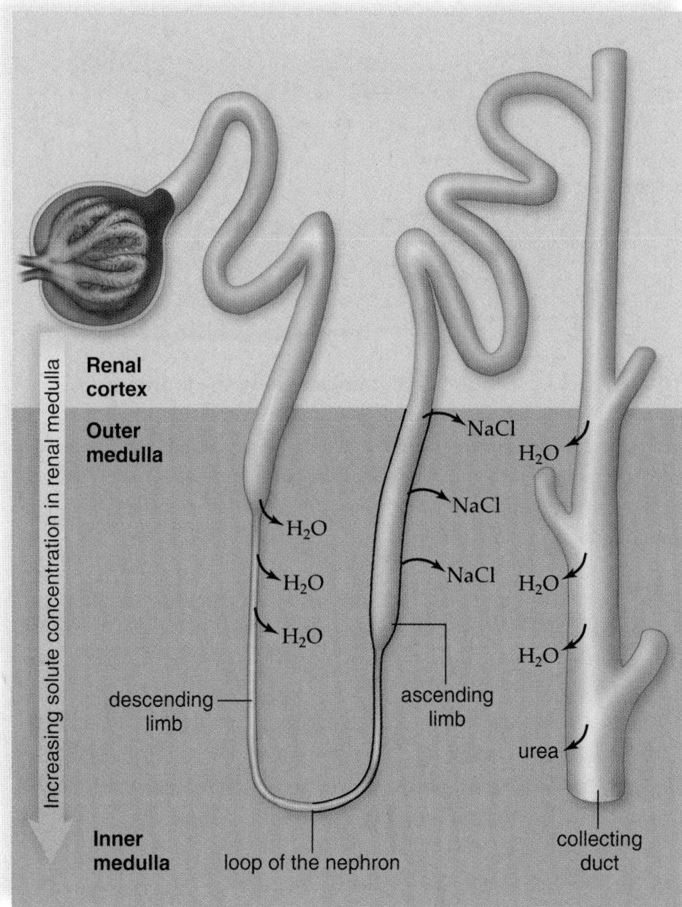

Figure 10.10 **Movement of salt and water within a nephron.** Salt (NaCl) diffuses and is actively transported out of the ascending limb of the loop of the nephron into the renal medulla; also, urea is believed to leak from the collecting duct and to enter the tissues of the renal medulla. This creates a hypertonic environment, which draws water out of the descending limb and the collecting duct. This water is returned to the cardiovascular system. (The thick black outline of the ascending limb means that it is impermeable to water.)

passively diffuses out of the lower portion of the *ascending limb*. Any remaining salt is actively transported from the thick upper portion of the limb into the tissue of the outer medulla (Fig. 10.10). In the end, the concentration of salt is greater in the direction of the inner medulla. Surprisingly, however, the inner medulla has an even higher concentration of solutes than expected. It is believed that urea leaks from the lower portion of the collecting duct, contributing to the high solute concentration of the inner medulla.

Water leaves the *descending limb* along its entire length via a countercurrent mechanism because of the osmotic gradient within the medulla. Although water is reabsorbed as soon as fluid enters the descending limb, the remaining fluid within the limb encounters an increasing osmotic concentration of solute. Therefore, water continues to be reabsorbed, even to the bottom of the descending limb. (The ascending limb does not reabsorb water—it has no aquaporins as indicated by the dark line in Fig. 10.10—its job is to help establish the solute concentration gradient.) Water is returned to the cardiovascular system when it is reabsorbed.

The Collecting Duct Fluid within the collecting duct encounters the same osmotic gradient established by the ascending limb of the nephron. Therefore, water diffuses from the entire length of the collecting duct into the blood if aquaporins are open, as they are if ADH is present.

To understand the action of ADH, consider its name, *antidiuretic hormone*. Diuresis means increased amount of urine, and antidiuresis means decreased amount of urine. When ADH is present, more water is reabsorbed (blood volume and pressure rise) and a decreased amount of urine results. ADH ultimately fine-tunes the tonicity of urine according to the needs of the body. For example, ADH is secreted at night when we are not drinking water, and this explains why the first urine of the day is more concentrated.

Interaction of Renin, Aldosterone, and ANH

If blood does not have the usual water–salt balance, blood volume and blood pressure are affected. Without adequate blood pressure, exchange across capillary walls cannot take place, nor is glomerular filtration possible in the kidneys.

What happens if you have insufficient sodium ions (Na^{+}) in your blood and tissue fluid? This can occur due to prolonged heavy sweating, as in athletes running a marathon. When blood Na^+ concentration falls too low, blood pressure falls and the renin–aldosterone sequence begins. Then the kidneys increase Na^+ reabsorption, conserving as much as possible. Subsequently, the osmolarity of the blood and the blood pressure return to normal.

The marathon runner should not drink too much water too fast. Quickly ingesting a large amount of pure water can dilute the body's remaining Na^+ and disrupt water–salt balance. Sports drinks preferred by athletes contain sodium and water, so both can be replaced simultaneously.

By contrast, think what happens if you eat a big tub of salty popcorn at the movies. When salt (NaCl) is absorbed from the digestive tract, the Na^+ content of the blood increases above normal. This results in increased blood volume. The atria of the heart are stretched by this increased blood volume, and the stretch triggers the release of ANH by the heart. ANH inhibits sodium and water reabsorption by the proximal convoluted tubule and collecting duct. Blood volume then decreases because more sodium and water are excreted in the urine.

Diuretics

Diuretics are chemicals that increase the flow of urine. Drinking alcohol causes diuresis because it inhibits the secretion of ADH. The dehydration that follows is believed to contribute to the symptoms of a hangover. Caffeine is a diuretic because it increases the glomerular filtration rate and decreases the tubular reabsorption of sodium ions (Na^+). Diuretic drugs developed to counteract high blood pressure also decrease the tubular reabsorption of Na^+. A decrease in water reabsorption and a decrease in blood volume and pressure follow.

Acid–Base Balance of Body Fluids

The pH scale, as discussed in Chapter 2, can be used to indicate the basicity (alkalinity) or the acidity of body fluids. A basic solution has a lesser hydrogen ion concentration $[H^+]$ than the neutral pH of 7.0. An acidic solution has a greater $[H^+]$ than neutral pH. The normal pH for body fluids is between 7.35 and 7.45. This is the pH at which our proteins, such as cellular enzymes, function properly. If the blood pH rises above 7.45, a person is said to have **alkalosis,** and if the blood pH decreases below 7.35, a person is said to have **acidosis.** Alkalosis and acidosis are abnormal conditions that may need medical attention.

The foods we eat add basic or acidic substances to the blood, and so does metabolism. For example, cellular respiration adds carbon dioxide that combines with water to form carbonic acid, and fermentation adds lactic acid. The pH of body fluids stays at just about 7.4 via several mechanisms, primarily acid–base buffer systems, the respiratory center, and the kidneys.

MP3 Acid-Base Balance

Acid–Base Buffer Systems

The pH of the blood stays near 7.4 because the blood is buffered. A **buffer** is a chemical or a combination of chemicals that can take up excess hydrogen ions (H^+) or excess hydroxide ions (OH^-). One of the most important buffers in the blood is a combination of carbonic acid (H_2CO_3) and bicarbonate ions (HCO_3^-). When hydrogen ions (H^+) are added to blood, the following reaction occurs:

$$H^+ + HCO_3^- \rightarrow H_2CO_3$$

When hydroxide ions (OH^-) are added to blood, this reaction occurs:

$$OH^- + H_2CO_3 \rightarrow HCO_3^- + H_2O$$

These reactions temporarily prevent any significant change in blood pH. A blood buffer, however, can be overwhelmed unless some more permanent adjustment is made. The next adjustment to keep the pH of the blood constant occurs at pulmonary capillaries.

Respiratory Center

As discussed in Chapter 9, the respiratory center in the medulla oblongata increases the breathing rate if the hydrogen ion concentration of the blood rises. Increasing the breathing rate rids the body of hydrogen ions because the following reaction takes place in pulmonary capillaries:

$$H^+ + HCO_3^- \rightleftharpoons H_2CO_3 \rightleftharpoons H_2O + CO_2$$

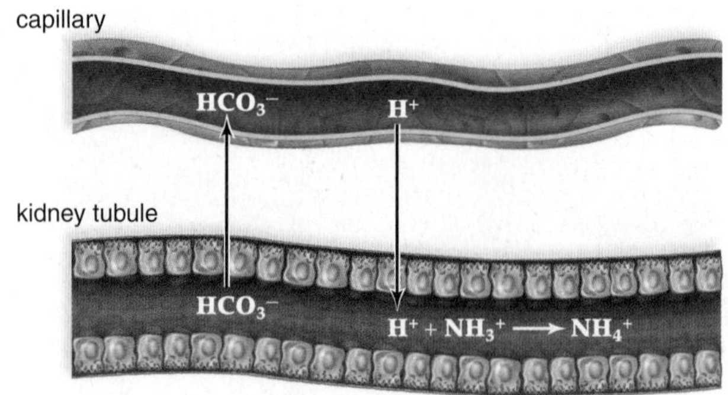

Figure 10.11 Blood pH is maintained by the kidneys. In the kidneys, bicarbonate ions (HCO_3^-) are reabsorbed, and hydrogen ions (H^+) are excreted as needed to maintain the pH of the blood. Excess hydrogen ions are buffered, for example, by ammonia (NH_3), which becomes ammonium (NH_4^+). Ammonia is produced in tubule cells by the deamination of amino acids.

In other words, when carbon dioxide is exhaled, this reaction shifts to the right and the amount of hydrogen ions is reduced.

It is important to have the correct proportion of carbonic acid and bicarbonate ions in the blood. Breathing readjusts this proportion so that this particular acid–base buffer system can continue to absorb H^+ and OH^- as needed.

The Kidneys

As powerful as the acid–base buffer and the respiratory center mechanisms are, only the kidneys can rid the body of a wide range of acidic and basic substances and, otherwise, adjust the pH. The kidneys are slower acting than the other two mechanisms, but they have a more powerful effect on pH. For the sake of simplicity, we can think of the kidneys as reabsorbing bicarbonate ions and excreting hydrogen ions as needed to maintain the normal pH of the blood (Fig. 10.11). If the blood is acidic, hydrogen ions are excreted and bicarbonate ions are reabsorbed. If the blood is basic, hydrogen ions are not excreted and bicarbonate ions are not reabsorbed. The urine is usually acidic, so it follows that an excess of hydrogen ions is usually excreted. Ammonia (NH_3) provides another means of buffering and removing the hydrogen ions in urine:

$$H^+ + NH_3 \rightarrow NH_4^+$$

Ammonia (whose presence is obvious in the diaper pail or kitty litter box) is produced in tubule cells by the deamination of amino acids. Phosphate provides another means of buffering hydrogen ions in urine.

The importance of the kidneys' ultimate control over the pH of the blood cannot be overemphasized. As mentioned, the enzymes of cells cannot continue to function if the internal environment does not have near-normal pH.

Connections and Misconceptions

Is it true that urine will neutralize the sting from a jellyfish?

You might remember the scene from the TV series *Survivor* when a player was stung by a jellyfish. The individual called for someone to come urinate on the sting. However, urine has not been shown to help jellyfish stings in scientific studies. Instead, experts recommend the application of vinegar to a sting to alleviate its effects.

The Kidneys Assist Other Systems

Aside from producing renin, the kidneys assist the endocrine system and also the cardiovascular system by producing erythropoietin (see Fig. 10.11). Erythropoietin is used to stimulate red bone marrow production in patients in renal failure or recovering from chemotherapy. The kidneys assist the skeletal, nervous, and muscular systems by helping to regulate the amount of calcium ions (Ca^{2+}) in the blood. The kidneys convert vitamin D to its active form needed for Ca^{2+} absorption by the digestive tract, and they regulate the excretion of electrolytes, including Ca^{2+}. The kidneys also regulate the sodium ion (Na^+) and potassium ion (K^+) content of the blood. These ions, needed for nerve conduction, are necessary to the contraction of the heart and other muscles in the body.

Connecting the Concepts

For additional information on water–salt balance and maintenance of the pH of the blood, refer to the following discussions.

Section 2.2 explains how buffers regulate pH levels.

Section 9.5 examines the role of the respiratory center in the medulla oblongata of the brain.

Section 15.4 explores the production of aldosterone by the adrenal glands.

Check Your Progress 10.4

1. Detail the differences between dilute and concentrated urine.
2. Describe the action of the three hormones used to influence urine production and discuss how they work together.
3. Summarize how the action of the kidneys regulates body fluid pH and clarify why this is important to homeostasis.

10.5 Kidney Function Disorders

Learning Outcomes

Upon completion of this section, you should be able to

1. List the major diseases of the urinary system and summarize their causes.
2. Describe how hemodialysis can help restore homeostasis of the blood in the event of kidney failure.

Many types of illnesses—especially diabetes, hypertension, and inherited conditions—cause progressive renal disease and renal failure. Infections are also contributory. If the infection is localized in the urethra, it is called **urethritis.** If the infection invades the urinary bladder, it is called **cystitis.** Finally, if the kidneys are affected, the infection is called **pyelonephritis.**

Urinary tract infections, an enlarged prostate gland, pH imbalances, or an intake of too much calcium can lead to kidney stones. Kidney stones are hard granules made of calcium, phosphate, uric acid, and protein. Kidney stones form in the renal pelvis and usually pass unnoticed in the urine flow. If they grow to several centimeters and block the renal pelvis or ureter, a reverse pressure builds up and destroys nephrons. When a large kidney stone passes, strong contractions within a ureter can be excruciatingly painful.

One of the first signs of nephron damage is albumin, white blood cells, or even red blood cells in the urine. As described on page 226, urinalysis can detect urine abnormalities rapidly. If damage is so extensive that more than two-thirds of the nephrons are inoperative, urea and other waste substances accumulate in the blood. This condition is called **uremia.** Although nitrogenous wastes can cause serious damage, the retention of water and salts is of even greater concern. The latter causes edema, fluid accumulation in the body tissues. Imbalance in the ionic composition of body fluids can lead to loss of consciousness and to heart failure.

Connections and Misconceptions

Does cranberry juice really prevent or cure a urinary tract infection?

Research has supported the use of cranberry juice to prevent urinary tract infections. It appears to prevent bacteria that would cause infection from adhering to the surfaces of the urinary tracts. However, cranberry juice has not been shown to be an effective treatment for an already existing urinary tract infection.

Video Cranberries vs. Bacteria

Hemodialysis

Patients with renal failure can undergo **hemodialysis,** using either an artificial kidney machine or continuous ambulatory peritoneal dialysis (CAPD). *Dialysis* is defined as the diffusion of dissolved molecules through a semipermeable natural or synthetic membrane that has pore sizes that allow only small molecules to pass through. In an artificial kidney machine (Fig. 10.12), the patient's blood is passed through a membranous tube that is in contact with a dialysis solution, or **dialysate.** Substances more concentrated in the blood diffuse into the dialysate, and substances more concentrated in the dialysate diffuse into the blood. The dialysate is continuously replaced to maintain favorable concentration gradients. In this way, the artificial kidney can be used either to extract substances from blood, including waste products or toxic chemicals and drugs, or to add substances to blood—for example, bicarbonate ions (HCO_3^-) if the blood is acidic. In the course of a 3- to 6-hour hemodialysis, from 50 to 250 g of urea can be removed from a patient, which greatly exceeds the amount excreted by normal kidneys. Therefore, a patient needs to undergo treatment only about twice a week.

CAPD is so named because the peritoneum is the dialysis membrane. A fresh amount of dialysate is introduced directly into the abdominal cavity from a bag that is temporarily attached to a permanently implanted plastic tube. The dialysate flows into the peritoneal cavity by gravity. Waste and salt molecules pass from the blood vessels in the abdominal wall into the dialysate before the fluid is collected 4 to 8 hours later. The solution is drained into a bag from the abdominal cavity by gravity, and then it is discarded. One advantage of CAPD over an artificial kidney machine is that the individual can go about his or her normal activities during CAPD.

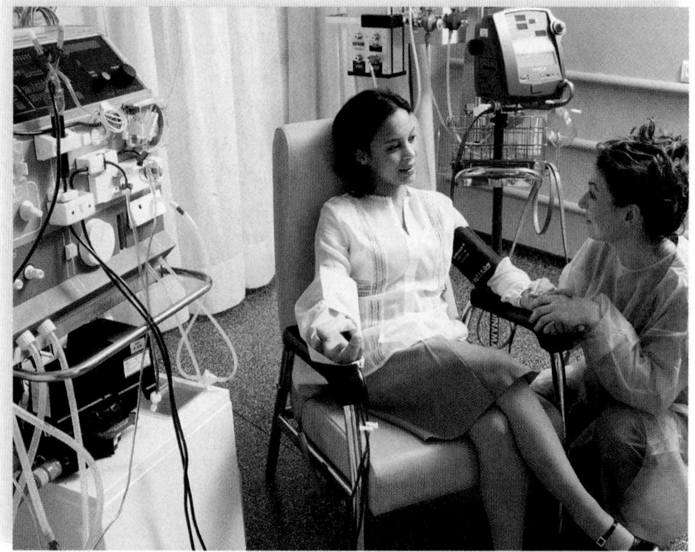

Figure 10.12 Hemodialysis using an artificial kidney machine.
As the patient's blood is pumped through dialysis tubing, it is exposed to a dialysate (dialysis solution). Wastes exit from blood into the solution because of a preestablished concentration gradient. In this way, not only is blood cleansed but its water–salt and acid–base balances can also be adjusted.

Replacing a Kidney

Patients with renal failure sometimes undergo a kidney transplant operation, during which a functioning kidney from a donor is received. As with all organ transplants, there is the possibility of organ rejection. Receiving a kidney from a close relative has the highest chance of success. The current one-year survival rate is 97% if the kidney is received from a relative and 90% if it is received from a nonrelative. In the future, transplantable kidneys may be created in a laboratory. Another option could be to use kidneys from specially bred pigs whose organs would not be antigenic to humans.

Connecting the Concepts

For more information on organ transplants, refer to the following discussions.

Section 5.7 describes the options available for individuals experiencing heart failure.

Section 7.6 examines how the immune system potentially interferes with organ transplants.

Section 19.4 provides an overview of how bone marrow transplants can be used to treat cancer.

Check Your Progress 10.5

1. List and detail a few common causes of renal disease.
2. Describe hemodialysis and relate its function to that of a kidney.
3. Explain why hemodialysis would need to be done frequently in a patient with renal failure.

Biology Matters Science Focus

Lab-Grown Bladders

You're probably familiar with organ transplants done with organs harvested from people who have recently died or even from living donors. Did you know that some organs can now be grown in a lab and used for transplantation?

Bladders grown in a lab have been successfully transplanted into a number of patients. Each patient in one particular study contributed cells from his or her own diseased bladder. Those cells were cultured in a lab and encouraged to form a new bladder by growing them on a collagen form shaped like a bladder. Eventually, the new bladders were attached to each patient's diseased bladder. This alleviated the incontinence problems the patients had experienced. The risk of kidney damage was also decreased in these individuals by lowering the pressure inside their enlarged bladders. The cells for the new bladders were taken from each patient, so there was no risk of rejection.

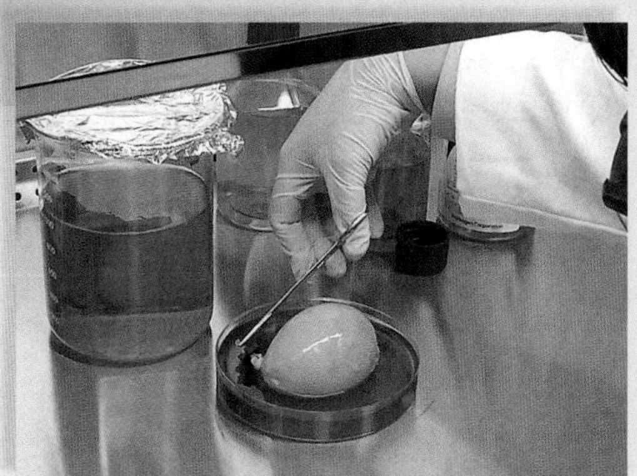

Figure 10C A lab-grown bladder is readied for transplantation.
There is less risk of rejection when a lab-grown bladder is transplanted.

Researchers hope this technique can be used to grow other types of tissues and organs that would be available for transplant. This could be the one potential solution to the limited number of organs available for transplant.

CASE STUDY CONCLUSION

Michael and Jada learned that there are two forms of polycystic kidney disease (PKD) and that both are genetic disorders. The majority of cases (over 90%) belong to a class called autosomal dominant PKD, named after the type of mutation that causes the disease (see Chap. 20). Individuals with this form of PKD typically develop symptoms in their 30s and 40s. The other, rarer form, is called autosomal recessive PKD, and it is caused by a defect in a gene called *PKHD1*. *PKHD1* contains the instructions for the manufacture of a protein called fibrocystin, an important molecule in ensuring that the kidneys function correctly. Unfortunately for Michael and Jada, the genetic test indicated that they were both carriers for a defective copy of *PKHD1*, and that Aiesha had inherited a defective copy from both parents. Without functioning fibrocystin protein, Aiesha's kidneys—and also possibly her liver—would eventually fail as the nephrons inside her kidneys struggled to perform their function and fluid accumulated inside the cysts. The only alternative for Aiesha would be a kidney transplant. Luckily for Aiesha, kidney transplants have a 90%+ success rate, so if a compatible donor could be found, the chances of Aiesha having a healthy childhood were very good.

Media Study Tools

www.mhhe.com/maderhuman12e

Enhance your study of this chapter with study tools and practice tests. Also ask your instructor about the resources available through ConnectPlus, including the media-rich eBook, interactive learning tools, and animations.

Summarizing the Concepts

10.1 The Urinary System

Organs of the Urinary System

Only the urinary system contains urine.

- Kidneys produce urine.
- Ureters take urine to the bladder.
- The urinary bladder stores urine.
- The urethra releases urine to the outside.

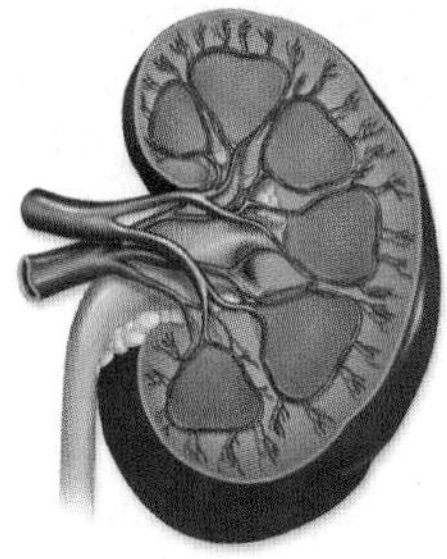

Functions of the Urinary System

The functions of the urinay system include:

- Excreting nitrogenous wastes, including urea, uric acid, and creatinine.
- Maintaining the normal water–salt balance of the blood.
- Maintaining the acid–base balance of the blood.
- Assisting the endocrine system in hormone secretion by secreting erythropoietin and releasing renin.

10.2 Kidney Structure

- Macroscopic structures are the renal cortex, renal medulla, and renal pelvis.
- Microscopic structures are the nephrons.

Anatomy of a Nephron

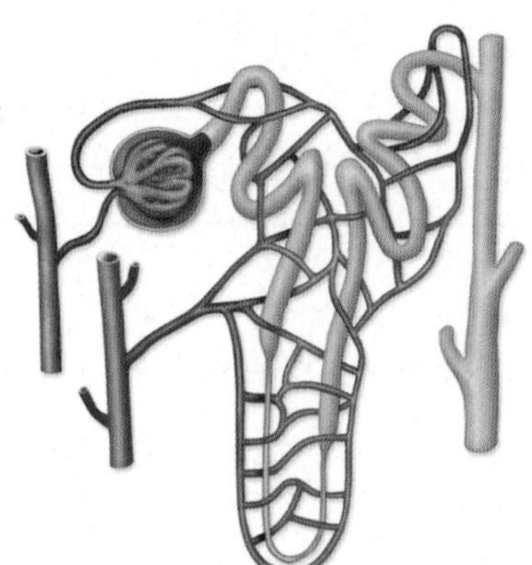

- Each nephron has its own blood supply. The afferent arteriole divides to become the glomerulus. The efferent arteriole branches into the peritubular capillary network.
- Each region of the nephron is anatomically suited to its task in urine formation.

10.3 Urine Formation

Urine is composed primarily of nitrogenous waste products and salts in water. The steps in urine formation include the following:.

Glomerular Filtration

Water, salts, nutrients, and wastes move from the glomerulus to the inside of the glomerular capsule.

Tubular Reabsorption

Nutrients and salt molecules are reabsorbed from the convoluted tubules into the peritubular capillary network; water follows.

Tubular Secretion

Certain molecules are actively secreted from the peritubular capillary network into the convoluted tubules.

10.4 Kidneys and Homeostasis

- The kidneys maintain the water–salt balance of blood.
- The kidneys also keep blood pH within normal limits.
- The urinary system works with the other systems of the body to maintain homeostasis.

Water–Salt Balance

- In the renal cortex, most of the reabsorption of salts and water occurs in the proximal convoluted tubule. Aldosterone controls the reabsorption of sodium, and antidiuretic hormone (ADH) controls the reabsorption of water in the distal convoluted tubule. Atrial natriuretic hormone (ANH) acts contrary to aldosterone.
- In the renal medulla, the ascending limb of the loop of the nephron establishes a solute gradient that increases toward the inner medulla. The solute gradient draws water from the descending limb of the loop of the nephron and from the collecting duct. When ADH is present, more water is reabsorbed from the collecting duct and a decreased amount of urine results.

Acid–Base Balance of Body Fluids

- The kidneys maintain the acid–base balance of blood (blood pH).
- They reabsorb HCO_3^- and excrete H^+ as needed to maintain the pH at about 7.4. Ammonia buffers H^+ in the urine.

10.5 Kidney Function Disorders

Various types of problems, including diabetes, kidney stones, and infections, can lead to renal failure, which necessitates either undergoing hemodialysis—by using a kidney machine or CAPD—or receiving a kidney transplant.

Understanding Key Terms

acidosis 232
aldosterone 230
alkalosis 232
antidiuretic hormone (ADH) 230
aquaporin 228
atrial natriuretic hormone (ANH) 230
buffer 232
collecting duct 224
creatinine 220
cystitis 233
diabetes mellitus 227
dialysate 234
distal convoluted tubule 224
diuretic 231
erythropoietin (EPO) 220
excretion 218
glomerular capsule 223
glomerular filtrate 227
glomerular filtration 225
glomerulus 223
gout 220
hemodialysis 234
juxtaglomerular apparatus 230
kidney 218
loop of the nephron 224
micturition 219
nephron 220
peritubular capillary network 223
proximal convoluted tubule 224
pyelonephritis 233
renal artery 218
renal cortex 220
renal medulla 220
renal pelvis 220
renal vein 218
renin 230
tubular reabsorption 227
tubular secretion 228
urea 219
uremia 233
ureter 218
urethra 219
urethritis 233
uric acid 220
urinary bladder 219

Match the key terms to these definitions.

a. ________ Drug used to counteract hypertension by causing the excretion of water.

b. ________ Removal of metabolic wastes from the body.

c. ________ Tubular structure that receives urine from the bladder and carries it to the outside of the body.

d. ________ Hollow chamber in the kidney that lies inside the renal medulla and receives freshly prepared urine from the collecting ducts.

e. ________ Filtered portion of blood contained within the glomerular capsule.

Testing Your Knowledge of the Concepts

1. Describe the path of urine and the structure and function of each organ mentioned. (pages 218–220)
2. In what ways do the four functions of the urinary system contribute to homeostasis? (pages 219–220)
3. Describe the macroscopic structure of the kidney. How does the structure aid in the function? (page 220)
4. Trace the path of blood into and out of the kidney. Why is the kidney so highly vascularized? (pages 220–221)
5. Name the parts of a nephron, and describe the structure of each part. (pages 223–224)
6. Why would it be proper to associate glucose with the first two processes involved in urine formation, glomerular filtration and tubular reabsorption, but not the third process, tubular secretion? (pages 225–227)
7. Explain how hypertonic urine can be formed, detailing where and how salt and water move, and the influence of hormones on the process. (pages 227–231)
8. Describe three ways that the body maintains the acid–base balance of body fluids. (pages 232–233)
9. Explain how an artificial kidney machine and continuous ambulatory peritoneal dialysis work to cleanse the blood. (pages 234)
10. How do the kidneys assist other body systems? (pages 233–234)
11. Which of these is found in the renal medulla?
    a. loop of the nephron
    b. collecting ducts
    c. peritubular capillaries
    d. All of these are correct.
12. Which of these functions of the kidneys are mismatched?
    a. excretes metabolic wastes—rids the body of urea
    b. maintains the water–salt balance—helps regulate blood pressure
    c. maintains the acid–base balance—rids the body of uric acid
    d. secretes hormones—secretes erythropoietin
    e. All of these are correct.
13. Which of the following is not correct?
    a. Uric acid is produced from the breakdown of amino acids.
    b. Creatinine is produced from breakdown reactions in the muscles.
    c. Urea is the primary nitrogenous waste of humans.
    d. Ammonia results from the deamination of amino acids.
14. When tracing the path of filtrate, the loop of the nephron follows which structure?
    a. collecting duct
    b. distal convoluted tubule
    c. proximal convoluted tubule
    d. glomerulus
    e. renal pelvis
15. When tracing the path of blood, the blood vessel that follows the renal artery is the
    a. peritubular capillary.
    b. efferent arteriole.
    c. afferent arteriole.
    d. renal vein.
    e. glomerulus.
16. The function of the descending limb of the loop of the nephron in the process of urine formation is
    a. reabsorption of water.
    b. production of filtrate.
    c. reabsorption of solutes.
    d. secretion of solutes.
17. Which of the following materials would not normally be filtered from the blood at the glomerulus?
    a. water
    b. urea
    c. protein
    d. glucose
    e. sodium ions
18. Which of the following materials would not be maximally reabsorbed from the filtrate?
    a. water
    b. glucose
    c. sodium ions
    d. urea
    e. amino acids
19. Reabsorption of the glomerular filtrate occurs primarily at the
    a. proximal convoluted tubule.
    b. distal convoluted tubule.
    c. loop of the nephron.
    d. collecting duct.
20. A countercurrent mechanism draws water from the
    a. proximal convoluted tubule.
    b. descending limb of the loop of the nephron.
    c. distal convoluted tubule.
    d. collecting duct.
    e. Both b and d are correct.
21. Sodium is actively extruded from which part of the nephron?
    a. descending portion of the proximal convoluted tubule
    b. ascending portion of the loop of the nephron
    c. ascending portion of the distal convoluted tubule
    d. descending portion of the collecting duct
22. Excretion of hypertonic urine in humans is best associated with the
    a. glomerular capsule and the tubules.
    b. proximal convoluted tubule only.
    c. loop of the nephron and collecting duct.
    d. distal convoluted tubule and peritubular capillary.
23. Which of these hormones is most likely to directly cause a drop in blood pressure?
    a. aldosterone
    b. antidiuretic hormone (ADH)
    c. erythropoietin
    d. atrial natriuretic hormone (ANH)
24. The presence of ADH (antidiuretic hormone) causes an individual to excrete
    a. sugars.
    b. less water.
    c. more water.
    d. Both a and c are correct.
25. To lower blood acidity,
    a. hydrogen ions are excreted and bicarbonate ions are reabsorbed.
    b. hydrogen ions are reabsorbed and bicarbonate ions are excreted.
    c. hydrogen ions and bicarbonate ions are reabsorbed.
    d. hydrogen ions and bicarbonate ions are excreted.
    e. urea, uric acid, and ammonia are excreted.

26. The function of erythropoietin is
    a. reabsorption of sodium ions.
    b. excretion of potassium ions.
    c. reabsorption of water.
    d. to stimulate red blood cell production.
    e. to increase blood pressure.

In questions 27–30, match the function of the urinary system to the human organ system in the key.

Key:

a. muscular system
b. nervous system
c. endocrine system
d. cardiovascular system
e. respiratory system
f. digestive system
g. reproductive system

27. Liver synthesizes urea.
28. Smooth muscular contraction assists voiding of urine.
29. ADH, aldosterone, and atrial natriuretic hormone regulate reabsorption of sodium ions (Na^{+}) by kidneys.
30. Blood vessels deliver waste to be excreted.
31. Label this diagram of a nephron.

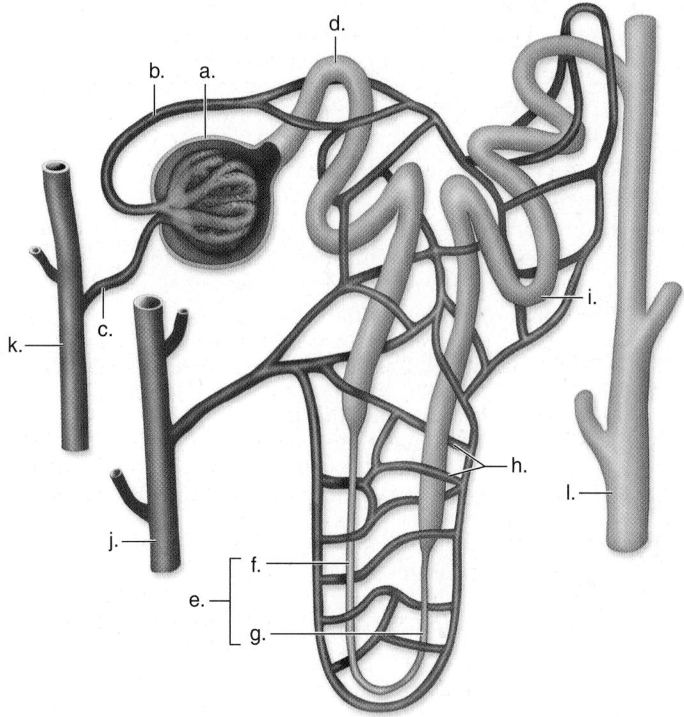

Thinking Critically About the Concepts

A form of diabetes that may be unfamiliar to most people is diabetes insipidus. When the word "diabetes" is used, everyone tends to think of the type related to insulin. That form of diabetes is diabetes mellitus, which can be type 1 or type 2. In diabetes insipidus, the kidneys are unable to perform their function of water and salt homeostasis due to a lack of ADH in the body. In both cases of diabetes, kidney function is affected.

1. Why would frequent urination be a symptom of both diabetes mellitus and diabetes insipidus?
2. Why would increasing the production of red blood cells in people in renal failure alleviate their symptom of fatigue?
3. A side effect of using medications to calm an overactive bladder is constipation. Look back to Chapter 8 and identify the type of muscle found throughout the digestive system. Speculate as to why constipation might occur after the use of Detrol LA.
4. Men who have prostate cancer often have some or all of the prostate removed. The surgeon tries very hard not to damage the sphincter muscles associated with the bladder. What problem might a man experience if the bladder's sphincter muscles were damaged during the surgery?
5. Recall the anatomical position of the kidneys, and consider the ways they might be damaged in sports. Describe special equipment designed specifically to protect the kidneys.

CHAPTER 11

Skeletal System

CASE STUDY KNEE REPLACEMENT

Jackie was an outstanding athlete in high school, and even now, in her early 50s, she tried to stay in shape. But during her customary 3-mile jogs, she was having an increasingly hard time ignoring the pain in her left knee. She had torn some ligaments in her knee playing intramural and intermural volleyball in college, and it had never quite felt the same. In her 40s, she was able to control the pain by taking over-the-counter medications; but two years ago she had had arthroscopic surgery to remove some torn cartilage and calcium deposits. Now that the pain was getting worse than before, she knew her best option might be a total knee replacement.

Although it sounds drastic, replacing old, arthritic joints with new artificial ones is becoming increasingly routine. About 500,000 artificial knees were installed in U.S. patients in 2006, which represents a 65% increase from 2000. During the procedure, a surgeon removes bone from the bottom of the femur and the top of the tibia and replaces each with caps made of metal or ceramic, held in place with bone cement. A plastic plate is installed to allow the femur and tibia to move smoothly against each other, and a smaller plate is attached to the kneecap (patella) so that it can function properly.

As you read through the chapter, think about the following questions.

1. What is the role of cartilage in the knee joint?
2. What specific portions of these long bones are being removed during knee replacement?
3. Why does Jackie's physical condition make her an ideal candidate for knee replacement?

CHAPTER CONCEPTS

11.1 Overview of the Skeletal System
Bones are the organs of the skeletal system. The tissues of the system are compact and spongy bone, various types of cartilage, and fibrous connective tissue in the ligaments that hold bones together.

11.2 Bone Growth, Remodeling, and Repair
Bone is a living tissue that grows, remodels, and repairs itself. In all of these processes, some bone cells break down bone and some repair bone.

11.3 Bones of the Axial Skeleton
The axial skeleton lies in the midline of the body and consists of the skull, the hyoid bone, the vertebral column, and the rib cage.

11.4 Bones of the Appendicular Skeleton
The appendicular skeleton consists of the bones of the pectoral girdle, upper limbs, pelvic girdle, and lower limbs.

11.5 Articulations
Joints are classified according to their degree of movement. Synovial joints are freely movable.

BEFORE YOU BEGIN

Before beginning this chapter, take a few moments to review the following discussions.

Section 4.1 What is the role of connective tissue in the body?

Section 4.9 How does the skeletal system contribute to homeostasis?

11.1 Overview of Skeletal System

Learning Outcomes

Upon completion of this section, you should be able to

1. State the functions of the skeletal system.
2. Describe the structure of a long bone and list the types of tissues it contains.
3. List the three types of cartilage found in the body and provide a function for each.

Connections and Misconceptions

Does everyone have 206 bones?

A newborn has nearly 300 bones, some of which fuse together as the child grows. The adult human skeleton has approximately 206 bones, but the number varies between individuals. Some people have extra bones, called Wormian bones, that help to fuse skull bones together. Others may have additional small bones in the ankles and feet.

The skeletal system consists of two types of connective tissue: bone and the cartilage found at **joints.** In addition, **ligaments** formed of fibrous connective tissue join the bones.

Functions of the Skeleton

The skeleton does more than merely provide a frame for the body. For example, consider the following functions:

The skeleton supports the body. The bones of the legs support the entire body when we are standing, and bones of the pelvic girdle support the abdominal cavity.

The skeleton protects soft body parts. The bones of the skull protect the brain; the rib cage protects the heart and lungs; and the vertebrae protect the spinal cord, which makes nervous connections to all the muscles of the limbs.

The skeleton produces blood cells. All bones in the fetus have red bone marrow that produces blood cells. In the adult, only certain bones produce blood cells.

The skeleton stores minerals and fat. All bones have a matrix that contains calcium phosphate, a source of calcium ions and phosphate ions in the blood. Fat is stored in yellow bone marrow.

The skeleton, along with the muscles, permits flexible body movement. Whereas articulations (joints) occur between all the bones, we associate body movement in particular with the bones of limbs.

Anatomy of a Long Bone

The bones of the body vary greatly in size and shape. To better understand the anatomy of a bone, we will use a *long bone* (Fig. 11.1), a type of bone common in the arms and legs. The shaft, or main portion of the bone, is called the diaphysis. The diaphysis has a large **medullary cavity,** whose walls are composed of compact bone. The medullary cavity is lined with a thin, vascular membrane (the endosteum) and is filled with yellow bone marrow that stores fat.

The expanded region at the end of a long bone is called an epiphysis (pl., epiphyses). The epiphyses are composed largely of spongy bone that contains red bone marrow, where blood cells are made. The epiphyses are coated with a thin layer of hyaline cartilage, which is called **articular cartilage** because it occurs at a joint.

Except for the articular cartilage on the bone's ends, a long bone is completely covered by a layer of fibrous connective tissue called the **periosteum.** This covering contains blood vessels, lymphatic vessels, and nerves. Note in Figure 11.1 how a blood vessel penetrates the periosteum and enters the bone. Branches of the blood vessel are found throughout the medullary cavity. Other branches can be found in hollow cylinders called central canals within the bone tissue. The periosteum is continuous with ligaments and tendons connected to a bone.

MP3 Bone Structure

Bone

Compact bone is highly organized and composed of tubular units called osteons. In a cross section of an osteon, bone cells called **osteocytes** lie in lacunae, tiny chambers arranged in concentric circles around a central canal (see Fig. 11.1). Matrix fills the space between the rows of lacunae. Tiny canals called canaliculi (sing., canaliculus) run through the matrix. These canaliculi connect the lacunae with one another and with the central canal. The cells stay in contact by strands of cytoplasm that extend into the canaliculi. Osteocytes nearest the center of an osteon exchange nutrients and wastes with the blood vessels in the central canal. These cells then pass on nutrients and collect wastes from the other cells via gap junctions (see Fig. 3.16).

Compared with compact bone, **spongy bone** has an unorganized appearance (see Fig. 11.1). It contains numerous thin plates (called trabeculae) separated by unequal spaces. Although this makes spongy bone lighter than compact bone, spongy bone is still designed for strength. Just as braces are used for support in buildings, the trabeculae follow lines of stress. The spaces of spongy bone are often filled with **red bone marrow,** a specialized tissue that produces all types of blood cells. The osteocytes of spongy bone are irregularly placed within the trabeculae. Canaliculi bring them nutrients from the red bone marrow.

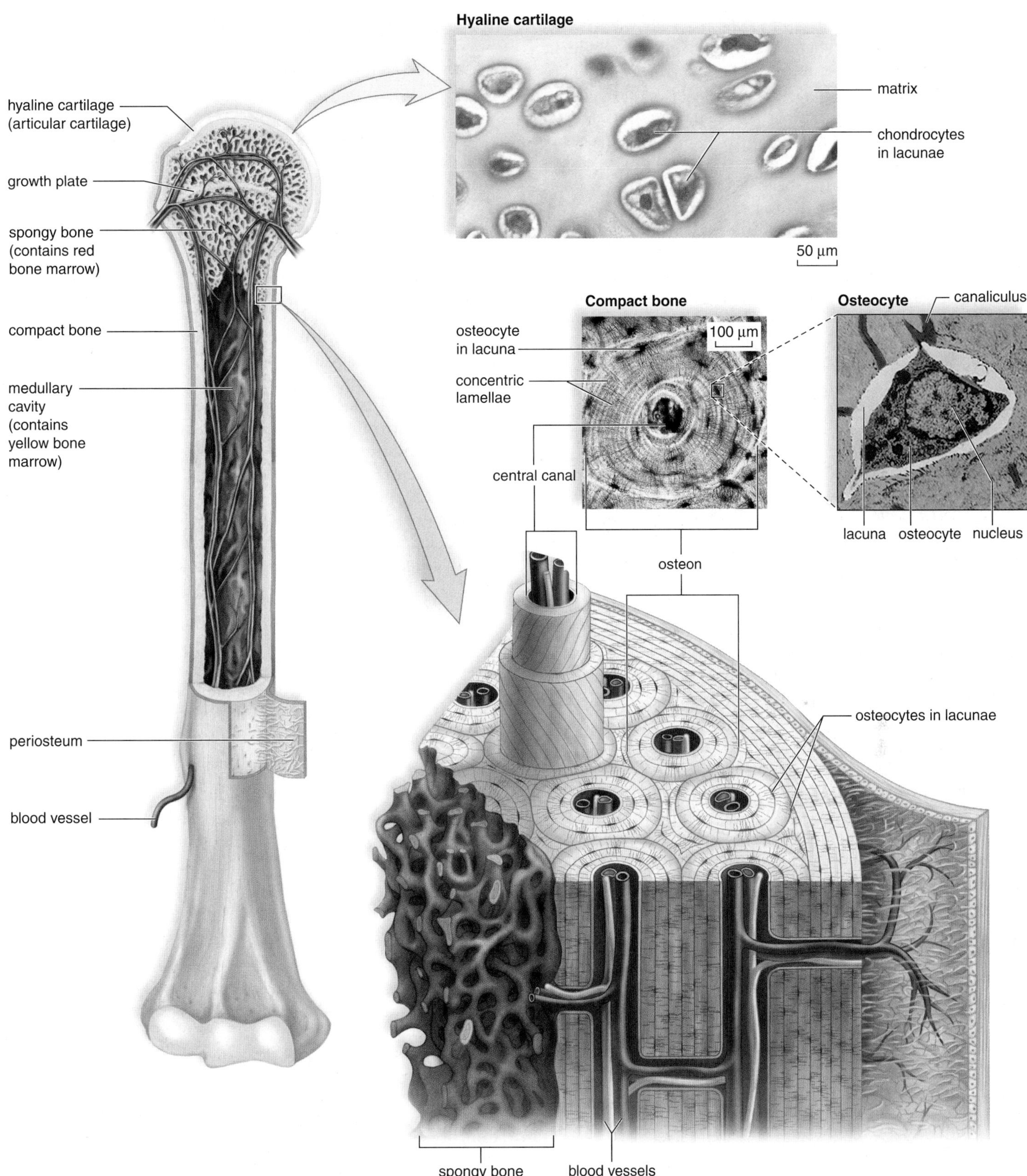

Figure 11.1 **The anatomy of a long bone.**
Left: A long bone is formed of an outer layer of compact bone, shown in the enlargement and micrograph (*right*). Spongy bone, which lies beneath compact bone, may contain red bone marrow. The central shaft of a long bone contains yellow marrow, a form of stored fat. Periosteum, a fibrous membrane, encases the bone except at its ends. Hyaline cartilage (see micrograph) covers the ends of bones.

Cartilage

Cartilage is not as strong as bone, but it is more flexible. Its matrix is gel-like and contains many collagenous and elastic fibers. The cells, called **chondrocytes,** lie within lacunae that are irregularly grouped. Cartilage has no nerves, making it well suited for padding joints where the stresses of movement are intense. Cartilage also has no blood vessels and relies on neighboring tissues for nutrient and waste exchange. This makes it slow to heal.

The three types of cartilage differ according to the type and arrangement of fibers in the matrix. *Hyaline cartilage* is firm and somewhat flexible. The matrix appears uniform and glassy, but actually it contains a generous supply of collagen fibers. Hyaline cartilage is found at the ends of long bones, in the nose, at the ends of the ribs, and in the larynx and trachea.

Fibrocartilage is stronger than hyaline cartilage because the matrix contains wide rows of thick, collagenous fibers. Fibrocartilage is able to withstand both tension and pressure and is found where support is of prime importance—in the disks located between the vertebrae and also in the cartilage of the knee.

Elastic cartilage is more flexible than hyaline cartilage because the matrix contains mostly elastin fibers. This type of cartilage is found in the ear flaps and the epiglottis.

Fibrous Connective Tissue

Fibrous connective tissue contains rows of cells called fibroblasts separated by bundles of collagenous fibers. This tissue makes up ligaments and tendons. Ligaments connect bone to bone. Tendons connect muscle to bone at a joint (also called an articulation).

Connecting the Concepts

For a better understanding of the types of connective tissue presented in this section, refer to the following discussions.

Section 4.2 examines the general structure and function of the connective tissues of the body.

Section 6.1 provides an overview of the blood cells formed in the bone marrow.

Section 12.1 provides additional information on the function of tendons in the body.

Check Your Progress 11.1

1. Describe the typical anatomy of compact and spongy bone.
2. List the three types of cartilage and detail their structure and where each can be found in the body.
3. Predict what might happen to homeostasis if the skeletal system did not exist. Give a few examples.

11.2 Bone Growth, Remodeling, and Repair

Learning Outcomes

Upon completion of this section, you should be able to

1. Summarize the process of ossification and list the types of cells involved.
2. Describe the process of bone remodeling.
3. Explain the steps in the repair of bone.

The importance of the skeleton to the human form is evident by its early appearance during development. The skeleton starts forming at about six weeks, when the embryo is only about 12 mm (0.5 in.) long. Most bones grow in length and width through adolescence, but some continue enlarging until about age 25. In a sense, bones can grow throughout a lifetime because they are able to respond to stress by changing size, shape, and strength. This process is called remodeling. If a bone fractures, it can heal by bone repair.

Bones are living tissues, as shown by their ability to grow, remodel, and undergo repair. Several different types of cells are involved in bone growth, remodeling, and repair:

Osteoblasts are bone-forming cells. They secrete the organic matrix of bone and promote the deposition of calcium salts into the matrix.

Osteocytes are mature bone cells derived from osteoblasts. They maintain the structure of bone.

Osteoclasts are bone-absorbing cells. They break down bone and assist in returning calcium and phosphate to the blood.

Throughout life, osteoclasts are removing the matrix of bone and osteoblasts are building it up. When osteoblasts are surrounded by calcified matrix, they become the osteocytes within lacunae.

Bone Development and Growth

The term **ossification** refers to the formation of bone. The bones of the skeleton form during embryonic development in two distinctive ways: intramembranous ossification and endochondral ossification (Fig. 11.2).

Intramembranous Ossification

Flat bones, such as the bones of the skull, are examples of intramembranous bones. In **intramembranous ossification,** bones develop between sheets of fibrous connective tissue. Here, cells derived from connective tissue cells become osteoblasts located in ossification centers. The osteoblasts secrete the organic matrix of bone. This matrix consists of mucopolysaccharides and collagen fibrils. Calcification occurs when calcium salts are added to the organic matrix.

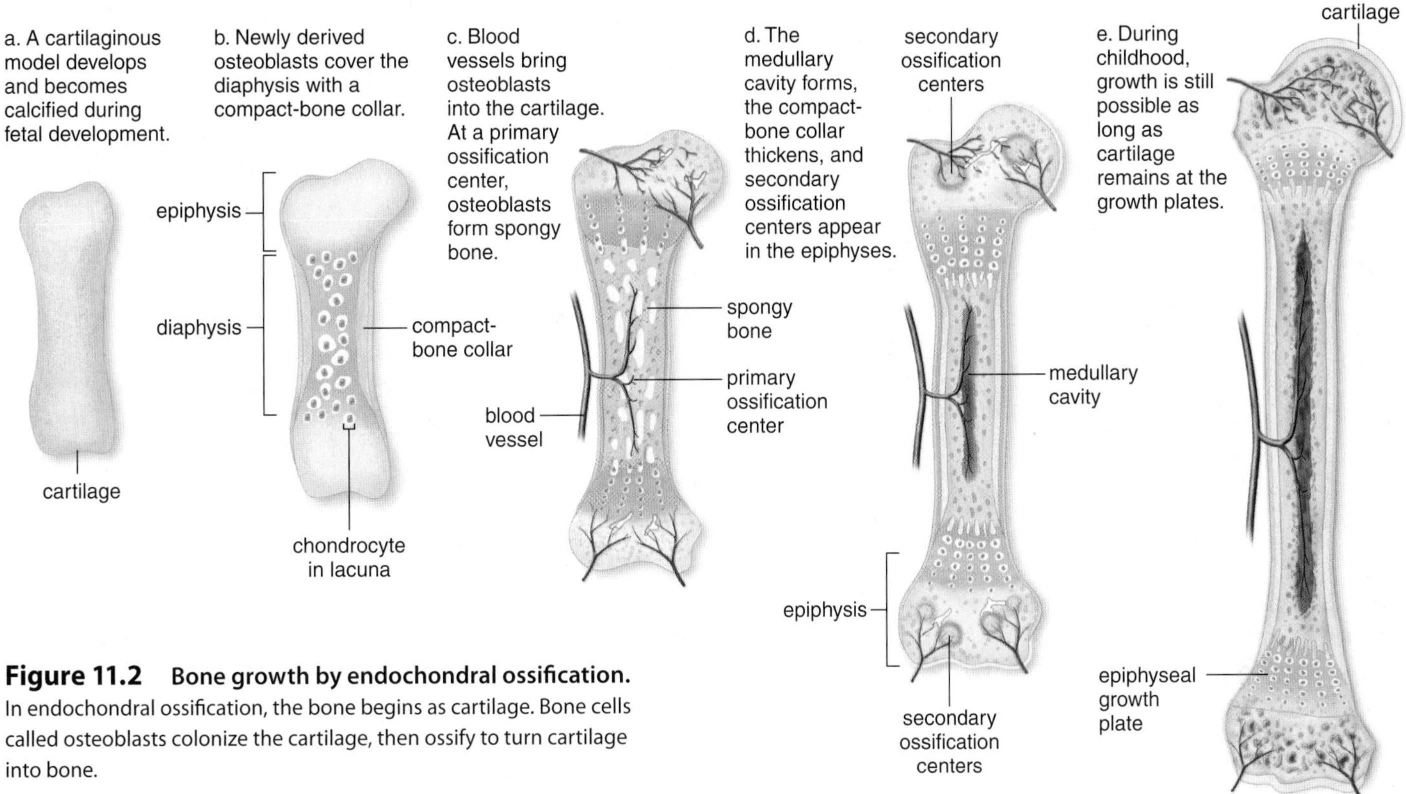
a. A cartilaginous model develops and becomes calcified during fetal development.
b. Newly derived osteoblasts cover the diaphysis with a compact-bone collar.
c. Blood vessels bring osteoblasts into the cartilage. At a primary ossification center, osteoblasts form spongy bone.
d. The medullary cavity forms, the compact-bone collar thickens, and secondary ossification centers appear in the epiphyses.
secondary ossification centers
e. During childhood, growth is still possible as long as cartilage remains at the growth plates.
articular cartilage
epiphysis
diaphysis
compact-bone collar
cartilage
chondrocyte in lacuna
blood vessel
spongy bone
primary ossification center
medullary cavity
epiphysis
secondary ossification centers
epiphyseal growth plate

Figure 11.2 **Bone growth by endochondral ossification.** In endochondral ossification, the bone begins as cartilage. Bone cells called osteoblasts colonize the cartilage, then ossify to turn cartilage into bone.

The osteoblasts promote calcification, or ossification, of the matrix. Ossification results in the trabeculae of spongy bone. Spongy bone remains inside a flat bone. The spongy bone of flat bones, such as those of the skull and clavicles (collarbones), contains red bone marrow.

A periosteum forms outside the spongy bone. Osteoblasts derived from the periosteum carry out further ossification. Trabeculae form and fuse to become compact bone. The compact bone forms a bone collar that surrounds the spongy bone on the inside.

Endochondral Ossification

Most of the bones of the human skeleton are formed by **endochondral ossification.** During endochondral ossification, bone replaces the cartilaginous models of the bones. Gradually, the cartilage is replaced by the calcified bone matrix that makes these bones capable of bearing weight.

Animation Bone Growth in Width

Inside, bone formation spreads from the center to the ends, and this accounts for the term used for this type of ossification. (*Endochondral* literally means "within cartilage.") The long bones, such as the tibia, provide examples of endochondral ossification (see Fig. 11.2).

1. *The cartilage model.* In the embryo, chondrocytes lay down hyaline cartilage, which is shaped like the future bones. Therefore, they are called cartilage models of the future bones. As the cartilage models calcify, the chondrocytes die off.
2. *The bone collar.* Osteoblasts are derived from the newly formed periosteum. Osteoblasts secrete the organic bone matrix, and the matrix undergoes calcification. The result is a bone collar, which covers the diaphysis (see Fig. 11.2). The bone collar is composed of compact bone. In time, the bone collar thickens.
3. *The primary ossification center.* Blood vessels bring osteoblasts to the interior, and they begin to lay down spongy bone. This region is called a primary ossification center because it is the first center for bone formation.
4. *The medullary cavity and secondary ossification sites.* The spongy bone of the diaphysis is absorbed by osteoclasts, and the cavity created becomes the medullary cavity. Shortly after birth, secondary ossification centers form in the epiphyses. Spongy bone persists in the epiphyses, and it persists in the red bone marrow for quite some time. Cartilage is present at two locations: the epiphyseal (growth) plate and articular cartilage, which covers the ends of long bones.
5. *The epiphyseal (growth) plate.* A band of cartilage called a **growth plate** remains between the primary ossification center and each secondary center. The limbs keep increasing in length as long as growth plates are still present.

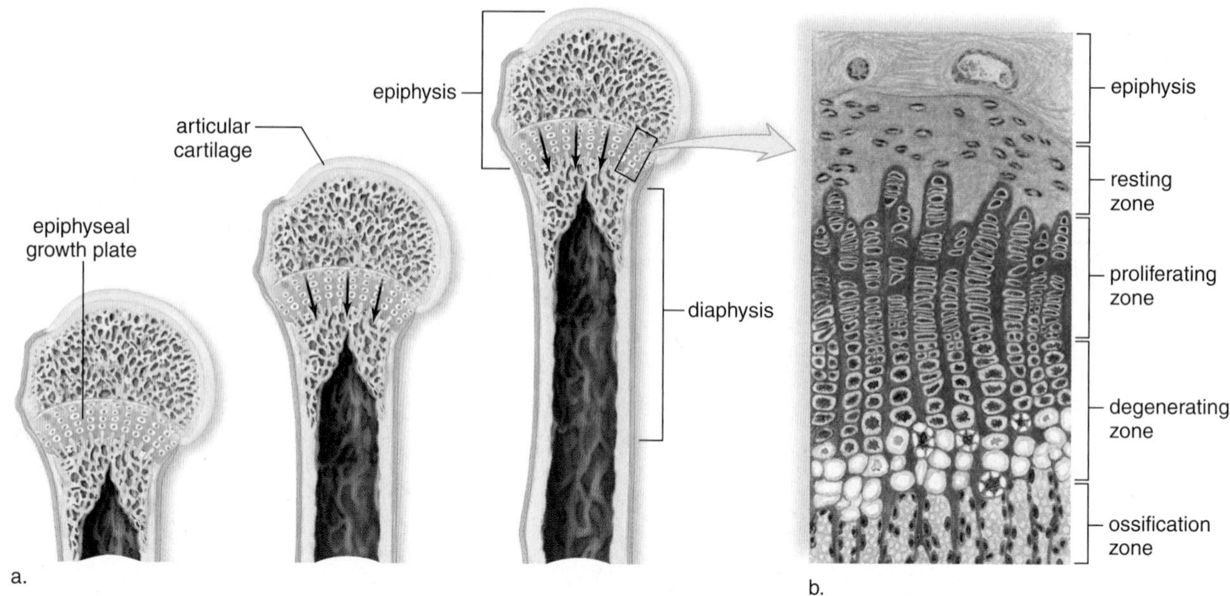

Figure 11.3 Increasing bone length.
a. Length of a bone increases when cartilage is replaced by bone at the growth plate. **b.** Chondrocytes produce new cartilage in the proliferating zone, and cartilage becomes bone in the ossification zone closest to the diaphysis. Arrows show the direction of ossification.

Figure 11.3 shows that the epiphyseal plate contains four layers. The layer nearest the epiphysis is the resting zone, where cartilage remains. The next layer is the proliferating zone, in which chondrocytes are producing new cartilage cells. In the third layer, the degenerating zone, the cartilage cells are dying off; and in the fourth layer, the ossification zone, bone is forming. Bone formation here causes the length of the bone to increase. The inside layer of articular cartilage also undergoes ossification in the manner described.

The diameter of a bone enlarges as a bone lengthens. Osteoblasts derived from the periosteum are active in new bone deposition as osteoclasts enlarge the medullary cavity from inside.

Final Size of the Bones When the epiphyseal plates close, bone lengthening can no longer occur. The epiphyseal plates in the arms and legs of women typically close at about age 16 to 18, and they do not close in men until about age 20. Portions of other types of bones may continue to grow until age 25. Hormones are chemical messengers secreted by the endocrine glands and distributed about the body by the bloodstream. Hormones control the activity of the epiphyseal plate, as is discussed next.

Hormones Affect Bone Growth

The importance of bone growth is signified by the involvement of several hormones in bone growth. A hormone, a chemical messenger, is produced by one part of the body and acts on a different part of the body.

Vitamin D is formed in the skin when it is exposed to sunlight, but it can also be consumed in the diet. Milk, in particular, is often fortified with vitamin D. In the kidneys, vitamin D is converted to a hormone that acts on the intestinal tract. The chief function of vitamin D is intestinal absorption of calcium. In the absence of vitamin D, children can develop rickets, a condition marked by bone deformities including bowed long bones (see Fig. 8.15*a*).

Growth hormone (GH) directly stimulates growth of the epiphyseal plate, as well as bone growth in general. However, growth hormone will be somewhat ineffective if the metabolic activity of cells is not promoted. Thyroid hormone, in particular, promotes the metabolic activity of cells. Too little growth hormone in childhood results in dwarfism. Too much growth hormone during childhood (prior to epiphyseal fusion) can produce excessive growth and even gigantism. Acromegaly results from excess GH in adults following epiphyseal fusion. This condition produces excessive growth of bones in the hands and face (see Figs. 15.7 and 15.8).

Adolescents usually experience a dramatic increase in height, called the growth spurt, due to an increased level of sex hormones. These hormones apparently stimulate osteoblast activity. Rapid growth causes epiphyseal plates to become "paved over" by the faster-growing bone tissue within one or two years of the onset of puberty.

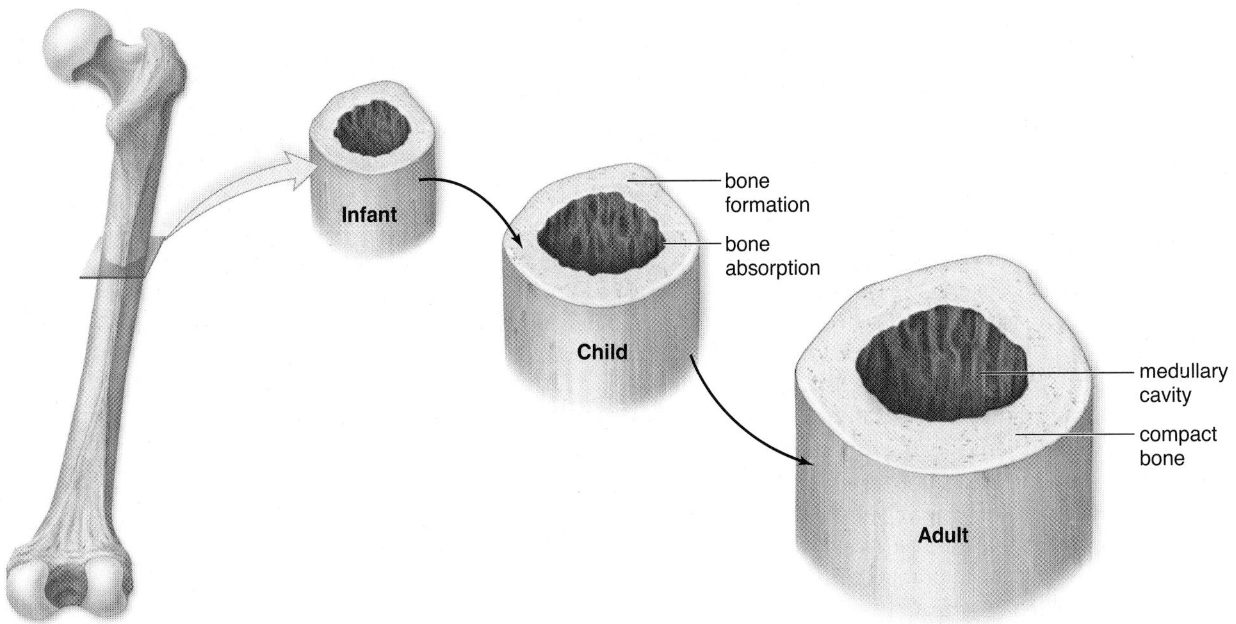

Figure 11.4 **Bone remodeling.**
Diameter of a bone increases as bone absorption occurs inside the shaft and is matched by bone formation outside the shaft.

Bone Remodeling and Its Role in Homeostasis

Bone is constantly being broken down by osteoclasts and re-formed by osteoblasts in the adult. As much as 18% of bone is recycled each year. This process of bone renewal, often called **bone remodeling,** normally keeps bones strong (Fig. 11.4). In Paget's disease, new bone is generated at a faster-than-normal rate. This rapid remodeling produces bone that's softer and weaker than normal bone and can cause bone pain, deformities, and fractures.

Bone recycling allows the body to regulate the amount of calcium in the blood. To illustrate that the blood calcium level is critical, recall that calcium is required for blood to clot. Also, if the blood calcium concentration is too high, neurons and muscle cells no longer function. If calcium falls too low, nerve and muscle cells become so excited that convulsions occur. The bones are the storage sites for calcium—if the blood calcium rises above normal, at least some of the excess is deposited in the bones. If the blood calcium dips too low, calcium is removed from the bones to bring it back up to the normal level.

Two hormones in particular are involved in regulating the blood calcium level. Parathyroid hormone (PTH) stimulates osteoclasts to dissolve the calcium matrix of bone. In addition, parathyroid hormone promotes calcium reabsorption in the small intestine and kidney. Thus, the blood calcium level increases. Calcitonin is a hormone that acts opposite to PTH. The female sex hormone estrogen can actually increase the number of osteoblasts; the reduction of estrogen in older women is often given as reason for the development of weak bones, called osteoporosis. Osteoporosis is discussed in the Health Focus, *You Can Avoid Osteoporosis.* In the young adult, the activity of osteoclasts is matched by the activity of osteoblasts, and bone mass remains stable until about age 45 in women. After that age, bone mass starts to decrease.

Video **Bear Bones**

Bone remodeling also accounts for why bones can respond to stress (see Fig. 11.4). If you engage in an activity that calls upon the use of a particular bone, the bone enlarges in diameter at the region most affected by the activity. During this process, osteoblasts in the periosteum form compact bone around the external bone surface and osteoclasts break down bone on the internal bone surface around the medullary cavity. Increasing the size of the medullary cavity prevents the bones from getting too heavy and thick. Today, exercises such as walking, jogging, and weight lifting are recommended. These exercises strengthen bone because they stimulate the work of osteoblasts instead of osteoclasts.

Biology Matters Health Focus

You Can Avoid Osteoporosis

Osteoporosis is a condition in which the bones are weakened due to a decrease in the bone mass that makes up the skeleton. The skeletal mass continues to increase until ages 20 to 30. After that, there is an equal rate of formation and breakdown of bone mass until ages 40 to 50. Then, reabsorption begins to exceed formation, and the total bone mass slowly decreases.

Over time, men are apt to lose 25% and women to lose 35% of their bone mass. But we have to consider that men—unless they have taken asthma medications that decrease bone formation—tend to have denser bones than women anyway. Whereas a man's testosterone (male sex hormone) level generally declines slowly after the age of 45, estrogen (female sex hormone) levels in women begin to decline significantly at about age 45. Sex hormones play an important role in maintaining bone strength, so this difference means that women are more likely than men to suffer a higher incidence of fractures, involving especially the hip, vertebrae, long bones, and pelvis. Although osteoporosis may at times be the result of various disease processes, it is essentially a disease that occurs as we age.

How to Avoid Osteoporosis

Everyone can take measures to avoid having osteoporosis when they get older. Adequate dietary calcium throughout life is an important protection against osteoporosis. The National Osteoporosis Foundation (www.nof.org) recommends that adults under the age of 50 take in 1,000 mg of vitamin D per day. After the age of 50, the daily intake should exceed 1,200 mg per day.

A small daily amount of vitamin D is also necessary for the body to use calcium correctly. Exposure to sunlight is required to allow skin to synthesize a precursor to vitamin D. If you reside on or north of a "line" drawn from Boston to Milwaukee, to Minneapolis, to Boise, chances are you're not getting enough vitamin D during the winter months. Therefore, you should take advantage of the vitamin D present in fortified foods such as low-fat milk and cereal. If you are under age 50, you should be receiving 400–800 IU of vitamin D per day. After age 50, this amount should increase to 800–1,000 IU of vitamin D daily.

Very inactive people, such as those confined to bed, lose bone mass 25 times faster than people who are moderately active. On the other hand, moderate weight-bearing exercise, such as regular walking or jogging, is another good way to maintain bone strength (Fig. 11A).

Diagnosis and Treatment

Postmenopausal women with any of the following risk factors should have an evaluation of their bone density:

- white or Asian race
- thin body type
- family history of osteoporosis
- early menopause (before age 45)
- smoking
- a diet low in calcium, or excessive alcohol consumption and caffeine intake
- sedentary lifestyle

Bone density is measured by a method called dual-energy X-ray absorptiometry (DEXA). This test measures bone density based on the absorption of photons generated by an X-ray tube. Soon there may be blood and urine tests to detect the biochemical markers of bone loss. Then it will be made easier for physicians to screen older women and at-risk men for osteoporosis.

If the bones are thin, it is worthwhile to take all possible measures to gain bone density because even a slight increase can significantly reduce fracture risk. Although estrogen therapy does reduce the incidence of hip fractures, long-term estrogen therapy is rarely recommended for osteoporosis. Estrogen is known to increase the risk of breast cancer, heart disease, stroke, and blood clots. Other medications are available, however. Calcitonin, a thyroid hormone, has been shown to increase bone density and strength, while decreasing the rate of bone fractures. Also, the bisphosphonates are a family of nonhormonal drugs used to prevent and treat osteoporosis. To achieve optimal results with calcitonin or one of the bisphosphonates, patients should also receive adequate amounts of dietary calcium and vitamin D.

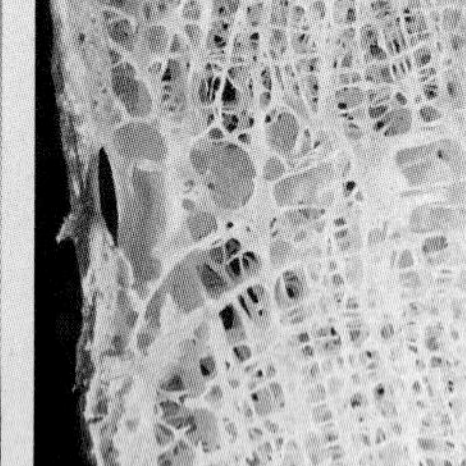

a. Normal bone

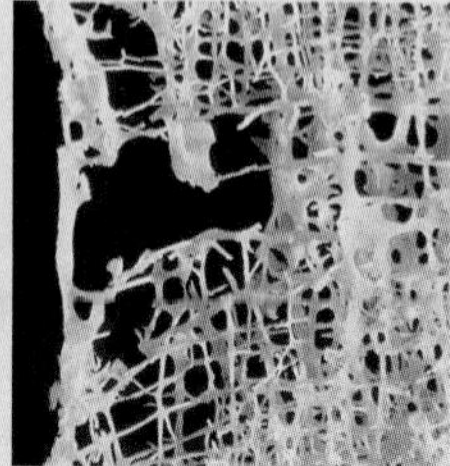

b. Osteoporosis

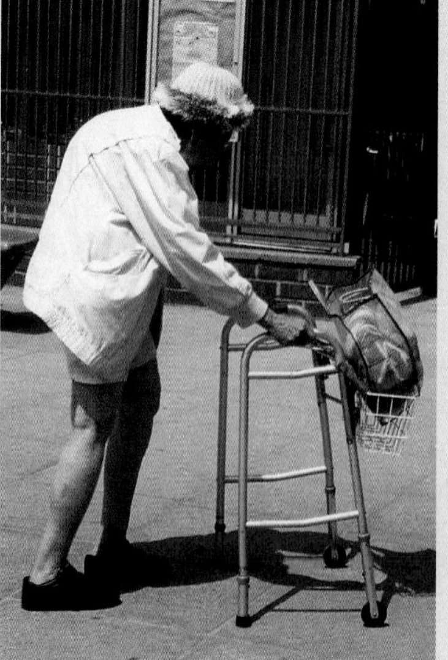

Figure 11A **Preventing osteoporosis.**
Weight-bearing exercise, when we are young, can help prevent osteoporosis when we are older. **a.** Normal bone. **b.** Bone from a person with osteoporosis.

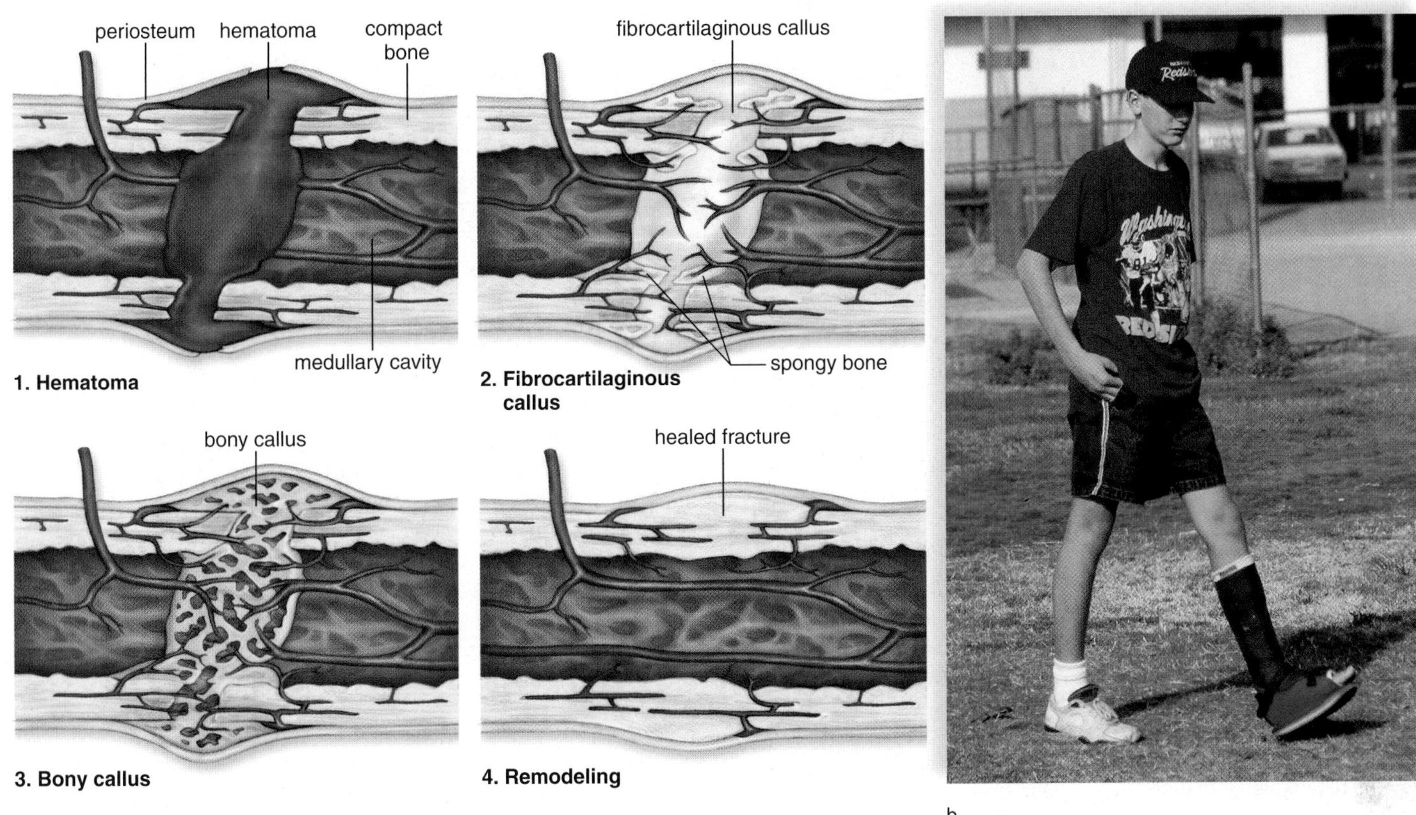

Figure 11.5 Bone repair following a fracture.
a. Steps in the repair of a fracture. **b.** A cast helps stabilize the bones while repair takes place. Fiberglass casts are now replacing plaster of paris as the usual material for a cast.

Bone Repair

Repair of a bone is required after it breaks or fractures. Fracture repair takes place over a span of several months in a series of four steps, shown in Figure 11.5 and listed here:

1. *Hematoma.* After a fracture, blood escapes from ruptured blood vessels and forms a hematoma (mass of clotted blood) in the space between the broken bones. The hematoma forms within 6 to 8 hours.
2. *Fibrocartilaginous callus.* Tissue repair begins, and a fibrocartilaginous callus fills the space between the ends of the broken bone for about three weeks.
3. *Bony callus.* Osteoblasts produce trabeculae of spongy bone and convert the fibrocartilage callus to a bony callus that joins the broken bones together. The bony callus lasts about three to four months.
4. *Remodeling.* Osteoblasts build new compact bone at the periphery. Osteoclasts absorb the spongy bone, creating a new medullary cavity.

In some ways, bone repair parallels the development of a bone except that the first step, hematoma, indicates that injury has occurred. Further, a fibrocartilaginous callus precedes the production of compact bone.

The naming of fractures tells you what type of break occurred. A fracture is complete if the bone is broken clear through and incomplete if the bone is not separated into two parts. A fracture is simple if it does not pierce the skin and compound if it does pierce the skin. Impacted means that the broken ends are wedged into each other. A spiral fracture occurs when the break is ragged due to twisting of a bone.

MP3 Remodeling and Repair

Connecting the Concepts

For more on bone development and the hormones that influence bone growth, refer to the following discussions.

Section 8.6 provides additional information on inputs of vitamin D and calcium in the diet.

Section 15.2 examines the role of growth hormones in the body.

Section 15.3 describes the action of the hormones calcitonin and PTH.

Check Your Progress 11.2

1. Classify cells of the skeletal system into ones involved in bone growth, remodeling, and repair.
2. Describe how bone growth occurs during development.
3. Summarize the stages in the repair of bone.

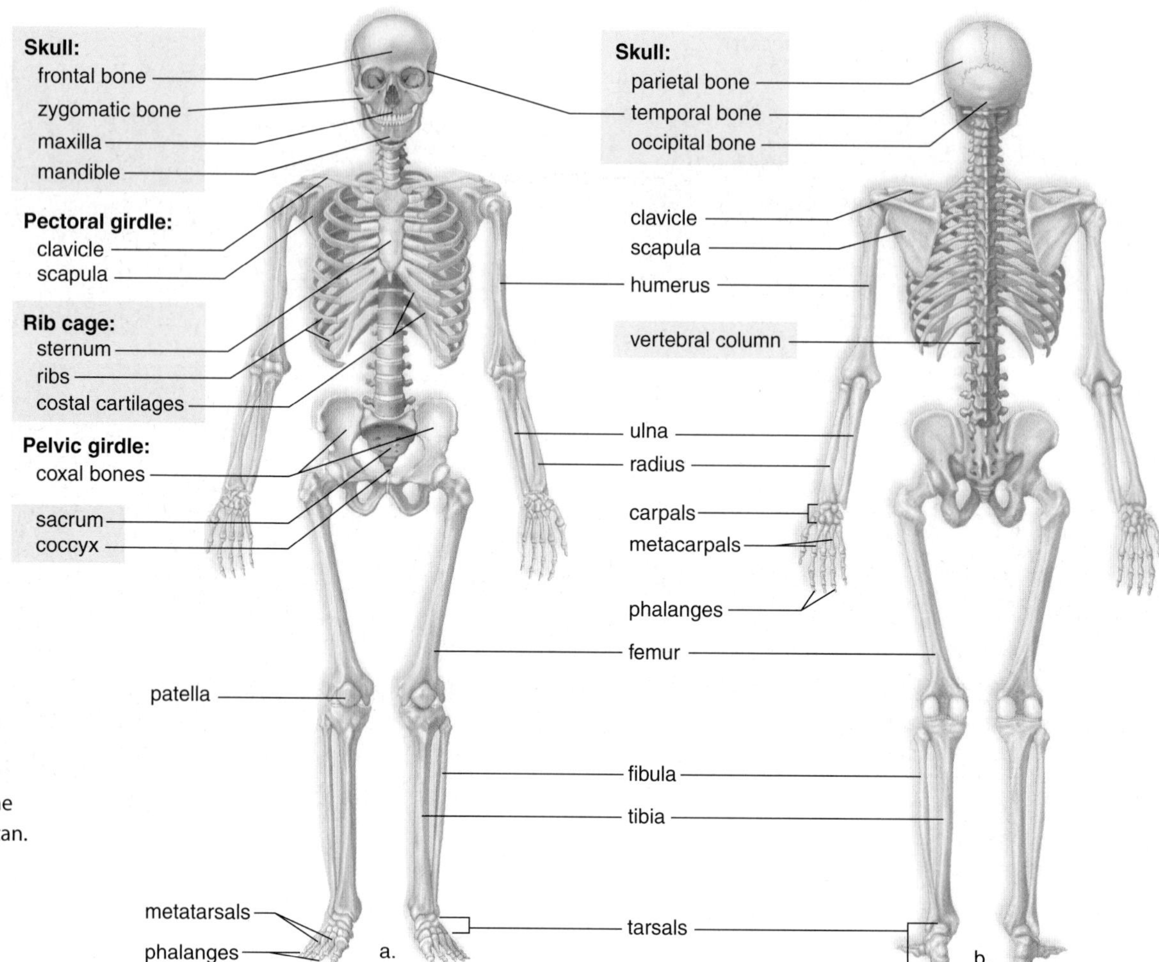

Figure 11.6 The axial and appendicular skeletons.
a. Axial skeleton bones are colored blue. **b.** Bones of the appendicular skeleton are tan.

11.3 Bones of the Axial Skeleton

Learning Outcomes

Upon completion of this section, you should be able to

1. Identify the bones of the skull, hyoid, vertebral column, and rib cage.
2. Identify the regions of the vertebral column.
3. Explain the function of the sinuses and intervertebral disks in relation to the axial skeleton.

The 206 bones of the skeleton are classified according to whether they occur in the axial skeleton or the appendicular skeleton (see Fig. 11.6). The **axial skeleton** lies in the midline of the body and consists of the skull, hyoid bone, vertebral column, and the rib cage.

The Skull

The **skull** is formed by the cranium (braincase) and the facial bones. However, some cranial bones contribute to the face.

The Cranium

The cranium protects the brain. In adults, it is composed of eight bones fitted tightly together. In newborns, certain cranial bones are not completely formed. Instead, these bones are joined by membranous regions called **fontanels.** The fontanels usually close by the age of 16 months by the process of intramembranous ossification.

Some of the bones of the cranium contain the **sinuses,** air spaces lined by mucous membrane. The sinuses reduce the weight of the skull and give a resonant sound to the voice. Two sinuses, called the mastoid sinuses, drain into the middle ear. **Mastoiditis,** a condition that can lead to deafness, is an inflammation of these sinuses.

The major bones of the cranium have the same names as the lobes of the brain: frontal, parietal, occipital, and temporal. On the top of the cranium (Fig. 11.7*a*), the *frontal bone* forms the forehead, the *parietal bones* extend to the sides, and the *occipital bone* curves to form the base of the skull. Here there is a large opening, the **foramen magnum** (Fig. 11.7*b*), through which the spinal cord passes and becomes the brain stem. Below the much larger parietal bones, each *temporal*

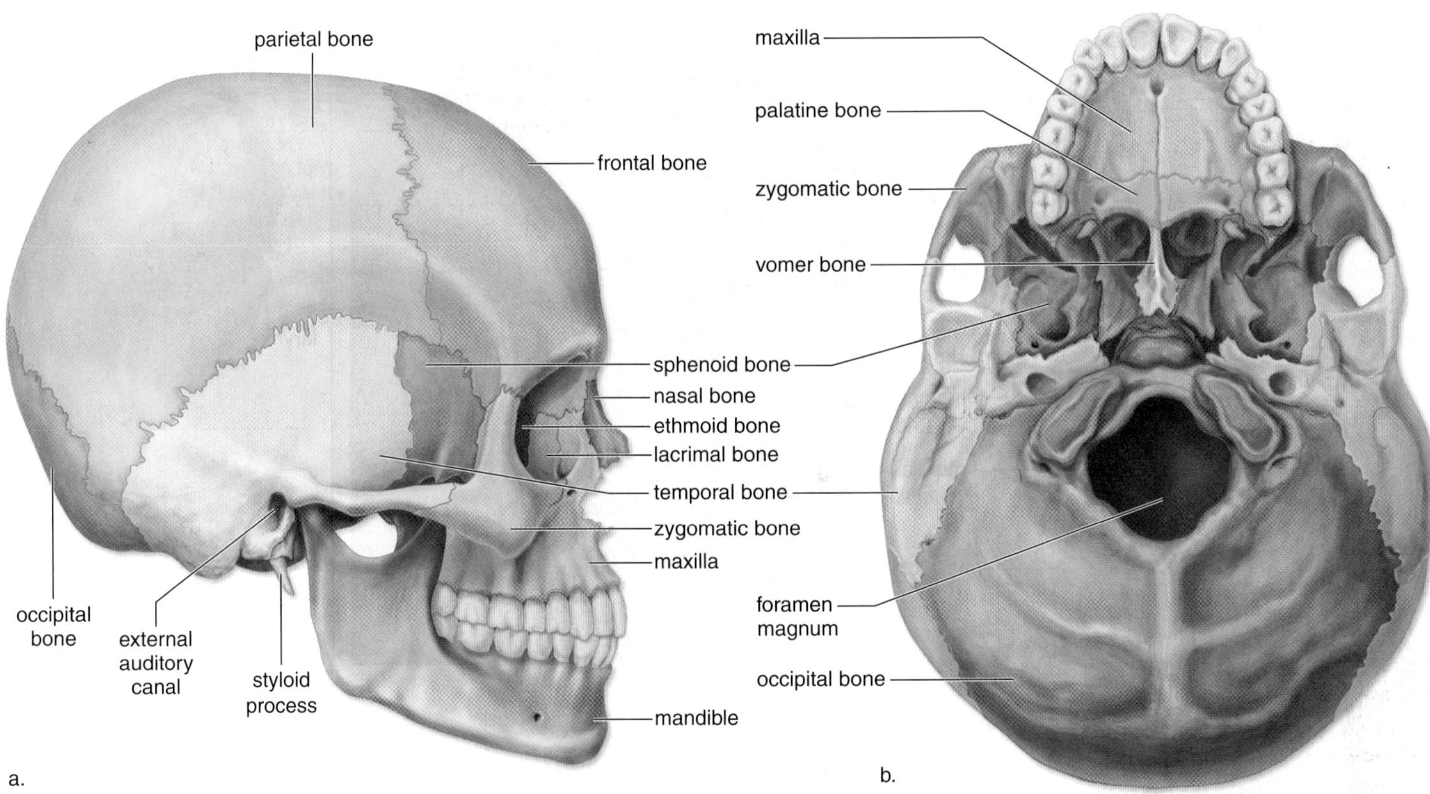

Figure 11.7 **The bones of the skull.**
a. Lateral view. **b.** Inferior view.

bone has an opening (external auditory canal) that leads to the middle ear.

The *sphenoid bone,* shaped like a bat with outstretched wings, extends across the floor of the cranium from one side to the other. The sphenoid is the keystone of the cranial bones because all the other bones articulate with it. The sphenoid completes the sides of the skull and also contributes to forming the orbits (eye sockets). The *ethmoid bone,* which lies in front of the sphenoid, also helps form the orbits and the nasal septum. The orbits are completed by various facial bones. The eye sockets are called orbits because we can rotate our eyes.

The Facial Bones

The most prominent of the facial bones are the mandible, the maxillae (sing., maxilla), the zygomatic bones, and the nasal bones.

The mandible, or lower jaw, is the only movable portion of the skull, and it also forms the chin (Figs. 11.7, 11.8). The maxillae form the upper jaw and a portion of the eye socket. Further, the hard palate and the floor of the nose are formed by the maxillae (anterior) joined to the palatine bones (posterior). Tooth sockets are located on the mandible and on the maxillae. The grinding action of the mandible and maxillae allows us to chew our food.

The lips and cheeks have a core of skeletal muscle. The *zygomatic bones* are the cheekbone prominences, and the *nasal bones* form the bridge of the nose. Other bones (e.g., ethmoid and vomer) are a part of the nasal septum, which divides the interior of the nose into two nasal cavities. The lacrimal bone (see Fig. 11.7*a*) contains the opening for the nasolacrimal canal, which drains tears from the eyes to the nose.

Certain cranial bones contribute to the face. The temporal bone and the wings of the sphenoid bone account for the flattened areas we call the temples. The frontal bone forms the forehead and has supraorbital ridges, where the eyebrows are located. Glasses sit where the frontal bone joins the nasal bones.

The exterior portions of ears are formed only by cartilage and not by bone. The nose is a mixture of bones, cartilages, and connective tissues. The cartilages complete the tip of the nose, and fibrous connective tissue forms the flared sides of the nose.

MP3 **The Skull**

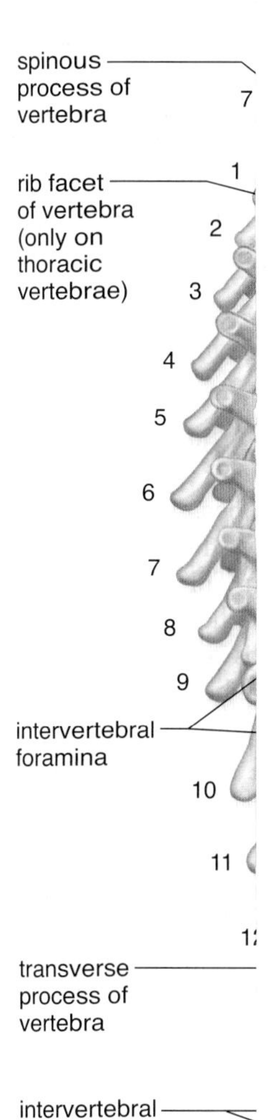

Figure 11.9 **The**
The vertebral column is
intervertebral disks. The
The vertebrae are name
example, the thoracic v
coccyx, which is also ca

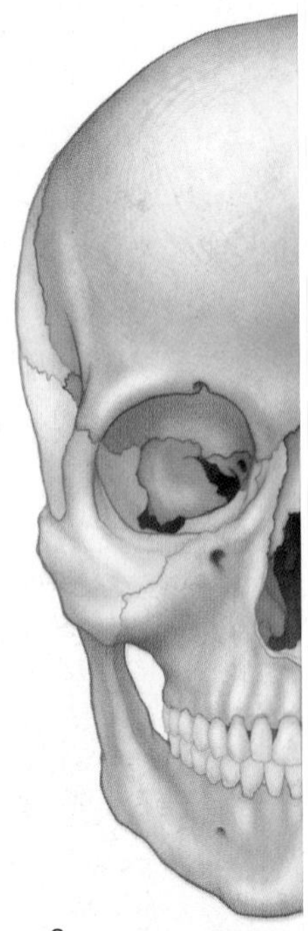

Figure 12.6 **The structure of a skeletal muscle fiber.** A muscle fiber contains many myofibrils divided into sarcomeres, which are contractile. When the myofibrils of a muscle fiber contract, the sarcomeres shorten. The actin (thin) filaments slide past the myosin (thick) filaments toward the center. The Z lines have moved and the H zone has gotten smaller, to the point of disappearing.

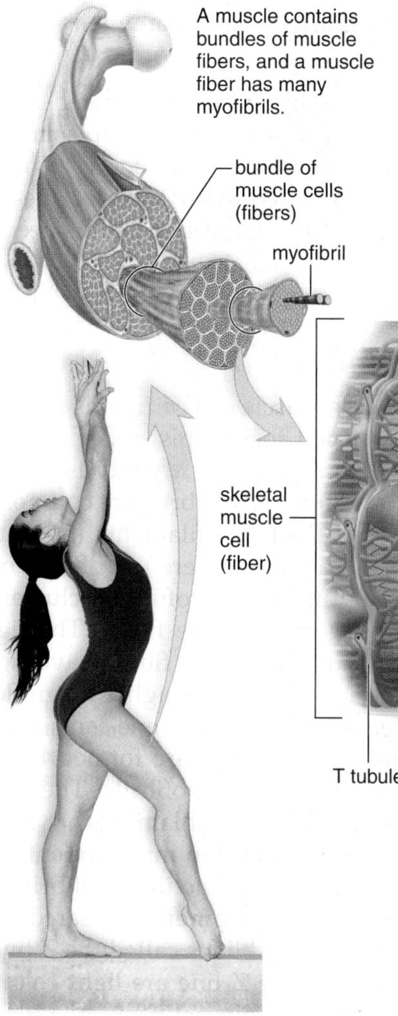

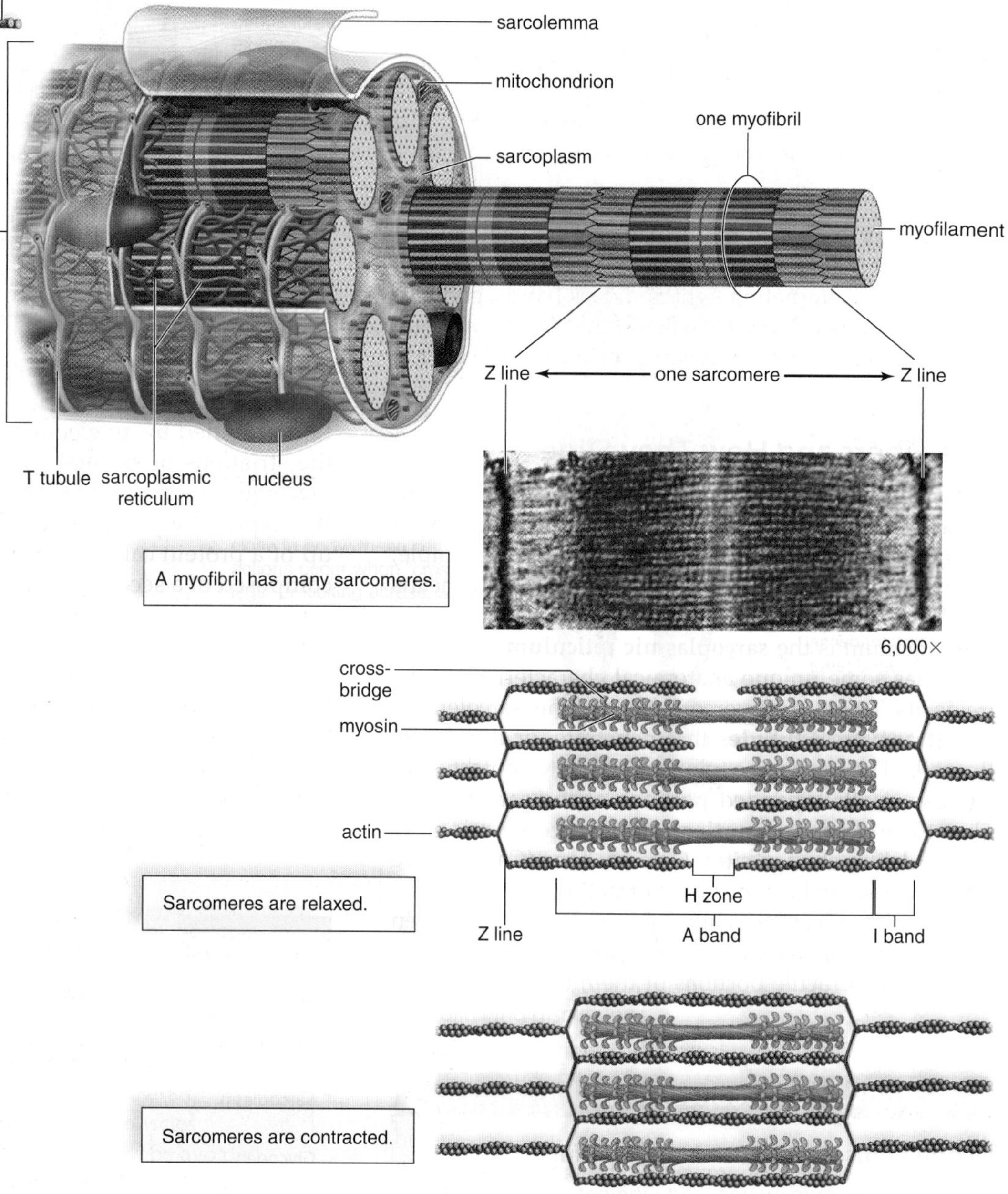

A muscle contains bundles of muscle fibers, and a muscle fiber has many myofibrils.

A myofibril has many sarcomeres.

Sarcomeres are relaxed.

Sarcomeres are contracted.

Myofilaments

The thick and thin filaments differ in the following ways:

Thick filaments. A thick filament is composed of several hundred molecules of the protein myosin. Each myosin molecule is shaped like a golf club, with the straight portion of the molecule ending in a globular head, or cross-bridge. The cross-bridges occur on each side of a sarcomere but not in the middle (Fig. 12.6).

Thin filaments. Primarily, a thin filament consists of two intertwining strands of the protein actin. Two other proteins, called tropomyosin and troponin, also play a role, as we will discuss later in this section.

Sliding filaments. We will also see that when muscles are stimulated, electrical signals travel across the sarcolemma and then down a T tubule. In turn, this signals calcium to be released from the sarcoplasmic reticulum. Now the muscle fiber contracts as the sarcomeres within the myofibrils

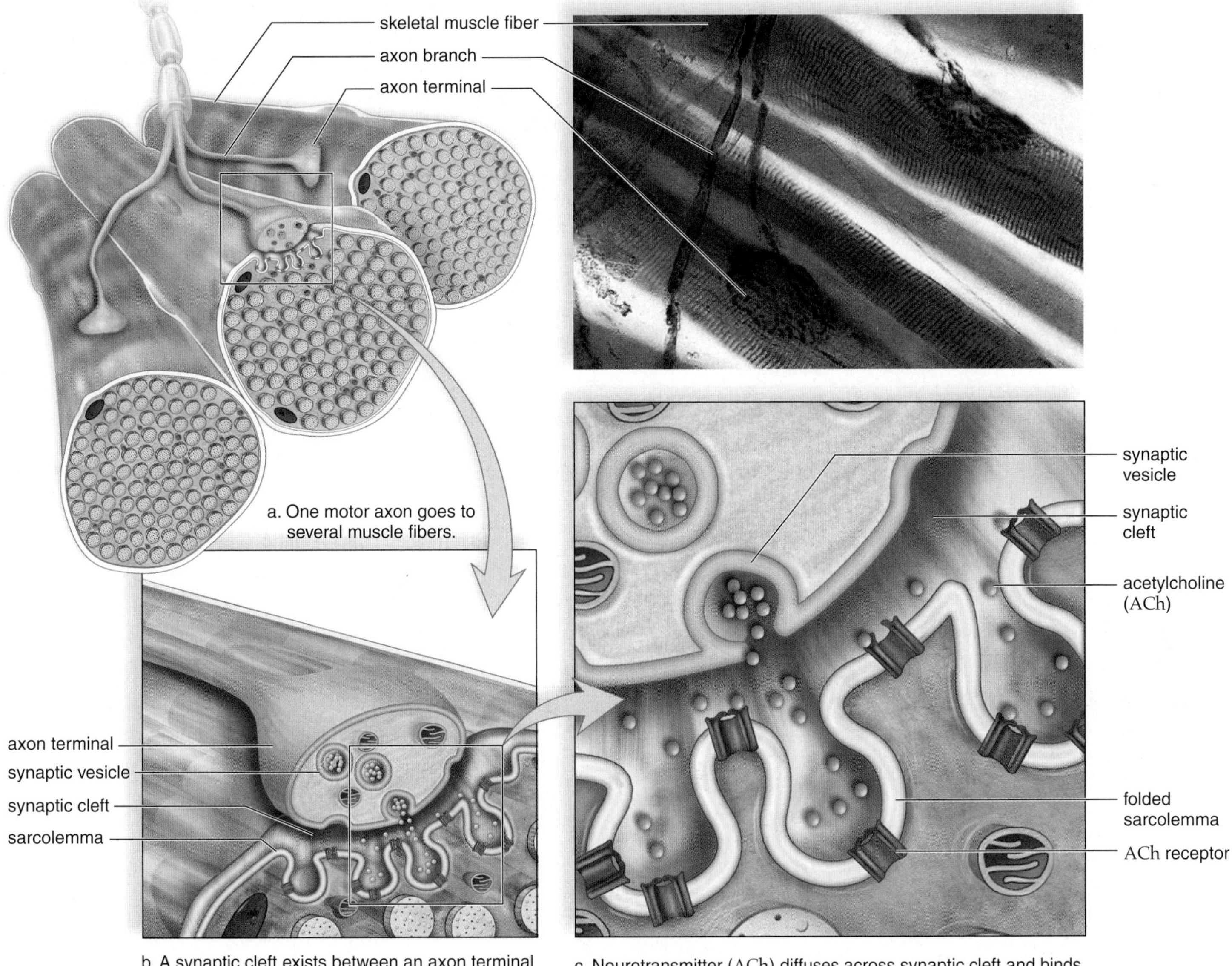
skeletal muscle fiber
axon branch
axon terminal
a. One motor axon goes to several muscle fibers.
axon terminal
synaptic vesicle
synaptic cleft
sarcolemma
b. A synaptic cleft exists between an axon terminal and a muscle fiber.
synaptic vesicle
synaptic cleft
acetylcholine (ACh)
folded sarcolemma
ACh receptor
c. Neurotransmitter (ACh) diffuses across synaptic cleft and binds to receptors in sarcolemma.

Figure 12.7 **Motor neurons and skeletal muscle fibers join neuromuscular junctions**
a. The branch of a motor nerve fiber terminates in an axon terminal. **b.** A synaptic cleft separates the axon terminal from the sarcolemma of the muscle fiber. **c.** Nerve impulses traveling down a motor fiber cause synaptic vesicles to discharge acetylcholine, which diffuses across the synaptic cleft and binds to ACh receptors. Impulses travel down the T system of a muscle fiber and the muscle fiber contracts.

shorten. As you compare the relaxed sarcomere (Fig. 12.6) with the contracted sarcomere (Fig. 12.6), note that the filaments themselves remain the same length. When a sarcomere shortens, the actin (thin) filaments approach one another as they slide past the myosin (thick) filaments. This causes the I band to shorten, the Z line to move inward, and the H zone to almost or completely disappear (see Fig. 12.6). The sarcomere changes from a rectangular shape to a square as it shortens. The movement of actin filaments in relation to myosin filaments is called the **sliding filament model** of muscle contraction. ATP supplies the energy for muscle contraction. Although the actin filaments slide past the myosin filaments, it is the myosin filaments that do the work. Myosin filaments break down ATP, and their cross-bridges pull the actin filament toward the center of the sarcomere.

Animation Sarcomere Contraction

As an analogy, think of yourself and a group of friends as myosin. Collectively, your hands are the cross-bridges, and you are pulling on a rope (actin) to get an object tied to the end of the rope (the Z line). As you pull the rope, you grab, pull, release, and then grab farther along on the rope.

Control of Muscle Fiber Contraction

Muscle fibers are stimulated to contract by motor neurons whose axons are grouped together to form nerves. The axon of one motor neuron can stimulate from a few to several muscle fibers of a muscle because each axon has several branches (Fig. 12.7*a*). Each branch of an axon ends in an axon terminal that lies in close proximity to the sarcolemma of a muscle fiber. A small gap, called a synaptic cleft, separates the axon terminal from the sarcolemma (Fig. 12.7*b*). This entire region is called a **neuromuscular junction.**

Animation Muscle Contraction

Axon terminals contain synaptic vesicles filled with the neurotransmitter acetylcholine (ACh). Nerve signals travel down the axons of motor neurons and arrive at an axon terminal. The signals trigger the synaptic vesicles to release ACh into the synaptic cleft (Fig. 12.7*c*). When ACh is released, it quickly diffuses across the cleft and binds to receptors in the sarcolemma. Now, the sarcolemma generates electrical signals that spread across the sarcolemma and down the T tubules. Recall that the T tubules lie adjacent to the sarcoplasmic reticulum, but the two structures are not connected. Nonetheless, signaling from the T tubules causes the release of Ca^{2+} from the sarcoplasmic reticulum, which leads to sarcomere contraction, as explained in Figure 12.7.

Animation Breakdown of ATP and Cross-Bridge Movement

Two other proteins are associated with an actin filament. Threads of **tropomyosin** wind about an actin filament, covering binding sites for myosin located on each actin molecule. **Troponin** occurs at intervals along the threads. When calcium ions (Ca^{2+}) are released from the sarcoplasmic reticulum, they combine with troponin. This causes the tropomyosin threads to shift their position, exposing myosin-binding sites. In other words, myosin can now bind to actin (see Fig.12.8*a*).

To fully understand muscle contraction, study Figure 12.8*b*. (1) The heads of a myosin filament have ATP-binding sites. At

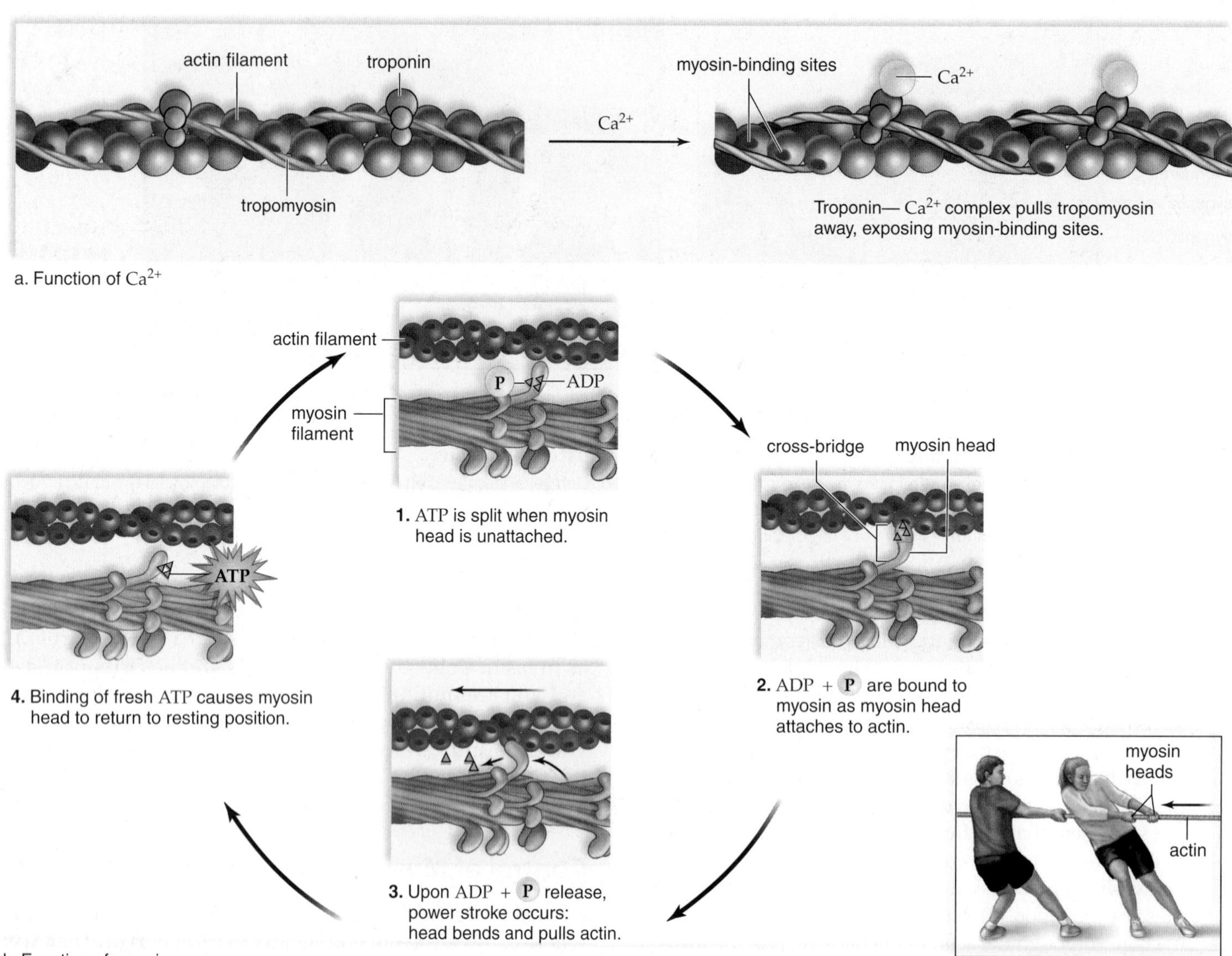

actin filament
troponin
tropomyosin
Ca^{2+}
myosin-binding sites
Ca^{2+}
Troponin— Ca^{2+} complex pulls tropomyosin away, exposing myosin-binding sites.
a. Function of Ca^{2+}
actin filament
myosin filament
P
ADP
1. ATP is split when myosin head is unattached.
cross-bridge
myosin head
2. ADP + P are bound to myosin as myosin head attaches to actin.
3. Upon ADP + P release, power stroke occurs: head bends and pulls actin.
ATP
4. Binding of fresh ATP causes myosin head to return to resting position.
myosin heads
actin
b. Function of myosin

Figure 12.8 **The role of calcium ions and ATP during muscular contraction.**
a. Calcium ions (Ca^{2+}) bind to troponin, exposing myosin-binding sites. **b.** Follow steps 1 through 4 to see how myosin uses ATP and does the work of pulling actin toward the center of the sarcomere, much as a group of people pulling a rope (*right*).

this site, ATP is hydrolyzed, or split, to form ADP and Ⓟ. (2) The ADP and Ⓟ remain on the myosin heads, and the heads attach to an actin-binding site. Joining myosin to actin forms temporary bonds called cross-bridges. (3) Now, ADP and Ⓟ are released and the cross-bridges bend sharply. This is the power stroke that pulls the actin filament toward the center of the sarcomere. (4) When ATP molecules again bind to the myosin heads, the cross-bridges are broken. Myosin heads detach from the actin filament. This is the step that does not happen during rigor mortis. Relaxing the muscle is impossible, because ATP is needed to break the bond between an actin-binding site and the myosin cross-bridge.

In living muscle, the cycle begins again and myosin reattaches farther along the actin filament. The cycle recurs until calcium ions are actively returned to the calcium storage sites. This step also requires ATP.

MP3 Sliding Filament Theory

Connections and Misconceptions

How does Botox remove wrinkles?

As described on page 270, sarcomere contraction can be prevented by the action of a potent neurotoxin. Botox is the trade name for botulinum toxin A, produced by a bacterium. Botox prevents wrinkling of the brow and skin about the eyes, because it blocks the release of ACh into the synaptic cleft. As a result, the sarcolemma is never activated and muscle contraction never occurs. However, the treatment is not permanent and must be repeated every few months.

Connecting the Concepts

For more information on ATP and how the nervous system controls the contraction of skeletal muscle, refer to the following discussions.

Figure 3.20 illustrates the ATP–ADP cycle.

Section 13.2 explains the role of neurotransmitters, such as acetylcholine, in the nervous system.

Figure 13.4 illustrates the action of neurotransmitters in the synaptic cleft.

Check Your Progress 12.2

1. Identify the myofibril, myofilament, and sarcomere in a muscle fiber.
2. Explain how the thin and thick filaments interact in the sliding filament model.
3. Describe the role of both ATP and calcium ions in muscle contraction.

12.3 Whole Muscle Contraction

Learning Outcomes

Upon completion of this section, you should be able to

1. List the stages of a muscle twitch and explain what is occurring in each stage.
2. Explain how summation and tetanus increase the strength of whole muscle contraction.
3. Summarize how muscle cells produce ATP for muscle contraction.
4. Distinguish between fast-twitch and slow-twitch muscle fibers.

In order for a whole muscle, such as a bicep or tricep, to contract, the individual muscle fibers must be activated by signals from the nervous system.

Muscles Have Motor Units

As already mentioned, each axon within a nerve stimulates a number of muscle fibers. A nerve fiber with all of the muscle fibers it innervates is called a **motor unit.** A motor unit obeys a principle called the **all-or-none law.** Why? Because all the muscle fibers in a motor unit are stimulated at once. They all either contract or do not contract. A variable of interest is the number of muscle fibers within a motor unit. For example, in the ocular muscles that move the eyes, the innervation ratio is one motor axon per 23 muscle fibers. By contrast, in the gastrocnemius muscle of the leg, the ratio is about one motor axon per 1,000 muscle fibers. Thus, moving the eyes requires finer control than moving the legs.

When a motor unit is stimulated by infrequent electrical impulses, a single contraction occurs. This response is called a **muscle twitch** and lasts only a fraction of a second. A muscle twitch is customarily divided into three stages. We can use our knowledge of muscle fiber contraction to understand these events. The latent period is the time between stimulation and initiation of contraction (Fig. 12.9*a*). During this time, we can imagine that the events that begin muscle contraction are occurring. The neurotransmitter ACh diffuses across the synaptic cleft, causing an electrical signal to spread across the sarcolemma and down the T tubules. The contraction period follows as calcium leaves the sarcoplasmic reticulum and myosin–actin cross-bridges form. As you know, the muscle shortens as it contracts. On the graph, the force increases as the muscle contracts. Finally, the relaxation period completes the muscle twitch. Myosin–actin cross-bridges are broken, and calcium returns to the sarcoplasmic reticulum. Force diminishes as the muscle returns to its former length.

If a motor unit is given a rapid series of stimuli, it can respond to the next stimulus without relaxing completely. **Summation** is increased muscle contraction until maximal sustained contraction, called **tetanus,** is achieved (Fig. 12.9*b*). Tetanus continues until the muscle fatigues due to depletion of energy reserves. Fatigue is apparent when a muscle relaxes, even though stimulation continues. The *tetanus* of muscle cells is not the same as the infection called tetanus. The infection called tetanus is caused by the bacterium *Clostridium tetani.* Death occurs because the muscles, including the respiratory muscles, become fully contracted and do not relax.

A whole muscle typically contains many motor units. As the intensity of nervous stimulation increases, more motor units in a muscle are activated. This phenomenon is known as *recruitment.* Maximum contraction of a muscle would require that all motor units be undergoing tetanic contraction. This rarely happens because they could all fatigue at the same time. Instead, some motor units are contracting maximally while others are resting, allowing sustained contractions to occur.

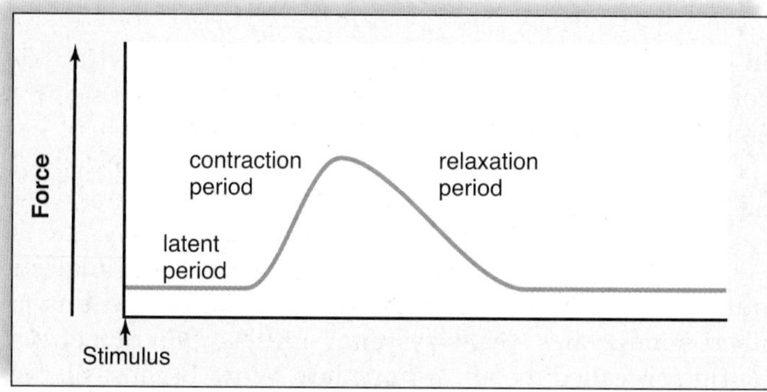

a.

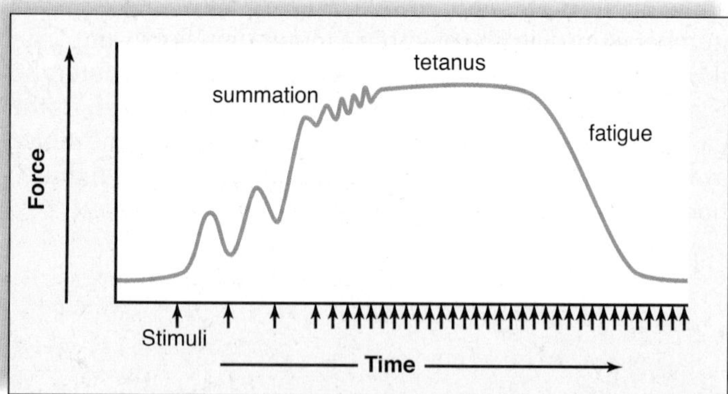

b.

Figure 12.9 The three phases of a single muscle twitch and how summation and tetanus increase the force of contraction.
a. Stimulation of a muscle by a single electrical signal results in a simple muscle twitch: first, a latent period, followed by contraction and relaxation. **b.** Repeated stimulation results in summation and tetanus, which creates greater force because the motor unit cannot relax between stimuli.

Muscle Tone

One desirable effect of exercise is to have good **muscle tone.** Muscles that have good tone are firm and solid, as opposed to being soft and flabby. The amount of muscle tone is dependent on muscle contraction. Some motor units are always contracted—but not enough to cause movement.

Energy for Muscle Contraction

Muscles can use various fuel sources for energy, and they have various ways of producing ATP needed for muscle contraction.

Fuel Sources for Exercise

A muscle has four possible energy sources (Fig. 12.10). Two of these are stored in muscle (glycogen and triglycerides), and two are acquired from blood (glucose and fatty acid). The amount that each of these is used depends on exercise intensity and duration. Figure 12.10 shows the percentage of energy derived from these sources due to submaximal exercise (65–75% of effort that an individual is capable of) over time. Notice that as the length of the exercise period is increased, use of muscle glycogen decreases and use of muscle triglyceride (fat) increases.

Muscles also make use of blood glucose and plasma fatty acids as energy sources. Both of these are delivered to muscles by circulating blood. Some of us specifically exercise to lose weight, and we are particularly interested in the use of plasma fatty acids by exercising muscles. Adipose tissue is the source of plasma fatty acids for muscle contraction, but it also tends to make us look fat. Figure 12.10 shows that the amount of fat burned increases when more time is spent in exercise. If we are on a diet that restricts the amount of fat eaten, exercise will decrease body fat. Submaximal exercise burns fat better than maximal exercise, for reasons we will now explore.

Animation Energy Sources for Prolonged Exercise

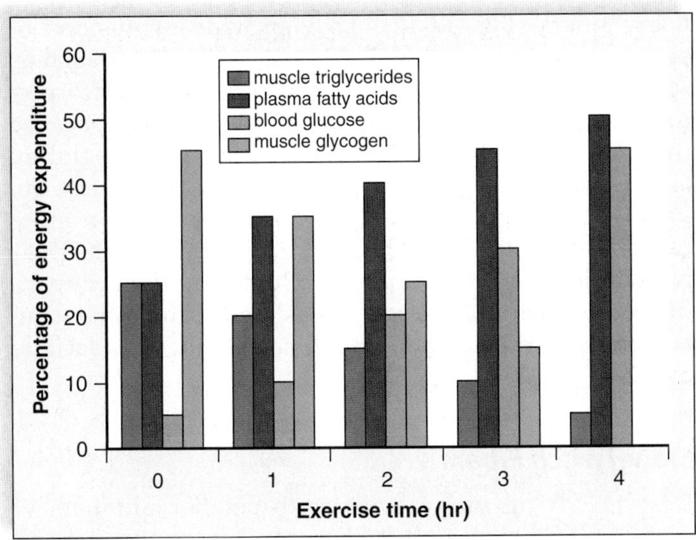

Figure 12.10 **The sources of energy for muscle contraction.** The percentage of energy derived from each of the four major fuel sources during submaximal exercise (65–75% of effort) is illustrated. The amount from plasma fatty acids increases during the time span shown.

Sources of ATP for Muscle Contraction

Muscle cells store limited amounts of ATP. Once stored ATP is used up, the cells have three ways to produce more ATP (Fig. 12.11). The three ways include (1) formation of ATP by the creatine phosphate (CP) pathway; (2) formation of ATP by fermentation (see Chapter 3); and (3) formation of ATP by cellular respiration, which involves the use of oxygen by mitochondria. Aerobic exercising depends on cellular respiration to supply ATP. Neither formation of ATP by the CP pathway nor by fermentation involves the need for oxygen. Both are anaerobic processes.

The CP Pathway The simplest and most rapid way for muscle to produce ATP is to use the CP pathway because it only consists of one reaction (Fig. 12.11*a*), as shown in the following graphic:

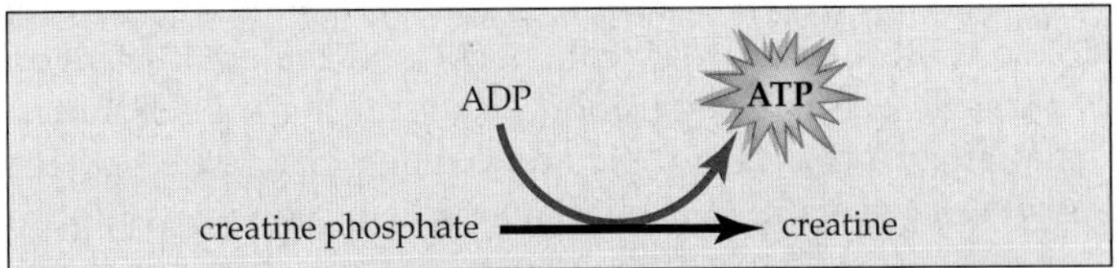

This reaction occurs in the midst of sliding filaments; therefore, this method of supplying ATP is the speediest energy source available to muscles. Creatine phosphate is formed only when a muscle cell is resting, and only a limited amount is stored. The CP pathway is used at the beginning of submaximal exercise and during short-term, high-intensity exercise that lasts less than 5 seconds. The energy to complete a single play in a football game comes principally from the CP system. Intense activities lasting longer than 5 seconds also make use of fermentation.

Fermentation Fermentation, as you know, produces two ATP molecules from the anaerobic breakdown of glucose to lactate. This pathway is the one most likely to begin with glycogen. Hormones provide the signal to muscle cells to break down glycogen, making glucose available as an energy source.

Fermentation, like the CP pathway, is fast-acting, but it results in the buildup of lactate (Fig. 12.11*b*). Formation of lactate is noticeable because it produces short-term muscle aches and fatigue upon exercising. We have all had the experience of needing to continue breathing following strenuous exercise. This continued intake of oxygen, called **oxygen debt,** is required, in part, to complete the metabolism of lactate and restore cells to their original energy state. The lactate is transported to the liver, where 20% of it is completely broken down to carbon dioxide and water. The ATP gained by this respiration is then used to reconvert 80% of the lactate to glucose and then glycogen. In persons who train, the number of mitochondria in individual muscles increases. There is a greater reliance on these additional mitochondria to produce ATP. Muscles rely less on fermentation as a result.

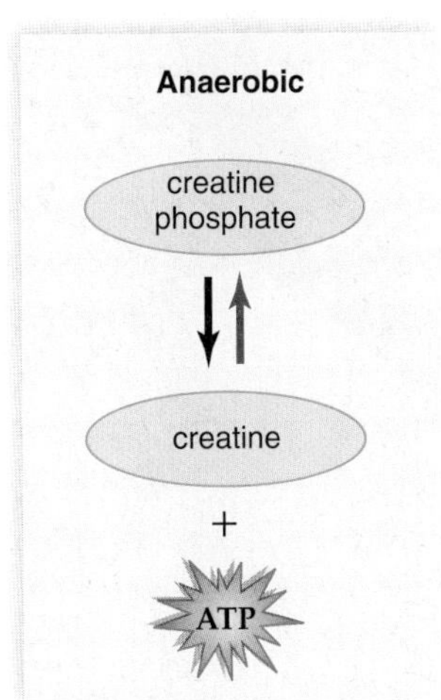

a.

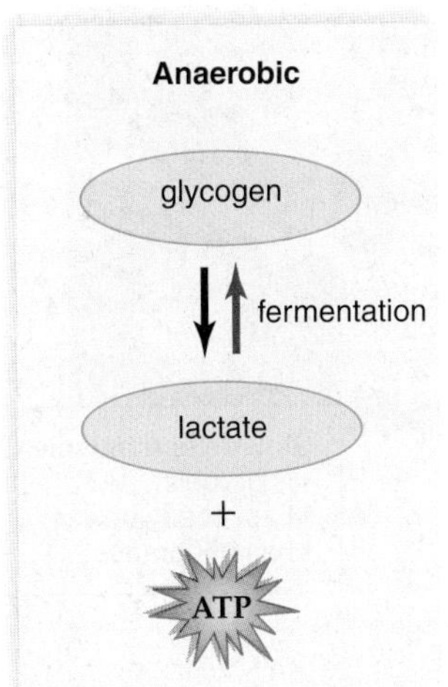

b.

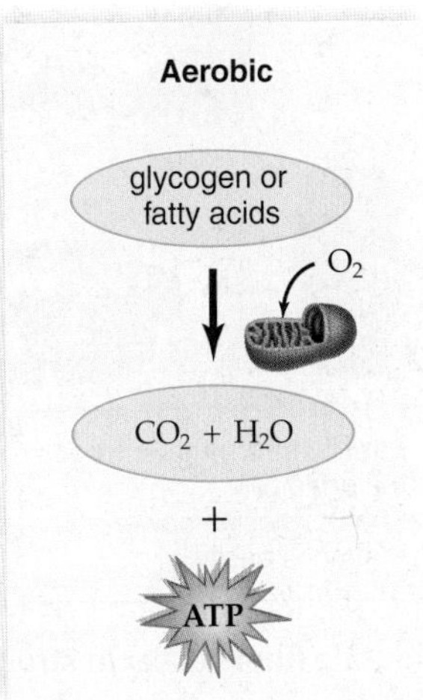

c.

Figure 12.11 **The three pathways by which muscle cells produce the ATP energy needed for contraction.** **a.** When contraction begins, muscle cells break down creatine phosphate to produce ATP. When resting, muscle cells rebuild their supply of creatine phosphate (red arrow). **b.** Muscle cells also use fermentation to produce ATP quickly. When resting, muscle cells metabolize lactate, reforming as much glucose and then glycogen as possible (red arrow). **c.** For the long term, muscle cells switch to cellular respiration to produce ATP aerobically.

Cellular Respiration Cellular respiration is the slowest of all three mechanisms used to produce ATP. However, it is also the most efficient, typically producing several dozen molecules of ATP from each food molecule. As you will recall, cellular respiration occurs in the mitochondria. Thus, the process is aerobic and oxygen is supplied by the respiratory system. In addition, a protein called **myoglobin** found within muscle cells delivers oxygen directly to the mitochondria. Cellular respiration can make use of glucose from the breakdown of stored muscle glycogen, glucose taken up from blood, and/or fatty acids from fat digestion (Fig. 12.11*c*). Also, cellular respiration is more likely to supply ATP when exercise is submaximal in intensity. According to Figure 12.9, if you are interested in exercising to lose weight, you should do so at a lower intensity and for a generous amount of time. This means that your aerobic exercise class at the local gym burns triglyceride, or fat, by making ATP using cellular respiration.

MP3 Energy Source for Muscle Contraction

Fast-Twitch and Slow-Twitch Muscle Fibers

We have seen that all muscle fibers metabolize aerobically and anaerobically. However, some muscle fibers use one method more than the other to provide myofibrils with ATP. Fast-twitch fibers tend to rely on the creatine phosphate pathway and fermentation, anaerobic means of supplying ATP to muscles. Slow-twitch fibers tend to prefer cellular respiration, which is aerobic.

Fast-Twitch Fibers

Fast-twitch fibers are usually anaerobic and seem to be designed for strength because their motor units contain many fibers (Fig. 12.12). They provide explosions of energy and are most helpful in sports activities such as sprinting, weight lifting, swinging a golf club, or throwing a shot. Fast-twitch fibers are light in color because they have fewer mitochondria, little or no myoglobin, and fewer blood vessels than slow-twitch fibers do. Fast-twitch fibers can develop maximum tension more rapidly than slow-twitch fibers can. In addition, their maximum tension is greater. However, their dependence on anaerobic energy leaves them vulnerable to an accumulation of lactate, which causes them to fatigue quickly.

Slow-Twitch Fibers

Despite having motor units with smaller numbers of muscle fibers, slow-twitch fibers have more stamina and a steadier "tug." These muscle fibers are most helpful in endurance sports, such as long distance running, biking, jogging, and swimming. They produce most of their energy aerobically, so they tire only when their fuel supply is gone. Slow-twitch fibers have many mitochondria and are dark in color because they contain myoglobin, the respiratory pigment found in muscles (Fig. 12.12). They are also surrounded by dense capillary beds and draw more blood and oxygen than fast-twitch fibers. Slow-twitch fibers have a low maximum tension, which develops slowly, but the muscle fibers are highly resistant to fatigue. Slow-twitch fibers have a substantial reserve of glycogen and fat, so their abundant mitochondria can maintain a steady, prolonged production of ATP when oxygen is available.

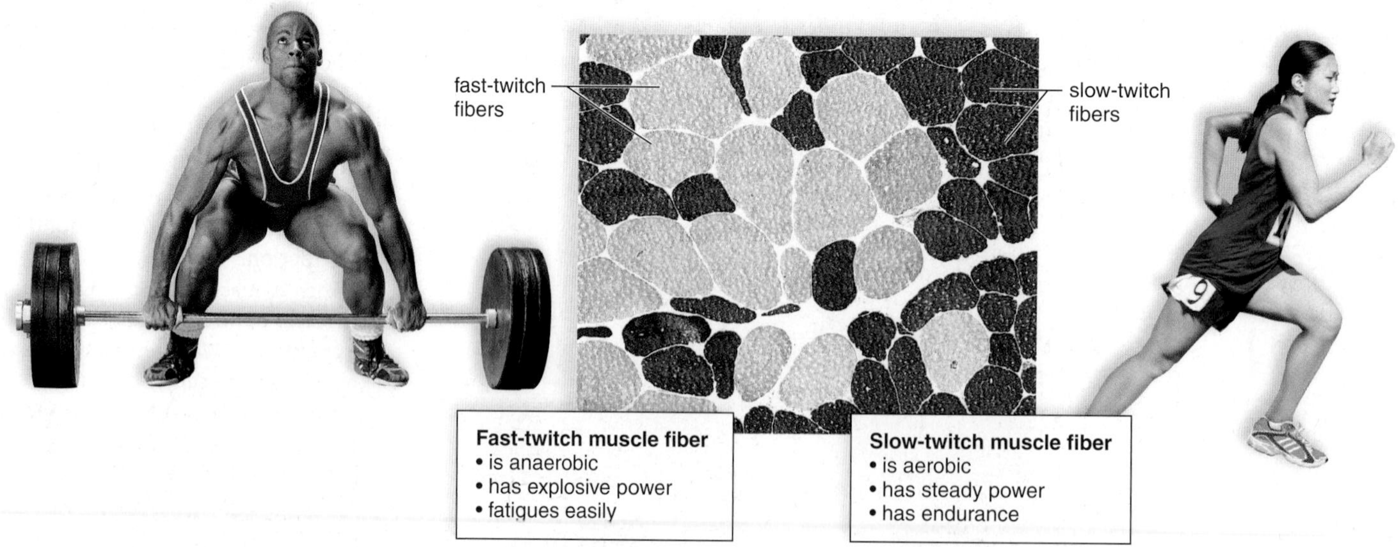
fast-twitch fibers
slow-twitch fibers
Fast-twitch muscle fiber
• is anaerobic
• has explosive power
• fatigues easily
Slow-twitch muscle fiber
• is aerobic
• has steady power
• has endurance

Figure 12.12 Fast-twitch and slow-twitch muscle fibers differ in structure.
If your muscles contain many fast-twitch fibers (light color), you would probably do better at a sport like weight lifting. If your muscles contain many slow-twitch fibers (dark color), you would probably do better at a sport like cross-country running.

Biology Matters **Health Focus**

Exercise, Exercise, Exercise

Exercise programs improve muscular strength, muscular endurance, and flexibility. Muscular strength is the force a muscle group (or muscle) can exert against a resistance in one maximal effort. Muscular endurance is judged by the ability of a muscle to contract repeatedly or to sustain a contraction for an extended period. Flexibility is tested by observing the range of motion about a joint.

Exercise also improves cardiorespiratory endurance. The heart rate and capacity increase, and the air passages dilate so that the heart and lungs are able to support prolonged muscular activity. The blood level of high-density lipoprotein (HDL) increases. HDL is the molecule that slows the development of artherosclerotic plaques in blood vessels (see Chap. 5). Also, body composition—the proportion of protein to fat—changes favorably when you exercise.

Exercise also seems to help prevent certain types of cancer. Cancer prevention involves eating properly, not smoking, avoiding cancer-causing chemicals and radiation, undergoing appropriate medical screening tests, and knowing the early warning signs of cancer. However, studies show that people who exercise are less likely to develop colon, breast, cervical, uterine, and ovarian cancers.

Physical training with weights can improve the density and strength of bones and the strength and endurance of muscles in all adults, regardless of age. Even men and women in their 80s and 90s can make substantial gains in bone and muscle strength that help them lead more independent lives. Exercise helps prevent osteoporosis, a condition in which the bones are weak and tend to break (see page 246). Exercise promotes the activity of osteoblasts in young as well as older people. The stronger the bones when a person is young, the less chance of osteoporosis as that person ages. Exercise helps prevent weight gain, not only because the level of activity increases but also because muscles metabolize faster than other tissues. As a person becomes more muscular, the body is less likely to accumulate fat.

Exercise relieves depression and enhances the mood. Some people report that exercise actually makes them feel more energetic. Further, after exercise, particularly in the late afternoon, people sleep better that night. Self-esteem rises because of improved appearance, as well as other factors that are not well understood. For example, vigorous exercise releases endorphins, hormonelike chemicals known to alleviate pain and provide a feeling of tranquility.

A sensible exercise program is one that provides all of these benefits without the detriments of a too-strenuous program. Overexertion can be harmful to the body and may result in sports injuries, such as lower back strains or torn ligaments of the knees. The beneficial programs suggested in Table 12A are tailored according to age.

Dr. Arthur Leon at the University of Minnesota performed a study involving 12,000 men, and the results showed that only moderate exercise is needed to lower the risk of a heart attack by one-third. In another study conducted by the Institute for Aerobics Research in Dallas, Texas, which included 10,000 men and more than 3,000 women, even a little exercise was found to lower the risk of death from cardiovascular diseases and cancer. Increasing daily activity by walking to the corner store instead of driving and by taking the stairs instead of the elevator can improve your health.

Table 12A **Staying Fit**

Exercise	Children, 7–12	Teenagers, 13–18	Adults, 19–55
Amount	Vigorous activity 1–2 hr daily	Vigorous activity 1 hr, 3–5 days a week; otherwise, ½ hr daily moderate activity	Vigorous activity 1 hr, 3 days a week; otherwise, ½ hr daily moderate activity
Purpose	Free play	Build muscle with calisthenics	Exercise to prevent lower back pain: aerobics, stretching, or yoga
Organized	Build motor skills through team sports, dancing, or swimming	Continue team sports: dancing, hiking, or swimming	Do aerobic exercise to control buildup of fat cells
Group	Enjoy more exercise outside of physical education classes	Pursue sports that can be enjoyed for a lifetime: tennis, swimming, or horseback riding	Find exercise partners: join a running club, bicycle club, or outing group
Family	Participate in family outings: bowling, boating, camping, or hiking	Take active vacations: hike, bicycle, or cross-country ski	Initiate family outings: bowling, boating, camping, or hiking

Biology Matters Science Focus

Rigor Mortis

When a person dies, the physiological events that accompany death occur in an orderly progression. Respiration ceases, the heart ultimately stops beating, and tissue cells begin to die. The first tissues to die are those with the highest oxygen requirement. Brain and nervous tissues have an extremely high requirement for oxygen. Deprived of oxygen, these cells typically die after only 6 minutes because of a lack of ATP. However, tissues that can produce ATP by fermentation (which does not require oxygen) can "live" for an hour or more before ATP is completely depleted. Muscle is capable of generating ATP by fermentation. Therefore, muscle cells can survive for a time after clinical death occurs. Muscle death is signaled by a process termed **rigor mortis,** the "stiffness of death."

Stiffness occurs because—for biochemical reasons we will discuss later—muscles cannot relax unless they have a supply of ATP. Without ATP, the muscles remain fixed in their last state of contraction. If, for example, a murder victim dies while sitting at a desk, the body in rigor mortis will be frozen in the sitting position. Rigor mortis resolves approximately 24 to 36 hours after death. Muscles lose their stiffness because lysosomes rupture. The lysosomes release enzymes that break the bonds between the muscle proteins actin and myosin.

Body temperature and the presence or absence of rigor mortis allow the time of death to be estimated. For example, the body of someone dead for 3 hours or less is still warm (close to body temperature, 98.6°F, 37°C) and rigor mortis is absent. After approximately 3 hours, the body is significantly cooler than normal and rigor mortis begins to develop. The corpse of an individual dead at least 8 hours is in full rigor mortis, and the temperature of the body is the same as the surroundings. Forensic pathologists know that a person has been dead for more than 24 hours if the body temperature is the same as the environment and there is no longer a trace of rigor mortis.

Connections and Misconceptions

What causes muscles to be sore a few days after exercising?

Many of us have experienced delayed onset muscle soreness (DOMS), which generally appears some 24 to 48 hours after strenuous exercise. It is thought that DOMS is due to tissue injury that takes several days to heal. Any movement you aren't used to can lead to DOMS, but it is especially associated with any activity that causes muscles to contract while they are lengthening. Examples include walking down stairs, running downhill, lowering weights, and the downward motion of squats and push-ups. To prevent DOMS, try warming up thoroughly and cooling down completely. Stretch after exercising. When beginning a new activity, start gradually and build up your endurance gradually. Avoid making sudden major changes in your exercise routine.

Connecting the Concepts

For more information on energy sources for muscle contraction and the pathways for generating ATP, refer to the following discussions.

Sections 2.4 and 2.5 describe the structure of carbohydrates and lipids and examine their function as energy nutrients.

Section 3.6 explores how ATP is generated by the cellular respiration and fermentation pathways.

Figure 3.20 illustrates the ATP–ADP cycle.

Check Your Progress 12.3

1. List the stages of a muscle twitch.
2. Contrast the activities of a single muscle twitch with the action of summation and tetanus.
3. Summarize how the CP pathway, fermentation, and aerobic respiration produce ATP for muscle contraction.
4. Explain why weight lifters are not well adapted for distance running.

12.4 Muscular Disorders

Learning Outcomes

Upon completion of this section, you should be able to

1. Distinguish between common muscle conditions such as strains and sprains.
2. Summarize the causes of fibromyalgia, muscular dystrophy, and myasthenia gravis.

Muscular disorders are common for most people. However, there are some disorders that can be life-threatening.

Common Muscular Conditions

Spasms are sudden and involuntary muscular contractions most often accompanied by pain. Spasms can occur in smooth and skeletal muscles. A spasm of the smooth muscle in the intestinal tract is a type of colic sometimes called a bellyache. Multiple spasms of skeletal muscles are called a seizure, or

Biology Matters Historical Focus

Iron Horse: Lou Gehrig (1903–1941)

It is truly unfortunate that over time, the name Lou Gehrig has become recognized first and foremost with a disease. Gehrig, the "Iron Horse" of baseball, was born in 1903. The son of German immigrants, Gehrig was a talented athlete who excelled in many sports.

Gehrig's dedication to baseball and love for the game were legendary. He won his nickname by playing 2,130 consecutive games. His record stood until 1995, when it was finally broken by Cal Ripken. Illness, back spasms, even broken bones in his hands—nothing stopped Gehrig from playing his game. Doctors who X-rayed his hands were amazed to discover 17 different fractures of the bones in Gehrig's hands. All had slowly healed as Gehrig continued to play baseball.

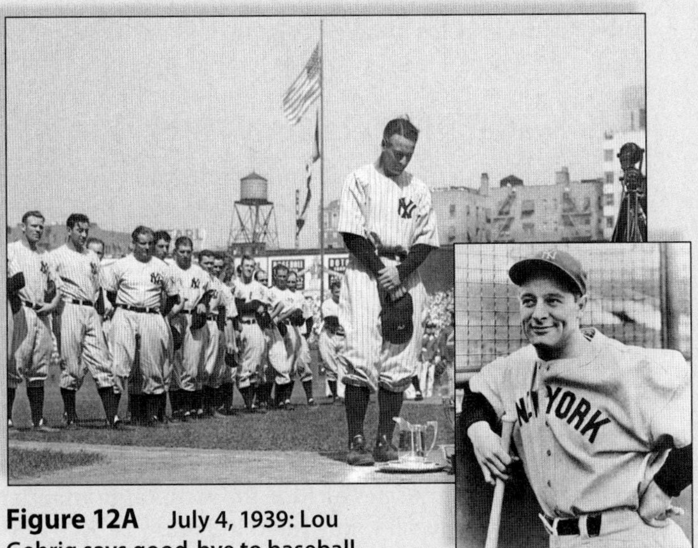

Figure 12A **July 4, 1939: Lou Gehrig says good-bye to baseball.**

During his years of playing, Gehrig set numerous records that stood for decades. His career included 493 home runs and set the record for the most home runs hit by any first baseman in history. (Mark McGwire broke Gehrig's record in 1998, but McGwire has admitted using performance-enhancing drugs during that time. See the Bioethical Focus, *Anabolic Steroid Use.*) Gehrig was the first American League player to hit four home runs in a single game. For 12 consecutive years, this quiet, humble man would have a batting average over .300. Gehrig would always play in the shadow of more famous players: first, home run slugger Babe Ruth, and later Joe DiMaggio. Yet, Gehrig was always recognized and respected for his achievements.

In 1938, Gehrig's performance went into a steady decline, and his consecutive game play ended on May 2, 1939. As the Yankee team captain, he presented the team lineup to the game umpire as always. But for the first time in 15 years, his name wasn't included. He had voluntarily pulled himself out of the lineup, and his baseball career was over. Gehrig was diagnosed with **amyotrophic lateral sclerosis (ALS)**, a rare degenerative neuromuscular disease that now bears his name. The cause of ALS is unknown. ALS sufferers gradually lose the ability to walk, talk, chew, and swallow. Mental abilities and sensation are not affected, however. The patient dies of respiratory complications, usually within three years of diagnosis. Today, drugs that slow the progression of the disease are available, but there remains no cure.

His farewell speech to team and fans was given at Yankee Stadium on July 4, 1939. As he said good-bye to the game he loved, Gehrig showed his extraordinary courage with the following words:

> *Yet today, I consider myself the luckiest man on the face of this earth. . . . So I close in saying that I may have had a tough break, but I have an awful lot to live for.*

ALS claimed Lou Gehrig's life on June 2, 1941.

convulsion. Cramps are strong, painful spasms, especially of the leg and foot, usually due to strenuous activity. Cramps can even occur when sleeping after a strenuous workout. **Facial tics,** such as periodic eye blinking, head turning, or grimacing, are spasms that can be controlled voluntarily, but only with great effort.

A **strain** is caused by stretching or tearing of a muscle. A **sprain** is a twisting of a joint leading to swelling and injury, not only of muscles but also of ligaments, tendons, blood vessels, and nerves. The ankle and knee are often subject to sprains. When a tendon is inflamed by a sprain, **tendinitis** results. Tendinitis may irritate the bursa underlying the tendon, causing **bursitis.**

Muscular Diseases

These conditions are more serious and always require close medical care.

Myalgia and Fibromyalgia

Myalgia refers to achy muscles. The most common cause for myalgia is either overuse or overstretching of a muscle or group of muscles. Myalgia without a traumatic history is often due to viral infections. Myalgia may accompany myositis (inflammation of the muscles), either in response to viral infection or as an immune system disorder. **Fibromyalgia** is a chronic condition whose symptoms include achy pain, tenderness, and stiffness of muscles. Its precise cause is not known, but it may also be due to an underlying infection that is not obvious at first.

Muscular Dystrophy

Muscular dystrophy is a broad term applied to a group of disorders characterized by a progressive degeneration and weakening of muscles. As muscle fibers die, fat and connective tissue take their place. **Duchenne muscular dystrophy,** the most common type, is inherited through a flawed gene on the X chromosome. It is now known that the lack of a protein called dystrophin causes the condition. When dystrophin is absent, calcium leaks into the cell and activates an enzyme that dissolves muscle fibers. In an attempt to treat the condition, muscles have been injected with immature muscle cells that do produce dystrophin.

Myasthenia Gravis

Myasthenia gravis is an autoimmune disease characterized by weakness that especially affects the muscles of the eyelids, face, neck, and extremities. Muscle contraction is impaired because the immune system mistakenly produces antibodies that destroy acetylcholine (ACh) receptors. In many cases, the first sign of the disease is a drooping of the eyelids and double vision. Treatment includes drugs that inhibit the enzyme that digests acetylcholine so that ACh accumulates in neuromuscular junctions.

Connecting the Concepts

For background information on the physiology of these diseases, refer to the following discussions.

Section 7.3 describes the inflammatory response.

Section 7.6 examines how an autoimmune response causes diseases in humans.

Section 20.4 provides additional information on sex-linked inheritance.

Check Your Progress 12.4

1. Distinguish between a strain and a sprain.
2. Compare and contrast the potential causes of myalgia and myasthenia gravis.
3. Discuss the potential symptoms of muscular dystrophy.

12.5 Homeostasis

Learning Outcomes

Upon completion of this section, you should be able to

1. Summarize the role of the muscular and skeletal systems in calcium homeostasis.
2. Summarize the role of the muscular and skeletal systems in body temperature homeostasis.
3. Describe how the muscular and skeletal systems aid in the homeostasis of the circulatory system.

In this section, our discussion centers on the contribution of the muscular and skeletal systems to homeostasis (Fig. 12.13).

Both Systems Produce Movement

Movement is essential to maintaining homeostasis. The skeletal and muscular systems work together to enable body movement (Fig. 12.14). This is most evidently illustrated by what happens when skeletal muscles contract and pull on the bones to which they are attached, causing movement at joints. Body movement of this sort allows us to respond to certain types of changes in the environment. For instance, if you are sitting in the sun and start to feel hot, you can get up and move to a shady spot.

The muscular and skeletal systems work for other types of movements that are just as important for maintaining homeostasis. Contraction of skeletal muscles associated with the jaw and tongue allow you to grind food with the teeth. The rhythmic smooth muscle contractions of peristalsis move ingested materials through the digestive tract. These processes are necessary for supplying the body's cells with nutrients. The ceaseless beating of your heart, which propels blood into the arterial system, is caused by the contraction of cardiac muscle. Contractions of skeletal muscles in the body, especially those associated with breathing and leg movements, aid in the process of venous return by pushing blood back toward the heart. This is why soldiers and members of marching bands are cautioned not to lock their knees when standing at attention. The reduction in venous return causes a drop in blood pressure that can result in fainting. The pressure exerted by skeletal muscle contraction also helps to squeeze tissue fluid into the lymphatic capillaries, where it is referred to as lymph.

Both Systems Protect Body Parts

The skeletal system plays an important role by protecting the soft internal organs of your body. The brain, heart, lungs, spinal cord, kidneys, liver, and most of the endocrine glands are shielded by the skeleton. In particular, the nervous and endocrine organs must be defended so they can carry out activities necessary for homeostasis.

The skeletal muscles pad and protect the bones, and the tendons and bursae associated with skeletal muscles reinforce and cushion the joints. Muscles of the abdominal wall offer additional protection to the soft internal organs. Examples of these muscles include the rectus abdominis and external oblique muscles illustrated in Figure 12.5.

Bones Store and Release Calcium

Under the direction of the endocrine system, the skeletal system performs vital tasks for calcium homeostasis. Calcium ions (Ca^{2+}) are needed for a variety of processes in your body, such as muscle contraction and nerve conduction. They are also necessary for the regulation of cellular metabolism by acting in cellular messenger systems. Thus, it is important to always maintain an adequate level of Ca^{2+} in the blood. When you have plenty of Ca^{2+} in your blood, the hormone calcitonin from the thyroid gland ensures that calcium salts are deposited in bone tissue. Thus, the skeleton acts as a reservoir for storage of this important mineral. If your blood Ca^{2+} level starts to fall, parathyroid hormone secretion stimulates osteoclasts to break down bone tissue and thereby make Ca^{2+} available to the blood. Vitamin D is needed for the absorption of Ca^{2+} from the digestive tract,

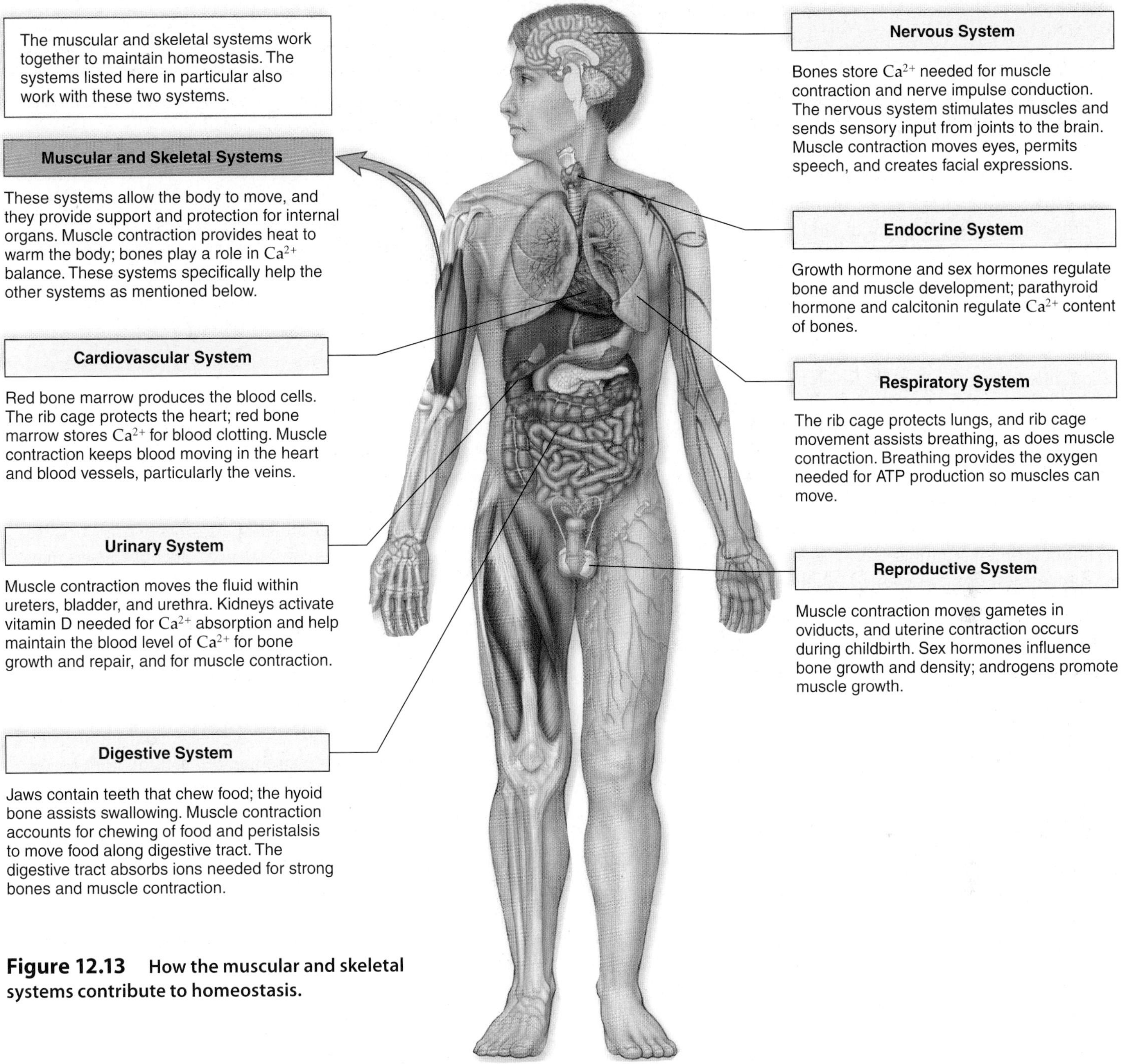

The muscular and skeletal systems work together to maintain homeostasis. The systems listed here in particular also work with these two systems.

Muscular and Skeletal Systems

These systems allow the body to move, and they provide support and protection for internal organs. Muscle contraction provides heat to warm the body; bones play a role in Ca^{2+} balance. These systems specifically help the other systems as mentioned below.

Nervous System

Bones store Ca^{2+} needed for muscle contraction and nerve impulse conduction. The nervous system stimulates muscles and sends sensory input from joints to the brain. Muscle contraction moves eyes, permits speech, and creates facial expressions.

Endocrine System

Growth hormone and sex hormones regulate bone and muscle development; parathyroid hormone and calcitonin regulate Ca^{2+} content of bones.

Cardiovascular System

Red bone marrow produces the blood cells. The rib cage protects the heart; red bone marrow stores Ca^{2+} for blood clotting. Muscle contraction keeps blood moving in the heart and blood vessels, particularly the veins.

Respiratory System

The rib cage protects lungs, and rib cage movement assists breathing, as does muscle contraction. Breathing provides the oxygen needed for ATP production so muscles can move.

Urinary System

Muscle contraction moves the fluid within ureters, bladder, and urethra. Kidneys activate vitamin D needed for Ca^{2+} absorption and help maintain the blood level of Ca^{2+} for bone growth and repair, and for muscle contraction.

Reproductive System

Muscle contraction moves gametes in oviducts, and uterine contraction occurs during childbirth. Sex hormones influence bone growth and density; androgens promote muscle growth.

Digestive System

Jaws contain teeth that chew food; the hyoid bone assists swallowing. Muscle contraction accounts for chewing of food and peristalsis to move food along digestive tract. The digestive tract absorbs ions needed for strong bones and muscle contraction.

Figure 12.13 **How the muscular and skeletal systems contribute to homeostasis.**

which is why vitamin D deficiency can result in weak bones. It is easy to get enough of this vitamin, because your skin produces it when exposed to sunlight, and the milk you buy at the grocery store is fortified with vitamin D.

Blood Cells Are Produced in Bones

The bones of your skeleton contain two types of marrow: yellow and red. Fat is stored in yellow bone marrow, thus making it part of the body's energy reserves.

Red bone marrow is the site of blood cell production. The red blood cells are the carriers of oxygen in the blood. Oxygen is necessary for the production of ATP by aerobic cellular respiration. White blood cells also originate in the red bone marrow. The white cells are involved in defending your body against pathogens and cancerous cells; without them, you would soon succumb to disease and die.

Muscles Help Maintain Body Temperature

The muscular system helps to regulate body temperature. When you are very cold, smooth muscle constricts inside the blood vessels supplying the skin. Thus, the amount of blood close to the surface of the body is reduced. This helps to conserve heat in the body's core, where vital organs lie. If you are cold enough, you may start to shiver. Shivering is caused by involuntary skeletal muscle contractions. This is initiated by temperature-sensitive neurons in the hypothalamus of the brain. Skeletal muscle contraction requires ATP, and using ATP generates heat. You may also notice that you

Biology Matters Bioethical Focus

Anabolic Steroid Use

They're called "performance-enhancing steroids," and their use is alleged to be widespread by athletes, both amateur and professional. Whether the sport is baseball, football, professional cycling, or track and field events—no activity seems to be safe from drug abuse. Steroid abuse admitted by Marion Jones forever changed Olympic history. Jones was the first female athlete to win five medals for track and field events during the 2000 Sydney Olympics. In 2008, she was stripped of all medals she had earned, as well as disqualified from a fifth-place finish in the 2004 Athens games. Future Olympic record books will not include her name. The records of her teammates in the relay events have also been tainted.

Baseball records will also likely require revisions. The exciting slugfest between Mark McGwire and Sammy Sosa in the summer of 1998 was largely credited with reviving national interest in baseball. However, the great home run competition drew unwanted attention to the darker side of professional sports when it was alleged that McGwire and Sosa were using anabolic steroids at the time. Since then, players such as Jose Canseco and McGwire have admitted using anabolic steroids to recover from baseball-related injuries. However, the controversy continues. Similar charges of drug abuse may prevent baseball great Roger Clemens from entering the Baseball Hall of Fame, despite holding the record for Cy Young awards. Likewise, because controversy continues to surround baseball legend Barry Bonds, this talented athlete may never achieve Hall of Fame status. Though he scored a record 715 home runs and won more Most Valuable Player awards than anyone in history, Bonds remains accused of steroid abuse.

Use of anabolic steroids in professional sports continues to be denied by most athletes and officials. However, many people from both inside and outside the industry maintain that such abuse has been going on for many years—and that it continues despite the negative publicity. Congress continues to investigate the controversy, yet the finger pointing and accusations steadily increase.

What Are Anabolic Steroids?

Steroids encompass a large category of substances, both beneficial and harmful. Anabolic steroids are a class of steroids that generally cause tissue growth by promoting protein production. They are naturally occurring hormones created by the body and commonly used to regulate many physiological processes, from growth to sexual function. Most anabolic steroids are closely related to male sex hormones, such as testosterone.

These metabolically potent drugs are controlled substances available only by prescription under the close supervision of a physician because of their many side effects and vast potential for abuse. However, a few anabolic steroids are still legal due to loopholes in drug laws, despite being banned by most professional sports organizations.

Figure 12B Jose Canseco wrote a book in 2006 about anabolic steroid abuse in major league baseball.

Robust muscle growth is not possible from prescription doses of anabolic steroids, so large doses must be used to obtain that effect. Athletes may take dangerously large amounts of anabolic steroids to enhance athletic performance or to increase strength, often with serious consequences. Steroid abusers may vary the type and quantity of drug taken (called "stacking") or may take the drugs and then stop for a time, only to resume later (called "cycling"). Stacking and cycling are done to minimize serious side effects while maximizing the desired effects of the drugs. Even so, dangerous health consequences can occur.

Dangerous Health Consequences

The most common health consequences include high blood pressure, jaundice (yellowing of the skin), acne, and a greatly increased risk of cancer. In women, anabolic steroid abuse may cause masculinization, including a deepened voice, excessive facial and body hair, coarsening of the hair, menstrual cycle irregularities, and enlargement of the clitoris. Anabolic steroid abuse is even more dangerous during adolescence. When taken prior to or during the teenage growth spurt, steroids may result in permanently shortened height or early onset of puberty. Ironically, whereas proper use of anabolic steroids has been helpful in treating many cases of impotence in males, abuse of these drugs may cause impotence and even shrinking of the testicles.

Perhaps the most frightening aspects of anabolic steroid abuse are the reports of increased aggressive behavior and violent mood swings. Furthermore, many users have reported extremely severe withdrawal symptoms upon quitting. Also, many anabolic steroids have also been identified as "gateway drugs," leading abusers to escalate their drug habit to more dangerous drugs such as heroin and cocaine.

Decide Your Opinion

1. Should recognitions such as admission to the Hall of Fame be denied to athletes if steroid abuse is alleged but cannot be proven?
2. Do you believe the techniques athletes use to train and enhance performance should be regulated? If so, who can or should enforce regulation?

Figure 12.14 Coordination of the skeletal and muscular system allows for complex movement.
Skeletal muscles and bones must cooperate to produce complex motion.

get goose bumps when you are cold. This is because arrector pili muscles contract. These tiny bundles of smooth muscle are attached to the hair follicles and cause the hairs to stand up. This is not very helpful in keeping humans warm, but it is quite effective in our furrier fellow mammals. Think of a cat or dog outside on a cold winter day. Its fur is a better insulator when standing up than when lying flat. Goose bumps can also be a sign of fear. Although a human with goose bumps may not look very impressive, a frightened or aggressive animal whose fur is standing on end looks bigger and (it is hoped) more intimidating to a predator or rival.

Connecting the Concepts

For more information on calcium and body temperature homeostasis, refer to the following discussions.

Section 4.8 explores how the body maintains homeostasis using feedback mechanisms.

Figure 4.17 examines how the hypothalamus is involved in body temperature regulation.

Section 15.3 describes how the thyroid and parathyroid glands are involved in calcium homeostasis.

Check Your Progress 12.5

1. List the functions of the muscular and skeletal system that contribute to calcium homeostasis.
2. Summarize how the muscular and skeletal systems work to maintain body temperature.
3. Explain how the muscular and skeletal systems interact with the cardiovascular system.

CASE STUDY CONCLUSION

There are nine different classes of muscular dystrophy. In the most common types of muscular dystrophy, symptoms occur very early in life. In Kyle's case, the relatively late onset of the disease suggested that he had a rarer form called Becker muscular dystrophy. For Kyle, the good news was that this was a much slower-progressing form of the disease, with most patients living well into their 30s without being confined to a wheelchair. Furthermore, many of the symptoms of Becker muscular dystrophy can be controlled with medication. Becker muscular dystrophy is also known to cause heart problems later in life. However, researchers are actively studying whether it may be possible to use gene therapy (see Chapter 21) to replace the defective dystrophin gene. In the interim, patients of Becker muscular dystrophy, such as Kyle, are recommended to regularly exercise to slow the loss of muscle tissue over time.

Media Study Tools

www.mhhe.com/maderhuman12e

Enhance your study of this chapter with study tools and practice tests. Also ask your instructor about the resources available through ConnectPlus, including the media-rich eBook, interactive learning tools, and animations.

Virtual Lab

The virtual lab "Muscle Stimulation" provides an interactive look at how skeletal muscles are stimulated by workload.

Summarizing the Concepts

12.1 Overview of the Muscular System

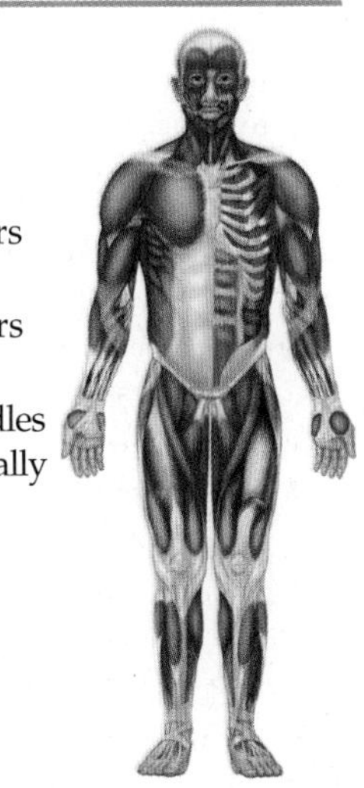

Humans have three types of muscle tissue:

- Smooth muscle is involuntary and occurs in walls of internal organs.
- Cardiac muscle is involuntary and occurs in walls of the heart.
- Skeletal muscle is voluntary, contains bundles of muscle fibers called fascicles, and is usually attached by tendons to the skeleton.

Skeletal muscle functions:

- Helps maintain posture
- Provides movement and heat
- Protects underlying organs

Skeletal Muscles of the Body

When achieving movement, some muscles are prime movers, some are synergists, and others are antagonists.

Names and Actions of Skeletal Muscles

Muscles are named for their size, shape, location, direction of fibers, number of attachments, and action.

12.2 Skeletal Muscle Fiber Contraction

Muscle fibers contain myofibrils, and myofibrils contain actin and myosin filaments. Muscle contraction occurs when sarcomeres shorten and actin filaments slide past myosin filaments.

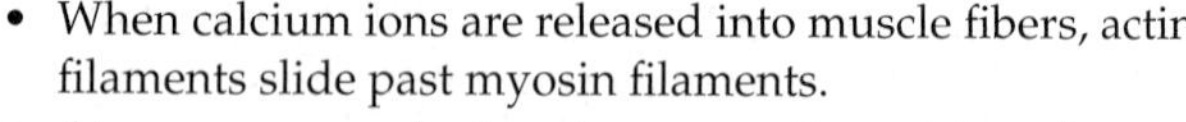

- Nerve impulses travel down motor neurons and stimulate muscle fibers at neuromuscular junctions.
- The sarcolemma of a muscle fiber forms T tubules that almost touch the sarcoplasmic reticulum, which stores calcium ions.
- When calcium ions are released into muscle fibers, actin filaments slide past myosin filaments.
- At a neuromuscular junction, synaptic vesicles release acetylcholine (neurotransmitter), which diffuses across the synaptic cleft.
- When acetylcholine (ACh) is received by the sarcolemma, electrical signals begin and lead to the release of calcium.
- Calcium ions bind to troponin, exposing myosin binding sites.
- Myosin filaments break down ATP and attach to actin filaments, forming cross-bridges.
- When ADP and Ⓟ are released, cross-bridges change their positions.
- This pulls actin filaments to the center of a sarcomere.

12.3 Whole Muscle Contraction

Muscles Have Motor Units

- A muscle contains motor units: several fibers under the control of a single motor axon.
- Motor unit contraction is described in terms of a muscle twitch, summation, and tetanus.
- The strength of muscle contraction varies according to recruitment of motor units.
- In the body, a continuous slight tension (called muscle tone) is maintained by muscle motor units that take turns contracting.

Energy for Muscle Contraction

A muscle fiber has three ways to acquire ATP for muscle contraction:

- Creatine phosphate (CP) transfers a phosphate to ADP, and ATP results. This CP pathway is the most rapid.

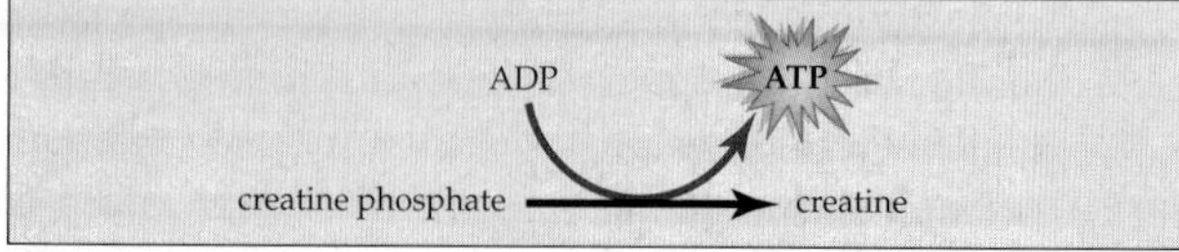

- Fermentation also produces ATP quickly. Fermentation is associated with an oxygen debt because oxygen is needed to metabolize the lactate that accumulates.

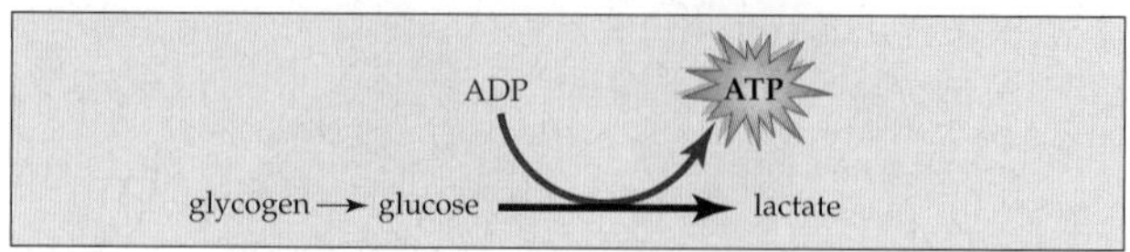

- Cellular respiration provides most of the muscle's ATP but takes longer because much of the glucose and oxygen must be transported in blood to mitochondria. Cellular respiration occurs during aerobic exercise and burns fatty acids in addition to glucose.

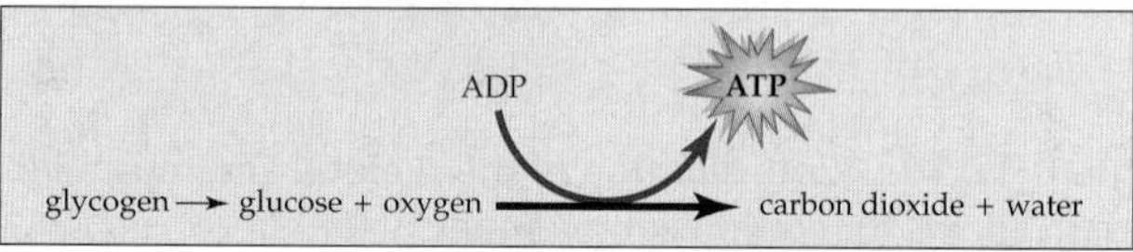

Fast-Twitch and Slow-Twitch Muscle Fibers

- Fast-twitch fibers, for sports like weight lifting, rely on an anaerobic means of acquiring ATP; have few mitochondria and myoglobin, but motor units contain more muscle fibers; and are known for explosive power but fatigue quickly.
- Slow-twitch fibers, for sports like running and swimming, rely on aerobic respiration to acquire ATP and have a plentiful supply of mitochondria and myoglobin, which gives them a dark color.

12.4 Muscular Disorders

Muscular disorders include spasms and injuries, as well as diseases such as muscular dystrophy and myasthenia gravis.

12.5 Homeostasis

- The muscles and bones produce movement and protect body parts.
- The bones produce red blood cells and are involved in the regulation of blood calcium levels.
- The muscles produce the heat that gives us a constant body temperature.

Understanding Key Terms

actin 267
all-or-none law 271
amyotrophic lateral sclerosis (ALS) 277
bursa 264
bursitis 277
cardiac muscle 263
convulsion 277
cramp 277
Duchenne muscular dystrophy 278
facial tic 277
fibromyalgia 277
insertion 264
intercalated disk 263
motor unit 271
muscle fiber 263
muscle tone 272
muscle twitch 272
muscular dystrophy 277
myalgia 277
myasthenia gravis 278
myofibril 267
myoglobin 274
myosin 267
neuromuscular junction 270
origin 264
oxygen debt 273
rigor mortis 276
sarcolemma 267

sarcomere 267
sarcoplasmic reticulum 267
skeletal muscle 264
sliding filament model 269
smooth (visceral) muscle 263
spasm 277
sprain 277
strain 277
T (transverse) tubule 267
tendinitis 277
tendon 264
tetanus 272
tropomyosin 270
troponin 270

Match the key terms to these definitions.

a. ________ Structural and functional unit of a myofibril; contains actin and myosin filaments.

b. ________ End of a muscle attached to a movable bone.

c. ________ Sustained maximal muscle contraction.

d. ________ Contraction of muscles at death due to lack of ATP.

e. ________ Stretching or tearing of a muscle.

Testing Your Knowledge of the Concepts

1. What are the characteristics of the three types of muscles in the human body? Where is each type found? (pages 263–264)
2. Give an example to show that skeletal muscles work in antagonistic pairs. Explain. (page 264)
3. What criteria are used to name muscles? Give an example of each one. (page 266)
4. What are the functions of a muscle fiber's components? (page 267)
5. Describe the sliding filament model of muscle contraction within a sarcomere. Begin with the nerve impulse and end with the relaxation of the muscle. (pages 267–270)
6. What are the four possible energy sources for a muscle and the three sources of ATP for muscle contraction? (pages 272–274)
7. Compare fast- and slow-twitch muscle fibers. (pages 274–275)
8. What are some common muscular disorders and some more serious muscular diseases? (pages 277–278)
9. How does the muscular system help maintain homeostasis? (pages 278–281)
10. Impulses that move down the T system of a muscle fiber most directly cause
    a. movement of tropomyosin.
    b. attachment of the cross-bridges to myosin.
    c. release of Ca^{2+} from the sarcoplasmic reticulum.
    d. splitting of ATP.
11. Which of the following statements about cross-bridges is false?
    a. They are composed of myosin.
    b. They bind to ATP after they detach from actin.
    c. They contain an ATPase.
    d. They split ATP before they attach to actin.
12. Which statement about sarcomere contraction is incorrect?
    a. The A bands shorten.
    b. The H zones shorten.
    c. The I bands shorten.
    d. The sarcomeres shorten.
13. Which of the following muscles would have motor units with the lowest innervation ratio?
    a. leg muscles
    b. arm muscles
    c. muscles that move the fingers
    d. muscles of the trunk
14. The thick filaments of a muscle fiber are made up of
    a. actin.
    b. troponin.
    c. fascia.
    d. myosin.
15. As ADP and Ⓟ are released from a myosin head,
    a. actin filaments move toward the H zone.
    b. myosin cross-bridges pull the thin filaments.
    c. a sarcomere shortens.
    d. Only a and c are correct.
    e. All of these are correct.
16. Which of these is a direct source of energy for muscle contraction?
    a. ATP
    b. creatine phosphate
    c. lactic acid
    d. glycogen
    e. Both a and b are correct.
17. When muscles contract,
    a. sarcomeres increase in length.
    b. actin breaks down ATP.
    c. myosin slides past actin.
    d. the H zone disappears.
    e. calcium is taken up by the sarcoplasmic reticulum.
18. Nervous stimulation of muscles
    a. occurs at a neuromuscular junction.
    b. results in an impulse that travels down the T system.
    c. causes calcium to be released from expanded regions of the sarcoplasmic reticulum.
    d. All of these are correct.
19. In a muscle fiber,
    a. the sarcolemma is connective tissue holding the myofibrils together.
    b. the T system causes release of Ca^{2+} from the sarcoplasmic reticulum.
    c. both actin and myosin filaments have cross-bridges.
    d. there is no endoplasmic reticulum.
    e. All of these are correct.
20. To increase the force of muscle contraction,
    a. individual muscle cells have to contract with greater force.
    b. motor units have to contract with greater force.
    c. motor units need to be recruited.
    d. All of these are correct.
    e. None of these is correct.
21. Lack of calcium in muscles
    a. results in no contraction.
    b. causes weak contraction.
    c. causes strong contraction.
    d. has no effect.
    e. None of these is correct.
22. Which of these energy relationships is mismatched?
    a. creatine phosphate—anaerobic
    b. cellular respiration—aerobic
    c. fermentation—anaerobic
    d. oxygen debt—anaerobic
    e. All of these are properly matched.

Propagation of an Action Potential

If an axon is unmyelinated, an action potential at one locale stimulates an adjacent part of the axon membrane to produce an action potential. Conduction along the entire axon in this fashion can be rather slow (approximately 1 meter/second in thin axons) because each section of the axon must be stimulated.

Animation Action Potential Propagation

In myelinated fibers, an action potential at one node of Ranvier causes an action potential at the next node, jumping over the entire myelin-coated portion of the axon. This type of conduction is called **saltatory conduction** (*saltatio* is a Latin word that means "to jump") and is much faster. In thick, myelinated fibers, the rate of transmission is more than 100 m/s. Regardless of whether an axon is myelinated or not, its action potentials are self-propagating. Each action potential generates another, along the entire length of the axon.

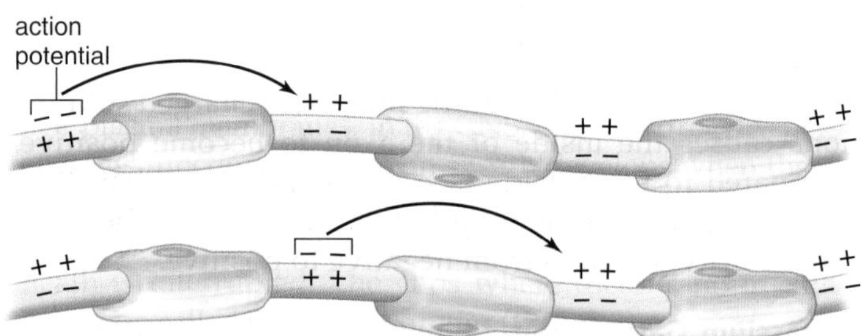
action
potential

Like the action potential itself, conduction of an action potential is an all-or-none event—either an axon conducts its action potential or it does not. The intensity of a message is determined by how many action potentials are generated within a given time. An axon can conduct a volley of action potentials very quickly, because only a small number of ions are exchanged with each action potential. Once the action potential is complete, the ions are rapidly restored to their proper place through the action of the sodium–potassium pump.

As soon as the action potential has passed by each successive portion of an axon, that portion undergoes a short **refractory period** during which it is unable to conduct an action potential. This ensures the one-way direction of a signal from the cell body down the length of the axon to the axon terminal.

It is interesting to observe that all functions of the nervous system, from our deepest emotions to our highest reasoning abilities, are dependent on the conduction of nerve signals.

The Synapse

Every axon branches into many fine endings, each tipped by a small swelling called an **axon terminal.** Each terminal lies very close to either the dendrite or the cell body of another neuron. This region of close proximity is called a **synapse** (Fig. 13.4). At a synapse, a small gap called the **synaptic cleft** separates the sending neuron from the receiving neuron. The nerve signal is unable to jump the cleft. Therefore, another means is needed to pass the nerve signal from the sending neuron to the receiving neuron.

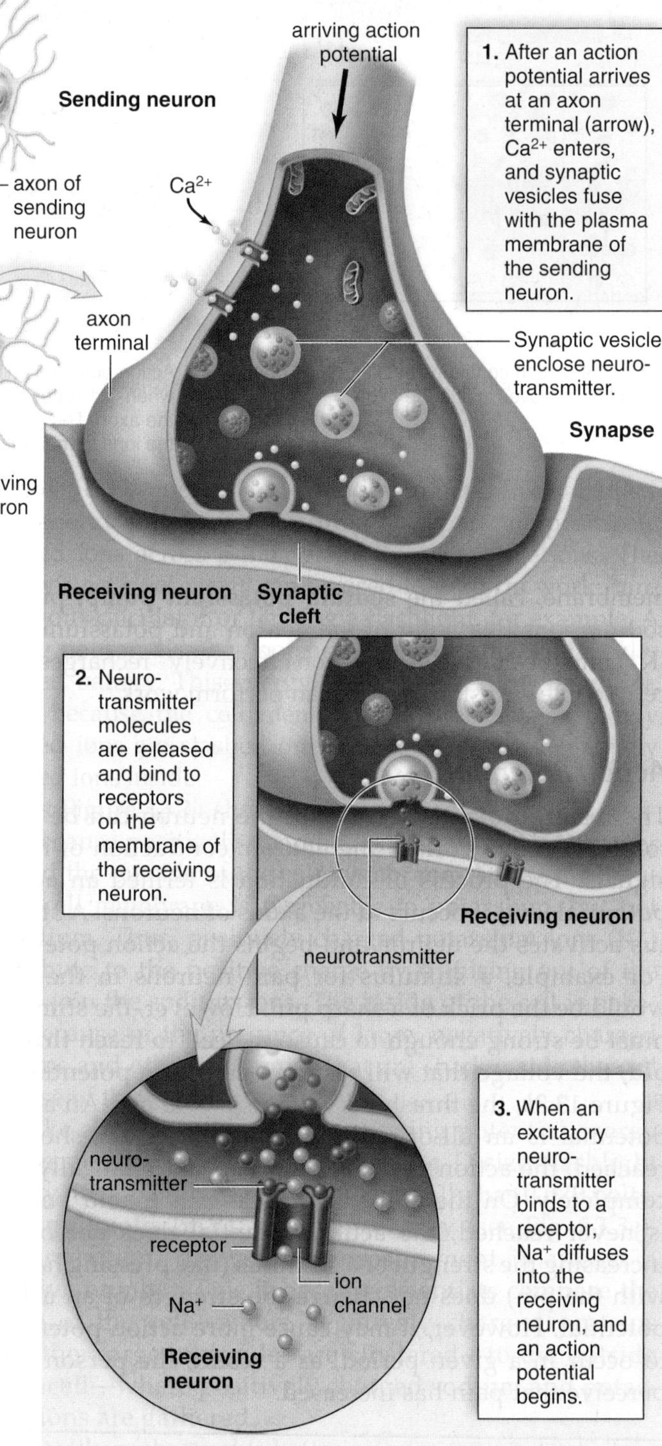
Sending neuron
axon of sending neuron
receiving neuron
arriving action potential
Ca^{2+}
axon terminal
1. After an action potential arrives at an axon terminal (arrow), Ca^{2+} enters, and synaptic vesicles fuse with the plasma membrane of the sending neuron.
Synaptic vesicles enclose neurotransmitter.
Synapse
Receiving neuron
Synaptic cleft
2. Neurotransmitter molecules are released and bind to receptors on the membrane of the receiving neuron.
Receiving neuron
neurotransmitter
neurotransmitter
receptor
ion channel
Na^+
Receiving neuron
3. When an excitatory neurotransmitter binds to a receptor, Na^+ diffuses into the receiving neuron, and an action potential begins.

Figure 13.4 Signal transmission at the synapse.
Transmission across a synapse from one neuron to another occurs when a neurotransmitter is released and diffuses across a synaptic cleft and binds to a receptor in the membrane of the receiving neuron.

Transmission across a synapse is carried out by molecules called **neurotransmitters,** stored in synaptic vesicles in the axon terminals. (Recall from Chap. 3 that a vesicle is a membranous sac that stores and transports substances.) The events at a synapse are (1) nerve signals traveling along an axon to reach an axon terminal; (2) calcium ions entering the terminal and stimulating synaptic vesicles to merge with the sending membrane; and (3) neurotransmitter molecules releasing into the synaptic cleft and diffusing across the cleft to the receiving membrane; there, neurotransmitter molecules bind with specific receptor proteins. Depending on the type of neurotransmitter, the response of the receiving neuron can be toward excitation or toward inhibition. In Figure 13.5, excitation occurs because the neurotransmitter, such as acetylcholine (ACh), has caused the sodium gate to open. Sodium ions diffuse into the receiving neuron. Inhibition would occur if a neurotransmitter caused potassium ions to exit the receiving neuron.

Once a neurotransmitter has been released into a synaptic cleft and has initiated a response, it is removed from the cleft. In some synapses, the receiving membrane contains enzymes that rapidly inactivate the neurotransmitter. For example, the enzyme **acetylcholinesterase (AChE)** breaks down acetylcholine. In other synapses, the sending membrane rapidly reabsorbs the neurotransmitter, possibly for repackaging in synaptic vesicles or for molecular breakdown.

Animation Chemical Synapse

The short existence of neurotransmitters at a synapse prevents continuous stimulation (or inhibition) of receiving membranes. The receiving cell needs to be able to respond quickly to changing conditions. If the neurotransmitter were to linger in the cleft, the receiving cell would be unable to respond to a new signal from a sending cell.

MP3 Synapses

Neurotransmitter Molecules

Among the more than 100 substances known or suspected to be neurotransmitters are **acetylcholine (ACh), norepinephrine (NE), dopamine, serotonin, glutamate,** and **GABA (gamma aminobutyric acid).** Neurotransmitters transmit signals between nerves. Nerve–muscle, nerve–organ, and nerve–gland synapses also communicate using neurotransmitters.

Acetylcholine and norepinephrine are active in both the CNS and PNS. In the PNS, these neurotransmitters act at synapses called neuromuscular junctions. Neuromuscular junctions are discussed in Chapter 12.

In the PNS, ACh excites skeletal muscle but inhibits cardiac muscle. It has either an excitatory or inhibitory effect on smooth muscle or glands, depending on their location.

Norepinephrine generally excites smooth muscle. In the CNS, norepinephrine is important to dreaming, waking, and mood. Serotonin is involved in thermoregulation, sleeping, emotions, and perception. Many drugs that affect the nervous system act at the synapse. Some interfere with the actions of neurotransmitters, and other drugs prolong the effects of neurotransmitters (see Section 13.5).

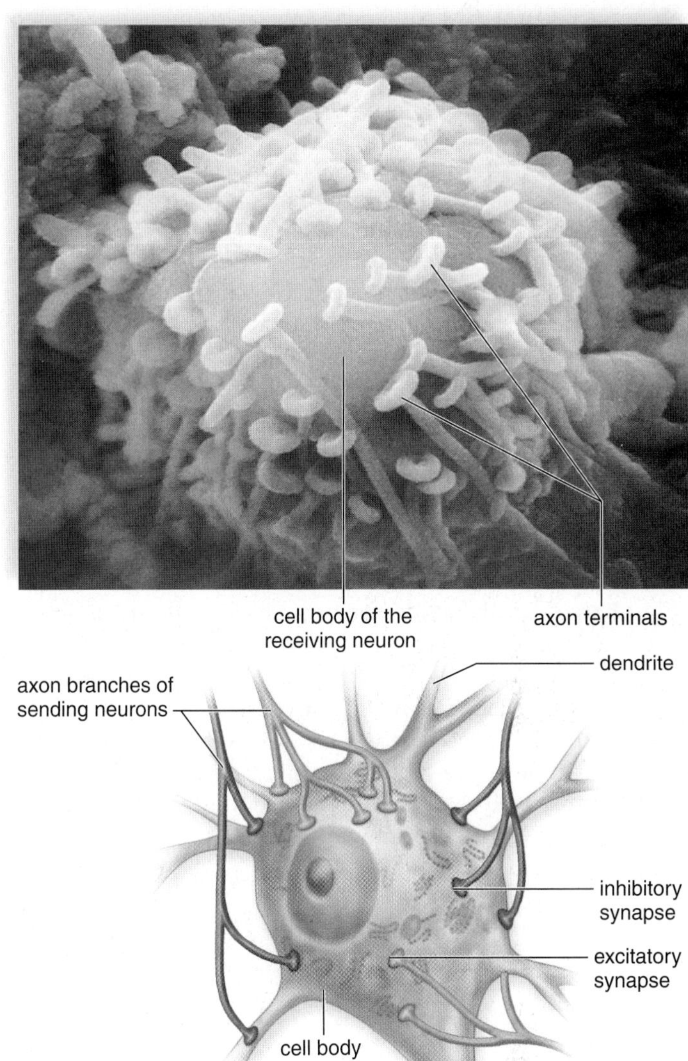

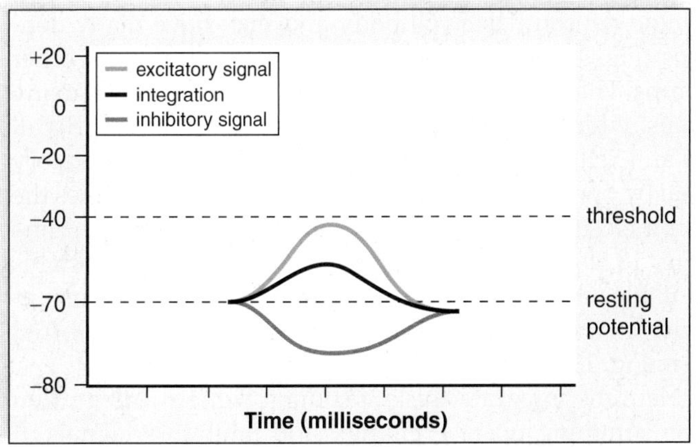

Figure 13.5 Integration of excitatory and inhibitory signal integrated at the synapse.
a. Inhibitory signals and excitatory signals are summed up in the dendrite and cell body of the postsynaptic neuron. Only if the combined signals cause the membrane potential to rise above threshold does an action potential occur. **b.** In this example, threshold was not reached.

Biology Matters Historical Focus

Artist and Scientist: Santiago Ramon y Cajal (1852–1934)

In any discipline—humanities, music and the arts, history, mathematics, education, social sciences—innovation and discovery happen because of the hard work of dedicated people. The contributions of the passionate scientist Dr. Santiago Ramon y Cajal established a foundation for ongoing studies of the nervous system.

In the late-nineteenth century, the brain was believed to be a continuous network of "filaments" and scientists were not convinced that the filaments were even cells. Using a new technique, Cajal stained samples of brain tissue with a dye containing metallic silver. Careful microscopic studies showed Cajal that the brain was composed of individual cells. A later researcher named nerve cells *neurons.* Cajal then discovered that the neurons were not directly connected to one another. This discovery allowed later scientists to research this gap—the synapse—as well as the neurotransmitters that allow nerve cells to communicate across a synapse. An artist as well as a researcher, Cajal illustrated his microscopic discoveries. His drawings were reproduced in textbooks for decades.

Cajal's theory of "dynamic polarization" described the idea of the resting and action potential. He proposed that neurons received signals at the cell body and dendrites and transmitted these signals via their axons to other neurons. Cajal described this basic principle of neuron function long before methods were devised to prove his theories.

Cajal was awarded the Nobel Prize in 1906 for his discoveries in the structure and function of the nervous system.

a.

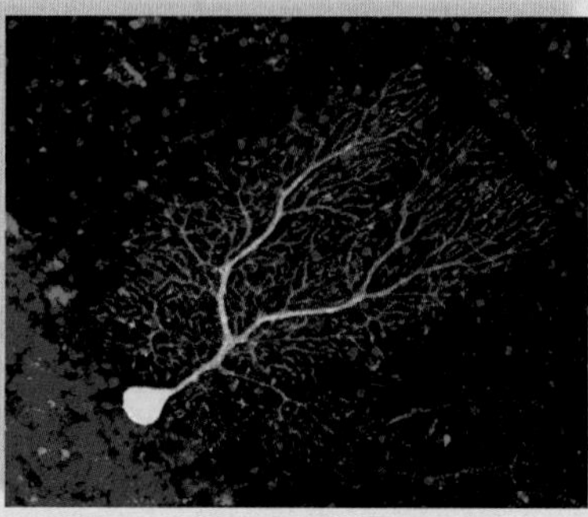
b.

Figure 13A Structure of a neuron.
a. Santiago Ramon y Cajal's sketch of neuron. **b.** Actual photomicrograph of the same neuron.

Synaptic Integration

A single neuron has a cell body and may have many dendrites (Fig. 13.5*a*). All can have synapses with many other neurons. Therefore, a neuron is on the receiving end of many signals, which can either be excitatory or inhibitory. Recall that an excitatory neurotransmitter produces an excitatory signal by opening sodium gates at a synapse. This drives the neuron closer to its threshold (illustrated by the green line in Fig. 13.5*b*). If threshold is reached, an action potential is inevitable. On the other hand, an inhibitory neurotransmitter drives the neuron farther from an action potential (red line in Fig. 13.5*b*) by opening the gates for potassium.

Neurons integrate these incoming signals. **Integration** is the summing up of excitatory and inhibitory signals. If a neuron receives enough excitatory signals (either from different synapses or at a rapid rate from a single synapse) to outweigh the inhibitory ones, chances are the axon will transmit a signal. On the other hand, if a neuron receives more inhibitory than excitatory signals, summing these signals may prohibit the axon from reaching threshold and then depolarizing (black solid line in Fig. 13.5*b*).

Connecting the Concepts

For more information on neurons and the nervous systems, refer to the following discussions.

Section 4.4 explores how stem cells may be used to regenerate nervous tissue.

Figure 12.7 illustrates the role of the synapse in the neuromuscular junction.

Section 14.1 explains how the peripheral nervous system sends information to and from the central nervous system.

Check Your Progress 13.1

1. Describe the three types of neurons and list the three main parts of a neuron.
2. Describe how a nerve impulse is propagated.
3. Summarize how a nerve impulse is transmitted from one neuron to the next.

13.2 The Central Nervous System

Learning Outcomes

Upon completion of this section, you should be able to

1. Identify the structures of the spinal cord and provide a function for each.
2. Identify the structures of the brain and provide a function for each.
3. Identify the lobes and major areas of the human brain.
4. Distinguish between the functions of the primary motor and the primary somatosensory areas of the brain.

The spinal cord and the brain make up the CNS, where sensory information is received and motor control is initiated. As mentioned previously, both the spinal cord and the brain are protected by bone. The spinal cord is surrounded by vertebrae, and the brain is enclosed by the skull. Also, both the spinal cord and the brain are wrapped in protective membranes known as **meninges** (sing., meninx). *Meningitis* is an infection of the meninges. The spaces between the meninges are filled with **cerebrospinal fluid,** which cushions and protects the CNS. A small amount of this fluid is sometimes withdrawn from around the spinal cord for laboratory testing when a spinal tap (lumbar puncture) is performed.

Cerebrospinal fluid is also contained within the ventricles of the brain and in the central canal of the spinal cord. The brain has four **ventricles,** interconnecting chambers that produce and serve as a reservoir for cerebrospinal fluid (Fig. 13.6). Normally, any excess cerebrospinal fluid drains away into the cardiovascular system. However, blockages can occur. In an infant, the brain can enlarge due to cerebrospinal fluid accumulation, resulting in a condition called hydrocephalus ("water on the brain"). If cerebrospinal fluid collects in an adult, the brain cannot enlarge. Instead, it is pushed against the skull. Such situations cause severe brain damage and can be fatal unless quickly corrected.

The CNS is composed of two types of nervous tissue—gray matter and white matter. **Gray matter** contains cell bodies and short, nonmyelinated fibers. **White matter** contains myelinated axons that run together in bundles called **tracts.**

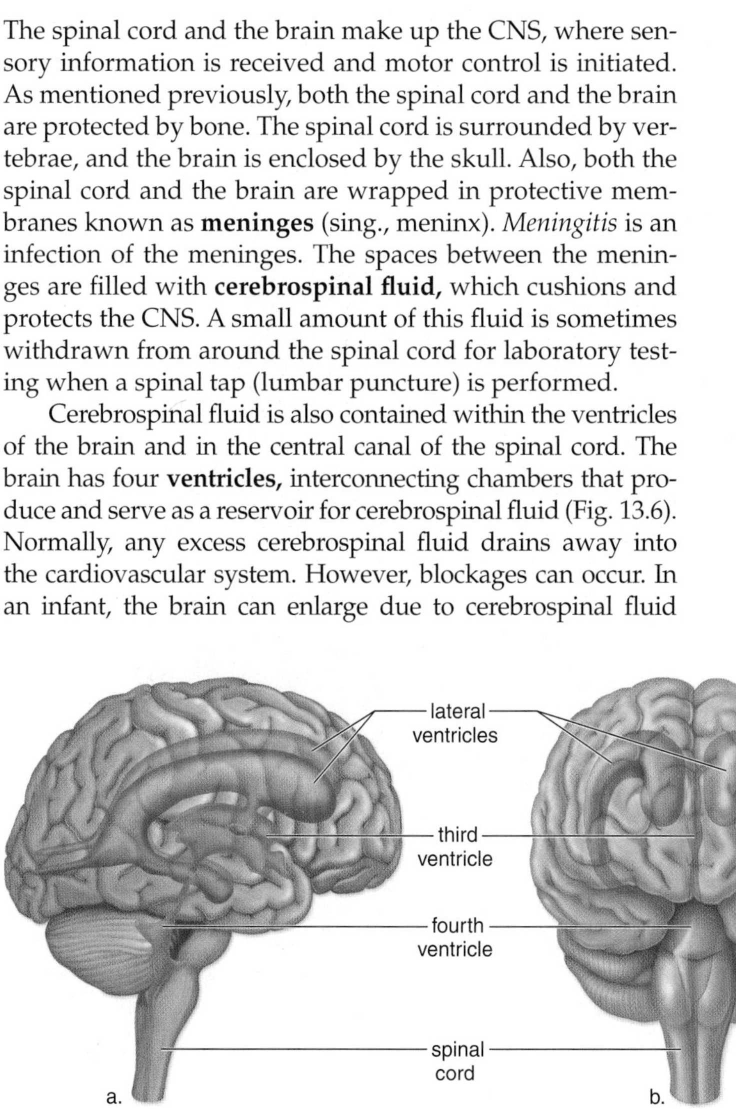

Figure 13.6 **The ventricles of the brain.**
The brain has four ventricles. A lateral ventricle is found on each side of the brain. They join at the third ventricle. The third ventricle connects with the fourth ventricle superiorly; the central canal of the spinal cord joins the fourth ventricle inferiorly. All structures are filled with cerebrospinal fluid. **a.** Lateral view of ventricles seen through a transparent brain. **b.** Anterior view of ventricles seen through a transparent brain.

The Spinal Cord

The **spinal cord** extends from the base of the brain through a large opening in the skull called the foramen magnum (see Fig. 11.7). From the foramen magnum, the spinal cord proceeds inferiorly in the vertebral canal.

Structure of the Spinal Cord

A cross section of the spinal cord shows a central canal, gray matter, and white matter (Fig. 13.7*a*). Figure 13.7*b* shows how an individual vertebra protects the spinal cord. The spinal nerves project from the cord through intervertebral foramina (see Fig. 11.9). Fibrocartilage intervertebral disks separate the vertebrae. If the disk ruptures or herniates, the vertebrae compress a spinal nerve. Pain and loss of mobility result.

The central canal of the spinal cord contains cerebrospinal fluid, as do the meninges that protect the spinal cord. The gray matter is centrally located and shaped like the letter H (Fig. 13.7*a*, *b*, *c*). Portions of sensory neurons and motor neurons are found in gray matter, as are interneurons that communicate with these two types of neurons. The dorsal root of a spinal nerve contains sensory fibers entering the gray matter. The ventral root of a spinal nerve contains motor fibers exiting the gray matter. The dorsal and ventral roots join before the spinal nerve leaves the vertebral canal, forming a mixed nerve (Fig. 13.7*c*, *d*). Spinal nerves are a part of the PNS.

The white matter of the spinal cord occurs in areas around the gray matter. The white matter contains ascending tracts taking information to the brain (primarily located posteriorly) and descending tracts taking information from the brain (primarily located anteriorly). Many tracts cross just after they enter and exit the brain, so the left side of the brain controls the right side of the body. Likewise, the right side of the brain controls the left side of the body.

Functions of the Spinal Cord

The spinal cord provides a means of communication between the brain and the peripheral nerves that leave the cord. When someone touches your hand, sensory receptors generate nerve signals that pass through sensory fibers to the spinal cord and up ascending tracts to the brain (see Fig. 13.1*b*, red arrows).

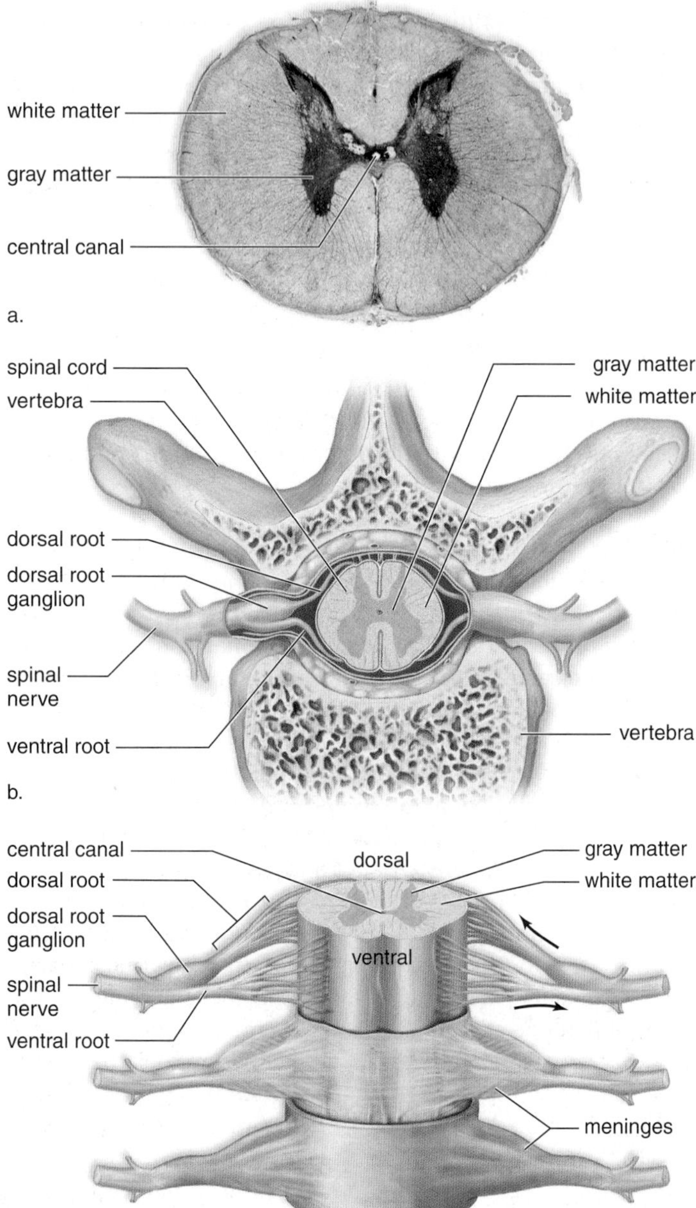

Figure 13.7 The organization of white and gray matter in the spinal cord and the spinal nerves.
a. Cross section of the spinal cord, showing arrangements of white and gray matter. **b.** Spinal nerves originating from the spinal cord. **c.** The spinal cord is protected by vertebrae. **d.** Spinal nerves emerging from the cord.

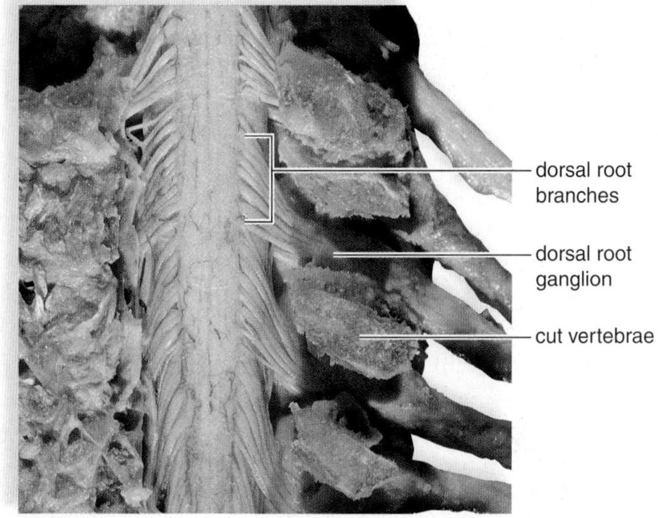

d. Dorsal view of spinal cord and dorsal roots of spinal nerves.

The gate control theory of pain proposes that the tracts in the spinal cord have "gates" and that these "gates" control the flow of pain messages from the peripheral nerves to the brain. Depending on how the gates process a pain signal, the pain message can be allowed to pass directly to the brain or can be prevented from reaching the brain. As mentioned earlier, endorphins can temporarily block pain messages and so can other messages, such as those received from touch receptors.

To touch the person back, the brain initiates voluntary control over our limbs. Motor signals originating in the brain pass down descending tracts to the spinal cord and out to our muscles by way of motor fibers (see Fig. 13.1*b*, black arrows). Therefore, if the spinal cord is severed, we suffer a loss of sensation and a loss of voluntary control—paralysis. If the cut occurs in the thoracic region, the lower body and legs are paralyzed, a condition known as *paraplegia.* If the injury is in the neck region, all four limbs are usually affected, a condition called *quadriplegia.*

Reflex Actions The spinal cord is the center for thousands of reflex arcs (see Fig. 13.16). A stimulus causes sensory receptors to generate signals that travel in sensory axons to the spinal cord. Interneurons integrate the incoming data and relay signals to motor neurons. A response to the stimulus occurs when motor axons cause skeletal muscles to contract. Motor neurons in a reflex arc may also affect smooth muscle, organs, or glands. Each interneuron in the spinal cord has synapses with many other neurons. Therefore, interneurons send signals to other interneurons and motor neurons.

Similarly, the spinal cord creates reflex arcs for the internal organs. For example, when blood pressure falls, internal receptors in the carotid arteries and aorta generate nerve signals that pass through sensory fibers to the cord and then up an ascending tract to a cardiovascular center in the brain. Thereafter, nerve signals pass down a descending tract to the spinal cord. Motor signals then cause blood vessels to constrict so that the blood pressure rises.

The Brain

The human **brain** has been called the last great frontier of biology. The goal of modern neuroscience is to understand the structure and function of the brain's various parts so well that it will be possible to prevent or correct the thousands of mental disorders that rob human beings of a normal life. This section gives only a glimpse of what is known about the brain and the modern avenues of research.

Video Brain Bank

We discuss the parts of the brain with reference to the cerebrum, the diencephalon, the cerebellum, and the brain stem. The brain's four ventricles (see Fig. 13.6) are called, in turn, the two lateral ventricles, the third ventricle, and the fourth ventricle. It may be helpful for you to associate the cerebrum with the two lateral ventricles, the diencephalon with the third ventricle, and the brain stem and the cerebellum with the fourth ventricle (Fig. 13.8*a*).

MP3 The Brain

The Cerebrum

The **cerebrum,** also called the telencephalon, is the largest portion of the brain in humans. The cerebrum is the last center to receive sensory input and carry out integration before commanding voluntary motor responses. It communicates with and coordinates the activities of the other parts of the brain.

Cerebral Hemispheres Just as the human body has two halves, so does the cerebrum. These halves are called the left and right **cerebral hemispheres** (Fig. 13.8*b*). A deep groove called the longitudinal fissure divides the left and right cerebral hemispheres. The two cerebral hemispheres communicate via the **corpus callosum,** an extensive bridge of nerve tracts.

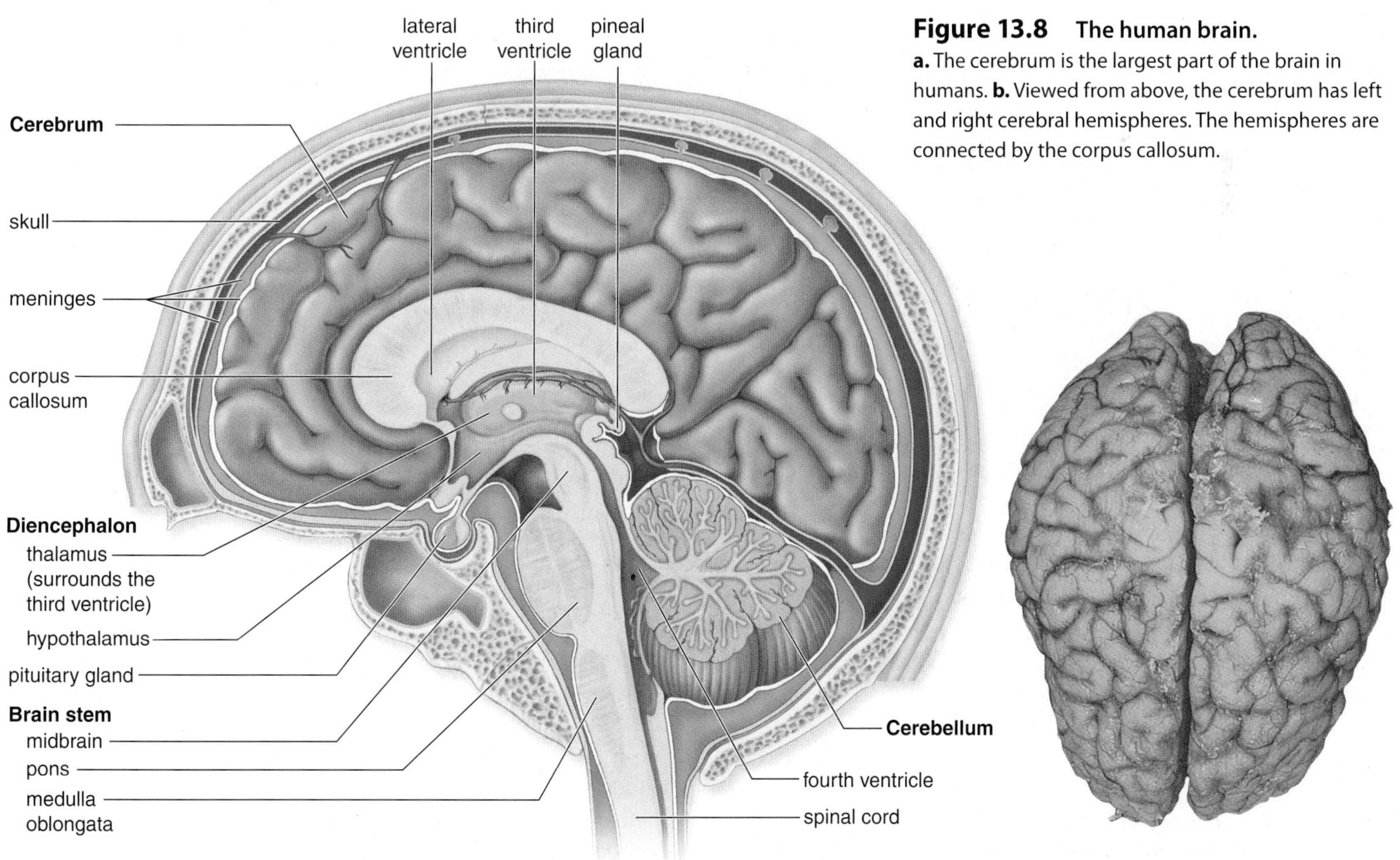

Figure 13.8 The human brain.
a. The cerebrum is the largest part of the brain in humans. **b.** Viewed from above, the cerebrum has left and right cerebral hemispheres. The hemispheres are connected by the corpus callosum.

Shallow grooves called *sulci* (sing., sulcus) divide each hemisphere into lobes (Fig. 13.9). The *frontal lobe* is the most anterior of the lobes (directly behind the forehead). The *parietal lobe* is posterior to the frontal lobe. The *occipital lobe* is posterior to the parietal lobe (at the rear of the head). The *temporal lobe* lies inferior to the frontal and parietal lobes (at the temple and the ear).

Each lobe is associated with particular functions, as indicated in Figure 13.9.

The Cerebral Cortex The **cerebral cortex** is a thin, highly convoluted outer layer of gray matter that covers the cerebral hemispheres. (Recall that gray matter consists of neurons whose axons are unmyelinated.) The cerebral cortex contains over 1 billion cell bodies and is the region of the brain that accounts for sensation, voluntary movement, and all the thought processes we associate with consciousness.

Primary Motor and Sensory Areas of the Cortex The cerebral cortex contains motor areas and sensory areas, as well as association areas. The **primary motor area** is in the frontal lobe just anterior to (before) the central sulcus. Voluntary commands to skeletal muscles begin in the primary motor area, and each part of the body is controlled by a certain section (Fig. 13.10*a*). Observe the illustration carefully. You'll see that large areas of cerebral cortex are devoted to controlling structures that carry out very fine, precise movements. Thus, the muscles that control facial movements—swallowing, salivation, expression—take up an especially large portion of the primary motor area. Likewise, hand movements require tremendous accuracy. Together, these two structures command nearly two-thirds of the primary motor area.

The **primary somatosensory area** is just posterior to the central sulcus in the parietal lobe. Sensory information from the skin and skeletal muscles arrives here, where each part of the body is sequentially represented (Fig. 13.10*b*). Like the primary motor cortex, large areas of the primary sensory cortex are dedicated to those body areas with acute sensation. Once again, the face and hands require the largest proportion of the sensory cortex.

Reception areas for the other primary sensations—taste, vision, hearing, and smell—are located in other areas of the cerebral cortex (see Fig. 13.9). The primary taste area in the parietal lobe (pink) accounts for taste sensations. Visual

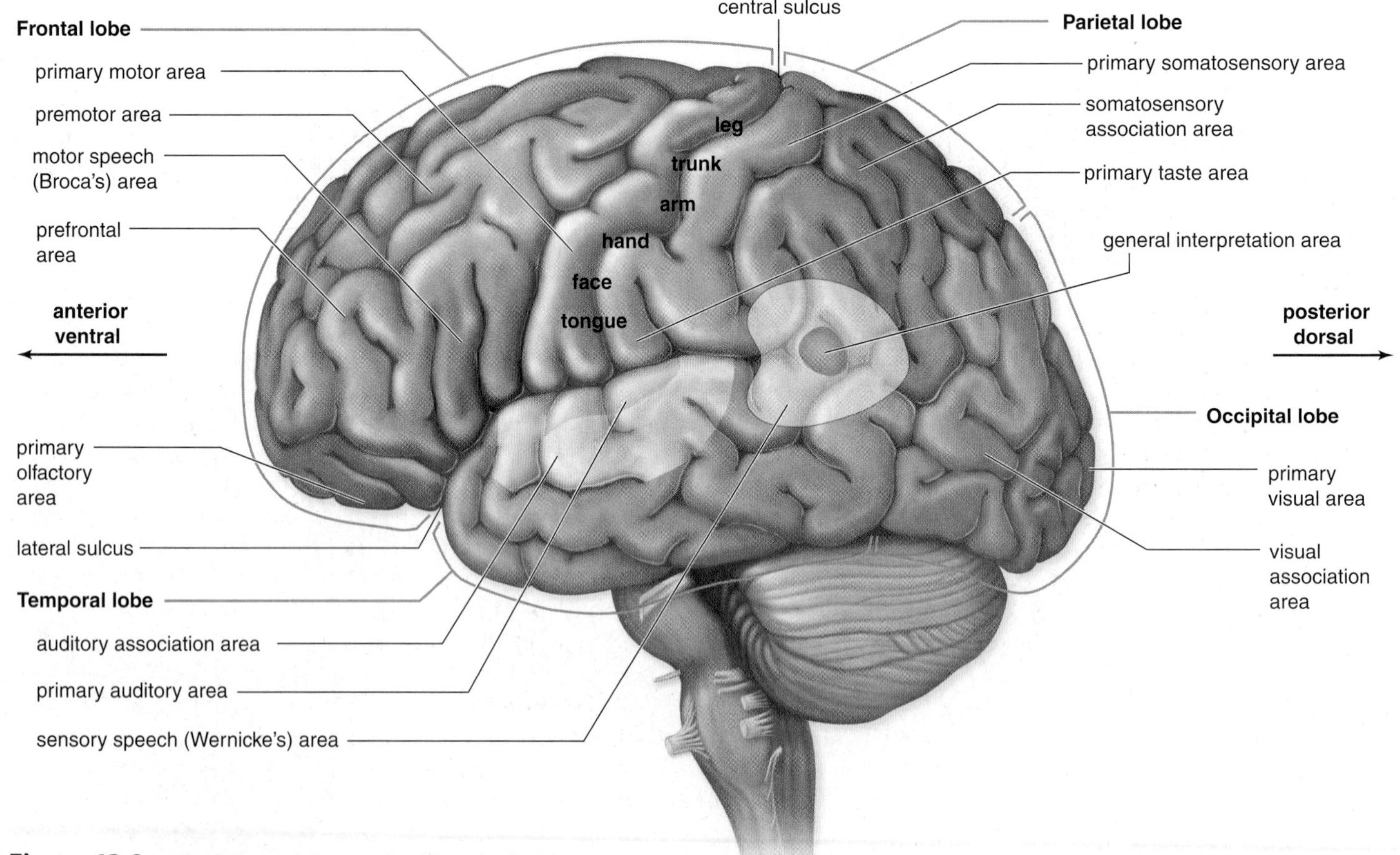

Figure 13.9 **The lobes of the cerebral hemispheres.**
Each cerebral hemisphere is divided into four lobes: frontal, parietal, temporal, and occipital. Centers in the frontal lobe control movement and higher reasoning, as well as the smell sensation. Somatic sensing is carried out by parietal lobe neurons, and those of the temporal lobe receive sound information. Visual information is received and processed in the occipital lobe.

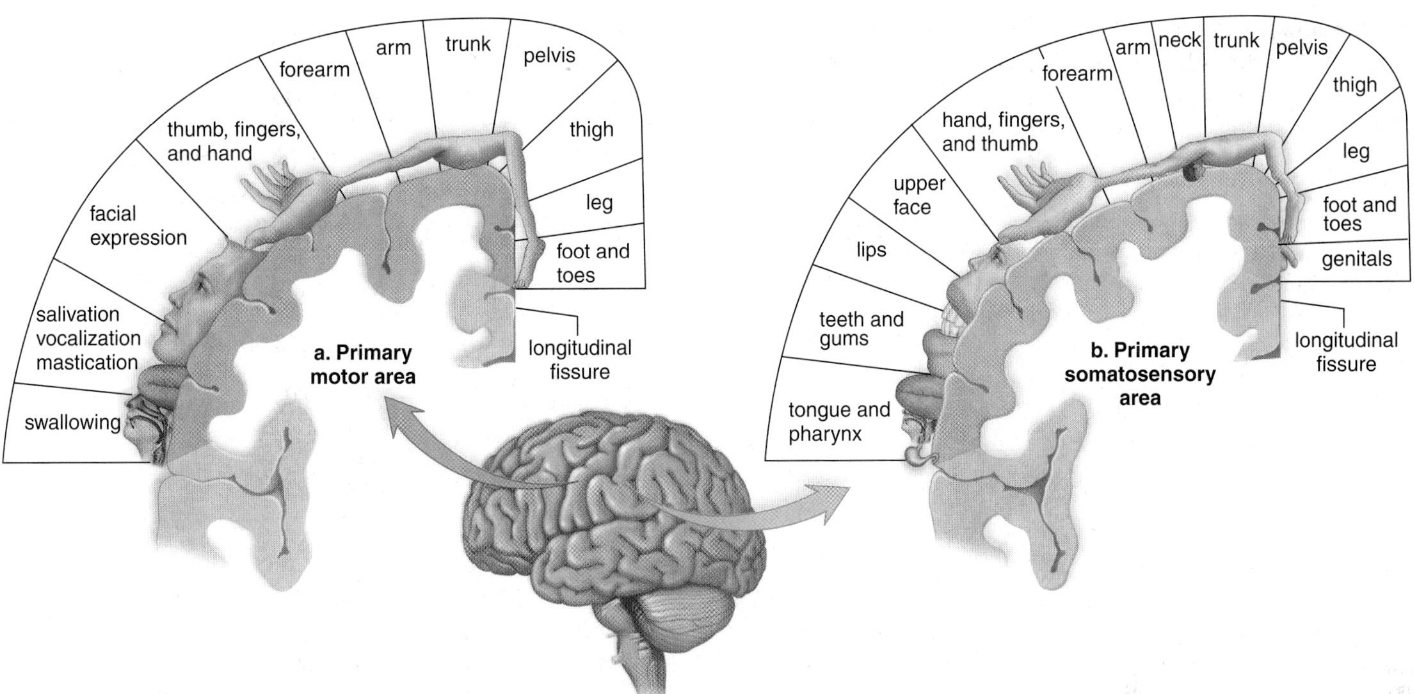

Figure 13.10 The primary motor and primary somatosensory areas of the brain.
a. The primary motor area (blue) is located in the frontal lobe, adjacent to (**b**) the primary somatosensory area in the parietal lobe. The primary taste area is colored pink. The size of each body region shown indicates the relative amount of cortex devoted to control of that body region.

information is received by the primary visual cortex (blue) in the occipital lobe. The primary auditory area in the temporal lobe (dark green) accepts information from our ears. Smell sensations travel to the primary olfactory area (yellow) found on the deep surface of the frontal lobe.

Association Areas **Association areas** are places where integration occurs and where memories are stored. Anterior to the primary motor area is a premotor area. The premotor area organizes motor functions for skilled motor activities, such as walking and talking at the same time. Next, the primary motor area sends signals to the cerebellum, which integrates them. A momentary lack of oxygen during birth can damage the motor areas of the cerebral cortex. Cerebral palsy, a condition characterized by a spastic weakness of the arms and legs, results. The *somatosensory association area,* located just posterior to the primary somatosensory area, processes and analyzes sensory information from the skin and muscles. The *visual association area* in the occipital lobe associates new visual information with stored visual memories. It might "decide," for example, if we have seen a face, scene, or symbol before. The *auditory association area* in the temporal lobe performs the same functions with regard to sounds.

Processing Centers Processing centers of the cortex receive information from the other association areas and perform higher-level analytical functions. The **prefrontal area,** an association area in the frontal lobe, receives information from the other association areas and uses this information to reason and plan our actions. Integration in this area accounts for our most cherished human abilities. Reasoning, critical thinking, and formulating appropriate behaviors are possible because of integration carried out in the prefrontal area.

The unique ability of humans to speak is partially dependent upon two processing centers found only in the left cerebral cortex. **Wernicke's area** is located in the posterior part of the left temporal lobe. **Broca's area** is located in the left frontal lobe. Broca's area is located just anterior to the portion of the primary motor area for speech musculature (lips, tongue, larynx, and so forth) (see Fig. 13.9). Wernicke's area helps us understand both the written and spoken word and sends the information to Broca's area. Broca's area adds grammatical refinements and directs the primary motor area to stimulate the appropriate muscles for speaking and writing.

Central White Matter Much of the rest of the cerebrum is composed of white matter. Myelination occurs and white matter develops as a child grows. Progressive myelination enables the brain to grow in size and complexity. For example, as neurons become myelinated within tracts designed for language development, children become more capable of speech. Descending tracts from the primary motor area communicate with lower brain centers, and ascending tracts from lower brain centers send sensory information

MP3
The Cerebrum

Connections and Misconceptions

Why does a stroke on the right side of the brain cause weakness or paralysis, as well as decreased sensation, on the left side of the body?

Descending motor tracts (from the primary motor area) and ascending sensory tracts (from the primary somatosensory area) cross over in the spinal cord and medulla. Motor neurons in the right cerebral hemisphere control the left side of the body and vice versa because of crossing-over. Likewise, sensation from the left half of the body travels to the right cerebral hemisphere. Destruction of brain tissue by a stroke interferes with outgoing motor signals to the opposite side of the body, as well as incoming sensory information.

up to the primary somatosensory area. Tracts within the cerebrum also take information between the different sensory, motor, and association areas pictured in Figure 13.9. As previously mentioned, the corpus callosum contains tracts that join the two cerebral hemispheres.

Video Brain Surgery

Basal Nuclei

Though the majority of each cerebral hemisphere is composed of tracts, there are masses of gray matter deep within the white matter. These **basal nuclei** integrate motor commands to ensure that the proper muscle groups are stimulated or inhibited. Integration ensures that movements are coordinated and smooth. **Parkinson disease** (see Chap. 17) is believed to be caused by degeneration of specific neurons in the basal nuclei.

The Diencephalon

The hypothalamus and the thalamus are in the **diencephalon,** a region that encircles the third ventricle. The **hypothalamus** forms the floor of the third ventricle. The hypothalamus is an integrating center that helps maintain homeostasis. It regulates hunger, sleep, thirst, body temperature, and water balance. The hypothalamus controls the pituitary gland and thereby serves as a link between the nervous and endocrine systems.

The **thalamus** consists of two masses of gray matter located in the sides and roof of the third ventricle. The thalamus is on the receiving end for all sensory input except the sense of smell. Visual, auditory, and somatosensory information arrives at the thalamus via the cranial nerves and tracts from the spinal cord. The thalamus integrates this information and sends it on to the appropriate portions of the cerebrum. The thalamus is involved in arousal of the cerebrum, and it also participates in higher mental functions such as memory and emotions.

The pineal gland, which secretes the hormone melatonin, is located in the diencephalon. Presently there is much popular interest in the role of melatonin in our daily rhythms. Some researchers believe it can help alleviate jet lag or insomnia. Scientists are also interested in the possibility that the hormone may regulate the onset of puberty.

The Cerebellum

The **cerebellum** lies under the occipital lobe of the cerebrum and is separated from the brain stem by the fourth ventricle. The cerebellum has two portions joined by a narrow median portion. Each portion is primarily composed of white matter. In alongitudinal section, the white matter has a treelike pattern called *arbor vitae*. Overlying the white matter is a thin layer of gray matter that forms a series of complex folds.

The cerebellum receives sensory input from the eyes, ears, joints, and muscles about the present position of body parts. It also receives motor output from the cerebral cortex about where these parts should be located. After integrating this information, the cerebellum sends motor signals by way of the brain stem to the skeletal muscles. In this way, the cerebellum maintains posture and balance. It also ensures that all of the muscles work together to produce smooth, coordinated, voluntary movements. The cerebellum assists the learning of new motor skills such as playing the piano or hitting a baseball.

The Brain Stem

The tracts cross in the **brain stem,** which contains the midbrain, the pons, and the medulla oblongata (see Fig. 13.8*a*). The **midbrain** acts as a relay station for tracts passing between the cerebrum and the spinal cord or cerebellum. It also has reflex centers for visual, auditory, and tactile responses. The word **pons** means "bridge" in Latin. True to its name, the pons contains bundles of axons traveling between the cerebellum and the rest of the CNS. In addition, the pons functions with the medulla oblongata to regulate breathing rate. Reflex centers in the pons coordinate head movements in response to visual and auditory stimuli.

The **medulla oblongata** contains a number of reflex centers for regulating heartbeat, breathing, and vasoconstriction (blood pressure). It also contains the reflex centers for vomiting, coughing, sneezing, hiccuping, and swallowing. The medulla oblongata lies just superior to the spinal cord, and it contains tracts that ascend or descend between the spinal cord and higher brain centers. Recall that tracts are groups of axons that travel together (page 293). Ascending tracts convey sensory information. Motor information is transmitted on descending tracts.

The Reticular Formation The **reticular formation** is a complex network of nuclei (sing., **nucleus**), which are masses of gray matter, and fibers that extend the length of the brain stem (Fig. 13.11). The reticular formation is a major component of the reticular activating system (RAS). The RAS receives sensory signals and sends them to higher centers. Motor signals received by the RAS are sent to the spinal cord.

The RAS arouses the cerebrum via the thalamus and causes a person to be alert. If you want to awaken the RAS,

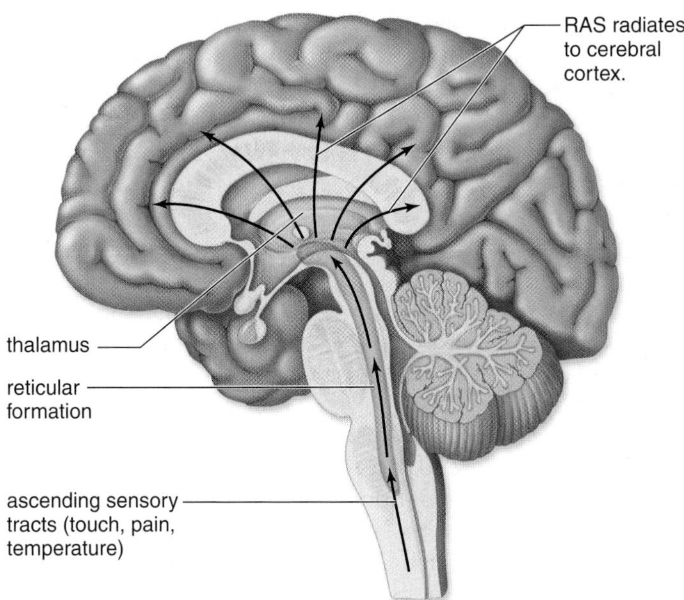

Figure 13.11 The reticular formation of the brain.
The reticular formation receives and sends on motor and sensory information to various parts of the CNS. One portion, the reticular activating system (RAS; see arrows), arouses the cerebrum and, in this way, controls alertness versus sleep.

surprise it with sudden stimuli, such as an alarm clock ringing, bright lights, smelling salts, or splashing cold water on your face. The RAS can filter out unnecessary sensory stimuli, explaining why you can study with the TV on. Similarly, the RAS allows you to take a test without noticing the sounds of the people around you—unless the sounds are particularly distracting. To inactivate the RAS, remove visual or auditory stimuli, allowing yourself to become drowsy and drop off to sleep. General anesthetics function by artificially suppressing the RAS. A severe injury to the RAS can cause a person to be comatose, from which recovery may be impossible.

Connecting the Concepts

For more information on the central nervous system, refer to the following discussions.

Section 9.5 examines how the central nervous system controls breathing.

Section 17.5 explores how aging and diseases such as Alzheimer disease influence the brain.

Sections 22.3 and 22.4 outline how the size of the brain has changed over the course of human evolution.

Check Your Progress 13.2

1. List the functions of the spinal cord.
2. Detail the four major parts of the brain and describe the general function of each.
3. Relate how the RAS aids in homeostasis.

13.3 The Limbic System and Higher Mental Functions

Learning Outcomes

Upon completion of this section, you should be able to

1. Identify the structures of the limbic system.
2. Explain how the limbic system is involved in memory, language, and speech.
3. Summarize the types of memory associated with the limbic system.

The limbic system integrates our emotions (fear, joy, sadness) with our higher mental functions (reason, memory). Because of the limbic system, activities such as sexual behavior and eating seem pleasurable and mental stress can cause high blood pressure.

Limbic System

The **limbic system** is an evolutionary ancient group of linked structures deep within the cerebrum. It is a functional grouping rather than an anatomical one (Fig. 13.12). The limbic system blends primitive emotions and higher mental functions into a united whole. As already noted, it accounts for why activities such as sexual behavior and eating seem pleasurable. Conversely, unpleasant sensations or emotions (pain, frustration, hatred, despair) are translated by the limbic system into a stress response.

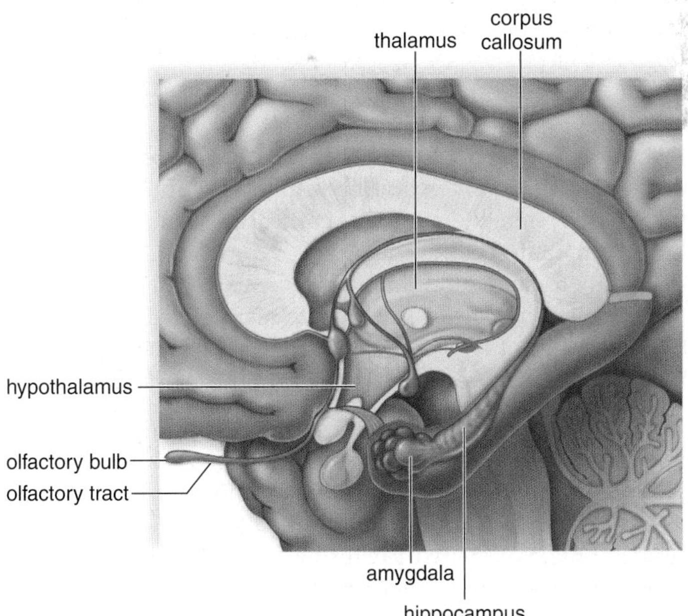

Figure 13.12 The regions of the brain associated with the limbic system.
In the limbic system (purple), structures deep within each cerebral hemisphere and surrounding the diencephalon join higher mental functions, such as reasoning, with more primitive feelings, such as fear and pleasure. Therefore, primitive feelings can influence our behavior, but reason can also keep them in check.

Two significant structures within the limbic system are the amygdala and the hippocampus. The **amygdala,** in particular, can cause experiences to have emotional overtones, and it creates the sensation of fear. This center can use past knowledge fed to it by association areas to assess a current situation. If necessary, the amygdala can trigger the fight-or-flight reaction. So if you are out late at night and you turn to see someone in a ski mask following you, the amygdala may immediately cause you to start running. The frontal cortex can override the limbic system and cause you to rethink the situation and prevent you from acting out strong reactions.

The **hippocampus** is believed to play a crucial role in learning and memory. The hippocampal region acts as an information gateway during the learning process. It determines what information about the world is to be sent to memory and how this information is to be encoded and stored by other regions in the brain. Most likely, the hippocampus can communicate with the frontal cortex because we know that memories are an important part of our decision-making processes.

It's hoped that research into the functioning of the hippocampus will help in understanding **Alzheimer disease**, a brain disorder characterized by gradual loss of memory (see Chap. 17).

Higher Mental Functions

As in other areas of biological research, brain research has progressed due to technological breakthroughs. Neuroscientists now have a wide range of techniques at their disposal for studying the human brain, including modern technologies that allow us to record its functioning.

Memory and Learning

Just as the connecting tracts of the corpus callosum are evidence that the two cerebral hemispheres work together, so the limbic system indicates that cortical areas may work with lower centers to produce learning and memory. **Memory** is the ability to hold a thought in mind or to recall events from the past, ranging from a word we learned only yesterday to an early emotional experience that has shaped our lives. **Learning** takes place when we retain and use past memories.

Types of Memory We have all tried to remember a seven-digit telephone number for a short time. If we say we are trying to keep it in the forefront of our brain, we are exactly correct. The prefrontal area, active during **short-term memory,** lies just posterior to our forehead! There are some telephone numbers that we have memorized. In other words, they have gone into **long-term memory.** Think of a telephone number you know by heart, and try to bring it to mind without also thinking about the place or person associated with that number. Most likely you cannot. Typically, long-term memory is a mixture of what is called **semantic memory** (numbers, words, etc.) and **episodic memory** (persons, events, etc.).

Skill memory is another type of memory that can exist independent of episodic memory. Skill memory is involved in performing motor activities such as riding a bike or playing ice hockey. When a person first learns a skill, more areas of the cerebral cortex are involved than after the skill is perfected. In other words, you have to think about what you are doing when you learn a skill, but later the actions become automatic. Skill memory involves all the motor areas of the cerebrum below the level of consciousness.

Long-Term Memory Storage and Retrieval Our long-term memories are apparently stored in bits and pieces throughout the sensory association areas of the cerebral cortex. Visions are stored in the vision association area, sounds are stored in the auditory association area, and so forth. As previously mentioned, the hippocampus serves as a bridge between the sensory association areas (where memories are stored) and the prefrontal area (where memories are used). The prefrontal area communicates with the hippocampus when memories are stored and when these memories are brought to mind. Some memories are emotionally charged because the amygdala seems to be responsible for fear conditioning and associating danger with sensory stimuli received from various parts of the brain.

Long-Term Potentiation Though it is helpful to know the memory functions of various portions of the brain, an important step toward curing mental disorders is understanding memory on the cellular level. After synapses have been used intensively for a short time, they release more neurotransmitters than before. This phenomenon, called **long-term potentiation (LTP)**, may be involved in memory storage.

Connections and Misconceptions

What is amnesia?

Amnesia results from disruption of the memory pathways and can be temporary or permanent. In anterograde amnesia, injury to the limbic system separates long-term memories of events that occurred prior to the injury from events that occur in the here and now. An affected person might carry on a conversation about past events (memories of a long-ago birthday) but cannot recall a breakfast menu from that morning. In retrograde amnesia, a blow to the head or similar injury abolishes all memories for a variable time before the injury. For example, a head injury occurring during a car accident may abolish all memories from hours to days prior to the accident.

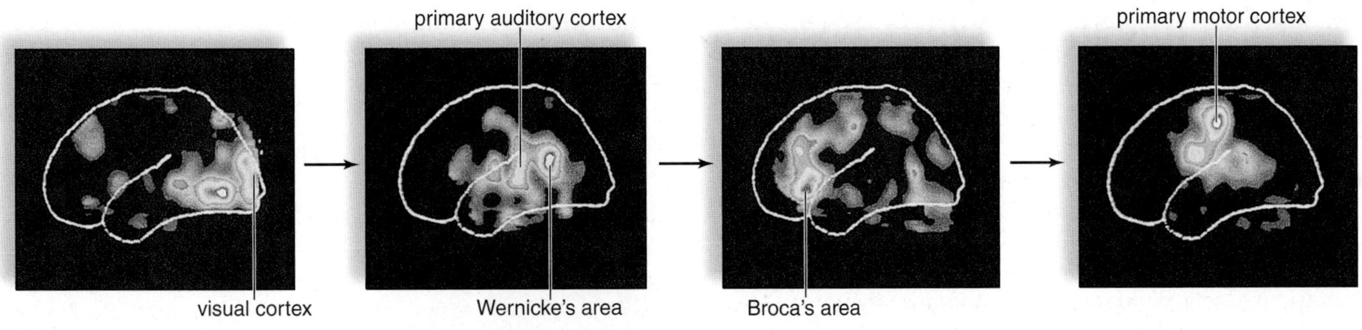

primary auditory cortex
primary motor cortex
visual cortex
Wernicke's area
Broca's area
1. The word is seen in the visual cortex.
2. Information concerning the word is interpreted in Wernicke's area.
3. Information from Wernicke's area is transferred to Broca's area.
4. Information is transferred from Broca's area to the primary motor area.

Figure 13.13 The areas of the brain involved in reading.
These functional images were captured by a high-speed computer during a PET (positron-emission tomography) scan of the brain. A radioactively labeled solution is injected into the subject, and then the subject is asked to perform certain activities. Cross-sectional images of the brain generated by the computer reveal where activity is occurring because the solution is preferentially taken up by active brain tissue and not by inactive brain tissue. These PET images show the cortical pathway for reading words and then speaking them. Red indicates the most active areas of the brain, and blue indicates the least active areas.

Language and Speech

Language depends on semantic memory. Therefore, we would expect some of the same areas in the brain to be involved in both memory and language. Any disruption of these pathways could contribute to an inability to comprehend our environment and use speech correctly.

Seeing and hearing words depends on sensory centers in the occipital and temporal lobes, respectively. Damage to Wernicke's area, discussed earlier, results in the inability to comprehend speech. Damage to Broca's area, on the other hand, results in the inability to speak and write. The functions of the visual cortex, Wernicke's area, and Broca's area are shown in Figure 13.13.

One interesting aside pertaining to language and speech is the recognition that the left brain and the right brain may have different functions. Recall that the left hemisphere contains both Broca's area and Wernicke's area. As you might expect, it appears that the left hemisphere plays a role of great importance in language functions. The role of the isolated left hemisphere can be studied in patients after surgery to sever the corpus callosum. This procedure is used for seizure control in patients with epilepsy. After surgery, the patient is termed "split brain," because there is no longer direct communication between the two cerebral hemispheres. If a split-brain individual views an object with only the right eye, its image will be sent only to the right hemisphere. This person will be able to choose the proper object for a particular use—scissors to cut paper, for example—but will be unable to name that object.

Research on the split brain is ongoing. In a very general way, the left brain can be contrasted with the right brain.

Left Hemisphere	Right Hemisphere
Verbal	Nonverbal, visuospatial
Logical, analytical	Intuitive
Rational	Creative

Researchers now believe that the hemispheres process the same information differently. The left hemisphere is more global, whereas the right hemisphere is more specific in its approach.

Connecting the Concepts

For more information on these topics, refer to the following discussions.

Section 17.5 examines the effects of aging on the body, including the nervous system.

Check Your Progress 13.3

1. Define the function of the limbic system.
2. List what limbic system structures are involved in the fight-or-flight reaction, learning, and long-term memory.
3. Describe the relationship between the left and right sides of the brain and language and speech.

13.4 The Peripheral Nervous System

Learning Outcomes

Upon completion of this section, you should be able to

1. Describe the series of events during a spinal reflex.
2. Distinguish between the somatic and autonomic divisions of the peripheral nervous system.
3. Distinguish between the sympathetic and parasympathetic divisions of the autonomic division.

The peripheral nervous system (PNS), which lies outside the central nervous system, contains the nerves. Nerves are designated as cranial nerves when they arise from the brain and spinal nerves when they arise from the spinal cord. In any case, all nerves carry signals to and from the CNS. So right now, your eyes are sending messages by way of a cranial nerve to the brain, allowing you to read a page. When you're finished, your brain will direct the muscles in your fingers to turn the page by way of the spinal cord and a spinal nerve.

Figure 13.14 illustrates the anatomy of a nerve. The cell body and the dendrites of neurons are in either the CNS or ganglia. **Ganglia** (sing., **ganglion**) are collections of nerve cell bodies outside the CNS. The axons of neurons project from the CNS and form the spinal cord. In other words, nerves, whether cranial or spinal, are composed of axons, the long part of neurons.

Humans have 12 pairs of **cranial nerves** attached to the brain. By convention, the pairs of cranial nerves are referred to by roman numerals (Fig. 13.15). Some cranial nerves are sensory nerves—they contain only sensory fibers; some are motor nerves that contain only motor fibers; and others are mixed nerves that contain both sensory and motor fibers. Cranial nerves are largely concerned with the head, neck, and facial regions of the body. However, the vagus nerve (X) has branches not only to the pharynx and larynx but also to most of the internal organs. From which part of the brain do you think the vagus arises? It arises from the brain stem, specifically, the medulla oblongata that communicates so well with the hypothalamus. These two parts of the brain control the internal organs.

As you know, the **spinal nerves** of humans emerge from either side of the spinal cord (see Fig. 13.7). There are 31 pairs of spinal nerves. The roots of a spinal nerve physically separate the axons of sensory neurons from the axons of motor neurons, forming an arrangement resembling a letter Y. The posterior root of a spinal nerve contains sensory fibers that direct sensory receptor information inward (toward the spinal cord). The cell body of a sensory neuron is in a posterior-root ganglion (also termed a **dorsal-root ganglion).** The anterior (also termed ventral) root of a spinal nerve contains motor fibers that conduct impulses outward (away from the cord) to the effectors. Observe in Figure 13.7 that the anterior and posterior roots rejoin to form a spinal nerve. All spinal nerves are called mixed nerves because they contain both sensory and motor fibers. Each spinal nerve serves the particular region of the body in which it is located. For example, the intercostal muscles of the rib cage are innervated by thoracic nerves.

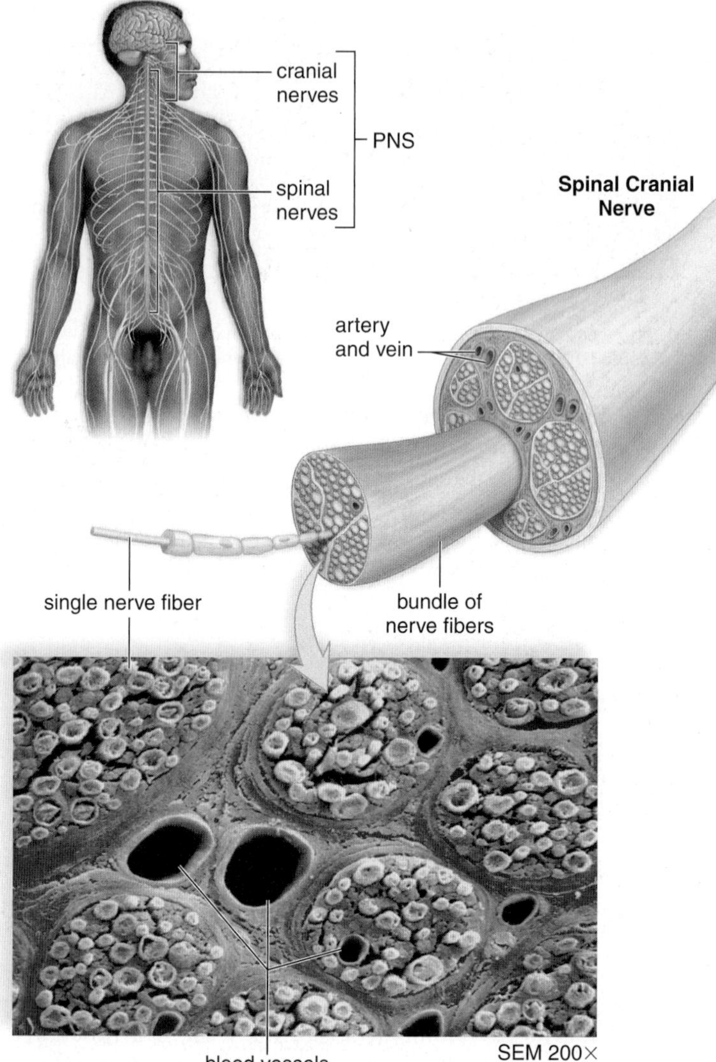

Figure 13.14 The structure of a nerve.
The peripheral nervous system consists of the cranial nerves and the spinal nerves. A nerve is composed of bundles of axons separated from one another by connective tissue.

Somatic System

The PNS has divisions, and right now we are going to consider the somatic system. The nerves in the **somatic system** serve the skin, skeletal muscles, and tendons (see Fig. 13.1). The somatic system sensory nerves take sensory information from external sensory receptors to the CNS. Motor commands leaving the CNS travel to skeletal muscles via somatic motor nerves.

Not all somatic motor actions are voluntary. Some actions are automatic. Automatic responses to a stimulus in the somatic system are called **reflexes.** A reflex occurs quickly, without your even having to think about it. For

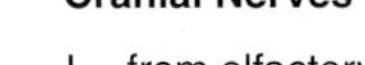

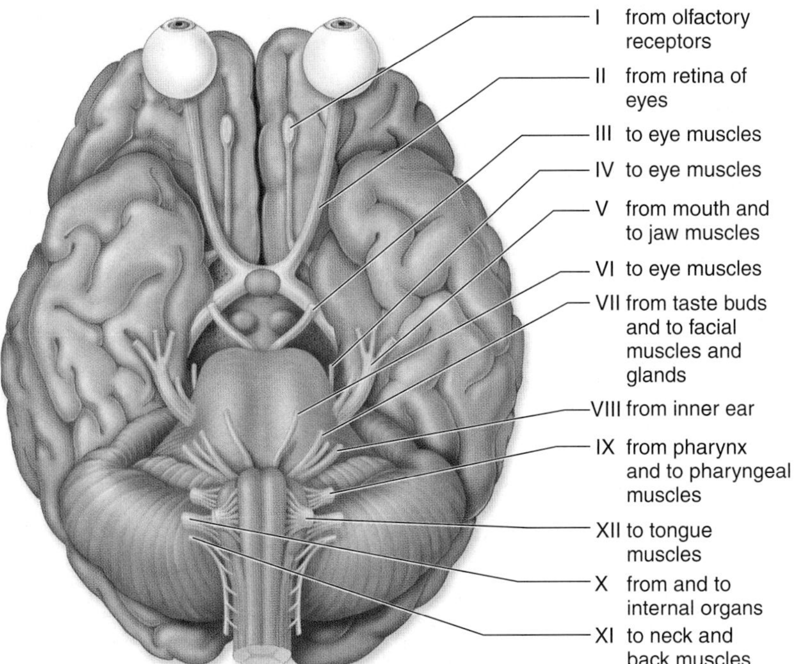

Figure 13.15 **The cranial nerves.**
Overall, cranial nerves receive sensory input from and send motor outputs to the head region. The spinal nerves receive sensory input from and send motor outputs to the rest of the body. Two important exceptions are the vagus nerve, X, which communicates with internal organs, and the spinal accessory nerve, XI, which controls neck and back muscles.

example, a reflex may cause you to blink your eyes in response to a flash of light, without your willing it. We will study the path of a reflex because it allows us to study in detail the path of nerve signals to and from the CNS.

The Reflex Arc

Figure 13.16 illustrates the path of a reflex that involves only the spinal cord. If your hand touches a sharp pin, sensory receptors in the skin generate nerve signals that move along sensory fibers through the dorsal-root ganglia toward the spinal cord. Sensory neurons that enter the cord dorsally (posteriorly) pass signals on to many interneurons. Some of these interneurons synapse with motor neurons whose short dendrites and cell bodies are in the spinal cord. Nerve signals travel along these motor fibers to an effector, which brings about a response to the stimulus. In this case, the effector is a muscle, which contracts so that you withdraw your hand from the pin. Various other reactions are also possible—you will most likely look at the pin, wince, and cry out in pain. This whole series of responses occur because some of the interneurons involved carry nerve signals to the brain.

Figure 13.16 **The events in a spinal reflex.**
A stimulus (e.g., sharp pin) causes sensory receptors in the skin to generate nerve signals that travel in sensory axons to the spinal cord. Interneurons integrate data from sensory neurons and then relay signals to motor neurons, causing contraction of a skeletal muscle and movement of the hand away from the pin.

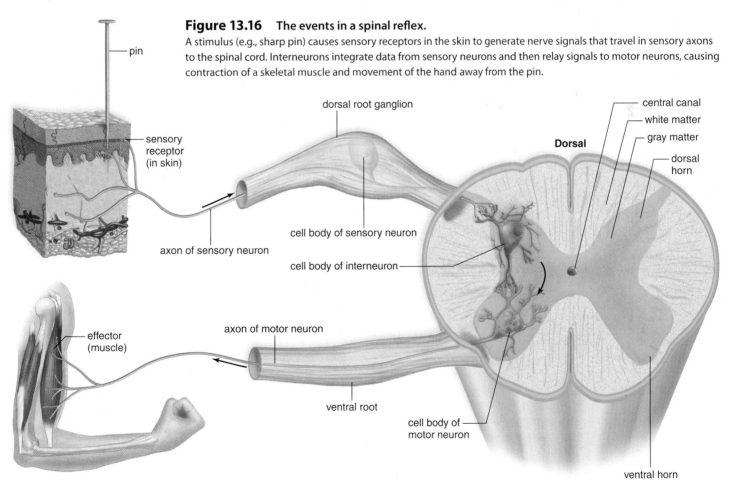

The brain makes you aware of the stimulus and directs these other reactions to it. You don't feel pain until the brain receives the information and interprets it!

Autonomic System

The **autonomic system** is also in the PNS (see Fig. 13.1). The autonomic system regulates the activity of cardiac and smooth muscles, organs, and glands. The system is divided into the sympathetic and parasympathetic divisions (Fig. 13.17). Activation of these two systems generally causes opposite responses.

Although their functions are different, the two divisions share some features: (1) They function automatically and usually in an involuntary manner; (2) they innervate all internal organs; and (3) they use two neurons and one ganglion for each impulse. The first neuron has a cell body within the CNS and a preganglionic fiber that enters the ganglion. The second neuron has a cell body within a ganglion and a postganglionic fiber that leaves the ganglion.

Reflex actions, such as those that regulate blood pressure and breathing rate, are especially important to the maintenance of homeostasis. These reflexes begin when the sensory neurons in contact with internal organs send messages to the CNS. They are completed by motor neurons within the autonomic system.

Sympathetic Division

Most preganglionic fibers of the **sympathetic division** arise from the middle, or thoracolumbar, portion of the spinal cord. They terminate almost immediately in ganglia that lie near the cord. Therefore, in this division, the preganglionic fiber is short, but the postganglionic fiber that contacts an organ is long.

The sympathetic division is especially important during emergency situations when you might be required to fight or take flight. It accelerates the heartbeat and dilates the bronchi—active muscles, after all, require a ready supply of glucose and oxygen. Sympathetic neurons inhibit the digestive organs, as well as the kidneys and urinary bladder. The activities of these organs—digestion, defecation, and urination—are not immediately necessary if you're under attack. The neurotransmitter released by the postganglionic axon is primarily norepinephrine (NE). The structure of NE is like that of epinephrine (adrenaline), an adrenal medulla hormone that usually increases heart rate and contractility.

Parasympathetic Division

The **parasympathetic division** includes a few cranial nerves (e.g., the vagus nerve) as well as fibers that arise from the sacral (bottom) portion of the spinal cord. Therefore, this division is often referred to as the craniosacral portion of the autonomic system. In the parasympathetic division, the preganglionic fiber is long, and the postganglionic fiber is short because the ganglia lie near or within the organ.

The parasympathetic division, sometimes called the housekeeper division, promotes all the internal responses we associate with a relaxed state. For example, it causes the pupil of the eye to contract, promotes digestion of food, and slows heart rate. It has been suggested that the parasympathetic system could be called the *rest-and-digest* system. The neurotransmitter used by the parasympathetic division is acetylcholine (ACh).

The Somatic Versus the Autonomic Systems

Recall that the PNS includes the somatic system and the autonomic system. Table 13.1 summarizes the features and functions of the somatic motor pathway with the motor pathways of the autonomic system.

Table 13.1 Comparison of Somatic Motor and Autonomic Motor Pathway

	Somatic Motor Pathway	Autonomic Motor Pathways	
		Sympathetic	Parasympathetic
Type of control	Voluntary/involuntary	Involuntary	Involuntary
Number of neurons per message	One	Two (preganglionic shorter than postganglionic)	Two (preganglionic longer than postganglionic)
Location of motor fiber	Most cranial nerves and all spinal nerves	Thoracolumbar spinal nerves	Cranial (e.g., vagus) and sacral spinal nerves
Neurotransmitter	Acetylcholine	Norepinephrine	Acetylcholine
Effectors	Skeletal muscles	Smooth and cardiac muscle, glands and organs	Smooth and cardiac muscle, glands and organs

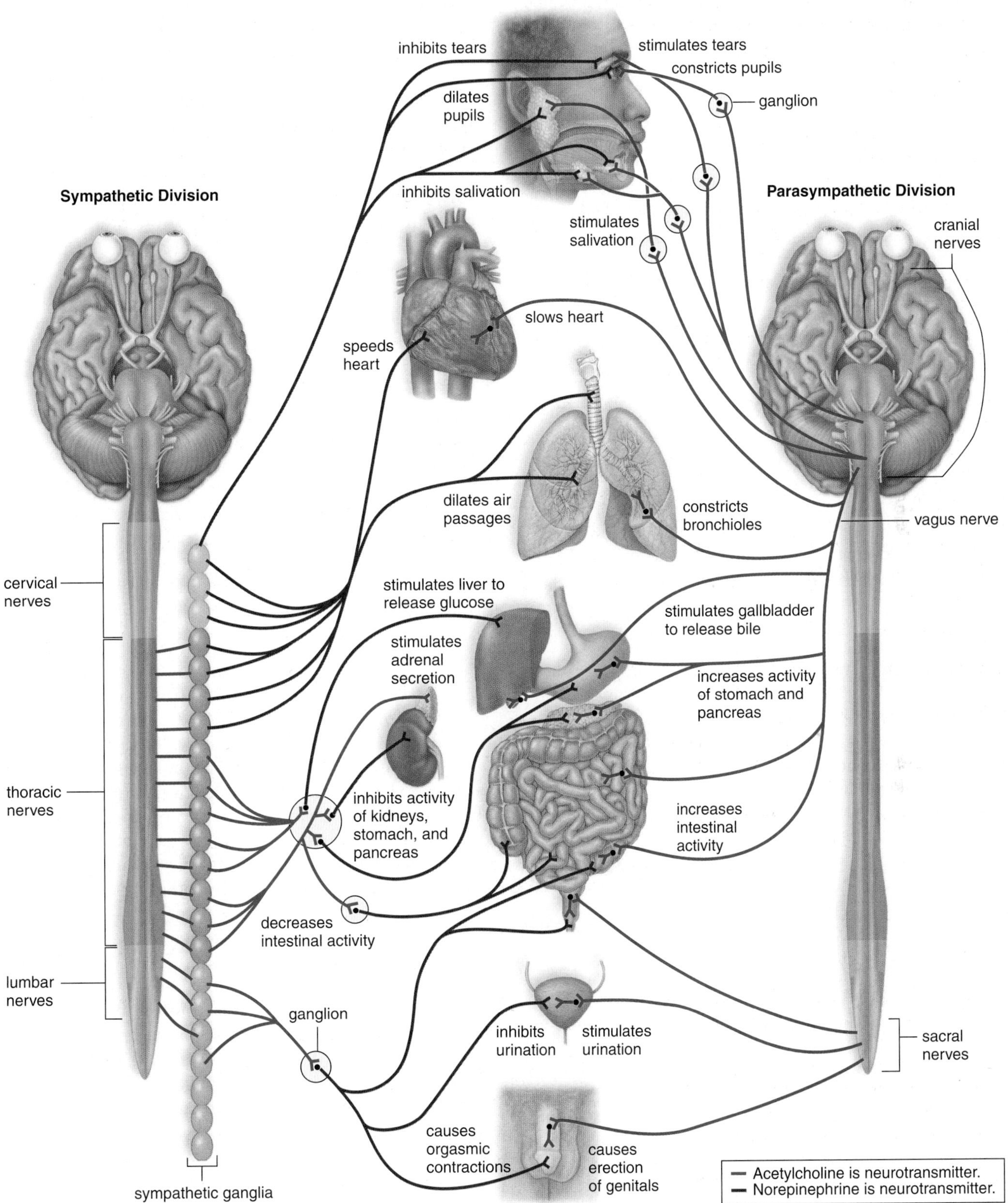

Figure 13.17 The two divisions of the autonomic nervous system.
Sympathetic preganglionic fibers (*left*) arise from the thoracic and lumbar portions of the spinal cord; parasympathetic preganglionic fibers (*right*) arise from the cranial and sacral portions of the spinal cord. Each system innervates the same organs but has contrary effects.

Connections and Misconceptions

How does aspirin work?

Aspirin is made of a chemical called acetylsalicylic acid (ASA). When tissue is damaged, it produces large amounts of a type of fatty acid called prostaglandin. Prostaglandin acts as a signal to the nervous system that tissue damage has occurred, which is interpreted by the brain as pain. Prostaglandins are manufactured in the cell by an enzyme called COX (cyclooxygenase). ASA reduces the capabilities of this enzyme, thus lowering the amount of prostaglandin produced and the perception of pain.

Connecting the Concepts

For more on the interaction of the PNS with the other systems of the body, refer to the following discussions.

Section 5.3 explores how the divisions of the autonomic system regulate the heart rate and help maintain homeostasis.

Section 9.5 examines how signals between the brain and the diaphragm control the rate of breathing.

Section 14.1 provides an overview of the types of sensory inputs processed by the peripheral nervous system.

Check Your Progress 13.4

1. Contrast cranial and spinal nerves.
2. Detail the fastest way for you to react to a stimulus.
3. Predict what might happen to homeostasis if the autonomic nervous system failed.

13.5 Drug Therapy and Drug Abuse

Learning Outcomes

Upon completion of this section, you should be able to

1. Explain the ways that drugs interact with the nervous system.
2. Classify drugs as to whether they have a depressant, stimulant, or psychoactive effect on the nervous system.
3. List the long-term effects of drug use on the body.

As you are reading these words, synapses throughout your brain are organizing, integrating, and cataloging the information you take in. Neurotransmitters at these synapses control the firing of countless action potentials, thus creating a network of neural circuits. It is amazing to realize that all thoughts, feelings, and actions of a human being are dependent on neurotransmitters in the CNS and PNS. By modifying or controlling synaptic transmission, a wide variety of drugs with neurological activity—both legal pharmaceuticals and illegal drugs of abuse—can alter mood, emotional state, behavior, and personality.

As mentioned previously (page 291), there are more than 100 known neurotransmitters. The most widely studied neurotransmitters to date are acetylcholine, norepinephrine, dopamine, serotonin, and GABA. Acetylcholine is an essential CNS neurotransmitter for memory circuits in the limbic system. Norepinephrine is important to dreaming, waking, and mood. The neurotransmitter dopamine plays a central role in the brain's regulation of mood. Dopamine is also the basal nuclei neurotransmitter that helps to organize coordinated movements. Serotonin is involved in thermoregulation, sleeping, emotions, and perception. GABA is an abundant inhibitory neurotransmitter in the CNS.

Neuromodulators are naturally occurring molecules that block the release of a neurotransmitter or modify a neuron's response to a neurotransmitter. Two well-known neuromodulators are substance P and endorphins. Substance P is released by sensory neurons when pain is present. Endorphins block the release of substance P and serve as natural painkillers. Endorphins are produced by the brain during times of physical and/or emotional stress. They are associated with the "runner's high" of joggers.

Both pharmaceuticals and illegal drugs have several basic modes of action:

- They promote the action of a neurotransmitter, usually by increasing the amount of neurotransmitter at a synapse. Examples include drugs such as Xanax and Valium, which increase GABA. These medications are used for panic attacks and anxiety. Reduced levels of norepinephrine and serotonin are linked to depression. Drugs such as Prozac, Paxil, and Cymbalta allow norepinephrine and/or serotonin to accumulate at the synapse, which explains their effectiveness as antidepressants. Alzheimer disease causes a slow, progressive loss of memory (Chapter 17). Drugs used for Alzheimer disease allow acetylcholine to accumulate at synapses in the limbic system.

  Video Apples for Alzheimers
- They interfere with or decrease the action of a neurotransmitter. For instance, antipsychotic drugs used for the treatment of schizophrenia decrease the activity of dopamine. The caffeine in coffee, chocolate, and tea keeps us awake by interfering with the effects of inhibitory neurotransmitters in the brain.
- They replace or mimic a neurotransmitter or neuromodulator. The opiates—namely, codeine, heroin, and morphine—bind to endorphin receptors and in this way reduce pain and produce a feeling of well-being.

Ongoing research into neurophysiology and neuropharmacology (the study of nervous system function and the way drugs work in the nervous system) continues to provide evidence that mental illnesses are caused by imbalances in neurotransmitters. These studies will undoubtedly improve treatments for mental illness, as well as provide insight into the problem of drug abuse.

Like mental illness, drug abuse is also linked to neurotransmitter levels. As mentioned previously, the neurotransmitter dopamine is essential for mood regulation. Dopamine plays a central role in the working of the brain's built-in *reward circuit*. The reward circuit is a collection of neurons that, under normal circumstances, promotes healthy, pleasurable activities, such as consuming food. It's possible to abuse behaviors such as eating, spending, or gambling because the behaviors stimulate the reward circuit and make us feel good. Drug abusers take drugs that artificially affect the reward circuit to the point that they neglect their basic physical needs in favor of continued drug use.

Drug abuse is apparent when a person takes a drug at a dose level and under circumstances that increase the potential for a harmful effect. Drug abusers are apt to display a psychological and/or physical dependence on the drug. Psychological dependence is apparent when a person craves the drug, spends time seeking the drug, and takes it regularly. With physical dependence, formerly called "addiction," the person has become tolerant to the drug. More is needed to get the same effect, and withdrawal symptoms occur when he or she stops taking the drug. This is true for not only teenagers and adults but also newborn babies of mothers who abuse and are addicted to drugs (Fig. 13.18). Alcohol, drugs, and tobacco can all adversely affect the developing embryo, fetus, or newborn.

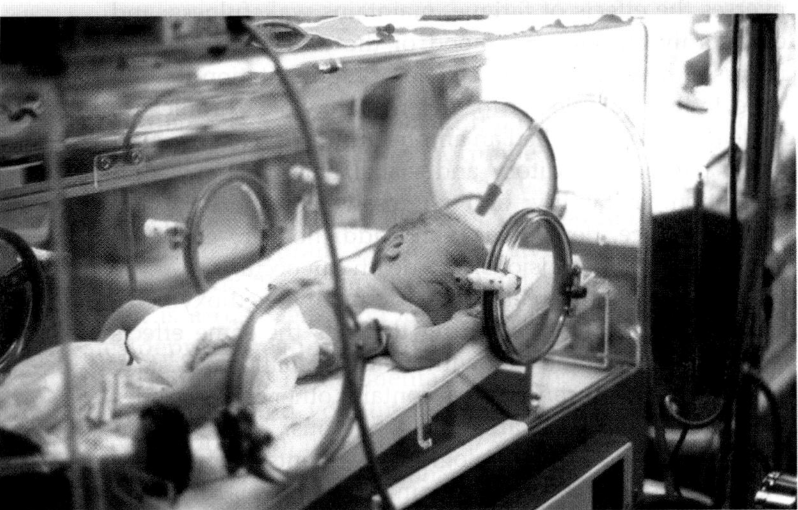

Figure 13.18 Drug abuse affects the fetus of a pregnant woman.
Many drugs can negatively effect developed of the fetus. This newborn is addicted to cocaine.

Alcohol

Alcohol consumption is the most socially accepted form of drug use worldwide. Its widespread use probably correlates with its ancient origin and its derivation from grains or fruits. The approximate number of adults that consume alcohol in the United States on a regular basis is 65%. Of those who are drinkers, 5% are heavy drinkers. Notably, 80% of college-age young adults drink.

Alcohol (ethanol) has known harmful effects on the body and brain. Alcohol readily crosses cell membranes, including the blood–brain barrier. It denatures protein structures, causing damage to several tissues, including vital organs of the body such as the liver and brain. The liver is the major detoxification organ of the body; prolonged alcohol consumption scars the liver and impairs its function. Depending on the amount consumed, the effects of alcohol on the brain can lead to a feeling of relaxation, a lowering of inhibitions, impaired concentration and coordination, slurred speech, and vomiting. If the alcohol level of blood becomes too high, coma or death can occur.

In the central nervous system (CNS), alcohol acts as a *depressant* (Table 13.2) and influences many brain regions and neurotransmitter systems. For example, alcohol increases the action of GABA and increases the release of endorphins in the hypothalamus. Chronic alcohol consumption can damage the frontal lobes, decrease overall brain size, and increase the size of the ventricles. Brain damage is manifested by permanent memory loss, amnesia, confusion, apathy, disorientation, or lack of motor coordination.

There is no effective treatment to cure alcoholism. However, some drugs have been approved by the FDA to help reduce cravings in those who seriously wish to quit drinking.

Nicotine

Nicotine is a small molecule not normally produced by the body, and it acts as a *stimulant*. During the smoking of tobacco, nicotine is rapidly delivered to the CNS, especially the midbrain. When nicotine binds to neurons in the CNS, dopamine is released. In the peripheral nervous system,

Table 13.2 Drug Influence on CNS and Route

Substance	Effect	Mode of Transmission
Alcohol	Depressant	Drink
Nicotine	Stimulant	Smoked or smokeless tobacco
Cocaine	Stimulant	Sniffed/snorted, injected, or smoked
Methamphetamine	Stimulant	Smoked or pill form
Heroin	Depressant	Sniffed/snorted, injected, or smoked
Marijuana	Psychoactive	Smoked or consumed

30. The sympathetic division of the autonomic system does not cause
    a. the liver to release glycogen.
    b. dilation of bronchioles.
    c. the gastrointestinal tract to digest food.
    d. an increase in the heart rate.
31. Label this diagram.

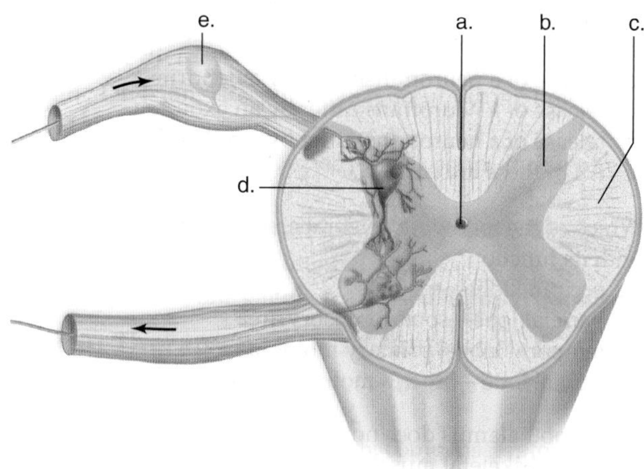

Thinking Critically About the Concepts

Demyelinating disorders, like the multiple sclerosis discussed in the chapter opener, are the subject of numerous research projects. Many investigations focus on the cells that create myelin: the Schwann cells of the PNS and oligodendrocytes in the CNS. Other studies focus on immune system cells that attack this myelin sheath. The goal of this research is to determine how to restore lost myelin, which might help (or possibly cure) folks living with MS and other demyelinating diseases. Investigations into the role played by the sheath in nerve regeneration may offer hope to victims of spinal cord injury.

1. Why are impulses transmitted more quickly down a myelinated axon than an unmyelinated axon?
2. A buildup of very long chain saturated fatty acids is believed to be the cause of myelin loss in adrenoleukodystrophy. This rare disease is a demyelinating disorder like MS. It is the subject of the film *Lorenzo's Oil*. This real-life drama focuses on Lorenzo Odone, whose parents successfully developed a diet that helped their son.
    a. From your study of chemistry in Chapter 2, a fatty acid is a part of what type of molecule?
    b. What distinguishes a saturated fatty acid from an unsaturated fatty acid?
    c. From your study of nutrition in Chapter 8, what types of foods would contain saturated fatty acids?
3. Why would you expect the motor skills of a child to improve as myelination continues during early childhood development?

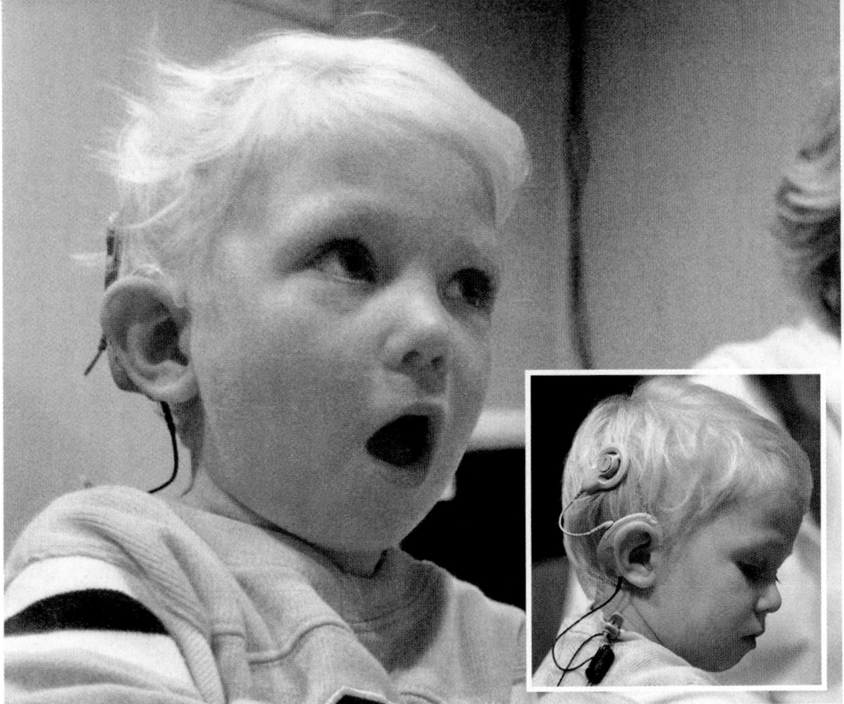

CHAPTER

14

Senses

CASE STUDY COCHLEAR IMPLANTS

Jacob and Marlene were married for almost a year when they learned that Marlene was pregnant with their first child. Almost immediately, they began discussing what they would do if their baby was born with hearing problems. Marlene had been born deaf because of a genetic condition called *nonsyndromic deafness.* The term *nonsyndromic* means that Marlene's deafness was not caused by another medical problem such as diabetes or a genetic syndrome such as Charcot–Marie–Tooth disease or Waardenburg syndrome. Researchers now know that many forms of nonsyndromic deafness are caused by a deletion in a gene on chromosome 13 called *GJB2*. *GJB2* produces a protein that is responsible for the formation of the cochlea, the key structure in hearing. However, when Marlene was a child, little was known as to the causes of deafness, so Marlene had compensated for her disability by learning sign language and had even become a teacher at a school for the deaf. Since then, advances in technology and medicine have made it possible to provide hearing for babies with many forms of deafness.

One option for treatment is cochlear implants. Cochlear implants are different from hearing aids, which simply amplify sounds. A cochlear implant consists of an external device that is surgically implanted under the skin. The external part picks up sounds from the environment and converts them to electrical impulses, which are sent directly to different regions of the auditory nerve and then to the brain, as we will see in this chapter. As of 2007, almost 120,000 people worldwide had received cochlear implants, including 17,800 children and 21,700 adults in the United States.

As you read through the chapter, think about the following questions.

1. Where is the cochlea located?
2. What are the roles of the cochlea and auditory nerve in hearing?
3. How might the technology associated with cochlear implants also be used to help people with vision problems?

CHAPTER CONCEPTS

14.1 Overview of Sensory Receptors and Sensations
Each type of sensory receptor detects a particular type of stimulus. When stimulation occurs, sensory receptors initiate nerve signals transmitted to the CNS. Sensation occurs when nerve impulses reach the cerebral cortex.

14.2 Proprioceptors, Cutaneous Receptors, and Pain Receptors
Proprioception, an awareness of the part of the body, is dependent on sensory receptors in muscles and joints. Touch, pressure, temperature, and pain are dependent on sensory receptors in the skin.

14.3 Senses of Taste and Smell
Taste and smell involve the activity of sensory receptors in the mouth and nose, respectively.

14.4 Sense of Vision
Vision depends on the sensory receptors in the eyes, the optic nerves, and the visual cortex.

14.5 Sense of Hearing
Hearing depends on sensory receptors in the ears, the cochlear nerve, and the auditory cortex.

14.6 Sense of Equilibrium
The inner ear contains sensory receptors for our sense of equilibrium.

BEFORE YOU BEGIN

Before beginning this chapter, take a few moments to review the following discussions.

Figure 13.1 What are the roles of the central and peripheral nervous systems in the body?

Section 13.2 What is the role of the primary somatosensory area of the cerebral cortex?

Section 13.2 How does the cerebellum help maintain balance?

14.1 Overview of Sensory Receptors and Sensations

Learning Outcomes

Upon completion of this section, you should be able to

1. List the four categories of sensory receptors and describe what each responds to.
2. Distinguish between perception and sensation.
3. Explain the purpose of integration and sensory adaptation.

Sensory receptors are dendrites specialized to detect certain types of stimuli (sing., **stimulus**). **Exteroceptors** are sensory receptors that detect stimuli from outside the body, such as those that result in taste, smell, vision, hearing, and equilibrium (Table 14.1). **Interoceptors** receive stimuli from inside the body. Interoceptors include pressoreceptors (sometimes referred to as baroreceptors) that respond to changes in blood pressure and osmoreceptors to monitor the body's water–salt balance. Chemoreceptors are interoceptors that monitor the pH of the blood.

Interoceptors are directly involved in homeostasis and are regulated by a negative feedback mechanism. For example, when blood pressure rises, pressoreceptors signal a regulatory center in the brain. The brain responds by sending out nerve signals to the arterial walls, causing their smooth muscle to relax. The blood pressure then falls. Once blood pressure is returned to normal, the pressoreceptors are no longer stimulated.

Exteroreceptors such as those in the eye and ear continually send messages to the central nervous system. In this way, they keep us informed regarding the surrounding environment.

Types of Sensory Receptors

Sensory receptors in humans can be classified into just four categories: chemoreceptors, photoreceptors, mechanoreceptors, and thermoreceptors.

Chemoreceptors respond to chemical substances in the immediate vicinity. As Table 14.1 indicates, taste and smell, which detect external stimuli, use chemoreceptors. However, so do various other organs sensitive to internal stimuli. Chemoreceptors that monitor blood pH are located in the carotid arteries and aorta. If the pH lowers, the breathing rate increases. As more carbon dioxide is exhaled, the blood pH rises. **Pain receptors** (nociceptors) are a type of chemoreceptor. They are naked dendrites that respond to chemicals released by damaged tissues. Pain receptors are protective because they alert us to possible danger. For example, without the pain of appendicitis, we might never seek the medical help needed to avoid a ruptured appendix.

Photoreceptors respond to light energy. Our eyes contain photoreceptors that are sensitive to light rays and thereby provide us with a sense of vision. Stimulation of the photoreceptors known as rod cells results in black-and-white vision. Stimulation of the photoreceptors known as cone cells results in color vision.

Mechanoreceptors are stimulated by mechanical forces, which most often result in pressure of some sort. When we hear, airborne sound waves are converted to fluid-borne pressure waves that can be detected by mechanoreceptors in the inner ear. Mechanoreceptors are responding to fluid-borne pressure waves when we detect changes in gravity and motion, helping us keep our balance. These receptors are in the vestibule and semicircular canals of the inner ear.

The sense of touch depends on pressure receptors sensitive to either strong or slight pressures. Pressoreceptors located in certain arteries detect changes in blood pressure, and stretch receptors in the lungs detect the degree of lung inflation. Proprioceptors respond to the stretching of muscle fibers, tendons, joints, and ligaments. Signals from proprioceptors make us aware of the position of our limbs.

Thermoreceptors located in the hypothalamus and skin are stimulated by changes in temperature. Those that respond when temperatures rise are called warmth receptors, and those that respond when temperatures lower are called cold receptors. On a hot summer day, rising body temperature will most likely have you seeking shade.

How Sensation Occurs

Sensory receptors respond to environmental stimuli by generating nerve signals. When the nerve signals arrive at the cerebral cortex of the brain, **sensation**, the conscious perception of stimuli, occurs.

MP3 Sensations and Receptors

Table 14.1 Exteroceptors

Sensory Receptor	Stimulus	Category	Sense	Sensory Organ
Taste cells	Chemicals	Chemoreceptor	Taste	Taste bud
Olfactory cells	Chemicals	Chemoreceptor	Smell	Olfactory epithelium
Rod cells and cone cells in retina	Light rays	Photoreceptor	Vision	Eye
Hair cells in spiral organ of the inner ear	Sound waves	Mechanoreceptor	Hearing	Ear
Hair cells in semicircular canals of the inner ear	Motion	Mechanoreceptor	Rotational equilibrium	Ear
Hair cells in vestibule of the inner ear	Gravity	Mechanoreceptor	Gravitational equilibrium	Ear

As discussed in Chapter 13, sensory receptors are the first element in a reflex arc. We are aware of a reflex action only when sensory information reaches the brain. At that time, the brain integrates this information with other information received from other sensory receptors. Consider what happens if you burn yourself and quickly remove your hand from a hot stove. The brain receives information not only from your skin but also from your eyes, nose, and all sorts of sensory receptors.

Some sensory receptors are free nerve endings or encapsulated nerve endings, and others are specialized cells closely associated with neurons. Often, the plasma membrane of a sensory receptor contains receptor proteins that react to the stimulus. For example, the receptor proteins in the plasma membrane of chemoreceptors bind to certain chemicals. When this happens, ion channels open, and ions flow across the plasma membrane. If the stimulus is sufficient, nerve signals begin and are carried by a sensory nerve fiber within the PNS to the CNS (Fig. 14.1). The stronger the stimulus, the greater the frequency of nerve signals. Nerve signals that reach the spinal cord first are conveyed to the brain by ascending tracts. If nerve signals finally reach the cerebral cortex, sensation and perception occur.

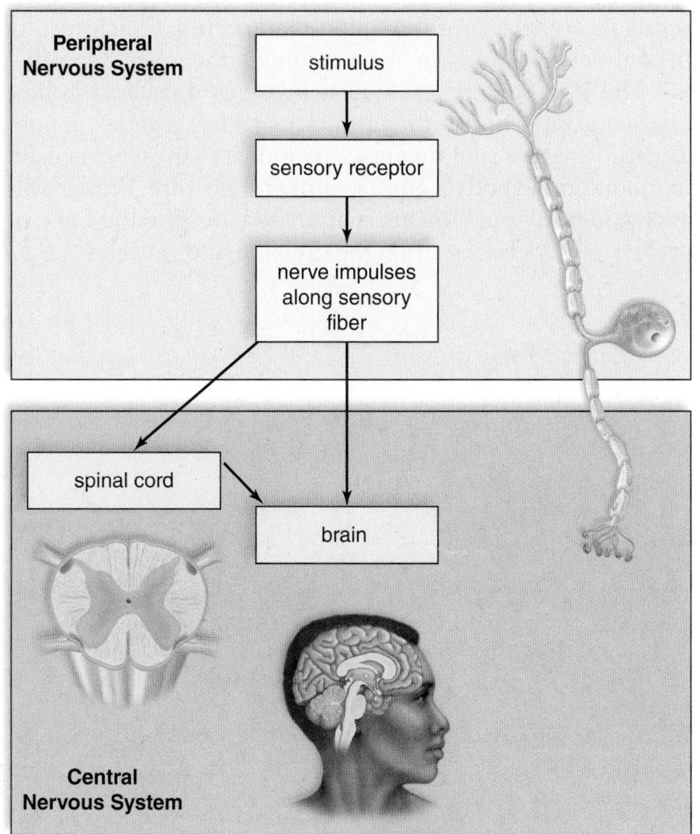

Figure 14.1 The role of the CNS and PNS in sensation and sensory perception.
After detecting a stimulus, sensory receptors initiate nerve signals within the PNS. These signals give the central nervous system (CNS) information about the external and internal environment. The CNS integrates all incoming information, and then initiates a motor response to the stimulus.

Connections and Misconceptions

Why does rubbing your closed eyes (as you might if you're tired or your eyes itch) produce a visual sensation?

As described in Section 14.4, the eye is a flexible container filled with fluid and a soft gelatinous material. Compressing the eyes by rubbing on them increases pressure in the eyes. In turn, the photoreceptors of the eye are stimulated by the increased eye pressure. When the nerve signals are conveyed to the brain, the brain senses "vision." We "see stars" because nerve signals from the eyes can only result in sight.

All sensory receptors initiate nerve signals. The sensation that results depends on the part of the brain receiving the nerve signals. Nerve signals that begin in the optic nerve eventually reach the visual areas of the cerebral cortex. Thereafter, we see objects. Nerve signals that begin in the auditory nerve eventually reach the auditory areas of the cerebral cortex. We hear sounds when the auditory cortex is stimulated. If it were possible to switch these nerves, stimulation of the eyes would result in hearing!

Before sensory receptors initiate nerve signals, they also carry out **integration,** the summing up of signals. One type of integration is called **sensory adaptation,** a decrease in response to a stimulus. We have all had the experience of smelling an odor when we first enter a room and then later not being aware of it. Some authorities believe that when sensory adaptation occurs, sensory receptors have stopped sending impulses to the brain. Others believe that the reticular activating system (RAS) has filtered out the ongoing stimuli. You will recall that sensory information is conveyed from the brain stem through the thalamus to the cerebral cortex by the RAS. The thalamus acts as a gatekeeper and passes on information only of immediate importance. Just as we gradually become unaware of particular environmental stimuli, we can suddenly become aware of stimuli that may have been present for some time. This can be attributed to the workings of the RAS, which has synapses with all the great ascending sensory tracts.

The functioning of our sensory receptors makes a significant contribution to homeostasis. Without sensory input, we would not receive information about our internal and external environment. This information leads to appropriate reflex and voluntary actions to keep the internal environment constant.

Connecting the Concepts

For more information on the regions of the brain that are associated with sensation, refer to the following discussions.

Section 13.2 describes the location and function of the reticular activating system (RAS).

Figure 13.10 illustrates the somatosensory regions of the cerebral cortex.

Figure 13.16 illustrates the portions of the peripheral nervous system that are involved in a reflex arc.

Check Your Progress 14.1

1. Describe the functions of the four different types of sensory receptors.
2. Explain how perception differs from sensation and give an example.
3. Summarize the importance of sensory receptors in the maintenance of homeostasis in the body.

14.2 Proprioceptors, Cutaneous Receptors, and Pain Receptors

Learning Outcomes

Upon completion of this section, you should be able to

1. Distinguish between proprioceptors and cutaneous receptors with regard to function.
2. State the location and general function of each type of cutaneous receptor.
3. Explain the role of nociceptors and summarize the type of sensory input that they detect.

Sensory receptors in the muscles, joints and tendons, other internal organs, and skin send nerve signals to the spinal cord. From there, they travel up the spinal cord in tracts to the somatosensory areas of the cerebral cortex (see Fig. 13.10). These general sensory receptors can be categorized into three types: proprioceptors, cutaneous receptors, and pain receptors.

Proprioceptors

Proprioceptors are mechanoreceptors involved in reflex actions that maintain muscle tone and, thereby, the body's equilibrium and posture. They help us know the position of our limbs in space by detecting the degree of muscle relaxation, the stretch of tendons, and the movement of ligaments. Muscle spindles built into a muscle act to increase muscle contraction. Golgi tendon organs found in tendons decrease muscle contraction. The result is a muscle that has the proper length and tension, or muscle tone.

Figure 14.2 illustrates the activity of a muscle spindle. In a muscle spindle, sensory nerve endings are wrapped around thin muscle cells within a connective tissue sheath. When the muscle relaxes and its length increases, the muscle spindle is stretched and nerve signals are generated. The more the muscle stretches, the faster the muscle spindle sends its signals. A reflex action then occurs, which results in contraction of muscle fibers adjoining the muscle spindle.

The knee-jerk reflex, which involves muscle spindles, offers an opportunity for physicians to test a reflex action. The information sent by muscle spindles to the CNS is used to maintain the body's equilibrium and posture. Proper balance and body position are maintained, despite the force of gravity always acting upon the skeleton and muscles.

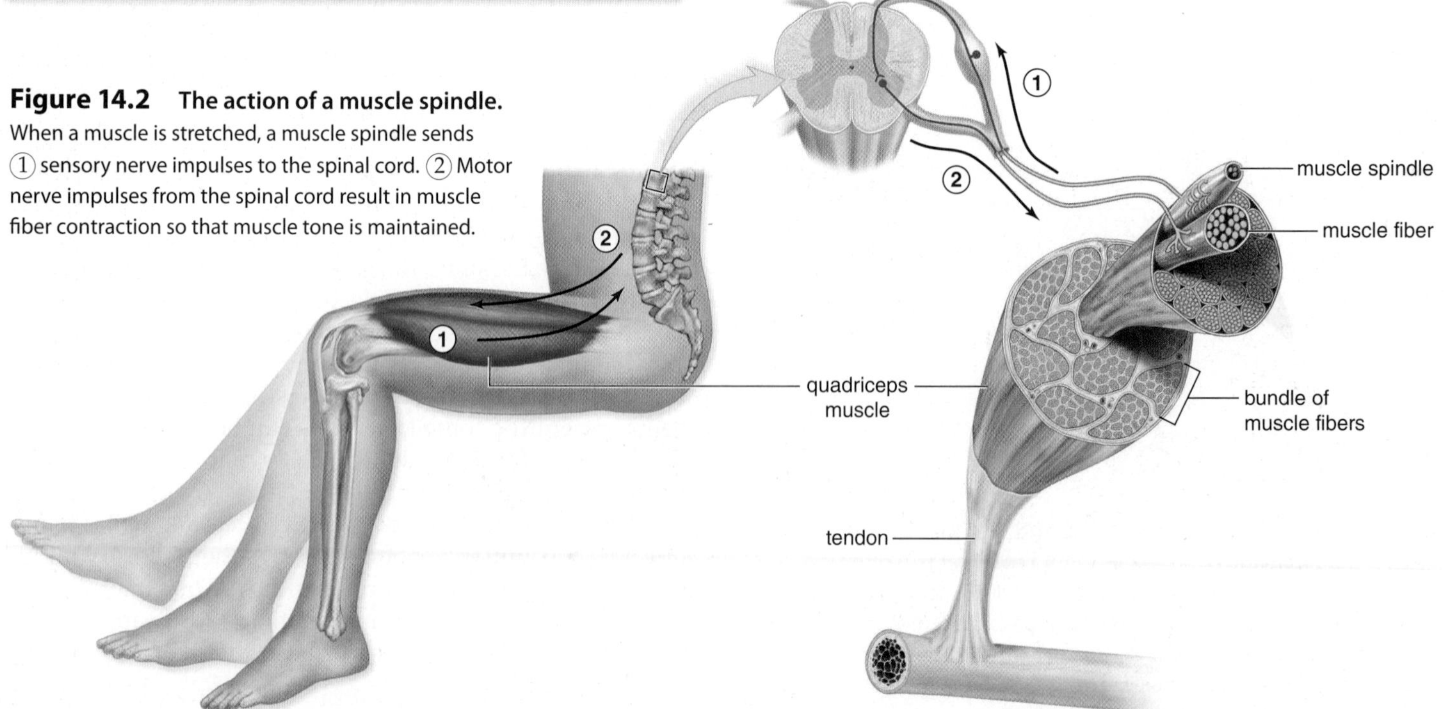

Figure 14.2 The action of a muscle spindle. When a muscle is stretched, a muscle spindle sends (1) sensory nerve impulses to the spinal cord. (2) Motor nerve impulses from the spinal cord result in muscle fiber contraction so that muscle tone is maintained.

Cutaneous Receptors

The skin is composed of two layers: the epidermis and the dermis (Fig. 14.3). The epidermis is stratified squamous epithelium (Chapter 4). Cells become keratinized as they rise to the surface, where they are sloughed off. The dermis is a thick connective tissue layer. The dermis contains **cutaneous receptors,** which make the skin sensitive to touch, pressure, pain, and temperature (warmth and cold). The dermis is a mosaic of these tiny receptors, as you can determine by slowly passing a metal probe over your skin. At certain points, you will feel touch or pressure; at others, you will feel heat or cold (depending on the probe's temperature).

Three types of cutaneous receptors are sensitive to fine touch. These receptors give a person specific information such as the location of the touch as well as its shape, size, and texture. *Meissner corpuscles* and *Krause end bulbs* are concentrated in the fingertips, the palms, the lips, the tongue, the nipples, the penis, and the clitoris. *Merkel disks* are found where the epidermis meets the dermis. A free nerve ending called a *root hair plexus* winds around the base of a hair follicle. This receptor responds if the hair is touched.

Two types of cutaneous receptors sensitive to pressure are Pacinian corpuscles and Ruffini endings. *Pacinian corpuscles* are onion-shaped sensory receptors that lie deep inside the dermis. *Ruffini endings* are encapsulated by sheaths of connective tissue and contain lacy networks of nerve fibers.

Temperature receptors are simply free nerve endings in the epidermis. Some free nerve endings are responsive to cold. Others respond to warmth. Cold receptors are far more numerous than warmth receptors, but the two types have no known structural differences.

Connections and Misconceptions

What are phantom sensation and phantom pain?

Suppose a person loses a foot and a leg due to an injury. In addition to dealing with loss of a limb, an amputee often must cope with the phenomenon of phantom sensation or phantom pain—or both. Phantom sensation is a painless awareness of the amputated limb. For example, a patient whose foot and lower leg have been removed may have an itchy or tingly sensation in the "foot," though the foot is no longer there. Similarly, pain can be sensed as originating from the absent body part. Researchers believe that any stimulus (such as a touch) to the stump will fool the brain into a perceived sensation, because the brain has received signals from the leg and foot for such a long time.

Phantom sensation may last for years but usually disappears without treatment. Phantom pain must be treated with a combination of medication, massage, and physical therapy.

Pain Receptors

Like the skin, many internal organs have pain receptors, also called *nociceptors*. These receptors are sensitive to chemicals released by damaged tissues. When inflammation occurs because of mechanical, thermal, or electrical stimuli or toxic substances, cells release chemicals that stimulate pain receptors. Aspirin and ibuprofen reduce pain by inhibiting the synthesis of one class of these chemicals.

Sometimes, stimulation of internal pain receptors is felt as pain from the skin, as well as the internal organs. This is

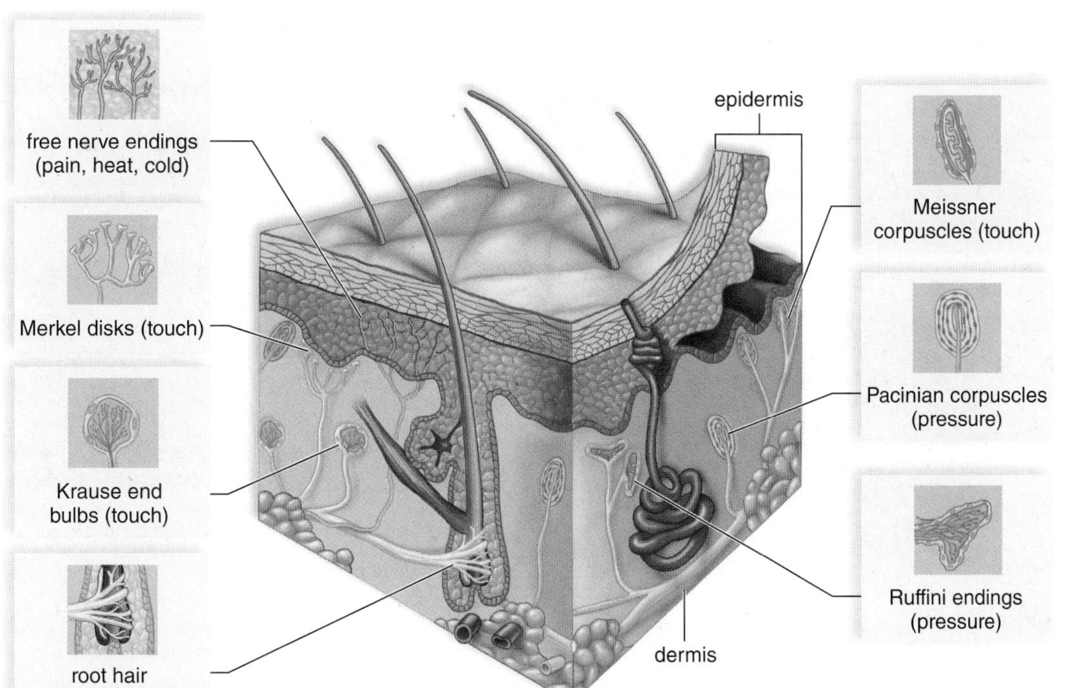

Figure 14.3 **Sensory receptors of the skin.**
The classical view is that each sensory receptor has the main function shown here. However, investigators report that matters are not so clear-cut. For example, microscopic examination of the skin of the ear shows only free nerve endings (pain receptors); yet, the skin of the ear is sensitive to all sensations. Therefore, it appears that the receptors of the skin are somewhat—but not completely—specialized.

called **referred pain.** Some internal organs have a referred pain relationship with areas located in the skin of the back, groin, and abdomen. For example, pain from the heart is often felt in the left shoulder and arm. This most likely happens when nerve impulses from the pain receptors of internal organs travel to the spinal cord and synapse with neurons also receiving impulses from the skin. Frequently, this type of referred pain is more common in men than in women. The nonspecific symptoms that women often experience during a heart attack may delay a diagnosis.

Connecting the Concepts

For more information on the material in this section, refer to the following discussions.

Figure 4.8 provides a more detailed look at the structure of human skin.

Section 12.2 provides an overview of muscle fiber contraction.

Section 13.2 presents the gate control theory of how the brain responds to input from pain receptors.

Check Your Progress 14.2

1. Describe how proprioceptors are used by the body to indicate the position of the arms and legs.
2. Summarize the role of each type of cutaneous receptor and classify it as to whether it is located in the epidermis or the dermis.
3. Explain why pain receptors are important for the maintenance of homeostasis.

14.3 Senses of Taste and Smell

Learning Outcomes

Upon completion of this section, you should be able to

1. Compare and contrast the senses of taste and smell.
2. Identify the structures of the tongue and olfactory areas of the nose.
3. Summarize how the brain receives taste and odor information.

Taste and smell are called chemical senses because their receptors are sensitive to molecules in the food we eat and the air we breathe.

Taste cells and olfactory cells bear chemoreceptors. Chemoreceptors are also in the carotid arteries and in the aorta, where they are primarily sensitive to the pH of the blood. These chemoreceptors are called carotid and aortic bodies. They communicate via sensory nerve fibers with the respiratory center in the medulla oblongata. When the pH drops, they signal this center. Immediately thereafter, the breathing rate increases. Exhaling CO_2 raises the pH of the blood.

Chemoreceptors are plasma membrane receptors that bind to particular molecules. They are divided into two types: those that respond to distant stimuli and those that respond to direct stimuli. Olfactory cells act from a distance and taste cells act directly. pH receptors also respond to direct stimuli.

Video Sex and the Senses

Sense of Taste

In adult humans, approximately 3,000 **taste buds** are located primarily on the tongue (Fig. 14.4). Many taste buds lie along

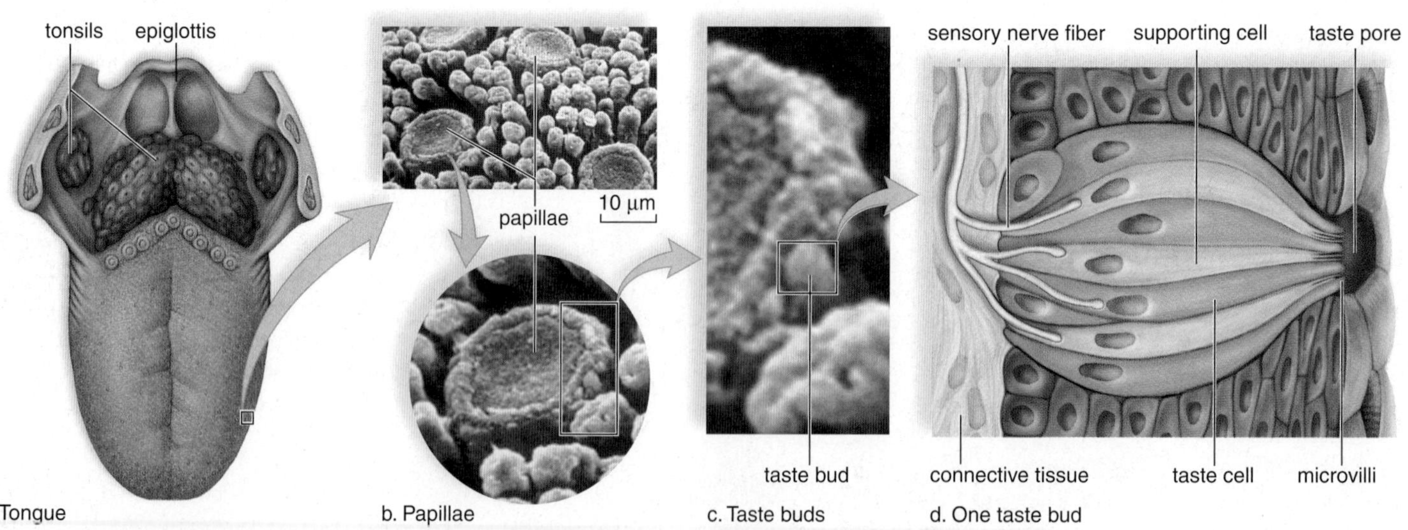

Figure 14.4 The tongue and the sense of taste.
a. Papillae on the tongue contain taste buds sensitive to sweet, sour, salty, bitter, and perhaps umami. **b.** Photomicrograph and enlargement of papilla. **c.** Taste buds occur along the walls of the papillae. **d.** Taste cells in microvilli that bear receptor proteins for certain molecules. When molecules bind to the receptor proteins, nerve signals are generated and go to the brain, where the sensation of taste occurs.

the walls of the papillae. These small elevations on the tongue are visible to the naked eye. Isolated taste buds are also present on the hard palate, the pharynx, and the epiglottis. There are at least four primary types of taste (sweet, sour, salty, and bitter). A fifth taste, called umami, allows us to enjoy the savory flavors of certain cheeses, beef, and mushrooms. Taste buds for each of these tastes are located throughout the tongue, although certain regions may be most sensitive to particular tastes. The tip of the tongue is most sensitive to sweet tastes, the margins of the tongue react to salty and sour tastes, and the rear of the tongue responds to bitter tastes.

How the Brain Receives Taste Information

Taste buds open at a taste pore. They have supporting cells and a number of elongated taste cells that end in microvilli. When molecules bind to receptor proteins of the microvilli, nerve signals are generated in sensory nerve fibers that go to the brain. Signals reach the gustatory (taste) cortex, located primarily in the parietal lobe, where they are interpreted as particular tastes.

Humans can respond to a range of sweet, sour, salty, and bitter tastes. As the signals are received, the brain appears to survey their overall pattern. A "weighted average" of all taste messages is used by the brain as the perceived taste. For example, a glass of lemonade transmits sour, sweet, and perhaps bitter sensations. If you remember the taste of lemonade, your brain will recognize this pattern, and you'll think, "Ahh, lemonade." Again, we can note that even though our senses depend on sensory receptors, the cortex integrates the incoming information and gives us our sensations.

Animation Taste

Sense of Smell

Approximately 80–90% of what we perceive as "taste" actually is due to the sense of smell. This accounts for how dull food tastes when we have a head cold or a stuffed-up nose. Our sense of smell depends on between 10 and 20 million **olfactory cells** located within olfactory epithelia high in the roof of the nasal cavity (Fig. 14.5). Olfactory cells are modified neurons. Each cell ends in a tuft of about five olfactory cilia, which bear receptor proteins for odor molecules.

How the Brain Receives Odor Information

Each olfactory cell has only one out of several hundred different types of receptor proteins. Nerve fibers from similar olfactory cells lead to the same neuron in the olfactory bulb (an extension of the brain). An odor contains many odor molecules, which activate a characteristic combination of receptor proteins. For

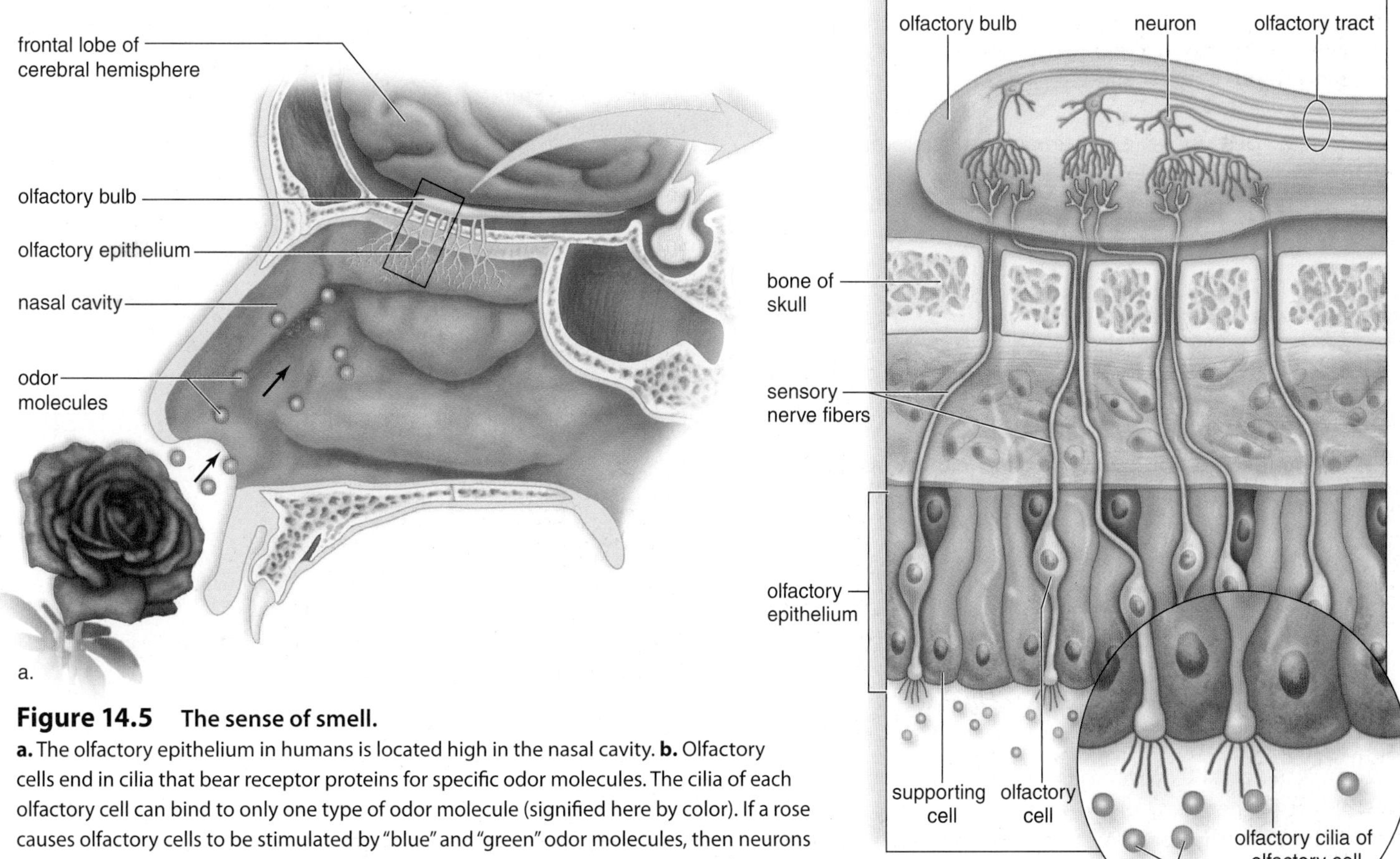

Figure 14.5 **The sense of smell.**
a. The olfactory epithelium in humans is located high in the nasal cavity. **b.** Olfactory cells end in cilia that bear receptor proteins for specific odor molecules. The cilia of each olfactory cell can bind to only one type of odor molecule (signified here by color). If a rose causes olfactory cells to be stimulated by "blue" and "green" odor molecules, then neurons designated by blue and green in the olfactory bulb are activated. The primary olfactory area of the cerebral cortex interprets the pattern of stimulation as the scent of a rose.

example, a rose may stimulate olfactory cells, designated by blue and green in Figure 14.5, whereas a carnation may stimulate a different combination. An odor's signature in the olfactory bulb is determined by which neurons are stimulated. When the neurons communicate this information via the olfactory tract to the olfactory areas of the cerebral cortex, we know we have smelled a rose or a carnation.

The olfactory cortex is located in the temporal lobe. Some areas of the olfactory cortex receive smell sensations, and other areas contain olfactory memories.

Have you ever noticed that a certain aroma vividly brings to mind a certain person or place and can re-create emotions you feel about that person or place? A person's cologne may depress you by reminding you of a failed relationship, whereas the smell of boxwood may create happier emotions by recalling your grandfather's farm. The olfactory bulbs have direct connections with the limbic system and its centers for emotion and memory. One investigator showed that when subjects smelled an orange while viewing a painting, memories of the painting were more vividly recalled and the subjects also had many deep feelings about the painting.

The number of olfactory cells declines with age. This can be dangerous if an older person can't smell smoke or a gas leak. Older people also tend to apply excessive amounts of perfume or cologne before they can detect its smell.

MP3 Taste and Smell

Connecting the Concepts

For more information on chemoreceptors, refer to the following discussions.

Section 9.5 describes the function of the respiratory center in the medulla oblongata.

Section 13.3 explains the role of the limbic system in maintaining memories such as smell and taste.

Check Your Progress 14.3

1. Identify the structures of the tongue and nose that are involved in the sense of taste and smell.
2. Compare and contrast the function of the chemoreceptors on the tongue and in the nose.
3. Summarize the pathway of sensory information on taste and smell from the receptors to the brain.

14.4 Sense of Vision

Learning Outcomes

Upon completion of this section, you should be able to

1. Identify the structures of the human eye.
2. Explain how the eye focuses on near and far objects.
3. Describe the role of photoreceptors in vision.
4. Summarize the abnormalities of the eye that produce vision problems.

Vision requires the work of the eyes and the brain. As we shall see, much processing of stimuli occurs in the eyes before nerve signals are sent to the brain. Still, researchers estimate that at least a third of the cerebral cortex takes part in processing visual information.

Anatomy and Physiology of the Eye

The eyeball is an elongated sphere about 2.5 cm in diameter. It has three layers, or coats: the sclera, the choroid, and the retina (Fig. 14.6, Table 14.2). The outer layer, the **sclera,** is white and fibrous except for the **cornea,** which is made of transparent collagen fibers. Therefore, the cornea is known as the window of the eye.

The **choroid** is the thin middle coat. It has an extensive blood supply, and its dark pigment absorbs stray light rays that photoreceptors have not absorbed. This helps visual acuity. Toward the front, the choroid becomes the doughnut-shaped **iris.** The iris regulates the size of the **pupil,** a hole in the center of the iris through which light enters the eyeball. The color of the iris (and therefore the color of the eyes) correlates with its pigmentation. Heavily pigmented eyes are brown, and lightly pigmented eyes are green or blue. Behind the iris, the choroid thickens and forms the circular ciliary body. The **ciliary body** contains the ciliary muscle, which controls the shape of the lens for near and far vision.

The **lens** is attached to the ciliary body by suspensory ligaments and divides the eye into two compartments. The anterior compartment is in front of the lens, and the posterior compartment is behind it. The anterior compartment is filled with a clear, watery fluid called the **aqueous humor.** A small amount of aqueous humor is continually produced each day. Normally, it leaves the anterior compartment by way of tiny ducts. When a person has **glaucoma,** these drainage ducts are blocked and aqueous humor builds up. If glaucoma is not treated, the resulting pressure compresses the arteries that serve the nerve fibers

Table 14.2 Structures of the Eye

Part	Function
Sclera	Protects and supports eyeball
Cornea	Refracts light rays
Pupil	Admits light
Choroid	Absorbs stray light
Ciliary body	Holds lens in place, accommodation
Iris	Regulates light entrance
Retina	Contains sensory receptors for sight
Rod cells	Make black-and-white vision possible
Cone cells	Make color vision possible
Fovea centralis	Makes acute vision possible
Other	
Lens	Refracts and focuses light rays
Humors	Transmit light rays and support eyeball
Optic nerve	Transmits impulse to brain

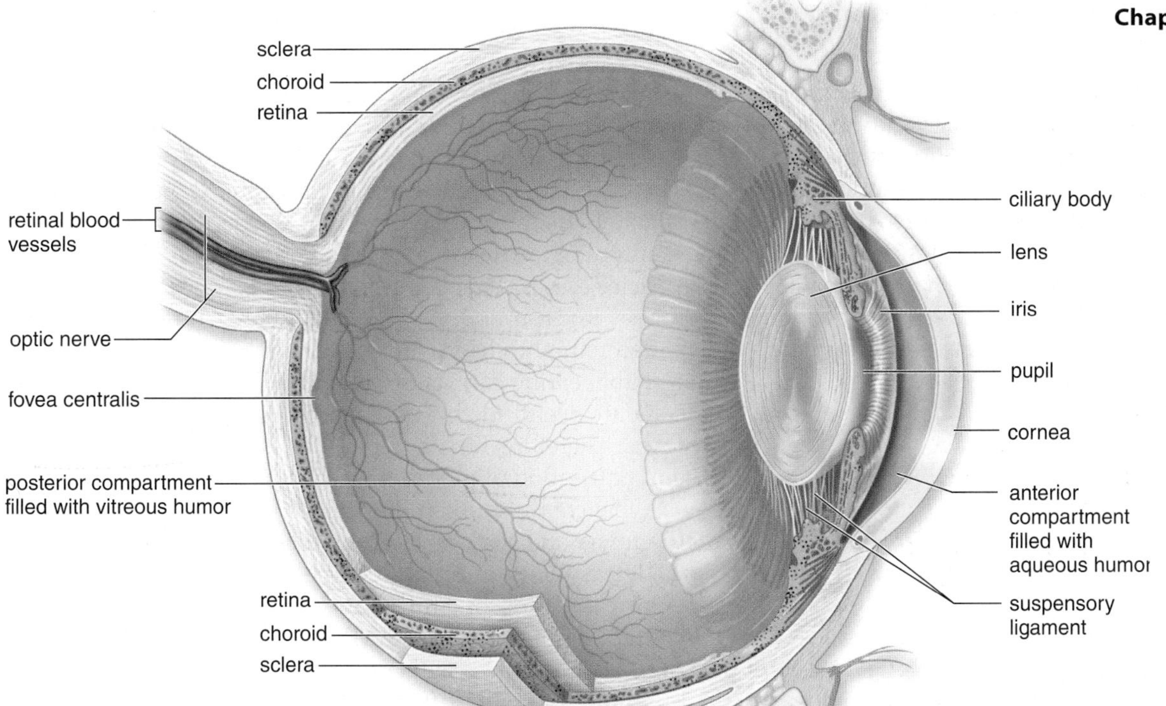

Figure 14.6
The structures of the human eye.
The sclera (the outer layer of the eye) becomes the cornea and the choroid (the middle layer) is continuous with the ciliary body and the iris. The retina (the inner layer) contains the photoreceptors for vision. The fovea centralis is the region where vision is most acute.

of the retina, where photoreceptors are located. The nerve fibers begin to die because of lack of nutrients, and the person becomes partially blind. Eventually, total blindness can result.

The third layer of the eye, the **retina,** is located in the posterior compartment. This compartment is filled with a clear, gelatinous material called the **vitreous humor.** The vitreous humor holds the retina in place and supports the lens. The retina contains photoreceptors called rod cells and cone cells. The rods are very sensitive to light, but they do not "see" color. Therefore, at night or in a darkened room, we see only shades of gray. The cones, which require bright light, are sensitive to different wavelengths of light. This sensitivity gives us the ability to distinguish colors. The retina has a very special region called the **fovea centralis** where cone cells are densely packed. Light is normally focused on the fovea when we look directly at an object. This is helpful because the sharpest images are produced by the fovea centralis. Sensory fibers from the retina form the **optic nerve,** which takes nerve signals to the visual cortex.

Function of the Lens

The cornea, assisted by the lens and the humors, focuses images on the retina. Focusing starts with the cornea and continues as the rays pass through the lens and the humors. The image produced is much smaller than the object because light rays are bent (refracted) when they are brought into **focus.** If the eyeball is too long or too short, the person may need corrective lenses to bring the image into focus. The image on the retina is inverted (upside down) and reversed from left to right.

Visual accommodation occurs for close vision. During visual accommodation, the lens changes its shape to bring the image into focus on the retina. The shape of the lens is controlled by the ciliary muscle, within the ciliary body. When we view a distant object, the ciliary muscle is relaxed, causing the suspensory ligaments attached to the ciliary body to be taut. The ligaments put tension on the lens and cause it to remain relatively flat (Fig. 14.7*a*). When we view a near object, the ciliary muscle

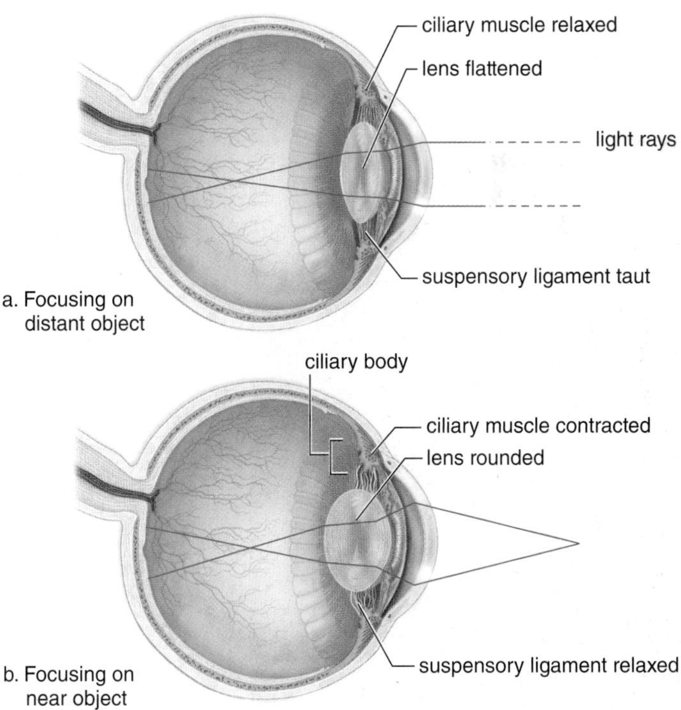

Figure 14.7 **Focusing light on the retina of the eye.**
Light rays from each point on an object are bent by the cornea and the lens in such a way that an inverted and reversed image of the object forms on the retina. **a.** When focusing on a distant object, the lens is flat because the ciliary muscle is relaxed and the suspensory ligament is taut. **b.** When focusing on a near object, the lens accommodates—it becomes rounded because the ciliary muscle contracts, causing the suspensory ligament to relax.

contracts, releasing the tension on the suspensory ligaments. The lens becomes round and thick due to its natural elasticity (Fig. 14.7*b*). Thus, contraction or relaxation of the ciliary muscle allows the image to be focused on the retina. Close work requires contraction of the ciliary muscle, so it often causes muscle fatigue, known as eyestrain. Usually, after the age of 40, the lens loses some of its elasticity and is unable to accommodate. For near vision, simple glasses with a magnifying lens solve this problem. Bifocal lenses may be necessary for those who already use corrective lenses.

Visual Pathway to the Brain

The pathway for vision begins once light has been focused on the photoreceptors in the retina. Some integration occurs in the retina, where nerve signals begin before the optic nerve transmits them to the brain.

Function of Photoreceptors Figure 14.8*a* illustrates the structure of the photoreceptors called **rod cells** and **cone cells.** Both rods and cones have an outer segment joined to an inner segment by a short stalk. Pigment molecules are embedded in the membrane of the many disks present in the outer segment. Synaptic vesicles are located at the synaptic endings of the inner segment.

The visual pigment in rods is a deep purple pigment called rhodopsin (Fig. 14.8*b*). **Rhodopsin** is a complex molecule made up of the protein opsin and a light-absorbing molecule called **retinal,** a derivative of vitamin A. When a rod absorbs light, rhodopsin splits into opsin and retinal. This leads to a cascade of reactions and the closure of ion channels in the rod cell's plasma membrane. The release of inhibitory transmitter molecules from the rod's synaptic vesicles ceases. Thereafter, signals go to other neurons in the retina. Rods are very sensitive to light and, therefore, are suited to night vision. (Carrots are rich in vitamin A, so it is true that eating carrots can improve your night vision.) Rod cells are plentiful throughout the entire retina, except the fovea. Therefore, rods also provide us with peripheral vision and perception of motion.

The cones, on the other hand, are located primarily in the fovea and are activated by bright light. They allow us to detect the fine detail and the color of an object. **Color vision** depends on three different types of cones, which contain pigments called the B (blue), G (green), and R (red) pigments. Each pigment is made up of retinal and opsin, but there is a slight difference in the opsin structure of each. This accounts for their individual absorption patterns. Various combinations of cones are believed to be stimulated by in-between shades of color.

Function of the Retina The retina has three layers of neurons (Fig. 14.9). The layer closest to the choroid contains

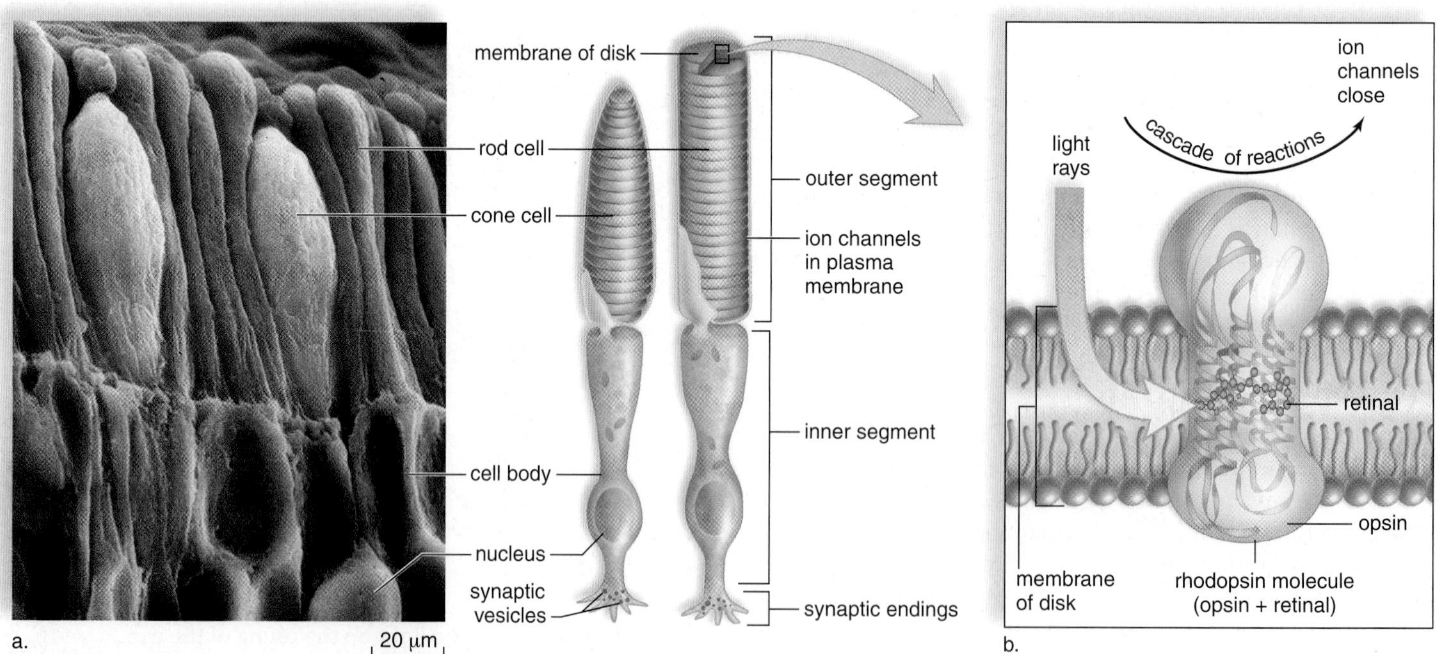

Figure 14.8 **The two types of photoreceptors in the eye.**
a. The outer segment of rods and cones contains stacks of membranous disks, which contain visual pigments. **b.** In rods, the membrane of each disk contains rhodopsin, a complex molecule containing the protein opsin and the pigment retinal. When rhodopsin absorbs light energy, it splits, releasing retinal, which sets in motion a cascade of reactions that cause ion channels in the plasma membrane to close. Thereafter, nerve signals go to the brain.

the rod cells and cone cells. A layer of bipolar cells covers the rods and cones. The innermost layer contains ganglion cells, whose sensory fibers become the optic nerve. Only the rod cells and the cone cells are sensitive to light; therefore, light must penetrate to the back of the retina before the rods and cones are stimulated.

The rod cells and the cone cells synapse with the bipolar cells. Next, signals from bipolar cells stimulate ganglion cells whose axons become the optic nerve. Notice in Figure 14.9 that there are many more rod cells and cone cells than ganglion cells. The retina has as many as 150 million rod cells and 6 million cone cells but only 1 million ganglion cells. The sensitivity of cones versus rods is mirrored by how directly they connect to ganglion cells. As many as 150 rods may activate the same ganglion cell. No wonder stimulation of rods results in vision that is blurred and indistinct. In contrast, some cone cells in the fovea centralis activate only one ganglion cell. This explains why cones, especially in the fovea centralis, provide us with a sharper, more detailed image of an object.

As signals pass to bipolar cells and ganglion cells, integration occurs. Therefore, considerable processing occurs in the retina before ganglion cells generate nerve signals. Ganglion cells converge to form the optic nerve, which transmits information to the visual cortex. Additional integration occurs in the visual cortex.

Blind Spot Figure 14.9 also provides an opportunity to point out that there are no rods and cones where the optic nerve exits the retina. Therefore, no vision is possible in this area. You can prove this to yourself by putting a dot to the right of center on a piece of paper. Use your right hand to move the paper slowly toward your right eye, and make sure you look straight ahead. The dot will disappear at one point—this is your right eye's **blind spot.** The two eyes together provide complete vision because the blind spot for the right eye is not the same as the blind spot for the left eye. The blind spot for the right eye is right of center and the blind spot for the left eye is left of center.

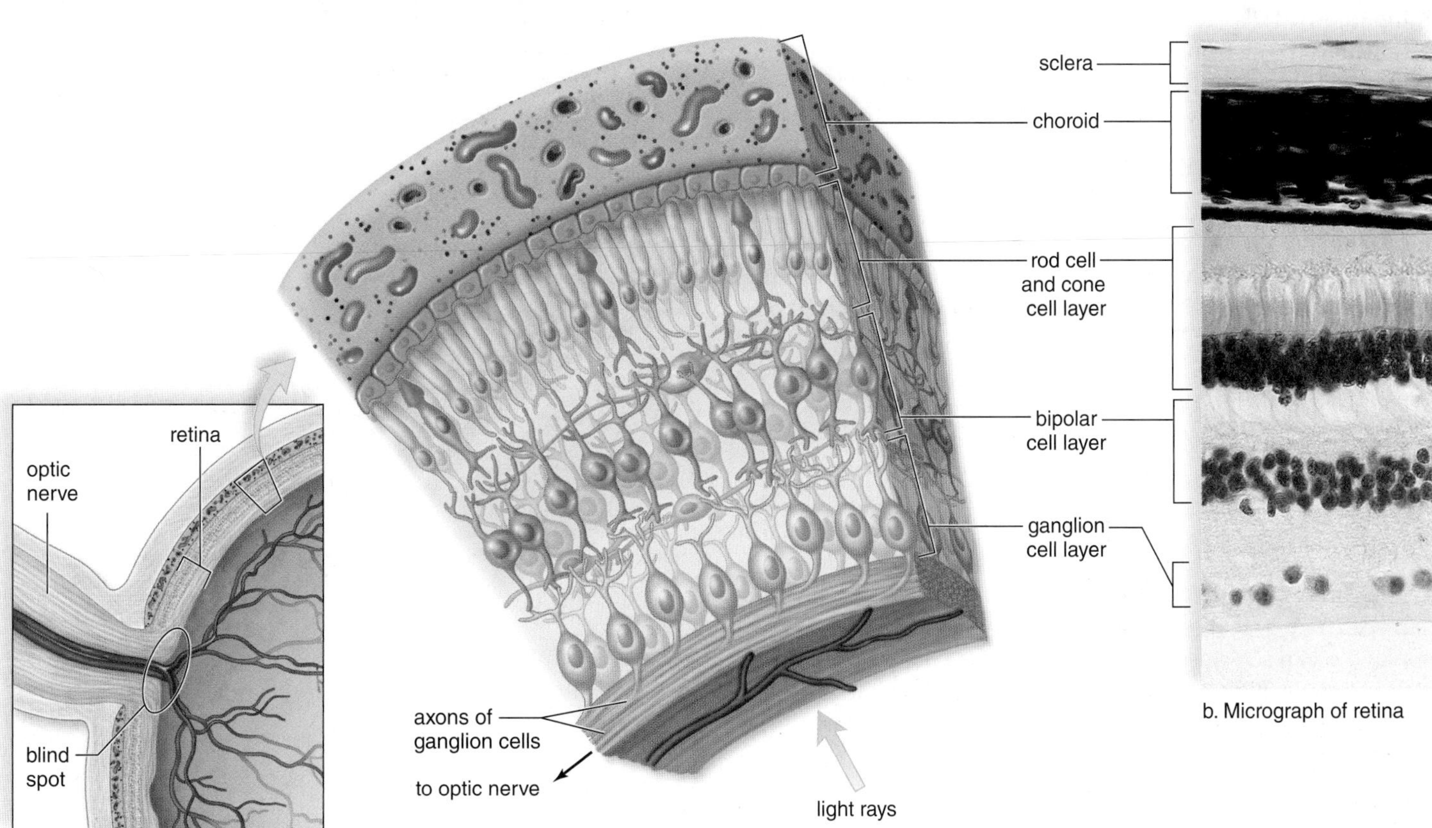

Figure 14.9 The structure of the retina.
a. The retina is the inner layer of the eyeball. Rod and cone cells, located at the back of the retina nearest the choroid, synapse with bipolar cells, which synapse with ganglion cells. Integration of signals occurs at these synapses; therefore, much processing occurs in bipolar and ganglion cells. Further, many rod cells share one bipolar cell but cone cells do not. Certain cone cells synapse with only one ganglion cell. Cone cells, in general, distinguish more detail than do rod cells. **b.** Micrograph shows that the sclera and choroid are relatively thin compared to the retina, which has several layers of cells.

From the Retina to the Visual Cortex To reach the visual cortex, the optic nerves carry nerve impulses from the eyes to the optic chiasma (Fig. 14.10). The **optic chiasma** has an X shape, formed by a crossing-over of optic nerve fibers. After exiting the optic chiasma, the optic nerves continue as **optic tracts.** Fibers from the right half of each retina converge and continue on together in the right optic tract. Similarly, the nerve fibers from the left half of each retina join to form the left optic tract, traveling together to the brain.

The optic tracts sweep around the hypothalamus, and most fibers synapse with neurons in nuclei (masses of neuron cell bodies) within the thalamus. Axons from the thalamic nuclei form optic radiations that take nerve impulses to the *visual cortex* within the occipital lobe. The image is split in the visual cortex. This division of incoming information happens because the right visual cortex receives information from the right optic tract, and the left visual cortex receives information from the left optic tract. For good depth perception, the right and left visual cortices communicate with each other. Also, because the image is inverted and reversed, it must be righted in the brain for us to correctly perceive the visual field.

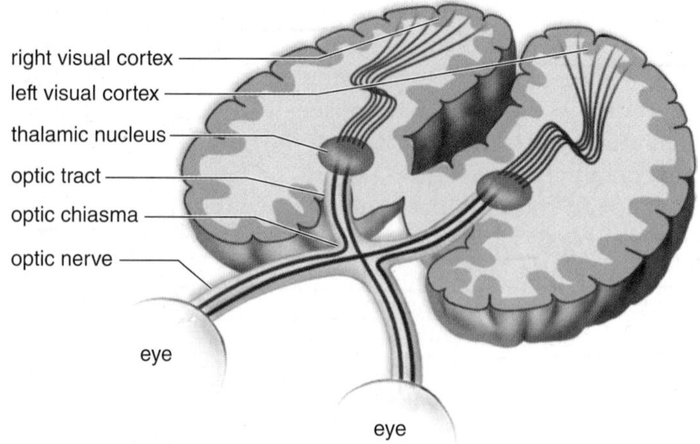

Figure 14.10 **The function of the optic chiasma.**
Because of the optic chiasma, data from the right half of each retina go to the right visual cortex, and data from the left half of the retina go to the left visual cortex. These data are then combined to allow us to see the entire visual field.

Abnormalities of the Eye

Color blindness and misshapen eyeballs are two common abnormalities of the eyes. Complete color blindness is extremely rare and is caused by a genetic mutation. In most instances, only one type of cone is defective or deficient in number. The most common mutation is the inability to see the colors red and green. This abnormality affects 5–8% of the male population. If the eye lacks cones that respond to red wavelengths, green colors are accentuated, and vice versa.

Distance Vision

If you can see from 20 feet what a person with normal vision can see from 20 feet, you are said to have 20/20 vision. Persons who can see close objects but can't see the letters on an optometrist's chart from 20 feet are said to be nearsighted. **Nearsighted** people can see close objects better than they can see objects at a distance. These individuals have an elongated eyeball, and when they attempt to look at a distant object, the image is brought to focus in front of the retina (Fig. 14.11*a*). They can see close objects because their lens can compensate for the long eyeball. To see distant objects, these people can wear concave lenses, which spread the light rays so that the image focuses on the retina.

Persons who can easily see the optometrist's chart but cannot see close objects well are **farsighted.** These individuals can see distant objects better than they can see close objects. They have a shortened eyeball, and when

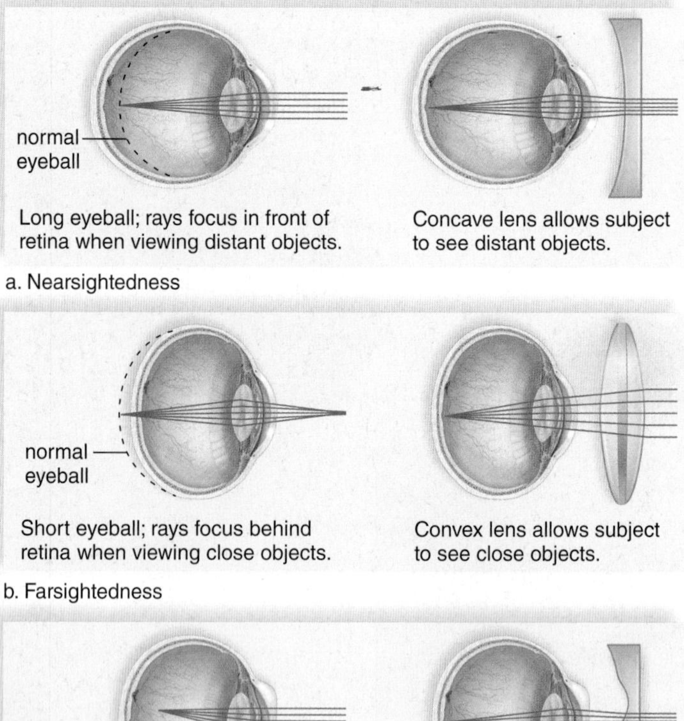

normal
eyeball
Long eyeball; rays focus in front of retina when viewing distant objects.
Concave lens allows subject to see distant objects.
a. Nearsightedness
normal
eyeball
Short eyeball; rays focus behind retina when viewing close objects.
Convex lens allows subject to see close objects.
b. Farsightedness

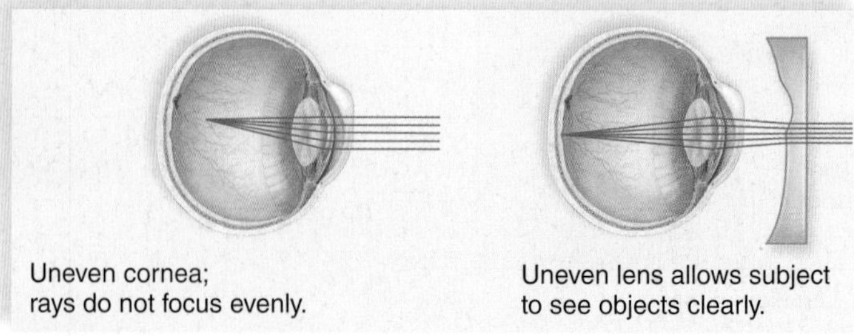

rays do not focus evenly.

Figure 14.11 **How corrective lenses correct vision problems.**
a. A concave lens in nearsighted persons focuses light rays on the retina.
b. A convex lens in farsighted persons focuses light rays on the retina.
c. An uneven lens in persons with astigmatism focuses light rays on the retina.

Biology Matters Health Focus

Correcting Vision Problems

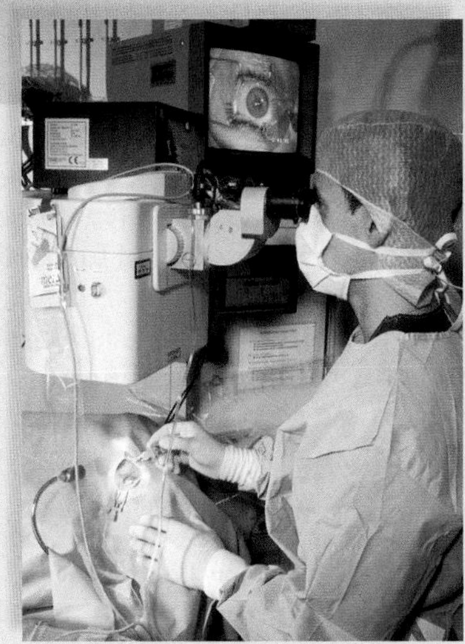

Figure 14A
LASIK surgery.

Poor vision can be due to a number of problems, some more serious than others. For example, retinal detachment, cataracts, and glaucoma are three conditions that need medical attention.

Cataracts and Glaucoma

Cataracts develop when the lens of the eye becomes cloudy. Normally, the lens is clear and allows light to pass through easily. A cloudy lens decreases light levels that reach the retina and slowly causes vision loss. Fortunately, a doctor can surgically remove the cloudy lens and replace it with a clear plastic lens, which often restores the light level passing through the lens and improves the patient's vision.

When fluid pressure builds up inside the eye, a patient is diagnosed with glaucoma. Glaucoma leads to a decrease in vision and can eventually cause blindness. Special eye drops and oral medications are often prescribed to help reduce the intraocular pressure. If eye drops and medications are not capable of controlling the pressure, surgery may be the only option. During glaucoma surgery, the doctor uses a laser to create tiny holes in the eye where the cornea and iris meet. It is hoped this will increase fluid drainage from the eye and decrease the pressure inside the eye.

The Benefits of LASIK Surgery

For many people, a sign of aging is the slow and steady decrease of their ability to see close-up. A difficulty in focusing on small print is usually the first sign of presbyopia, a condition that usually begins in the late 30s. By age 55, nearly 100% of the population is affected. Reading in low-light situations becomes more difficult, and letters begin to look fuzzy when reading close-up. Many people who suffer from presbyopia experience headaches while reading. To accommodate for their deteriorating vision, presbyopia sufferers hold reading materials at arm's length. One solution to presbyopia is to wear a simple magnifying lens (purchased at a pharmacy, for example). Those already wearing glasses may need bifocal lenses. Bifocals are designed to correct vision at a distance of 12 to 18 inches. These lenses work well for most people while reading but pose a problem for people who use a computer. Computer monitors are usually 19 to 24 inches away. This forces bifocal lens wearers to constantly move their head up and down in an attempt to switch between the close and distant viewing sections of the bifocals. Another solution is to wear contact lenses with one eye corrected to see close objects and the other eye corrected to see distant objects. The same type of correction can be done with LASIK surgery (Fig. 14A). LASIK, which stands for *laser-assisted in situ keratomileusis*, is generally a safe and effective treatment option for a wide array of vision problems.

LASIK is a quick and painless procedure that involves the use of a laser to permanently change the shape of the cornea. For the majority of patients, LASIK improves their vision and reduces their dependency on corrective lenses. The ideal LASIK candidate is over 18 years of age and has had a stable contact or glasses prescription for at least two years. Patients need to have a cornea thick enough to allow the surgeon to safely create a clean corneal flap of appropriate depth. Typically, patients affected by common vision problems (nearsightedness, astigmatism, or farsightedness) respond well to LASIK. Anyone who suffers from any disease that decreases his or her ability to heal properly after surgery is not a candidate for LASIK. Candidates should thoroughly discuss the procedure with their eye-care professional before electing to have LASIK. People need to realize that the goal of LASIK is to reduce their dependency on glasses or contact lenses, not completely eliminate them.

Individuals who suffer from cataracts, advanced glaucoma, or corneal or other eye diseases are not considered for LASIK. Patients who expect LASIK to completely correct their visual problems and make them completely independent of their corrective lenses are not good candidates either.

The LASIK Procedure

During the LASIK procedure, a small flap of tissue (the conjunctiva) is cut away from the front of the eye. The flap is folded back exposing your cornea, allowing the surgeon to remove a defined amount of tissue from your cornea. Each pulse of the laser removes a small amount of corneal tissue, allowing the surgeon to flatten or increase the steepness of the curve of your cornea. After the procedure, the flap of tissue is put back into place and allowed to heal on its own. LASIK patients receive eye drops or medications to help relieve the pain of the procedure. Improvements to your vision begin as early as the day after the surgery but typically take two to three months. Most patients will have vision close to 20/20, but your chances for improved vision are based in part on how good your eyes were before the surgery.

As with any surgery, complications are possible. Adverse effects include a sensation of having something in your eye or having blurred vision. You might also see halos around objects or be very sensitive to glare. In addition, dryness can cause eye irritation. Typically, these effects are temporary, and the rate of complications following surgery is very low. Always consult with your doctor before considering any type of surgery.

they try to see close objects, the image is focused behind the retina (Fig. 14.11*b*). When the object is distant, the lens can compensate for the short eyeball. When the object is close, these persons can wear convex lenses to increase the bending of light rays so that the image can be focused on the retina.

When the cornea or lens is uneven, the image is fuzzy. The light rays cannot be evenly focused on the retina. This condition, called **astigmatism,** can be corrected by an unevenly ground lens to compensate for the uneven cornea (Fig. 14.11*c*).

Many people today opt to have LASIK surgery instead of wearing lenses. LASIK surgery is discussed in the Health Focus, *Correcting Vision Problems.*

Video
Artificial Eye

Connections and Misconceptions

What is pinkeye?

At some point in their lives, most people have suffered from conjunctivitis, or pinkeye. Conjunctivitis is the inflammation of a mucous membrane called the conjunctiva, which covers the eyeball (except the cornea) and the inner part of the eyelid. The purpose of the conjunctiva is to lubricate the eye and keep it from drying out. In the case of viral conjunctivitis, the most common type, this membrane becomes inflamed as part of an immune response against the viral pathogens. Viral conjunctivitis is highly contagious; individuals with the condition must be careful not to spread the disease. However, not all conjunctivitis is contagious; allergies and other medical conditions may cause pinkeyelike symptoms. Treatment usually involves the use of eyedrops that help lubricate the eye and reduce inflammation.

Connecting the Concepts

For more information on the information presented in this section, refer to the following discussions.

Table 8.7 describes the function and dietary sources of vitamin A.

Section 13.2 describes the function of the visual association area in the cerebral cortex of the brain.

Section 20.4 explores the pattern of inheritance associated with color blindness.

Check Your Progress 14.4

1. Identify the structures of the eye and provide a function of each.
2. Describe the two types of photoreceptors and state the function of each.
3. Summarize the movement of sensory information from the photoreceptors to the visual cortex.

14.5 Sense of Hearing

Learning Outcomes

Upon completion of this section, you should be able to

1. Identify the structures of the ear that are involved in hearing.
2. Summarize how sound waves are converted into nerve signals.
3. Describe the pathway of sensory information from the ear to the brain.

The ear has two sensory functions: hearing and balance (equilibrium). The sensory receptors for both of these are located in the inner ear. Each consists of **hair cells** with **stereocilia** (sing., stereocilium)—long microvilli—that are sensitive to mechanical stimulation. They are mechanoreceptors.

Anatomy and Physiology of the Ear

Figure 14.12 shows that the ear has three divisions: outer, middle, and inner. The **outer ear** consists of the **pinna** (external flap) and the **auditory canal.** The opening of the auditory canal is lined with fine hairs and sweat glands. Modified sweat glands are located in the upper wall of the canal. They secrete earwax, a substance that helps guard the ear against the entrance of foreign materials, such as air pollutants.

The **middle ear** begins at the **tympanic membrane** (eardrum) and ends at a bony wall containing two small openings covered by membranes. These openings are called the **oval window** and the **round window.** Three small bones are found between the tympanic membrane and the oval window. Collectively, they are called the **ossicles.** Individually, they are the **malleus** (hammer), the **incus** (anvil), and the **stapes** (stirrup) because their shapes resemble these

Connections and Misconceptions

Why does that annoying song you hear seem to replay itself in your head all day?

Scientists who study the brain and special senses call such phenomena *earworms.* These are most likely caused by songs with repetitive, silly, or catchy lyrics. Their cause is not fully understood. Scientists have documented that the auditory cortex—the site for storing memories of sounds you've heard—becomes very active when earworms run through your mind. Earworms may be a way to keep the brain "idling," much as a car idles when you're at a stoplight. When you're ready for active thinking and problem solving, your brain is ready to go.

The good news is, you can make an earworm go away. Distract yourself with a complex task, such as reading or solving a puzzle. Eat something very spicy—the taste sensation will stimulate other areas of the brain. If all else fails, just think of a song that is equally annoying and it will most likely replace the first!

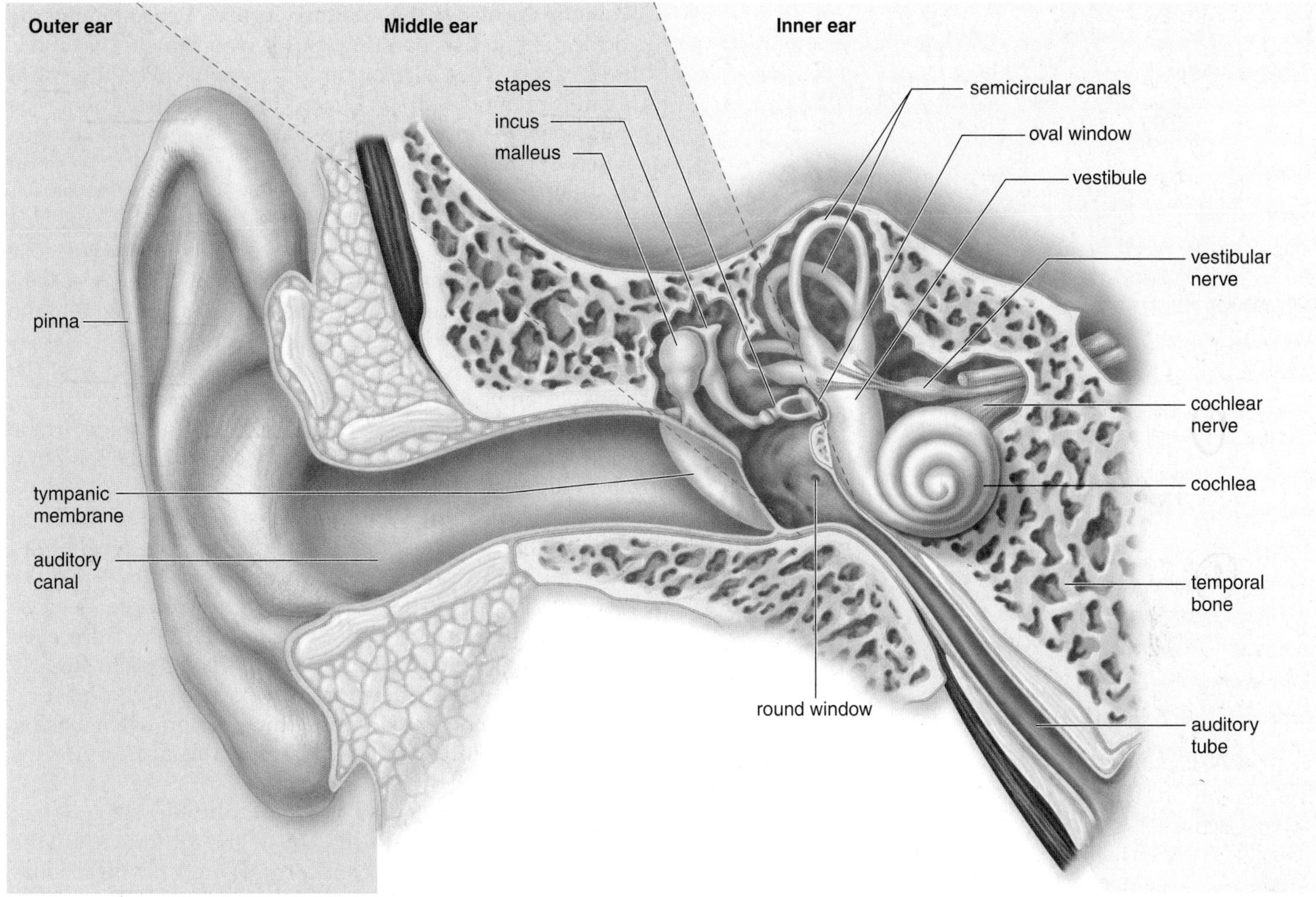

Figure 14.12 **The three divisions of the human ear.**
The external ear consists of the pinna (the structure commonly referred to as the "ear") and the auditory canal. The tympanic membrane separates the external ear from the middle ear. In the middle ear, the malleus (hammer), the incus (anvil), and the stapes (stirrup) amplify sound waves. In the inner ear, the mechanoreceptors for equilibrium are in the semicircular canals and the vestibule. The mechanoreceptors for hearing are in the cochlea.

objects. The malleus adheres to the tympanic membrane, and the stapes touches the oval window. An **auditory (eustachian) tube**, which extends from the middle ear to the nasopharynx, permits equalization of air pressure. Chewing gum, yawning, and swallowing in elevators and airplanes help move air through the auditory tubes upon ascent and descent. As this occurs, we often feel the ears "pop."

Whereas the outer ear and the middle ear contain air, the inner ear is filled with fluid. The **inner ear** has three areas: The **semicircular canals** and the **vestibule** are concerned with equilibrium; the **cochlea** is concerned with hearing. The cochlea resembles the shell of a snail because it spirals.

Auditory Pathway to the Brain

The auditory pathway begins with the auditory canal. Thereafter, hearing requires the other parts of the ear, the cochlear nerve, and the brain.

Through the Auditory Canal and Middle Ear The process of hearing begins when sound waves enter the auditory canal. Just as ripples travel across the surface of a pond, sound waves travel by the successive vibrations of molecules. Ordinarily, sound waves do not carry much energy. However, when a large number of waves strike the tympanic membrane, it moves back and forth (vibrates) ever so slightly. As you know, the auditory ossicles attach to one another: malleus to incus, incus to stapes. The malleus is attached to the inner wall of the tympanic membrane. Thus, vibrations of the tympanic membrane cause vibration of the malleus and, in turn, the incus and stapes. The magnitude of the original pressure wave increases significantly as the vibrations move along the auditory ossicles. The pressure is multiplied about 20 times. Finally, the stapes strikes the membrane of the oval window, causing it to vibrate. In this way, the pressure is passed to the fluid within the cochlea.

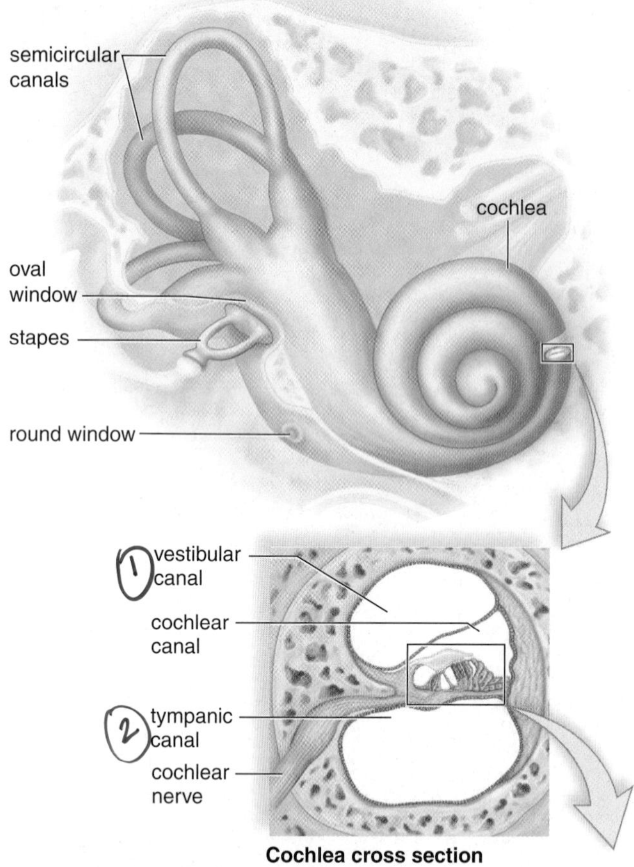

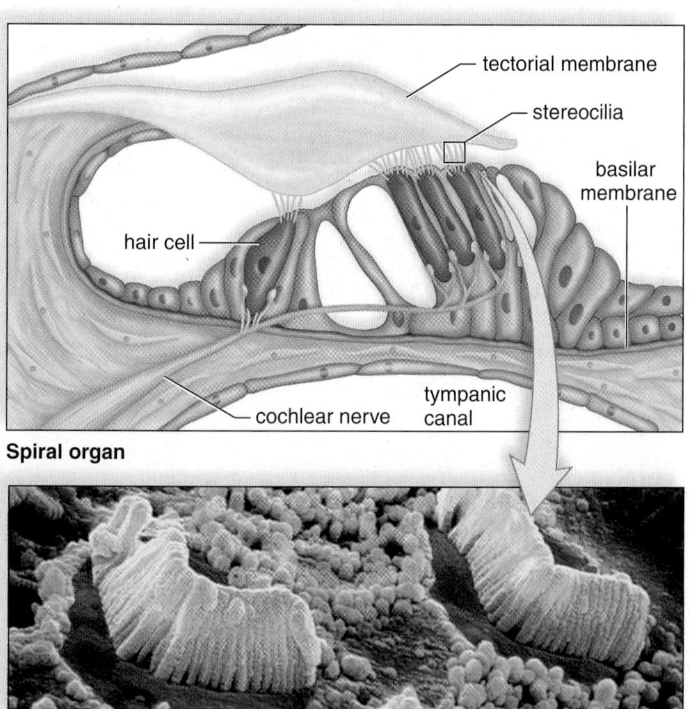

Figure 14.13 **How the spiral organ (organ of Corti) translates sound waves into nerve signals.**
The spiral organ (organ of Corti) is located within the cochlea. The spiral organ consists of hair cells resting on the basilar membrane, with the tectorial membrane above. Pressure waves moving through the canals cause the basilar membrane to vibrate. This causes the stereocilia embedded in the tectorial membrane to bend. Nerve impulses traveling in the cochlear nerve result in hearing.

From the Cochlea to the Auditory Cortex By examining the cochlea in cross section (Fig. 14.13), you can see that it has three canals. The sense organ for hearing, called the **spiral organ** (organ of Corti), is located in the cochlear canal. The spiral organ consists of little hair cells and a gelatinous material called the **tectorial membrane.** The hair cells sit on the basilar membrane, and their stereocilia are embedded in the tectorial membrane.

Animation Effect of Sound Waves on Cochlear Structures

When the stapes strikes the membrane of the oval window, pressure waves move from the vestibular canal to the tympanic canal across the basilar membrane. The basilar membrane moves up and down, and the stereocilia of the hair cells embedded in the tectorial membrane bend. Then, nerve signals begin in the **cochlear nerve** and travel to the brain. When they reach the auditory cortex in the temporal lobe, they are interpreted as a sound.

Each part of the spiral organ is sensitive to different wave frequencies, or pitch. Near the tip, the spiral organ responds to low pitches, such as those of a tuba. Near the base (beginning), it responds to higher pitches, such as those of a bell or a whistle. The nerve fibers from each region along the length of the spiral organ lead to slightly different areas in the auditory cortex. The pitch sensation we experience depends upon which region of the basilar membrane vibrates and which area of the auditory cortex is stimulated.

Volume is a function of the amplitude (strength) of sound waves. Loud noises cause the fluid within the vestibular canal to exert more pressure and the basilar membrane to vibrate to a greater extent. The resulting increased stimulation is interpreted by the brain as volume.

Animation Hearing

Connecting the Concepts

For more information on the material in this section, refer to the following discussions.

Section 13.2 describes the function of the cerebral cortex area of the brain in hearing.

Figure 13.14 illustrates the structure of a nerve.

Check Your Progress 14.5

1. Identify the structures of the ear involved in hearing, and provide a function for each.
2. Describe the role of mechanoreceptors in the sense of hearing.
3. Summarize how the spiral organ translates sound waves to nerve impulses.

Biology Matters Bioethical Focus

Noise Pollution

Though we can sometimes tune its presence out, unwanted noise is all around us. Noise pollution is noise from the environment that is annoying, distracting, and potentially harmful. It comes from airplanes, cars, lawn mowers, machinery, and our own loud music and that of our neighbors. It is present at our workplaces, in public spaces like amusement parks, and at home. Its prevalence allows loud noise to have a potentially high impact on our welfare.

How Does Noise Affect Us?

How does noise affect human health? Perhaps the greatest worry about noise pollution is that exposure to loud (>85 decibels) or chronic noises can damage cells of the inner ear and cause hearing loss (Fig. 14B). When we are young, we often do not consider the damage that noise may be doing to our spiral organ. The stimulation of loud music is often sought by young people at rock concerts without regard to the possibility that their hearing may be diminished as a result. Over the years, loud noises can bring deafness and accompanying depression when we are seniors.

Noise can affect well-being by other means, too. Data from studies of environmental noise can be difficult to interpret because of the presence of other confounding factors, including physical or chemical pollution. The tolerance level for noise

(continued)

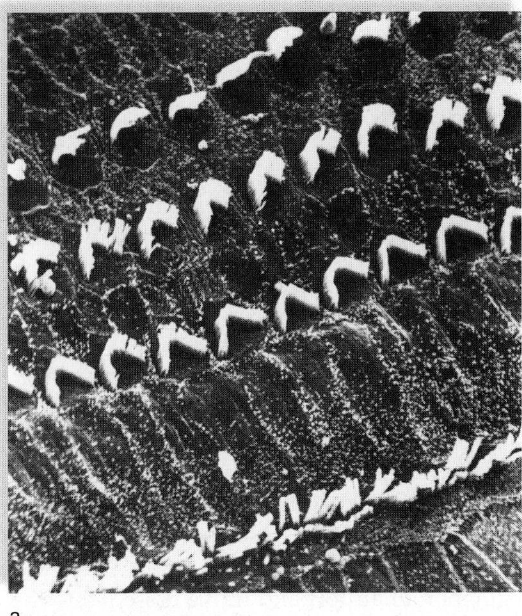

a.

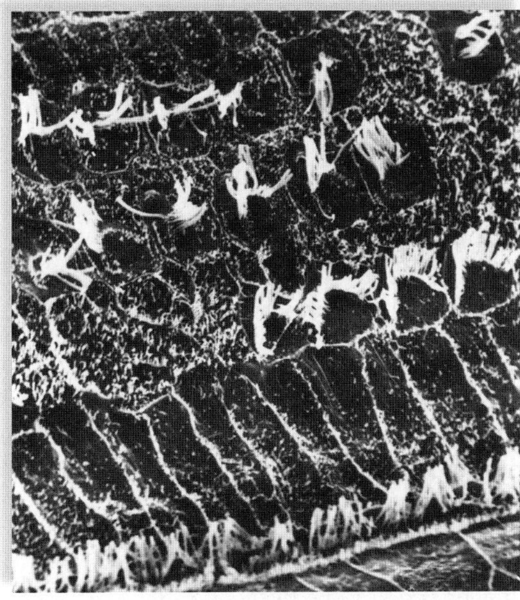

b.

Figure 14B Loud noise damages the hair cells in the spiral organ. **a.** Normal hair cells in the spiral organ of a guinea pig. **b.** Damaged cells. This damage occurred after 24-hour exposure to a noise level equivalent to that at a heavy-metal rock concert (see Table 14A). Hearing is permanently impaired because lost cells will not be replaced, and damaged cells may also die.

Table 14A Noises That Affect Hearing

Type of Noise	Sound Level (Decibels)	Effect
"Boom car," jet engine, shotgun, rock concert	Over 125	Beyond threshold of pain; potential for hearing loss high
Nightclub, thunderclap	Over 120	Hearing loss likely
Earbuds in external ear canal	110–120	Hearing loss likely
Chain saw, pneumatic drill, jackhammer, symphony orchestra, snowmobile, garbage truck, cement mixer	100–200	Regular exposure of more than 1 min risks permanent hearing loss.
Farm tractor, newspaper press, subway, motorcycle	90–100	Fifteen minutes of unprotected exposure; potentially harmful
Lawn mower, food blender	85–90	Continuous daily exposure for more than 8 hr can cause hearing damage.
Diesel truck, average city traffic noise	80–85	Annoying; constant exposure may cause hearing damage.

(continued from p. 331)

also varies from person to person. Nonetheless, laboratory and field studies show that noise may be detrimental in nonauditory ways. Its effects on mental health include annoyance, inability to concentrate, and increased irritability. Long-term noise exposure from air or car traffic may impair cognitive ability, language learning, and memory in children. Noise often causes loss of sleep and reduced productivity and can induce stress. Additionally, several studies have suggested that noise pollution negatively affects cardiovascular health, though more research needs to be done to confirm this link.

Federal and Local Control

Noise pollution has been a concern for several decades. In 1972, the Noise Control Act was passed as a means for coordinating federal noise control and research and to develop noise emission standards. The aim was to protect Americans from "noise that jeopardizes their health or welfare." The Environmental Protection Agency (EPA) had federal authority to regulate noise pollution, and their Office of Noise Abatement and Control (ONAC) worked on establishing noise guidelines. However, funding for ONAC was cut off in 1981. Today, there is no national noise policy, though several federal government departments have some oversight.

Workplace noise exposure is controlled by the Occupational Safety and Health Administration (OSHA). OSHA has set guidelines for workplace noise. OSHA regulations require that protective gear be provided if sound levels exceed certain values. This may include noise-reducing earmuffs and other equipment for people who work around big equipment. However, OSHA guidelines don't cover things like telephone ringing and computer or typewriter noise that may be present in a nonindustrial environment such as an open-plan office. Aviation noise and traffic noise reduction plans are overseen by the Department of Transportation, the Federal Aviation Administration (FAA), and the Federal Highway Administration (FHA), respectively.

Some cities and local governments have taken steps to decrease unwanted or annoying noise. Many communities have noise ordinances and control noise levels during particular times of day. Persons throwing loud house parties might get a knock on their door from the police, and owners of barking dogs, a visit from animal control. Business enterprises may also be regulated. For example, concerts at outdoor venues may be required to finish by a certain time in the evening. However, there is often opposition to this type of control on the basis that it impedes commerce.

Many people believe that much more should be done to curb noise pollution. Yet there are difficulties in regulating noise pollution generated by private persons or business owners. Individuals often feel that they should be free to do as they wish on private property, including playing music at full volume or running machinery. For example, club and café owners may believe that their livelihood depends on having reasonably loud sound systems. The same may hold true of sports fields and stadiums. There are also challenges to regulating airport and traffic noise. The main responsibilities of the FAA and FHA—promoting aviation industry growth and maintaining highway infrastructure—can be at odds with controlling noise for the public's sake. Therefore, noise abatement isn't necessarily their top priority.

Finding neighbor- and commerce-friendly solutions to noise problems will likely require individual citizens, communities, and governments to work together. Stopping the source of noise may not be the only answer to this issue. The implementation of new technology could also make things quieter by preventing sounds from traveling too far away from their sources.

Decide Your Opinion

1. Do you think that more should be done to curb noise pollution in your neighborhood? If so, why?
2. Do you believe that the government has the right to regulate what someone does on his or her own property? Why or why not?
3. Should regulation of airport and traffic noise fall to the Department of Transportation? If not, who may better control airport or highway noise?
4. If technology is available to reduce noise on highways, who do you think should pay for implementing it—tax payers, car owners, or others?

14.6 Sense of Equilibrium

Learning Outcomes

Upon completion of this section, you should be able to

1. Explain how mechanoreceptors are involved in the sense of equilibrium.
2. Identify the structures of the ear involved in the sense of equilibrium.
3. Distinguish between rotational and gravitational equilibrium.

The vestibular nerve originates in the semicircular canals, saccule, and utricle. It takes nerve signals to the brain stem and cerebellum (Fig. 14.14). Through its communication with the brain, the vestibular nerve helps us achieve equilibrium, but other structures in the body are also involved. For example, we already mentioned that proprioceptors are necessary for maintaining our equilibrium. Vision, if available, usually provides extremely helpful input the brain can act upon. To explain, let's take a look at the two sets of mechanoreceptors for equilibrium.

Rotational Equilibrium Pathway

Mechanoreceptors in the **semicircular canals** detect rotational and/or angular movement of the head—**rotational equilibrium** (Fig. 14.14*a*). The three semicircular canals are arranged so that there is one in each dimension of space. The base, or **ampulla,** of each of the three canals is slightly enlarged. Little hair cells, whose stereocilia are embedded within a gelatinous material called a cupula, are found within the ampullae. Each ampulla responds to head rotation in a different plane of space because of the way the semicircular canals are arranged. As fluid within a semicircular canal flows over and displaces a cupula, the

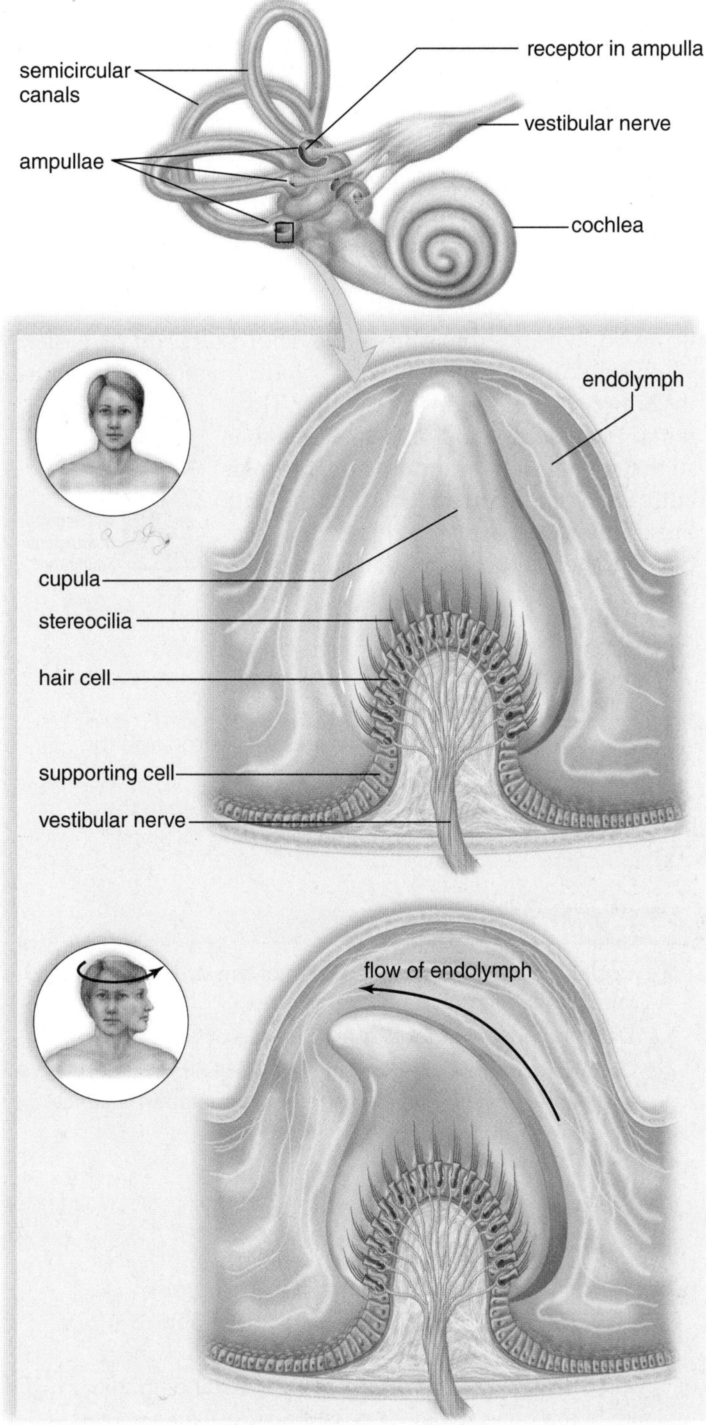

a. Rotational equilibrium: receptors in ampullae of semicircular canal

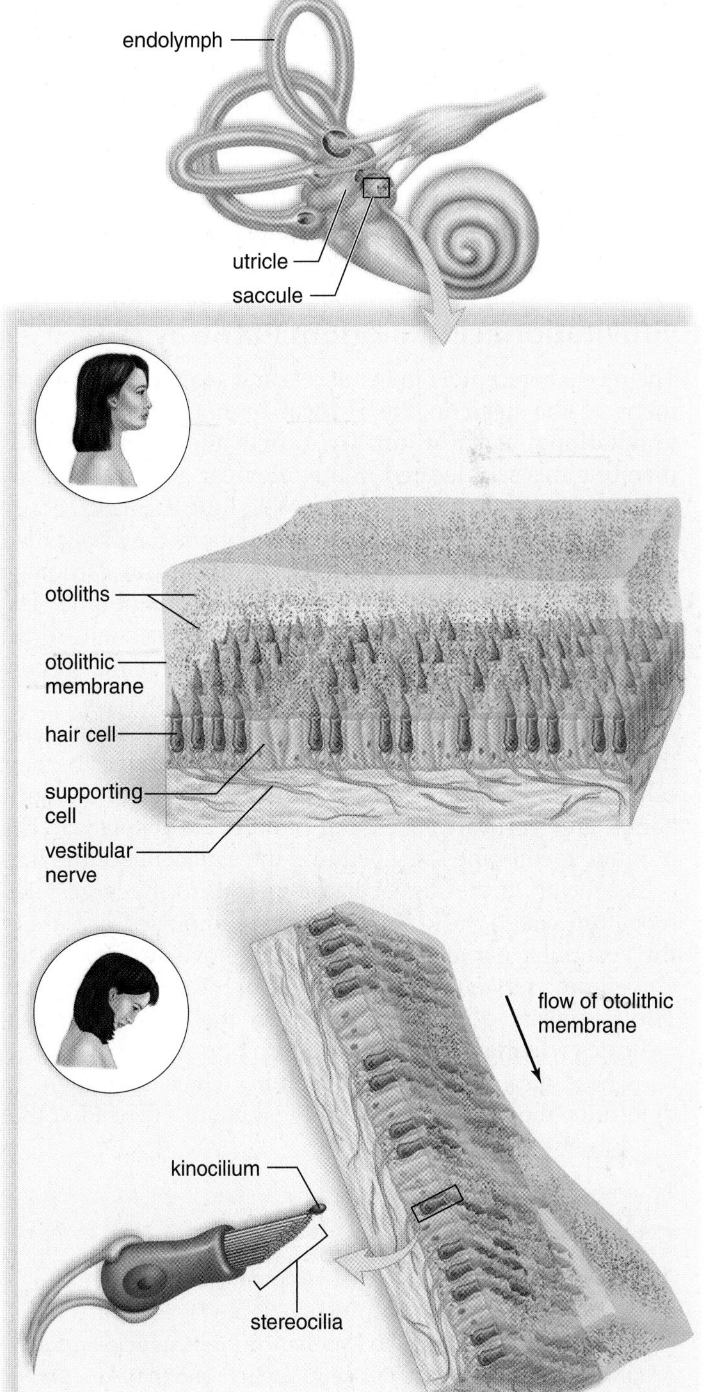

b. Gravitational equilibrium: receptors in utricle and saccule of vestibule

Figure 14.14 The mechanoreceptors of the inner ear and the sense of balance.
a. Rotational equilibrium is coordinated by receptors in the ampullae of the semicircular canals. **b.** Gravitational equilibrium is coordinated by receptors in the utricule and saccule located near the semicircular canals.

stereocilia of the hair cells bend. This causes a change in the pattern of signals carried by the vestibular nerve to the brain. The brain uses information from the hair cells within each ampulla of the semicircular canals to maintain equilibrium. Appropriate motor output to various skeletal muscles can correct our present position in space as needed.

Why does spinning around cause you to become dizzy? When we spin, the cupula slowly begins to move in the same direction we are spinning, and bending of the stereocilia causes hair cells to send messages to the brain. As time goes by, the cupula catches up to the rate we are spinning, and the hair cells no longer send messages to the brain. When we stop spinning, the slow-moving cupula continues to move in the direction of the spin and the stereocilia bend again, indicating that we are moving. Yet, the eyes know we have stopped. The mixed messages sent to the brain cause us to feel dizzy.

Gravitational Equilibrium Pathway

The mechanoreceptors in the utricle and saccule detect movement of the head in the vertical or horizontal planes—**gravitational equilibrium.** The **utricle** and **saccule** are two membranous sacs located in the inner ear near the semicircular canals. Both of these sacs contain little hair cells whose stereocilia are embedded within a gelatinous material called an *otolithic membrane* (see Fig. 14.14*b*). Calcium carbonate ($CaCO_3$) granules, or **otoliths,** rest on this membrane. The utricle is especially sensitive to horizontal (back-and-forth) movements and the bending of the head, and the saccule responds best to vertical (up-and-down) movements.

When the body is still, the otoliths in the utricle and the saccule rest on the otolithic membrane above the hair cells. When the head bends or the body moves in the horizontal and vertical planes, the otoliths are displaced. The otolithic membrane sags, bending the stereocilia of the hair cells beneath. If the stereocilia move toward the largest stereocilium, called the **kinocilium,** nerve impulses increase in the vestibular nerve. If the stereocilia move away from the kinocilium, nerve impulses decrease in the vestibular nerve. The frequency of nerve impulses in the vestibular nerve indicates whether you are moving up or down.

These data reach the cerebellum, which uses them to determine the direction of the movement of the head at that moment. Remember that the cerebellum (Chapter 13) is vital to maintaining balance and gravitational equilibrium. The cerebellum processes information from the inner ear (the semicircular canals, utricle, and saccule) as well as visual and proprioceptive inputs. In addition, the motor cortex in the frontal lobe of the brain signals where the limbs should be located at any particular moment. After integrating all these nerve inputs, the cerebellum coordinates skeletal muscle contraction to correct our position in space if necessary.

Continuous stimulation of the stereocilia can contribute to motion sickness, especially when messages reaching the brain conflict with visual information from the eyes. Imagine that you are standing inside a ship that is tossing up and down on the waves. Your visual inputs signal that you are standing still, because you can see the wall in front of you and that wall isn't moving. However, the inputs from all three sensory areas of the inner ear tell your brain that you are moving up and down and from side to side. The result? You're seasick!

If you can match the two sets of information coming into the brain, you will begin to feel better. Thus, it makes sense to stand on deck if possible, so that visual signals and inner-ear signals both tell your brain that you're moving. Prescription and over-the-counter medications can help alleviate motion sickness.

Animation Sense and Balance

MP3 The Sense of Hearing and Equilibrium

Connecting the Concepts

For more information on the sense of equilibrium, refer to the following discussions.

Section 13.1 examines the structure of a neuron and the generation of a nerve impulse.

Section 13.2 explains the role of the cerebellum in the processing of sensory information regarding balance.

Check Your Progress 14.6

1. Explain the location and function of the structures that are involved in maintaining balance.
2. Describe how rotational equilibrium is achieved.
3. Contrast rotational and gravitational equilibrium and explain how the two work together to maintain balance.

CASE STUDY CONCLUSION

The sounds that people hear with cochlear implants are different from what hearing people are familiar with. However, with training and experience, implant recipients can develop the ability to understand speech, and they function very much like a person with normal hearing. Several recent studies have shown that deaf babies that receive implants as young as six months of age develop language and speech skills very much like those of hearing children. By identifying the gene responsible for many forms of nonsyndromic deafness (*GJB2*), genetic counselors will soon be able to inform parents with a family history of deafness as to the chances of having a deaf child. Furthermore, by understanding the function of this gene, it may be possible to develop a greater understanding as to how the cochlea develops in the fetus and how environmental factors and genetics play a role in some forms of deafness.

Media Study Tools

www.mhhe.com/maderhuman12e

Enhance your study of this chapter with study tools and practice tests. Also ask your instructor about the resources available through ConnectPlus, including the media-rich eBook, interactive learning tools, and animations.

Summarizing the Concepts

14.1 Overview of Sensory Receptors and Sensations

There are four types of sensory receptors: chemoreceptors; photoreceptors; mechanoreceptors; and thermoreceptors.

- Sensory receptors initiate nerve signals transmitted to the spinal cord and/or brain.
- Sensation occurs when nerve signals reach the cerebral cortex.
- Perception is an interpretation of sensations.

14.2 Proprioceptors, Cutaneous Receptors, and Pain Receptors

Proprioceptors

- are mechanoreceptors involved in reflex actions.
- help maintain equilibrium and posture.

Cutaneous Receptors

- are found in the skin.
- sense touch, pressure, temperature, and pain.

Pain Receptors

- are also known as nociceptors.
- respond to chemical signals from damaged tissues.

14.3 Senses of Taste and Smell

Taste and smell are due to chemoreceptors stimulated by molecules in the environment.

Sense of Taste

Microvilli of taste cells have receptor proteins for molecules that cause the brain to distinguish sweet, sour, salty, and bitter tastes.

Sense of Smell

The cilia of olfactory cells have receptor proteins for molecules that cause the brain to distinguish odors.

14.4 Sense of Vision

Vision depends on the eye, the optic nerves, and the visual areas of the cerebral cortex.

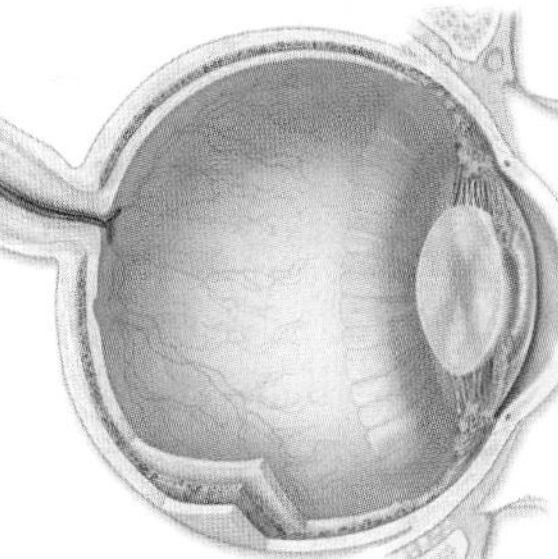

Anatomy and Physiology of the Eye

The eye has three layers:

- The sclera (outer layer) protects and supports the eyeball.
- The choroid (middle, pigmented layer) absorbs stray light rays.
- The retina (inner layer) contains the rod cells (sensory receptors for dim light) and cone cells (sensory receptors for bright light and color).

  Function of the Lens The lens (assisted by the cornea and the humors) brings the light rays to focus on the retina. To see a close object, visual accommodation occurs as the lens becomes round and thick.

  Visual Pathway to the Brain The visual pathway begins when light strikes photoreceptors (rod cells and cone cells) in the retina. The optic nerves carry nerve impulses from the eyes to the optic chiasma, then pass through the thalamus before reaching the primary vision area in the occipital lobe of the brain.

Abnormalities of the Eye

- color blindness
- misshapen eyeballs (cause of nearsightedness, farsightedness, or astigmatism)

14.5 Sense of Hearing

Hearing depends on the ear, the cochlear nerve, and the auditory areas of the cerebral cortex.

Anatomy and Physiology of the Ear

The ear has three parts:

- In the outer ear, the pinna and the auditory canal direct sound waves to the middle ear.
- In the middle ear, the tympanic membrane and the ossicles (malleus, incus, and stapes) amplify sound waves.
- In the inner ear, the semicircular canals detect rotational equilibrium; the utricle and saccule detect gravitational equilibrium; and the cochlea houses the spiral organ, which contains mechanoreceptors for hearing.

Auditory Pathway to the Brain The auditory pathway begins when the outer ear receives and the middle ear amplifies sound waves that then strike the oval window membrane.

- The mechanoreceptors for hearing are hair cells on the basilar membrane of the spiral organ.
- Nerve signals begin in the cochlear nerve and are carried to the primary auditory area in the temporal lobe of the cerebral cortex.

14.6 Sense of Equilibrium

The ear also contains mechanoreceptors for equilibrium.

Rotational Equilibrium Pathway

- Mechanoreceptors (hair cells) in the semicircular canals detect rotational and/or angular movement of the head.

Gravitational Equilibrium Pathway

- Mechanoreceptors (hair cells) in the utricle and saccule detect head movement in the vertical or horizontal planes.

Understanding Key Terms

ampulla 333
aqueous humor 322
astigmatism 328
auditory canal 328
auditory (eustachian) tube 329
blind spot 325
chemoreceptor 316
choroid 322
ciliary body 322
cochlea 328
cochlear nerve 330
color vision 324
cone cell 324
cornea 322
cutaneous receptor 319
exteroceptor 316
farsighted 326
focus 323
fovea centralis 323
glaucoma 322
gravitational equilibrium 334
hair cell 328
incus 328
inner ear 328
integration 317
interoceptor 316
iris 322
kinocilium 334
lens 322
malleus 328
mechanoreceptor 316
middle ear 328
nearsighted 326
olfactory cell 321
optic chiasma 326
optic nerve 323
optic tracts 326
ossicle 328
otolith 334
outer ear 328
oval window 328
pain receptor 316
photoreceptor 316
pinna 328
proprioceptor 318
pupil 322
referred pain 360
retina 323
retinal 324
rhodopsin 324
rod cell 324
rotational equilibrium 333
round window 328
saccule 334
sclera 322
semicircular canal 328, 333
sensation 316
sensory adaptation 317
sensory receptor 316
spiral organ 330
stapes 328
sterocilia 328
stimulus 316
taste bud 320
tectorial membrane 330
thermoreceptor 316
tympanic membrane 328
utricle 334
vestibule 329
visual accommodation 323
vitreous humor 323

Match the key terms to these definitions.

a. ________ Structure that receives sensory stimuli and is a part of a sensory neuron or transmits signals to a sensory neuron.

b. ________ Inner layer of the eyeball containing the photoreceptors—rod cells and cone cells.

c. ________ Outer, white, fibrous layer of the eye that surrounds the eye except for the transparent cornea.

d. ________ Receptor sensitive to chemical stimulation—for example, receptors for taste and smell.

e. ________ Specialized region of the cochlea containing the hair cells for sound detection and discrimination.

Testing Your Knowledge of the Concepts

1. Contrast exteroceptors and interoceptors. (page 316)
2. What is sensory adaptation? (page 317)
3. What is proprioception, and what is the role of muscle spindles? (page 318)
4. List the cutaneous receptors and the type of stimulus to which each responds. (pages 319–320)
5. Describe the structure of a taste bud and explain how a taste cell functions. (pages 320–321)
6. Describe the structure and function of the olfactory epithelium. How does the sense of smell come about? (pages 321–322)
7. Describe the anatomy of the eye and the function of each part. (pages 322–323)
8. How does the eye respond when viewing an object far away? When viewing a close object? (pages 323–324)
9. Describe the sequence of events after rhodopsin absorbs light. (pages 324–325)
10. What are three abnormalities caused by a misshapen eyeball, and what type lens can be helpful? (page 326)
11. Describe the anatomy of the ear and the function of each part. (pages 328, 329)
12. Explain the pathway of sound and how sound is produced. (pages 329–330)
13. What structures are involved in equilibrium, their structures, and functions? (pages 332–334)
14. A sensory receptor
    a. is the first portion of a reflex arc.
    b. initiates nerve impulses.
    c. can be internal or external.
    d. All of these are correct.
15. Receptors sensitive to changes in blood pressure are
    a. interoceptors.
    b. exteroceptors.
    c. proprioceptors.
    d. nociceptors.
16. Conscious interpretation of changes in the internal and external environment is called
    a. responsiveness.
    b. perception.
    c. sensation.
    d. accommodation.
17. Which of these is an incorrect difference between proprioceptors and cutaneous receptors?

	Proprioceptors	**Cutaneous Receptors**
a.	located in muscles and tendons	located in the skin
b.	chemoreceptors	mechanoreceptors
c.	respond to tension	respond to pain, hot, cold, touch, pressure
d.	interoceptors	exteroceptors

    e. All of these contrasts are correct.
18. Pain perceived as coming from another location is known as
    a. intercepted pain.
    b. phantom pain.
    c. referred pain.
    d. parietal pain.
19. Tasting something "sweet" versus "salty" is a result of activating
    a. different sensory receptors.
    b. many versus few sensory receptors.
    c. no sensory receptors.
    d. None of these are correct.

20. Which structure of the eye is incorrectly matched with its function?
    a. lens—focusing
    b. cones—color vision
    c. iris—regulation of amount of light
    d. choroid—location of cones
    e. sclera—protection
21. Which of the following gives the correct path for light rays entering the human eye?
    a. sclera, retina, choroid, lens, cornea
    b. fovea centralis, pupil, aqueous humor, lens
    c. cornea, pupil, lens, vitreous humor, retina
    d. cornea, fovea centralis, lens, choroid, rods
    e. optic nerve, sclera, choroid, retina, humors
22. The thin, darkly pigmented layer that underlies most of the sclera is the
    a. conjunctiva.
    b. cornea.
    c. retina.
    d. choroid.
23. Adjustment of the lens to focus on objects close to the viewer is called
    a. convergence.
    b. visual accommodation.
    c. focusing.
    d. constriction.
24. To focus on objects that are close to the viewer, the
    a. suspensory ligaments must be pulled tight.
    b. lens needs to become more rounded.
    c. ciliary muscle will be relaxed.
    d. image must focus on the area of the optic nerve.
25. Which abnormality of the eye is incorrectly matched with its cause?
    a. astigmatism—either the lens or cornea is not even
    b. farsightedness—eyeball is shorter than usual
    c. nearsightedness—image focuses behind the retina
    d. color blindness—genetic disorder in which certain types of cones may be missing
26. Which of these associations is incorrectly matched?
    a. semicircular canals—inner ear
    b. utricle and saccule—outer ear
    c. auditory canal—outer ear
    d. cochlea—inner ear
    e. ossicles—middle ear
27. Which of the following is not involved in the sense of hearing?
    a. auditory canal
    b. tympanic membrane
    c. ossicles
    d. semicircular canals
    e. cochlea
28. The middle ear is separated from the inner ear by the
    a. oval window.
    b. tympanic membrane.
    c. round window.
    d. Both a and c are correct.
29. Which one of these correctly describes the location of the spiral organ?
    a. between the tympanic membrane and the oval window in the inner ear
    b. in the utricle and saccule within the vestibule
    c. between the tectorial membrane and the basilar membrane in the cochlear canal
    d. between the nasal cavities and the throat
    e. between the outer and inner ear within the semicircular canals
30. Which of the following structures would allow you to know that you were upside down, even if you were in total darkness?
    a. utricle and saccule
    b. cochlea
    c. semicircular canals
    d. tectorial membrane
31. Which of these is an incorrect difference between olfactory receptors and equilibrium receptors?

	Olfactory Receptors	Equilibrium Receptors
a.	located in nasal cavities	located in the inner ear
b.	chemoreceptors	mechanoreceptors
c.	respond to molecules in the air	respond to movements of the body
d.	communicate with brain via a tract	communicate with brain via vestibular nerve

    e. All of these contrasts are correct.
32. Both olfactory receptors and sound receptors
    a. are chemoreceptors.
    b. are a part of the brain.
    c. are mechanoreceptors.
    d. initiate nerve impulses.
    e. All of these are correct.
33. Label this diagram of a human eye.

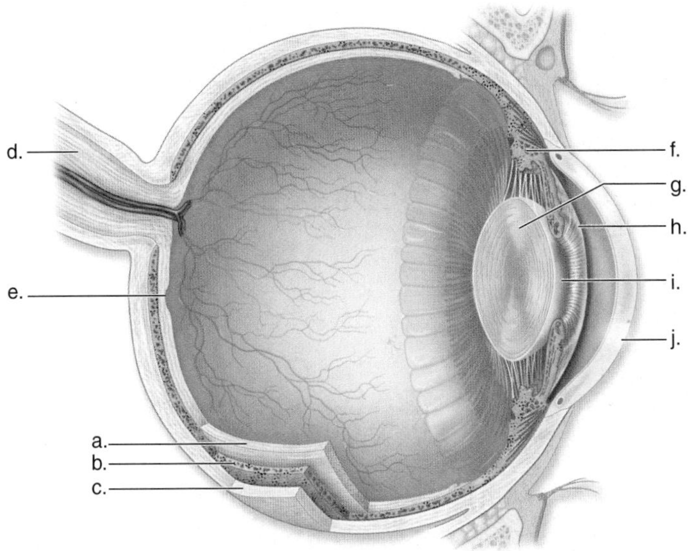

34. Label this diagram of the human ear.

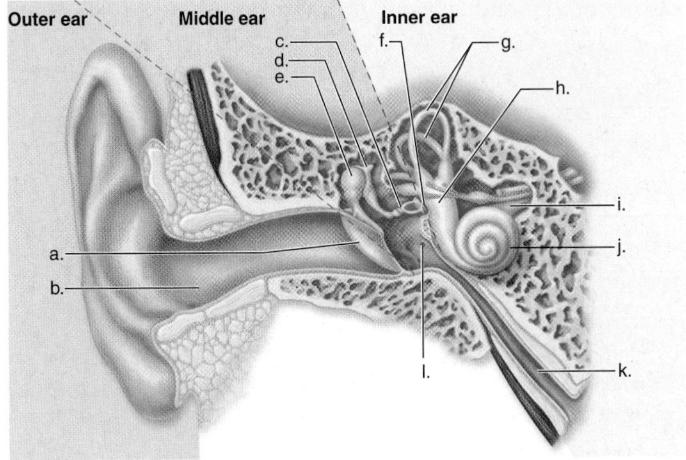

Thinking Critically About the Concepts

Both exteroreceptors and interoreceptors are essential to provide information to the central nervous system. In the chapter opener, Marlene is deaf due to a birth defect. Loss of one of the special senses is immediately debilitating and can seriously affect homeostasis over time. Without vision, hearing, taste, or smell, we are vulnerable to dangerous conditions in our surroundings. Loss of cutaneous senses—touch, pressure, temperature, pain—can immediately impact survival. Imagine, for example, being unaware that you've been seriously cut because you can't feel pain. In that circumstance, it would be possible to bleed to death.

Keep in mind the hugely important role of interoreceptors in maintaining homeostasis. Any unacceptable change in internal homeostatic conditions is reported to the brain by some type of interoreceptor. Pressoreceptors of the cardiovascular system help to regulate blood pressure. Osmoreceptors in the hypothalamus help to control water–salt balance, and chemoreceptors monitor pH in the blood. Our lives depend on proper function of all components of the sensory system.

1. What receptors are activated when we enjoy supper in a pizza restaurant?
2. Besides the blood pH mentioned, what other homeostatic conditions are monitored by chemoreceptors?
3. If a person takes a blow to the back of the head, what sense is most likely to be affected?
4. Airport and construction workers are likely to be exposed to continuous, loud noises. What would you predict the long-term effect on their hearing to be? Why?
5. The acoustic and vestibular nerves travel together to the brain. If a tumor grows on this combined nerve, what sensations will be affected?
6. Explain how a brain tumor in the cerebral cortex might be diagnosed by a doctor.

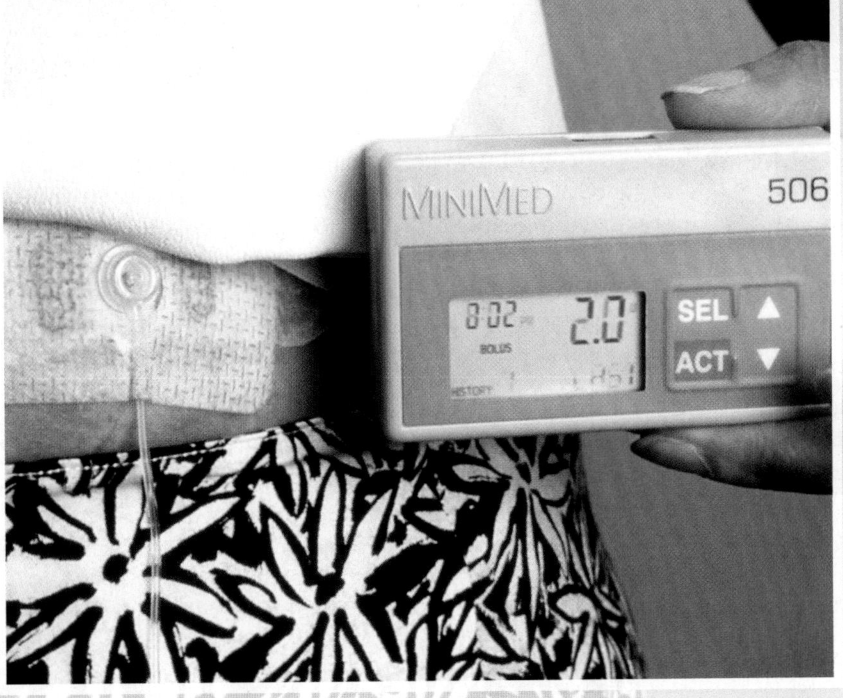

C H A P T E R

15

Endocrine System

CASE STUDY DIABETES

For some time, Hank had been feeling very sluggish and had been losing weight. At first, he attributed this to his very active lifestyle. Between school, work, and his social activities, Hank had very little time for sleep. However, he was beginning to notice that he was always thirsty and was urinating much more frequently than usual. Concerned about his health, Hank visited the local health clinic where he discussed his health history and symptoms with the physician. The doctor mentioned that his symptoms were consistent with many disorders, including both viral infections and diabetes. As a quick test, the doctor ordered a urinalysis test to see if there was any glucose in his urine, which would indicate that Hank's symptoms were caused by diabetes mellitus, a disease that affects over 23.6 million Americans. The results of the urinalysis indicated that there were small amounts of glucose in Hank's urine, a sign that Hank's body may not be adequately maintaining its blood glucose levels. The doctor scheduled Hank for a blood glucose test the following morning and instructed him to not eat or drink anything for 8 hours prior to the test.

During a blood glucose test, a small vial of blood is removed and the amount of glucose measured. Normally, after 8 hours of fasting, the blood glucose level should be between 70 and 100 mg per deciliter of blood (mg/dL). Hank's value was slightly above this, but it was not high enough for the doctor to conclude that diabetes was the cause of Hank's symptoms. The next test was an oral glucose tolerance test (OGTT). In this test, Hank drank a solution containing 75 milligrams of glucose. Then, over the next 3 hours, five additional vials of blood were drawn and tested for glucose levels. In a normal individual participating in this test, blood glucose levels rise rapidly and then fall to below 140 mg/dL within 2 hours. In Hank's case, the response was much slower, and his 2-hour blood glucose level was 150 mg/dL. The physician told Hank that the cause of his symptoms was most likely type 2 diabetes mellitus, a disease of the endocrine system, the organ system that is responsible for the long-term homeostasis of the body.

As you read through the chapter, think about the following questions.

1. What hormones control the level of glucose in the blood?
2. What is the difference between type 1 and type 2 diabetes?
3. How do feedback mechanisms help control blood glucose levels?

CHAPTER CONCEPTS

15.1 Endocrine Glands
Endocrine glands produce hormones that are secreted into the bloodstream and distributed to target cells where they alter cellular metabolism.

15.2 Hypothalamus and Pituitary Gland
The hypothalamus controls the secretions of the pituitary gland, which, in turn, controls the secretion of certain other glands.

15.3 Thyroid and Parathyroid Glands
The hormones of the thyroid and parathyroid glands stimulate cellular metabolism and help maintain blood calcium homeostasis.

15.4 Adrenal Glands
The adrenal glands release hormones to respond to both long-term and short-term stress.

15.5 Pancreas
The pancreas secretes insulin and glucagon, which together keep the blood glucose level fairly constant.

15.6 Other Endocrine Glands
Other endocrine glands include the testes and ovaries, the thymus gland, and the pineal gland.

15.7 Homeostasis
The nervous and endocrine systems work together to control the other organ systems and maintain homeostasis throughout the body.

BEFORE YOU BEGIN

Before beginning this chapter, take a few moments to review the following discussions.

Section 2.5 What is the structure of a steroid?

Section 4.8 How are negative feedback mechanisms involved in homeostasis?

Section 13.2 What is the role of the hypothalamus in the nervous system?

15.1 Endocrine Glands

Learning Outcomes

Upon completion of this section, you should be able to

1. Distinguish between the mode of action of a neurotransmitter and that of a hormone.
2. Distinguish between endocrine and exocrine glands.
3. Identify the organs and glands of the endocrine system.
4. Compare the action of peptide and steroid hormones.

The nervous system and the endocrine system work together to regulate the activities of the other systems. Both systems use chemical signals when they respond to changes that might threaten homeostasis. However, they have different means of delivering these signals (Fig. 15.1). As discussed in Chapter 13, the nervous system is composed of neurons. In this system, sensory receptors detect changes in the internal and external environment. The central nervous system (CNS) then integrates the information and responds by stimulating muscles and glands. Communication depends on nerve signals, conducted in axons, and neurotransmitters, which cross synapses. Axon conduction occurs rapidly and so does diffusion of a neurotransmitter across the short distance of a synapse. In other words, the nervous system is organized to respond rapidly to stimuli. This is particularly useful if the stimulus is an external event that endangers our safety—we can move quickly to avoid being hurt.

The endocrine system functions differently. The endocrine system is largely composed of glands (Fig. 15.2). These glands secrete **hormones,** carried by the bloodstream to target cells throughout the body. It takes time to deliver hormones, and it takes time for cells to respond. The effect initiated by the endocrine system is longer lasting. In other words, the endocrine system is organized for a slow but prolonged response. Table 15.1 summarizes the hormones of the endocrine system and provides the function and target of these hormones in the body.

MP3
Major Endocrine Glands

Endocrine glands can be contrasted with **exocrine glands.** Exocrine glands have ducts and secrete their products into these ducts. The glands' products are carried to the lumens of other organs or outside the body. For example, the salivary glands send saliva into the mouth by way of the salivary ducts. Endocrine glands, as stated, secrete their products into the bloodstream, which delivers them throughout the body. It must be stressed that only certain cells, called target cells,

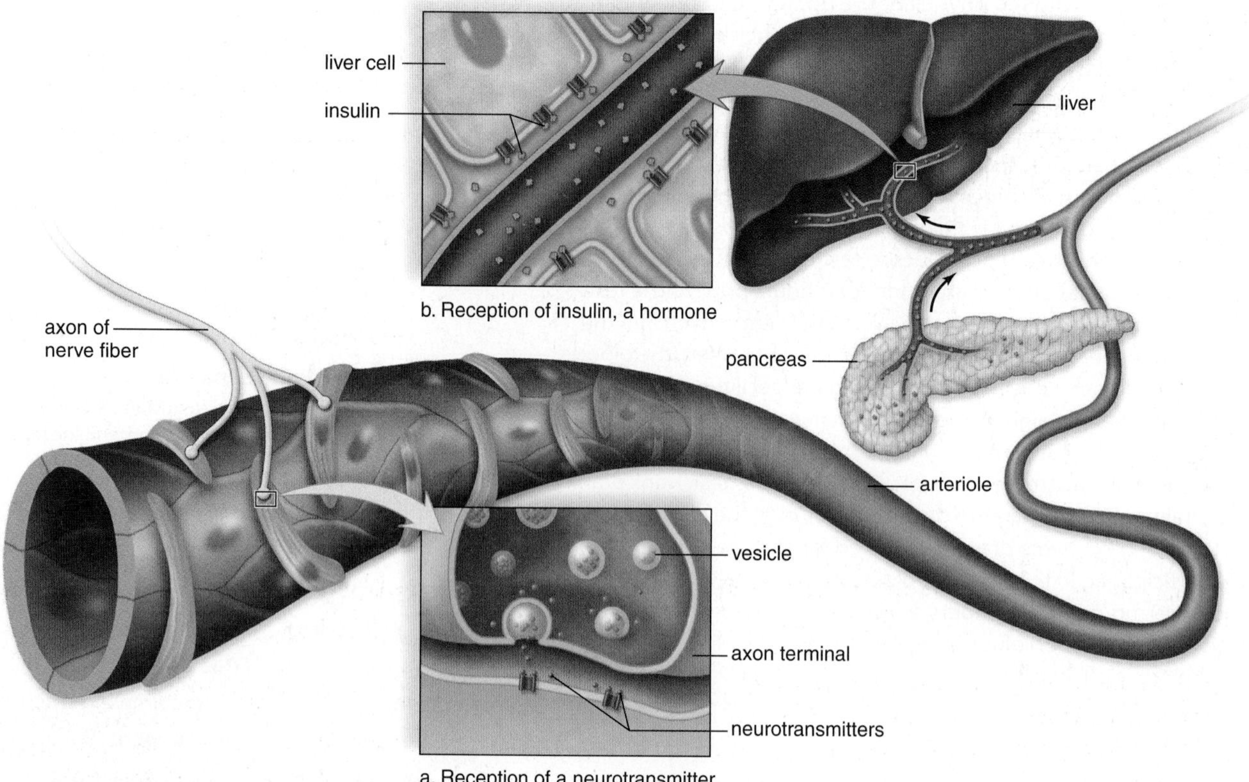

Figure 15.1 The action of a neurotransmitter differs from that of a hormone.
a. Nerve impulses passing along an axon cause the release of a neurotransmitter. The neurotransmitter, a chemical signal, causes the wall of an arteriole to constrict. **b.** The hormone insulin, a chemical signal, travels in the cardiovascular system from the pancreas to the liver, where it causes liver cells to store glucose as glycogen.

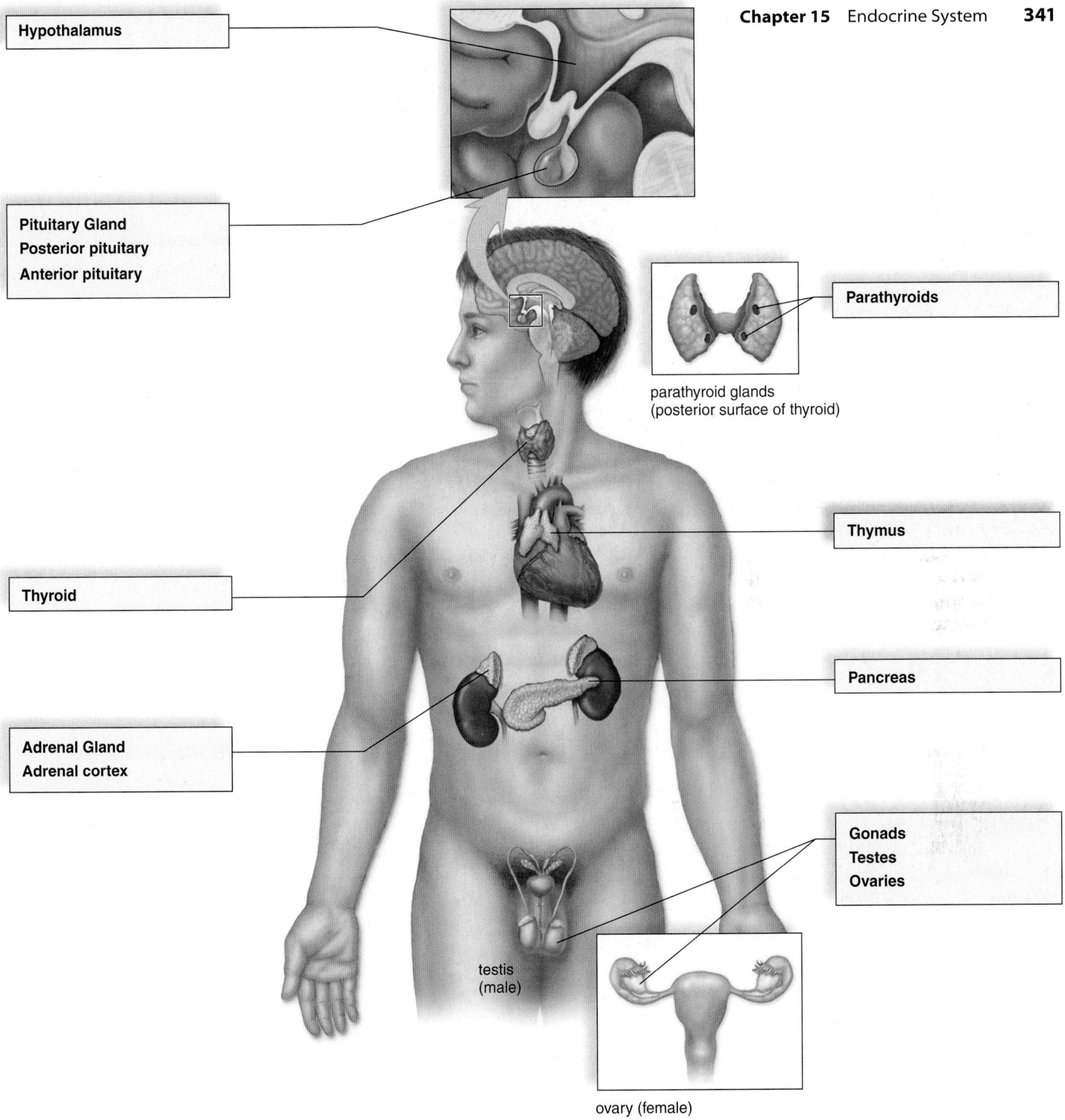

Hypothalamus

Pituitary Gland
Posterior pituitary
Anterior pituitary

Parathyroids

parathyroid glands
(posterior surface of thyroid)

Thymus

Thyroid

Pancreas

Adrenal Gland
Adrenal cortex

Gonads
Testes
Ovaries

testis
(male)

ovary (female)

Figure 15.2 The endocrine system.
The major endocrine glands in the body. Other organs such as the kidneys, gastrointestinal tract, and the heart also produce hormones, but this is not the primary function of these organs.

can respond to certain hormones. A target cell for a particular hormone will have a receptor protein for that hormone. The receptor protein and hormone bind together like a key that fits a lock. The target cell then responds to that hormone.

Both the nervous system and the endocrine system make use of negative feedback mechanisms. If the blood pressure falls, sensory receptors signal a control center in the brain. This center sends out nerve signals to the arterial walls so that they constrict, and blood pressure rises. Now the sensory receptors are no longer stimulated, and the feedback mechanism is inactivated. Similarly, a rise in blood glucose level causes the pancreas to release insulin. This, in turn, promotes glucose uptake by the liver, muscles, and other cells of the body (see Fig. 15.1). When the blood glucose level falls, the pancreas no longer secretes insulin.

MP3
Endocrine System

Table 15.1 Principal Endocrine Glands and the Hormones They Produce

Endocrine Gland	Hormone Released	Target Tissues/Organs	Chief Functions of Hormone
Hypothalamus	Hypothalamic-releasing	Anterior pituitary	Regulate anterior pituitary hormones and inhibiting hormones
Pituitary gland			
Posterior pituitary	Antidiuretic (ADH)	Kidneys	Stimulates water reabsorption by kidneys
	Oxytocin	Uterus, mammary glands	Stimulates uterine muscle contraction, release of milk by mammary glands
Anterior pituitary	Thyroid-stimulating (TSH)	Thyroid	Stimulates thyroid
	Adrenocorticotropic (ACTH)	Adrenal cortex	Stimulates adrenal cortex
	Gonadotropic (FSH, LH)	Gonads	Egg and sperm production; sex hormone production
	Prolactin (PRL)	Mammary glands	Milk production
	Growth (GH)	Soft tissues, bones	Cell division, protein synthesis, and bone growth
	Melanocyte-stimulating (MSH)	Melanocytes in skin	Unknown function in humans; regulates skin color in lower vertebrates
Thyroid	Thyroxine (T_4) and triiodothyronine (T_3)	All tissues	Increases metabolic rate; regulates growth and development
	Calcitonin	Bones, kidneys, intestine	Lowers blood calcium level
Parathyroids	Parathyroid (PTH)	Bones, kidneys, intestine	Raises blood calcium level
Adrenal gland			
Adrenal cortex	Glucocorticoids (cortisol)	All tissues	Raise blood glucose level; stimulate breakdown of protein
	Mineralocorticoids (aldosterone)	Kidneys	Reabsorb sodium and excrete potassium
	Sex hormones	Gonads, skin, muscles, bones	Stimulate reproductive organs and bring about sex characteristics
Adrenal medulla	Epinephrine and norepinephrine	Cardiac and other muscles	Released in emergency situations; raise blood glucose level
Pancreas	Insulin	Liver, muscles, adipose tissue	Lowers blood glucose level; promotes formation of glycogen
	Glucagon	Liver, muscles, adipose tissue	Raises blood glucose level
Gonads			
Testes	Androgens (testosterone)	Gonads, skin, muscles, bones	Stimulate male sex characteristics
Ovaries	Estrogens and progesterone	Gonads, skin, muscles, bones	Stimulate female sex characteristics
Thymus	Thymosins	T lymphocytes	Stimulate production and maturation of T lymphocytes
Pineal gland	Melatonin	Brain	Controls circadian and circannual rhythms; possibly involved in maturation of sexual organs

Hormones Are Chemical Signals

Like other **chemical signals,** hormones are a means of communication between cells, between body parts, and even between individuals. They affect the metabolism of cells that have receptors to receive them (Fig. 15.3). In a condition called *androgen insensitivity,* an individual has X and Y sex chromosomes. The testes, which remain in the abdominal cavity, produce the sex hormone testosterone. However, the body cells lack receptors that are able to combine with testosterone. Therefore, the individual appears to be a normal female.

Like testosterone, most hormones act at a distance between body parts. They travel in the bloodstream from the gland that produced them to their target cells. Also considered to be hormones are the secretions produced by neurosecretory cells in the hypothalamus, a part of the brain. They travel in the capillary network that runs between the hypothalamus and the pituitary gland. Some of these secretions stimulate the pituitary to secrete its hormones, and others prevent it from doing so.

Not all hormones act between body parts. As we shall see, prostaglandins are a good example of *local hormones.* After prostaglandins are produced, they are not carried elsewhere in the bloodstream. Instead, they affect neighboring cells, sometimes promoting pain and inflammation. Also, growth factors are local hormones that promote cell division and mitosis.

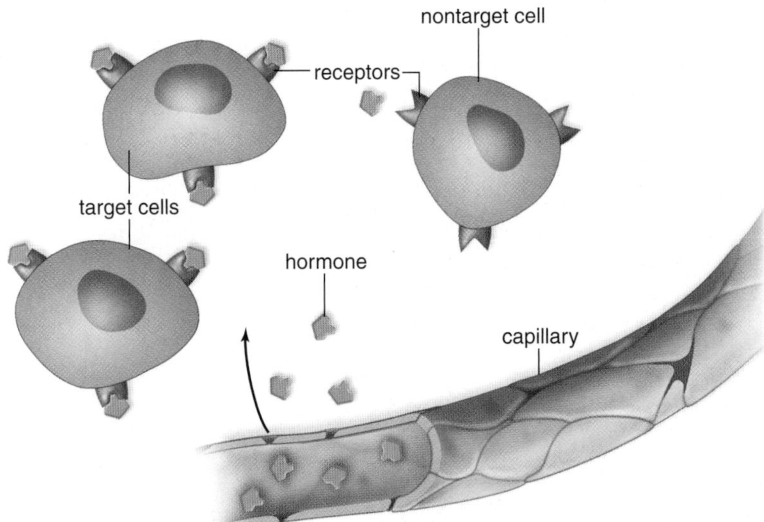

Figure 15.3 Hormones target specific cells.
Most hormones are distributed by the bloodstream to target cells. Target cells have receptors for the hormones, and a hormone combines with a receptor as a key fits a lock.

Chemical signals that influence the behavior of other individuals are called **pheromones.** Other animals rely heavily on pheromones for communication. Pheromones are used to mark one's territory and to attract a mate. Humans produce pheromones too. A researcher has isolated one released by men that reduces premenstrual nervousness and tension in women. Women who live in the same household often have menstrual cycles in synchrony. This is likely caused by the armpit secretions of a woman who is menstruating affecting the menstrual cycle of other women in the household.

The Action of Hormones

Hormones have a wide range of effects on cells. Some of these effects induce a target cell to increase its uptake of particular substances (such as glucose) or ions (such as calcium). Other effects bring about an alteration of the target cell's structure in some way. A few hormones simply influence cell metabolism. Growth hormone is a peptide that influences cell metabolism leading to a change in the structure of bone. The term **peptide hormone** is used to include hormones that are peptides, proteins, glycoproteins, and modified amino acids. Growth hormone is a protein produced and secreted by the anterior pituitary. **Steroid hormones** have the same complex of four carbon rings because they are all derived from cholesterol (see Fig. 2.17).

The Action of Peptide Hormones Most endocrine glands secrete peptide hormones. The actions of peptide hormones can vary. We will concentrate on what happens in muscle cells after the hormone epinephrine binds to a receptor in the plasma membrane (Fig. 15.4). In muscle cells, the reception of epinephrine leads to the breakdown of glycogen to glucose, which provides energy for ATP production. The immediate result of binding is the formation of **cyclic adenosine monophosphate (cAMP).** Cyclic AMP contains one phosphate group attached to adenosine at two locations. Therefore, the molecule is cyclic. Cyclic AMP activates a protein kinase enzyme in the cell. This enzyme, in turn, activates another enzyme, and so forth. The series of enzymatic reactions that follows cAMP formation is called an enzyme cascade. Each enzyme can be used over and over at every step of the cascade, so more enzymes are involved. Finally, many molecules of glycogen are broken down to glucose, which enters the bloodstream.

Animation Peptide Hormone Action

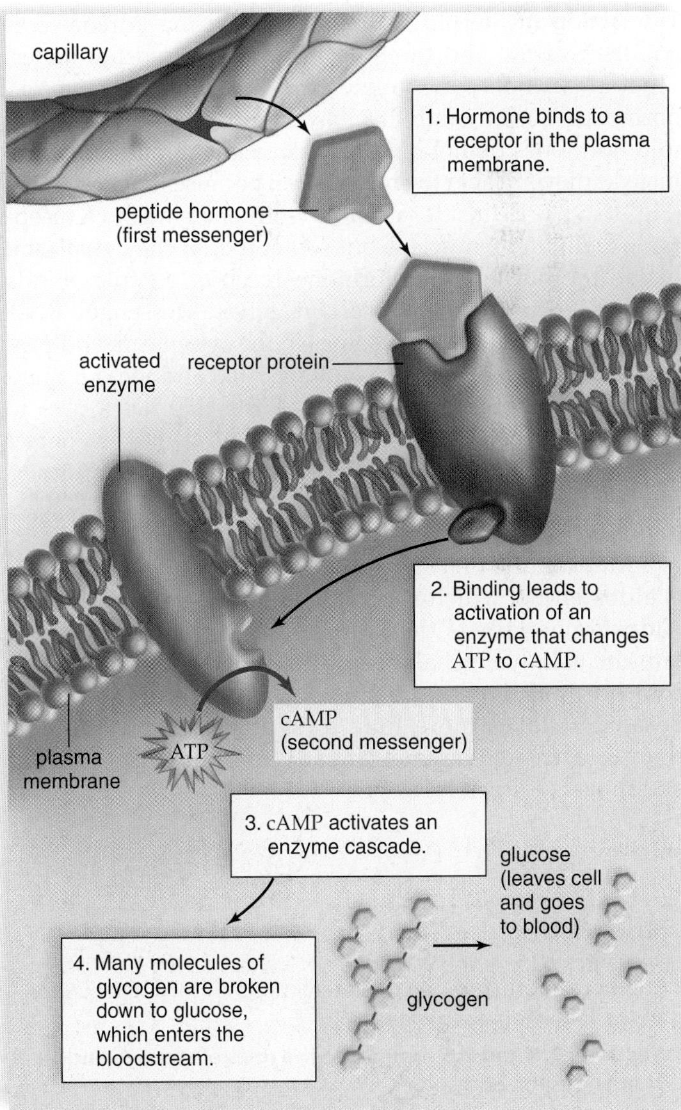

Figure 15.4 Action of a peptide hormone.
A peptide hormone (first messenger) binds to a receptor in the plasma membrane. Thereafter, cyclic AMP (second messenger) forms and activates an enzyme cascade.

Typical of a peptide hormone, epinephrine never enters the cell. Therefore, the hormone is called the **first messenger;** cAMP, which sets the metabolic machinery in motion, is called the **second messenger.** To explain this terminology, let's imagine that the adrenal medulla, which produces epinephrine, is like the home office that sends out a courier (i.e., the hormone epinephrine is the first messenger) to a factory (the cell). The courier doesn't have a pass to enter the factory, so when he arrives at the factory, he tells a supervisor through the screen door that the home office wants the factory to produce a particular product. The supervisor (i.e., cAMP, the second messenger) walks over and flips a switch that starts the machinery (the enzymatic pathway), and a product is made.

Animation Second Messengers: cAMP

The Action of Steroid Hormones Only the adrenal cortex, the ovaries, and the testes produce steroid hormones. Thyroid hormones are amines and act similarly to steroid hormones, even though they have a different structure. Steroid hormones do not bind to plasma membrane receptors. Instead they are able to enter the cell because they are lipids (Fig. 15.5). Once inside, a steroid hormone binds to a receptor, usually in the nucleus but sometimes in the cytoplasm. Inside the nucleus, the hormone–receptor complex binds with DNA and activates certain genes. Messenger RNA (mRNA) moves to the ribosomes in the cytoplasm, and protein (e.g., enzyme) synthesis follows. To continue our analogy, a steroid hormone is like a courier that has a pass to enter the factory (the cell). Once inside, it makes contact with the plant manager (DNA), who sees to it that the factory (cell) is ready to produce a product.

Animation Mechanism of Steroid Hormone Action

An example of a steroid hormone is aldosterone, which is produced by the adrenal glands. Aldosterone targets the kidneys where it helps to regulate the salt–water balance of the blood. In general, steroid hormones act more slowly than peptide hormones because it takes more time to synthesize new proteins than to activate enzymes already present in cells. Their action lasts longer, however.

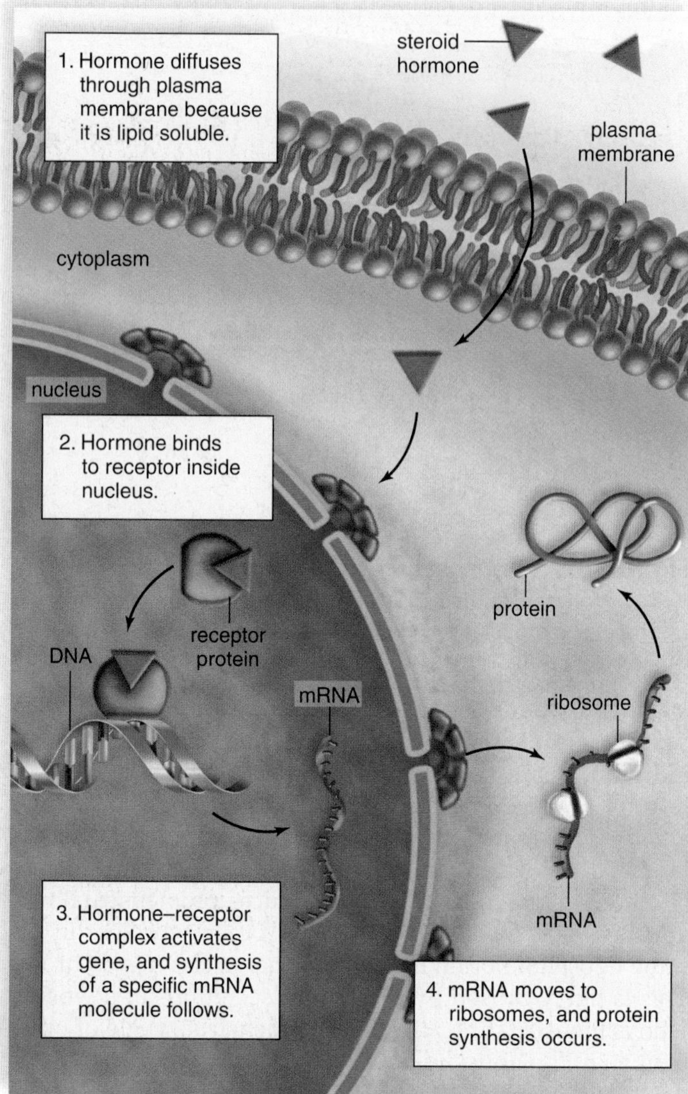

1. Hormone diffuses through plasma membrane because it is lipid soluble.
steroid hormone
plasma membrane
cytoplasm
nucleus
2. Hormone binds to receptor inside nucleus.
receptor protein
DNA
mRNA
protein
ribosome
mRNA
3. Hormone–receptor complex activates gene, and synthesis of a specific mRNA molecule follows.
4. mRNA moves to ribosomes, and protein synthesis occurs.

Figure 15.5 **Action of a steroid hormone.**
A steroid hormone passes directly through the target cell's plasma membrane before binding to a receptor in the nucleus or cytoplasm. The hormone–receptor complex binds to DNA, and gene expression follows.

Connecting the Concepts

For more information on the interactions in this section, refer to the following discussions.

Sections 2.5 and 2.6 summarize the roles of steroids and proteins in the body.

Figure 3.6 illustrates the structure of the plasma membrane and the proteins that are associated with it.

Section 13.2 describes the location and function of the hypothalamus, which integrates the nervous and endocrine systems.

Check Your Progress 15.1

1. State the role of a hormone.
2. Compare and contrast the nervous and endocrine systems with regard to function and the types of signals used.
3. Summarize the differences between a peptide and steroid hormone.
4. Explain why second-messenger systems are needed for peptide hormones.

15.2 Hypothalamus and Pituitary Gland

Learning Outcomes

Upon completion of this section, you should be able to

1. Explain the role of the hypothalamus in the endocrine system.
2. List the hormones produced by the anterior and posterior pituitary glands, and provide a function for each.
3. Summarize the conditions produced by excessive and inadequate levels of growth hormone.

The **hypothalamus** acts as the link between the nervous and endocrine systems. It regulates the internal environment through communications with the autonomic nervous system. For example, it helps control body temperature and water–salt balance. The hypothalamus also controls the glandular secretions of the **pituitary gland.** The pituitary, a small gland about 1 cm in diameter, is connected to the hypothalamus by a stalklike structure. The pituitary has two portions: the posterior and the anterior pituitary.

Posterior Pituitary

Neurons in the hypothalamus called neurosecretory cells produce the hormones **antidiuretic hormone (ADH)** and **oxytocin** (Fig. 15.6, *next page*). These hormones pass through axons into the **posterior pituitary** where they are stored in axon endings. Certain neurons in the hypothalamus are sensitive to the water–salt balance of the blood. When these cells determine that the blood is too concentrated, ADH is released from the posterior pituitary. Upon reaching the kidneys, ADH causes more water to be reabsorbed into kidney capillaries. As the blood becomes dilute, ADH is no longer released. This is an example of control by negative feedback because the effect of the hormone (to dilute blood) acts to shut down the release of the hormone. Negative feedback maintains stable conditions and homeostasis.

Inability to produce ADH causes *diabetes insipidus.* A person with this type of diabetes produces copious amounts of urine. Excessive urination results in severe dehydration and loss of important ions from the blood. The condition can be corrected by the administration of ADH.

Oxytocin, the other hormone made in the hypothalamus, causes uterine contraction during childbirth and milk letdown when a baby is nursing. The more the uterus contracts during labor, the more nerve signals reach the hypothalamus, causing oxytocin to be released. Similarly, as a baby suckles while being breast-fed, nerve signals from breast tissue reach the hypothalamus. As a result, oxytocin is produced by the hypothalamus and released from the posterior pituitary. The hormone causes the woman's breast milk to be released. The sound of a baby crying (even someone else's baby!) may also stimulate the release of oxytocin and milk letdown, much to the chagrin of women who are nursing. In both instances, the release of oxytocin from the posterior pituitary is controlled by **positive feedback.** The stimulus continues to bring about an effect that ever increases in intensity. Positive feedback terminates due to some external event, as when a baby is full and stops suckling. Positive feedback is not a way to maintain stable conditions and homeostasis.

Connections and Misconceptions

How is labor induced if a woman's pregnancy extends past her due date?

After medication to prepare the birth canal for delivery, pitocin (a synthetic version of oxytocin) is used to induce labor. During labor, pitocin may also be given to increase the strength of contractions. Stronger contractions speed the labor process if necessary (for example, if the woman's uterus is contracting poorly or if the health of the mother or child is at risk during delivery). Pitocin is routinely used following delivery to minimize postpartum bleeding by ensuring that strong uterine contractions continue.

Use of pitocin must be monitored carefully, because it may cause excessive uterine contractions. Should this occur, the uterus could potentially tear itself. Further, reduced blood supply to the fetus caused by very strong contractions may be fatal to the baby. Though it reduces the actual duration of labor, inducing labor with pitocin can be very painful for the mother. Whenever possible, gentler and more natural methods should be used to induce labor and/or strengthen contractions. Walking slowly, while stopping to breathe through contractions, is a time-tested way to facilitate the birthing process. Stimulating the nipples very gently will release oxytocin, and that will help too. Assisted by her birthing coach, the woman may also try rocking in a rocking chair, changing positions, or rolling on a birthing ball.

Anterior Pituitary

A portal system, consisting of two capillary systems connected by a vein, lies between the hypothalamus and the **anterior pituitary**. The hypothalamus controls the anterior pituitary by producing **hypothalamic-releasing** and **hypothalamic-inhibiting hormones,** which pass from the hypothalamus to the anterior pituitary by way of the portal system. For example, there is a thyroid-releasing hormone (TRH) and a thyroid-inhibiting hormone (TIH). The TRH stimulates the anterior pituitary to secrete thyroid-stimulating hormone, and the TIH inhibits the pituitary from secreting thyroid-stimulating hormone.

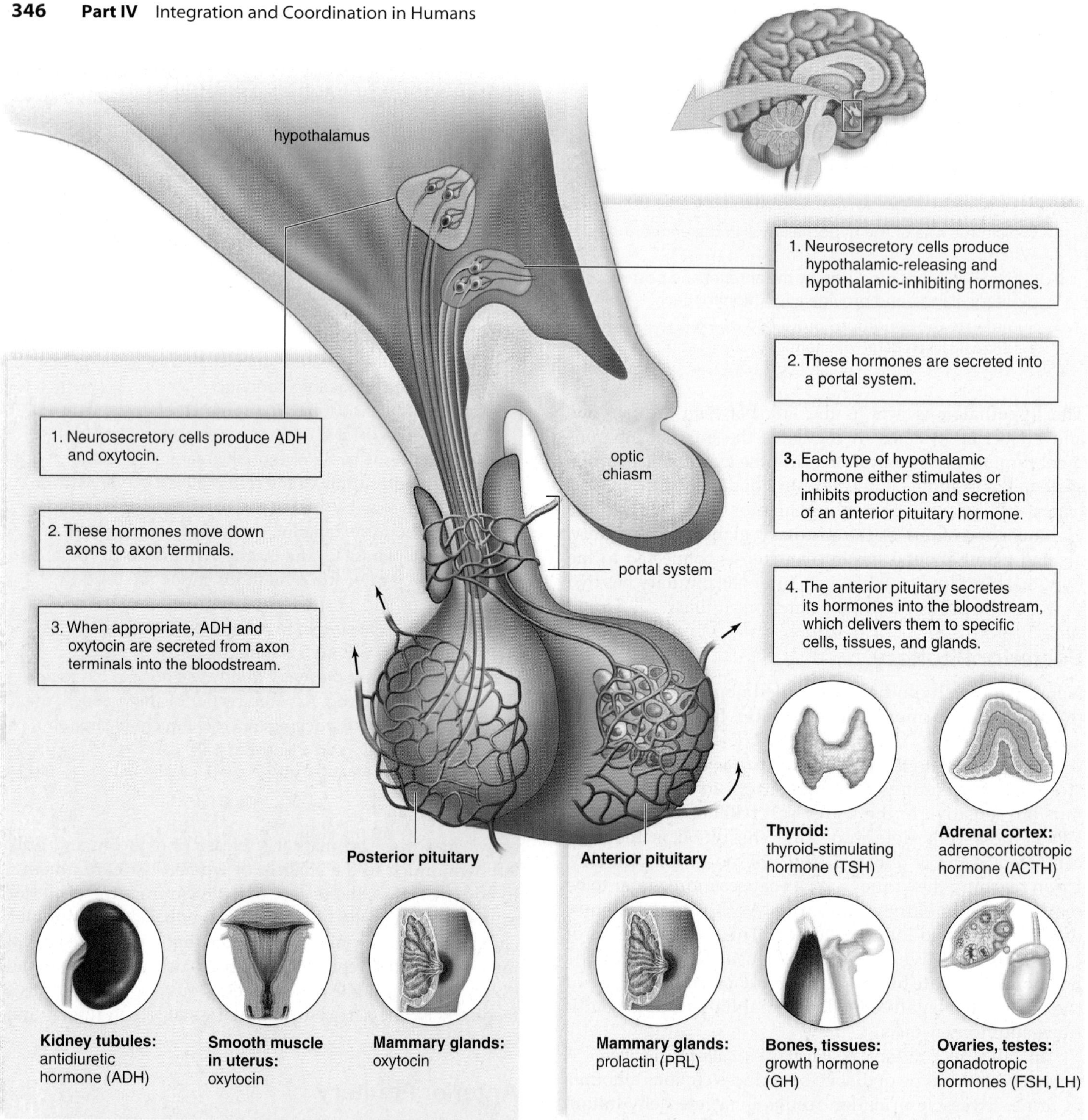

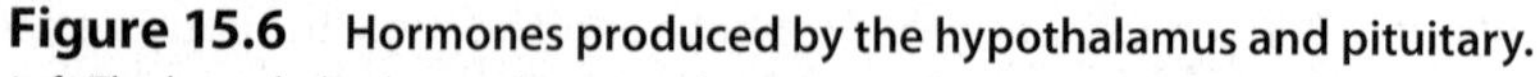
hypothalamus
1. Neurosecretory cells produce ADH and oxytocin.
2. These hormones move down axons to axon terminals.
3. When appropriate, ADH and oxytocin are secreted from axon terminals into the bloodstream.
1. Neurosecretory cells produce hypothalamic-releasing and hypothalamic-inhibiting hormones.
2. These hormones are secreted into a portal system.
3. Each type of hypothalamic hormone either stimulates or inhibits production and secretion of an anterior pituitary hormone.
4. The anterior pituitary secretes its hormones into the bloodstream, which delivers them to specific cells, tissues, and glands.
optic chiasm
portal system
Posterior pituitary
Anterior pituitary
Thyroid: thyroid-stimulating hormone (TSH)
Adrenal cortex: adrenocorticotropic hormone (ACTH)
Kidney tubules: antidiuretic hormone (ADH)
Smooth muscle in uterus: oxytocin
Mammary glands: oxytocin
Mammary glands: prolactin (PRL)
Bones, tissues: growth hormone (GH)
Ovaries, testes: gonadotropic hormones (FSH, LH)

Figure 15.6 Hormones produced by the hypothalamus and pituitary.
Left: The hypothalamus produces two hormones, ADH and oxytocin, stored and secreted by the posterior pituitary. *Right:* The hypothalamus controls the secretions of the anterior pituitary, and the anterior pituitary controls the secretions of the thyroid, adrenal cortex, and gonads, which are also endocrine glands.

Four of the seven hormones produced by the anterior pituitary (Fig. 15.6, *right*) have an effect on other glands. **Thyroid-stimulating hormone** (**TSH**) stimulates the thyroid to produce the thyroid hormones. **Adrenocorticotropic hormone** (**ACTH**) stimulates the adrenal cortex to produce cortisol. The **gonadotropic hormones, follicle-stimulating hormone** (**FSH**) and **luteinizing hormone** (**LH**), stimulate the gonads (the testes in males and the ovaries in females) to produce gametes and sex hormones. In each instance, the blood level of the last hormone in the sequence exerts negative feedback control over the secretion of the first two hormones (Fig. 15.7).

Animation Positive and Negative Feedback

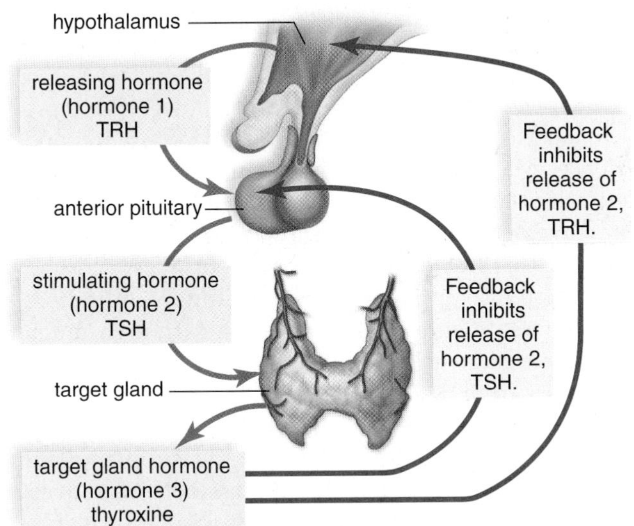

Figure 15.7 Negative feedback mechanisms in the endocrine system.
Feedback mechanisms (red arrows) provide a mechanism of controlling the amount of hormones produced (blue arrows) by the hypothalamus and pituitary glands.

The other three hormones produced by the anterior pituitary do not affect other endocrine glands. **Prolactin (PRL)** is produced in quantity only after childbirth. It causes the mammary glands in the breasts to develop and produce milk. It also plays a role in carbohydrate and fat metabolism.

Melanocyte-stimulating hormone (MSH) causes skin-color changes in many fishes, amphibians, and reptiles having melanophores, special skin cells that produce color variations. The concentration of this hormone in humans is very low.

Growth hormone (GH), or somatotropic hormone, promotes skeletal and muscular growth. It stimulates the rate at which amino acids enter cells and protein synthesis occurs. It also promotes fat metabolism as opposed to glucose metabolism. The production of insulin-like growth factor 1 (IGF-1) by the liver is stimulated by growth hormone as well. IGF-1 is often measured as a means of determining GH level. Growth and development are also stimulated by IGF-1, and it may well be the means by which GH truly influences growth and development.

In the 1980s, growth hormone became a biotechnology product, and it was possible to treat short children and those diagnosed as pituitary dwarfs. A growth hormone blood test can be done to tell if a child is able to produce the normal amount of growth hormone. If not, growth hormone can be injected as a medication.

Animation Hormonal Communication

Effects of Growth Hormone

Growth hormone is produced by the anterior pituitary. The quantity is greatest during childhood and adolescence, when most body growth is occurring. If too little GH is produced during childhood, the individual has **pituitary dwarfism,** characterized by perfect proportions but small stature. If too much GH is secreted, a person can become a giant (Fig. 15.8). Giants usually have poor health, primarily because GH has a secondary effect on the blood sugar level, promoting an illness called *diabetes mellitus* (see pages 354–56).

On occasion, GH is overproduced in the adult and a condition called **acromegaly** results. Long bone growth is no longer possible in adults, so only the feet, hands, and face (particularly the chin, nose, and eyebrow ridges) can respond, and these portions of the body become overly large (Fig. 15.9).

Connecting the Concepts

For more information on the hormones presented in this section, refer to the following discussions.

Section 11.2 examines the influence of growth hormone on bone growth.

Section 16.2 describes the role of pituitary hormones in the production of sperm cells in males.

Section 16.4 describes the role of pituitary hormones in the female ovarian cycle.

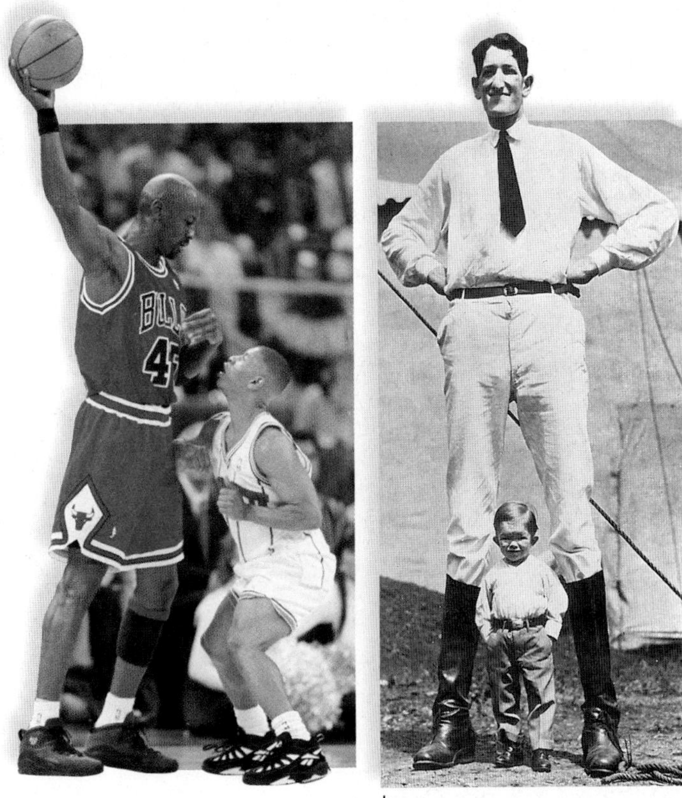

Figure 15.8 Growth hormone influences height.
a. The amount of growth hormone produced by the anterior pituitary during childhood affects the height of an individual. Plentiful growth hormone produces very tall basketball players. **b.** Too much growth hormone can lead to gigantism, whereas an insufficient amount results in limited stature and even pituitary dwarfism.

Biology Matters Bioethical Focus

Using Growth Hormones to Treat Pituitary Dwarfism

Without treatment, children with a deficiency of growth hormone (GH) experience pituitary dwarfism: slow growth, short stature, and in some cases failure to begin puberty. Prior to the advent of biotechnology in the 1980s, treating these children was incredibly difficult and expensive. The GH needed to treat deficiencies had to be obtained from cadaver pituitaries. Whereas the treatment was generally very successful, the use of cadaveric GH caused Creutzfeldt–Jakob disease (a neurological disease similar to "mad cow" disease) in a small number of treated individuals.

Thanks to biotechnology, technologists are now able to synthesize human GH (HGH) using bacteria. These bacteria have had the gene for HGH inserted into their genetic information. The altered bacteria are then grown in laboratories and make unlimited amounts of GH. Children with insufficient GH can be treated more safely and inexpensively with this GH. Recombinant HGH can also be used to treat other disorders such as the chromosomal deficiency known as Turner syndrome (discussed in Chap. 18). It may even be possible to slow or reverse the aging process with HGH treatments.

There is some controversy surrounding treating short children without HGH deficiency for essentially cosmetic reasons. Unfortunately, Americans are obsessed with height. Shorter children are often bullied and teased by their peers. Some data suggest that shorter individuals are discriminated against at their jobs. Their salaries are often lower than those of their taller counterparts with equivalent education and experience. Many people of short stature report having greater self-esteem problems than individuals of average to above-average height. Treatment with HGH could be the solution to these problems.

Although the supply of HGH is seemingly unlimited, the cost of treatments is still quite high (though much cheaper than cadaveric GH). The cost of a year's treatment ranges from \$13,000 to \$30,000. Many insurance companies will not cover these costs. Of greater concern, however, are the potential side effects of supplemental HGH therapy, which are not well understood. Moreover, it is not clear whether HGH treatment will result in a significant increase in the final height of short children.

Decide Your Opinion

1. Now that HGH is easier to obtain, what potential abuses would you predict?
2. Do you think insurance companies should be expected to pay for HGH treatment if a child shows no hormone deficiency and is simply short?
3. What will be the overall influence on society if HGH therapy becomes widely available?

Age 9

Age 16

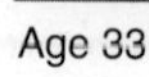

Age 33

Age 52

Figure 15.9 Overproduction of growth hormone in adults leads to acromegaly.
Acromegaly is caused by overproduction of GH in the adult. It is characterized by enlargement of the bones in the face, the fingers, and the toes as a person ages.

Check Your Progress 15.2

1. Explain how the endocrine system and nervous system communicate with one another.
2. List the hormones produced by the posterior pituitary and provide a function for each.
3. List the hormones produced by the anterior pituitary and provide a function for each.
4. Discuss what symptoms can occur if a person's anterior pituitary is inactivated by a tumor.

15.3 Thyroid and Parathyroid Glands

Learning Outcomes

Upon completion of this section, you should be able to

1. List the hormones produced by the thyroid and parathyroid glands and provide a function for each.
2. Describe the negative feedback mechanism that is involved in the maintenance of blood calcium homeostasis.
3. Summarize the diseases and conditions associated with the thyroid and parathyroid glands.

The **thyroid gland** is a large gland located in the neck, where it is attached to the trachea just below the larynx (see Fig. 15.2). The parathyroid glands are embedded in the posterior surface of the thyroid gland.

Thyroid Gland

The thyroid gland is composed of a large number of follicles. Each follicle is a small spherical structure made of thyroid cells filled with **triiodothyronine** (T_3), which contains three iodine atoms, and **thyroxine (T_4),** which contains four.

Effects of Thyroid Hormones

To produce triiodothyronine (T_3) and thyroxine (T_4), the thyroid gland actively requires iodine. The concentration of iodine in the thyroid gland can increase to as much as 25 times that of the blood. If iodine is lacking in the diet, the thyroid gland is unable to produce the thyroid hormones. In response to constant stimulation by TSH from the anterior pituitary, the thyroid enlarges, resulting in a **simple goiter** (Fig. 15.10*a*). Some years ago, it was discovered that the use of iodized salt allows the thyroid to produce the thyroid hormones and, therefore, helps prevent simple goiter.

Thyroid hormones increase the metabolic rate. They do not have a target organ. Instead, they stimulate all cells of the body to metabolize at a faster rate. More glucose is broken down, and more energy is used.

If the thyroid fails to develop properly, a condition called **congenital hypothyroidism** results (Fig. 15.10*b*). Individuals with this condition are short and stocky and have had extreme hypothyroidism (undersecretion of thyroid hormone) since infancy or childhood. Thyroid hormone therapy can initiate growth, but unless treatment is begun within the first two months of life, mental retardation results. The occurrence of hypothyroidism in adults produces the condition known as **myxedema.** Lethargy, weight gain, loss of hair, slower pulse rate, lowered body temperature, and thickness and puffiness of the skin are characteristics of myxedema. The administration of adequate doses of thyroid hormones restores normal function and appearance.

In the case of hyperthyroidism (oversecretion of thyroid hormone), the thyroid gland is overactive and a goiter forms. This type of goiter is called **exophthalmic goiter** (Fig. 15.10*c*). The eyes protrude because of edema in eye socket tissues and swelling of the muscles that move the eyes. The patient usually becomes hyperactive, nervous, and irritable and suffers

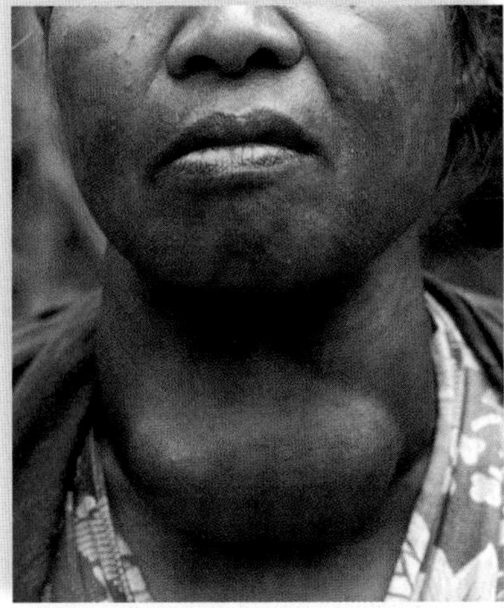

a. Simple goiter

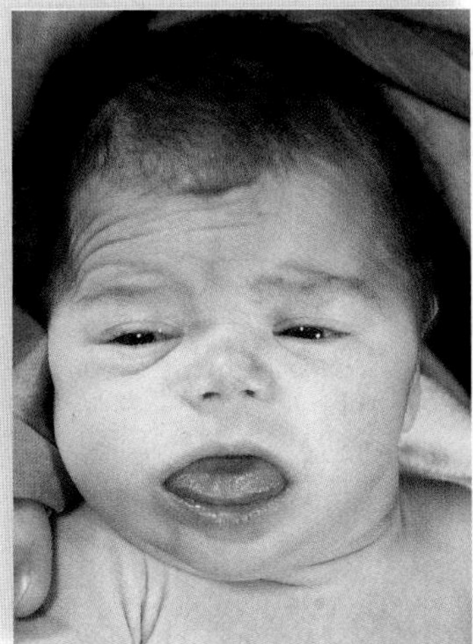

b. Congenital hypothyroidism

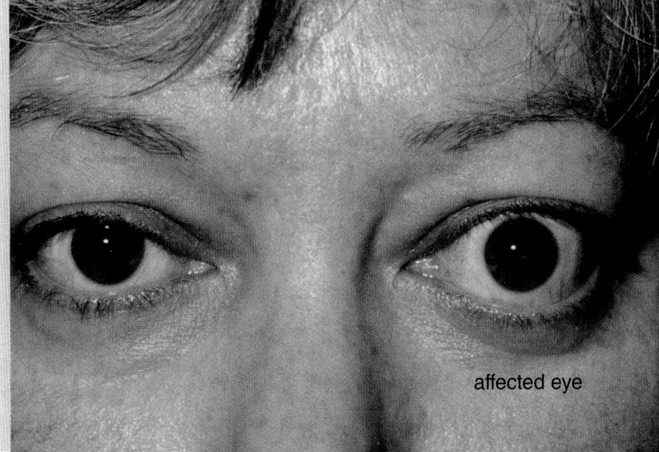

c. Exophthalmic goiter

Figure 15.10 **Effects of insufficient dietary iodine, hypothyroidism, and hyperthyroidism.**
a. An enlarged thyroid gland is often caused by a lack of iodine in the diet. Without iodine, the thyroid is unable to produce its hormones, and continued anterior pituitary stimulation causes the gland to enlarge. **b.** Individuals who develop hypothyroidism during infancy or childhood do not grow and develop as others do. Unless medical treatment is begun, the body is short and stocky; mental retardation is also likely. **c.** In exophthalmic goiter, a goiter is due to an overactive thyroid and the eyes protrude because of edema in eye socket tissue.

from insomnia. Removal or destruction of a portion of the thyroid by means of radioactive iodine is sometimes effective in curing the condition. Hyperthyroidism can also be caused by a thyroid tumor, usually detected as a lump during physical examination. Again, the treatment is surgery in combination with administration of radioactive iodine. The prognosis for most patients is excellent.

Animation Mechanism of Thyroxine Action

Calcitonin

Calcium ions (Ca^{2+}) play a significant role in both nervous conduction and muscle contraction. They are also necessary for blood clotting. The blood calcium level is regulated in part by **calcitonin,** a hormone secreted by the thyroid gland when the blood calcium level rises (Fig. 15.11). The primary effect of calcitonin is to bring about the deposit of calcium ions in the bones. It also temporarily reduces the activity and number of osteoclasts. When the blood calcium level lowers to normal, the release of calcitonin by the thyroid is inhibited.

Parathyroid Glands

Parathyroid hormone (**PTH**), produced by the **parathyroid glands,** causes the blood calcium level to increase. A low blood calcium level stimulates the release of PTH, which promotes the activity of osteoclasts and the release of calcium from the bones. PTH also activates vitamin D in the kidneys. Activated vitamin D, a hormone sometimes called *calcitriol,* then promotes calcium reabsorption by the kidneys. The absorption of calcium ions from the intestine is also stimulated by calcitriol. These effects bring the blood calcium level back to the normal range, and PTH secretion stops.

Many years ago, the four parathyroid glands were sometimes mistakenly removed during thyroid surgery because of their size and location. Gland removal caused insufficient PTH production, which resulted in *hypoparathyroidism.* Hypoparathyroidism causes a dramatic drop in blood calcium, followed by excessive nerve excitability. Nerve signals happen spontaneously and without rest, causing a phenomenon called tetany. In **tetany,** the body shakes from continuous muscle contraction. Without treatment, severe hypoparathyroidism causes seizures, heart failure, and death.

Untreated *hyperparathyroidism* (oversecretion of PTH) can result in osteoporosis because of continuous calcium release from the bones. Hyperparathyroidism may also cause formation of calcium kidney stones.

When a bone is broken, homeostastis is disrupted. For the fracture to heal, osteoclasts will have to destroy old bone, and osteoblasts will have to lay down new bone. Many factors influence the formation of new bone, including parathyroid hormone, calcitonin, and vitamin D. The calcium needed to repair the fracture is made readily available as new blood capillaries penetrate the fractured area.

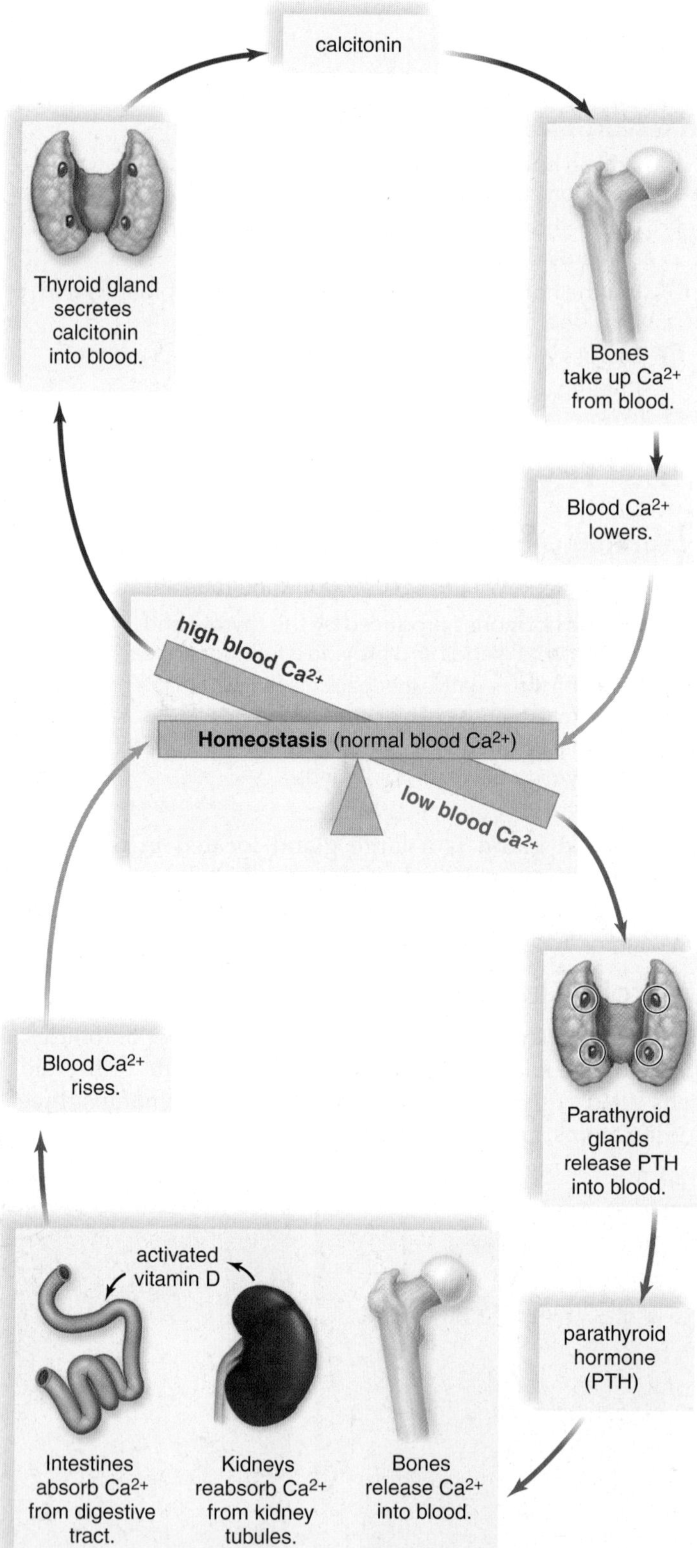

Figure 15.11 Blood calcium homeostasis.
Top: When the blood calcium level is high, the thyroid gland secretes calcitonin. Calcitonin promotes the uptake of calcium ions (Ca^{2+}) by the bones; therefore, the blood calcium level returns to normal. *Bottom:* When the blood calcium level is low, the parathyroid glands release parathyroid hormone (PTH). PTH causes the bones to release calcium ions (Ca^{2+}). It also causes the kidneys to reabsorb Ca^{2+} and activate vitamin D; thereafter, the intestines absorb Ca^{2+}. Therefore, the blood calcium level returns to normal.

Connecting the Concepts

For more information on the importance of calcium, refer to the following discussions.

Section 12.2 examines how calcium ions are involved in muscle contraction.

Section 12.5 explains the role of the bones in maintaining calcium homeostasis.

Section 13.1 explores how calcium ions are involved in the activity of a neural synapse.

Check Your Progress 15.3

1. Explain how the hormones of the thyroid gland influence the metabolic rate.
2. Describe how calcitonin and parathyroid hormone interact to regulate blood calcium levels.
3. Distinguish between hyperthyroidism and hyperparathyroidism with regard to the effects on the body.

15.4 Adrenal Glands

Learning Outcomes

Upon completion of this section, you should be able to

1. List the hormones produced by the adrenal medulla and adrenal cortex and provide a function for each.
2. Explain how the adrenal cortex is involved in the stress response.
3. Distinguish between mineralocorticoid and glucocorticoid hormones.

The **adrenal glands** sit atop the kidneys (see Fig. 15.2). Each adrenal gland consists of an inner portion called the **adrenal medulla** and an outer portion called the **adrenal cortex.** These portions, like the anterior and the posterior pituitary, are two functionally distinct endocrine glands. The adrenal medulla is under nervous control. Portions of the adrenal cortex are under the control of corticotropin-releasing hormone (CRH) from the hypothalamus and ACTH, an anterior pituitary hormone. Stress of all types, including emotional and physical trauma, prompts the hypothalamus to stimulate a portion of the adrenal glands (Fig. 15.12).

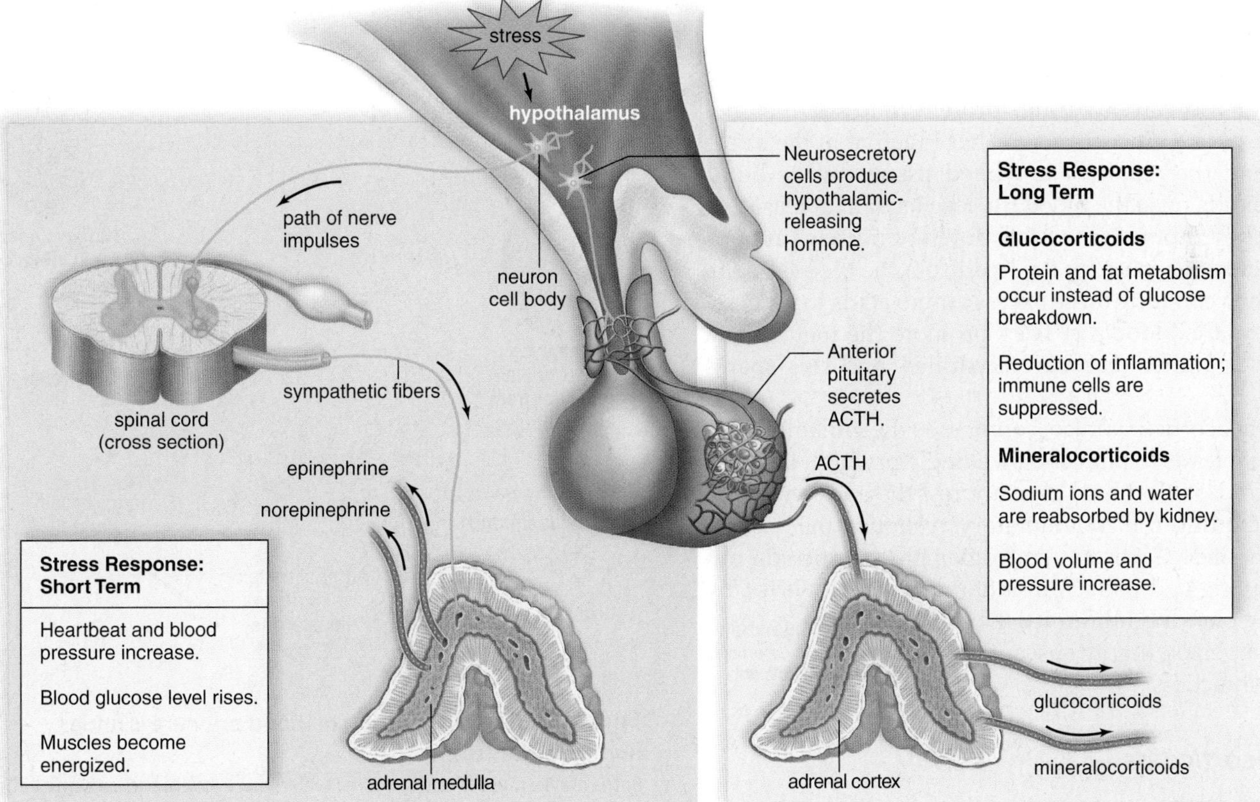

Figure 15.12 **Response of the adrenal medulla and the adrenal cortex to stress.**
Both the adrenal medulla and the adrenal cortex are under the control of the hypothalamus when they help us respond to stress. *Left:* Nervous stimulation causes the adrenal medulla to provide a rapid but short-term stress response. *Right:* The adrenal cortex provides a slower but long-term stress response. ACTH causes the adrenal cortex to release glucocorticoids. Independently, the adrenal cortex releases mineralocorticoids.

Adrenal Medulla

The hypothalamus initiates nerve signals that travel by way of the brain stem, spinal cord, and preganglionic sympathetic nerve fibers to the adrenal medulla. These signals stimulate the adrenal medulla to secrete its hormones. The cells of the adrenal medulla are thought to be modified postganglionic neurons.

Epinephrine (adrenaline) and **norepinephrine** (**NE**) (noradrenaline) are the hormones produced by the adrenal medulla. These hormones rapidly bring about all the body changes that occur when an individual reacts to an emergency situation in a fight-or-flight manner. These hormones provide a short-term response to stress.

Adrenal Cortex

In contrast, the hormones produced by the adrenal cortex provide a long-term response to stress (see Fig. 15.12). The two major types of hormones produced by the adrenal cortex are the glucocorticoids and the mineralocorticoids. The adrenal cortex also secretes a small amount of male sex hormone and a small amount of female sex hormone. This is the case in both males and females.

Glucocorticoids

The **glucocorticoids,** whose secretion is controlled by ACTH, regulate carbohydrate, protein, and fat metabolism. **Cortisol** is a glucocorticoid that is active in the stress response and the repair of damaged tissues in the body. Glucocorticoids raise the blood glucose level in at least two ways. (1) They promote the breakdown of muscle proteins to amino acids, taken up by the liver from the bloodstream. The liver then converts these excess amino acids to glucose, which enters the blood. (2) They promote the metabolism of fatty acids rather than carbohydrates, and this spares glucose.

The glucocorticoids also counteract the inflammatory response that leads to pain and swelling. Very high levels of glucocorticoids in the blood can suppress the body's defense system, including the inflammatory response that occurs at infection sites. Cortisone and other glucocorticoids can relieve swelling and pain from inflammation. However, by suppressing pain and immunity, they can also make a person highly susceptible to injury and infection.

Animation Glucocorticoid Hormones

Mineralocorticoids

Aldosterone is the most important of the **mineralocorticoids.** Aldosterone primarily targets the kidney, where it promotes renal absorption of sodium ions (Na^+) and renal excretion of potassium ions (K^+).

The secretion of mineralocorticoids is not controlled by the anterior pituitary. When the blood sodium level and pressure are low, the kidneys secrete **renin** (Fig. 15.13).

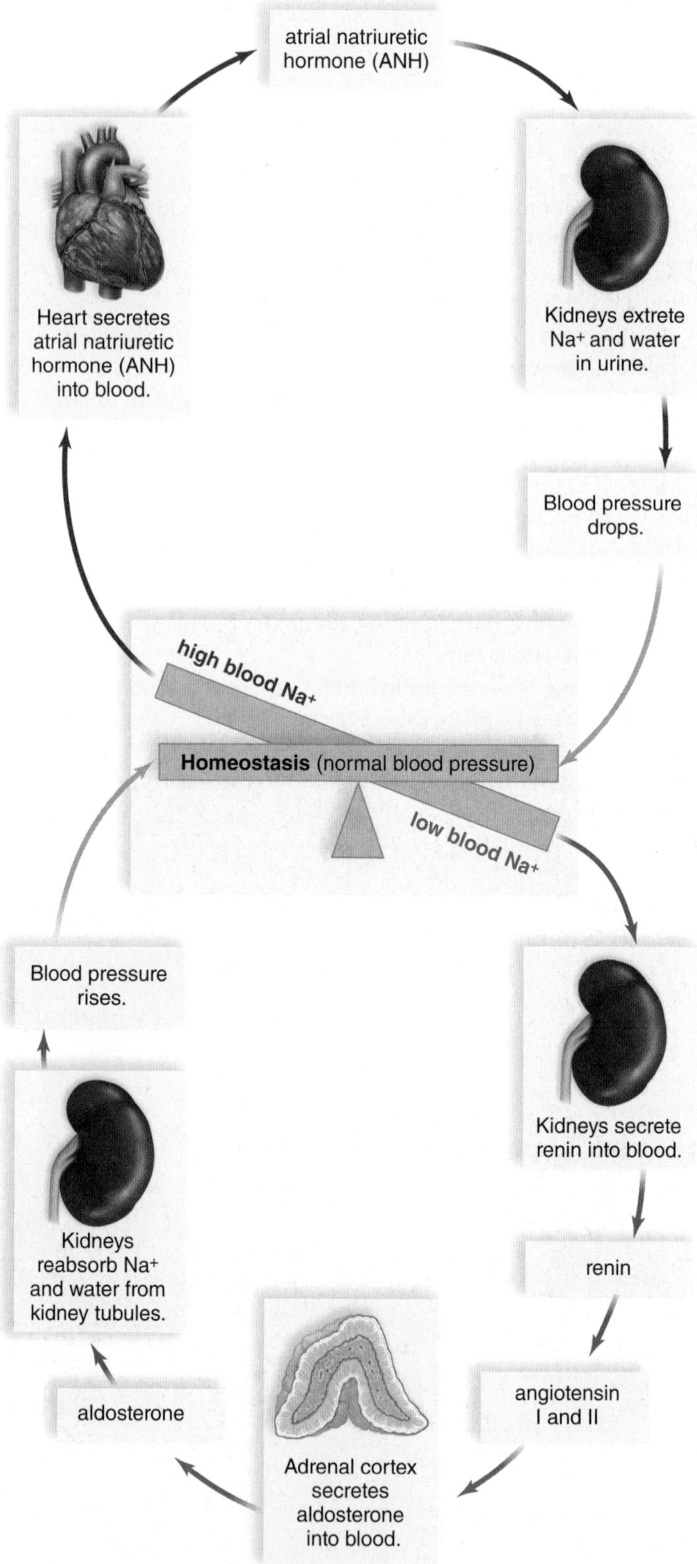

Figure 15.13 Regulation of blood pressure is under hormonal control.
Bottom: When the blood sodium level is low, a low blood pressure causes the kidneys to secrete renin. Renin leads to the secretion of aldosterone from the adrenal cortex. Aldosterone causes the kidneys to reabsorb sodium ions (Na^+), and water follows, so that blood volume and pressure return to normal. *Top*: When a high blood sodium level accompanies a high blood volume, the heart secretes atrial natriuretic hormone (ANH). ANH causes the kidneys to excrete sodium ions (Na^+), and water follows. The blood volume and pressure return to normal.

Renin is an enzyme that converts the plasma protein angiotensinogen to angiotensin I. Angiotensin I is changed to angiotensin II by a converting enzyme found in lung capillaries. Angiotensin II stimulates the adrenal cortex to release aldosterone. The effect of this system, called the renin–angiotensin–aldosterone system, is to raise blood pressure in two ways. Angiotensin II constricts the arterioles, and aldosterone causes the kidneys to reabsorb sodium ions (Na^+). When the blood sodium level rises, water is reabsorbed, in part, because the hypothalamus secretes ADH (see page 345). Reabsorption means that water enters kidney capillaries and, thus, the blood. Then blood pressure increases to normal.

Recall that we studied the role of the kidneys in maintaining blood pressure (see Chapter 10). At that time, we mentioned that if the blood pressure rises due to the reabsorption of sodium ions (Na^+), the atria of the heart are apt to stretch. Due to a great increase in blood volume, cardiac cells release a chemical called **atrial natriuretic hormone (ANH)**, which inhibits the secretion of aldosterone from the adrenal cortex. In other words, the heart is among various organs in the body that release a hormone but obviously not as the major function. (Therefore, the heart is not included as an endocrine gland in Fig. 15.2.) The effect of this ANH is to cause *natriuresis*, the excretion of sodium ions (Na^+). When sodium ions are excreted, so is water; therefore, blood pressure lowers to normal.

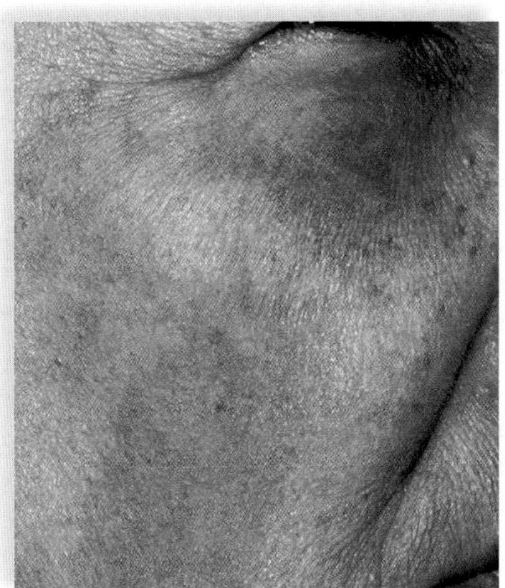

a.

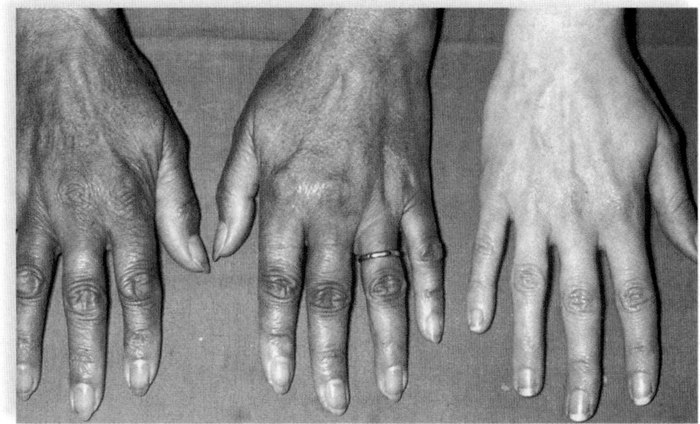

b.

Figure 15.14 Addison disease. Addison disease is characterized by a peculiar bronzing of the skin, particularly noticeable in these light-skinned individuals. Note the color of (**a**) the face and (**b**) the hands compared with the hand of an individual without the disease.

Malfunction of the Adrenal Cortex

When the blood level of glucocorticoids is low due to hyposecretion, a person develops **Addison disease.** The presence of excessive but ineffective ACTH causes a bronzing of the skin because ACTH, like MSH, can lead to a buildup of melanin (Fig. 15.14). Without the glucocorticoids, glucose cannot be replenished when a stressful situation arises. Even a mild infection can lead to death. In some cases, hyposecretion of aldosterone results in a loss of sodium and water. Low blood pressure and, possibly, severe dehydration can develop as a result. Left untreated, Addison disease can be fatal.

When the level of glucocorticoids is high due to hypersecretion, a person develops **Cushing syndrome.** The excess glucocorticoids result in a tendency toward diabetes mellitus as muscle protein is metabolized and subcutaneous fat is deposited in the midsection. The result is a swollen "moon" face (Fig. 15.15*a*) and an obese trunk, with arms and legs of normal size. Children show obesity and poor growth in height. Depending on the cause and duration of the Cushing syndrome, some people may have more dramatic changes. These include masculinization with increased blood pressure and weight gain. However, Cushing syndrome may be treated by the use of cortisol-inhibiting drugs (Fig 15.15*b*).

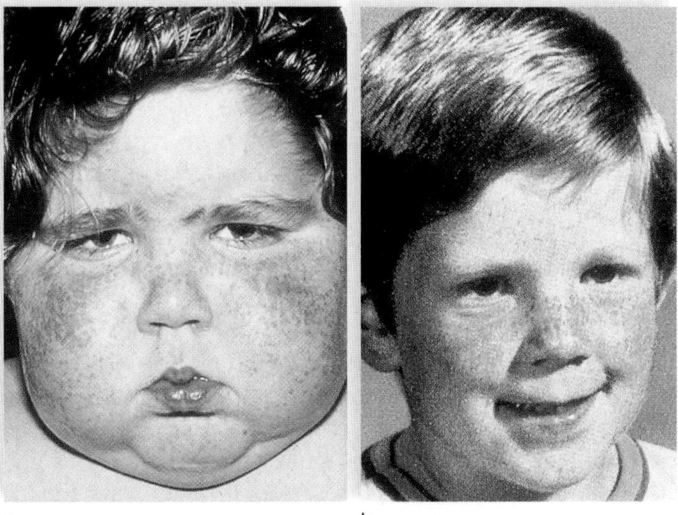

a. b.

Figure 15.15 Cushing syndrome.
Cushing syndrome results from hypersecretion of adrenal cortex hormones. **a.** Patient first diagnosed with Cushing syndrome. **b.** Four months later, after therapy.

Connecting the Concepts

For more information on the hormones produced by the adrenal glands, refer to the following discussions.

Section 5.3 describes how epinephrine and norepinephrine influence the heart rate.

Section 10.4 examines how aldosterone is involved in maintaining the water–salt balance of the body fluids.

Check Your Progress 15.4

1. List the hormones produced by the adrenal glands and indicate whether they are produced by the adrenal cortex or the adrenal medulla.
2. Summarize the involvement of the adrenal glands during a stress response.
3. Explain why maintaining the correct balance of water and salt is important in homeostasis.

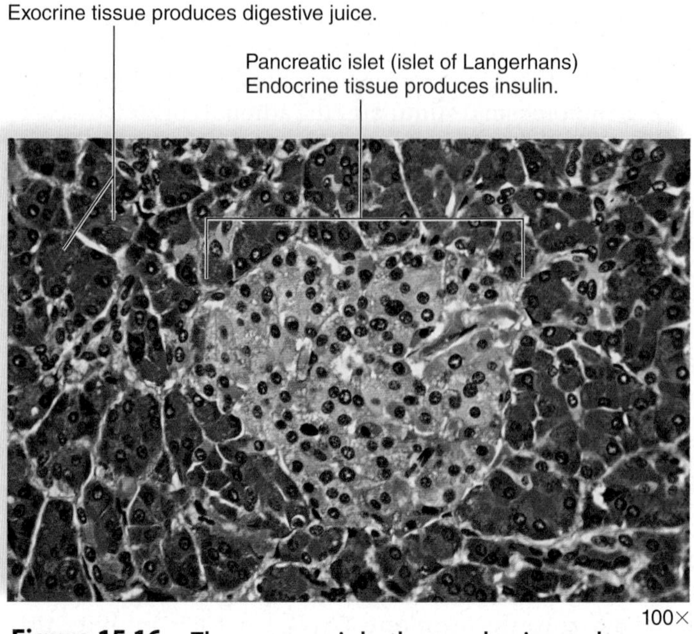

Exocrine tissue produces digestive juice.

Pancreatic islet (islet of Langerhans) Endocrine tissue produces insulin.

100×

Figure 15.16 The pancreas is both an endocrine and exocrine gland.

This light micrograph shows that the pancreas has two types of cells. The exocrine tissue produces a digestive juice, and the endocrine tissue produces the hormones insulin and glucagon.

15.5 Pancreas

Learning Outcomes

Upon completion of this section, you should be able to

1. Explain why the pancreas is both an endocrine and exocrine gland.
2. List the hormones produced by the pancreas and provide a function for each.
3. Describe how the pancreatic hormones help maintain blood glucose homeostasis.
4. Distinguish between type 1 and type 2 diabetes mellitus.

The **pancreas** is a fish-shaped organ that stretches across the abdomen behind the stomach and near the duodenum of the small intestine. It is composed of two types of tissue. Exocrine tissue produces and secretes digestive juices that go by way of ducts to the small intestine. Endocrine tissue, called the **pancreatic islets** (islets of Langerhans), produces and secretes the hormones **insulin** and **glucagon** directly into the blood (Fig. 15.16).

The pancreas is not under pituitary control. Insulin is secreted when the blood glucose level is high, which usually occurs just after eating. Insulin stimulates the uptake of glucose by cells, especially liver cells, muscle cells, and adipose tissue cells. In liver and muscle cells, glucose is then stored as glycogen. In muscle cells, the glucose supplies energy for muscle contraction. Glucose enters the metabolic pool in fat cells and thereby supplies glycerol for the formation of fat. In these various ways, insulin lowers the blood glucose level (Fig. 15.17, *top*).

Glucagon is secreted from the pancreas, usually between eating, when the blood glucose level is low. The major target tissues of glucagon are the liver and adipose tissue. Glucagon stimulates the liver to break down glycogen to glucose. It also promotes the use of fat and protein in preference to glucose as energy sources. Adipose tissue cells break down fat to glycerol and fatty acids. The liver takes these up and uses them as substrates for glucose formation. In these ways, glucagon raises the blood glucose level (Fig. 15.17, *bottom*).

Diabetes Mellitus

Over 23.6 million Americans (7.8% of the population) have **diabetes mellitus.** Of these, it is estimated that almost 5.7 million are unaware that they have the disease. This disease is characterized by the inability of the body's cells to take up glucose, especially liver and muscle cells. This causes blood glucose to be higher than normal, and cells rely on other "fuels" like fatty acids for energy. A common symptom of diabetes mellitus is glucose in the urine (*mellitus,* from Greek, refers to "honey" or "sweetness"). As blood glucose increases, more glucose and water are excreted in the urine. This results in frequent urination and complaints of great thirst by affected people. Other symptoms include fatigue, unusual hunger, and weight changes. Diabetics often experience blurred vision and vision loss caused by the swelling in the lens of the eye due to high blood sugar levels. If untreated, diabetics develop serious and often fatal complications. Sores that don't heal result in severe infections. Blood vessel changes cause kidney failure, nerve destruction, heart attack, or stroke.

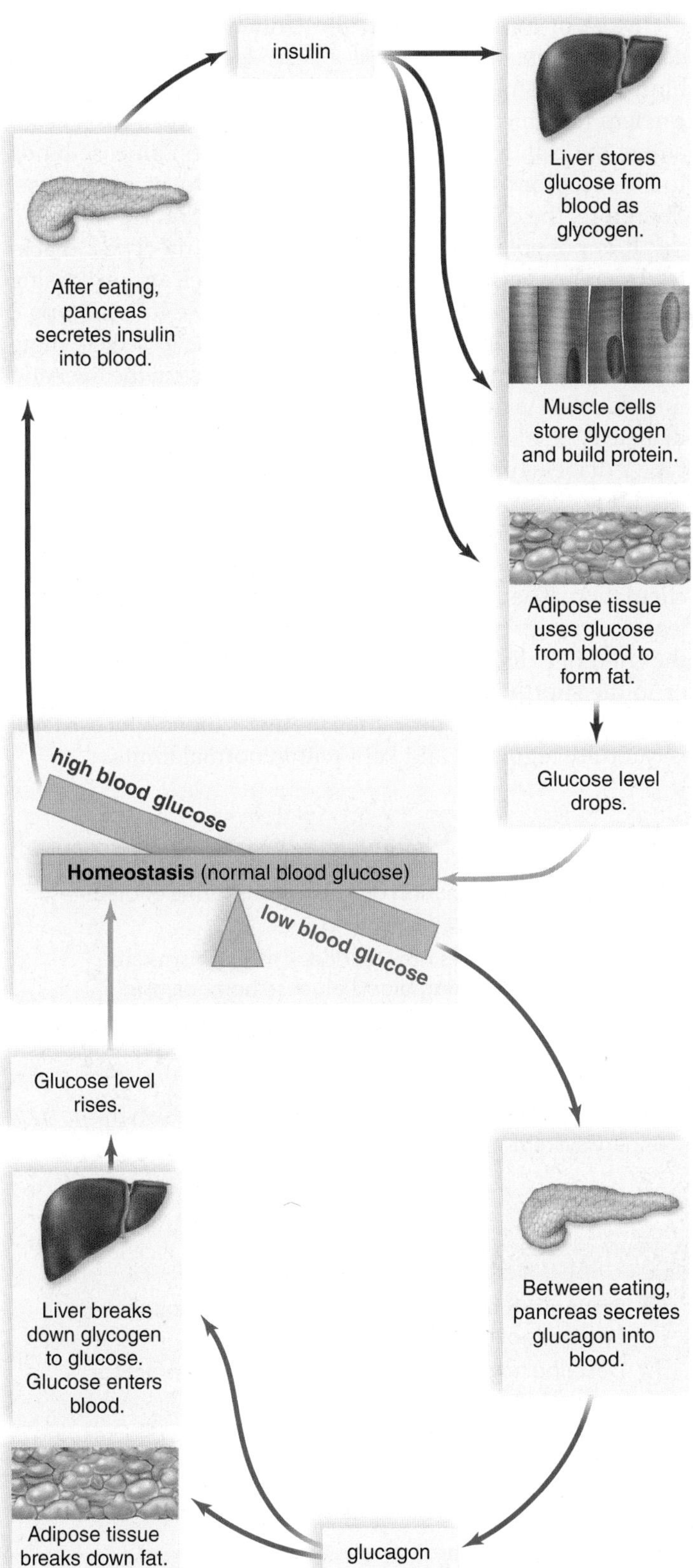

Figure 15.17 Blood glucose homeostasis.
Top: When the blood glucose level is high, the pancreas secretes insulin. Insulin promotes the storage of glucose as glycogen and the synthesis of proteins and fats. Therefore, insulin lowers the blood glucose level. *Bottom:* When the blood glucose level is low, the pancreas secretes glucagon. Glucagon acts opposite to insulin; therefore, glucagon raises the blood glucose level to normal.

The glucose tolerance test assists in the diagnosis of diabetes mellitus. After the patient is given 100 grams of glucose, the blood glucose concentration is measured at intervals. In a diabetic, the blood glucose level rises greatly and remains elevated for several hours (Fig. 15.18). In the meantime, glucose appears in the urine. In a nondiabetic person, the blood glucose level rises somewhat and then returns to normal after about two hours.

Animation Blood Sugar Regulation in Diabetics

Types of Diabetes

There are two types of diabetes mellitus. In *type 1 diabetes,* the pancreas is not producing insulin. This condition is believed to be brought on by exposure to an environmental agent, most likely a virus, whose presence causes cytotoxic T cells to destroy the pancreatic islets. The body turns to the metabolism of fat, which leads to the buildup of ketones in the blood and, in turn, to acidosis (acid blood), which can lead to coma and death. As a result, the individual must have daily insulin injections. These injections control the diabetic symptoms but can still cause inconveniences because the blood sugar level may swing between hypoglycemia (low blood glucose level) and hyperglycemia (high blood glucose level). Without testing the blood glucose level, it is difficult to be certain which of these is present because the symptoms are similar. The symptoms include perspiration, pale skin, shallow breathing, and anxiety. Whenever these symptoms appear, immediate attention is required to bring the blood glucose level to its proper level. If the problem is hypoglycemia, the treatment is

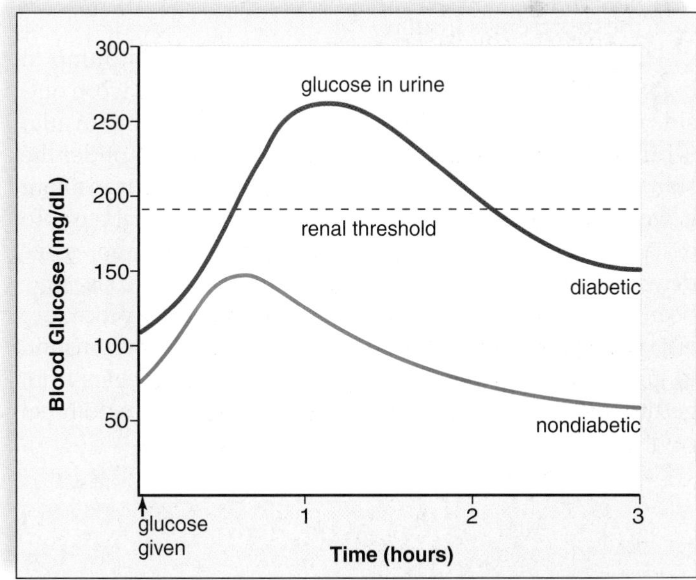

Figure 15.18 The results of a glucose tolerance test for diabetes.
Following the administration of 100 grams of glucose, the blood glucose level rises dramatically in the diabetic, and glucose appears in the urine. Also, the blood glucose level at 2 hours is equal to more than 200 mg/dL.

Connections and Misconceptions

What is gestational diabetes, and what causes it?

Women who were not diabetic prior to pregnancy but have high blood glucose during pregnancy have *gestational* diabetes. Gestational diabetes affects a small percentage of pregnant women. This form of diabetes is caused by insulin resistance—body insulin concentration is normal, but cells fail to respond normally. Gestational diabetes and insulin resistance generally develop later in the pregnancy. Carefully planned meals and exercise often control this form of diabetes, but insulin injections may be necessary.

If the woman is not treated, additional glucose crosses the placenta, causing high blood glucose in the fetus. The extra energy in the fetus is stored as fat, resulting in macrosomia or a "fat" baby. Delivery of a very large baby can be dangerous for both the infant and the mother. Cesarean section is often necessary. Complications after birth are common for these babies. Further, there is a greater risk that the child will become obese and develop type 2 diabetes mellitus later in life.

Gestational diabetes usually goes away after the birth of the child. However, once a woman has experienced gestational diabetes, she has a greater chance of developing it again during future pregnancies. These women also tend to develop type 2 diabetes later in life.

two glucose tablets or two doses of glucose gel. Hard candy or orange juice would work too. If the problem is hyperglycemia, the treatment is insulin.

Some diabetics have learned to use an insulin pump to better regulate their blood sugar level. The pump is worn outside the body, usually attached to a belt or waistband. Insulin is pumped from a reservoir through a tube inserted under the skin of the abdominal wall. It is also possible to transplant a working pancreas into patients with type 1 diabetes. To do away with the necessity of taking immunosuppressive drugs after the transplant, fetal pancreatic islet cells have been injected into patients. Another experimental procedure is to place pancreatic islet cells in a capsule that allows insulin to get out but prevents antibodies and T lymphocytes from getting in. This artificial organ is implanted in the abdominal cavity.

Most of the diabetics in the United States have *type 2 diabetes.* Often, the patient is obese. Usually, after insulin binds to a plasma membrane receptor, the number of protein carriers for glucose increases and more glucose than usual enters the cell. In the case of type 2 diabetes, glucose binds to the receptor but the number of carriers does not increase. Therefore, the cell is said to be insulin resistant.

It is possible to prevent or at least control type 2 diabetes by adhering to a low-fat, low-sugar diet and exercising regularly. If this fails, oral drugs are available to treat type 2 diabetes. These drugs stimulate the pancreas to secrete more insulin and enhance the metabolism of glucose in the liver and muscle cells. As many as 7 million Americans may have type 2 diabetes without being aware of it; yet, the effects of untreated type 2 diabetes are as serious as those of type 1 diabetes.

Long-term complications of both types of diabetes are blindness; kidney disease; and cardiovascular disorders, including atherosclerosis, heart disease, stroke, and reduced circulation. The latter can lead to gangrene in the arms and legs. Pregnancy carries an increased risk of diabetic coma, and the child of a diabetic is somewhat more likely to be stillborn or to die shortly after birth. These complications of diabetes are not expected to appear if the mother's blood glucose level is carefully regulated and kept within normal limits.

Connecting the Concepts

For additional information on the various forms of diabetes, refer to the following discussions.

Section 4.8 examines how feedback mechanisms are involved in maintaining blood glucose homeostasis.

Section 8.6 explains the body mass index (BMI), an indicator that is used to determine obesity and the subsequent risk of diabetes.

Section 10.3 examines the influence of diabetes on the urinary system.

Check Your Progress 15.5

1. Distinguish between the exocrine and endocrine functions of the pancreas.
2. Describe how the pancreatic hormones interact to regulate blood glucose levels.
3. Explain the difference in the function of the pancreas in type 1 and type 2 diabetes.

Biology Matters Historical Focus

Surviving Diabetes Mellitus

Millions of diabetics owe an enormous debt of gratitude to two men that most have probably never heard of: Frederick Banting (1891–1941) and Charles Best (1899–1978). Prior to their research, those unfortunates diagnosed with diabetes mellitus wasted away until death claimed them. Banting and Best were the team who first successfully isolated insulin, the hormone that allows diabetic patients to survive.

In 1920, Banting began work to isolate secretions from the pancreas that could treat the high blood sugar associated with diabetes. Earlier researchers had already determined that diabetes mellitus was caused by lack of a protein hormone produced by the islets of Langerhans, clumps of cells found in the pancreas. Referring to its origin, the hormone was termed *insulin* (after the Latin *insula,* or "island"). Feeding diabetic patients pancreatic tissue failed to produce insulin activity. Attempts to extract insulin from the islets of Langerhans were also unsuccessful. Banting theorized that the unproductive attempts to recover insulin occurred because pancreatic enzymes digested the protein. If he could prevent insulin's destruction, Banting speculated, he might have a treatment for diabetes.

Banting presented his idea to a University of Toronto physiologist named John MacLeod, who offered Banting the use of lab space and experimental dogs. A medical student named Charles Best worked as Banting's assistant. The pair made amazing progress toward their goal of isolating insulin during the summer of 1921. Their achievements in such a short time are unparalleled. By fall, they were able to extend the lives of diabetic dogs with the material they'd isolated from the islets of Langerhans. Further experimentation allowed them to collect much larger insulin samples from cattle. Finally, here was a source that would provide enough insulin for human testing.

Early in 1922, they treated their first human patient, a 14-year-old boy whose life was saved by their extract. After insulin treatment, the boy's blood sugar decreased, glucose was absent from his urine, and he was freed from other signs of diabetes. In February of 1922, Banting and Best published their first paper about the treatment of diabetes with insulin.

Figure 15A Charles Best and Frederick G. Banting. Best and Banting photographed at the University of Toronto in the summer of 1921. The dog was the first one to be kept alive with insulin treatment.

The Nobel Prize in Physiology or Medicine was awarded in 1923 to Banting and MacLeod, though MacLeod had never taken part in the original research. Best was not included in the award, because he was only a medical student. Banting was stung by the injustice to Best and divided his prize money with Best.

The work begun by Banting and Best continues to evolve. Today, human insulin is produced by genetically engineered bacteria, eliminating the need for an animal source. Several forms of human insulin are available, each with its own distinctive action. Combinations of insulins, delivered by high-tech insulin pumps, give diabetics stable insulin concentration throughout the day.

15.6 Other Endocrine Glands

Learning Outcomes

Upon completion of this section, you should be able to

1. List the hormones produced by the sex organs, thymus, and pineal gland, and provide a function for each.
2. List the hormones that are produced by glands and organs outside of the endocrine system.

The **gonads** are the testes in males and the ovaries in females. The gonads are endocrine glands. Other lesser-known glands and some tissues also produce hormones.

Testes and Ovaries

The activity of the testes and ovaries is controlled by the hypothalamus and pituitary. The **testes** are located in the scrotum, and the **ovaries** are located in the pelvic cavity. The testes produce **androgens** (e.g., **testosterone**), the male sex hormones. The ovaries produce **estrogens** and **progesterone,** the female sex hormones. These hormones feed back to control the hypothalamic secretion of gonadotropin-releasing hormone (GnRH). The pituitary gland secretion of **follicle-stimulating hormone** (FSH) and **luteinizing hormone** (LH), the gonadotropic hormones (Fig. 15.19), is controlled by feedback from the sex hormones, too. The activities of FSH and LH are discussed in Chapter 16.

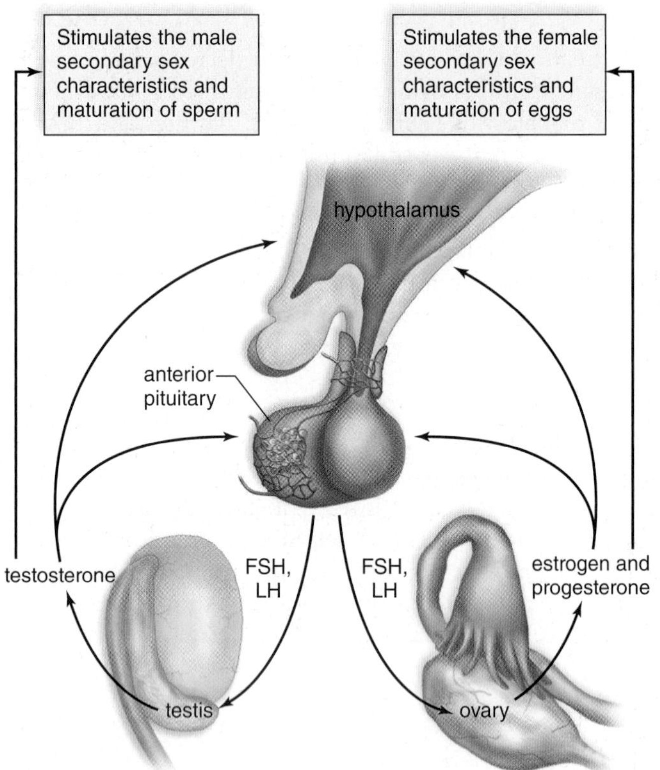

Figure 15.19 The hormones produced by the testes and the ovaries.
The testes and ovaries secrete the sex hormones. The testes secrete testosterone, and the ovaries secrete estrogens and progesterone. In each sex, secretion of GnRH from the hypothalamus and secretion of FSH and LH from the pituitary are controlled by their respective hormones.

Under the influence of the gonadotropic hormones, the testes begin to release increased amounts of testosterone at the time of puberty. Testosterone stimulates the growth of the penis and the testes. Testosterone also brings about and maintains the male secondary sex characteristics that develop during puberty. These include the growth of facial, axillary (underarm), and pubic hair. It prompts the larynx and the vocal cords to enlarge, causing the voice to lower. Testosterone also stimulates oil and sweat glands in the skin. It is largely responsible for acne and body odor. Another side effect of testosterone is baldness. Although females, like males, do inherit genes for baldness, baldness is seen more often in males because of the presence of testosterone. Testosterone is partially responsible for the muscular strength of males, and this is why some athletes take supplemental amounts of **anabolic steroids,** which are either testosterone or related chemicals. The Bioethical Focus, *Anabolic Steroid Use,* in Chapter 12 discusses the detrimental effect anabolic steroids can have on the body.

The female sex hormones, estrogens (often referred to in the singular) and progesterone, have many effects on the body. In particular, estrogen secreted at the time of puberty stimulates the growth of the uterus and the vagina. Estrogen is necessary for egg maturation and is largely responsible for the secondary sex characteristics in females. These include female body hair and fat distribution. In general, females have a more rounded appearance than males because of a greater accumulation of fat beneath the skin. Also, the pelvic girdle is wider in females than in males, resulting in a larger pelvic cavity. Both estrogen and progesterone are required for breast development and for regulation of the uterine cycle. This includes monthly menstruation (discharge of blood and mucosal tissues from the uterus).

Thymus Gland

The lobular **thymus gland** lies just beneath the sternum (see Fig. 15.2). This organ reaches its largest size and is most active during childhood. With aging, the organ gets smaller and becomes fatty. Lymphocytes that originate in the bone marrow and then pass through the thymus are transformed into T lymphocytes. The lobules of the thymus are lined by epithelial cells that secrete hormones called **thymosins.** These hormones aid in the differentiation of lymphocytes packed inside the lobules. Although thymosins ordinarily work in the thymus, there is hope that these hormones could be injected into AIDS or cancer patients, where they would enhance T- lymphocyte function.

Video
Lymphocytes

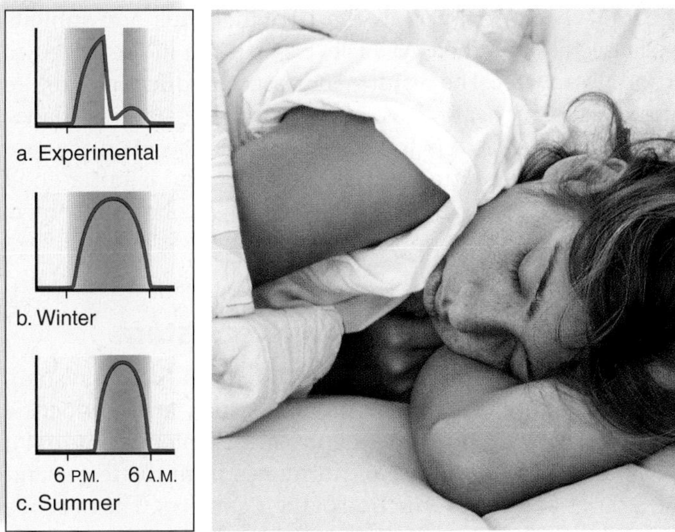

Figure 15.20 **Melatonin production changes by season.** Melatonin production is greatest at night when we are sleeping. Light suppresses melatonin production (**a**), so it is secreted for a longer time in the winter (**b**) than in the summer (**c**).

Pineal Gland

The **pineal gland,** located in the brain (see Fig. 15.2), produces the hormone **melatonin,** primarily at night. Melatonin is involved in our daily sleep–wake cycle. Normally, we grow sleepy at night when melatonin levels increase and awaken once daylight returns and melatonin levels are low (Fig. 15.20). Daily 24-hour cycles such as this are called **circadian rhythms.** These rhythms are controlled by a biological clock located in the hypothalamus.

Video Winter Mood

Animal research suggests that melatonin also regulates sexual development. In keeping with these findings, it has been noted that children whose pineal glands have been destroyed due to brain tumors experience early puberty.

Hormones from Other Organs or Tissues

Some organs not usually considered endocrine glands do secrete hormones. We have already mentioned that the kidneys secrete renin and that the heart produces atrial natriuretic hormone (see page 353); recall also that the stomach and the small intestine produce peptide hormones that regulate digestive secretions. A number of other types of tissues produce hormones.

Erythropoietin

In response to a low oxygen blood level, the kidneys secrete erythropoietin. Erythropoietin stimulates red blood cell formation in the red bone marrow. A greater number of red blood cells results in increased blood oxygen. A number of different types of organs and cells also produce peptide growth factors, which stimulate cell division and mitosis. Growth factors can be considered hormones because they act on cell types with specific receptors to receive them. Some are released into the blood; others diffuse to nearby cells.

Leptin

Leptin is a protein hormone produced by adipose tissue. Leptin acts on the hypothalamus, where it signals satiety or fullness. Strange to say, the blood of obese individuals may be rich in leptin. It is possible that the leptin they produce is ineffective because of a genetic mutation or because their hypothalamic cells lack a suitable number of receptors for leptin.

Prostaglandins

Prostaglandins are potent chemical signals produced within cells from arachidonate, a fatty acid. Prostaglandins are not distributed in the blood. They act locally, quite close to where they were produced. In the uterus, prostaglandins cause muscles to contract. Therefore, they are implicated in the pain and discomfort of menstruation in some women. Also, prostaglandins mediate the effects of pyrogens, chemicals believed to reset the temperature regulatory center in the brain. Aspirin reduces body temperature and controls pain because of its effect on prostaglandins.

Certain prostaglandins reduce gastric secretion and have been used to treat gastric reflux. Others lower blood pressure and have been used to treat hypertension. Still others inhibit platelet aggregation and have been used to prevent thrombosis. However, different prostaglandins have contrary effects, and it has been very difficult to successfully standardize their use. Therefore, prostaglandin therapy is still considered experimental.

Connecting the Concepts

For more information on the hormones presented in this section, refer to the following discussions.

Figure 6.4 illustrates the role of erythropoietin in the manufacture of new red blood cells.

Section 8.4 examines the role of the digestive hormones.

Sections 16.2 and 16.3 explain the role of the male and female sex hormones.

Check Your Progress 15.6

1. Summarize the role of testosterone and estrogen in the body.
2. Explain the relationship between melatonin and the sleep–wake cycle.
3. Describe the response of the body to low levels of oxygen in the blood.

15.7 Homeostasis

Learning Outcomes

Upon completion of this section, you should be able to

1. Summarize how the endocrine and nervous systems respond to external changes in the body.
2. Summarize how the endocrine and nervous systems respond to internal changes in the body.

The nervous and endocrine systems exert control over the other systems and thereby maintain homeostasis (Fig. 15.21).

Responding to External Changes

The nervous system is particularly able to respond to changes in the external environment. Some responses are automatic as you can verify by trying this: Take a piece of clear plastic and hold it just in front of your face. Get someone to gently toss a soft object, such as a wadded-up piece of paper, at the plastic. Can you prevent yourself from blinking? This reflex protects your eyes.

The eyes and other organs that have sensory receptors provide us with valuable information about the external environment. The central nervous system, on the receiving end of millions of bits of information, integrates information, compares it with previously stored memories, and "decides" on the proper course of action. The nervous system often responds to changes in the external environment through body movement. It gives us the ability to stay in as moderate an environment as possible. Otherwise, we test the ability of the nervous system to maintain homeostasis despite extreme conditions.

Responding to Internal Changes

The governance of internal organs usually requires that the nervous and endocrine systems work together. This usually occurs below the level of consciousness. Subconscious control often depends on reflex actions that involve the hypothalamus and the medulla oblongata. Let's take blood pressure as an example. You've just run 3 miles to raise money for hunger relief, and you decide to sit down under a tree to rest a bit. When you stand up to push off again, you feel faint. The feeling quickly passes because the medulla oblongata responds to input from the baroreceptors in the aortic arch and carotid arteries. The sympathetic system immediately acts to increase heart rate and constrict the blood vessels so that your blood pressure rises. Sweating may have upset the water–salt balance of your blood. If so, the hormone aldosterone from the adrenal cortex will act on the kidney tubules to conserve sodium ions (Na^+), and water reabsorption will follow. The hypothalamus can also help by sending antidiuretic hormone (ADH) to the posterior pituitary gland, which releases it into the blood. ADH actively promotes water reabsorption by the kidney tubules.

Recall from Chapter 13 that certain drugs, such as alcohol, can affect ADH secretion. When you consume alcohol, it is quickly absorbed across the stomach lining into the bloodstream, where it travels to the hypothalamus and inhibits ADH secretion. When ADH levels fall, the kidney tubules absorb less water. The result is increased production of dilute urine. Excessive water loss, or dehydration, is a disturbance of homeostasis. This is why drinking alcohol when you are exercising or perspiring heavily on a hot day is not a good idea. Instead of keeping you hydrated, an alcoholic beverage, such as beer, has the opposite effect.

Controlling the Reproductive System

Few systems intrigue us more than the reproductive system, which couldn't function without nervous and endocrine control. The hypothalamus controls the anterior pituitary, which controls the release of hormones from the testes and the ovaries and the production of their gametes. The nervous system directly controls the muscular contractions of the ducts that propel the sperm. Contractions of the oviducts, which move a developing embryo to the uterus where development continues, are stimulated by the nervous system, too. Without the positive feedback cycle involving oxytocin produced by the hypothalamus, birth might not occur.

The Neuroendocrine System

The nervous and endocrine systems work so closely together that they form what is sometimes called the neuroendocrine system. As we have seen, the hypothalamus certainly bridges the regulatory activities of the nervous and endocrine systems. In addition to producing the hormones released by the posterior pituitary, the hypothalamus produces hormones that control the anterior pituitary. The nerves of the autonomic system, which control other organs, are acted upon directly by the hypothalamus. The hypothalamus truly belongs to both the nervous and endocrine systems. Indeed, it is often and appropriately referred to as a neuroendocrine organ.

Connecting the Concepts

For more information on the organ systems presented in this section, refer to the following discussions.

Section 5.3 examines the factors that regulate heart rate.

Section 10.4 explains the role of aldosterone and ADH on the function of the kidney.

Section 13.2 explores the roles of the hypothalamus and medulla oblongata in the CNS.

Check Your Progress 15.7

1. Summarize the role of the nervous system in the monitoring of the internal and external environments.
2. Explain how the body restores its water–salt balance after it has lost water and salt through sweating.
3. Explain why the nervous and endocrine systems are integrated with one another.

The nervous and endocrine systems work together to maintain homeostasis. The systems listed here in particular also work with these two systems.

Nervous and Endocrine Systems

The nervous and endocrine systems coordinate the activities of the other systems. The brain receives sensory input and controls the activity of muscles and various glands. The endocrine system secretes hormones that influence the metabolism of cells, the growth and development of body parts, and homeostasis.

Urinary System

Nerves stimulate muscles that permit urination. Hormones (ADH and aldosterone) help kidneys regulate the water–salt balance and the acid–base balance of the blood.

Digestive System

Nerves stimulate smooth muscle and permit digestive tract movements. Hormones help regulate digestive juices that break down food to nutrients for neurons and glands.

Muscular System

Nerves stimulate muscles, whose contractions allow us to move out of danger. Androgens promote growth of skeletal muscles. Sensory receptors in muscles and joints send information to the brain. Muscles protect neurons and glands.

Cardiovascular System

Nerves and epinephrine regulate contraction of the heart and constriction/dilation of blood vessels. Hormones regulate blood glucose and ion levels. Growth factors promote blood cell formation. Blood vessels transport hormones to target cells.

Respiratory System

The respiratory center in the brain regulates the breathing rate. The lungs carry on gas exchange for the benefit of all systems, including the nervous and endocrine systems.

Reproductive System

Nerves stimulate contractions that move gametes in ducts, and uterine contraction that occurs during childbirth. Sex hormones influence the development of the secondary sex characteristics.

Integumentary System

Nerves activate sweat glands and arrector pili muscles. Sensory receptors in skin send information to the brain about the external environment. Skin protects neurons and glands.

Skeletal System

Growth hormone and sex hormones regulate the size of the bones; parathyroid hormone and calcitonin regulate their Ca^{2+} content and therefore bone strength. Bones protect nerves and glands.

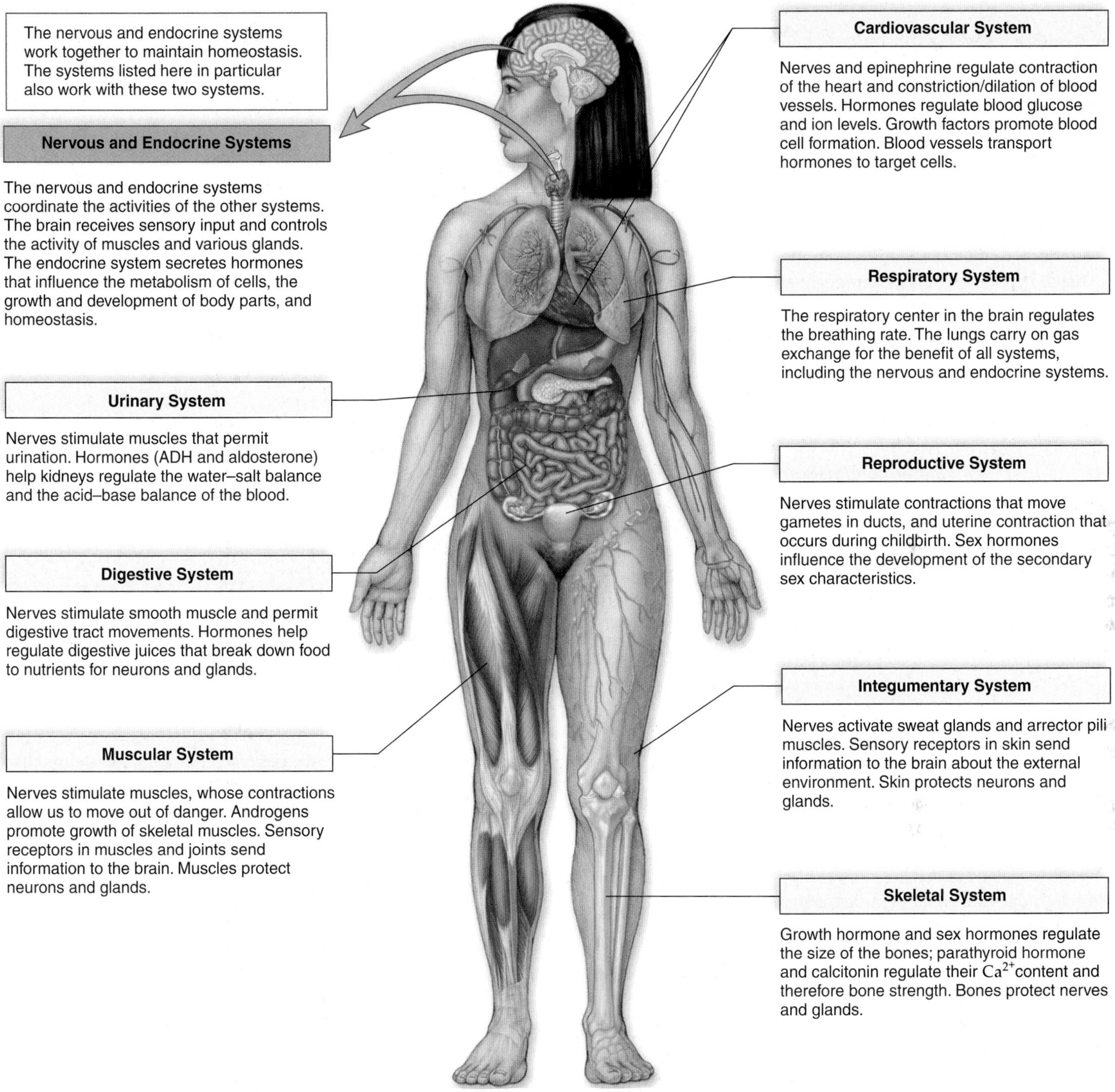

Figure 15.21 **Homeostasis is maintained through cooperation of multiple organ systems.**
The nervous and endocrine systems work together to regulate and control the other systems.

CASE STUDY CONCLUSION

For diabetics, the prospects of controlling their blood glucose levels and living a healthy life are better today than in the past. Prior to the development of recombinant DNA technology, which now allows human insulin to be produced in large quantities, insulin was derived from the pancreases of pigs or cows. This required laborious purification, and because the animal insulins were not identical to the human form, sometimes immunologic reactions occurred. Another recent advance is the insulin pump, a device a little bigger than a cell phone, which can deliver precise amounts of insulin under the skin using a small plastic catheter. The insulin pump more accurately mimics the pancreas's natural release of the correct amount of insulin needed by the body. Studies have shown that insulin pumps are more effective than traditional injections of insulin in controlling blood sugar levels. Currently, around 150,000 people in the United States are using an insulin pump. In the near future, it may be possible to implant a device—sometimes called an "artificial pancreas"—into patients with diabetes that will not only monitor the blood sugar level but will also provide the appropriate doses of insulin.

Media Study Tools

www.mhhe.com/maderhuman12e

Enhance your study of this chapter with study tools and practice tests. Also ask your instructor about the resources available through ConnectPlus, including the media-rich eBook, interactive learning tools, and animations.

Summarizing the Concepts

15.1 Endocrine Glands

Endocrine glands secrete hormones into the bloodstream; from there, they are distributed to target organs or tissues.

- Hormones, a type of chemical signal, usually act at a distance between body parts.
- Hormones are either peptides or steroids.
- Reception of a peptide hormone at the plasma membrane activates an enzyme cascade inside the cell.
- Steroid hormones combine with a receptor, and the complex attaches to and activates DNA. Protein synthesis follows.

15.2 Hypothalamus and Pituitary Gland

Neurosecretory cells in the hypothalamus produce antidiuretic hormone and oxytocin, which are stored in axon endings in the posterior pituitary until released.

- The hypothalamus produces hypothalamic-releasing and hypothalamic-inhibiting hormones, which pass to the anterior pituitary by way of a portal system.
- The anterior pituitary produces seven types of hormones, and some of these stimulate other hormonal glands to secrete hormones.

15.3 Thyroid and Parathyroid Glands

The thyroid gland requires iodine to produce triiodothyronine and thyroxine, which increase the metabolic rate.

- If iodine is available in limited quantities, a simple goiter develops.
- If the thyroid is overactive, an exophthalmic goiter develops.
- The thyroid gland produces calcitonin, which helps lower the blood calcium level.
- The parathyroid glands secrete parathyroid hormone, which raises the blood calcium level.

15.4 Adrenal Glands

The adrenal glands respond to stress.

Adrenal Medulla

The adrenal medulla immediately secretes epinephrine and norepinephrine. Heartbeat and blood pressure increase, blood glucose level rises, and muscles become energized.

Adrenal Cortex

The adrenal cortex produces the glucocorticoids (e.g., cortisol) and the mineralocorticoids (e.g., aldosterone). The glucocorticoids regulate carbohydrate, protein, and fat metabolism and also suppress the inflammatory response. Mineralocorticoids regulate water and salt balance, leading to increases in blood volume and blood pressure.

15.5 Pancreas

The pancreatic islets secrete the hormones insulin and glucagon.

- Insulin lowers the blood glucose level.
- Glucagon raises the blood glucose level.
- Diabetes mellitus is due to the failure of the pancreas to produce insulin or the failure of the cells to take it up.

15.6 Other Endocrine Glands

Other endocrine glands produce hormones.

- The testes and ovaries produce the sex hormones. Male sex hormones are the androgens (e.g., testosterone); female sex hormones are the estrogens and progesterone.
- The thymus gland secretes thymosins, which stimulate T- lymphocyte production and maturation.
- The pineal gland produces melatonin, which may be involved in circadian rhythms and the development of the reproductive organs.

Some organs and tissues also produce hormones.

- Kidneys produce erythropoietin.
- Adipose tissue produces leptin, which acts on the hypothalamus.
- Prostaglandins are produced within cells and act locally.

15.7 Homeostasis

The nervous and endocrine systems exert control over the other systems and thereby maintain homeostasis.

- The nervous system is able to respond to the external environment after receiving data from the sensory receptors. Sensory receptors are present in such organs as the eyes and ears.
- The nervous and endocrine systems work together to govern the subconscious control of internal organs. This control often depends on reflex actions involving the hypothalamus and medulla oblongata.
- The nervous and endocrine systems work so closely together that they form what is sometimes called the neuroendocrine system.

Understanding Key Terms

acromegaly 347
Addison disease 353
adrenal cortex 351
adrenal gland 351
adrenal medulla 351
adrenocorticotropic hormone (ACTH) 346
aldosterone 357
anabolic steroid 358
androgen 358
anterior pituitary 345
antidiuretic hormone (ADH) 345
atrial natriuretic hormone (ANH) 353
calcitonin 350
chemical signal 342
circadian rhythm 359
congenital hypothyroidism 349
cortisol 352
Cushing syndrome 353
cyclic adenosine monophosphate (cAMP) 343
diabetes mellitus 354
endocrine gland 340
epinephrine 352
estrogen 358
exocrine gland 340
exophthalmic goiter 349
first messenger 344
follicle-stimulating hormone (FSH) 346, 358
glucagon 354
glucocorticoid 352
gonad 358
gonadotropic hormone 346
growth hormone (GH) 347
hormone 340
hypothalamic-inhibiting hormone 345
hypothalamic-releasing hormone 345
hypothalamus 345
insulin 354
leptin 359
luteinizing hormone (LH) 346, 358
melanocyte-stimulating hormone (MSH) 347
melatonin 359
mineralocorticoid 352
myxedema 349
norepinephrine (NE) 352
ovary 358
oxytocin 345
pancreas 354
pancreatic islets 354
parathyroid gland 350
parathyroid hormone (PTH) 350
peptide hormone 343
pheromone 343
pineal gland 359
pituitary dwarfism 347
pituitary gland 345
positive feedback 345
posterior pituitary 345
progesterone 358
prolactin (PRL) 347
prostaglandin 359
renin 352
second messenger 344
simple goiter 349
steroid hormone 343
testes 358
testosterone 358
tetany 350
thymosin 358
thymus gland 358
thyroid gland 349
thyroid-stimulating hormone (TSH) 346
thyroxine (T_4) 349
triiodothyronine (T_3) 349

Match the key terms to these definitions.

a. ________ Organ in the neck that secretes several important hormones, including thyroxine and calcitonin.

b. ________ Condition characterized by high blood glucose level and the appearance of glucose in the urine.

c. ________ Hormone secreted by the anterior pituitary that stimulates portions of the adrenal cortex.

d. ________ Type of hormone that causes the activation of an enzyme cascade in cells.

e. ________ Hormone released by the posterior pituitary that causes contraction of the uterus and milk letdown.

Testing Your Knowledge of the Concepts

1. Compare and contrast the nervous and endocrine systems. (page 340)
2. How does the action of a peptide hormone differ from that of a steroid hormone? (pages 343–344)
3. Explain the relationship between the hypothalamus and the posterior pituitary gland and to the anterior pituitary gland. List the hormones secreted by the anterior pituitary and the posterior pituitary glands and their actions. (pages 345–346)
4. Give an example of the negative feedback relationship among the hypothalamus, the anterior pituitary, and other endocrine glands. (page 345)
5. Discuss the action of growth hormone on the body. What occurs if there is too much or too little GH during the growing years? What occurs if there is too much GH in an adult? (page 347)
6. What types of conditions are associated with a malfunctioning thyroid gland? Explain each type. (page 349)
7. Explain how the thyroid and parathyroid glands work together to maintain blood calcium homeostasis. (page 350)
8. What hormones are secreted by the adrenal cortex and adrenal medulla, and what are their actions? (pages 351–352)
9. What are the causes and symptoms of Addison disease and Cushing syndrome? (page 353)
10. Explain how insulin and glucagon maintain blood glucose homeostasis. What are the two types of diabetes mellitus, and what are the major symptoms? (pages 354–355)
11. How does the neuroendocrine system work with other systems to maintain homeostasis? (pages 360–361)
12. Which type of glands are ductless?
    a. exocrine
    b. endocrine
    c. Both a and b are correct.
    d. Neither a nor b is correct.
13. Which hormones can cross cell membranes?
    a. peptide hormones
    b. steroid hormones
    c. Both a and b are correct.
    d. Neither a nor b is correct.
14. The anterior pituitary controls the secretions of both
    a. the adrenal medulla and the adrenal cortex.
    b. the thyroid and adrenal cortex.
    c. the ovaries and testes.
    d. Both b and c are correct.

15. Growth hormone is produced by the
    a. posterior adrenal gland.
    b. posterior pituitary.
    c. anterior pituitary.
    d. kidneys.
    e. None of these is correct.

16. ________ is released through positive feedback and causes ________ contractions.
    a. Insulin, stomach
    b. Oxytocin, stomach
    c. Oxytocin, uterine
    d. None of these is correct.

17. PTH causes the blood level of calcium to ________, and calcitonin causes it to ________.
    a. increase, increase
    b. increase, decrease
    c. decrease, increase
    d. decrease, decrease

18. Bodily response to stress includes
    a. water reabsorption by the kidneys.
    b. blood pressure increase.
    c. increase in blood glucose levels.
    d. heart rate increase.
    e. All of these are correct.

19. Lack of aldosterone will cause a blood imbalance of
    a. sodium.
    b. potassium.
    c. water.
    d. All of these are correct.
    e. None of these is correct.

20. Long-term complications of diabetes include
    a. blindness.
    b. kidney disease.
    c. circulatory disorders.
    d. All of these are correct.
    e. None of these is correct.

21. Diabetes mellitus is associated with
    a. too much insulin in the blood.
    b. too high a blood glucose level.
    c. blood that is too dilute.
    d. All of these are correct.

22. Which of these is not a pair of antagonistic hormones?
    a. insulin—glucagon
    b. calcitonin—parathyroid hormone
    c. cortisol—epinephrine
    d. aldosterone—atrial natriuretic hormone (ANH)

23. Which hormone and condition are mismatched?
    a. growth hormone—acromegaly
    b. thyroxine—goiter
    c. parathyroid hormone—tetany
    d. cortisol—myxedema
    e. insulin—diabetes

In questions 24–28, match the hormones to the correct gland in the key.

Key:

a. glucagon
b. prostaglandin
c. melatonin
d. insulin
e. leptin

24. Raises blood glucose levels
25. Conversion of glucose to glycogen
26. Hunger control
27. Controls circadian rhythms
28. Causes uterine contractions
29. Complete the following diagram.

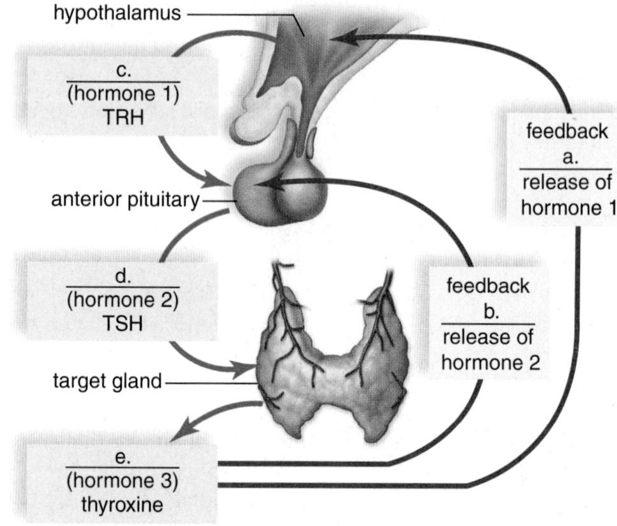

Thinking Critically About the Concepts

Blood tests are a way to diagnose any number of endocrine disorders because hormones are transported by the circulatory system. GH and IGF-1 can be checked to determine if deficiencies are the reason for a child's slow growth. Blood levels of TSH, T_3, and T_4 provide information about thyroid function. Some tests, such as the glucose tolerance test from the chapter opener, do not directly measure the level of the glucose-regulating hormones (in this case, insulin) but rather indirectly monitor whether an endocrine gland is performing correctly by measuring specific compounds in the blood.

1. How is follicle-stimulating hormone similar to growth hormone with regard to how their target cells respond to their signals?
2. It is possible to diagnose hypothyroidism by high levels of TSH in the blood. Explain what would cause a high TSH level. (Hint: You may want to consider what happens to TSH when the activity of the thyroid is normal.)
3. Why would a diabetic urinate frequently and always be thirsty?
4. Many diets advertise that they are specifically designed for diabetics. How would these diets be different from a "normal" diet?

CHAPTER 16

Reproductive System

CASE STUDY CERVICAL CANCER

Ann always dreaded her visits to the gynecologist. At each visit, her doctor had warned her to quit smoking. However, for the past 20 years, she had been a regular smoker, and even though she had cut back considerably in the past few years, she was still smoking around a pack of cigarettes a day. Ann's annual Pap smears had always been normal, so Ann was beginning to view her annual trip to the gynecologist as just a formality. Since she had turned 40, her visits had become more sporadic; and over the past few years, she had stopped the visits completely.

Until recently, Ann had felt fine. However, starting just a few months ago, she had started to experience some abnormal vaginal bleeding, usually shortly after sexual intercourse with her partner. This was often accompanied by small amounts of pain. Concerned about these recent changes, Ann scheduled an appointment with her doctor.

At the appointment, the doctor performed a complete physical exam, which included a Pap smear. As the doctor expected, the results of her test were abnormal. Her doctor sent the results to an oncologist, a cancer specialist, who confirmed that Ann's symptoms were being caused by an early stage of cervical cancer. To check the extent of the cancer and to see if it had spread to any additional organs, the oncologist ordered a computed tomography (CT) scan of Ann's pelvis and abdomen, as well as a series of blood tests to look for evidence of cancer. Both the CT scan and the blood tests indicated that Ann was lucky—they had caught the cancer at an early stage of development. Her oncologist was convinced that a hysterectomy might be avoided, but Ann would have to immediately begin both chemotherapy and radiation treatment to stop the spread of the cancer.

As you read through the chapter, think about the following questions.

1. What is the role of the cervix in the female reproductive system?
2. What is a Pap smear test used to detect?
3. What is a hysterectomy?

CHAPTER CONCEPTS

16.1 Human Life Cycle
The male reproductive system produces sperm, and the female reproductive system produces eggs, each of which have only 23 chromosomes due to meiosis.

16.2 Male Reproductive System
The male reproductive system produces sperm and the male sex hormones.

16.3 Female Reproductive System
In the female, the ovaries produce eggs and the sex hormones.

16.4 The Ovarian Cycle
The sex hormones fluctuate in monthly cycles, resulting in ovulation once a month followed by menstruation if pregnancy does not occur.

16.5 Control of Reproduction
Numerous birth control methods are available for those who wish to prevent pregnancy. Infertile couples may use assisted reproductive technologies to have a child.

16.6 Sexually Transmitted Diseases
Medications have been developed to control AIDS and genital herpes, but these STDs are not curable. STDs caused by bacteria are curable with antibiotic therapy, but resistance is making this more and more difficult.

BEFORE YOU BEGIN

Before beginning this chapter, take a few moments to review the following discussions.

Section S.1 What factors contribute to an increased risk of an HIV infection?

Section 15.6 Where is testosterone produced and what is its function in the male body?

Section 15.6 Where are estrogen and progesterone produced and what are their functions in the female body?

16.1 Human Life Cycle

Learning Outcomes

Upon completion of this section, you should be able to

1. List the functions of the reproductive system in humans.
2. Describe the human life cycle and explain the role of mitosis and meiosis in this cycle.

Unlike the other systems of the body, the reproductive system is quite different in males and females. *Puberty* is the sequence of events by which a child becomes a sexually competent young adult. The reproductive system does not begin to fully function until puberty is complete. Sexual maturity occurs between the ages of 10 and 14 in girls and 12 and 16 in boys. At the completion of puberty, the individual is capable of producing children.

The reproductive organs (genitals) have the following functions:

1. Males produce sperm within testes, and females produce eggs within ovaries.
2. Males nurture and transport the sperm in ducts until they exit the penis. Females transport the eggs in uterine tubes to the uterus.
3. The male penis functions to deliver sperm to the female vagina, which functions to receive the sperm. The vagina also transports menstrual fluid to the exterior and acts as the birth canal.
4. The uterus of the female allows the fertilized egg to develop within her body. After birth, the female breast provides nourishment in the form of milk.
5. The testes and ovaries produce the sex hormones. The sex hormones have a profound effect on the body because they bring about masculinization or feminization of various features. In females, the sex hormones also allow a pregnancy to continue.

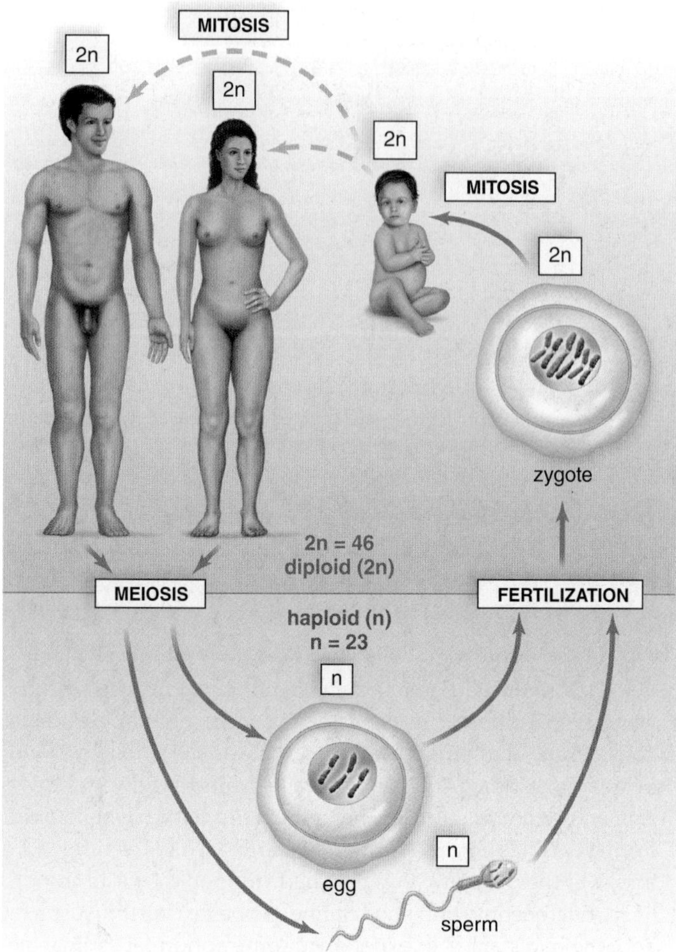

Figure 16.1 **The human life cycle.**
The human life cycle has two types of cell divisions: mitosis, in which the chromosome number stays constant, and meiosis, in which the chromosome number is reduced. During growth or cell repair, mitosis ensures that each new cell has 46 chromosomes. During production of sex cells, the chromosome number is reduced from 46 to 23. Therefore, an egg and a sperm each have 23 chromosomes so that when the sperm fertilizes the egg, the new cell, called a zygote, has 46 chromosomes.

Mitosis and Meiosis

Our DNA is distributed among 46 chromosomes within the nucleus. The majority of the cell types in the body have 46 chromosomes. Ordinarily, when a cell divides by a process called mitosis, the new cells also have 46 chromosomes. Mitosis is *duplication division.* (As an analogy, imagine the cell producing exact copies of itself during mitosis, much like a duplicating machine does with a page of notes.) In the life cycle of a human being, mitosis is the type of cell division that takes place during growth and repair of tissues (Fig. 16.1).

In addition to mitosis, human cells undergo a type of cell division called meiosis, which is *reduction division.* Meiosis takes place only in the testes of males during the production of sperm and in the ovaries of females during the production of eggs. During meiosis, the chromosome number is reduced from the normal 46 chromosomes, called the diploid or 2n number, down to 23 chromosomes, called the haploid or

Connections and Misconceptions

What types of cells do not have 46 chromosomes?

In humans, the cell types that do not have the standard 23 pairs of chromosomes are the red blood cells and the cells of the liver. Recall from Section 6.2 that red blood cells lack a nucleus; therefore, they do not have any chromosomes. Cells in the liver, called hepatocytes, typically have more than 3 copies of each chromosome (giving them 69 or more chromosomes). This condition is called polyploidy, and it is believed to provide the liver with its ability to degrade toxic compounds.

n number of chromosomes. Meiosis requires two successive divisions, called meiosis I and meiosis II.

The flagellated sperm is small compared to the egg. It is specialized to carry only chromosomes as it swims to the egg. The egg is specialized to await the arrival of a sperm and to provide the new individual with cytoplasm in addition to chromosomes. The first cell of a new human being is called the zygote. A sperm has 23 chromosomes and the egg has 23 chromosomes, so the zygote has 46 chromosomes altogether. Without meiosis, the chromosome number in each generation of human beings would double, and the cells would no longer be able to function.

Connecting the Concepts

For more information on the topics presented in this section, refer to the following discussions.

Section 15.6 provides an introduction to the sex hormones.

Sections 18.2 and 18.3 examine the stages of mitosis and meiosis.

Section 18.4 compares the processes of mitosis and meiosis.

Check Your Progress 16.1

1. Compare the functions of the reproductive system in males and females.
2. Contrast the two types of cell division in the human life cycle.
3. Explain the location of meiosis in males and females.

16.2 Male Reproductive System

Learning Outcomes

Upon completion of this section, you should be able to

1. Identify the structures of the male reproductive system and provide a function for each.
2. Describe the location and stages of spermatogenesis.
3. Summarize how hormones regulate the male reproductive system.

The male reproductive system includes the organs depicted in Figure 16.2 and listed in Table 16.1. The male gonads, or primary sex organs, are paired **testes** (sing., testis), suspended within the sacs of the **scrotum.**

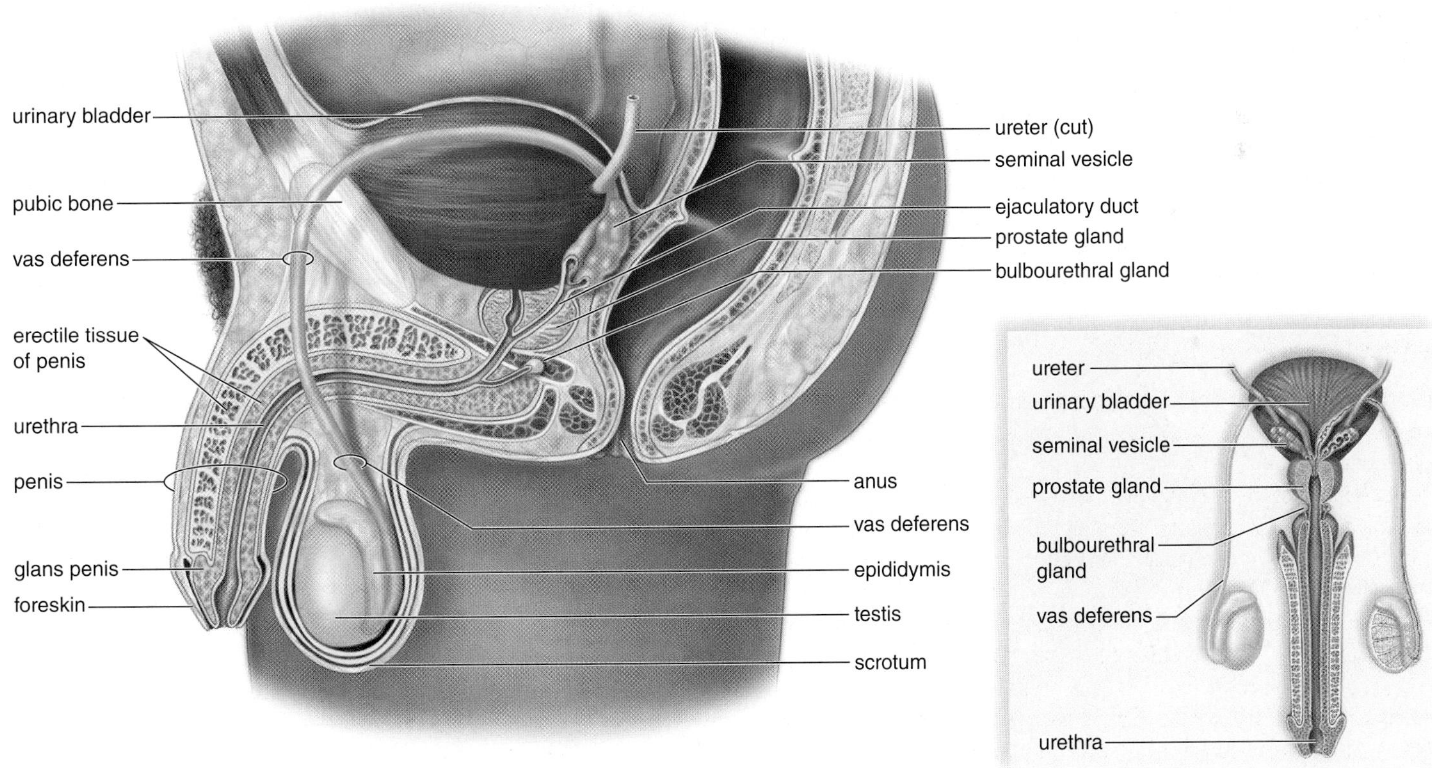

Figure 16.2 The male reproductive system.
The testes produce sperm. The seminal vesicles, the prostate gland, and the bulbourethral glands provide a fluid medium for the sperm, which move from the vas deferens through the ejaculatory duct to the urethra in the penis. The foreskin (prepuce) is removed when a penis is circumcised.

Table 16.1 Male Reproductive Organs

Organ	Function
Testes	Produce sperm and sex hormones
Epididymides	Ducts where sperm mature and some sperm are stored
Vasa deferentia	Conduct and store sperm
Seminal vesicles	Contribute nutrients and fluid to semen
Prostate gland	Contributes fluid to semen
Urethra	Conducts sperm
Bulbourethral glands	Contribute mucus-containing fluid to semen
Penis	Organ of sexual intercourse

Sperm produced by the testes mature within the **epididymis** (pl., epididymides), a tightly coiled duct lying just outside each testis. Maturation seems to be required for sperm to swim to the egg. When sperm leave an epididymis, they enter a **vas deferens** (pl., vasa deferentia), also called the ductus deferens. The sperm may be stored for a time in the vas deferens. Each vas deferens passes into the abdominal cavity, where it curves around the bladder and empties into an ejaculatory duct. The ejaculatory ducts enter the **urethra.**

At the time of ejaculation, sperm leave the penis in a fluid called **semen.** The seminal vesicles, the prostate gland, and the bulbourethral glands (Cowper glands) add secretions to seminal fluid. The **seminal vesicles** (a pair) lie at the base of the bladder, and each has a duct that joins with a vas deferens. The **prostate gland** is a single, doughnut-shaped gland that surrounds the upper portion of the urethra just below the bladder. In older men, the prostate can enlarge and squeeze off the urethra, making urination painful and difficult. The condition can be treated medically. **Bulbourethral glands** are pea-sized organs that lie posterior to the prostate on either side of the urethra. Their secretion makes the seminal fluid gelatinous.

Each component of seminal fluid seems to have a particular function. Sperm are more viable in a basic solution; seminal fluid, milky in appearance, has a slightly basic pH (about 7.5). Swimming sperm require energy; and seminal fluid contains the sugar fructose, which presumably serves as an energy source. Semen also contains prostaglandins, chemicals that cause the uterus to contract. Some investigators believe that uterine contractions help propel the sperm toward the egg.

The Penis and Male Orgasm

The **penis** (Fig. 16.3) is the male organ of sexual intercourse. The penis has a long shaft and an enlarged tip called the glans penis. The glans penis is normally covered by a layer of skin called the foreskin. **Circumcision,** the surgical removal of the foreskin, if done, is usually performed soon after birth.

Spongy, erectile tissue containing distensible blood spaces extends through the shaft of the penis. During sexual arousal, autonomic nerves release nitric oxide, NO. This stimulus leads to the production of cGMP (cyclic guanosine monophosphate), a high-energy compound similar to ATP. The cGMP causes the smooth muscle of incoming arterial walls to relax and the erectile tissue to fill with blood. The veins that take blood away from the penis are compressed, and the penis becomes erect. **Erectile dysfunction** (formerly called *impotency* but now commonly called *ED*) is the inability to achieve or maintain an erection suitable for sexual intercourse. ED may be caused by a number of factors, including poor blood flow, certain medications, and many illnesses.

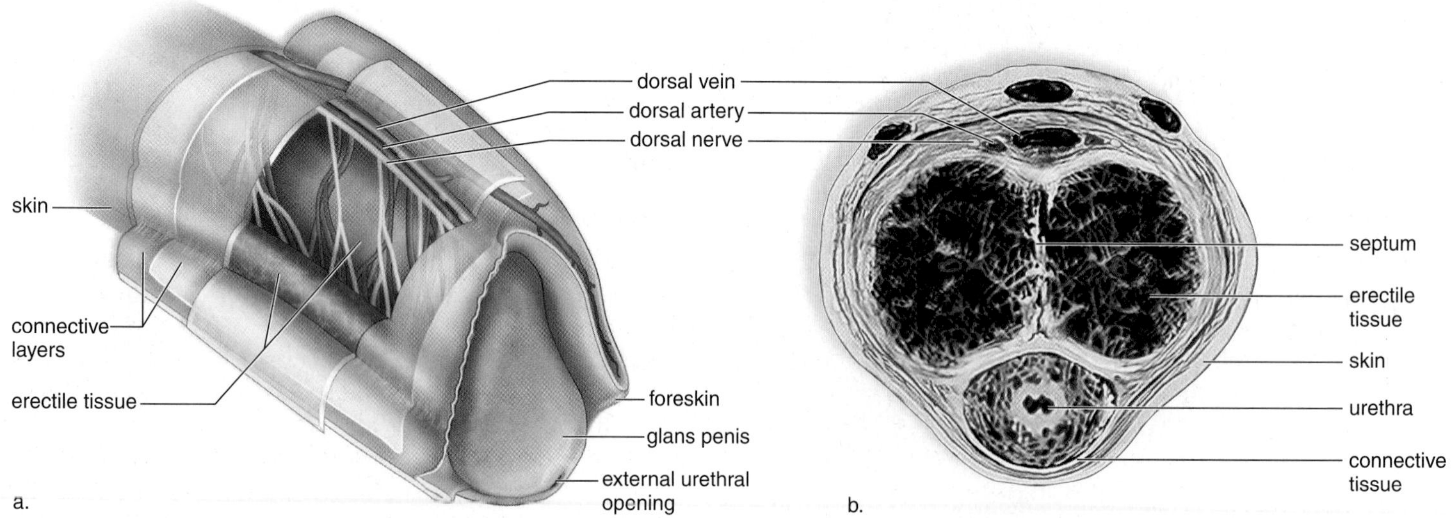

Figure 16.3 The structure of the penis.
a. Penis shaft. The shaft of the penis ends in an enlarged tip called the glans penis, which, in uncircumsized males, is partially covered by a foreskin (prepuce). The penis contains columns of erectile tissue. **b.** Micrograph of shaft in cross section showing location of erectile tissue. One column surrounds the urethra.

Medications for the treatment of erectile dysfunction inhibit the enzyme that breaks down cGMP, ensuring that a full erection will take place. Some of these medications can cause vision problems because the same enzyme occurs in the retina. During an erection, a sphincter closes off the bladder so that no urine enters the urethra. (The urethra carries either urine or semen at different times.)

As sexual stimulation intensifies, sperm enter the urethra from each vas deferens, and the glands contribute secretions to the seminal fluid. Once seminal fluid is in the urethra, rhythmic muscle contractions cause it to be expelled from the penis in spurts (ejaculation).

The contractions that expel seminal fluid from the penis are a part of male orgasm, the physiological and psychological sensations that occur at the climax of sexual stimulation. The psychological sensation of pleasure is centered in the brain. However, the physiological reactions involve the genital (reproductive) organs and associated muscles, as well as the entire body. Marked muscular tension is followed by contraction and relaxation. Following ejaculation and/or loss of sexual arousal, the penis returns to its normal flaccid state. Usually a period of time, called the refractory period, follows during which stimulation does not bring about an erection. The length of the refractory period increases with age.

There may be in excess of 400 million sperm in the 3.5 mL of semen expelled during ejaculation. The sperm count can be much lower than this, however, and fertilization of the egg by a sperm can still take place.

Male Gonads: The Testes

The testes, which produce sperm and also the male sex hormones, lie outside the abdominal cavity of the male, within the scrotum. The testes begin their development inside the abdominal cavity. They descend into the scrotal sacs through the inguinal canal during the last two months of fetal development. If the testes do not descend and the male is not treated or operated on to place the testes in the scrotum, sterility (the inability to produce offspring) usually follows. This is because the internal temperature of the body is too high to produce viable sperm. The scrotum helps regulate the temperature of the testes by holding them closer to or farther away from the body.

Seminiferous Tubules

A longitudinal section of a testis shows that it is composed of compartments called lobules, each of which contains one to three tightly coiled **seminiferous tubules** (Fig. 16.4*a*). A microscopic cross section of a seminiferous tubule reveals that it is packed with cells undergoing **spermatogenesis** (Fig. 16.4*b*), the production of sperm.

During the production of sperm, spermatogonia divide to produce primary spermatocytes (2n). Primary spermatocytes move away from the outer wall, increase in size, and undergo meiosis I to produce secondary spermatocytes. Each secondary spermatocyte has only 23 chromosomes (Fig. 16.4*c*). Secondary spermatocytes (n) undergo meiosis II to produce four spermatids, each of which also has 23 chromosomes. Spermatids then develop into sperm (Fig. 16.4*d*). Note the presence of **Sertoli cells** (purple), which support, nourish, and regulate the process of spermatogenesis. It takes approximately 74 days for sperm to undergo development from spermatogonia to sperm.

Animation Spermatogenesis

Mature **sperm,** or spermatozoa, have three distinct parts: a head, a middle piece, and a tail (Fig. 16.4*e*). Mitochondria in the middle piece provide energy for the movement of the tail, which is a flagellum. The head contains a nucleus covered by a cap called the **acrosome,** which stores enzymes needed to penetrate the egg. The ejaculated semen of a normal human male contains several hundred million sperm, but only one sperm normally enters an egg. Sperm usually do not live more than 48 hours in the female genital tract.

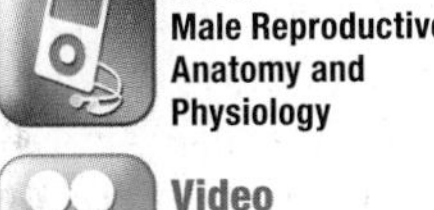

Video Human Sperm

Interstitial Cells

The male sex hormones, the androgens, are secreted by cells that lie between the seminiferous tubules. These cells are called **interstitial cells.** The most important of the androgens is testosterone, whose functions are discussed next.

Hormonal Regulation in Males

The hypothalamus has ultimate control of the testes' sexual function because it secretes a hormone called **gonadotropin-releasing hormone (GnRH).** GnRH stimulates the anterior pituitary to secrete the gonadotropic hormones. There are two gonadotropic hormones, **follicle-stimulating hormone (FSH)** and **luteinizing hormone (LH),** which are present in both males and females. In males, FSH promotes the production of sperm in the seminiferous tubules. LH in males controls the production of testosterone by the interstitial cells.

All these hormones are involved in a negative feedback relationship that maintains the fairly constant production

> **Connections and Misconceptions**
>
> **Boxers or briefs?**
>
> Most people seem to know that the scrotum's role is to keep the temperature of the testes lower than body temperature. The lower temperature is necessary for normal sperm production. It might follow that the man's type of underwear might change that temperature, affecting sperm production. However, research has not supported this assumption. The style of underwear worn by a man, loose or close fitting, has not been shown to affect sperm count or fertility significantly. So, boxers or briefs? It's up to you!

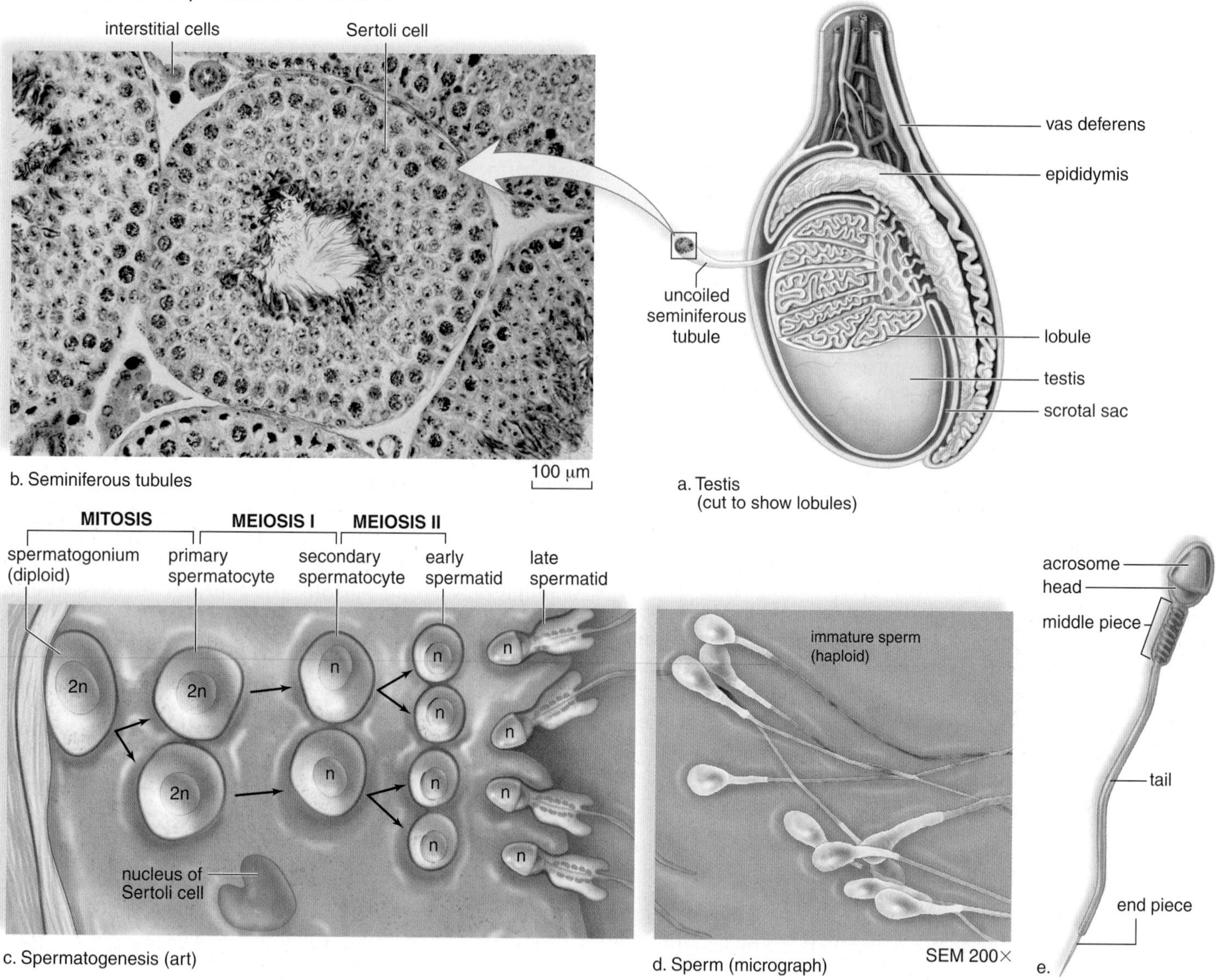

Figure 16.4 Spermatogenesis produces sperm cells.
a. The lobules of a testis contain seminiferous tubules. **b.** Electron micrograph of a cross section of the seminiferous tubules, where spermatogenesis occurs. Note the location of interstitial cells in clumps among the seminiferous tubules. **c.** Diagrammatic representation of spermatogenesis, which occurs in wall of tubules. **d.** Micrograph of sperm. **e.** A sperm has a head, a middle piece, and a tail. The nucleus is in the head, capped by the enzyme-containing acrosome.

of sperm and testosterone (Fig. 16.5). When the amount of testosterone in the blood rises to a certain level, it causes the hypothalamus and anterior pituitary to decrease their respective secretion of GnRH and LH. As the level of testosterone begins to fall, the hypothalamus increases its secretion of GnRH and the anterior pituitary increases its secretion of LH. These stimulate the interstitial cells to produce testosterone. A similar feedback mechanism maintains the continuous production of sperm. The Sertoli cells in the wall of the seminiferous tubules produce a hormone called *inhibin* that blocks GnRH and FSH secretion when appropriate (Fig. 16.4).

Testosterone, the main sex hormone in males, is essential for the normal development and functioning of the organs listed in Table 16.1. Testosterone also brings about and maintains the male secondary sex characteristics that develop at the time of puberty. Males are generally taller than females and have broader shoulders and longer legs relative to trunk length. The deeper voices of males compared with those of females are due to a larger larynx with longer vocal cords. The so-called Adam's apple, part of the larynx, is usually more prominent in males than in females. Testosterone causes males to develop noticeable hair on the face, chest, and occasionally other regions of the body, such as the back. A related chemical also leads to the receding hairline and male-pattern baldness that occur in males.

Testosterone is responsible for the greater muscular development in males. Knowing this, both males and females sometimes take anabolic steroids, either testosterone or related steroid hormones resembling testosterone. Health problems involving the kidneys, the cardiovascular system, and hormonal imbalances can arise from such use.

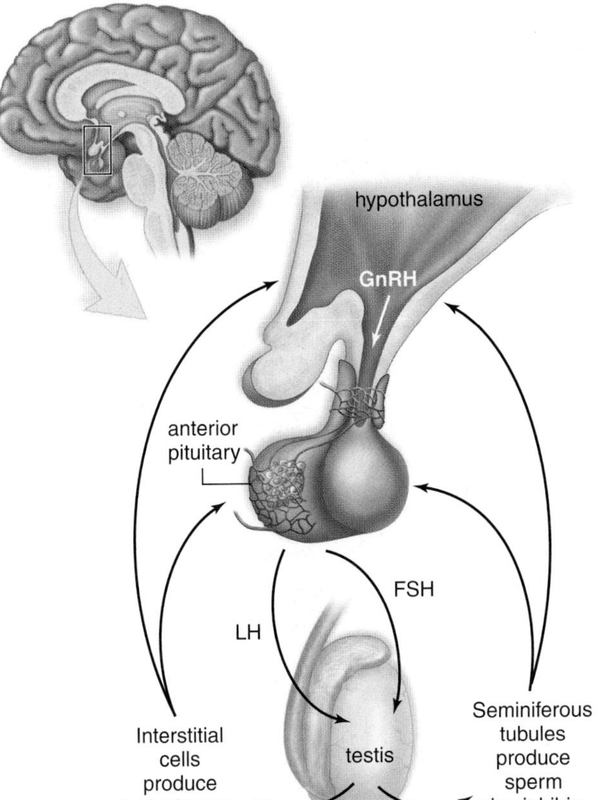

Figure 16.5 The hormones that control the production of sperm and testosterone by the testes.
Gonadotropin-releasing hormone (GnRH) stimulates the anterior pituitary to secrete the gonadotropic hormones: Follicle-stimulating hormone (FSH) stimulates the production of sperm, and luteinizing hormone (LH) stimulates the production of testosterone. Testosterone and inhibin exert negative feedback control over the hypothalamus and the anterior pituitary, and this regulates the level of testosterone in the blood and the production of sperm by the testes.

Connecting the Concepts

For more information on the topics presented in this section, refer to the following discussions.

Figure 2.20 provides the chemical structure of testosterone.

Section 15.6 provides additional information on the male sex hormones.

Section 18.3 examines how meiosis reduces the chromosome number during spermatogenesis.

Check Your Progress 16.2

1. Identify the structures of the male reproductive system, and then trace the movement of sperm through the system.
2. Describe the process of spermatogenesis.
3. Explain the importance of testoterone to the male reproductive system.

16.3 Female Reproductive System

Learning Outcome

Upon completion of this section, you should be able to

1. Identify the structures of the female reproductive system and provide a function for each.

The female reproductive system includes the organs depicted in Figure 16.6 and listed in Table 16.2. The female gonads are paired **ovaries** that lie in shallow depressions, one on each side of the upper pelvic cavity. The ovaries produce **eggs** (technically referred to as *ova*) and the female sex hormones estrogen and progesterone.

The Genital Tract

The **oviducts,** also called the uterine or fallopian tubes, extend from the uterus to the ovaries. However, the oviducts are not attached to the ovaries. Instead, they have fingerlike projections called fimbriae (sing., **fimbria**) that sweep over the ovaries. When an egg (ovum) bursts from an ovary during ovulation, it usually is swept into an oviduct by the combined action of the fimbriae and the beating of cilia that line the oviducts.

Once in the oviduct, the egg is propelled slowly by ciliary movement and tubular muscle contraction toward the uterus. An egg lives approximately 6 to 24 hours, unless fertilization occurs. Fertilization, and therefore **zygote** formation, usually takes place in the oviduct. A developing embryo normally arrives at the uterus after several days, and then **implantation** occurs. During implantation, the embryo embeds in the uterine lining, which has been prepared to receive it.

The **uterus** is a thick-walled, muscular organ about the size and shape of an inverted pear. Normally, it lies above and is tipped over the urinary bladder. The oviducts join the uterus at its upper end; at its lower end, the **cervix** enters the vagina nearly at a right angle.

Cancer of the cervix is a common form of cancer in women (see chapter opener). Early detection is possible by means of a **Pap test,** which requires the removal of a few

Table 16.2 Female Reproductive Organs

Organ	Function
Ovaries	Produce eggs and sex hormones
Oviducts	Conduct eggs; location of fertilization (uterine or fallopian tubes)
Uterus (womb)	Houses developing fetus
Cervix	Contains opening to uterus
Vagina	Receives penis during sexual intercourse; serves as birth canal and as an exit for menstrual flow

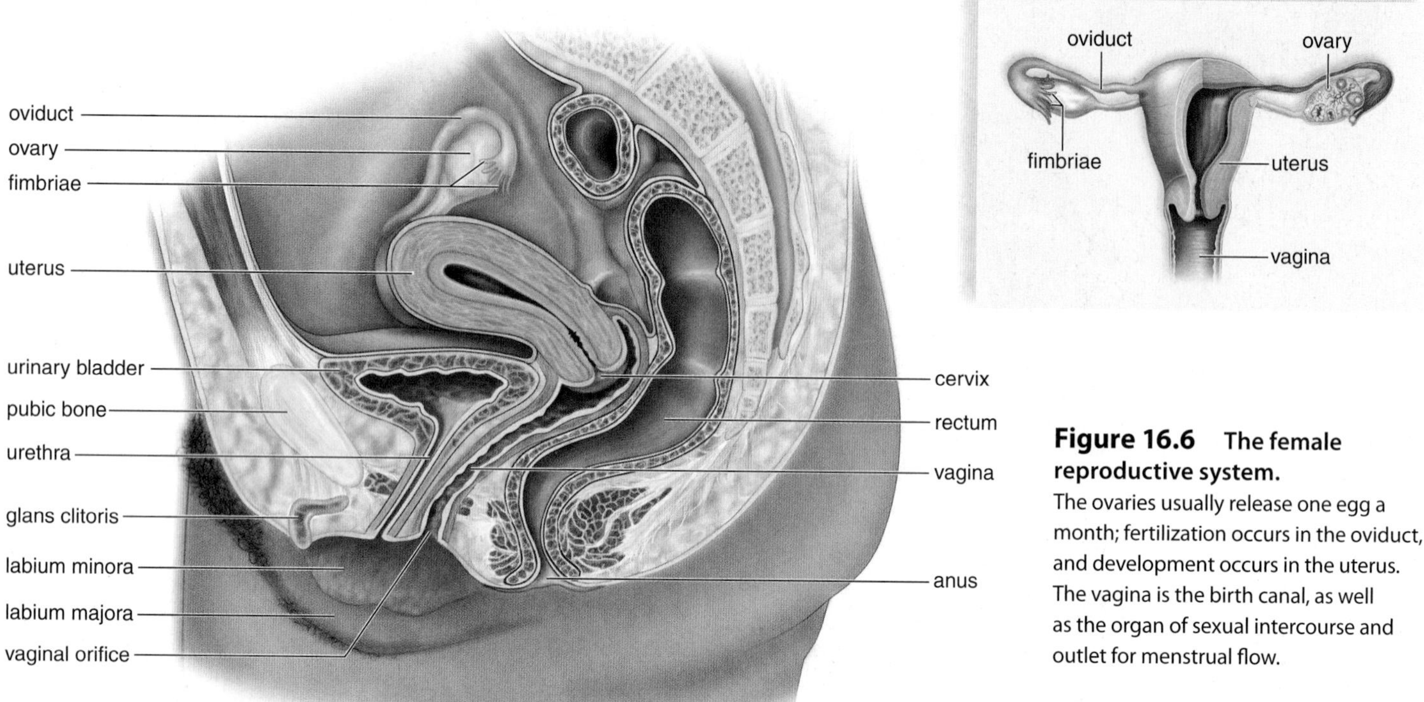

Figure 16.6 **The female reproductive system.** The ovaries usually release one egg a month; fertilization occurs in the oviduct, and development occurs in the uterus. The vagina is the birth canal, as well as the organ of sexual intercourse and outlet for menstrual flow.

cells from the region of the cervix for microscopic examination. If the cells are cancerous, a physician may recommend a hysterectomy. A hysterectomy is the removal of the uterus, including the cervix. Removal of the ovaries in addition to the uterus is technically termed an *ovariohysterectomy* (radical hysterectomy). The vagina remains, so the woman can still engage in sexual intercourse.

Development of the embryo and fetus normally takes place in the uterus. This organ, sometimes called the womb, is approximately 5 cm wide in its usual state. It is capable of stretching to over 30 cm wide to accommodate a growing fetus. The lining of the uterus, called the **endometrium,** participates in the formation of the placenta (see page 378). The endometrium supplies nutrients needed for embryonic and fetal development. The endometrium has two layers: a functional layer that is shed during each menstrual period (see Section 16.4) and a basal layer of reproducing cells. In the nonpregnant female, the functional layer of the endometrium varies in thickness according to a monthly reproductive cycle called the uterine cycle.

A small opening in the cervix leads to the vaginal canal. The **vagina** is a tube that lies at a 45° angle to the small of the back. The mucosal lining of the vagina lies in folds and can extend. This is especially important when the vagina serves as the birth canal, and it facilitates sexual intercourse when the vagina receives the penis. The vagina also acts as an exit for menstrual flow. Several different types of bacteria normally reside in the vagina and create an acidic environment. This environment is protective against the possible growth of pathogenic bacteria, but sperm prefer the basic environment provided by seminal fluid.

External Genitals

The external genital organs of the female are known collectively as the **vulva** (Fig. 16.7). The vulva includes two large, hair-covered folds of skin called the labia majora. The labia majora extend backward from the mons pubis, a fatty prominence underlying the pubic hair. The labia minora are two small folds lying just inside the labia majora. They extend forward from the vaginal opening to encircle and form a foreskin for the glans clitoris. The glans clitoris is the organ of sexual arousal in females and, like the penis, contains a shaft of erectile tissue that becomes engorged with blood during sexual stimulation.

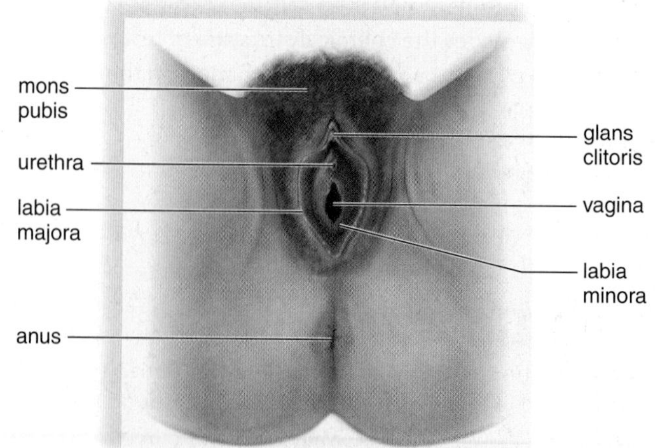

Figure 16.7 **The external genitals of a female.** The external genitals of the female include the labia majora, labia minora, and glans clitoris. These organs are also referred to as the vulva.

Biology Matters Bioethical Focus

Male and Female Circumcision

At birth, a layer of skin (the foreskin) covers the end of a male baby's penis. In the United States, more than 50% of infant males are circumcised shortly after birth. During circumcision, the glans penis is exposed when the foreskin is removed during a surgical procedure. This procedure is done before babies go home from the hospital or during religious ceremonies in the home.

The decision to circumcise a baby boy is made by the parents and is often based on the religious or cultural beliefs of the parents. Circumcision may be done so that male children will resemble their father. Some choose circumcision because of concerns about cleanliness. Claims that circumcision increases or decreases sexual pleasure later in life have not been supported by research.

There is some evidence that suggests urinary tract infections are less common in circumcised infants. Research also shows that circumcision reduces the spread of HIV during heterosexual contact. In areas of the world where HIV infections are prevalent, circumcision may become an important means of limiting the spread of AIDS.

As with any type of surgery, there are risks associated with circumcision. The most common complications are minor bleeding and localized infections that can be treated easily. One of the biggest concerns is about the pain experienced by the baby during circumcision. The American Academy of Pediatrics (AAP) now recommends using a form of local anesthesia during the procedure. The AAP does not recommend nor argue against circumcision of male babies.

However, the circumcision of females is a highly controversial topic. Female circumcision (also referred to as female genital cutting [FGC], or female genital mutilation) is done strictly for cultural or religious reasons, though no religion specifically calls for its practice. The procedure involves partially or totally cutting away the external genitalia of a female. Cultures that practice FGC believe it to be a necessary rite of passage for girls. In the views of these cultures, FGC must be done to preserve the virginity of females and to prevent promiscuity. It is also done for aesthetic reasons, because the clitoris is thought to be an unhealthy and unattractive organ. Moreover, FGC is seen as an essential prerequisite for marriage. Females with an intact clitoris are believed to be unclean. Such women are considered to be potentially harmful to a man during intercourse or to a baby during childbirth if either is touched by the clitoris. Many believe that FGC enhances a husband's sexual pleasure and a woman's fertility.

Many girls die from infection after FGC. FGC also causes life-long urinary and reproductive tract infections, infertility, and pelvic pain. Victims report an absent or greatly diminished pleasurable response to sexual intercourse.

FCG is most commonly performed on girls between the ages of 4 and 12 and in countries in central Africa. It is also performed in some Middle Eastern countries and among Muslim groups in various other locations. With increasing immigration from these countries, there are also greater numbers of women who have been subjected to FGC. Likewise, there are more girls in the United States who are at risk for FGC.

Figure 16A Waris Dirie, Somalian-born supermodel, victim of FGC, advocates against female genital circumcision in her book *Desert Flower.*

Thanks to the efforts of mutilation victim Waris Dirie and others like her, the need to eliminate FGC is now discussed openly, and action is being taken in many countries to outlaw the practice. FGC is considered to be a violation of human rights by the United Nations, UNICEF, and the World Health Organization. It is illegal to perform FGC in many African and Middle Eastern countries, but the practice continues because the laws are not enforced. In the United States, FGC is a criminal practice. In 1996, the United States granted asylum to a woman from Togo, who was trying to escape an arranged marriage and the FGC that would accompany it. Unfortunately, many immigrants to the United States continue the practice of FGC by sending their daughters abroad for the procedure or by importing someone to perform it. A number of educational approaches to eliminate FGC have been tried. These include community education that teaches about the harm done by FGC and substituting alternative rituals for the rite of passage to womanhood. Education may do even more to halt FGC, because more highly educated women are less likely to support having their daughters mutilated in this fashion.

Decide Your Opinion

1. In your view, is male circumcision unjustifiable? Why or why not?
2. Should families who accept the idea of FGC be allowed to immigrate?
3. How should the United States prosecute parents who have subjected their daughters to FGC?

The cleft between the labia minora contains the openings of the urethra and the vagina. The vagina may be partially closed by a ring of tissue called the hymen. The hymen is ordinarily ruptured by sexual intercourse or by other types of physical activities. If remnants of the hymen persist after sexual intercourse, they can be surgically removed.

The urinary and reproductive systems in the female are entirely separate. For example, the urethra carries only urine, and the vagina serves only as the birth canal and the organ for sexual intercourse.

MP3 Female Reproductive Anatomy and Physiology

Orgasm in Females

Upon sexual stimulation, the labia minora, the vaginal wall, and the clitoris become engorged with blood. The breasts also swell, and the nipples become erect. The labia majora enlarge, redden, and spread away from the vaginal opening.

The vagina expands and elongates. Blood vessels in the vaginal wall release small droplets of fluid that seep into the vagina and lubricate it. Mucus-secreting glands beneath the labia minora on either side of the vagina also provide lubrication for entry of the penis into the vagina. Although the vagina is the organ of sexual intercourse in females, the clitoris plays a significant role in the female sexual response. The extremely sensitive clitoris can swell to two or three times its usual size. The thrusting of the penis and the pressure of the pubic symphyses of the partners act to stimulate the clitoris.

Orgasm occurs at the height of the sexual response. Blood pressure and pulse rate rise, breathing quickens, and the walls of the uterus and oviducts contract rhythmically. A sensation of intense pleasure is followed by relaxation when organs return to their normal size. Females have no refractory period, and multiple orgasms can occur during a single sexual experience.

Connecting the Concepts

For more information on the topics presented in this section, refer to the following discussions.

Section 17.1 outlines the steps in the fertilization of an egg by a sperm cell.

Section 17.2 examines the stages of fetal development in the uterus.

Sections 19.1 and 19.2 examine the characteristics of cancer cells and the causes of cancer.

Check Your Progress 16.3

1. Distinguish the structures of the female reproductive system that (a) produce the egg, (b) transport the egg, (c) house a developing embryo, and (d) serve as the birth canal.
2. Explain the purpose of the vagina and uterus in the female reproductive system.
3. Discuss why the urinary and reproductive systems are separate in a female.

16.4 The Ovarian Cycle

Learning Outcomes

Upon completion of this section, you should be able to

1. List the stages of the ovarian cycle and explain what is occurring in each stage.
2. Describe the process of oogenesis.
3. Summarize how estrogen and progesterone influence the ovarian cycle.

Hormone levels cycle in the female on a monthly basis, and the ovarian cycle drives the uterine cycle, as discussed in this section.

Ovarian Cycle: Nonpregnant

An ovary contains many **follicles,** and each one contains an immature egg called an oocyte. A female is born with as many as 2 million follicles, but the number is reduced to 300,000 to 400,000 by the time of puberty. Only a small number of follicles (about 400) ever mature because a female usually produces only one egg per month during her reproductive years. As the follicle matures during the **ovarian cycle,** it changes from a primary to a secondary to a vesicular (Graafian) follicle (Fig. 16.8). Epithelial cells of a primary follicle surround a primary oocyte. Pools of follicular fluid bathe the oocyte in a secondary follicle. In a vesicular follicle, the fluid-filled cavity increases to the point that the follicle wall balloons out on the surface of the ovary.

Animation Maturation of the Follicle and Oocyte

Figure 16.9 traces the steps of **oogenesis.** A primary oocyte undergoes meiosis I, and the resulting cells are haploid with 23 chromosomes each. One of these cells is called a **polar body**. A polar body is a sort of cellular "trash can" because its function is simply to hold discarded chromosomes. The secondary oocyte undergoes meiosis II but only if it is first fertilized by a sperm cell. If the secondary oocyte remains unfertilized, it never completes meiosis and will die shortly after being released from the ovary.

Connections and Misconceptions

Do women make testosterone?

The adrenal glands and ovaries of women make small amounts of testosterone. Women's small testosterone levels may affect libido, or sex drive. The use of supplemental testosterone to restore a woman's libido has not been well researched.

By the way, men make estrogen, too. Androgens are converted to estrogen by an enzyme in the gonads and peripheral tissues. Estrogen may prevent osteoporosis in males.

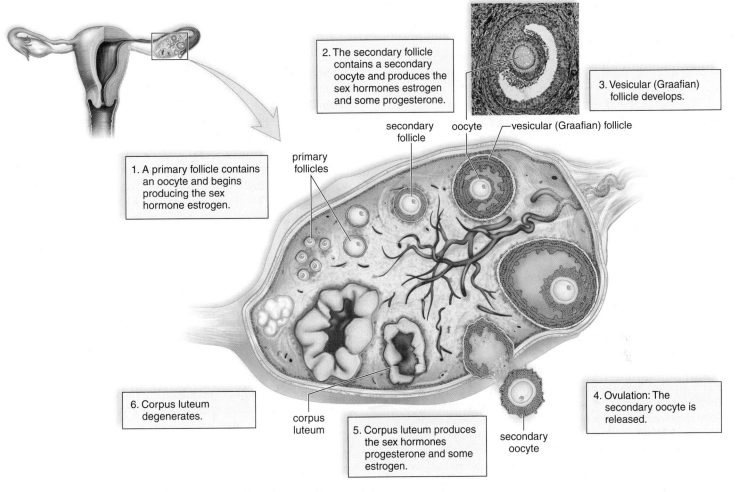

1. A primary follicle contains an oocyte and begins producing the sex hormone estrogen.
2. The secondary follicle contains a secondary oocyte and produces the sex hormones estrogen and some progesterone.
3. Vesicular (Graafian) follicle develops.
4. Ovulation: The secondary oocyte is released.
5. Corpus luteum produces the sex hormones progesterone and some estrogen.
6. Corpus luteum degenerates.

Figure 16.8 The ovarian cycle.
A single follicle goes through all stages (1–6) in one place within the ovary. As the follicle matures, layers of follicle cells surround a secondary oocyte. Eventually, the mature follicle ruptures, and the secondary oocyte is released. The follicle then becomes the corpus luteum, which eventually disintegrates.

When appropriate, the vesicular follicle bursts, releasing the oocyte (often called an egg) surrounded by a clear membrane. This process is referred to as **ovulation.** Once a vesicular follicle has lost the oocyte, it develops into a **corpus luteum,** a glandlike structure. If the egg is not fertilized, the corpus luteum disintegrates.

As mentioned previously, the ovaries produce eggs and also the female sex hormones estrogen and progesterone. A primary follicle produces estrogen, and a secondary follicle produces estrogen and some progesterone. The corpus luteum produces progesterone and some estrogen.

Phases of the Ovarian Cycle

Similar to the testes, the hypothalamus has ultimate control of the ovaries' sexual function because it secretes gonadotropin-releasing hormone, or GnRH. GnRH stimulates the anterior pituitary to produce FSH and LH, and these hormones control the ovarian cycle. The gonadotropic hormones are not present in constant amounts. Instead they are secreted at different rates during the cycle. For simplicity's sake, it is convenient to emphasize that during the first half, or *follicular phase,* FSH promotes the development of

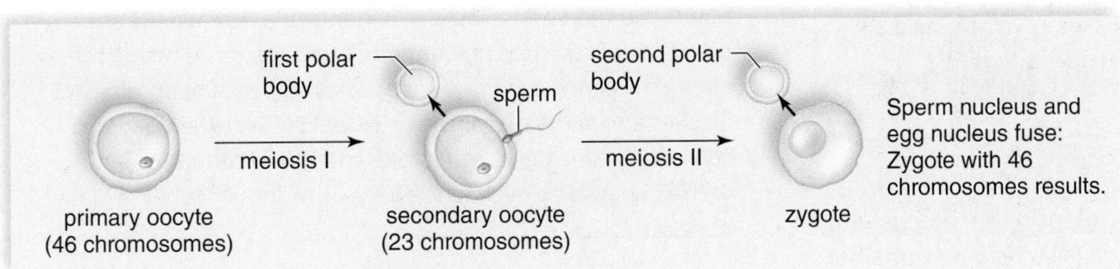

Figure 16.9 Oogenesis produces egg cells.
During oogenesis, the chromosome number is reduced from 46 to 23. Oogenesis produces one functional egg cell and three nonfunctional polar bodies.

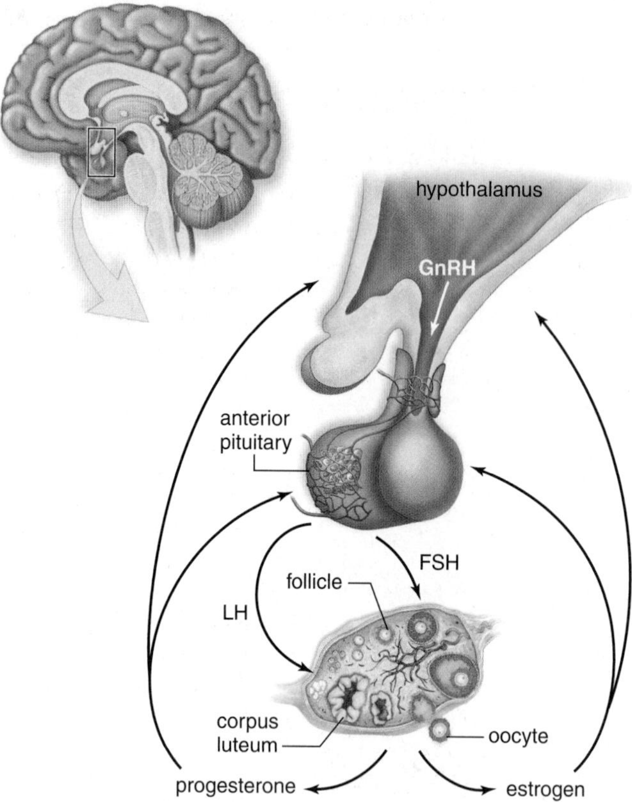

Figure 16.10 The hormones that control the production of estrogen and progesterone by the ovaries.
The hypothalamus produces gonadotropin-releasing hormone (GnRH). GnRH stimulates the anterior pituitary to produce follicle-stimulating hormone (FSH) and luteinizing hormone (LH). FSH stimulates the follicle to produce primarily estrogen, and LH stimulates the corpus luteum to produce primarily progesterone. Estrogen and progesterone maintain the sexual organs (e.g., uterus) and the secondary sex characteristics, and they exert feedback control over the hypothalamus and the anterior pituitary. Feedback control regulates the relative amounts of estrogen and progesterone in the blood.

follicles that primarily secrete estrogen (Fig. 16.10). As the estrogen level in the blood rises, it exerts negative feedback control over the anterior pituitary secretion of FSH. The follicular phase now comes to an end.

The estrogen spike at the end of the follicular phase has a positive feedback effect on the hypothalamus and pituitary gland. As a result, GnRH from the hypothalamus increases. There is a corresponding surge of LH released from the anterior pituitary. The LH surge triggers ovulation at about day 14 of a 28-day cycle.

Next, the *luteal phase* begins. During the luteal phase of the ovarian cycle, LH promotes the development of the corpus luteum. The corpus luteum secretes high levels of progesterone and some estrogen. When pregnancy does not occur, the corpus luteum regresses and a new cycle begins with menstruation (Fig. 16.11).

Animation Female Reproductive System

Estrogen and Progesterone

Estrogen and progesterone affect not only the uterus but other parts of the body as well. Estrogen is largely responsible for the secondary sex characteristics in females, including body hair and fat distribution. In general, females have a more rounded appearance than males because of a greater accumulation of fat beneath the skin. Like males, females develop axillary and pubic hair during puberty. In females, the upper border of pubic hair is horizontal, but in males, it tapers toward the navel. Both estrogen and progesterone are also required for breast development. Other hormones are involved in milk production (prolactin) following pregnancy and milk letdown (oxytocin) when a baby begins to nurse.

The pelvic girdle is wider and deeper in females, so the pelvic cavity usually has a larger relative size compared with that of males. This means that females have wider hips than males and their thighs converge at a greater angle toward the knees. The female pelvis tilts forward, so females tend to have more of a lower back curve than males, an abdominal bulge, and protruding buttocks.

Menopause, the period in a woman's life during which the ovarian cycle ceases, is likely to occur between ages 45 and 55. The ovaries are no longer responsive to the gonadotropic hormones produced by the anterior pituitary, and the ovaries no longer secrete estrogen or progesterone. At the onset of menopause, menstruation becomes irregular, but as long as it occurs, it is still possible for a woman to conceive. Therefore, a woman is usually not considered to have completed menopause until menstruation is absent for a year.

Uterine Cycle: Nonpregnant

The female sex hormones, **estrogen** and **progesterone,** have numerous functions. One function of these hormones affects the endometrium, causing the uterus to undergo a cyclical series of events known as the **uterine cycle** (see Fig. 16.11). Twenty-eight-day cycles are divided as follows:

During *days 1–5,* a low level of estrogen and progesterone in the body causes the endometrium to disintegrate and its blood vessels to rupture. On day 1 of the cycle, a

Connections and Misconceptions

What is hormone replacement therapy?

Hormone replacement therapy (HRT) is used to put hormones back into the body that are not made during and after menopause. The fluctuation of hormone levels during menopause can cause symptoms such as hot flashes, mood swings, trouble sleeping, increased abdominal fat, and thinning hair; and HRT can alleviate these symptoms. The use of HRT has its pros and cons. There is evidence that use of HRT after menopause can help prevent bone loss, decrease the risk of colorectal cancer, and decrease certain types of heart disease. Studies also show that some types of HRT in certain patients can increase incidences of stroke and blood clots. Women on HRT should be evaluated every six months by their physician.

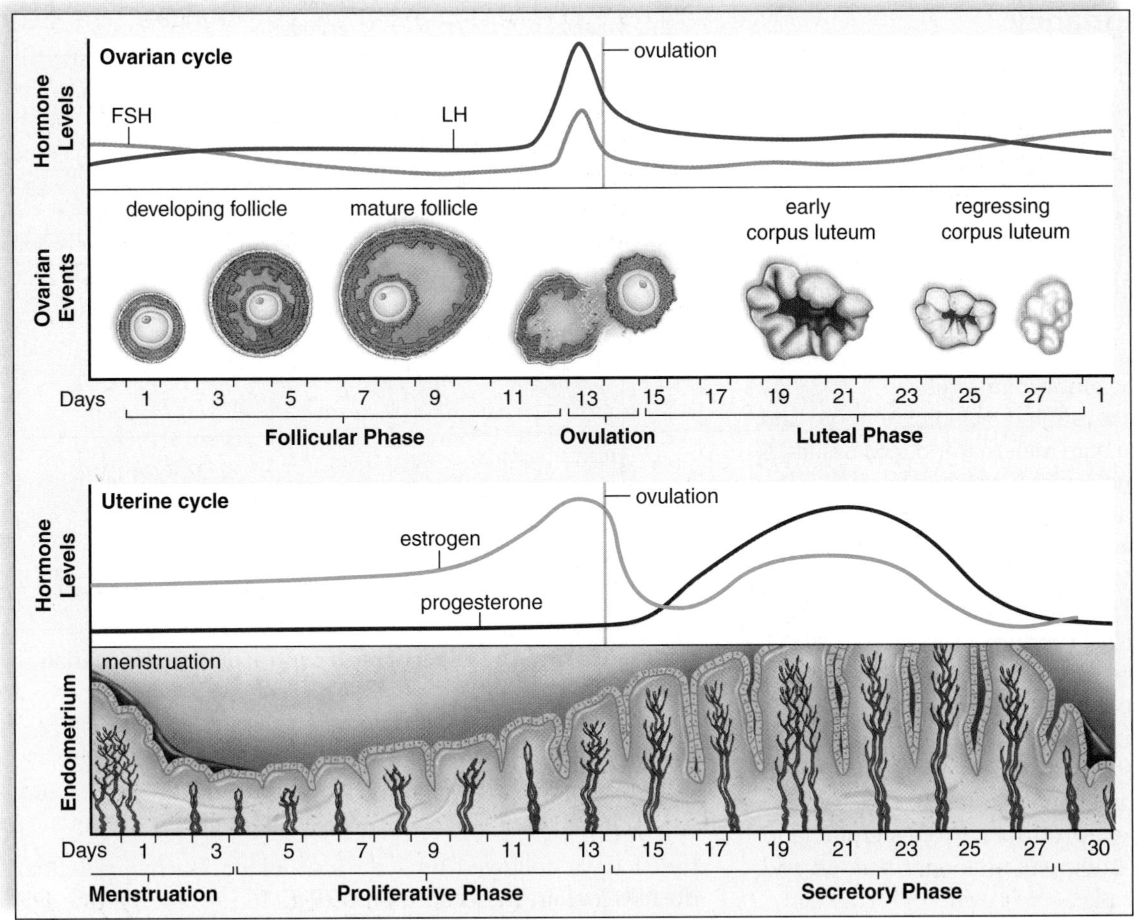

Figure 16.11 The effects of estrogen and progesterone on the endometrium during the uterine cycle. During the follicular phase, FSH released by the anterior pituitary promotes the maturation of a follicle in the ovary. The ovarian follicle produces increasing levels of estrogen, which causes the endometrium to thicken during the proliferative phase of the uterine cycle. After ovulation and during the luteal phase of the ovarian cycle, LH promotes the development of the corpus luteum. Progesterone in particular causes the endometrial lining to become secretory. Menses, due to the breakdown of the endometrium, begins when progesterone production declines to a low level.

flow of blood and tissues, known as the menses, passes out of the vagina during **menstruation,** also called the menstrual period.

During *days 6–13,* increased production of estrogen by a new ovarian follicle in the ovary causes the endometrium to thicken and become vascular and glandular. This is called the proliferative phase of the uterine cycle.

On *day 14* of a 28-day cycle, ovulation usually occurs.

During *days 15–28,* increased production of progesterone by the corpus luteum in the ovary causes the endometrium of the uterus to double or triple in thickness (from 1 mm to 2–3 mm). The uterine glands mature and produce a thick mucoid secretion in response to increased progesterone. This is called the secretory phase of the uterine cycle. The endometrium is now prepared to receive the developing embryo. If this does not occur, the corpus luteum in the ovary regresses. The low level of progesterone in the female body results in the endometrium breaking down during menstruation.

Table 16.3 compares the stages of the uterine cycle with those of the ovarian cycle when pregnancy does not occur.

Table 16.3 Ovarian and Uterine Cycles: Nonpregnant

Ovarian Cycle	Events	Uterine Cycle	Events
Follicular phase—days 1–13	FSH secretion begins.	Menstruation—days 1–5	Endometrium breaks down.
	Follicle maturation occurs.	Proliferative phase—days 6–13	Endometrium rebuilds.
	Estrogen secretion is prominent.		
Ovulation—day 14*	LH spike occurs.		
Luteal phase—days 15–28	LH secretion continues.	Secretory phase—days 15–28	Endometrium thickens, and glands are secretory.
	Corpus luteum forms.		
	Progesterone secretion is prominent.		

*Assuming a 28-day cycle.

Fertilization and Pregnancy

Following unprotected sexual intercourse, many sperm make their way into the oviduct, where the egg is located following ovulation. Only one sperm is needed to fertilize the egg, which is then called a zygote. Development begins even as the zygote travels down the oviduct to the uterus. The endometrium is now prepared to receive the developing embryo. The embryo implants in the endometrial lining several days following fertilization. Implantation signals the beginning of a pregnancy. An abortion may be spontaneous (referred to as a miscarriage) or induced. Each type of abortion ends with the loss of the embryo or fetus.

The **placenta,** which sustains the developing embryo and later the fetus, originates from both maternal and fetal tissues. It is the region of exchange of molecules between fetal and maternal blood, although the two rarely mix. At first, the placenta produces **human chorionic gonadotropin (hCG)**, which maintains the corpus luteum in the ovary. (A pregnancy test detects the presence of hCG in the blood or urine.) Rising amounts of hCG stimulate the corpus luteum to produce increasing amounts of progesterone. This progesterone shuts down the hypothalamus and anterior pituitary so that no new follicles begin in the ovary. The progesterone maintains the uterine lining where the embryo now resides. The absence of menstruation is a signal to the woman that she may be pregnant (Fig. 16.12).

Eventually, the placenta produces progesterone and some estrogen. The corpus luteum is no longer needed and it regresses.

Many women use birth control pills to prevent pregnancy (see Section 16.5). The most commonly used pills include active pills, containing a synthetic estrogen and progesterone, taken for 21 days, followed by 7 days of taking inactive pills that do not contain these hormones (Fig. 16.13). The uterine lining builds up to some degree while the active pills are being taken. Progesterone decreases when the last of the active pills are taken, causing a minimenstruation to occur. Some women

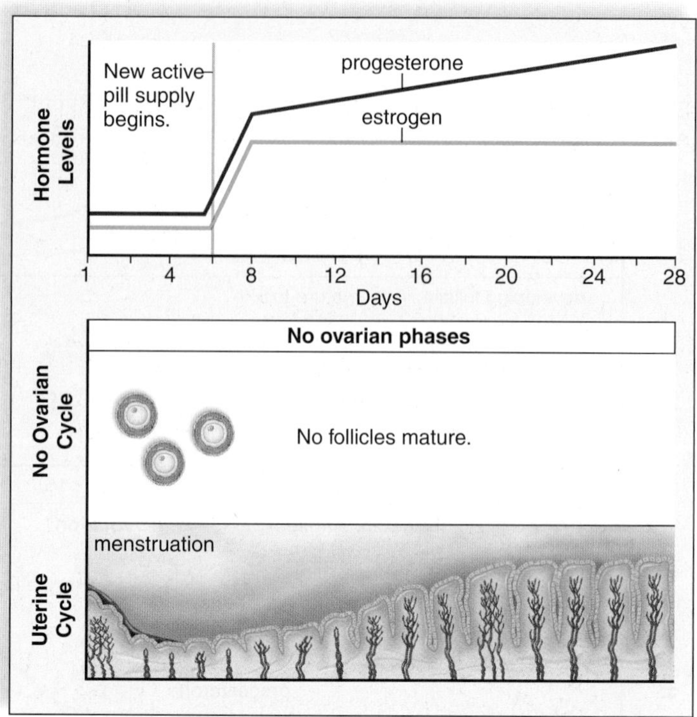

Figure 16.13 **The effect of birth control pills on the ovarian cycle.**
Active pills cause the uterine lining to build up, and this lining is shed when inactive pills are taken. Feedback inhibition of the hypothalamus and anterior pituitary means that the ovarian cycle does not occur.

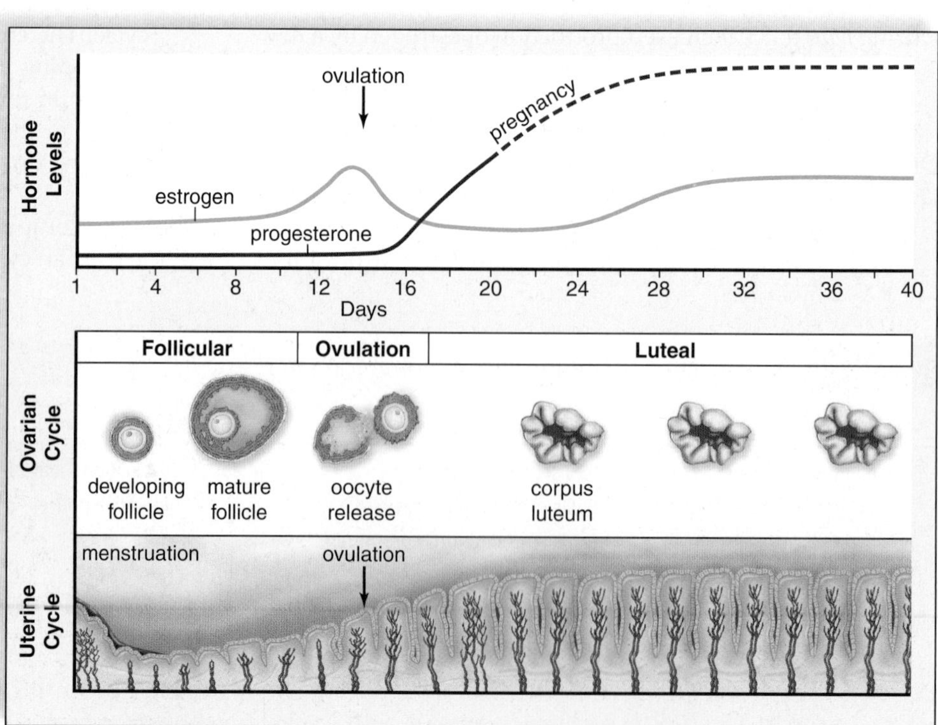

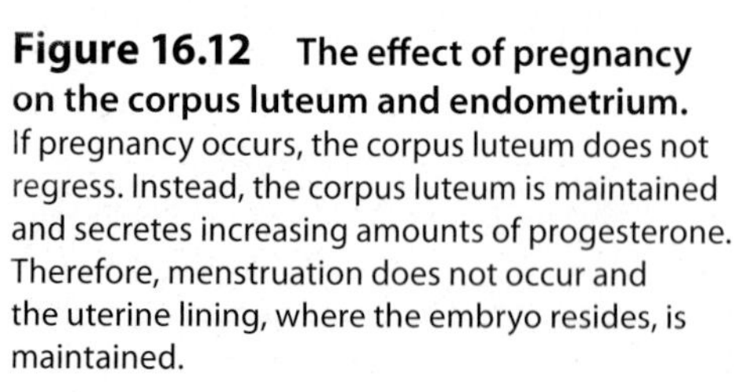

Figure 16.12 **The effect of pregnancy on the corpus luteum and endometrium.**
If pregnancy occurs, the corpus luteum does not regress. Instead, the corpus luteum is maintained and secretes increasing amounts of progesterone. Therefore, menstruation does not occur and the uterine lining, where the embryo resides, is maintained.

skip taking the inactive pills and start taking a new pack of active pills right away to skip menstruation (a period). A new form of birth control pills consists of three months of active pills. Women taking them have only four menstrual periods a year.

Connecting the Concepts

For more information on the topics presented in this section, refer to the following discussions.

Section 15.6 provides additional information on the male sex hormones.

Section 17.1 outlines the steps in the fertilization of an egg by a sperm cell.

Section 18.3 examines how meiosis reduces the chromosome number during oogenesis.

Check Your Progress 16.4

1. Summarize the roles of estrogen and progesterone in the ovarian and uterine cycles.
2. Describe the changes that occur in the ovarian and uterine cycles during pregnancy.
3. Describe the changes that occur in the ovarian and uterine cycles when birth control pills are used.

16.5 Control of Reproduction

Learning Outcomes

Upon completion of this section, you should be able to

1. List the forms of birth control and summarize how each reduces the chances of fertilization of an egg by a sperm cell.
2. Explain the causes of infertility.
3. Describe how the use of assisted reproductive technologies can increase the chances of conceiving a child.

Several means are available to reduce or enhance our reproductive potential. **Birth control methods** are used to regulate the number of children an individual or couple has. For individuals who are experiencing infertility, or an inability to achieve pregnancy, a number of assisted reproductive technologies may be used to increase the chances of conceiving a child.

Birth Control Methods

The most reliable method of birth control is abstinence—not engaging in sexual intercourse. This form of birth control has the added advantage of preventing transmission of sexually transmitted diseases. Table 16.4 lists other means of birth control used in the United States and rates their effectiveness. For example, with the birth control pill, we expect 98% effectiveness. Only 2% of sexually active women will get pregnant within the year. On the other hand, with the withdrawal method, we expect that 75% of women will not get pregnant. That makes the withdrawal method one of the least effective methods of contraception.

Figure 16.14 features some of the most effective and commonly used means of birth control. **Contraceptives** are medications and devices that reduce the chance of pregnancy. Oral contraception—**birth control pills**—contains a combination of estrogen and progesterone for the first 21 days, followed by seven inactive pills (Fig. 16.14*a*). The estrogen and progesterone in the birth control pill or a patch (Fig. 16.14*b*) applied to the skin effectively shuts down the pituitary production of both FSH and LH. Follicle development in the ovary is prevented. Because ovulation does not occur, pregnancy cannot take place. Women taking birth control pills or using a patch should see a physician regularly because of possible side effects.

An **intrauterine device (IUD)** is a small piece of molded plastic inserted into the uterus by a physician (Fig. 16.14*c*). IUDs are believed to alter the environment of the uterus and oviducts so that fertilization probably will not occur. If fertilization should occur, implantation cannot take place.

The **diaphragm** is a soft latex cup with a flexible rim that lodges behind the pubic bone and fits over the cervix (Fig. 16.14*d*). Each woman must be properly fitted by a physician, and the diaphragm can be inserted into the vagina no more than 2 hours before sexual relations. Also, it must be used with spermicidal jelly or cream and should be left in place at least 6 hours after sexual relations. The cervical cap is a minidiaphragm.

There has been a renewal of interest in barrier methods of birth control, because these methods offer some protection against sexually transmitted diseases. A **female condom,** now available, consists of a large polyurethane tube with a flexible ring that fits onto the cervix (Fig. 16.14*e*). The open end of the tube has a ring that covers the external genitals. A **male condom** is most often a latex sheath that fits over the erect penis (Fig. 16.14*f*). The ejaculate is trapped inside the sheath and thus does not enter the vagina. When used in conjunction with a spermicide, the protection is better than with the condom alone.

Contraceptive implants use a synthetic progesterone to prevent ovulation by disrupting the ovarian cycle. The older version of the implant consists of six match-sized, time-release capsules surgically implanted under the skin of a woman's upper arm. The newest version consists of a single capsule that remains effective for about three years (Fig. 16.14*g*).

Contraceptive injections are available as progesterone only (Fig. 16.14*h*) or a combination of estrogen and progesterone. The length of time between injections can vary from one to several months.

Contraceptive vaccines are now being developed. For example, a vaccine intended to immunize women against hCG, the hormone so necessary to maintaining the implantation of the embryo, was successful in a limited clinical trial. Because hCG is not normally present in the body, no autoimmune reaction is expected but the immunization does wear off with time. Others believe that it would also be possible to develop a safe antisperm vaccine that could be used in women.

Table 16.4 **Common Methods of Contraception**

Name	Procedure	How Does It Work?	Effectiveness?	Health Risk
Abstinence	Refrain from sexual intercourse	No sperm in vagina	100%	None; also protects against STDs
Natural family planning	Determine day of ovulation by keeping records	Intercourse avoided during the time that ovum is viable	80%	None
Withdrawal method	Penis withdrawn from vagina just before ejaculation	Ejaculation outside the woman's body; no sperm in vagina	75%	None
Douching	Vagina cleansed after intercourse	Washes sperm out of vagina	≥ 70%	May cause inflammation
Male condom	Sheath of latex, polyurethane, or natural material fitted over erect penis	Prevents entry of sperm into vagina; latex and polyurethane forms protect against STDs	89%	Latex allergy with latex forms; no protection against STDs with natural-material condoms
Female condom	Polyurethane liner fitted inside vagina	Prevents entry of sperm into vagina; some protection against STDs	79%	Possible allergy or irritation, urinary tract infection
Spermicide: jellies, foams, creams	Spermicidal products inserted into vagina before intercourse	Spermicide nonoxynol-9 kills large numbers of sperm cells.	50–80%	Irritation, allergic reaction, urinary tract infection
Contraceptive sponges	Sponge containing spermicide inserted into vagina and placed against cervix	Spermicide nonoxynol-9 kills large numbers of sperm cells.	72–86%	Irritation, allergic reaction, urinary tract infection, toxic shock syndrome
Combined hormone vaginal ring	Flexible plastic ring inserted into vagina; releases hormones absorbed into the bloodstream	Combined hormonal methods suppress ovulation by the combined actions of the hormones estrogen and progestin.	98%	Combined hormonal methods can cause dizziness; nausea; changes in menstruation, mood, and weight; rarely, cardiovascular disease, including high blood pressure, blood clots, heart attack, and strokes.
Combined hormone pill	Pills are swallowed daily; chewable form also available		98%	
Combined hormone 91-day regimen	Pills are swallowed daily; user has three to four menstrual periods a year		98%	
Combined hormone injection (Lunelle)	Injection of long-acting hormone given once a month		99%	
Combined hormone patch	Patch is applied to skin and left in place for 1 week; new patch applied		98%	
Progestin-only minipill	Pills are swallowed daily.	Thickens cervical mucus, preventing sperm from contacting egg	98%	Irregular bleeding, weight gain, breast tenderness
Progesterone-only injection (Depo-Provera)	Injection of progestin once every three months	Inhibits ovulation; prevents sperm from reaching the egg; prevents implantation	99%	Irregular bleeding, weight gain, breast tenderness, osteoporosis possible
Emergency contraception	Must be taken within 72 hours after unprotected intercourse	Suppresses ovulation by the combined actions of the hormones estrogen and progestin; prevents implantation	80%	Nausea, vomiting, abdominal pain, fatigue, headache
Diaphragm	Latex cup, placed into vagina to cover cervix before intercourse	Blocks entrance of sperm into uterus, spermicide kills sperm	90% with spermicide	Irritation, allergic reaction, urinary tract inflection, toxic shock syndrome
Cervical cap	Latex cap held over cervix	Blocks entrance of sperm into uterus, spermicide kills sperm	90% with spermicide	Irritation, allergic reaction, toxic shock syndrome, abnormal Pap smear
Cervical shield	Latex cap in upper vagina, held in place by suction	Blocks entrance of sperm into uterus, spermicide kills sperm	90% with spermicide	Irritation, allergic reaction, urinary tract infection, toxic shock syndrome
Intrauterine device Copper T	Placed in uterus	Causes cervical mucus to thicken; fertilized embryo cannot implant	99%	Cramps, bleeding, infertility, perforation of uterus
Intrauterine device progesterone-releasing type	Placed in uterus	Prevents ovulation; causes cervical mucus to thicken; fertilized embryo cannot implant	99%	Cramps, bleeding, infertility, perforation of uterus

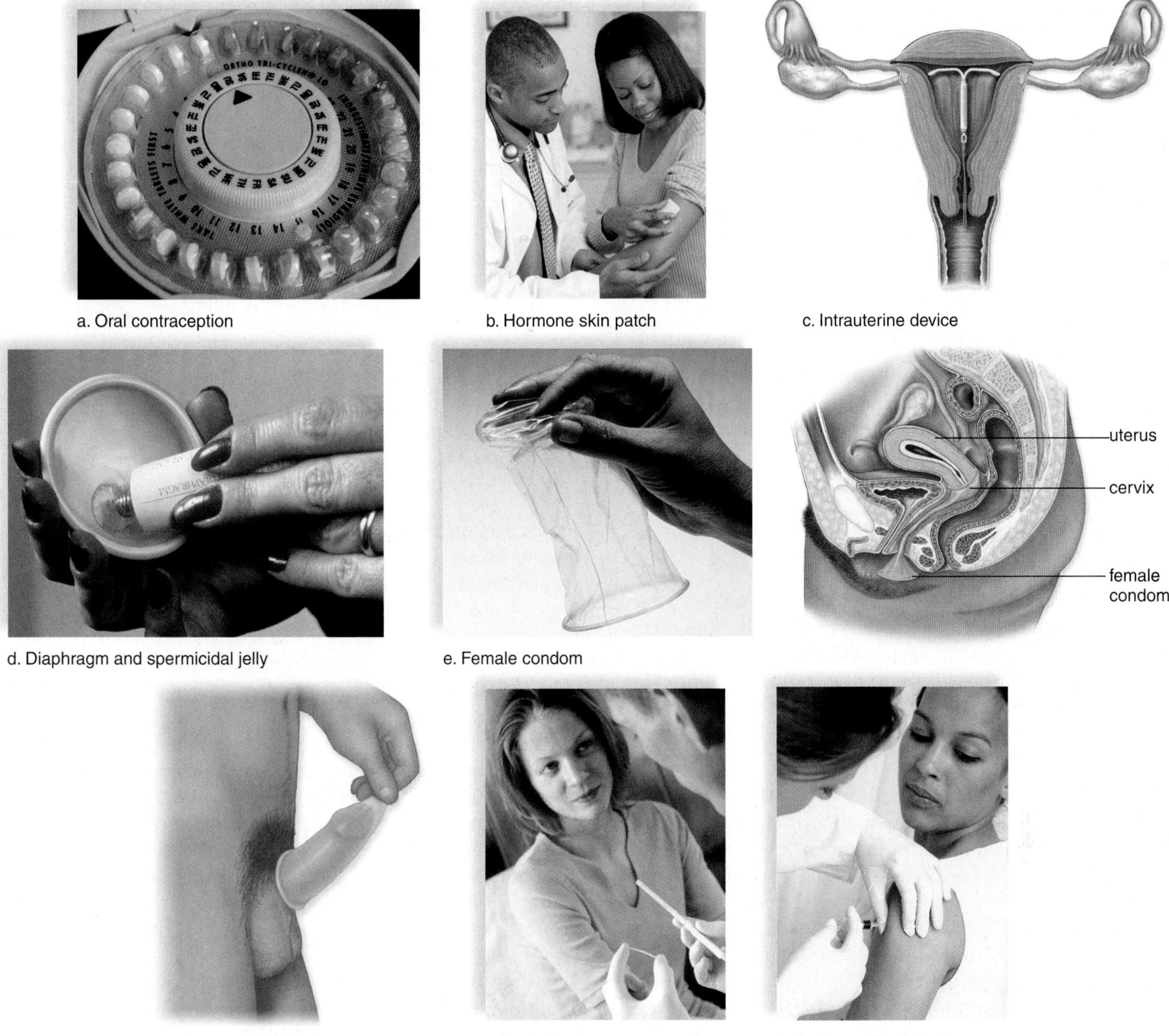

Figure 16.14 Some of the different forms of birth control that are currently available.
a. Oral contraception (birth control pills). **b.** Hormone skin patch. **c.** Intrauterine device. **d.** Diaphragm and spermicidal jelly. **e.** Female condom. **f.** Male condom. **g.** Implant. **h.** Contraceptive injection.

Vasectomy and Tubal Ligation

Vasectomy and tubal ligation are two methods used to bring about sterility, the inability to reproduce (Fig. 16.15). **Vasectomy** consists of cutting and sealing the vas deferens from each testis so that the sperm are unable to reach the seminal fluid ejected at the time of orgasm. The sperm are then largely reabsorbed. Following this operation, which can be done in a doctor's office, the amount of ejaculate remains normal because sperm account for only about 1% of the volume of semen. Also, there is no effect on the secondary sex characteristics because testosterone continues to be produced by the testes.

Tubal ligation consists of cutting and sealing the oviducts. Pregnancy rarely occurs because the passage of the egg through the oviducts has been blocked. Using a method called laparoscopy, which requires only two small incisions, the surgeon inserts a small, lighted telescope to view the oviducts and a small surgical blade to sever them.

It is best to view a vasectomy or tubal ligation as permanent. Even following successful reconnection, fertility is usually reduced by about 50%.

Connections and Misconceptions

What are emergency contraceptive pills?

Emergency contraceptive pills, or morning-after pills as they are sometimes called, are designed to prevent pregnancy after having unprotected sex. These pills contain levonorgestrel, a synthetic progestin that both inhibits the release of an unfertilized egg and stops the implantation of a fertilized egg. The chemical works by thinning the lining of the endometrium and slowing the movement of the fertilized egg through the oviducts. It is most effective if taken within the first 48 hours after unprotected sexual intercourse but may be taken within the next five days. The chemical does not protect against sexually transmitted diseases and will not abort an existing pregnancy. Emergency contraceptive pills are available without a prescription to women over 17 years of age and with a prescription for women under 17.

Infertility

Infertility is the failure of a couple to achieve pregnancy after one year of regular, unprotected intercourse. The American Medical Association (AMA) estimates that 15% of all couples are infertile. The cause of infertility can be attributed to the male (40%), the female (40%), or both (20%).

Causes of Infertility

The most frequent cause of infertility in males is low sperm count and/or a large proportion of abnormal sperm, which can be due to environmental influences. It appears that a sedentary lifestyle coupled with smoking and alcohol consumption is most often the cause of male infertility. When males spend most of the day sitting in front of a computer or the TV or driving, the testes temperature remains too high for adequate sperm production.

Body weight appears to be the most significant factor in causing female infertility. In women of normal weight, fat cells produce a hormone called leptin that stimulates the hypothalamus to release GnRH. FSH release and normal follicle development follow. In overweight women, leptin levels are higher, which impacts GnRH and FSH. The ovaries of overweight women often contain many small follicles that fail to ovulate. Other causes of infertility in females are blocked oviducts due to pelvic inflammatory disease (see page 387) and endometriosis. Endometriosis is the presence of uterine tissue outside the uterus, particularly in the oviducts and on the abdominal organs. Backward flow of menstrual fluid allows living uterine cells to establish themselves in the abdominal cavity. The cells go through the usual uterine cycle, causing pain and structural abnormalities that make it more difficult for a woman to conceive.

Sometimes the causes of infertility can be corrected by medical intervention so that couples can have children. If no obstruction is apparent and body weight is normal, it is possible to give females fertility drugs. These drugs are gonadotropic hormones that stimulate the ovaries and bring about ovulation. Such hormone treatments may cause multiple ovulations and multiple births.

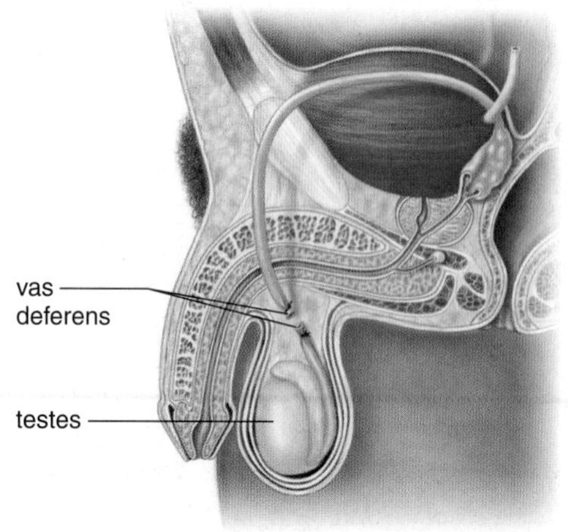

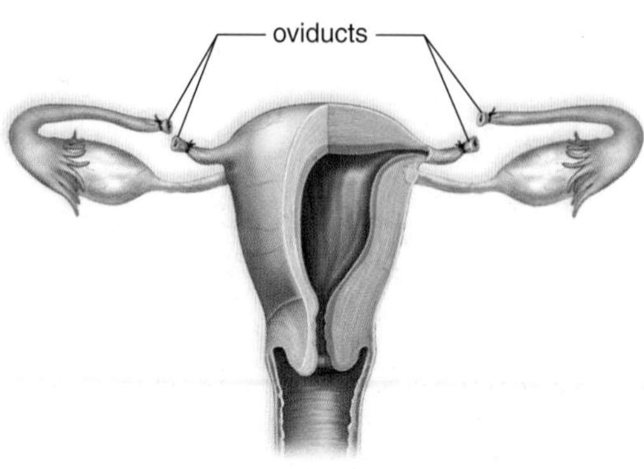

Figure 16.15 Vasectomies and tubal ligations.
a. Vasectomy involves making two small cuts in the skin of the scrotum. Each vas deferens is lifted out and cut. The cut ends are tied or sealed with an electrical current. The openings in the scrotum are closed with stitches. **b.** During tubal ligation, one or two small incisions are made in the abdomen. Using instruments inserted through the incisions, the oviducts are coagulated (burned), sealed shut with cautery, or cut and tied. The skin incision is then stitched closed.

Biology Matters Bioethical Focus

Should Infertility Be Treated?

Every day, couples make plans to start or expand their families. Yet, for many, their dreams might not be realized because conception is difficult or impossible. Before seeking medical treatment for infertility, a couple might want to decide how far they are willing to go to have a child. Here are some of the possible risks.

Some of the Procedures Used

If a man has low sperm count or motility, artificial or intrauterine insemination of his partner with a large number of specially selected sperm may be done to stimulate pregnancy. Although there are dangers in all medical procedures, artificial insemination is generally safe.

If a woman is infertile because of physical abnormalities in her reproductive system, she may be treated surgically. Whereas surgeries are now very sophisticated, they nonetheless have risks, including bleeding, infection, organ damage, and adverse reactions to anesthesia. Similar risks are associated with collecting eggs for in vitro fertilization (IVF). To ensure the collection of several eggs, a woman may be placed on hormone-based medications that stimulate egg production. Such medications may cause ovarian hyperstimulation syndrome—enlarged ovaries and abdominal fluid accumulation. In mild cases, the only symptom is discomfort; but in severe cases (though rare), a woman's life may be endangered. In any case, the fluid has to be drained.

Usually, IVF involves the creation of many embryos; the healthiest-looking ones are transferred into the woman's body. Others may be frozen for future attempts at establishing pregnancy, given to other infertile couples, donated for research, or destroyed. Of those that are transferred, none, one, or all might develop into fetuses. The significant increase of multifetal pregnancies in the United States in the last 15 years has been largely attributed to fertility treatment (Fig. 16B). Though the number of triplet and higher-number multiple pregnancies started to level off in 1999, twin pregnancies continue to climb.

They may seem like a dream come true, but multifetal pregnancies are difficult. The mother is more likely to develop complications such as gestational diabetes and high blood pressure than are women carrying single babies. Positioning of the babies in the uterus may make vaginal delivery less likely, and there is likely a chance of preterm labor. Babies born prematurely face numerous hardships. Infant death and long-term disabilities are also more common with multiple births. This is true even of twins. Even if all babies are healthy, parenting "multiples" poses unique challenges.

Figure 16B Reproductive technologies may lead to mutliple births.

What Happens to Frozen Embryos?

Despite potential trials, thousands of people undergo fertility treatment every year. Its popularity has brought a number of ethical issues to light. For example, the estimated high numbers of stored frozen embryos (a few hundred thousand in the United States) has generated debate about their fate, complicated by the fact that the long-term viability of frozen embryos is not well understood. Scientists may worry that embryos donated to other couples are ones screened out from one implantation and not likely to survive. Some religious groups strongly oppose destruction of these embryos or their use in research. Patients for whom the embryos were created generally feel that they should have sole rights to make decisions about their fate. However, a fertility clinic may no longer be receiving monetary compensation for the storage of frozen embryos and may be unable to contact the couples for whom they were produced. The question then becomes whether the clinic now has the right to determine their fate.

Who Should Be Treated?

Additionally, because fertility treatment is voluntary, is it ever acceptable to turn some people away? What if the prospect of a satisfactory outcome is very slim or almost nonexistent? This may happen when one of the partners is ill or the woman is at an advanced age. Should a physician go ahead with treatment even if it might endanger a woman's (or baby's) health? Those in favor of limiting treatment argue that a physician has a responsibility to prevent potential harm to a patient. On the other hand, there is concern that if certain people are denied fertility for medical reasons, might they be denied for other reasons also, such as race, religion, sexuality, or income?

Decide Your Opinion

1. Should couples go to all lengths to have children even if it could endanger the life of one or both spouses?
2. Should couples with "multiples" due to infertility treatment receive assistance from private and public services?
3. To what lengths should society go to protect frozen embryos?
4. Do you think that anyone should be denied fertility treatment? If so, what factors do you think a doctor should take into consideration when deciding whether to provide someone fertility treatment?

When reproduction does not occur in the usual manner, many couples adopt a child. Others sometimes try one of the assisted reproductive technologies discussed in the following paragraphs.

Assisted Reproductive Technologies

Assisted reproductive technologies (ART) consist of techniques used to increase the chances of pregnancy. Often, sperm and/or eggs are retrieved from the testes and ovaries, and fertilization takes place in a clinical or laboratory setting.

Artificial Insemination by Donor (AID) During artificial insemination, sperm are placed in the vagina by a physician. Sometimes a woman is artificially inseminated by her partner's sperm. This is especially helpful if the partner has a low sperm count, because the sperm can be collected over time and concentrated so that the sperm count is sufficient to result in fertilization. Often, however, a woman is inseminated by sperm acquired from a donor who is a complete stranger to her. At times, a combination of partner and donor sperm is used.

A variation of AID is *intrauterine insemination* (*IUI*). In IUI, fertility drugs are given to stimulate the ovaries. Then the donor's sperm are placed in the uterus rather than in the vagina.

If the prospective parents wish, sperm can be sorted into those believed to be X-bearing or Y-bearing to increase the chances of having a child of the desired sex. Fertilization of an egg with an X-bearing sperm results in a female child. Fertilization by a Y-bearing sperm yields a male child.

In Vitro Fertilization (IVF) During IVF, conception occurs in laboratory glassware. Ultrasound machines can now spot follicles in the ovaries that hold immature eggs; therefore, the latest method is to forgo the administration of fertility drugs and retrieve immature eggs by using a needle. The immature eggs are then brought to maturity in glassware, and then concentrated sperm are added (Fig. 16.16). After about two to four days, the embryos are ready to be transferred to the uterus of the woman, who is now in the secretory phase of her uterine cycle. If desired, the embryos can be tested for a genetic disease, and only those found to be free of disease will be used. If implantation is successful, development is normal and continues to term.

Gamete Intrafallopian Transfer (GIFT) The term **gamete** refers to a sex cell, either a sperm or an egg. Gamete intrafallopian transfer (GIFT) was devised to overcome the low success rate (15–20%) of in vitro fertilization. The method is exactly the same as in vitro fertilization, except the eggs and the sperm are placed in the oviducts immediately after they have been brought together. GIFT has the advantage of being a one-step procedure for the woman—the eggs are removed and reintroduced all at the same time. A variation on this procedure is to fertilize the eggs in the laboratory and then place the zygotes in the oviducts.

Connections and Misconceptions

How many babies are born annually in the United States using ART?

The very first IVF baby born in the United States was Elizabeth Carr on December 28, 1981. Since that time, assisted reproductive technologies have improved along with their success rates. In 2009, more than 4 million babies were born in the United States with more than 1% of that total conceived from assisted reproductive technologies.

Surrogate Mothers In some instances, women are contracted and paid to have babies. These women are called surrogate mothers. The sperm and even the egg can be contributed by the contracting parents.

Intracytoplasmic Sperm Injection (ICSI) In this highly sophisticated procedure, a single sperm is injected into an egg. It is used effectively when a man has severe infertility problems.

Connecting the Concepts

For more information on the hormones presented in this section, refer to the following discussions.

Section 15.2 explains the role of the gonadotropic hormones.

Section 15.6 describes the hormones produced by the female reproductive system.

Check Your Progress 16.5

1. List the major forms of birth control in order of effectiveness.
2. Explain why vasectomies and tubal ligations represent a permanent form of birth control.
3. Distinguish between an IVF and an ICSI procedure to compensate for infertility.

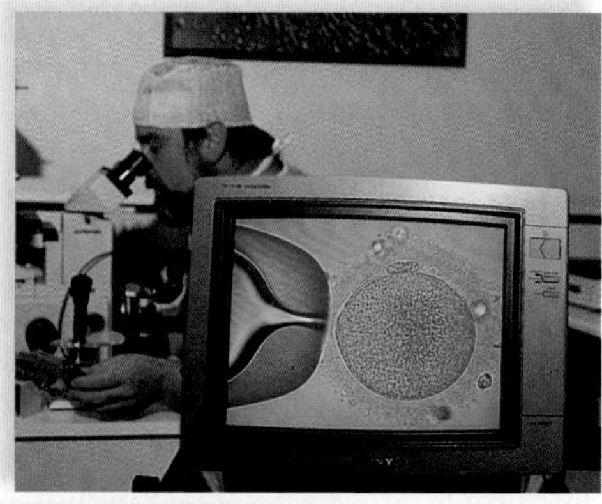

Figure 16.16 Intracytoplasmic sperm injection. A microscope connected to a television screen is used to carry out intracytoplasmic sperm injection. A pipette holds the egg steady while a needle (not visible) introduces the sperm into the egg.

16.6 Sexually Transmitted Diseases

Learning Outcomes

Upon completion of this section, you should be able to

1. Distinguish between sexually transmitted diseases (STDs) caused by viruses and those caused by bacteria.
2. Describe the causes and treatments of selected STDs.

Sexually transmitted diseases (STDs) are caused by viruses, bacteria, fungi, and parasites.

STDs Caused by Viruses

Among those STDs caused by viruses, effective treatment is available for AIDS (acquired immunodeficiency syndrome) and genital herpes. However, treatment for HIV/AIDS and genital herpes cannot presently eliminate the virus from the person's body. Drugs used for treatment can merely slow replication of the viruses. Thus, neither viral disease is presently curable. Further, antiviral drugs have serious, debilitating side effects on the body.

HIV Infections

The Infectious Diseases Supplement, which begins on page 155, discusses HIV infections at greater length than this brief summary. At present, there is no vaccine to prevent an HIV infection nor is there a cure for AIDS. The best course of action is to follow the guidelines for preventing transmission of STDs outlined in the Health Focus, *Preventing Transmission of STDs.*

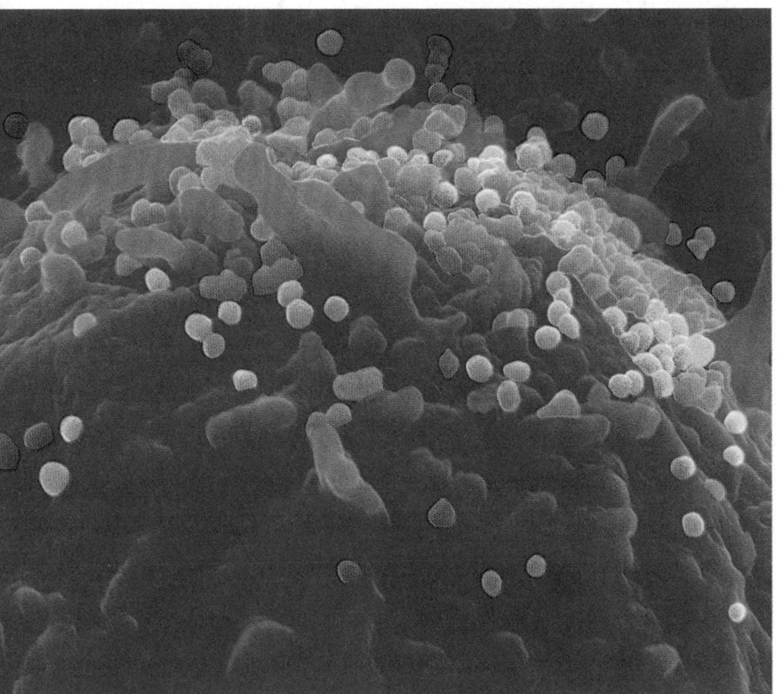

Figure 16.17 **Cells infected by the HIV virus.**
HIV viruses (yellow) can infect helper T cells (blue) and also macrophages, which work with helper T cells to stem the infection.

Connections and Misconceptions

Can you catch an STD from a toilet seat?

When HIV/AIDS was first identified in the mid-1980s, many people were concerned about being infected by the virus on toilet seats. Toilet seats are plastic and inert, so they're not very hospitable to disease-causing organisms. So if you're deciding whether to hover or sit, remember sitting on a toilet seat will not give you an STD.

The primary host for HIV is a helper T lymphocyte (Fig. 16.17). These are the very cells that stimulate an immune response, so the immune system becomes severely impaired in persons with AIDS. During the first stage of an HIV infection, symptoms are few, but the individual is highly contagious. Several months to several years after infection, the helper T lymphocyte count falls. Following this decrease, infections, such as other sexually transmitted diseases, begin to appear. In the last stage of infection, called AIDS, the helper T cell count falls way below normal. At least one opportunistic infection is present. Such diseases have the opportunity to occur only because the immune system is severely weakened. Persons with AIDS typically die from an opportunistic disease, such as *Pneumocystis* pneumonia.

Animation How the HIV Infection Cycle Works

There is no cure for AIDS. A treatment called highly active antiretroviral therapy (HAART) is usually able to stop HIV reproduction to the extent that the virus becomes undetectable in the blood. The medications must be continued indefinitely because as soon as HAART is discontinued, the virus rebounds.

Genital Warts

Genital warts are caused by the human papillomaviruses (HPVs). Many times, carriers either do not have any sign of warts or merely have flat lesions. When present, the warts commonly are seen on the penis and foreskin of men and near the vaginal opening in women. A newborn can become infected while passing through the birth canal.

Individuals currently infected with visible growth may have those growths removed by surgery, freezing, or burning with lasers or acids. However, visible warts that are removed may recur. A vaccine has been released for the human papillomaviruses that most commonly cause genital warts. This development is an extremely important step in the prevention of cancer, as well as in the prevention of warts themselves. Genital warts are associated with cancer of the cervix, as well as tumors of the vulva, vagina, anus, and penis. Researchers believe that these viruses may be involved in up to 90% of all cases of cancer of the cervix. Vaccination might make such cancers a thing of the past (see the chapter opener).

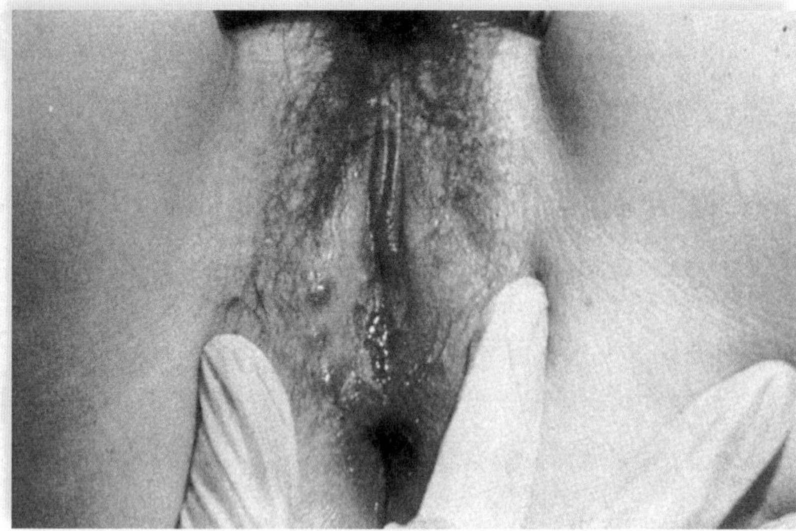

a.

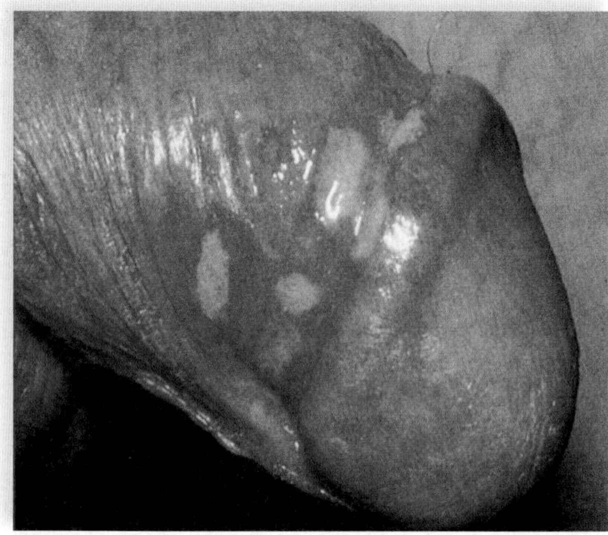

b.

Figure 16.18 **Herpes simplex virus 2 and genital herpes.**
There are several types of herpes viruses, and usually the ones called herpes simplex virus 2 cause genital herpes. About 1 million persons become infected each year, most of them teens and young adults. Symptoms of genital herpes include an outbreak of blisters, which can be present on the labia of females (**a**) or on the penis of males (**b**).

Genital Herpes

Genital herpes is caused by herpes simplex virus. Type 1 usually causes cold sores and fever blisters, while type 2 more often causes genital herpes (Fig. 16.18).

Persons usually get infected with herpes simplex virus 2 when they are adults. Some people exhibit no symptoms. Others may experience a tingling or itching sensation before blisters appear on the genitals. Once the blisters rupture, they leave painful ulcers that may take between five days and three weeks to heal. The blisters may be accompanied by fever; pain on urination; swollen lymph nodes in the groin; and, in women, a copious discharge. At this time, the individual has an increased risk of acquiring an HIV infection.

After the ulcers heal, the disease is only latent, and blisters can recur, although usually at less frequent intervals and with milder symptoms. Fever, stress, sunlight, and menstruation are associated with recurrence of symptoms. Exposure to herpes in the birth canal can cause an infection in the newborn, which leads to neurological disorders and even death. Birth by cesarean section prevents this possibility. There are antiviral drugs available that reduce the number and length of outbreaks. However, these drugs are not a cure for genital herpes. Latex or polyurethane condoms are recommended by the FDA to prevent the transmission of the virus to sexual partners.

Hepatitis

Hepatitis infects the liver and can lead to liver failure, liver cancer, and death. There are six known viruses that cause hepatitis, designated A, B, C, D, E, and G. Hepatitis A is usually acquired from sewage-contaminated drinking water, but this infection can also be sexually transmitted through oral–anal contact. Hepatitis B is spread through sexual contact and by blood-borne transmission (accidental needlestick on the job, receiving a contaminated blood transfusion, a drug abuser sharing infected needles while injecting drugs, from mother to fetus, etc.). Simultaneous infection with hepatitis B and HIV is common, because both share the same routes of transmission. Fortunately, a combined vaccine is available for hepatitis A and B. It is recommended that all children receive the vaccine to prevent infection (see Chapter 7). Hepatitis C (also called *non-A, non-B hepatitis*) causes most cases of posttransfusion hepatitis. Hepatitis D and G are sexually transmitted, and hepatitis E is acquired from contaminated water. Screening of blood and blood products can prevent transmission of hepatitis viruses during a transfusion. Proper water-treatment techniques can prevent contamination of drinking water.

Video
Halting Hepatitis

STDs Caused by Bacteria

Only STDs caused by bacteria are curable with antibiotics. Antibiotic resistance acquired by these bacteria may require treatment with extremely strong drugs for an extended period to achieve a cure.

Chlamydia

Chlamydia is named for the tiny bacterium that causes it, *Chlamydia trachomatis*. The incidence of new chlamydia infections has steadily increased since 1984.

Chlamydia infections of the lower reproductive tract are usually mild or asymptomatic, especially in women. About 18 to 21 days after infection, men may experience a mild burning sensation on urination and a mucoid discharge. Women may have a vaginal discharge along with the symptoms of a urinary tract infection. Chlamydia also causes cervical ulcerations, which increase the risk of acquiring HIV.

If the infection is misdiagnosed or if a woman does not seek medical help, there is a particular risk of the infection spreading from the cervix to the uterine tubes so that pelvic inflammatory disease (PID) results. This very painful condition can result in blockage of the uterine tubes with the possibility of sterility and infertility. If a baby comes in contact with chlamydia during birth, inflammation of the eyes or pneumonia can result.

Gonorrhea

Gonorrhea is caused by the bacterium *Neisseria gonorrhoeae.* Diagnosis in the male is not difficult, because typical symptoms are pain upon urination and a thick, greenish-yellow urethral discharge. In males and females, a latent infection leads to pelvic inflammatory disease (PID), which can also cause sterility in males. If a baby is exposed during birth, an eye infection leading to blindness can result. All newborns are given eyedrops to prevent this possibility.

Gonorrhea proctitis, an infection of the anus characterized by anal pain and blood or pus in the feces, also occurs in patients. Oral–genital contact can cause infection of the mouth, throat, and tonsils. Gonorrhea can spread to internal parts of the body, causing heart damage or arthritis. If, by chance, the person touches infected genitals and then touches his or her eyes, a severe eye infection can result. Up to now, gonorrhea was curable by antibiotic therapy. However, resistance to antibiotics is becoming more and more common, and 40% of all strains are now known to be resistant to therapy.

Syphilis

Syphilis is caused by a bacterium called *Treponema pallidum* (Fig. 16.19). As with many other bacterial diseases, penicillin is an effective antibiotic. Syphilis has three stages, often separated by latent periods, during which the bacteria are resting before multiplying again. During the primary stage, a hard **chancre** (ulcerated sore with hard edges) indicates the site of infection. The chancre usually heals spontaneously, leaving little scarring. During the secondary stage, the victim breaks out in a rash that does not itch and is seen even on the palms of the hands and the soles of the feet. Hair loss and infectious gray patches on the mucous membranes may also occur. These symptoms disappear of their own accord.

The tertiary stage lasts until the patient dies. During this stage, syphilis may affect the cardiovascular system by causing aneurysms, particularly in the aorta. In other instances, the disease may affect the nervous system, resulting in psychological disturbances. Also, gummas, large destructive ulcers, may develop on the skin or within the internal organs.

Congenital syphilis is caused by syphilitic bacteria crossing the placenta. The child is born blind and/or with numerous anatomical malformations. Control of syphilis depends on prompt and adequate treatment of all new cases. Therefore, it is crucial for all sexual contacts to be traced so they can be treated. Diagnosis of syphilis can be made by blood tests or by microscopic examination of fluids from lesions.

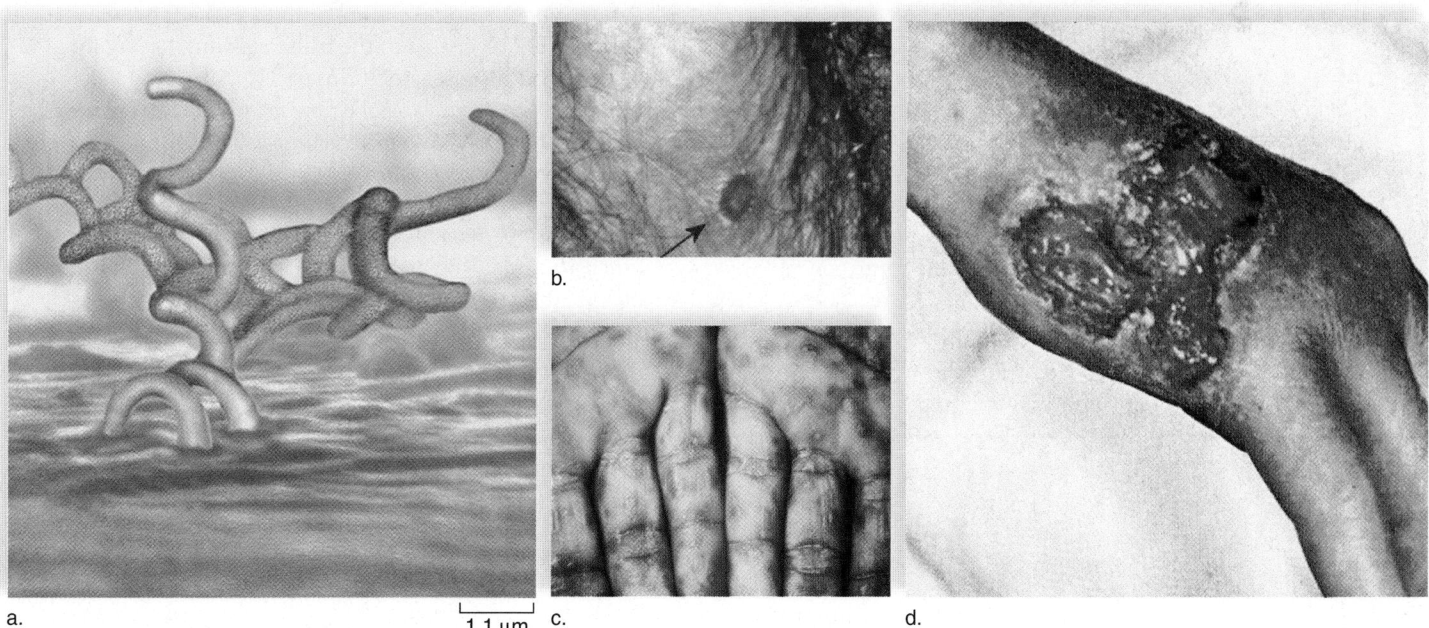

Figure 16.19 The three stages of syphilis.
a. Scanning electron micrograph of *Treponema pallidum,* the cause of syphilis. **b.** The three stages of syphilis. The primary stage of syphilis is a chancre at the site where the bacterium enters the body. **c.** The secondary stage is a body rash that occurs even on the palms of the hands and soles of the feet. **d.** In the tertiary stage, gummas may appear on the skin or internal organs.

Biology Matters **Health Focus**

Preventing Transmission of STDs

Sexual Activities Transmit STDs

Abstain from sexual intercourse or develop a long-term monogamous (always the same partner) sexual relationship with a partner who is free of STDs (Fig. 16C).

Refrain from having multiple sex partners or having relations with someone who has multiple sex partners. If you have sex with two other people and each of these has sex with two people, and so forth, the number of people who are relating is quite large.

Be aware that having relations with an intravenous drug user is risky because the behavior of this group risks AIDS and hepatitis B. Be aware that anyone who already has another sexually transmitted disease is more susceptible to an HIV infection.

Avoid anal intercourse (in which the penis is inserted into the rectum) because this behavior increases the risk of an HIV infection. The lining of the rectum is thin, and infected CD4 T cells can easily enter the body there. Also, the rectum is supplied with many blood vessels, and insertion of the penis into the rectum is likely to cause tearing and bleeding that facilitate the entrance of HIV. The vaginal lining is thick and difficult to penetrate, but the lining of the uterus is only one cell thick at certain times of the month and does allow CD4 T cells to enter.

Uncircumcised males are more likely to become infected than circumcised males because vaginal secretions can remain under the foreskin for a long time.

Practice Safer Sex

Always use a latex condom during sexual intercourse if you are not in a monogamous relationship. Be sure to follow the directions supplied by the manufacturer for the use of a condom. At one time, condom users were advised to use nonoxynol-9 in conjunction with a condom, but testing shows that this spermicide has no effect on viruses, including HIV.

Avoid fellatio (kissing and insertion of the penis into a partner's mouth) and cunnilingus (kissing and insertion of the tongue into the vagina) because they may be a means of transmission. The mouth and gums often have cuts and sores that facilitate catching an STD.

Figure 16C Sexual activities transmit STDs.

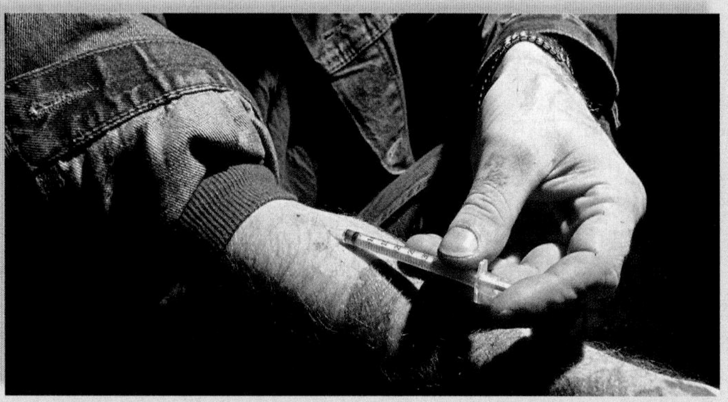

Figure 16D Sharing needles transmits STDs.

Practice penile, vaginal, oral, and hand cleanliness. Be aware that hormonal contraceptives make the female genital tract receptive to the transmission of sexually transmitted diseases, including HIV.

Be cautious about using alcohol or any drug that may prevent you from being able to control your behavior.

Drug Use Transmits HIV

Stop, if necessary, or do not start the habit of injecting drugs into your veins. Be aware that HIV and hepatitis B can be spread by blood-to-blood contact.

Always use a new sterile needle for injection or one that has been cleaned in bleach if you are a drug user and cannot stop your behavior (Fig. 16D).

Vaginal Infections

The term *vaginitis* is used to describe any vaginal infection or inflammation. It is the most commonly diagnosed gynecologic condition. Bacterial vaginosis (BV) is believed to cause 40–50% of the cases of vaginitis in the United States. Overgrowth of certain bacteria inhabiting the vagina causes vaginosis. A common culprit is the bacterium *Gardnerella vaginosis.* Overgrowth of this organism and subsequent symptoms can occur for nonsexual reasons. However, symptomless males may pass on the bacterium to women, who do experience symptoms.

The symptoms of BV are vaginal discharge that has a strong odor, a burning sensation during urination, and/or itching or pain in the vulva. Some women with BV have no signs of the infection. How women acquire these infections is not well understood. Having a new sex partner or multiple sex partners seems to increase the risk of getting BV, but females who are not sexually active get BV as well. Douching also appears to increase the incidence of BV. Women with BV are more susceptible to infection by other STDs, including HIV, herpes, chlamydia, and gonorrhea. Pregnant women with BV are at greater risk of premature delivery.

The yeast *Candida albicans,* and a protozoan, *Trichomonas vaginalis,* are two other causes of vaginitis. *Candida albicans* is normally found living in the vagina. Under certain circumstances, its growth increases above normal, causing vaginitis. For example, women taking birth control pills or antibiotics may be prone to yeast infections. Both can alter the normal balance of vaginal organisms, causing a yeast infection. A yeast infection causes a thick, white, curdlike vaginal discharge and is accompanied by itching of the vulva and/or vagina. Antifungal medications inserted into the vagina are used to treat yeast infections. Trichomoniasis, caused by *Trichomonas vaginalis,* affects both males and females. The urethra is usually the site of infection in males. Infected males are often asymptomatic and pass the parasite to their partner during sexual intercourse. Symptoms of trichomoniasis in females are a foul-smelling, yellow-green frothy discharge and itching of the vulva/vagina. Having trichomoniasis greatly increases the risk of infection by HIV. Prescription drugs are used to treat trichomoniasis, but if one partner remains infected, reinfection will occur. It is recommended that both partners in a sexual relationship be treated and abstain from having sex until the treatment is completed.

Connecting the Concepts

For more information on the topics presented in this section, refer to the following discussions.

Section 7.1 explores the structure of both viruses and bacteria.

Section S.1 provides a detailed examination of the HIV virus, its replication, and the disease AIDS.

Section S.3 examines how antibiotic resistance occurs and its consequences in the treatment of disease.

Check Your Progress 16.6

1. Explain what condition can occur due to both a chlamydial and gonorrheal infection.
2. Describe the medical conditions in women that are associated with genital warts.
3. Discuss the cause of most STDs.

CASE STUDY CONCLUSION

The good news for Ann is that early detection of cervical cancer is critical to successful treatment of the disease. For individuals with cervical cancer, the survival rate for those who have early detection is almost 100%, versus a less than 5% survival rate for those in whom the cancer has begun to spread, or metastasize, to other organs. In Ann's case, the years of smoking cigarettes probably were a major factor in her development of cervical cancer. However, for many cases of cervical cancer, the cause is the human papillomavirus, or HPV. There are over 15 forms of HPV that have been linked to cervical cancers. In 2006, the Food and Drug Administration (FDA) approved an HPV vaccine for women. The vaccine is designed to be administered at three doses starting between the ages of 11 and 12 years. Recently, the vaccine was also approved for men ages 9 to 26 to prevent genital warts and to reduce the chances that men will transmit HPV to their sexual partners. With the development of the HPV vaccine, it is hoped that the rates of cervical cancer in women will drop drastically over the next few decades.

Media Study Tools

www.mhhe.com/maderhuman12e

Enhance your study of this chapter with study tools and practice tests. Also ask your instructor about the resources available through ConnectPlus, including the media-rich eBook, interactive learning tools, and animations.

Summarizing the Concepts

16.1 Human Life Cycle

The life cycle of higher organisms requires two types of cell division:

- **Mitosis** growth and repair of tissues
- **Meiosis** gamete production

16.2 Male Reproductive System

The external genitals of males are:

- the penis (organ of sexual intercourse) and
- the scrotum (contains the testes).

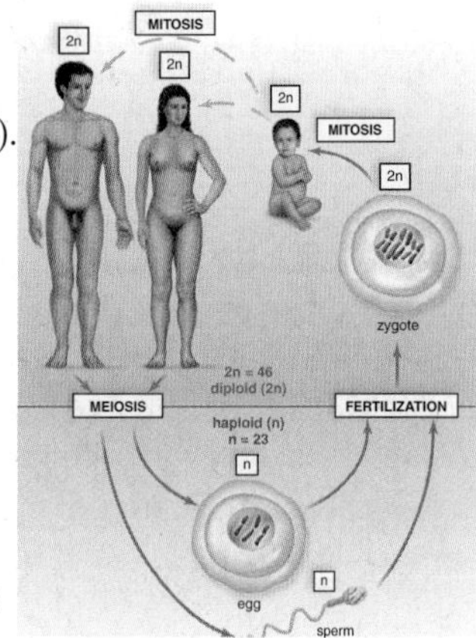

Spermatogenesis, occurring in seminiferous tubules of the testes, produces sperm.

- Mature sperm are stored in the epididymides.
- Sperm pass from the vasa deferentia to the urethra.
- The seminal vesicles, prostate gland, and bulbourethral glands add fluids (by secretion) to sperm.
- Sperm and secretions are called semen or seminal fluid.

Orgasm in males results in ejaculation of semen from the penis.

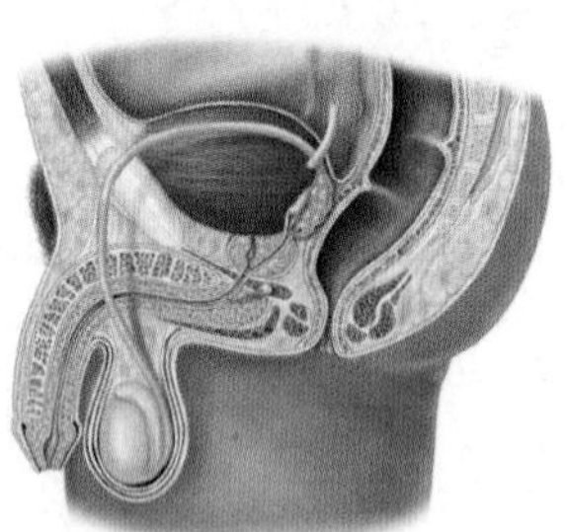

Hormonal Regulation in Males

- Hormonal regulation, involving secretions from the hypothalamus, the anterior pituitary, and the testes, maintains a fairly constant level of testosterone.
- FSH from the anterior pituitary promotes spermatogenesis.
- LH from the anterior pituitary promotes testosterone production by interstitial cells.

16.3 Female Reproductive System

Oogenesis occurring within the ovaries typically produces one mature follicle each month.

- This follicle balloons out of the ovary and bursts, releasing an egg that enters an oviduct.

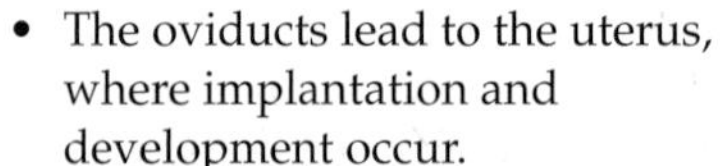

- The oviducts lead to the uterus, where implantation and development occur.

The female external genital area includes the vaginal opening, the clitoris, the labia minora, and the labia majora.

- The vagina is the organ of sexual intercourse and the birth canal in females.

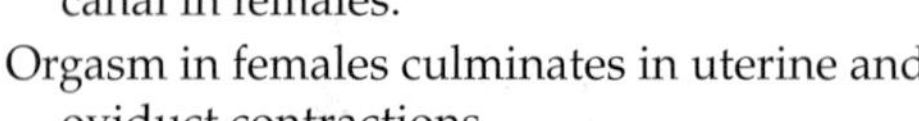

Orgasm in females culminates in uterine and oviduct contractions.

16.4 The Ovarian Cycle

Ovarian Cycle: Nonpregnant

- The ovarian cycle is under the hormonal control of the hypothalamus and the anterior pituitary.
- During the cycle's first half, FSH from the anterior pituitary causes maturation of a follicle that secretes estrogen and some progesterone.
- After ovulation and during the cycle's second half, LH from the anterior pituitary converts the follicle into the corpus luteum.
- The corpus luteum secretes progesterone and some estrogen.

Uterine Cycle: Nonpregnant

Estrogen and progesterone regulate the uterine cycle.

- Estrogen causes the endometrium to rebuild.
- Ovulation usually occurs on day 14 of a 28-day cycle.
- Progesterone produced by the corpus luteum causes the endometrium to thicken and become secretory.
- A low level of hormones causes the endometrium to break down as menstruation occurs.

Fertilization and Pregnancy

If fertilization takes place, the embryo implants in the thickened endometrium.

- The corpus luteum is maintained because of hCG production by the placenta; therefore, progesterone production does not cease.
- Menstruation usually does not occur during pregnancy.

16.5 Control of Reproduction

Numerous birth control methods and devices are available.

- A few of these are the birth control pill, diaphragm, and condom.
- Effectiveness varies.

Assisted reproductive technologies may help infertile couples to have children. Some of these technologies are:

- artificial insemination by donor (AID),
- in vitro fertilization (IVF),
- gamete intrafallopian transfer (GIFT), and
- intracytoplasmic sperm injection (ICSI).

16.6 Sexually Transmitted Diseases

STDs are caused by viruses, bacteria, fungi, and parasites.

STDs Caused by Viruses

- AIDS is caused by HIV (human immunodeficiency virus).
- Genital warts are caused by human papillomaviruses; these viruses cause warts or lesions on genitals and are associated with certain cancers.

- Genital herpes is caused by herpes simplex virus 2; causes blisters on genitals.
- Hepatitis is caused by hepatitis viruses A, B, C, D, E, and G. A and E are usually acquired from contaminated water; hepatitis B and C are from bloodborne transmission; and B, D, and G are sexually transmitted.

STDs Caused by Bacteria

- Chlamydia is caused by *Chlamydia trachomatis;* PID can result.
- Gonorrhea is caused by *Neisseria gonorrhoeae;* PID can result.
- Syphilis is caused by *Treponema pallidum.* It has three stages, with the third stage resulting in death.

Vaginal Infections

- Bacterial vaginosis commonly results from bacterial overgrowth. *Gardnerella vaginosis* often causes such infections.
- Infection with the yeast *Candida albicans* also occurs because of overgrowth, and antibiotics or hormonal contraceptives trigger this condition.
- The parasite *Trichomonas vaginalis* also causes vaginosis. This type affects both men and women, though men are often asymptomatic.

Understanding Key Terms

acrosome 369
birth control method 379
birth control pill 379
bulbourethral gland 368
cervix 370
chancre 387
chlamydia 386
circumcision 368
contraceptive 379
contraceptive implant 379
contraceptive injection 379
contraceptive vaccine 375
corpus luteum 375
diaphragm 379
egg 370
endometrium 372
epididymis 368
erectile dysfunction 368
estrogen 376
female condom 379
fimbria 370
follicle 374
follicle-stimulating hormone (FSH) 375
gamete 384
gonadotropin-releasing hormone (GnRH) 369
human chorionic gonadotropin (hCG) 378
implantation 370
infertility 382
interstitial cell 369
intrauterine device (IUD) 379
luteinizing hormone (LH) 369, 376
male condom 379
menopause 376
menstruation 377
oogenesis 374
ovarian cycle 374
ovary 370
oviduct 371
ovulation 375
Pap test 370
penis 368
placenta 378
polar body 374
progesterone 376
prostate gland 368
scrotum 367
semen 368
seminal vesicle 368
seminiferous tubule 369
Sertoli cell 369
sperm 369
spermatogenesis 369
testes 367
testosterone 370
tubal ligation 382
urethra 368
uterine cycle 376
uterus 370
vagina 372
vas deferens 368
vasectomy 382
vulva 372
zygote 370

Match the key terms to these definitions.

a. ________ Release of an oocyte from the ovary.

b. ________ Female sex hormone that causes the endometrium of the uterus to become secretory during the uterine cycle; along with estrogen, it maintains secondary sex characteristics in females.

c. ________ Thick, whitish fluid consisting of sperm and secretions from several glands of the male reproductive tract.

d. ________ Narrow end of the uterus, which projects into the vagina.

e. ________ Cap at the anterior end of a sperm that partially covers the nucleus and contains enzymes that help the sperm penetrate the egg.

Testing Your Knowledge of the Concepts

1. What type of cell division produces the gametes? Why is this type of cell division necessary? (page 366)
2. Trace the path of sperm. What glands contribute fluids to semen? (page 367)
3. Where are sperm produced in the testes? What is the process called? Where is testosterone produced in the testes? (pages 368–370)
4. Name the hormones involved in maintaining the sex characteristics of the male, and tell what each does. (pages 369–370)
5. What are the organs of the female reproductive tract and their functions? (pages 371–372)
6. Describe the ovarian cycle in a nonpregnant female and the hormones involved. (pages 374–375)
7. Describe the uterine cycle in a nonpregnant female, and relate it to the ovarian cycle. (pages 376–377)
8. Describe the hormonal role of the placenta. (page 378)
9. Briefly describe various birth control methods, along with their effectiveness. (pages 379–380)
10. Briefly describe various types of assisted reproductive technologies. (page 384)
11. What STDs are caused by viruses? List the causative agent, symptoms, and treatments. What STDs are caused by bacteria? List the causative agent, symptoms, and treatments. (pages 385–388)
12. Label this diagram of the male reproductive system.

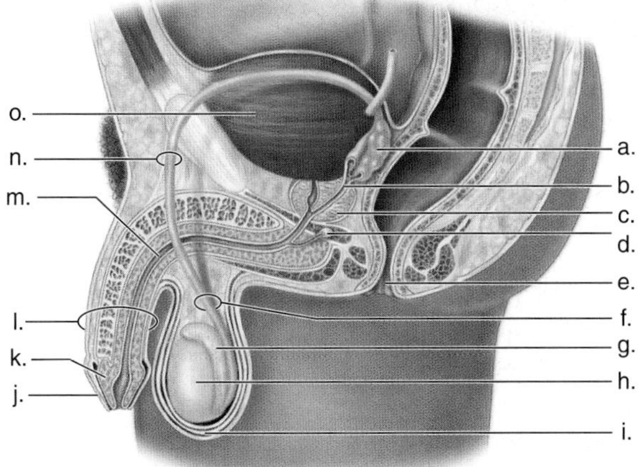

13. Follicle-stimulating hormone (FSH)
    a. is secreted by females but not by males.
    b. stimulates the seminiferous tubules to produce sperm.
    c. secretion is controlled by gonadotropin-releasing hormone (GnRH).
    d. Both b and c are correct.
14. Semen does not contain
    a. prostate fluid.
    b. urine.
    c. fructose.
    d. prostaglandins.
    e. Both b and d are correct.
15. Label this diagram of the female reproductive system.

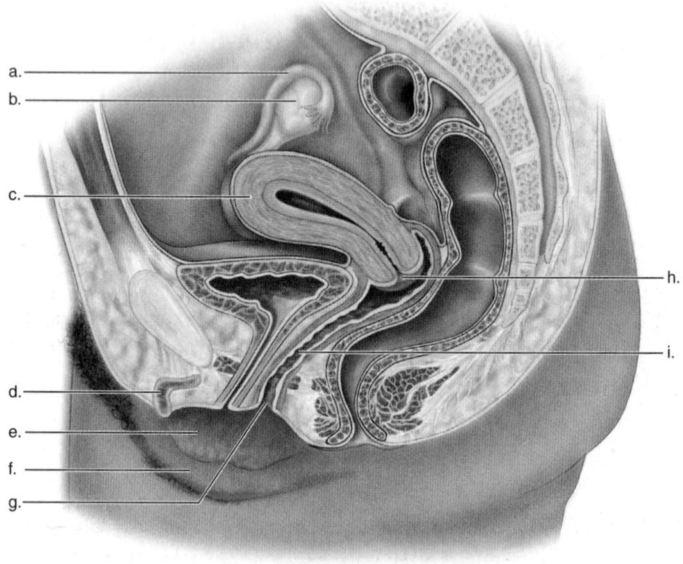

16. Testosterone is produced and secreted by
    a. spermatogonia.
    b. sustentacular cells.
    c. seminiferous tubules.
    d. interstitial cells.
17. The release of the oocyte from the follicle is caused by
    a. a decreasing level of estrogen.
    b. a surge in the level of follicle-stimulating hormone.
    c. a surge in the level of luteinizing hormone.
    d. progesterone released from the corpus luteum.
18. Which of the following is not an event of the ovarian cycle?
    a. FSH promotes the development of a follicle.
    b. The endometrium thickens.
    c. The corpus luteum secretes progesterone.
    d. Ovulation of an egg occurs.
19. An oocyte is fertilized in the
    a. vagina.
    b. uterus.
    c. oviduct.
    d. ovary.
20. During pregnancy,
    a. the ovarian and uterine cycles occur more quickly than before.
    b. GnRH is produced at a higher level than before.
    c. the ovarian and uterine cycles do not occur.
    d. the female secondary sex characteristics are not maintained.
21. Female oral contraceptives prevent pregnancy because
    a. the pill inhibits the release of luteinizing hormone.
    b. oral contraceptives prevent the release of an egg.
    c. follicle-stimulating hormone is not released.
    d. All of these are correct.

In questions 22–24, match each method of protection with a means of birth control in the key.

Key:

a. vasectomy
b. oral contraception
c. intrauterine device (IUD)
d. diaphragm
e. male condom

22. Blocks entrance of sperm to uterus
23. Traps sperm and also prevents STDs
24. Prevents implantation of an embryo

Thinking Critically About the Concepts

Male hormonal contraception (MHC) would be similar to hormonal contraception used by women. The primary goal of hormonal contraception is to inhibit or lower the gonadotropins, which in turn prevents gamete production or maturation. A successful product must have minimal side effects, and fertility should be restored shortly after the contraception is discontinued. There are numerous clinical trials being conducted to determine hormone dosages and an effective delivery method for MHC. Ideally, an MHC will be available to the general public in the next decade. That is good news for couples who would like an option for male contraception in addition to condoms and vasectomies.

1. Review Figures 16.5 and 16.13. See if you can figure out how giving males androgens and progestin (a synthetic progesterone) would inhibit the gonadotropins.
2. Refer to Figures 16.11 and 16.12.
   a. Redraw the illustration and add progesterone early in the cycle (to represent use of birth control pills) and draw new lines to show how FSH/LH will be affected by progesterone's presence early in the cycle.
   b. Redraw the illustration and show the effects of hCG (produced when the female gets pregnant) on the corpus luteum, progesterone levels, and endometrium.
3. Women who use birth control pills appear to have a lower risk of developing ovarian cancer. Women who use fertility enhancing drugs (which increase the number of follicles that develop) may increase their risk of developing ovarian cancer. Speculate about how these therapies may affect a woman's risk of developing ovarian cancer.
4. It is fairly common for very serious female athletes to not have their periods. These athletes might have a 10–15% body fat composition. Explain the connection between low body fat composition and the absence of a menstrual cycle in these athletes.

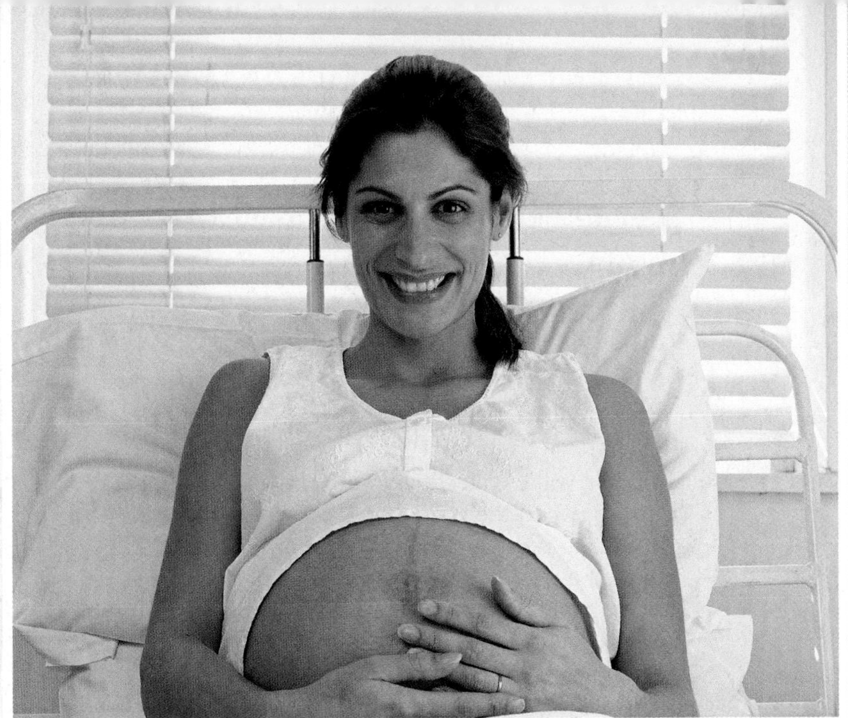

CHAPTER

17

Development and Aging

CASE STUDY PREGNANCY TESTING

For several months, Amber and Kent had been trying to conceive a child. Both Amber and Kent had put off having children for several years while they pursued their careers. Now, as they both approached the age of 40, they were beginning to feel the pressures of time. As a precautionary method, Amber had begun taking prenatal vitamins; additionally, she was much more aware of the contents of her diet. Although neither of them was ever really into physical exercise, both began walking several times a week in preparation for what they hoped would be news that Amber was pregnant. Finally, after two months, Amber proudly announced that the home pregnancy test was positive! They immediately scheduled an appointment with their regular physician to prepare for the next stage of their lives.

Both Amber and Kent were very satisfied with their choice of a doctor. At the first visit following the positive results of the home pregnancy test, their physician had performed a complete physical of Amber, as well as a blood test to confirm pregnancy. The physician informed the new parents that a blood test was much more accurate in detecting levels of the pregnancy hormone, human chorionic gonadotropin (hCG), than over-the-counter (OTC) urine tests. The results of the blood test confirmed what Amber and Kent suspected, that in a period of just 40 weeks, Amber and Kent would be parents.

Their physician promptly gave the parents-to-be a list of items to avoid. Amber was told to increase her level of exercise and watch her diet more closely. She needed to drink eight to ten glasses of water per day, as well as eating plenty of fruits and vegetables. The doctor informed them that it was crucial for Amber to inform her physician of any over-the-counter drugs she may want to take, especially in the first trimester. Her doctor told Amber that the first trimester was a period when critical organ systems developed in her baby and that most OTC medications and alcohol were now forbidden. Both Amber and Kent were up to the challenge and excited about the prospects of finally being parents.

As you read through the chapter, think about the following questions.

1. What is the role of the hCG hormone in pregnancy?
2. Why would the doctor request that Amber check before taking over-the-counter drugs?
3. What physiological changes should Amber expect over the course of her pregnancy?

CHAPTER CONCEPTS

17.1 Fertilization
During fertilization, a sperm nucleus fuses with the egg nucleus. Once one sperm penetrates the plasma membrane, the egg undergoes changes that prevent any more sperm from entering.

17.2 Pre-Embryonic and Embryonic Development
Pre-embryonic development occurs between the time of fertilization and implantation in the uterine lining. By the end of embryonic development, all organ systems are established and there is a mature and functioning placenta. The embryo is only about 38 mm (1.5 in.) long.

17.3 Fetal Development
During fetal development, the gender becomes obvious, the skeleton continues to ossify, fetal movement begins, and the fetus gains weight.

17.4 Pregnancy and Birth
During pregnancy, the mother gains weight as the uterus comes to occupy most of the abdominal cavity. A positive feedback mechanism that involves uterine contractions and oxytocin explains the onset and continuation of labor so that the child is born.

17.5 Development After Birth
Development after birth consists of infancy, childhood, adolescence, and adulthood. Aging is influenced by our genes, whole-body changes, and extrinsic factors.

BEFORE YOU BEGIN

Before beginning this chapter, take a few moments to review the following discussions.

Figure 16.4 How does spermatogenesis produce sperm cells?

Figure 16.9 What are the differences between oogenesis and spermatogenesis?

Section 16.4 What are the roles of estrogen and progesterone in the female reproductive system?

17.1 Fertilization

Learning Outcomes

Upon completion of this section, you should be able to

1. Identify the structures of an egg and sperm cell and provide a function for each.
2. Describe the steps in the fertilization of an egg cell by a sperm.

Connections and Misconceptions

How many sperm compete to fertilize the egg?

Studies indicate that there are between 150 and 200 million sperm in a man's ejaculation. Of that number, only 20 to 30 or so make it near the egg. Out of those, only a single sperm fertilizes the egg.

Fertilization is the union of a sperm and egg to form a **zygote,** the first cell of the new individual (Fig. 17.1).

Steps of Fertilization

The tail of a sperm is a flagellum, which allows it to swim toward the egg. The middle piece contains energy-producing mitochondria. The head contains a nucleus capped by a membrane-bound acrosome. The acrosome is an organelle containing digestive enzymes. Only the nucleus from the sperm head fuses with the egg nucleus. Therefore, the zygote receives cytoplasm and organelles only from the mother.

Video Human Sperm

The plasma membrane of the egg is surrounded by an extracellular matrix termed the *zona pellucida.* In turn, the zona pellucida is surrounded by a few layers of adhering follicular cells, collectively called the *corona radiata.* These cells nourished the egg when it was in a follicle of the ovary.

During fertilization, several sperm penetrate the corona radiata. Several sperm attempt to penetrate the zona pellucida, but only one sperm enters the egg. The acrosome plays a role in allowing sperm to penetrate the zona pellucida. After a sperm head binds tightly to the zona pellucida, the acrosome releases digestive enzymes that forge a pathway for the sperm through the zona pellucida. When a sperm binds to the egg, their plasma membranes fuse. This sperm (the head, the middle piece, and usually the tail) enters the egg. Fusion of the sperm nucleus and the egg nucleus follows.

To ensure proper development, only one sperm should enter an egg. Prevention of polyspermy (entrance of more than one sperm) depends on changes in the egg's plasma membrane and in the zona pellucida. As soon as a sperm touches an egg, the egg's plasma membrane depolarizes (from 265 mV to

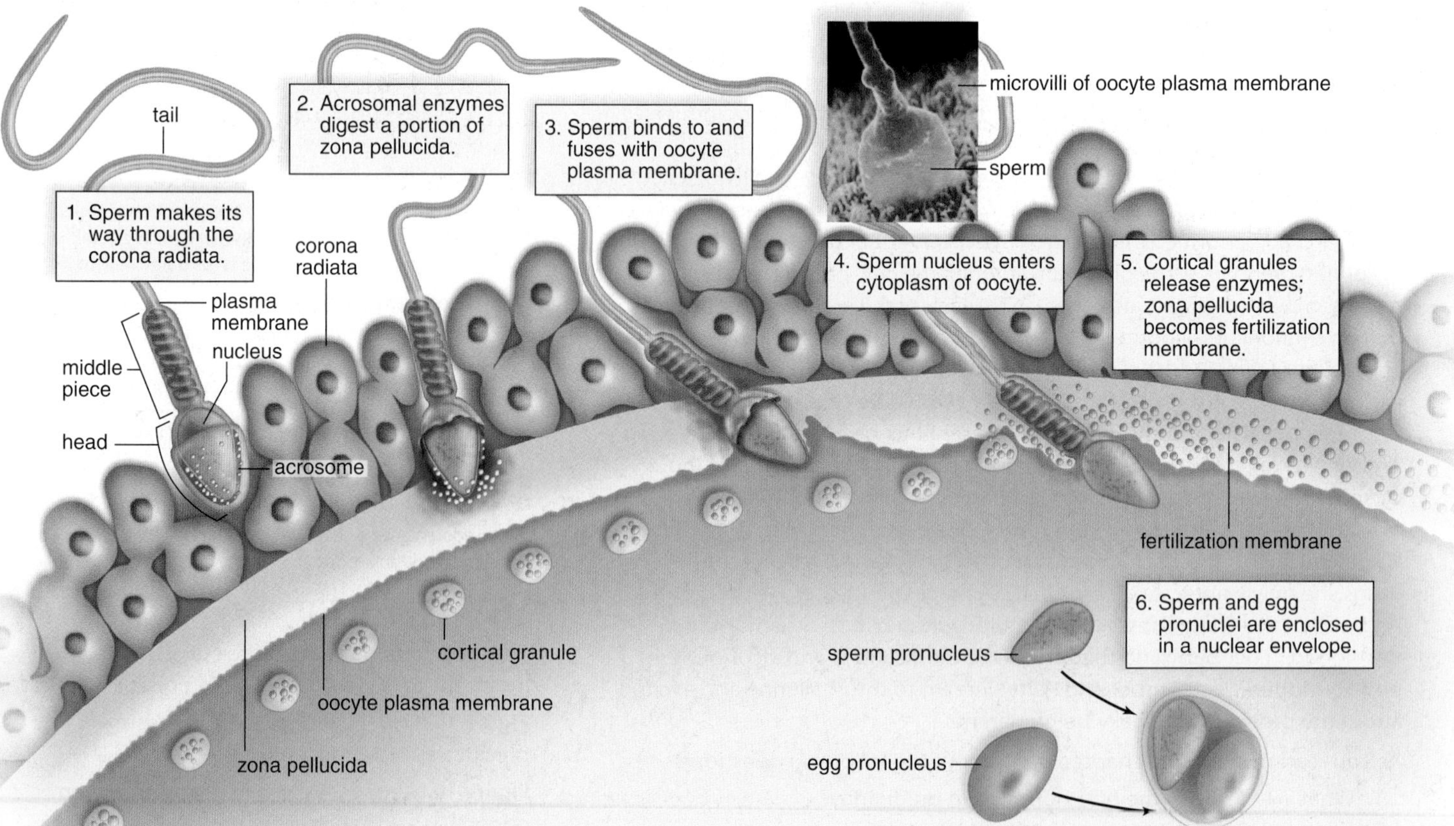

Figure 17.1 The steps in the fertilization of an egg.
During fertilization, a single sperm is drawn into the egg by microvilli of its plasma membrane (micrograph). With the help of enzymes from the acrosome, a sperm makes its way through the zona pellucida. After a sperm binds to the plasma membrane of the egg, changes occur that prevent other sperm from entering the egg. Fertilization is complete when the sperm pronucleus and the egg pronucleus contribute chromosomes to the zygote.

10 mV), and this prevents the binding of any other sperm. Then, vesicles called cortical granules release enzymes that cause the zona pellucida to become an impenetrable fertilization membrane. Now sperm cannot bind to the zona pellucida either.

Connecting the Concepts

For more information on egg and sperm cells, refer to the following discussions.

Section 16.2 explains how sperm are produced by spermatogenesis.

Section 16.4 explains how egg cells are produced by oogenesis.

Check Your Progress 17.1

1. Describe the steps in fertilization.
2. Distinguish between the function of the corona radiata and the zona pellucida.
3. Explain what prevents multiple sperm from fertilizing the same egg.

17.2 Pre-Embryonic and Embryonic Development

Learning Outcomes

Upon completion of this section, you should be able to

1. Recognize how cleavage, growth, morphogenesis, and differentiation all play a role in development.
2. Identify the extraembryonic membranes and provide a function for each.
3. Identify the organ systems that are formed from each of the primary germ layers.
4. Summarize the key events that occur at each stage of pre-embryonic and embryonic development.

Human development proceeds from pre-embryonic to embryonic development and then through fetal development. Table 17.1 outlines the major events during development.

Table 17.1 Human Development

Time	Events for Mother	Events for Baby
Pre-Embryonic Development		
First week	Ovulation occurs.	Fertilization occurs. Cell division begins and continues. Chorion appears.
Embryonic Development		
Second week	Symptoms of early pregnancy (nausea, breast swelling, and fatigue) are present.	Implantation occurs. Amnion and yolk sac appear. Embryo has tissues. Placenta begins to form. Blood pregnancy test is positive.
Third week	First menstruation is missed. Urine pregnancy test is positive. Symptoms of early pregnancy continue.	Nervous system begins to develop. Allantois and blood vessels are present. Placenta is well formed.
Fourth week		Limb buds form. Heart is noticeable and beating. Nervous system is prominent. Embryo has tail. Other systems form.
Fifth week	Uterus is the size of a hen's egg. Mother feels frequent need to urinate due to pressure of growing uterus on bladder.	Embryo is curved. Head is large. Limb buds show divisions. Nose, eyes, and ears are noticeable.
Sixth week	Uterus is the size of an orange.	Fingers and toes are present. Skeleton is cartilaginous.
Two months	Uterus can be felt above the pubic bone.	All systems are developing. Bone is replacing cartilage. Facial features are becoming refined. Embryo is about 38 mm (1.5 in.) long.
Fetal Development		
Third month	Uterus is the size of a grapefruit.	Gender can be distinguished by ultrasound. Fingernails appear.
Fourth month	Fetal movement is felt by a mother who has previously been pregnant.	Skeleton is visible. Hair begins to appear. Fetus is about 150 mm (6 in.) long and weighs about 170 grams (6 oz.).
Fifth month	Fetal movement is felt by a mother who has not previously been pregnant. Uterus reaches up to level of umbilicus, and pregnancy is obvious.	Protective cheesy coating called vernix caseosa begins to be deposited. Heartbeat can be heard.
Sixth month	Doctor can tell where baby's head, back, and limbs are. Breasts have enlarged, nipples and areolae are darkly pigmented, and colostrum is produced.	Body is covered with fine hair called lanugo. Skin is wrinkled and reddish.
Seventh month	Uterus reaches halfway between umbilicus and rib cage.	Testes descend into scrotum. Eyes are open. Fetus is about 300 mm (12 in.) long and weighs about 1,350 grams (3 lb).
Eighth month	Weight gain is averaging about a pound a week. Standing and walking are difficult for the mother because her center of gravity is thrown forward.	Body hair begins to disappear. Subcutaneous fat begins to be deposited.
Ninth month	Uterus is up to rib cage, causing shortness of breath and heartburn. Sleeping becomes difficult.	Fetus is ready for birth. It is about 530 mm (20.5 in.) long and weighs about 3,400 grams (7.5 lb).

Processes of Development

As a human being develops, these processes occur:

Cleavage. Immediately after fertilization, the zygote begins to divide so that there are first 2; then 4, 8, 16, and 32 cells; and so forth. Increase in size does not accompany these divisions (Fig. 17.2). Cell division during cleavage is mitotic, and each cell receives a full complement of chromosomes and genes.

Growth. During embryonic development, cell division is accompanied by an increase in size of the daughter cells.

Morphogenesis. Morphogenesis refers to the shaping of the embryo and is first evident when certain cells are seen to move, or migrate, in relation to other cells. By these movements, the embryo begins to assume various shapes.

Differentiation. When cells take on a specific structure and function, differentiation occurs. The first system to become visibly differentiated is the nervous system.

Stages of Development

Pre-embryonic development encompasses the events of the first week; **embryonic development** begins with the second week and lasts until the end of the second month.

Pre-Embryonic Development

The events of the first week of development are shown in Figure 17.2.

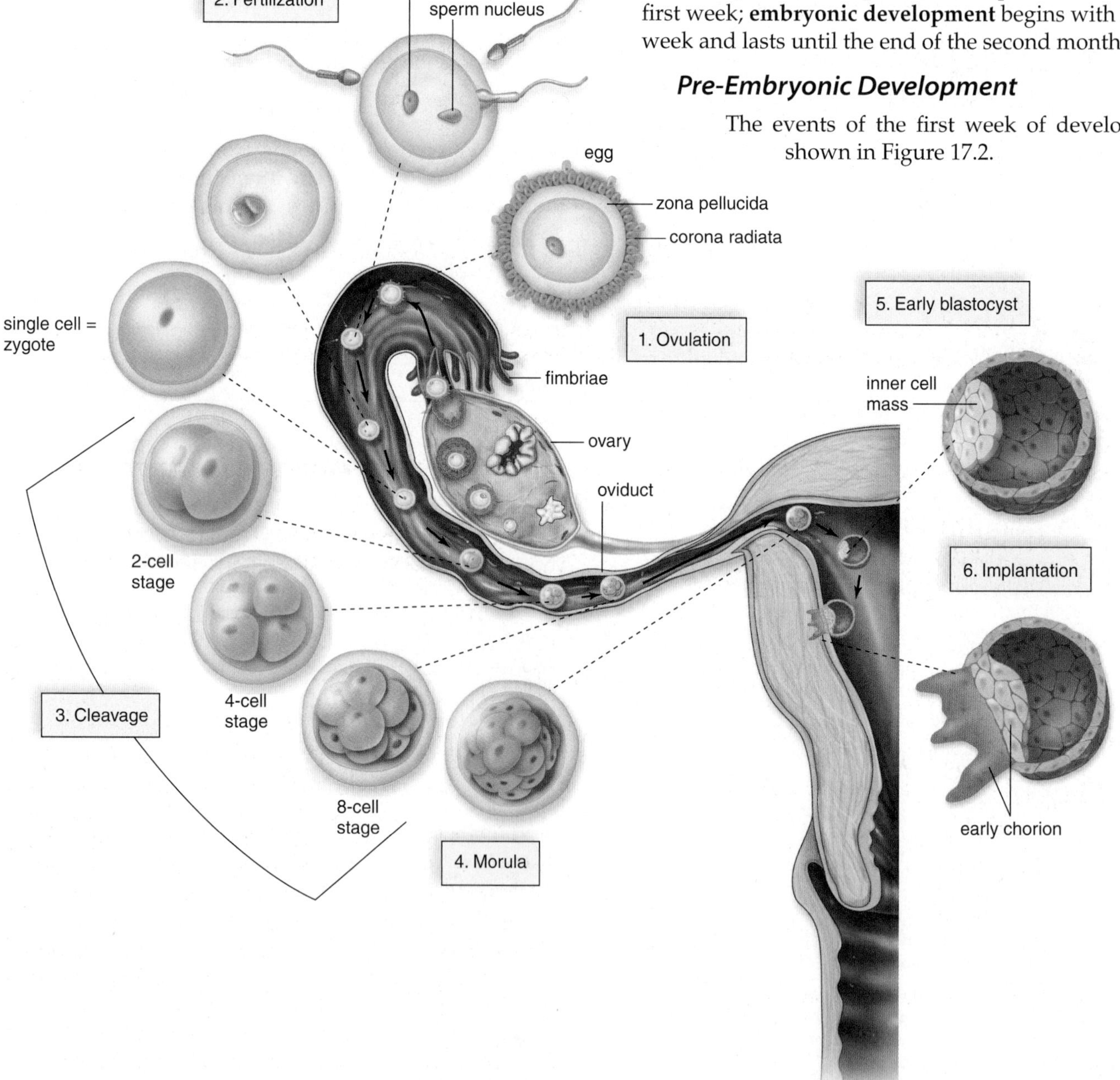

Figure 17.2 The stages of pre-embryonic development.
Structures and events proceed counterclockwise. **1.** At ovulation, the secondary oocyte leaves the ovary. A single sperm nucleus enters the egg, and (**2**) fertilization occurs in the oviduct. As the zygote moves along the oviduct, it undergoes (**3**) cleavage to produce (**4**) a morula. **5.** The blastocyst forms and (**6**) implants in the uterine lining.

Immediately after fertilization, the zygote divides repeatedly as it passes down the oviduct to the uterus. A **morula** is a compact ball of embryonic cells that becomes a **blastocyst.** The many cells of the blastocyst arrange themselves so that there is an **inner cell mass** surrounded by an outer layer of cells. The inner cell mass will become the embryo, and the layer of cells will become the chorion. The early appearance of the chorion emphasizes the complete dependence of the developing embryo on this extraembryonic membrane.

Video Blastocyst Formation

Each cell within the inner cell mass has the genetic capability of becoming any type of tissue. Sometimes, during development, the cells of the morula separate or the inner cell mass splits and two pre-embryos are present rather than one. If all goes well, these two pre-embryos will be identical twins because they have inherited exactly the same chromosomes. Fraternal twins arise when two different eggs are fertilized by two different sperm. They do not have identical chromosomes.

Extraembryonic Membranes

The **extraembryonic membranes** are not part of the embryo and fetus. Instead, as implied by their name, they are outside the embryo (Fig. 17.3). The names of the extraembryonic membranes in humans are strange to us because they are named for their function in animals, such as birds, that produce eggs with shells. In these animals, the chorion lies next to the shell and carries on gas exchange. The amnion contains the protective amniotic fluid, which bathes the developing embryo. The allantois collects nitrogenous wastes. The yolk sac surrounds the yolk, which provides nourishment.

The functions of the extraembryonic membranes are different in humans because humans develop inside the uterus. The extraembryonic membranes and their functions in humans follow.

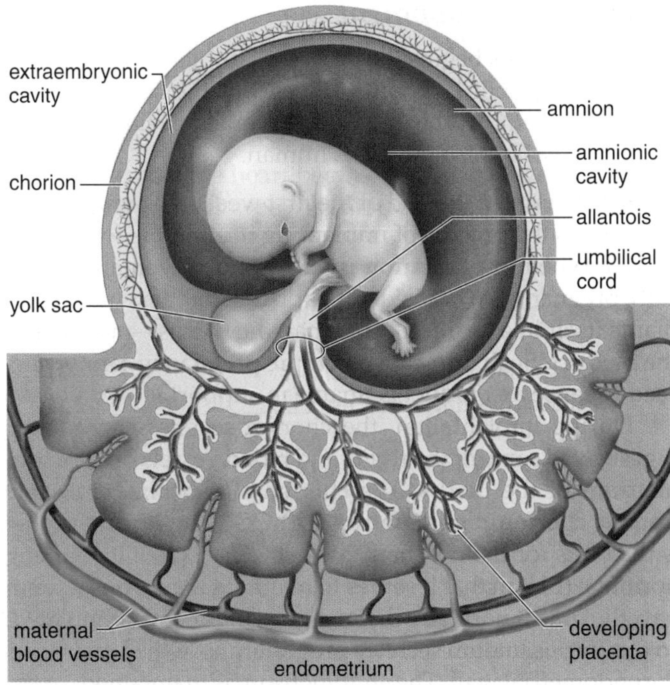

Figure 17.3 The extraembryonic membranes.
The chorion and amnion surround the embryo. The two other extraembryonic membranes, the yolk sac and allantois, contribute to the umbilical cord.

1. The **chorion** develops into the fetal half of the **placenta,** the organ that provides the embryo/fetus with nourishment and oxygen and takes away its waste. The chorionic villi are fingerlike projections of the chorion that increase the absorptive area of the chorion. Blood vessels within the chorionic villi are continuous with the umbilical blood vessels.
2. The **allantois,** like the yolk sac, extends away from the embryo. It accumulates the small amount of urine produced by the fetal kidneys and later gives rise to the urinary bladder. For now, its blood vessels become the umbilical blood vessels, which take blood to and from the fetus. The umbilical arteries carry oxygen-poor blood from the fetus to the placenta, and the umbilical veins carry oxygen-rich blood from the placenta to the fetus.
3. The **yolk sac** is the first embryonic membrane to appear. In shelled animals such as birds, the yolk sac contains yolk, food for the developing embryo. In mammals such as humans, this function is taken over by the placenta and the yolk sac contains little yolk. But the yolk sac contains plentiful blood vessels. It is the first site of blood cell formation.
4. The **amnion** enlarges as the embryo and then the fetus enlarges. It contains fluid to cushion and protect the embryo, which develops into a fetus.

Connections and Misconceptions

How is a baby's due date calculated?

The due date for arrival of the baby is calculated from the first day of the woman's last menstrual cycle before pregnancy. From conception to birth is approximately 266 days. Conception occurs approximately 14 days after the menstrual cycle begins (assuming ovulation occurs in the middle of the menstrual cycle). This gives a total of 280 days until the due date, or approximately 40 weeks.

To estimate the actual date, a calculation called Naegele's rule is often used:

1. Use the first day of the last period as a starting point.
2. Subtract three months from the month in which the period occurred.
3. Add seven days to the first day of the last period.

For example, if a woman's last period started on January 1, her baby's approximate due date is October 8. Ultrasound exams are also frequently used to tell the baby's due date.

Biology Matters Bioethical Focus

In March 1997, Scottish investigators announced that they had cloned a sheep called Dolly, and their procedure is now routinely used (Fig. 17A). A donor 2n nucleus is substituted for the n nucleus of an egg. A stimulus is applied that triggers cell division, and the resulting embryo is implanted into a surrogate mother where it develops to term. Using the procedure developed by the Scottish researchers, it is now common practice to clone all sorts of farm animals (horses, cows, sheep, goats, pigs) and also cats and monkeys.

Success of Cloning

Even so, cloning of animals is still in its infancy, and many problems still exist. (1) The vast majority of pregnancies involving clones are not successful. To clone Dolly the sheep, it took 247 tries before one was successful. In many cases, the clone grows abnormally large and the uterus enlarges with fluid to the point at which it can rip apart. Almost all clone pregnancies spontaneously abort. (2) Of the small number of animal clones born, most have severe abnormalities; malfunctioning livers, abnormal blood vessels and heart problems, underdeveloped lungs, diabetes, and immune system deficiencies are seen in newborns. Several cow clones had head deformities—none survived very long. (3) Even if the newborn clone appears healthy, it usually soon develops diseases seen in older animals. Dolly was euthanized in 2003 because she was suffering from lung cancer and crippling arthritis. She had lived only half the normal life span for a Dorset sheep.

Reproductive Cloning Versus Therapeutic Cloning

Animal cloning is a form of reproductive cloning. In **reproductive cloning,** the desired end is to create an individual. Even if such a feat were accomplished, a clone would not be identical to the person being cloned. Recall that mitochondria have genes, and these genes are contributed by the egg even if the egg nucleus is removed. The clone could be identical to its genetic parent only if the donor nucleus and the donor egg came from the same female. Further, the clone would be subject to different environmental factors and a different upbringing from his/her genetic parent.

In **therapeutic cloning,** the desired end is *not* an individual. Rather, it is the embryonic stem cells that could possibly be "coaxed" into becoming other types of cells. The goal is twofold. Stem-cell research will undoubtedly yield useful information about how cells develop and become specialized. However, the ultimate goal of therapeutic cloning is to provide cells and tissues that could treat human illnesses: insulin-secreting cells for diabetics, nerve cells for stroke patients or those with Parkinson disease, cardiac cells for those with heart disease, and so forth. Yet, ethical concerns about therapeutic cloning remain—after all, any pre-embryo is potentially a living, breathing human.

Animation
Heart Stem Cells

Anticipating intense interest in therapeutic cloning, the U.S. National Academy of Science proposed strict new guidelines for federally funded research in 2005. As a result of the Academy's recommendations, Embryonic Stem Cell Research Oversight (ESCRO) committees must approve embryonic stem-cell research before it is begun. Thus, this type of research is subject to two reviews: (1) an ESCRO committee analysis and (2) reviews already required by an Institutional Review Board. Why scrutinize research so carefully? You don't need to look any farther than the Tuskegee scandal (see Chapter 1) to understand.

ESCRO committees consist of bioethicists and legal experts, as well as members of the general public. All research requires informed consent from the donors of ova or sperm prior to beginning the research. Stem cells created for therapeutic cloning can never be used for reproductive purposes under the proposed guidelines. The guidelines also require that the embryos created cannot be grown in culture for longer than 14 days. At that point, the primitive streak of the developing nervous system begins to form.

Decide Your Opinion

1. Do you approve of the restrictions currently in place for therapeutic cloning? If not, how would you see these restrictions changed?
2. Scientific researchers have made great strides in stem-cell research, successfully converting both adult cells and umbilical cord cells back to stem-cell forms. In light of these successes, should therapeutic cloning be funded at all?
3. A commercial company, now out of business, once offered the opportunity to clone people's cats—for around $50,000. Should the cloning of pets be allowed when there are so many unwanted pets in the United States?
4. Should scientists be allowed to bring extinct species back to life (shades of Jurassic Park) using frozen tissue samples and reproductive cloning?

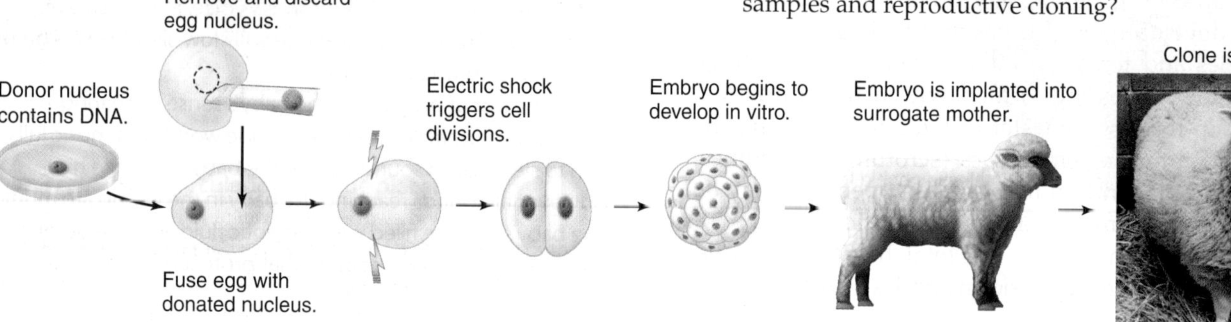
Remove and discard egg nucleus.
Donor nucleus contains DNA.
Fuse egg with donated nucleus.
Electric shock triggers cell divisions.
Embryo begins to develop in vitro.
Embryo is implanted into surrogate mother.
Clone is born.

Figure 17A The process of reproductive cloning.

Eighth Through Ninth Months

At the end of nine months, the fetus is about 530 mm (20.5 in.) long and weighs about 3,400 g (7.5 lb). Weight gain is due largely to an accumulation of fat beneath the skin. Full-term babies have the best chance of survival. Premature babies are subject to various challenges, such as respiratory distress syndrome, because their lungs are underdeveloped (see Chapter 9), jaundice (see Chapter 8), and infections.

As the end of development approaches, the fetus usually rotates so that the head is pointed toward the cervix. However, if the fetus does not turn, a **breech birth** (rump first) is likely. It is very difficult for the cervix to expand enough to accommodate this form of birth, and asphyxiation of the baby is more likely to occur. Thus, a **cesarean section** (incision through the abdominal and uterine walls) may be prescribed for delivery of the fetus.

Development of Male and Female Genitals

The sex of an individual is determined at the moment of fertilization. Males have a pair of chromosomes designated as X and Y, and females have two X chromosomes.

Normal Development of the Genitals

Development of the internal and external genitals is shown in Figure 17.9.

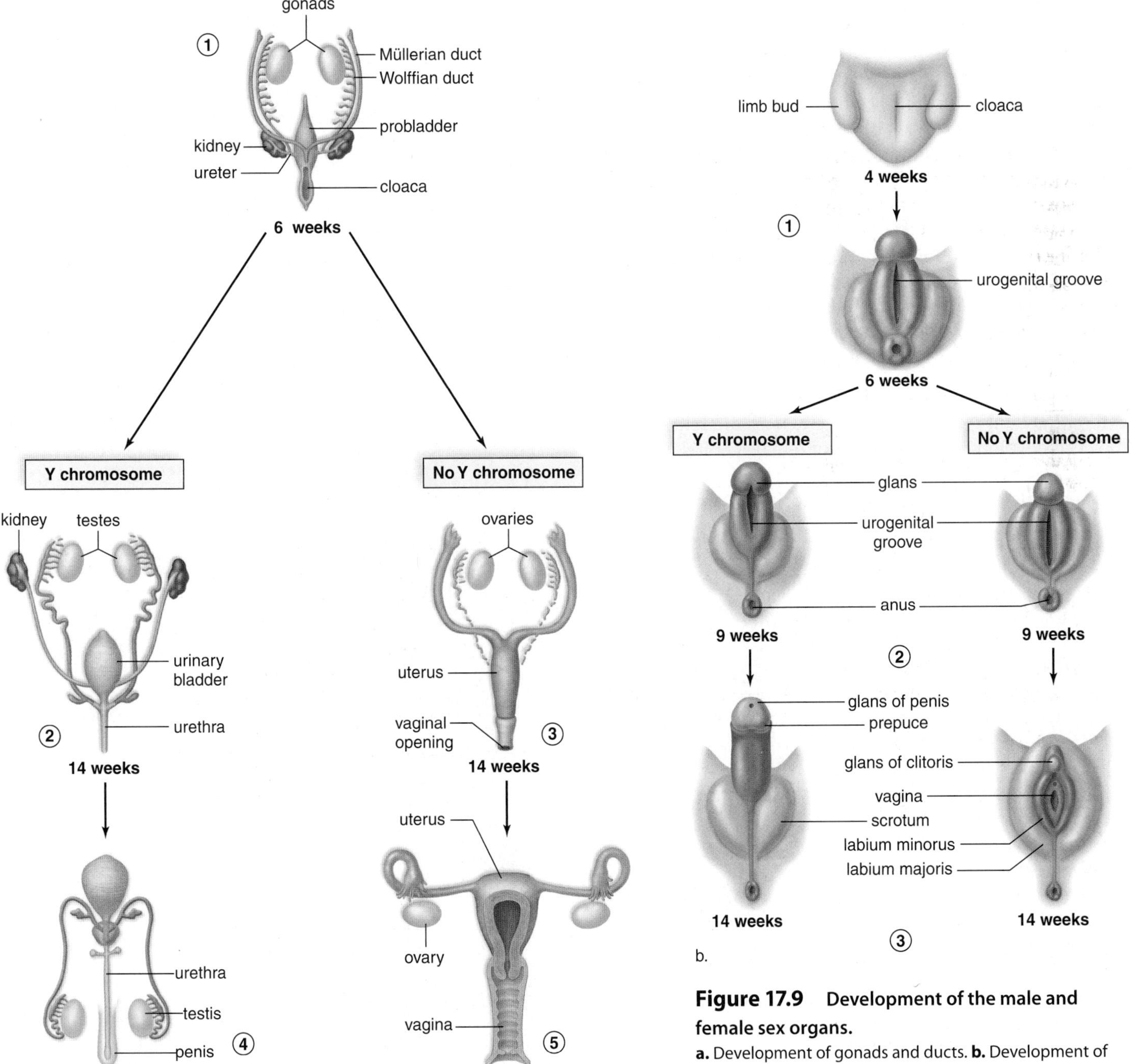

Figure 17.9 Development of the male and female sex organs.
a. Development of gonads and ducts. **b.** Development of external genitals.

Internal Genitals During the first several weeks of development, it is impossible to tell by external inspection whether the unborn child is a boy or a girl. Gonads don't start developing until the seventh week of development. The tissue that gives rise to the gonads is called *indifferent* because it can become testes or ovaries, depending on the action of hormones.

In Figure 17.9*a*, notice that ① at six weeks, both males and females have the same types of ducts. During this indifferent stage, an embryo has the potential to develop into a male or a female. If a gene called *SRY* is present, a protein called *testis-determining factor* is produced that regulates the initial development of the testes. The **testosterone** produced by the testes stimulates the Wolffian ducts to become male genital ducts. ② The Wolffian ducts enter the urethra, which belongs to both the urinary and reproductive systems in males. An anti-Müllerian hormone causes the Müllerian ducts to regress. In the absence of an *SRY* gene, ovaries develop instead of testes from the same indifferent tissue. ③ Now the Wolffian ducts regress, and because of an absence of testosterone, the Müllerian ducts develop into the uterus and oviducts. Estrogen has no effect on the Wolffian duct, which degenerates in females. A developing vagina also extends from the uterus. There is no connection between the urinary and genital systems in females.

At 14 weeks, both the primitive testes and ovaries are located deep inside the abdominal cavity. An inspection of the interior of the ovaries would indicate that they already contain large numbers of tiny follicles, each having an ovum. ④ Toward the end of development, the testes descend into the scrotal sac; ⑤ the ovaries remain in the abdominal cavity.

External Genitals Figure 17.9*b* shows the development of the external genitals. These tissues are also indifferent at first—they can develop into either male or female genitals. ① At six weeks, a small bud appears between the legs; this can develop into the male penis or the female clitoris. ② At nine weeks, a urogenital groove bordered by two swellings appears. ③ By 14 weeks, this groove has disappeared in males, and the scrotum has formed from the original swellings. In females, the groove persists and becomes the vaginal opening. Labia majora and labia minora are present instead of a scrotum. These changes are due to the presence or absence of the hormone dihydrotestosterone (DHT), which is manufactured in the adrenal glands and prostate glands from testosterone.

Connections and Misconceptions

Why is the female gender sometimes referred to as the "default sex"?

The term "default sex" has to do with the presence or absence of the Y chromosome in the fetus. On the Y chromosome, there is a gene called *SRY* (sex determining region Y) that produces a protein that will cause Sertoli cells in the testes to produce Müllerian-inhibiting substance (MIS). This causes Leydig cells in the testes to produce testosterone, which signals the development of the male sex organs (vasa deferentia, epididymis, penis, etc.). Without the *SRY* gene and this hormone cascade effect, female structures (uterus, fallopian tubes, ovaries, etc.) will begin to form.

Abnormal Development of the Genitals

It's not correct to say that all XY individuals develop into males. Some XY individuals become females (XY female syndrome). Similarly, some XX individuals develop into males (XX male syndrome). In individuals with the XY female syndrome, a piece of the Y chromosome is missing. In individuals with the XX male syndrome, this same small piece is present on an X chromosome. The piece of a Y chromosome that causes male genitals to develop is called the *SRY* (sex determining region of the Y) gene. The *SRY* gene causes testes to form, and then the testes secrete these hormones: (1) Testosterone stimulates development of the epididymides, vasa deferentia, seminal vesicles, and ejaculatory duct. (2) Anti-Müllerian hormone prevents further development of female structures and instead causes them to degenerate. (3) Dihydrotestosterone (DHT) directs the development of the urethra, prostate gland, penis, and scrotum.

Ambiguous Sex Determination The absence of any one or more of these hormones results in ambiguous sex determination. The individual has the external appearance of a female, although the gonads of a female are absent.

In *androgen insensitivity syndrome,* these three types of hormones are produced by testes during development, but the individual develops as a female because the receptors for testosterone are ineffective (Fig. 17.10). The external genitalia

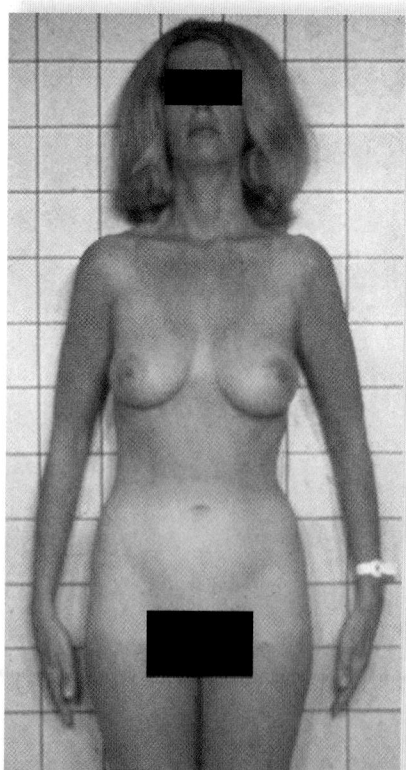

Figure 17.10 Androgen insensitivity affects sexual development. This individual has a female appearance but the XY chromosomes of a male. She developed as a female because her receptors for testosterone are ineffective. Underdeveloped testes are in the abdominal cavity, instead of a uterus and ovaries.

Chapter 17 Development and Aging 407

develop as female, and the Wolffian duct degenerates internally. The individual does not develop a scrotum, so the testes fail to descend and instead remain deep within the body. The individual develops the secondary sex characteristics of a female, and no abnormality is suspected until the individual fails to menstruate.

Connecting the Concepts

For more information on the topics presented in this section, refer to the following discussions.

Figures 16.2 and 16.3 illustrate the structures of the male reproductive system.

Figures 16.6 and 16.7 illustrate the structures of the female reproductive system.

Check Your Progress 17.3

1. Describe the path of blood flow in the fetus starting with the placenta and name the structures unique to fetal circulation.
2. Summarize major events by month during fetal development.
3. Explain how the development of the genitals differs in males and females.

17.4 Pregnancy and Birth

Learning Outcomes

Upon completion of this section, you should be able to

1. Explain the influence of progesterone and estrogen on female physiology during pregnancy.
2. Summarize the events that occur during each stage of birth.

Pregnancy

Major changes that take place in the mother's body during pregnancy are due to the hormones progesterone and estrogen.

The Energy Level Fluctuates

When first pregnant, the mother may experience nausea and vomiting, loss of appetite, and fatigue. These symptoms subside, and some mothers report increased energy levels and a general sense of well-being despite an increase in weight. During pregnancy, the mother gains weight due to breast and uterine enlargement; weight of the fetus; amount of amniotic fluid; size of the placenta; her own increase in total body fluid; and an increase in storage of proteins, fats, and minerals. The increased weight can lead to lordosis (swayback) and lower back pain.

The Uterus Relaxes

Aside from an increase in weight, many of the physiological changes in the mother are due to the presence of the placental hormones that support fetal development (Table 17.2). Progesterone decreases uterine motility by relaxing smooth muscle, including the smooth muscle in the walls of arteries. The arteries expand, and this leads to a low blood pressure that sets in motion the renin–angiotensin–aldosterone mechanism, promoted by estrogen. Aldosterone activity promotes sodium and water retention, and blood volume increases until it reaches its peak sometime during weeks 28 to 32 of pregnancy. Altogether, blood volume increases from 5 L to 7 L—a 40% rise. An increase in the number of red blood cells follows. With the rise in blood volume, cardiac output increases by 20–30%. Blood flow to the kidneys, placenta, skin, and breasts rises significantly. Smooth muscle relaxation also explains the common gastrointestinal effects of pregnancy. The heartburn experienced by many is due to relaxation of the esophageal sphincter and reflux of stomach contents into the esophagus. Constipation is caused by a decrease in intestinal tract motility.

The Pulmonary Values Increase

Of interest is the increase in pulmonary values in a pregnant woman. The bronchial tubes relax, but this alone cannot explain the typical 40% increase in vital capacity and tidal volume. The increasing size of the uterus from a nonpregnant weight of 60–80 g to 900–1,200 g contributes to an improvement in respiratory functions. The uterus comes to occupy most of the abdominal cavity, reaching nearly to the xiphoid process of the sternum. This increase in size not only pushes the intestines, liver, stomach, and diaphragm superiorly but it also widens the thoracic cavity. Compared with nonpregnant values, the maternal oxygen level changes little. Blood carbon dioxide levels fall by 20%, creating a concentration gradient favorable to the flow of carbon dioxide from fetal blood to maternal blood at the placenta.

Still Other Effects

The enlargement of the uterus does result in some problems. In the pelvic cavity, compression of the ureters and urinary bladder can result in stress **incontinence**, or the involuntary

Table 17.2 Effects of Placental Hormones on Mother

Hormone	Chief Effects
Progesterone	Relaxation of smooth muscle; reduced uterine motility; reduced maternal immune response to fetus
Estrogen	Increased uterine blood flow; increased renin–angiotensin–aldosterone activity; increased protein biosynthesis by the liver
Peptide hormones	Increased insulin resistance

leakage of urine from the urinary tract. Compression of the inferior vena cava, especially when lying down, decreases venous return, and the result is edema and varicose veins.

Aside from the steroid hormones progesterone and estrogen, the placenta also produces some peptide hormones. One of these makes cells resistant to insulin, and the result can be gestational diabetes. Some of the integumentary changes observed during pregnancy are also due to placental hormones. **Striae gravidarum,** commonly called "stretch marks," typically form over the abdomen and lower breasts in response to increased steroid hormone levels rather than stretching of the skin. Melanocyte activity also increases during pregnancy. Darkening of certain areas of the skin, including the face, neck, and breast areolae, is common.

Birth

The uterus has contractions throughout pregnancy. At first, these are light, lasting about 20 to 30 seconds and occurring every 15 to 20 minutes. Near the end of pregnancy, the contractions may become stronger and more frequent so that a woman thinks she is in labor. "False labor" contractions are called **Braxton Hicks contractions.** However, the onset of true labor is marked by uterine contractions that occur regularly every 15 to 20 minutes and last for 40 seconds or longer.

A positive feedback mechanism can explain the onset and continuation of labor. Uterine contractions are induced by a stretching of the cervix, which also brings about the release of oxytocin from the posterior pituitary gland. Oxytocin stimulates the uterine muscles, both directly and through the action of prostaglandins. Uterine contractions push the fetus downward, and the cervix stretches even more. This cycle keeps repeating itself until birth occurs.

Prior to or at the first stage of **parturition,** the process of giving birth to an offspring, there can be a "bloody show" caused by expulsion of a mucous plug from the cervical canal. This plug prevented bacteria and sperm from entering the uterus during pregnancy.

Stage 1

During the first stage of labor, the uterine contractions of labor occur in such a way that the cervical canal slowly disappears as the lower part of the uterus is pulled upward

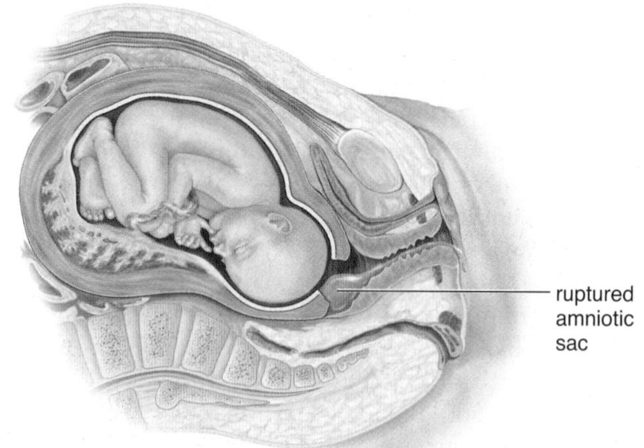

a. First stage of birth: Cervix dilates.

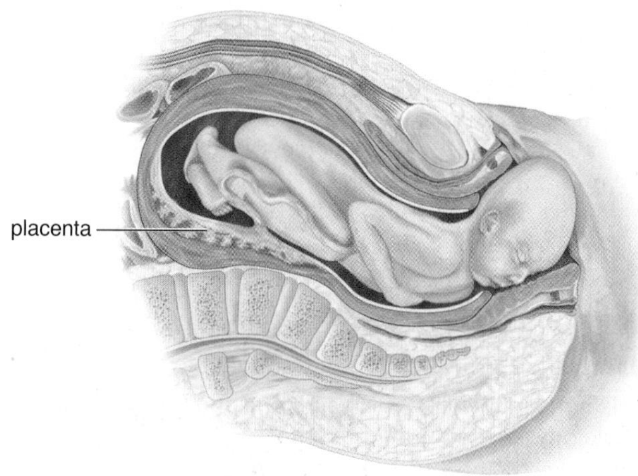

b. Second stage of birth: Baby emerges.

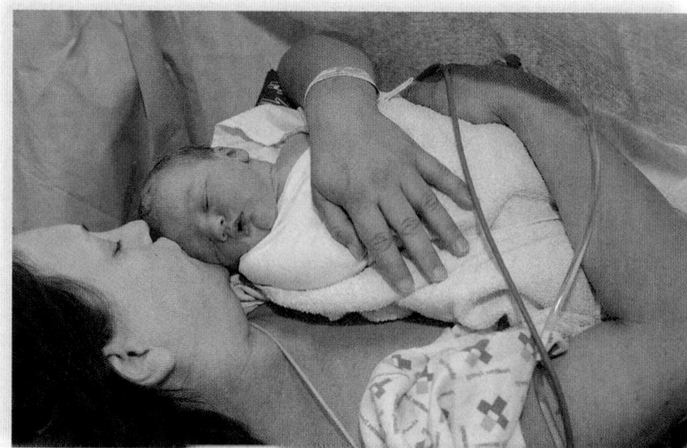

c. Baby has arrived.

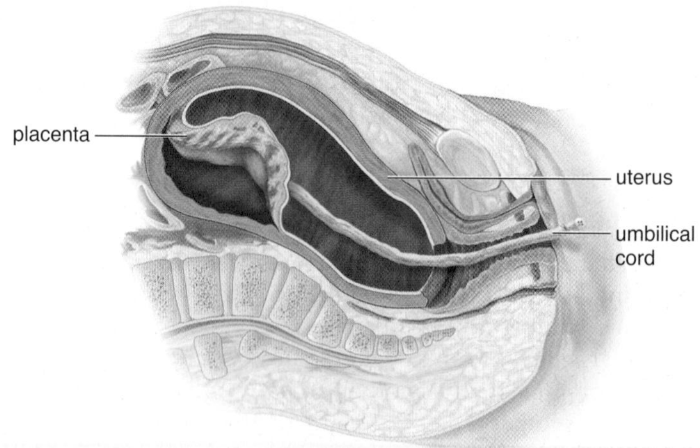

d. Third stage of birth: expelling afterbirth

Figure 17.11 **The stages of birth.**
Birth proceeds in three stages. The baby emerges in stage two.

Biology Matters Historical Focus

An End to "Laudable Pus"

Imagine that it is the nineteenth century. Now imagine that you or your loved one is expecting a baby. You are a little down on your luck, and you cannot seem to scrape together enough money for private childbirth at home. Instead, you will be forced to go to a hospital, and you are terrified. You fear for your life.

In fact, this scenario occurred all across Europe. In 1847, almost 20% of women whose babies were delivered by a doctor in a hospital died of so-called "childbed fever." A day or so after delivery, the woman began to show signs of infection: high fever, low blood pressure, and discharge of pus from her reproductive tract. Physicians of the time accepted these almost-routine infections as part of the delivery process. Even the discharge was thought to be beneficial and an indication of healing and was thus referred to as "laudable (praiseworthy) pus."

By contrast, a woman whose delivery was performed at home by a midwife had a very low incidence of childbed fever, and the Hungarian physician Ignaz Semmelweis (1818–1865) wondered why. At the time, there was no concept of germs or infection, but Semmelweis began to suspect that the doctors were killing their patients. The increasing urbanization of society had resulted in more and more women going to hospitals built to care for the poor. Urban hospitals were training grounds for physicians, and part of their education involved performing autopsies. Semmelweis began to understand the source of childbed fever when one of his mentors died following an accidental stab wound inflicted during an autopsy. When Semmelweis autopsied his mentor, the body looked like those of women with childbed fever.

Semmelweis proposed that something contagious was being carried from the autopsies to the delivery wards. He placed basins of chlorinated lime at the entrance to each ward and required that all who entered wash their hands. He also instituted the use of nailbrushes for cleaning the fingernails. The incidence of childbed fever dropped from 18% to 1.3% during the first seven months after the hand-washing protocol was instituted and continued to drop thereafter.

Unfortunately, despite his dramatic success, Semmelweis was neither liked nor respected by his colleagues. Suggesting that dirty, careless doctors were to blame for the mothers' deaths earned only hostility. Presentations and publications of his work were largely ignored. In 1865, he began to deteriorate mentally and was committed to a mental institution. He died two weeks later at the age of 47. Ironically, his own death was most likely from an infection similar to "childbed fever."

toward the baby's head. This process is called **effacement,** or "taking up the cervix." With further contractions, the baby's head acts as a wedge to assist cervical dilation (Fig. 17.11*a*). If the amniotic membrane has not already ruptured, it is apt to do so during this stage, releasing the amniotic fluid, which leaks out of the vagina (an event sometimes referred to as "breaking water"). The first stage of parturition ends once the cervix is dilated completely.

Stage 2

During the second stage of parturition, the uterine contractions occur every 1 to 2 minutes and last about 1 minute each. They are accompanied by a desire to push, or bear down. As the baby's head gradually descends into the vagina, the desire to push becomes greater. When the baby's head reaches the exterior, it turns so that the back of the head is uppermost (Fig. 17.11*b*). To enlarge the vaginal orifice, an **episiotomy** is often performed. This incision, which enlarges the opening, is sewn together later. As soon as the head is delivered, the physician may hold the head and guide it downward, while one shoulder and then the other emerges. The rest of the baby follows easily (Fig. 17.11*c*).

Once the baby is breathing normally, the umbilical cord is cut and tied, severing the child from the placenta. The stump of the cord shrivels and leaves a scar, the umbilicus (belly button).

Stage 3

The placenta, or **afterbirth,** is delivered during the third stage of parturition (Fig. 17.11*d*). About 15 minutes after delivery of the baby, uterine muscular contractions shrink the uterus and dislodge the placenta. The placenta then is expelled into the vagina. As soon as the placenta and its membranes are delivered, the third stage of parturition is complete.

Video
Triggering Birth

Connecting the Concepts

For more information on the content of this section, refer to the following discussions.

Section 4.9 explains positive feedback mechanisms.

Section 10.4 examines how aldosterone regulates urine formation in the kidneys.

Section 16.4 examines the role of the female sex hormones estrogen and progesterone.

Check Your Progress 17.4

1. Identify the hormonal changes that occur in a female during pregnancy.
2. Explain the following: Maternal blood carbon dioxide levels fall by 20% during pregnancy. How does this benefit the fetus?
3. Describe the three stages of labor.

17.5 Development After Birth

Learning Outcomes

Upon completion of this section, you should be able to

1. Explain the three hypotheses on why humans age.
2. Summarize the effects of aging on the organ systems of the body.

Development does not cease once birth has occurred but continues throughout the stages of life: infancy, childhood, adolescence, and adulthood. Infancy, the toddler years, and preschool years are times of remarkable growth. During the birth to 5-year-old stage, humans acquire gross motor and fine motor skills. These include the ability to sit up and then to walk, as well as being able to hold a spoon and manipulate small objects. Language usage begins during this time and will become increasingly sophisticated throughout childhood. As infants and toddlers explore their environment, their senses—vision, taste, hearing, smell, and touch—mature dramatically. Socialization is very important as a child forms emotional ties with its caregivers and learns to separate self from others. Babies do not all develop at the same rate, and there is a large variation in what is considered normal.

The preadolescent years, from 6 to 12 years of age, are a time of continued rapid growth and learning. Preadolescents form identities apart from parents, and peer approval becomes very important. Adolescence begins with the onset of puberty as the young person achieves sexual maturity. For girls, puberty begins between 10 and 14 years of age, whereas for boys it generally occurs between ages 12 and 16. During this time, the sex-specific hormones (see Chapter 15) cause the secondary sexual characteristics to appear. Profound social and psychological changes are also associated with the transition from childhood to adulthood.

Aging encompasses the progressive changes from infancy until eventual death. Today, **gerontology,** the study of aging, is of great interest because there are now more older individuals in our society than ever before. The number is expected to rise dramatically; in the next half-century, the number of people over age 65 will increase 147%. The human life span is judged to be a maximum of 120 to 125 years. The present goal of gerontology is not necessarily to increase the life span but to increase the health span, the number of years that an individual enjoys the full functions of all body parts and processes (Fig. 17.12).

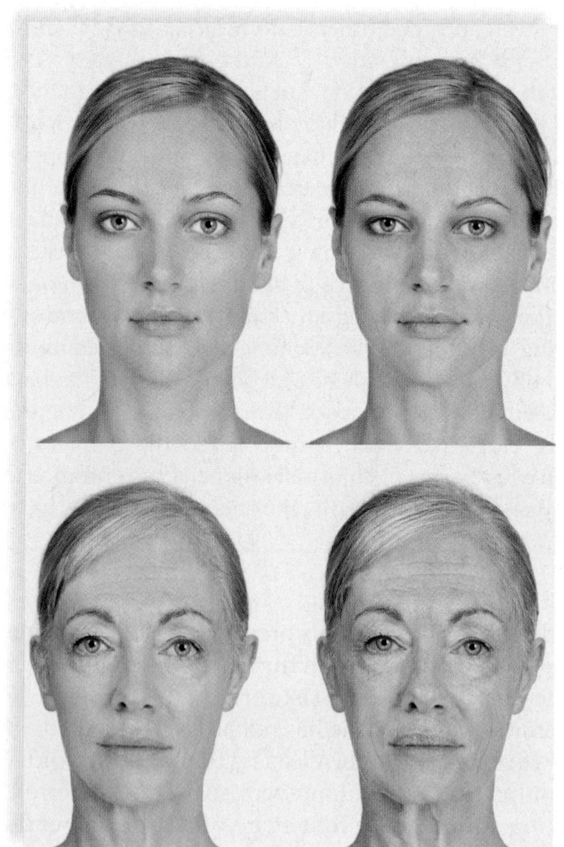

Figure 17.12 **The effects of aging.**
Aging is a slow process during which the body undergoes changes that eventually bring about death, even if no marked disease or disorder is present. Medical science is trying to extend the human life span and the health span, the length of time the body functions normally.

Hypotheses of Aging

Of the many hypotheses about the cause of aging, three are considered here.

Genetic in Origin

Several lines of research indicate that aging has a genetic basis. Researchers working with simple organisms, such as yeast, roundworms and fruit flies, have identified a host of genes whose expression decreases the life span. If these genes are silenced through mutations or restricted food intake, the organism lives longer. What do these genes have in common? Apparently, when these genes are inactive, mitochondria do not produce energy—the cell uses alternative pathways. The current *mitochondrial hypothesis of aging* has been supported by engineering mice that had a defective DNA polymerase. (Recall that mitochondria have their own DNA.) These mice aged much faster than their peers. Why? Possibly because their defective mitochondria produced more free radicals than usual. Free radicals (see Chapter 8) are unstable molecules that carry an extra electron. To become stable, free radicals donate an electron to another molecule, such as DNA or proteins (e.g., enzymes) or lipids, found in plasma membranes. Eventually, these molecules are unable to function and the cell is destroyed. The well-known observation that a low-calorie diet can expand the life span is consistent with the mitochondrial hypothesis of aging. Caloric restriction also shuts down the genes that decrease the life span—the genes that turn on the activity of mitochondria!

Whole-Body Process

A decline in the hormonal system can affect many different organs of the body. For example, type 2 diabetes is common in older individuals. The pancreas makes insulin, but the cell's receptors are ineffective. Menopause in women and andropause in men occur for similar reasons. The bloodstream contains adequate amounts of anterior pituitary hormones, but the ovaries and testes do not respond. The ovaries do not produce adequate amounts of estrogen and progesterone. Likewise, testosterone secretion diminishes. Perhaps aging results from the loss of hormonal activities and a decline in the functions they control.

The immune system, too, no longer performs as it once did, and this can affect the body as a whole. The thymus gland gradually decreases in size, and eventually most of it is replaced by fat and connective tissue. The incidence of cancer increases among the elderly, which may signify that the immune system is no longer functioning as it should. This idea is also substantiated by the increased incidence of autoimmune diseases in older individuals.

It is possible, though, that aging is due not to the failure of a particular system that can affect the body as a whole but to a specific type of tissue change that affects all organs. It has been noticed for some time that proteins such as the collagen fibers present in many support tissues become increasingly cross-linked as people age. Undoubtedly, this cross-linking contributes to the stiffening and loss of elasticity characteristic of aging tendons and ligaments. It may also account for the inability of organs such as the blood vessels, heart, and lungs to function as they once did.

Connections and Misconceptions

What genes may be associated with longevity?

In the past several years, researchers have begun to use new molecular techniques to unravel some of the mysteries as to why some individuals live past the age of 100—the centenarians. A variation of one gene, *FOX03A*, has been identified to be more prevalent in centenarians than the general population. This gene regulates the insulin pathways of the body and also appears to control the genetic mechanisms that protect cells against free radicals. Both insulin regulation and protection against free radicals have been known for some time to enhance longevity in a variety of organisms. Interestingly, *FOX03A* is also believed to be involved in the process of apoptosis and may help protect the body against cancer. Although researchers do not think that *FOX03A* is the sole gene for longevity, it is providing an starting point for larger studies on human aging.

Extrinsic Factors

The current data about the effects of aging are often based on comparisons of elderly people to people in younger age groups. When today's elderly people were young, they may not have been aware of the benefits of such things as diet and exercise to general health. It is possible, then, that much of what we attribute to aging is instead due to years of poor health habits.

Consider osteoporosis. This condition is associated with a progressive decline in bone density in both males and females, so fractures are more likely to occur after only minimal trauma. Osteoporosis is common in elderly people. By age 65, one-third of women will have vertebral fractures. By age 81, one-third of women and one-sixth of men will have suffered a hip fracture. Although there is no denying that a decline in bone mass occurs as a result of aging, certain extrinsic factors are also important. The occurrence of osteoporosis is associated with cigarette smoking, heavy alcohol intake, and inadequate calcium intake. It is possible not only to eliminate these negative factors by personal choice but also to add a positive factor. A moderate exercise program has been found to slow down the progressive loss of bone mass.

Even more important, a sensible exercise program and a proper diet that includes at least five servings of fruits and vegetables a day will most likely help eliminate cardiovascular disease. Experts no longer believe that the cardiovascular system necessarily suffers a large decrease in functioning ability with age. Persons 65 years of age and older can have well-functioning hearts and open coronary arteries if their health habits are good and they continue to exercise regularly.

Biology Matters Health Focus

Alzheimer Disease

In 1900, the average life span in the United States was 47 years of age. Today, it is 75 years of age. Normal aging does involve some changes in mental faculties, but many of the changes we associate with old age are related to disease, not aging. Two of the more common diseases are Alzheimer disease (AD) and Parkinson disease.

AD is characterized by the presence of abnormally structured neurons and a reduced amount of acetylcholine (ACh). The AD neuron has two characteristic features. Bundles of fibrous protein, called neurofibrillary tangles, extend from the axon to surround the nucleus of the neuron. Tangles form when the supporting protein, called tau, becomes malformed (Fig. 17B) and twists the neurofibrils, which are normally straight. In addition, protein-rich accumulations called amyloid plaques envelop branches of the axon. The plaques grow so dense that they trigger an inflammatory reaction that causes neuron death.

Video Prion Disease

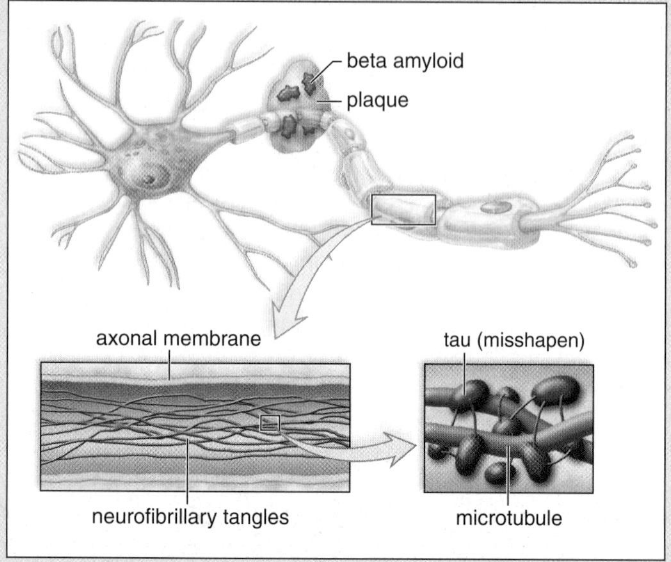

Figure 17B **Tau protein and Alzheimer disease.** Some of the neurons of Alzheimer disease (AD) patients have beta amyloid plaques and neurofibrillary tangles. AD neurons are present throughout the brain but concentrated in the hippocampus and amygdala.

Treatment for Alzheimer Disease

Treatment for AD involves using one of two categories of drugs. Cholinesterase inhibitors work at neuron synapses in the brain, slowing the activity of acetylcholinesterase, the enzyme that breaks down acetylcholine (ACh). Allowing ACh to accumulate in synapses keeps memory pathways in the brain functional for a longer period. A second drug, memantine, blocks *excitotoxicity,* the tendency of diseased neurons to self-destruct. This recently approved medication is used only in moderately to severely affected patients. Using the drug allows neurons involved in memory pathways to survive longer in affected patients. Successes with these medications indicates that treatment for AD patients should begin as soon as possible after diagnosis and continue indefinitely. However, neither type of medication cures AD; both merely slow the progress of disease symptoms, allowing the patient to function independently for a longer time. Additional research is currently underway to test the effectiveness of anticholesterol *statin* drugs, as well as anti-inflammatory medications, in slowing the progress of the disease.

Much of current research on AD focuses on the prevention and cure of the disease. Scientists believe that curing AD will require an early diagnosis, because it is thought that the disease may begin in the brain 15 to 20 years before symptoms ever develop. Currently, diagnosis can't be made with absolute certainty until the brain is examined at autopsy. A new test on the cerebrospinal fluid may allow early detection of amyloid proteins and a much earlier diagnosis of the disease. Researchers are also testing vaccines for AD, which would target the patient's immune system to destroy amyloid plaques.

Early findings have shown that risk factors for cardiovascular disease—heart attacks and stroke—also contribute to an increased incidence of AD. Risk factors for cardiovascular disease include elevated blood cholesterol and blood pressure, smoking, obesity, sedentary lifestyle, and diabetes mellitus. Thus, evidence suggests that a lifestyle tailored for good cardiovascular health may also prevent AD.

Video Apples for Alzheimers

Effect of Age on Body Systems

Data about how aging affects body systems are necessarily based on past events. It is possible that, in the future, age will not have these effects—or at least not to the same degree as those described here.

Skin

As aging occurs, skin becomes thinner and less elastic because the number of elastic fibers decreases. The collagen fibers undergo cross-linking, as discussed previously. Also, there is less adipose tissue in the subcutaneous layer. Therefore, older people are more likely to feel cold. The loss of thickness partially accounts for sagging and wrinkling of the skin.

Homeostatic adjustment to heat is also limited because there are fewer sweat glands for sweating to occur. The hair on the scalp and the extremities thins out because of fewer hair follicles. The number of oil (sebaceous) glands is reduced, and the skin tends to crack. Older people also experience a decrease in the number of melanocytes, making their hair gray and their skin pale. In contrast, some of the remaining pigment cells are larger, and pigmented blotches appear on the skin.

Processing and Transporting

Cardiovascular disorders are the leading cause of death today. The heart shrinks because of a reduction in cardiac muscle cell size. This leads to loss of cardiac muscle strength and reduced cardiac output. Still, the heart, in the absence of disease, is able to meet the demands of increased activity. It can double its rate or triple the amount of blood pumped each minute even though the maximum possible output declines.

The middle layer of arteries contains elastic fibers, most likely subject to cross-linking, so the arteries become more rigid with time. Their size is further reduced by plaque, a buildup of fatty material. Therefore, blood pressure readings gradually rise. Such changes are common in individuals living in Western industrialized countries but not in agricultural societies. A diet low in cholesterol and saturated fatty acids has been suggested as a way to control degenerative changes in the cardiovascular system.

Blood flow to the liver is reduced, and this organ does not metabolize drugs as efficiently as before. This means that as a person gets older, less medication is needed to maintain the same level of a drug in the bloodstream.

Cardiovascular problems are often accompanied by respiratory disorders, and vice versa. Growing inelasticity of lung tissue means that ventilation is reduced. We rarely use the entire vital capacity, so these effects are not noticed unless the demand for oxygen increases.

Blood supply to the kidneys is also reduced. The kidneys become smaller and less efficient at filtering wastes. Salt and water balance is difficult to maintain, and elderly people dehydrate faster than young people do. Difficulties involving urination include incontinence (lack of bladder control) and the inability to urinate. In men, the prostate gland may enlarge and reduce the diameter of the urethra. This can make urination so difficult that surgery is needed.

Integration and Coordination

Though most tissues of the body regularly replace their cells, some at a faster rate than others, the brain and the muscles ordinarily do not. However, contrary to previous opinion, studies show that few neural cells of the cerebral cortex are lost during the normal aging process. This means that cognitive skills remain unchanged even though a loss in short-term memory characteristically occurs. Although elderly people learn more slowly than young people, they can acquire and remember new material. The results of tests indicate that when more time is given for the subject to respond, age differences in learning decrease.

Neurons are extremely sensitive to oxygen deficiency; if neuron death does occur, it may be due not to aging but to reduced blood flow in narrowed blood vessels. Reaction time, however, does slow, and more stimulation is needed for hearing, taste, and smell receptors to function as before. After age 50, the ability to hear tones at higher frequencies decreases gradually; this can make it difficult to identify individual voices and to understand conversation in a group. The lens of the eye does not accommodate as well and also may develop a cataract. Glaucoma, the buildup of pressure due to increased fluid, is more likely to develop because of a reduction in the size of the anterior cavity of the eye.

Loss of skeletal muscle mass is not uncommon, but it can be controlled by following a regular exercise program. The capacity to do heavy labor decreases, but routine physical work should be no problem. Lung function may decrease because of decreases in both rib cage flexibility and respiratory muscle strength. Still, many healthy senior citizens will be able to continue everyday activities well into the eighth or ninth decade. Reduced muscularity of the urinary bladder contributes to an inability to empty the bladder completely during urination. Occasional urinary incontinence may result. Risk of urinary tract infection is also increased.

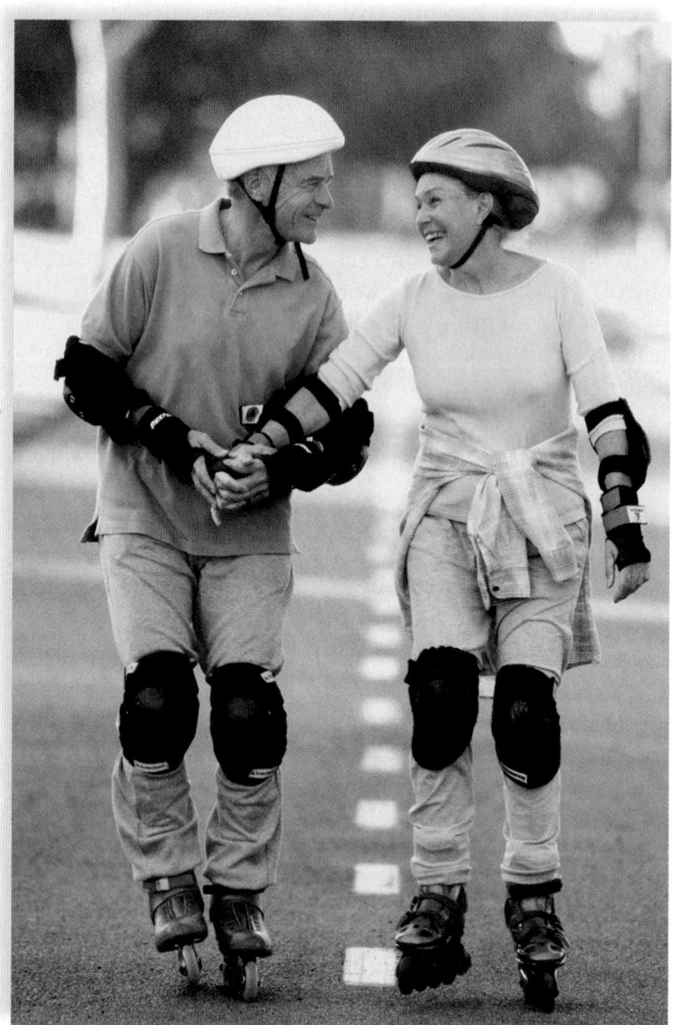

Figure 17.13 **What steps can an individual take to increase health span?**
Gerontology research has shown that regular physical exercise, as well as staying engaged both mentally and socially, can slow the progress of aging and lengthen the health span.

The Reproductive System

As mentioned, females undergo menopause; thereafter, the level of female sex hormones in the blood falls markedly. The uterus and the cervix decrease in size, and the walls of the oviducts and the vagina become thinner. The external genitals become less pronounced. Males undergo andropause, and the level of androgens falls gradually over the age span of 50 to 90, but some sperm production continues until death.

As a group, females live longer than males. Males suffer a marked increase in heart disease in their 40s. An increase is not noted in females until after menopause, when women lead men in the incidence of stroke. Men are still more likely than women to have a heart attack, however. At one time, it was thought that estrogen offers women some protection against cardiovascular disorders. This hypothesis is called into question because postmenopausal administration of estrogen has been shown to increase the risk of cardiovascular disorders in women.

Conclusion

We have listed many adverse effects of aging; however, though such effects are seen, they are not inevitable (Fig. 17.13). We must discover any extrinsic factors that precipitate these adverse effects and guard against them. Just as it is wise to make the proper preparations to remain financially independent when older, it is also wise to realize that, biologically, successful old age begins with the health habits developed when we are younger.

Connecting the Concepts

For more information on the organ systems presented in this section, refer to the following discussions.

Section 5.1 provides an overview of the cardiovascular system.

Section 10.1 provides an overview of the urinary system.

Section 13.1 provides an overview of the nervous system.

Check Your Progress 17.5

1. Distinguish between the three different hypotheses of aging presented in this section.
2. Summarize the effect of aging on the various body systems.
3. Discuss the best way to keep healthy, even though aging occurs.

CASE STUDY CONCLUSION

Over the course of the next few office visits, Amber's doctor performed a variety of tests. Because Amber was over 35 years old when she conceived her first child, the doctor recommended that a chorionic villus sampling test (see next chapter) be done to test for birth defects, such as Down syndrome. In addition, the doctor performed a glucose tolerance test to determine if Amber was experiencing gestational diabetes. An additional blood test, called an alpha-fetoprotein (AFP) test, screened for neural tube defects and additional evidence of Down syndrome. Much to the relief of Amber and Kent, all of these tests came back normal. Then, around week 20 of Amber's pregnancy, the doctor made an appointment for Amber to get an ultrasound exam, which is frequently used to confirm the due date of the baby but can also be used to look for birth defects such as cleft palate. Everything once again appeared normal. Finally, the technician asked the parents the question that they had been waiting for: were they interested in knowing the sex of their baby? Within a few minutes, Amber and Kent found out that, in just a few short months, they would be the parents of a baby girl.

Media Study Tools

www.mhhe.com/maderhuman12e

Enhance your study of this chapter with study tools and practice tests. Also ask your instructor about the resources available through ConnectPlus, including the media-rich eBook, interactive learning tools, and animations.

Summarizing the Concepts

17.1 Fertilization

The acrosome of a sperm releases enzymes that digest a pathway for the sperm through the zona pellucida. The sperm nucleus enters the egg and fuses with the egg nucleus.

17.2 Pre-Embryonic and Embryonic Development

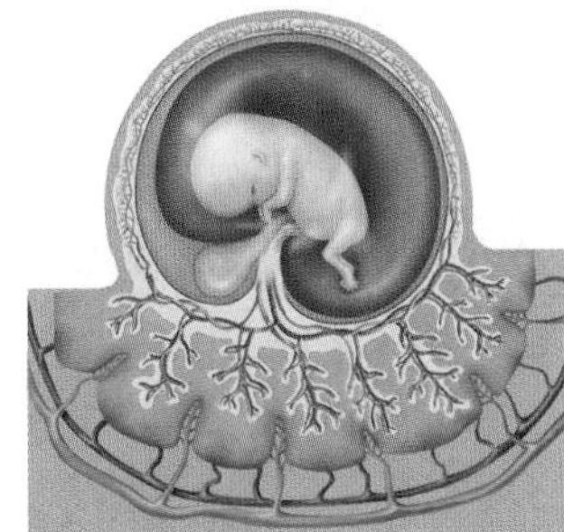

- Cleavage, growth, morphogenesis, and differentiation are the processes of development.
- The extraembryonic membranes (chorion, allantois, yolk sac, and amnion) function in internal development.

17.3 Fetal Development

- At the end of the embryonic period, all organ systems are established and there is a mature and functioning placenta. The umbilical arteries and umbilical vein take blood to and from the placenta, where exchanges take place.

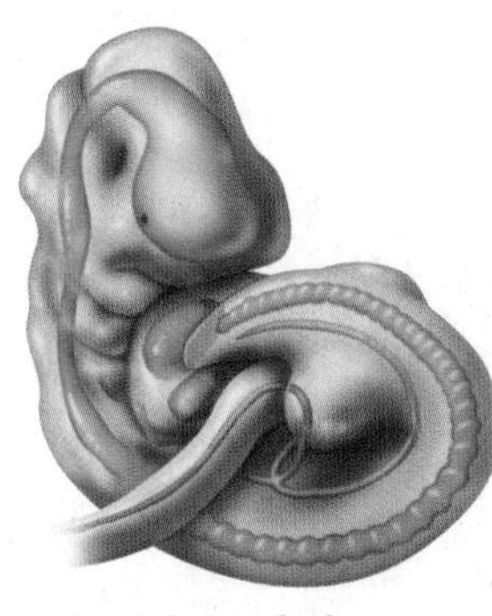

- Exchanges supply the fetus with oxygen and nutrients and rid the fetus of carbon dioxide and wastes.
- The venous duct joins the umbilical vein to the inferior vena cava.
- The oval duct and arterial duct allow the blood to pass through the heart without going to the lungs.

  Fetal development extends from the third through the ninth months.
- During the third and fourth months, the skeleton is becoming ossified.
- The sex of the fetus becomes distinguishable. If an *SRY* gene is present, testes and male genitals develop. Otherwise, ovaries and female genitals develop.
- During the fifth through the ninth months, the fetus continues to grow and to gain weight.

17.4 Pregnancy and Birth

Pregnancy

Major changes take place in the mother's body during pregnancy.

- Weight gain occurs as the uterus occupies most of the abdominal cavity.
- Many complaints, such as constipation, heartburn, darkening of certain skin areas, and pregnancy-induced diabetes, are due to the presence of placental hormones.

Birth

A positive feedback mechanism that involves uterine contractions and oxytocin explains the onset and continuation of labor.

- During stage 1 of parturition (birth), the cervix dilates.
- During stage 2, the child is born.
- During stage 3, the afterbirth is expelled.

17.5 Development After Birth

Development after birth consists of infancy, childhood, adolescence, and adulthood.

- Aging encompasses progressive changes from about age 20 on that contribute to an increased risk of infirmity, disease, and death.

Hypotheses of Aging

- Aging may have a genetic basis.
- Aging may be due to changes that affect the whole body (e.g., decline of hormonal system).
- Aging may be due to extrinsic factors (e.g., diet and exercise).

Effect of Age on Body Systems

- Deterioration of organ systems can possibly be prevented or reduced in part by using good health habits.

Understanding Key Terms

afterbirth 410
aging 410
allantois 397
amnion 397
blastocyst 397
Braxton Hicks contraction 408
breech birth 405
cesarean section 405
chorion 397
chorionic villi 400
cleavage 396
differentiation 396
ectopic pregnancy 398
effacement 409
embryo 398
embryonic development 396
embryonic disk 398
episiotomy 409
extraembryonic membrane 397
fertilization 394
fetal development 403
fontanel 403
gastrulation 398
gerontology 410
growth 396
human chorionic gonadotropin (hCG) 398
implantation 398
incontinence 407
inner cell mass 397
lanugo 403
morphogenesis 396
morula 397
parturition 408
placenta 397
pre-embryonic development 396
primary germ layer 398
reproductive cloning 404
striae gravidarum 408
testosterone 406
therapeutic cloning 404
umbilical cord 400
vernix caseosa 403
yolk sac 397
zygote 394

Match the key terms to these definitions.

a. ________ Short, fine hair present during the later portion of fetal development.

b. ________ The placenta delivered during the third stage of parturition.

c. ________ The study of aging.

d. ________ Union of a sperm nucleus and an egg nucleus, which creates a zygote with the diploid number of chromosomes.

e. ________ Mitotic cell division of the zygote with no increase in cell size.

Testing Your Knowledge of the Concepts

1. Describe how polyspermy is prevented during fertilization. (page 394)
2. Name the four embryonic membranes and give a human function for each one. (page 397)
3. Justify the division of development into pre-embryonic, embryonic, and fetal development. (pages 396–398)
4. What are the three primary germ layers, and what body structures come from each germ layer? (pages 398–399)
5. Briefly summarize the weekly events of embryonic development. (pages 396–400)
6. Briefly summarize the monthly events of fetal development. (pages 401–405)
7. Explain how blood circulates to and from the placenta and the fetus. How is blood shunted away from the lungs? (pages 402, 403)
8. List the hormones involved in the development of the male and female internal and external sex organs and state their functions. (pages 405–406)
9. Describe some of the changes that occur in the mother during pregnancy. (page 407)
10. What event marks the end of each stage of birth? (pages 408–410)
11. Discuss three hypotheses concerning aging. How can you prevent the major changes that can occur in the body as we age? (pages 411–414)
12. Only one sperm enters an egg because
    a. sperm have an acrosome.
    b. the corona radiata gets larger.
    c. changes occur in the zona pellucida.
    d. the cytoplasm hardens.
    e. All of these are correct.
13. When all three germ layers are present (ectoderm, endoderm, and mesoderm), what event has occurred?
    a. blastulation
    b. limb formation
    c. gastrulation
    d. morulation
14. Which of these is not a process of development?
    a. cleavage
    b. parturition
    c. growth
    d. morphogenesis
    e. differentiation
15. Which of these is mismatched?
    a. chorion—sense perception
    b. yolk sac—first site of blood cell formation
    c. allantois—umbilical blood vessels
    d. amnion—contains fluid that protects embryo

16. In human development, which part of the blastocyst will develop into a embryo?
    a. trophoblast
    b. inner cell mass
    c. chorion
    d. yolk sac
17. Which primary germ layer is not correctly matched to an organ system or organ that develops from it?
    a. ectoderm—the nervous system
    b. endoderm—lining of the digestive tract
    c. mesoderm—skeletal system
    d. endoderm—cardiovascular system
18. Human chorionic gonadotropin is a
    a. hormone.
    b. basis of pregnancy test.
    c. cause of ectopic pregnancy.
    d. Both a and b are correct.
19. Which is a correct sequence that ends with the stage that implants?
    a. morula, blastocyst, embryonic disk, gastrula
    b. ovulation, fertilization, cleavage, morula, early blastocyst
    c. embryonic disk, gastrula, primitive streak, neurula
    d. primitive streak, neurula, extraembryonic membranes, chorion
20. Differentiation is equivalent to which term?
    a. morphogenesis
    b. growth
    c. specialization
    d. gastrulation
21. Which process refers to the shaping of the embryo and involves cell migration?
    a. cleavage
    b. differentiation
    c. growth
    d. morphogenesis
22. Which association is not correct?
    a. third and fourth months—fetal heart has formed, but it does not beat
    b. fifth through seventh months—mother feels movement
    c. eighth through ninth months—usually head is now pointed toward the cervix
    d. All of these are correct.
23. At three months, the embryo has
    a. become a fetus.
    b. body systems already.
    c. a head, arms, and legs.
    d. ears and eyes, which don't function.
    e. All but b are correct.
24. Which of these structures is not a circulatory feature unique to the fetus?
    a. arterial duct
    b. oval opening
    c. umbilical vein
    d. pulmonary trunk
25. Which of these statements is correct?
    a. Fetal circulation, like adult circulation, takes blood equally to a pulmonary circuit and a systemic circuit.
    b. Fetal circulation shunts blood away from the lungs but makes full use of the systemic circuit.
    c. Fetal circulation includes exchange of substances between fetal blood and maternal blood at the placenta.
    d. Unlike adult circulation, fetal blood always carries oxygen-rich blood and therefore has no need for the pulmonary circuit.
    e. Both b and c are correct.
26. Which of these is a hormone involved in development of male and female sex organs?
    a. estrogen
    b. anti-Müllerian hormone
    c. dihydrotestosterone
    d. testosterone
    e. All of these hormones are involved.
27. Which hormone can be administered to begin the process of childbirth?
    a. estrogen
    b. oxytocin
    c. prolactin
    d. testosterone
    e. Both b and d are correct.
28. Label this diagram illustrating the placement of the extraembryonic membranes, and give a function for each membrane in humans.

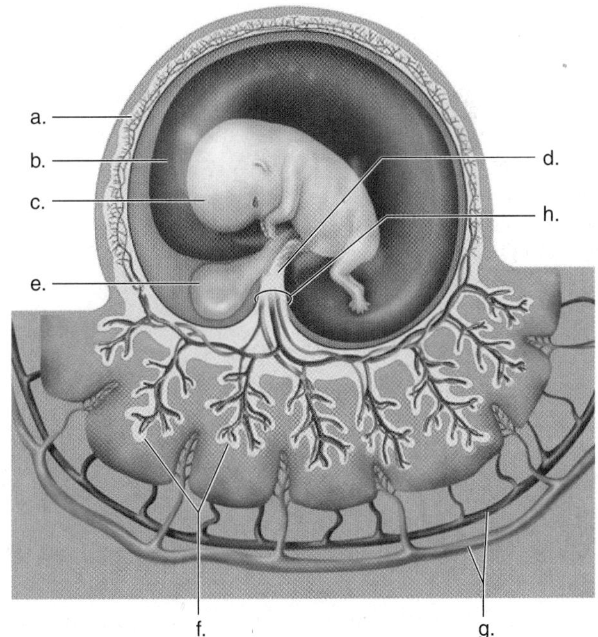

Thinking Critically About the Concepts

Amber and Kent used a home pregnancy test to determine if she was pregnant. These tests detect the level of hCG (human chorionic gonadotropin; see page 398) in the urine. This hormone is released following implantation of the embryo into the uterus, usually around six days after fertilization. Some tests claim that they are sensitive enough to detect hCG on the date that menstruation is expected to begin. However, doctors recommend waiting until menstruation is one week late. If pregnant, a woman's level of hCG rises with each passing day, and testing is more likely to be accurate. However, even with a negative test result, the woman may still be pregnant if hCG levels are too low to be detected at the time of the first test. The test should be repeated later if menstruation doesn't begin. The home pregnancy tests contain a positive control. This is a visual sign (usually a line or a +) that appears if the test is working correctly. If this line does not appear, the test is not valid and must be repeated.

1. At home, pregnancy tests check for the presence of hCG in a female's urine. Where does hCG come from? Why is hCG found in a pregnant woman's urine?
2. A blood test at a doctor's office can also check for the presence of hCG in a female's blood.
   a. Why would you expect to find hCG circulating in a pregnant female's blood?
   b. hCG is a protein; so how does hCG affect its target cells?
3. a. What pituitary hormone is checked with a blood test to diagnose menopause?
   b. Will levels of this hormone be increased or decreased if the female is in menopause?
   c. How does the changed (increased or decreased) level of this pituitary hormone cause the onset of menopause (cessation of menses)?

CHAPTER 18

Patterns of Chromosome Inheritance

CASE STUDY CELL CYCLE CONTROL

In August 2008, actress Christina Applegate, then 36 years old, revealed to shocked fans that she had been diagnosed with breast cancer. Every year, over 192,000 American women are diagnosed with breast cancer. Christina's mother is a breast cancer survivor, and Christina tested positive for a *BRCA1* (breast cancer predisposition gene 1) gene mutation, which is linked to both breast and ovarian cancer. Cancer results from a failure to control the cell cycle, a series of steps that all cells go through prior to initiating cell division. *BRCA1* is an important component of that control mechanism. *BRCA1* is a tumor suppressor gene, meaning that at specific points in the cell cycle, called checkpoints, *BRCA1* will slow down the cell cycle if it detects damage in the DNA. However, in Christina's case, the version of *BRCA1* that she inherited from her parents was not acting as a gatekeeper. Instead of preventing damaged cells from progressing into cell division, Christina's genes were allowing them to multiply and divide. The result of her body's failure to control the cell was a tumor in one of her breasts. It is estimated that 1 in 833 people possesses the mutant *BRCA1* allele. Although it is not the only genetic contribution to breast and ovarian cancers, it does play a major role. Several medical approaches, included increased surveillance and surgery, are available to those that test positive for a mutated *BRCA1* gene.

As you read through the chapter, think about the following questions.

1. What are the roles of checkpoints in the cell cycle?
2. At what checkpoint in the cell cycle would you think that *BRCA1* would normally be active?

CHAPTER CONCEPTS

18.1 Chromosomes
A karyotype is a picture of chromosomes about to divide.

18.2 The Cell Cycle
The cell cycle consists of interphase and mitosis and is regulated by a series of checkpoints.

18.3 Mitosis
Mitosis is duplication division in which the daughter cells have the same number and types of chromosomes as the mother cell.

18.4 Meiosis
Meiosis is reduction division in which the daughter cells have half the number of chromosomes, and different combinations of genes, as the parent cell.

18.5 Comparison of Meiosis and Mitosis
Meiosis I uniquely pairs and separates the paired chromosomes so that the daughter cells have half the number of chromosomes. Meiosis II is exactly like mitosis, except the cells have half the number of chromosomes.

18.6 Chromosome Inheritance
Abnormalities in chromosome inheritance occur due to changes in chromosome number and changes in chromosome structure.

BEFORE YOU BEGIN

Before beginning this chapter, take a few moments to review the following discussions.

Section 2.7 What is the structure of a DNA molecule? How does this structure differ from that of RNA?

Section 3.4 What is chromatin?

Section 3.5 What is the function of microtubules and actin filaments in a cell?

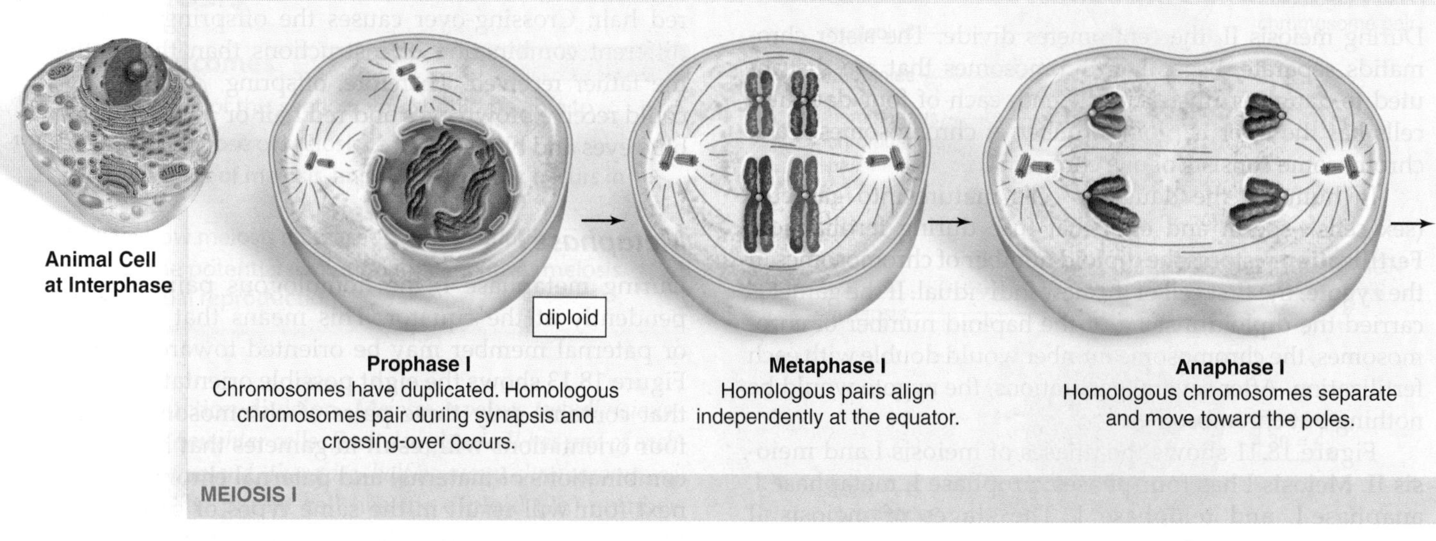
Animal Cell
at Interphase
diploid
Prophase I
Chromosomes have duplicated. Homologous chromosomes pair during synapsis and crossing-over occurs.
Metaphase I
Homologous pairs align independently at the equator.
Anaphase I
Homologous chromosomes separate and move toward the poles.
MEIOSIS I

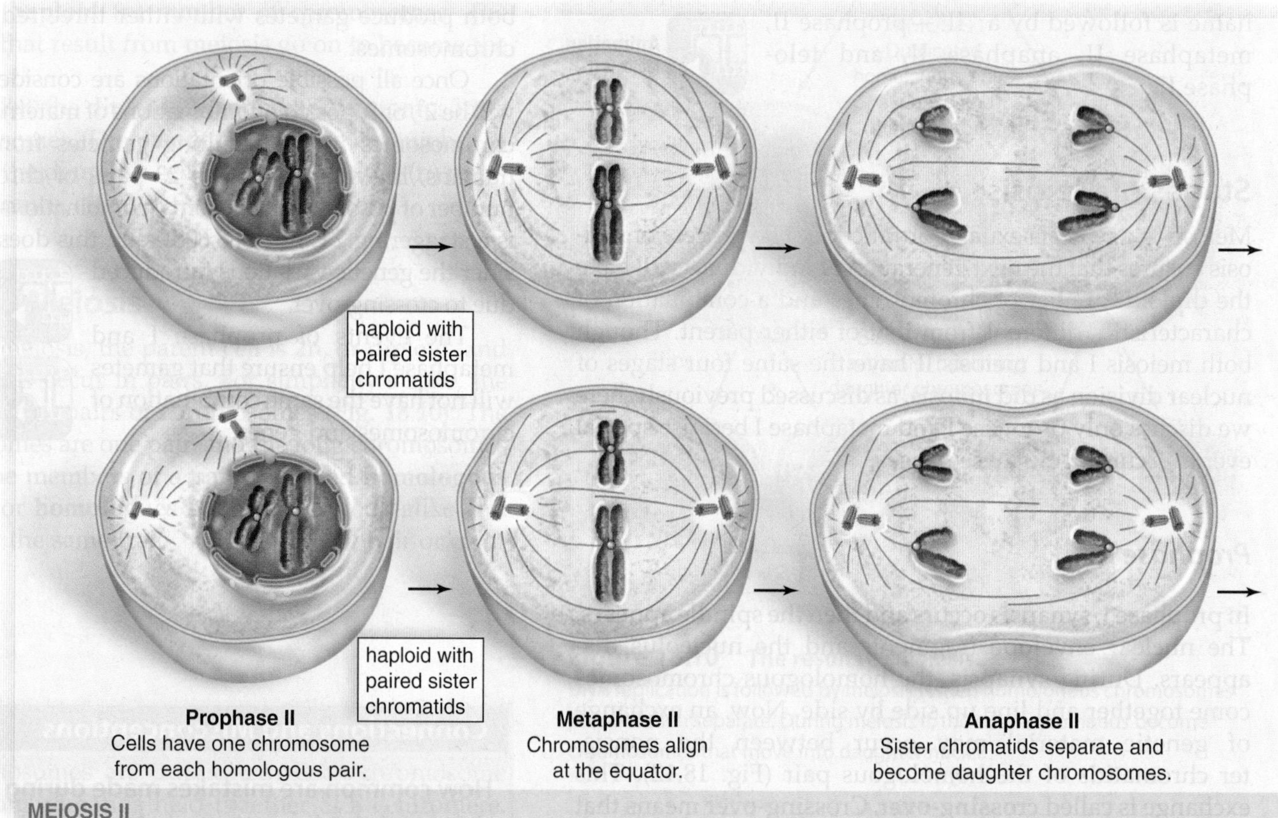
haploid with paired sister chromatids
haploid with paired sister chromatids
Prophase II
Cells have one chromosome from each homologous pair.
Metaphase II
Chromosomes align at the equator.
Anaphase II
Sister chromatids separate and become daughter chromosomes.
MEIOSIS II

Figure 18.11 The phases of meiosis.
Homologous chromosomes pair and then separate during meiosis I. Crossing-over, which occurs during meiosis I, is discussed more fully on page 432. Chromatids separate, becoming daughter chromosomes during meiosis II. Following meiosis II, there are four haploid daughter cells.

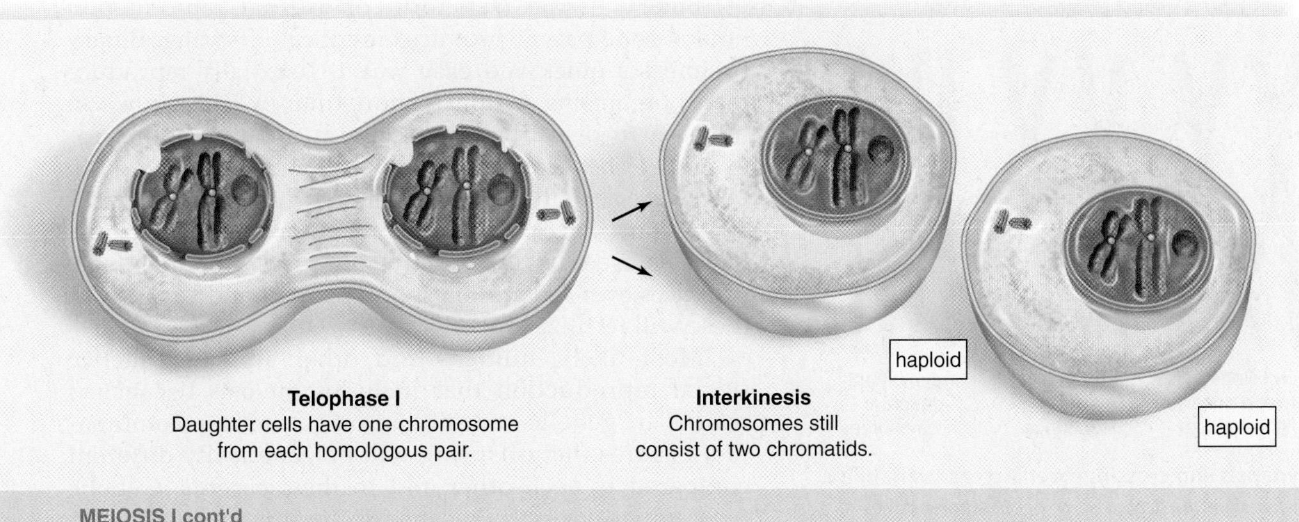
haploid
haploid
Telophase I
Daughter cells have one chromosome from each homologous pair.
Interkinesis
Chromosomes still consist of two chromatids.
MEIOSIS I cont'd

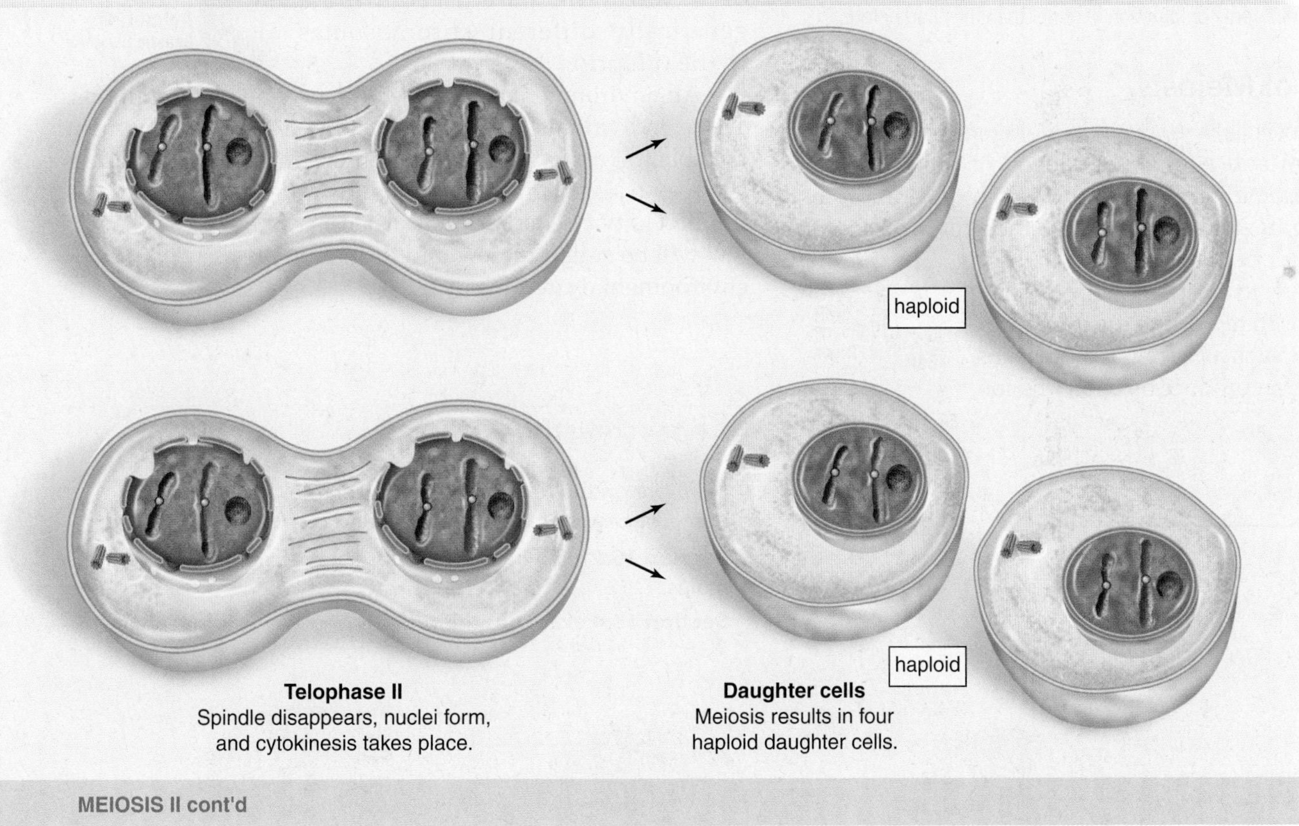
haploid
haploid
Telophase II
Spindle disappears, nuclei form, and cytokinesis takes place.
Daughter cells
Meiosis results in four haploid daughter cells.
MEIOSIS II cont'd

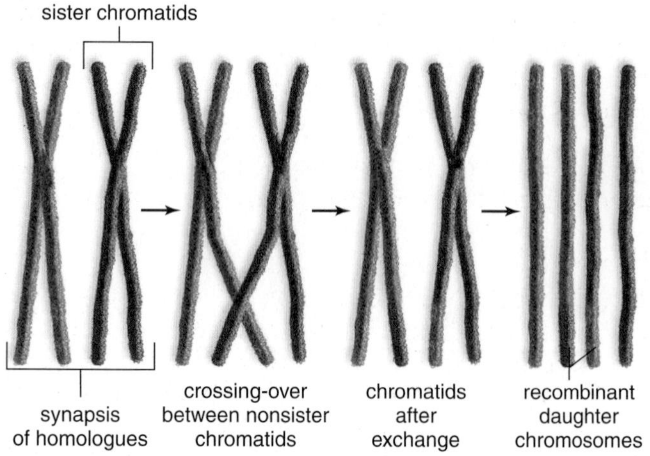

Figure 18.12 **Synapsis and crossing-over increase variability.** During meiosis I, *from left to right,* duplicated homologous chromosomes undergo synapsis and line up with each other. During crossing-over, nonsister chromatids break and then rejoin in the manner shown. Two of the resulting chromosomes have a different combination of genes than they had before.

Significance of Meiosis

In animals, meiosis is a part of gametogenesis, production of the sperm and egg. One function of meiosis is to keep the chromosome number constant from generation to generation. The gametes are haploid, so the zygote has only the diploid number of chromosomes.

An easier way to keep the chromosome number constant is to reproduce asexually. Unicellular organisms such as bacteria, protozoans, and yeasts (a fungi) reproduce by binary fission:

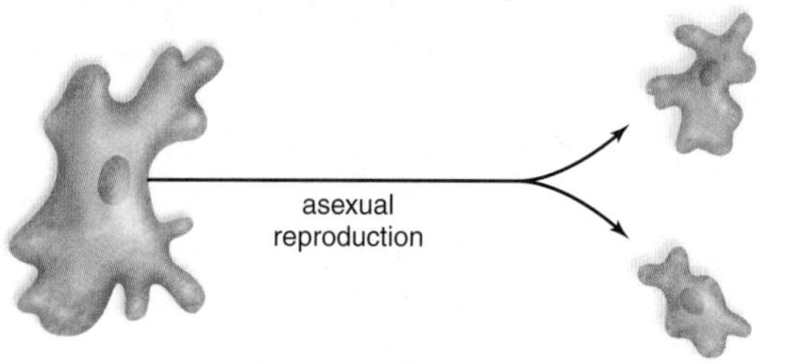

Binary fission is a form of asexual reproduction because one parent produces identical offspring. Binary fission is a quick and easy way to asexually reproduce many organisms within a short time. A bacterium can increase to over 1 million cells in about 7 hours, for example. Then why do organisms expend the energy to reproduce sexually? It takes energy to find a mate, carry out a courtship, and produce eggs or sperm that may never be used for reproductive purposes. A human male produces over 400 million sperm per day, and very few of these will fertilize an egg.

Most likely, humans and other animals practice sexual reproduction that includes meiosis because it results in genetic recombination. Genetic recombination ensures that offspring will be genetically different compared to each other and to their parents. Genetic recombination occurs because of crossing-over and independent alignment of chromosomes. Also, at the time of fertilization, parents contribute genetically different chromosomes to the offspring.

Animation Genetic Diversity

All environments are subject to a change in conditions. Those individuals able to survive in a new environment are able to pass on their genes. Environments are subject to change, so sexual reproduction is advantageous. It generates the diversity needed so that at least a few will be suited to new and different environmental circumstances.

MP3 Meiosis

Connecting the Concepts

For more on the relationship between meiosis and gametogenesis, refer to the following discussions.

Section 16.2 examines the process of spermatogenesis in males.

Section 16.4 examines the process of oogenesis in females.

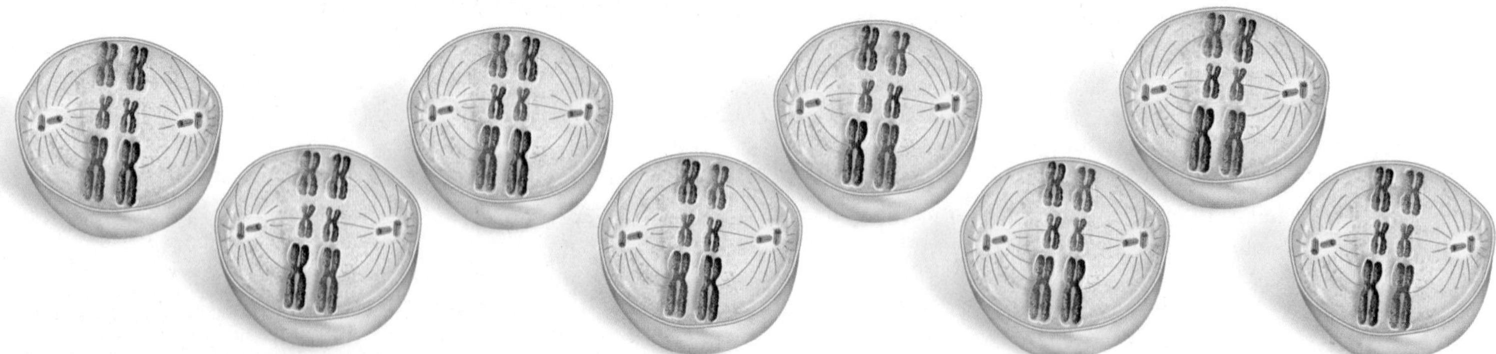

Figure 18.13 **Independent alignment at metaphase I increases variability.** When a parent cell has three pairs of homologous chromosomes, there are eight possible chromosome alignments at the equator due to independent assortment. Among the 16 daughter nuclei resulting from these alignments, there are eight different combinations of chromosomes.

Check Your Progress 18.4

1. Explain how, following meiosis the chromosome number of the daughter cells compares to the chromosome number of the parent cell.
2. Explain how meiosis reduces the likelihood that gametes will have the same combination of chromosomes and genes.
3. Summarize the events during the two cell divisions of meiosis.

18.5 Comparison of Meiosis and Mitosis

Learning Outcomes

Upon completion of this section, you should be able to

1. Distinguish between meiosis and mitosis with regard to the number of divisions and the number and chromosome content of the resulting cells.
2. Contrast the events of meiosis I and meiosis II with the events of mitosis.
3. List the stages of spermatogenesis and oogenesis.
4. Differentiate between spermatogenesis and oogenesis with regard to occurrence and the number of functional gametes produced by each process.

Meiosis and mitosis are both nuclear divisions, but there are several differences between them. You should be able to easily recognize how they differ. To this end, Figure 18.14 compares meiosis to mitosis. Study each of the differences listed here until you are familiar with them.

- DNA replication takes place only once prior to both meiosis and mitosis. Meiosis requires two nuclear divisions, but mitosis requires only one.
- Four daughter nuclei are produced by meiosis; following cytokinesis, there are four daughter cells. Mitosis followed by cytokinesis results in two daughter cells.
- The four daughter cells following meiosis are haploid (n) and have half the chromosome number of the parent cell (2n). The daughter cells following mitosis have the same chromosome number as the parent cell—the 2n, or diploid, number.
- The daughter cells from meiosis are not genetically identical to each other or to the parent cell. The daughter cells from mitosis are genetically identical to each other and to the parent cell.

The specific differences between these nuclear divisions can be categorized according to occurrence and process.

Occurrence

Meiosis occurs only at certain times in the life cycle of sexually reproducing organisms. In humans, meiosis occurs only in the reproductive organs and produces the gametes. Mitosis is more common because it occurs in all tissues during growth and repair. Which type of cell division can lead to cancer? Mitosis can result in a proliferation of body cells. Abnormal mitosis can lead to cancer.

Process

Comparison of Meiosis I with Mitosis

These events distinguish meiosis I from mitosis:

- Homologous chromosomes pair and undergo crossing-over during prophase I of meiosis but not during mitosis.
- Paired homologous chromosomes align at the equator during metaphase I in meiosis. These paired chromosomes have four chromatids altogether. Individual chromosomes align at the equator during metaphase in mitosis. They each have two chromatids.
- This difference makes it easy to tell whether you are looking at mitosis, meiosis I, or meiosis II. For example, if a cell has 16 chromosomes, then 16 chromosomes are at the equator during mitosis but only 8 chromosomes during meiosis II. Only meiosis I has paired duplicated chromosomes at the equator.
- Homologous chromosomes (with centromeres intact) separate and move to opposite poles during anaphase I of meiosis. Centromeres split, and sister chromatids, now called chromosomes, move to opposite poles during anaphase in mitosis.

Comparison of Meiosis II with Mitosis

The events of meiosis II are just like those of mitosis except that, in meiosis II, the nuclei contain the haploid number of chromosomes. If the parent cell has 16 chromosomes, then the cells undergoing meiosis II have 8 chromosomes, and the daughter cells have 8 chromosomes, for example.

Summary

To summarize the process, Tables 18.1 and 18.2 separately compare meiosis I and meiosis II with mitosis.

Animation Comparison of Mitosis and Meiosis

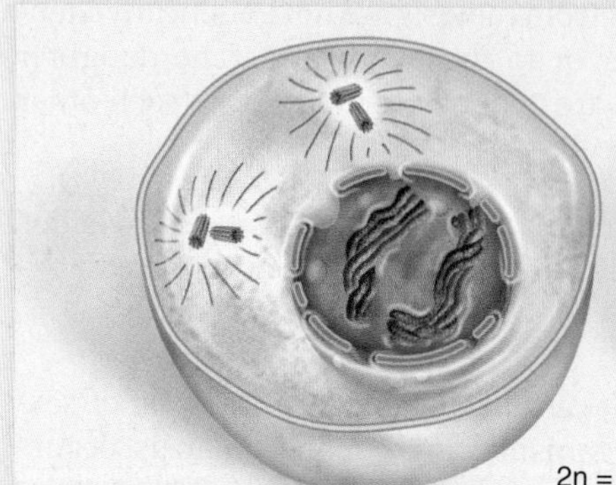

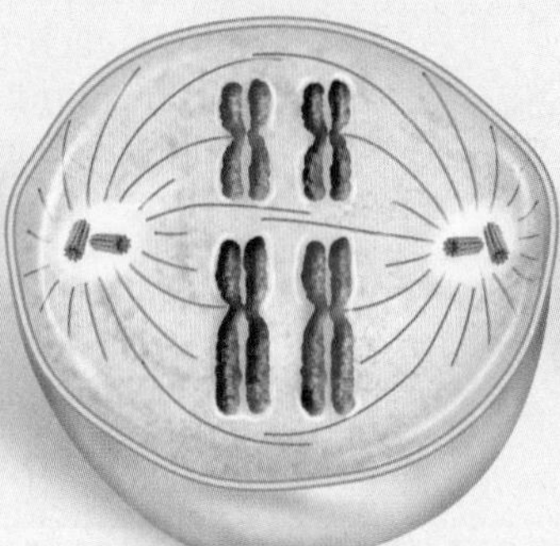

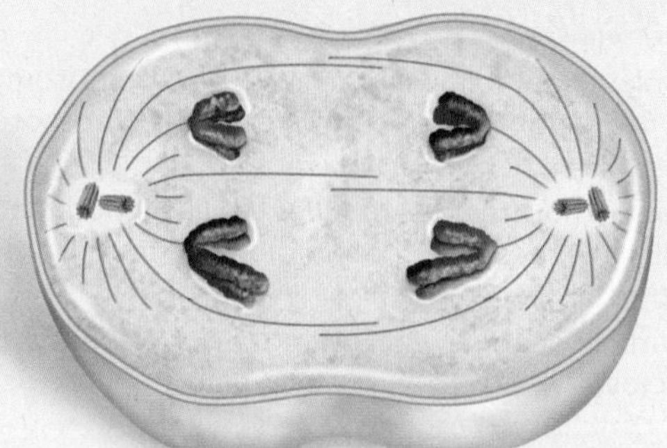

2n = 4

Prophase I
Synapsis and crossing-over occur.

Metaphase I
Homologous pairs align independently at the equator.

Anaphase I
Homologous chromosomes separate and move towards the poles.

MEIOSIS I

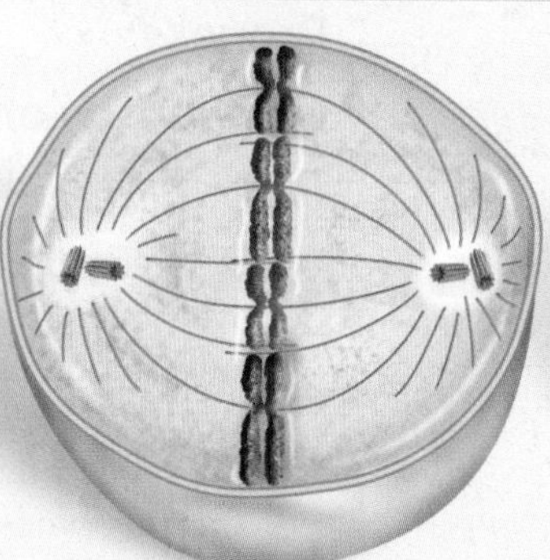

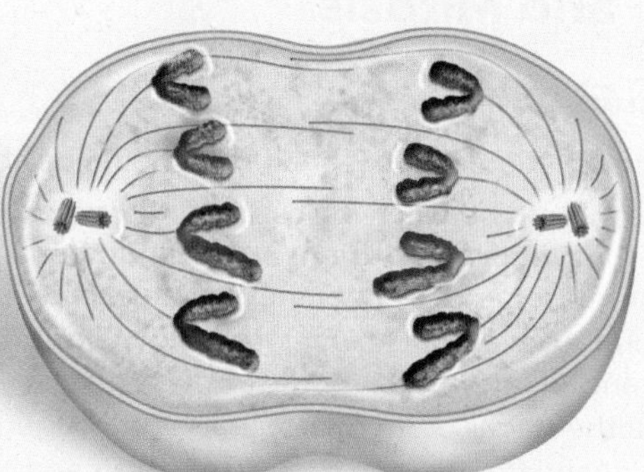

2n = 4

Prophase

Metaphase
Chromosomes align at the equator.

Anaphase
Sister chromatids separate and become daughter chromosomes.

MITOSIS

Table 18.1 Comparison of Meiosis I with Mitosis

Meiosis I	Mitosis
Prophase I	*Prophase*
Pairing of homologous chromosomes	No pairing of chromosomes
Metaphase I	*Metaphase*
Homologous duplicated chromosomes at equator	Duplicated chromosomes at equator
Anaphase I	*Anaphase*
Homologous chromosomes separate	Sister chromatids separate, becoming daughter chromosomes that move to the poles
Telophase I	*Telophase*
Two haploid daughter cells	Two daughter cells, identical to the parent cell

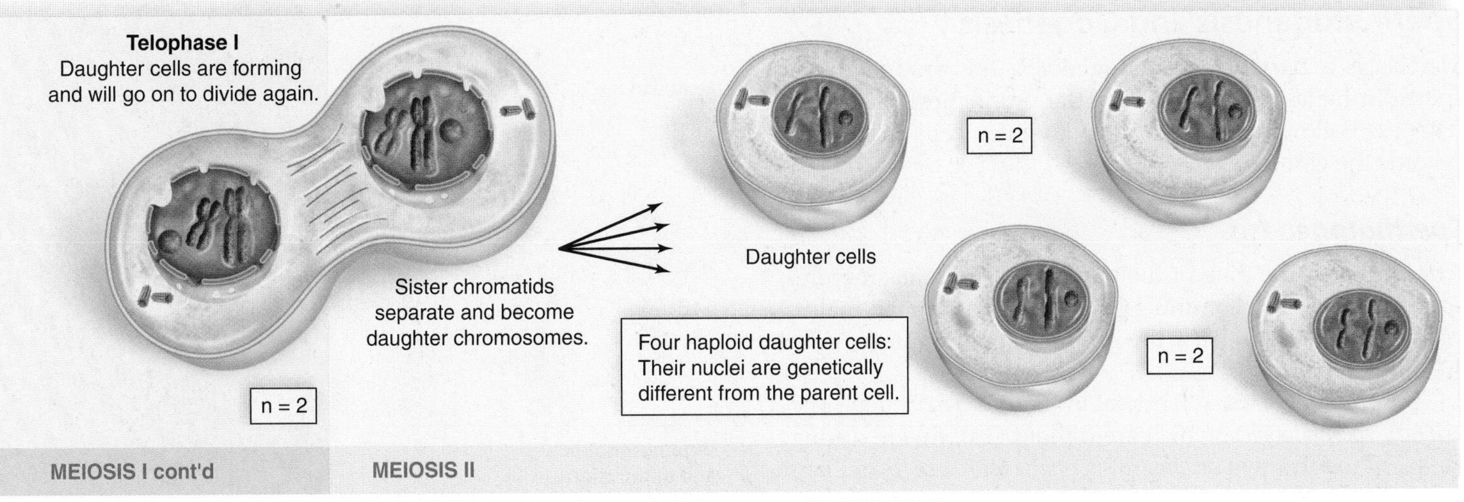

Telophase I
Daughter cells are forming and will go on to divide again.
Sister chromatids separate and become daughter chromosomes.
n = 2
Daughter cells
n = 2
n = 2
Four haploid daughter cells: Their nuclei are genetically different from the parent cell.
MEIOSIS I cont'd
MEIOSIS II

Daughter cells
Telophase
Daughter cells are forming.
Two diploid daughter cells: Their nuclei are genetically identical to the parent cell.
MITOSIS cont'd

Figure 18.14 **A comparison of meiosis and mitosis.**
Why does meiosis produce daughter cells with half the number and mitosis produces daughter cells with the same number of chromosomes as the parent cell? Compare metaphase I of meiosis to metaphase of mitosis. Only in metaphase I are the homologous chromosomes paired at the equator. Members of homologous chromosome pairs separate during anaphase I; therefore, the daughter cells are haploid. The blue chromosomes were inherited from the paternal parent, and the red chromosomes were inherited from the maternal parent. The exchange of color between nonsister chromatids represents the crossing-over that occurs during meiosis I.

Table 18.2 **Comparison of Meiosis II with Mitosis**

Meiosis II	Mitosis
Prophase II	*Prophase*
No pairing of chromosomes	No pairing of chromosomes
Metaphase II	*Metaphase*
Haploid number of duplicated chromosomes at equator	Duplicated chromosomes at equator
Anaphase II	*Anaphase*
Sister chromatids separate, becoming daughter chromosomes that move to the poles	Sister chromatids separate, becoming daughter chromosomes that move to the poles
Telophase II	*Telophase*
Four haploid daughter cells	Two daughter cells, identical to the parent cell

Spermatogenesis and Oogenesis

Meiosis is a part of **spermatogenesis,** the production of sperm in males, and **oogenesis,** the production of eggs in females. Following meiosis, the daughter cells mature to become the gametes.

Spermatogenesis

After puberty, the time of life when the sex organs mature, spermatogenesis is continual in the testes of human males. As many as 300,000 sperm are produced per minute, or over 400 million per day.

Spermatogenesis is shown in Figure 18.15, *top*. The *primary spermatocytes,* which are diploid (2n), divide during meiosis I to form two *secondary spermatocytes,* which are haploid (n). Secondary spermatocytes divide during meiosis II to produce four *spermatids,* which are also haploid (n). What's the difference between the chromosomes in haploid secondary spermatocytes and those in haploid spermatids? The chromosomes in secondary spermatocytes are duplicated and consist of two chromatids, whereas those in spermatids consist of only one. Spermatids mature into sperm (spermatozoa). In human males, sperm have 23 chromosomes, the haploid number. The process of meiosis in males always results in four cells that become sperm. In other words, all four daughter cells—the spermatids—become sperm.

Oogenesis

As you know, the ovary of a female contains many immature follicles (see Fig. 16.8). Each of these follicles contains a primary oocyte arrested in prophase I. As shown in Figure 18.15, *bottom,* a primary oocyte, which is diploid (2n), divides during meiosis I into two cells, each of which is haploid. The chromosomes are duplicated. One of these cells, termed the **secondary oocyte,** receives almost all the cytoplasm. The other is the first polar body. A **polar body** acts like a trash can to hold discarded chromosomes. The first polar body contains duplicated chromosomes and occasionally completes meiosis II. The secondary oocyte begins meiosis II but stops at metaphase II and doesn't complete it unless a sperm enters during the fertilization process.

The secondary oocyte (for convenience, called the egg) leaves the ovary during ovulation and enters an oviduct, where it may be fertilized by a sperm. If so, the oocyte is activated to complete the second meiotic division. Following meiosis II, there is one egg and two or possibly three polar bodies. The mature egg has 23 chromosomes. The polar bodies disintegrate, which is a way to discard unnecessary chromosomes while retaining much of the cytoplasm in the egg.

One egg can be the source of identical twins if after one division of the fertilized egg during development, the cells separate and each one becomes a complete individual. On the other hand, the occurrence of fraternal twins requires that two eggs be ovulated and then fertilized separately.

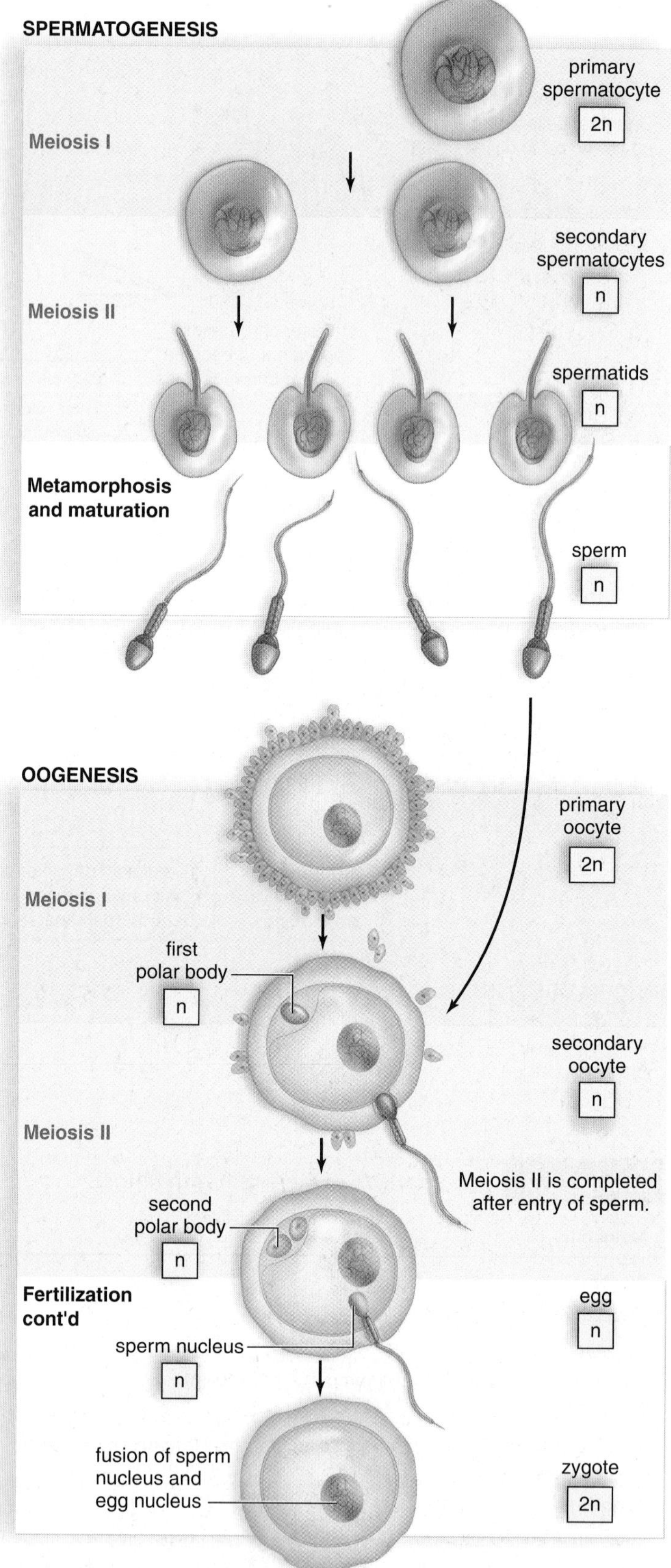

Figure 18.15 A comparison of spermatogenesis and oogenesis in mammals.
Spermatogenesis produces four viable sperm, whereas oogenesis produces one egg and at least two polar bodies. In humans, both sperm and egg have 23 chromosomes each; therefore, following fertilization, the zygote has 46 chromosomes.

Connecting the Concepts

For more on the role of gametogenesis in the reproductive system, refer to the following discussions.

Section 16.2 examines the process of spermatogenesis and the male reproductive system.

Section 16.4 examines the process of oogenesis and the female reproductive system.

Check Your Progress 18.5

1. List the similarities and differences between meiosis I and mitosis.
2. List the similarities and differences between meiosis II and mitosis.
3. Compare and contrast the stages of oogenesis and spermatogenesis.
4. Discuss the potential reasons that only one egg cell is produced by oogenesis.

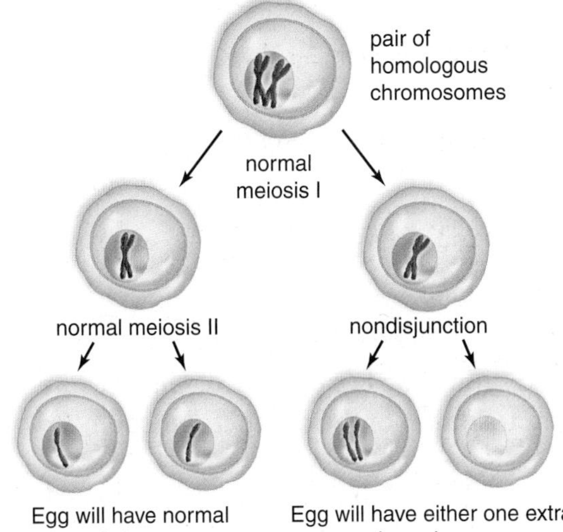

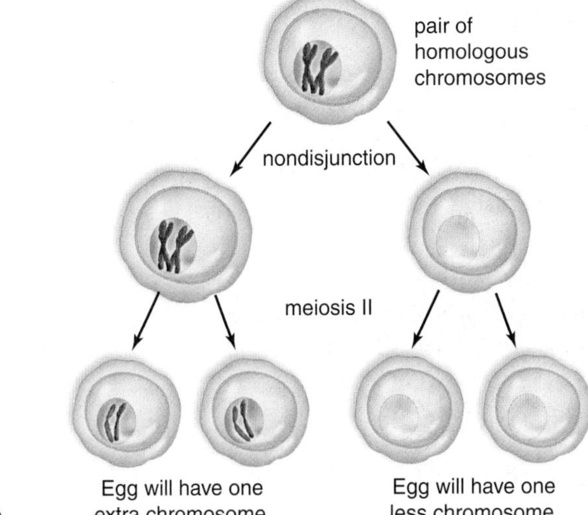

Figure 18.16 The consequences of nondisjunction of chromosomes during oogenesis.
a. Nondisjunction can occur during meiosis II if the sister chromatids separate, but the resulting chromosomes go into the same daughter cell. Then the egg will have one more (24) or one less (22) than the usual number of chromosomes. **b.** Nondisjunction can also occur during meiosis I and result in abnormal eggs that also have one more or one less than the normal number of chromosomes.

18.6 Chromosome Inheritance

Learning Outcomes

Upon completion of this section, you should be able to

1. Explain how nondisjunction produces monosomy and trisomy chromosome conditions.
2. Describe the causes and consequences of trisomy 21.
3. List the major syndromes associated with changes in the number of sex chromosomes.
4. Describe the effects of deletions, duplications, translocations, and inversions on chromosome structure.

Normally, an individual receives 22 pairs of autosomes and two sex chromosomes. Each pair of autosomes carries alleles for particular traits. The alleles can be different, as when one calls for freckles and one does not.

Changes in Chromosome Number

Sometimes individuals are born with either too many or too few autosomes or sex chromosomes, most likely due to nondisjunction during meiosis. **Nondisjunction** is the failure of the homologous chromosomes, or daughter chromosomes, to separate correctly during meiosis I and meiosis II, respectively. Nondisjunction may occur during meiosis I, when both members of a homologous pair go into the same daughter cell. It can also occur during meiosis II, when the sister chromatids fail to separate and both daughter chromosomes go into the same gamete. Figure 18.16 assumes that nondisjunction has occurred during oogenesis. Some abnormal eggs have 24 chromosomes, whereas others have only 22 chromosomes. If an egg with 24 chromosomes is fertilized with a normal sperm, the result is a **trisomy,** so called because one type of chromosome is present in three copies. If an egg with 22 chromosomes is fertilized with a normal sperm, the result is a **monosomy,** so called because one type of chromosome is present in a single copy.

Normal development depends on the presence of exactly two of each type of chromosome. An abnormal number of autosomes causes a developmental abnormality. Monosomy of all but the X chromosome is fatal. The affected infant rarely develops to full term. Trisomy is usually fatal, though there are some exceptions. Among autosomal trisomies, only trisomy 21 (Down syndrome) has a reasonable chance of survival after birth.

The chances of survival are greater when trisomy or monosomy involves the sex chromosomes. In normal XX females, one of the X chromosomes becomes a darkly staining mass of chromatin called a **Barr body** (named after the person who discovered it). A Barr body is an inactive X chromosome.

We now know that the cells of females function with a single X chromosome just as those of males do. This is most likely the reason that a zygote with one X chromosome (Turner syndrome) can survive. Then, too, all extra X chromosomes beyond a single one become Barr bodies; this explains why poly-X females and XXY males are seen fairly frequently. An extra Y chromosome, called Jacobs syndrome, is tolerated in humans, most likely because the Y chromosome carries few genes. Jacobs syndrome (XYY) is due to nondisjunction during meiosis II of spermatogenesis. We know this because two Ys are present only during meiosis II in males.

Connections and Misconceptions

Down syndrome is a trisomy of chromosome 21. Are there trisomies of the other chromosomes?

There are other chromosomal trisomies. However, because most chromosomes are much larger than chromosome 21, the abnormalities associated with three copies of these other chromosomes are much more severe than those found in Down syndrome. The extra genetic material causes profound congenital defects, resulting in fatality. Trisomies of the X and Y chromosomes appear to be exceptions to this, as noted in the text.

Chromosome 8 trisomy occurs rarely. Affected fetuses generally do not survive to birth or die shortly after birth. There are also trisomies of chromosomes 13 (Patau syndrome) and 18 (Edwards syndrome). Again, these babies usually die within the first few days of life.

Down Syndrome: An Autosomal Trisomy

The most common autosomal trisomy seen among humans is Down syndrome, also called trisomy 21. Persons with Down syndrome usually have three copies of chromosome 21 because the egg had two copies instead of one. (In 23% of the cases studied, however, the sperm had the extra chromosome 21.) The chances of a woman having a Down syndrome child increase rapidly with age, starting at about age 40. The reasons for this are still being investigated.

Although an older woman is more likely to have a Down syndrome child, most babies with Down syndrome are born to women younger than age 40 because this is the age group having the most babies. Karyotyping can detect a Down syndrome child. However, young women are not routinely

Biology Matters Science Focus

Mosaics, Barr Bodies, and Breast Cancer

Most people are familiar with calico cats, whose fur contains patches of orange, black, and white. These cats are genetic mosaics. A mosaic is formed by combination of different pieces to form a whole (a stained glass window is one example). Likewise, in genetics, a mosaic refers to an individual whose cells have at least two—and sometimes more—different types of genetic expression. In the case of the calico cat, the fur colors are due to expression of different genes. Some of the hair cells of these cats express the paternal copy of the gene. If an orange-haired father's copy of the gene is activated, a patch of orange hair develops. In other cells, the maternal gene is activated. A calico kitten with a black mother will grow black patches of hair scattered among the orange. Were you aware that human females are also mosaics?

The nucleus of human cells contains 46 chromosomes arranged into a set of 23 pairs. One chromosome from each pair is maternal, and the other is paternal. Each of the chromosomes in the first 22 pairs resembles its mate. Further, each member of a pair contains the same genes as the other member. Sex chromosomes that determine a person's gender are the last pair. Females have two X chromosomes, and males have one X and one Y chromosome. The Y chromosome is very small and contains far fewer genes than the X chromosome. Almost all of the genes on the X chromosome lack a corresponding gene on the Y chromosome. Thus, females have two copies of X genes, whereas males have only one. The body compensates for this extra dose of genetic material by inactivating one of the X chromosomes in each cell of the female embryo. Inactivation occurs early in development (at approximately the 100-cell stage). The inactivated X chromosome is called a Barr body, named after its discoverer. Barr bodies are highly condensed chromatin that appear as dark spots in the nucleus. Which X chromosome is inactivated in a given cell appears to be random. But every cell that develops from the original group of 100 cells will have the same inactivated X chromosome as its parent cell. Some of a woman's cells have inactivated the maternal X chromosome and other cells have inactivated the paternal X chromosome—she is a mosaic. Problems with inactivation of the X chromosome in humans could be linked to the development of cancer. For example, women who contain one defective copy of the breast cancer gene *BRCA1* have a greatly increased risk of developing breast and ovarian cancer. The BRCA1 protein produced from the gene is called a tumor suppressor. When the protein is functioning normally, it suppresses the development of cancer. This same protein is involved in X chromosome inactivation, although its exact role is uncertain. Presumably, increased cancer risk occurs because abnormal BRCA1 protein can neither inactivate the X chromosome nor function as a tumor suppressor.

encouraged to undergo the procedures necessary to get a sample of fetal cells (i.e., amniocentesis or chorionic villi sampling) because the risk of complications is greater than the risk of having a Down syndrome child. Fortunately, a test based on substances in maternal blood can help identify fetuses who may need to be karyotyped.

Down syndrome is easily recognized by these common characteristics: short stature; an eyelid fold; a flat face; stubby fingers; a wide gap between the first and second toes; a large, fissured tongue; a round head; and a palm crease, the so-called simian line. Unfortunately, mental retardation, which can vary in intensity, is also a characteristic. Chris Burke (Fig. 18.17*a*) was born with Down syndrome, and his parents were advised to put him in an institution. But Chris's parents didn't do that. They gave him the same loving care and attention they gave their other children, and it paid off. Chris is remarkably talented. He is a playwright, actor, and musician. He starred in *Life Goes On* (1989–1993), a TV series written just for him, and he is sometimes asked to be a guest star in other TV shows. His love of music and collaboration with other musicians have led to the release of several albums—like Chris, the songs are uplifting and inspirational. You can read more about this remarkable individual in his autobiography, *A Special Kind of Hero.*

The genes that cause Down syndrome are located on the bottom third of chromosome 21 (Fig. 18.17*b*). Extensive investigative work has been directed toward discovering the specific genes responsible for the characteristics of the syndrome. Thus far, investigators have discovered several genes that may account for various conditions seen in persons with Down syndrome. For example, they have located genes most likely responsible for the increased tendency toward leukemia, cataracts, accelerated rate of aging, and mental retardation. The gene for mental retardation, dubbed the *Gart* gene, causes an increased level of purines in the blood, a finding associated with mental retardation. One day, it may be possible to control the expression of the *Gart* gene even before birth so that at least this symptom of Down syndrome does not appear.

Connections and Misconceptions

Why is the age of a female a factor in Down syndrome?

One reason may be due to a difference in the timing of meiosis between males and females. Following puberty, males produce sperm continuously throughout their entire lives. In contrast, meiosis for females begins about five months after being conceived. However, the process is paused at prophase I of meiosis. Only after puberty are a selected few number of these cells allowed to continue meiosis as part of the female menstrual cycle. Because long periods of time may occur between the start and completion of meiosis, there is a greater chance that nondisjunction will occur; thus, as a female ages, there is a greater chance of producing a child with Down syndrome.

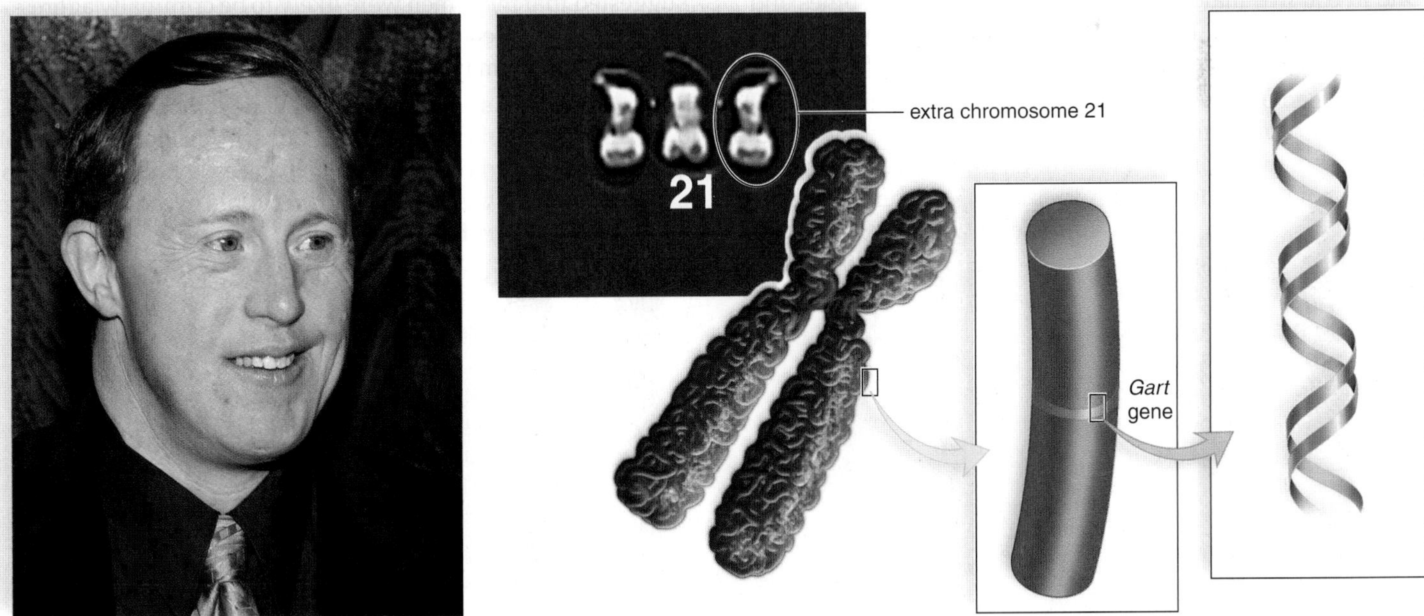

Figure 18.17 Down syndrome.
a. Chris Burke was born with Down syndrome. Common characteristics of the syndrome include a wide, rounded face, and a fold on the upper eyelids. Mental retardation, along with an enlarged tongue, makes it difficult for a person with Down syndrome to speak distinctly. **b.** Karotype of an individual with Down syndrome shows an extra chromosome 21. More sophisticated technologies allow investigators to pinpoint the location of specific genes associated with the syndrome. An extra copy of the *Gart* gene, which leads to a high level of purines in the blood, may account for the mental retardation seen in persons with Down syndrome.

Cri du chat (cat's cry) syndrome is seen when chromosome 5 is missing an end piece. The affected individual has a small head, is mentally retarded, and has facial abnormalities. Abnormal development of the glottis and larynx results in the most characteristic symptom—the infant's cry resembles that of a cat.

Translocation Syndromes A person who has both of the chromosomes involved in a translocation has the normal amount of genetic material and is healthy, unless the chromosome exchange broke an allele into two pieces. The person who inherits only one of the translocated chromosomes will no doubt have only one copy of certain alleles and three copies of certain other alleles. A genetics counselor begins to suspect a translocation has occurred when spontaneous abortions are commonplace and family members suffer from various syndromes.

Figure 18.21 shows a daughter and father who have a translocation between chromosomes 2 and 20. Although they have the normal amount of genetic material, they have the distinctive face (broad, prominent forehead and small, pointed chin), abnormalities of the eyes and internal organs, and severe itching characteristic of Alagille syndrome. People with this syndrome ordinarily have a deletion on chromosome 20. Therefore, it can be deduced that the translocation disrupted an allele on chromosome 20 in the father. The symptoms of Alagille syndrome range from mild to severe, so some people may not be aware they have the syndrome. This father did not realize it until he had a child with the syndrome.

Translocations can also be responsible for a variety of other disorders including certain types of cancer. In the 1970s, new staining techniques identified that a translocation from a portion of chromosome 22 to chromosome 9 was responsible for chronic myelogenous leukemia. This translocated chromosome was called Philadelphia chromosome. In Burkett lymphoma, a cancer common in children in equatorial Africa, a large tumor develops from lymph glands in the region of the jaw. This disorder involves a translocation from a portion of chromosome 8 to chromosome 14.

Connecting the Concepts

For more information on the topics presented in this section, refer to the following discussions.

Section 17.3 explains how the *SRY* gene directs the formation of the male reproductive system.

Section 20.2 explores how chromosomes are involved in patterns of inheritance.

Check Your Progress 18.6

1. Explain what causes an individual to have an abnormal number of chromosomes.
2. Describe the specific chromosome abnormality of a person with Down syndrome.
3. List some syndromes that result from inheritance of an abnormal sex chromosome number.
4. Describe other types of chromosome mutations, aside from abnormal chromosome number.

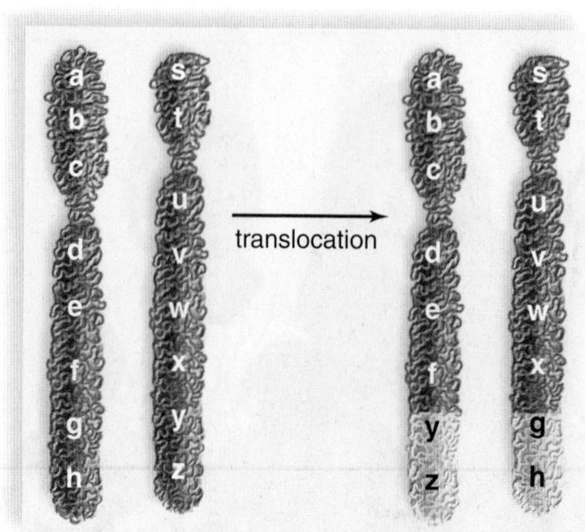

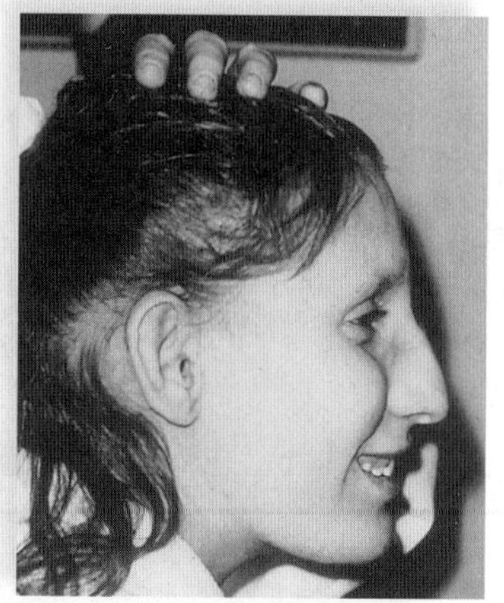

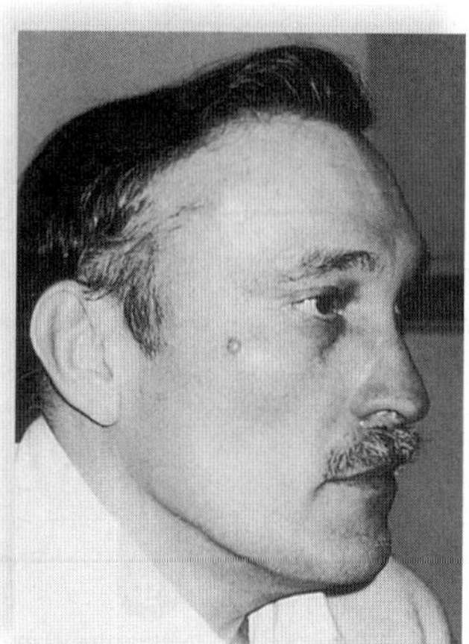

Figure 18.21 A chromosomal translocation. **a.** When chromosomes 2 and 20 exchange segments, (**b**) Alagille syndrome, with distinctive facial features, sometimes results because the translocation disrupts an allele on chromosome 20.

Biology Matters Bioethical Focus

Human beings have always attempted to influence the characteristics of their children. For example, couples have attempted to determine the sex of their children for centuries through a variety of methods. Amniocentesis has allowed us to test fetuses for chromosomal abnormalities and debilitating developmental defects before birth. Modern genetic testing technology enables parents to directly select children bearing desired traits, even at the very earliest stages of development. See the Chapter 20 Health Focus, *Preimplantation Genetic Diagnosis,* for an explanation of this technology.

Fanconi's Anemia

Recently, preimplantation genetic diagnosis selected an embryo for a couple because the newborn could save the life of his sister (Fig. 18A). The couple, Jack and Lisa Nash, had a daughter with Fanconi's anemia, a rare inherited disorder in which affected persons cannot properly repair DNA damage that results from certain toxins. The disease primarily afflicts the bone marrow and, therefore, results in a reduction of all types of blood cells. Anemia occurs, due to a deficiency of red blood cells. Patients are also at high risk of infection, because of low white blood cell numbers, and of leukemia, because white blood cells cannot properly repair any damage to their DNA.

Fanconi's anemia may be treated by a traditional bone marrow transplant or by an adult stem-cell transplant. The donor should be preferably a parent or sibling, because the risk of rejection is lower. Adult stem cells are almost always the preferred treatment option because stem cells are hardier and much less likely to be rejected than a bone marrow transplant. Recall that the umbilical cord of a newborn is a rich source of adult stem cells for all types of blood cells. (These are called adult because they are not embryonic stem cells.)

Testing the Embryo

During preimplantation genetic diagnosis, an embryonic cell is removed and tested for a disorder that runs in the family. In this case, doctors wanted to find an embryo that would become a healthy individual and also who would be able to benefit his or her sister. The parents underwent in vitro fertilization, and the 15 resulting embryos were screened to see which ones would not produce an individual with Fanconi's anemia and also had similar recognition proteins to those of their daughter. Two embryos met these requirements; only one implanted in the uterus, and it developed into a healthy baby boy. Adult stem cells were harvested from the umbilical cord of the newborn and were successfully used to treat his sister's anemia. The physician who performed the genetic screening stated that he has received numerous inquiries about performing the procedure for other couples with diseased children. Molly Nash, now 11 years old, is still doing well.

Ethical Dilemma

This case and other related cases have raised a number of ethical issues surrounding prenatal selection of children based on genetic traits. The American Medical Association (AMA) insists that selection based on traits not related to disease is unethical. However, the AMA's chair of the Council on Ethical and Judicial Affairs made an exception for this case because the child was selected for medical reasons. Dr. Jacques Montagut, who helped develop in vitro fertilization, believes that it is dangerous to bear children for the purpose of curing others and compares it with "a new form of biological slavery." Still others think that children will be selected for less altruistic reasons, such as for their height, physical prowess, or intellectual abilities.

Decide Your Opinion

1. In general, explain whether you think it is ethical to have children to cure medically related conditions, regardless of how fertilization has occurred.
2. The brother was created ostensibly as a treatment for his sister's disease. Discuss whether you believe that there is a moral obligation to provide him with compensation.
3. Discuss whether the use of embryonic stem cells derived from an aborted fetus and cultured in the laboratory would be an acceptable substitute.
4. Would you willingly donate sperm or eggs for in vitro fertilization to produce a healthy child for a couple who could not have one because of the risk of an inherited disease, such as Fanconi's anemia? Explain your answer.

Figure 18A The Nash family.
Jack, Molly holding baby Adam, and Lisa Nash. Adam was genetically selected as an embryo because the stem cells of his umbilical cord would save the life of his sister Molly.

CASE STUDY CONCLUSION

Christina was lucky. She had been undergoing routine mammograms since age 30, so the cancer was caught at an early stage. However, most women may be unaware that they carry the mutated *BRCA1* gene. A blood test can easily screen for mutations in the *BRCA1* gene; depending on the mutation, age of the patient, and health history of the patient, a genetics counselor can advise on the probability of breast or ovarian cancer in the future. In Christina's case, she made the decision to have a double mastectomy, removing both breasts even though the tumor was only located in one, instead of undergoing longer-term treatment such as chemotherapy. The next chapter will explore some of the options that cancer patients, such as Christina, face when they learn that they carry a potentially dangerous combination of alleles that make them suspectible to cancer.

Media Study Tools

www.mhhe.com/maderhuman12e

Enhance your study of this chapter with study tools and practice tests. Also ask your instructor about the resources available through ConnectPlus, including the media-rich eBook, interactive learning tools, and animations.

Virtual Lab

The virtual lab "The Cell Cycle and Cancer" provides a more detailed look at how a failure in the control system of the cell cycle can lead to cancer.

Summarizing the Concepts

18.1 Chromosomes

- Chromosomes occur in pairs in body cells.
- A karyotype is a visual display of a person's chromosomes.

A normal human karyotype shows 22 homologous pairs of autosomes and one pair of sex chromosomes.

- Normal sex chromosomes in males: XY
- Normal sex chromosomes in females: XX

18.2 The Cell Cycle

- The cell cycle occurs continuously and has four stages: G_1, S, G_2 (the interphase stages), and M (the mitotic stage, which includes cytokinesis and the stages of mitosis).
- In G_1, a cell doubles organelles and accumulates materials for DNA synthesis.
- In S, DNA replication occurs.
- In G_2, a cell synthesizes proteins needed for cell division.
- Checkpoints and external signals control the progression of the cell cycle.

18.3 Mitosis

Mitosis is duplication division that ensures that all body cells have the diploid number and the same types of chromosomes as the cell that divides.

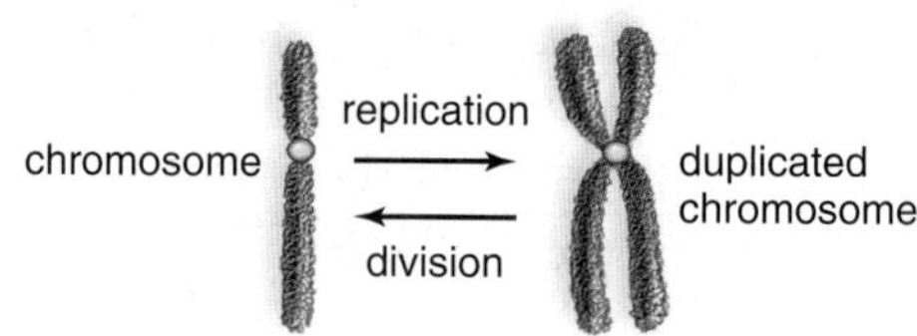

The phases of mitosis are prophase, metaphase, anaphase, and telophase.

- **Prophase** Chromosomes attach to spindle fibers.
- **Metaphase** Chromosomes align at the equator.
- **Anaphase** Chromatids separate, becoming chromosomes that move toward the poles.
- **Telophase** Nuclear envelopes form around chromosomes; cytokinesis begins.

Cytokinesis is the division of cytoplasm and organelles following mitosis.

- The proper workings of the cell cycle and mitosis are critical to growth and tissue repair.

18.4 Meiosis

Meiosis involves two cell divisions: meiosis I and meiosis II.

- **Meiosis I** Homologous chromosomes pair and then separate.
- **Meiosis II** Sister chromatids separate, resulting in four cells with the haploid number of chromosomes that move into daughter nuclei.

Meiosis results in genetic recombination due to crossing-over; gametes have all possible combinations of chromosomes. Upon fertilization, recombination of chromosomes occurs.

18.5 Comparison of Meiosis and Mitosis

- In prophase I, homologous chromosomes pair; there is no pairing in mitosis.
- In metaphase I, homologous duplicated chromosomes align at equator.
- In anaphase I, homologous chromosomes separate.

Spermatogenesis and Oogenesis

- **Spermatogenesis** In males, produces four viable sperm.
- **Oogenesis** In females, produces one egg and two or three polar bodies. Oogenesis goes to completion if the sperm fertilizes the developing egg.

18.6 Chromosome Inheritance

Meiosis is a part of gametogenesis (spermatogenesis in males and oogenesis in females) and contributes to genetic diversity.

Changes in Chromosome Number

- Nondisjunction changes the chromosome number in gametes, resulting in trisomy or monosomy.
- Autosomal syndromes include trisomy and Down syndrome.

Changes in Sex Chromosome Number

- Nondisjunction during oogenesis or spermatogenesis can result in gametes that have too few or too many X or Y chromosomes.
- Syndromes include Turner, Klinefelter, poly-X, and Jacobs.

Changes in Chromosome Structure

- Chromosomal mutations can produce chromosomes with deleted, duplicated, inverted, or translocated segments.
- These result in various syndromes such as Williams and cri du chat (deletion) and Alagille and certain cancers (translocation).

Understanding Key Terms

anaphase 426
apoptosis 432
aster 425
Barr body 437
cell cycle 422
centriole 425
centromere 421
centrosome 425
checkpoints 422
cleavage furrow 427
crossing-over 429
cytokinesis 422
daughter cell 424
deletion 441
diploid (2n) 424
duplication 441
fertilization 429
gamete 429
haploid (n) 428
homologous chromosome 428
homologue 428
interkinesis 428
interphase 422
inversion 441
meiosis 428
metaphase 426
mitosis 421, 422, 424
mitotic spindle 425
monosomy 437
nondisjunction 437
oogenesis 436
parent cell 424
polar body 436
prophase 425
secondary oocyte 436
sister chromatids 421
spermatogenesis 436
synapsis 428
syndrome 421
telophase 427
translocation 441
trisomy 437
zygote 429

Match the key terms to these definitions.

a. ________ Dark-staining nuclei of females that contain a condensed, inactive X chromosome.

b. ________ Member of a pair of chromosomes that are alike and come together in synapsis of prophase of meiosis I.

c. ________ Having one less chromosome than usual.

d. ________ A change in chromosome structure in which a segment is turned around 180°.

e. ________ Having the n number of chromosomes; for humans, n = 23.

Testing Your Knowledge of the Concepts

1. Describe the two parts of the cell cycle. (pages 422–423)
2. Explain how the checkpoints regulate the cell cycle. (pages 422–423)
3. How do external signals, such as hormones, influence the cell cycle? (page 423)
4. What is the function of the mitotic spindle? (page 425)
5. What is the importance of mitosis? (page 424)
6. How do the terms *diploid* (2n) and *haploid* (n) relate to meiosis? (page 428)
7. List and describe the events in meiosis I and meiosis II. (pages 428–431)
8. What is the significance of meiosis? (page 432)
9. Contrast mitosis and meiosis I and meiosis II. (pages 433–436)
10. How does spermatogenesis differ from oogenesis? (page 436)
11. What is nondisjunction, when can it occur, and what are the results? (page 437)
12. What are some syndromes caused by changes in chromosome number, and what is the change for each? (pages 437–439)
13. Describe three changes that can occur in chromosome structure and some syndromes caused by such changes. (pages 440–442)
14. The point of attachment for two sister chromatids is the
    a. centriole.
    b. centromere.
    c. chromosome.
    d. karyotype.
15. Label the drawing of the cell cycle; then tell the main event of each stage.

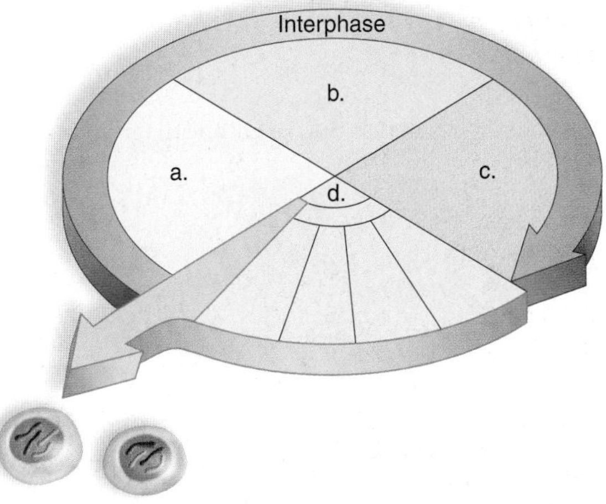

In questions 16–22, match the statement to interphase or the phase of mitosis in the key.

Key:

a. metaphase
b. interphase
c. telophase
d. prophase
e. anaphase

16. Spindle fibers begin to appear.
17. DNA synthesis occurs.
18. Chromosomes line up at the equator.
19. Duplicated chromosomes become visible.
20. Centromere splits and sister chromosomes move to opposite poles.
21. Cytokinesis occurs during this phase.
22. Chromosomes duplicate.
23. If a parent cell has 18 chromosomes before mitosis, how many chromosomes will the daughter cells have?
    a. 18
    b. 36
    c. 9
    d. 27
24. Crossing-over occurs between
    a. sister chromatids of the same chromosome.
    b. chromatids of nonhomologous chromosomes.
    c. nonsister chromatids of a homologous pair.
    d. Both b and c are correct.
25. The products of ________ are ________ cells.
    a. mitosis, diploid
    b. meiosis, haploid
    c. meiosis, diploid
    d. Both a and b are correct.
26. If a parent cell has 22 chromosomes, the daughter cells following meiosis II will have
    a. 22 chromosomes.
    b. 44 chromosomes.
    c. 11 chromosomes.
    d. Any of these could be correct.
27. Which of these drawings represents metaphase of meiosis I?

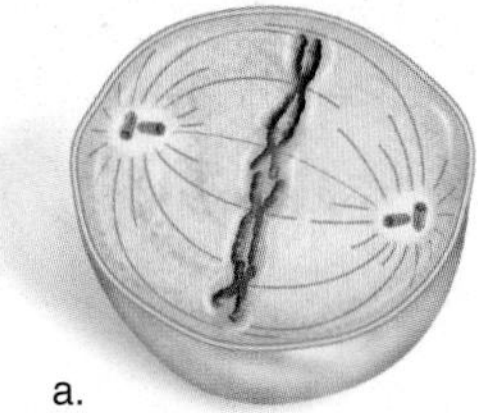
a.

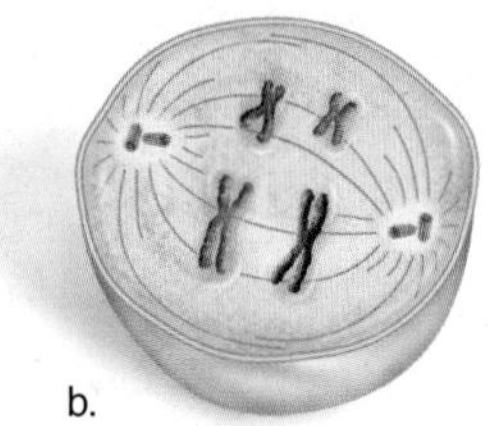
b.

In questions 28–32, match the part of the diagram to the correct label.

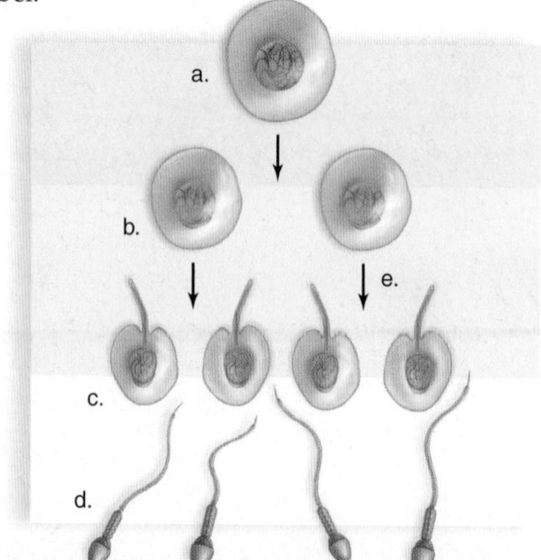

28. Meiosis II
29. Primary spermatocyte
30. Sperm
31. Secondary spermatocyte
32. Spermatids
33. Which of these helps to ensure that genetic diversity will be maintained?
    a. independent alignment during metaphase I
    b. crossing-over during prophase I
    c. fusion of sperm and egg nuclei during fertilization
    d. All of these are correct.
34. How many viable cells are produced by oogenesis?
    a. four
    b. three
    c. one
    d. two
35. Monosomy or trisomy occurs because of
    a. crossing-over.
    b. inversion.
    c. translocation.
    d. nondisjunction.
36. A person with Klinefelter syndrome is ________ and has ________ sex chromosomes.
    a. male, XYY
    b. male, XXY
    c. female, XXY
    d. female, XO

Thinking Critically About the Concepts

1. Benign and cancerous tumors occur when the cell cycle control mechanisms no longer operate correctly. What types of genes may be involved in these cell cycle control mechanisms?
2. Mitosis goes on continually in your body: in your blood cells, skin cells, and the cells that line the respiratory and digestive tracts. Why might Michelle have problems with wound healing, but not with these other sites where mitosis occurs continuously?
3. Explain how the separation of homologous chromosomes during meiosis affects the appearance of siblings (that some resemble each other and others look very different from one another).
4. a. What would you conclude about the ability of nervous and muscle tissue to repair themselves if nerve and muscle cells are typically arrested in G_1 of interphase?
   b. What are the implications of this arrested state to someone who suffers a spinal cord injury or heart attack?

CHAPTER

19

Cancer

CASE STUDY NEPHROBLASTOMA

Cody had all of the appearances of being a healthy 3-year-old boy. He liked to play with his toys, fought with his sister, and chased the cat around the house. However, shortly after his last birthday, Cody had begun to complain of stomach aches and often said that he felt sick. At first, his parents paid little attention to his condition, until they noticed that his appetite was decreasing and that he kept running low fevers without any apparent cause. One morning, Cody yelled to his Mom that his pee was pink. His mother rushed into the bathroom and discovered that Cody had blood in his urine. She immediately called their pediatrician and scheduled an appointment for that afternoon.

At the office, the doctor performed a complete physical exam and had Cody drink a couple of glasses of water to get a urine sample. During the exam, the doctor noticed a small lump in Cody's abdomen, just around the location of his kidney. The symptoms suggested that Cody may be having a kidney problem, so the doctor scheduled a magnetic resonance imaging (MRI) test the next morning for Cody.

The results of the MRI indicated that Cody had a mass (or tumor) on his kidney, which was a sign of nephroblastoma, or Wilms disease, a rare form of kidney cancer that affects only 500 children annually. The good news was that a follow-up computed tomography (CT) scan of Cody's abdomen did not suggest that the cancer had spread to any other organs. If this was the case, Cody's chances of survival were greater than 92%, but he was going to have to undergo surgery to remove the kidney, followed by several weeks of chemotherapy to make sure that no cancer cells were present outside the kidney. The doctor was very optimistic that they had caught Cody's cancer in time.

As you read through the chapter, think about the following questions

1. What are the characteristics of cancer cells that distinguish them from normal cells?
2. Why do cancer cells form tumors?
3. Why would the doctor recommend both surgery and chemotherapy?

CHAPTER CONCEPTS

19.1 Cancer Cells
Cancer cells have a number of abnormal characteristics that prevent them from functioning in the same manner as normal cells. They divide repeatedly and form tumors in the place of origin and in other parts of the body.

19.2 Causes and Prevention of Cancer
Whether cancer develops is partially due to inherited genes, but exposure to carcinogens such as UV radiation, tobacco smoke, pollutants, industrial chemicals, and certain viruses also plays a significant role.

19.3 Diagnosis of Cancer
Cancer is usually diagnosed by certain screening procedures and by imaging the body and tissues using various techniques.

19.4 Treatment of Cancer
Surgery followed by radiation and/or chemotherapy has now become fairly routine. Immune therapy, bone marrow transplants, and other methods are under investigation.

BEFORE YOU BEGIN

Before beginning this chapter, take a few moments to review the following discussions.

Section 8.6 What are antioxidants?

Section 18.1 What is the role of the checkpoints in the cell cycle?

Section 18.1 What external factors may regulate cell division?

19.1 Cancer Cells

Learning Outcomes

Upon completion of this section, you should be able to

1. Describe the characteristics of cancer cells that distinguish them from normal cells.
2. Distinguish between a proto-oncogene and a tumor suppressor gene with regard to their effects on the cell cycle.
3. Identify the common types of cancers and the body system with which each is associated.

Cancer is over a hundred different diseases, and each type of cancer can vary from another. However, some characteristics are common to cancer cells.

Characteristics of Cancer Cells

Cancer is a cellular disease, and cancer cells share traits that distinguish them from normal cells.

Cancer Cells Lack Differentiation

Differentiation is the process of cellular development by which a cell acquires a specific structure and function. Red blood cells are examples of differentiated cells in the circulatory system. In comparison, cancer cells are nonspecialized and do not contribute to the functioning of a body part. A cancer cell does not look like a differentiated epithelial, muscle, nervous, or connective tissue cell. Instead, it looks distinctly abnormal.

Cancer Cells Have Abnormal Nuclei

In Figure 19.1, you can compare the appearance of normal cervical cells (*a*) with that of precancerous (*b*) and cancerous (*c*) cervical cells. The nuclei of cancer cells are enlarged and may contain an abnormal number of chromosomes. The nuclei of the cervical cancer cells (Fig. 19.1*c*) have increased to the point that they take up most of the cell.

In addition to nuclear abnormalities, cancer cells have defective chromosomes. Some portions of the chromosomes may be duplicated, and/or some may be deleted. In addition, gene amplification (extra copies of specific genes) is seen much more frequently than in normal cells. Ordinarily, cells with damaged DNA undergo **apoptosis,** or programmed cell death. Cancer cells fail to undergo apoptosis, even though they are abnormal cells.

Tissues that divide frequently, such as those that line the respiratory and digestive tracts, are more likely to become cancerous. Cell division gives them the opportunity to undergo genetic mutations, each one making the cell more abnormal and giving it the ability to produce more of its own type.

Cancer Cells Have Unlimited Potential to Replicate

Ordinarily, cells divide about 60 to 70 times and then just stop dividing and die. Cancer cells are immortal and keep on dividing for an unlimited number of times.

Just as shoelaces are capped by small pieces of plastic, chromosomes in human cells end with special repetitive DNA sequences called **telomeres.** Specific proteins bind to telomeres in both normal and cancerous cells. These telomere proteins protect the ends of chromosomes from DNA repair enzymes. Though the enzymes effectively repair DNA in the center of the chromosome, they always tend to bind together the naked ends of chromosomes. In a normal cell, the telomeres get shorter after each cell cycle and protective telomere proteins gradually decrease. In turn, repair enzymes eventually cause the chromosomes' ends to bind together, causing the cell to undergo apoptosis and die. Telomerase is an enzyme that can rebuild telomere sequences and in that way prevent a cell from ever losing its potential to divide. The gene that codes for telomerase is constantly turned on in cancer cells, and telomeres are continuously rebuilt. The telomeres remain at a constant length, and the cell can keep dividing over and over.

Animation Telomerase Function

Cancer Cells Form Tumors

Normal cells anchor themselves to a substratum and/or adhere to their neighbors. They exhibit *contact inhibition*—when they come in contact with a neighbor, they stop dividing. Cancer cells have lost all restraint. They pile on top of

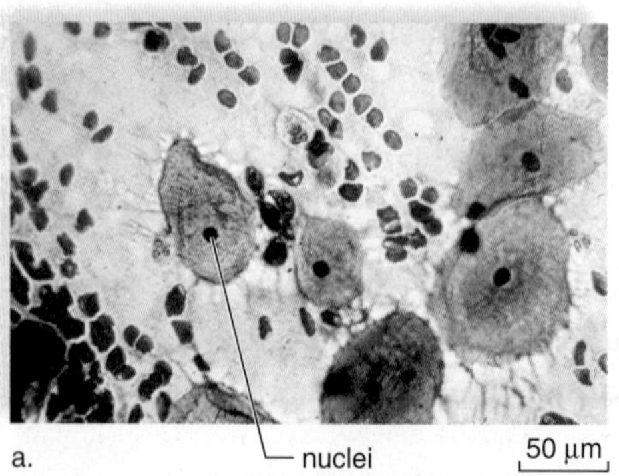

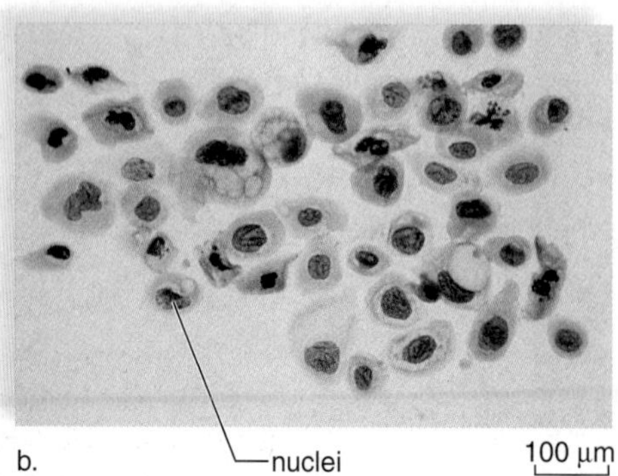

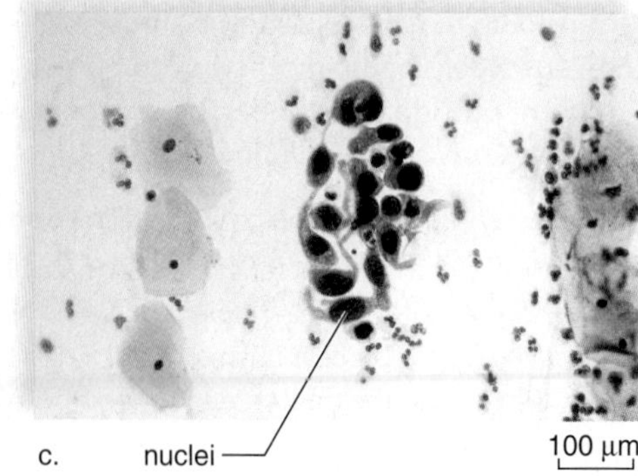

Figure 19.1 A comparison of normal tissue cells and cancer cells.
a. Normal cervical cells. **b.** Precancerous cervical cells. **c.** Cancerous cervical cells.

Connections and Misconceptions

capsule

benign tumor

cancer in situ

What's the difference between a benign tumor and a malignant tumor?

A *benign* tumor is usually surrounded by a connective tissue capsule. A benign tumor does not invade adjacent tissue because of the capsule. The cells of a benign tumor resemble normal cells fairly closely. Chances are you have one of these tumors. If you have a mole, or nevus, on your body, you have a benign tumor of the skin melanocytes. Cells from a mole closely resemble other melanocytes.

However, despite the term benign, this type of tumor isn't necessarily harmless. If it presses on normal tissue or restricts the normal tissue's blood supply, a benign tumor can be fatal. For example, a benign neuroma (nerve cell tumor) growing near the brainstem eventually affects control centers for heartbeat and respiration.

On the other hand, a malignant tumor is able to invade surrounding tissues and its cells don't resemble normal cells. By traveling in blood or lymph vessels, the tumor can spread throughout the body. As a general rule, badly deformed cells are most likely to spread. Thus, they are the most malignant.

Cancer in situ (in place) is a malignant tumor found in its place of origin. As yet, the cancer has not spread beyond the basement membrane, the nonliving material that anchors tissues to one another. If recognized and treated early, cancer in situ is usually curable.

one another and grow in multiple layers, forming a **tumor.** As cancer develops, the most aggressive cell becomes the dominant cell of the tumor (Fig. 19.2).

Cancer Cells Have No Need for Growth Factors

Chemical signals between cells tell them whether or not they should be dividing. These chemical signals, called growth factors, are of two types: stimulatory growth factors and inhibitory growth factors. Cancer cells keep on dividing even when stimulatory growth factors are absent, and they do not respond to inhibitory growth factors.

Cancer Cells Gradually Become Abnormal

Figure 19.2 illustrates that **carcinogenesis,** the development of cancer, is a multistage process that can be divided into these three phases:

Initiation: A single cell undergoes a mutation that causes it to begin to divide repeatedly (Fig. 19.2*a*).

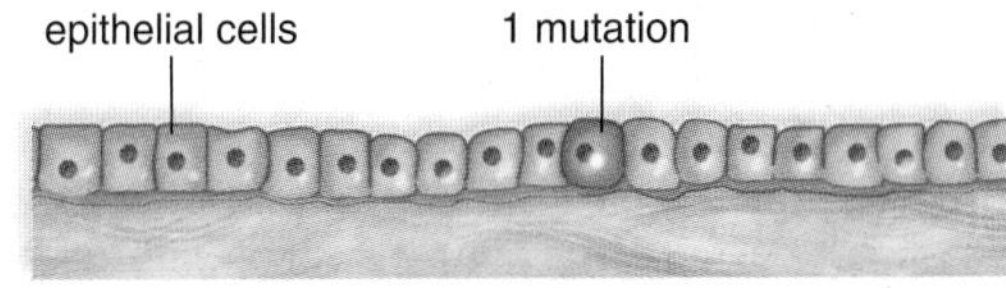
a. Cell (dark pink) acquires a mutation for repeated cell division.
epithelial cells
1 mutation

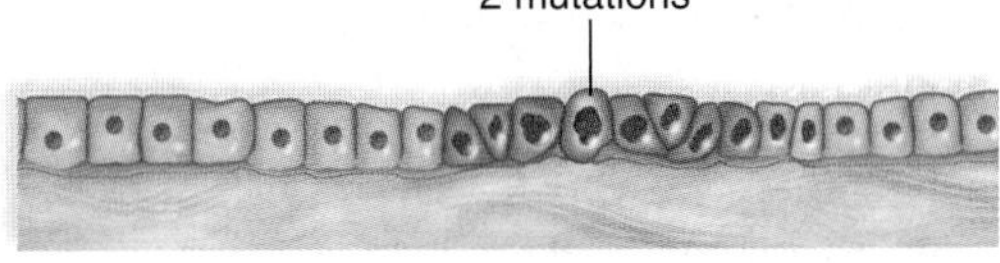
b. New mutations arise, and one cell (brown) has the ability to start a tumor.
2 mutations

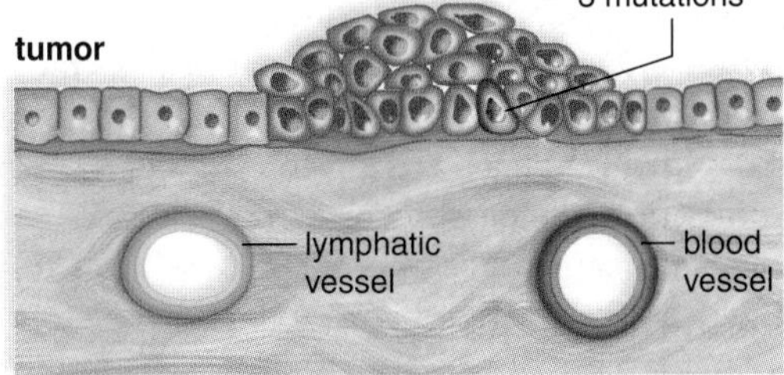
c. Cancer in situ. The tumor is at its place of origin. One cell (purple) mutates further.
tumor
3 mutations
lymphatic vessel
blood vessel

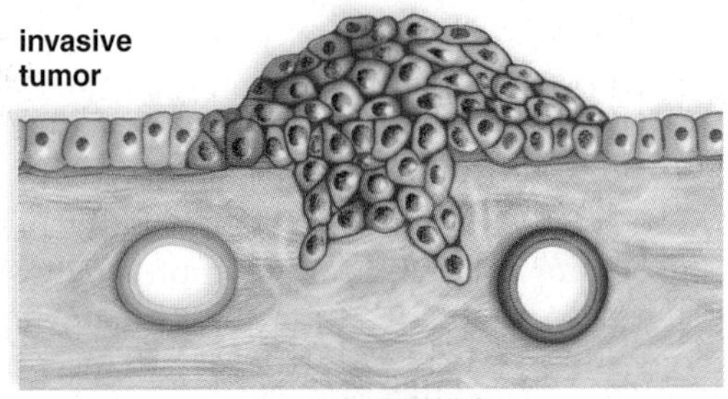
d. Cells have gained the ability to invade underlying tissues by producing a proteinase enzyme.
invasive tumor

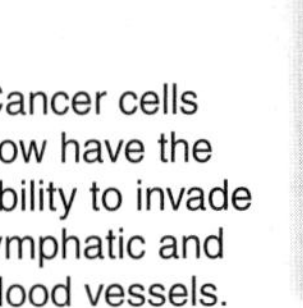

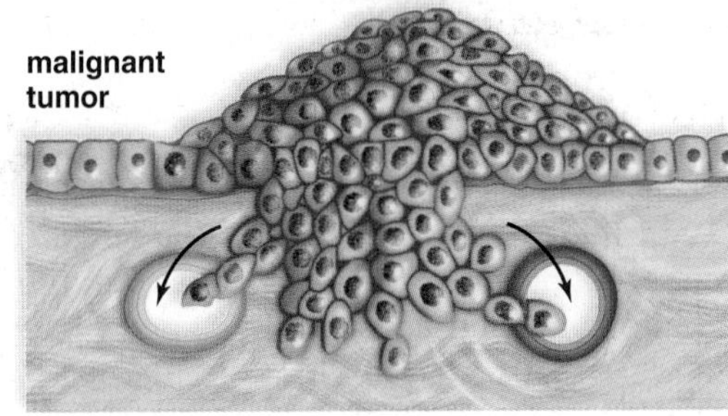
e. Cancer cells now have the ability to invade lymphatic and blood vessels.
malignant tumor

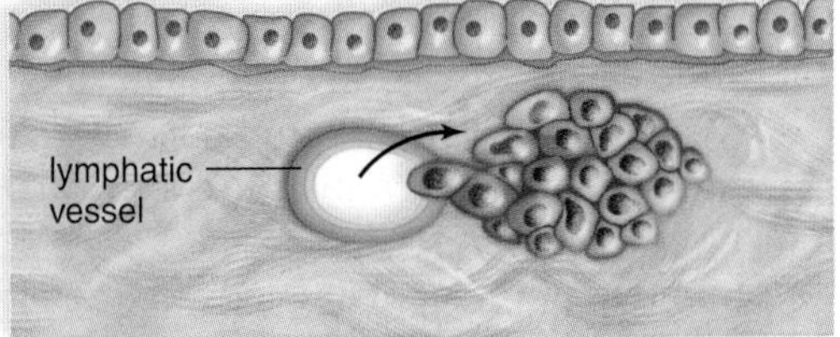
f. New metastatic tumors are found some distance from the original tumor.
distant tumor
lymphatic vessel

Figure 19.2 **Progression from a single mutation to a tumor.**

a. One cell (dark pink) in a tissue mutates. **b.** This mutated cell divides repeatedly and a cell (brown) with two mutations appears. **c.** A tumor forms and a cell with three mutations (purple) appears. **d.** This cell (purple), which takes over the tumor, can invade underlying tissue. **e.** Tumor cells invade lymphatic and blood vessels. **f.** A new tumor forms at a distant location.

Promotion: A tumor develops, and the tumor cells continue to divide. As they divide, they undergo mutations (Fig. 19.2*b*, *c*).

Progression: One cell undergoes a mutation that gives it a selective advantage over the other cells (Fig. 19.2*c*). This process is repeated several times; eventually, there is a cell that has the ability to invade surrounding tissues (Fig. 19.2*d*, *e*).

Cancer Cells Undergo Angiogenesis and Metastasis

To grow larger than about a billion cells (about the size of a pea), a tumor must have a well-developed capillary network to bring it nutrients and oxygen. **Angiogenesis** is the formation of new blood vessels. The low oxygen content in the middle of a tumor may turn on genes coding for angiogenic growth factors that diffuse into the nearby tissues and cause new vessels to form.

Due to mutations, cancer cells tend to be motile. They have a disorganized internal cytoskeleton and lack intact actin filament bundles. To metastasize, cancer cells must make their way across the basement membrane and invade a blood vessel or lymphatic vessel. Invasive cancer cells are sperm-shaped (see Fig. 19.1*c*) and don't look at all like the normal cell seen nearby. Cancer cells produce proteinase enzymes that degrade the basement membrane and allow them to invade underlying tissues. Malignancy is present when cancer cells are found in nearby lymph nodes. When these cells begin new tumors far from the primary tumor, **metastasis** has occurred (Fig. 19.2*f*). Not many cancer cells achieve this ability (maybe 1 in 10,000), but those that successfully metastasize to various parts of the body lower the prognosis (the predicted outcome of the disease) for recovery.

Cancer Results from Gene Mutation

Recall that the cell cycle (see Chapter 8) consists of interphase, followed by mitosis. **Checkpoints** in the cell cycle monitor the condition of the cell and regulate its ability to divide. **Cyclin** is a protein molecule that has to be present for a cell to proceed from interphase to mitosis. When cancer develops, the cell cycle occurs repeatedly, in large part, due to mutations in two types of genes.

Animation Control of the Cell Cycle

1. **Proto-oncogenes** code for proteins that promote the cell cycle and prevent apoptosis. They are often likened to the gas pedal of a car because they cause acceleration of the cell cycle.
2. **Tumor suppressor genes** code for proteins that inhibit the cell cycle and promote apoptosis. They are often likened to the brakes of a car because they inhibit acceleration.

Proto-Oncogenes Become Oncogenes

When proto-oncogenes mutate, they become cancer-causing genes called **oncogenes.** These mutations can be called "gain-of-function" mutations because overexpression is the result (Fig. 19.3). Whatever a proto-oncogene does, an oncogene does it better.

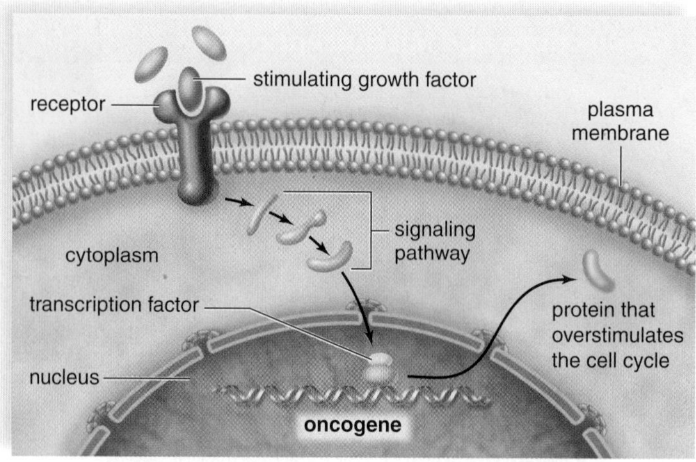

Figure 19.3 Mutations in tumor suppressor genes cause a loss of cell cycle control.
Normally, tumor suppressor genes code for a protein that inhibits the cell cycle. If mutations occur, the resulting protein loses the function of controlling the cell cycle.

A **growth factor** is a signal that activates a cell signaling pathway, resulting in cell division. Some proto-oncogenes code for a growth factor or for a receptor protein that receives a growth factor. When these proto-oncogenes become oncogenes, receptor proteins are easy to activate and may even be stimulated by a growth factor produced by the receiving cell. Several proto-oncogenes code for Ras proteins that promote mitosis by activating cyclin. *Ras* oncogenes are typically found in many different types of cancers. *Cyclin D* is a proto-oncogene that codes for cyclin directly. When this gene becomes an oncogene, cyclin is readily available all the time.

Another protein, p53, activates repair enzymes. At the same time, p53 turns on genes that stop the cell cycle from proceeding. If repair is impossible, the p53 protein promotes *apoptosis*, programmed cell death. Apoptosis is an important way for carcinogenesis to be prevented. One proto-oncogene codes for a protein that functions to make p53 unavailable. When this proto-oncogene becomes an oncogene, no p53 will be available, regardless of how much is made. Many tumors are lacking in p53 activity.

Tumor Suppressor Genes Become Inactive

When tumor suppressor genes mutate, their products no longer inhibit the cell cycle nor promote apoptosis. Therefore, these mutations can be called "loss-of-function" mutations (Fig. 19.4).

Mutation of the tumor suppressor gene *Bax* is a good example. Its product, the protein Bax, promotes apoptosis. When *Bax* mutates, Bax protein is not present and apoptosis is less likely to occur. The *Bax* gene contains a line of eight consecutive G bases in its DNA (see Chapter 2). When the same base molecules are lined up in this fashion, the gene is more likely to be subject to mutation.

Animation How Tumor Supressor Genes Block Cell Division

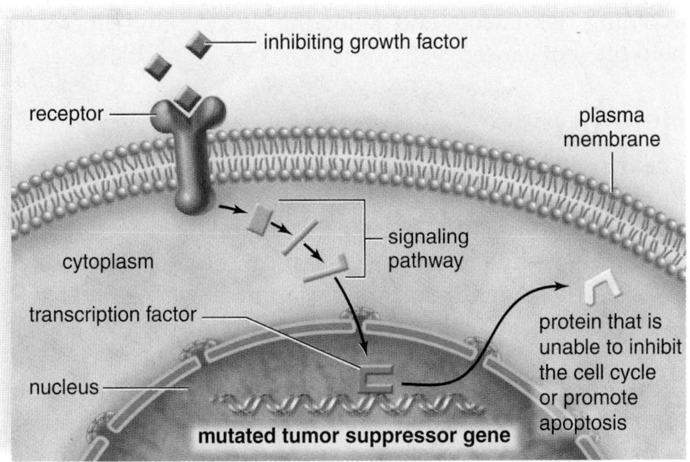

Figure 19.4 **Mutations in proto-oncogenes produce oncogenes that stimulate the cell cycle.**
Normally, proto-oncogenes stimulate cell division but are typically off in differentiated cells. A mutation may convert a proto-oncogene to an oncogene, which then produces a protein that overstimulates the cell cycle.

Types of Cancer

Statistics indicate that one in three Americans will deal with cancer in their lifetime. Therefore, this is a topic of considerable importance to the health and well-being of every individual. **Oncology** is the study of cancer. A medical specialist in cancer is, therefore, known as an **oncologist.** The patient's prognosis (probable outcome) depends on (1) whether the tumor has invaded surrounding tissues and (2) whether there are metastatic tumors in distant parts of the body.

Tumors are classified according to their place of origin. **Carcinomas** are cancers of the epithelial tissues, and adenocarcinomas are cancers of glandular epithelial cells. Carcinomas include cancer of the skin, breast, liver, pancreas, intestines, lung, prostate, and thyroid. **Sarcomas** are cancers that arise in muscles and connective tissue, such as bone and fibrous connective tissue. **Leukemias** are cancers of the blood, and **lymphomas** are cancers of lymphatic tissue. A blastoma is a cancer composed of immature cells. Recall that the embryo is formed from three primary germ layers: ectoderm, mesoderm, and endoderm (see Chap. 17). Each blastoma cell resembles the cells in its original primary germ layer. For example, a nephroblastoma (like Cody's in the case study) has cells similar to mesoderm cells, because the kidney grows from mesoderm.

Common Cancers

Cancer occurs in all parts of the body, but some organs are more susceptible than others (Fig. 19.5). In the respiratory system, lung cancer is the most common type. Overall, this is one of the most common types of cancer, and smoking is known to increase a person's risk for this disease. Smoking also increases the risk for cancer in the oral cavity.

In the digestive system, colorectal cancer (colon/rectum) is another common tumor. Other cancers of the digestive system include those of the pancreas, stomach, esophagus,

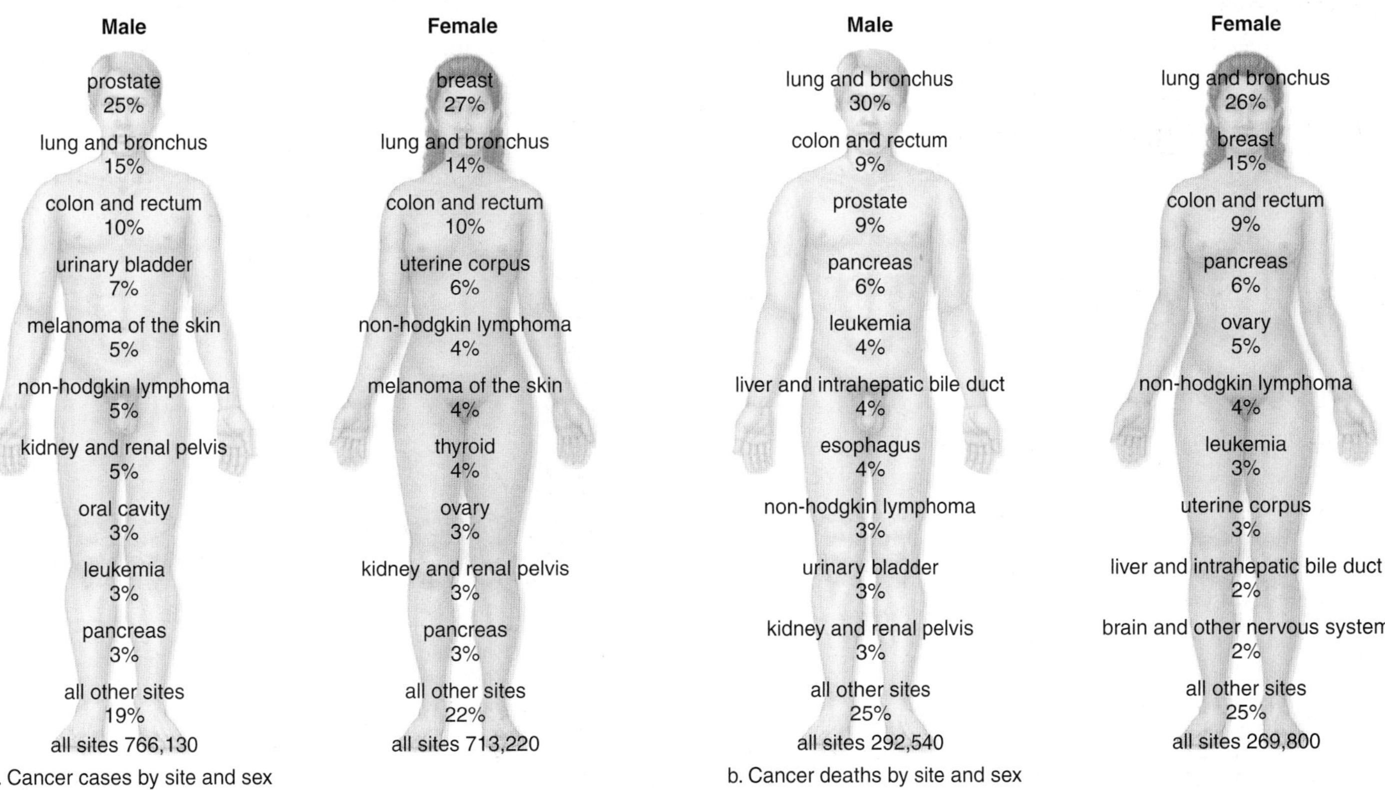

Figure 19.5 **Estimated cases of cancer and cancer deaths in the United States.**
a. Estimated new cancer cases by sex in 2009. **b.** Estimated deaths from cancer by sex in 2009.

and other organs. In the cardiovascular system, cancers include leukemia and plasma cell tumors. In the lymphatic system, cancers are classified as either Hodgkin or non-Hodgkin lymphoma. Hodgkin lymphomas develop from mutated B cells. Non-Hodgkin lymphomas can arise from B cells or T cells. Thyroid cancer is the most common type of tumor in the endocrine system, whereas brain and spinal tumors are found in the central nervous system.

Breast cancer is one of the most common types of cancer. Although predominantly found in women, occasionally it is also found in men. Cancers of the cervix, ovaries, and other reproductive structures also occur in women. In males, prostate cancer is one of the most common cancers. Other cancers of the male reproductive system include cancer of the testis and the penis. Bladder and kidney cancers are associated with the urinary system. Skin cancers include melanoma and basal cell carcinoma.

Connecting the Concepts

For more information on the topics in this section, refer to the following discussions.

Section 18.2 describes the role of the checkpoints of the cell cycle.

Section 21.2 describes how the information in a gene is expressed to form a protein.

Check Your Progress 19.1

1. List the characteristics of cancer cells that allow them to grow uncontrollably.
2. Describe how mutations in tumor suppressor genes and proto-oncogenes have opposite effects on the cell cycle.
3. Summarize some of the more common types of cancer.

19.2 Causes and Prevention of Cancer

Learning Outcomes

Upon completion of this section, you should be able to

1. Explain how heredity and the environment may both contribute to cancer.
2. Identify the genetic mechanisms of select forms of cancer.
3. Summarize how protective behaviors and diet can help prevent cancer.

Our current understanding of the causes of cancer is incomplete. However, by studying patterns of cancer development in populations, scientists have determined that both heredity and environmental risk factors come into play. Obviously, one's genetic inheritance can't be changed; but avoiding risk factors and following dietary guidelines can help prevent cancer.

Heredity

The first gene associated with breast cancer was discovered in 1990. Scientists named this gene *br*east *ca*ncer 1 or *BRCA1* (pronounced "brak-uh"). Later, they found that breast cancer could also be due to another breast cancer gene they called *BRCA2*. These genes are tumor suppressor genes that follow a particular inheritance pattern. We inherit two copies of every gene, one from each parent. If a mutated copy of *BRCA1* or *BRCA2* is inherited from either parent, a mutation in the other copy is required before predisposition to cancer is increased. Each cell in the body already has the single mutated copy, but cancer develops wherever the second mutation occurs. If the second mutation occurs in the breast, breast cancer may develop. Ovarian cancer may develop if a second cancer-causing mutation occurs in an ovary.

The *RB* gene is also a tumor suppressor gene. It takes its name from its association with retinoblastoma, a rare eye cancer that occurs almost exclusively in early childhood. In an affected child, a single mutated copy of the *RB* gene is inherited. A second mutation, this time to the other copy of the *RB* gene, causes the cancer to develop (Fig. 19.6). Retinoblastoma affecting a single eye is treated by removal of the eye, because total removal is more likely to result in a cure. Radiation, laser, and chemotherapy are options when both eyes are involved, but these treatments are less likely to produce a cure.

An abnormal *RET* gene, which predisposes an individual to thyroid cancer, can be passed from parent to child. *RET* is a proto-oncogene known to be inherited in an autosomal dominant manner—only one mutation is needed to increase a predisposition to cancer (see Chap. 20). The remainder of the mutations necessary for a thyroid cancer to develop are acquired (not inherited).

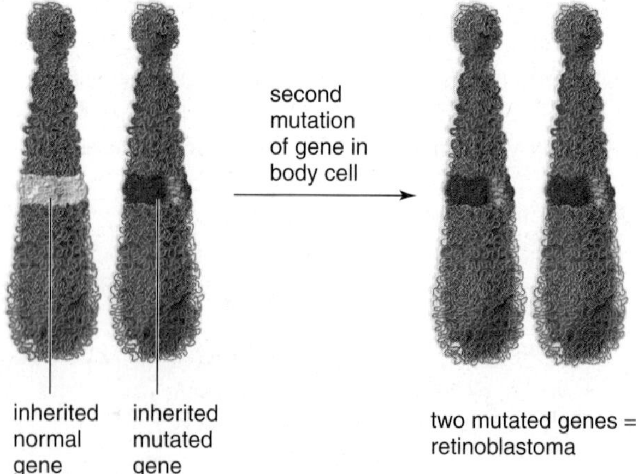

Figure 19.6 Inheritance of retinoblastoma.
A child is at risk for an eye tumor when a mutated copy of the *RB* gene is inherited, even though a second mutation in the normal copy is required before the tumor develops.

Environmental Carcinogens

A **mutagen** is an agent that causes mutations. A simple laboratory test—an Ames test—is used to determine whether a substance is mutagenic. A **carcinogen** is a chemical that causes cancer by being mutagenic. Some carcinogens cause only initiation. Others cause initiation and promotion.

Heredity can predispose a person to cancer, but whether it develops or not depends on environmental mutagens, such as the ones we will be discussing.

Radiation

Ionizing radiation, such as in radon gas, nuclear fuel, and X-rays, is capable of affecting DNA and causing mutations. Though not a form of ionizing radiation, ultraviolet light also causes mutations. Ultraviolet radiation in sunlight and tanning lamps is most likely responsible for the dramatic increases seen in skin cancer in the past several years. Today, at least six cases of skin cancer occur for every one case of lung cancer. Nonmelanoma skin cancers are usually curable through surgery. However, melanoma skin cancer tends to metastasize and is responsible for 1–2% of cancer deaths in the United States.

Another natural source of ionizing radiation is radon gas, which comes from the natural (radioactive) breakdown of uranium in soil, rock, and water. The Environmental Protection Agency recommends that every home be tested for radon because it is the second leading cause of lung cancer in the United States. The combination of inhaling radon gas and smoking cigarettes can be particularly dangerous. A vent pipe system and fan, which pull radon from beneath the house and vent it to the outside, are the most common tools to rid a house of radon after the house is constructed.

We are well aware of the damaging effects of a nuclear bomb explosion or accidental emissions from nuclear power plants. For example, more cancer deaths have occurred in the vicinity of the Chernobyl Power Station (in Ukraine), which suffered a terrible accident in 1986. Usually, however, diagnostic X-rays account for most of our exposure to artificial sources of radiation. The benefits of these procedures can far outweigh the possible risk, but it is still wise to avoid X-ray procedures that are not medically warranted. When X-rays are necessary for therapy, nearby noncancerous tissues should be shielded carefully.

Recently, there has been a great deal of public concern regarding the presumed danger of nonionizing radiation. This energy form is given off by cell phones, electrical lines, and appliances. No evidence has been found that links nonionizing radiation to cancer, however.

Organic Chemicals

Certain chemicals, particularly synthetic organic chemicals, have been found to be risk factors for cancer. We will mention only two examples: the organic chemicals in tobacco and those pollutants in the environment. There are many other chemical carcinogens.

Tobacco Smoke Tobacco smoke contains a number of organic chemicals that are known mutagens, including *N*-nitrosonornicotine, vinyl chloride, and benzo[a]pyrene (a known suppressor of p53). The greater the number of cigarettes smoked per day and the earlier the habit starts, the more likely it is that cancer will develop. On the basis of data, such as those shown in Figure 19.7, scientists estimate that about 80% of all cancers, including oral cancer and cancers of the larynx, esophagus, pancreas, bladder, kidney, and cervix, are related to the use of tobacco products. When smoking is combined with drinking alcohol, the risk of these cancers increases.

Passive smoking, or inhalation of someone else's tobacco smoke, is also dangerous. Researchers continue to collect evidence that confirms the link between passive smoking and cancer.

Pollutants Being exposed to substances such as metals, dust, chemicals, or pesticides at work can increase the risk of cancer. Asbestos, nickel, cadmium, uranium, radon, vinyl chloride, benzidine, and benzene are well-known examples of carcinogens in the workplace. For example, inhaling asbestos fibers increases the risk of lung diseases, including cancer. The cancer risk is especially high for asbestos workers who smoke.

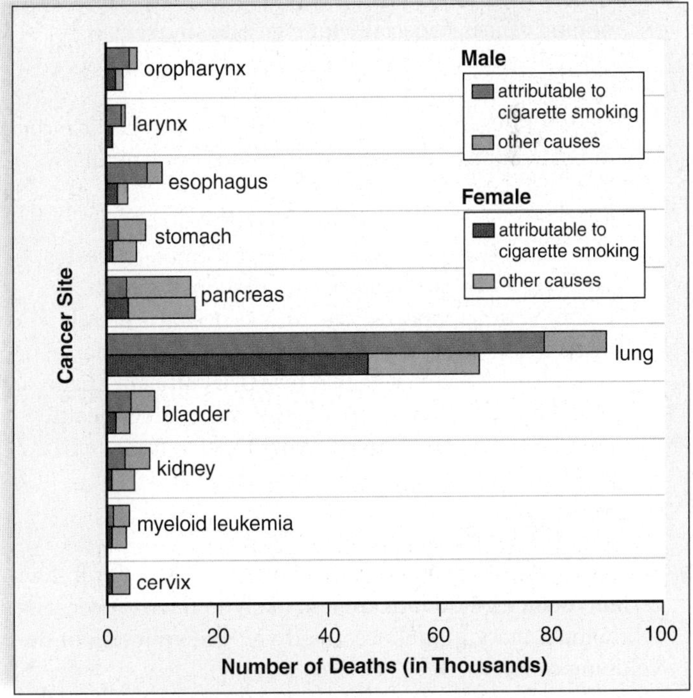

Figure 19.7 **Cancer deaths attributed to smoking.** These data from the Centers for Disease Control (CDC) show that, overall, the majority of cancer deaths for (**a**) men and (**b**) women may be attributed to cigarette smoking.

Biology Matters **Health Focus**

Prevention of Cancer

Protective Behaviors

These behaviors help prevent cancer:

Don't use tobacco. Cigarette smoking accounts for over 30% of cancer deaths. Smoking is responsible for 90% of lung cancer cases among men and 79% among women—about 87% on average. People who smoke two or more packs of cigarettes a day have lung cancer mortality rates 15 to 25 times greater than those of nonsmokers. Smokeless tobacco (chewing tobacco or snuff) increases the risk of cancers of the mouth, larynx, throat, and esophagus.

Don't sunbathe or use a tanning booth. Almost all cases of basal cell and squamous cell skin cancers are considered sun-related. Further, sun exposure is a major factor in the development of melanoma, and the incidence of this cancer increases for people living near the equator.

Avoid radiation. Excessive exposure to ionizing radiation can increase cancer risk. Even though most medical and dental X-rays are adjusted to deliver the lowest dose possible, unnecessary X-rays should be avoided. Excessive radon exposure in homes increases the risk of lung cancer, especially for cigarette smokers. It is best to test your home for radon and take the proper remedial actions.

Be tested for cancer. Do the shower check for breast cancer or testicular cancer. Have other exams done regularly by a physician. (See Table 19.1.)

Be aware of occupational hazards. Exposure to several different industrial agents (nickel, chromate, asbestos, vinyl chloride, etc.) and/or radiation increases the risk of various cancers. Risk from asbestos is greatly increased when combined with cigarette smoking.

Get vaccinated. Get vaccinated for HPV and hepatitis A and B. Consult with your physician or other health professional for advice regarding these vaccines and whether they are appropriate for you.

The Right Diet

Statistical studies have suggested that people who follow certain dietary guidelines are less likely to have cancer. The following dietary guidelines greatly reduce your risk of developing cancer:

Avoid obesity. Obesity increases the risk for many different types of cancer, especially those related to the reproductive systems in both men and women. Cancers of the colon, rectum, esophagus, breast, kidney, and prostate are all associated with being overweight.

Eat plenty of high-fiber foods. Studies have indicated that a high-fiber diet (whole grain cereals, fruits, and vegetables) may protect against colon cancer, a frequent cause of cancer deaths. Foods high in fiber also tend to be low in fat!

Increase consumption of foods that are rich in vitamins A and C. Beta-carotene, a precursor of vitamin A, is found in dark green, leafy vegetables; carrots; and various fruits. Vitamin C is present in citrus fruits. These vitamins are called antioxidants because in cells they prevent the formation of free radicals (organic ions having an unpaired electron) that can possibly damage DNA. Vitamin C also prevents the conversion of nitrates and nitrites into carcinogenic nitrosamines in the digestive tract.

Reduce consumption of salt-cured, smoked, or nitrite-cured foods. Salt-cured or pickled foods may increase the risk of stomach and esophageal cancers. Smoked foods, such as ham and sausage, contain chemical carcinogens similar to those in tobacco smoke. Nitrites are sometimes added to processed meats (e.g., hot dogs and cold cuts) and other foods to protect them from spoilage. As mentioned previously, nitrites are converted to nitrosamines in the digestive tract.

Include vegetables from the cabbage family in the diet. The cabbage family includes cabbage, broccoli, brussels sprouts, kohlrabi, and cauliflower. These vegetables may reduce the risk of gastrointestinal and respiratory tract cancers.

Drink alcohol in moderation. Risks of cancer development rise as the level of alcohol intake increases. The risk increases still further for those who smoke or chew tobacco while drinking. Heavy drinkers face a much greater risk for oral, pharyngeal, esophageal, and laryngeal cancer. Higher rates for cancers of the breast and liver are also linked to alcohol abuse. Men should limit daily alcohol consumption to two drinks or fewer. Women should have only one alcoholic drink daily. (A drink is defined as 12 oz. of beer, 5 oz. of wine, or 1.5 oz. of distilled spirits.)

Data show the incidence in soft tissue sarcomas (STS), malignant lymphomas, and non-Hodgkin lymphomas has increased in farmers living in Nebraska and Kansas. All used 2,4-D, a commonly used herbicidal agent, on crops and to clear weeds along railroad tracks.

Viruses

At least four types of DNA viruses—hepatitis B and C viruses, Epstein–Barr virus, and human papillomavirus (HPV)—are directly believed to cause human cancers.

In China, almost all the people have been infected with the hepatitis B virus. This correlates with the high incidence of liver cancer in that country. A combined vaccine is now available for hepatitis A and B. For a long time, circumstances suggested that cervical cancer was a sexually transmitted disease. Now, human papillomaviruses (HPVs) are routinely isolated from cervical cancers. Burkitt lymphoma occurs frequently in Africa, where virtually all children are infected with the Epstein–Barr virus. In China, the Epstein–Barr virus is isolated in nearly all nasopharyngeal cancer specimens.

RNA-containing retroviruses, in particular, are known to cause cancers in animals. In humans, the retrovirus HTLV-1 (human T-cell lymphotropic virus, type 1) has been shown to cause hairy cell leukemia. This disease occurs frequently in parts of Japan, the Caribbean, and Africa, particularly in regions where people are known to be infected with the virus. HIV, the virus that causes AIDS, and also Kaposi's sarcoma-associated herpes virus (KSHV) are responsible for the development of Kaposi's sarcoma and certain lymphomas. This occurs due to the suppression of proper immune system functions.

Animation Replication Cycle of a Retrovirus

Dietary Choices

Nutrition is emerging as a way to help prevent cancer. The incidence of breast and prostate cancer parallels a high-fat diet, as does obesity. The Health Focus, *Prevention of Cancer,* discusses how to protect yourself from cancer. The American Cancer Society recommends consumption of fruits and vegetables, whole grains instead of processed (refined) grains, and limited consumption of red meats (especially high-fat and processed meats). Moderate to vigorous activity for 30 to 45 minutes a day, five or more days a week, is also recommended.

Connecting the Concepts

For more information on cancer and its prevention, refer to the following discussions.

Section 4.7 describes methods of protecting yourself against skin cancer.

Section 8.6 summarizes how the components of a healthy diet reduce the risk of cancer.

Section 9.7 examines the influence of smoking on lung cancer.

Check Your Progress 19.2

1. Summarize the evidence that there is a hereditary component to cancer.
2. List the environmental carcinogens that are known to play a role in the development of cancer.
3. Discuss the proactive steps that you can take to reduce your risk of cancer.

19.3 Diagnosis of Cancer

Learning Outcomes

Upon completion of this section, you should be able to

1. List the seven warning signs of cancer.
2. Describe the tests that may be used to diagnose cancer in an individual.

The earlier a cancer is detected, the more likely it can be effectively treated. At present, physicians have ways to detect several types of cancer before they become malignant. Researchers are always looking for new and better detection methods. A growing number of researchers now believe the future of early detection lies in testing for the "fingerprints"—molecular changes caused by cancer. Several teams of scientists are working on blood, saliva, and urine tests to catch cancerous gene and protein patterns in these bodily fluids before a tumor develops. In the meantime, cancer is usually diagnosed by the methods discussed here.

Seven Warning Signs

At present, diagnosis of cancer before metastasis is difficult, although treatment at this stage is usually more successful.

Connections and Misconceptions

Can transposons cause cancer?

Transposons are small, mobile sequences of DNA that have the ability to move throughout the genome. From an evolutionary perspective, they are closely related to the retroviruses. Also known as "jumping genes," transposons are known to cause mutation as they move throughout the genome. Although the chances of a transposon disrupting the activity of a proto-oncogene or a tumor suppressor gene in a specific cell are very small (our genome has over 3.4 billion nucleotides), there have been cases when a transposon has caused a loss of cell cycle control and been a factor in the development of cancer. Transposon activity has also been associated with the development of other diseases, such as some forms of hemophilia and muscular dystrophy.

Animation Transposons: Shifting Segments of the Genome

The American Cancer Society publicizes seven warning signals, which spell out the word CAUTION and of which everyone should be aware:

C change in bowel or bladder habits
A a sore that does not heal
U unusual bleeding or discharge
T thickening or lump in breast or elsewhere
I indigestion or difficulty in swallowing
O obvious change in wart or mole
N nagging cough or hoarseness

Keep in mind that these signs do not necessarily mean that you have cancer. However, they are an indication that something is wrong and a medical professional should be consulted. Unfortunately, some of these symptoms are not obvious until cancer has progressed to one of its later stages.

Routine Screening Tests

Self-examination, followed by examination by a physician, can help detect the presence of cancer. For example, the letters ABCDE can help your self-exam of the skin for melanoma, the most serious form of skin cancer (Fig. 19.8). Breast cancer in women and testicular cancer in men can often be detected during a monthly shower check. The technique is discussed in the Health Focus, *Shower Check for Cancer.*

The ideal tests for cancer are relatively easy to do, cost little, and are fairly accurate. The Pap test for cervical cancer fulfills these three requirements. A physician merely takes a sample of cells from the cervix, which are then examined microscopically for signs of abnormality (see Fig. 19.1). Any woman who chooses to receive the new HPV vaccine should still get regular Pap tests because (1) the vaccine will not protect against all types of HPV that cause cervical cancer, and (2) the vaccine does not protect against any HPV infections she may already have if she is sexually active. Regular Pap tests are credited with preventing over 90% of deaths from cervical cancer. Screening for colon cancer also depends upon three types of testing. A digital rectal examination performed by a physician is of limited value because only a portion of the rectum can be reached. With flexible sigmoidoscopy, the second procedure, a much larger portion of the colon can be examined by using a thin, pliable, lighted tube. Finally, a stool blood test (fecal occult blood test) consists of examining a stool sample to detect any hidden blood. The sample is smeared on a slide, and a chemical is added that changes color in the presence of hemoglobin. This procedure is based on the supposition that a cancerous polyp bleeds, although some polyps do not bleed. Moreover, bleeding is not always because of a polyp. Therefore, the percentage of false negatives and false positives is high. All positive tests are followed up by a colonoscopy, an examination of the entire colon, or by X-ray after a barium enema. If the colonoscopy detects polyps, they can be destroyed by laser therapy. Blood tests are used to detect leukemia. Urinalysis aids in the diagnosis of bladder cancer.

Breast cancer is not as easily detected, but three procedures are recommended. First, every woman should do a monthly breast self-examination. But this is not a sufficient screen for breast cancer. Therefore, all women should have an annual physical examination, especially women over 40, when a physician will do the same procedure. Even then, this type of examination may not detect lumps before metastasis has already taken place. That is the goal of the third recommended procedure, *mammography*, an X-ray study of the breast

A = Asymmetry: one-half of mole does not look like the other half.

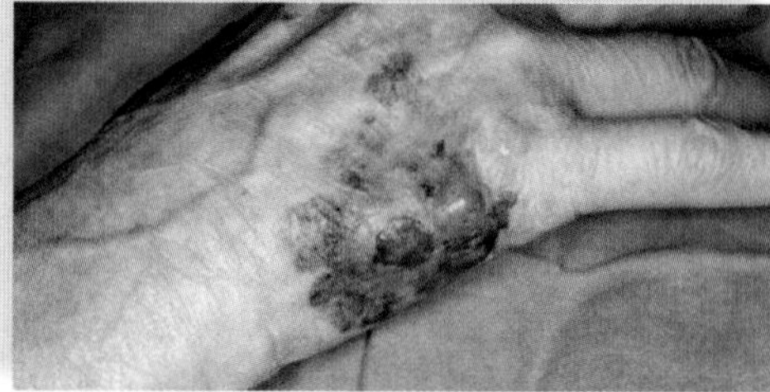

B = Border: irregular scalloped or poorly circumscribed border.

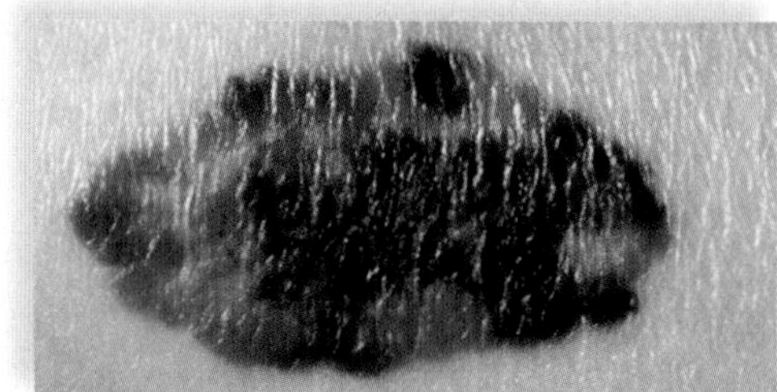

C = Color: varied from one area to another; shades of tan, brown, black, or sometimes white, red, or blue.

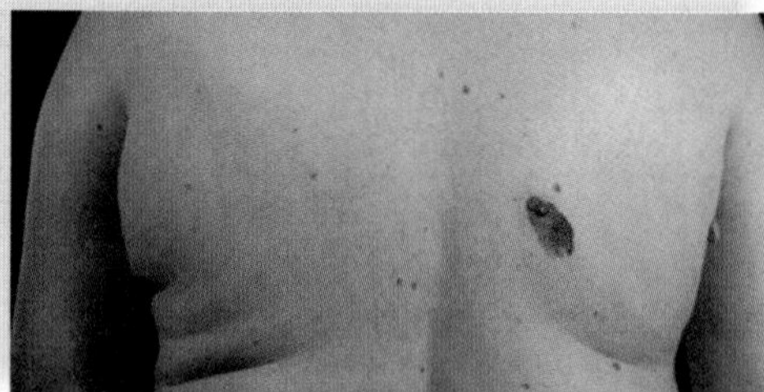

D = Diameter: larger than 6 mm (the diameter of a pencil eraser).

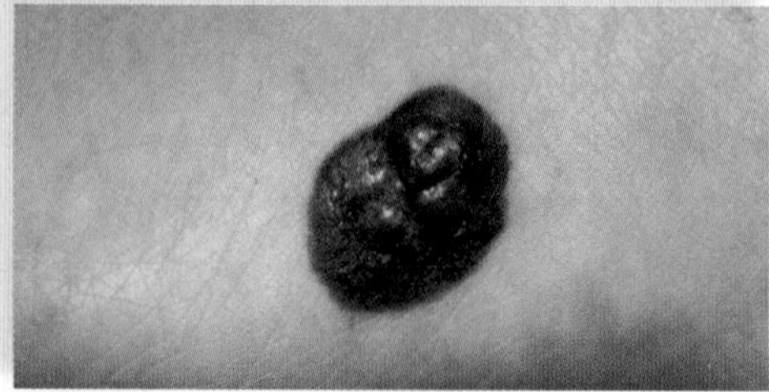

E = Elevated: above skin surface, and **evolving,** or changing over time

Figure 19.8 The ABCDE test for melanoma.
Suspicion of melanoma can begin by discovering a mole that has one or more of these characteristics.

Biology Matters Health Focus

Shower Check for Cancer

The American Cancer Society urges women to do a breast self-exam and men to do a testicle self-exam every month. Breast cancer and testicular cancer are far more curable if found early, and we must all take on the responsibility of checking for one or the other.

Shower Self-Exam for Women

1. Check your breasts for any lumps, knots, or changes about one week after your period.
2. Place your right hand behind your head. Move your *left* hand over your *right* breast in a circle. Press firmly with the pads of your fingers (Fig. 19A). Also check the armpit.
3. Now place your left hand behind your head and check your *left* breast with your *right* hand in the same manner as before. Also check the armpit.
4. Check your breasts while standing in front of a mirror right after you do your shower check. First, put your hands on your hips and then raise your arms above your head (Fig. 19B). Look for any changes in the way your breasts look: dimpling of the skin, changes in the nipple, or redness or swelling.
5. If you find any changes during your shower or mirror check, see your doctor right away.

You should know that the best check for breast cancer is a mammogram. When your doctor checks your breasts, ask about getting a mammogram.

Shower Self-Exam for Men

1. Check your testicles once a month.
2. Roll each testicle between your thumb and finger as shown in Figure 19C. Feel for hard lumps or bumps.
3. If you notice a change or have aches or lumps, tell your doctor right away so he or she can recommend proper treatment.

Cancer of the testicles can be cured if you find it early. You should also know that prostate cancer is the most common cancer in men. Men over age 50 should have an annual health checkup that includes a prostate examination.

Information provided by the American Cancer Society. Used by permission.

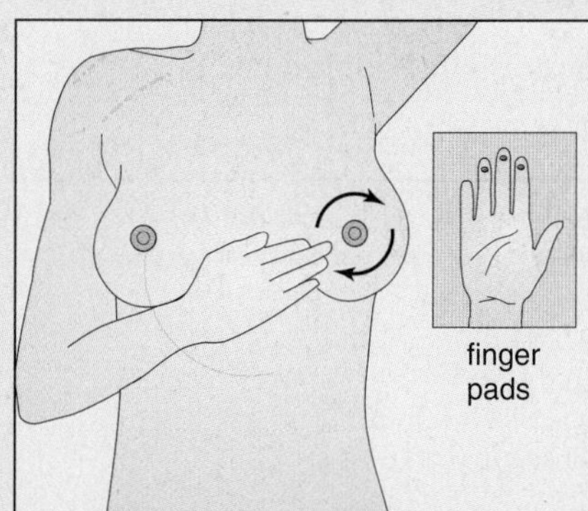

Figure 19A Shower check for breast cancer.

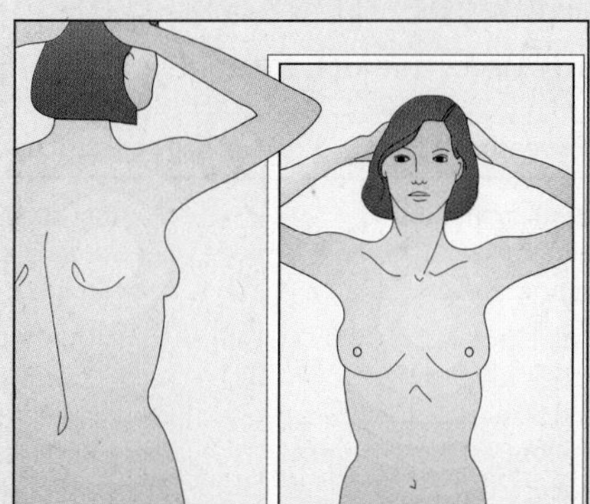

Figure 19B Mirror check for breast cancer.

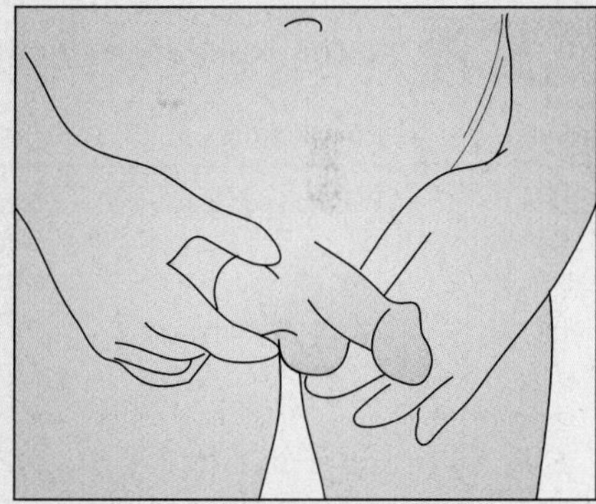

Figure 19C Shower check for testicular cancer.

(Fig. 19.9). However, mammograms do not show all cancers, and new tumors may develop in the interval between mammograms. The hope is that a mammogram will reveal a lump too small to be felt, at a time when the cancer is still highly curable. Table 19.1 outlines when routine screening tests should be done for various cancers, including breast cancer.

The diagnosis of cancer in other parts of the body may involve other types of imaging. A computerized axial tomography (CAT) scan uses computer analysis of scanning X-ray images to create cross-sectional pictures that portray a tumor's size and location (see Fig. 2.3). Magnetic resonance imaging (MRI) is another type of imaging technique that depends on computer analysis. MRI is particularly useful for analyzing tumors in tissues surrounded by bone, such as tumors of the brain or spinal cord. A radioactive scan obtained after a radioactive isotope is administered can reveal any abnormal isotope accumulation due to a tumor. During ultrasound, echoes of high-frequency sound waves directed at a part of the body are used to reveal the size, shape, and location of tissue masses. Ultrasound can confirm tumors of the stomach, prostate, pancreas, kidney, uterus, and ovary.

Aside from various imaging procedures, a diagnosis of cancer can be confirmed without major surgery by performing a *biopsy* or viewing body parts. Needle biopsies allow removal of a few cells for examination. Sophisticated techniques, such as laparoscopy, permit viewing of body parts.

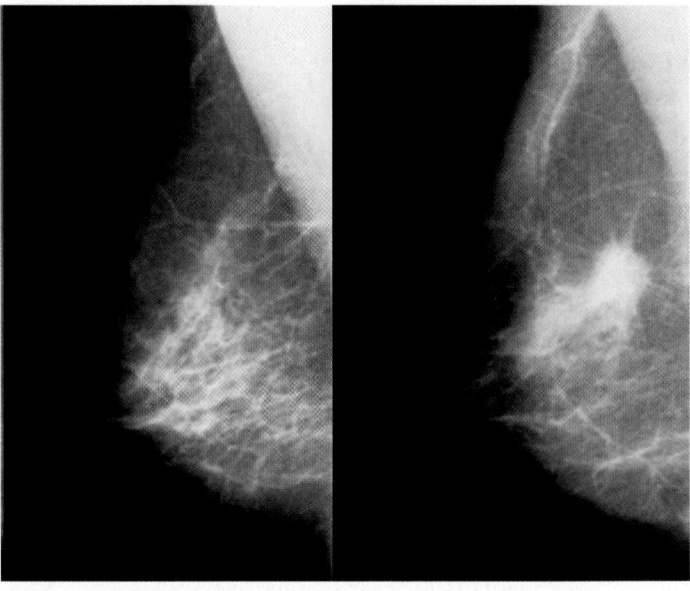

Figure 19.9 **Mammograms can detect breast cancer.** An X-ray image of the breast can find tumors too small to be felt. **a.** An image of a normal breast; (**b**) a breast with a tumor.

Tumor Marker Tests

Tumor marker tests are blood tests for antigens and/or antibodies. Blood tests are possible because tumors release substances that provoke an antibody response in the body.

Table 19.1 **Recommendations for the Early Detection of Cancer in Average-Risk Asymptomatic People**

Cancer	Population	Test or Procedure	Frequency
Breast	Women, age ≥ 20 years	Breast self-examination Clinical breast examination and mammography	Monthly, starting at age 20 years; see the Health Focus, p. 457. For women in 20s and 30s, every 3 years; aged 40 years and over, preferably annually
Colorectal	Men and women, age ≥ 50 years	Fecal occult blood test (FOBT) or fecal immunochemical test (FIT),* *or* FOBT (or FIT) and flexible sigmoidoscopy *or* double-contrast barium enema *or* colonoscopy	Annually, starting at age 50 years. Annually, starting at age 50 years Every 5 years, starting at age 50 years Every 10 years, starting at age 50 years
Prostate	Men, age ≥ 50 years	Digital rectal examination and prostate-specific antigen test (PSA)	Discussion with physician annually, starting at age 50 years, for men who have a life expectancy of at least 10 more years
Cervical	Women, age ≥ 21 years	Pap test	3 years after vaginal intercourse begins but no later than 21 years; after age 30 years, every 2 to 3 years if three normal pap tests in a row. Women age 70 years may choose to stop cervical cancer screening if three normal pap tests in a row in the past 10 years.
Endometrial	Women, after menopause	Report any unexpected bleeding or spotting to a physician.	
Testicular	Men, age 20 years	Testicle, self-examination	Monthly, starting at age 20 years, see Health Focus, p. 457.
Other cancers	Men and women, age 20 years	On the occasion of a periodic health examination, a cancer-related checkup should include examination for other cancers. See Health Focus, p. 457, for counseling about tobacco, sun exposure, diet and nutrition, risk factors, and environmental and occupational exposures.	

*FOBT, as it is sometimes done in physicians' offices, with the single stool sample collected on a fingertip during a digital rectal examination, is not an adequate substitute for the recommended at-home procedure of collecting two samples from three consecutive specimens. Toilet-bowl FOBT tests also are not recommended. FIT is more patient-friendly and is likely to be equal or better in sensitivity and specificity. *Source:* American Cancer Society website: www.cancer.org.

For example, if an individual has already had colon cancer, it is possible to use the presence of an antigen called *carcinoembryonic antigen (CEA)* to detect any relapses. When the CEA level rises, additional tumor growth has occurred.

There are also tumor marker tests that can be used for early cancer diagnosis. They are not reliable enough to count on solely, but in conjunction with physical examination (see page 457) and ultrasound, they are considered useful. For example, there is a *prostate-specific antigen (PSA) test* for prostate cancer, a *CA-125* test for ovarian cancer, and an *alpha-fetoprotein (AFP) test* for liver tumors.

Genetic Tests

Tests for genetic mutations in proto-oncogenes and tumor suppressor genes are making it possible to detect the likelihood of cancer before the development of a tumor. Tests are available that signal the possibility of colon, bladder, breast, and thyroid cancers, as well as melanoma. Physicians now believe that a mutated *RET* gene means that thyroid cancer is present or may occur in the future, and a mutated *p16* gene appears to be associated with melanoma. Genetic testing can also be used to determine if cancer cells still remain after a tumor has been removed.

A genetic test is also available for the presence of *BRCA1* (breast cancer gene 1). A woman who has inherited this gene can choose to either have prophylactic surgery or to be frequently examined for signs of breast cancer. Physicians can use microsatellites to detect chromosomal deletions that accompany bladder cancer. Microsatellites are small regions of DNA that always have two (di-), three (tri-), or four (tetra-) nucleotide repeats. They compare the number of nucleotide DNA repeats in a lymphocyte microsatellite with the number in a microsatellite of a cell found in urine. When the number of repeats is less in the cell from urine, a bladder tumor is suspected.

Telomerase, you will recall, is the enzyme that keeps telomeres a constant length in cells. The gene that codes for telomerase is turned off in normal cells but is active in cancer cells. Therefore, if the test for the presence of telomerase is positive, the cell is cancerous.

Video Melanoma Marker

Video New Cancer Clue

Connecting the Concepts

For more information on the information listed in this section, refer to the following discussions.

Section 7.4 describes how the body generates antibodies against antigens.

Section 8.5 examines how polyps in the large intestine (colon) may lead to colon cancer.

Section 16.3 examines the structure of the female reproductive system and illustrates the location of the cervix.

Check Your Progress 19.3

1. List the seven warning signs of cancer.
2. Discuss the routine screening tests that are available to detect and diagnose cancer.
3. Discuss how genetic tests and tumor marker tests may be used to prevent cancer.

19.4 Treatment of Cancer

Learning Outcomes

Upon completion of this section, you should be able to

1. Describe how radiation therapy, chemotherapy, and surgery may all be used to treat cancer.
2. Summarize some of the new advances in the treatment of cancer.

Certain therapies for cancer have been available for some time. Other methods of therapy are in clinical trials. If successful, these new therapies will become more generally available in the future.

Standard Therapies

Surgery, radiation therapy, and chemotherapy are the standard methods of cancer therapy.

Surgery

With the exception of cancers of the blood, surgery may be sufficient for cancer in situ. But because there is always the danger that some cancer cells have been left behind, surgery is often preceded by and/or followed by radiation therapy and chemotherapy.

Radiation Therapy

Ionizing radiation causes chromosomal breakage and cell cycle disruption. Therefore, rapidly dividing cells, such as cancer cells, are more susceptible to its effects than other cells. Powerful X-rays or gamma rays can be administered through an externally applied beam. In some instances, tiny radioactive sources can be implanted directly into the patient's body. Cancer of the cervix and larynx, early stages of prostate cancer, and Hodgkin disease are often treated with radiation therapy alone.

Although X-rays and gamma rays are the mainstays of radiation therapy, protons and neutrons also work well. Proton beams can be aimed at the tumor like a rifle bullet hitting the bull's-eye of a target.

Side effects of radiation therapy greatly depend upon which part of the body is being irradiated and how much radiation is used. Weakness and fatigue are very typical. Dry mouth, nausea, and diarrhea often affect the digestive tract. Dry, red, or irritated skin or even blistering burns may occur at the treatment site. Hair loss at the treatment site, which in some situations can be permanent, can be very depressing to the patient. Fortunately, most side effects of radiation treatment are temporary.

Chemotherapy

Radiation is localized therapy, whereas chemotherapy is a way to catch cancer cells that have spread (Figs. 19.10 and 19.11). Unlike radiation, which treats only the part of the body exposed to the radiation, chemotherapy treats the entire body. As a result, any cells that may have escaped from where the cancer originated are treated. Most chemotherapeutic drugs kill cells by damaging their DNA or interfering with DNA synthesis. The hope is that all cancer cells will be killed, while leaving enough normal cells untouched to allow the body to keep functioning. Combining drugs that have different actions at the cellular level may help destroy a greater number of cancer cells. Combinations may also reduce the risk of the cancer developing resistance to one particular drug. What chemicals are used is generally based on the type of cancer and the patient's age, general health, and perceived ability to tolerate potential side effects. Some of the types of chemotherapy medications commonly used to treat cancer include the following.

Alkylating agents. These medications interfere with the growth of cancer cells by blocking the replication of DNA.

Antimetabolites. These drugs block the enzymes needed by cancer cells to live and grow.

Antitumor antibiotics. These antibiotics—different from those used to treat bacterial infections—interfere with DNA, blocking certain enzymes and cell division and changing cell membranes.

Connections and Misconceptions

What can be done to lessen the side effects of radiation and chemotherapy?

Victims of cancer will often describe the side effects of treatment as the worst part of the disease. Nausea, vomiting, diarrhea, weight loss and hair loss, anxiety and/or depression, and extreme fatigue are common symptoms that may be caused by chemotherapy or radiation therapy. However, strategies now exist to help sufferers deal with treatment side effects, and others continue to be devised. Antiemetic drugs alleviate nausea and vomiting. Genetically engineered erythropoietin will help to stimulate red blood cell production and reduce fatigue (see Chap. 6). Marinol, a medication closely related to marijuana, stimulates appetite and helps with weight loss. Hair loss can be minimized by cryotherapy, which involves applying cold packs on the scalp during a treatment. Cold temperatures slow the metabolic rate of hair follicles, and more follicles survive treatment. Antidepressants and antianxiety medications will help the patient to deal with the psychological effects of the disease.

Increasingly, cancer patients are turning to alternative therapies as well for help in coping with the disease. Calming techniques, including yoga, meditation, and tai chi, help the sufferer to relax. Hypnosis can have the same restful effect. Aromatherapy, massage therapy, and music therapy seem to ease tension and stress, perhaps by stimulating pleasure centers in the brain. Exercise is known to produce endorphins, the brain's pain-relieving neurotransmitters (see Chap. 13).

Figure 19.10 Radiation treatment for cancer.
Most people who receive radiation therapy for cancer have external radiation on an outpatient basis. The patient's head is immobilized so that radiation can be delivered to a very precise area.

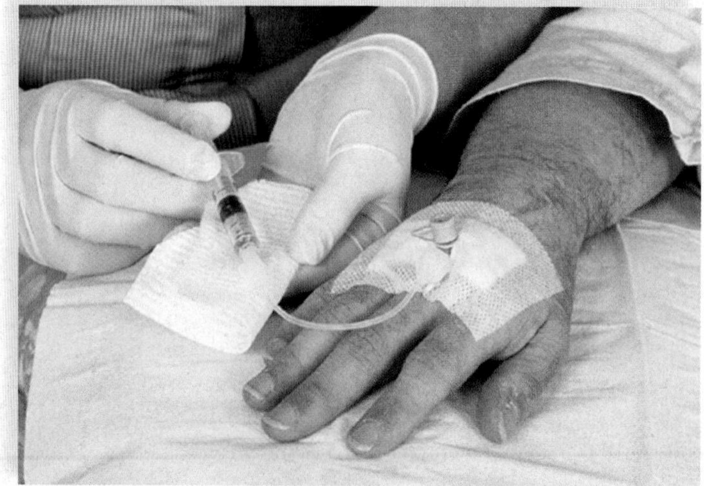

Figure 19.11 Treating cancer with chemotherapy.
The intravenous route is the most common, allowing chemotherapy drugs to spread quickly throughout the entire body by way of the bloodstream.

Mitotic inhibitors. These drugs inhibit cell division or hinder certain enzymes necessary in the cell reproduction process.
Nitrosoureas. These medications impede the enzymes that help repair DNA.

Whenever possible, chemotherapy is specifically designed for the particular cancer. For example, in some cancers, a small portion of chromosome 9 is missing. Therefore, DNA metabolism differs in the cancerous cells compared with normal cells. Specific chemotherapy for the cancer can exploit this metabolic difference and destroy the cancerous cells.

One drug, *taxol*, extracted from the bark of the Pacific yew tree, was found to be particularly effective against advanced ovarian cancers, as well as breast, head, and neck tumors. Taxol interferes with microtubules needed for cell division. Now, chemists have synthesized a family of related drugs, called taxoids, which may be more powerful and have fewer side effects than taxol.

Certain types of cancer, such as leukemias, lymphomas, and testicular cancer, are now successfully treated by combination chemotherapy alone. The survival rate for children with childhood leukemia is 80%. Hodgkin disease, a lymphoma, once killed two out of three patients. Combination therapy, using four different drugs, can now wipe out the disease in a matter of months. Three out of four patients achieve a cure, even when the cancer is not diagnosed immediately. In other cancers—most notably, breast and colon cancer, chemotherapy can reduce the chance of recurrence after surgery has removed all detectable traces of the disease.

Chemotherapy sometimes fails because cancer cells become resistant to one or several chemotherapeutic drugs, a phenomenon called multidrug resistance. This occurs because a plasma membrane carrier pumps the drug (or drugs) out of the cancer cell before it can be harmed. Researchers are testing drugs known to poison the pump in an effort to restore the effectiveness of the drugs. Another possibility is to use combinations of drugs with different toxic activities, because cancer cells can't become resistant to many different types at once.

Bone marrow transplants are sometimes done in conjunction with chemotherapy. The red bone marrow contains large populations of dividing cells. Therefore, red bone marrow is particularly prone to destruction by chemotherapeutic drugs. In bone marrow autotransplantation, a patient's stem cells are harvested and stored before chemotherapy begins. High doses of radiation or chemotherapeutic drugs are then given within a relatively short time. This prevents multidrug resistance from occurring, and the treatment is more likely to catch every cancer cell. The stored stem cells can then be returned to the patient by injection. They automatically make their way to bony cavities, where they resume blood cell formation.

Newer Therapies

Several therapies are now in clinical trials and are expected to be increasingly used to treat cancer.

Immunotherapy

When cancer develops, the immune system has failed to dispose of cancer cells, even though they bear antigens that make them different from the body's normal cells. A vaccine, called Melacine, which contains broken melanoma cells from two different sources, is under investigation for use against melanoma.

Another idea is to use immune cells, genetically engineered to bear the tumor's antigens (Fig. 19.12). When these

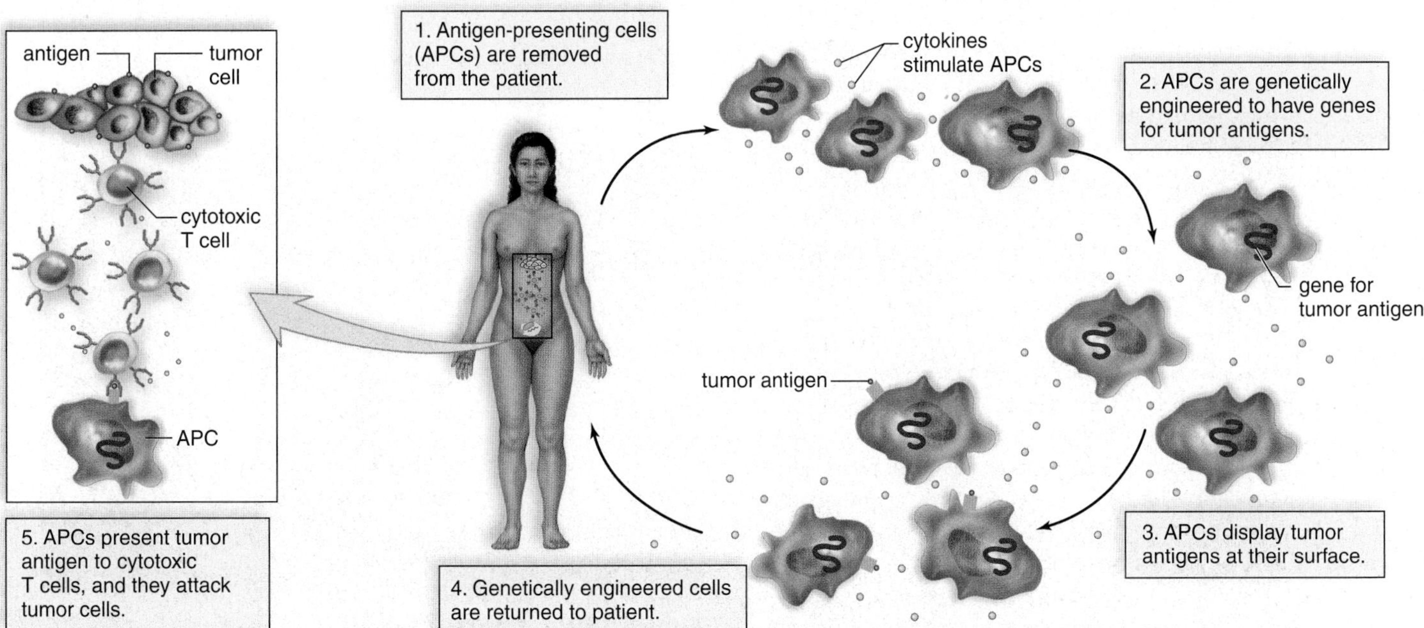

Figure 19.12 Use of immunotherapy to treat cancer.
1. Antigen-presenting cells (APCs) are removed and (**2**) are genetically engineered to (**3**) display tumor antigens. **4.** After these cells are returned to the patient, (**5**) they present the antigen to cytotoxic T cells, which then kill tumor cells.

Biology Matters Historical Focus

The Immortal Henrietta Lacks

If you've ever had food spoil in your refrigerator, you already know that it's very easy to grow mold and even easier to grow bacteria. Not so for healthy human tissue cells. These cells can be coaxed to reproduce using a technique called tissue culture. Unfortunately, healthy cells typically all die very rapidly, usually after dividing only a few times. However, metastatic cells—cancer cells—can survive and thrive in tissue culture. The techniques used to grow human cells successfully in tissue culture are part of the legacy of Henrietta Lacks (Fig. 19D). The cancer that claimed her life ensured that a part of Henrietta will live forever.

Lacks was a young African-American woman and the mother of five children. In February of 1951, she experienced unexplained vaginal bleeding and sought help from a Baltimore hospital. Physicians found a quarter-sized tumor on Henrietta's cervix. Samples of the tumor were quickly obtained, then sent to the tissue culture laboratories at Johns Hopkins University. Though radiation treatment was attempted, the tumor ravaged Henrietta's body, appearing on all her major organs within months. After eight months of suffering, she died in October 1951. She was survived by her husband and five children, three of whom were still in diapers.

Henrietta's tumor cells ended up in the laboratory of George and Margaret Gey at Johns Hopkins. The couple had been trying to culture human cells, with little success, for more than two decades. The cells from Henrietta Lacks' tumor, now termed HeLa cells (Fig. 19E), not only lived but multiplied like wildfire. At last, here was a source of human cells that grew rapidly and seemed almost indestructible.

Using these cells, the Geys directed their research toward curing polio. Infantile paralysis, or polio, occurred in epidemics that damaged the brain and spinal cord in a small percentage of patients. Paralysis of major muscle groups resulted. If the diaphragm and other respiratory muscles were affected, the disease could be fatal. For the first time, the virus could be grown in human cells—HeLa cells—so that its characteristics could be studied. The results of these studies enabled Dr. Jonas Salk to develop a vaccine for polio. Today, polio is almost unheard of in the Western world.

Research with HeLa cells did not stop with the Geys. For over 50 years, these durable cells have been used worldwide to study many types of viruses, as well as leukemia and other cancers. The cells are human in origin, so they have also been the test subjects to determine the harmful effects caused by drugs or radiation. Effective testing was developed for chromosome abnormalities and hereditary diseases using HeLa. Samples of HeLa cells have even been launched in the space shuttle for experiments involving a zero- gravity environment. HeLa cell colonies can be found around the world. The cells can even be purchased from biological supply companies.

HeLa cells are almost too sturdy. In 1974, researchers were shocked and dismayed to learn that cells thought to originate from other body tissues were HeLa. Countless hours and millions of dollars had been wasted by scientists who thought they were studying the brain, for example. HeLa cells, like bacteria, could be transmitted through the air, on a researcher's glove, or on contaminated glassware. This discovery led to better sterile techniques not only in the laboratory but also in operating rooms where cancerous tumors were removed. Once again, the cells of a poor woman, buried in an unmarked grave, served humanity by living on.

Figure 19D Henrietta Lacks, 1920–1951.

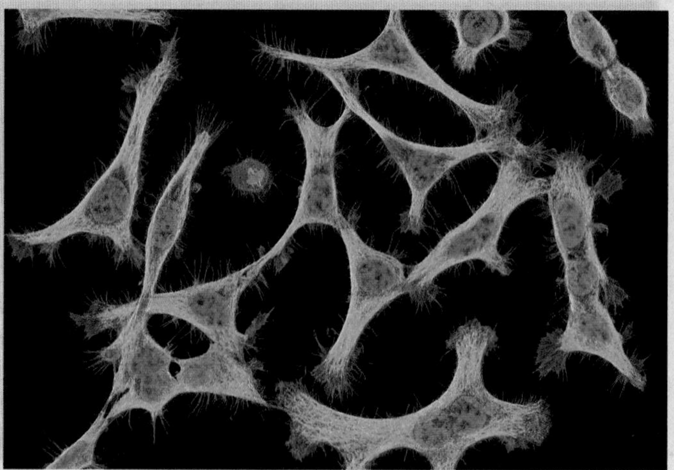

Figure 19E SEM of HeLa cells.

cells are returned to the body, they produce cytokines. Recall that cytokines are chemical messengers for immune cells (see Chap. 7). Cytokines stimulate the body's immune cells to attack the tumor. Further, the altered immune cells present the tumor antigen to cytotoxic T cells, which then go forth and destroy tumor cells in the body.

Passive immunotherapy is also possible. Monoclonal antibodies have the same structure because they are produced by the same plasma cell (see Fig. 7.17). Some monoclonal antibodies are designed to zero in on the receptor proteins of cancer cells. To increase the killing power of monoclonal antibodies, they are linked to radioactive isotopes or chemotherapeutic drugs. It is expected that soon they will be used as initial therapies, in addition to chemotherapy.

p53 *Gene Therapy*

Researchers believe that *p53* gene expression is needed for only 19 hours to trigger apoptosis, programmed cell death. And the *p53* gene seems to trigger cell death only in cancer cells—elevating the *p53* level in a normal cell doesn't do any harm, possibly because apoptosis requires extensive DNA damage.

Ordinarily, when adenoviruses infect a cell, they first produce a protein that inactivates *p53*. In a cleverly designed procedure, investigators genetically engineered an adenovirus that lacks the gene for this protein. Now, the adenovirus can infect and kill only cells that lack a *p53* gene. Which cells are those? Tumor cells, of course. Another plus to this procedure is that the injected adenovirus spreads through the cancer, killing tumor cells as it goes. This genetically engineered virus is now in clinical trials.

Other Therapies

Many other therapies are now being investigated. Among them, drugs that inhibit angiogenesis are a proposed therapy under investigation. Antiangiogenic drugs confine and reduce tumors by breaking up the network of new capillaries in the vicinity of a tumor. A number of antiangiogenic compounds are currently being tested in clinical trials. Two highly effective drugs, called angiostatin and endostatin, have been shown to inhibit angiogenesis in laboratory animals and are expected to do the same in humans.

Connecting the Concepts

For more information on the topics presented in this section, refer to the following discussions.

Section 7.3 examines the role of chemical signals, such as cytokines, in protecting the body from infection.

Section 7.4 explores the function of cytotoxic T cells in the immune response.

Section 21.4 provides an overview on how genetic engineering can alter the genes within a virus.

Check Your Progress 19.4

1. Describe the three types of therapy that are presently the standard ways to treat cancer.
2. Discuss why cancer treatment may involve a combination of chemotherapy, radiation therapy, and surgery.
3. Discuss why a cancer treatment that targets p53 may be successful.
4. Explain the rationale for antiangiogenesis therapy.

CASE STUDY CONCLUSION

In Cody's case, his treatment involved a process called a simple nephrectomy. During this procedure, the surgeon removes the entire kidney. Typically, nephroblastoma does not occur in both kidneys, and because the MRI and CT scan did not indicate any problems with Cody's other kidney, it most likely did not have the nephroblastoma. Because a single kidney is all that is needed to maintain blood homeostasis and water–salt balance in the body, Cody did not need a transplant. Initially, the chemotherapy treatment caused Cody to lose his hair, and his appetite decreased, but the doctors informed the concerned parents that this was normal, and very soon Cody would be back to his normal self.

Although Wilms disease is a rare form of cancer, researchers have identified a number of genes that appear to be associated with the condition. One of these is *WT1*, a gene that plays a major role in the differentiation of renal (kidney) tissue and the development of the kidneys. In Cody's case, he probably inherited one copy of a defective gene from one of his parents. Sometime during development, the other gene in one of his kidney cells acquired a mutation that initiated the formation of the tumor. This explained why Cody's other kidney was unaffected. Because *WT1* is active early in development, it was unlikely that Cody would have any additional complications from his cancer.

Media Study Tools

www.mhhe.com/maderhuman12e

Enhance your study of this chapter with study tools and practice tests. Also ask your instructor about the resources available through ConnectPlus, including the media-rich eBook, interactive learning tools, and animations.

Summarizing the Concepts

19.1 Cancer Cells

Characteristics of Cancer Cells

malignant tumor

Certain characteristics are common to cancer cells. Cancer cells

- lack differentiation and do not contribute to function.
- do not undergo apoptosis—they enter the cell cycle an unlimited number of times.
- form tumors and do not need growth factors to signal them to divide.
- gradually become abnormal—carcinogenesis is comprised of initiation, promotion, and progression.
- undergo angiogenesis (the growth of blood vessels to support them) and can spread throughout the body (metastasis).

Cancer Results from Gene Mutation

Cells become increasingly abnormal due to mutations in proto-oncogenes and tumor suppressor genes.

In normal cells, the cell cycle functions normally.

- Proto-oncogenes promote cell cycle activity and restrain apoptosis. Proto-oncogenes can mutate into oncogenes.
- Tumor suppressor genes restrain the cell cycle and promote apoptosis.

In cancer cells, the cell cycle is accelerated and occurs repeatedly.

- Oncogenes cause an unrestrained cell cycle and prevent apoptosis.
- Mutated tumor suppressor genes cause an unrestrained cell cycle and prevent apoptosis.

Types of Cancer

Cancers are classified according to their places of origin.

- Carcinomas originate in epithelial tissues.
- Sarcomas originate in muscle and connective tissues.
- Leukemias originate in blood.
- Lymphomas originate in lymphatic tissue.

Certain body organs are more susceptible to cancer than others.

19.2 Causes and Prevention of Cancer

Development of cancer is determined by a person's genetic profile, plus exposure to environmental carcinogens.

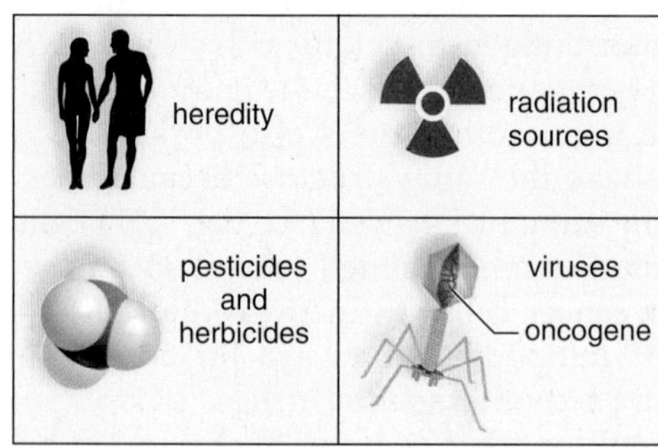

some sources contributing to development of cancer

- Cancers that run in families are most likely due to the inheritance of mutated genes (e.g., breast cancer, retinoblastoma tumor).
- Certain environmental factors are carcinogens (e.g., UV radiation, tobacco smoke, pollutants).
- Industrial chemicals (e.g., pesticides and herbicides) are carcinogenic.
- Certain viruses (e.g., hepatitis B and C, human papillomavirus, and Epstein–Barr virus) cause specific cancers.

19.3 Diagnosis of Cancer

The earlier a cancer is diagnosed, the more likely it can be effectively treated. Tests for cancer include

- Pap test for cervical cancer.
- mammogram for breast cancer.
- tumor marker tests—blood tests that detect tumor antigens/antibodies.
- tests for genetic mutations of oncogenes and tumor suppressor genes.
- biopsy and imaging—used to confirm the diagnosis of cancer.

C hange in bowel or bladder habits
A sore that does not heal
U nusual bleeding or discharge
T hickening or lump in breast or elsewhere
I ndigestion or difficulty in swallowing
O bvious change in wart or mole
N agging cough or hoarseness

19.4 Treatment of Cancer

Surgery, radiation, and chemotherapy are traditional methods of treating cancer. Other methods include

- chemotherapy involving bone marrow transplants.
- immunotherapy.
- *p53* gene therapy (one type of *p53* gene therapy ensures that cancer cells undergo apoptosis).
- other therapies, such as inhibitory drugs for angiogenesis and metastasis, which are being investigated.

Understanding Key Terms

angiogenesis 450
apoptosis 448
bone marrow transplant 461
cancer 448
carcinogen 453
carcinogenesis 449
carcinoma 451
checkpoints 450
cyclin 450
differentiation 448
growth factor 450
initiation 449
leukemia 451
lymphoma 451
metastasis 450
mutagen 453
oncogene 450
oncologist 451
oncology 451
progression 450
promotion 450
proto-oncogene 450
sarcoma 451
telomere 448
tumor 449
tumor marker test 458
tumor suppressor gene 450

Match the key terms to these definitions.

a. ________ End of chromosome that prevents it from binding to another chromosome; normally gets shorter with each cell division.

b. ________ Environmental agent that contributes to the development of cancer.

c. ________ Normal gene involved in cell growth and differentiation that becomes an oncogene through mutation.

d. ________ Formation of new blood vessels, such as a capillary network.

e. ________ Spread of cancer from the place of origin to throughout the body; caused by the ability of cancer cells to migrate and invade tissues.

Testing Your Knowledge of the Concepts

1. List and briefly discuss the seven characteristics of cancer cells that cause them to be abnormal. (pages 448–450)
2. What are the roles of the two genes that mutate, causing cancer to develop? (pages 450–451)
3. What are the four leading types of cancer in men and women? (page 451)
4. What are the general causes of cancer? What are several types of environmental mutagens? (pages 452–455)
5. What are the seven warning signals of cancer? (pages 455–456)
6. List and describe three tests designed to detect cancer. (pages 456–458)
7. Describe three standard therapies for cancer treatment. (pages 459–460)
8. Describe some newer therapies that may be successful in cancer treatment. (pages 461–463)
9. Whereas ________ stimulate the cell cycle, ________ inhibit the cell cycle.
   a. tumor suppressor genes, oncogenes
   b. oncogenes, tumor suppressor genes
   c. proto-oncogenes, oncogenes
   d. proto-oncogenes, tumor suppressor genes
10. Growth factors lead to
    a. increased cell division.
    b. the functioning of cyclin proteins.
    c. progression through the cell cycle.
    d. All of these are correct.
11. Which of these is not true of the gene *p53*?
    a. Mutations of both proto-oncogenes and tumor suppressor genes lead to inactivity of *p53*.
    b. Normally, *p53* functions to stop the cell cycle and initiate repair enzymes, when necessary.
    c. Normally, *p53* restores the length of telomeres.
    d. Cancer cells shut down the activity of *p53*, even though it may be present.
12. Which association is incorrect?
    a. proto-oncogenes—code for cyclin and proteins that inhibit the activity of *p53*
    b. oncogenes—"gain-of-function" genes
    c. mutated tumor suppressor genes—code for cyclin and proteins that inhibit the activity of *p53*
    d. mutated tumor suppressor genes—"loss-of-function" mutations
    e. Both a and c are incorrect.
13. Following each cell cycle, telomeres
    a. get longer.
    b. get shorter.
    c. return to the same length.
    d. bind to cyclin proteins.
14. Bone marrow transplants are done in conjunction with chemotherapy for the following reason.
    a. Smoking and pollutants can kill bone marrow stem cells.
    b. Blood cells will be engineered to have cancer antigens.
    c. This procedure is no longer done.
    d. Chemotherapy sometimes kills bone marrow stem cells.
    e. All of these are correct.
15. Angiogenic growth factors function to
    a. stimulate the development of new blood vessels.
    b. activate tumor suppressor genes.
    c. change proto-oncogenes into oncogenes.
    d. promote metastasis.
16. A tumor with cells that spread to secondary locations is referred to as
    a. benign.
    b. cancer in situ.
    c. malignant.
    d. encapsulated.
17. Which of the following is not a type of carcinogen?
    a. tobacco smoke
    b. radiation
    c. pollutants
    d. viruses
    e. All of these are carcinogens.
18. What type of cancer is associated with human papilloma viruses?
    a. breast
    b. cervical
    c. kidney
    d. lymphatic

19. Why is cancer called a genetic disease?
   a. Cancer is always inherited.
   b. Carcinogenesis is accompanied by mutations.
   c. Cancer causes mutations that are passed on to offspring.
   d. All of these are correct.
20. Leukemia is a form of cancer that affects
   a. lymphatic tissue.
   b. bone tissue.
   c. blood-forming cells.
   d. nervous system structures.
21. Which is the name of the tumor suppressor gene that causes retinoblastoma?
   a. *p21*
   b. *RB*
   c. *ras*
   d. *TGF-b*
22. Concerning the causes of cancer, which one is incorrect?
   a. Genetic mutations cause cancer.
   b. Genetic mutations can be caused by environmental influences, such as radiation, organic chemicals, and viruses.
   c. An active immune system, diet, and exercise can help prevent cancer.
   d. Heredity cannot be a cause of cancer.
23. Which of the following is not a warning signal for cancer?
   a. a sore that does not heal
   b. change in bowel or bladder habits
   c. nagging cough or hoarseness
   d. shortness of breath or fatigue
24. Which of these tests for the particular cancer is mismatched?
   a. breast cancer—mammogram
   b. lung cancer—X-ray
   c. cervical cancer—Pap test
   d. prostate cancer—*CA-125* test
25. Following a biopsy, what does a doctor look for to diagnose cancer?
   a. abnormal-appearing cells
   b. whether the cells can divide
   c. whether the cells will respond to growth factors
   d. whether the chromosomes have telomeres
26. Most chemotherapeutic drugs kill cells by
   a. producing pores in plasma membranes.
   b. interfering with protein synthesis.
   c. interfering with cellular respiration.
   d. interfering with DNA and/or enzymes.
27. Multidrug resistance to chemotherapeutic drugs occurs because
   a. cancer cells can use plasma membrane carriers to pump drugs out of the cell.
   b. one drug may interfere with the activity of another drug.
   c. using several drugs at once will overtax the patient's immune system.
   d. using several drugs at once decreases the effectiveness of radiation therapy.
28. *p53* gene therapy
   a. triggers cytotoxic T cells to destroy tumor cells.
   b. triggers apoptosis in cancer cells.
   c. produces monoclonal antibodies against the tumor cells.
   d. reduces tumors by breaking up their blood vessels.

Thinking Critically About the Concepts

Today, cancer is often treated with a combination of chemotherapy, radiation treatment, and surgery. In Cody's case, his cancer required surgery to remove the cancerous kidney, followed by chemotherapy treatment to ensure that no cancer cells had metastasized to other locations in his body. From what you have learned about cancer in this chapter, answer the following questions.

1. After his initial treatments in the hospital, Cody had to avoid public exposure. Based on what you know about the immune system, explain.
2. Cody's cancer was called a nephroblastoma. The suffix –blast refers to an immature cell. Would this type of cancer be more or less likely to spread throughout the body?
3. Cody's nephroblastoma caused blood to be found in his urine. What large blood vessels supplying the kidney might have been damaged by cancer? (See Chap. 5.)
4. In addition to blood in his urine, what other abnormal symptoms did Cody show?
5. Many types of cancer (ovarian cancer, for example) show no symptoms until they are well advanced. What might be the consequence for the patient?
6. The vaccine for HPV is recommended for females ages 11–26. There is some controversy over this vaccine. Some believe vaccination for sexually transmitted disease might lull these young women into a false sense of security regarding safety during sexual intercourse. Others fear that vaccination will encourage girls to engage in sex.
   a. Why would the vaccine be recommended for young girls who may not be sexually active yet?
   b. What type of immunity would women develop from getting vaccinated for HPV? You may need to revisit Chapter 7 on immunity to answer this question.
   c. If women are not vaccinated against HPV, how else could they protect themselves from infection?
7. Why are lymph nodes surrounding a breast with an invasive cancer often removed when a mastectomy is performed?
8. What are some healthy lifestyle choices you could make that might prevent cancer later in your life?

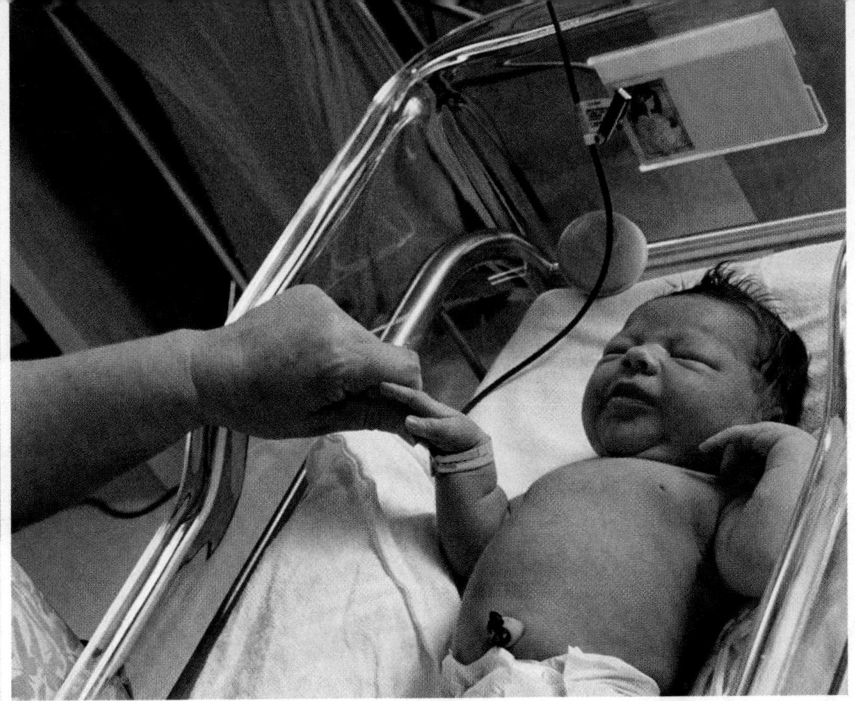

CHAPTER

20

Patterns of Genetic Inheritance

CASE STUDY PHENYLKETONURIA

As part of the routine newborn screening test performed on most children born in the United States and other developed countries, a high-performance liquid chromatography (HPLC) test was performed by taking a small amount of blood from a heel prick of 12-hour-old Patrick. The test came back positive for phenylketonuria (PKU). Dr. Preston explained to Patrick's parents that PKU was a disorder in which the body has a deficiency in the hepatic enzyme phenylalanine hydroxylase (PAH). PAH is an important enzyme that the body needs to properly metabolize the amino acid phenylalanine into the amino acid tyrosine. When phenylalanine is not metabolized, it accumulates in the body and is converted into the compound phenylpyruvate. An accumulation of phenylpyruvate can lead to impaired brain development, mental retardation, and seizure disorders. Patrick's parents wondered how he had developed this disorder. Dr. Preston explained that it was a genetic disorder; the cause of the disorder was found in the DNA Patrick had acquired from both his parents. Neither of Patrick's parents was affected by PKU and wondered how their son could be if neither of them showed any signs. Dr. Preston explained that this kind of disorder was inherited in an autosomal recessive manner.

As you read through the chapter, think about the following questions.

1. What is an autosomal recessive disorder? What are the different types of genetic inheritance?
2. How can people pass on conditions they do not show any signs of?
3. Can you always tell from looking at people what genes they carry in their DNA?

CHAPTER CONCEPTS

20.1 Genotype and Phenotype
The genotype refers to the genes for a particular trait. The phenotype refers to physical characteristics, such as hairline, blood type, color blindness, and the operation of cellular pathways.

20.2 One- and Two-Trait Inheritance
In humans, each trait is controlled by two alleles. In most cases, dominant alleles mask recessive alleles. It is possible to predict patterns of inheritance using Punnett squares and pedigrees. Many disorders, including some cellular disorders, are inherited in a dominant or recessive manner.

20.3 Inheritance of Genetic Disorders
Inheritance of traits can be traced throughout generations of a family using a pedigree chart.

20.4 Beyond Simple Inheritance Patterns
There are other inheritance patterns beyond simple dominant or recessive ones. The environment and other alleles may both influence the phenotype.

20.5 Sex-Linked Inheritance
Not all traits on the sex chromosomes are associated with sex. Because males only have one X chromosome, the alleles on that chromosome are always expressed. Therefore, males are more apt to have an X-linked disorder than are females.

BEFORE YOU BEGIN

Before beginning this chapter, take a few moments to review the following discussions.

Section 2.7 What is the role of DNA in a cell?

Figures 16.4 and 16.8 How are sperm and egg cells produced?

Section 18.4 How does meiosis produce new combinations of alleles?

20.1 Genotype and Phenotype

Learning Outcomes

Upon completion of this section, you should be able to

1. Distinguish between a genotype and a phenotype.
2. Define allele, gene, dominant, and recessive as they relate to patterns of inheritance.
3. Given the genotype of an individual, identify the phenotype.

Genotype

Genotype refers to the genes of the individual. Recall that genes are segments of DNA on a chromosome that code for a trait (characteristic). These genes are units of heredity existing in specific positions, or loci (sing., **locus**), on a chromosome. An **allele** can be easily described as an alternate form of a gene. So if the trait the gene coded for was eye color, an allele would be the code for blue eyes or the code for brown eyes. Alleles are classified as either dominant or recessive. If an allele is dominant, only one copy of that allele needs to be present in both chromosomes for that trait to appear. If an allele is recessive, both alleles need to code for the recessive trait for it to appear in the individual.

Phenotype

The physical appearance of a trait is called the **phenotype.** Alleles are designated by a one-letter abbreviation. A **dominant allele** is assigned an uppercase letter, and the lowercase letter is used for **recessive alleles.** In humans, for example, unattached (free) earlobes are dominant over attached earlobes. A suitable key would be *E* for unattached earlobes and *e* for attached earlobes.

For humans, we receive one chromosome from each parent; therefore, we inherit one allele from each parent, resulting in a pair of alleles for each trait. Figure 20.1 shows three possible fertilizations between different alleles for the earlobe trait and the resulting offspring of those fertilizations. In the first instance, the chromosomes of both the sperm and the egg carry the dominant trait, designated *E*, resulting in an individual with the alleles *EE*. This type of genotype is called **homozygous dominant** and results in the appearance of the dominant phenotype—in this example, unattached earlobes. Notice in Figure 20.1 that the genotype (letters) and then the phenotype (description) are given after the drawing.

In the second fertilization, the zygote has received two recessive alleles (*ee*), a genotype called **homozygous recessive.** An individual with this genotype has the recessive phenotype, attached earlobes. In the third fertilization, the resulting individual has the alleles *Ee*, called a **heterozygous** genotype. A heterozygote shows the dominant phenotype because the dominant allele only needs one copy to appear in the phenotype. Therefore, this individual has unattached earlobes.

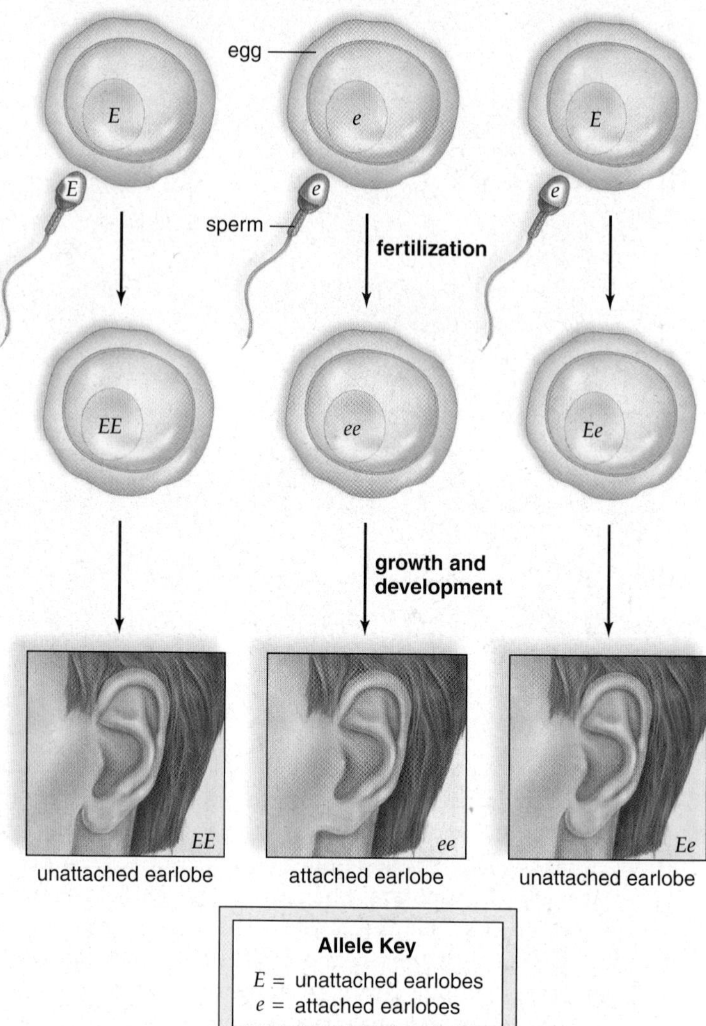

Figure 20.1 **Genetic inheritance affects our characteristics.** For humans, we receive one chromosome from each parent; therefore, we inherit one allele from each parent, resulting in a pair of alleles for each trait. The inheritance of a single dominant allele (*E*) causes an individual to have unattached earlobes; two recessive alleles (*ee*) cause an individual to have attached earlobes. Each individual receives one allele from the father (by way of a sperm) and one allele from the mother (by way of an egg).

In our example, the phenotypes were attached or unattached earlobes. From this, you may get the impression that the phenotype has to be an easily observable trait. However, the phenotype can be any characteristic of the individual. This could include color blindness or a metabolic disorder such as the lack of an enzyme to metabolize the amino acid phenylalanine.

Connecting the Concepts

The genotype of an individual is based on the information in the DNA of its cells. For more information on DNA, refer to the following discussions.

Sections 2.7 and 21.1 describe the basic structure of the DNA molecule.

Section 21.2 examines how the information in the DNA is expressed as a protein.

Check Your Progress 20.1

1. Define the following terms: *gene, locus, chromosome, dominant, recessive,* and *heterozygous.*
2. Describe the difference between genotype and phenotype.
3. Summarize the three possible genotypes and their corresponding phenotypes.

20.2 One- and Two-Trait Inheritance

Learning Outcomes

Upon completion of this section, you should be able to

1. Understand the relationship between probability and one- and two-trait crosses.
2. Calculate the probability of a specific genotype or phenotype in an offspring of a genetic cross.

In one-trait crosses (one-characteristic crosses), the inheritance of only one set of alleles is being considered. In two-trait crosses (two-characteristic crosses), the inheritance of two different alleles is being considered. For both types of crosses, it will be necessary to determine the gametes of both individuals who are reproducing.

Forming the Gametes

When sperm and eggs are formed, the chromosome number is divided in half. An individual has 46 chromosomes, but the sperm or egg only has 23 chromosomes. If reduction of the chromosome number didn't happen, the new individual would have twice as many chromosomes after fertilization, which would not result in a viable embryo. This division of chromosome numbers occurs during meiosis, when homologous chromosomes separate. One of these homologous chromosomes was originally from the mother, and the other was donated by the father. The alleles are on the homologous chromosomes, so they also separate during meiosis.

For example, let's say that the gene for unattached or attached earlobes is on a particular chromosome. On the homologous chromosome originally from the mother, the allele is an *E*. On the father's paired chromosome, the allele is also an *E*. Therefore, *E* is the only option for alleles on either chromosome, so every gamete formed will carry an *E*. This occurs whether the gamete gets the mother's or the father's original chromosome. Similarly, if the homologous chromosome originally from the mother carries an *e* and the father's chromosome also bears an *e*, then all gametes will have the *e* allele. What if the mother's homologous chromosome has an *E* allele and the father's has an *e*? Then, half of the gametes formed will receive an *E* allele from the mother's homologue. The other half will get the father's homologue with the *e*.

Figure 20.2 shows the genotypes and phenotypes for certain other traits in humans. You can practice deciding

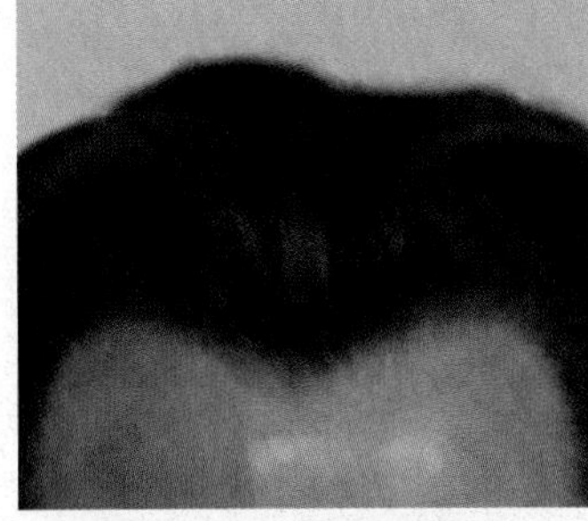

a. Widow's peak: *WW* or *Ww*

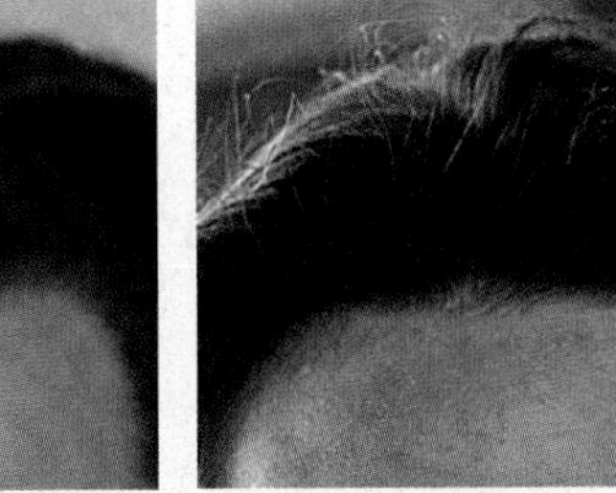

b. Straight hairline: *ww*

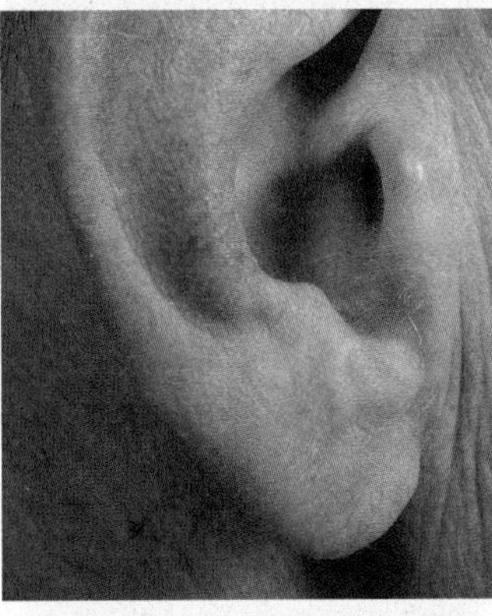

c. Unattached earlobes: *EE* or *Ee*

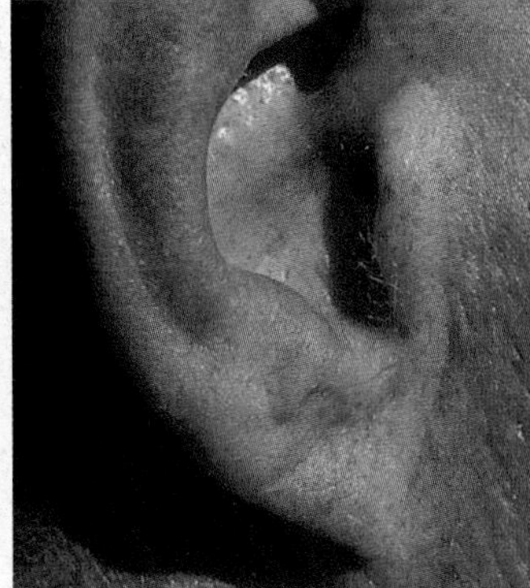

d. Attached earlobes: *ee*

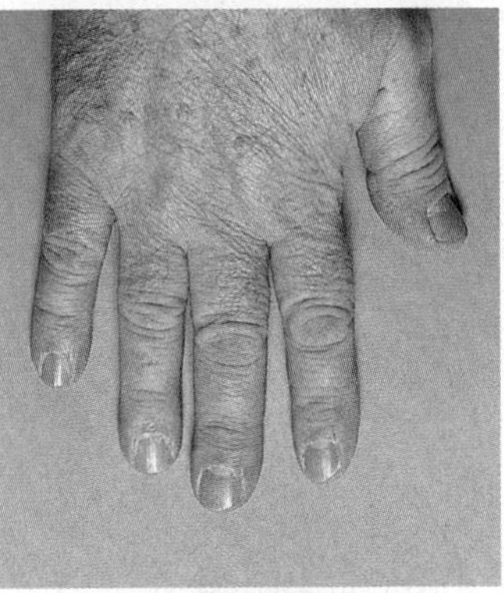

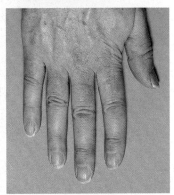

e. Short fingers: *SS* or *Ss*

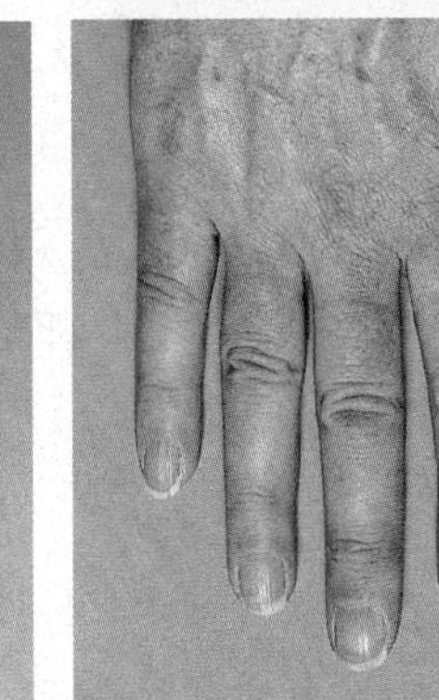

f. Long fingers: *ss*

g. Freckles: *FF* or *Ff*

h. No freckles: *ff*

Figure 20.2 Common inherited traits in humans. The allele keys indicate which traits are dominant and which are recessive.

what alleles the gametes would carry in order to produce these genotypes. If the genotype is *EeSs*, all combinations of any two different letters can be present. Therefore, *ES*, *eS*, *Es*, and *es* are all possible gametes.

One-Trait Crosses

Parents often like to know the chances of having a child with a certain genotype and, therefore, a certain phenotype. To illustrate, let us consider a cross involving freckles. What happens when two parents without freckles have children? Will the children of this couple have freckles? In solving the problem, we will (1) use the key provided in Figure 20.3 to indicate the genotype of each parent; (2) determine the possible gametes for each parent; (3) combine all possible gametes; and (4) finally, determine the genotypes and the phenotypes of all the offspring. If both parents do not have freckles, then their genotypes are both *ff*. The only gametes they can produce contain the *f* allele. All of the children will therefore be *ff* and will not have freckles. In the following diagram, the letters in the first row give the genotypes of the parents. Each parent has only one type of gamete with regard to freckles; therefore, all the children have a similar genotype and phenotype.

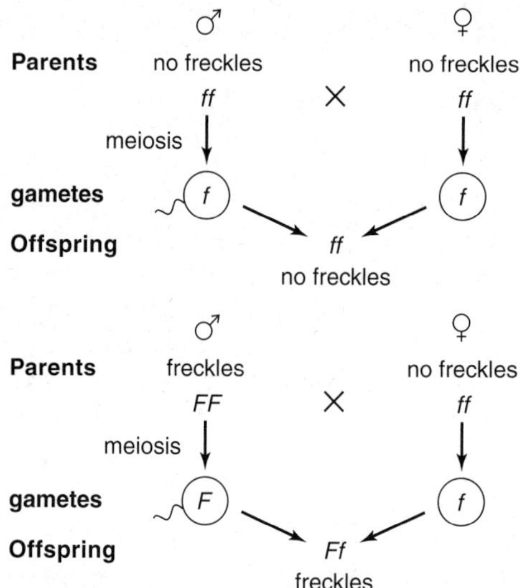

To challenge you a bit more, let's consider the results when a homozygous dominant man with freckles has children with a woman with no freckles. Will the children of this couple have freckles? The children are heterozygous (*Ff*) and have freckles. When writing a heterozygous genotype, always put the capital letter (for the dominant allele) first to avoid confusion.

These children are **monohybrids.** They are heterozygous with regard to one pair of alleles. If they have children with someone else of the same genotype, will their children have freckles? In this problem (*Ff* × *Ff*), each parent has two possible types of gametes (*F* or *f*), and we must ensure that all types of sperm have an equal chance to fertilize all possible types of eggs. One way to do this is to use a **Punnett square** (Fig. 20.3), in which all possible types of sperm are lined up vertically and all possible types of eggs are lined up horizontally (or vice versa). Every possible combination of gametes occurs within the squares.

After we determine the genotypes and the phenotypes of the offspring, we can determine the genotypic and then the phenotypic ratio. The genotypic ratio is 1 *FF*: 2 *Ff*: 1 *ff* or simply 1:2:1, but the phenotypic ratio is 3:1. Why? Three individuals will have freckles (the *FF* and the two *Ff*) and one will not have freckles (the *ff*).

This 3:1 phenotypic ratio is always expected for a monohybrid cross when one allele is completely dominant over the other. The exact ratio is more likely to be observed if a large number of matings take place and if a large number of offspring result. Only then do all possible types of sperm have an equal chance of fertilizing all possible types of eggs. Naturally, we do not routinely observe hundreds of offspring from a single type of cross in humans. The best interpretation of Figure 20.3 in humans is to say that each child has three chances out of four to have freckles, or one chance out of four to not have freckles.

Every fertilization has the exact same chance for combinations as the previous fertilization. For example, if two heterozygous parents already have three children with freckles and are expecting a fourth child, this child still has a 75% chance of having freckles and a 25% chance of not having freckles, just like each of its siblings did. The chance of achieving a new phenotype not previously shown does not increase with each fertilization; it stays the exact same with every fertilization. Every new fertilization is not influenced by any previous fertilizations. Each one is considered an individual occurrence, each time subject to the probabilities of the gametes of the parents.

Connections and Misconceptions

Why are some alleles dominant to other alleles?

In a simple example, the dominant allele (A) codes for a particular protein. Let's assume this gene codes for an enzyme (protein) responsible for brown eye color. The recessive allele (a) is a mutated allele that no longer codes for the enzyme. At the molecular level, having the enzyme is dominant to the lack of the enzyme. In other words, having brown in the eye is dominant over not having dark pigment in the eye. Using this example, a homozygous dominant individual (AA) may have twice the amount of enzyme as a heterozygous individual (Aa). However, it may not be possible to detect a difference between homozygous and heterozygous, as long as there is enough enzyme to bring about the dominant phenotype. Half the amount of enzyme may still turn the eyes completely brown. So a person with the genotype Aa would have eyes just as brown as a person with the genotype AA. A homozygous recessive individual (aa) makes no enzyme, so the brown phenotype is absent. Gene expression is discussed further in Chapter 21.

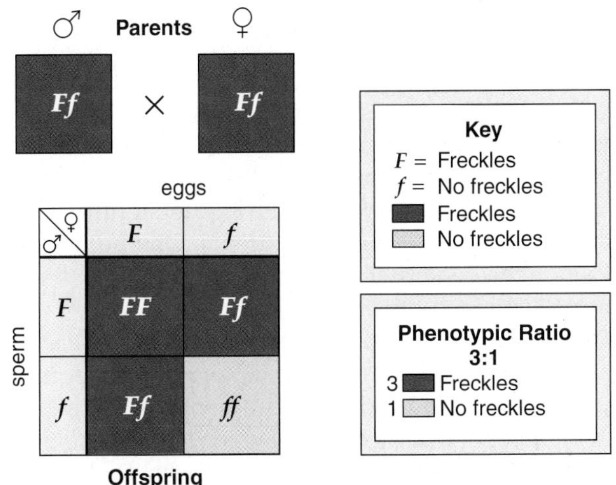

Figure 20.3 Expected results of a monohybrid cross.
A Punnett square diagrams the results of a cross. When the parents are heterozygous, each child has a 75% chance of having the dominant phenotype and a 25% chance of having the recessive phenotype.

Connections and Misconceptions

Why is it called a Punnett square?

The Punnett square was first proposed by the English geneticist Reginald Punnett (1875–1967). Punnett first used the diagram as a teaching tool to explain basic patterns of inheritance to his introductory genetics classes in 1909. This teaching tool has remained basically unchanged for 100 years!

Punnett squares are useful for visualizing crosses that involve one or two traits, but they quickly become cumbersome with complex traits involving many different factors. Most geneticists rely on statistical analysis to predict the outcomes of complex crosses.

Other One-Trait Crosses

It is not possible to tell by inspection if a person expressing a dominant allele is homozygous dominant or heterozygous. However, it is sometimes possible to tell by the results of a cross. For example, Figure 20.4 shows two possible results when a man with a widow's peak reproduces with a woman who has a straight hairline. If the man is homozygous dominant, all his children will have a widow's peak. If the man is heterozygous, each child has a 50% chance of having a straight hairline. The birth of just one child with a straight hairline indicates that the man is heterozygous.

Consider, also, that a person's parentage sometimes tells you that he or she is heterozygous. In Figure 20.4, each of the offspring with a widow's peak has to be heterozygous. Why? One of the parents was homozygous recessive and therefore had to give each offspring a *w*. Be sure to do all the practice problems in the Testing Your Knowledge section at the end of the chapter to ensure that you can do one-trait genetics problems.

The Punnett Square and Probability

Two laws of probability apply to genetics. The first is the product rule. According to this rule, the chance of two different events occurring simultaneously is equal to the multiplied probabilities of each event occurring separately. For example, what is the probability that a coin toss will be "heads"? There are only two options, so the probability of "heads" is 1 out of 2 or 50%. If we wanted to know what the probability was of a first coin toss being "heads" *and* a second coin toss being "heads," we use the product rule. *The*

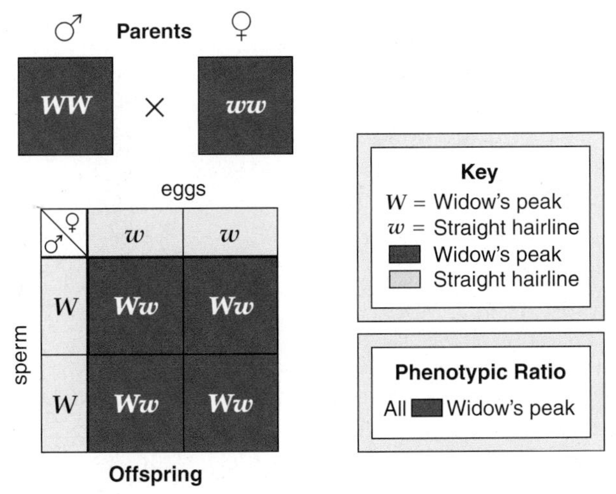

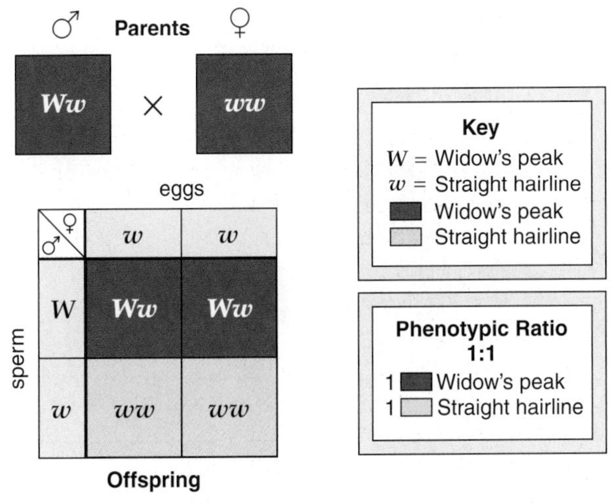

Figure 20.4 Determining if a dominant phenotype is homozygous or heterozygous.
This cross will determine if an individual with the dominant phenotype is homozygous or heterozygous. **a.** All offspring show the dominant characteristic, so the individual is most likely homozygous, as shown. **b.** The offspring show a 1:1 phenotypic ratio, so the individual is heterozygous, as shown.

product rule is often applied to cases in which the word "and" is used. This would be $\frac{1}{2} \times \frac{1}{2} = \frac{1}{4}$, or 25%. The Punnett square allows you to determine the probability that an offspring will have a particular genotype or phenotype. When you bring the alleles donated by the sperm and egg together into the same square, you are using the product rule. Both father and mother each give a chance of having a particular allele, so the probability for that allele in an offspring is the multiple of these two separate chances.

The second law of probability is the sum rule. Using this rule, individual probabilities can be added to determine total probability for an event. If we toss a coin and we want to know the probability that it will be either heads *or* tails, we use the sum rule. *The sum rule is often applied to cases in which the word "or" is used.* The probability of heads is $\frac{1}{2}$. The probability of tails is $\frac{1}{2}$. Therefore, the probability of the coin toss giving heads or tails is $\frac{1}{2} + \frac{1}{2} = 1$ (meaning it's a sure thing). In the Punnett square, you use the sum rule when you add up the results of each square to determine the final genotype and phenotype ratios.

Two-Trait Crosses

Figure 20.5 allows you to relate the events of meiosis to the formation of gametes when a cross involves two traits. In the example given, a cell has two pairs of homologues, recognized by length. One pair of homologues is short, and the other is long. The color in the figure signifies that we inherit chromosomes from our parents. One homologue of each pair is the "paternal" chromosome, and the other is the "maternal" chromosome.

The homologues separate during meiosis I, so each gamete receives one member from each pair of homologues. The homologues separate independently so it does not matter which member of a pair goes into which gamete. All possible combinations of alleles occur in the gametes. In the simplest of terms, a gamete in Figure 20.5 will *receive one short and one long chromosome of either color.* Therefore, all possible combinations of chromosomes and alleles are in the gametes.

Specifically, assume that the alleles for two genes are on these homologues. The alleles *S* and *s* are on one pair of

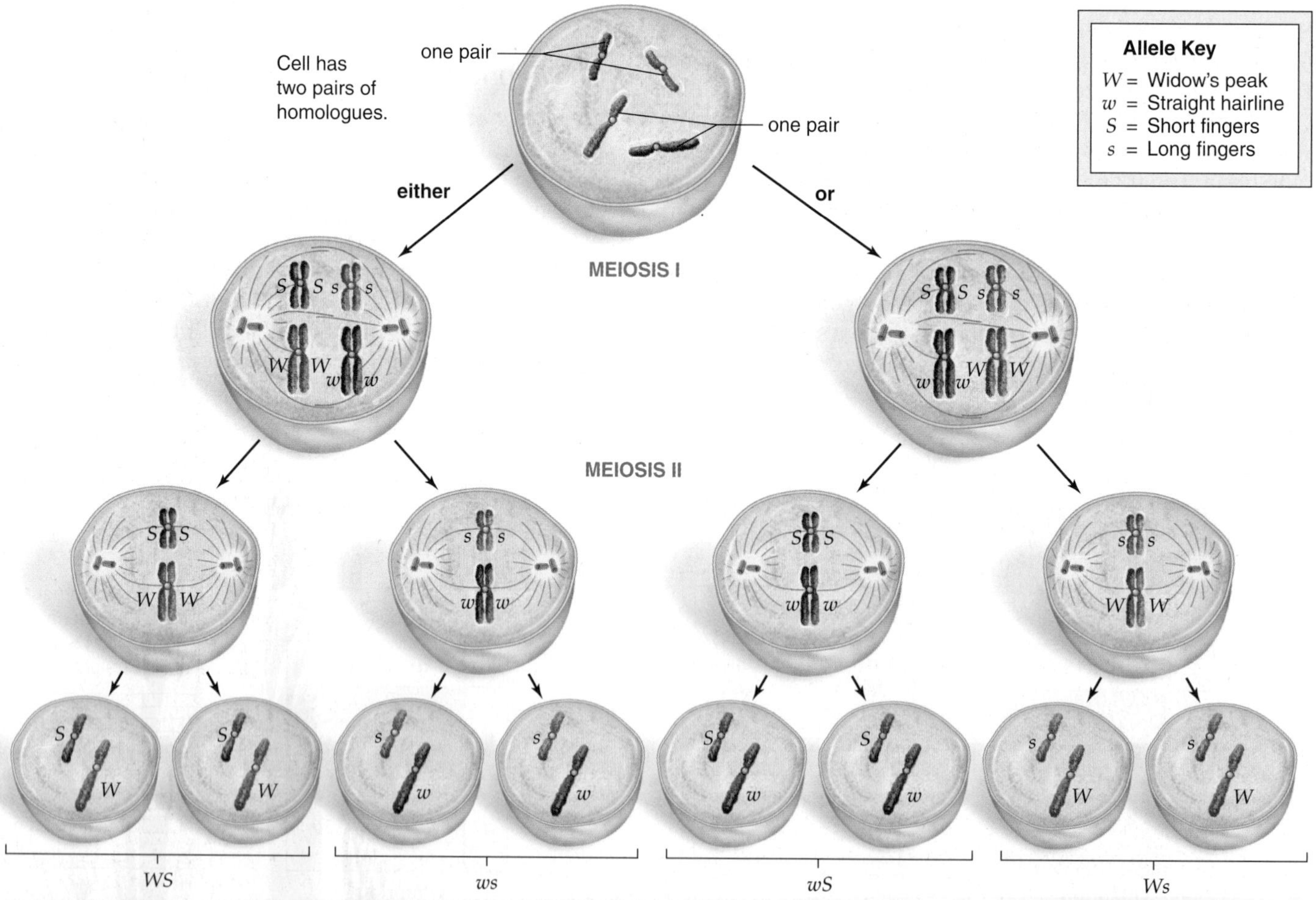

Figure 20.5 Meiosis results in genetic diversity of gametes.
A cell has two pairs of homologous chromosomes (homologues), recognized by length, not color. The long pair of homologues carries alleles for type of hairline and the short pair of homologues carries alleles for finger length. The homologues, and the alleles they carry, align independently during meiosis. Therefore, all possible combinations of chromosomes and alleles occur in the gametes as shown in the last row of cells.

homologues, and the alleles *W* and *w* are on the other pair of homologues. The homologues separate, so a gamete will have either an *S* or an *s* and either a *W* or a *w*. They will never have two of the same letter of the alphabet. Also, because the homologues align independently at the equator, either the paternal or maternal chromosome of each pair can face either pole.

Therefore, there are no restrictions as to which homologue goes into which gamete. A gamete can receive either an *S* or an *s* and either a *W* or a *w* in any combination. In the end, the gametes will collectively have all possible combinations of alleles. You should be able to transfer this information to any cross that involves two traits. In other words, the process of meiosis explains why a person with the genotype *EeFf* would produce the gametes *EF, ef, Ef,* and *eF* in equal number.

The Dihybrid Cross

In the two-trait cross depicted in Figure 20.6, a person homozygous for widow's peak and short fingers (*WWSS*) reproduces with one who has a straight hairline and long fingers (*wwss*). The gametes for the *WWSS* parent must be *WS* and the gametes for the *wwss* parent must be *ws*. Therefore, the offspring will all have the genotype *WwSs* and the same phenotype (widow's peak with short fingers). This genotype is called a **dihybrid** because the individual is heterozygous in two regards: hairline and fingers.

When a dihybrid *WwSs* has children with another dihybrid that is *WwSs,* what gametes are possible? Each gamete can have only one letter of each type in all possible combinations. Therefore, these are the gametes for both dihybrids: *WS, Ws, wS,* and *ws*.

A Punnett square makes sure that all possible sperm fertilize all possible eggs. If so, these are the expected phenotypic results:

9 widow's peak and short fingers:
3 widow's peak and long fingers:
3 straight hairline and short fingers:
1 straight hairline and long fingers

This 9:3:3:1 phenotypic ratio is always expected for a dihybrid cross when simple dominance is present. We can use this expected ratio to predict the chances of each child receiving a certain phenotype. For example, the chance of getting the two dominant phenotypes together is 9 out of 16. The chance of getting the two recessive phenotypes together is 1 out of 16.

Two-Trait Crosses and Probability

It is also possible to use the rules of probability we discussed on pages 471-72 to predict the results of a dihybrid cross. For example, we know the probable results for two separate monohybrid crosses are as follows:

1. Probability of widow's peak = $\frac{3}{4}$
   Probability of straight hairline = $\frac{1}{4}$
2. Probability of short fingers = $\frac{3}{4}$
   Probability of long fingers = $\frac{1}{4}$

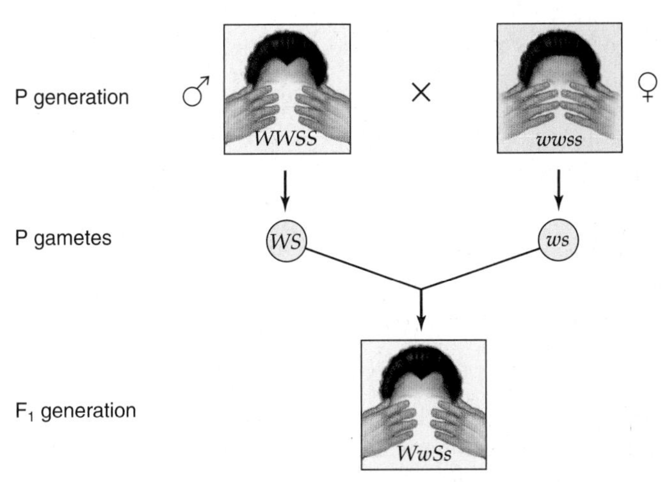

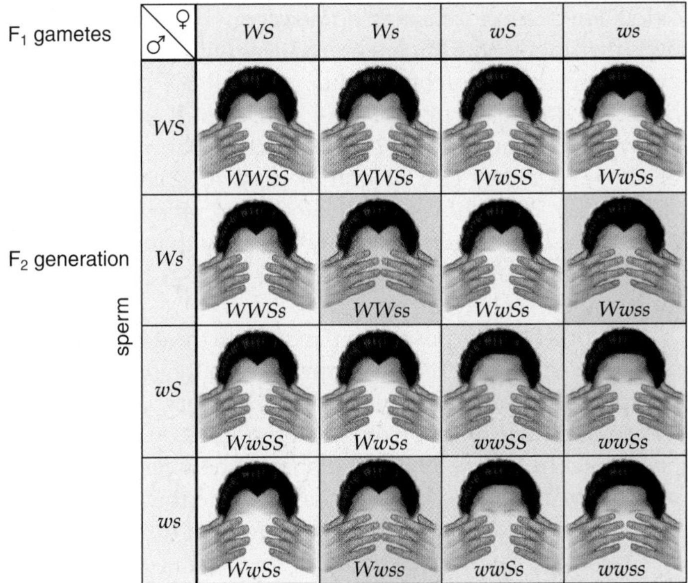

Allele Key
W = Widow's peak
w = Straight hairline
S = Short fingers
s = Long fingers

Phenotypic Ratio
9:3:3:1
9 Widow's peak, short fingers
3 Widow's peak, long fingers
3 Straight hairline, short fingers
1 Straight hairline, long fingers

Figure 20.6 Expected results of a dihybrid cross.
Each dihybrid can form four possible types of gametes, so four different phenotypes occur among the offspring in the proportions shown.

Using the product rule, we can calculate the probable outcome of a dihybrid cross as follows:

Probability of widow's peak and short fingers =
$\frac{3}{4} \times \frac{3}{4} = \frac{9}{16}$

Probability of widow's peak and long fingers =
$\frac{3}{4} \times \frac{1}{4} = \frac{3}{16}$

Probability of straight hairline and short fingers =
$\frac{1}{4} \times \frac{3}{4} = \frac{3}{16}$

Probability of straight hairline and long fingers =
$\frac{1}{4} \times \frac{1}{4} = \frac{1}{16}$

In this way, the rules of probability tell us that the expected phenotypic ratio when all possible sperm fertilize all possible eggs is 9:3:3:1.

Other Two-Trait Crosses

It is not possible to tell by inspection whether an individual expressing the dominant alleles for two traits is homozygous dominant or heterozygous. But, if the individual has children with the homozygous recessive, it may be possible to tell. For example, if a man homozygous dominant for widow's peak and short fingers reproduces with a female homozygous recessive for both traits, then all his children will have the dominant phenotypes. However, if a man is heterozygous for both traits, then each child has a 25% chance of showing either one or both recessive traits. A Punnett square (Fig. 20.7) shows that the expected ratio is 1 widow's peak with short fingers: 1 widow's peak with long fingers: 1 straight hairline with short fingers: 1 straight hairline with long fingers, or 1:1:1:1.

For practical purposes, if a parent with the dominant phenotype in either trait has an offspring with the recessive phenotype, the parent has to be heterozygous for that trait. Also, it is possible to tell if a person is heterozygous by knowing the parentage. In Figure 20.7, no offspring showing a dominant phenotype is homozygous dominant for either trait. Why? The mother is homozygous recessive for that trait.

Table 20.1 gives the phenotypic results for certain crosses we have been studying. These crosses always give these phenotypic results. Therefore, it is not necessary to do a Punnett square to arrive at the results. To facilitate doing crosses, study Table 20.1 so that you can understand why these are the results expected for these crosses.

Table 20.1 Phenotypic Ratios of Common Crosses

Genotypes	Phenotypes
Monohybrid *Gg* × monohybrid *Gg*	3:1 (dominant to recessive)
Monohybrid *Gg* × recessive *gg*	1:1 (dominant to recessive)
Dihybrid *GgRr* × dihybrid *GgRr*	9:3:3:1 (9 both dominant: 3 one dominant: 3 other dominant: 1 both recessive)
Dihybrid *GgRr* × recessive *ggrr*	1:1:1:1 (all possible combinations in equal number)

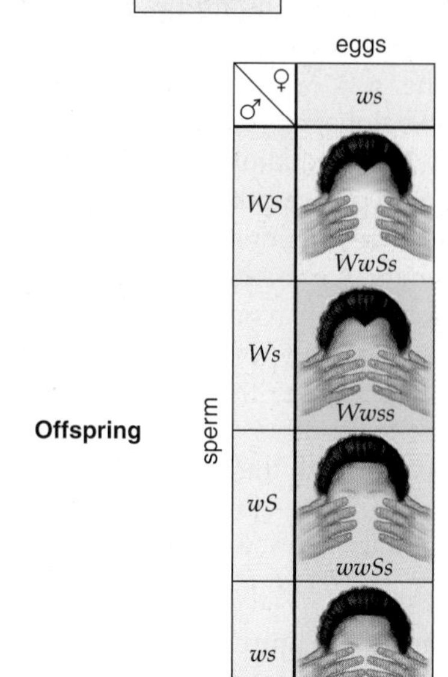

Figure 20.7 Two-trait cross.
The results of this cross indicate that the individual with the dominant phenotypes is heterozygous for both traits because some of the children are homozygous recessive for one or both traits. The chance of receiving any possible phenotype is 25%.

Connecting the Concepts

For more information on the relationship between meiosis and patterns of inheritance, refer to the following discussions.

Section 16.2 explains gamete production in males.

Section 16.3 explores the process of gamete formation in females.

Section 18.4 describes how meiosis produces genetic variation.

Check Your Progress 20.2

1. A man with a widow's peak has a mother with a straight hairline. Widow's peak (*W*) is dominant over straight hairline (*w*). State the genotype of the man.
2. Predict what genotype the children will have if one parent is homozygous recessive for earlobes and homozygous dominant for hairline (*eeWW*) and the other parent is homozygous dominant for unattached earlobes and homozygous recessive for hairline (*EEww*).
3. A child who does not have dimples or freckles is born to a man who has dimples and freckles (both dominant traits) and a woman who does not. Distinguish the genotypes of all persons concerned.

Biology Matters Bioethical Focus

Preimplantation Genetic Diagnosis

If prospective parents are heterozygous for one of the genetic disorders discussed on pages 476–79, they may want the assurance that their offspring will be free of the disorder. Determining the genotype of the embryo will provide this assurance. For example, if both parents are *Aa* for a recessive disorder, the embryo will develop normally if it has the genotype *AA* or *Aa*. On the other hand, if one of the parents is *Aa* for a dominant disorder, the embryo will develop normally only if it has the genotype *aa*.

Following in vitro fertilization (IVF), the zygote (fertilized egg) divides. When the embryo has eight cells (Fig. 20A*a*), removal of one of these cells for testing purposes has no effect on normal development. Only embryos that will not have the genetic disorders of interest are placed in the uterus to continue developing.

It is estimated that over 10,000 children have been born worldwide with normal genotypes following preimplantation embryo analysis for genetic disorders that run in their families. No American agency currently tracks these statistics, however. In the future, it's possible that embryos that test positive for a disorder could be treated by gene therapy so that they, too, would be allowed to continue to term.

Testing the egg is possible if the condition of concern is recessive. Recall that meiosis in females results in a single egg and at least two polar bodies (see Fig. 16.9). Polar bodies later disintegrate. They receive very little cytoplasm, but they do receive a haploid number of chromosomes. When a woman is heterozygous for a recessive genetic disorder, about half of the first polar bodies have received the mutated allele. In these instances, the egg received the normal allele. Therefore, if a polar body tests positive for a recessive mutated allele, the egg received the normal dominant allele. Only normal eggs are then used for IVF. Even if the sperm should happen to carry the mutation, the zygote will, at worst, be heterozygous. But the phenotype will appear normal.

Decide Your Opinion

1. Of the two diagnostic procedures described, does either seem more ethically responsible? Why?
2. Caring for an individual with a genetic condition can be very costly. Should society require preimplantation studies for the carriers of a genetic disease?

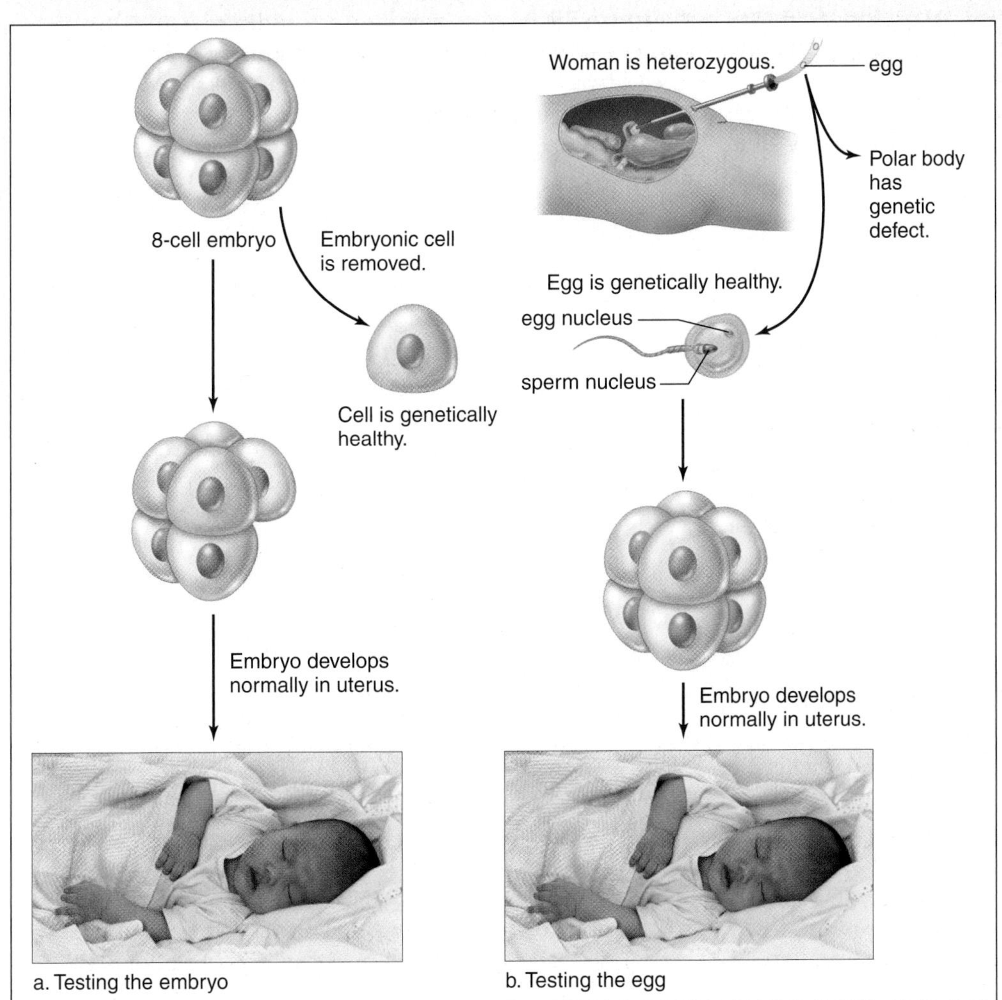

Figure 20A **The process of preimplantation genetic diagnosis.** **a.** Following IVF and cleavage, genetic analysis is performed on one cell removed from an eight-cell embryo. If it is found to be free of the genetic defect of concern, the seven-cell embryo is implanted in the uterus and develops into a newborn with a normal phenotype. **b.** Chromosomal and genetic analysis is performed on a polar body attached to an egg. If the egg is free of a genetic defect, it is used for IVF and the embryo is implanted in the uterus for further development.

20.3 Inheritance of Genetic Disorders

Learning Outcomes

Upon completion of this section, you should be able to

1. Interpret a human pedigree to identify the pattern of inheritance for a trait.
2. Understand the genetic basis of select human autosomal dominant and autosomal recessive genetic disorders.

We inherit many different traits from our parents—not only our hair and eye color but also traits for diseases and disorders. Many of these diseases occur as a result of changes, or mutations, in our parents' genetic codes. The abnormal gene could be present in each of your parents' cells and thus passed down in the sperm or egg. Your parent may or may not have been affected by this genetic mutation. Alternatively, the genetic mutation might have occurred only in the sperm or egg that became a part of you. Some genetic diseases require two damaged alleles for the disease to manifest itself. Others need only one. When a genetic disorder is autosomal dominant, an individual with the alleles *AA* or *Aa* will have the disorder. When a genetic disorder is autosomal recessive, only individuals with the alleles *aa* will have the disorder. Genetic counselors often construct pedigrees to determine whether a condition that runs in the family is dominant or recessive. A pedigree shows the pattern of inheritance for a particular condition. Consider these two possible patterns of inheritance:

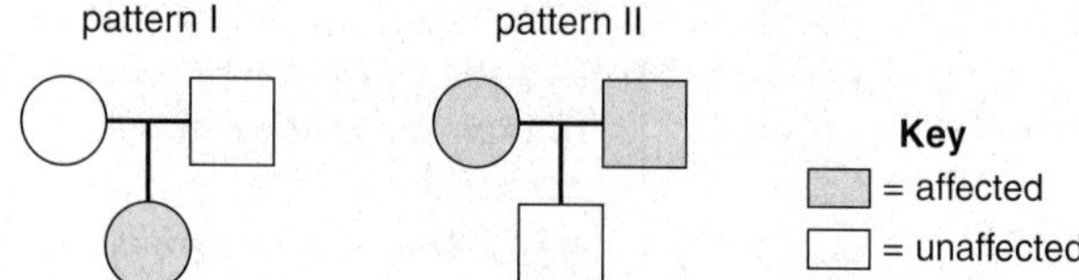
pattern I
pattern II
Key
= affected
= unaffected

In both patterns, males are designated by squares and females by circles. Shaded circles and squares are affected individuals. A line between a square and a circle represents a mating. A vertical line going downward leads, in these patterns, to a single child. (If there are more children, they are placed off a horizontal line.) Which pattern of inheritance do you suppose represents an autosomal recessive characteristic? Which represents an autosomal dominant characteristic?

Autosomal recessive disorder: In pattern I, the child is affected but neither parent is. This can happen if the condition is recessive and the parents are *Aa*. The parents are **carriers** because they carry the recessive trait in their DNA but their phenotype is dominant. Figure 20.8 shows a typical pedigree chart for a recessive genetic disorder. Other ways to recognize an autosomal recessive pattern of inheritance are also listed in the figure. If both parents are affected, all the children are affected. Why? The parents can pass on only recessive alleles for this condition. All children will be homozygous recessive, just like the parents.

Autosomal dominant disorder: In pattern II, the child is unaffected but the parents are affected. Of the two patterns, this one shows a dominant pattern of inheritance. The condition is dominant, so the parents can be *Aa* (heterozygous). The child inherited a recessive allele from each parent and, therefore, is unaffected. Figure 20.9 shows a typical pedigree for a dominant disorder. Other ways to recognize an autosomal dominant pattern of inheritance are also listed. When a disorder is dominant, an affected child must have at least one affected parent.

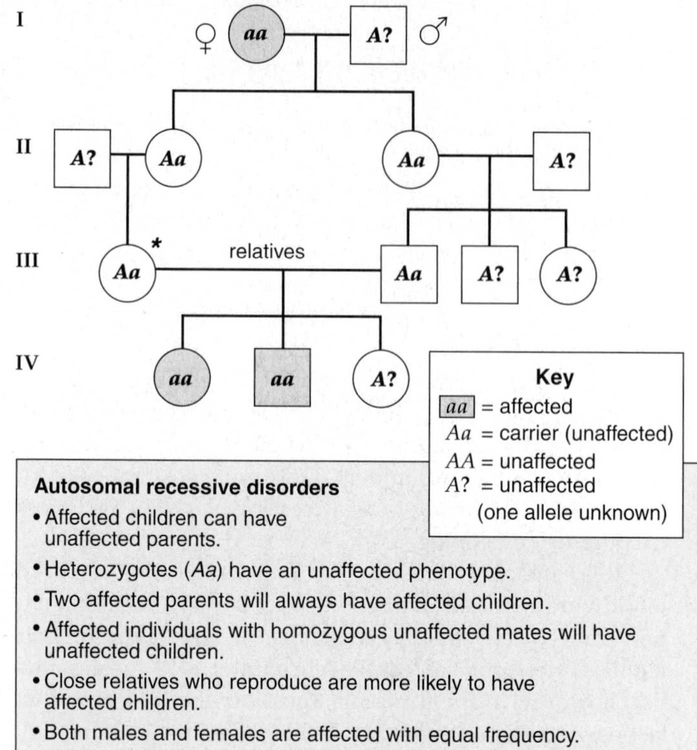
I
aa
A?
II
A?
Aa
Aa
A?
III
*
relatives
Aa
Aa
A?
A?
IV
aa
aa
A?
Key
aa = affected
Aa = carrier (unaffected)
AA = unaffected
A? = unaffected
(one allele unknown)
Autosomal recessive disorders
• Affected children can have unaffected parents.
• Heterozygotes (Aa) have an unaffected phenotype.
• Two affected parents will always have affected children.
• Affected individuals with homozygous unaffected mates will have unaffected children.
• Close relatives who reproduce are more likely to have affected children.
• Both males and females are affected with equal frequency.

Figure 20.8 **Autosomal recessive disorder pedigree.** The list gives ways to recognize an autosomal recessive disorder. How would you know that the individual at the asterisk is heterozygous? (Answer is provided in Appendix B.)

Genetic Disorders of Interest

Medical genetics has traditionally focused on disorders caused by single gene mutations; they are well understood due to their straightforward patterns of inheritance. Presently, it is estimated that there are over 4,000 identified disorders caused by single gene mutations in humans. Here, we will focus on only a few.

Autosomal Recessive Disorders

Inheritance of two recessive alleles is required for an autosomal recessive disorder to be the expressed phenotype.

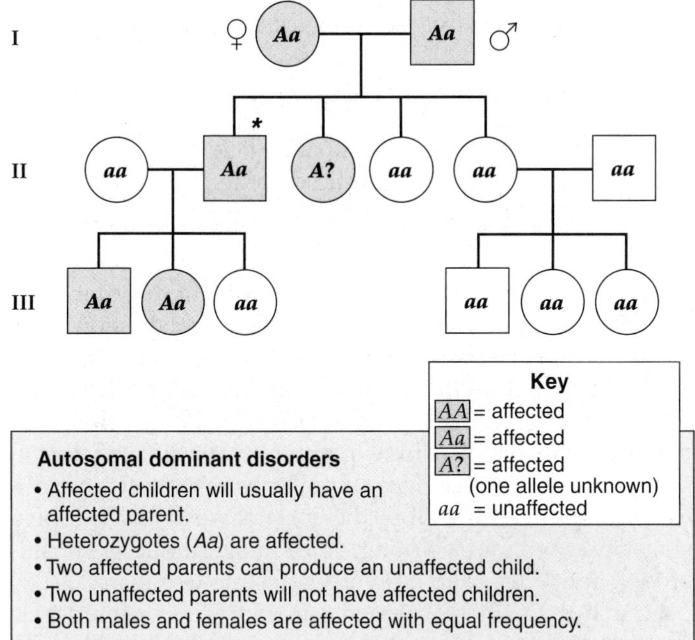

I
Aa
Aa
II
aa
Aa
A?
aa
aa
aa
III
Aa
Aa
aa
aa
aa
aa
Key
AA = affected
Aa = affected
A? = affected (one allele unknown)
aa = unaffected
Autosomal dominant disorders
• Affected children will usually have an affected parent.
• Heterozygotes (Aa) are affected.
• Two affected parents can produce an unaffected child.
• Two unaffected parents will not have affected children.
• Both males and females are affected with equal frequency.

Figure 20.9 Autosomal dominant disorder pedigree.
The list gives ways to recognize an autosomal dominant disorder. How would you know that the individual at the asterisk is heterozygous? (Answer is provided in Appendix B.)

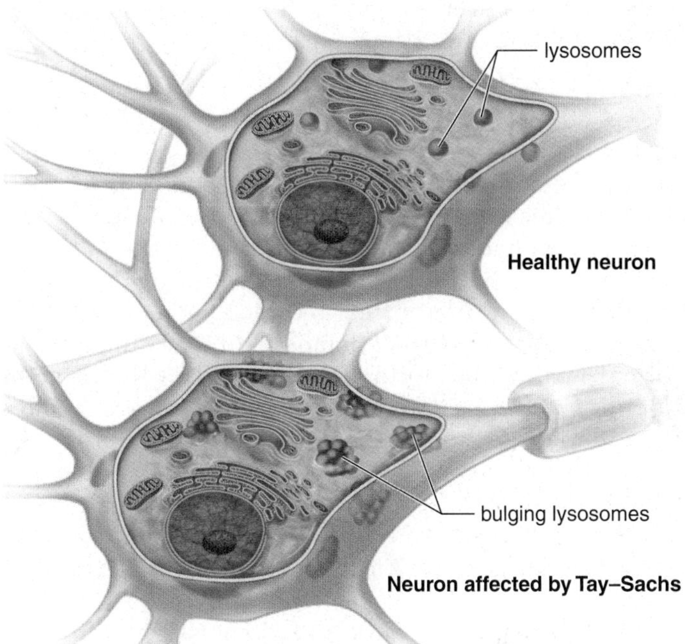

Figure 20.10 Neuron affected by Tay–Sachs disease.
In Tay–Sachs disease, a lysosomal enzyme is missing. This causes the substrate of that enzyme to accumulate within the lysosomes.

Tay–Sachs Disease **Tay–Sachs disease** is a well-known autosomal recessive disorder that occurs usually among Ashkenazic Jewish people (those from Central and Eastern Europe) and their descendants. Tay–Sachs disease results from a lack of a lysosome enzyme, hex A, which clears out fatty acid proteins that build up in cells of the brain. Without this enzyme, the buildup will interfere with proper brain development and growth and cause malfunctions in vision, movement, hearing, and overall mental development. This impairment leads to blindness, seizures, and paralysis. Currently, there is no cure for Tay–Sachs disease; affected children normally die by the age of 5 (Fig. 20.10).

Animation Lysosomes

Cystic Fibrosis **Cystic fibrosis** (**CF**) is an autosomal recessive disorder that occurs among all ethnic groups but is most prevalent in Caucasians. It is estimated that 1 in 29 Caucasians in the United States carries a CF gene. Research has demonstrated that chloride ions (Cl^-) fail to pass through a plasma membrane channel protein in the cells of these patients (Fig. 20.11). Ordinarily, after chloride ions have passed through the membrane, sodium ions (Na^+) and water follow. It is believed that lack of water is the cause of abnormally thick mucus in the bronchial tubes, GI tract, and pancreatic ducts. In these patients, the mucus in the bronchial tubes and pancreatic ducts is particularly thick and viscous, interfering with their functions. The past few decades have seen advances in the treatment of CF to loosen the mucus and ease breathing, as well as medications that improve digestion. These

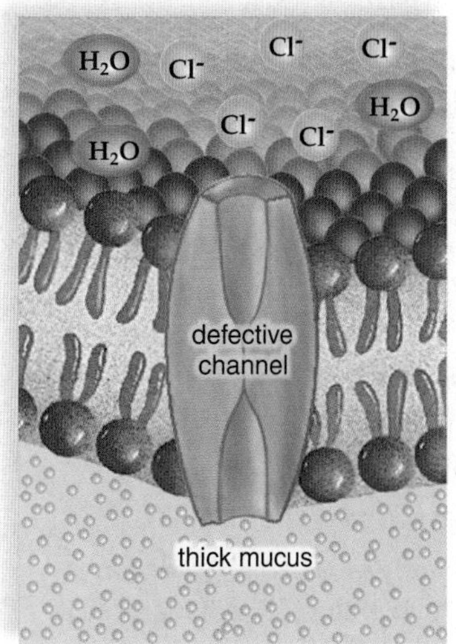

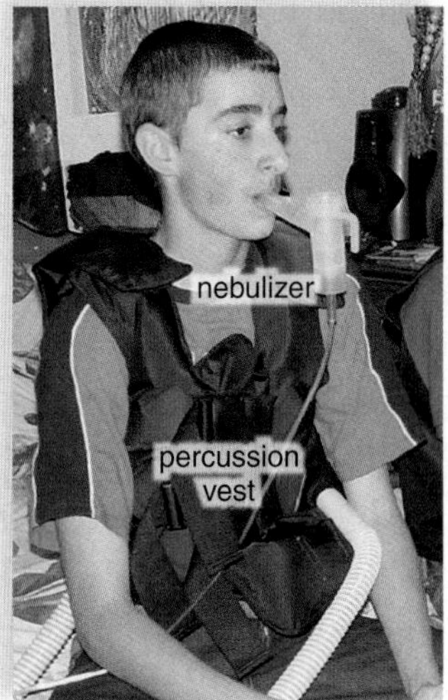

Figure 20.11 Cystic fibrosis disease.
Cystic fibrosis is due to a faulty protein that is supposed to regulate the flow of chloride ions into and out of cells through a channel protein. The nebulizer delivers drugs in aerosol form to the bronchial passages.

advances have not only improved the quality of life of those affected but increased the average life span of CF patients to their late 30s, whereas the mid-20s were the ages of the average life span about 20 years ago.

Video Good Poison

Sickle-Cell Disease **Sickle-cell disease** is an autosomal recessive disorder in which the red blood cells are not biconcave disks like normal red blood cells. Many are sickle- or boomerang-shaped, and these red blood cells live for only about two weeks, unlike the average four-month life span of a normal red blood cell. The defect is caused by an abnormal hemoglobin that differs from normal hemoglobin by one amino acid in the protein globin. The single amino acid change causes hemoglobin molecules to stack up and form insoluble rods. This causes the red blood cells to become sickle-shaped. This single gene defect can affect any race but is prevalent among African Americans; it is estimated that 1 in every 625 African Americans is affected by sickle-cell disease.

Sickle-shaped cells can't pass along narrow capillary passageways as disk-shaped cells can, so they clog the vessels preventing adequate circulation. This results in anemia, tissue damage, jaundice, joint pain, and gallstones. Those affected by sickle-cell disease are also susceptible to many types of bacterial infections and have a higher incidence of stroke. Many treatment options, including blood transfusions and bone marrow transplants, are highly effective. The most common medicinal treatment, hydroxyurea, has been on the market for over a decade and is considered one of the most effective daily treatments for the reduction in sickle- cell-related anemia, joint pain, and tissue damage.

The sickle-cell gene is one of only a few autosomal recessive disorders in which the heterozygote can express variations of the recessive phenotype. Sickle-cell heterozygotes have sickle-cell traits in which the blood cells are normal unless they experience dehydration or mild oxygen deprivation. Intense exertion may cause sickling of some red blood cells for a short period of time. These patients can experience episodes and symptoms very similar to those patients with the autosomal recessive genotype.

Connections and Misconceptions

Why do diet sodas carry the warning, "Phenylketonurics: Contains Phenylalanine"?

The sweetener used in diet sodas is aspartame. Aspartame is formed by the combination of two amino acids: aspartic acid and phenylalanine. When aspartame is broken down by the body, phenylalanine is released. Phenylalanine can be toxic for those with PKU and must be avoided. Other foods high in phenylalanine include eggs, meat, milk, and bananas.

Autosomal Dominant Disorders

Inheritance of only one dominant allele is necessary for an autosomal dominant genetic disorder to be displayed. Here, we discuss only two of the many autosomal dominant disorders.

Marfan Syndrome The autosomal dominant disorder **Marfan syndrome** is caused by a defect in the production of an elastic connective tissue protein called fibrillin. This protein is normally abundant in the lens of the eye; the bones of limbs, fingers, and ribs; and also in the wall of the aorta and the blood vessels. This explains why the affected person often has a dislocated lens, long limbs and fingers, and a caved-in chest. The wall of the aorta is weak and can possibly burst without warning. Marfan syndrome is considered a "rare" disorder, currently affecting less than 200,000 people in the United States or 1 in every 2,000 Americans. Treatments for Marfan syndrome include beta blockers to control the cardiovascular symptoms, corrective lenses or eye surgery, and braces or orthopedic surgery for musculoskeletal symptoms.

Huntington Disease An autosomal dominant neurological disorder that leads to progressive degeneration of brain cells (Fig. 20.12), **Huntington disease** is caused by a mutated copy of the gene for a protein called huntingtin. The defective gene contains segments of DNA in which the base sequence CAG repeats again and again. This type of structure, called a trinucleotide repeat, causes the huntingtin protein to have too many copies of the amino acid glutamine. The normal version of huntingtin has stretches of between 10 and 25 glutamines. If huntingtin has more than 36 glutamines, as is seen in the mutated Huntington gene, it changes shape and forms clumps inside neurons. These clumps attract other proteins to clump, rendering them inactive. One of these proteins that attaches itself to the clumps, called CBP, helps nerve cells survive.

The onset of symptoms for Huntington disease is normally not first seen until later in life (average age of onset is late 30s to late 40s) although it is not unusual for an affected person to develop symptoms as early as their late teens. Huntington disease has a range of symptoms but is normally characterized by uncontrolled movements, unsteady gait, dementia, and speech impairment. On average, patients live 15 to 20 years after onset of symptoms, as the disease progresses rapidly. Current effective treatments include medications that slow the progression of the disease. Additionally, dopamine blockers have been found to be very effective in reducing uncontrolled movements and improving eye–hand coordination and steadiness during walking.

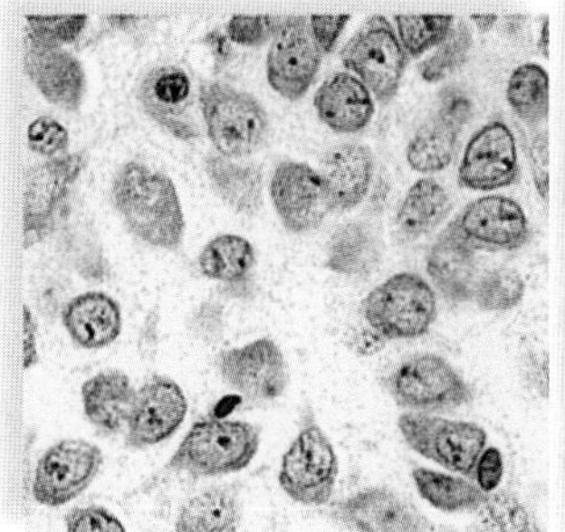
many neurons in normal brain

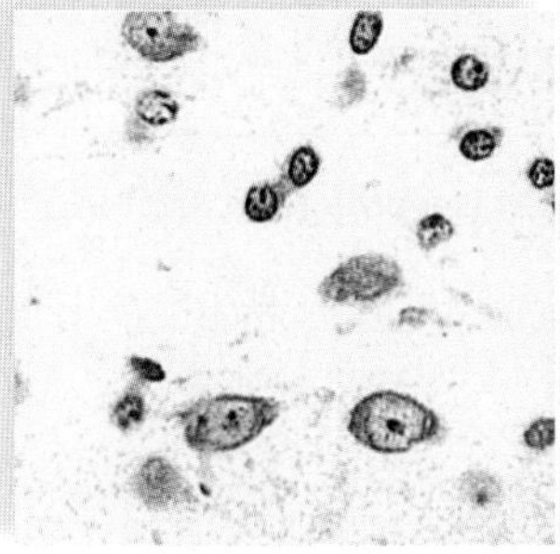
loss of neurons in Huntington brain

Figure 20.12 **Huntington disease.**
Huntington disease is characterized by increasingly serious psychomotor and mental disturbances because of a loss of nerve cells.

Connecting the Concepts

Many of the diseases discussed in this section are associated with specific aspects of human physiology. For more information on these diseases (indicated in parentheses), refer to the following discussions.

Section 3.3 describes the function of proteins in the cell membrane (cystic fibrosis).

Section 3.4 contains descriptions of the lysosomes (Tay–Sachs disease).

Section 4.2 reviews the role of connective tissue in the body (Marfan syndrome).

Section 6.2 overviews the role of the red blood cells (sickle-cell anemia).

Check Your Progress 20.3

1. Solve the following: In a pedigree, all the members of one family are affected. Based on this knowledge, list the genotypes of the parents (a) if the trait is recessive and (b) if the trait is dominant.
2. Predict the chances that homozygous normal parents for cystic fibrosis will have a child with cystic fibrosis.
3. Explain why some incidences of autosomal recessive disorders are higher in one race or culture.

20.4 Beyond Simple Inheritance Patterns

Learning Outcomes

Upon completion of this chapter, you should be able to

1. Summarize how polygenic inheritance, codominance, and incomplete dominance differ from simple one-trait crosses.
2. Explain how a combination of genetics and the environment can influence a phenotype.
3. Predict a person's blood type based upon his or her genotype.

Certain traits, such as those studied in Section 20.2, are controlled by one set of alleles that follows a simple dominant or recessive inheritance. We now know of many other types of inheritance patterns.

Polygenic Inheritance

Polygenic traits, such skin color and height, are governed by several sets of alleles. The individual has a copy of all allelic pairs, possibly located on many different pairs of chromosomes. Each dominant allele codes for a product; therefore, the dominant alleles have a quantitative effect on the phenotype, that is, these effects are additive. The result is a *continuous variation* of phenotypes, resulting in a distribution of these phenotypes that resembles a bell-shaped curve. The more genes involved, the more continuous the variations and distribution of the phenotypes. Also, environmental effects cause many intervening phenotypes. In the case of height, differences in nutrition bring about a bell-shaped curve (Fig. 20.13).

Skin Color

Skin color is an example of a polygenic trait likely controlled by many pairs of alleles. Even so, we will use the simplest model and assume that skin has only two pairs of alleles (*Aa* and *Bb*) and that each capital letter contributes pigment to the skin. When a very dark person has children with a very light person, the children have medium brown skin. When two people with the genotype *AaBb* have children with one another, the children may range in skin color from very dark to very light:

Genotypes	Phenotypes
AABB	Very dark
AABb or *AaBB*	Dark
AaBb or *AAbb* or *aaBB*	Medium brown
Aabb or *aaBb*	Light
aabb	Very light

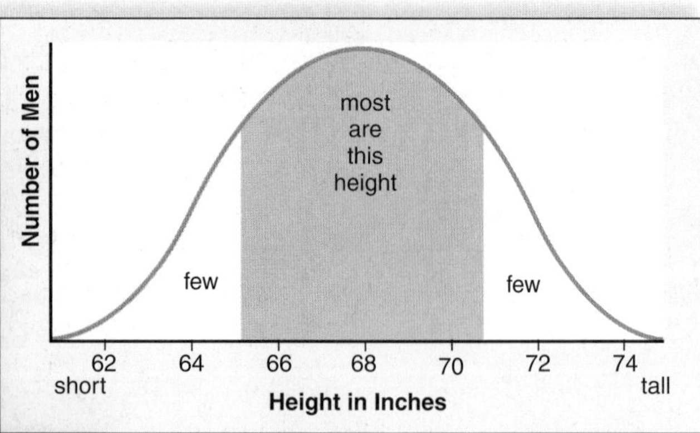

Figure 20.13 **Height is a polygenic trait in humans.**
When you record the heights of a large group of people chosen at random, the values follow a bell-shaped curve. Such a continuous distribution is because of control of a trait by several sets of alleles. Environmental effects are also involved.

A range of phenotypes exists, and several possible phenotypes fall between the two extremes. Therefore, the distribution of these phenotypes is expected to follow a bell-shaped curve. It is important to understand that in a polygenic inheritance, like skin color, the offspring do not always have a phenotype exactly in between the two parents. Because we are dealing with many genes, each with its own trait, the offspring can have phenotypes anywhere in the bell curve of options. Skin color is also a **multifactorial trait,** a polygenic trait commonly influenced by the environment. For skin color, one environmental influence is sun exposure.

Connections and Misconceptions

Is skin color a good indication of a person's race?

The simple answer to this question is, No, an individual's skin color is not a true indicator of his or her genetic heritage. People of the same skin color are not necessarily genetically related. Because only a few genes control skin color, a person's skin color may not indicate his or her ancestors' origins. Scientists are actively investigating the genetic basis of race and have found some interesting relationships between a person's genetic heritage and medicine. This will be explored in greater detail in Chapter 22.

Multifactorial Disorders

Many human disorders, such as cleft lip and/or palate, clubfoot, schizophrenia, diabetes, phenylketonuria (see chapter opener), and even allergies and cancers, are most likely controlled by polygenes subject to environmental influences. The coats of Siamese cats and Himalayan rabbits are darker in color at the ears, nose, paws, and tail (Fig. 20.14). Himalayan rabbits are *homozygous* for the allele *ch,* involved in the production of melanin. Experimental evidence suggests that the enzyme coded for by this gene is active only at a low temperature. Therefore, black fur occurs only at the extremities where body heat is lost to the environment.

Recent studies have reported that all sorts of behavioral traits, such as alcoholism, phobias, and even suicide, can be associated with particular genes. However, in almost all cases, the environment plays an important role in the severity of the phenotype. Therefore, they must be multifactorial traits. Current research focuses on determining what percentage of the trait is due to nature (inheritance) and what percentage is due to nurture (the environment). Some studies use identical and fraternal twins separated at birth and raised in different environments. The supposition is that, if identical twins in different environments share the same trait, that trait is most likely inherited. Identical twins are more similar in their intellectual

Figure 20.14 **Himalayan rabbit with temperature-susceptible coat color.**
The dark ears, nose, and feet of this rabbit are believed to be because of a lower body temperature in these areas.

talents, personality traits, and levels of lifelong happiness than are fraternal twins separated from birth. This substantiates the belief that behavioral traits are partly heritable. It also supports the belief that genes exert their effects by acting together in complex combinations susceptible to environmental influences.

Incomplete Dominance and Codominance

Incomplete dominance occurs when the heterozygote is intermediate between the two homozygotes. For example, when a curly-haired individual has children with a straight-haired individual, their children have wavy hair. When two wavy-haired persons have children, the expected phenotypic ratio among the offspring is 1:2:1—one curly-haired child to two with wavy hair to one with straight hair. We can explain incomplete dominance by assuming that only one allele codes for a product and the single dose of the product gives the intermediate result.

Codominance occurs when alleles are equally expressed in a heterozygote. A familiar example is the human blood type AB, in which the red blood cells have the characteristics of both type A and type B blood. We can explain codominance by assuming that both genes code for a product, and we observe the results of both products being present. Blood type inheritance is said to be an example of multiple alleles, described in the next section.

Incompletely Dominant Disorders

The prognosis in **familial hypercholesterolemia (FH)** parallels the number of LDL-cholesterol receptor proteins in the plasma membrane. A person with two mutated alleles lacks LDL-cholesterol receptors. A person with only one mutated allele has half the normal number of receptors, and a person with two normal alleles has the usual number of receptors. People with the full number of receptors do not have familial hypercholesterolemia. When receptors are completely absent, excessive cholesterol is deposited in various places in the body, including under the skin (Fig. 20.15).

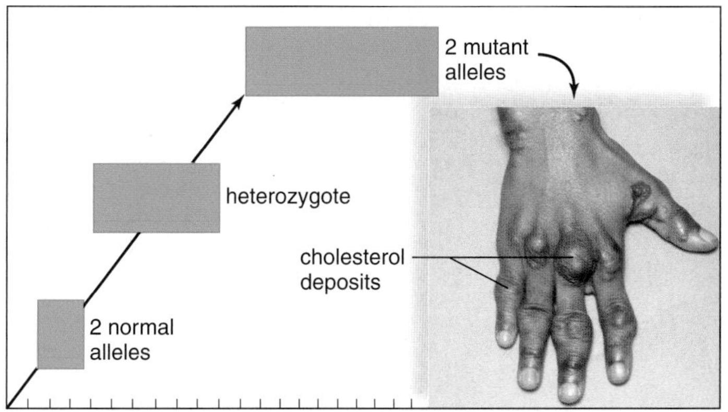

Figure 20.15 The inheritance of familial hypercholesterolemia.

Familial hypercholesterolemia is incompletely dominant. Persons with one mutated allele have an abnormally high level of cholesterol in the blood, and those with two mutated alleles have a higher level still.

The presence of excessive cholesterol in the blood causes cardiovascular disease. Therefore, those with no receptors die of cardiovascular disease as children. Individuals with half the number of receptors may die when young or after they have reached middle age.

Multiple-Allele Inheritance

When a trait is controlled by **multiple alleles,** the gene exists in several allelic forms. However, each person can only have two of the possible alleles.

ABO Blood Types

Three alleles for the same gene control the inheritance of ABO blood types. These alleles determine the presence or absence of antigens on red blood cells:

I^A = A antigen on red blood cells
I^B = B antigen on red blood cells
i = neither A nor B antigen on red blood cells

Each person has only two of the three possible alleles, and both I^A and I^B are dominant over i. Therefore, there are two possible genotypes for type A blood and two possible genotypes for type B blood. On the other hand, I^A and I^B are fully expressed in the presence of the other. Therefore, if a person inherits one of each of these alleles, that person will have type AB blood. Type O blood can result only from the inheritance of two i alleles.

The possible genotypes and phenotypes for blood type are as follows:

Phenotype	Genotype
A	I^AI^A, I^Ai
B	I^BI^B, I^Bi
AB	I^AI^B
O	ii

Blood typing was traditionally used in paternity suits. However, a blood test of a supposed father can only suggest that he *might* be the father, not that he definitely *is* the father. For example, it is possible, but not definite, that a man with type A blood (genotype I^Ai) is the father of a child with type O blood. On the other hand, a blood test sometimes can definitely prove that a man is not the father. For example, a man with type AB blood cannot possibly be the father of a child with type O blood. Therefore, blood tests can be used in legal cases only to try to exclude a man from possible paternity. Modern identification relies more heavily on analysis of patterns in the DNA, called DNA fingerprinting (see Chapter 21, Section 21.4).

Figure 20.16 shows that matings between certain genotypes can have surprising results in terms of blood type. Parents with type A and type B blood can have offspring with all four possible blood types.

As a point of interest, the Rh factor is inherited separately from A, B, AB, or O blood types. When you are Rh-positive, your red blood cells have a particular antigen; when you are

Blology Matters Historical Focus

Hemophilia: The Royal Disease

The pedigree in Figure 20B shows why hemophilia is often referred to as "The Royal Disease." Queen Victoria of England, who reigned from 1837 to 1901, was the first of the royals to carry the gene. From her, the disease eventually spread to the Prussian, Spanish, and Russian royal families. In that era, monarchs arranged marriages between their children to consolidate political alliances. This practice allowed the gene for hemophilia to spread throughout the royal families. It is assumed that a spontaneous mutation arose either in Queen Victoria after her conception or in one of the gametes of her parents. However, in the book *Queen Victoria's Gene* by D. M. Potts, the author postulates that Edward Augustus, Duke of Kent, may not have been Queen Victoria's father. Potts suggests that Victoria may have instead been

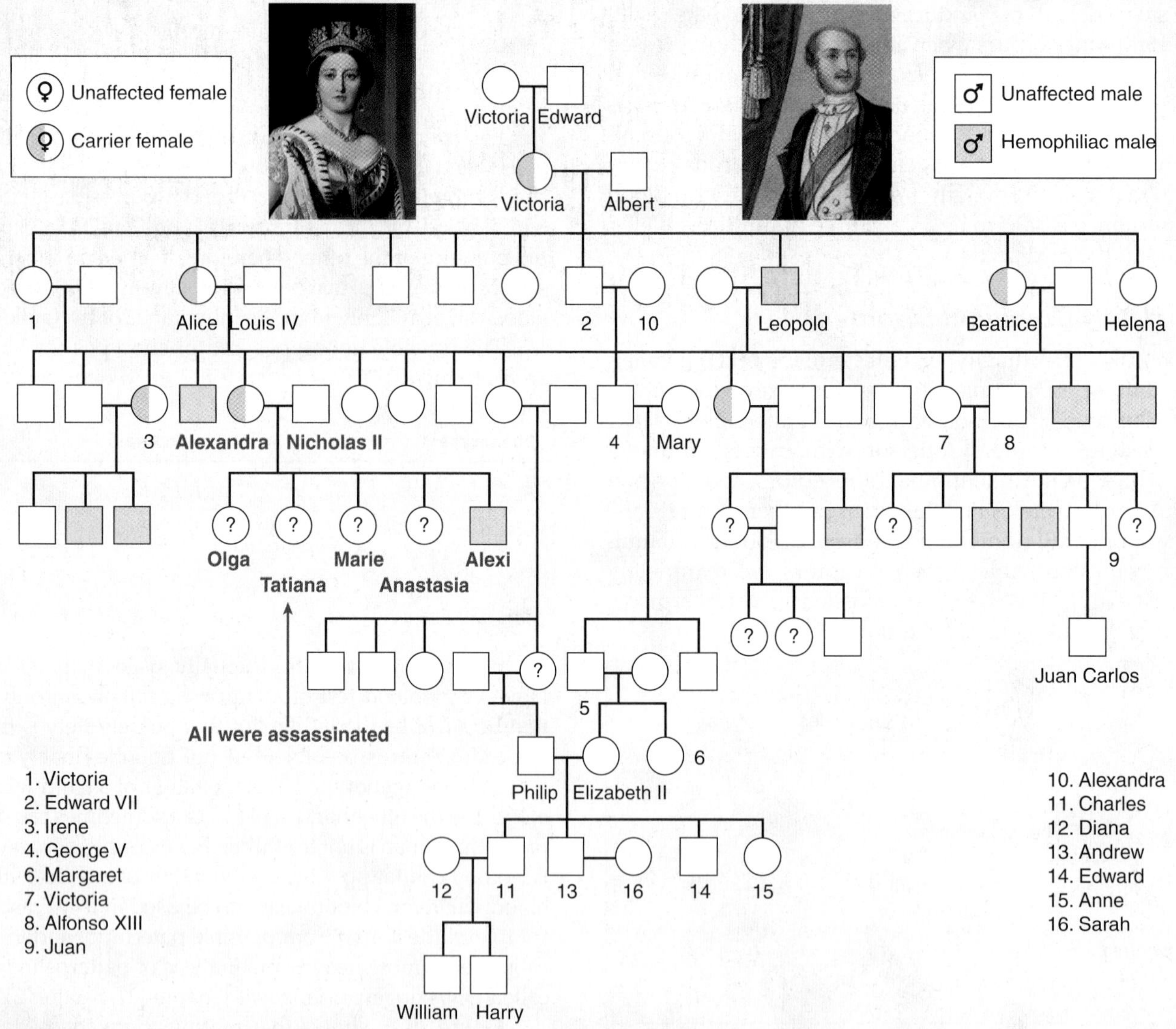

Figure 20B The royal families' X-linked pedigree.
Queen Victoria was a carrier, so each of her sons had a 50% chance of having the disorder, and each of her daughters had a 50% chance of being a carrier. This pedigree shows only the affected descendants. Many others are unaffected, including the members of the present British royal family.

the illegitimate child of a hemophiliac male. Regardless of her parentage, had Victoria not been crowned, the fate of the various royal households may have been very different. Further, the history of Europe could have been changed dramatically as well.

However, Victoria did become queen. Queen Victoria and her husband Prince Albert had nine children. Fortunately, only one son, Leopold, suffered from hemophilia. He experienced severe hemorrhages and died in 1884 at the age of 31 as the result of a minor fall. He left behind a daughter Alice, a carrier for the disease. Her son Rupert also suffered from hemophilia and in 1928 died of a brain hemorrhage as a result of a car accident. Queen Victoria's eldest son Edward VII, the heir to the throne, did not have the disease; thus, the current British royal family is free of the disease.

Two of Queen Victoria's daughters, Alice and Beatrice, were carriers of the disease. Alice married Louis IV, the Grand Duke of Hesse. Of her six children, three were affected by hemophilia. Her son Frederick died of internal bleeding from a fall. Alice's daughter Irene married Prince Henry of Prussia, her first cousin. Two of their three sons suffered from hemophilia. One of these sons, Waldemar, died at age 56 due to the lack of blood-transfusion supplies during World War I. Henry, Alice's other son, bled to death at the age of 4.

Alice's daughter Alexandra married Nicholas II of Russia. Alexandra gave birth to four daughters before giving birth to Alexei, the heir to the Russian throne. It was obvious almost from birth that Alexei had hemophilia. Every fall caused bleeding into his joints, which led to crippling of his limbs and excruciating pain. The best medical doctors could not help Alexei. Desperate to relieve his suffering, his parents turned to the monk Rasputin. Rasputin was able to relieve some of Alexei's suffering by hypnotizing him and putting him to sleep. Alexandra and Nicholas, the czar and czarina, put unlimited trust in Rasputin. The illness of the only heir to the czar's throne, the strain Alexei's illness placed on the czar and czarina, and the power of Rasputin were all factors leading to the Russian Revolution of 1917. The czar and czarina, as well as their children, were all murdered during the revolution.

Queen Victoria's other carrier daughter, Beatrice, married Prince Henry of Battenberg. Her son Leopold was a hemophiliac, dying at 32 during a knee operation. Beatrice's daughter Victoria Eugenie married Alfonso XII of Spain. Queen Ena, as Victoria Eugenie came to be known, was not popular with the Spanish people. Her firstborn son Alfonso, the heir to the Spanish throne, did not stop bleeding upon his circumcision. When it became obvious that she had given her son hemophilia, it is alleged that her husband never forgave her. Like his cousin Rupert, Alfonso died in 1938 from internal bleeding after a car accident. Victoria's youngest son Gonzalo was also a hemophiliac whose life was claimed by a car accident in 1934.

Today, no members of any European royal family are known to have hemophilia. Individuals with the disease gene born in the late 1800s and early 1900s have all died, eliminating the gene from the current royal houses.

Rh-negative, that antigen is absent. There are multiple recessive alleles for an Rh– status, but they are all recessive to an Rh+ status.

Connecting the Concepts

For more information on the topics presented in this section, refer to the following discussions.

Section 4.7 provides more information on the melanocyte cells that control skin color.

Section 6.2 describes the role of red blood cells in the human body.

Section 22.5 explores the evolution of humans and human races.

Check Your Progress 20.4

1. Detail why polygenic inheritance follows a bell-shaped curve.
2. Describe a multifactorial trait that could have diet and nutrition as environmental influences.
3. Discuss the potential the evolutionary advantages of having multiple alleles for a trait.

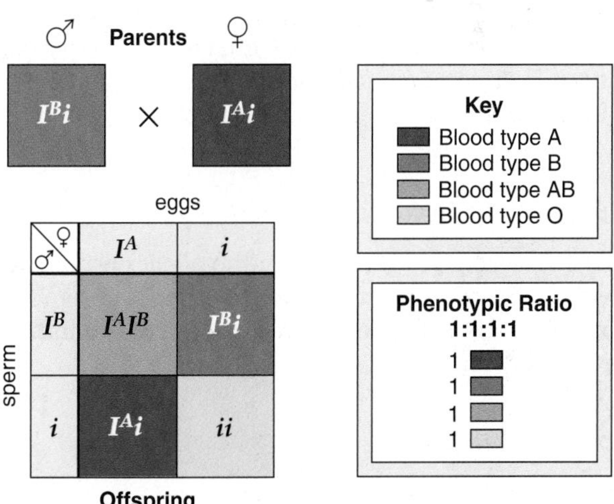

Figure 20.16 The inheritance of ABO blood types. Blood type exemplifies multiple allele inheritance. The *I* gene has two codominant alleles, designated as I^A and I^B, and one recessive allele, designated by *i*. Therefore, a mating between individuals with type A blood and type B blood can result in any one of the four blood types. Why? The parents are I^Ai and I^Bi. If both parents were type AB blood, no child would have what blood type? See Appendix B for answers.

20.5 Sex-Linked Inheritance

Learning Outcomes

Upon completion of this section, you should be able to

1. Understand the differences between autosomal and sex-linked patterns of inheritance.
2. Interpret a human pedigree to determine sex-linked inheritance of a trait.

Normally, both males and females have 23 pairs of chromosomes; 22 pairs are called **autosomes,** and one pair is the sex chromosomes. These are called the **sex chromosomes** because they differ between the sexes. In humans, males have the sex chromosomes X and Y, and females have two X chromosomes. The Y chromosome contains the gene responsible for determining male gender.

Traits controlled by genes on the sex chromosomes are said to be **sex-linked.** An allele on an X chromosome is **X-linked,** and an allele on the Y chromosome is Y-linked. Most sex-linked genes are only on the X chromosomes. The Y chromosome is lacking these. Very few Y-linked alleles have been found on the much smaller Y chromosome.

Many of the genes on the X chromosomes, such as those that determine normal as opposed to red–green color blindness, are unrelated to the gender of the individual. In other words, the X chromosome carries genes that affect both males and females. It would be logical to suppose that a sex-linked trait is passed from father to son or from mother to daughter, but this is not the case. A male always receives an X-linked allele from his mother, from whom he inherited an X chromosome. *The Y chromosome from the father does not carry an allele for the trait.* Usually, a sex-linked genetic disorder is recessive. Therefore, a female must receive two alleles, one from each parent, before she has the disorder.

X-Linked Alleles

When considering X-linked traits, the allele on the X chromosome is shown as a letter attached to the X chromosome. For example, this is the key for red–green color blindness, a well-known X-linked recessive disorder:

X^B = normal vision
X^b = color blindness

The possible genotypes and phenotypes in both males and females follow.

Genotypes	Phenotypes
X^BX^B	Female who has normal color vision
X^BX^b	Carrier female who has normal color vision
X^bX^b	Female who is color-blind
X^BY	Male who has normal vision
X^bY	Male who is color-blind

The second genotype is a carrier female. Although a female with this genotype appears normal, she is capable of passing on an allele for color blindness. Color-blind females are rare because they must receive the allele from both parents. Color-blind males are more common because they need only one recessive allele to be color-blind. The allele for color blindness must be inherited from their mother because it is on the X chromosome. Males only inherit the Y chromosome from their father (Fig. 20.17).

Now let us consider a mating between a man with normal vision and a heterozygous woman (Fig. 20.17). What is the chance that this couple will have a color-blind daughter? A color-blind son? All daughters will have normal color vision because they all receive an X^B from their father. The sons, however, have a 50% chance of being color-blind, depending on whether they receive an X^B or an X^b from their mother. The inheritance of a Y chromo-

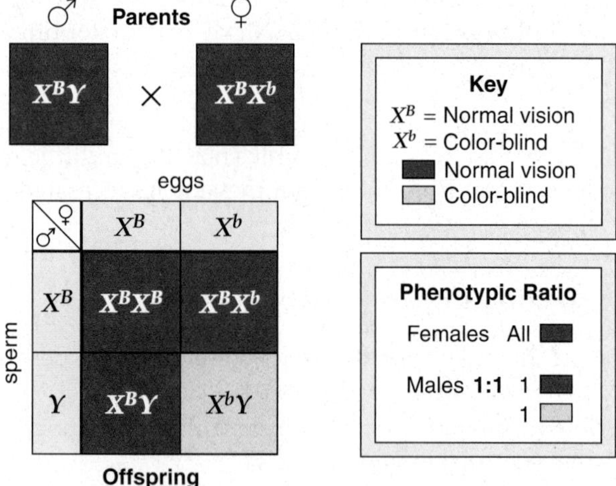

Figure 20.17 Results of an X-linked cross.
The male parent is normal, but the female parent is a carrier—an allele for color blindness is located on one of her X chromosomes. Therefore, each son has a 50% chance of being color-blind. The daughters will appear normal, but each one has a 50% chance of being a carrier.

some from their father cannot offset the inheritance of an X^b from their mother. The Y chromosome doesn't have an allele for the trait, so it can't possibly prevent color blindness in a son. Notice in Figure 20.17 that the phenotypic results for sex-linked traits are given separately for males and females.

Pedigree for X-Linked Disorders

Like color blindness, most sex-linked disorders are usually carried on the X chromosome. Figure 20.18 gives a pedigree for an *X-linked recessive disorder.* More males than females have the disorder because recessive alleles on the X chromosome are always expressed in males. The Y chromosome lacks an allele for the disorder. X-linked recessive conditions often pass from grandfather to grandson because the daughters of a male with the disorder are carriers. Figure 20.18 lists various ways to recognize a recessive X-linked disorder.

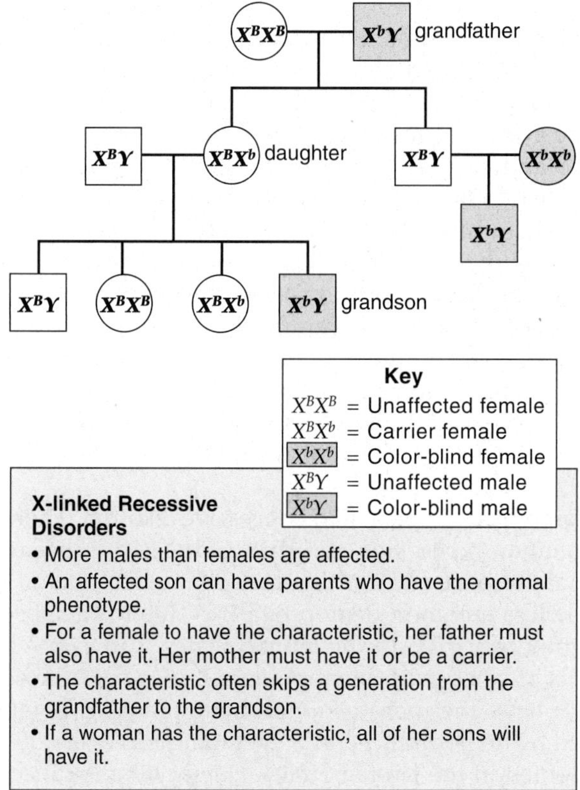

X^BX^B
X^bY grandfather
X^BY
X^BX^b daughter
X^BY
X^bX^b
X^bY
X^BY
X^BX^B
X^BX^b
X^bY grandson
Key
X^BX^B = Unaffected female
X^BX^b = Carrier female
X^bX^b = Color-blind female
X^BY = Unaffected male
X^bY = Color-blind male
X-linked Recessive Disorders
• More males than females are affected.
• An affected son can have parents who have the normal phenotype.
• For a female to have the characteristic, her father must also have it. Her mother must have it or be a carrier.
• The characteristic often skips a generation from the grandfather to the grandson.
• If a woman has the characteristic, all of her sons will have it.

Figure 20.18 **X-linked recessive disorder pedigree.** This pedigree for color blindness exemplifies the inheritance pattern of an X-linked recessive disorder. The list gives various ways of recognizing the X-linked recessive pattern of inheritance.

Only a few known traits are *X-linked dominant.* If a disorder is X-linked dominant, affected males pass the trait *only* to daughters, who have a 100% chance of having the condition. Females can pass an X-linked dominant allele to both sons and daughters. If a female is heterozygous and her partner is normal, each child has a 50% chance of escaping an X-linked dominant disorder. This depends on the maternal X chromosome that is inherited.

X-Linked Recessive Disorders of Interest

Color blindness, an X-linked recessive disorder, is the inability to differentiate certain color perceptions. There is no treatment for color blindness, but the disorder itself does not lead to a significant disability for the patient. Color blindness affects about 8–12% of Caucasian males and 0.5% of Caucasian females in the United States. Most people affected see brighter greens as tans, olive greens as browns, and reds as reddish-browns. A few cannot tell reds from greens at all. They see only yellows, blues, blacks, whites, and grays.

Duchenne muscular dystrophy is an X-linked recessive disorder characterized by a degeneration of the muscles. Symptoms, such as waddling gait, toe walking, frequent falls, and difficulty in rising, may appear as soon as the child starts to walk. Muscle weakness intensifies and respiratory and cardiovascular conditions progress until the individual is confined to a wheelchair, normally by ages 7 to 10. Death usually occurs by ages 20 to 25. Therefore, affected males are rarely fathers. The recessive allele remains in the population by passage from carrier mother to carrier daughter.

The absence of a protein, now called dystrophin, is the cause of Duchenne muscular dystrophy. Much investigative work determined that dystrophin is involved in the release of calcium from the sarcoplasmic reticulum in muscle fibers. The lack of dystrophin causes calcium to leak into the cell, which promotes the action of an enzyme that dissolves muscle fibers. (Refer to Chap. 12 to review this process.) When the body attempts to repair the tissue, fibrous tissue forms (Fig. 20.19). This cuts off the blood supply so that more and more cells die. Immature muscle cells can be injected into muscles, but it takes 100,000 cells for dystrophin production to increase by 30–40%.

Fragile X syndrome is the most common cause of inherited mental impairment. These impairments can range from mild learning disabilities to more severe intellectual disabilities. It is also the most common

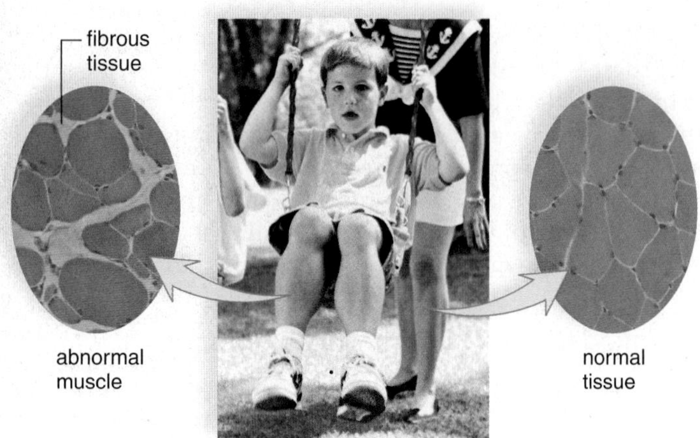

Figure 20.19 Muscular dystrophy.
In muscular dystrophy, an X-linked recessive disorder, calves enlarge because fibrous tissue develops as muscles waste away, due to lack of the protein dystrophin.

Connections and Misconceptions

Why is this disease called "fragile" X syndrome?

The name of fragile X syndrome may make you think that the disease has something to do with a breakable X chromosome. Actually, the name refers to the appearance of the chromosome under a microscope. In extreme cases of fragile X syndrome, in which there are many copies of the trinucleotide repeat, a portion of the X chromosome appears to be hanging from a thread. However, the appearance of the chromosome is not a cause of the disease. Fragile X syndrome is a result of a malfunction in the fragile X mental retardation protein (FMRP).

known cause of *autism,* a class of social, behavioral, and communication disorders. Fragile X syndrome affects 1 out of every 4,000 males and 1 out of every 8,000 females on average. Males with full symptoms of the condition have characteristic physical abnormalities—a long face, prominent jaw, large ears, joint laxity (excessively flexible joints), and genital abnormalities. Other common characteristics include tactile defensiveness (dislike of being touched), poor eye contact, repetitive speech patterns, hand flapping, and distractibility. Females with the condition present with variable symptoms. Most of the same traits seen in males with fragile X syndrome have been reported in females as well, but often the symptoms are milder in females and present with lower frequency.

A person with fragile X syndrome does not make the fragile X mental retardation protein (FMRP). Lack of this protein in the brain results in the various manifestations of the disease. As the name implies, a gene defect is found on the X chromosome. The genetic basis of fragile X syndrome is similar to that seen in Huntington disease (see pages 478-79). In both cases, the gene in question has too many repeated copies of a DNA sequence containing three nucleotides (called a trinucleotide repeat). In fragile X syndrome, the DNA sequence is CGG. (Recall that in Huntington disease, the repeated sequence is CAG.) Fewer than 59 copies of the repeated sequence is considered normal. Between 59 and 200 copies is considered "premutation." Generally, both males and females with the premutation genotype have normal intellect and appearance, although they can have subtle intellectual or behavioral symptoms. Persons whose DNA has over 200 copies of the repeat have "full mutation" and show the physical and behavioral traits of fragile X syndrome.

The trinucleotide-repeat disorders, like Huntington disease and fragile X syndrome, exhibit what is called *anticipation.* This means that the number of repeats in the gene can increase in each successive generation. For example, a female with 100 copies of the repeat is considered to have a premutation. When this female passes her X chromosome with the premutation to her offspring, the number of repeats may expand to over 200. This would result in a full mutation in her child. In a recent study, maternal premutations of between 90 and 200 repeats resulted in an expansion to full mutation in 80–100% of the offspring.

Hemophilia is an X-linked recessive disorder. There are two common types. Hemophilia A is due to the absence or minimal presence of a clotting factor known as factor VIII, and the less common hemophilia B is due to the absence of clotting factor IX. In the United States, 1 in every 5,000 males is affected with hemophilia. Of those, 80–85% specifically have the more common hemophilia A. A female affected by hemophilia is rare, averaging 1 in every 50 million women in the United States. Hemophilia is called the bleeder's disease because the affected person's blood either does not clot or clots very slowly. Although hemophiliacs bleed externally after an injury, they also bleed internally, particularly around joints. Hemorrhages can be stopped with transfusions of fresh blood (or plasma) or concentrates of the clotting protein. Also, factors VIII and IX are now available as biotechnology products.

Connecting the Concepts

The sex-linked conditions described in this section are based on other systems of the human body. For more information, refer to the following discussions.

Section 6.4 lists the various types of hemophilia.

Section 12.4 describes a variety of muscular disorders, including muscular dystrophy.

Section 18.1 outlines how genetics counselors obtain chromosomes to screen for diseases such as fragile X syndrome.

Check Your Progress 20.5

1. Solve the following: In a given family, a man and woman have two children, a boy and a girl, and both are color-blind. List the possible genotypes of the parents if both parents have normal vision.
2. Predict if a woman affected by an X-linked dominant disorder can have a child that is not affected. Why or why not?
3. Discuss why X-linked disorders are more common than Y-linked disorders.

CASE STUDY CONCLUSION

Dr. Preston referred Patrick's parents to the hospital's genetics counselor Ms. Wolfe. A genetics counselor works in the medical field as a liaison between patients and members of all health-care disciplines. Their job descriptions include identifying couples at risk for passing on genetic disorders and working with patients affected by disorders and their families to better understand diagnoses, treatment options, and counseling.

Ms. Wolfe explained that PKU was autosomal recessive in inheritance. She drew out a pedigree chart depicting both Patrick and his parents. Patrick was affected and neither of his parents was, which meant that both of his parents were carriers of the disorder. She explained how the diagnosis of PKU was made on Patrick by use of the HPLC test and what the disorder might mean to his health and development. The characteristics of PKU—reduced brain development, seizures, and so on—can be avoided; because the problem is the buildup of phenylalanine levels, the main goal during developmental years is to eliminate or reduce the intake of phenylalanine. She explained that a diet restricting phenylalanine intake starting right away and continuing until Patrick finished puberty would result in little to no adverse effects. This diet, combined with prescribed protein supplements, would help Patrick avoid the buildup of phenylalanine in his neurons and would allow his brain to develop normally.

Media Study Tools

www.mhhe.com/maderhuman12e

Enhance your study of this chapter with study tools and practice tests. Also ask your instructor about the resources available through ConnectPlus, including the media-rich eBook, interactive learning tools, and animations.

Virtual Labs

For additional assistance in understanding monohybrid, dihybrid, and sex-linked crosses, use the virtual labs "Punnett Square" and "Sex-Linked Traits."

Summarizing the Concepts

For an overall review of this chapter, play the MP3 file "Genetics and Inheritance."

20.1 Genotype and Phenotype

Genotype refers to the alleles of the individual, and phenotype refers to the physical characteristics associated with these alleles.

- Homozygous dominant individuals (*EE*) have the dominant phenotype (i.e., unattached earlobes).
- Homozygous recessive individuals (*ee*) have the recessive phenotype (i.e., attached earlobes).
- Heterozygous individuals (*Ee*) have the dominant phenotype (i.e., unattached earlobes).

20.2 One- and Two-Trait Inheritance

One-Trait Crosses

The first step in doing one-trait problems is to determine the genotype and then the gametes.

- An individual has two alleles for every trait, but a gamete has one allele for every trait.

The next step is to combine all possible sperm with all possible eggs. If there is more than one possible sperm and/or egg, a Punnett square is helpful in determining the genotypic and phenotypic ratio among the offspring.

- For a monohybrid–monohybrid cross, a 3:1 ratio is expected among the offspring.
- For a monohybrid–recessive cross, a 1:1 ratio is expected among the offspring.
- The expected ratio can be converted to the chance of a particular genotype or phenotype. For example, a 3:1 ratio = a 75% chance of the dominant phenotype and a 25% chance of the recessive phenotype.

Two-Trait Crosses

- If an individual is heterozygous for two traits, four gamete types are possible as can be substantiated by knowledge of meiosis.
- For a dihybrid–dihybrid cross (*AaBb* × *AaBb*), a 9:3:3:1 ratio is expected among the offspring.
- For a dihybrid–recessive cross (*AaBb* × *aabb*), a 1:1:1:1 ratio is expected among the offspring.

20.3 Inheritance of Genetic Disorders

A pedigree shows the pattern of inheritance for a trait from generation to generation of a family. This first pattern appears in a family pedigree for a recessive disorder—both parents are carriers. The second pattern appears in a family pedigree for a dominant disorder. Both parents are again heterozygous.

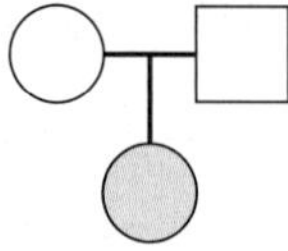

Trait is recessive.

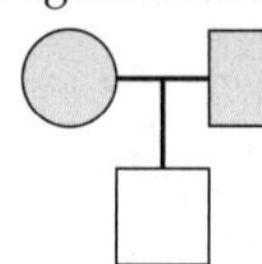

Trait is dominant.

Genetic Disorders of Interest

- Tay–Sachs, cystic fibrosis, and sickle-cell disease are autosomal recessive disorders.
- Marfan syndrome and Huntington disease are autosomal dominant disorders.

20.4 Beyond Simple Inheritance Patterns

In some patterns of inheritance, the alleles are not just dominant or recessive.

Polygenic Inheritance

Polygenic traits, such as skin color and height, are controlled by more than one set of alleles. The dominant alleles have an additive effect on the phenotype.

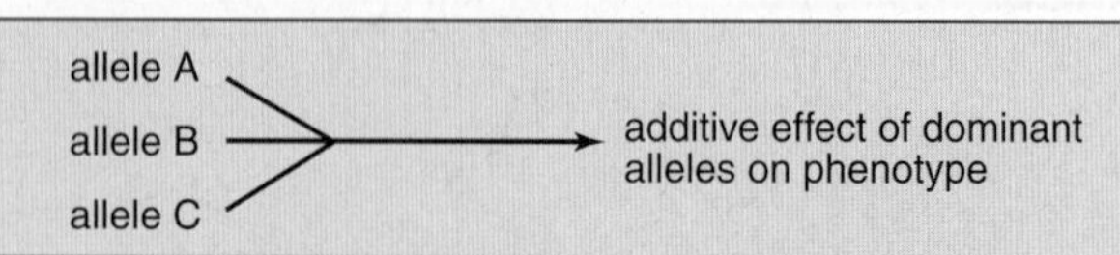

Incomplete Dominance and Codominance

In incomplete dominance (e.g., familial hypercholesterolemia), the heterozygote is intermediate between the two homozygotes. In codominance (e.g., blood type AB), both dominant alleles are expressed equally.

Multiple-Allele Inheritance

The multiple-allele inheritance pattern is exemplified in humans by blood type inheritance. Every individual has two out of three possible alleles: I^A, I^B, or *i*. Both I^A and I^B are expressed. Therefore, this is also a case of codominance.

20.5 Sex-Linked Inheritance

X-Linked Alleles

Many genes on the X chromosomes, such as those that determine normal vision as opposed to color blindness, are unrelated to the gender of the individual. Common X-linked genetic crosses are

- $X^BX^b \times X^BY$ All daughters will be normal, even though they have a 50% chance of being carriers, but sons have a 50% chance of being color-blind.
- $X^BX^B \times X^bY$ All children are normal (daughters will be carriers).

Pedigree for X-Linked Disorders

- A pedigree for an X-linked recessive disorder shows that the trait often passes from grandfather to grandson by way of a carrier daughter. Also, more males than females have the characteristic.
- Like most X-linked disorders, color blindness, muscular dystrophy, fragile X syndrome, and hemophilia are recessive.

Understanding Key Terms

allele 468
autosome 484
carrier 476
codominance 481
color blindness 485
cystic fibrosis (CF) 477
dihybrid 473
dominant allele 468
Duchenne muscular dystrophy 485
familial hypercholesterolemia (FH) 481
fragile X syndrome 485
genotype 468
hemophilia 486
heterozygous 468
homozygous dominant 468
homozygous recessive 468
Huntington disease 478
incomplete dominance 481
locus 468
Marfan syndrome 478
monohybrid 470
multifactorial trait 480
multiple allele 481
phenotype 468
polygenic trait 479
Punnett square 470
recessive allele 468
sex chromosome 484
sex-linked 484
sickle-cell disease 478
Tay–Sachs disease 477
X-linked 484

Match the key terms to these definitions.

a. ________ Gridlike device used to calculate the expected results of simple genetic crosses.

b. ________ Alternate forms of a gene that occur at the same site on homologous chromosomes.

c. ________ Allele that exerts its phenotype effect in the heterozygote; it masks the expression of the recessive allele.

d. ________ Particular site where a gene is found on a chromosome.

e. ________ Alleles of an individual for a particular trait or traits, expressed such as *BB*, *Aa*, or *BBAa*.

Testing Your Knowledge of the Concepts

1. Parents both have unattached earlobes, but some of their children have attached earlobes. Explain in terms of phenotype and genotype. (page 468)
2. Explain why the gametes have only one allele for a trait. (pages 469–470)
3. What is the chance of producing a child with the dominant phenotype from each of the following crosses? (pages 470–471)
   a. *AA* × *AA*
   b. *Aa* × *AA*
   c. *Aa* × *Aa*
   d. *aa* × *aa*
4. Which of the crosses in question 3 can result in an offspring with the recessive phenotype? Explain. (page 470)
5. What are the expected genotypic and phenotypic results of the following crosses? (pages 470–474)
   a. monohybrid × monohybrid
   b. monohybrid × recessive
   c. dihybrid × dihybrid
   d. dihybrid × recessive in both traits
6. Using a pedigree, show an autosomal recessive disorder pattern and an autosomal dominant disorder pattern. (pages 476–478)
7. What is required for an autosomal recessive disorder to appear? Name four autosomal recessive disorders. What is required for an autosomal dominant disorder to appear? Name two autosomal dominant disorders. (pages 476–478)
8. Define and give an example of each of the following inheritance patterns: polygenic inheritance, multifactorial trait, incomplete dominance, codominance, and multiple alleles. (pages 480–481)
9. What are the phenotypes and genotypes for the ABO blood groups? If both parents are type AB blood, what are the possible genotypes and phenotypes for their children, including the expected percentages? (pages 481, 483)
10. How is an X-linked trait different from an autosomal trait? Show a pedigree of an X-linked trait. (pages 484–485)
11. A woman with normal vision reproduces with a man with normal vision. Three sons are color-blind, and one son has normal vision. Explain how this occurred, using the genotypes and gametes of all individuals. (pages 484–485)
12. Which of these is a correct statement?
    a. Each gamete contains two alleles for each trait.
    b. Each individual has one allele for each trait.
    c. Fertilization gives each new individual one allele for each trait.
    d. All of these are correct.
    e. None of these is correct.
13. What possible gametes can be produced by *AaBb*?
    a. *Aa*, *Bb*
    b. *A*, *a*, *B*, *b*
    c. *AB*, *ab*
    d. *AB*, *Ab*, *aB*, *ab*
14. A straight hairline is recessive. If two parents with a widow's peak have a child with a straight hairline, then what is the chance that their next child will have a straight hairline?
    a. no chance
    b. ¼
    c. $\frac{3}{16}$
    d. ½
    e. $\frac{1}{16}$
15. What is the chance that an *Aa* individual will be produced from an *Aa*–*Aa* cross?
    a. 50%
    b. 75%
    c. 0%
    d. 25%
    e. 100%
16. The genotype of an individual with the dominant phenotype can be determined best by reproduction with
    a. the recessive genotype or phenotype.
    b. a heterozygote.
    c. the dominant phenotype.
    d. the homozygous dominant.
    e. Both a and b are correct.
17. The homologous chromosomes align independently at the equator during meiosis so
    a. all possible combinations of alleles can occur in the gametes.
    b. only the parental combinations of gametes can occur in the gametes.
    c. only the nonparental combinations of gametes can occur in the gametes.
18. Which of the following is not a feature of multifactorial inheritance?
    a. Effects of dominant alleles are additive.
    b. Genes affecting the trait may be on multiple chromosomes.
    c. Environment influences phenotype.
    d. Recessive alleles are harmful.
19. The ABO blood system exhibits
    a. codominance.
    b. multiple alleles.
    c. incomplete dominance.
    d. Both a and b are correct.
20. If a child has type O blood and the mother is type A, then which of the following could be the blood type of the child's father?
    a. A only
    b. B only
    c. O only
    d. A or O
    e. A, B, or O

Thinking Critically About the Concepts

In the case study, Patrick was affected with an autosomal recessive disorder, PKU. The chapter detailed many other autosomal recessive disorders, other types of genetic inheritance, and one- and two-trait inheritance patterns. Think about the basics learned from this chapter when answering the following questions.

1. In the pedigree for an autosomal recessive disorder on page 476, the chart depicts a skip in the generations of affected individuals. Is this always the case for autosomal recessive disorders? Why or why not?
2. In two-trait crosses, does one trait being dominant have an effect on the inheritance of the second trait?
3. Would you think an X-linked disorder or an autosomal disorder would appear more in a family pedigree? Explain your reasoning.

CHAPTER 21

DNA Biology and Technology

CASE STUDY DIABETES

At 8 years old, Kaya was diagnosed with type 1 diabetes mellitus. For months before her diagnosis, she had been fatigued. She had also had noticeable weight loss; an almost insatiable thirst; and dry, itchy skin that was slow to heal after a cut or injury. Her pediatrician ran a series of diagnostic tests over several days. She had a fasting plasma glucose (FPG) test and a random plasma glucose test performed. Kaya's doctor was administering these tests to detect her blood glucose levels at different times during the day and in response to fasting and nonfasting situations. Type 1 diabetes results from the body's failure to either produce insulin or produce enough insulin for the body's needs. Insulin is a hormone that helps move glucose from the blood stream into the cells of the body where it can be used to make the fuel, ATP, for our cells. Kaya's pediatrician explained to her and her parents that this type of diabetes was a chronic disease that she will have to control throughout her life. Left untreated, type 1 diabetes can lead to blindness, kidney failure, nerve damage, cardiovascular disease, and death. Dr. Weber explained that 17 million people in the United States are afflicted with diabetes and 5–10% of them have type 1, like Kaya. In fact, diabetes is the third leading major cause of death in the United States behind heart disease and cancer. Dr. Weber explained how Kaya would have to treat her condition. She would need to monitor her blood glucose level several times per day and use insulin injections multiple times per day when needed. Kaya's parents asked where the insulin she would be injecting came from. Dr. Weber explained that at one time it was derived from the pancreas of cows and pigs, but now with the advances in biotechnology, human insulin is used to treat diabetes patients. The gene for human insulin can be inserted into a bacteria cell and the bacteria can produce the insulin Kaya will inject.

As you read through the chapter, think about the following questions:

1. What advances in biotechnology make the production of human insulin in bacterial cells possible?
2. What is gene expression? What processes are involved to express a gene in a cell?
3. How do scientists clone a gene? Why is this done?

CHAPTER CONCEPTS

21.1 DNA and RNA Structure and Function

DNA is a double helix composed of two polynucleotide strands. When DNA replicates, each strand serves as a template for a new strand. RNA is involved with processing the information contained within the DNA molecule.

21.2 Gene Expression

Gene expression results in an RNA or protein product. Each protein has a sequence of amino acids. The blueprint for building a protein is coded in the sequence of nucleotides in the DNA.

21.3 DNA Technology

DNA technology allows us to clone a portion of DNA for various purposes, including DNA fingerprinting and transfer of DNA to other organisms. Transgenic (genetically modified) organisms receive and express foreign DNA. During gene therapy, humans receive foreign DNA to cure some particular condition.

21.4 Genomics

Genomics is the study of genetic information in a particular cell or organism, including humans and other organisms.

BEFORE YOU BEGIN

Before beginning this chapter, take a few moments to review the following discussions.

Section 2.7 What are the roles of nucleic acids in a cell?

Section 3.4 What is the structure of the nucleus?

Section 18.1 When are the chromosomes replicated in the cell cycle?

21.1 DNA and RNA Structure and Function

Learning Outcomes

Upon completion of this section, you should be able to

1. Describe the structure of a DNA molecule.
2. Explain the process of DNA replication.
3. Distinguish between the structures of DNA and RNA.
4. State the roles of RNA in a cell.

For life on Earth, **DNA (deoxyribonucleic acid)** is the genetic material. DNA is organized into structures called chromosomes. In eukaryotic cells, including those of humans, the majority of the DNA is located in the nucleus. A small amount of DNA is also found in the mitochondria, but for this chapter we will focus on the nuclear DNA. Genetic material must be able to do three things: (1) replicate so that it can be transmitted to the next generation, (2) store information, and (3) undergo change (mutation) to provide genetic variability. DNA meets all three of these criteria.

Structure of DNA

DNA is a **double helix.** It is composed of two strands that spiral about each other (Fig. 21.1*a*). Each strand is a polynucleotide because it is composed of a series of nucleotides. A nucleotide is a molecule composed of three subunits—phosphoric acid (phosphate), a pentose sugar (deoxyribose), and a nitrogen-containing base (either adenine [A], cytosine [C], guanine [G], or thymine [T]; see Fig.2.24). Looking at just one strand of DNA, notice that the phosphate and sugar molecules make up a backbone and the bases project to one side. Put the two strands together, and DNA resembles a ladder (Fig. 21.1*b*). The phosphate–sugar backbones make up the supports of the ladder. The rungs of the ladder are the paired bases. The bases are held together by hydrogen bonding: A pairs with T by forming two hydrogen bonds, and G pairs with C by forming three hydrogen bonds, or vice versa. This is called *complementary base pairing* (Fig. 21.1*c*).

These **complementary paired bases** are important to the functioning of DNA. Remember that adenine and guanine are purines (purine is a structure with two rings), and cytosine and thymine are pyrimidines (pyrimidine has one ring) (Fig 21.1*c*).

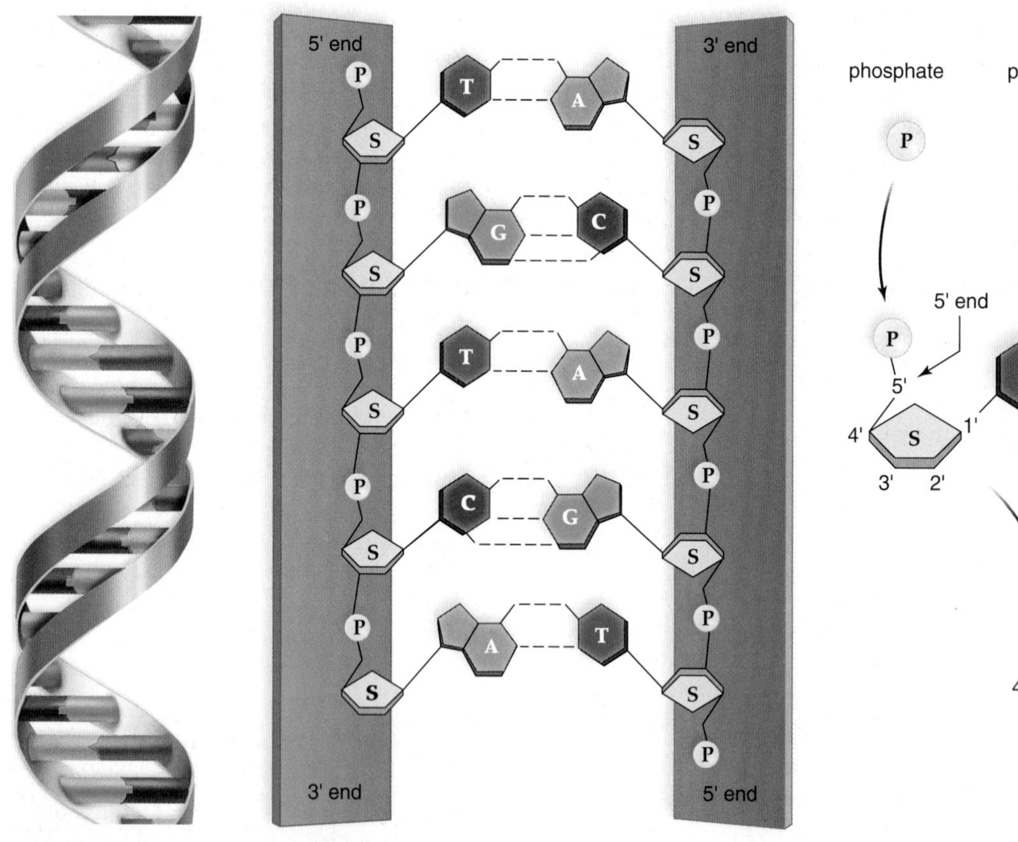

Figure 21.1 **The structure of DNA.**
a. DNA double helix. **b.** When the helix is unwound, a ladder configuration shows that the supports are composed of sugar (S) and phosphate (P) molecules and the rungs are complementary bases. The bases in DNA pair in such a way that the phosphate–sugar backbones are oriented in different directions. **c.** The DNA strands are antiparallel, which is apparent by numbering the carbon atoms in deoxyribose (5′–3′).

Biology Matters Historical Focus

Overlooked Genius: Rosalind Franklin

A brilliant scientist and researcher, Dr. Rosalind Franklin made one of the most significant contributions to understanding the structure of DNA. Her X-ray "Photograph 51" was crucial in revealing DNA's makeup. But when the Nobel Prize for the discovery of the double helix was awarded in 1962, the honor went to James Watson, Francis Crick, and Maurice Wilkins. Rosalind Franklin was not listed as one of the recipients.

Rosalind Franklin was born in 1920 in London, England. At the age of 15, she decided to become a scientist, against her father's wishes. She went on to earn a doctorate in physical chemistry from Cambridge University by the age of 26. Her first job was in X-ray diffraction in a cutting-edge laboratory in Paris. In this process, a substance is first formed into crystals. X-rays shone through the crystals provide information about the molecule's structure. In 1951, Rosalind joined a laboratory at King's College, London. There, she began working on the X-ray diffraction of DNA. She refined an X-ray machine and perfected her photographs of crystallized DNA, some of which required 100-hour exposures to complete. By studying these photographs, she discovered that the linked sugar–phosphate strands in DNA were located on the outside of the molecule. She also discovered that the DNA helix was composed of two separate strands.

Yet Rosalind faced intense gender discrimination in her work at King's College. Women were not allowed to eat in the male-only dining rooms at the university. Her colleague, Maurice Wilkins, treated her as a research assistant, not as a peer. There was friction between the two. The atmosphere was so uncomfortable that Franklin left King's College to accept a position at Brikbeck College in London.

At that time, there was intense competition worldwide to be the first to determine the structure of DNA. In America, Linus Pauling had discovered the alpha helix structure of proteins and was using these data to investigate DNA. Francis Crick and James Watson at Cambridge University were also studying DNA. Wilkins showed Watson one of Franklin's X-ray photographs, without her permission. With the information from the photograph, Watson and Crick were finally able to determine the structure of DNA. They published their findings in the journal *Nature* in April 1953. Although Franklin's work was published in the same issue, she was not given credit in the publication describing the arrangement of DNA.

Figure 21A Rosalind Franklin (1920–1958).

Would Rosalind Franklin's work ever have received the acclaim she earned? Might she have been included in the 1962 Nobel Prize? The answers will never be known. Tragically, Franklin developed ovarian cancer in 1956, and she died in 1958 at the age of 37. The Nobel Prize is not awarded posthumously, and so Franklin was ineligible.

Video
Dark Lady of DNA

The two strands of DNA are antiparallel, meaning they run in opposite directions. Notice in Figure 21.1*b* that in one strand, the sugar molecules appear right side up, and in the other, they appear upside down. This is due to the position of certain carbon molecules on the deoxyribose sugar molecules. When looking at a double helix, one side will have the 5′ carbon at one end, and the other side will have the 3′ carbon (Fig. 21.1*b*). This orientation becomes important when the DNA is replicated.

Animation
DNA Structure

Replication of DNA

When cells divide, each new cell gets an exact copy of DNA. The process of copying a DNA helix is called **DNA replication.** During the S phase of interphase during mitosis when DNA is replicated, the double-stranded structure of DNA allows each original strand to serve as a **template** (mold) for the formation of a complementary new strand. DNA replication is termed *semiconservative* because each new double helix has one original strand

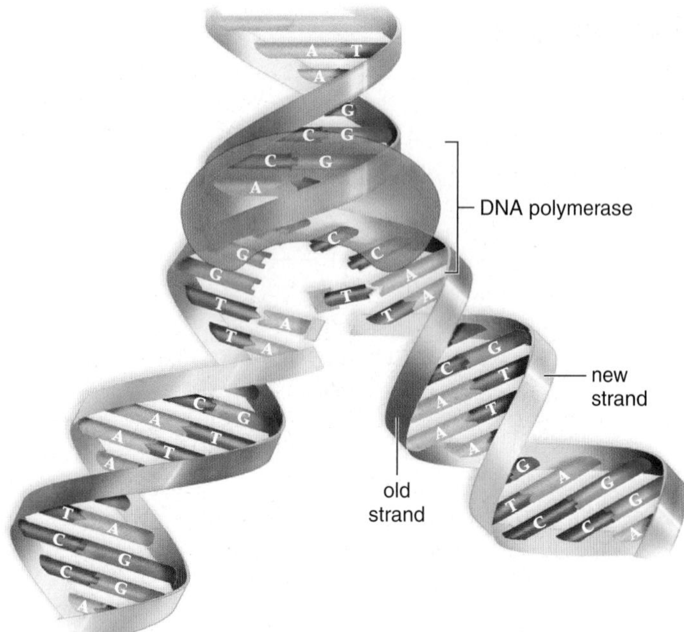

Figure 21.2 **DNA replication is "semiconservative."** Replication is called semiconservative because each new double helix is composed of an original strand and a new strand.

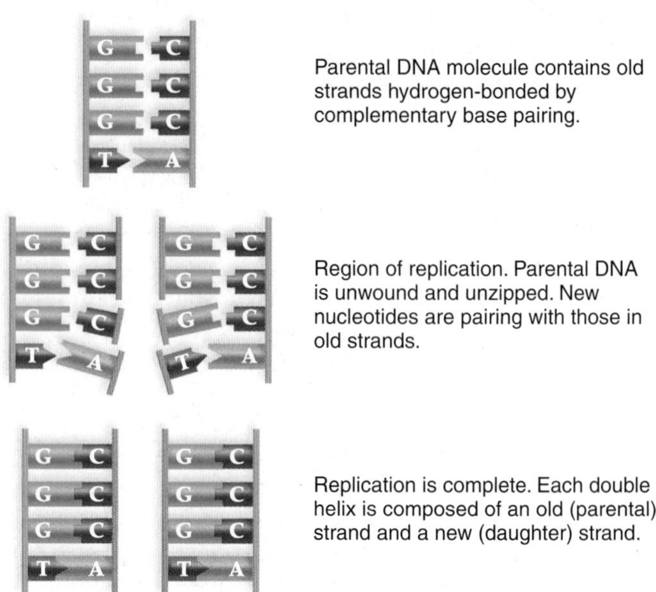

Parental DNA molecule contains old strands hydrogen-bonded by complementary base pairing.
Region of replication. Parental DNA is unwound and unzipped. New nucleotides are pairing with those in old strands.
Replication is complete. Each double helix is composed of an old (parental) strand and a new (daughter) strand.

Figure 21.3 **The process of DNA replication.** Use of the ladder configuration better illustrates how complementary nucleotides, available in the cell, pair with those of each old strand before they are joined together to form a daughter strand.

and one new strand. In other words, one of the original strands is conserved, or present, in each new double helix. Each original strand has produced a new strand through complementary base pairing, so there are now two DNA helices identical to each other and to the original molecule (Fig. 21.2).

Figure 21.3 shows how complementary nucleotides pair in the daughter strand.

1. Before replication begins, the two strands that make up parental DNA are hydrogen-bonded to each other.
2. An enzyme (DNA helicase) unwinds and "unzips" double-stranded DNA (i.e., the weak hydrogen bonds between the paired bases break).
3. New complementary DNA nucleotides (composed of a sugar, phosphate, and one nitrogen-containing base), always present in the nucleus, fit into place by the process of complementary base pairing. These are positioned and joined by the enzyme *DNA polymerase.*
4. To complete replication, the enzyme ligase seals any breaks in the sugar–phosphate backbone. The DNA returns to its coiled structure.
5. The two double-helix molecules are identical to each other and to the original DNA molecule.

Connections and Misconceptions

How long does it take to copy the DNA in one human cell?

The enzyme DNA polymerase in humans can copy approximately 50 bases per second. If only one DNA polymerase were used to copy human DNA, it would take almost three weeks! However, multiple DNA polymerases copy the human genome by starting at many different places. The entire 3 billion base pairs can be copied in 8 hours in a rapidly dividing cell.

Rarely, a replication error occurs, making the sequence of the bases in the new strand different from the parental strand. But, if an error does occur, the cell has repair enzymes that usually fix it. A replication error that persists is a **mutation,** a permanent change in the sequence of bases. A mutation is not always a bad thing; a mutation can possibly cause a change in the phenotype and introduce variability. Such variabilities make you different from your neighbor and humans different from other animals.

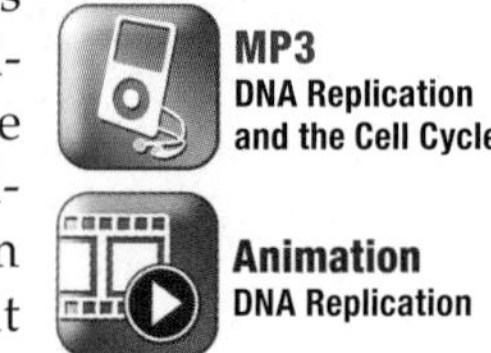

The Structure and Function of RNA

RNA (ribonucleic acid) is made up of nucleotides containing the sugar ribose. This sugar accounts for the scientific name of this polynucleotide. The four nucleotides that make up the RNA molecule have the following bases: adenine (A), uracil (U), cytosine (C), and guanine (G) (Fig. 21.4). One of the differences between RNA and DNA is the replacement of thymine with uracil. As with DNA, complementary base pairing may occur

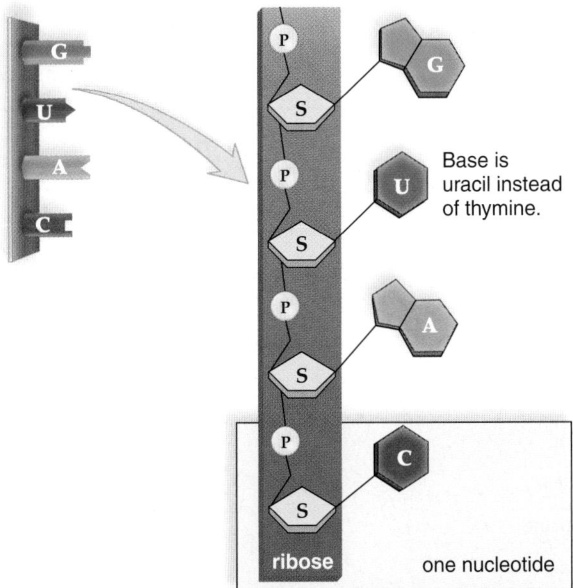

Figure 21.4 **The structure of RNA.**
Like DNA, RNA is a polymer of nucleotides. In an RNA nucleotide, the sugar ribose is attached to a phosphate molecule and to a base: G, U, A, or C. In RNA, the base uracil replaces thymine as one of the pyrimidine bases. RNA is single-stranded, whereas DNA is double-stranded.

in RNA; cytosine pairs with guanine and adenine pairs with uracil. However, unlike the double-helix structure of DNA, most RNA is single-stranded (Fig. 21.4). Table 21.1 lists the similarities and differences between DNA and RNA.

RNA is divided into coding and noncoding RNAs. The coding RNA is messenger RNA (mRNA), which is translated into protein. Noncoding RNAs are divided into ribosomal RNA (rRNA), transfer RNA (tRNA), and the small RNAs. The small RNAs are involved in the expression of the genes that code for mRNA and rRNA.

Table 21.1 DNA–RNA Similarities and Differences

DNA–RNA Similarities	
Nucleic acids	
Composed of nucleotides	
Sugar–phosphate backbone	
Four different types of bases	
DNA–RNA Differences	
DNA	**RNA**
Found in nucleus and mitochondria	Found in nucleus and cytoplasm
The genetic material	Helper to DNA
Sugar is deoxyribose	Sugar is ribose
Bases are A, T, C, G	Bases are A, U, C, G
Double-stranded	Single-stranded
Is transcribed (to give RNA)	Can be translated (to give proteins)

Messenger RNA

Messenger RNA (mRNA) is produced in the nucleus, where DNA serves as a template for its formation. This type of RNA carries genetic information from DNA to the ribosomes in the cytoplasm, where protein synthesis occurs. Messenger RNA is a linear molecule, as shown in Figure 21.4.

Ribosomal RNA

In eukaryotes, **ribosomal RNA (rRNA)** is produced using a DNA template in the nucleolus of the nucleus. Ribosomal RNA joins with specific proteins to form the large and small subunits of ribosomes. The subunits then leave the nucleus and either attach themselves to the endoplasmic reticulum or remain free within the cytoplasm. During the process of protein synthesis, the large and small ribosomal subunits combine to form a complex (the ribosome) that acts as a workbench for the manufacture of proteins (see Chap. 3).

Transfer RNA

Transfer RNA (tRNA) is produced in the nucleus, and a portion of DNA also serves as a template for its production. Appropriate to its name, tRNA transfers amino acids to the ribosomes. At the ribosomes, the amino acids are bonded together in the correct order to form a protein. There are 20 different types of amino acids used to make proteins. Each type of tRNA carries only one type of amino acid; therefore, at least 20 different tRNA molecules must be functioning in the cell to properly make a protein.

Small RNAs

Small RNAs are divided into several classes. Small nuclear RNAs (snRNAs) are involved in splicing the mRNA (see page 498–99) before it is exported from the nucleus to the cytoplasm for translation. Small nucleolar RNAs (snoRNAs) modify the ribosomal RNAs within the nucleolus of the cell. MicroRNAs (miRNAs) attach to mRNAs in the cytoplasm. Messenger RNAs are thus prevented from being translated unnecessarily. Small interfering RNAs (siRNAs) also bind to mRNAs. Attachment of an siRNA prepares the mRNA for degradation.

Connecting the Concepts

For more information on DNA and RNA, refer to the following discussions.

Section 2.7 describes the role of DNA and RNA as organic molecules.

Section 18.6 examines the role of chromosomes in inheritance, as well as the consequences of changes in chromosome number and structure.

Section 22.3 explores how mitochondrial DNA is being used to study human evolution.

Connections and Misconceptions

What is a microRNA and how is it used?

A microRNA (miRNA) is a small, noncoding gene that plays a role in developmental biology specifically by acting to regulate the events of gene expression. Historically, miRNAs were commonly used in early detection of various forms of cancer. Researchers currently are using them to determine evolutionary relationships between organisms. They have discovered that once an miRNA is fixed in a genome, it is rarely lost, so organisms with similar miRNA sequences are closely related. This was discovered by studying the miRNA sequences in the annelids (segmented worms). Earthworms, leeches, and bristle worms from all over the globe had retained similar sequences despite differences in speciation over millions of years.

Check Your Progress 21.1

1. Compare and contrast the structure and function of DNA and RNA.
2. Describe the function of the different types of RNA.
3. Explain how the structure of DNA allows it to be replicated.

21.2 Gene Expression

Learning Outcomes

Upon completion of this section, you should be able to

1. Summarize how the information contained within the DNA is expressed as a protein.
2. Interpret a sequence of DNA nucleotides.

As we shall see, DNA provides the cell with a blueprint for synthesizing proteins. In simplest terms, DNA acts as a template for making RNA, which in turn acts as a template for the manufacture of proteins. In gene expression, also known as *protein synthesis,* the process of transcription makes an RNA copy of DNA and the process of translation makes protein from the RNA.

Before discussing the mechanics of gene expression, let's review the structure of proteins.

Structure and Function of Proteins

Proteins are composed of subunits called amino acids (Table 21.2). Twenty different amino acids are commonly found in proteins. Proteins differ because the number and order of their amino acids differ. The sequence of amino acids in a protein leads to its particular shape. (See Chap. 2 for more detail on the levels of protein structure.) Proteins

Table 21.2 Amino Acids

Amino Acid	Abbreviation
alanine	ala
arginine	arg
asparagine	asn
aspartic acid	asp
cysteine	cys
glutamine	gln
glutamic acid	glu
glycine	gly
histidine	his
isoleucine	ile
leucine	leu
lysine	lys
methionine	met
phenylalanine	phe
proline	pro
serine	ser
threonine	thr
tryptophan	trp
tyrosine	tyr
valine	val

have many different functions in the body as they determine the structure and function of various cells in the body. Proteins are used as structural and regulatory components of cells. They are used as enzymes to catalyze chemical reactions, neurotransmitters to aid in the function of the nervous system, antibodies for the immune system, and hormones to change activities of certain cells. Proteins have this great diversity of functions due to the arrangement of these 20 different amino acids in their individual structures.

Gene Expression: An Overview

The first step in gene expression is called *transcription* and the second step is called *translation* (Fig. 21.5). During **transcription,** a strand of mRNA forms that is complementary to a portion of DNA. The mRNA molecule that forms is a *transcript* of a gene. *Transcription* means "to make a faithful copy." In this case, a sequence of nucleotides in DNA is copied to a sequence of nucleotides in mRNA.

Protein synthesis requires the process of **translation.** *Translation* means "to put information into a different language." In this case, a sequence of nucleotides (the mRNA)

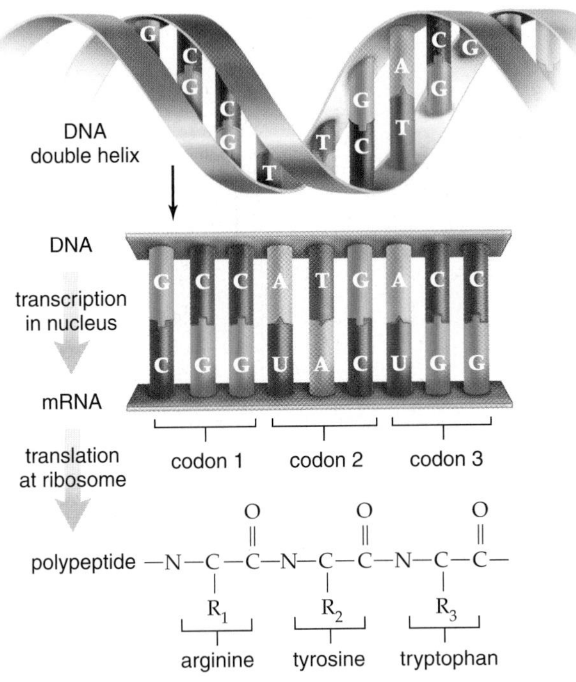

Figure 21.5 **Summary of gene expression.**
Transcription occurs when DNA acts as a template for RNA (e.g., mRNA) synthesis. (Uracil [U] in RNA takes the place of thymine [T] in DNA.) Translation occurs when the sequence of codons of mRNA specify the sequence of amino acids in a polypeptide.

First Base	Second Base: U	Second Base: C	Second Base: A	Second Base: G	Third Base
U	UUU phenylalanine	UCU serine	UAU tyrosine	UGU cysteine	U
U	UUC phenylalanine	UCC serine	UAC tyrosine	UGC cysteine	C
U	UUA leucine	UCA serine	UAA *stop*	UGA *stop*	A
U	UUG leucine	UCG serine	UAG *stop*	UGG tryptophan	G
C	CUU leucine	CCU proline	CAU histidine	CGU arginine	U
C	CUC leucine	CCC proline	CAC histidine	CGC arginine	C
C	CUA leucine	CCA proline	CAA glutamine	CGA arginine	A
C	CUG leucine	CCG proline	CAG glutamine	CGG arginine	G
A	AUU isoleucine	ACU threonine	AAU asparagine	AGU serine	U
A	AUC isoleucine	ACC threonine	AAC asparagine	AGC serine	C
A	AUA isoleucine	ACA threonine	AAA lysine	AGA arginine	A
A	**AUG (*start*) methionine**	ACG threonine	AAG lysine	AGG arginine	G
G	GUU valine	GCU alanine	GAU aspartate	GGU glycine	U
G	GUC valine	GCC alanine	GAC aspartate	GGC glycine	C
G	GUA valine	GCA alanine	GAA glutamate	GGA glycine	A
G	GUG valine	GCG alanine	GAG glutamate	GGG glycine	G

Figure 21.6 **The genetic code chart.**
In this chart, each of the codons (white rectangles) is composed of three letters representing the first base, second base, and third base. For example, find the rectangle where C for the first base and A for the second base intersect. You will see that U, C, A, or G can be the third base. CAU and CAC are codons for histidine; CAA and CAG are codons for glutamine.

is translated into the sequence of amino acids (the protein). This is possible only if the bases in DNA and mRNA code for amino acids. This code is called the genetic code.

The Genetic Code

The genetic code (Fig. 21.6) corresponds to a three-base sequence in the mRNA molecule called a **codon.** Each codon represents a specific amino acid. The reason that each codon contains three bases instead of one or two is a matter of mathematics. There are 20 different amino acids that are used to build proteins. If a codon consisted of just a single base, then 16 amino acids could not be coded for. If each codon contained two bases, only 16 amino acids would be covered. However, the use of three bases allows for 64 possible codons, more than enough to code for the 20 amino acids. A closer examination of the genetic code (Fig. 21.6.) shows that 61 codons correspond to a particular amino acid. The remaining three are stop codons, which signal polypeptide termination. One of the codons (AUG) stands for the amino acid methionine, the amino acid that is used to signal the beginning of the polypeptide. Notice as well that most amino acids have more than one codon. For example. leucine, serine, and arginine each have six different codons. This redundancy in the genetic code offers some protection against possibly harmful mutations that change the sequence of the bases.

The genetic code is almost universal in living things, that is, it is the same for most living organisms. This suggests that the code dates back to the very first organisms on Earth and that all living things are related. In other words, all organisms share an evolutionary heritage.

Connections and Misconceptions

Are all mutations bad?

A mutation simply means a change from the original structure of DNA during replication or division. In some cases, these changes can result in a faulty or detrimental product that can negatively affect the organism. In other cases, all the diversity in our traits can be considered mutations. Without the ability to change, we would not have been able to adapt to changing environments; thus, evolution would not have been possible.

Transcription

During transcription, a segment of the DNA serves as a template for the production of an RNA molecule. Although all classes of RNA are formed by transcription, we will focus on transcription to form mRNA.

Forming mRNA

Transcription begins in the nucleus when the enzyme **RNA polymerase** opens up the DNA helix so that complementary base pairing can occur. Recall that RNA contains uracil instead of thymine, so base pairing between DNA and the mRNA strand will be A–U and C–G (Fig. 21.7). Then, RNA polymerase joins the RNA nucleotides, and an mRNA molecule with a sequence of bases complementary to the DNA segment results.

Processing mRNA

Before the transcribed mRNA leaves the nucleus for translation in the cytoplasm, it undergoes a series of processing steps.

The newly synthesized *primary* mRNA molecule becomes a *mature* mRNA molecule after processing (Fig. 21.8). Most

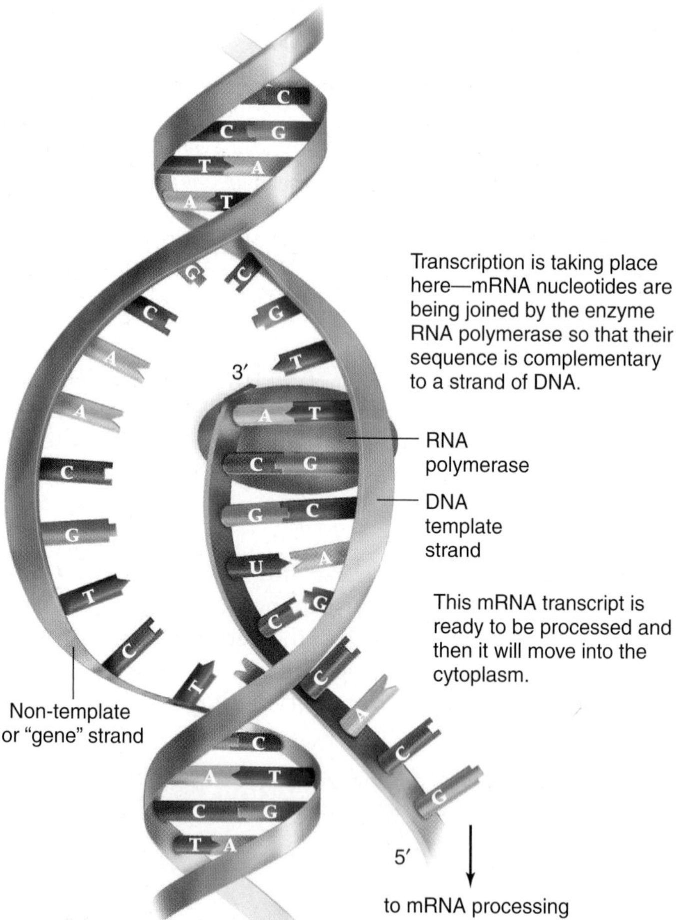

Figure 21.7 **Transcription of DNA into mRNA.**
During transcription, complementary RNA is made from a DNA template. A portion of DNA unwinds and unzips at the point of attachment of RNA polymerase. A strand of mRNA is produced when complementary bases join in the order dictated by the sequence of bases in template DNA.

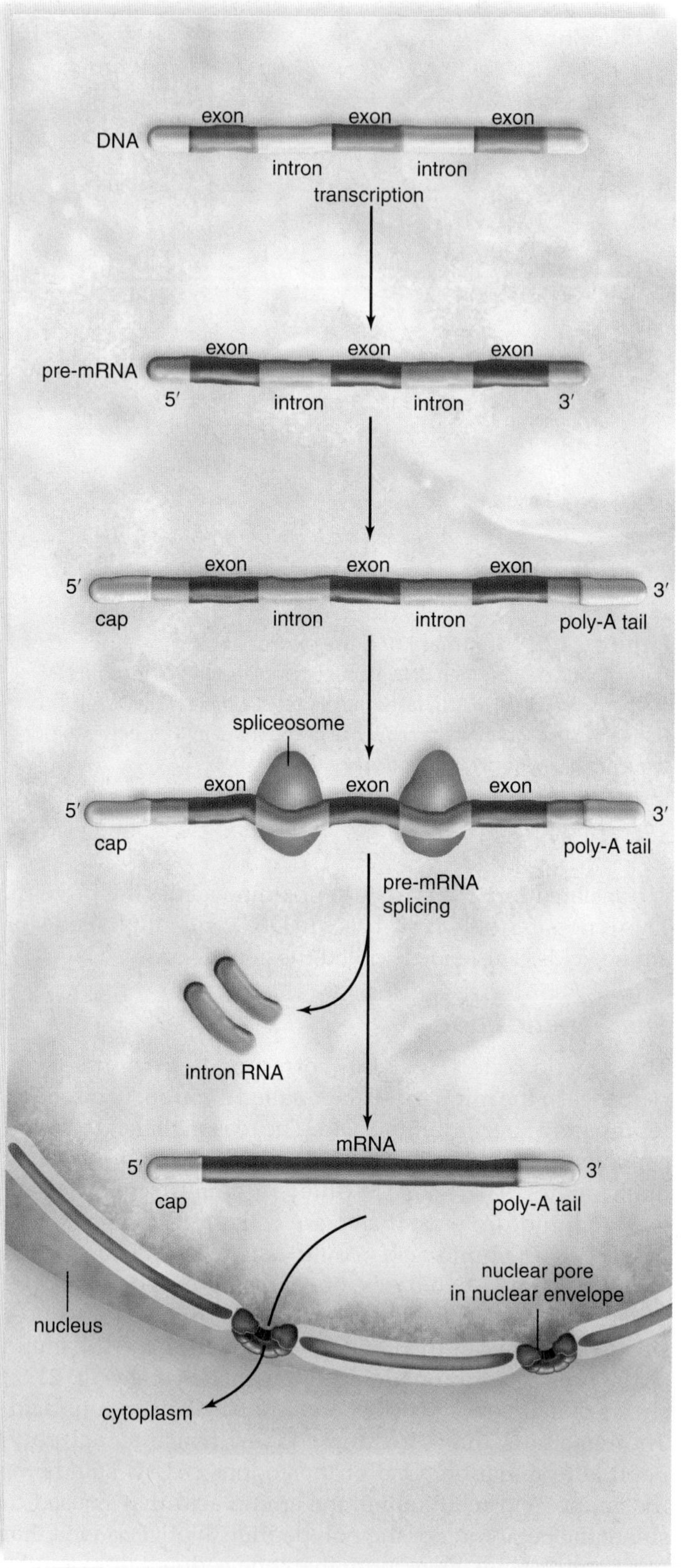

Figure 21.8 **mRNA processing.**
During processing, a cap and tail are added to mRNA, and the introns are removed so that only exons remain. Then the mRNA molecule is ready to leave the nucleus.

genes in humans are interrupted by segments of DNA that are not part of the gene. These portions are called *introns* because they are intragene segments and do not code for a functional protein. The other portions of the gene are called *exons* because they are ultimately expressed. Only exons result in a protein product.

Primary mRNA contains bases complementary to both exons and introns, but during processing, (1) one end of the mRNA is capped by the addition of an altered guanine nucleotide. The other end is given a tail, by the addition of multiple adenosine nucleotides. (2) The introns are removed, and the exons are joined to form a mature mRNA molecule consisting of continuous exons. This *splicing* of mRNA is done by a complex composed of both RNA (a small RNA molecule) and protein. Surprisingly, the RNA portion, not the protein, is functioning as the enzyme, and so it is called a *ribozyme*.

Animation How Spliceosomes Process RNA

Ordinarily, processing brings together all the exons of a gene. In some instances, cells use only certain exons rather than all of them to form a mature RNA transcript. Alternate mRNA splicing is believed to account for the ability of a single gene to result in different proteins in a cell and the different complexities between all living organisms despite the universal genetic code. Increasingly, small RNA molecules have been found that regulate not only mRNA processing but also transcription and translation. DNA codes for proteins, but RNA orchestrates the outcome.

Translation

During translation, transfer RNA (tRNA) molecules bring amino acids to the ribosomes (Fig. 21.9), where polypeptide synthesis occurs. Recall that a ribosome consists of a small and a large subunit that will join together during translation and bind to the mRNA strand. This creates a translation complex (Fig. 21.9*a*). The ribosome contains special binding sites called the A and P sites where individual tRNA molecules can bind with the mRNA strand. A tRNA molecule has an almost cloverleaf shape with an area for binding onto an amino acid and a region called an **anticodon** (Fig. 21.9*c*). The anticodon is a three-base sequence of RNA that will complementary base pair with the codons of mRNA. This complementary base pairing between an anticodon and a codon is how the tRNA brings the correct amino acid into the correct order instructed by the mRNA strand. Each amino acid, coded by specific codons, has a tRNA molecule with a specific anticodon that carries it to the translation complex. The tRNA molecules attach to the translation complex at the A site where each anticodon pairs with the complementary codon (Fig. 21.9*b*).

The order in which tRNA molecules link to the ribosome is directed by the sequence of the mRNA codons. In this way, the order of codons in mRNA brings about a particular order of amino acids in a protein. The tRNA attached to the growing polypeptide moves from the A site to the P site of a ribosome, as shown in Figure 21.9*b*.

If the codon sequence in a portion of the mRNA is ACC, GUA, and AAA, what will be the sequence of amino acids in a portion of the polypeptide? The genetic code chart in Figure 21.6 allows us to determine this:

Codon	Anticodon	Amino Acid
ACC	UGG	Threonine
GUA	CAU	Valine
AAA	UUU	Lysine

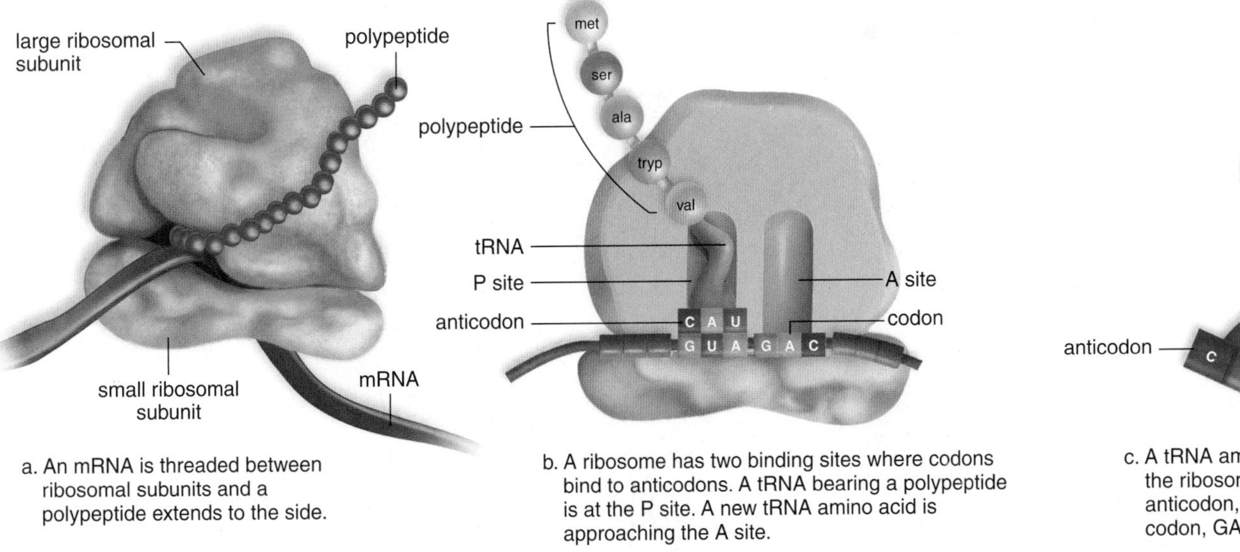

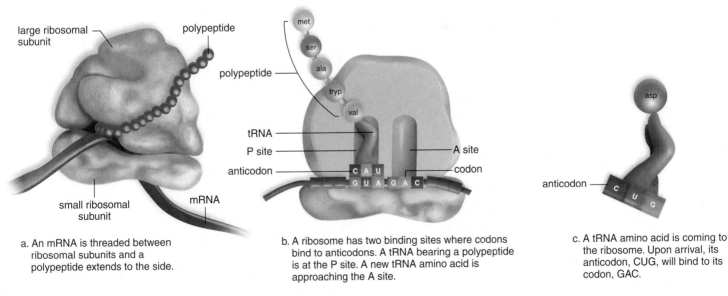

a. An mRNA is threaded between ribosomal subunits and a polypeptide extends to the side.

b. A ribosome has two binding sites where codons bind to anticodons. A tRNA bearing a polypeptide is at the P site. A new tRNA amino acid is approaching the A site.

c. A tRNA amino acid is coming to the ribosome. Upon arrival, its anticodon, CUG, will bind to its codon, GAC.

Figure 21.9 The roles of all three forms of RNA in translation.
Protein synthesis occurs at a ribosome. **a.** Side view of a ribosome showing mRNA and a growing polypeptide. **b.** The large ribosomal subunit contains two binding sites for tRNAs. **c.** tRNA structure and function.

Polypeptide synthesis requires three steps: initiation, elongation, and termination (Fig. 21.10).

1. During *initiation,* mRNA binds to the smaller of the two ribosomal subunits. Then the larger ribosomal subunit associates with the smaller one, forming the translation complex.
2. During *elongation,* the polypeptide lengthens, one amino acid at a time, about five amino acids per second. An incoming tRNA arrives at the A site and then receives the growing peptide chain from the outgoing tRNA. The ribosome moves laterally down the mRNA strand one codon at a time so that again the P site is filled by a tRNA–peptide complex (Fig. 21.10*b*). The A site is now available to receive another incoming tRNA as the complex has moved down one codon. In this manner, the peptide grows, and the linear structure of a polypeptide is made. The particular shape of a polypeptide is formed later (see Chap. 2).
3. Then *termination* of synthesis occurs when one of the three stop codons is reached by the A site. The ribosome dissociates into its two subunits and falls off the mRNA molecule. The individual portions of the translation complex can then re-form at the beginning of the mRNA strand to repeat this process and make another polypeptide.

Additionally, many ribosomes are at work forming the same polypeptide at the same time. As soon as the initial portion of mRNA has been translated by one ribosome and the ribosome has begun to move down the mRNA, another ribosome attaches to the mRNA to begin translation. Therefore, several ribosomes, collectively called a **polyribosome,** can move along one mRNA at a time. Several polypeptides of the same type can be synthesized using one mRNA molecule (Fig. 21.11). This, in addition to the recycling of the

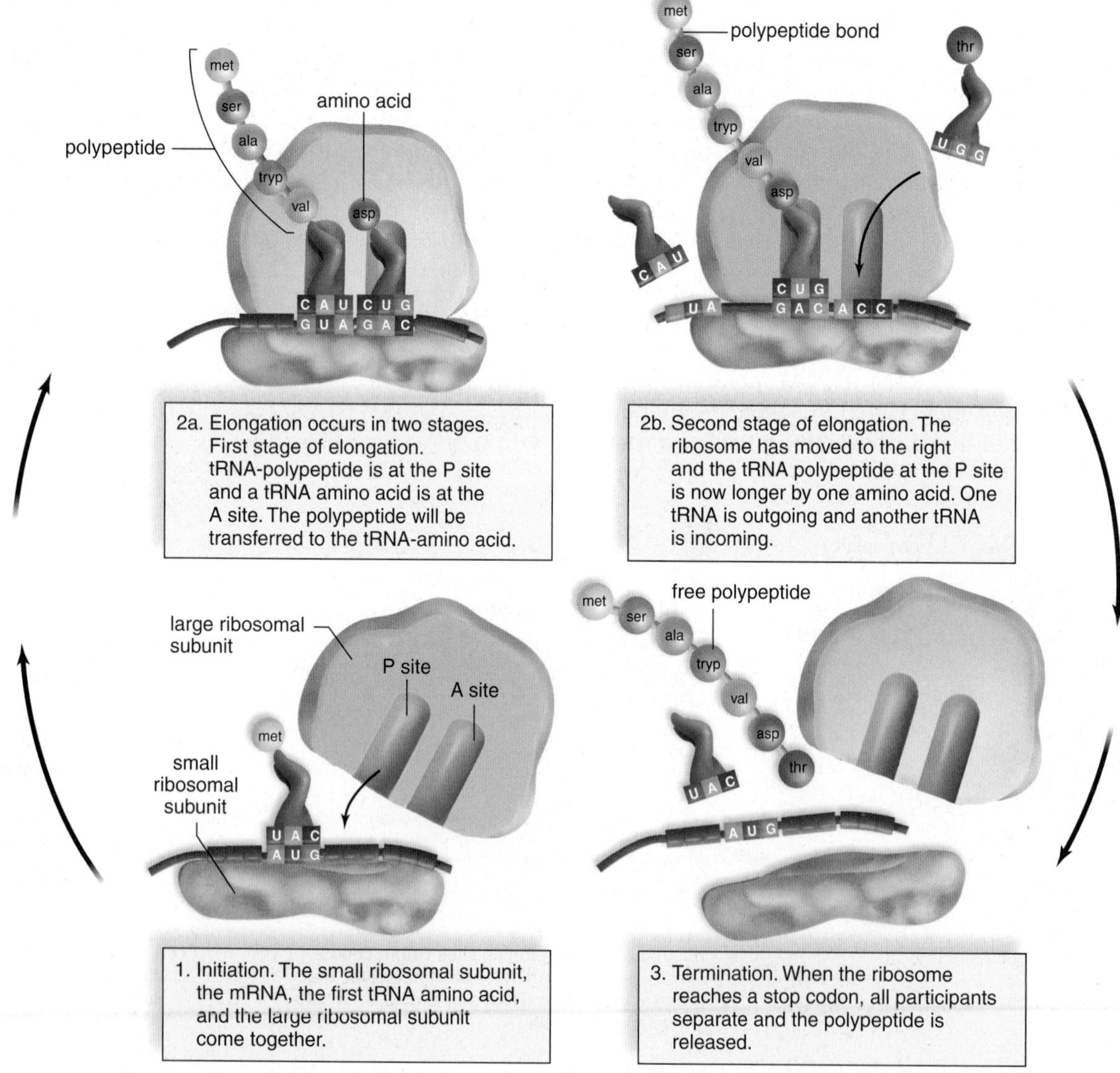
polypeptide
met
ser
ala
tryp
val
asp
amino acid
C A U C U G
G U A G A C
2a. Elongation occurs in two stages. First stage of elongation. tRNA-polypeptide is at the P site and a tRNA amino acid is at the A site. The polypeptide will be transferred to the tRNA-amino acid.
polypeptide bond
thr
U G G
C A U
U A
C U G
G A C A C C
2b. Second stage of elongation. The ribosome has moved to the right and the tRNA polypeptide at the P site is now longer by one amino acid. One tRNA is outgoing and another tRNA is incoming.
large ribosomal subunit
P site
A site
small ribosomal subunit
U A C
A U G
1. Initiation. The small ribosomal subunit, the mRNA, the first tRNA amino acid, and the large ribosomal subunit come together.
free polypeptide
A U G
3. Termination. When the ribosome reaches a stop codon, all participants separate and the polypeptide is released.

Figure 21.10 Formation of the polypeptide during translation.
Polypeptide synthesis takes place at a ribosome and has three steps: (**1**) initiation, (**2**) elongation, and (**3**) termination.

translation complex, gives every cell the ability to make sufficient amounts of proteins.

Review of Gene Expression

DNA in the nucleus contains genes that are transcribed into RNAs. Some of these RNAs are mRNAs that will then be translated into proteins. During transcription, a segment of a DNA strand (a gene) serves as a template for the formation of RNA. The bases in RNA are complementary to those in DNA. In mRNA, every three bases is a *codon* for a certain amino acid (Fig. 21.12 and Table 21.3). Messenger RNA is processed before it leaves the nucleus. During

Table 21.3 Participants in Gene Expression

Name of Molecule	Special Significance	Definition
DNA	Genetic information	Sequence of DNA bases
mRNA	Codons	Sequence of three RNA bases complementary to DNA
tRNA	Anticodon	Sequence of three RNA bases complementary to codon
rRNA	Ribosome	Site of protein synthesis
Amino acid	Building block for protein	Transported to ribosome by tRNA
Protein	Enzyme, structural protein, or secretory product	Amino acids joined in a predetermined order

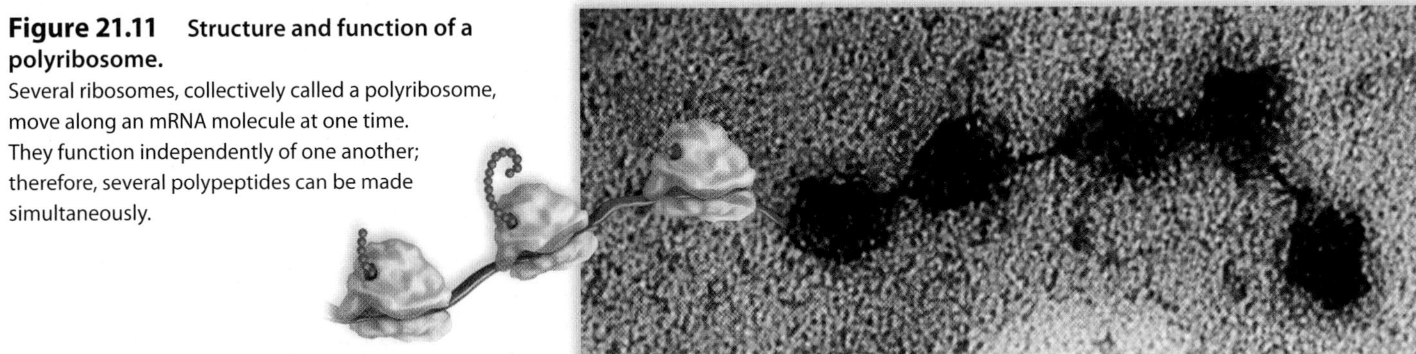

Figure 21.11 Structure and function of a polyribosome.
Several ribosomes, collectively called a polyribosome, move along an mRNA molecule at one time. They function independently of one another; therefore, several polypeptides can be made simultaneously.

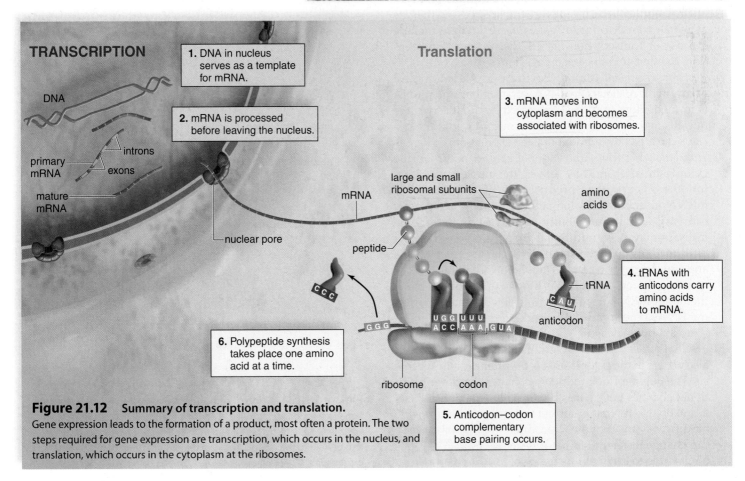

TRANSCRIPTION
1. DNA in nucleus serves as a template for mRNA.
2. mRNA is processed before leaving the nucleus.
DNA
introns
primary mRNA
exons
mature mRNA
nuclear pore
Translation
3. mRNA moves into cytoplasm and becomes associated with ribosomes.
mRNA
large and small ribosomal subunits
amino acids
peptide
4. tRNAs with anticodons carry amino acids to mRNA.
tRNA
anticodon
6. Polypeptide synthesis takes place one amino acid at a time.
ribosome
codon
5. Anticodon–codon complementary base pairing occurs.

Figure 21.12 Summary of transcription and translation.
Gene expression leads to the formation of a product, most often a protein. The two steps required for gene expression are transcription, which occurs in the nucleus, and translation, which occurs in the cytoplasm at the ribosomes.

processing, the introns are removed and the ends are modified. The mRNA carries a sequence of codons to the *ribosomes,* composed of rRNA and proteins. A tRNA bonded to a particular amino acid has an *anticodon* that pairs with a codon in mRNA. During translation, tRNAs and their attached amino acids arrive at the ribosomes. The linear sequence of codons of mRNA determines the order in which amino acids become incorporated into a protein.

The Regulation of Gene Expression

All cells receive a copy of all genes. However, cells differ as to which genes are actively expressed. Muscle cells, for example, have a different set of genes that are turned on in the nucleus and different proteins that are active in the cytoplasm than do nerve cells. A variety of mechanisms regulate gene expression, from transcription to protein activity, in our cells (Fig. 21.13). These mechanisms can be grouped under five primary levels of control—three that pertain to transcription in the nucleus and two that pertain to translation in the cytoplasm:

1. *Pretranscriptional control:* In the nucleus, the DNA must be available to the enzymes necessary for transcription. The chromosome in the region must decondense, or uncoil. Proteins and chemical modifications that protect the DNA must be removed before transcription can begin.
2. *Transcriptional control:* In the nucleus, a number of mechanisms regulate which genes are transcribed and the rate at which transcription of genes occurs. These include the use of transcription factors that initiate transcription, the first step in gene expression.
3. *Posttranscriptional control:* Posttranscriptional control occurs in the nucleus after DNA is transcribed and mRNA is formed. How mRNA is processed before it leaves the nucleus and also how fast mature mRNA leaves the nucleus can affect the amount of gene expression.
4. *Translational control:* Translational control occurs in the cytoplasm after mRNA leaves the nucleus and before there is a protein product. The life expectancy of mRNA molecules (how long they exist in the cytoplasm) can vary, as can their ability to bind ribosomes. Some mRNAs may need additional changes before they are translated. Two of the small RNA classes, microRNAs and small interfering RNAs, are involved at this level of gene expression. MicroRNAs bind to mRNAs and block the target mRNA from being translated. Small interfering RNAs bind to mRNAs, marking the mRNA for destruction by nucleases.
5. *Posttranslational control:* Posttranslational control, which also occurs in the cytoplasm, occurs after protein synthesis. The polypeptide product may have to undergo additional changes before it is biologically functional. Also, a functional enzyme is subject to feedback control—the binding of an enzyme's product can change its shape so that it is no longer able to carry out its reaction.

Animation Control of Gene Expression in Eukaryotes

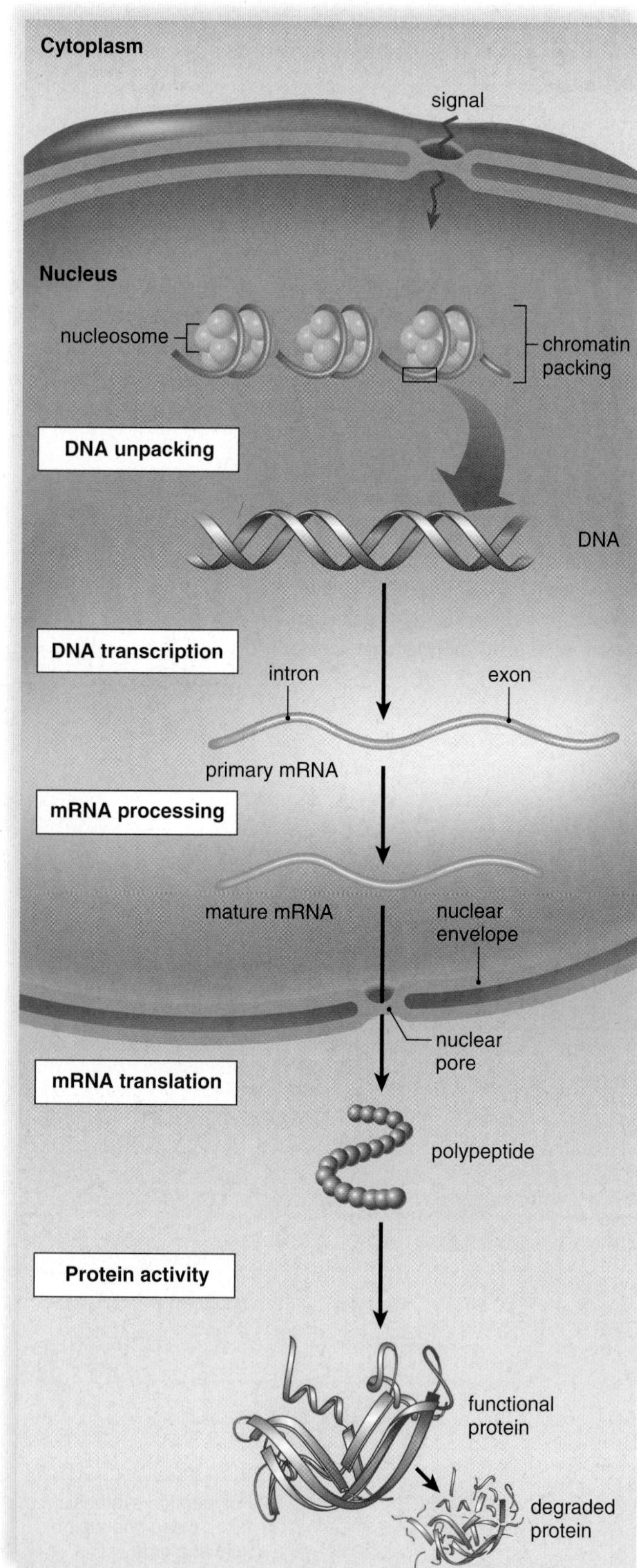

Figure 21.13 Control of gene expression in eukaryotic cells. Gene expression is controlled at various levels in eukaryotic cells. There are three mechanisms pertaining to the nucleus and two mechanisms that pertain to the cytoplasm. An external signal (red) may also alter gene expression.

Connections and Misconceptions

Why is the genetic code being universal important?

With a few exceptions, the genetic code is universal, meaning that the amino acid a particular codon codes for in a human is the same amino acid it will code for in a monkey, a fern, or a flea. A notable exception is the codon UGA, which in some protists and in the mitochondria of the cell, codes for the amino acid tryptophan, instead of a stop codon. This universality among living organisms suggests a common evolutionary heritage. The genetic code supports the underlying concept in biology that all living organisms are related.

Transcription Factors

In human cells, **transcription factors** are DNA-binding proteins. Every cell contains many different types of transcription factors. A specific combination of transcription factors and modifiers is believed to regulate the activity of any particular gene. After the right combination binds to DNA, an RNA polymerase attaches to the DNA and begins the process of transcription.

Animation Transcription Factors

As cells mature, they differentiate and become specialized. Specialization is determined by which genes are active and, therefore, perhaps, by which transcription factors are active in that cell. Signals received from inside and outside the cell could turn on or off genes that code for certain transcription factors. For example, in an embryo, the hand is flattened and paddle-shaped. The five fingers are joined by webs of skin and connective tissue. The gene for apoptosis (programmed cell death) is turned on in the embryo's hand. Separated fingers are formed when the webs of tissue die off.

Connecting the Concepts

For more information on DNA, RNA, and proteins, refer to the following discussions.

Section 2.6 examines the structure of proteins and their role in the body.

Section 2.7 describes the role of DNA and RNA as organic molecules.

Check Your Progress 21.2

1. Describe the processes of transcription and translation.
2. Discuss the genetic code and explain how it works with different types of RNA to make a protein. Evaluate the significance of the genetic code being universal.
3. Identify the various means of gene regulation and tell why they are important to homeostasis.

21.3 DNA Technology

Learning Outcomes

Upon completion of this section, you should be able to

1. Recognize the importance of DNA sequencing to the study of biology.
2. State the purpose of the polymerase chain reaction and DNA cloning.
3. Summarize some of the products that have been produced using biotechnology.

DNA Sequencing

The Human Genome Project was made possible by a research procedure called DNA sequencing. In this procedure, the order of nucleotides in a segment of DNA is determined. When DNA technology was in its inception back in the early 1970s, this technique was performed manually using dye-terminator substances or radioactive tracer elements attached to each of the four nucleotides during DNA replication, with results being deciphered from their pattern on a gel plate. Modern day sequencing normally involves dyes attached to the nucleotides and use of a laser to detect the different dyes by an automated sequencing machine, which shows the order of nucleotides on a grid called an electropheragram (Fig. 21.14). DNA sequencing is currently used to aid in medical research, pharmaceutical developments, molecular biology research, and forensic biology. To begin

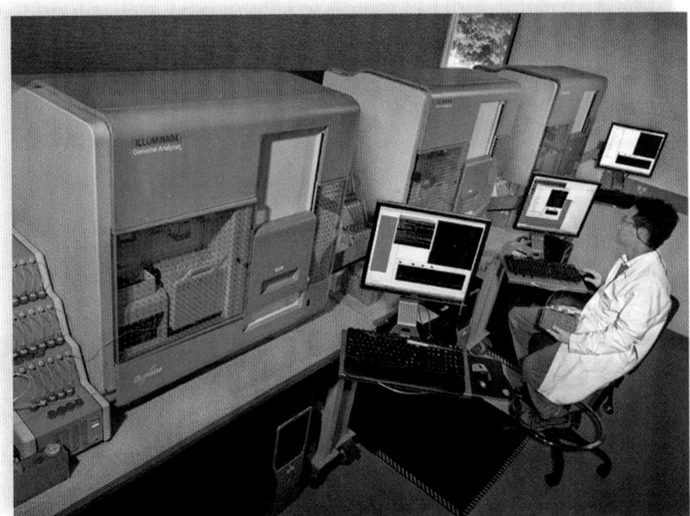

240 250 260
TTAAGTGAATTTAGGTGGACAAGACACAAGTCTA
TTAAGTGAATTTAGGTGGACAAGACACAAGTCTA

Small section of *Arabidopsis* genome

Figure 21.14 Automated DNA sequencer and an electropheragram. This typical automated DNA sequencer can sequence 1,000 base-pair sections of DNA in a matter of several hours.

for example. So, the quest is on to possibly discover how the regulation of genes explains why we have one set of traits and mice have another set, despite the similarity of our base sequences. One possibility is alternative gene splicing. We may differ from mice in the types of proteins we manufacture and/or by when and where certain proteins are present.

Proteomics and Bioinformatics

Proteomics is the study of the structure, function, and interaction of cellular proteins. Many of our genes are translated into proteins at some time, in some of our cells. The translation of all coding genes results in a collection of proteins called the human proteome. The analysis of proteomes is more challenging than the analysis of genomes. A single gene can code for more than 1,000 different proteins, and protein concentrations can differ widely in cells. Researchers have to be able to identify proteins, regardless of whether there is one or thousands of copies of a protein in a cell. Any particular protein differs minute by minute in concentration, interactions, cellular location, and chemical modifications, among other features. Yet, to understand a protein, all these features must be analyzed. Computer modeling of the three-dimensional shape of cellular proteins is an important part of proteomics. The study of cellular proteins and how they function is essential to understanding the causes of certain diseases and disorders. This study is also important in the discovery of better drugs, because most drugs are proteins or molecules that affect the function of proteins.

Bioinformatics is the application of computer technologies to the study of the genome. Specifically, it is the process of creating databases of information, then mapping and analyzing the information gained from DNA sequencing and proteomics. Genomics and proteomics produce raw data. These fields depend on computer analysis to find significant patterns in the data. As a result of bioinformatics, scientists hope to find cause-and-effect relationships between various genetic profiles and genetic disorders caused by multifactorial genes. By correlating any sequence changes with resulting phenotypes, one current focus of bioinformatics research is discovering if noncoding regions of the genome do have functions and, if so, what effect those functions may have on homeostasis.

A Person's Genome Can Be Modified

Gene therapy is the insertion of genetic material into human cells for the treatment of a disorder. Gene therapy has been used to cure inborn errors of metabolism and also to treat more generalized disorders such as cardiovascular disease and cancer. Most recently, a 2008 clinical trial showed that gene therapy could successfully treat a type of inherited blindness. Both ex vivo (outside the body) and in vivo (inside the body) gene therapy methods are used.

Ex Vivo Gene Therapy

Figure 21.24 describes the methodology for treating children who have severe combined immunodeficiency disease

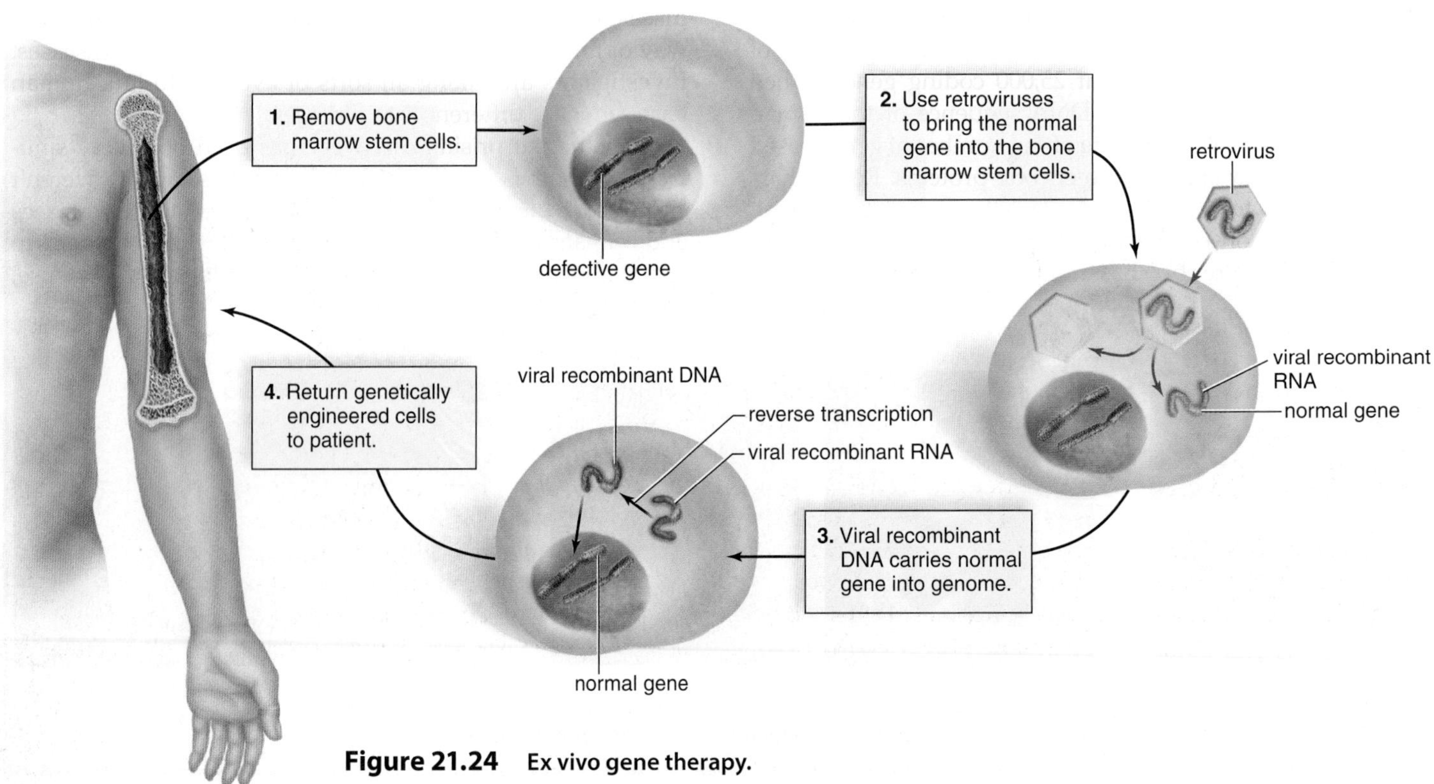
1. Remove bone marrow stem cells.
defective gene
2. Use retroviruses to bring the normal gene into the bone marrow stem cells.
retrovirus
viral recombinant RNA
normal gene
3. Viral recombinant DNA carries normal gene into genome.
viral recombinant DNA
reverse transcription
viral recombinant RNA
normal gene
4. Return genetically engineered cells to patient.

Figure 21.24 Ex vivo gene therapy.

(SCID, see Chapter 7). These children lack the enzyme adenosine deaminase (ADA), which is involved in the maturation of T and B cells. To carry out ex vivo gene therapy, stem cells are removed from bone marrow. These cells are mixed with a normal gene for the enzyme that is contained in a vector, or "carrier" molecule. The most successful vectors are found to be viruses that have been altered for safety. The combined cells are then returned to the patient. Bone marrow stem cells are preferred for this procedure because they divide to produce more cells with the same genes. SCID patients who have undergone this procedure show significantly improved immune function and a sustained rise in ADA enzyme activity in the blood. Ex vivo gene therapy has also been proven effective for hemophilia A, Alzheimer disease, Parkinson disease, Crohn's disease, and certain cancers.

In Vivo Gene Therapy

Cystic fibrosis patients lack a gene that codes for the transmembrane carrier of the chloride ion. They often suffer from numerous and potentially deadly infections of the respiratory tract. For in vivo gene therapy, the therapeutic DNA is injected straight into the body cells. As in ex vivo therapy, the use of a vector, or carrier molecule, is needed. For in vivo therapy, a retrovirus or adenovirus is used as the vector that carries the corrective gene. For CF patients, the adenovirus with the functioning gene is either sprayed into the nose or injected into the lower respiratory tract. This type of gene therapy is minimally invasive and has quick results, but the effects are not as long lasting as ex vivo therapy. In vivo gene therapy has been used for cardiovascular diseases, endocrine disorders, and Huntington disease.

Problems

Gene therapy is still considered experimental. It has not yet been approved by the U.S. Food and Drug Administration (FDA). Research and clinical trials are currently being done to gain the overall FDA approval for widespread use. Most of the issues come from guaranteeing the safety of the vectors used on the patients. The use of a virus, even altered with no reproductive viral DNA remaining, can sometimes stimulate the immune system and cause an infection. Some patients have also had severe allergic reactions to both ex vivo and in vivo gene therapy. Researchers are continuing to modify vectors to make gene therapy a safe and effective treatment option.

Connecting the Concepts

The study of genomics and proteomics plays an important role in our understanding of evolutionary change. For more information, refer to the following discussions.

Section 22.2 examines how biochemical evidence, including the analysis of DNA, helps us to understand biological evolution.

Section 22.3 explores how DNA evidence enables a deeper understanding of human evolution.

Check Your Progress 21.4

1. Describe how the information collected in the Human Genome Project affects everyday life.
2. Discuss what the genetic code being universal has to do with proteomics.
3. Explain why gene therapy would be used over medication and list some advantages and disadvantages.

CASE STUDY CONCLUSION

Dr. Weber explained that insulin for diabetes patients has been made since the late 1970s in large vats called bioreactors. Here, a non-disease-producing strain of *Escherichia coli* has been made that contains the human gene for the production of insulin through recombinant DNA technology. In the bioreactor with the recombinant cells is also a medium that contains a food source for the bacteria. The bacteria stay alive and make billions of copies of themselves, while at the same time producing human insulin. The insulin is retrieved from the medium and used for injections. Humalog and Humulin N are two very common types of medical insulin made in this manner. Kaya will have to monitor her blood sugar level several times per day using a glucose meter and, when appropriate, give herself injections of insulin to help regulate her levels. Insulin is injected because if it were in pill form and used orally, it would be inactivated by the digestive system. Instead, insulin is administered by intramuscular injection so it can enter directly into the blood stream. Advances in the treatment of diabetes have also included the use of an insulin pump. The pump is a small device worn on a belt or pocket that delivers fast-acting insulin into the body via an infusion set—a thin plastic tube ending in a small, flexible plastic cannula, or a very thin needle. The cannula is inserted beneath the skin at the infusion site, usually in the person's abdomen or upper buttocks. The infusion set is in place for two to three days (sometimes more), and then it is moved to a new location. The insulin pump is not an artificial pancreas; it is a computer-driven device that delivers fast-acting insulin in precise amounts at preprogrammed times. Under her doctor's supervision, Kaya will monitor her diabetes to discover the appropriate treatment options for her lifestyle and medical needs.

Media Study Tools

www.mhhe.com/maderhuman12e

Enhance your study of this chapter with study tools and practice tests. Also ask your instructor about the resources available through ConnectPlus, including the media-rich eBook, interactive learning tools, and animations.

Virtual Labs

The following virtual labs explore different uses of biotechnology in answering questions in the life sciences.

"Classifying Using Biotechnology" explores how biotechnology can be used to classify bacteria.

"Knocking Out Genes" examines how knocking out specific genes can create a better understanding of how plants adapt to specific environmental conditions.

"Gene Splicing" uses biotechnology techniques to generate transgenic organisms.

Summarizing the Concepts

21.1 DNA and RNA Structure and Function

DNA is the genetic material found in the chromosomes. It replicates, stores information, and mutates for genetic variability.

Structure of DNA

- DNA is a double helix composed of two polynucleotide strands. Each nucleotide is composed of a deoxyribose sugar, a phosphate, and a nitrogen-containing base (A, T, C, G).
- The base A is bonded to T, and G is bonded to C.

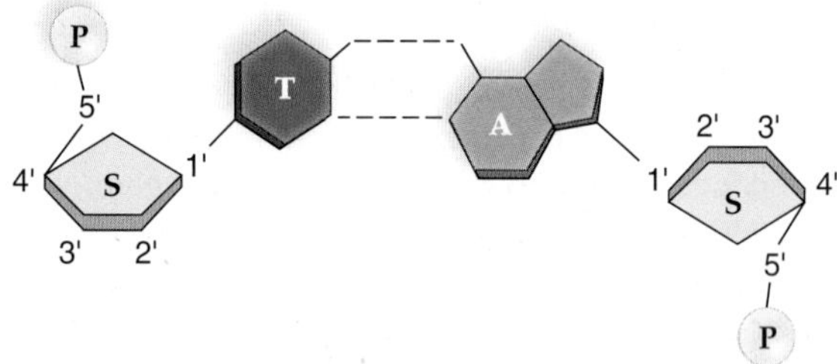

Replication of DNA

- DNA strands unzip, and a new complementary strand forms opposite each old strand, resulting in two identical DNA molecules:

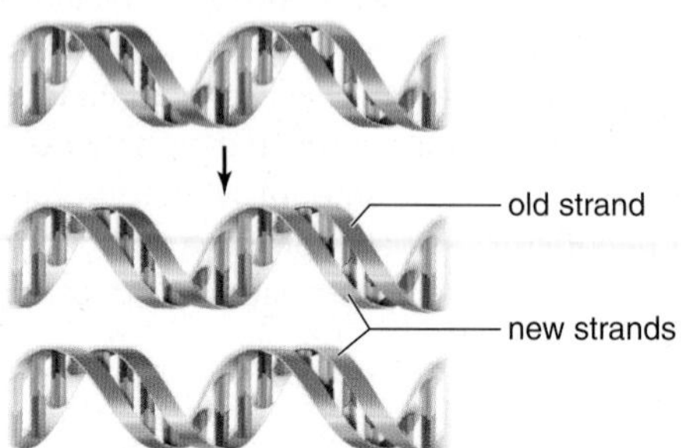

The Structure and Function of RNA

- RNA is a single-stranded nucleic acid in which the base U (uracil) occurs instead of T (thymine).
- The four forms of RNA are rRNA (found in the ribosomes); mRNA (carries the DNA message to the ribosomes); tRNA (transfers amino acids to the ribosomes where protein synthesis occurs); and small RNAs (control of gene expression).

21.2 Gene Expression

Gene expression leads to the formation of a product, either an RNA or a protein. Proteins differ by the sequence of their amino acids. Gene expression for proteins requires transcription and translation.

Transcription

Transcription occurs in the nucleus. The three-base DNA codon is passed to an mRNA that contains codons. Introns are removed from mRNA during mRNA processing.

Translation

Translation occurs in the cytoplasm at the ribosomes. tRNA molecules bind to their amino acids, and then their anticodons pair with mRNA codons.

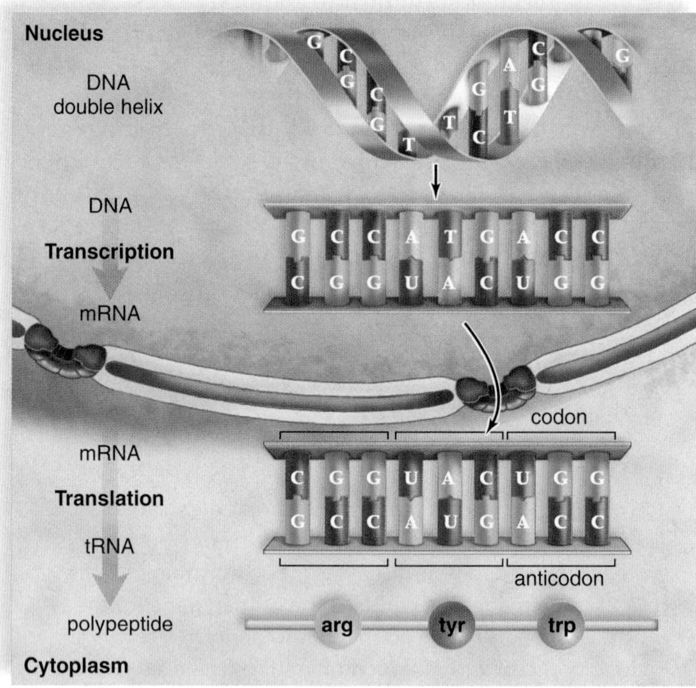

The Regulation of Gene Expression

Regulation of gene expression occurs at five levels in a human cell:

- *Pretranscriptional control:* In the nucleus; the DNA is made available to transcription factors and enzymes.
- *Transcriptional control:* In the nucleus; the degree to which a gene is transcribed into mRNA determines the amount of gene product.
- *Posttranscriptional control:* In the nucleus; involves mRNA processing and how fast mRNA leaves the nucleus.

- *Translational control:* In the cytoplasm; affects when translation begins and how long it continues. Includes inactivation and degradation of mRNA.
- *Posttranslational control:* In the cytoplasm; occurs after protein synthesis.

21.3 DNA Technology

- Recombinant DNA contains DNA from two different sources. The foreign gene and vector DNA are cut by the same restriction enzyme and then the foreign gene is sealed into vector DNA. Bacteria take up recombinant plasmids.
- PCR uses DNA polymerase to make multiple copies of a specific piece of DNA. Following PCR, DNA can be subjected to DNA fingerprinting.
- Transgenic organisms (bacteria, plants, and animals that have had a foreign gene inserted into them) can produce biotechnology products, such as hormones and vaccines.
- Transgenic bacteria can promote plant health, remove sulfur from coal, clean up toxic waste and oil spills, extract minerals, and produce chemicals.
- Transgenic crops can resist herbicides and pests.
- Transgenic animals can be given growth hormone to produce larger animals; can supply transplant organs; and can produce pharmaceuticals.

21.4 Genomics

The human genome has now been sequenced via the 13-year-long Human Genome Project.

- Genomes of other organisms have also been sequenced.

Functional and Comparative Genomics

- Functional genomics is the study of how the 20,500 genes in a human genome function.
- Comparative genomics is a way to determine how species have evolved and how genes and noncoding regions of the genome function.

Proteomics and Bioinformatics

- Proteomics is the study of the structure, function, and interaction of cellular proteins.
- Bioinformatics is the application of computer technologies to the study of the genome.

Understanding Key Terms

anticodon 499
bioinformatics 512
biotechnology product 506
cloning 506
codon 497
complementary DNA (cDNA) 506
complementary paired bases 492
DNA (deoxyribonucleic acid) 492
DNA ligase 506
DNA replication 493
double helix 492
functional genomics 511
gene cloning 506
genetic engineering 507
messenger RNA (mRNA) 495
mutation 494
plasmid 506
polymerase chain reaction (PCR) 504
polyribosome 500
proteomics 512
recombinant DNA (rDNA) 506
restriction enzyme 506
ribosomal RNA (rRNA) 495
RNA (ribonucleic acid) 494
RNA polymerase 498
small RNA 495
template 493
transcription 496
transcription factor 503
transfer RNA (tRNA) 495
transgenic organism 506
translation 496
vector 506
xenotransplantation 509

Match the key terms to these definitions.

a. ________ Cluster of ribosomes attached to the same mRNA molecule; each ribosome is producing a copy of the same polypeptide.

b. ________ Free-living organism in the environment that has had a foreign gene inserted into it.

c. ________ Pattern or guide used to make copies.

d. ________ The study of the structure, function, and interactions of cellular proteins.

e. ________ A means to transfer foreign genetic material into a cell.

Testing Your Knowledge of the Concepts

1. Explain why it is important that each new DNA helix is complementary to the parent DNA helix. (pages 492–494)
2. Describe the structure and function of the three types of RNA. (page 495)
3. Explain why the terms *transcription* and *translation* are appropriate for these processes. (pages 496–497)
4. If a DNA strand is TAC AAT AAA CGT GTC ATT, what are the codons of mRNA, the anticodons of tRNA, and the amino acid sequence? (pages 497–499)
5. What are the different types of genetic control and where does each one occur in the cell? (page 502)
6. Compare and contrast in vivo and ex vivo gene therapy. (pages 512–513)
7. What is the methodology for producing transgenic bacteria? (pages 506–509)
8. What is the polymerase chain reaction (PCR), and how is it used to produce multiple copies of a DNA segment? (page 504)
9. What are some biotechnology products from bacteria, plants, and animals; what are they used for? (pages 506–509)
10. The double-helix model of DNA resembles a twisted ladder in which the rungs of the ladder are
    a. complementary base pairs.
    b. A paired with G and C paired with T.
    c. A paired with T and G paired with C.
    d. a sugar–phosphate paired with a sugar–phosphate.
    e. Both a and c are correct.
11. The enzyme or enzymes responsible for adding new nucleotides to a growing DNA chain during DNA replication is/are
    a. helicase.
    b. RNA polymerase.
    c. DNA polymerase.
    d. ribozymes.

12. RNA processing
    a. removes the introns, leaving only the exons.
    b. is the same as transcription.
    c. is an event that occurs after RNA is transcribed.
    d. is the rejection of old, worn-out RNA.
    e. Both a and c are correct.
13. During protein synthesis, an anticodon of a tRNA pairs with
    a. amino acids in the polypeptide.
    b. DNA nucleotide bases.
    c. rRNA nucleotide bases.
    d. mRNA nucleotide bases.
14. Which of the following is involved in controlling gene expression?
    a. the occurrence of transcription
    b. activity of the polypeptide product
    c. life expectancy of the mRNA molecule in the cell
    d. All of these are involved.
15. Restriction enzymes found in bacterial cells are ordinarily used
    a. during DNA replication.
    b. to degrade the bacterial cell's DNA.
    c. to degrade viral DNA that enters the cell.
    d. to attach pieces of DNA together.
16. Which of the following is a benefit to having insulin produced by biotechnology?
    a. It is just as effective.
    b. It can be mass-produced.
    c. It is nonallergenic.
    d. It is less expensive.
    e. All of these are correct.
17. Following is a segment of a DNA molecule. (Remember that only one strand is transcribed.) What are (a) the mRNA codons, (b) the tRNA anticodons, and (c) the sequence of amino acids?

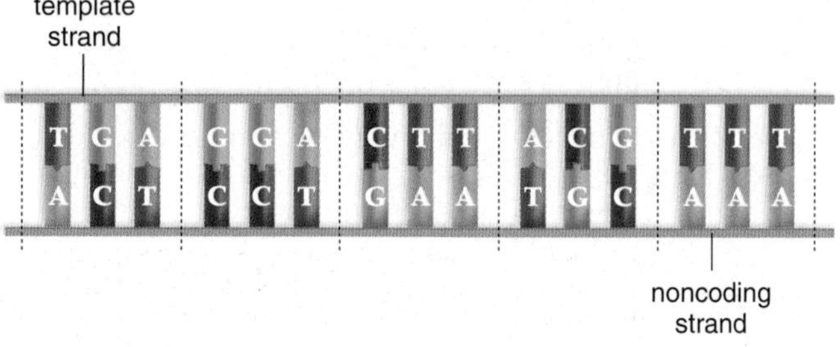

Thinking Critically About the Concepts

The advances in biotechnology since the 1970s have made a huge impact on medicine and the treatment and management of certain diseases. The ability to combine DNA from two different organisms has given us the means to produce substances the body needs, as seen in the case study with Kaya using *E.coli*–derived human insulin. It has also given doctors the ability to insert properly functioning genes into patients lacking those genes. With certain advances in biotechnology, ethical issues arise. Think about both sides when answering the following questions.

1. What are the advantages of using a recombinant DNA human product instead of a product isolated from another organism? (For example, what advantage would there be to using human recombinant insulin versus insulin produced from cows or pigs?)
2. Are there any disadvantages to using a recombinant DNA product? (Hint: You may want to think about how the product is produced and then purified.)
3. Recombinant human growth hormone is available for children and adults with growth hormone deficiency.
    a. When would use of the hormone be appropriate? Under what circumstances would hormone use be improper?
    b. How would a physician know that the hormone treatment was completely effective? What further information might you need to answer this question?
4. In general, diseases caused by a single protein deficiency are more easily treated with a recombinant DNA-produced product. Yet certain cancers can be treated in this manner. What aspects of cancer could be treated with a recombinant DNA-produced product?
5. What types of regulations should be placed on advances in DNA technology and who should be responsible for these regulations?

CHAPTER 22

Human Evolution

CASE STUDY THE NEANDERTALS

For years, Neandertals have been perceived as an offshoot of human evolution, a branch of the evolutionary tree that was only distantly connected with the evolution of modern *Homo sapiens*. However, that changed in 2010 when researchers from the Plank Institute for Evolutionary Anthropology released their first-draft sequence of the Neandertal genome. To obtain the DNA, the researchers ground up bone fragments from three Neandertal skeletons that were between 38,000 and 45,000 years old. Recall from Chapter 11 that bone consists of cells (osteocytes) surrounded by a nonliving matrix. By grinding the bone, the researchers were able to get at the small amounts of DNA located within the osteocytes. These DNA samples were then very carefully processed so as to avoid contamination by DNA from the human research team. The team then compared the Neandertal sequences to five genomes of modern humans living across the globe. The results surprised even the researchers. Despite the fact that Neandertals and modern *Homo sapiens* diverged from a common ancestor a brief 500,000 years ago, the comparison indicated that there were at least 15 areas of the *Homo sapiens* genome that showed evidence of selection following the Neandertal–*Homo sapiens* split. Some of the areas of the genome that showed the most change were in cranial development (size of the skull), metabolic functions, and cognitive (thinking) abilities. In many ways, this landmark study provided some of the first direct evidence of selection and evolution in our species, and the results are already changing the way that scientists view our species' place among our ancestors.

As you read through the chapter, think about the following questions.

1. What are some of the known differences between Neandertals and the early Cro-Magnons?
2. What is the current thinking on the evolutionary relationship of the Neandertals to modern humans?
3. What is the difference between the "out-of-Africa" and the "multiregional" hypotheses on the evolution of modern humans?

CHAPTER CONCEPTS

22.1 Origin of Life
Data suggest that chemical evolution produced the first cell.

22.2 Biological Evolution
Descent from a common ancestor explains the unity of living things. For example, all living things have a cellular structure and a common chemistry from their common ancestor. Adaptation to different environments explains the great diversity of living things.

22.3 Classification of Humans
The classification of humans can be used to trace the ancestry of humans.

22.4 Evolution of Hominins
Modern humans evolved from a group of hominins called the Australopithecines.

22.5 Evolution of Humans
Among the members of the genus *Homo, Homo habilis* was the first to make tools. *Homo erectus* was the first to have the use of fire. Neandertals were highly adapted to life in cold climates. Cro-Magnons resembled modern humans in appearance.

BEFORE YOU BEGIN

Before beginning this chapter, take a few moments to review the following discussions.

Section 1.1 What is the definition of *evolution*?

Section 2.7 What are the differences between RNA and DNA?

Section 3.1 What are the basic principles of the cell theory?

22.1 Origin of Life

Learning Outcomes

Upon completion of this section, you should be able to

1. Describe the conditions of the early Earth's atmosphere.
2. Discuss how the first organic materials may have arisen on the planet.
3. Distinguish between the RNA-first and protein-first hypotheses.

Our study of evolution begins with the origin of life. A fundamental principle of biology, the **cell theory**, states that all living things are made of cells and that every cell comes from a pre-existing cell. But if this is so, how did the first cell come about? It was the very first living thing, so it had to come from nonliving chemicals. Could there have been a slow increase in the complexity of chemicals? Could a **chemical evolution** have produced the first cells or cells on the primitive Earth?

The Primitive Earth

The sun and the planets, including Earth, probably formed over a 10-billion-year period from aggregates of dust particles and debris. At 4.6 billion years ago (BYA), the solar system was in place. Dense silicate minerals became the semiliquid mantle.

The Earth's mass is such that the gravitational field is strong enough to have an atmosphere. If the Earth had had less mass, atmospheric gases would have escaped into outer space. The early Earth's atmosphere was not the same as today's atmosphere. Most likely, the first atmosphere was formed by gases escaping from volcanoes. If so, the primitive atmosphere would have consisted mostly of water vapor (H_2O), nitrogen (N_2), and carbon dioxide (CO_2), with only small amounts of hydrogen (H_2) and carbon monoxide (CO). The primitive atmosphere had little, if any, free oxygen.

At first, the Earth and its atmosphere were extremely hot. Water, existing only as a gas, formed dense, thick clouds. Then, as the Earth cooled, water vapor condensed to liquid water, and rain began to fall. Rain fell in such enormous quantities over hundreds of millions of years that the oceans of the world were produced.

Small Organic Molecules

The rain washed the other gases, such as N_2 and CO_2, into the oceans (Fig. 22.1*a*). The primitive Earth had many sources of

a. The primitive atmosphere contained gases, including H_2O, CO_2, and N_2, that escaped from volcanoes. As the water vapor cooled, some gases were washed into the oceans by rain.

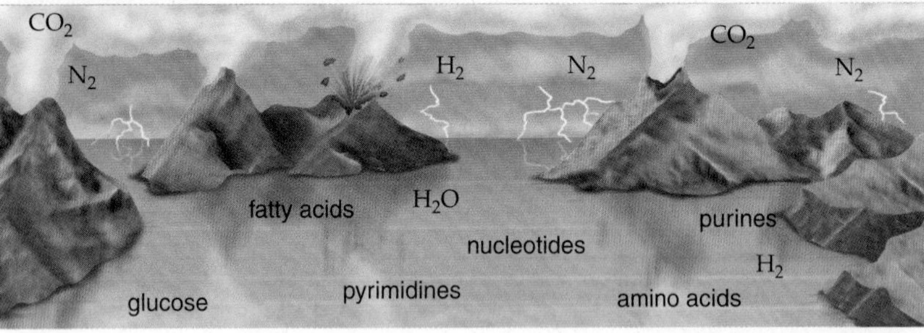

b. The availability of energy from volcanic eruption and lightning allowed gases to form small organic molecules, such as nucleotides and amino acids.

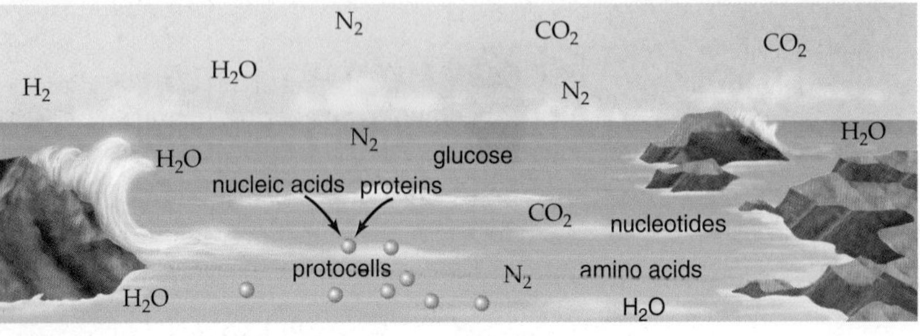

c. Small organic molecules could have joined to form proteins and nucleic acids, which became incorporated into membrane-bound spheres. The spheres became the first cells, called protocells. Later protocells became true cells that could reproduce.

Figure 22.1 Origins of early life on Earth.
A chemical evolution could have produced the protocell, which became a true cell once it had genes composed of DNA and could reproduce.

energy. These included volcanoes, meteorites, radioactive isotopes, lightning, and ultraviolet radiation. In the presence of so much available energy, the primitive gases may have reacted with one another. This may have produced small organic compounds, such as nucleotides and amino acids (Fig. 22.1*b*). In 1953, Stanley Miller (1930–2007) performed an experiment (illustrated in Fig. 22.2). To simulate the Earth's early environment, Miller placed the inorganic materials believed to have been present on the early Earth in a closed system, heated it, and circulated it past an electrical spark. After a week, the solution contained a variety of amino acids and organic compounds. This and other similar experiments support the hypothesis that inorganic chemicals may form organic molecules in the presence of a strong energy source, even if oxygen is not present.

Animation
Miller-Urey Experiments

Macromolecules

The newly formed small organic molecules likely joined to produce organic macromolecules (Fig. 22.1*c*). There are two hypotheses of special interest concerning this stage in the origin of life. One is the **RNA-first hypothesis.** This hypothesis suggests that only the macromolecule RNA (ribonucleic acid) was needed at this time to progress toward formation of the first cell or cells. This hypothesis was formulated after the discovery that RNA can sometimes be both a substrate and an enzyme during RNA processing (see Fig. 21.9). At that time, the splicing of mRNA to remove introns was done by a complex composed of both RNA and protein. The RNA, not the protein, is the enzyme. RNA enzymes are called ribozymes. Then, too, ribosomes where protein synthesis occurs contain rRNA. Perhaps RNA, then, could have carried out the processes of life commonly associated today with DNA (deoxyribonucleic acid) and proteins. Scientists who support this hypothesis are fond of saying that it was an "RNA world" some 3.5 BYA. However, DNA, being a double helix, is more stable than RNA. In the RNA-first hypothesis, RNA would have served as the first genetic material, with a transition to DNA occurring later. As far as we currently know, all organisms on the planet use DNA as their genetic material, but scientists continue to look for evidence to support the RNA-first hypothesis.

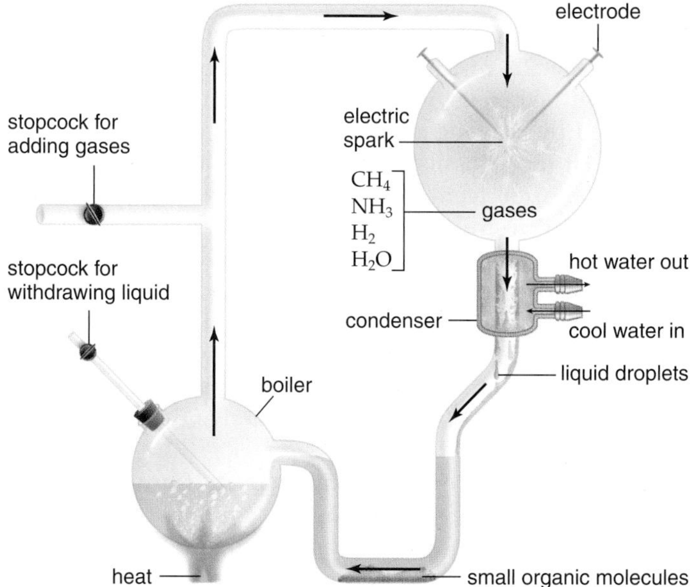

Figure 22.2 The Stanley Miller experiment.
Gases thought to be present early in the Earth's atmosphere were admitted to the apparatus, circulated past an energy source (electrical spark), and cooled to produce a liquid that could be withdrawn. Upon chemical analysis, the liquid was found to contain various small organic molecules.

Another hypothesis is termed the **protein-first hypothesis.** Sidney Fox (1912–1998), an American biochemist, demonstrated that amino acids join together when exposed to dry heat. He suggested that amino acids collected in shallow puddles along the rocky shore. The heat of the sun caused them to form proteinoids, small polypeptides that have some catalytic properties. When proteinoids are returned to water, they form microspheres. Microspheres are structures composed only of protein that have many of the properties of a cell.

The Protocell

A cell has a lipid–protein membrane. Fox demonstrated that if lipids are made available to microspheres, the two tend to become associated, producing a lipid–protein membrane. A **protocell,** which could carry on metabolism but could not reproduce, could have come into existence in this manner.

The protocell would have been able to use the still-abundant small organic molecules in the ocean as food. Therefore, the protocell was, most likely, a **heterotroph,** an organism that takes in preformed food. Further, the protocell would have been a fermenter, because there was no free oxygen.

The True Cell

A true cell can reproduce. In today's cells, DNA replicates before cell division occurs. Enzymatic proteins carry out the replication process.

How did the first cell acquire both DNA and enzymatic proteins? Scientists who support the RNA-first hypothesis propose a series of steps. According to this hypothesis, the first cell had RNA genes that, like messenger RNA, could have specified protein synthesis. Some of the proteins formed would have been enzymes. Perhaps one of these enzymes, such as reverse transcriptase found in retroviruses, could use RNA as a template to form DNA. Replication of DNA would then proceed normally.

By contrast, supporters of the protein-first hypothesis suggest that some of the proteins in the protocell would have evolved the enzymatic ability to synthesize DNA from nucleotides in the ocean. Then, DNA would have gone on to specify protein synthesis; in this way, the cell could have acquired all of its enzymes, even the ones that replicated DNA.

Connecting the Concepts

For more information on the topics presented in this section, refer to the following discussions.

Section 3.1 describes the principles of the cell theory.

Section 3.3 examines the structure of the plasma membrane of cells.

Section 21.1 describes the structure and function of RNA.

Check Your Progress 22.1

1. Compare and contrast the composition of the early Earth's atmosphere with that of the modern atmosphere.
2. Explain what the Miller experiments demonstrated about the formation of the first organic molecules.
3. Discuss the importance of RNA in the formation of the first protocells.

22.2 Biological Evolution

Learning Outcomes

Upon completion of this section, you should be able to

1. Explain the relationship between adaptation and the process of biological evolution.
2. Discuss how the fossil record, biogeography, and anatomical and biochemical evidence all support the concept of biological evolution.
3. Distinguish between homologous and analogous structure.
4. Describe how the process of natural selection supports the concept of biological evolution.

The first true cells were the simplest of life-forms. Therefore, they must have been **prokaryotic cells,** which lack a nucleus. Later, **eukaryotic cells** (protists), which have nuclei, evolved. Then, multicellularity and the other kingdoms (fungi, plants, and animals) evolved (see Fig. 1.5). Obviously, all these types of organisms—even prokaryotic cells—are alive today. Each type of organism has its own evolutionary history that is traceable back to the first cell or cells.

Biological evolution is the process by which a species changes through time. Biological evolution has two important aspects: descent from a common ancestor and adaptation to the environment. Descent from the original cell or cells explains why all living things have a common chemistry and a cellular structure. An **adaptation** is a characteristic that makes an organism able to survive and reproduce in its environment. Adaptations to different environments help explain the diversity of life—why there are so many different types of living things.

Common Descent

Charles Darwin was an English naturalist who first formulated the theory of evolution that has since been supported by so much independent data. At the age of 22, Darwin sailed around the world as the naturalist on board the HMS *Beagle.* Between 1831 and 1836, the ship sailed in the tropics of the Southern Hemisphere. There, life-forms are more abundant and varied than in Darwin's native England.

Even though it was not his original intent, Darwin began to gather evidence that life-forms change over time and from place to place. The types of evidence that convinced Darwin that common descent occurs were fossil, anatomical, and biogeographical.

Fossil Evidence Supports Evolution

Fossils are the best evidence for evolution. They are the actual remains of species that lived on Earth at least 10,000 years ago and up to billions of years ago. Fossils can be the traces of past life or any other direct evidence that past life existed. Traces include trails, footprints, burrows, worm casts, or even preserved droppings. Fossils can also be such items as pieces of bone, impressions of plants pressed into shale, and even insects trapped in tree resin (which we know as amber). Most fossils, however, are found embedded in or recently eroded from sedimentary rock. Sedimentation, a process that has been going on since the Earth was formed, can take place on land or in bodies of water. Weathering and erosion of rocks produce an accumulation of particles. These particles vary in size and nature and are called sediment. Sediment becomes a stratum (pl., strata), a recognizable layer in a sequence of layers. Any given stratum is older than the one above it and younger than the one immediately below it (Fig. 22.3*a*). This allows fossils to be dated.

Usually, when an organism dies, the soft parts are either consumed by scavengers or decomposed by bacteria. This means that most fossils consist only of hard parts such as shells, bones, or teeth (Fig. 22.3*b*). These are usually not consumed or destroyed. When a fossil is found encased by rock, the remains were first buried in sediment. The hard parts were preserved by a process called mineralization. Finally, the surrounding sediment hardened to form rock. Subsequently, the fossil has to be found by a human. Most estimates suggest that less than 1% of past species have been preserved as fossils. Only a small fraction of these have been found.

More and more fossils have been found because researchers, called paleontologists, and their assistants have been out in the field looking for them. Usually, paleontologists remove fossils from the strata to study them in the laboratory. Then they may decide to exhibit them. The **fossil record** is the history of life recorded by fossils. *Paleontology* is the science of discovering the fossil record. Decisions about the history of life, ancient climates, and environments can be made using the fossil record. The

Figure 22.3 Fossils and strata.
a. Due to erosion, it is often possible to see a number of strata, layers of rock or sedimentary material that contain fossils. **b.** A fossil snake dated 90 MYA.

fossil record is the most direct evidence we have that evolution has occurred. The species found in ancient sedimentary rock are not the species we see today.

Darwin relied on fossils to formulate his theory of evolution. Today, we have a far more complete record than was available to Darwin. The record is complete enough to tell us that, in general, life has progressed from the simple to the complex. Unicellular prokaryotes are the first signs of life in the fossil record. Unicellular eukaryotes and then multicellular eukaryotes followed. Among the latter, fishes evolved before terrestrial plants and animals. On land, nonflowering plants preceded the flowering plants. Amphibians preceded the reptiles, including the dinosaurs. Dinosaurs are directly linked to the birds, but they are only indirectly linked to the evolution of mammals—including humans.

Transitional fossils are those that have characteristics of two different groups. In particular, they tell us who is related to whom and how evolution occurred. Even in Darwin's day, scientists knew of the *Archaeopteryx* fossils, which are intermediate to reptiles and birds. The dinosaur-like skeleton of these fossils had reptilian features, including jaws with teeth, and a long, jointed tail. But *Archaeopteryx* also had feathers and wings. Figure 22.4 not only shows a fossil

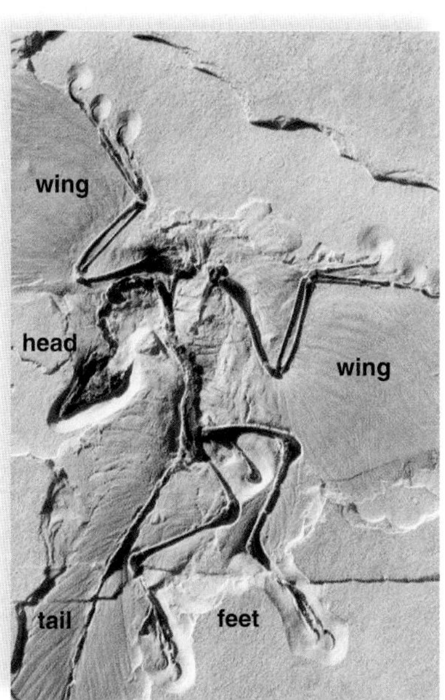

Archaeopteryx fossil

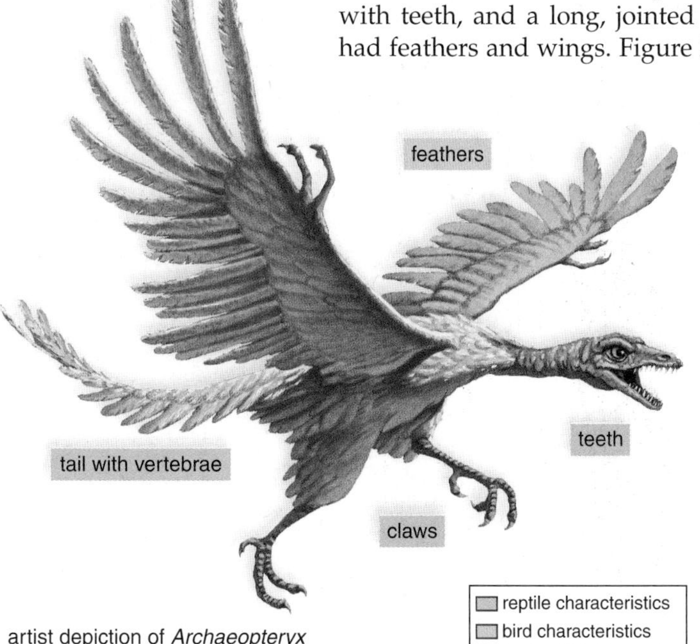

artist depiction of *Archaeopteryx*

Figure 22.4 ***Archaeopteryx* is an example of a transitional fossil.**
Archaeopteryx had a combination of reptilian and bird characteristics.

of *Archaeopteryx* but also gives us an artist's representation of the animal based on the fossil remains. Many more transitional fossils have been discovered in China.

Another example of how transitional fossils can be used to trace the evolutionary history of an organism is the whale. It had always been thought that whales had terrestrial ancestors. Now, fossils have been discovered that support this hypothesis (Fig. 22.5). *Ambulocetus natans* ("the walking whale that swims") was the size of a large sea lion, with broad webbed feet on both fore- and hindlimbs. This animal could both walk and swim. It also had tiny hooves on its toes and the primitive skull and teeth of early whales. It is believed that *Ambulocetus* was a predator that patrolled freshwater streams looking for prey.

The origin of land mammals is also well documented. The synapsids are mammal-like reptiles whose descendants were wolflike and bearlike predators, as well as several types of piglike herbivores. Slowly, mammalian-like fossils acquired features, such as a palate, that would have enabled them to breathe and eat at the same time. They also acquired a muscular diaphragm and rib cage that would have helped them breathe efficiently. The earliest true mammals were shrew-size creatures found in fossil beds about 200 million years old.

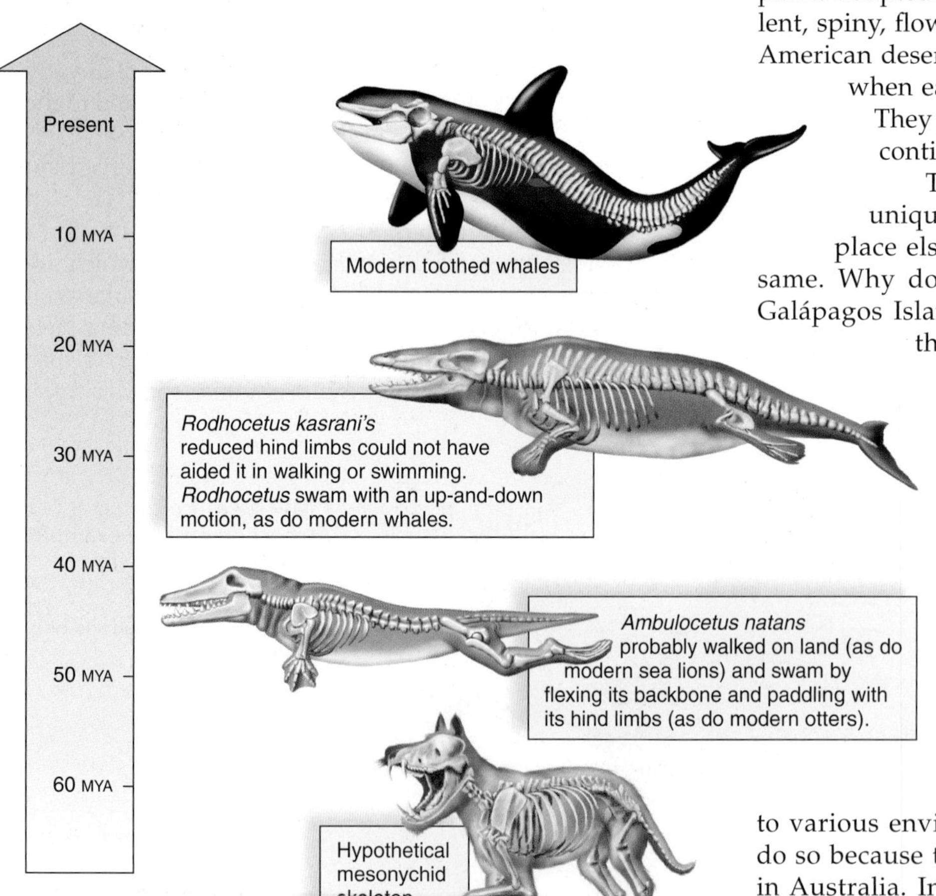

Figure 22.5 Evolution of the whales.
The discovery of *Ambulocetus* has filled in the gaps in the evolution of whales.

Other Evidence Supports Evolution

Many different types of evidence support the hypothesis that organisms are related through common descent. Darwin cited much of the evidence to support his theory of evolution. However, he had no knowledge of the genetic and biochemical data that became available after his time.

Biogeographical Evidence **Biogeography** is the study of the distribution of plants and animals in different places throughout the world. Such distributions are consistent with the hypothesis that life-forms evolved in a particular locale before they spread out. Therefore, you would expect a different mix of plants and animals whenever geography separates continents, islands, or seas. For example, Darwin noted that South Africa lacked rabbits, even though the environment was suitable for them. He concluded that no rabbits lived in South America because rabbits evolved somewhere else and had no means of reaching South America. Instead, the Patagonian hare lives in South America. The Patagonian hare resembles a rabbit in anatomy and behavior, but it has the face of a guinea pig, from which it probably evolved (Fig. 22.6).

As another example, both cacti and euphorbia are plants adapted to a hot, dry environment. Both are succulent, spiny, flowering plants. Why do cacti grow in North American deserts and euphorbia grow in African deserts, when each would do well on the other continent? They just happened to evolve on their respective continents.

The islands of the world are home to many unique species of animals and plants found no place else, even when the soil and climate are the same. Why do so many species of finches live on the Galápagos Islands, when these same species are not on the mainland? The reasonable explanation is that finches from the ancestral species migrated to all the different islands. Then, geographic isolation allowed the ancestral finches to evolve into a different species on each island.

Video Finches Adaptive Radiation

In the history of the Earth, South America, Antarctica, and Australia were originally connected. Marsupials (pouched mammals) arose at this time and today are found in both South America and Australia. But when Australia separated and drifted away, the marsupials diversified into many different forms suited to various environments of Australia. They were free to do so because there were few, if any, placental mammals in Australia. In South America, where there are placental mammals, marsupials are not as diverse. This supports the hypothesis that evolution is influenced by the mix of plants and animals in a particular continent—by biogeography.

a.

b.

Figure 22.6 The Patagonian hare and the European rabbit.
a. The Patagonian hare, a native of South America. **b.** A European rabbit, which does not occur naturally in South America.

Anatomical Evidence Darwin was able to show that a common-descent hypothesis offers a plausible explanation for anatomical similarities among organisms. Vertebrate forelimbs are used for flight (birds and bats), orientation during swimming (whales and seals), running (horses), climbing (arboreal lizards), or swinging from tree branches (monkeys). Yet all vertebrate forelimbs contain the same sets of bones organized in similar ways, despite their dissimilar functions (Fig. 22.7). The most plausible explanation for this unity is that the basic forelimb plan belonged to a common ancestor. The basic plan was then modified in the succeeding groups as each continued along its own evolutionary pathway. Structures that are anatomically similar because they are inherited from a common ancestor are called **homologous structures.** In contrast, **analogous structures** serve the same function but are not constructed similarly; nor do they share a common ancestry. The wings of birds and insects and the jointed appendages of a lobster and humans are analogous structures. The presence of homology, not analogy, is evidence that organisms are related.

Figure 22.7 Vertebrate forelimbs are homologous structures.
Despite differences in function, vertebrate forelimbs have the same bones.

Vestigial structures are anatomical features that are fully developed in one group of organisms but that are reduced and may have no function in similar groups. Modern whales have a vestigial pelvic girdle and legs. The ancestors of whales walked on land, but whales are totally aquatic animals today. Most birds have well-developed wings used for flight. Some bird species (e.g., ostrich), however, have greatly reduced wings

Connections and Misconceptions

What are some examples of vestigial organs in humans?

The human body is littered with vestigial organs from our evolutionary past. One example is the tiny muscles (called piloerectors) that surround each hair follicle. During times of stress, these muscles cause the hair to stand straight up—a useful protection mechanism for small mammals trying to escape predators but one that has little function in humans. Wisdom teeth are also considered to be vestigial organs, because most people now retain their original teeth for the majority of their lives.

and do not fly. Similarly, snakes have no use for hindlimbs; yet some have remnants of a pelvic girdle and legs. Humans have a tailbone but no tail. The presence of vestigial structures can be explained by common descent. Vestigial structures occur because organisms inherit their anatomy from their ancestors. They are traces of an organism's evolutionary history.

The homology shared by vertebrates extends to their embryological development (Fig. 22.8). At some time during development, all vertebrates have a postanal tail and exhibit paired pharyngeal pouches. In fish and amphibian larvae, these pouches develop into functioning gills. In humans, the first pair of pouches becomes the cavity of the middle ear and the auditory tube. The second pair becomes the tonsils. The third and fourth pairs become the thymus and parathyroid glands, respectively. Why should terrestrial vertebrates develop and then modify structures like pharyngeal pouches that have lost their original function? The most likely explanation is that fish are ancestral to other vertebrate groups.

Biochemical Evidence Almost all living organisms use the same basic biochemical molecules, including DNA (deoxyribonucleic acid), ATP (adenosine triphosphate), and many identical or nearly identical enzymes. Further, organisms use the same DNA triplet code and the same 20 amino acids in their proteins. The sequences of DNA bases in the genomes of many organisms are now known, so it has become clear that humans share a large number of genes with much simpler organisms. Evolutionists who study development have also found that many developmental genes (called *Hox* genes) are shared in animals ranging from worms to humans. It appears that life's vast diversity has come about by only a slight difference in the regulation of genes. The result has been widely divergent types of bodies.

When the degree of similarity in DNA base sequences or the degree of similarity in amino acid sequences of proteins is examined, the data are as expected, assuming common descent. Cytochrome *c* is a molecule used in the electron transport chain of many organisms. Data regarding differences in the amino acid sequence of cytochrome *c* show that the sequence in a human differs from that in a monkey by only two amino acids. The human sequence differs from that in a duck by 11 amino acids, and from that in a yeast by 51 amino acids (Fig. 22.9). These data are consistent with other data regarding the anatomical similarities of these organisms and, therefore, their relatedness.

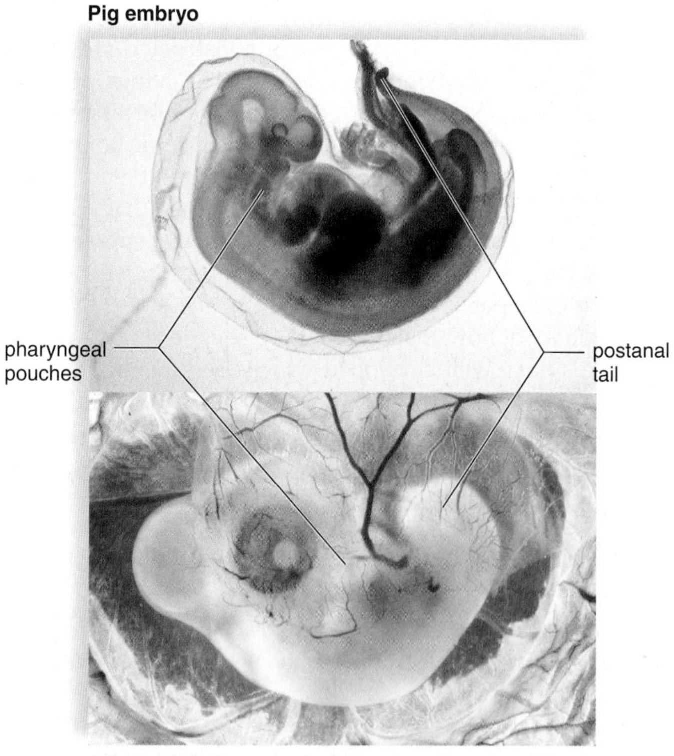

Pig embryo

pharyngeal pouches

postanal tail

Chick embryo

Figure 22.8 Homologous structures in vertebrate embryos. Vertebrate embryos have features in common, such as pharyngeal pouches, despite different ways of life as adults.

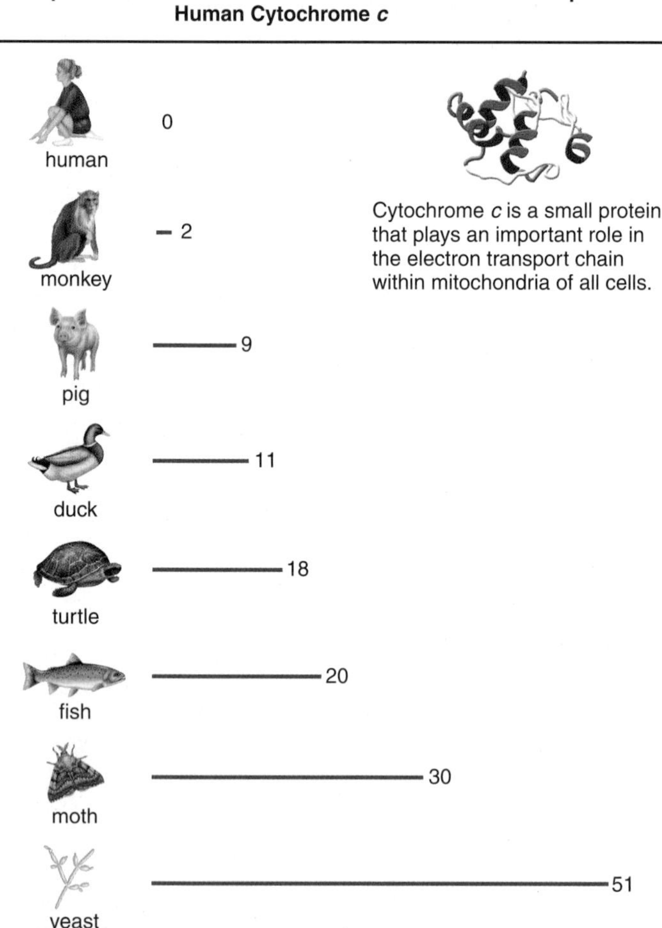

Figure 22.9 Biochemical evidence describes evolutionary relationships. The number of amino acid differences in cytochrome *c* between humans and other species is indicated.

Connections and Misconceptions

What is intelligent design?

Evolution is a scientific theory. Sometimes we use the word *theory* when we mean a hunch or a guess. But in science, the term *theory* is reserved for those ideas that scientists have found to be all-encompassing because they are based on evidence (data) collected in a number of different fields. In other words, evolutionary theory has been supported by repeated scientific experiments and observations.

Some persons advocate the teaching of ideas that run contrary to the theory of evolution in schools. The emphasis is currently placed on intelligent design, a belief system that maintains that the diversity of life could never have arisen without the involvement of an "intelligent agent." Many scientists, and even religions, argue that intelligent design is faith-based and not science-based. It would not be possible to test in a scientific way whether an intelligent agent exists. If it were possible to structure such an experiment, scientists would be the first to do it.

Natural Selection

When Darwin returned home, he spent the next 20 years gathering data to support the principle of biological evolution. Darwin's most significant contribution was to describe a mechanism for adaptation—**natural selection.** During adaptation, a species becomes suited to its environment. On his trip, Darwin visited the Galápagos Islands. He saw a number of finches that resembled one another but had different ways of life. Some were seed-eating ground finches, some cactus-eating ground finches, and some insect-eating tree finches. A warbler-type finch had a beak that could take honey from a flower. A woodpecker-type finch lacked the long tongue of a woodpecker but could use a cactus spine or twig to pull insects from cracks in the bark of a tree. Darwin thought the finches were all descended from a mainland ancestor whose offspring had spread out among the islands and become adapted to different environments.

Video Finches Natural Selection

To emphasize the nature of Darwin's natural selection process, it is often contrasted with a process espoused by Jean-Baptiste Lamarck, another nineteenth-century naturalist. Lamarck's explanation for the long neck of the giraffe was based on the assumption that the ancestors of the modern giraffe were trying to reach into the trees to browse on high-growing vegetation (Fig. 22.10). Continual stretching of

Lamarck's proposal
Darwin's proposal
Originally, giraffes had short necks.
Originally, giraffe neck length varied.
Giraffes stretched their necks in order to reach food.
Competition for resources causes long-necked giraffes to have the most offspring.
With continual stretching, most giraffes now have long necks.
Due to natural selection, most giraffes now have long necks.

Figure 22.10 The two major mechanisms for evolutionary change in the nineteenth century.
This diagram contrasts Jean-Baptiste Lamarck's process of acquired characteristics with Charles Darwin's process of natural selection.

Connections and Misconceptions

When did Darwin publish his book on natural selection? What was the reaction to his work?

Darwin published his book on natural selection in 1859. It was entitled *On the Origin of Species by Means of Natural Selection, or The Preservation of Favoured Races in the Struggle for Life.* The title is usually shortened to *On the Origin of Species*. It was instantly controversial, even though he avoided the use of the word "evolution." He only alludes to humans with the statement that, by his theory, ". . . light will be thrown on the origin of man and his history."

The book is still in print and you can obtain a copy.

Connections and Misconceptions

When did humans first start to practice artificial selection?

Almost all animals that are currently used in modern agriculture are the result of thousands of years of artificial selection by humans. But perhaps the longest-running experiment in artificial selection is the modern dog. Analysis of canine DNA indicates that dogs (*Canis familiaris*) are a direct descendant of the gray wolf (*Canis lupus*). This domestication, and the subsequent selection for desirable traits, appear to have begun over 130,000 years ago. Artificial selection of dogs continues to this day, with over 150 variations (or breeds) currently known.

the neck caused it to become longer, and this acquired characteristic was passed on to the next generation. Lamarck's mechanism will not work because acquired characteristics cannot be inherited (Fig. 22.10).

The critical elements of the natural selection process are

- *Variation.* Individual members of a species vary in physical characteristics. Physical variations can be passed from generation to generation. (Darwin was never aware of genes, but we know today that the inheritance of the genotype determines the phenotype.)
- *Competition for limited resources.* Even though each individual could eventually produce many descendants, the number in each generation usually stays about the same. Why? Resources are limited and competition for resources results in unequal reproduction among members of a population.
- *Adaptation.* Those members of a population with advantageous traits capture more resources and are more likely to reproduce and pass on these traits. Thus, over time, the environment "selects" for the better-adapted traits. Each subsequent generation includes more individuals adapted in the same way to the environment.

Darwin noted that when humans help carry out **artificial selection**, they breed selected animals with particular traits to reproduce. For example, prehistoric humans probably noted desirable variations among wolves and selected particular individuals for breeding. Therefore, the desired traits increased in frequency in the next generation. This same process was repeated many times, resulting in today's numerous varieties of dogs, all descended from the wolf. In a similar way, several varieties of vegetables can be traced to a single ancestor. Chinese cabbage, brussels sprouts, and kohlrabi are all derived from a single species, *Brassica oleracea*.

Natural selection can account for the great diversity of life. Environments differ widely; therefore, adaptations are varied. From vampire bats to sea turtles to the many finches observed by Darwin, all the different organisms are adapted to their way of life.

Connecting the Concepts

For more information on the information provided in this section, refer to the following discussions.

Section 1.1 explains why evolutionary change is the core concept of the study of biology.

Section 18.4 describes how meiosis introduces the variation that is the basis of evolutionary change.

Section 21.3 explores how scientists study changes in DNA and proteins.

Check Your Progress 22.2

1. Define *biological evolution,* and explain what its two most important aspects are.
2. Describe the types of evidence that convinced Charles Darwin that biological evolution does occur.
3. Discuss why natural selection is the mechanism for biological evolution.

22.3 Classification of Humans

Learning Outcomes

Upon completion of this section, you should be able to

1. Describe how DNA analysis is used to study primate evolution.
2. Describe the evolutionary trends that occur in the primates.
3. Compare the structure of chimpanzee and human skeletons and list the adaptations in humans that make upright walking possible.

To begin a study of human evolution, we turn to the classification of humans because biologists classify organisms according to their evolutionary relatedness. The **binomial name** of an organism gives its genus and species. Organisms in the same domain have only general characteristics in common. Those in the same genus have specific characteristics in common. Table 22.1 lists some of the characteristics that help classify humans. The dates in the first column of the table tell us when these groups of animals first appear in the fossil record.

DNA Data and Human Evolution

We are accustomed to using the characteristics given in Table 22.1 to determine evolutionary relationships, but researchers are just as apt to use DNA data. Researchers are depending more and more on DNA data to trace the history of life. DNA data are particularly useful when anatomical differences are unavailable.

For example, in the late 1970s, Carl Woese and his colleagues at the University of Illinois decided to use ribosomal RNA (rRNA) sequence data to decide how prokaryotes are related. They knew that the DNA coding for rRNA changes slowly during evolution. Ribosomal RNA genes may change only when there is a major evolutionary event. Woese reported, on the basis of rRNA sequence data, that there are three domains of life and members of Archaea are more closely related to members of Eukarya than to those of Bacteria (Fig. 22.11). (See Fig. 1.6 for a description of these domains.) In other words, major decisions regarding the history of life are now being made on the basis of DNA/rRNA/protein sequencing data. (Fig. 22.9 gave an example of the use of protein data to show evolutionary relationships.) For example, studies of rRNA sequences indicate that among the major groups of eukaryotes, animals are more closely related to fungi than they are to plants.

Animation Three Domains

Later in this chapter, you will learn that DNA sequencing data were used to decide when the last common ancestor for the apes and humans must have existed. As yet, the fossil record has not revealed this ancestor. Comparative DNA data between apes and humans tell us this ancestor must have existed about 7 MYA. Mitochondrial DNA (mtDNA) is used to decide the timing of recent evolutionary events because mtDNA changes occur frequently. Mitochondrial DNA data indicate that modern humans arose in Africa and later migrated to Eurasia. (See page 536 for a discussion of this study.)

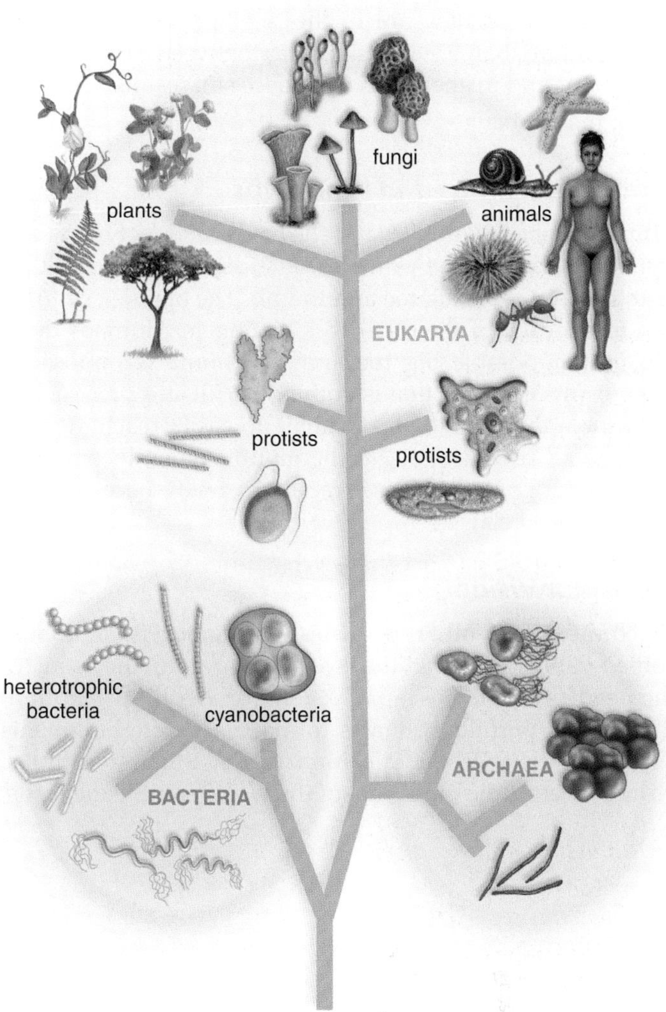

Figure 22.11 The three domains of life.
Representatives of each domain are depicted in the ovals. The evolutionary tree of life shows that domain Archaea is more closely related to domain Eukarya than either is to domain Bacteria.

Table 22.1 Evolution and Classification of Humans

BYA/MYA*	Classification Category	Characteristics
2 BYA	Domain Eukarya	Membrane-bound nucleus
600 MYA	Kingdom Animalia	Multicellular, motile, heterotrophic
540 MYA	Phylum Chordata	Sometime in life history: dorsal tubular nerve cord, notochord, pharyngeal pouches
120 MYA	Class Mammalia	Vertebrates with hair, mammary glands
60 MYA	Order Primates	Well-developed brain, adapted to live in trees
7 MYA	Family Hominidae	Adapted to upright stance and bipedal locomotion
3 MYA	Genus *Homo*	Most developed brain, made and used tools
0.1 MYA	Species *Homo sapiens*†	Modern humans; speech centers of brain well-developed

*BYA = billions of years ago; MYA = millions of years ago.
†To specify an organism, you must use the full binomial name, such as *Homo sapiens*.

Humans Are Primates

In contrast to the other orders of placental mammals, **primates** are adapted to an arboreal life—for living in trees. Primates have mobile limbs; grasping hands; a flattened face; binocular vision; a large, complex brain; and a reduced reproductive rate. The order Primates may be divided into two major groups—the **prosimians,** which include lemurs tarsiers, and lorises; and the **anthropoids,** which include monkeys, apes, and humans. This classification tells us that humans are more closely related to the monkeys and apes (Fig. 22.12) than they are to the prosimians. Remarkably, there is more than 95–98% similarity between related genes in humans and in apes. This difference still results in a number of major changes.

Mobile Forelimbs and Hindlimbs

Primate limbs are mobile, and the hands and feet both have five digits each. Many primates, such as chimpanzees, have both an opposable big toe and thumb. The big toe or thumb can touch each of the other toes or fingers. Humans don't have an opposable big toe, but the thumb is opposable. This results in a grip that is both powerful and precise. The opposable thumb allows a primate to easily reach out and bring food, such as fruit, to the mouth. When locomoting, primates grasp and release tree limbs freely because nails have replaced claws.

Binocular Vision

In chimps, like other primates, the snout is shortened considerably, allowing the eyes to move to the front of the head. The stereoscopic vision (or depth perception) that results permits primates to make accurate judgments about the distance and position of adjoining tree limbs. Humans and the apes have three different cone cells, which are able to discriminate between greens, blues, and reds. (See Chap. 14 for a review.) Cone cells require bright light, but the image is sharp and in color. The lens of the eye focuses light directly on the fovea, a region of the retina, where cone cells are concentrated.

Large, Complex Brain

The evolutionary trend among primates is toward a larger and more complex brain. The brain size is smallest in prosimians and largest in modern humans. The cerebral cortex, with many association areas, expands so much that it becomes extensively folded in humans. The portion of the brain devoted to smell is smaller. The portions devoted to sight have increased in size and complexity during primate evolution. Also, more of the brain is involved in controlling and processing information received from the hands and the thumb. The result is good hand–eye coordination in chimpanzees and humans.

Reduced Reproductive Rate

It is difficult to care for several offspring while moving from limb to limb, and one birth at a time is the norm in primates. The juvenile period of dependency is extended, and there is an emphasis on learned behavior and complex social interactions.

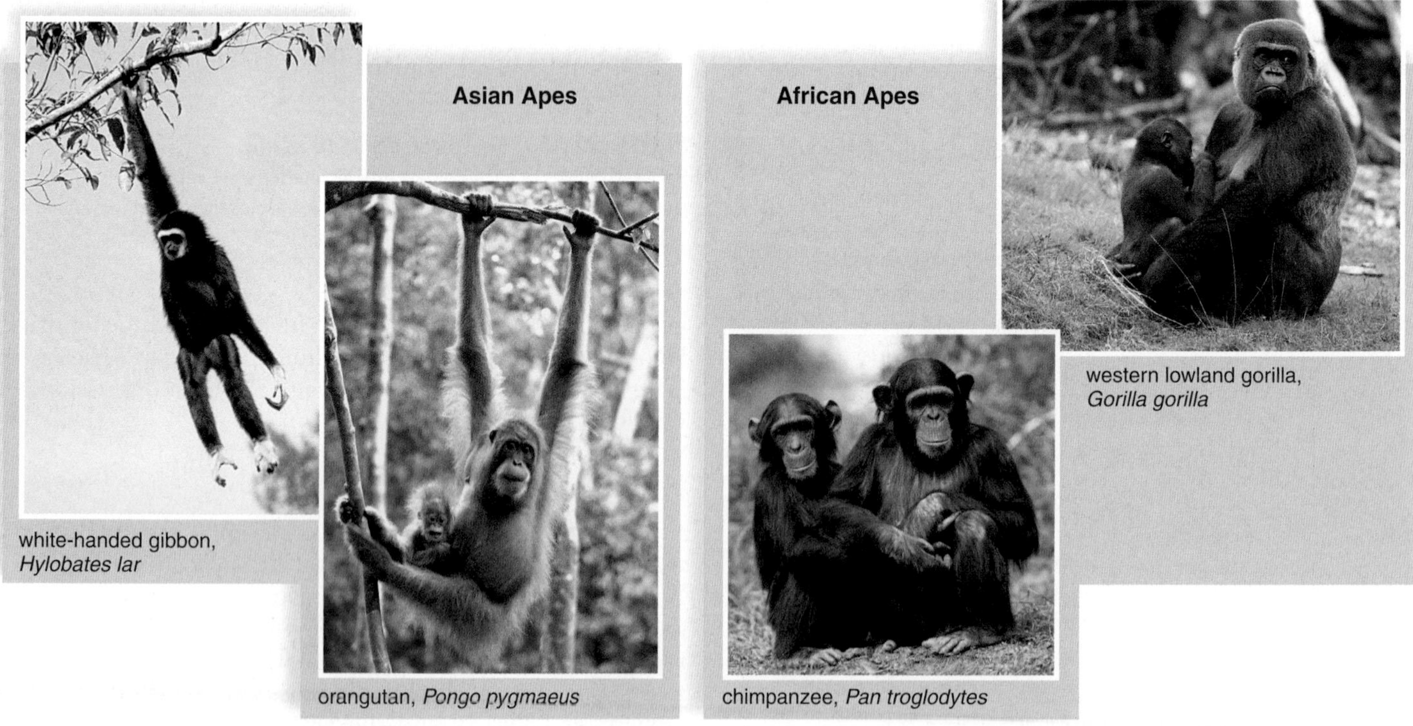

Figure 22.12 **Asian and African apes.**
The apes can be divided into the Asian apes (gibbons and orangutans) and the African apes (chimpanzees and gorillas). Molecular data and the location of early fossil remains tell us that we are more closely related to the African than to the Asian apes.

Comparing the Human Skeleton to the Chimpanzee Skeleton

Figure 22.13 compares anatomical differences between chimpanzees and humans, which relate to the upright stance of humans when they walk compared to the chimpanzees' practice of knuckle-walking. When chimpanzees walk, their forearms rest on their knuckles.

These differences in anatomy between chimpanzees and humans determine that humans—but not chimps—are adapted for an upright stance. (1) In humans, the spine exits inferior to the center of the skull, and this places the skull in the midline of the body. (2) The longer, S-shaped spine of humans places the trunk's center of gravity squarely over the feet. (3) The broader pelvis and hip joint of humans keep them from swaying when they walk. (4) The longer neck of the femur in humans causes the femur to angle inward at the knees. (5) The human knee joint is modified to support the body's weight; the femur is larger at the bottom, and the tibia is larger at the top. (6) Finally, the human toe is not opposable; instead, the foot has an arch. The arch enables humans to walk long distances and run with less chance of injury.

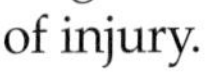

Human spine exits from the skull's center; ape spine exits from rear of skull.
Human spine is S-shaped; ape spine has a slight curve.
Human pelvis is bowl-shaped; ape pelvis is longer and more narrow.
Human femurs angle inward to the knees; ape femurs angle out a bit.
Human knee can support more weight than ape knee.
Human foot has an arch; ape foot has no arch.
a.
b.

Figure 22.13 Adaptations in the human skeleton allow upright locomotion.
a. Human skeleton compared to (**b**) chimpanzee skeleton.

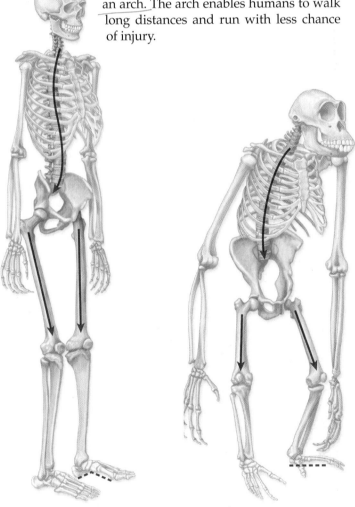

Connecting the Concepts

For more information on the material in this section, refer to the following discussions.

Section 3.6 discusses the structure and function of the mitochondria in a eukaryotic cell.

Section 13.2 examines the function of the cerebral cortex in the human brain.

Section 21.1 explores the structure and function of DNA and RNA molecules.

Check Your Progress 22.3

1. List the general characteristics of primates.
2. Discuss the benefits of binocular vision and a complex brain structure.
3. Summarize the major differences between the chimpanzee skeleton and the human skeleton.

22.4 Evolution of Hominins

Learning Outcomes

Upon completion of this section, you should be able to

1. Describe the major events in the evolution of the hominins.
2. Summarize the significance of the australopithecines in the study of human evolution.

Once biologists have studied the characteristics of a group of organisms, they can construct an **evolutionary tree** that is a working hypothesis of their past history. The evolutionary tree in Figure 22.14 shows that all primates share one common ancestor and that the other types of primates diverged from the human line of descent over time. When any two lines of descent, called a **lineage,** first diverge from a common ancestor, the genes and proteins of the two lineages are nearly identical. As time goes by, each lineage accumulates genetic changes, which lead to RNA and protein changes. Many genetic changes are neutral (not tied to adaptation) and accumulate at a fairly constant rate. Such changes can be used as a type of **molecular clock** to indicate the relatedness of two groups and when they diverged from each other. Molecular data also suggest that hominids split from the ape line of descent about 7 MYA.

The names of the various classifications of primates has changed rapidly over the past several years as new discoveries and biochemical analyses have unveiled more information on primate evolution. For example, as previously mentioned, the *prosimians* now include the lemurs, tarsiers, and lorises; the *anthropoids* include the monkeys, apes, and humans. The designation ***hominid*** includes the apes (gorillas and orangutans), chimpanzees, humans, and the closest extinct relatives of humans. The term ***hominine*** is now used to include only the gorillas, chimpanzees, and humans and their closest extinct relatives. The designation ***hominin*** refers to all members of the genus *Homo* and their close relatives.

One of the most unfortunate misconceptions concerning human evolution is the belief that Darwin and others suggested that humans evolved from apes. On the contrary, humans and apes are thought to have shared a common apelike ancestor. Today's apes are our distant cousins, and we couldn't have evolved from our cousins because we are contemporaries—living on Earth at the same time. Humans and apes have been evolving separately from a common ancestor for about 7 million years. Following the split between humans and apes, different environments selected for the different traits that apes and humans have now.

The First Hominins

Paleontologists use certain anatomical features when they try to determine if a fossil is a hominin. One of these features is **bipedal posture** (walking on two feet). Until recently, many scientists thought that hominins began to walk upright on two feet because of a dramatic change in climate that caused forests to be replaced by grassland. Now, some biologists suggest that the first hominin began to assume a bipedal posture even while it lived in trees. Why? They cannot find evidence of a dramatic shift in vegetation about 7 MYA. The first hominin's environment is now thought to have included some forest, some woodland, and some grassland. While still living in trees, the first hominins may have walked upright on large branches as they collected fruit from overhead. Then, when they began to forage on the ground among bushes, it would have been easier to shuffle along on their hindlimbs. Bipedalism would also have prevented them from getting heatstroke because an upright stance exposes more of the body to breezes. Bipedalism may have been an advantage in still another way. Males may have acquired food far afield. If they could carry it back to females, they would have been more assured of having sexual intercourse.

Two other hominin features of importance are the shape of the face and brain size. Today's humans have a flatter face and a more pronounced chin than do the apes because the human jaw is shorter than that of the apes. Then, too, our teeth are generally smaller and less specialized. We don't have the sharp canines of an ape, for example. Chimpanzees have a brain size of about 400 cm^3, and modern humans have a brain size of about 1,300 cm^3.

It's hard to decide which fossils are hominins because human features evolved gradually and they didn't evolve at the same rate. Most investigators rely first and foremost on bipedal posture as the hallmark of a hominin, regardless of the size of the brain.

Earliest Fossil Hominins

Fossils have been found that can be dated at the time the ape and human lineages split. The oldest of these fossils, called *Sahelanthropus tchadensis*, dated at 7 MYA, was found in Chad, located in central Africa, far from eastern and southern Africa where other hominid fossils were excavated. The only find, a skull, appears to be that of a hominin because it has smaller canines and thicker tooth enamel than an ape. The braincase, however, is very apelike. It is impossible to tell if this hominin walked upright. Some suggest this fossil is ancestral to the gorilla.

Orrorin tugenensis, dated at 6 MYA and found in eastern Africa, is thought to be another early hominin, especially because the limb anatomy suggests a bipedal posture. However, the canine teeth are large and pointed, and the arm and finger bones retain adaptations for climbing. Some suggest this fossil is ancestral to the chimpanzee.

Ardipithecus kadabba, found in eastern Africa and dated between 5.8 and 5.2 MYA, is closely related to the later-appearing *Ardipithecus ramidus.* This ardipithecine is thought to be closely related to the australopithecines, discussed next.

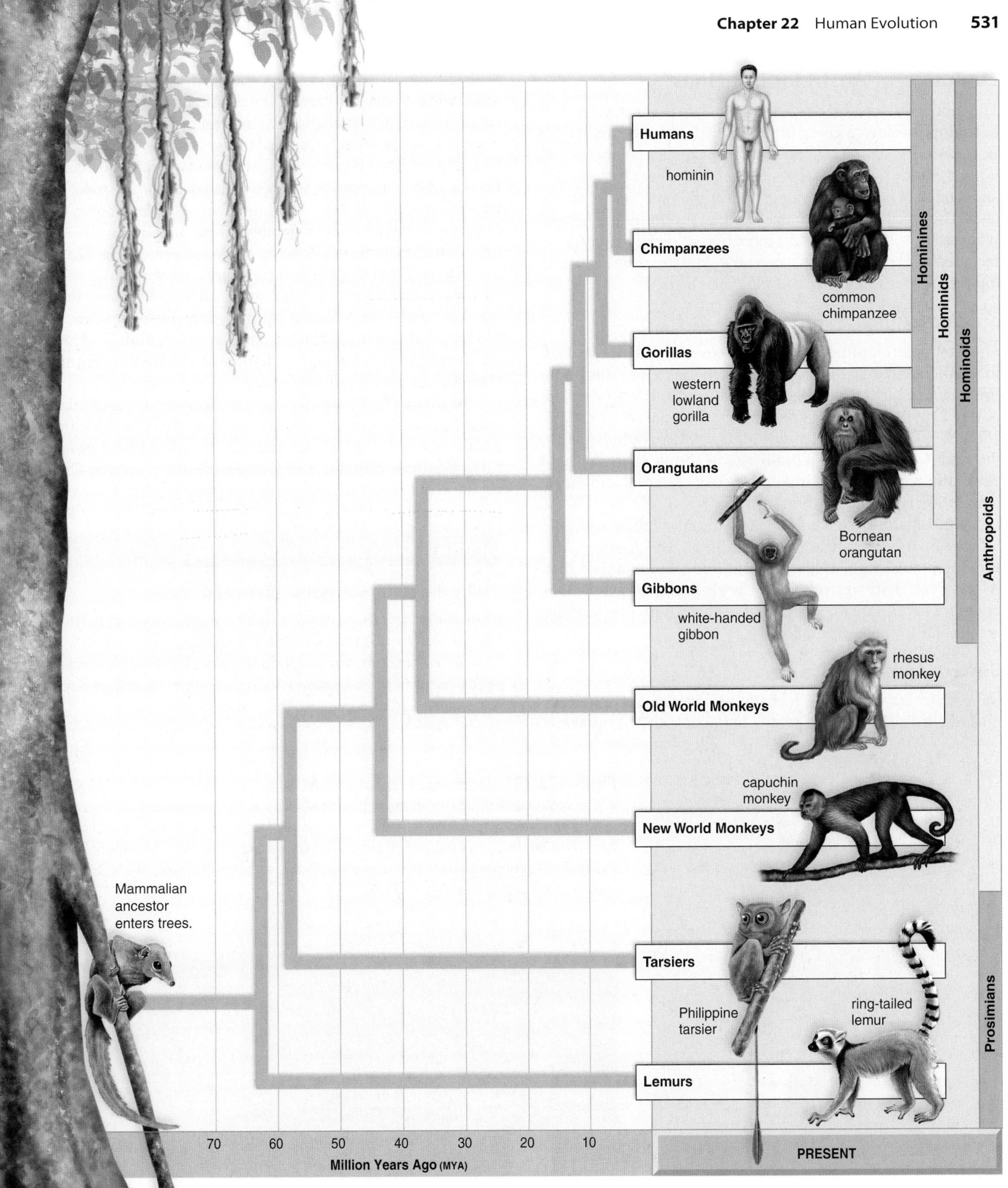

Figure 22.14 **The evolutionary tree of the primates.**
Humans are related to all other primates through a common ancestor.

Evolution of Australopithecines

The hominin line of descent begins in earnest with the **australopithecines,** a group of species that evolved and diversified in Africa. Some australopithecines were slight of frame and termed *gracile* ("slender") types. Some were *robust* ("powerful") and tended to have strong upper bodies and especially massive jaws. The gracile types most likely fed on soft fruits and leaves, whereas the robust types had a more fibrous diet that may have included hard nuts. In other words, the skull structure of australopithecines was suited to their particular diets.

Southern Africa

The first australopithecine to be discovered was unearthed in southern Africa by Raymond Dart in the 1920s. This hominin, named *Australopithecus africanus,* is a gracile type dated about 2.8 MYA. *Australopithecus robustus,* dated from 2 to 1.5 MYA, is a robust type from southern Africa. Both *A. africanus* and *A. robustus* had a brain size of about 500 cm^3. Their skull differences are essentially because of dental and facial adaptations to different diets.

Limb anatomy suggests these hominids walked upright. However, the proportions of the limbs are apelike. The forelimbs are longer than the hindlimbs. Some argue that *A. africanus,* with its relatively large brain, is a possible ancestral candidate for early *Homo,* whose limb proportions are similar to those of this fossil.

Eastern Africa

More than 20 years ago, a team led by Donald Johanson unearthed nearly 250 fossils of a hominin called *A. afarensis.* A now-famous female skeleton dated at 3.18 MYA is known worldwide by its field name, Lucy. Although her brain was small (400 cm^3), the shapes and relative proportions of her limbs indicate that Lucy stood upright and walked bipedally (Fig. 22.15*a*). Even better evidence of bipedal locomotion comes from a trail of footprints in Laetoli dated about 3.7 MYA. The larger prints are double, as though a smaller-sized being was stepping in the footfalls of another. There are additional small prints off to the side, within hand-holding distance (Fig. 22.15*b*).

The fact that the australopithecines were apelike above the waist (small brain) and humanlike below the waist (walked erect) shows that human characteristics did not evolve all at one time. The term **mosaic evolution** is applied when different body parts change at different rates and, therefore, at different times.

Australopithecus afarensis, a gracile type, is most likely ancestral to the robust types found in eastern Africa: *A. aethiopicus* and *A. boisei. Australopithecus boisei* had a powerful upper body and the largest molars of any hominin. These robust types died out; therefore, it is possible that *A. afarensis* is ancestral to both *A. africanus* and early *Homo* (Fig. 22.16).

Connections and Misconceptions

Why did Johanson name his fossil *Lucy*?

The evening of the discovery, there was a party in camp. The team gathered to celebrate the find of what appeared to be an almost complete hominid skeleton. The Beatles' song "Lucy in the Sky with Diamonds" was played over and over, and the name *Lucy* was given to the skeleton.

a.

b.

Figure 22.15 ***Australopithecus afarensis.***
a. A reconstruction of Lucy on display at the St. Louis Zoo. **b.** These fossilized footprints occur in ash from a volcanic eruption some 3.7 MYA. The larger footprints are double (one followed behind the other), and a third, smaller individual was walking to the side. (A female holding the hand of a youngster may have been walking in the footprints of a male.) The footprints suggest that *A. afarensis* walked bipedally.

Connecting the Concepts

For additional information on the topics in this section, refer to the following discussions.

Section 8.2 describes the types of teeth found in modern humans.

Figure 11.6 illustrates the skeletal structure of a modern human.

Section 13.2 examines the brain structure of modern humans.

Check Your Progress 22.4

1. Distinguish between a hominid and a hominin and give an example of each.
2. Name three features characteristic of hominins.
3. Discuss why the discovery of Lucy was an important event in the study of human evolution.

22.5 Evolution of Humans

Learning Outcomes

Upon completion of this section, you should be able to

1. Explain the adaptations of *Homo erectus.*
2. Distinguish between the multiregional and out-of-Africa hypotheses of *Homo sapiens* evolution.
3. Describe the differences between Neandertals and Cro-Magnons.

Fossils are assigned to the genus *Homo* if (1) the brain size is 600 cm^3 or greater, (2) the jaw and teeth resemble those of humans, and (3) tool use is evident. In this section, we will discuss early *Homo: Homo habilis* and *Homo erectus;* and later *Homo:* the Neandertals and the Cro-Magnons, the first modern humans.

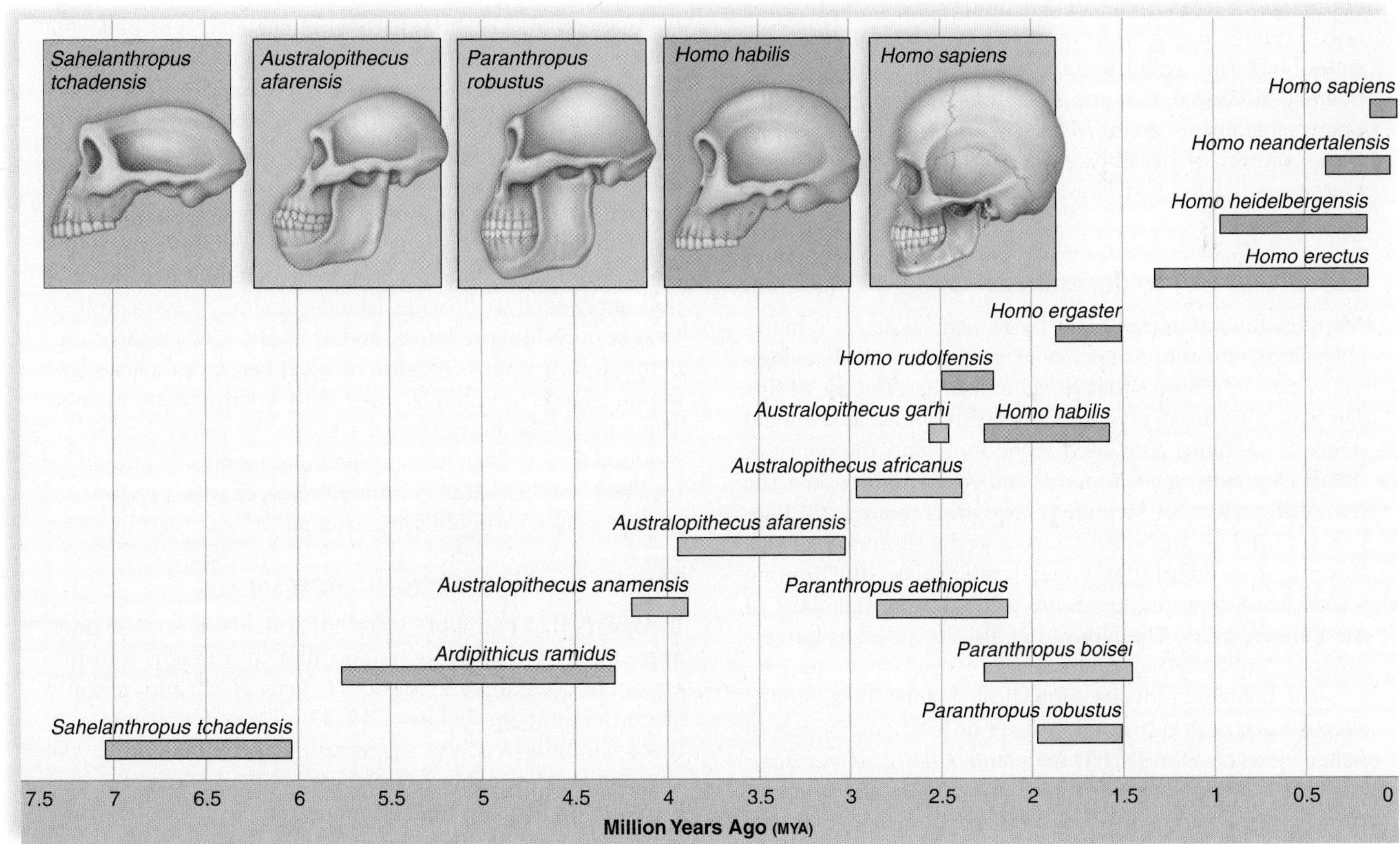

Figure 22.16 Human evolution.

Several groups of extinct hominins preceded the evolution of modern humans. The groups have been divided into the early humanlike hominins (orange), later humanlike hominins (green), early *Homo* species (lavender), and finally the later *Homo* species (blue).

Biology Matters Science Focus

Homo floresiensis

In 2003, scientists made one of the most spectacular discoveries in evolutionary history. Nine skeletons were discovered in a cave on the isle of Flores. Flores is an Indonesian island east of Bali, located midway between Asia and Australia. This new species of humans grew no taller than a modern 3-year-old child (Fig. 22A). One skeleton was that of an adult female who died when she was approximately 30 years old. She stood 1 meter tall (3.3 ft) and weighed approximately 25 kilograms (55 lb). Scientists estimate that she died around 18,000 years ago.

After examining the first skeleton, the research team concluded that they had discovered a new human species. They named the species *Homo floresiensis* after the island where it was found. The workers at the excavation site nicknamed the tiny creatures "hobbits" after the fictional creatures in the *Lord of the Rings* books by J. R. R. Tolkien.

a. *Homo floresiensis*, artist's impression

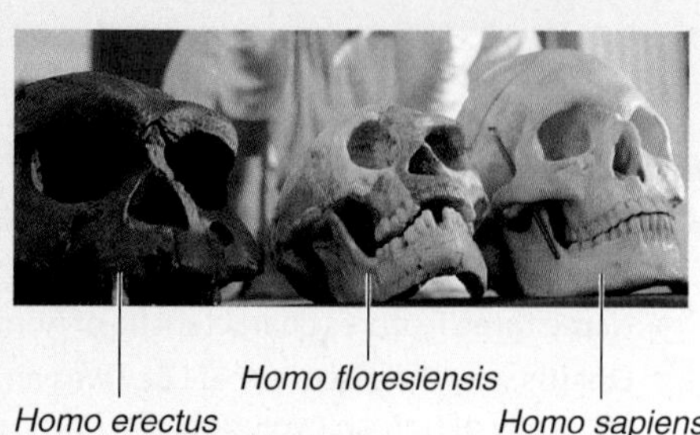

b. Comparison of skulls

Figure 22A *Homo floresiensis.*
a. Artist's re-creation of *H. floresiensis.* **b.** The *H. floresiensis* skull is smaller than that of *H. erectus* and that of *H. sapiens.*

Classification of *Homo floresiensis*

Homo floresiensis have skulls the size of a grapefruit and a brain size of approximately 417 cc. The teeth are humanlike, the eyebrow ridges are thick, the forehead slopes sharply, and the face lacks a chin. Even though the hobbits stand only 3 ft tall, they are not classified as pygmies. Despite the small body size, small brain size, and a mixture of primitive and advanced anatomical features, *H. floresiensis* is distinctly a member of genus *Homo.* The researchers believe that *Homo floresiensis* possibly evolved from a population of *H. erectus* that reached Flores approximately 840,000 years ago.

Culture of *Homo floresiensis*

Many of the habits exhibited by *H. floresiensis* are remarkably similar to those of other *Homo* species. Archaeological evidence indicates that *H. floresiensis* had the use of fire. The skeletons discovered on Flores were found in sediment deposits that also contained stone tools and the bones of dwarf elephants, giant rodents, and Komodo dragons. The dwarf elephants, or stegodons, weighed about 1,000 kilograms (2,200 lb) and would have posed a serious challenge to men who were only 1 meter tall. Successful hunting would have required communication among members of the hunting party. The Flores diet also included fish, frogs, birds, rodents, snakes, and tortoises.

The hobbits produced sophisticated stone tools, hunted successfully in groups, and crossed at least two bodies of water to reach Flores from mainland Asia. And yet, their brain was about one-third the size of modern humans. *Homo floresiensis* is the smallest species of human ever discovered. Pound for pound, they outcompete every other member of genus *Homo.*

Further Research Needed

Researchers are interested in determining why the hobbits were so small. The first-discovered skeleton was believed to be that of a small child. There is no evidence of any other 1-meter-tall adults in genus *Homo.* Modern pygmies are 1.4 to 1.5 meters (4.6 to nearly 5 ft) tall. Over thousands of years, it is possible that a population of *Homo erectus* evolved into *H. floresiensis.* If so, a smaller body size would have been favored by natural selection. The members of each generation could have been smaller than the previous generation. Dwarfing of mammals on islands is a well-known process that can be seen worldwide. Islands generally have a limited food supply, few predators, and at least a few species competing for the same ecological niche. It behooves species living on islands to minimize their daily energy requirements. The smaller the body size, the fewer the calories per day required for survival. At this point, there is no substantiation for this hypothesis, but continued research may produce an answer as to why hobbits are so small.

The Extinction of *Homo floresiensis*

It appears that many of Flores' inhabitants became extinct approximately 12,000 years ago due to a major volcanic eruption. Researchers found *H. floresiensis* and pygmy stegodon remains below a 12,000-year-old volcanic ash layer. Hobbits reached the island approximately 11,000 years ago and possibly intermingled with modern humans. Rumors, myths, and legends among the indigenous tribes of Flores about "the tiny people who lived in the forest" have persisted.

Early *Homo*

Homo habilis, dated between 2.0 and 1.9 MYA, may be ancestral to modern humans. Some of these fossils have a brain size as large as 775 cm^3, about 45% larger than that of *A. afarensis.* The cheek teeth are smaller than even those of the gracile australopithecines. Therefore, it is likely that these early members of the genus *Homo* were omnivores who ate meat in addition to plant material. Bones at their campsites bear cut marks, indicating that they used tools to strip them of meat.

The stone tools made by *H. habilis,* whose name means "handyman," are rather crude. It's possible that these are the cores from which they took flakes sharp enough to scrape away hide, cut tendons, and easily remove meat from bones.

Early *Homo* skulls suggest that the portions of the brain associated with speech were enlarged. We can speculate that the ability to speak may have led to hunting cooperatively. Other members of the group may have remained plant gatherers. If so, both hunters and gatherers most likely ate together and shared their food. In this way, society and culture could have begun.

Culture, which encompasses human behavior and products (e.g., technology and the arts), depends upon the capacity to speak and transmit knowledge. We can further speculate that the advantages of a culture to *H. habilis* may have hastened the extinction of the australopithecines.

Homo erectus

Homo erectus and like fossils are found in Africa, Asia, and Europe and dated between 1.9 and 0.3 MYA. A Dutch anatomist named Eugene Dubois was the first to unearth *H. erectus* bones in Java in 1891. Since that time, many other fossils have been found in the same area. Although all fossils assigned the name *H. erectus* are similar in appearance, enough discrepancy exists to suggest that several different species have been included in this group. In particular, some experts suggest that the Asian form is *Homo erectus* and the African form is *Homo ergaster* (Fig. 22.17).

Compared with *H. habilis, H. erectus* had a larger brain (about 1,000 cm^3) and a flatter face. The nose projected, however. This type of nose is adaptive for a hot, dry climate because it permits water to be removed before air leaves the body. The recovery of an almost complete skeleton of a 10-year-old boy indicates that *H. ergaster* was much taller than the hominids discussed thus far. Males were 1.8 meters tall (about 6 feet), and females were 1.55 meters (approaching 5 feet). Indeed, these hominids were erect and most likely had a striding gait like ours. The robust and most likely heavily muscled skeleton still retained some australopithecine features. Even so, the size of the birth canal indicates that infants were born in an immature state that required an extended period of care.

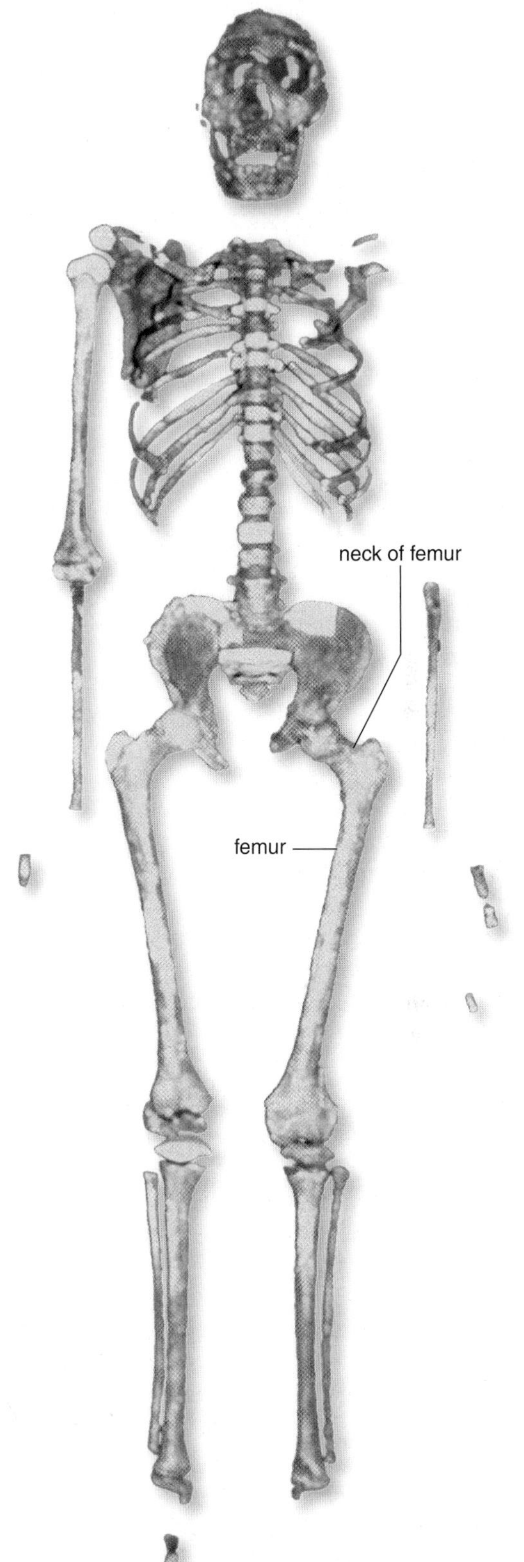

Figure 22.17 ***Homo ergaster.***
This skeleton of a ten-year-old boy who lived 1.6 MYA in eastern Africa shows angled femurs because the femur neck is quite long.

Homo ergaster may have first appeared in Africa and then migrated into Asia and Europe. At one time, the migration was thought to have occurred about 1 MYA. Recently, *H. erectus* fossil remains in Java and the Republic of Georgia have been dated at 1.9 and 1.6 MYA, respectively. These remains push the evolution of *H. erectus* in Africa to an earlier date than has yet been determined. In any case, such an extensive population movement is a first in the history of humankind and a tribute to the intellectual and physical skills of the species.

Homo erectus was the first hominid to use fire and also fashioned more advanced tools than early *Homos*. These hominids used heavy, teardrop-shaped axes and cleavers. Flake tools were probably used for cutting and scraping. It could be that *H. ergaster* was a systematic hunter and brought kills to the same site over and over. In one location, researchers have found over 40,000 bones and 2,647 stones. These sites could have been "home bases," where social interaction occurred and a prolonged childhood allowed time for learning. Perhaps a language evolved and a culture more like our own developed.

Evolution of Modern Humans

Most researchers accept the idea that ***Homo sapiens*** (modern humans) evolved from *H. erectus*, but they differ as to the details. Perhaps *Homo sapiens* evolved from *H. erectus* separately in Asia, Africa, and Europe. The hypothesis that *Homo sapiens* evolved in several different locations is called the **multiregional continuity hypothesis** (Fig. 22.18*a*). This hypothesis proposes that evolution to modern humans was essentially similar in several different places. If so, each region should show a continuity of its own anatomical characteristics from the time when *H. erectus* first arrived in Europe and Asia.

Opponents argue that it seems highly unlikely that evolution would have produced essentially the same result in these different places. They suggest, instead, the **out-of-Africa hypothesis,** which proposes that *H. sapiens* evolved from *H. erectus* only in Africa and, thereafter, *H. sapiens* migrated to Europe and Asia about 100,000 years BP (before present) (Fig. 22.18*b*). If this were the case, there would be no continuity of characteristics between fossils dated 200,000 years BP and 100,000 years BP in Europe and Asia.

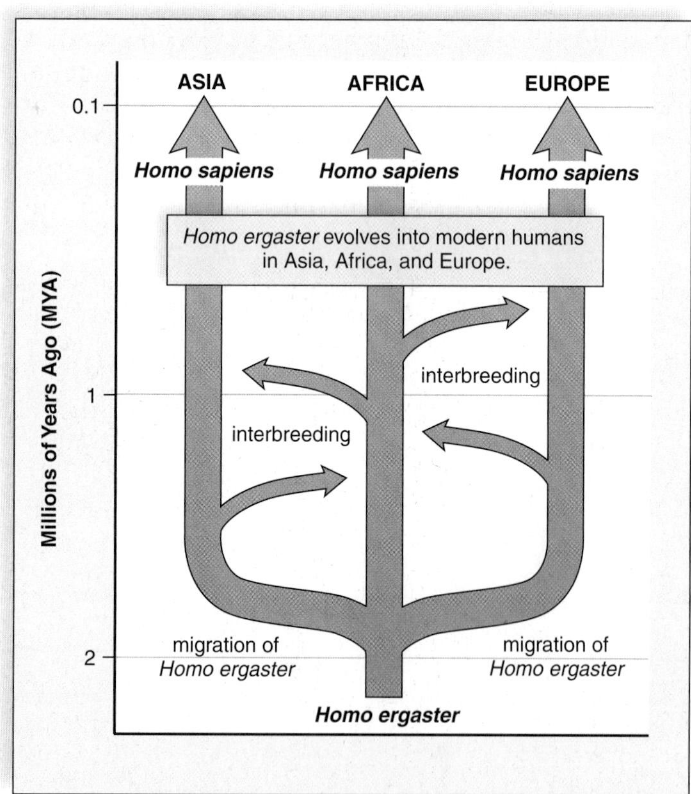

a. Multiregional continuity

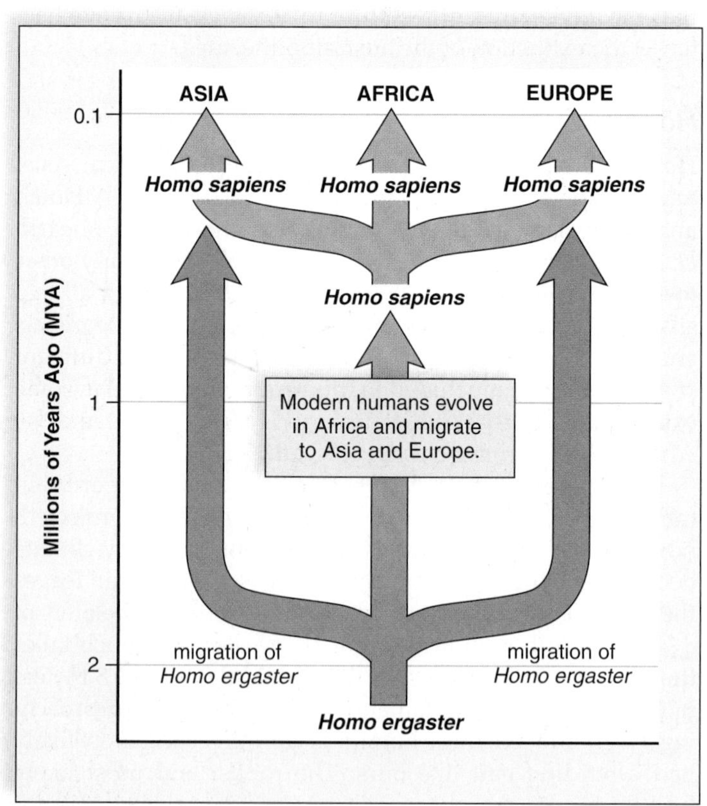

b. Out of Africa

Figure 22.18 **Two theories on the evolution of modern humans.**
a. The multiregional continuity hypothesis proposes that *Homo sapiens* evolved separately in at least three different places: Asia, Africa, and Europe. Therefore, continuity of genotypes and phenotypes is expected in each region but not between regions. **b.** The out-of-Africa hypothesis proposes that *Homo sapiens* evolved only in Africa and then migrated out of Africa, as *H. ergaster* did many years before. *Homo sapiens* would have supplanted populations of *Homo* in Asia and Europe about 100,000 years BP.

According to which hypothesis would modern humans be most genetically alike? The multiregional continuity hypothesis states that human populations have been evolving separately for a long time. Therefore, genetic differences are expected. The out-of-Africa hypothesis states that we are all descended from a few individuals from about 100,000 years BP. Therefore, the out-of-Africa hypothesis suggests that we are more genetically similar.

A few years ago, a study attempted to show that all the people of Europe (and the world, for that matter) have essentially the same mitochondrial DNA. Called the "mitochondrial Eve" hypothesis by the press (this is a misnomer because no single ancestor is proposed), the statistics that calculated the date of the African migration were found to be flawed. Still, the raw data—which indicate a close genetic relationship among all Europeans—support the out-of-Africa hypothesis.

These opposing hypotheses have sparked many other innovative studies to test them. Most scientists have come to the conclusion that the out-of-Africa hypothesis is supported.

Neandertals

Neandertals *(H. neandertalensis)* take their name from Germany's Neander Valley, where one of the first Neandertal skeletons, dated some 200,000 years BP, was discovered. The Neandertals had massive brow ridges; and their nose, jaws, and teeth protruded far forward. The forehead was low and sloping, and the lower jaw lacked a chin. New fossils show that the pubic bone was long compared with ours.

According to the out-of-Africa hypothesis, Neandertals were eventually supplanted by modern humans. Surprisingly, however, the Neandertal brain was, on the average, slightly larger than that of *H. sapiens* (1,400 cm^3, compared with 1,360 cm^3 in most modern humans). The Neandertals were heavily muscled, especially in the shoulders and neck (Fig. 22.19). The bones of the limbs were shorter and thicker than those of modern humans. It is hypothesized that a larger brain than that of modern humans was required to control the extra musculature. The Neandertals lived in Europe and Asia during the last Ice Age, and their sturdy build could have helped conserve heat.

The Neandertals give evidence of being culturally advanced. Most lived in caves, but those living in the open may have built houses. They manufactured a variety of stone tools, including spear points, which could have been used for hunting. Scrapers and knives could have helped in food preparation. They most likely successfully hunted bears, woolly mammoths, rhinoceroses, reindeer, and other contemporary animals. They used and could control fire, which probably helped them cook meat and keep themselves warm. They even buried their dead with flowers and tools and may have had a religion. Perhaps they believed in life after death. If so, they were capable of thinking symbolically.

Cro-Magnons

Cro-Magnons are the oldest fossils to be designated *Homo sapiens*. Cro-Magnons are named for a fossil location in France. In keeping with the out-of-Africa hypothesis, the Cro-Magnons were the modern humans who entered Asia and Europe from Africa 100,000 years BP or even earlier. Cro-Magnons had a thoroughly modern appearance (Fig. 22.20). Analysis of Neandertal DNA indicates that it is so different from Cro-Magnon DNA that these two groups of people did

Figure 22.19 The Neandertals.
This drawing shows that the nose and mouth of the Neandertals protruded from their faces, and their muscles were massive. They made stone tools and were most likely excellent hunters.

Figure 22.20 The Cro-Magnons.
Cro-Magnon people are one of the earliest groups to be designated *Homo sapiens*. Their tool-making ability and other cultural attributes, including artistic talents, are legendary.

Biology Matters **Bioethical Focus**

Effects of Biocultural Evolution on Population Growth

Human beings today undergo biocultural evolution because culture has developed to the point that adaptation to the environment is not dependent on genes but on the passage of culture from one generation to the next.

Tool Use and Language Began

The first step toward biocultural evolution began when *Homo habilis* made primitive stone tools. *Homo erectus* continued the tradition and most likely was a hunter of sorts. It's possible that the campsites of *H. erectus* were "home bases," where the women stayed behind with the children while the men went out to hunt. Hunting was an important event in the development of culture, especially because it encourages the development of language. If *H. erectus* didn't have the use of language, certainly Cro-Magnons did. People who have the ability to speak a language would have been able to cooperate better as they hunted and even as they sought places to gather plants. Among animals, only humans have a complex language that allows them to communicate their experiences. Words are not objects and events. They stand for objects and events that can be pictured in the mind.

Agriculture Began

About 10,000 years ago, people gave up being full-time hunter–gatherers and became at least part-time farmers. What accounts for the rise of agriculture? The answer is not known, but several explanations have been put forth. About 12,000 years ago, a warming trend occurred as the Ice Age came to a close. A variety of big-game animals became extinct, including the saber-toothed cats, mammoths, and mastodons. This may have made hunting less productive. As the weather warmed, the glaciers retreated and left fertile valleys, where rivers and streams were full of fish and the soil was good. The Fertile Crescent in Mesopotamia is one such example. Here, fishing villages may have sprung up, causing people to settle down.

The people were probably already knowledgeable about what crops to plant. Most likely, as hunter–gatherers, people had already selected seeds with desirable characteristics for propagation. Then, a chance mutation may have made these plants particularly suitable as a source of food. So now they began to till the good soil where they had settled, and they began to systematically plant crops (Fig. 22B).

As people became more sedentary, they may have had more children, especially because the men were home more often. Population increases may have tipped the scales and caused them to adopt agriculture full time, especially if agriculture could be counted on to provide food for hungry mouths. The availability of agricultural tools must have contributed to making agriculture worthwhile. The digging stick, the hoe, the sickle, and the plow were improved when iron tools replaced bronze in the stone–bronze–iron sequence of ancient tools. Irrigation began as a way to control water supply, especially in semiarid areas and regions of periodic rainfall.

If evolutionary success is judged by population size, agriculture was extremely beneficial to our success because it caused a rapid increase of human numbers all over the Earth. Also, agriculture ushered in civilization as we know it. When crops became bountiful, some people were freed from raising their own food. They began to specialize for other ways of life in towns and then cities. Some people became traders, shopkeepers, bakers, and teachers, to name a few possibilities. Others became the nobility, priests, and soldiers. Today, farming is highly mechanized and cities are extremely large. However, we are on a treadmill. As the human population increases, we need new innovations to produce greater amounts of food. As soon as food production increases, populations grow once again, and the demand for food becomes still greater. Will there be a point when the population is greater than the food capacity? Is that time already upon us?

Industrial Revolution Began

The Industrial Revolution began in England during the eighteenth century and with it a demand for energy in the form of coal and oil that today seems unlimited. Our ability to construct any number of tools, including high-tech computers, is not stored in our genes. We learn it from the previous generation. Our modern civilization that began due to the advent of agriculture is now altering the global environment in a way that affects the evolution of other species. Species are becoming extinct unless they are able to adapt to the presence of our civilization. It could be that biocultural evolution will be so harmful to the biosphere that the human species also will eventually be driven to extinction.

Decide Your Opinion

1. Can technology be used to help us not pollute the environment? How?
2. Should the extinction of other species be prevented? How can it be prevented?
3. Should the human population be reduced in size? How could this occur?

Figure 22B Primitive agriculture.
Agriculture began in several locations across the globe about 10,000 years ago.

not interbreed. Instead, Cro-Magnons seem to have replaced the Neandertals in the Middle East and then spread to Europe 40,000 years ago. There, they lived side by side with the Neandertals for several thousand years. If so, the Neandertals are cousins and not ancestors to us.

Cro-Magnons made advanced stone tools, including compound tools, as when stone flakes were fitted to a wooden handle. They may have been the first to make knifelike blades and throw spears, enabling them to kill animals from a distance. They were such accomplished hunters that some researchers suggest they were responsible for the extinction of many larger mammals, such as the giant sloth, the mammoth, the saber-toothed tiger, and the giant ox, during the late Pleistocene epoch.

Cro-Magnons hunted cooperatively, and perhaps they were the first to have a language. Most likely, they lived in small groups, with the men hunting by day while the women remained at home with the children. It's possible that this hunting way of life among prehistoric people influences our behavior today. The Cro-Magnon culture included art. They sculpted small figurines out of reindeer bones and antlers. They also painted beautiful drawings of animals on cave walls in Spain and France (Fig. 22.20).

Human Variation

Human beings have been widely distributed about the globe since they evolved. As with any other species that has a wide geographic distribution, phenotypic and genotypic variations are noticeable between populations. Today, we say that people have different ethnicities (Fig. 22.21*a*).

It has been hypothesized that human variations evolved as adaptations to local environmental conditions. One obvious difference among people is skin color. A darker skin is protective against the high ultraviolet (UV) intensity of bright sunlight. On the other hand, a white skin ensures vitamin D production in the skin when the UV intensity is low. Harvard University geneticist Richard Lewontin points out, however, that this hypothesis concerning the survival value of dark and light skin has never been tested.

Two correlations between body shape and environmental conditions have been noted since the nineteenth century. The first, Bergmann's rule, states that animals in colder regions of their range have a bulkier body build. The second, Allen's rule, states that animals in colder regions of their range have shorter limbs, digits, and ears. Both of these effects help regulate body temperature by increasing the surface area-to-volume ratio in hot climates and decreasing the ratio in cold climates. For example, Figure 22.21, *b* and *c*, shows that the Massai of East Africa tend to be very tall and slender, with elongated limbs. By contrast, the Eskimos, who live in northern regions, are bulky with short limbs.

Other anatomical differences among ethnic groups, such as hair texture, a fold on the upper eyelid (common in Asian peoples), or the shape of lips, cannot be explained as adaptations to the environment. Perhaps these features became fixed in different populations due to genetic drift. As far as intelligence is concerned, no significant disparities have been found among different ethnic groups.

a.

b.

c.

Figure 22.21 **Ethnic variations in modern humans.**
a. Some of the differences between the three prevalent ethnic groups in the United States may be due to adaptations to the original environment. **b.** The Maasai live in East Africa. **c.** The Kuna live on the San Blas Islands near Panama.

Genetic Evidence for a Common Ancestry

The two hypotheses regarding the evolution of humans, discussed on pages 536-37, pertain to the origin of ethnic groups. The multiregional hypothesis suggests that different human populations came into existence as long as a million years ago, giving time for significant ethnic differences to accumulate despite some gene flow. The out-of-Africa hypothesis, on the other hand, proposes that all modern humans have a relatively recent common ancestor who evolved in Africa and then spread into other regions. Paleontologists tell us that the variation among modern populations is considerably less than among human populations some 250,000 years ago. This would mean that all ethnic groups evolved from the same single, ancestral population.

A comparative study of mitochondrial DNA shows that the differences among human populations are consistent with their having a common ancestor no more than a million years ago. Lewontin, mentioned previously, found that the genotypes of different modern populations are extremely similar. He examined variations in 17 genes, including blood groups and various enzymes, among seven major geographic groups: Caucasians, black Africans, mongoloids, south Asian Aborigines, Amerinds, Oceanians, and Australian Aborigines. He found that the great majority of genetic variation—85%—occurs within ethnic groups, not among them. In other words, the amount of genetic variation between individuals of the same ethnic group is greater than the variation between ethnic groups.

Connecting the Concepts

For more information on the topics presented in this section, refer to the following discussions.

Section 3.6 describes the structure and function of the mitochondria.

Section 23.1 examines the major terrestrial ecosystems that are occupied by humans.

Check Your Progress 22.5

1. Name the fossil hominid that is the oldest of the australopithecines.
2. Name the hominin that was the first to migrate out of Africa.
3. Discuss how *Homo habilis* might have differed from the australopithecines.
4. Discuss how *Homo erectus* might have differed from *Homo habilis.*
5. Describe how the out-of-Africa hypothesis influenced the evolution of modern humans.
6. Describe how Cro-Magnons differ from the other species in the genus *Homo.*

CASE STUDY CONCLUSION

In a follow-up study of the Neandertal genomes, the researchers unveiled another startling discovery. The comparison of the Neandertal and human genomes suggests that the two groups may have interbred long after they had diverged from their common ancestor. The Neandertal genome shares more commonality with modern humans from outside Africa than with modern African populations. Statistical studies have suggested that almost 4% of the genomes of modern Europeans may be Neandertal in nature, suggesting that these two groups interbred freely in the past. If more samples support these findings, scientists agree that they may have to go back and revise some of their basic ideas of how modern humans evolved, and what our species' interactions may have been with the other hominins.

Media Study Tools

www.mhhe.com/maderhuman12e

Enhance your study of this chapter with study tools and practice tests. Also ask your instructor about the resources available through ConnectPlus, including the media-rich eBook, interactive learning tools, and animations.

Summarizing the Concepts

22.1 Origin of Life

A chemical evolution could have produced the protocell.

- Using an outside energy source, small organic molecules were produced by reactions between early Earth's atmospheric gases.
- Macromolecules evolved and interacted.
- The RNA-first hypothesis proposed that only macromolecule RNA was needed for the first cell or cells.

- The protein-first hypothesis proposed that amino acids join to form polypeptides when exposed to dry heat.
- The protocell, a heterotrophic fermenter, lived on preformed organic molecules in the ocean.

The protocell eventually became a true cell once it had genes composed of DNA and could reproduce.

22.2 Biological Evolution

Biological evolution explains both the unity and diversity of life.

- Descent from a common ancestor explains the unity (sameness) of living things.
- Adaptation to different environments explains the great diversity of living things.
- *Fossil evidence supports evolution:* The fossil record gives us the history of life in general and allows us to trace the descent of a particular group.

Darwin discovered much evidence for common descent.

- *Biogeographical evidence:* The distribution of organisms on Earth is explainable by assuming that organisms evolved in one locale.
- *Anatomical evidence:* The common anatomies and development of a group of organisms are explainable by descent from a common ancestor.

	Observation	Result	Conclusion
1	a. Organisms have variations.	b. New adaptations to the environment arise.	Organisms become more adapted with each generation.
2	a. Organisms struggle to exist.	b. More organisms are present than can survive.	
3	a. Organisms differ in fitness.	b. Organisms best suited to the environment survive and reproduce.	

- *Biochemical evidence:* All organisms have similar biochemical molecules.

Darwin developed a mechanism for adaptation known as natural selection:

- The result of natural selection is a population adapted to its local environment.

22.3 Classification of Humans

The classification of humans can be used to trace their ancestry.

- Humans are primates.
- A primate evolutionary tree shows that humans share a common ancestor with African apes.

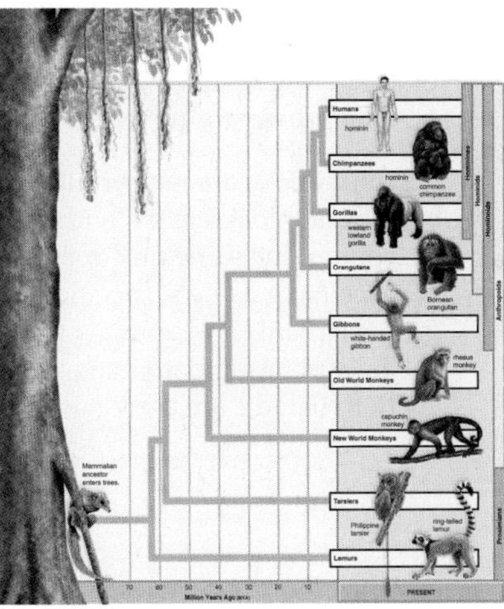

22.4 Evolution of Hominins

- The first hominin (includes humans) most likely lived about 6–7 MYA.
- Certain features (bipedal posture, flat face, and brain) identify fossil hominins.
- Ardipithecines were most likely hominins.

Evolution of Australopithecines

The evolutionary tree of hominids resembles a bush (not a straight line of fossils leading to modern humans).

- Australopithecines (a hominid) lived about 3 MYA.
- They could walk erect, but they had a small brain.
- This testifies to a mosaic evolution for humans (not all advanced features evolved at the same time).

22.5 Evolution of Humans

Fossils are classified as *Homo* with regard to brain size (over 600 cm^3), jaws and teeth (resemble modern humans), and evidence of tool use.

- *Homo habilis* made and used tools.
- *Homo erectus* was the first *Homo* to have a brain size of more than 1,000 cm^3.
- *Homo erectus* migrated from Africa into Europe and Asia.
- *Homo erectus* used fire and may have been big-game hunters.

Evolution of Modern Humans

Two hypotheses of modern human evolution are being tested.

- The multiregional continuity hypothesis suggests that modern humans evolved separately in Europe, Africa, and Asia.
- The out-of-Africa hypothesis says that *H. sapiens* evolved in Africa but then migrated to Asia and Europe.

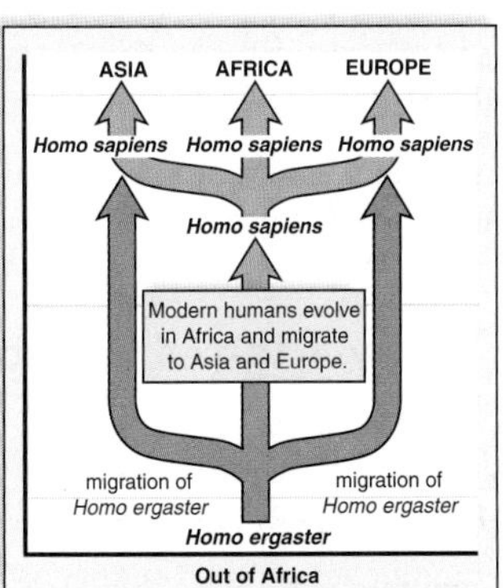

Neandertals

- The Neandertals were already living in Europe and Asia before modern humans arrived.
- They had a culture but did not have the physical traits of modern humans.

Cro-Magnons

- Cro-Magnons are the oldest fossil to be designated *H. sapiens.*
- Their tools were sophisticated, and they had a culture.

Understanding Key Terms

adaptation 520
analogous structure 523
anthropoid 527
artificial selection 526
australopithecine 532
binomial name 526
biocultural evolution 538
biogeography 522
biological evolution 520
bipedal posture 530
cell theory 518
chemical evolution 518
Cro-Magnon 537
culture 535
eukaryotic cell 520
evolutionary tree 530
fossil 520
fossil record 520
heterotroph 519
hominid 530
hominin 530
hominine 530
Homo erectus 535
Homo habilis 535
homologous structure 523
Homo sapiens 536
lineage 530
molecular clock 530
mosaic evolution 532
multiregional continuity hypothesis 536
natural selection 525
Neandertal 537
out-of-Africa hypothesis 536
primate 527
prokaryotic cell 520
prosimian 527
protein-first hypothesis 519
protocell 519
RNA-first hypothesis 519
vestigial structure 523

Match the key terms to these definitions.

a. ________ Process by which populations become adapted to their environment.

b. ________ Organism's modification in structure, function, or behavior suitable to the environment.

c. ________ Structure that is similar in two or more species because of common ancestry.

d. ________ Increase in the complexity of chemicals over time that could have led to the first cells.

e. ________ Any remains of an organism that have been preserved in the Earth's crust.

Testing Your Knowledge of the Concepts

1. List and discuss the steps by which a chemical evolution could have produced a protocell. (pages 518–519)
2. You are studying finches on the Galápagos Islands. What evidence do you need to show common descent and adaptation to the environment? (page 520)
3. Describe the evidence that supports descent from a common ancestor. (pages 520–524)
4. Describe the critical elements of Darwin's natural selection process. (pages 525–526)
5. List a characteristic of each category of human classification. (page 527)
6. What type of life are primates adapted to? Discuss several primate characteristics that aid them in their adaptation. (pages 527–528)
7. How is the human skeleton like that of a chimpanzee skeleton? What's the major difference? (page 529)
8. Why do most investigators use bipedal locomotion, not size of brain, to designate a hominid? (page 530)
9. Describe the characteristics of australopithecines, early Homo, Homo erectus, Neandertals, and Cro-Magnons. (pages 535–537, 539)
10. Contrast the multiregional continuity hypothesis with the out-of-Africa hypothesis for the evolution of Homo sapiens. (page 536)
11. Which of these did Stanley Miller place in his experimental system to show that organic molecules could have arisen from inorganic molecules on the early Earth?
    a. microspheres
    b. purines and pyrimidines
    c. gases in the atmosphere of early Earth
    d. only RNA
    e. All of these are correct.
12. Evolution of the DNA → RNA → protein system was a milestone because the protocell could now
    a. be a heterotrophic fermenter.
    b. pass on genetic information.
    c. use energy to grow.
    d. take in preformed molecules.
    e. All of these are correct.
13. According to Darwin,
    a. the adapted individual is the one who survives and passes on its genes to offspring.
    b. changes in phenotype are passed on by way of the genotype to the next generation.
    c. organisms are able to bring about a change in their phenotype.
    d. evolution is striving toward particular traits.
    e. All of these are correct.
14. If evolution occurs, we would expect different biogeographical regions with similar environments to
    a. contain the same mix of plants and animals.
    b. each have its own specific mix of plants and animals.
    c. have plants and animals with similar adaptations.
    d. have plants and animals with different adaptations.
    e. Both b and c are correct.

15. The fossil record offers direct evidence for evolution because you can
    a. see that the types of fossils change over time.
    b. sometimes find common ancestors.
    c. trace the ancestry of a particular group.
    d. trace the biological history of living things.
    e. All of these are correct.

16. Organisms such as whales and sea turtles adapted to an aquatic way of life
    a. will probably have homologous structures.
    b. will have similar adaptations but not necessarily homologous structures.
    c. may very well have analogous structures.
    d. will have the same degree of fitness.
    e. Both b and c are correct.

17. Which of these gives the correct order of divergence from the main line of descent leading to humans?
    a. prosimians, monkeys, Asian apes, African apes, humans
    b. gibbons, baboons, prosimians, monkeys, African apes, humans
    c. monkeys, gibbons, prosimians, African apes, baboons, humans
    d. African apes, gibbons, monkeys, baboons, prosimians, humans
    e. *H. habilis, H. erectus, H. neandertalensis,* Cro-Magnon

18. Lucy is a member of what species?
    a. *Homo erectus*
    b. *Australopithecus afarensis*
    c. *H. habilis*
    d. *A. robustus*
    e. *A. anamensis* and *A. afarensis* are alternative forms of Lucy.

19. A hominid includes all of the following, except
    a. chimpanzees
    b. orangutans
    c. Old World monkeys
    d. gorillas

20. What possibly may have influenced the evolution of bipedalism?
    a. A larger brain developed.
    b. Food gathering was easier.
    c. The climate became colder.
    d. Both b and c are correct.
    e. Both a and c are correct.

21. *H. ergaster* could have been the first to
    a. use and control fire.
    b. migrate out of Africa.
    c. make tools.
    d. Both a and b are correct.
    e. a, b, and c are correct.

22. Which of these characteristics is not consistent with the others?
    a. opposable thumb
    b. learned behavior
    c. multiple births
    d. well-developed brain
    e. stereoscopic vision

23. The last common ancestor for African apes and hominids
    a. has been found, and it resembles a gibbon.
    b. has not yet been identified, but it is expected to be dated from about 6–7 MYA.
    c. has been found, and it has been dated at 30 MYA.
    d. is not expected to be found because there was no such common ancestor.
    e. most likely lived in Asia, not Africa.

24. If the multiregional continuity hypothesis is correct, then
    a. hominid fossils in China after 100,000 BP are not expected to resemble earlier fossils.
    b. hominid fossils in China after 100,000 BP are expected to resemble earlier fossils.
    c. the mitochondrial Eve study must be invalid.
    d. Both a and c are correct.
    e. Both b and c are correct.

25. A primate evolutionary tree
    a. exists only for humans.
    b. shows the evolutionary relationship among the different types of primates.
    c. should not include extinct forms.
    d. indicates that the ape lineage and human lineage are still joined.
    e. All of these are correct.

26. Classify humans by filling in the missing lines.

Domain	Eukarya
Kingdom	Animalia
Phylum	a. _____
b. _____	Mammalia
c. _____	Primates
Family	d. _____
e. _____	*Homo*
Species	f. _____

In questions 27–30, match each description to a type of evolutionary evidence in the key.

Key:

a. biogeography
b. fossil record
c. comparative biochemistry
d. comparative anatomy

27. Species change over time.
28. Forms of life are variously distributed.
29. A group of related species have homologous structures.
30. The same types of molecules are found in all living things.

Thinking Critically About the Concepts

Much of the work done in archaeology is very tedious. For example, the Science Focus, *Homo floresiensis,* describes the discovery of *H. floresiensis* in a limestone cave on the island of Flores. Researchers Michael Morwood, of the University of New England in Australia, and Radien Soejono, of the Indonesian Center for Archaeology in Jakarta, were in charge of the dig. They began the work in July 2001. It was not until the end of three seasons of field work that they found evidence of hominins—a single tooth. Once the first female skeleton had been found, colleague Peter Brown spent three months analyzing it.

1. In archaeology, why is it critically important to know where to dig? What clues would you look for to determine where to dig?
2. What artifacts do archaeologists use to determine a population's bioculture?
3. What types of information can scientists derive about an ancient people, using only a single human skull from that group?

CHAPTER 23

Global Ecology and Human Interferences

CASE STUDY THREATS TO ECOSYSTEMS

It is the year 2100—the future Earth. Pristine ecosystems, such as the one pictured here, may be a thing of the past. People in the northeastern United States no longer have shovels or snowblowers as winters barely get below-freezing temperatures. Corn is now being grown in the middle latitudes of Canada, as opposed to Iowa. New Orleans is completely under water, and Boston's sewage system no longer functions because of the rise in sea level. Several island countries in the Pacific and Indian oceans no longer exist, and many coastal regions are plagued by flooding. Skin cancer has reached an all-time high and affects half of the people in the world. Malaria is now a problem for over half of the population of the United States, and many houses now use mosquito nets that historically were only found in tropical regions. Nearly all water bodies are overgrown with algae and pond scum, and freshwater fish are disappearing. Freshwater is limited and costs over $50 a bottle. Hurricanes are more frequent and severe than ever before . . .

These scenarios are based on the fact that there are consequences to humans when ecosystems fail. This chapter will take a look at the structure of an ecosystem and of the cycling of nutrients and energy within ecosystems.

As you read through the chapter, think about the following questions.

1. What is the difference between a biotic and an abiotic component of an ecosystem?
2. How does the cycling of energy and chemicals differ in an ecosystem?
3. How is human activity disrupting the normal cycling of carbon and water?

CHAPTER CONCEPTS

23.1 The Nature of Ecosystems
The biosphere encompasses that part of Earth where living things live in ecosystems. Populations interact among themselves and with the physical environment in ecosystems. Ecosystems are characterized by energy flow and chemical cycling.

23.2 Energy Flow
Ecosystems contain food webs, in which the various populations are connected by predator and prey. Food chains have a limited length. As demonstrated by food pyramids, only about 10% of energy is passed from one feeding level to the other. Eventually, all the energy dissipates but the chemicals cycle back to the photosynthesizers.

23.3 Global Biogeochemical Cycles
Biogeochemical cycles contain reservoirs, which retain nutrients; exchange pools, where nutrients are readily available; and the biotic community, which passes nutrients from one population to the next. The water, carbon, and nitrogen cycles are gaseous because the exchange pool is the atmosphere. The phosphorus cycle is a sedimentary cycle.

BEFORE YOU BEGIN

Before beginning this chapter, take a few moments to review the following discussions.

Section 1.1 What is the relationship between populations and ecosystems in the levels of biological organization?

Section 1.1 Why must living organisms acquire materials and energy?

Section 1.2 What are some of the threats from humans to the biosphere?

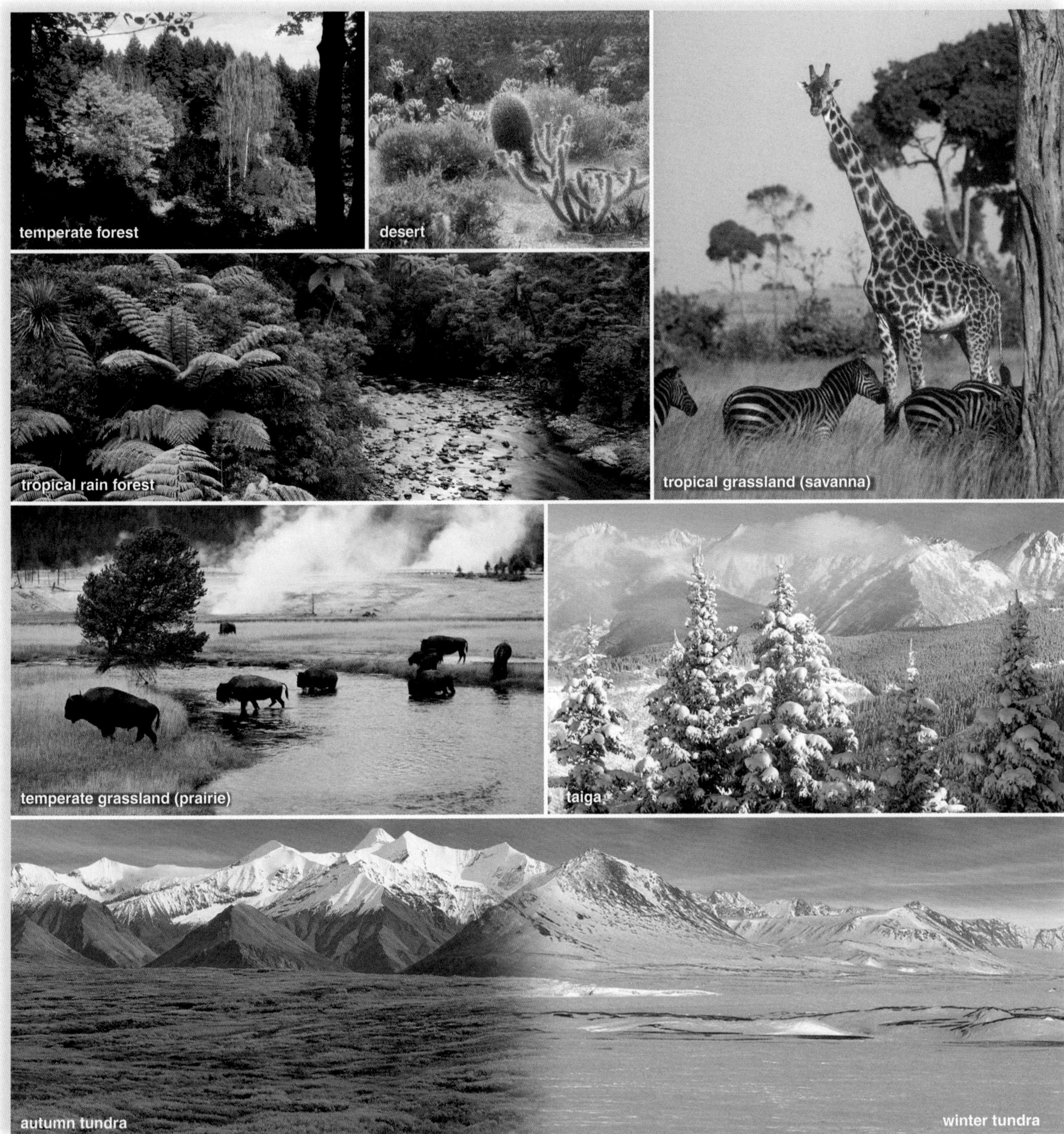

Figure 23.1 The temperature and rainfall characteristics of the major terrestrial ecosystems. Temperate forests have mild temperatures and occur where rainfall is moderate, yet sufficient to support trees. Deserts have changeable temperatures with minimal rainfall. Tropical rain forests, which generally occur near the equator, have a high average temperature and the greatest amount of rainfall of all the terrestrial ecosystems. A tropical grassland (savanna) has high temperatures and moderate/seasonal rainfall. A temperate grassland (prairie) has low to high temperatures, with low annual rainfall. The taiga, a coniferous forest that encircles the globe, has a low average temperature but moderate rainfall. The tundra is the northernmost terrestrial ecosystem and has the lowest average temperature of all the terrestrial ecosystems, with minimal to moderate rainfall.

a.

c.

d.

Figure 23.2 Examples of freshwater and saltwater ecosystems.
Aquatic ecosystems are divided into those that have salt water, such as (**a**) the ocean, and (**b**) those that have freshwater, such as a river. Saltwater, or marine, ecosystems also include (**c**) coral reefs and (**d**) salt marshes.

23.1 The Nature of Ecosystems

Learning Outcomes

Upon completion of this section, you should be able to

1. State the relationship between ecosystems and the biosphere.
2. Explain the differences between the roles of autotrophs and heterotrophs in an ecosystem.
3. Define the term *niche*.
4. Distinguish between the paths of energy and chemicals in an ecosystem.

The **biosphere** is where organisms are found on planet Earth, from the atmosphere above to the depths of the oceans below and everything in between. Specific areas of the biosphere where organisms interact among themselves and the physical and chemical environment are called **ecosystems.** The interactions that occur in an ecosystem maintain balance in that specific area, which in turn affects the balance of the biosphere. Human activities can alter the interactions between organisms and their environments in ways that reduce the abundance and diversity of life in an ecosystem. It is important to understand how ecosystems function so that we can repair past damage and predict how human activities might change normal conditions.

Types of Ecosystems

Scientists recognize several distinctive major types of terrestrial ecosystems, also called **biomes** (Fig. 23.1). Temperature and rainfall define the biomes. A variety of organisms adapted to the regional climate characterize each biome. The tropical rain forest, which occurs at the equator, is dominated by large, evergreen, broad-leaved trees. The savanna is a tropical grassland that supports many types of grazing animals. Temperate grasslands receive less rainfall than temperate forests (where many trees lose their leaves during the winter). Deserts receive scant rainfall and therefore lack trees. The taiga is a very cold northern forest of conifers such as pine, spruce, hemlock, and fir. Bordering the North Pole is the frigid tundra, which has long winters and a short growing season. A permafrost persists even during the summer in the tundra and prevents large plants from becoming established.

Animation Biomes

Aquatic ecosystems are divided into those composed of freshwater and those composed of salt water (marine ecosystems). The ocean is a marine ecosystem that covers 70% of the Earth's surface. Two types of freshwater ecosystems are those with standing water, such as lakes and ponds, and those with running water, such as rivers and streams. The richest marine ecosystems lie near the coasts. Coral reefs are located offshore, and salt marshes occur where rivers meet the sea (Fig. 23.2).

Video Coral Reef Ecosystems

Biotic Components of an Ecosystem

The biotic components of an ecosystem are living things that can be categorized according to their food source (Fig. 23.3). The abiotic components are the nonliving components, such as soil type, water, and weather. Some biotic species are autotrophs, and some are heterotrophs.

Autotrophs

Autotrophs require only inorganic nutrients and an outside energy source to produce organic nutrients for their own use and for all the other members of a community. Therefore, they are called **producers,** meaning they produce food (Fig. 23.3*a*). Photosynthetic organisms produce most of the organic nutrients for the biosphere. Algae of all types possess chlorophyll and carry on photosynthesis in freshwater and marine habitats. Green plants are the dominant photosynthesizers on land.

Video Plants

Heterotrophs

Heterotrophs need a source of organic nutrients. They are **consumers,** meaning they consume food. **Herbivores** are organisms that graze directly on plants or algae (Fig. 23.3*b*). You're probably familiar with deer, rabbits, and cows that are herbivores in terrestrial habitats. Some insects are herbivores, too. In aquatic ecosystems, some protists are herbivores. **Carnivores** feed on other animals. Spiders that feed on insects are carnivores, as are osprey that feed on fish (Fig. 23.3*c*). This example allows us to mention that there are primary consumers (e.g., plant-eating insects), secondary consumers (e.g., insect-eating spiders), and tertiary consumers (e.g., osprey, hawks). Sometimes tertiary consumers are called top predators. **Omnivores** are animals that feed on plants and animals. Humans are omnivores.

Video Snake Eating

Detritus feeders are organisms that feed on detritus, decomposing particles of organic matter. Marine fan worms take detritus from the water, and clams take it from the sub-

Connections and Misconceptions

How do producers at the bottom of the ocean floor make food without sunlight?

Producers at the bottom of the ocean floor use chemical energy instead of sunlight to make food. They are called *chemoautotrophs.* Volcanoes at the bottom of the ocean floor release hydrogen sulfide gas through cracks called hydrothermal vents. (Hydrogen sulfide is the nasty smelling gas we associate with rotten eggs.) Some chemoautotrophs split hydrogen sulfide to obtain the energy needed to link carbon atoms together to form glucose.

The glucose contained in these chemoautotrophs sustains a variety of bizarre organisms such as giant tube worms, anglerfish, and giant clams.

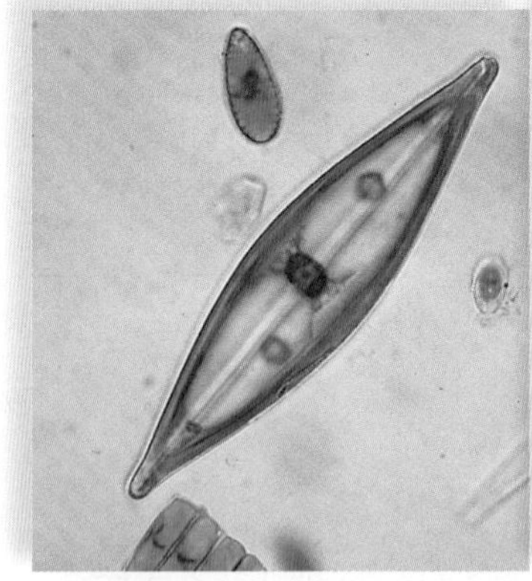

a. Producers

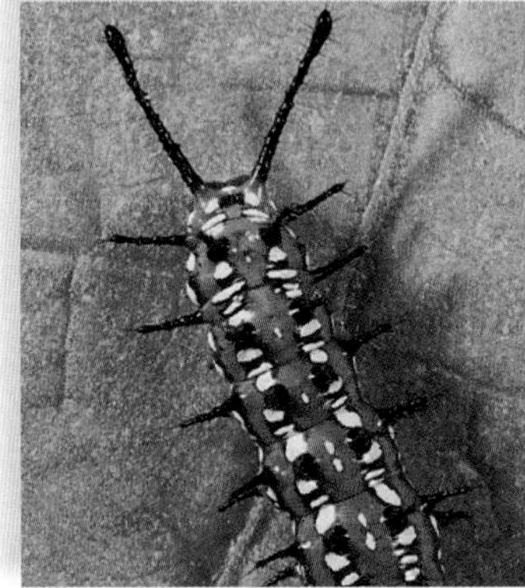

b. Herbivores

c. Carnivores

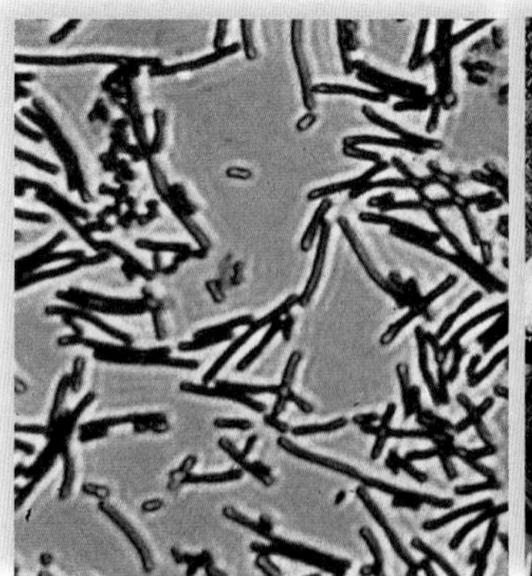

d. Decomposers

Figure 23.3 The biotic components of an ecosystem.
a. Diatoms and green plants are autotrophs. **b.** Caterpillars and rabbits are herbivores. **c.** Spiders and osprey are carnivores. **d.** Some bacteria and some mushrooms are decomposers.

stratum. Earthworms and some beetles, termites, and ants are terrestrial detritus feeders. Bacteria and fungi, including mushrooms, are decomposers. They acquire nutrients by breaking down dead organic matter, including animal wastes. Decomposers perform a valuable service because they release inorganic substances that are taken up by plants once more (Fig. 23.3*d*). Otherwise, plants would be completely dependent only on physical processes, such as the release of minerals from rocks, to supply them with inorganic nutrients.

Video Decomposers

Niche

A **niche** is the role of an organism in an ecosystem. Descriptions of a niche include how an organism gets its food, how it interacts with other populations in the same community, and what eats it. Human beings are also part of the cycle of life. The chemicals making up our bodies must also return to nature.

Video Cichlid Specialization

Energy Flow and Chemical Cycling

When we diagram the interactions of all the populations in an ecosystem, it is possible to illustrate two phenomena that characterize every ecosystem. One of these phenomena is energy flow, which begins when producers absorb solar (and in some cases, chemical) energy. The second, nutrient cycling, occurs when producers take in inorganic chemicals from the physical environment. Thereafter, producers make organic nutrients (food) directly for themselves and indirectly for the other populations of the ecosystem. Energy flow occurs when nutrients pass from one population to another. Eventually, all the energy content is converted to heat. The heat then dissipates in the environment. Therefore, most ecosystems cannot exist without a continual supply of solar energy. Chemicals cycle when inorganic nutrients are returned to the producers from the atmosphere or soil (Fig. 23.4).

Only a portion of the organic nutrients made by autotrophs is passed on to heterotrophs because plants use organic molecules to fuel their cellular respiration (see Fig. 3.19). Similarly, only a small percentage of nutrients taken in by heterotrophs is available to higher-level consumers. Figure 23.5 shows why. Some of the food eaten by an herbivore is never digested and is eliminated as feces. Metabolic wastes are excreted in urine. Of the assimilated energy, a large portion is used during cellular respiration and thereafter becomes heat. Only the remaining food energy, converted into increased body weight (or additional offspring), becomes available to carnivores. Plants carry on cellular respiration, too, so only about 55% of the original energy absorbed by plants is available to an ecosystem. Further, as organisms feed on one another, less and less of this 55% is available in a usable form.

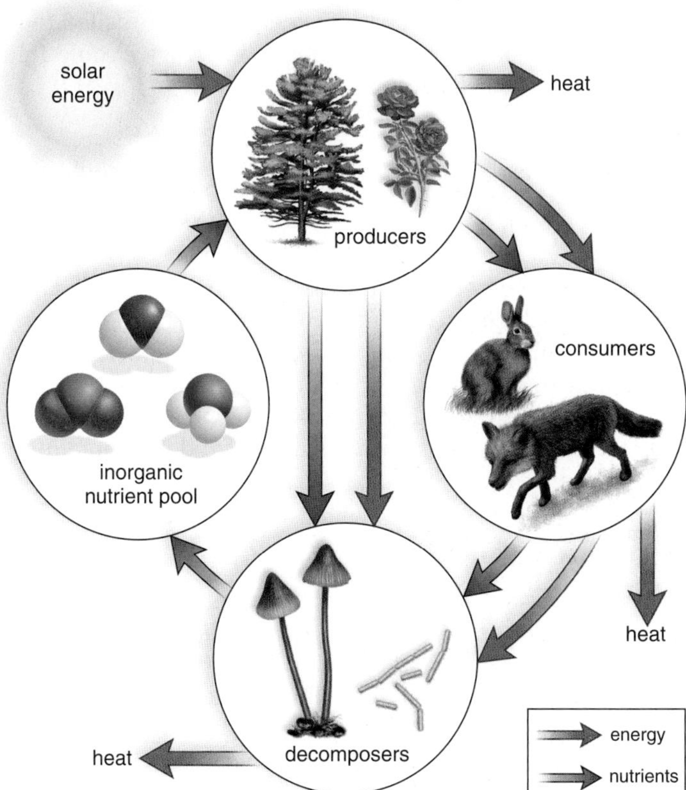

Figure 23.4 Energy flow and chemical cycling in an ecosystem.
Chemicals cycle, but energy flows through an ecosystem. As energy transformations repeatedly occur, all the energy derived from the sun eventually dissipates as heat.

Figure 23.5 The fate of food energy taken in by an herbivore.
Only about 10% of the food energy taken in by an herbivore is passed on to carnivores. A large portion goes to detritus feeders via defecation, excretion, and death; another large portion is used for cellular respiration.

The elimination of wastes and the deaths of all organisms do not mean that substances are lost to an ecosystem. Instead they are nutrients made available to decomposers. Decomposers convert organic nutrients, such as glucose, back into inorganic chemicals, such as carbon dioxide and water. The inorganic chemicals are then released to the soil or atmosphere. Chemicals complete their cycle within an ecosystem when inorganic chemicals are absorbed by the producers from the atmosphere or soil.

Connecting the Concepts

For more information on the topics presented in this section, refer to the following discussions.

Section 1.1 describes the relationship between populations, the ecosystems on the planet, and the biosphere.

Section 3.6 examines how organisms release the energy stored in organic nutrients by cellular respiration.

Check Your Progress 23.1

1. Discuss why it could be said that the biosphere is a giant ecosystem.
2. Explain why autotrophs are called producers and heterotrophs are called consumers.
3. List the different types of consumers that may be present in an ecosystem.
4. Distinguish between the movement of energy and chemicals in an ecosystem.

23.2 Energy Flow

Learning Outcomes

Upon completion of this section, you should be able to

1. Summarize the differences between a grazing and a detrital food web.
2. Explain the relationship between the trophic level of an ecological pyramid and the biomass at each level.

The principles we have been discussing can now be applied using a forest ecosystem. The various interconnecting paths of energy flow are represented by a **food web,** which is a diagram that describes **trophic** (feeding) **relationships.** Figure 23.6*a* is a **grazing food web** because it begins with an oak tree and grass. Caterpillars feed on oak leaves; mice, rabbits, and deer feed on leaves and grass at or near the ground. Birds, chipmunks, and mice feed on seeds and nuts, but they are omnivores because they also feed on caterpillars. These herbivores and omnivores are then food for a number of different carnivores.

Figure 23.6*b* is a **detrital food web.** This type of food web begins with wastes and the remains of dead organisms. Detritus is food for decomposers and soil organisms like earthworms. Earthworms may be food for carnivorous invertebrates. In turn, salamanders and shrews may consume the carnivorous insects. The members of detrital food webs may become food for aboveground carnivores, so the detrital and grazing food webs are connected.

Video
Decomposers

We naturally tend to think that aboveground plants, such as trees, are the largest storage form of organic matter and energy. This is not necessarily the case. In this particular forest, the organic matter lying on the forest floor and mixed into the soil contains over twice as much energy as the leaves of living trees. Therefore, more energy in a forest may be funneling through the detrital food web than through the grazing food web.

Trophic Levels

The arrangement of the species in Figure 23.6 suggests that organisms are linked to one another in a straight line, according to feeding or predator–prey relationships. Diagrams that show a single path of energy flow are called **food chains.** For example, in the grazing food web, we could find this **grazing food chain:**

leaves → caterpillars → birds → hawks

And in the detrital food web (Fig. 23.6*b*), we could find this **detrital food chain:**

detritus → earthworms → beetles → shrews

A **trophic level** is composed of all the organisms that feed at a particular link in a food chain. In the grazing food web in Figure 23.6*a*, going from left to right, the trees are producers (first trophic level); the first series of animals are primary consumers (second trophic level); and the next group of animals are secondary consumers (third trophic level).

Ecological Pyramids

The shortness of food chains can be attributed to the loss of energy between trophic levels. As mentioned, only about 10% of the energy of one trophic level is available to the next trophic level. Therefore, if an herbivore population consumes 1,000 kg of plant material, only about 100 kg is converted to herbivore tissue, 10 kg to first-level carnivores, and 1 kg to second-level carnivores. This 10% rule of thumb explains why so few carnivores can be supported in a food

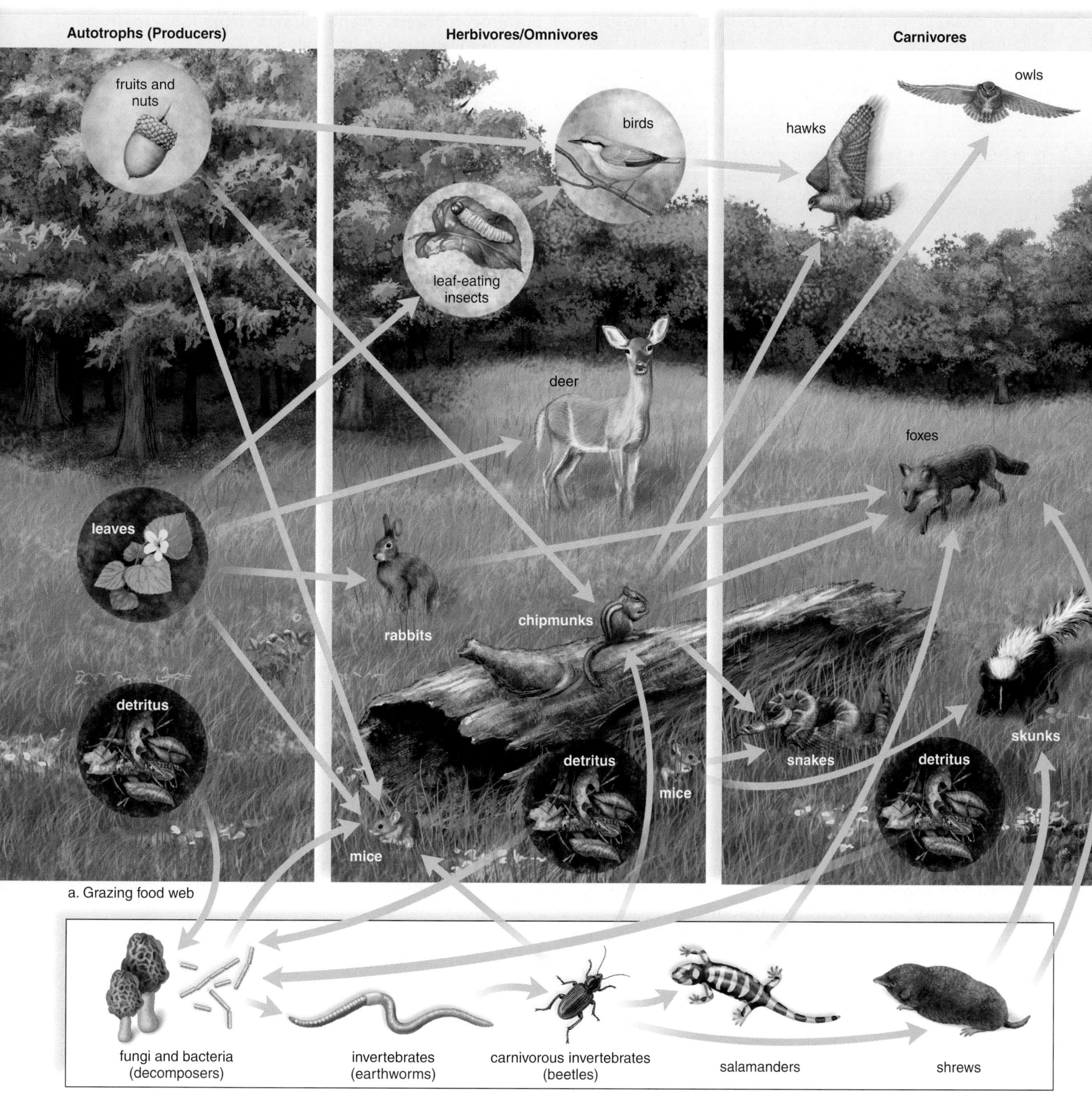

Figure 23.6 **Food webs illustrate ecological relationships.**
Food webs are descriptions of who eats whom. **a.** Tan arrows illustrate possible grazing food webs. The tree and other organisms that convert sun energy into food energy are producers (first trophic level). Mice and other animals that eat the producers are primary consumers (second trophic level). Carnivores that rely on primary consumer animals for energy (the hawk, fox, skunk, snake, and owl) are secondary consumers (third trophic level). **b.** Green arrows illustrate possible detrital food webs. These begin with detritus: organic waste and remains of dead organisms. Decomposers and detritus feeders recycle these organic nutrients. The organisms in the detrital food web may be prey for animals in the grazing food web, as when chipmunks feed on bugs. Thus, the grazing food web and detrital food web are interconnected.

Figure 23.7 The influence of trophic level on biomass. The biomass, or dry weight (g/m^2), for trophic levels in a grazing food web in a bog at Silver Springs, Florida. There is a sharp drop in biomass between the producer level and herbivore level. This is consistent with the common knowledge that the detrital food web plays a significant role in bogs.

web. The flow of energy with large losses between successive trophic levels is sometimes depicted as an **ecological pyramid** (Fig. 23.7).

Energy losses between trophic levels also result in pyramids based on the number of organisms in each trophic level. When constructing a pyramid based on number of organisms, problems arise, however. For example, in Figure 23.6*a*, each tree would contain numerous caterpillars. Therefore, there would be more herbivores than autotrophs. The explanation has to do with size. An autotroph can be as tiny as a microscopic alga or as big as a beech tree. Similarly, an herbivore can be as small as a caterpillar or as large as an elephant.

Pyramids of biomass eliminate size as a factor because **biomass** is the number of organisms multiplied by the weight of organic matter contained in one organism. You would certainly expect the biomass of the producers to be greater than the biomass of the herbivores and that of the herbivores to be greater than that of the carnivores. In aquatic ecosystems, such as lakes and open seas, the herbivores may have a greater biomass than the producers. This is because algae are the only producers. Over time, algae reproduce rapidly but are also consumed at a high rate.

These types of problems are making some ecologists hesitant about using pyramids to describe ecological relationships. Another issue concerns the role played by decomposers. These organisms are rarely included in pyramids, even though a large portion of energy becomes detritus in many ecosystems.

Connecting the Concepts

For more information on the topics presented in this section, refer to the following discussions.

Section 24.1 examines how human population growth is threatening the structure of many ecological pyramids.

Check Your Progress 23.2

1. Explain the difference between a grazing food web and a detrital food web and give an example of each.
2. Describe the type of diagram that represents a single path of energy flow in an ecosystem.
3. Calculate the amount of biomass available to a second-level carnivore if the producer biomass in the ecosystem is 1,500 g/m^2.

23.3 Global Biogeochemical Cycles

Learning Outcomes

Upon completion of this section, you should be able to

1. Identify whether a given biogeochemical cycle is gaseous or sedimentary.
2. Describe the water cycle and discuss how human activity interferes with this cycle.
3. Describe the carbon cycle and discuss how human activity is altering the carbon cycle.
4. Summarize the nitrogen cycle and discuss how human activity is creating problems with this cycle.
5. Summarize the phosphorous cycle and its relationship to eutrophication.

In this section, we will examine in more detail how chemicals cycle through ecosystems. All organisms require a variety of organic and/or inorganic nutrients. For example, carbon dioxide and water are necessary nutrients for photosynthesizers. Nitrogen is a component of all the structural and functional proteins and nucleic acids that sustain living tissues. Phosphorus is essential for ATP and nucleotide production.

The pathways by which chemicals circulate through ecosystems involve both living (biotic) and nonliving (abiotic) components. Therefore, they are known as **biogeochemical cycles.** A biogeochemical cycle can be gaseous or sedimentary. In a gaseous cycle, such as the carbon and nitrogen cycles, the element returns to and is withdrawn from the atmosphere as a gas. The phosphorus cycle is a sedimentary cycle. Phosphorous is absorbed from the soil by plant roots, passed to heterotrophs, and eventually returned to the soil by decomposers.

Chemical cycling involves the components of ecosystems shown in Figure 23.8. A *reservoir* is a source normally unavailable to producers, such as carbon in calcium carbonate shells on ocean bottoms. An *exchange pool* is a source from which organisms do generally take chemicals, such as the atmosphere or soil. Chemicals move along food chains in a *biotic community.*

Human activities (purple arrows) remove chemicals from reservoirs and exchange pools and make them available to the biotic community. In this way, human activities result in pollution because it upsets the normal balance of nutrients for producers in the environment.

Video Salmon Farming

Connections and Misconceptions

How much water is required to produce your food?

Growing a single serving of lettuce takes about 6 gallons of water. Producing an 8 oz. glass of milk requires 49 gallons of water. That includes the amount of water the cow drinks, the water used to grow the cow's food, and water needed to process the milk. Producing a single serving of steak consumes more than 2,600 gallons of water.

Figure 23.8 The cycling of nutrients between biotic communities and biogeochemical reservoirs. Reservoirs, such as fossil fuels, minerals in rocks, and sediments in oceans, are normally relatively unavailable sources of nutrients for the biotic community. Nutrients in exchange pools, such as the atmosphere, soil, and water, are available sources of chemicals for the biotic community. When human activities (purple arrows) remove chemicals from a reservoir or an exchange pool and make them available to the biotic community, pollution can result. This is because not all the nutrients are used. For example, when humans burn fossil fuels, CO_2 increases in the atmosphere and contributes to global warming.

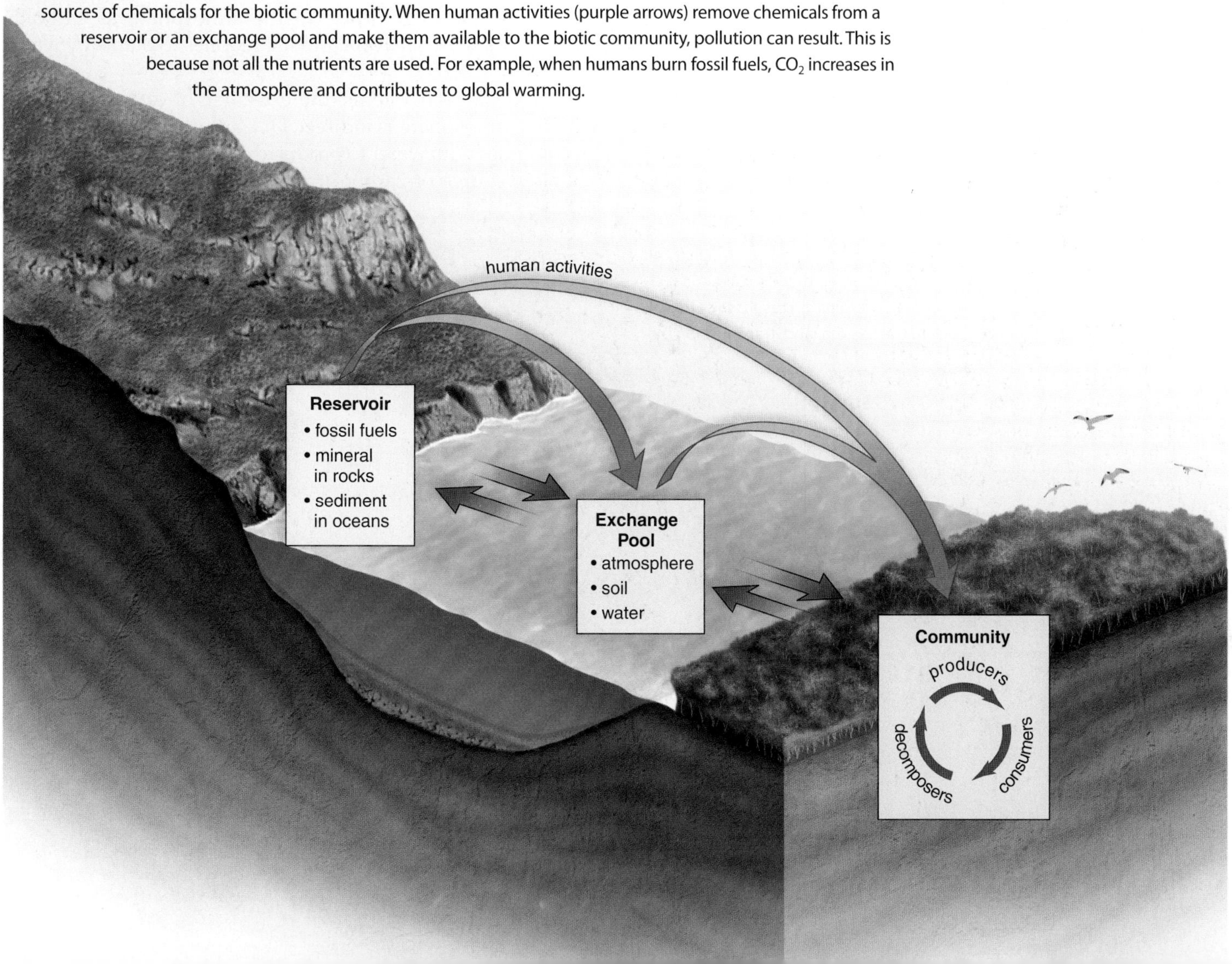

The Water Cycle

The **water (hydrologic) cycle** is described in Figure 23.9. The width of the arrows in this and the other cycles that will be examined indicates the transfer rate of water between components of an ecosystem.

① During **evaporation** in the water cycle, the sun's rays cause freshwater to evaporate from seawater, leaving the salts behind. Net condensation occurs. During condensation, a gas is changed into a liquid. ② Vaporized freshwater rises into the atmosphere, condenses, and then falls as **precipitation** (e.g., rain, snow, sleet, hail, and fog) over the oceans and the land.

③ Water also evaporates from land and from plants (evaporation from plants is called transpiration). ④ Land lies above sea level, so gravity eventually returns all freshwater to the sea. In the meantime, much water is contained within standing waters (lakes and ponds), flowing water (streams and rivers), and groundwater. ⑤ **Runoff** is water that flows directly into nearby streams, lakes, wetlands, or the ocean.

Instead of running off, some precipitation sinks, or percolates, into the ground. This saturates the earth to a certain level. The top of the saturation zone is called the groundwater table, or the water table. ⑥ Sometimes, groundwater is also located in **aquifers,** rock layers that contain water and release it in appreciable quantities to wells or springs. Aquifers are recharged when rainfall and melted snow percolate into the soil.

Human Activities

Humans interfere with the water cycle in three ways. First, they withdraw water from aquifers. Second, they clear vegetation from land and build roads and buildings that prevent percolation and increase runoff. Third, they interfere with the natural processes that purify water and instead add pollutants like sewage and chemicals to water.

Video Thames River

In some parts of the United States, especially the arid West and southern Florida, withdrawals from aquifers exceed any possibility of recharge. This is called "groundwater mining." In these locations, the groundwater is dropping. Residents may run out of groundwater, at least for irrigation purposes, within a few short years. Freshwater, which makes up only about 3% of the world's supply of water, is called a renewable resource because a new supply is always being produced. However, it is possible to run out of freshwater when the available supply runs off instead of entering bodies of freshwater and aquifers. Freshwater may also become so polluted that it is not usable.

The Carbon Cycle

The carbon dioxide (CO_2) in the atmosphere is the exchange pool for the carbon cycle. In this cycle, organisms in both terrestrial and aquatic ecosystems exchange carbon dioxide with the atmosphere (Fig.23.10). ① On land, plants take up carbon dioxide from the air. Through photosynthesis, they incorporate carbon into nutrients used by autotrophs and heterotrophs. ② When organisms, including plants, respire, carbon is returned to the atmosphere as carbon dioxide. Therefore, carbon dioxide recycles to plants by way of the atmosphere.

In aquatic ecosystems, the exchange of carbon dioxide with the atmosphere is indirect. ③ Carbon dioxide from the air combines with water to produce bicarbonate ion (HCO_3^-). This is a source of carbon for algae that produce food for

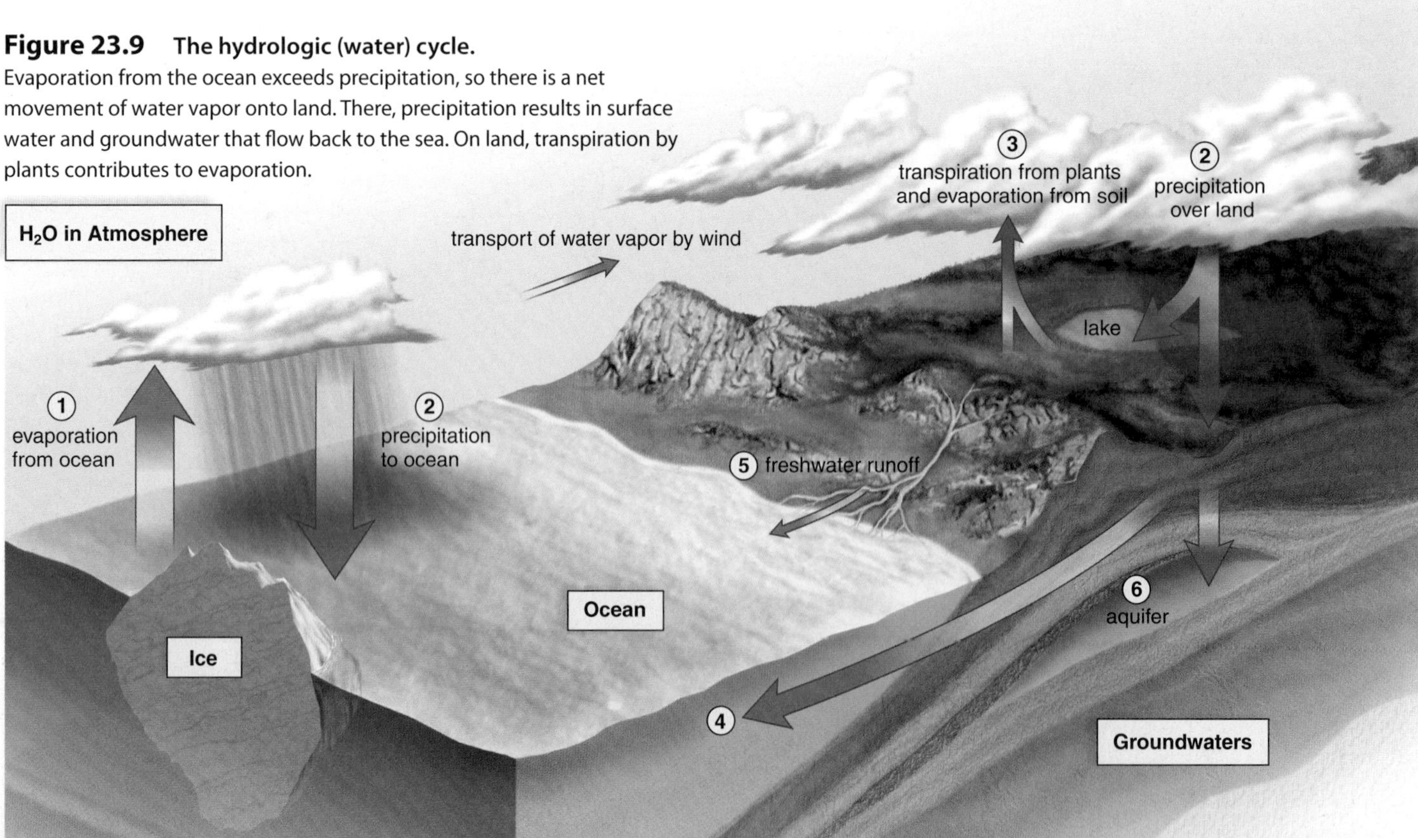

Figure 23.9 **The hydrologic (water) cycle.**
Evaporation from the ocean exceeds precipitation, so there is a net movement of water vapor onto land. There, precipitation results in surface water and groundwater that flow back to the sea. On land, transpiration by plants contributes to evaporation.

themselves and for heterotrophs. Similarly, when aquatic organisms respire, the carbon dioxide they give off becomes bicarbonate ion. ④ The amount of bicarbonate in the water is in equilibrium with the amount of carbon dioxide in the air.

Reservoirs Hold Carbon

⑤ Living and dead organisms contain organic carbon and serve as one of the reservoirs for the carbon cycle. The world's biotic components, particularly trees, contain over 800 billion tons of organic carbon. An additional 1,000–3,000 billion metric tons are estimated to be held in the remains of plants and animals in the soil. Ordinarily, decomposition of plants and animals returns CO_2 to the atmosphere.

⑥ In the history of the Earth, some 300 MYA, plant and animal remains were transformed into coal, oil, and natural gas. We call these materials the **fossil fuels.** Another reservoir for carbon is the inorganic carbonate that accumulates in limestone and in calcium carbonate shells. Many marine organisms have calcium carbonate shells that remain in bottom sediments long after the organisms have died. Geologic forces change these sediments into limestone.

Human Activities

The transfer rates of carbon dioxide because of photosynthesis and cellular respiration, which include the work of decomposers, are just about even. However, more carbon dioxide is being deposited in the atmosphere than is being removed. ⑦ This increase is largely due to the burning of fossil fuels and the destruction of forests to make way for farmland and pasture. When we do away with forests,

Figure 23.10 **The carbon cycle.**
The carbon cycle is a gaseous biogeochemical cycle. Producers take in carbon dioxide from the atmosphere and convert it to organic molecules that feed all organisms. The transfer rate of carbon into the atmosphere due to respiration approximately matches the rate due to withdrawal by plants for photosynthesis. Fossil fuels arise when organisms die but do not decompose. When humans burn fossil fuels and destroy vegetation (purple arrows), more carbon dioxide is added to the atmosphere than is withdrawn. This causes environmental pollution.

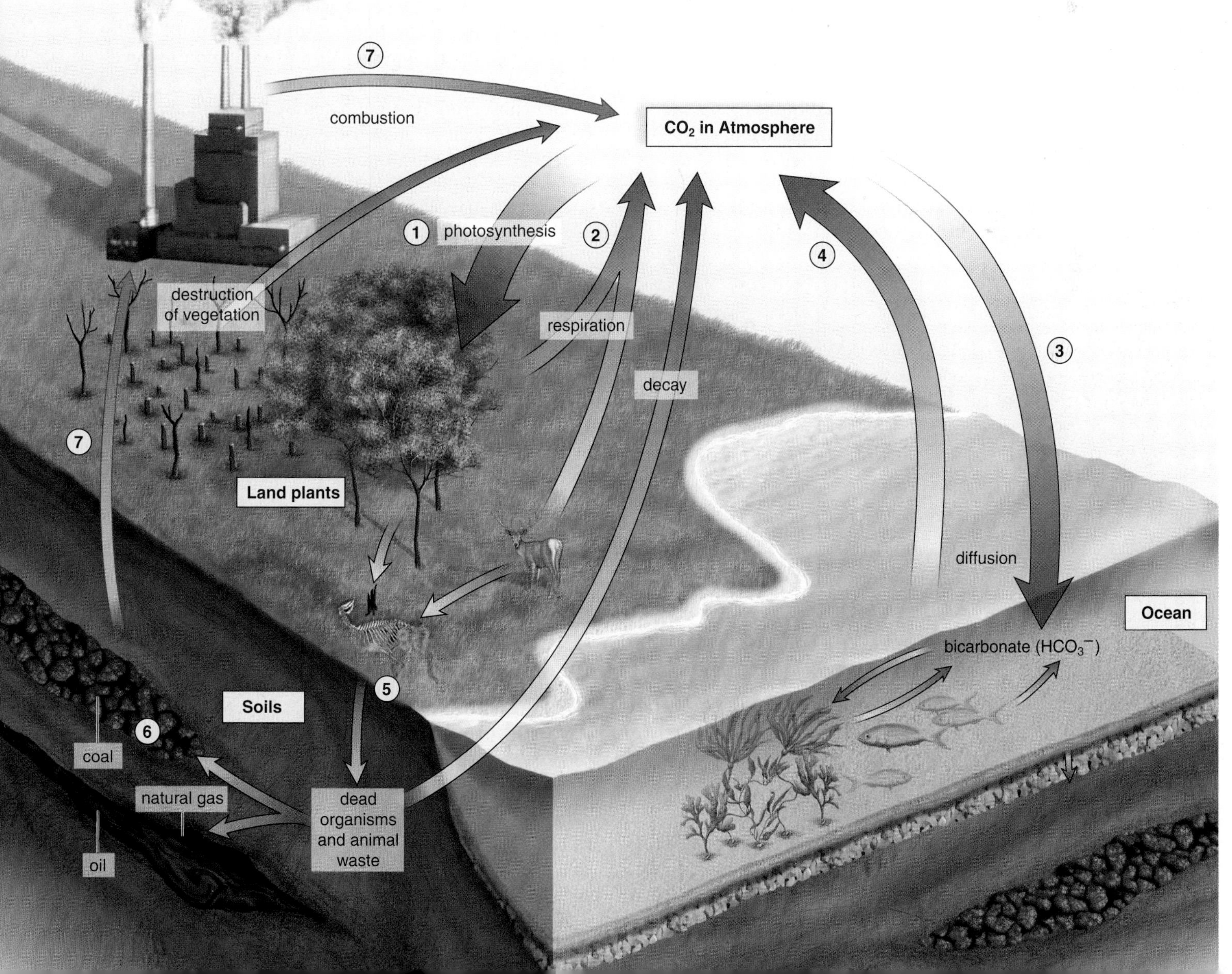

Biology Matters Bioethical Focus

Guaranteeing Access to Safe Drinking Water

The ability to get safe drinking water may not be something you've ever thought about. In the United States and other developed countries, safe, clean water usually pours out of your home faucets every time you turn them on. When water is plentiful, it's taken for granted and a great deal of it is wasted. However, water is a precious and often scarce resource in many parts of the world.

Almost 20% of the Earth's population doesn't have access to safe drinking water (Fig. 23A). Further, basic sanitation (proper disposal of human and animal waste) is unavailable to 40%. Most people lacking safe drinking water and/or basic sanitation live in China, the Middle East, or Africa. These are some of the world's poorest people. If they have to pay for water, they are often charged much more than wealthy people living in the same area. They may have to walk miles to collect water and then return to their homes.

Having safe drinking water and basic sanitation would impact people's lives and health more than any other intervention. More than 80% of diseases, including typhoid fever and cholera, are associated with foul water and improper sanitation. Each year, more than 5 million people die from water-pollution-related illnesses. This is the leading cause of death for children under age 5.

Solving the problem of making safe water available to more people is no easy task. One of the Millennium Development Goals (agreed upon by members of the United Nations in 2000) is to halve the number of people without access to safe water by 2015. The cost has been projected to be between $4 billion and $11.3 billion per year. An unusual twist to the water crisis is the interest in water as an investment and commodity to be traded or sold. The Goldman Sachs Group, Inc. (often referred to as *Goldman Sachs*), is a large global banking company involved in investment banking. Goldman Sachs estimates the value of the water industry at $425 billion. At a recent conference hosted by the company, the water shortage was listed among the world's top threats to global stability.

Each day, an average American uses 100–176 gallons of water. By contrast, the typical African family uses about 5 gallons. We may need to consider how much of the water we use daily is necessary. More effort will be needed to ensure that all people have the ability to get safe drinking water. A more stable global community may depend on it.

Decide Your Opinion

1. Should developed nations assume responsibility for providing safe drinking water and sanitation to developing countries? If so, to what extent?
2. Water conservation is mandated by law in many states. Should legislation requiring water conservation be extended to include all states? Why or why not?
3. What form might this legislation take? How should it be applied?

Figure 23A **Many people in developing nations lack access to safe drinking water.**
Individuals in developing countries may have little or no access to safe drinking water. Providing a universal clean water supply for all is a global necessity.

Connections and Misconceptions

What are some other greenhouse gases?

In addition to carbon dioxide, these gases also play a role in the greenhouse effect:

- *Methane (CH_4):* A single molecule of methane has 21 times the warming potential of a molecule of carbon dioxide, making it a powerful greenhouse gas. Methane is a natural by-product of the decay of organic material but is also released by landfills and the production of coal, natural gas, and oil.
- *Nitrous oxide (N_2O):* Nitrous oxide is released from the combustion of fossil fuels and as gaseous waste from many industrial activities.
- *Hydrofluorocarbons:* These were initially produced to reduce the levels of ozone-depleting compounds in the upper atmosphere. Unfortunately, although present in very small quantities, they are potent greenhouse gases.

we reduce a reservoir and lose the organisms that take up excess carbon dioxide. Today, the amount of carbon dioxide released into the atmosphere is about twice the amount that remains in the atmosphere. It's believed that much of this has been dissolving into the ocean.

CO_2 and Global Warming Carbon dioxide and other gases are being emitted due to human activities. The other gases include nitrous oxide (N_2O) and methane (CH_4). Fertilizers and animal wastes are sources of nitrous oxide. In the digestive tracts of animals, bacterial decomposition produces methane. Decaying sediments and flooded rice paddies are methane sources, as well. In the atmosphere, nitrous oxide and methane are known as **greenhouse gases** because they act just like the panes of a greenhouse. They allow solar radiation to penetrate to Earth but hinder the escape of infrared rays (heat) back into space—a phenomenon called the **greenhouse effect.** The greenhouse gases are contributing significantly to an overall rise in the Earth's ambient temperature. This phenomenon is called **global warming.**

Video Global Warming

Figure 23.11 shows the Earth's radiation balances. One thing to be learned from this diagram is that water vapor is a greenhouse gas. Clouds (composed of water vapor) also reradiate heat back to Earth. If the Earth's temperature rises, more water will evaporate, forming more clouds. This sets up a positive feedback effect that could increase global warming still more. The global climate has already warmed about 0.6°C since the Industrial Revolution. Computer models are unable to consider all possible variables, but the Earth's temperature may rise 1.5–4.5°C by 2100 if greenhouse emissions continue at the current rates.

Video Warming Hurts Rice

Global warming will bring about other effects, which computer models attempt to forecast. It is predicted that, as the oceans warm, temperatures in the polar regions will rise to a greater degree than in other regions. As a result, sea levels will rise because glaciers will melt, and water expands as it warms. Water evaporation will increase, and most likely there will be increased rainfall along the coasts and dryer conditions inland. The occurrence of droughts will reduce agricultural yields and also cause trees to die off. Expansion of forests into Arctic areas might not offset the loss of forests in the temperate zones. Coastal agricultural lands, such as the deltas

Video Green Concrete

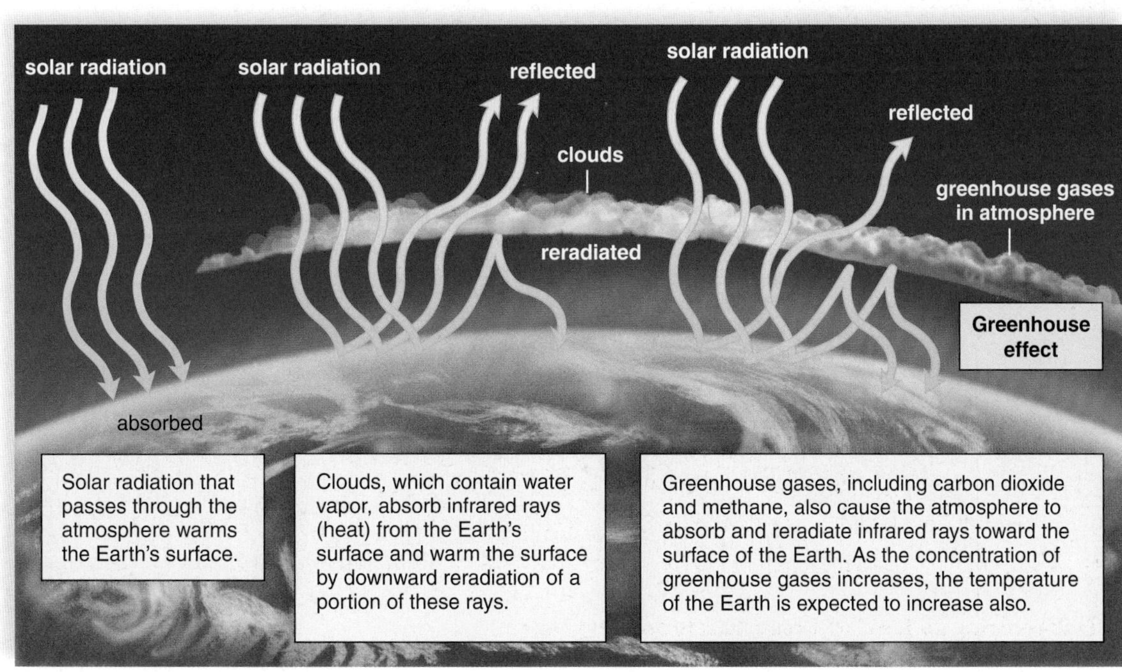

solar radiation
solar radiation
reflected
solar radiation
reflected
clouds
greenhouse gases in atmosphere
reradiated
Greenhouse effect
absorbed
Solar radiation that passes through the atmosphere warms the Earth's surface.
Clouds, which contain water vapor, absorb infrared rays (heat) from the Earth's surface and warm the surface by downward reradiation of a portion of these rays.
Greenhouse gases, including carbon dioxide and methane, also cause the atmosphere to absorb and reradiate infrared rays toward the surface of the Earth. As the concentration of greenhouse gases increases, the temperature of the Earth is expected to increase also.

Figure 23.11 The greenhouse effect. The effect of the greenhouse gases (*far right*) is contributing to global warming. A positive feedback cycle is predicted. As the Earth's temperature rises, more water will evaporate, causing clouds to thicken and absorb more solar radiation. Water vapor in clouds and the other greenhouse gases prevent the escape of heat (infrared rays), and it is reradiated back to the Earth, causing even more evaporation, and so forth.

of Bangladesh and China, will be inundated with water. Billions of dollars will have to be spent to keep coastal cities such as New Orleans, New York, Boston, Miami, and Galveston from disappearing into the sea.

Video Karoo Global Warming

The Nitrogen Cycle

Nitrogen gas (N_2) makes up about 78% of the atmosphere, but plants cannot make use of nitrogen in this form. Therefore, nitrogen can be a nutrient that limits the amount of growth in an ecosystem.

Video Dung Beetles

Ammonium (NH_4^+) Formation and Use

① In the nitrogen cycle, **nitrogen fixation** occurs when nitrogen gas (N_2) is converted to ammonium (NH_4^+), a form plants can use (Fig. 23.12). Some cyanobacteria in aquatic ecosystems and some free-living bacteria in soil are able to fix atmospheric nitrogen in this way. Other nitrogen-fixing bacteria live in nodules on the roots of legumes, such as beans, peas, and clover. They make organic compounds containing nitrogen available to the host plants so that the plant can form proteins and nucleic acids.

Figure 23.12 **The nitrogen cycle.**
Nitrogen is primarily made available to biotic communities by internal cycling of the element. Without human activities, the amount of nitrogen returned to the atmosphere (denitrification in terrestrial and aquatic communities) exceeds withdrawal from the atmosphere (N_2 fixation and nitrification). Human activities (dark purple arrow) result in an increased amount of NO_3^- in terrestrial communities, with resultant runoff to aquatic biotic communities.

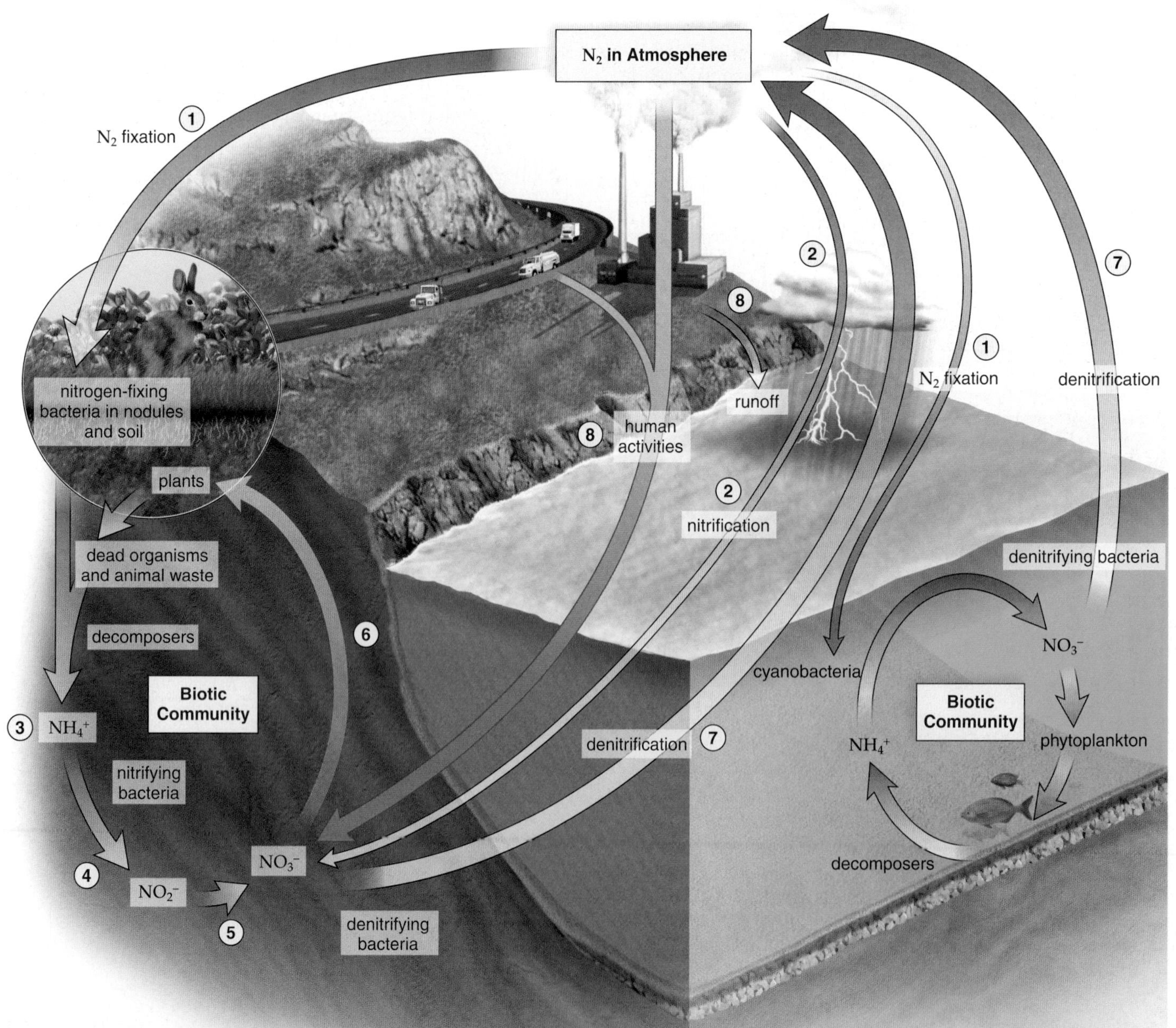

Connections and Misconceptions

How acidic is acid rain?

The pH of normal rain varies from 4.5 to 5.6 (see Fig. 2.9). That's about the acidity of tomatoes or black coffee, and neither of these is particularly corrosive or harmful to the body. Rain collected in the eastern United States during the summer has an average pH of 3.6. Vinegar has a pH similar to this rain—can you imagine watering trees or house plants with vinegar? Fog with a pH of 2 has been measured in some locations. A solution with pH 2 is only slightly less acid than stomach acid or battery acid. It's easy to imagine the damage such a solution could do to the environment.

Nitrate (NO_3^-) Formation and Use

Plants can also use nitrates (NO_3^-) as a source of nitrogen. The production of nitrates during the nitrogen cycle is called **nitrification.** Nitrification can occur in various ways: ② Nitrogen gas (N_2) is converted to nitrate (NO_3^-) in the atmosphere, where high energy is available for nitrogen to react with oxygen. This energy may be supplied by cosmic radiation, meteor trails, or lightning. ③ Ammonium (NH_4^+) in the soil from various sources, including decomposition of organisms and animal wastes, is converted to nitrate by soil bacteria. ④ Nitrite-producing bacteria convert ammonium to nitrite (NO_2^-). ⑤ Nitrate-producing bacteria convert nitrite to nitrate. ⑥ During the process of **assimilation,** plants take up ammonia and nitrate from the soil and use these ions to produce proteins and nucleic acids.

In Figure 23.12, observe the biotic community subcycles occurring on land and in water. Those subcycles do not depend on the presence of nitrogen gas.

Formation of Nitrogen Gas from Nitrate

⑦ **Denitrification** is the conversion of nitrate back to nitrogen gas, which enters the atmosphere. Denitrifying bacteria living in the anaerobic mud of lakes, bogs, and estuaries carry out this process during their metabolism. In the nitrogen cycle, denitrification would counterbalance nitrogen fixation except for human activities.

Human Activities

⑧ Human activities significantly alter the transfer rates in the nitrogen cycle by producing fertilizers from N_2. They nearly double the fixation rate. Fertilizer, which also contains phosphate, runs off into lakes and rivers. This results in an overgrowth of algae and rooted aquatic plants. As discussed on page 561, the result is cultural eutrophication (overenrichment), which can lead to an algal boom. When the algae die off, enlarged decomposer populations use up all the oxygen in the water. The result is a massive fish kill.

Acid deposition occurs because nitrogen oxides (NO_x) and sulfur dioxide (SO_2) enter the atmosphere from the burning of fossil fuels (Fig. 23.13*a*). Both of these gases combine with water vapor to form acids that eventually return to the Earth in acid

a.

b.

c.

Figure 23.13 The effects of acid deposition.
a. Many forests in higher elevations of northeastern North America and northern Europe are dying due to acid deposition.
b. Air pollution due to fossil-fuel burning in factories and modes of transportation is the major cause of acid deposition.
c. Acid deposition damages architectural features.

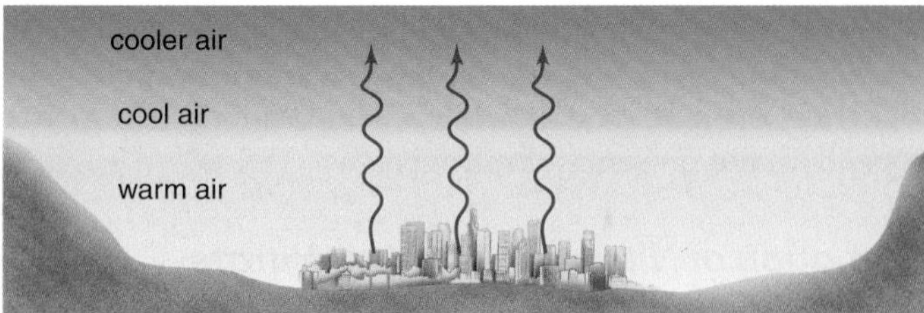

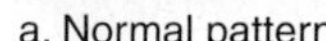
a. Normal pattern

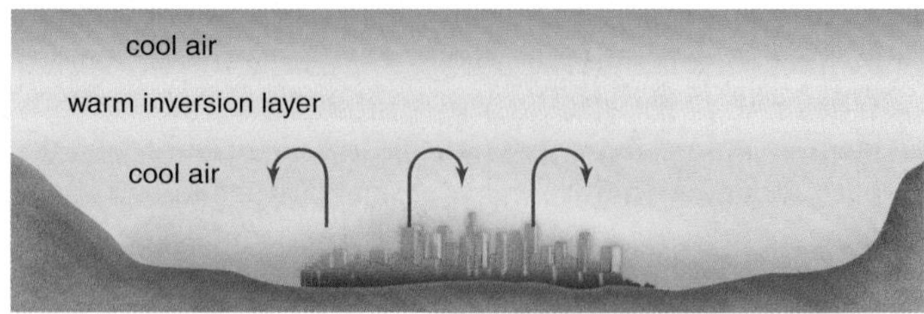

b. Thermal inversion

Figure 23.14 **Thermal inversions.**
a. Normally, pollutants escape into the atmosphere when warm air rises. **b.** During a thermal inversion, a layer of warm air (warm inversion layer) overlies and traps pollutants in cool air below.

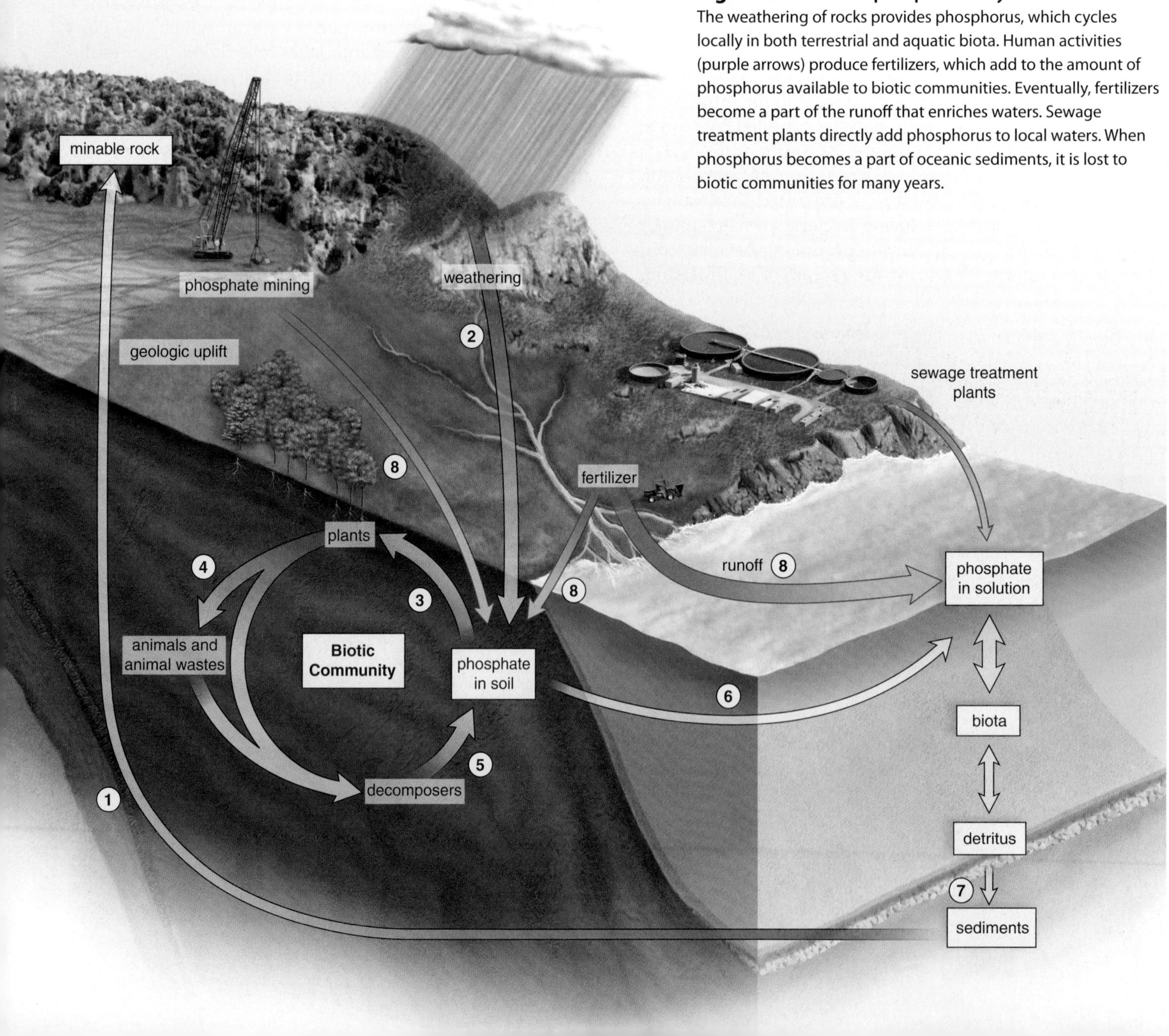

minable rock
phosphate mining
weathering
geologic uplift
sewage treatment plants
fertilizer
plants
runoff 8
phosphate in solution
animals and animal wastes
Biotic Community
phosphate in soil
biota
decomposers
detritus
sediments

Figure 23.15 **The phosphorous cycle.**
The weathering of rocks provides phosphorus, which cycles locally in both terrestrial and aquatic biota. Human activities (purple arrows) produce fertilizers, which add to the amount of phosphorus available to biotic communities. Eventually, fertilizers become a part of the runoff that enriches waters. Sewage treatment plants directly add phosphorus to local waters. When phosphorus becomes a part of oceanic sediments, it is lost to biotic communities for many years.

rain. Acid deposition has drastically affected forests and lakes in northern Europe, Canada, and the northeastern United States. The soil in these forests is naturally acidic, whereas the surface water is only mildly alkaline (basic). Soil pH is made even more acidic by acid rain. The increased acidity kills trees (Fig. 23.13*b*) and reduces agricultural yields. Marble, metal, and stonework are corroded by acid deposition, which results in the loss of architectural features (Fig. 23.13*c*).

Nitrogen oxides and hydrocarbons (HC) from the burning of fossil fuels react with one another in the presence of sunlight to produce smog. Smog contains dangerous pollutants. Warm air near the Earth usually escapes into the atmosphere, taking pollutants with it. However, during a thermal inversion, pollutants are trapped near the Earth beneath a layer of warm, stagnant air. The air does not circulate, so pollutants can build up to dangerous levels. Areas surrounded by hills are particularly susceptible to the effects of a thermal inversion. This is because the air tends to stagnate and little turbulent mixing can occur (Fig. 23.14).

The Phosphorus Cycle

In the phosphorus cycle (Fig. 23.15), ① phosphorus trapped in oceanic sediments moves onto land after a geologic upheaval. ② On land, the very slow weathering of rocks places phosphate ions (PO_4^{3-} and HPO_4^{2-}) in the soil. ③ Some of this soil becomes available to plants, which use phosphate in a variety of molecules. Molecules requiring phosphate include phospholipids, ATP, and the nucleotides that become a part of DNA and RNA. ④ Animals eat producers and incorporate some of the phosphate into teeth, bones, and shells, which take many years to decompose. ⑤ However, the eventual death and decomposition of all organisms and their wastes do make phosphate ions available to producers once again. The available amount of phosphate is already being used within food chains, so phosphate is usually a limiting inorganic nutrient for plants. In other words, the lack of it limits the size of populations in ecosystems.

⑥ Some phosphate naturally runs off into aquatic ecosystems. There, algae acquire phosphate from the water before it becomes trapped in sediments. ⑦ Phosphate in marine sediments does not become available to producers on land again until a geologic upheaval exposes sedimentary rocks on land. Now, the cycle begins again. Phosphorus does not enter the atmosphere. Therefore, the phosphorus cycle is called a sedimentary cycle.

Phosphorus and Water Pollution

⑧ Human beings boost the supply of phosphate by mining phosphate ores for fertilizer and detergent production. As mentioned previously, runoff of phosphate and nitrogen into water occurs due to fertilizer use. This contamination, as well as that from animal wastes in livestock feedlots and discharge from sewage treatment plants, results in **cultural eutrophication** (overenrichment) of waterways.

Figure 23.16 lists the various sources of water pollution. *Point* sources of pollution are specific, and *nonpoint*

Sources of Water Pollution	
Leading to Cultural Eutrophication	
Oxygen-demanding waste	Biodegradable organic compounds (e.g., sewage, wastes from food-processing plants, paper mills, and tanneries)
Plant nutrients	Nitrates and phosphates from detergents, fertilizers, and sewage treatment plants
Sediments	Enriched soil in water due to soil erosion
Thermal discharges	Heated water from power plants
Health Hazards	
Disease-causing agents	Bacteria and viruses from sewage and barnyard waste (causing, for example, cholera, food poisoning, and hepatitis)
Synthetic organic compounds	Pesticides, industrial chemicals (e.g., PCBs)
Inorganic chemicals and minerals	Acids from mines and air pollution; dissolved salts; heavy metals (e.g., mercury) from industry
Radiation	Radioactive substances from nuclear power plants, medical and research facilities

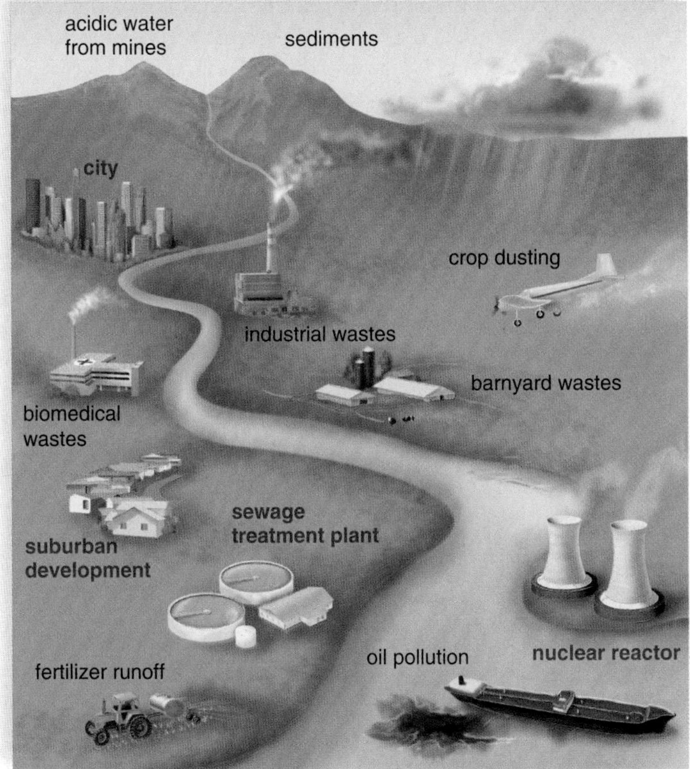

Figure 23.16 Some sources of surface water pollution. Many bodies of water are dying due to the introduction of pollutants from point sources, which are easily identifiable, and nonpoint sources, which cannot be specifically identified.

Biology Matters **Science Focus**

Ozone Shield Depletion

Ozone is a gas that can be found in the troposphere (the layer of atmosphere closest to the ground) and in the stratosphere (a layer of the Earth's upper atmosphere). Exhaust from our cars and emissions from various industries contribute to the formation of ozone in the troposphere. It's more likely to form during the hot, sunny days associated with late spring through early fall. This type of ozone is "bad" ozone and is considered a pollutant. A large component of smog is ozone. You may have heard ozone levels being discussed in your area during daily air-quality reports. When ozone levels around us are high, it adversely affects our ability to breathe and may cause permanent lung damage. Children and people who already suffer from lung diseases or conditions should stay indoors when tropospheric ozone levels are high.

In the stratosphere, ozone forms when ultraviolet (UV) radiation from the sun splits molecules of oxygen gas (O_2). Single oxygen atoms (O) combine with molecules of O_2 to form ozone (O_3). This ozone—often called the **ozone shield**— is considered "good" ozone because of its ability to absorb most of the sun's UV rays. The absorption of UV radiation by the ozone shield is critical for living things. In humans, UV radiation causes mutations that can lead to skin cancer and can make the lens of the eye develop cataracts. A U.N. Environmental Program report predicts a 26% rise in cataracts and nonmelanoma skin cancers for every 10% drop in the ozone level. Further, the immune system is damaged by UV radiation, making us more susceptible to infectious diseases. The growth of crops and trees is also adversely affected by UV radiation. In water ecosystems, algae and krill (tiny shrimplike animals) that sustain other aquatic life are killed by UV radiation. Therefore, an inadequate ozone shield that allows more UV rays to strike the Earth would threaten both our health and our food sources.

In the 1980s, it became apparent that loss of ozone in the shield had occurred worldwide. The depletion was most severe over Antarctica in the spring. There, the so-called **ozone hole** covered an area two and a half times the size of Europe (Fig. 23B). This bare area did not expose only Antarctica to harmful UV rays. The southern tip of South America and vast areas of the Pacific and Atlantic oceans were exposed as well. Australia is also impacted by the Antarctica hole. A second ozone hole formed above the Arctic. Other holes have been detected over areas where large populations of people live.

Scientists discovered that chlorine atoms are a major cause of ozone depletion. These chlorine atoms come primarily from the breakdown of **chlorofluorocarbons (CFCs)**. The CFC you may be most familiar with is Freon, a coolant found in refrigerators and air conditioners. CFCs are also used as a foaming agent during the production of Styrofoam coffee cups, egg cartons, insulation, and padding. At one time, CFCs were used as propellants in spray cans, but this use is now banned in the United States and several European countries. A pesticide called methyl bromide significantly decreases ozone levels as well.

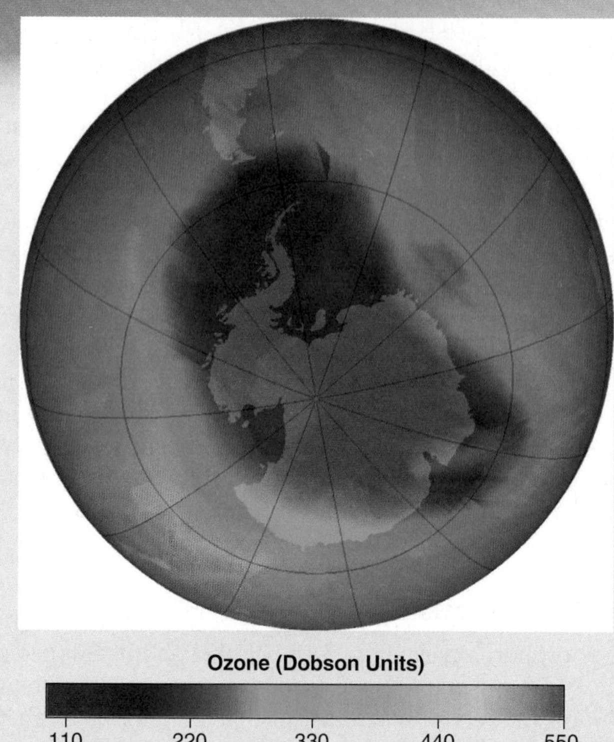

Figure 23B Ozone shield depletion.
Map of ozone levels in the atmosphere of the Southern Hemisphere on September 17, 2010. The ozone depletion, often called an ozone hole, is larger than the size of Europe. The purple regions indicate the lowest levels of ozone.

The Montreal Protocol was drafted in 1987 and was initially signed by 43 countries. Agreement to phase out the use of CFCs by 1995 was reached (though developing countries have until 2010 to complete their phase-out). In 2007, China closed five of its six remaining plants that produced CFCs, ahead of schedule. Methyl bromide was also later included in the Montreal Protocol. Its phase-out in developed countries occurred in 2005. The Montreal Protocol has now been signed by at least 183 countries.

Recent satellite measurements indicate that the amount of harmful chlorine pollution in the stratosphere has started to decline. However, recovery of the ozone shield may take many more years (the Antarctic hole is expected to be present until 2050 even if the Montreal Protocol is followed). Other pollution-fighting approaches may be required in addition to lowering chlorine pollution. The global warming phenomenon may also adversely impact ozone restoration. As greenhouse gases increase, less heat is reflected from the Earth's surface, and the stratosphere becomes colder. Ozone depletion increases when the stratosphere is very cold.

sources are those caused by runoff from the land. Industrial wastes can include heavy metals and organochlorides, such as DDT and PCBs. These materials are not readily degraded under natural conditions or in conventional sewage treatment plants. **Biological magnification** occurs as these toxic chemicals pass along a food chain. Pollutants become increasingly concentrated with each higher consumer level, because they remain in the body and are not excreted. Aquatic food chains are more likely to experience biological magnification because there are more links than in a terrestrial food chain. Mercury levels can be high in some of the fish we eat (top consumers like shark, swordfish, and tuna) for this reason.

Coastal regions are the immediate receptors for local pollutants and the final receptors for pollutants carried by rivers that empty at a coast. Waste dumping occurs at sea. However, ocean currents sometimes transport both trash and pollutants back to shore. Offshore mining and shipping add pollutants to the oceans. Some 5 million metric tons of oil a year end up in the oceans. Large oil spills kill plankton, fish, and shellfishes, as well as birds and marine mammals.

In the last 50 years, humans have polluted the seas and exploited their resources to the point that many species are on the brink of extinction. Fisheries once rich and diverse, such as George's Bank off the coast of New England, are in severe decline. Haddock was once the most abundant species in this fishery, but now it accounts for less than 2% of the total catch. Cod and bluefin tuna have suffered a 90% reduction in population size. In warm, tropical regions, many areas of coral reefs are now overgrown with algae. This is because the fish that normally keep the algae under control have been killed off.

Video Ocean Fishing Ban

Connecting the Concepts

For more information on the material presented in this section, refer to the following discussions.

Section 24.2 describes some of the problems associated with the overexploitation of natural resources.

Section 24.4 explores some of the ways that humans can move towards a sustainable society.

Check Your Progress 23.3

1. List three components that are involved in the cycling of chemicals in the biosphere.
2. Identify whether the carbon, water, nitrogen, and phosphorous cycles are examples of gaseous or sedimentary biogeochemical cycles.
3. Discuss the roles of bacteria in the nitrogen cycle.
4. Summarize the ecological problems that are associated with water, carbon, nitrogen, and phosphorus cycles.
5. Discuss the potential consequences of global warming.

CASE STUDY CONCLUSION

Although the scenario in the chapter opener may appear extreme, it is possible given the influence of modern humans on ecocystems and chemical cycling. An ecosystem is characterized by the interactions of its biotic (living) and abiotic (nonliving) components. The biotic community depends on the cycling of chemicals, such as carbon, nitrogen, and phosphorous.

Media Study Tools

www.mhhe.com/maderhuman12e

Enhance your study of this chapter with study tools and practice tests. Also ask your instructor about the resources available through ConnectPlus, including the media-rich eBook, interactive learning tools, and animations.

Virtual Lab

The virtual lab "Model Ecosystems" provides an interactive look at some ofthe factors that influence the structure of an ecosystem.

Summarizing the Concepts

23.1 The Nature of Ecosystems

Ecology is the study of the interactions of organisms with each other and with the physical environment.

- Organisms interact with the physical and chemical environment, and the result is an ecosystem.

Types of Ecosystems

- Terrestrial ecosystems are forests (tropical rain forests, coniferous forests, temperate deciduous forests); grasslands (savannas and prairies); and deserts, including the tundras.
- Aquatic ecosystems are either salt water (i.e., seashores, oceans, coral reefs, estuaries) or freshwater (i.e., lakes, ponds, rivers, and streams).

Biotic Components of an Ecosystem

- In a community, each population has a habitat (residence) and a niche (its role in the community).
- Autotrophs (producers) produce organic nutrients for themselves and others from inorganic nutrients and an outside energy source.
- Heterotrophs (consumers) consume organic nutrients.
- Consumers are herbivores (eat plants/algae), carnivores (eat other animals), and omnivores (eat both plants/algae and animals).
- Decomposers feed on detritus, releasing inorganic substances back into the ecosystem.

Energy Flow and Chemical Cycling

Ecosystems are characterized by energy flow and chemical cycling.

- Energy flows through the populations of an ecosystem.
- Chemicals cycle within and among ecosystems.

23.2 Energy Flow

Various interconnecting paths of energy flow are called a food web.

- A food web is a diagram showing how various organisms are connected by eating relationships.
- Grazing food webs begin with vegetation eaten by an herbivore that becomes food for a carnivore.
- Detrital food webs begin with detritus, food for decomposers and for detritivores.
- Members of detrital food webs can be eaten by aboveground carnivores, joining the two food webs.

Trophic Levels

A trophic level is all the organisms that feed at a particular link in a food chain.

Ecological Pyramids

- Ecological pyramids illustrate that biomass and energy content decrease from one trophic level to the next because of energy loss.

23.3 Global Biogeochemical Cycles

Chemicals circulate through ecosystems via biogeochemical cycles, pathways involving both biotic and geologic components. Biogeochemical cycles:

- can be gaseous or sedimentary.
- have reservoirs (e.g., ocean sediments, the atmosphere, and organic matter) that contain inorganic nutrients available to living things on a limited basis.

Exchange pools are sources of inorganic nutrients.

- Nutrients cycle among the biotic communities (producers, consumers, decomposers) of an ecosystem.

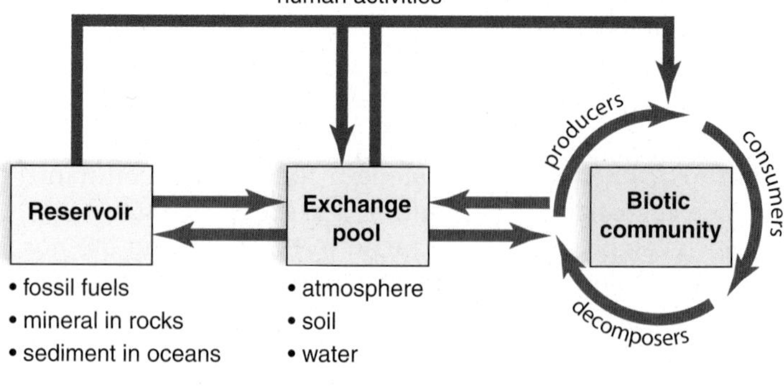

The Water Cycle

- The reservoir of the water cycle is freshwater that evaporates from the ocean.
- Water that falls on land enters the ground, surface waters, or aquifers and evaporates again.
- All water returns to the ocean.

The Carbon Cycle

- The reservoirs of the carbon cycle are organic matter (e.g., forests and dead organisms for fossil fuels), limestone, and the ocean (e.g., calcium carbonate shells).
- The exchange pool is the atmosphere.
- Photosynthesis removes carbon dioxide from the atmosphere.
- Respiration and combustion add carbon dioxide to the atmosphere.

The Nitrogen Cycle

- The reservoir of the nitrogen cycle is the atmosphere.
- Nitrogen gas must be converted to a form usable by plants (producers).
- Nitrogen-fixing bacteria (in root nodules) convert nitrogen gas to ammonium, a form producers can use.
- Nitrifying bacteria convert ammonium to nitrate.
- Denitrifying bacteria convert nitrate back to nitrogen gas.

The Phosphorus Cycle

- The reservoir of the phosphorus cycle is ocean sediments.
- Phosphate in ocean sediments becomes available through geologic upheaval, which exposes sedimentary rocks to weathering.
- Weathering slowly makes phosphate available to the biotic community.
- Phosphate is a limiting nutrient in ecosystems.

Understanding Key Terms

acid deposition 559
aquifer 554
assimilation 559
autotroph 548
biogeochemical cycle 553
biological magnification 563
biomass 552
biome 547
biosphere 547
carnivore 548
chlorofluorocarbon (CFC) 562
consumer 548
cultural eutrophication 561
denitrification 559
detrital food chain 550
detrital food web 550
detritus feeder 548

ecological pyramid 552
ecosystem 547
evaporation 554
food chain 550
food web 550
fossil fuel 555
global warming 557
grazing food chain 550
grazing food web 550
greenhouse effect 557
greenhouse gases 557
herbivore 548
heterotroph 548
niche 549
nitrification 559
nitrogen fixation 558
omnivore 548
ozone hole 562
ozone shield 562
precipitation 554
producer 548
runoff 554
trophic level 550
trophic relationship 550
water (hydrologic) cycle 554

Match the key terms to these definitions.

a. ________ Animals that feed on both plants and animals.

b. ________ The organisms that feed at a particular link in a food chain.

c. ________ Remains of once-living organisms that are burned to release energy, such as coal, oil, and natural gas.

d. ________ Process by which atmospheric nitrogen gas is changed to forms that plants can use.

e. ________ Photosynthetic organism at the start of a grazing food chain that makes its own food.

Testing Your Knowledge of the Concepts

1. What are the major types of terrestrial ecosystems? Describe each with its temperature, rainfall, and type of vegetation. (pages 546–550)
2. What are the major types of aquatic ecosystems? (page 547)
3. What is a niche? (page 549)
4. Name the four different types of consumers (heterotrophs) found in ecosystems. (page 548)
5. Explain why energy flows but chemicals cycle through an ecosystem. (page 549)
6. Describe the two types of food webs and two types of food chains in terrestrial ecosystems. Which of these typically moves more energy through an ecosystem? (pages 550–552)
7. What is a trophic level? An ecological pyramid? (pages 550–552)
8. What is a biochemical cycle? What is a reservoir and an exchange pool? Give an example of each. (page 553)
9. Draw a diagram to illustrate the water, carbon, nitrogen, and phosphorus biochemical cycles. Include how human activities can change a particular cycle's balance. (pages 554–555, 557)

In questions 10–13, match each description to a population in the key.

Key:

a. producer
b. consumer
c. decomposer
d. herbivore

10. Heterotroph that feeds on plant material.
11. Autotroph that manufactures organic nutrients.
12. Any type of heterotroph that feeds on plant material or on other animals.
13. Heterotroph that breaks down detritus as a source of nutrients.
14. Of the total amount of energy that passes from one trophic level to another, about 10% is
    a. respired and becomes heat.
    b. passed out as feces or urine.
    c. stored as body tissue.
    d. recycled to autotrophs.
    e. All of these are correct.
15. Compare this food chain:
    algae → water fleas → fish → green herons
    with this food chain:
    trees → tent caterpillars → red-eyed vireos → hawks
16. Both water fleas and tent caterpillars are
    a. carnivores.
    b. primary consumers.
    c. detritus feeders.
    d. present in grazing and detrital food webs.
    e. Both a and b are correct.
17. In what way are decomposers like producers?
    a. Either may be the first member of a grazing or a detrital food chain.
    b. Both produce oxygen for other forms of life.
    c. Both require nutrient molecules and energy.
    d. Both are present only on land.
    e. Both produce organic nutrients for other members of ecosystems.
18. Why are ecosystems dependent on a continual supply of solar energy?
    a. Carnivores have a greater biomass than producers.
    b. Decomposers process the greatest amount of energy in an ecosystem.
    c. Energy transformation results in a loss of usable energy to the environment.
    d. Energy cycles within and between ecosystems.
19. Nutrient cycles always involve
    a. rocks as a reservoir.
    b. movement of nutrients through the biotic community.
    c. the atmosphere as an exchange pool.
    d. loss of the nutrients from the biosphere.
20. Which of the following contribute/s to the carbon cycle?
    a. respiration
    b. photosynthesis
    c. fossil fuel combustion
    d. decomposition of dead organisms
    e. All of these are correct.
21. How do plants contribute to the carbon cycle?
    a. When plants respire, they release CO_2 into the atmosphere.
    b. When plants photosynthesize, they consume CO_2 from the atmosphere.
    c. When plants photosynthesize, they provide oxygen to heterotrophs.
    d. When plants emigrate, they transport carbon molecules between ecosystems.
    e. Both a and b are correct.

22. How do nitrogen-fixing bacteria in the soil contribute to the nitrogen cycle?
    a. They return nitrogen to the atmosphere.
    b. They change ammonium to nitrate.
    c. They change nitrogen to ammonium.
    d. They withdraw nitrate from the soil.
    e. They decompose and return nitrogen to autotrophs.
23. What is the reservoir in the phosphorus cycle?
    a. oceans
    b. marine sediments
    c. plants
    d. animals

For questions 23-27, match each characteristic to the cycles listed in the key. More than one answer can be used, and answers can be used more than once.

Key:

a. water cycle
b. carbon cycle
c. nitrogen cycle
d. phosphorus cycle
e. none of these
f. all of these

23. Occurs on land but not in the water
24. Can occur without the participation of humans
25. Always involves the participation of decomposers
26. The atmosphere is involved.
27. Rocks are the reservoir in this cycle.
28. Label the following diagram of an ecosystem.

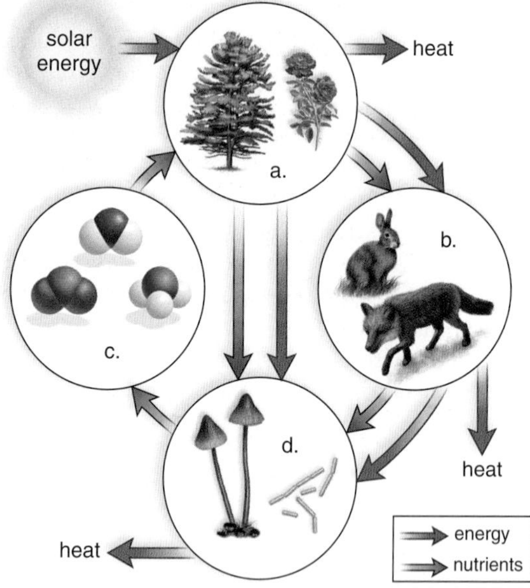

Thinking Critically About the Concepts

Throughout this chapter, we have seen that human activity has a negative impact on many of the world's ecosystems. In response, many people are making efforts to reduce the environmental impact of their lifestyles. Across the country, state agencies, businesses, local organizations, and educational institutions are developing environmental projects and/or increasing their conservation methods. Many colleges and universities have initiatives that make the campuses more environmentally friendly. Our own contributions help, too. Just a few of the many ideas you can consider are water conservation, recycling trash, replacing paper or plastic containers with cloth, conserving electricity, starting a compost pile, and collecting rainwater for plants in rain barrels.

1. a. What impact might fungicides and pesticides have on detrital food webs?
   b. How do you think the nutrient cycles would be affected by the use of fungicides and pesticides?
2. Why would an agricultural extension agent recommend planting a legume such as clover in a pasture?
3. a. What types of things could you do at home to facilitate the cycling of nutrients such as carbon and nitrogen?
   b. What types of things could you do at home to conserve water or improve the quality of a nearby body of water?
4. If farmers use parasitoids to control flies on their farm, they may be able to stop using pesticides that contain organochlorides. What impact would this have on food chains involving wildlife around the farms?

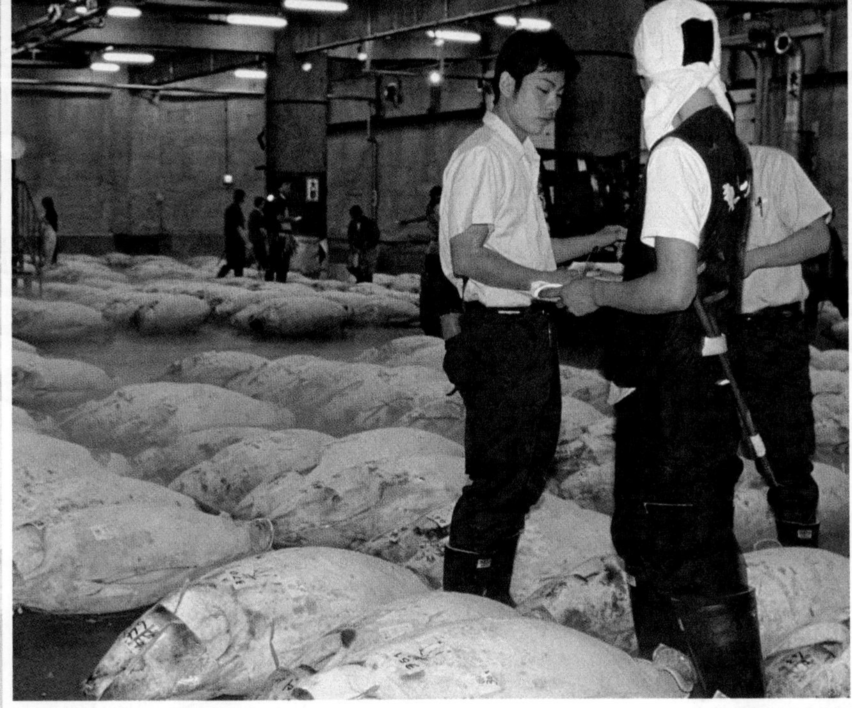

CHAPTER

24

Human Population, Planetary Resources, and Conservation

CASE STUDY OVERFISHING

Overfishing threatens the world's fish populations, such as that of the swordfish. Global fisheries' production has increased steadily over the past 50 years, from about 18 million tons per year in 1950 to over 100 million tons in 2007. The U.N. Food and Agricultural Organization estimates that by 2030 we will need to catch an additional 37 million tons per year to feed the growing human population. These numbers lead some to predict a global fishery collapse by 2048. Overfishing occurs when the catch of a particular fish species exceeds its reproduction, resulting in declining numbers. In other words, the death rate exceeds the birthrate. Normally, fish populations exist at the carry capacity of the environment (the numbers the environment can support). However, if humans reduce fish populations well below the carrying capacity, a lag in population growth occurs. If fishing continues at a high rate during this lag phase, it can drive fish populations to extinction.

As you read through the chapter, think about the following questions.

1. How are increases in the human population influencing the availability of natural resources?
2. Are fish populations considered to be a renewable or nonrenewable resource?
3. What may be done to make commercial fishing a sustainable business?

CHAPTER CONCEPTS

24.1 Human Population Growth
The present growth rate for the world's population has decreased to around 1.2%; still 78 million more people are expected within a year because the human population is so large.

24.2 Human Use of Resources and Pollution
Human beings use land, water, food, energy, and minerals to meet their basic needs. Use of these resources leads to pollution.

24.3 Biodiversity
A biodiversity crisis is upon us because of habitat loss. The introduction of alien species, pollution, overexploitation, and disease all contribute to the crisis. Yet, wildlife has both a direct value and an indirect value for us.

24.4 Working Toward a Sustainable Society
Our present-day society is not sustainable, however, there are ways to make it more sustainable.

BEFORE YOU BEGIN

Before beginning this chapter, take a few moments to review the following discussions.

Section 21.4 How has biotechnology been applied to plants such as corn and potatoes to produce increases in food production?

Section 23.1 How does the cycling of energy and chemicals differ in an ecosystem?

Section 23.3 How are human activities influencing the water cycle?

24.1 Human Population Growth

Learning Outcomes

Upon completion of this section, you should be able to

1. Define the terms *exponential growth* and *carrying capacity* and explain how each relates to human population growth.
2. Explain the relationship between birthrate, death rate, and the annual growth rate of a population.
3. Distinguish between more-developed countries (MDCs) and less-developed countries (LDCs) with regard to population growth.

The world's population has risen steadily to a present size of close to 7 billion people (Fig. 24.1). Prior to 1750, the growth of the human population was relatively slow. As more reproducing individuals were added, population growth increased. The curve began to slope steeply upward, indicating that the population was undergoing **exponential growth.** The number of people added annually to the world population peaked at about 87 million around 1990. Currently it is a little over 78 million per year.

The **growth rate** of a population is determined by considering the difference between the number of persons born per year (birthrate, or natality) and the number who die per year (death rate, or mortality). It is customary to record these rates per 1,000 persons. For example, the world at the present time has a birthrate of 19.9 per 1,000 per year, but it has a death rate of 8.2 per 1,000 per year. This means that the world's population growth, or its growth rate, is

$$\frac{19.9 - 8.2}{1{,}000} = \frac{11.7}{1{,}000} = 0.0117 \times 100 = 1.17\%.$$

Note that whereas the birthrate and death rate are expressed in terms of 1,000 persons, the growth rate is expressed per 100 persons, or as a percentage.

After 1750, the world population growth rate steadily increased, until it peaked at 2% in 1965. It has since fallen to its present level of around 1.2%. Yet, the world population is still steadily growing because of its past exponential growth.

In the wild, exponential growth indicates that a population is enjoying its **biotic potential.** This is the maximum growth rate under ideal conditions. Growth begins to decline because of limiting factors such as food and space. Finally, the population levels off at the carrying capacity. The **carrying capacity** is the maximum population that the environment can support for an indefinite period. The carrying capacity of the Earth for humans has not been determined. Some authorities think the Earth may be able to sustain 50–100 billion people. Others think we already have more humans than the Earth can adequately support.

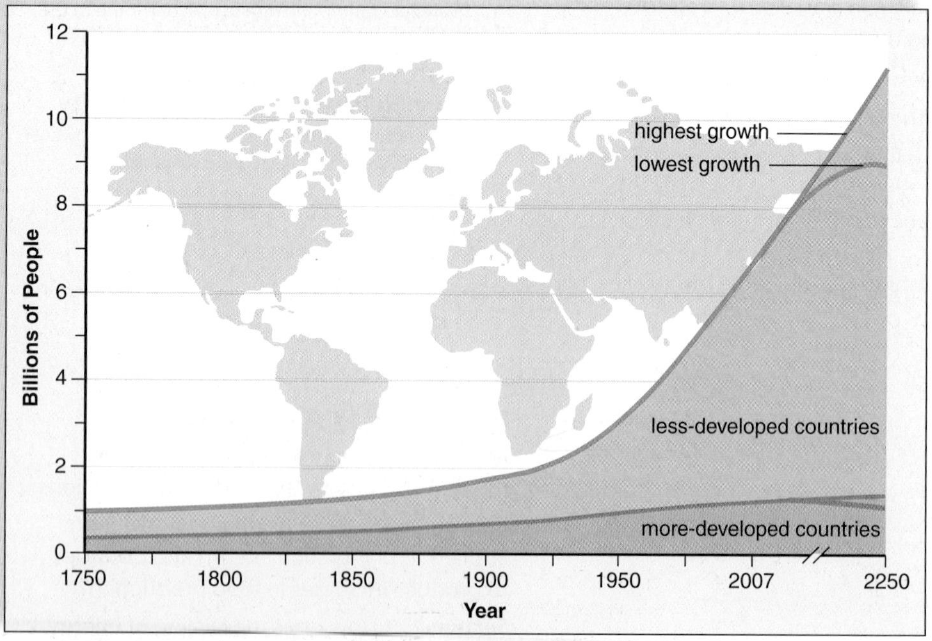

Figure 24.1 Projections for human population growth.
The world's population of humans is now close to 7 billion. It is predicted that the world's population size may level off at 9 billion or increase to more than 11 billion by 2250, depending on the speed with which the growth rate declines.

The MDCs Versus the LDCs

The countries of the world can be divided into two groups. The more-developed countries (MDCs), typified by countries in North America and Europe, are those in which population growth is modest. The people in these countries enjoy a good standard of living. The less-developed countries (LDCs), typified by some countries in Asia, Africa, and Latin America, are those in which population growth is dramatic. The majority of people in these countries live in poverty.

The MDCs

The MDCs did not always have low population increases. Between 1850 and 1950, they doubled their populations. This was largely because of a decline in the death rate due to development of modern medicine and improvements in public health and socioeconomic conditions. The decline in the death rate was followed shortly thereafter by a decline in the birthrate. As a result, the MDCs have experienced only modest growth since 1950 (Fig. 24.1).

The growth rate for the MDCs as a whole is about 0.1%. In some countries, population is not increasing or is decreasing in size. The MDCs are expected to increase by 52 million between 2002 and 2050, but this amount will still keep their total population at

Connections and Misconceptions

How do birthrates and death rates vary between countries?

In 2009, the country with the highest birthrate was Niger (51.6%), followed by Mali (49.1%) and Uganda (47.8%), whereas Japan was the country with lowest birthrate (7.64%). For death rates, Swaziland had the highest rate (30.8%) and the United Arab Emirates had the lowest (2.1%). For overall annual growth rates, the fastest-growing country in 2009 was the United Arab Emirates (3.7%). Several countries experienced declines, including Montenegro (–0.8%) and Bulgaria (–0.8%). In comparison, in the United States, the 2009 birthrate was 13.82%, the death rate was 8.38%, and the annual growth rate was 0.98%.
Source: www.globalhealthfacts.com

just about 1.2 billion. In contrast to the other MDCs, growth in the United States has not leveled off. The population of the United States is now greater than 300 million and continues to increase. Though the birthrate in the United States has increased slightly, much of the continued population growth is due to immigration.

The LDCs

The death rate began to decline steeply in the LDCs following World War II with the introduction of modern medicine. However, the birthrate remained high. The growth rate of the LDCs peaked at 2.5% between 1960 and 1965. Since that time, the collective growth rate for the LDCs has declined. However, the growth rate has not declined in all LDCs. In many countries in Sub-Saharan Africa, women give birth to more than five children each.

Between 2002 and 2050, the population of the LDCs may jump from 5 billion to at least 8 billion. Some of this increase will occur in Africa, but most will occur in Asia. Many deaths from AIDS are slowing the growth of the African population. Continued growth in Asia is expected to cause acute water scarcity, a significant loss of biodiversity, and more urban pollution. Twelve of the world's 15 most polluted cities are in Asia.

Comparing Age Structure

The LDCs are experiencing a population momentum because they have more women entering the reproductive years than older women leaving them. Populations have three age groups: prereproductive, reproductive, and postreproductive. This is best visualized by plotting the proportion of individuals in each group on a bar graph. This produces an age-structure diagram (Fig. 24.2).

Laypeople are sometimes under the impression that if each couple has two children, zero population growth will take place immediately. However, **replacement reproduction,** as this practice is called, will still cause most LDCs

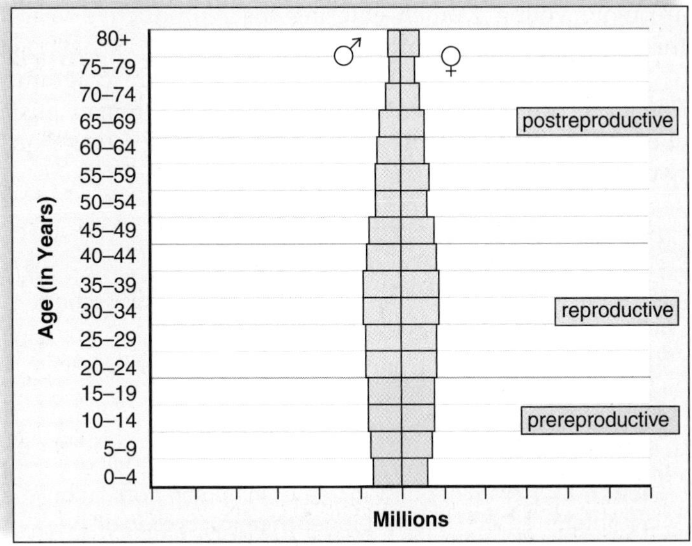

a. More-developed countries (MDCs)

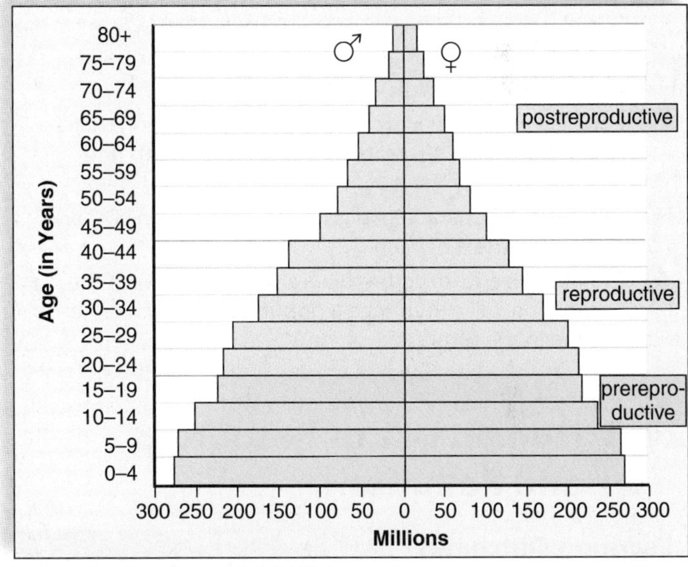

b. Less-developed countries (LDCs)

c.

Figure 24.2 **Age-structure diagrams of MDCs and LDCs.** The shape of these age-structure diagrams allows us to predict that (**a**) the populations of MDCs are approaching stabilization, and (**b**) the populations of LDCs will continue to increase for some time. **c.** Improved women's rights and increasing contraceptive use could change this scenario. Here, a community health worker is instructing women in Bangladesh about the use of contraceptives.

today to have a positive growth rate. This is because there are more young women entering the reproductive years than older women leaving them.

Most MDCs have a stabilized age-structure diagram. Therefore, their populations are expected to remain just about the same or decline if couples are having fewer than two children each.

Connecting the Concepts

For more information on the topics presented in this section, refer to the following discussions.

Figure S.1 illustrates the number of cases worldwide of HIV in 2009.

Section 16.5 describes how assisted reproductive technologies are used to treat individuals with infertility.

Section 23.3 examines how increases in human population have interfered with natural biogeochemical cycling of water, carbon, nitrogen, and phosphorous.

Check Your Progress 24.1

1. Calculate the annual growth rate of a population that is experiencing a birthrate of 20.5% and a death rate of 9.8%.
2. Compare the characteristics of an MDC with an LDC and give an example of each.
3. Evaluate an age-structure diagram to determine if the population will experience a population growth or decline in the future.

24.2 Human Use of Resources and Pollution

Learning Outcomes

Upon completion of this section, you should be able to

1. Distinguish between renewable and nonrenewable resources and give an example of each.
2. Explain how human activity is influencing the natural resources of land, water, food, and energy.
3. Provide an example of biological magnification.

Human beings have certain basic needs. A resource is anything from the biotic or abiotic environment that helps meet these needs. Land, water, food, energy, and minerals are the maximally used resources that will be discussed in this chapter (Fig. 24.3). The total amount of resources used by an individual to meet his/her needs is sometimes referred to as an **ecological footprint**. A person can make his or her ecological footprint smaller by driving an energy-efficient car, living in a smaller house, owning fewer possessions, eating vegetables as opposed to meat, and so forth.

Figure 24.3 The five classes of resources needed to meet basic human needs.
Human beings use land, water, food, energy, and minerals to meet their basic needs, such as a place to live, food to eat, and products that make their lives easier.

Some resources are nonrenewable, and some are renewable. **Nonrenewable resources** are limited in supply. For example, the amount of land, fossil fuels, and minerals is finite and can be exhausted. Efficient use, recycling, or substitution can make the supply last longer, but eventually these resources will run out.

Renewable resources are capable of being naturally replenished. We can use water and certain forms of energy (e.g., solar energy) or harvest plants and animals for food. A new supply will always be forthcoming. However, even with renewable resources, we have to be careful not to squander them.

Unfortunately, a side effect of resource consumption can be pollution. **Pollution** is any undesired alteration of the environment and is often caused by human activities. The effect of humans on the environment is proportional to the population size. As the population grows, so does the need for resources and the amount of pollution caused by using these resources. Seven people adding waste to the ocean may not be alarming, but 7 billion people doing so would certainly affect its cleanliness. In modern times, the consumption of mineral and energy resources has grown faster than population size. This has occurred as people in LDCs have increased their use of resources.

Land

People need a place to live. Naturally, land is also needed for a variety of uses aside from homes. Land is used for agriculture, electrical power plants, manufacturing plants, highways, hospitals, schools, and so on.

Beaches and Human Habitation

At least 40% of the world population lives within 100 km (60 mi) of a coastline, and this number is expected to increase. Living right on the coast is an unfortunate choice because it leads to beach erosion. Loss of habitat for marine organisms and loss of a buffer zone for storms also occur. Figure 24.4 shows how severe the problem can be in the United States. Coastal wetlands are also impacted when people fill them in. One reason to protect coastal wetlands is that they are spawning areas for fish and other forms of marine life. They are

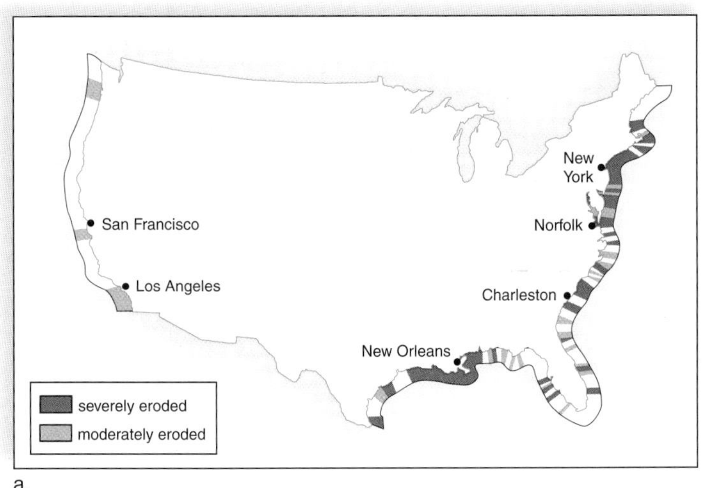

Figure 24.4 **Beach erosion and coastal development.**
a. Most of the U.S. coastline is subject to beach erosion. **b.** Therefore, people who choose to live near the coast may eventually lose their homes.

also habitats for certain terrestrial species, including many types of birds. Wetlands also protect coastal areas from storms. The coast is particularly subject to pollution because toxic substances placed in freshwater lakes, rivers, and streams may eventually find their way to the coast.

Semiarid Lands and Human Habitation

Forty percent of the Earth's lands are already deserts. Land adjacent to a desert is in danger of becoming unable to support human life if it is improperly managed by humans (Fig. 24.5). **Desertification** is the conversion of semiarid land to desertlike conditions.

Often, desertification begins when humans allow animals to overgraze the land. The soil can no longer hold rainwater, and it runs off instead of keeping the remaining plants alive or replenishing wells. Humans then remove whatever vegetation they can find to use as fuel or fodder for their animals. The result is a lifeless desert. That area is then abandoned as people move on to continue the process someplace else. Some estimate that nearly three-quarters of all rangelands worldwide are in danger of desertification. Many famines are due, at least in part, to degradation of the land to the point that it can no longer support human beings and their livestock.

Tropical Rain Forests and Human Habitation

Deforestation, the removal of trees, has long allowed humans to live in areas where forests once covered the land (Fig. 24.6). This land, too, is subject to desertification. Soil in the tropics is often thin and nutrient-poor because all the nutrients are tied up in the trees and other vegetation. When the trees are felled and the land is used for agriculture or grazing, it quickly loses its fertility. Then it is subject to desertification.

Water

In the water-poor areas of the world, people may not have ready access to drinking water and, if they do, the water

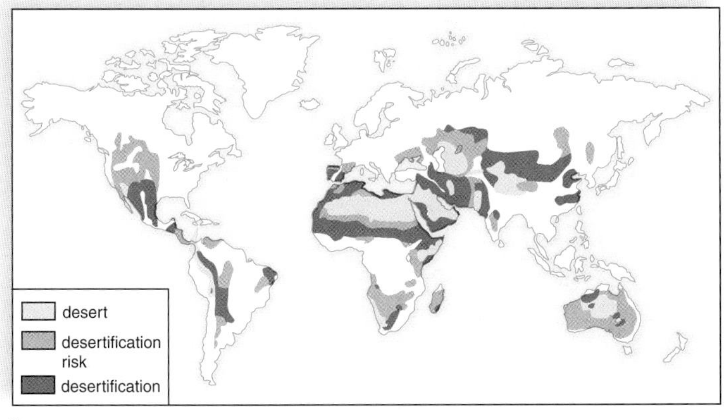

Figure 24.5 **Desertification.**
a. Desertification is a worldwide occurrence that (**b**) reduces the amount of land suitable for human habitation.

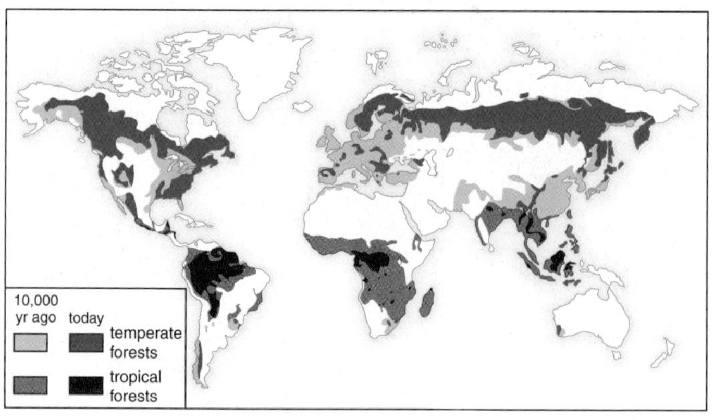

a.

b.

Figure 24.6 Deforestation.
a. Nearly half of the world's forestlands have been cleared for farming, logging, and urbanization. **b.** The soil of tropical rain forests is not suitable for long-term farming.

may be impure. It's considered a human right for people to have clean drinking water. In reality, most freshwater is used by industry and agriculture (Fig. 24.7). Worldwide, 70% of freshwater is used to irrigate crops! Much of a recent surge in demand for water stems from increased industrial activity and irrigation-intensive agriculture. This type of agriculture now supplies about 40% of the world's food crops. In the MDCs, more water is usually used for bathing, flushing toilets, and watering lawns than for drinking and cooking.

Increasing Water Supplies

The needs of the human population overall do not exceed the renewable supply. However, this is not the case in certain regions of the United States and the world. About 40% of the world's land is desert, and deserts are bordered by semiarid land. When needed, humans increase the supply of freshwater by damming rivers and withdrawing water from aquifers.

Dams The world's 45,000 large dams catch 14% of all precipitation runoff and provide water for up to 40% of irrigated land. They also give some 65 countries more than half their electricity. Damming of certain rivers has been so extensive that they no longer flow as they once did. The Yellow River in China fails to reach the sea most years. The Colorado River barely makes it to the Gulf of California. Even the Rio Grande dries up before it can merge with the

a. Agriculture uses most of the freshwater consumed.

b. Industrial use of water is about half that of agricultural use.

c. Domestic use of water is about half that of industrial use.

Figure 24.7 Water use by agriculture, industries, and households.
a. Agriculture primarily uses water for irrigation. **b.** Industry uses water variously. **c.** Households use water to drink, shower, flush toilets, and water lawns.

Gulf of Mexico. The Nile in Egypt and the Ganges in India are also so overexploited they hardly make it to the ocean at some times of the year.

Dams have other drawbacks as well, not the least of which is losing water by evaporation and seepage into underlying rock beds. The amount of water lost sometimes equals the amount made available! The salt left behind by evaporation and agricultural runoff increases salinity and can make a river's water unusable farther downstream. Dams hold back less water with time because of sediment buildup. Sometimes a reservoir becomes so full of silt that it is no longer useful for storing water.

Aquifers To meet their freshwater needs, people are pumping vast amounts of water from **aquifers.** These are reservoirs found just below or as much as 1 km below the surface of the Earth. Aquifers hold about 1,000 times the amount of water that falls on land as precipitation each year. This water accumulates from rain that has fallen in far-off regions. In the past 50 years, groundwater depletion has become a problem in many areas of the world. In substantial portions of the High Plains Aquifer (stretching from South Dakota to Texas), more than half of the water has been pumped out.

Consequences of Groundwater Depletion Removal of water from aquifers is causing land **subsidence,** a settling of the soil as it dries out. An area in California's San Joaquin valley has subsided at least 30 cm due to groundwater depletion. In the worst spot, the surface of the ground has dropped more than 9 meters! Subsidence damages canals, buildings, and underground pipes. Withdrawal of groundwater can cause **sinkholes.** These form when an underground cavern collapses because water no longer holds up its roof (Fig. 24.8).

Figure 24.8 Sinkholes may be caused by groundwater depletion.
Sinkholes occur when an underground cavern collapses after groundwater has been withdrawn.

a.

b.

c.

Figure 24.9 Measures that can be taken to conserve water.
a. Planting drought-resistant crops in the fields and drought-resistant plants in parks and gardens cuts down on the need to irrigate. **b.** When irrigation is necessary, drip irrigation is preferable to using sprinklers. **c.** Wastewater can be treated and reused instead of withdrawing more water from a river or aquifer.

Saltwater intrusion is another consequence of aquifer depletion. The flow of water from streams and aquifers usually keeps them fairly free of seawater. As water is withdrawn, the water table can lower to the point that seawater backs up into streams and aquifers. Saltwater intrusion reduces the supply of freshwater along the coast.

Conservation of Water

By 2025, two-thirds of the world's population may be living in countries facing serious water shortages. Some solutions for expanding water supplies have been suggested. Planting drought- and salt-tolerant crops would help a lot. Using drip irrigation delivers more water to crops and increases crop yields as well (Fig. 24.9). Although the first drip systems were developed in 1960, they're used on only less than 1% of irrigated land. Most governments subsidize irrigation so heavily that farmers have little incentive to invest in drip systems or other water-saving methods. Reusing water and adopting conservation measures could help the world's industries cut their water demands by more than half.

Food

In 1950, the human population numbered 2.5 billion. Only enough food was produced to provide less than 2,000 calories per person per day. Now, with almost 7 billion people on Earth, the world food supply provides more calories per person per day. Generally speaking, food comes from three activities: growing crops, raising animals, and fishing the seas. The increase in the food supply has largely been possible because of modern farming methods. Unfortunately, many of these methods include some harmful practices:

1. *Planting of a few genetic varieties.* The majority of farmers practice monoculture. Wheat farmers plant the same type of wheat, and corn farmers plant the same type of corn. *Monoculture* means that a single type of parasite can destroy entire crops.
2. *Heavy use of fertilizers, pesticides, and herbicides.* Fertilizer production is energy intensive, and fertilizer runoff contributes to water pollution. Pesticides reduce soil fertility because they kill off beneficial soil organisms as well as pests. Some pesticides and herbicides are linked to the development of cancer. **Agricultural runoff** places these chemicals in our water supply.
3. *Generous irrigation.* As already discussed, water is sometimes taken from aquifers. In the future, the water content of these aquifers may become so reduced that it could be too expensive to pump out any more.
4. *Excessive fuel consumption.* Irrigation pumps remove water from aquifers. Large farming machines are used to spread fertilizers, pesticides, and herbicides, as well as to sow and harvest the crops. In effect, modern farming methods transform fossil fuel energy into food energy.

Figure 24.10 shows ways to minimize the harmful effects of modern farming practices. *Polyculture* is the planting of two or more different crops in the same area. In Figure 24.10*a*, a farmer has planted alfalfa in between strips of corn. The alfalfa replenishes the nitrogen content of the soil so that fertilizer doesn't have to be added. In Figure 24.10*b*, contour farming with no-till conserves topsoil because it reduces agricultural runoff. Contour farming is planting and plowing according to the slope of the land. No-till farming allows the previous crop to remain on the land, recycling nutrients and preventing soil erosion. Biological control (Fig. 24.10*c*) relies on the use of natural predators to destroy organisms that harm crops. This reduces the need for pesticides.

Soil Loss

Land suitable for farming and grazing animals is being degraded worldwide. Topsoil is the richest in organic matter and the most capable of supporting grass and crops. When bare soil is acted on by water and wind, soil erosion occurs

a. Polyculture

b. Contour farming

c. Biological pest control

Figure 24.10 Methods that make farming more friendly to the environment.
a. Polyculture reduces the ability of one parasite to wipe out an entire crop and reduces the need to use a herbicide to kill weeds. This farmer has planted alfalfa in between strips of corn, which also replenishes the nitrogen content of the soil (instead of adding fertilizers). Alfalfa, a legume, has root nodules that contain nitrogen-fixing bacteria. **b.** Contour farming with no-till conserves topsoil because water has less tendency to run off. **c.** Instead of pesticides, it is possible to use a natural predator. Here, ladybugs are feeding on cottony-cushion scales (insects) on citrus trees.

and topsoil is lost. As a result, marginal rangeland becomes desertized, and farmland loses its productivity.

The custom of planting the same crop in straight rows, which facilitates the use of large farming machines, has caused the United States and Canada to have one of the highest rates of soil erosion in the world. Conserving the nutrients now being lost could save farmers billions of dollars annually in fertilizer costs. Much of the eroded sediment ends up in lakes and streams, where it reduces the ability of aquatic species to survive.

Green Revolutions

About 50 years ago, researchers began to breed tropical wheat and rice varieties specifically for farmers in the LDCs. The dramatic increase in yield due to the introduction of these new varieties around the world was called "the green revolution." These plants helped the world food supply keep pace with the rapid increase in world population. Unfortunately, most green revolution plants are called "high responders" because they need high levels of fertilizer, water, and pesticides to produce a high yield. They require the same subsidies and create the same ecological problems as do modern farming methods.

Genetic Engineering As discussed in Chapter 21, genetic engineering can produce transgenic plants with new and different traits. For example, resistance to both insects and herbicides are traits that can be introduced into plant DNA. When herbicide-resistant crops are planted, weeds are easily controlled, less tillage is needed, and soil erosion is minimized. Researchers also want to produce crops that tolerate salt, drought, and cold. Some progress has also been made in increasing the food quality of crops so that they will supply more of the proteins, vitamins, and minerals people need. Genetically engineered crops could result in still another green revolution.

Some are opposed to the use of genetically engineered crops. It is feared that these crops will damage the environment and lead to health problems in humans. The Chapter 21 Health Focus, *Are Genetically Engineered Foods Safe?*, discusses this issue.

Video
Pesticide
Plants

Domestic Livestock

A low-protein, high-carbohydrate diet consisting only of grains such as wheat, rice, or corn can lead to malnutrition. In the LDCs, kwashiorkor, caused by a severe protein deficiency, is seen in infants and children ages 1 to 3. It usually occurs after a new arrival in the family and the older children are no longer fed milk. The diet then consists of protein-poor starches. Such children are lethargic, irritable, and have bloated abdomens. Mental retardation is expected.

In the MDCs, many people tend to have more than enough protein in their diet. Almost two-thirds of United States cropland is devoted to producing livestock feed. This means that a

Figure 24.11 The effects on the environment of raising livestock.
Raising livestock requires the use of more fossil fuels and water than raising crops. Livestock waste often washes into nearby bodies of water, creating water pollution.

large percentage of the fossil fuel, fertilizer, water, herbicides, and pesticides used are for the purpose of raising livestock. Typically, cattle are range-fed for about four months. Then they are brought to crowded feedlots where they may receive growth hormone and antibiotics. At the feedlots, they feed on grain or corn. Most pigs and chickens spend their entire lives cooped up in crowded pens and cages (Fig. 24.11).

If livestock eat a large proportion of the crops in the United States, then raising livestock accounts for much of the pollution associated with farming. Fossil fuel energy is needed not just to produce herbicides and pesticides and to grow food but also to make the food available to the livestock. Raising livestock is extremely energy-intensive in the MDCs. In addition, water is used to wash livestock wastes into nearby bodies of water, where they add significantly to water pollution. Whereas human wastes are sent to sewage treatment plants, raw animal wastes are not.

For these reasons, it is prudent to recall the ecological energy pyramid (see Fig. 23.7), which shows that as you move up the food chain, energy is lost. As a rule of thumb, for every 10 calories of energy from a plant, only 1 calorie is available for the production of animal tissue in an herbivore. A great deal of energy is wasted when the human diet contains more protein than is needed to maintain good health. It is possible to feed ten times as many people on grain as on meat.

Energy

Modern society runs on various sources of energy. Some of these energy sources are nonrenewable, and others are renewable. The consumption of nonrenewable energy supplies results in environmental degradation. Renewable energy is expected to be used more in the future.

Nonrenewable Sources

Presently, about 6% of the world's energy supply comes from nuclear power and 75% comes from fossil fuels. Both of these are finite, nonrenewable sources. It was once predicted that the nuclear power industry would fulfill a significant portion of the world's energy needs. However, this has not happened for two reasons. One is that people are very concerned about nuclear power dangers, such as the meltdown at the Chernobyl nuclear power plant in Russia in 1986. Further, radioactive wastes from nuclear power plants remain a threat to the environment for thousands of years. We still have not decided how best to safely store them.

Fossil fuels (oil, natural gas, and coal) are derived from the compressed remains of plants and animals that died thousands of years ago. Of the fossil fuels, oil burns more cleanly than coal, which may contain a lot of sulfur. When the use of coal releases sulfur, acid rain forms. Thus, despite the fact that coal is plentiful in the United States, imported oil is our preferred fossil fuel. Regardless of which fossil fuel is used, all contribute to environmental problems because of the pollutants released when they're burned. Current research into clean-coal technology may make U.S. coal less polluting.

Fossil Fuels and Global Climate Change In 1850, the level of carbon dioxide in the atmosphere was about 280 parts per million (ppm); today, it is over 350 ppm. This increase is largely due to the burning of fossil fuels and the burning and clearing of forests. Human activities are causing the emission of other gases as well. These gases are known as **greenhouse gases.** Just like the panes of a greenhouse, they allow solar radiation to pass through but hinder the escape of infrared heat back into space.

Video Global Warming

Computer models predict the Earth may warm to temperatures never before experienced by living things. The global climate has already warmed about 0.6°C, and it may rise as much as 1.5–4.5°C by 2100. If so, sea levels will rise as glaciers melt and warm water expands. Major coastal cities of the United States could eventually be threatened. The present wetlands will be inundated. Great losses of aquatic habitat will occur wherever wetlands cannot move inward because of coastal development and levees. Coral reefs, which prefer shallow waters, will most likely "drown" as the waters rise.

On land, regions of suitable climate for various species will shift toward the poles and higher elevations. Plants migrate when seeds disperse and growth occurs in a new locale. The present assemblages of species in ecosystems will be disrupted as some species migrate northward faster than others. Trees, for example, cannot migrate as fast as nonwoody plants. Also, too many species of organisms are confined to relatively small habitat patches surrounded by agricultural or urban areas. Even if such species have the capacity to disperse to new sites, suitable habitats may not be available.

Renewable Energy Sources

Renewable types of energy include hydropower, geothermal, wind, and solar.

Hydropower Hydroelectrical plants convert the energy of falling water into electricity (Fig. 24.12*a*). Hydropower accounts for about 10% of the electrical power generated in the United States and almost 98% of the total renewable energy used. Worldwide, hydropower presently generates 19% of all electricity used. This percentage is expected to rise because of increased use in certain countries.

Much of the hydropower development in recent years has been due to the construction of enormous dams. These are known to have detrimental environmental effects (see pages 571–72). The better choice is believed to be small-scale dams that generate less power per dam but do not have the same environmental impact.

Geothermal Energy Elements such as uranium, thorium, radium, and plutonium undergo radioactive decay below the Earth's surface. This heats the surrounding rocks to hundreds of degrees Celsius. When the rocks are in contact with underground streams or lakes, huge amounts of steam and hot water are produced. This steam can be piped up to the surface to supply hot water for home heating or to run steam-driven turbogenerators. The California's Geysers project is the world's largest geothermal electricity-generating complex.

Wind Power Wind power is expected to account for a significant percentage of our energy needs in the future. A common belief is that a huge amount of land is required for the "wind farms" that produce commercial electricity. The amount of land needed for a wind farm compares favorably with the amount of land required by a coal-fired power plant or a solar thermal energy system (Fig. 24.12*b*).

A community generating its own electricity by using wind power can solve the problem of uneven energy production. Electricity can be sold to a local public utility when an excess is available. Then electricity is bought from the same facility when wind power is in short supply.

Energy and the Solar–Hydrogen Revolution Solar energy is diffuse energy that must be collected, converted to another form, and stored if it is to compete with other available forms of energy. Passive solar heating of a house is successful when the windows of the house face the sun and the building is well insulated. Successful heating also requires that heat can be stored in water tanks, rocks, bricks, or some other suitable material.

In a **photovoltaic** (**solar**) **cell,** a wafer of the electron-emitting metal is in contact with another metal that collects the electrons. Electrons are then passed along into wires in a steady stream. Spurred by the oil shocks of the 1970s, the U.S. government has been supporting the development of photovoltaics ever since. As a result, the price of buying a photovoltaic cell has dropped from about $100 per watt to around $4. Photovoltaic cells placed on roofs generate elec-

a.

c.

b.

d.

Figure 24.12 Sources of renewable energy.
a. Hydropower dams provide a clean form of energy but can be ecologically disastrous in other ways. **b.** Wind power requires land on which to place enough windmills to generate energy. **c.** Photovoltaic cells on rooftops and (**d**) sun-tracking mirrors on land can collect diffuse solar energy more cheaply than could be done formerly.

tricity that can be used inside a building and/or sold back to a power company (Fig. 24.12*c*).

Several types of solar power plants are now operational in California. In one type, huge reflectors focus sunlight on a pipe containing oil. The heated pipes boil water, generating steam that drives a conventional turbogenerator. In another type, 1,800 sun-tracking mirrors focus sunlight onto a molten salt receiver mounted on a tower. The hot salt generates steam that drives a turbogenerator (Fig. 24.12*d*).

Scientists are working on the possibility of using solar energy to extract hydrogen from water via electrolysis. The hydrogen can then be used as a clean-burning fuel. When it burns, water is produced. Presently, cars have internal combustion engines that run on gasoline. In the future, vehicles are expected to be powered by fuel cells, which use

a.

b.

Figure 24.13 Hydrogen fuel cells may reduce air pollution and fossil fuel consumption.

a. This bus is powered by hydrogen fuel. **b.** The use of fuel-cell hybrid vehicles, such as this prototype, will reduce air pollution and dependence on fossil fuels.

hydrogen to produce electricity (Fig. 24.13). The electricity runs a motor that propels the vehicle. Fuel cells are now powering buses in Vancouver and Chicago. Additional buses are planned.

Hydrogen fuel can be produced locally or in central locations, using energy from photovoltaic cells. The fuel produced in central locations can be piped to filling stations using the natural gas pipes already plentiful in the United States. However, two major hurdles are storage and production. Advantages of a solar-hydrogen revolution are decreased dependence on oil and fewer environmental problems.

Video Spinach Battery

Minerals

Minerals are nonrenewable raw materials in the Earth's crust that can be mined (extracted) and used by humans. Nonmetallic raw materials such as sand, gravel, and phosphate are considered minerals. Metals, such as aluminum, copper, iron, lead, and gold fall into this category, as well.

One of the greatest threats to the maintenance of ecosystems and biodiversity is surface mining, called strip mining. In the United States, huge machines can go as far as removing mountaintops to reach a mineral. The land devoid of vegetation takes on a surreal appearance, and rain washes toxic waste deposits into nearby streams and rivers.

The most dangerous metals to human health are the heavy metals. These include lead, mercury, arsenic, cadmium, tin, chromium, zinc, and copper. They are used to produce batteries, electronics, pesticides, medicines, paints, inks, and dyes. In the ionic form, they enter the body and inhibit vital enzymes. That's why these items should be discarded carefully and taken to hazardous waste sites.

Hazardous Wastes

The consumption of minerals and use of synthetic organic chemicals contribute to the buildup of hazardous waste in the environment. Every year, countries around the world discard billions of tons of solid waste on land and in freshwater and salt water. The Environmental Protection Agency (EPA) oversees the cleanup of hazardous waste disposal sites in the United States. An EPA program called the Superfund provides the funds that helps pay for this cleanup. Commonly found contaminants include heavy metals (such as lead, mercury, and arsenic) and chlorine-containing organic chemicals (such as chloroform and polychlorinated biphenyls, or PCBs). Some of these contaminants interfere with hormone activity and proper endocrine system functioning.

Humanmade organic chemicals play a role in the production of plastics, pesticides, herbicides, cosmetics, and hundreds of other products. For example, the so-called *halogenated hydrocarbons* are compounds made from carbon and hydrogen and also include halogen atoms such as chlo-

Connections and Misconceptions

What is methylmercury and why is it dangerous?

Methylmercury is a form of the element mercury that has been bound to a methyl (CH_3) group. Because of this methyl group, methylmercury easily accumulates in the food chain by biological magnification or the concentration of chemicals in a food chain. Methylmercury is released into the environment by the burning of coal, the mining of certain metals, and the incineration of medical waste. Methylmercury is a powerful neurotoxin that also inhibits the activity of the immune system. Because of this, the Food and Drug Administration (FDA) and the Environmental Protection Agency (EPA) recommend that pregnant women and small children not eat shark, swordfish, tilefish, or king mackerel and limit the amount of albacore tuna to less than 6 ounces per week. Most states have also posted warnings on eating local fish that have been caught from mercury-contaminated waters. For more information, visit the EPA website, www.epa.gov/waterscience/fish.

rine and fluorine. These compounds, **chlorofluorocarbons (CFCs)**, have been shown to damage the Earth's ozone shield. Recall from Chapter 23 that this shield protects terrestrial life from harmful UV radiation and has recently been depleted by the use of CFCs.

Further, these types of synthetic compounds pose a threat to the health of living things, including humans, because they undergo **biological magnification.** Such chemicals accumulate in fat and are not excreted, so they remain in an organism's body. Therefore, they become more concentrated as they pass from organism to organism along a food chain. Biological magnification is more apt to occur in aquatic food chains, which have more links than terrestrial food chains. Rachel Carson's book *Silent Spring,* published in 1962, made the public aware of the harmful effects of pesticides such as DDT. These substances accumulate in the mud of deltas and estuaries of highly polluted rivers and cause environmental problems if disturbed. After working its way up the food chain, high concentrations of DDT in predatory birds like bald eagles and pelicans interfered with their ability to reproduce (Fig. 24.14). There are health advisories about eating certain types of fish due to the high levels of mercury they contain. Humans are often the final consumers in a variety of food chains and are affected by biological magnification as well. The breast milk of humans has been found to contain significant levels of DDT, PCBs, solvents, and heavy metals.

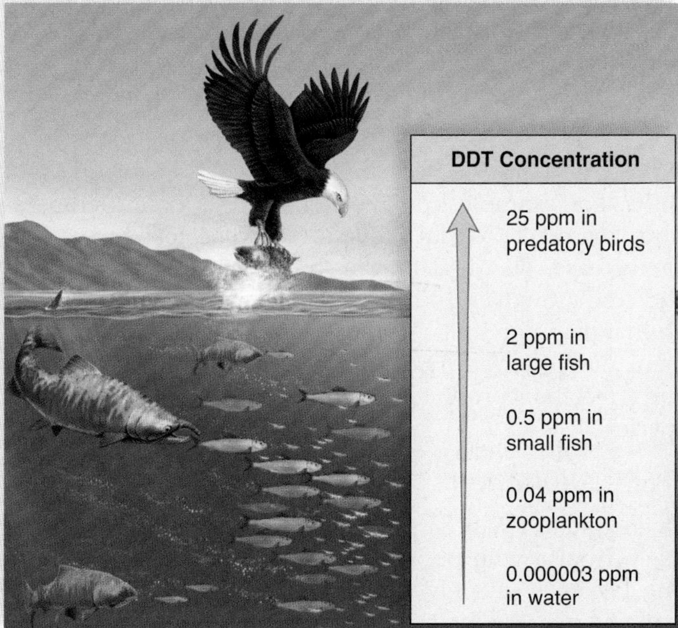

Figure 24.14 Biological magnification concentrates chemicals in the food chain.
Various synthetic organic chemicals, such as DDT, accumulate in animal fat. Therefore, the chemicals become increasingly concentrated at higher trophic levels. By the time DDT was banned in the United States, it had interfered with predatory bird reproduction by causing eggshell thinning.

Raw sewage causes oxygen depletion in lakes and rivers. As the oxygen level decreases, the diversity of life is greatly reduced. Also, human feces can contain pathogenic microorganisms that cause cholera, typhoid fever, and dysentery. In regions of the LDCs where sewage treatment is practically nonexistent, many children die each year from these diseases. Typically, sewage treatment plants use bacteria to break down organic matter to inorganic nutrients, such as nitrates and phosphates, which then enter surface waters. The result can be cultural eutrophication (see Chapter 23).

Connecting the Concepts

For more information on the topics in this section, refer to the following discussions.

Section 23.2 explores the flow of energy in an ecosystem.

Section 23.3 examines the water and carbon cycles and explores the influence of human activity on these cycles.

Check Your Progress 24.2

1. List the five main classes of resources.
2. Describe how humans have increased the availability of groundwater and food resources, and summarize the potential problems that these activities may be creating.
3. Summarize why is it better for humans to use renewable rather than nonrenewable sources of energy.
4. Discuss the consequences of using fossil fuels as an energy source.
5. Describe how hazardous wastes may interfere with natural environmental processes.

24.3 Biodiversity

Learning Outcomes

Upon completion of this section, you should be able to

1. Describe the factors that are contributing to the current biodiversity crisis.
2. Summarize the direct values to society for conserving biodiversity.
3. Discuss the indirect values to society for conserving biodiversity.

Biodiversity can be defined as the variety of life on Earth and described in terms of the number of different species. We are presently in a biodiversity crisis. The number of extinctions (loss of species) expected to occur in the near future is unparalleled in the history of the Earth.

Loss of Biodiversity

Figure 24.15 identifies the major causes of extinction.

Habitat Loss

Human occupation of the coastline, semiarid lands, tropical rain forests, and other areas have contributed to the loss of biodiversity. Scientists are especially concerned about the tropical rain forests and coral reefs because they are particularly rich in species. Already, tropical rain forests have been reduced from their original 14% of landmass to the present 6%. Also, 60% of coral reefs have been destroyed or are on the verge of destruction. It's possible that all coral reefs may disappear during the next 40 years.

Video Coral Reef Ecosystems

Alien Species

Alien species, sometimes called exotics, are nonnative members of an ecosystem. Humans have introduced alien species into new ecosystems via colonization, horticulture and agriculture, and accidental transport. For example, the pilgrims brought the dandelion to the United States as a familiar salad green. Kudzu is a vine from Japan that the U.S. Department of Agriculture thought would help prevent soil erosion. The plant now covers much landscape in the South. The zebra mussel from the Caspian Sea was accidentally introduced into the Great Lakes in 1988. It is now found in many U.S. rivers where it forms dense beds that squeeze out native mussels. Alien species that crowd out native species are termed **invasive.** One way to counteract alien species is to replant native (original to the area) species.

Video Alien Invasion

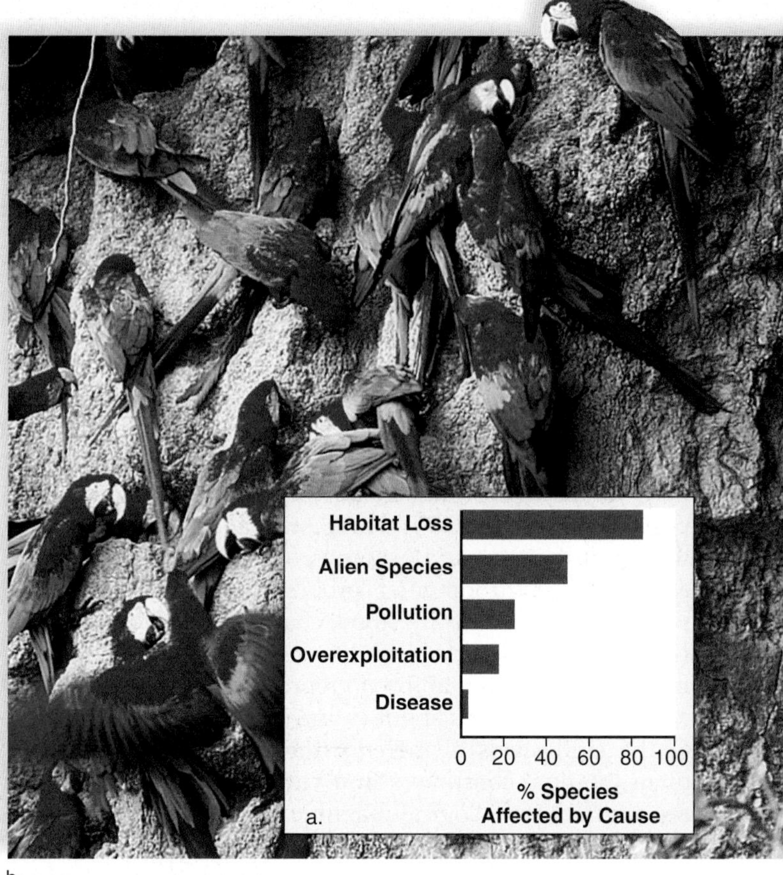

b.

Figure 24.15 Causes for the loss of biodiversity.
a. Habitat loss, alien species, pollution, overexploitation, and disease have been identified as causes of extinction of organisms. **b.** Macaws that reside in South American tropical rain forests are endangered for the reasons listed in the graph.

Pollution

Pollution brings about environmental change that adversely affects the lives and health of living things. Acid deposition weakens trees and increases their susceptibility to disease and insects. This can decimate a forest. Global warming is predicted to be the cause of habitat loss due to temperature shifts. For example, coral reefs die off as ocean temperatures increase. Coastal wetlands may be lost as sea levels increase. The depletion of the ozone shield limits crop and tree growth and kills plankton that sustain ocean life. Humanmade organic chemicals released into the environment may interfere with hormone function and affect the reproductive ability of different species.

Video Ocean Garbage

> **Connections and Misconceptions**
>
> **Why are there beagle dogs at the customs area in an international airport?**
>
> The beagles working at an airport's customs checkpoint are trained to detect agricultural items (foodstuffs, plants, animals, and the like) that may not be brought into the United States. Importing foreign species is prohibited by law. The goal of this law is to prevent the introduction of exotic life and/or diseases that might affect crops or animals in the United States. About $2 million worth of illegal articles are seized yearly. Dogs assist in about 10% of those cases. Beagles are used because of their keen sense of smell and good-natured temperament.

Overexploitation

Overexploitation occurs when the number of individuals taken from a wild population is so great that the population becomes severely reduced. A positive feedback cycle explains overexploitation. The smaller the population, the more valuable its members and the greater the incentive to exploit the few remaining organisms.

Markets for decorative plants and exotic pets support both legal and illegal trade in wild species. Rustlers dig up rare cacti, such as the single-crested saguaro, and sell them to gardeners. Parakeets and macaws are among the birds taken from the wild for sale to pet owners. For every bird delivered alive, many more have died in the process. The same holds true for tropical fish, which often come from the coral reefs of Indonesia and the Philippines. Divers dynamite reefs or use plastic squeeze bottles of cyanide to stun them. In the process, many fish die.

Declining species of mammals are still hunted for their hides, tusks, horns, or bones. A single Siberian tiger is now worth more than $500,000 because of its rarity. Its bones are pulverized and used as a medicinal powder. The horns of rhinoceroses become ornate carved daggers, and their bones are ground up to sell as a medicine. The ivory of an elephant's tusk is used to make art objects, jewelry, or piano keys. The fur of a Bengal tiger sells for as much as $100,000 in Tokyo.

Fish are a renewable resource if harvesting does not exceed the ability of the fish to reproduce. Today, larger and more efficient fishing fleets decimate fishing stocks (Fig. 24.16). Tuna and similar fish are captured by purse seining. A very large net surrounds a school of fish, and then the net is closed in the same manner as a drawstring purse. Dolphins that accompany the tuna are killed by this type of net. Other fishing boats drag huge trawling nets, large enough to accommodate 12 jumbo jets, along the seafloor to capture bottom-dwelling fish. Trawling has been called the marine equivalent of clear-cutting trees because after the net goes by, the sea bottom is devastated (Fig. 24.16*b*). Only large fish are kept. Undesirable small fish and sea turtles are discarded, dying, back into the ocean. Cod and haddock were once the most abundant bottom-dwelling fish along the Northeast Coast. Now, they are often outnumbered by dogfish and skate.

A marine ecosystem can be disrupted by overfishing, as exemplified on the U.S. West Coast. When sea otters began to decline in numbers, investigators found that they were being eaten by orcas (killer whales). Usually, orcas prefer seals and sea lions to sea otters, but they began eating sea otters when seals and sea lions could not be found. The decline in seals

Connections and Misconceptions

How does one choose the best fish to eat?

Before you place your order for fish, take a moment to think about your choice. You may want to consider whether it's a sustainable species of fish, one that's being overfished, or one that contains high levels of mercury. Pacific halibut is a wild fish that is a good choice. The Pacific halibut is usually caught with bottom longlines, rather than trawling nets. This fishing technique won't damage the surrounding environment or catch unwanted fish or animals. Tilapia that is farm grown in the United States is another good choice. It's grown in closed inland systems that prevent exposure of the fish to pollutants. But avoid tilapia from Chile or Taiwan—fish there are raised in open systems where pollutants, especially mercury, can affect the fish. You might also want to consider the levels of healthy omega-3 fatty acids in the fish you eat. Farm-raised salmon or trout supply high concentrations of omega-3 acids (thought to protect against heart disease) and contain little or no mercury or other pollutants.

Shark, sole, haddock, and swordfish are poor choices. These species have been severely overfished and/or contain high levels of mercury.

Video Fishing for Trouble

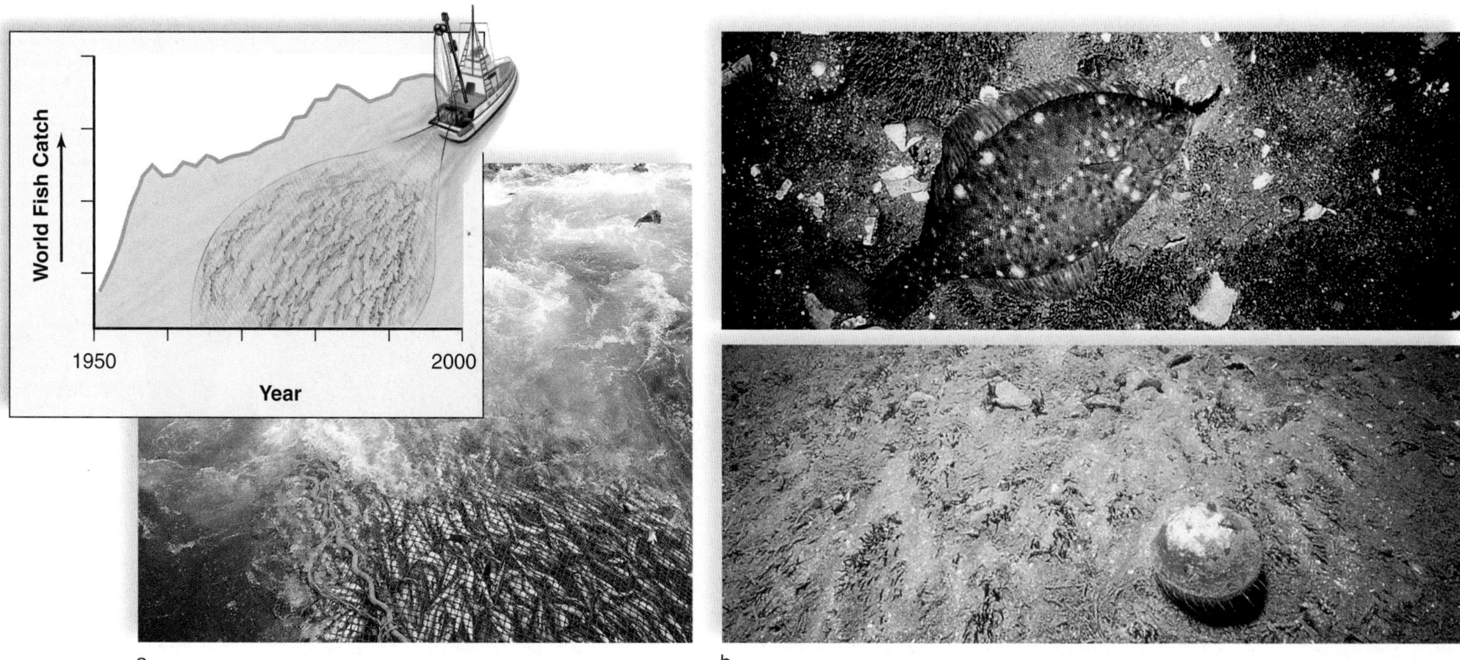

Figure 24.16 The impact of modern fishing practices.
a. The world fish catch has declined in recent years (insert). **b.** Devastation of the seafloor after trawling (lower photo).

and sea lions occurred because their preferred food sources (perch and herring) were no longer plentiful due to overfishing. Ordinarily, sea otters keep the population of sea urchins, which feed on kelp, under control. But with fewer sea otters around, the sea urchin population exploded and decimated the kelp beds. Thus, overfishing set in motion a chain of events that adversely affected the food web of an ecosystem.

Video Ocean Fishing Ban

Disease

Wildlife is subject to emerging diseases just as humans are. Exposure to domestic animals and their pathogens occurs due to the encroachment of humans on wildlife habitats. Wildlife can also be infected by animals not ordinarily encountered. For example, African elephants carry a strain of herpesvirus that is fatal to Asian elephants. Deaths can result if the two types of elephants are housed together.

The significant effect of diseases on biodiversity is underscored by a National Wildlife Health Center study. The study found that almost half of sea otter deaths along the coast of California are due to infectious diseases. Scientists tell us that the number of pathogens that cause disease are on the rise. Just as human health is threatened, so is that of wildlife. Extinctions due to disease may occur.

Direct Value of Biodiversity

Various individual species perform useful services for human beings and contribute greatly to the value of biodiversity. The direct value of wildlife species is related to their medicinal value, agricultural value, and consumptive use value. These are the most obvious values that are discussed here and illustrated in Figure 24.17.

Medicinal Value

Most of the prescription drugs used in the United States were originally derived from living organisms. The rosy periwinkle from Madagascar is a tropical plant that has provided us with useful medicines. Potent chemicals from this plant are now used to treat leukemia and Hodgkin disease. The survival rate for childhood leukemia has gone from 10% to 90%, and Hodgkin disease is usually curable because of these drugs. Although the value of saving a life cannot be calculated, it is still sometimes easier for us to appreciate the worth of a resource if it is explained in monetary terms. Based on past success, it has been estimated that more than 300 types of drugs may yet be found in tropical rain forests. The value of this resource could be in excess of $140 billion.

You may already know that the antibiotic penicillin is derived from a fungus. Certain species of bacteria pro-

Figure 24.17 The direct value of biodiversity. The direct services of wild species benefit human beings immensely, and it is sometimes possible to calculate the monetary value, which is always surprisingly large.

Wild species, like the rosy periwinkle, are sources of many medicines.

Wild species, like many marine species, provide us with food.

Wild species, like the nine-banded armadillo, play a role in medical research.

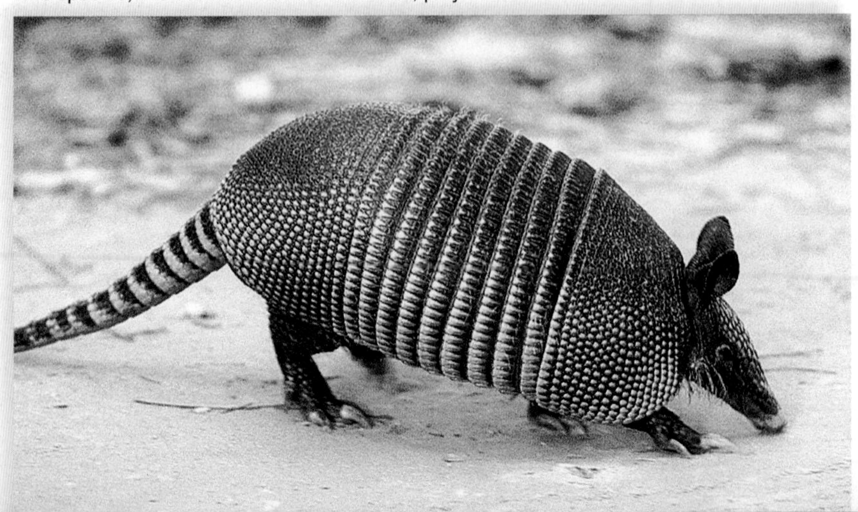

duce the antibiotics tetracycline and streptomycin. These drugs have proven to be indispensable in the treatment of diseases.

Leprosy is a disease for which there is no cure as of yet. The bacterium that causes leprosy will not grow in the laboratory. However, scientists discovered that it grows naturally in the nine-banded armadillo. Having a source for the bacterium may make it possible to find a cure for leprosy. The blood of horseshoe crabs contains a substance called limulus amoebocyte lysate. This chemical is used to ensure that medical devices, such as pacemakers, surgical implants, and prosthetic devices, are free of bacteria. Blood is taken from 250,000 crabs a year, and then they are returned to the sea unharmed.

Video Good Poison

Agricultural Value

Crops, such as wheat, corn, and rice, are derived from wild plants that have been modified to be high producers. The same high-yield, genetically similar strains tend to be grown worldwide. At one time, rice crops in Africa were being devastated by a virus. Researchers grew wild rice plants from thousands of seed samples until they found one that contained a gene for resistance to the virus. These wild plants were then used in a breeding program to transfer the gene into high-yield rice plants. If this variety of wild rice had become extinct before being discovered, rice cultivation in Africa might have collapsed.

The use of biological pest controls (natural predators and parasites) is often preferable to using chemical pesticides. When a rice pest, called the brown planthopper, became resistant to pesticides, farmers began to use natural brown planthopper enemies instead. The economic savings were calculated at well over $1 billion. Similarly, cotton growers in Cañete Valley, Peru, found that the cotton aphid was resistant to the pesticides being used. Research identified natural predators that are now being used to an ever greater degree by cotton farmers. Again, savings have been enormous.

Most flowering plants are pollinated by animals, such as bees, wasps, butterflies, beetles, birds, and bats. The honeybee *Apis mellifera* has been domesticated. It now pollinates almost $10 billion worth of food crops annually in the United States. The value of wild bee pollinators to the U.S. agricultural economy has been calculated at $4.1 to $6.7 billion a year. And yet, modern agriculture often kills wild bees by spraying fields with pesticides.

Consumptive Use Value

We have had much success cultivating crops, keeping domesticated animals, growing trees in plantations, and so forth. However, aquaculture, the growing of fish and shellfish for human consumption, has contributed only minimally to human welfare. Instead, most freshwater and marine harvests depend on

Wild species, like the lesser long-nosed bat, are pollinators of agricultural and other plants.

Wild species, like ladybugs, play a role in biological control of agricultural pests.

Wild species, like rubber trees, can provide a product indefinitely if the forest is not destroyed.

Biology Matters Science Focus

Mystery of the Vanishing Bees

Imagine standing in the produce section of your supermarket. You're shocked to see that there are no apples, cucumbers, broccoli, onions, pumpkins, squash, carrots, blueberries, avocados, almonds, or cherries. This could happen at grocery stores in the future. All the crops mentioned, as well as many others, are dependent on honeybees for pollination. Your diet and $15 billion worth of crops could suffer if the honeybees aren't available to perform their important job of moving pollen. Although there are wild bee populations that pollinate crops, domestic honeybees are easily managed and transported from place to place when their pollination services are needed.

Colonies of honeybees have experienced a number of health problems since the 1980s. Mites—animals similar to ticks—have always been a danger for bees. Varroa mites and tracheal mites were early causes of colony stress and bee deaths (Fig. 24A). However, beekeepers were very alarmed in 2006 when entire colonies of bees began to vanish. Researchers started referring to the phenomenon as colony collapse disorder (CCD). There doesn't appear to be one single factor that causes seemingly healthy bees to vanish from their hives. Scientists now believe multiple factors may stress bees, causing them to be vulnerable to infection by a parasite or pathogen. The indiscriminate use of pesticides, the strain of being moved from place to place to pollinate crops, and/or poor nutrition (because genetically engineered plants don't provide as much food for the bees) may contribute to CCD.

CCD is also occurring in the honeybee populations in other countries. Ideally, a cause-and-effect treatment for CCD will be found soon, thanks to worldwide research dedicated to solving the problem, as well as improved funding from agricultural agencies. Until then, there are things you can do to keep bees in your area healthy. Research and then plant native plants in your yard and garden. These typically require less fertilizer and water than other plants, and they will provide more pollen and nectar for the bees. In the southern and midwestern regions of the United States, bees enjoy red clover, foxglove, bee balm, and joe-pye weed. Desert willow and manzanita will attract desert bees. Choose palms for tropical areas. In addition, native plants that flower at different times of the year will provide a constant food source. Midday is typically when bees are out foraging, so if you have to use pesticides, apply them late in the day. Plants that rely on the honeybees for pollination—as well as your body—will thank you!

Figure 24A **Parasitic mites may destroy entire bee colonies.**

the catching of wild fish (e.g., trout and tuna), crustaceans (e.g., shrimps and crabs), and mammals (e.g., whales). These aquatic organisms are an invaluable biodiversity resource.

The environment provides all sorts of other products that are sold in the marketplace worldwide. Wild fruits and vegetables, skins, fibers, beeswax, and seaweed are only a few examples. Also, some people obtain their meat directly from the environment. The economic value of wild pig in the diet of native hunters in Sarawak, East Malaysia, has been calculated to be approximately $40 million per year.

Similarly, many trees are still felled in the natural environment for their wood. Researchers have calculated that a species-rich forest in the Peruvian Amazon is worth far more if the forest is used for fruit and rubber production than for timber production. Fruit and the latex needed to produce rubber can be brought to market for an unlimited number of years. Once the trees are gone, no timber can be harvested until new trees replace those that have been taken.

Indirect Value of Biodiversity

To bring about the preservation of wildlife, it is necessary to make all people aware that biodiversity is a resource of immense value. If we want to preserve wildlife, it is more economical to save ecosystems than individual species. Ecosystems perform many useful services for modern humans, who increasingly live in cities. These services are said to be indirect because they are pervasive and not easily discernible. Our very survival depends on the functions that ecosystems perform for us. The indirect value of biodiversity can be associated with the following services.

Waste Disposal

Decomposers break down dead organic matter and other types of wastes to inorganic nutrients. The nutrients are then used by the producers within ecosystems. This function aids humans immensely because we dump millions of tons of waste material into natural ecosystems each year. Waste would soon cover the entire surface of our planet without decomposition. We can build expensive sewage treatment plants, but few of them break down solid wastes completely to inorganic nutrients. It is less expensive and more efficient to water plants and trees with partially treated wastewater and let soil bacteria cleanse it completely.

Biological communities are also capable of breaking down and immobilizing pollutants. These include heavy metals and pesticides that humans release into the environment.

Provision of Freshwater

Few terrestrial organisms are adapted to living in a salty environment. They need freshwater. The water cycle continually supplies freshwater to terrestrial ecosystems. Humans use freshwater in innumerable ways, including drinking it and irrigating their crops. Freshwater ecosystems, such as rivers and lakes, also provide us with fish and other types of organisms for food.

Unlike other commodities, there is no substitute for freshwater. We can remove salt from seawater to obtain freshwater. However, the cost of desalination is about four to eight times the average cost of freshwater acquired via the water cycle.

Forests and other natural ecosystems exert a "sponge effect." They soak up water and then release it at a regular rate. When rain falls in a natural area, plant foliage and dead leaves lessen its impact. The soil then slowly absorbs it, especially if the soil has been aerated by organisms. The water-holding capacity of forests reduces the possibility of flooding. Forests release water slowly for days or weeks after the rains have ceased. Rivers flowing through forests in West Africa release between three and five times as much at the end of the dry season, as do rivers from coffee plantations.

Prevention of Soil Erosion

Intact ecosystems naturally retain soil and prevent soil erosion. The importance of this ecosystem attribute is especially observed following deforestation. In Pakistan, the world's largest dam, the Tarbela Dam, is losing its storage capacity of 12 billion cubic meters many years sooner than expected. Deforestation is causing silt to build up behind the dam, decreasing its storage capacity. At one time, the Republic of the Philippines was exporting $100 million worth of oysters, mussels, clams, and cockles each year. Now, silt carried down rivers following deforestation is smothering the mangrove ecosystem that serves as a nursery for these shellfish. Most coastal ecosystems are not as bountiful as they once were because of deforestation and a myriad of other assaults.

Biogeochemical Cycles

You'll recall from Chapter 23 that ecosystems are characterized by energy flow and chemical cycling. The biodiversity within ecosystems contributes to the workings of the water, phosphorus, nitrogen, carbon, and other biogeochemical cycles. We depend on these cycles for freshwater, provision of phosphate, uptake of excess soil nitrogen, and removal of carbon dioxide from the atmosphere. When human activities upset the usual workings of biogeochemical cycles, the dire environmental consequences include the release of excess pollutants that are harmful to us. Technology is unable to substitute for any of the biogeochemical cycles.

Video
Dung Beetles

Regulation of Climate

At the local level, trees provide shade and reduce the need for fans and air conditioners during the summer. In a rain forest, trees maintain area rainfall. Forests would become arid without the trees.

Globally, forests stabilize the climate because they take up carbon dioxide. The leaves of trees use carbon dioxide when they photosynthesize, and the bodies of the trees store carbon. When trees are cut and burned, carbon dioxide is released into the atmosphere. Carbon dioxide makes a significant contribution to global warming, which is expected to be stressful for many plants and animals. Only a small percentage of wildlife may be able to move northward, where the weather will be suitable for them.

Ecotourism

Almost everyone prefers to vacation in the natural beauty of an ecosystem. In the United States, nearly 100 million people enjoy vacationing in a natural setting. To do so, they spend $4 billion each year on fees, travel, lodging, and food. Many tourists want to go sport fishing, whale watching, boat riding, hiking, bird watching, and the like.

Connecting the Concepts

For more information on the topics presented in this section, refer to the following discussions.

Section 23.3 examines the water, carbon, nitrogen, and phosphorous cycles.

Check Your Progress 24.3

1. Describe the factors that are contributing to the current extinction crisis.
2. Summarize the direct and indirect benefits of preserving wildlife.
3. Discuss the importance of biodiversity to human society.

24.4 Working Toward a Sustainable Society

Learning Outcomes

Upon completion of this section, you should be able to

1. Describe the characteristics of a sustainable society.
2. Distinguish methods of developing sustainability between rural and urban environments.
3. List the methods of determining economic well-being and quality of life.

A **sustainable** society would always be able to provide the same amount of goods and services for future generations as it does at present. At the same time, biodiversity would be preserved.

To achieve a sustainable society, resources cannot be depleted and must be preserved. In particular, future generations need clean air, water, an adequate amount of food, and enough space in which to live. This goal is not possible unless we carefully regulate our consumption of resources today, taking into consideration that the human population is still increasing.

Today's Unsustainable Society

We are quick to realize that population growth in the LDCs creates an environmental burden. However, we also need to consider that the excessive resource consumption of the MDCs also stresses the environment. Sustainability is incompatible with the current level of consumption plus the generation of wastes practiced by the MDCs. Overpopulation of the LDCs and overconsumption by the MDCs account for the increasing amount of worldwide pollution and the extinction of wildlife observable today (Fig. 24.18).

At present, a considerable proportion of land is being used for human purposes (homes, agriculture, factories, etc.). Agriculture uses large inputs of fossil fuels, fertilizer, and pesticides. These create much pollution. More freshwater is used for agriculture than in homes. Almost half of the agricultural yield in the United States goes toward feeding animals. According to the ten-to-one rule of thumb, it takes 10 lb of grain to grow 1 lb of meat. Therefore, it is wasteful for citizens in MDCs to eat as much meat as they do.

Farm animals and crops require freshwater from surface water and groundwater, and so do humans. Available supplies are dwindling, and what groundwater remains is in danger of being contaminated. Sewage and animal wastes wash into bodies of surface water and cause overenrichment, which robs aquatic animals of the oxygen they need to survive.

Our society primarily uses nonrenewable fossil fuel energy, which leads to global warming, acid deposition, and smog. The result is weakened ecosystems. The demand for goods has increased to the point that facilities to meet the demand are strained. Construction of improved infrastructure to support increased transportation needs only increases the use of nonrenewable energy resources. LDCs have increased needs for energy, making it imperative for the MDCs to develop renewable energy sources.

Figure 24.18 **Characteristics of an unsustainable society.**
Arrows point outward to signify that these types of activities reduce the carrying capacity of the Earth.

The human population is expanding into all regions on the face of the planet, so habitats for other species are being lost. A wildlife extinction crisis is expected.

Characteristics of a Sustainable Society

A natural ecosystem can offer clues about how to make today's society sustainable. A natural ecosystem makes use of only renewable solar energy. Its materials cycle through the various populations back to the producer once again. For example, coral reefs have been sustaining themselves for millions of years. At the same time, the reefs have provided sustenance to the LDCs. The value of coral reefs has been assessed at over $300 billion a year. Their aesthetic value is immeasurable.

It is clear that if we want to develop a sustainable society, we too should use renewable energy sources and recycle materials. We should protect natural ecosystems that help sustain our modern society. At least a quarter of the coral reefs exist close to the shores of an MDC country, and the chances are good that these coral reefs will be protected. Unfortunately, other coral reefs are threatened by unsustainable practices. The good news is that reefs are remarkably regenerative and will return to their former condition if left alone. The message of today's environmentalists is about what can be done to improve matters and make the environment sustainable (Fig. 24.19). There is still time to make changes and improvements.

Sustainability should be practiced in various areas of human endeavor, from agriculture to business enterprises. Efficiency is the key to sustainability. For example, an efficient car would be ultralight and gas thrifty. Efficient cars could be just as durable and speedy as the inefficient ones of today. Only through efficiency can we meet the challenges of limited resources and finances in the future.

People generally live in either the country or the city, but the two regions depend on one another. Achieving sustainability requires that we understand how the two regions are interdependent. It would be impossible to have one sustainable and not the other because the two regions are linked. What happens within one ultimately affects the other. Let's consider, therefore, the importance of both rural and urban sustainability.

Rural Sustainability

In rural areas, we must put the emphasis on preservation. We need to preserve ecosystems, including terrestrial ecosystems (such as forests and prairies) and aquatic ecosystems (freshwater and brackish ones along the coast). We should also preserve agricultural land, groves of fruit trees, and other areas that provide us with renewable resources.

It is imperative that we take all possible steps to preserve what remains of our topsoil and replant areas with native plants. Native grasses stabilize the soil, rebuild soil nutrients, and also serve as a source of renewable biofuel. Native trees can be planted to break the wind, protect the soil from erosion, and provide a product as well. Creative solutions to today's ecological problems are very much needed.

Here are some other possible ways to help make rural areas sustainable:

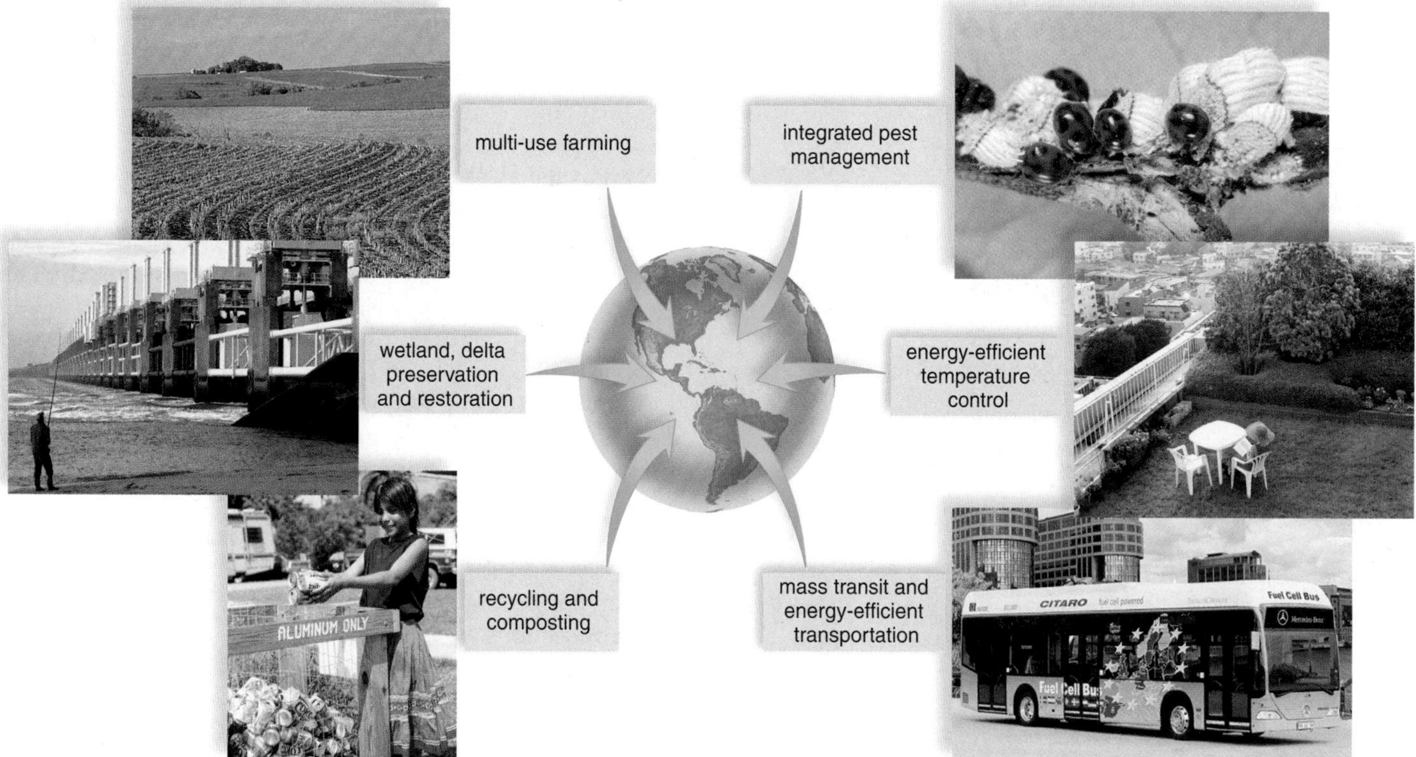

Figure 24.19 Characteristics of a sustainable society.
Arrows point inward to signify that these types of activities increase the carrying capacity of the Earth.

- Plant *cover crops,* which often are a mixture of legumes and grasses, to stabilize the soil between rows of cash crops or between seasonal plantings of cash crops.
- Use *multiuse farming* by planting a variety of crops, and use a variety of farming techniques to increase the amount of organic matter in the soil.
- Replenish soil nutrients through composting, organic gardening, or other self-renewable methods.
- Use low-flow or trickle irrigation, retention ponds, and other water-conserving methods.
- Increase the planting of *cultivars* (plants propagated vegetatively), which are resistant to blight, rust, insect damage, salt, drought, and encroachment by noxious weeds.
- Use *precision farming* (*PF*) techniques that rely on accumulated knowledge to reduce habitat destruction and improve crop yields.
- Use *integrated pest management* (*IPM*), which encourages the growth of competitive beneficial insects and uses biological controls to reduce the abundance of a pest.
- Plant a variety of species, including native plants, to reduce dependence on traditional crops.
- Plant *multipurpose trees*—trees with the ability to provide numerous products and perform a variety of functions, in addition to serving as windbreakers (Fig. 24.20). Remember that mature trees can provide many different types of products. For example, mature rubber trees provide us with rubber, and tagua nuts are an excellent substitute for ivory.
- Maintain and restore wetlands, especially in hurricane- or tsunami-prone areas. Protect deltas from storm damage. By protecting wetlands, we protect the spawning ground for many valuable fish nurseries.
- Use renewable forms of energy, such as wind and biofuel.

Connections and Misconceptions

What are some simple things you can do to conserve energy and/or water and help solve environmental problems like global warming?

A few things you could easily do include the following:

- Change the lightbulbs in your home to compact fluorescent bulbs. They use 75–80% less electricity than incandescent bulbs.
- Walk, ride your bike, carpool, or use mass transit. It will save you a lot of money, too!
- Get cloth or mesh bags for groceries and other purchases. Plastic bags may take 10–20 years to degrade. They're also dangerous to wildlife if mistaken for food and consumed.
- Turn off the water while you brush your teeth. If you don't finish a bottle of water, use it to water your plants. In dry climates, plant native plants that won't require frequent watering.

Figure 24.20 The roles of trees in a sustainable society. Trees planted by a farmer to break the wind and prevent soil erosion can also have other purposes, such as supplying nuts and fruits.

- Support local farmers, those who fish, and feed stores by buying food products produced close to home.

Urban Sustainability

More and more people are moving to the city. Much thought needs to be given about how to serve the needs of new arrivals without overexpansion of the city. Resources need to be shared in a way that will allow urban sustainability.

Here are some other possible ways to help make a city sustainable:

- Design an energy-efficient transportation system to rapidly move people about.
- Use solar or geothermal energy to heat buildings. Cool them with an air-conditioning system that uses seawater. In general, use conservation methods to regulate the temperature of buildings.
- Use *green roofs.* Grow a wild garden of grasses, herbs, and vegetables on the tops of buildings. This will assist temperature control, supply food, reduce the amount of rainwater runoff, and be visually appealing (Fig. 24.21).
- Improve storm-water management by using sediment traps for storm drains, artificial wetlands, and holding ponds. Increase use of porous surfaces for walking paths, parking lots, and roads. These surfaces reflect less heat and soak up rainwater runoff.
- Instead of traditional grasses, plant native species that attract bees and butterflies. These require less water and fewer fertilizers.
- Create *greenbelts* that suit the particular urban setting. Include plentiful walking and bicycle paths.
- Revitalize old sections of a city before developing new sections.
- Use lighting fixtures that hug the walls or ground and send light down. Control noise levels by designing quiet motors.
- Promote sustainability by encouraging recycling of business equipment. Use low-maintenance building materials rather than wood.

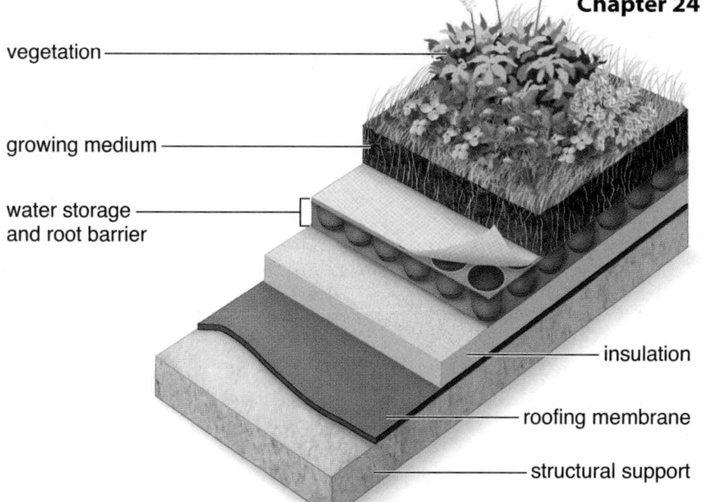

Figure 24.21 A green roof.
A green roof has plants growing on it that help control temperature, supply food, and reduce water runoff.

Assessing Economic Well-Being and Quality of Life

The gross national product (GNP) is a measure of the flow of money from consumers to businesses in the form of goods and services purchased. It can also be considered the total costs of all manufacturing, production, and services. Costs include salaries and wages, mortgage and rent, interest and loans, taxes, and profit within and outside the country. In other words, GNP pertains solely to economic activities.

When calculating GNP, economists do not necessarily consider whether an activity is environmentally or socially harmful. For example, destruction of forests due to clear-cutting, strip mining, or land development is not a part of the GNP. In the same way, the cost of medical services does not include the pain or suffering caused by illness, for example.

Measures that include noneconomic indicators are most likely better at revealing our quality of life than is the GNP. The index of sustainable economic welfare (ISEW) includes real per capita income, distributional equity, natural resources depletion, environmental damage, and the value of unpaid labor. The ISEW *does* take into account other forms of value, beyond the purely monetary value of goods and services. Another such index is called the genuine progress indicator (GPI). This indicator attempts to consider the quality of life, an attribute that does not necessarily depend on worldly goods. For example, the quality of life might depend on how much respect we give other human beings. The Grameen Bank in Bangladesh decided that if women were loaned small amounts of money, they would pay it back after starting up small businesses. The loans give women the opportunity to make choices that can improve the quality of their lives. For these women, a loan is a way to sustain their lives while, in part, fulfilling their dreams. It is difficult to assign a value to well-being or happiness. However, economists are trying to devise a way to measure these values. The following criteria, among others, can be used.

Use value: actual price we pay to use or consume a resource, such as the entrance fees into national parks.
Option value: preserving options for the future, such as saving a wetland or a forest.
Existence value: saving things we might not realize exist yet. This might be flora and fauna in a tropical rain forest that, one day, could be the source of new drugs.
Aesthetic value: appreciating an area or creature for its beauty and/or contribution to biodiversity.
Cultural value: factors such as language, mythology, and history that are important for cultural identity.
Scientific and educational value: valuing the knowledge of naturalists, or even an experience of nature, as a type of rational facts.

Development of the environment will always continue. Still, we can use these values to help us direct future development. Growth creates increases in demand, but development includes the direction of growth. If we permit unbridled growth, resources will become depleted. However, if development restrains resource consumption and still promotes economic growth, perhaps a balance can be reached. We can then retain sources for future generations.

Each person has a particular comfort level, and human beings do not like to make sacrifices that reduce their particular comfort level. So, despite our knowledge of the need to protect fisheries and forests, we continue to exploit them. People from LDCs directly depend on these resources to survive and so have much to lose. Even so, it is difficult for them to sacrifice today for the sake of the future. Yet, there is still hope because nature is incredibly resilient. One solution to deforestation is reforestation. Costa Rica has been successfully reforesting since the early 1980s. Also, declining fisheries can be restocked and then managed for sustainability. It will take an informed citizenry, creativity, and a desire to bring about change for the better to move toward sustainability.

Connecting the Concepts

For more information on the material presented in this section, refer to the following discussions.

Section 23.2 describes the how the 10% rule of thumb relates to energy flow in an ecosystem.

Section 23.3 examines human influence on the major biogeochemical cycles.

Check Your Progress 24.4

1. Describe the characteristics of today's society that make it unsustainable.
2. Discuss what changes are needed to convert today's society into to one that is sustainable.
3. Summarize how scientists assess economic well-being and quality of life.

Media Study Tools

www.mhhe.com/maderhuman12e

Enhance your study of this chapter with study tools and practice tests. Also ask your instructor about the resources available through ConnectPlus, including the media-rich eBook, interactive learning tools, and animations.

Summarizing the Concepts

24.1 Human Population Growth

- Populations have a biotic potential for increase in size.
- Biotic potential is normally held in check by environmental resistance.
- Population size usually levels off at carrying capacity.

The MDCs Versus the LDCs

- The MDCs have had a 0.1% growth rate since 1950.
- The LDC growth rate is presently 1.6% after peaking at 2.5% in the 1960s.

Age-structure diagrams can be used to predict population growth.

- MDCs are approaching a stable population size.
- LDC populations will continue to increase in size.

24.2 Human Use of Resources and Pollution

Five resources are maximally used by humans:

Resources are either nonrenewable or renewable.

- Nonrenewable resources are not replenished and are limited in quantity (e.g., land, fossil fuels, minerals).
- Renewable resources are replenished but still are limited in quantity (e.g., water, solar energy, food).

Land

Human activities, such as habitation, farming, and mining, contribute to erosion, pollution, desertification, deforestation, and loss of biodiversity.

Water

Industry and agriculture use most of the freshwater supply. Water supplies are increased by damming rivers and drawing from aquifers. As aquifers are depleted, subsidence, sinkhole formation, and saltwater intrusion can occur. If used by industries, water conservation methods could cut world water consumption by half.

Food

Food comes from growing crops, raising animals, and fishing.

- Modern farming methods increase the food supply, but some methods harm the land, pollute water, and consume fossil fuels excessively.
- Genetically engineered plants increase the food supply and reduce the need for chemicals.
- Raising livestock contributes to water pollution and uses fossil fuel energy.
- The increased number and high efficiency of fishing boats have caused the world fish catch to decline.

Energy

Fossil fuels (oil, natural gas, coal) are nonrenewable sources. Burning fossil fuels and burning to clear land for farming cause pollutants and gases to enter the air.

- Greenhouse gases include CO_2 and other gases. Greenhouse gases cause global warming because solar radiation can pass through, but infrared heat cannot escape back into space.
- Renewable resources include hydropower, geothermal, wind, and solar power.

Minerals

Minerals are nonrenewable resources that can be mined. These raw materials include sand, gravel, phosphate, and metals. Mining causes destruction of the land by erosion, loss of vegetation, and toxic runoff into bodies of water. Some metals are dangerous to health. Land ruined by mining can take years to recover.

Hazardous Wastes

Billions of tons of solid waste are discarded on land and in water.

- Heavy metals (lead, arsenic, cadmium, chromium)
- Synthetic organic chemicals include chlorofluorocarbons (CFCs), which are involved in the production of plastics, pesticides, herbicides, and other products.
- Ozone shield destruction is associated with CFCs.
- Other synthetic organic chemicals enter the aquatic food chain, where the toxins become more concentrated (biological magnification).

24.3 Biodiversity

Biodiversity is the variety of life on Earth.

Loss of Biodiversity

The five major causes of biodiversity loss and extinction are

- habitat loss,
- introduction of alien species,
- pollution,
- overexploitation of plant and animals, and
- disease.

Direct Value of Biodiversity

Direct values of biodiversity are

- medicinal value (medicines derived from living organisms),
- agricultural value (crops derived from wild plants; biological pest controls; animal pollinators), and
- consumptive use values (food production).

Indirect Value of Biodiversity

Biodiversity in ecosystems contributes to

- waste disposal (through the action of decomposers and the ability of natural communities to purify water and take up pollutants);
- freshwater provision through the water biogeochemical cycle;
- prevention of soil erosion, which occurs naturally in intact ecosystems;
- function of biogeochemical cycles;
- climate regulation (plants take up carbon dioxide); and
- ecotourism (human enjoyment of a beautiful ecosystem).

24.4 Working Toward a Sustainable Society

A sustainable society would use only renewable energy sources, would reuse heat and waste materials, and would recycle almost everything. It would also provide the same goods and services presently provided and would preserve biodiversity.

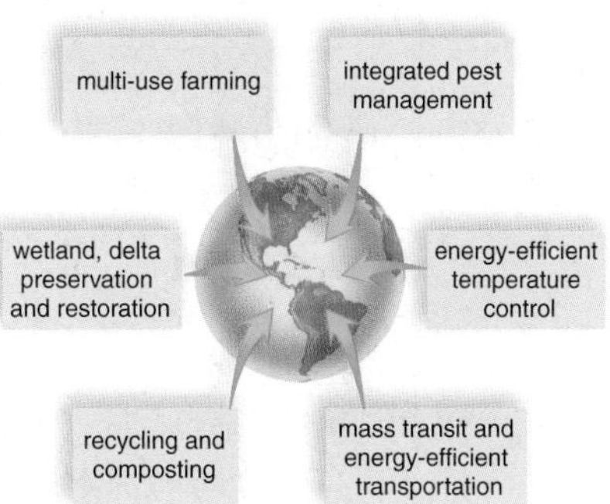

Understanding Key Terms

agricultural runoff 574	greenhouse gases 576
alien species 580	growth rate 568
aquifer 573	invasive 580
biodiversity 579	mineral 578
biological magnification 579	nonrenewable resource 570
biotic potential 568	photovoltaic (solar) cell 576
carrying capacity 568	pollution 570
chlorofluorocarbon (CFC) 579	renewable resource 570
deforestation 571	replacement reproduction 569
desertification 571	saltwater intrusion 573
ecological footprint 570	sinkhole 573
exponential growth 568	subsidence 573
fossil fuel 576	sustainable 586

Match the key terms to these definitions.

a. ________ The ability of a society or ecosystem to maintain itself while also providing services to human beings.

b. ________ Largest number of organisms of a particular species that can be maintained indefinitely in an ecosystem.

c. ________ Gases such as carbon dioxide and methane in the atmosphere that trap heat.

d. ________ Concentration of a synthetic organic chemical as it passes along a food chain.

e. ________ Water reservoir below the Earth's surface.

Testing Your Knowledge of the Concepts

1. Define carrying capacity and discuss the relevancy of this concept to the size of today's human population. (page 568)
2. Distinguish between MDCs and LDCs. Why are most LDCs, but not most MDCs, increasing in size? (pages 568–569)
3. Name three locales where humans have settled with unfortunate environmental consequences. What are those consequences? (pages 571–572)
4. What steps can be taken to conserve water? (pages 573–574)
5. What are the environmental benefits of reducing the amount of meat in the American diet? (page 575)
6. What are the types of fossil fuels, and what environmental problems are associated with burning fossil fuels? (page 576)
7. What renewable energy sources are available? What are the benefits of the solar-hydrogen revolution? (pages 576–577)
8. What are minerals, and what are the drawbacks of mining them? (page 578)
9. Exponential growth is best described by
   a. steep unrestricted growth.
   b. an S-shaped growth curve.
   c. a constant rate of growth.
   d. growth that levels off after rapid growth.
   e. Both b and d are correct.
10. When the carrying capacity of the environment is exceeded, the population will typically
    a. increase, but at a slower rate.
    b. stabilize at the highest level reached.
    c. decrease.
    d. die off entirely.
11. Decreased death rate followed by decreased birthrate has occurred in
    a. MDCs.
    b. LDCs.
    c. MDCs and LDCs.
    d. neither MDCs nor LDCs.
12. Renewable resources
    a. are in limited supply compared to nonrenewable resources.
    b. are always forthcoming but still may be inadequate for human needs.
    c. include such energy sources as wind, solar, and biomass.
    d. All of these are correct.
13. Desertification is often caused by
    a. overuse of aquifers.
    b. urban sprawl.
    c. air pollution.
    d. overgrazing.
14. Most freshwater is used for
    a. domestic purposes, such as bathing, flushing toilets, and watering lawns.
    b. domestic purposes, such as cooking and drinking.
    c. agriculture.
    d. industry.

15. Removal of groundwater from aquifers may cause
    a. pollution.
    b. subsidence.
    c. mineral depletion.
    d. soil erosion.
    e. All of the above are correct.
16. Which of the following is not a component of modern agriculture?
    a. dependency on chemical inputs
    b. frequent irrigation
    c. high fuel consumption
    d. high diversity of cultivars planted
17. The raising of domestic livestock
    a. consumes large amounts of fossil fuels.
    b. leads to water pollution.
    c. is energetically wasteful.
    d. All of the above are correct.
18. The best way to maintain fish supplies is to
    a. limit harvesting to the ability of fish to reproduce.
    b. do away with all the other animals that feed on fish.
    c. use larger and better types of nets.
    d. All of these are good ways to maintain fish supplies.

For questions 19–23, match the description to the type of fuel in the key. Each answer can be used more than once. Each question may have more than one answer.

Key:

a. nuclear power
b. fossil fuels
c. hydropower
d. solar power
e. wind power
f. geothermal power

19. Renewable source of power
20. Waste products are harmful
21. Detrimental environmental effects
22. More available in certain geographic locations
23. Contribute/s to global warming
24. The Earth's ozone shield has been damaged by
    a. heavy metals.
    b. strip mining.
    c. chlorofluorocarbons.
    d. fossil fuels.
25. Which of these are indirect values of species?
    a. participation in biogeochemical cycles
    b. participation in waste disposal
    c. provision of freshwater
    d. prevention of soil erosion
    e. All of these are indirect values.
26. Which of the following is not a function that ecosystems can perform for humans?
    a. purification of water
    b. immobilization of pollutants
    c. reduction of soil erosion
    d. removal of excess soil nitrogen
    e. breakdown of heavy metals
27. In which of the following is biological magnification most pronounced?
    a. aquatic food chains
    b. terrestrial food chains
    c. long food chains
    d. energy pyramids
    e. Both a and c are correct.
28. Complete the following graph by labeling each bar with a cause of extinction, from the most influential to the least.

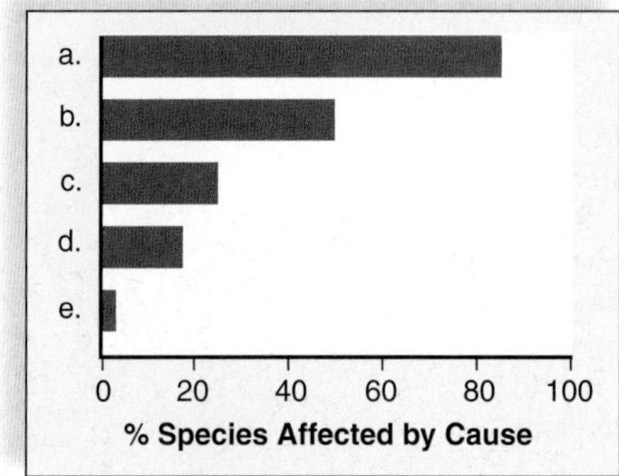

Thinking Critically About the Concepts

1. What environmental reasons would you give a friend for becoming a vegetarian?
2. a. What size footprint do you currently have on Earth? Visit www.myfootprint.org to determine the size of your footprint.
   b. What types of things could you do to decrease the size of your footprint on Earth?
3. How would failure to recycle items that can be recycled (they end up in a landfill) affect the nutrient cycles, such as the carbon cycle covered in Chapter 23?
4. a. What type of environmental activities/initiatives (Earth Day festivities, recycling program, etc.) are in place at your school?
   b. How could you increase awareness of environmental issues on your campus?

Appendix A

Periodic Table of the Elements

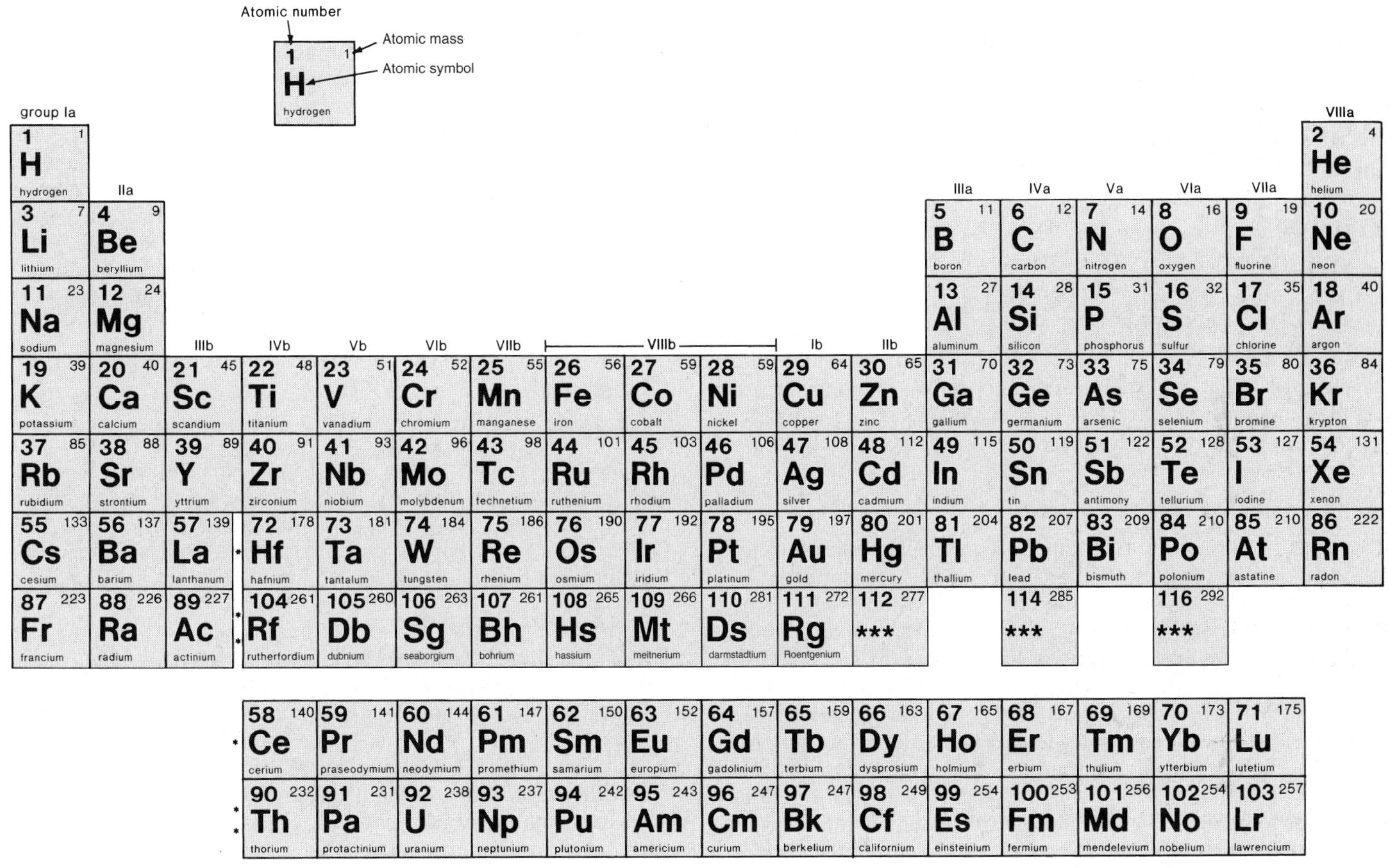

Each cell: atomic number, atomic symbol, name, atomic mass.

group Ia	IIa	IIIb	IVb	Vb	VIb	VIIb	VIIIb			Ib	IIb	IIIa	IVa	Va	VIa	VIIa	VIIIa
1 H hydrogen 1																	2 He helium 4
3 Li lithium 7	4 Be beryllium 9											5 B boron 11	6 C carbon 12	7 N nitrogen 14	8 O oxygen 16	9 F fluorine 19	10 Ne neon 20
11 Na sodium 23	12 Mg magnesium 24											13 Al aluminum 27	14 Si silicon 28	15 P phosphorus 31	16 S sulfur 32	17 Cl chlorine 35	18 Ar argon 40
19 K potassium 39	20 Ca calcium 40	21 Sc scandium 45	22 Ti titanium 48	23 V vanadium 51	24 Cr chromium 52	25 Mn manganese 55	26 Fe iron 56	27 Co cobalt 59	28 Ni nickel 59	29 Cu copper 64	30 Zn zinc 65	31 Ga gallium 70	32 Ge germanium 73	33 As arsenic 75	34 Se selenium 79	35 Br bromine 80	36 Kr krypton 84
37 Rb rubidium 85	38 Sr strontium 88	39 Y yttrium 89	40 Zr zirconium 91	41 Nb niobium 93	42 Mo molybdenum 96	43 Tc technetium 98	44 Ru ruthenium 101	45 Rh rhodium 103	46 Pd palladium 106	47 Ag silver 108	48 Cd cadmium 112	49 In indium 115	50 Sn tin 119	51 Sb antimony 122	52 Te tellurium 128	53 I iodine 127	54 Xe xenon 131
55 Cs cesium 133	56 Ba barium 137	57 La lanthanum 139	72 Hf hafnium 178	73 Ta tantalum 181	74 W tungsten 184	75 Re rhenium 186	76 Os osmium 190	77 Ir iridium 192	78 Pt platinum 195	79 Au gold 197	80 Hg mercury 201	81 Tl thallium 204	82 Pb lead 207	83 Bi bismuth 209	84 Po polonium 210	85 At astatine 210	86 Rn radon 222
87 Fr francium 223	88 Ra radium 226	89 Ac actinium 227	104 Rf rutherfordium 261	105 Db dubnium 260	106 Sg seaborgium 263	107 Bh bohrium 261	108 Hs hassium 265	109 Mt meitnerium 266	110 Ds darmstadtium 281	111 Rg Roentgenium 272	112 *** 277		114 *** 285		116 *** 292		

58 Ce cerium 140	59 Pr praseodymium 141	60 Nd neodymium 144	61 Pm promethium 147	62 Sm samarium 150	63 Eu europium 152	64 Gd gadolinium 157	65 Tb terbium 159	66 Dy dysprosium 163	67 Ho holmium 165	68 Er erbium 167	69 Tm thulium 169	70 Yb ytterbium 173	71 Lu lutetium 175
90 Th thorium 232	91 Pa protactinium 231	92 U uranium 238	93 Np neptunium 237	94 Pu plutonium 242	95 Am americium 243	96 Cm curium 247	97 Bk berkelium 247	98 Cf californium 249	99 Es einsteinium 254	100 Fm fermium 253	101 Md mendelevium 256	102 No nobelium 254	103 Lr lawrencium 257

Appendix B

Answer Key

This appendix contains the answers to the Case Study, Understanding Key Terms, Testing Your Knowledge of the Concepts, and Thinking Critically About the Concepts questions, which appear at the end of each chapter, and the Check Your Progress questions, which appear within each chapter.

Chapter 1

Case Study

1. Organization, acquiring materials and energy, reproducing, growing and developing, being homeostatic, responding to stimuli, evolution; **2.** Look for any clues that correspond to the characteristics of life, for example, fossils of carbon-based life forms; **3.** Similar environments might give us insight on varying diversity; Different environments might allow us to examine variations we do not possess.

Check Your Progress

(1.1) 1. Organization, acquiring materials and energy, reproducing, growing and developing, being homeostatic, responding to stimuli, evolution; **2.** Responding to stimuli allows organisms to detect changes and bring internal conditions back to normal, organization allows living organisms to perform various functions required for homeostasis, living organisms cannot maintain organization without acquiring materials and energy, growth and development are needed in order to maintain internal conditions, organisms have evolved adaptations that make homeostasis possible in particular environments, and reproduction allows these adaptations to be passed on to offspring; **3.** Organization leads to structure and structure dictates function. **(1.2) 1.** From the surface of the Earth through the atmosphere and down to the sea and soil; Areas where life can be found; **2.** To be able to study adaptation and genetic diversity through similarities and differences; **3.** Larger populations put a strain on resources and modify ecosystems. **(1.3) 1.** Observation—watch and take note; Hypothesis—testable statement, experiment-testing, conclusion-data supported; **2.** A controlled study gives a basis of comparison for a certain variable; **3.** Pro—Primary author is researcher; Current research; Con—Technical; difficult to understand; limited data. **(1.4) 1.** Anecdotal data are testimonials by individuals, not results from research; **2.** To fully understand how the researcher obtained the data and came to the conclusion; **3.** They show relationship between quantities and summarize data in a clear, concise manner. **(1.5). 1.** Nuclear physics (cancer therapy, nuclear bomb), biotechnology (GM plants/bacteria, possibly endangering biosphere), gene technology (stem cells, destroying embryos); **2.** Human ethics, values, and morals set guidelines for science and technology; **3.** All members.

Understanding Key Terms

a. biosphere; **b.** cell; **c.** scientific theory; **d.** homeostasis; **e.** experiment

Testing Your Knowledge of the Concepts

9. c; **10.** b; **11.** e; **12.** a; **13.** a; **14.** c; **15.** b; **16.** c; **17.** c; **18.** a; **19.** d;

Thinking Critically About the Concepts

1. Currently the smallest living things are single-celled organisms. These organisms have a cell membrane, genetic information, and cytoplasm (cell contents). Viruses lack cell membranes and contents are considered by most people to be nonliving. Evolution or genetic change is the only characteristic of living things displayed by viruses; **2.** It is not reliable in the terms of research data, numbers studied, or control groups, but it can be useful as participants give their opinion of the conclusion of the research through first-hand knowledge; **3.** It would alter our basic definition of life if the environment was different than ours; that would also change the definition of biosphere for those places.

Chapter 2

Case study

1. Cholesterol can cause build-ups and blockages in the cardiovascular system; **2.** No, HDLs are beneficial; **3.** Proteins are used as enzymes, hormones, signalling, transport in the body. Proteins make up a certain percentage of the cholesterol molecule which relates to David's values.

Check Your Progress

(2.1) 1. An atom is composed of a single nucleus that contains its protons and neutrons. Electrons circle the atoms in orbitals. Orbitals are then arranged into groups called *shells*; **2.** The first isotope has $40 - 20 = 20$ neutrons. The second isotope has $48 - 20 = 28$ neutrons; **3.** Isotopes are the forms of an element that differ in the number of neutrons. A radioisotope decays over time, releasing rays and subatomic particles. Radioisotopes can be used for medical purposes and for sterilizing objects, including food; **4.** The two basic types of bonding found between atoms are the ionic bond and the covalent bond. In ionic bonding, one or more atoms lose electrons and other atoms gain electrons. In this way, ions are formed. In covalent bonding, the ions are shared between atoms. **(2.2) 1.** Water is a liquid at room temperature, its temperature changes very slowly, and it requires a great deal of heat to become vapor. These characteristics keep water available for supporting life and enable it to help in cooling. Frozen water is less dense than water, so ice floats. Ice helps to insulate the water underneath, allowing marine life to survive the cold. Water molecules are cohesive yet flow freely, allowing water to fill vessels. Water is a solvent for polar molecules and facilitates chemical reactions; **2.** Acids contain greater numbers of H^+ ions than OH^- ions. Their pH will be less than 7. Bases contain greater numbers of OH^- ions than H^+ ions. Their pH will be greater than 7; **3.** The pH of soil and the body remain relatively constant because of the action of buffers. **(2.3) 1.** An organic molecule contains carbon and hydrogen, an inorganic molecule does not contain carbon; **2.** Carbohydrates— quick and short term energy storage; lipids—energy storage; proteins—support, enzymes, transport, defense, hormones, motion; nucleic acids—stores genetic information; **3.** A type of synthesis reaction called dehydration reaction where a hydroxyl group and a hydrogen group are removed as the molecule forms. **(2.4) 1.** Simple carbohydrate is used for quick and short-term energy; Polysaccharides are used for long-term energy storage; **2.** Simple carbohydrates contain low numbers of carbon atoms (3–7), complex carbohydrates contain high numbers of carbon atoms in chains of sugar (glucose) units; **3.** Insoluble fiber stimulates movements of the large intestine, soluble fiber prevents the absorption of cholesterol. **(2.5) 1a.** Energy storage; **2.** Phospholipids—cellular membranes; Steroids—sex hormones; **3.** There would be a shortage of long term energy storage and possibly humans would have to eat at more regular intervals for energy; Cell structure would change without phospholipid bilayer. **(2.6) 1.** Support, enzymes, transport, defense, hormones, motion; **2.** A central carbon bonded to a hydrogen, a —NH_2 (amino group) a —COOH (acid group), and an R group; **3.** Once a protein loses its normal shape, it is no longer able to perform its usual function. **(2.7) 1.** ATP consists of an adenosine and three phosphate groups; **2.** Hydrogen bonds are used because individually they are easy to break but in great numbers they are very strong; **3.** DNA—deoxyribose sugar, phosphate, nitrogen containing bases (adenine, thymine, cytosine, guanine), double stranded; RNA—ribose sugar, phosphate, nitrogen containing bases (adenine, guanine, uracil, cytosine), single stranded. DNA is used to store genetic information and RNA is used to synthesize proteins.

Understanding Key Terms

a. emulsification; **b.** ion; **c.** covalent bond; **d.** hydrophilic; **e.** hydrogen bond

Testing Your Knowledge of the Concepts

14. b; **15.** c; **16.** b; **17.** c; **18.** d; **19.** a; **20.** b; **21.** d; **22.** c; **23.** b; **24.** a; **25.** d; **26.** b; **27.** d; **28.** a; **29.** c; **30.** a; **31.** c; **32.** d; **33.** b; **34. a.** subunits; **b.** dehydration reaction; **c.** macromolecule; **d.** hydrolysis reaction

Thinking Critically About the Concepts

1. Because cholesterol is in the food we eat and the body makes a certain amount of cholesterol as well; **2.** The different types of cholesterol include some that are more beneficial than others. Some are good and aid in transport, others build-up in the lining of the blood vessels causing blockages; **3.** Carbohydrates, lipids, oils, proteins; **4.** Altering the pH of blood will result in changes in the structures of molecules in the blood such as proteins, which will affect their function; **5.** Storage is important so the body does not have to constantly intake certain molecules.

Chapter 3

Case Study

1. Golgi apparatus; **2.** Lysosomes contain hydrolytic enzymes that digest macromolecules and cell parts; **3.** The lysosome digests excess amounts of fatty acids, if it were to malfunction, an excess of fatty acids would build up.

Check Your Progress

(3.1) 1. A cell is the basic unit of life, all living things are made up of cells, new cells arise only from preexisting cells; this is important in evolutionary history and diversity; **2.** Small cells have a greater surface-area-to-volume ratio, thus a greater ability to get material in and out of the cell; **3.** Light microscopes use light rays to magnify objects and can be used to view living specimens; electron microscopes use a stream of electrons to magnify objects and have a higher resolving power than light microscopes. **(3.2) 1a.** Nucleus, plasma membrane, cytoplasm; **2.** Eukaryotic cells have a nucleus, prokaryotic cells lack a nucleus; **3.** Nucleus—through invagination of the plasma membrane. Mitochondria and chloroplast—by engulfing prokaryotic cells; **(3.3) 1.** Structure is a phospholipid bilayer with attached or embedded proteins (fluid-mosaic model), function is to keep cells intact and selectively allow passage of molecules and ions; **2.** Diffusion is the random movement of particles from the area of high to low concentration until they are equally distributed. Osmosis is the net movement of water from an area of high to low concentration across a semipermeable membrane; **3.** Diffusion, osmosis, facilitated transport—molecules are transported across the plasma membrane from high to low concentration, active transport—molecules are moving from low to high concentration using energy (ATP). **(3.4) 1.** Smooth ER—synthesizes lipids; Rough ER—synthesizes proteins and packages them in vesicles; Golgi apparatus modifies lipids and proteins from the ER, sorts and packages them in vesicles; Lysosomes—contains digestive enzymes that break down cell parts or substances entering by vesicles; **2.** Nucleus contains the DNA which directs the synthesis of proteins; ribosomes aid the mRNA in the production of proteins; Rough ER is where proteins are synthesized and folded; **3.** Endomembrane system consists of most organelles needed for the function of the cell. Transport vesicles move molecules from the ER to the Golgi. **(3.5) 1.** Microtubules, actin filaments, intermediate filaments; **2.** Cilia and flagella both contain an inner core of microtubules within a covering of plasma membrane. Cilia function by moving cells slowly in their environment and by sweeping debris. Flagella function in moving a cell very quickly; **3.** Adhesion junctions mechanically attach adjacent cells and prevent overstretching; tight junctions provide a zipperlike barrier between adjacent cells and do not allow movement of molecules between the two cells; Gap junctions are channels that are used for communication between adjacent cells. **(3.6) 1.** Enzymes speed up chemical reactions; **2.** Glycolysis begins with glucose and ends with two molecules of pyruvate; the citric acid cycle continues the breakdown of glucose and ends with two molecules of ATP; the electron transport chain takes NADH molecules and ends with 32 ATP molecules; **3.** If electron transport chain or citric acid cycle were missing, then glycolysis would be the only means of creating ATP, instead of more than 32 ATP molecules being made, only 2 would be made and the cell could not keep up with its energy demands.

Understanding Key Terms

a. fluid-mosaic model; **b.** osmosis; **c.** selectively permeable; **d.** fermentation; **e.** cellular respiration

Testing Your Knowledge of the Concepts

11. b; **12.** e; **13.** a; **14.** c; **15.** c; **16.** a; **17.** d; **18.** b; **19.** c; **20.** e; **21.** b; **22.** c; **23. a.** carbohydrate chain; **b.** hydrophilic heads; **c.** phospholipid bilayer; **d.** hydrophobic tails; **e.** filaments of cytoskeleton; **f.** protein; **g.** protein; **24. a.** active site; **b.** substrates; **c.** product; **d.** enzyme; **e.** enzyme-substrate complex; **f.** enzyme.

Thinking Critically About the Concepts

1. A malfunctioning mitochondria would result in the malfunction of cellular respiration, thus the cell would only produce a small amount of ATP by glycolysis; **2.** Chemical reactions would not occur as quickly; **3.** Lysosomes might start digesting needed substances in the cell, or autodigesting a functioning cell.

Chapter 4

Case Study

1. All four types of tissues; **2.** Skin burns affect homeostasis of fluid levels and pH levels in the body; **3.** The dermis contains blood vessels whereas the epidermis does not.

Check Your Progress

(4.1) 1. Connective tissue, muscular tissue, nervous tissue, and epithelial tissue; 2. Cancers are classified according to the type of tissue from which they arise; **3.** Because they are related to the rate of cell division. **(4.2) 1.** Fibrous connective tissue (adipose tissue and dense fibrous connective tissue) and specialized connective tissue (cartilage and bone); **2.** Loose fibrous connective tissue contains fibroblasts separated from each other, it supports epithelium and many internal organs; Dense fibrous connective tissue also contains fibroblasts separated from each other, as well as many collagen fibers, it is found in tendons and ligaments; Bone is the most compact of the connective tissues and contains inorganic salts, it makes up the bones of the skeletal system; Cartilage contains cells separated by a solid, yet flexible matrix, it is found at the ends of long bones, in the outer ear, and in joints; **3.** Blood transports nutrients and oxygen to tissue fluid. Lymph absorbs excess tissue fluids. **(4.3) 1.** Skeletal muscle—voluntary movement of body, striated cells with multiple nuclei; smooth muscle—movement of substances in lumens of body, spindle-shaped cells each with a single nuclei; cardiac muscle—pumping blood, branching striated cells with a single nucleus; **2.** Skeletal muscles—attached to the skeleton, smooth muscle—blood vessels and walls of the digestive tract, cardiac muscle—walls of the heart; **3.** These muscles are involuntary because they directly relate to overall homeostasis and sustaining life. They work without our needing to be conscious of their activity. **(4.4)** Dendrites—receive signals from sensory receptors or other neurons; cell body—contains most of the cell's cytoplasm and the nucleus; axon—conducts the nerve impulses; **2.** Microglia—support and nourish neurons, engulf bacterial and cellular debris, astrocytes—produce hormones, oligodendrocytes—form myelin sheaths; **3.** Both work in sensory input, integration of data, and motor output. **(4.5) 1.** Its main function is protection, lining and covering; **2.** Simple—one layer of cells; stratified—two or more layers of cells; pseudostratified—one layer, but looks like more; squamous—thin, flat cell; cuboidal- cube shaped cell; columnar—column shape cell; transitional changes shape according to function; **3.** Simple squamous— quick diffusion; stratified squamous—thick for protection. **(4.6) 1.** Epidermis is composed of stratified squamous epithelium that functions in protection. Dermis is a region of dense fibrous connective tissue beneath the dermis with collagen and elastic fibers that functions in sensory reception; **2.** Nails are protective covering; hair is used for protection and covering; sweat glands aid in regulating temperature; **3.** Because the epidermis layer is routinely shed. **(4.7) 1.** Integumentary—protects body, receives sensory input, regulates temperature, synthesizes vitamin D; cardiovascular—transports nutrients and gases; lymphatic and immune—controls fluid balance, defends against infectious disease; digestive—ingest and digest food, absorb nutrients and eliminates waste; respiratory—gas exchange, regulate pH; urinary—excrete metabolic wastes, control fluid and pH balance; skeletal—support and protects, stores minerals, locomotion; muscular—maintain posture, moves body and internal organs, produces heat; nervous—receives, stores, and integrates sensory input, initiates motor output, coordinates organ systems; endocrine—produces hormones, coordinates organ systems, regulates metabolism; reproductive—produce and transport gametes, produces sex hormones; **2.** Dorsal cavity includes the cranial and vertebral cavities. Ventral cavity contains the thoracic, abdominal, and pelvic cavities; **3.** Mucous membrane is composed of epithelium over loose fibrous connective tissue and is used for protection from bacteria and virus; Serous membrane secrete a watery fluid that supports organs and lungs; Synovial membrane is composed of loose connective tissue and lines cavities of freely movable joints; Meninges are membranes of connective tissue in the dorsal cavity. **(4.8) 1.** The body's ability to maintain a relative constancy of its internal environment by adjusting its physiological processes. Fluctuation in internal conditions can result in illness; **2.** Issue electrochemical signals, release hormones, supply oxygen, maintain body temperature, remove waste, maintain adequate nutrient levels, adjust the water-salt and acid-base balance of the blood; **3.** Negative feedback keeps a variable close to a particular value, positive feedback brings about an ever greater change in the same direction, which assists the body in completing a process.

Understanding Key Terms

a. ligament; **b.** epidermis; **c.** carcinoma; **d.** homeostasis; **e.** spongy bone

Testing Your Knowledge of the Concepts

11. a; **12.** d; **13.** d; **14.** d; **15.** c; **16.** b; **17.** b; **18.** c; **19.** d; **20.** b; **21.** d; **22.** d; **23.** b; **24.** b; **25.** b.

Thinking Critically About the Concepts

1. c; **2.** Sweat glands and hair follicles are located in the dermis region; **3.** Protection and a decrease in fluid loss; **4.** Fluid loss and pH imbalance; **5.** Sweat glands, hair follicles, collagen, elastic fibers, sensory receptors. Hair follicles and sweat glands may never grow back once they have been destroyed; **6.** No protection from bacteria and viruses, no regulation/coordination of other systems.

Chapter 5

Case Study

1. Materials will adhere to the lining of the inside of the vessel causing a blockage; or large particles will get stuck inside a vessel blocking flow; **2.** Tissue will die without proper circulation; **3.** Atherosclerosis can develop, even in a healthy person, due to an accumulation of plaque inside the walls of an artery.

Check Your Progress

(5.1) 1. Heart and blood vessels; **2.** Generate blood pressure, transport blood, create exchange at the capillaries, and regulate blood flow; **3.** Lymphatic vessels collect excess tissue fluid and return it to the cardiovascular system. **(5.2) 1.** Arteries, capillaries, veins; **2.** Arteries carry blood away from the heart and have strong walls; capillaries have thin walls, where exchange can occur; veins carry blood to the heart, have thinner walls than arteries, and contain valves; **3.** Areas that need more gas exchange than others, like organs and skeletal muscles in limbs. **(5.3) 1.** Oxygen poor blood begins in the right atrium, through the tricuspid valve into the right ventricle, through the pulmonary trunk into the lungs, drops off carbon dioxide and picks up oxygen, then through pulmonary veins into left atrium, through bicuspid valve into left ventricle then out the aortic valve; **2.** "Lub" occurs when increasing pressure of blood inside a ventricle forces the cusps of the AV valve to slam shut, "dup" occurs when the ventricles relax and blood in the arteries pushes back, causing the semilunar valves to close; **3.** External is cardiac control center of the medulla oblongata and inside the heart is the conduction system of the heart. **(5.4) 1.** Heart rate; **2.** Blood pressure accounts for blood flow in arteries and capillaries, in veins it is skeletal muscle pump, respiratory pump, and valves; **3.** Valves are needed to prevent the backflow of blood in the veins. **(5.5) 1.** Right ventricle→pulmonary trunk→pulmonary arteries→pulmonary capillaries→pulmonary veins→left atrium; **2.** Left ventricle→aorta→mesenteric arteries→digestive tract capillary bed→hepatic portal vein→liver capillary bed→hepatic vein→inferior vena cava→right atrium; **3.** The pulmonary arteries carry oxygen-poor blood while the pulmonary veins carry oxygen-rich blood. **(5.6) 1.** Excess fluid goes into tissue fluid and eventually into the lymphatic vessels; **2.** Simple diffusion due to the capillary membrane; **3.** Osmotic pressure is greater than blood pressure and fluid moves back into the capillary. **(5.7) 1.** Hypertension, stroke, heart attack, aneurysm, heart failure; **2.** Common treatments include drugs to lower blood pressure, nitroglycerin given at onset of heart attack, replace diseased/damaged portion of the vessel, open clogged arteries, dissolve clots, surgery, heart transplant; **3.** Mainly due to diet, exercise, and lifestyle choices.

Understanding Key Terms

a. diastole; **b.** inferior vena cava; **c.** pulse; **d.** venule; **e.** systemic circuit

Testing Your Knowledge of the Concepts

15. a; **16.** e; **17.** b; **18.** d; **19.** c; **20.** c; **21.** c; **22.** e; **23.** d; **24.** e; **25.** c; **26. a.** jugular vein; **b.** pulmonary artery; **c.** superior vena cava; **d.** inferior vena cava; **e.** hepatic vein; **f.** hepatic portal vein; **g.** renal vein; **h.** iliac vein; **i.** carotid artery; **j.** pulmonary vein; **k.** aorta; **l.** mesenteric arteries; **m.** renal artery; **n.** iliac artery; **27. a.** blood pressure; **b.** osmotic pressure; **c.** blood pressure; **d.** osmotic pressure

Thinking Critically About the Concepts

1. He misdiagnosed his limb numbness as muscle fatigue from exercising; **2.** Diabetes; **3.** To determine the chemical composition of the blood; **4.** Aorta is the largest diameter blood vessel; if blocked no oxygenated blood gets delivered to the body tissues from the heart; **5.** Contractions are occurring too quickly and forcefully; This might cause the heart to beat irregularly and disturb blood flow to the body; **6.** Yes, lymphatic vessels have valves that prevent backflow of lymph.

Chapter 6

Case Study

1. Function in immunity; **2.** Bone marrow is where blood cells originate and mature; **3.** Because the cells are not responding to normal cell cycle signals.

Check Your Progress

(6.1) 1. Transport, defense and regulation; **2.** Plasma proteins include albumins (contribute to osmotic pressure and transport organic molecules), globulins (transport and antibodies), and fibrinogens (when activated form clots); **3.** Because it consists of cells within a matrix. **(6.2) 1.** The RBCs contain hemoglobin chains with an iron-containing heme group that binds to oxygen; **2.** Pro—they can hold more oxygen molecules, and they are more malleable to move through vessels; Con—they can not go through mitosis to repair themselves; **3.** Anemia—iron, B_{12}, or folic acid deficiency, tired, run down feeling; sickle-cell disease—hereditary disease, hemolysis of red blood cells. **(6.3) 1.** To protect against different types of pathogens and immune system disorders; **2.** Neutrophils—granular with a multilobed nucleus, first responders in bacterial infection; Eosinophils—granular with a bilobed nucleus, increase in number during a parasitic worm infection or allergic reaction; Basophils—granular with a U-shaped nucleus, release histamine; Lymphocytes—do not have granules and have nonlobular nuclei, responsible for specific immunity to particular pathogens and their toxins; monocytes—do not have granules and are the largest of the white blood cells, phagocytize pathogens, and stimulate other white blood cells; **3.** Severe combined immunodeficiency—stem cells of white blood cells lack adenosine deaminase; Leukemia—uncontrolled white blood cell proliferation; Infectious mononucleosis—EBV infection of lymphocytes. **(6.4) 1.** Platelets—clump at the site of puncture; Thrombin—activates fibrinogen, forming long threads of fibrin; Fibrin threads—wind around platelets, plasmin—destroys the fibrin network; Serum (released after blood clots)—contains all the components of plasma; **2.** Damaged platelets release prothrombin activator, thrombin forms fibrin which winds around the platelet plug trapping RBCs forming a clot; **3.** Thrombocytopenia is low platelet count, hemophilia is a deficiency in clotting factors. **(6.5) 1a.** A, B, AB, and O. Blood types are based on the presence or absence of type A antigen and type B antigen; **2.** Type A can give to type A or AB; **3.** When an Rh^- mother is pregnant with her second Rh^+ baby. The mother's antibodies cross the placenta and destroy the fetal red blood cells. **(6.6) 1.** Delivers oxygen from the lungs and nutrients from the digestive system, and removes metabolic wastes; **2.** Digestive system—provides molecules for plasma protein and blood cell formation; Urinary system—helps maintain water-salt balance of blood; Muscular system—moves blood; Nervous system—regulates heart contraction and constriction/dilation of blood vessels; Endocrine system—produces hormones that regulate blood pressure and volume; Respiratory system—gas exchange; Lymphatic system—helps maintain blood volume; Skeletal system—protects heart and red bone marrow; **3.** Diabetes—blood sugar levels rise above normal; Decrease in blood calcium levels—problems with clotting; Kidney not producing erythropoietin—decrease in the number of red blood cells.

Understanding Key Terms

a. hemoglobin; **b.** formed element; **c.** plasma; **d.** leukemia; **e.** prothrombin

Testing Your Knowledge of the Concepts

10. a; **11.** a; **12.** b; **13.** d; **14.** c; **15.** c; **16.** a; **17.** e; **18.** d; **19.** b; **20.** a; **21.** d; **22.** d; **23.** e; **24.** b; **25.** e; **26.** d; **27.** e.

Thinking Critically About the Concepts

1a. Carbon monoxide is found in automobile exhaust. Burning charcoal, or wood in some cases, will also give off CO. Fumes from all of these sources must be properly vented to the outside; **1b.** Without oxygen, the cell mitochondria can't metabolize food to form ATP; **2a.** iron, vitamin B_{12}, and folic acid; **2b.** Good sources of protein include beef, turkey, spinach, and fortified foods. Foods rich in B_{12} include organ meats, fish, shellfish, and dairy products. Folic acid can be found in cereals, baked goods, leafy vegetables, fruits (bananas, melons, lemons), and organ meat; **3.** Hemoglobin is a protein, and you might recall from Chapter 2 that it is composed of smaller subunits. Iron is necessary for hemoglobin synthesis; **4.** After blood loss for any reason, erythropoietin will be needed to grow erythrocytes. At high altitude, erythropoietin will help a person to acclimate to the lower concentration of oxygen in the air. Remember that it will take several weeks to replace lost red blood cells; **5a.** Blood packed with too many red blood cells becomes too dense for the heart to pump; **5b.** The athlete is more likely to die at night because the person isn't drinking fluids, and blood pressure falls during sleep. Blood becomes increasingly dense during sleep, further increasing the workload on the heart; **6.** Any parasitic disease can cause the student's symptoms. The most common parasitic disease is malaria; **7.** The respiratory system, the skeletal system, and the urinary system are examples. Lungs oxygenate the blood, bone marrow within the bones produces blood cells, and the kidneys produce erythropoietin to stimulate erythrocyte production.

Chapter 7

Case Study

1. Disease that results when the immune system mistakenly attacks the body's tissues; **2.** Yes, fatigue and muscle soreness are common; **3.** There appears to be a genetic predisposition but infections, stress, and hormones may play a role.

Check Your Progress

(7.1) 1. Bacteria grow in host tissues and cause damage. They can also release toxins. Viruses infect host cells and cause damage. **2.** Capsule, flagella, fimbriae, pilus, and toxins; **3.** Viruses must replicate inside a living cell. **(7.2) 1.** Lymphatic capillaries absorb excess tissue fluid and return it to the bloodstream; In the small intestine, lymphatic capillaries called lacteals absorb fat in the form of lipoproteins and transport them to the bloodstream; The lymphatic system is responsible for the production, maintenance, and distribution of lymphocytes; The lymphatic system helps defend the body against pathogens; **2.** Primary include red bone marrow and thymus and are the sites where white blood cells are produced and mature; Secondary includes lymph nodes and spleen and are the sites where white blood cells react to pathogens and blood and lymph are purified; **3.** A fluid imbalance in the cardiovascular system may result and that would lead to possible pH differences and improper circulation. **(7.3) 1.** Skin and mucous membranes, chemical barriers, resident bacteria, inflammatory response, and protective proteins; **2.** Neutrophils and macrophages engulf pathogens by phagocytosis; **3.** They "complement" certain immune responses. **(7.4) 1.** Innate defenses act indiscriminately against all pathogens, acquired defenses respond to antigens; **2.** Antibody-mediated immunity, cell-mediated immunity; **3.** B cells produce plasma and memory cells, plasma cells produce antibodies, memory cells produce antibodies in the future, T cells regulate immune responses and produce cytotoxic T cells and helper T cells, cytotoxic T cells kill virus infected cells and cancer cells, helper T cells regulate immunity, and memory T cells kill in the future. **(7.5) 1.** Immunity that occurs naturally through infection or is brought about artificially by medical intervention; Active immunity—infection with a pathogen, immunization; Passive immunity—transfer of IgG antibodies across the placenta, breast feeding, gamma globulin injection; **2.** Active immunity develops after an exposure to a pathogen and passive immunity occurs when an individual is given antibodies or immune cells to combat a disease; **3.** Monoclonal antibodies are all of the same type because they are derived from the same B cell; Cytokine therapy regulates white blood cell production. **(7.6) 1.** Allergies—hypersensitivities to substances that ordinarily would do no harm to the body; Tissue rejection—the recipient's immune system recognizes that the transplanted tissue is not self; autoimmune disease—Cytotoxic T cells or antibodies mistakenly attack the body's own cells, as if they bear foreign antigens; **2.** When an allergen attaches to the IgE antibodies on mast cells they release histamines that make an allergic reaction occur; **3.** Because the antibodies produces sometimes react with other tissue causing inflammation or damage.

Understanding Key Terms

a. vaccine; **b.** complement system; **c.** antigen; **d.** apoptosis; **e.** T cell

Testing Your Knowledge of the Concepts

17. b; **18.** d; **19.** e; **20.** b; **21.** a; **22.** c; **23.** d; **24.** d; **25.** b; **26.** c; **27.** c; **28.** c; **29.** a; **30.** e; **31.** d; **32.** d; **33.** c

Thinking Critically About the Concepts

1. B cells produce the IgE necessary for the allergic reaction; **2.** Injection of small doses of the allergen to help a patient build immunity to the allergen; **3.** Due to the diversity of our genes giving us different immune system strengths and weaknesses; **4.** Barrier defenses are supposed to prevent the entrance of pathogens to someone's body, similar as a fence keeping intruders off your property; **5.** They should get another shot of anti-venom, because the first shot gave them passive immunity.

Infectious Diseases Supplement

Check Your Progress

(S.1) 1. Outbreak—when an epidemic is confined to a local area—cold; Epidemic—a disease with more cases than expected in a certain area during a certain period—flu; Pandemic- global epidemics—HIV/AIDS; **2.** CD4 cells; Attachment, fusion, entry, reverse transcription, integration, biosynthesis and cleavage, assembly, budding; **3.** The mosquito carries the parasite to the human. The parasite completes the sexual part of its life cycle within the mosquito and the asexual part within the human; **4.** By exposure to antivirals and antibiotics. **(S.2) 1.** Diseases that are newly recognized or that have reappeared after a significant decline in incidence, avian influenza, SARS, tuberculosis; **2.** New or increased exposure to animals/insects, changes in human behavior, mutations in pathogens; **3.** Careful use of antibiotics, education, research, surveillance, and quarantine. **(S.3) 1.** Through mutations in their genetic material, often as a result of the misuse of antibiotics; **2.** Take the entire dose of antibiotics for the entire time prescribed, do not skip doses or discontinue treatment early, do not share antibiotics, do not treat viral infections with antibiotics; **3.** Both are drug-resistant strains of pathogens.

Chapter 8

Case Study

1. Obesity and a sedentary lifestyle contribute to many diseases including gastrointestinal disease; **2.** Heartburn occurs when some of the stomach contents escape into the esophagus. Sedentary lifestyle, obesity, over-eating, and gastrointestinal disease can cause heartburn; **3.** PillCam and colonoscopy can aid in the detection of polyps, cancer, IBS, diverticulitis, and IBD.

Check Your Progress

(8.1) 1. Organs—mouth, pharynx, esophagus, stomach, small intestine, large intestine, rectum, anus; Accessory organs—salivary glands, liver, gallbladder, pancreas; **2.** Ingestion—the mouth takes in food; Digestion—divides food into pieces and hydrolyzes food to molecular nutrients; Movement—food is passed along from one organ to the next and indigestible remains are expelled; absorption—unit molecules produced by digestion cross the wall of the GI tract and enter the blood for delivery to cells, elimination—removal of indigestible wastes through the anus; **3.** Mucosa—diverticulitis, submucosa—inflammatory bowel disease, muscularis—irritable bowel syndrome, serosa—appendicitis. **(8.2) 1.** Mouth contains a hard and soft palate, salivary glands, and a tongue; Teeth have two main divisions, a crown with a layer of enamel and dentin and a root filled with dentin and pulp; Pharynx is a hollow tube at the back of the throat connecting the mouth to the esophagus; Esophagus is a hollow tube connecting the pharynx to the stomach; **2.** Mechanical digestion—teeth chew food into pieces convenient for swallowing and the tongue moves food around the mouth; Chemical digestion—salivary amylase begins the process of digesting starch; 3. The soft palate moves back to close off the nasal passages, and the trachea moves up under the epiglottis to cover the glottis. **(8.3) 1.** Store food, initiate the digestion of protein, and control the movement of chyme into the small intestine; The muscularis contains an oblique layer that allows the stomach to stretch and mechanically break food down, the mucosa has millions of gastric pits, which lead into gastric glands that produce gastric juice; **2.** Complete digestion using enzymes, which digest all types of food and absorb the products of the digestive process; It contains villi that have an outer layer of columnar epithelial cells, each containing thousands of microvilli; **3.** Nutrients are not absorbed through the stomach lining due to the acidity of the gastric juices found in the stomach; once in the intestines, the gastric juices are neutralized. **(8.4) 1.** Pancreas—secretes pancreatic juice into the small intestine and insulin into the blood; liver—filters the blood, removes toxic substances, stores iron and vitamins, functions in sugar homeostasis, regulates blood cholesterol; gallbladder—stores bile; **2.** If the pancreas is not functioning, it will not make pancreatic juices to aid in digestion; If the liver is not functioning, bile will not be produced to emulsify fats; if the gallbladder is not functioning bile does not have a storage organ; **3.** Regulation of digestive secretions is important to have the correct product at the correct time to digest the correct nutrient or chemical. Without these, the process of digestion could not occur properly and nutrients would not be able to get into our circulatory system and into our cells. **(8.5) 1.** Cecum and colon—absorb water and vitamins; rectum—form feces; anal canal—defecation; **2.** Large intestine contributes to homeostasis by regulating fluid balance and removing waste; 3. Without the functions of the large intestine water and vitamin absorption would be deficient. **(8.6) 1.** Carbohydrates are simple or complex ringed structures and are used as an energy source and include products made from refined grains, beans, peas, nuts, fruits, and whole grain products; Proteins are long chains of amino acids the body breaks down to make other proteins and are found in meat, eggs, and milk; Lipids are fats and oils and contain either saturated or unsaturated fatty acid chains. They are used for energy storage; **2.** Because they are consumed in excess and with our sedentary lifestyle are not burned off in exercise but stored; **3.** To obtain all the chemicals and nutrients required for all cells to function correctly.

Understanding Key Terms

a. vitamin; **b.** lipase; **c.** lacteal; **d.** esophagus; **e.** gallbladder

Testing Your Knowledge of the Concepts

9. d; **10.** b; **11.** e; **12.** c; **13.** b; **14.** d; **15.** e; **16.** c; **17.** e; **18.** b; **19.** c; **20.** d; **21.** a; **22.** d; **23.** a, b; **24.** d; **25.** e; **26.** e; **27.** a, b; **28.** c; **29.** d; **30.** e; **31.** a; **32.** c; **33.** b

Thinking Critically About the Concepts

1a. To prepare the food for enzymes to perform chemical digestion, such as chewing does. The smaller stomach from bariatric surgery no longer performs a significant amount of mechanical digestion; **1b.** Because the stomach now holds only a few ounces at a time; **2.** If stomach is overfilled, stomach contents would flow into the esophagus causing heartburn; if chronic, serious problems can result.

Chapter 9

Case Study

1. Nasal cavity, nasopharynx, oropharynx, laryngopharynx, glottis, larynx; **2.** Because the brain controls breathing and heart rate; **3.** Diet and alcohol affect sleep patterns, as well as excess weight affecting breathing rate and the functions of the upper respiratory system during sleep.

Check Your Progress

(9.1) 1. Nasal cavity→pharynx→glottis→larynx→trachea→bronchus→bronchioles→lungs; **2.** Inspiration is inhaling air from the atmosphere into the lungs through a series of cavities, tubes and openings; Expiration is exhaling air form the lungs to the atmosphere through the same structures; **3.** Ensure that oxygen enters the body and carbon dioxide leaves the body. **(9.2) 1.** Nose—filter, warm, and moisten the air; Pharynx—connect nasal and oral cavities to the larynx; Larynx—sound production; **2.** Nasal cavity (filters and warms incoming air), oral cavity (where food is received), and larynx (voice box); **3.** Air passage and food passage. **(9.3) 1.** Trachea—keeps lungs clean by sweeping mucus upwards and connects the larynx to the primary bronchi; Bronchial tree—passage of air to the lungs; Lungs—site of gas exchange between air in the alveoli and blood in the capillaries; **2.** Alveoli are composed of alveolar sacs surrounded by blood capillaries used for gas exchange; **3.** If alveoli do not function properly gas exchange cannot happen and cells will not receive oxygen or be able to remove carbon dioxide and will cease to function. **(9.4) 1.** As the volume (size) of the thoracic cavity increases (when you inhale), the pressure in the lungs decreases. As the volume decreases (when you exhale), the pressure in the lungs increases; **2.** Tidal volume is achieved through normal breathing, vital capacity is the maximum volume that can be achieved, inspiratory and expiratory reserve volume is achieved through forced breathing, and residual volume is the air that remains in the lungs after exhalation. **3.** The body would not be able to get rid of excess carbon dioxide, which would affect pH of the blood, energy metabolism, and other important functions. **(9.5) 1.** The rhythm of ventilation is controlled by a respiratory control center, located in the medulla oblongata. If you breathe any less than that, correct gas exchange would not occur; **2.** The rhythm of ventilation is controlled by a respiratory control center located in the medulla oblongata. This center automatically sends out nerve signals to the diaphragm and the external intercostal muscles of the rib cage, causing inspiration. When signals are not longer sent, the muscles relax and expiration occurs; **3.** As you hold your breath, blood CO_2 increases, which makes the blood more acidic. The respiratory center initiates exhalation in response to the increased acidity. **(9.6) 1.** External respiration refers to the exchange of gases between air in the alveoli and blood in the pulmonary capillaries; Internal respiration refers to the exchange of gases between the blood in systemic capillaries and the tissue fluid; **2.** Takes up oxygen and becomes oxyhemoglobin, Takes up carbon dioxide and becomes carbaminohemoglobin; **3.** P_{O_2} allows the gases to diffuse through the membranes and move through circulation. For example, oxygen moves out of the blood into the tissues because the P_{O_2} of the tissues is lower than that of the blood. **(9.7) 1.** Common infections include strep throat (sore throat and high fever), sinusitis (postnasal discharge and facial pain), otitis media (ear pain), laryngitis (sore throat and hoarseness), bronchitis (deep cough with mucus), pneumonia (chest pain, fever, and chills), and tuberculosis (nonproductive cough). Common disorders include pulmonary fibrosis, chronic bronchitis, asthma, and emphysema. Symptoms of disorders generally include difficulty breathing and increased exertion required for breathing. **2.** When the common respiratory infections are caused by bacteria, they are treated with antibiotics. **3.** Chronic bronchitis, emphysema, lung cancer.

Understanding Key Terms

a. pharynx; **b.** surfactant; **c.** vocal cord; **d.** bicarbonate ion; **e.** expiration

Testing Your Knowledge of the Concepts

10. d; **11.** d; **12.** c; **13.** f; **14.** d; **15.** a; **16.** b; **17.** e; **18.** d; **19.** b; **20.** d; **21.** a; **22.** b; **23.** e; **24.** d; **25.** b; **26. a.** nasal cavity; **b.** nostril; **c.** pharynx; **d.** epiglottis; **e.** glottis; **f.** larynx; **g.** trachea; **h.** bronchus; **i.** bronchiole

Thinking Critically About the Concepts

1. The enlarged tonsils or adenoids physically obstruct the airway, inhibiting airflow. Surgical removal may be required to restore sufficient passage of air. **2.** The expression describes the movement of food particles past the epiglottis, through the glottis, and into the trachea; dust, bacteria, and other airborne contaminants from reaching the lungs; **3a.** Ciliated cells are damaged from smoking, so coughing prevents dust, bacteria, and other airborne contaminants from reaching the lungs; **3b.** Their ciliated epithelium is fully functioning; **4.** Diaphragm and abdominal muscles; **5.** The blood O_2 of someone who has nearly drowned is low. When hemoglobin is not bound to O_2, blood is a much darker color and appears bluish because of the diffusion of light by the skin.

Chapter 10

Case Study

1. The kidneys produce urine, excrete metabolic wastes, maintain a water-salt balance, an acid-base balance, secrete hormones, and reabsorb filtered nutrients and convert vitamin D; **2.** Because urine would not be delivered from the nephrons to the renal pelvis so it could exit the kidney through the ureters; **3.** Blood would not be filtered and reabsorbed but removed as waste causing a fluid imbalance that would lead to changes in blood pressure.

Check Your Progress

(10.1) 1. Kidneys—primary organs of excretion, ureters—conduct urine from the kidneys to the bladder, urinary bladder—stores urine until it is expelled, urethra—small tube that extends from the urinary bladder to an external opening; **2.** Excretion of metabolic wastes, maintenance of water-salt balance, maintenance of acid-base balance, and secretion of hormones; **3.** Waste products would build up in the system impeding the correct functioning of the cells. **(10.2) 1.** Renal cortex, renal medulla, and renal pelvis; **2.** The glomerulus structure facilitates easy passage of small molecules to the glomerular capsule. The proximal convoluted tubule has a large surface area for the reabsorption of filtrate components; The loop of the nephron facilitates reabsorption of water by the nephron. The distal convoluted tubules function in ion exchange. The collecting ducts carry urine to the renal pelvis. **3.** Due to the function of filtering blood. **(10.3) 1.** Glomerular filtration, tubular reabsorption, and tubular secretion. Due to glomerular blood pressure, water and small blood molecules move from the glomerulus to the inside of the glomerular capsule; Tubular reabsorption occurs as molecules and ions are both passively and actively reabsorbed from the nephron into the blood of the peritubular capillary network; Certain molecules are then actively secreted from the peritubular capillary network into the convoluted tubules; **2.** Glomerular filtration moves water, salts, nutrients, and waste molecules from the glomerulus to the glomerular capsule. Tubular reabsorption reabsorbs nutrients and salt molecules from the convoluted tubules into the peritubular capillary network. Tubular secretion secretes certain molecules into the convoluted tubules; **3.** If glomerular filtration did not occur, the blood would not get filtered hindering proper circulation and respiration. **(10.4) 1.** Dilute urine has more water than solutes; Concentrated urine has less water than solutes; **2.** Aldosterone promotes the excretion of potassium ions and the reabsorption of sodium ions, leading to a reabsorption of water; Atrial natriuretic hormone promotes the excretion of sodium, leading to a decrease in water reabsorption; Antidiuretic hormone increases water reabsorption; **3.** If the blood is acidic, hydrogen ions are excreted and bicarbonate ions are reabsorbed; If the blood is basic, hydrogen ions are not excreted and bicarbonate ions are not reabsorbed. The blood must be at the proper pH for all of the body systems to function properly. **(10.5) 1.** Infections, diabetes, hypertension, and inherited conditions cause renal disease; **2.** Hemodialysis uses either an artificial kidney machine or CAPD to filter the blood, removing wastes and reabsorb needed nutrients and water just like the functions of a kidney; **3.** Because blood needs to be continuously filtered for homeostasis.

Understanding Key Terms

a. diuretic; **b.** excretion; **c.** urethra; **d.** renal pelvis; **e.** glomerular filtrate

Testing Your Knowledge of the Concepts

11. d; **12.** c; **13.** a; **14.** c; **15.** c; **16.** a; **17.** c; **18.** d; **19.** a; **20.** e; **21.** b; **22.** c; **23.** d; **24.** b; **25.** a; **26.** d; **27.** f; **28.** a; **29.** c; **30.** d; **31. a.** glomerular capsule; **b.** efferent arteriole; **c.** afferent arteriole; **d.** proximal convoluted tubule; **e.** loop of the nephron; **f.** descending limb; **g.** ascending limb; **h.** peritubular capillaries; **i.** distal convoluted tubule; **j.** renal vein; **k.** renal artery; **l.** collecting duct

Thinking Critically About the Concepts

1. Symptoms common to diabetes mellitus and diabetes insipidus are great thirst and frequent urination due to increased filtration and removal of waste products; **2.** If there are more red blood cells, more oxygen will be delivered to the cells that will use the oxygen to make more ATP; **3.** It calms the smooth muscle of the bladder, which the large intestine also has. When the large intestine's contractions are lessened, constipation may result; **4.** The man is likely to be incontinent; **5.** Damage to the kidneys may result from football tackles or blows from a hard ball (like one used in lacrosse). Players often wear pads designed to protect the ribs and kidneys from damage.

Chapter 11

Case Study

1. It provides flexibility and support; **2.** The epiphysis; **3.** She is active and exercises before surgery which can make recovery easier.

Check Your Progress

(11.1) 1. Compact bone is highly organized and composed of tubular units called osteons; Spongy bone has an unorganized appearance and is composed of trabeculae; **2.** Hyaline cartilage—uniform and glassy matrix with abundant collagen fibers (the ends of long bones, in the nose, at the ends of the ribs, and in the larynx and trachea); Fibrocartilage—wide rows of thick, collagenous fibers (the disks located between the vertebrae and also in the cartilage of the knee); Elastic cartilage—matrix contains mostly elastin fibers (the ear flaps and the epiglottis); **3.** Without the skeletal system, organs would not be protected, blood cells would not be produced, and there would be no storage of certain materials. **(11.2) 1.** Osteoblasts are bone-forming cells, osteocytes are mature bone cells that maintain bone structure, and osteoclasts are bone-absorbing cells. Osteoblasts are involved in bone growth, osteoblasts and osteoclasts are both involved in bone remodeling and repair; **2.** Through intramembranous ossification, in which bone develops between sheets of fibrous connective tissue, and endochondral ossification, in which bone replaces a cartilage model; **3.** A hematoma is formed, next tissue repair begins, and a fibrocartilaginous callus is formed between the ends of the broken bone, then the fibrocartilaginous callus is converted into a bony callus and remodeled. **(11.3) 1.** Skull, hyoid bone, vertebral column, rib cage, and the ear ossicles; **2.** The frontal bone forms the forehead, the parietal bones extend to the sides, and the occipital bone curves to form the base of the skull, each temporal bone is located below the parietal bones, the sphenoid bone extends across the floor of the cranium from one side to the other, and the ethmoid bone lies in front of the sphenoid. Mandible forms the lower jaw and chin, maxillae forms the upper jaw and the anterior portion of the hard palate, zygomatic bones are the cheekbone prominences, and the nasal bones form the bridge of the nose; **3.** Cervical vertebrae—located in the neck and allow movement of the head; Thoracic vertebrae—form the thoracic curvature and have long, thin, spinous processes and articular facets for the attachment of the ribs; Lumbar vertebrae—form the lumbar curvature and have a large body and thick processes;

Sacral vertebrae—fused together, forming the pelvic curvature; Coccyx—fused vertebrae that form the tailbone. **(11.4) 1.** Scapula, clavicle; humerus, radius, ulna, carpals, metacarpals, phalanges; **2.** Two coxal bones; femur, patella, tibia, fibula, tarsals, metatarsals, phalanges; **3.** In the female the iliac bones are more flared, the pelvic cavity is more shallow, but the outlet is wider. **(11.5) 1.** Fibrous, cartilaginous, synovial; **2.** Cartilaginous are slightly movable and found in the elbow and ankle. Fibrous joints are immovable and are found in the skull; **3.** Flexion and extension—knee; adduction and abduction—hip and shoulder; rotation and circumduction—hip and shoulder; inversion and eversion—ankle.

Understanding Key Terms

a. ligament; **b.** compact bone; **c.** sinus; **d.** fontanel; **e.** osteocyte

Testing Your Knowledge of the Concepts

18. c; **19.** d; **20.** a; **21.** g; **22.** f; **23.** e; **24.** e; **25.** a; **26.** b; **27.** c; **28.** f; **29.** d; **30.** b; **31.** b; **32.** a; **33.** b; **34.** c; **35.** e; **36.** c; **37.** c; **38.** b; **39.** d; **40.** c; **41.** c; **42.** F; **43.** T; **44.** F; **45.** T; **46.** T; **47.** b; **48.** a; **49.** e; **50.** c; **51.** d; **52.** **a.** frontal bone; **b.** zygomatic bone; **c.** maxilla; **d.** mandible; **e.** clavicle; **f.** scapula; **g.** sternum; **h.** ribs; **i.** costal cartilages; **j.** coxal bones; **k.** sacrum; **l.** coccyx; **m.** patella; **n.** metatarsals; **o.** phalanges; **p.** temporal bone; **q.** vertebral column; **r.** humerus; **s.** ulna; **t.** radius; **u.** carpals; **v.** metacarpals; **w.** femur; **x.** fibula; **y.** tibia; **z.** tarsals

Thinking Critically About the Concepts

1. A balanced diet rich in calcium and containing the proper amount of protein will speed bone repair. Weight-bearing exercise stimulates bond growth at any age; **2.** Without sunlight, a person produces less vitamin D. Thus, less calcium can be absorbed, and bones become weakened. Supplemental vitamin D can help to prevent this; **3.** The growth rate for all cells becomes slower as a person ages; **4.** The typical fast-food diet is deficient in calcium, vitamin D, and other key nutrients. It may also be low in protein; **5.** Believe it or not, it is the player whose fibula is fractured. Bone cells reproduce faster than cells found in cartilage, such as the ligaments that may be torn in a sprain; **6a.** Low blood calcium triggers parathyroid hormone release; **6b.** In hyperparathyroidism, the bones lose too much calcium and become fragile. Calcium deposits form elsewhere in the body, especially in the kidneys; **7.** If the cervical vertebrae are fractured, they can no longer support the weight of the head. If the head falls forward or backward, the spinal cord will stretch and be damaged.

Chapter 12

Case Study

1. An EMG records the electrical signals that control the skeletal muscles of the body; **2.** So the muscle can perform all its functions properly; **3.** Uncontrollable muscle spasms, muscle degeneration, muscle atrophy.

Check Your Progress

(12.1) 1. Smooth, cardiac, and skeletal; **2.** Make bones move, help maintain a constant body temperature, assist movement in cardiovascular and lymphatic vessels, protect internal organs and stabilize joints; **3.** Orbicularis oculi, orbicularis oris, pectoralis major, serratus anterior, external oblique, quadriceps femoris, tibialis anterior, extensor digitorum longus, masseter, deltoid, biceps brachii, rectus abdominus, flexor carpi group, adductor longus, sartorius, trapezius, lattisimus dorsi, triceps brachii, extensor carpi group, extensor digitorum, gluteus maximus, biceps femoris, gastrocnemis; **4.** In opposite pairs, for example, one flexes and the other extends. **(12.2) 1.** Myofibril is a bundle of myofilaments that contract. Myofilament is actin filaments or myosin filaments whose structure and functions account for muscle striations and contractions. A muscle fiber contains many myofibrils divided into sarcomeres which are contractile; **2.** The actin filaments (thin filaments) slide past the myosin filaments (thick filaments) toward the center. The Z lines have moved and the H zone has gotten smaller, to the point of disappearing; **3.** Calcium binds to troponin exposing myosin binding sites. Myosin uses ATP and does the work of pulling actin towards the center of the sarcomere. **(12.3) 1.** First a latent period followed by contraction then relaxation; **2.** Stimulation of a muscle by a single electrical signal results in a simple muscle twitch (latent period, contraction, relaxation). Repeated stimulation results in summation and tetanus, which creates greater force because the motor unit cannot relax between stimuli; **3.** CP pathway converts ADP to ATP while creatine phosphate is converted to creatine. In fermentation, ATP is produced while glucose is broken down to lactic acid. Aerobic respiration uses the processes of cellular respiration, glycolysis, citric acid cycle, and electron transport chain, to produce ATP; **4.** Due to their weight lifting, their muscles contain more fast twitch fibers than slow twitch. **(12.4) 1.** In a strain, the muscle at a joint is stretched or torn. A sprain results in stretching and tearing of tendons and ligaments at a joint, as well as possible blood vessel and nerve damage; **2.** Myasthenia gravis is an autoimmune disease that results in impaired muscle contraction. Myalgia is achy muscles which usually results from overuse but may result from an immune disorder. **3.** Progressive degeneration and weakening of muscles. **(12.5) 1.** The skeletal system stores and releases calcium and muscle tissue stores calcium in the sarcoplasmic reticulum; **2.** By involuntary shivering to produce heat by friction; **3.** Blood cells are produced in bone; maintenance of internal body temperature; storage of calcium.

Understanding Key Terms

a. sarcomere; **b.** insertion; **c.** tetanus; **d.** rigor mortis; **e.** strain

Testing Your Knowledge of the Concepts

10. c; **11.** b; **12.** a; **13.** c; **14.** d; **15.** e; **16** a; **17.** d; **18.** d; **19.** b; **20.** c; **21.** a; **22.** e; **23.** c; **24.** e; **25.** d; **26.** a; **27.** b; **28.** c; **29.** c; **30.** a; **31.** d; **32.** b; **33.** c; **34.** a; **35.** b; **36.** a; **37.** c; **38.** b; **39.** b; **40.** a; **41.** c; **42.** **a.** T tubule; **b.** sarcoplasmic reticulum; **c.** myofibril; **d.** Z line; **e.** sarcomere; **f.** sarcolemma

Thinking Critically About the Concepts

1a. When dystrophin is absent, calcium leaks into the cell and activates an enzyme that dissolves muscle fibers; **1b.** As muscle fibers die, fat and connective tissue take their place; **2a.** Rapid cooling would slow the progression of rigor mortis, because it would slow cellular metabolism. **2b.** As cold can delay the onset of rigor mortis, heat can accelerate the onset, as can extreme exercise right before death, which depletes ATP stores in the muscles. **3.** Rigor mortis diminishes because the decay process will cause the body to lose its stiffness; **4.** If the diaphragm and external intercostal muscles fail, a person can no longer breathe.

Chapter 13

Case Study

1. The nerves will no longer conduct impulses correctly; **2.** These tests look at the structure and function of nervous tissue; **3.** The nerves from the brain and spinal cord to the muscles of the legs are some of the longest in the body and can deteriorate quickly.

Check Your Progress

(13.1) 1. Sensory neurons (take nerve signals from a sensory receptor to the CNS), interneurons (lie entirely within the CNS and communicate with other neurons), and motor neurons (take nerve impulses away from the CNS to an effector); Cell body, dendrites, and axon; **2.** Action potential traveling along a neuron; Exchange of ions generated along the length of an axon; **3.** Neurotransmitters. **(13.2) 1.** Provide a means of communication between the brain and the peripheral nerves, the center for reflex actions; **2.** Cerebrum—main part of the brain, communicates and coordinates the activities of the other parts of the brain; Diencephalon—contains the hypothalamus and thalamus, maintains homeostasis, receives sensory input; Cerebellum—sends out motor impulses by way of the brain stem to the skeletal muscles, produces smooth, coordinated voluntary movements; Brain stem—contains the midbrain, pons, and medulla oblongata; Acts as a relay station and medulla has reflex centers; **3.** The reticular activating system is responsible for a person being awake, alert, asleep and comatose. **(13.3) 1.** Blend primitive emotions and higher mental functions into a united whole; **2.** Amygdala—fight-or-flight, and hippocampus—learning and memory; **3.** Wernicke's area and Broca's area in the left hemisphere are related to speech. It appears that the left and right hemispheres have different functions pertaining to language and speech. **(13.4) 1.** Cranial nerves arise from the brain and there are 12 pairs. Spinal nerves arise from the spine and there are 31 pairs; **2.** Reflex; **3.** Without the autonomic nervous system, activities of the cardiac muscles, smooth muscles, and glands, could not be regulated involuntarily. **(13.5) 1.** Drug therapy is used to treat a disease or disorder. Drug abuse is using drugs without symptoms of disease or disorder; **2.** Affects the limbic system and either promotes or decreases the action of a particular neurotransmitter; **3.** Depressive, stimulant, psychoactive.

Understanding Key Terms

a. reflex; **b.** neurotransmitter; **c.** autonomic system; **d.** ganglion; **e.** acetylcholine

Testing Your Knowledge of the Concepts

18. b; **19.** d; **20.** c; **21.** a; **22.** a; **23.** b; **24.** c; **25.** c; **26.** a; **27.** c; **28.** b; **29.** d; **30.** c; **31. a.** central canal, **b.** dorsal horn; **c.** white matter; **d.** cell body of interneuron; **e.** dorsal root ganglion.

Thinking Critically About the Concepts

1. Myelin enables the signal to jump from node to node quickly, because the depolarization process occurs only at the node of Ranvier; **2a.** Triglyceride (fat or oil); **2b.** Unsaturated fatty acids are characterized by one or more double bonds between carbons, while saturated fatty acids are all single bonds; **2c.** Animal fat, such as butter and fatty cuts of meat; **3.** Myelination enables signals to travel quickly down an axon, which helps motor skills.

Chapter 14

Case Study

1. Inner ear; **2.** Pressure waves moving through the canals cause the basilar membrane, which lies underneath the spiral organ of the cochlea, to vibrate. This causes the stereocilia in the tectorial membrane to bend, which activate nerve impulses traveling in the cochlea nerve to travel to the brain; **3.** The technology in a cochlear implant picks up signals from the environment and converts them to electrical impulses that can be sent to the brain. This type of technology could work on vision if it picked up signals and sent electrical impulses to the nerves controlling vision.

Check Your Progress

(14.1) 1. Chemoreceptors respond to chemical substances. Photoreceptors respond to light energy. Mechanoreceptors are stimulated by mechanical forces that result in pressure. Thermoreceptors are stimulated by changes in temperature; **2.** Sensation is the conscious perception of stimuli that occurs after sensory receptors generate a nerve impulse that arrives at the cerebral cortex. Perception is the process of using the senses to acquire information about the environment or situation. For example, sensation is when nerve impulses from our ears reach the brain. Perception is when we recognize that sound as singing; **3.** Sensory receptors pick up changes in the internal and external environment so the body can maintain homeostasis. **(14.2) 1.** By detecting the degree of muscle relaxation, the stretch of the tendons, and the movement of ligaments; **2.** Meissner corpuscles (dermis), Krause end bulbs (dermis), Merkel discs (where epidermis meets dermis), and root hair plexus (dermis) are sensitive to fine touch. Pacinian corpuscles (dermis) and Ruffini endings (dermis) are sensitive to pressure. Temperature receptors are free nerve endings (epidermis). **3.** Pain receptors are sensitive to chemicals released by damaged tissues; in that way, the body will know there is a change in the internal environment that needs adjusting by homeostasis. **(14.3) 1.** In the tongue, taste buds in the papillae are used for taste. In the nose, olfactory cells within the olfactory epithelium are used for smell; **2.** They both respond to chemical stimuli, in the tongue the stimulus is direct, in the nose it can be distant; **3.** In the tongue, the microvilli in the taste cells bear receptor proteins for certain molecules and when they bind nerve signals are generated and go to the brain where the sensation of taste occurs. In the nose, the cilia of the olfactory cells bind a molecule that activates the connected neuron sending the information to the cerebral cortex where the sensation is interpreted. **(14.4) 1.** Sclera—protects and supports eyeball; Cornea—refracts light rays; Pupil—admits light; Choroid—absorbs stray light; Ciliary body—accommodation; Iris—regulates light entrance; Retina—contains sensory receptors; Rod cells—black and white vision; Cone cells—color vision; Fovea centralis—acute vision; Lens—refracts and focuses light; Humors—supports eyeball; Optic nerve—transmits impulse to brain; **2.** Rod cells are used for black and white vision and cone cells are used for color vision; **3.** Once photoreceptors initiate a visual signal, rods and cones synapse with bipolar cells, which synapse with ganglion cells whose axons become the optic nerve. **(14.5) 1.** Outer ear directs sound into middle ear where air from the outer ear vibrates the tympanic membrane and all ossicles which attach to the inner ear, where fluid is used to stimulate receptors that attach to nerves sending signals to the brain; **2.** Mechanoreceptors are sensitive to mechanical stimulation. Hair cells with stereocilia located in the spiral organ of the cochlea are mechanoreceptors; **3.** Pressure waves move through the canals causing the basilar membrane to vibrate. This causes the stereocilia embedded in the tectorial membrane to bend. Nerve impulses traveling in the cochlear nerve result in hearing. **(14.6) 1.** For rotational equlibrium—semicircular canals, ampullae, vestibular nerve, cochlea, cupula, stereocilia, hair cells, supporting cells, endolymph. For gravitational equilibrium—utricle, saccule, otoliths, otolithic membrane, hair cells, supporting cells, vestibular nerve; **2.** As fluid within the semicircular canal flows over and displaces a cupula, the stereocilia of the hair cells bend. This causes a change in the pattern of signals sent to the brain by the vestibular nerve. **3.** Rotational equilibrium occurs when the head is moved side to side and gravitational equilibrium occurs when the head is moved up and down. They work together to keep the head, and body, in position according to gravity.

Understanding Key Terms

a. sensory receptor; **b.** retina; **c.** sclera; **d.** chemoreceptor; **e.** spiral organ

Testing Your Knowledge of the Concepts

14. d; **15.** a; **16.** b; **17.** b; **18.** c; **19.** a; **20.** d; **21.** c; **22.** d; **23.** b; **24.** b; **25.** c; **26.** b; **27.** d; **28.** d; **29.** c; **30.** a; **31.** e; **32.** d; **33. a.** retina; **b.** choroid; **c.** sclera; **d.** optic nerve; **e.** fovea centralis; **f.** ciliary body; **g.** lens; **h.** iris; **i.** pupil; **j.** cornea; **34. a.** tympanic membrane; **b.** auditory canal; **c.** stapes; **d.** incus; **e.** malleus; **f.** oval window; **g.** semicircular canals; **h.** vestibule; **i.** cochlear nerve; **j.** cochlea; **k.** auditory tube; **l.** round window

Thinking Critically About the Concepts

1. Just about the entire sensory system: taste, smell, vision (seeing your pizza!), as well as receptors for temperature and texture in your mouth; **2.** Chemoreceptors also monitor the oxygen and carbon dioxide in the blood. Taste and smell receptors are also chemoreceptors; **3.** Vision, because the visual sensory and memory areas are found in the occipital lobe; **4.** Hearing receptors are severely damaged by continual loud noise. Without ear protection, the workers will go deaf; **5.** Both hearing and balance will be affected, sometimes severely; **6.** By seeing what senses are affected in the areas of the tumor in the cerebral cortex.

Chapter 15

Case Study

1. Insulin and glucagon; **2.** Type 1 occurs when the pancreas does not produce insulin. Type 2 occurs when cells become insulin-resistant; **3.** When there is not enough glucose, glucagon releases more from the body's storage areas. When there is too much glucose, insulin stores excess glucose.

Check Your Progress

(15.1) 1. A hormone is a chemical signal that affects the metabolism of a target cell; **2.** Endocrine system uses hormones to affect homeostasis and the nervous system uses neurons and neurotransmitters; Both regulate the activities of other systems. **3.** Peptide hormones cause the formation of cyclic AMP, which results in a series of enzymatic reactions, steroid hormones promote the synthesis of a specific enzyme; **4.** To regulate the hormones. **(15.2) 1.** Through hormones and neurotransmitters; **2.** Posterior pituitary does not produce any hormones; **3.** TSH—stimulates the thyroid to produce the thyroid hormones; ACTH—stimulates the adrenal cortex to produce cortisol; Gonadotropic hormones—stimulate the gonads; PRL—causes milk development in the breasts; MSH—skin color changes; GH—affects growth; **4.** Abnormal growth, malfunctioning kidneys and thyroid. **(15.3) 1.** By increasing the metabolic rate; **2.** When blood calcium level is high, the thyroid gland secretes calcitonin which promotes the uptake of calcium by the bones, returning the blood calcium level to normal. When blood calcium is low, the parathyroid releases parathyroid hormone causing bones to release calcium, the kidneys to reabsorb calcium and activate vitamin D so the intestines can absorb calcium until the blood calcium levels return to normal; **3.** In hypothyroidism (undersecretion of thyroid hormone) individuals do not grow and develop normally. In hyperthyroidism (oversecretion of thyroid hormones) goiters form. **(15.4) 1.** Adrenal medulla- epinephrine and norepinephrine; Adrenal cortex- cortisol, aldosterone, atrial natriuretic hormone; **2.** Short term stress response- heartbeat and blood pressure increase; blood glucose level rises; muscles become energized. Long term stress response- protein and fat metabolism instead of glucose breakdown; reduction of inflammation; sodium ions and water are reabsorbed by kidney; blood volume and pressure increase; **3.** Salt balance maintains fluid levels, and keeping proper water balance not only maintains fluid balance, but pH balance. **(15.5) 1.** Exocrine—produces and secretes digestive juices. Endocrine- produces and secretes insulin and glucagon; **2.** When blood glucose levels are high, insulin aids in storing excess glucose. When blood glucose levels are low, glucagon releases glucose from storage and delivers to bloodstream; **3.** Type 1 diabetes the pancreas does not produce insulin. In Type 2 diabetes, the cells are insulin-resistant. **(15.6) 1.** Estrogen maintains the secondary sexual characteristics in the female, and testosterone maintains the secondary sexual characteristics in males; **2.** Increased levels of melatonin occur at night and decrease by morning; **3.** The kidneys will secrete erythropoietin. **(15.7) 1.** For both internal and external stimuli, the nervous system detects and responds to maintain homeostasis; **2.** Aldosterone acts on kidney tubules to conserve sodium and water reabsorption will follow; **3.** Because of the role of the hypothalamus in each system.

Understanding Key Terms

a. thyroid gland; **b.** diabetes mellitus; **c.** adrenocorticotropic hormone (ACTH); **d.** peptide hormone; **e.** oxytocin

Testing Your Knowledge of the Concepts

14. d; **15.** b; **16.** b; **17.** d; **18.** c; **19.** c; **20.** b; **21.** e; **22.** e; **23.** d; **24.** d; **25.** d; **26.** b; **27.** c; **28.** d; **29.** a; **30.** d; **31.** e; **32.** c; **33.** b; **34.** f; **35.** b; **36.** c; **37.** a; **38.** e; **39.** d; **40. a.** inhibits; **b.** inhibits; **c.** releasing hormone; **d.** stimulating hormone; **e.** target gland hormone

Thinking Critically About the Concept

1. Follicle stimulating hormone and growth hormone are both protein hormones. They both bind to a receptor in the plasma membrane and activate the cAMP second messenger system; **2.** When thyroxine is produced, negative feedback occurs to stop TSH. Without thyroxine, there is no negative feedback to stop TSH. Therefore TSH levels remain high in someone with hypothyroidism; **3.** An increase of blood glucose will cause fluid to move from the cells into the bloodstream causing frequent urination and thirst; **4.** The diet would regulate the intake of glucose.

Chapter 16

Case Study

1. The cervix is the opening to the uterus; **2.** Irregularities in the cells of the cervix and uterus; **3.** The removal of the uterus and/or ovaries.

Check Your Progress

(16.1) 1. The male reproductive system functions in sperm production and the female in ovum production; **2.** Mitosis is duplication division (number of chromosomes stays the same), meiosis is reduction division (number of chromosomes is reduced); **3.** Meiosis in males occurs in the testes and in ovaries in females. **(16.2) 1.** Organs/structure of the male reproductive system include testes, epididymis, vas deferens, seminal vesicles, prostate gland, urethra, bulbourethral glands, ejaculatory duct, penis. Sperm flows from epididymis to vas deferens to ejaculatory duct to the urethra; **2.** During spermatogenesis, a diploid spermatogonium turns into a primary spermatocyte, then into a secondary spermatocyte, then into a spermatid, then a mature sperm; **3.** Testosterone maintains the secondary sexual characteristics in males. **(16.3) 1a.** Ovary; **1b.** Oviducts; **1c.** Uterus; **1d.** Vagina; **2.** Vagina is the birth canal, exit site for menstruation, and used for insertion of the penis during intercourse. The uterus is the site of menstruation and development of the embryo and fetus; **3.** Unlike the male, the female does not need to expel her gamete from her body for fertilization. **(16.4) 1.** Estrogen and progesterone control the uterine cycle; FSH and LH control the ovarian cycle. **2.** Corpeus luteum is maintained in the ovary and produces increasing concentrations of progesterone; progesterone shuts down the hypothalamus and anterior pituitary so that no new follicles begin in the ovary during pregnancy. The uterine lining is maintained; **3.** The hormones in birth control pills feedback to inhibit the hypothalamus and the anterior pituitary; therefore, no new follicles begin in the ovary. **(16.5) 1.** Abstinence, vasectomy, tubal ligation, IUD, contraceptive implants, contraceptive injections, birth control pills, diaphragm, condoms, morning-after pills; **2.** Vasectomy prevents sperm from entering the semen. Tubal ligation prevents the egg from getting into the fallopian tubes. In both cases the tubes are surgically closed; **3.** IVF conception occurs outside the body and the embryos are transferred to the uterus. In ICSI a single sperm is injected into an egg. **(16.6) 1.** Pelvic inflammatory disease; **2.** Cancer of the cervix; **3.** Unprotected sex.

Understanding Key Terms

a. ovulation; **b.** progesterone; **c.** semen; **d.** cervix; **e.** acrosome

Testing Your Knowledge of the Concepts

12a. Seminal vesicle; **b.** Ejaculatory duct; **c.** Prostate gland; **d.** Bulbourethral gland; **e.** Anus; **f.** Vas deferens; **g.** Epididymis; **h.** Testis; **i.** Scrotum; **j.** Foreskin; **k.** Glans penis; **l.** Penis; **m.** Urethra; **n.** Vas deferens; **o.** Urinary bladder. **13.** c; **14.** d; **15.** c; **16.** b; **17. a.** oviduct; **b.** ovary; **c.** uterus; **d.** glans clitoris; **e.** labium minora; **f.** labium majora; **g.** vaginal orifice; **h.** cervix; **i.** vagina; **18.** d; **19.** d; **20.** c; **21.** b; **22.** c; **23.** d; **24.** c; **25.** d; **26.** a; **27.** d; **28.** e; **29.** c; **30.** d.

Thinking Critically About the Concepts

1. Increased testosterone and progesterone inhibit GnRH. The inhibition of GnRH inhibits FSH production. Lower FSH decreases sperm production; **2a.** Drawing should show the introduction of progesterone early in the cycle, inhibiting FSH and LH; **2b.** Drawing should show that hCG prevents the corpus luteum from regressing, which means progesterone remains high and the uterine lining remains thick; **3.** Birth control pills inhibit FSH, which in turn inhibit follicle development. Fewer follicles may be the cause of a lower risk of ovarian cancer. Fertility drugs increase the number of follicles, which may be the cause of the higher risk of ovarian cancer; **4.** A low body fat composition may prevent the female athlete from producing estrogen and progesterone, which prevents the occurrence of a menstrual cycle.

Chapter 17

Case Study

1. Regulates the production of progesterone from the corpus luteum to maintain pregnancy; **2.** Some medications have negative effects on the development of the baby; **3.** Changes in size and shape of uterus, weight gain, breast swelling, increased urination, shortness of breath, difficulty sleeping.

Check Your Progress

(17.1) 1. A sperm makes its way through the corona radiata, acrosome releases digestive enzymes to digest zona pellucida, sperm binds to the egg, their plasma membranes fuse, sperm enters the egg, egg and sperm nuclei fuse; **2.** Corona radiata nourish the egg when it is a follicle in the ovary. The zona pellucida lies outside the plasma membrane and protects the egg from polyspermy; **3.** The changes in the egg's plasma membrane and the zona pellucida as soon as the first sperm penetrates the membrane. **(17.2) 1.** Morphogenesis refers to the shaping of the embryo; differentiation is when cells take on a specific structure and function; **2.** Chorion—develops into fetal half of the placenta; Allantois—forms umbilical blood vessels; Yolk sac—produces blood cells; Amnion—contains fluid to cushion and protect embyo; **3.** In pre-embryonic development the zygote divides repeatedly, develops into a morula, and then a blastocyst; In embryonic development the embryo implants in the uterine wall, gastrulation occurs and primary germ layers are formed, organ systems appear and develop. **(17.3) 1.** The path of blood is *chorionic villi of placenta→umbilical vein→venous ducts*→inferior vena cava→heart→blood is shunted into the left atrium by way of the *oval opening* and some enters the aorta by way of the *arterial duct*→ aorta→*umbilical arteries;* Structures unique to fetus in italics; **2.** Third and fourth months—skeleton becomes ossified, sex of fetus is distinguishable, fetus grows and gains weight; Fifth through seventh months—fetal limbs grow, fetus gains weight; Eighth through ninth months—fetus gains weight, fetus rotates in preparation for birth; **3.** The presence of an *SRY* gene, on the Y chromosome, leads to the development of testes and male genitals; otherwise, ovaries and female genitals develop. **(17.4) 1.** Due to placental hormones, energy levels fluctuate, the uterus relaxes, the pulmonary values increase; **2.** Creates a concentration gradient favorable to the flow of carbon dioxide from fetal blood to maternal blood at the placenta; **3.** Stage 1—cervix dilates; Stage 2—baby is born; Stage 3—placenta/afterbirth is delivered. **(17.5) 1.** genetic in origin (certain genes can increase life span), whole-body process (many different organs in the body deteriorate), extrinsic factors (poor health habits affect various systems); **2.** Skin—becomes thinner and less elastic, homeostatic adjustment to heat is limited; Processing and transporting—heart shrinks because of a reduction in cardiac muscle size, arteries become rigid, cardiovascular problems are often accompanied by respiratory disorders, blood supply to kidney and liver is reduced; Integration and coordination—reaction time slows, greater stimulation is needed for sense receptors to function, decline in bone density, loss of skeletal muscle mass; Reproductive system—females undergo menopause, males undergo andropause; **3.** Good health habits.

Understanding Key Terms

a. lanugo; **b.** afterbirth; **c.** gerontology; **d.** fertilization; **e.** cleavage

Testing Your Knowledge of the Concepts

12. c; **13.** c; **14.** b; **15.** a; **16.** b; **17.** d; **18.** d; **19.** b; **20.** c; **21.** d; **22.** a; **23.** e; **24.** d; **25.** e; **26.** e; **27.** b; **28. a.** chorion—fetal portion of the placenta, for nutrient, waste, and gas exchange; **b.** amnion—contains fluid to protect and cushion the embryo; **c.** embryo; **d.** allantois—becomes the umbilical vessels; **e.** yolk sac—first site of blood cell formation; **f.** fetal portion of placenta where fetal blood makes exchanges with maternal blood; **g.** maternal portion of placenta; **h.** umbilical cord—connects the embryo to the placenta.

Thinking Critically About the Concepts

1. It is secreted by the chorion. hCG is filtered from the blood by the nephron during glomerular filtration. It is not reabsorbed during tubular reabsorption, so it remains in the tubule and is a urine component; **2a.** hCG is a hormone secreted by the chorion that enters the blood; **2b.** It binds to a membrane receptor and activates the second messenger, cAMP; **3a.** FSH; **3b.** Decreased; **3c.** FSH stimulates follicle development and estrogen production by the follicle. Without FSH, follicle development doesn't occur and estrogen secretion does not cause the development of the uterine lining that is shed during menstruation.

Chapter 18

Case Study

1. They delay the continuation of the cell cycle until certain criteria are met to ensure proper cell division; **2.** During interphase.

Check Your Progress

(18.1) 1. Karyotype arranges the chromosomes of a cell into pairs so all chromosomes can be observed; **2.** Because they are exact copies of each other; **3.** CVS is performed between the 8th and 12th week and fetal cells are obtained from the villi by suction. Amniocentesis is performed between the 15th and 17th week and amniotic fluid is obtained. **(18.2) 1.** The cell cycle includes interphase (G1,S, G2) and mitosis (prophase, metaphase, anaphase, telophase) and is how a cell divides. Checkpoints occur at G1, G2 in interphase and during mitosis between metaphase and anaphase; **2.** DNA is duplicated; **3.** Checkpoints prevent the cell from dividing unless the proper signals are present and the DNA is not damaged; **4.** To stimulate/inhibit the cell cycle due to the body's needs. **(18.3) 1.** They are the same; **2.** Prophase—chromosomes attach to the spindle; Metaphase—chromosomes align at the equator; Anaphase—chromatids separate and chromosomes move toward poles; Telophase—nuclear envelopes form around chromosomes; **3.** During cytokinesis, a contractile ring pinches the cell in two. **(18.4) 1.** Each daughter cell contains half as many chromosomes as the parent cell; **2.** In prophase I, crossing-over occurs, in metaphase I, independent alignment occurs; **3.** Meiosis I—homologous chromosomes pair and then separate; Meiosis II—sister chromatids separate, resulting in four cells with a haploid number of chromosomes. **(18.5) 1.** Meiosis I—Prophase—pairing of homologous chromosomes; Metaphase I—homologous duplicated chromosomes at equator; Anaphase I—homologous chromosomes separate; Telophase I—two haploid daughter cells. Mitosis- Prophase- no pairing of chromosomes; metaphase- duplicated chromosomes at equator; anaphase- sister chromatids separate becoming daughter chromosomes that move to the poles; telophase- two daughter cells identical to the parent; **2.** Meiosis II- Prophase II- no pairing of chromosomes; metaphase II- haploid number of chromosomes at equator; anaphase II- sister chromatids separate becoming daughter chromosomes that move to the poles; telophase II- four haploid daughter cells; Mitosis- Prophase- no pairing of chromosomes; metaphase- duplicated chromosomes at equator; anaphase- sister chromatids separate becoming daughter chromosomes that move to the poles; telophase- two daughter cells identical to parent cell; **3.** Oogenesis- during meiosis I a primary oocyte turns into a secondary oocyte and a polar body. During meiosis II a secondary oocyte, if fertilized, turns into an egg and second polar body. Spermatogenesis—During meiosis I a primary spermatocyte turns into two secondary spermatocytes. During meiosis II secondary spermatocytes turn into four haploid spermatids; **4.** Human females do not need to produce as many gametes as human males. **(18.6) 1.** Nondisjunction; **2.** Trisomy 21; **3.** Turner syndrome, Klinefelter syndrome, poly-X females, Jacobs syndrome; **4.** Changes in chromosome structure, including deletions, duplications, inversions, and translocations.

Understanding Key Terms

a. Barr body; **b.** homologous; chromosome; **c.** monosomy; **d.** inversion; **e.** haploid

Testing Your Knowledge of the Concepts

14. b; **15. a.** G_1 phase—cells grow, organelles double; **b.** S phase—DNA synthesis; **c.** G_2 phase—cell prepares to divide; **d.** mitosis—nuclear division and cytokinesis; **16.** d; **17.** b; **18.** a; **19.** d; **20.** e; **21.** c; **22.** b; **23.** a; **24.** c; **25.** c; **26.** d; **27.** c; **28.** e; **29.** a; **30.** d; **31.** b; **32.** c; **33.** c; **34.** d; **35.** c; **36.** d; **37.** a; **38.** b

Thinking Critically About the Concepts

1. Cell signaling genes that send a cell into the cell cycle or prevent a cell from entering the cell cycle; **2.** It could be that the body responds abnormally only to injury, but that normal mitosis functions well. Injury initiates a number of responses besides just mitosis; **3.** Homologous chromosomes separate randomly. Sibilings that look similar inherited similar homologous chromosomes; sibilings that are dissimilar inherited different combinations of homologous chromosomes; **4a.** They do not undergo the phases of mitosis to form new cells that would replace the damaged cells; **4b.** They are not likely to recover fully.

Chapter 19

Case Study

1. Cancer cells lack differentiation, have abnormal nuclei, have unlimited potential to replicate, form tumors, have no need for growth factors, gradually become abnormal, undergo angiogenesis and metastasis; **2.** They do not have contact inhibition; **3.** To remove the existing tumor from the kidney and make sure no other tumor cells existed outside the kidney.

Check Your Progress

(19.1) 1. Cancer cells do not undergo apoptosis, have unlimited replicative potential due to telomerase, do not exhibit contact inhibition, have no need for growth factors, undergo angiogenesis and metastasis; **2.** Proto-oncogenes result in gain-of-function mutations that continuously promote the cell cycle, and tumor suppressors genes are loss-of-function mutations that can no longer inhibit the cell cycle; **3.** Lung cancer, colorectal cancer, prostate cancer, breast cancer. **(19.2) 1.** DNA-linkage studies have revealed breast cancer genes (*BRCA1* and *BRCA2*), a tumor suppressor gene has been associated with retinoblastoma, and an abnormal RET gene predisposes an individual to thyroid cancer; **2.** Ionizing radiation, tobacco smoke, pollutants, certain viruses; **3.** Good nutrition, avoid radiation, do not sunbathe, do not smoke, etc.; **(19.3) 1.** Change in bowel or bladder habits; A sore that does not heal; Unusual bleeding or discharge; Thickening or lump in breast or elsewhere; Indigestion or difficulty in swallowing; Obvious change in wart or mole; nagging cough or hoarseness; **2.** Self-examination, Pap test, mammography, CAT scan, MRI, radioactive scan, ultrasound, biopsy; **3.** Tumor marker tests detect cancer when in the very early stages when it is treatable. Genetic tests alert the patient to the predisposition to develop cancer and so indicate earlier and more frequent screening and preventative behaviors. **(19.4) 1.** Surgery (removal of tumor), radiation therapy (localized exposure to ionizing radiation), chemotherapy (whole body treatment with drugs that inhibit DNA synthesis or damage DNA); **2.** To remove an existing tumor, inhibit growth of new tumors, and make sure no other tumor cells have developed outside of tumor origin; **3.** Because it raises the level of p53 in all cells and uses an adenovirus to kill cells that lack a p53 gene; **4.** Antiangiogenic drugs confine and reduce tumors by breaking up the network of new capillaries in the vicinity of a tumor.

Understanding Key Terms

a. telomere; **b.** carcinogen; **c.** proto-oncogene; **d.** angiogenesis; **e.** metastasis

Testing Your Knowledge of the Concepts

9. d; **10.** d; **11.** c; **12.** c; **13.** b; **14.** d; **15.** a; **16.** c; **17.** e; **18.** b; **19.** b; **20.** c; **21.** b; **22.** d; **23.** d; **24.** d; **25.** a; **26.** d; **27.** a; **28.** b;

Thinking Critically About the Concepts

1. Chemotherapy and radiation therapy attack all rapidly reproducing cells, including immune cells. The immune response will be diminished; **2.** These cancers are more likely to spread, because the cells are immature. The greater the degree of cell immaturity, the more likely the cancer is to be malignant; **3.** The renal artery, renal vein, and/or any of the arteries within the kidney; **4.** Low-grade fever, abdominal pain, and "fussiness"—irritability in a baby like Cody; **5.** The tumors are more likely to be large and to have spread throughout the body; **6a.** To prevent the virus from ever infecting the girl's body; **6b.** She would develop artificial active immunity; **6c.** Abstinence, form and remain in a monogamous relationship, and practice safer sex using a condom; **7.** The lymphatic system is often a route used by cancer cells as they spread. Removing lymph nodes will remove cancer cells that may have spread there; **8.** A well-balanced diet, exercise, avoid smoking and excessive alcohol, practice safer sex in a monogamous relationship.

Chapter 20

Case Study

1. One that is caused by two recessive alleles. Autosomal recessive, autosomal dominant, sex-linked; **2.** They can be carriers (heterozygous) for a recessive disorder; **3.** No. You can tell the genotype from the phenotype homozygous recessive, but you cannot tell the genotype from the phenotype dominant (could be heterozygous or homozygous dominant).

Check Your Progress

(20.1) 1. Gene—unit of heredity existing in alleles on the chromosomes; Locus—particular site where a gene is found on a chromosome; Chromosome—chromatin condensed into a compact structure; Dominant—displays phenotype with only one allele copy; Recessive—needs both alleles to express phenotype; heterozygous-having one dominant and one recessive allele; **2.** Genotype refers to the genes of an individual; Phenotype is a characteristic of an individual; **3.** The three genotypes are homozygous dominant, heterozygous (which both display the dominant phenotype) and homozygous recessive which displays the recessive phenotype. **(20.2) 1.** *W; WS, Ws; T, t; Tg, tg; AB, Ab, aB, ab;* **2.** The mother is homozygous recessive and the son is heterozygous; **3.** *EeSs; eeSS;* **4.** EeWw **5.** father—DdFf, mother—ddff, child—ddff; **6.** phenotypic ratio—½ freckles, short fingers: ½ freckles, long fingers; genotypic ratio—1 FFSs:1 FfSs: 1 FFss: 1 Ffss **(20.3) 1a.** *aa;* **1b.** *A?*, the question mark means that each could be *AA* or *Aa;* **2.** 0%; **3.** Due to cultural and religious beliefs governing marriage and procreating with members of the same culture or religion. **(20.4) 1.** Control of trait by several sets of alleles with an additive effect; **2.** Height—genetics controls the potential height, but without proper diet and nutrition the full potential height would not be reached. **3.** Diversity in alleles. **(20.5) 1.** Mother is $X^{+}X$ and father is XY; **2.** Yes, assuming she is heterozygous, she can give her normal X chromosome to her offspring. **3.** Y chromosome has very few genes, mostly ones that determine male gender.

Understanding Key Terms

a. Punnett square; **b.** allele; **c.** dominant allele; **d.** locus; **e.** genotype

Testing Your Knowledge of the Concepts

12. e; **13.** d; **14.** b; **15.** a; **16.** a; **17.** a; **18.** d; **19.** d; **20.** e

Thinking Critically About the Concepts

1. Yes, unless both parents are homozygous recessive, this is because affected children can have unaffected parents and heterozygotes have an unaffected phenotype; **2.** No; **3.** X-linked would occur more often than autosomal due to the fact that males only have one X chromosome and will thus exhibit the disorder every time they inherit the X chromosome containing the defective allele.

Figure Questions

Fig. 20.8: Because she passed on the *a* allele to her first two children. **Fig. 20.9:** Because he passed on the *a* allele to his third child. **Fig. 20.16:** No child could have type O blood.

Chapter 21

Case Study

1. Recombinant DNA technology, PCR; **2.** Gene expression is the production of a protein from DNA which involves the processes of transcription and translation; **3.** Through recombinant DNA technology. Genes are cloned for research, diagnostic testing, and advances in technology.

Check Your Progress

(21.1) 1. Function of DNA is to store genetic information. Function of RNA is to aid DNA in the production of proteins. Structure of both include a sugar phosphate backbone and nitrogen containing bases. Differences include: the sugar in DNA is deoxyribose, the sugar in RNA is ribose, the nitrogen containing bases in DNA are cytosine, thymine, guanine, adenine. The bases in RNA are cytosine, uracil, guanine, adenine; **2.** rRNA composes the subunits of the ribosomes. tRNA transfers the amino acids to the ribosomes during translation. mRNA is the complementary copy of DNA produced by transcription. Small RNAs are involved in several processes including splicing and regulation; **3.** The double stranded structure of DNA allows each original strand to serve as a template for the formation of a complementary strand. **(21.2) 1.** Transcription occurs in the nucleus and results in a mRNA strand complementary to the DNA gene. Translation occurs in the cytoplasm and is how ribosomes, tRNA molecules and the mRNA strand make a protein; **2.** The genetic code is how 64 codons can code for the 20 amino acids which are used to make proteins. The tRNA molecule will complementary base pair its anticodon region to the codon region of the mRNA when bringing in its correct amino acid. The universal characteristic of the genetic code demonstrates an evolutionary relationship history among different organisms; **3.** Pretranscriptional, transcriptional control, posttranscriptional control, translational control, posttranslational control. They are important to homeostasis because they regulate protein production and aid in diversity of proteins produced. **(21.3) 1.** A desired gene is identified and isolated. It is then introduced into a fertilized ovum producing an offspring that synthesizes the desired gene product; **2.** Disease protection, herbicide resistance, salt tolerance, drought tolerance, cold tolerance, improved yield, modified wood pulp, fatty acid/oil content, protein/starch content, amino acid content; **3.** Safety regulation for population health and well being should be instituted and regulation is the responsibility of all citizens. **(21.4) 1.** Researchers are studying the functions of our 20,500 genes and how they form a human being, as well as what goes wrong in certain disorders. Proteomics and bioinformatics are involved; **2.** Through similarities and differences in the proteins used by different organisms even though they share a common genetic code to produce these proteins; **3.** Gene therapy has a longer lasting result than medication but is limited in capability and very expensive.

Understanding Key Terms

a. polyribosome; **b.** transgenic organism; **c.** template; **d.** proteomics; **e.** vector

Testing Your Knowledge of the Concepts

10. e; **11.** c; **12.** e; **13.** d; **14.** d; **15.** c; **16.** e; **17a.** ACU'CCU'GAA'UGC'AAA; **b.** UGA'GGA'CUU'ACG'UUU; **c.** thr-pro-glu-cys-lys

Thinking Critically About the Concepts

1. It would have the same amino acid sequence as the natural human product; **2.** The product could be contaminated by the growing process, in other words, with bacteria or bacterial products if grown in bacteria. Also, the organism it is grown in may modify the product in a manner that the human body may not; **3a.** It would be appropriate to use growth hormone in those who are deficient in the natural production. It would be inappropriate to use growth hormone to help normal people grow taller. **3b.** The addition of growth hormone may allow the patient to grow to normal stature. However, it treats the symptoms and not the cause. The addition of growth hormone does not correct the reason the patient cannot make growth hormone in the first place. If the growth hormone could be administered via gene therapy, then perhaps the patient could be cured; **4.** Recombinant proteins that attack the tumor or enhance the immune system, products that inhibit the growth of the tumor in various ways (such as preventing the tumor from developing a blood supply), vaccines to prevent infection by agents that cause cancer; **5.** Regulations in ethics, health and safety should be considered by all citizens.

Chapter 22

Case Study

1. Neandertals were muscular and had a slightly larger brain than modern humans. They manufactured a variety of stone tools. Cro-Magnons had a modern appearance, made advanced tools, and may have been the first to make knife-like blades and spears; **2.** Analysis of Neandertal DNA suggests that Neandertals are cousins and not ancestors to us; **3.** Out-of-Africa hypothesis proposes that modern humans originated only in Africa then migrated out of Africa and supplanted populations of early *Homo* in Asia and Europe about 100,000 years ago. The multiregional continuity hypothesis proposes that evolution to modern humans was essentially similar in several different places with each region showing continuity of its own anatomical characteristics.

Check Your Progress

(22.1) 1. Both contain gases including water vapor, carbon dioxide, and nitrogen, and a small amount of carbon monoxide and hydrogen. Primitive Earth's atmosphere, however, had little or no oxygen. **2.** Miller's experiment demonstrated that the primitive Earth's atmospheric gases could combine to form organic molecules leading to the formation of the first cell; **3.** RNA is the only macromolecule needed at that time to progress towards formation of the first cell. **(22.2) 1.** Change in a population or species over time. Its two most important aspects are descent from a common ancestor and adaptation to the environment; **2.** Fossil, biogeographical, and anatomical; **3.** Because it accounts for the great diversity of living things. **(22.3) 1.** Mobile forelimbs and hindlimbs, binocular vision, large complex brains, reduced reproductive rate, grasping hands, flattened face; **2.** Binocular vision allows for depth perception and the ability to make accurate judgments about distance and position. Complex brain with many foldings allows for many association areas; **3.** Human spine exits from the skull's center while ape spine exits from rear of skull. Human spine is S-shaped while ape spine has a slight curve. Human pelvis is bowl-shaped while ape pelvis is longer and more narrow. Human femurs angle inward to the knees while ape femurs angle out. Human knee can support more weight than ape knee. Human foot has an arch while ape foot does not. **(22.4) 1.** Hominin includes all members of genus *Homo* and their close relative. Hominids include apes, chimpanzees, humans, and all close extinct relatives; **2.** Bipedal posture, shape of face and size of brain; **3.** Lucy was important to evolution because she gave important information regarding the Australopithecines structure and function. **(22.5) 1.** *Australopithecus afarensis* (Lucy); **2.** *Homo ergaster;* **3.** Larger brain size, smaller teeth, enlarged portions of the brain associated with speech, culture; **4.** Larger brain, flatter face, used fire, fashioned advanced tools; **5.** *H.sapiens* evolved from *H. ergaster* only in Africa and thereafter *H.sapiens* migrated to Europe and Asia. These findings suggest that we all descend from a few individuals and are more genetically similar; **6.** A thoroughly modern appearance, advanced stone tools, possibly first with a language, and culture that included art.

Understanding Key Terms

a. natural selection; **b.** adaptation; **c.** homologous structure; **d.** chemical evolution; **e.** fossil

Testing Your Knowledge of the Concepts

11. c; **12.** b; **13.** a; **14.** e; **15.** e; **16.** e; **17.** a; **18.** b; **19.** c; **20.** b; **21.** d; **22.** c; **23.** b; **24.** e; **25.** b; **26. a.** chordata; **b.** class; **c.** order; **d.** hominidae; **e.** genus; **f.** *Homo sapiens;* **27.** b; **28.** a; **29.** d; **30.** c

Thinking Critically About the Concepts

1. There are many places to dig on the surface of the Earth where people have inhabited, and without some clues on the surface, it would be difficult to determine where to begin. Human habitations usually are built near water sources, near food sources, on high places (for protection), or near shelter (like a cave). Archaeologists look for unusual dirt structures, such as mounds or evidence of digging/holes. Sometimes aerial surveys and infrared photographs provide evidence of unusual structures that are not visible from the ground; **2.** Type of pottery, decorative items, jewelry, burial artifacts, monuments and memorials, seeds, tools; **3.** Brain case size can help determine intellectual capacity, location of the spine's exit from the skull determines whether upright stance, location of the eye sockets determines whether forward vision, size and shape of the jaw determine food source.

Chapter 23

Case Study

1. Abiotic is non-living and biotic is living organisms; **2.** Energy flows through an ecosystem from the sun to the Earth and chemicals are recycled throughout the ecosystem through the biotic and abiotic; **3.** Water cycling is interfered with by withdrawal from aquifers, clearing vegetation from land, and adding of pollutants. Carbon cycling is interfered with by the burning of fossil fuels and destruction of forests.

Check Your Progress

(23.1) 1. Because it is a place where organisms interact among themselves and with the physical and chemical environment; **2.** Autotrophs require only inorganic nutrients and energy to produce food. Heterotrophs need a source of organic nutrients and they must consume food; **3.** Herbivores, carnivores, omnivores, detritus feeders; **4.** Energy flow—occurs because, as nutrients pass from one population to another, all the energy is eventually converted into heat; Chemical cycling—inorganic nutrients are returned to the producers from the atmosphere or soil. **(23.2) 1.** Grazing food web begins with producers such as trees and grass, detrital food web begins with detritus such as fungi and bacteria; **2.** Food chain; **3.** 1.5g. **(23.3) 1.** Reservoir, exchange pool, biotic community; **2.** Carbon, water, and nitrogen are examples of gaseous biogeochemical cycles and phosphorous is an example of a sedimentary biogeochemical cycle; **3.** Convert nitrogen gas to ammonium; **4.** Ground water shortage, global warming, acid deposition, cultural eutrophication; **5.** Increased climate warming, sea levels increase, glaciers melt, increased rainfall along coastlines, dryer inland conditions.

Understanding Key Terms

a. omnivore; **b.** trophic level; **c.** fossil fuel; **d.** nitrogen fixation; **e.** producer

Testing Your Knowledge of the Concepts

10. d; **11.** a; **12.** b; **13.** c; **14.** c; **15.** b; **16.** c; **17.** c; **18.** b; **19.** e; **20.** e; **21.** c; **22.** b; **23.** e; **24.** f; **25.** b, c, d; **26.** a, b, c; **27.** d; **28. a.** producers; **b.** consumers; **c.** inorganic nutrient pool; **d.** decomposers

Thinking Critically About the Concepts

1a. Fungicides will kill the fungi and pesticides will kill the insects that are part of the detritus food webs; **1b.** Fungicides kill fungi that decompose dead/discarded tissues and pesticides kill many insects that are detritivores that recycle dead/discarded tissues. The nutrients in the dead/discarded tissues will not be released without the activities of the fungi and insects causing adverse effects to the cycles; **2.** Legumes like clover and soybeans have nitrogen-fixing bacteria living in their roots. These bacteria supply the plant with a form of nitrogen the plant can use. Nitrogen that is not needed by the plants (extra) is added to the soil around the plant, which enriches the soil for other plants. Fertilizer also adds nitrogen to the soil, but it may run off during heavy rainfalls, which contributes to overgrowth in aquatic habitats; **3a.** Answers will vary, but may include recycling newspapers, cans, and so forth or starting a compost pile for food scraps. The compost can be used to enrich the soil used for gardens; **3b.** Answers will vary, but may include turning off the water while brushing your teeth or running the dishwasher/washing machine with full loads only. Others might include minimizing the use of fertilizer and pesticides that can run off into bodies of water; **4.** Organochlorides accumulate in organisms higher up on a food chain, discontinuing the use of pesticides that contain organochlorides would improve the health of all organisms in food chains.

Chapter 24

Case Study

1. The availability of natural resources are declining; **2.** Renewable; **3.** Decrease pollution, overexploitation, habitat loss, and increase biodiversity.

Check Your Progress

(24.1) 1. 1.07%; **2.** MDC—Increased medicine, public health, socioeconomic conditions. USA. LDC—Increased growth, birth rates, deaths, urban populations, decreased availability of water and decreased biodiversity. Africa; **3.** Compare the number of young women entering the reproductive years to older women leaving them. If the number of young women is greater, the population will grow. If the number of older women is greater, the population will decline. **(24.2) 1.** Land, water, food, energy, and minerals; **2.** Groundwater—damming rivers and withdrawing water from aquifers; Land subsidence and saltwater intrusion; Food resources—planting a few genetic varieties, heavy use of fertilizers, pesticides and herbicides, generous irrigation, and excessive fuel consumption. Soil loss, agricultural runoff, energy consumption, susceptibility to pathogens. **3.** Renewable means more can be produced over time, nonrenewable resources cannot be replaced once used; **4.** They contribute to environmental problems because of the pollutants released when they're burned; **5.** They damage the Earth's ozone shield, pose a threat to the health of living things, and raw sewage causes oxygen depletion in lakes and rivers. **(24.3) 1.** Habitat loss, alien species, pollution, overexploitation, and disease; **2.** Direct value—medicinal value, agricultural value, consumptive use value; Indirect value—waste disposal, provision of fresh water, prevention of soil erosion, biogeochemical cycles, regulation of climate, ecotourism; **3.** Biodiversity is important to maintain a large variety of alleles present in a gene pool to maintain populations. **(24.4) 1.** Large portion of land used for human purposes, agriculture uses large amounts of nonrenewable energy and creates pollution, more freshwater is used in agriculture than used in homes, almost half of the agriculture yield goes toward feeding animals, decrease in surface water, use of nonrenewable fossil energy, expansion of the population into all regions of the planet; **2.** To make rural areas sustainable: plant cover crops, plant multiuse crops, use low flow irrigation, use precision farming to reduce habitat destruction, use integrated pest management, plant multipurpose trees, restore wetlands, use renewable forms of energy. To make urban areas sustainable: use energy-efficient modes of transport, use solar or geothermal energy to heat buildings, use green roofs, improve storm-water management, plant native grasses for lawns, create greenbelts, revitalize old sections of cities, use more efficient light fixtures, recycle business equipment, use low-maintenance building materials. **3.** Criteria include use value, option value, existence value, aesthetic value, cultural value, scientific and educational value.

Understanding Key Terms

a. sustainable; **b.** carrying capacity; **c.** greenhouse gases; **d.** biological magnification; **e.** aquifer

Testing Your Knowledge of the Concepts

9. a; **10.** c; **11.** a; **12.** d; **13.** d; **14.** a; **15.** c; **16.** c; **17.** b; **18.** d; **19.** d; **20.** a; **21.** c, d, e, f; **22.** a, b; **23.** a, b, c; **24.** c, d, e, f; **25.** b; **26.** a; **27.** c; **28.** c; **29.** e; **30.** e; **31.** b; **32.** e; **33. a.** habitat loss; **b.** alien species; **c.** pollution; **d.** overexploitation; **e.** disease

Thinking Critically About the Concepts

1. Less energy is needed to grow plant material than animals for human consumption; **2a.** Answers will vary; **2b.** Answers will vary. Example: drive less or drive an energy efficient vehicle, use alternative means of heating/cooling a home, purchase locally grown food; **3.** Most items in a landfill are not exposed to decomposers, so the nutrients do not cycle back; **4a.** Answers will vary; **4b.** Answers will vary. Examples: recycling programs, ride sharing/bus use, water/electricity use, events associated with Earth Day.

Glossary

A

absorption Taking in of substances by cells or membranes. 171

acetylcholine (ACh) (uh-seet-ul-koh-leen) Neurotransmitter active in both the peripheral and central nervous systems. 291

acetylcholinesterase (AChE) (uh-seet-ul-koh-luh-nes-tuh-rays) Enzyme that breaks down acetylcholine bound to postsynaptic receptors within a synapse. 291

acid Molecules that raise the hydrogen ion concentration in a solution and lower its pH numerically. 26

acid deposition The return to Earth in rain or snow of sulfate or nitrate salts of acids produced by commercial and industrial activities. 559

acidosis Excessive accumulation of acids in body fluids. 232

acquired immunodeficiency syndrome (AIDS) (im-yuh-noh-dih-fish-un-see) Disease caused by HIV and transmitted via body fluids; characterized by failure of the immune system. 156

acromegaly (ak-roh-meg-uh-lee) Condition resulting from an increase in growth hormone production after adult height has been achieved. 347

acrosome (ak-ruh-sohm) Cap at the anterior end of a sperm that partially covers the nucleus and contains enzymes that help the sperm penetrate the egg. 369

actin (ak-tin) One of two major proteins of muscle; makes up thin filaments in myofibrils of muscle fibers. See myosin. 267

actin filament Cytoskeletal filaments of eukaryotic cells composed of the protein actin; also refers to the thin filaments of muscle cells. 55

action potential Electrochemical changes that take place across the axomembrane; the nerve impulse. 289

active immunity Resistance to disease due to the immune system's response to a microorganism or a vaccine. 147

active site Region on the surface of an enzyme where the substrate binds and where the reaction occurs. 57

active transport Use of a plasma membrane carrier protein and energy to move a substance into or out of a cell from lower to higher concentration. 51

acute bronchitis (brahn-ky-tis) Infection of the primary and secondary bronchi. 210

adaptation Organism's modification in structure, function, or behavior suitable to the environment. 6, 520

Addison disease Condition resulting from a deficiency of adrenal cortex hormones; characterized by low blood glucose, weight loss, and weakness. 353

adenine (A) (ad-uh-neen) One of four nitrogen bases in nucleotides composing the structure of DNA and RNA. 38

adipose tissue (ah-duh-pohs) Connective tissue in which fat is stored. 67

ADP (adenosine diphosphate) (ah-den-ah-seen dy-fahs-fayt) Nucleotide with two phosphate groups that can accept another phosphate group and become ATP. 38

adrenal cortex (uh-dree-nul kor-teks) Outer portion of the adrenal gland; secretes mineralocorticoids, such as aldosterone, and glucocorticoids, such as cortisol. 351

adrenal gland (uh-dree-nul) An endocrine gland that lies atop a kidney; consisting of the inner adrenal medulla and the outer adrenal cortex. 351

adrenal medulla (uh-dree-nul muh-dul-uh) Inner portion of the adrenal gland; secretes the hormones epinephrine and norepinephrine. 351

adrenocorticotropic hormone (ACTH) (uh-dree-noh-kawrt-ih-koh-troh-pik) Hormone secreted by the anterior lobe of the pituitary gland that stimulates activity in the adrenal cortex. 346

aerobic Requiring oxygen. 59

afterbirth Placenta and the extraembryonic membranes, which are delivered (expelled) during the third stage of parturition. 410

agglutination (uh-gloot-un-ay-shun) Clumping of red blood cells due to a reaction between antigens on red blood cell plasma membranes and antibodies in the plasma. 125

aging Progressive changes over time, leading to loss of physiological function and eventual death. 410

agranular leukocyte White blood cell that does not contain distinctive granules. 122

agricultural runoff Water from precipitation and irrigation that flows over fields into bodies of water or aquifers. 574

albumin (al-byoo-mun) Plasma protein of the blood having transport and osmotic functions. 118

aldosterone (al-dahs-tuh-rohn) Hormone secreted by the adrenal cortex that decreases sodium and increases potassium excretion; raises blood volume and pressure. 230

alien species Nonnative species that migrate or are introduced by humans into a new ecosystem; also called exotics. 580

alkalosis Excessive accumulation of bases in body fluids. 232

allantois (uh-lan-toh-is) Extraembryonic membrane that contributes to the formation of umbilical blood vessels in humans. 397

allele (uh-leel) Alternative form of a gene; alleles occur at the same locus on homologous chromosomes. 468

allergen (al-ur-jun) Foreign substance capable of stimulating an allergic response. 150

allergy Immune response to substances that usually are not recognized as foreign. 150

all-or-none law Law that states that muscle fibers contract either maximally or not at all and that neurons conduct a nerve impulse either completely or not at all. 271

alveolus (pl., alveoli) (al-vee-uh-lus) Air sac of a lung. 200

Alzheimer disease (AD) Brain disorder characterized by a general loss of mental abilities. 300, 306

amino acid Organic molecule having an amino group and an acid group, which covalently bonds to produce peptide molecules. 35

amnion (am-nee-ahn) Extraembryonic membrane that forms an enclosing, fluid-filled sac. 397

ampulla (am-pool-uh, -pul-uh) Base of a semicircular canal in the inner ear. 333

amygdala (uh-mig-duh-luh) Portion of the limbic system that functions to add emotional overtones to memories. 300

amyotrophic lateral sclerosis (ALS) Chronic, progressive motor neuron disease characterized by the gradual degeneration of the nerve cells resulting in death. 277

anabolic steroid (a-nuh-bahl-ik) Synthetic steroid that mimics the effect of testosterone. 358

anaerobic Growing or metabolizing in the absence of oxygen. 58

analogous structure Structure that has a similar function in separate lineages but differs in anatomy and ancestry. 523

anaphase Mitotic phase during which daughter chromosomes move toward the poles of the spindle. 426

anaphylactic shock Severe systemic form of allergic reaction involving bronchiolar constriction, impaired breathing, vasodilation, and a rapid drop in blood pressure with a threat of circulatory failure. 150

androgen (an-druh-jun) Male sex hormone (e.g., testosterone). 358

anemia (uh-nee-mee-uh) Inefficiency in the oxygen-carrying ability of blood due to a shortage of hemoglobin. 120

aneurysm Saclike expansion of a blood vessel wall. 105

angina pectoris (an-jy-nuh pek-tuh-ris) Condition characterized by thoracic pain resulting from occluded coronary arteries; precedes a heart attack. 106

angiogenesis (an-jee-oh-jen-uh-sis) Formation of new blood vessels; one mechanism by which cancer spreads. 450

angioplasty (an-jee-uh-plas-tee) Surgical procedure for treating clogged arteries in which a plastic tube is threaded through a major blood vessel toward the heart and then a balloon at the end of the tube is inflated, forcing open the vessel. A stent is then placed in the vessel. 108

anorexia nervosa (a-nuh-rek-see-uh nur-voh-suh) Eating disorder characterized by a morbid fear of gaining weight. 191

anterior pituitary (pih-too-ih-tair-ee) Portion of the pituitary gland controlled by the hypothalamus and that produces seven types of hormones, some of which control other endocrine glands. 345

anthropoid Group of primates that includes monkeys, apes, and humans. 527

antibiotic resistance A characteristic of pathogens that causes them to survive treatment with chemicals (antibiotics) that normally would kill them. 386

antibody (an-tih-bahd-ee) Protein produced in response to the presence of an antigen; each antibody combines with a specific antigen. 116

antibody-mediated immunity Specific mechanism of defense in which plasma cells derived from B cells produce antibodies that combine with antigens. 144

antibody titer Amount of antibody present in a sample of blood serum. 148

anticodon (an-tih-koh-dahn) Three-base sequence in a tRNA molecule base that pairs with a complementary codon in mRNA. 499

antidiuretic hormone (ADH) (an-tih-dy-uh-ret-ik) Hormone secreted by the posterior pituitary that increases the permeability of the collecting ducts in a kidney. 230, 345

antigen (an-tih-jun) Foreign substance, usually a protein or a polysaccharide, that stimulates the immune system to produce antibodies. 121

antigen-presenting cell (APC) Cell that displays the antigen to the cells of the immune system so they can defend the body against that particular antigen. 145

anus Outlet of the digestive tract. 180

aorta (ay-or-tuh) Major systemic artery that receives blood from the left ventricle. 103

apoptosis (ap-uh-toh-sis) Programmed cell death involving a cascade of specific cellular events leading to death and destruction of the cell. 144, 422, 448

appendicular skeleton (ap-un-dik-yuh-lur) Portion of the skeleton forming the pectoral girdles and upper extremities and the pelvic girdle and lower extremities. 253

appendix In humans, small, tubular appendage that extends outward from the cecum of the large intestine. 171

aquaporin Protein membrane channel through which water can diffuse. 228

aqueous humor (ay-kwee-us, ak-wee-) Clear, watery fluid between the cornea and lens of the eye. 322

aquifer (ahk-wuh-fur) Rock layers that contain water released in appreciable quantities to wells or springs. 554, 573

arteriole (ar-teer-ee-ohl) Vessel that takes blood from an artery to capillaries. 93

arteriovenous shunt A pathway, usually abnormal, that connects an artery directly to a vein. 94

articular cartilage (ar-tik-yuh-lur) Hyaline cartilaginous covering over the articulating surface of the bones of synovial joints. 248

artificial selection The intentional breeding of organisms for the selection of certain traits. 526

assimilation Action of chemically changing absorbed substances. 559

association area One of several regions of the cerebral cortex related to memory, reasoning, judgment, and emotional feelings. 297

aster Short, radiating fibers about the centrioles at the poles of a spindle. 425

asthma (az-muh, as-muh) Condition in which bronchioles constrict and cause difficulty in breathing. 211

astigmatism (uh-stig-muh-tiz-um) Blurred vision due to an irregular curvature of the cornea or the lens. 328

atherosclerosis (ath-uh-roh-skluh-roh-sis) Condition in which fatty substances accumulate abnormally beneath the inner linings of the arteries. 106

atom Smallest particle of an element that displays the properties of the element. 2, 20

atomic mass Mass of an atom equal to the number of protons plus the number of neutrons with the nucleus. 20

atomic number Number of protons within the nucleus of an atom. 38

ATP (adenosine triphosphate) (uh-den-uh-seen try-fahs-fayt) Nucleotide with three phosphate groups. The breakdown of ATP into ADP + Ⓟ makes energy available for energy-requiring processes in cells. 38

atrial natriuretic hormone (ANH) (ay-tree-ul nay-tree-yoo-ret-ik) Hormone secreted by the heart that increases sodium excretion and, therefore, lowers blood volume and pressure. 230, 353

atrioventricular (AV) bundle (ay-tree-oh-ven-trik-yuh-lur) Group of specialized fibers that conduct impulses from the atrioventricular node to the ventricles of the heart; also called AV bundle. 98

atrioventricular (AV) valve Valve located between the atrium and the ventricle. 94

atrium (ay-tree-um) One of the upper chambers of the heart, either the left atrium or the right atrium, that receives blood. 94

auditory canal Curved tube extending from the pinna to the tympanic membrane. 328

auditory (eustachian) tube Extension from the middle ear to the nasopharynx that equalizes air pressure on the eardrum. 198, 329

australopithecine (aw-stray-loh-pith-uh-syn) Any of the first evolved hominids; classified into several species of *Australopithecus*. 532

autoimmune disease Disease that results when the immune system mistakenly attacks the body's tissues. 151

autonomic system (aw-tuh-nahm-ik) Branch of the peripheral nervous system that has control over the internal organs; consists of the sympathetic and parasympathetic systems. 304

autosome (aw-tuh-sohm) Any chromosome other than the sex chromosomes. 484

autotroph Organism that can capture energy and synthesize organic nutrients from inorganic nutrients. 548

AV (atrioventricular) node Small region of neuromuscular tissue that transmits impulses received from the sinoatrial node to the ventricles. 98

axial skeleton (ak-see-ul) Portion of the skeleton that supports and protects the organs of the head, the neck, and the trunk. 248

axon (ak-sahn) Elongated portion of a neuron that conducts nerve impulses typically from the cell body to the synapse. 287

axon terminal Small swelling at the tip of one of many endings of the axon. 290

B

bacillus A rod-shaped bacterial cell. 134

bacteria Members of one of three domains of life; prokaryotic cells other than archaea with unique genetic, biochemical, and physiological characteristics. 134

Barr body Dark-staining body (discovered by M. Barr) in the nuclei of female mammals that contains a condensed, inactive X chromosome. 437

basal nuclei Nerve cells that integrate motor commands to ensure balance and coordination. 298

base Molecules that lower the hydrogen ion concentration in a solution and raise the pH numerically. 26

basement membrane Layer of nonliving material that anchors epithelial tissue to underlying connective tissue. 72

basophil (bay-suh-fil) White blood cell with granular cytoplasm; able to be stained with a basic dye. 122

B cell (B lymphocyte) Lymphocyte that matures in the bone marrow and, when stimulated by the presence of a specific antigen, gives rise to antibody-producing plasma cells. 139

B-cell receptor (BCR) Molecule on the surface of a B lymphocyte to which an antigen binds. 143

bicarbonate ion Ion that participates in buffering the blood; the form in which carbon dioxide is transported in the bloodstream. 206

bile Secretion of the liver temporarily stored and concentrated in the gallbladder before being released into the small intestine, where it emulsifies fat. 176, 178

bilirubin The yellow-orange bile pigment produced from the breakdown of hemoglobin. 180

binge-eating disorder Condition characterized by overeating episodes that are not followed by purging. 192

binomial name Two-part scientific name of an organism. The first part designates the genus, the second part the specific epithet. 526

biocultural evolution Passage of culture from one generation to the next. 538

biodiversity Total number of species, the variability of their genes, and the communities in which they live. 8, 579

biogeochemical cycle (by-oh-jee-oh-kem-ih-kul) Circulating pathway of elements such as carbon and nitrogen, involving exchange pools, storage areas, and biotic communities. 533

biogeography Study of the geographic distribution of organisms. 522

bioinformatics Computer technologies used to study the genome. 512

biological evolution Change in life-forms that has taken place in the past and will take place in the future; includes descent from a common ancestor and adaptation to the environment. 520

biological magnification Process by which substances become more concentrated in organisms in the higher trophic levels of a food web. 563, 579

biology Scientific study of life. 2

biomass The number of organisms multiplied by their weight. 552

biome A major regional community of organisms, characterized primarily by climate and the types of plant species that are present. 547

biosphere (by-oh-sfeer) Zone of air, land, and water at the surface of the Earth in which living organisms are found. 2

biotechnology product Product created by using biotechnology techniques. 506

biotic potential Maximum reproductive rate of an organism, given unlimited resources and ideal environmental conditions. 568

bipedal posture Ability to walk upright on two feet. 530

birth control method Prevents either fertilization or implantation of an embryo in the uterine lining. 379

birth control pill Oral contraceptive containing estrogen and progesterone. 379

blastocyst (blas-tuh-sist) Early stage of human embryonic development that consists of a hollow, fluid-filled ball of cells. 397

blind spot Region of the retina lacking rods or cones where the optic nerve leaves the eye. 325

blood Type of connective tissue in which cells are separated by a liquid called plasma. 68

blood doping Practice of boosting the number of red blood cells in the blood to enhance athletic performance. 120

blood pressure Force of blood pushing against the inside wall of a vessel. 100

blood transfusion Introduction of whole blood or a blood component directly into the bloodstream. 125

body mass index (BMI) Calculation used to determine whether a person is overweight or obese. 183

bolus Small lump of food that has been chewed and swallowed. 173

bone marrow transplant A cancer patient's stem cells are harvested and stored before chemotherapy begins. Then, the stored cells are returned to the patient by injection. 461

bone remodeling Ongoing mineral deposits and withdrawals from bone that adjust bone strength and maintain levels of calcium and phosphorus in blood. 245

brain Enlarged superior portion of the central nervous system located in the cranial cavity of the skull. 295

brain stem Portion of the brain consisting of the medulla oblongata, pons, and midbrain. 298

Braxton Hicks contraction Strong, late-term uterine contractions prior to cervical dilation; also called false labor. 408

breech birth Birth in which the baby is positioned rump first. 405

Broca's area Region of the frontal lobe that coordinates complex muscular actions of the mouth, tongue, and larynx, making speech possible. 297

bronchiole (brahng-kee-ohl) Smaller air passages in the lungs that begin at the bronchi and terminate in alveoli. 200

bronchus (pl., bronchi) (brahng-kus) One of two major divisions of the trachea leading to the lungs. 200

buffer Substance or group of substances that tend to resist pH changes of a solution, thus stabilizing its relative acidity and basicity. 27, 232

bulbourethral gland (bul-boh-yoo-ree-thrul) Either of two small structures located below the prostate gland in males; each adds secretions to semen. 368

bulimia nervosa (byoo-lee-mee-uh, -lim-ee-, nur-voh-suh) Eating disorder characterized by binge eating followed by purging via self-induced vomiting or use of a laxative. 191

bursa (bur-suh) Saclike, fluid-filled structure, lined with synovial membrane, that occurs near a joint. 264

bursitis (bur-sy-tis) Inflammation of any of the friction-easing sacs called bursae within the knee joint. 277

C

calcitonin (kal-sih-toh-nin) Hormone secreted by the thyroid gland that increases the blood calcium level. 350

calorie Amount of heat energy required to raise the temperature of 1 g of water 1°C. 25

cancer Malignant tumor whose nondifferentiated cells exhibit loss of contact inhibition, uncontrolled growth, and the ability to invade tissue and metastasize. 448

capsule Gelatinous layer surrounding the cells of blue-green algae and certain bacteria. 134

carbaminohemoglobin Hemoglobin carrying carbon dioxide. 248

carbohydrate Class of organic compounds that includes monosaccharides, disaccharides, and polysaccharides. 29

carbonic anhydrase (kar-bahn-ik an-hy-drays, -drayz) Enzyme in red blood cells that speeds the formation of carbonic acid from the reactants water and carbon dioxide. 206

carcinogen (kar-sin-uh-jun) Environmental agent that causes mutations leading to the development of cancer. 453

carcinogenesis (kar-suh-nuh-jen-uh-sis) Development of cancer. 449

carcinoma (kar-suh-noh-muh) Cancer arising in epithelial tissue. 451

cardiac cycle One complete cycle of systole and diastole for all heart chambers. 97, 263

cardiac muscle Striated, involuntary muscle found only in the heart. 69

cardiovascular system (kar-dee-oh-vas-kyuh-lur) Organ system in which blood vessels distribute blood powered by the pumping action of the heart. 79

carnivore (kar-nuh-vor) Consumer in a food chain that eats other animals. 548

carrier Heterozygous individual who has no apparent abnormality but can pass on an allele for a recessively inherited genetic disorder. 476

carrying capacity Maximum number of individuals of any species that can be supported by a particular ecosystem on a long-term basis. 568

cartilage (kar-tul-ij, kart-lij) Connective tissue in which the cells lie within lacunae separated by a flexible proteinaceous matrix. 241, 242

cecum (see-kum) Small pouch that lies below the entrance of the small intestine and is the blind end of the large intestine. 180

cell Smallest unit that displays the properties of life; always contains cytoplasm surrounded by a plasma membrane. 2

cell body Portion of a neuron that contains a nucleus and from which dendrites and an axon extend. 287

cell cycle Repeating sequence of cellular events that consists of interphase, mitosis, and cytokinesis. 422

cell-mediated immunity Specific mechanism of defense in which T cells destroy antigen-bearing cells. 146

cell theory One of the major theories of biology; states that all organisms are made up of cells and cells come only from pre-existing cells. 44, 518

cellular respiration Metabolic reactions that use the energy primarily from carbohydrates but also from fatty acid or amino acid breakdown to produce ATP molecules. 57

cellulose (sel-yuh-lohs, -lohz) Polysaccharide that is the major complex carbohydrate in plant cell walls. 29

central nervous system (CNS) Portion of the nervous system consisting of the brain and spinal cord. 286

centriole (sen-tree-ohl) Cellular structure, existing in pairs, that possibly organizes the mitotic spindle for chromosomal movement during mitosis and meiosis. 425

centromere (sen-truh-meer) Constriction where sister chromatids of a chromosome are held together. 421

centrosome Central microtubule organizing center of cells. In animal cells, it contains two centrioles. 55, 425

cerebellum (ser-uh-bel-um) Part of the brain located posterior to the medulla oblongata and pons that coordinates skeletal muscles to produce smooth, graceful motions. 298

cerebral cortex (suh-ree-brul, ser-uh-brul kor-teks) Outer layer of cerebral hemispheres; receives sensory information and controls motor activities. 296

cerebral hemisphere One of the large, paired structures that together constitute the cerebrum of the brain. 295

cerebrospinal fluid (sair-uh-broh-spy-nul, suh-ree-broh-) Fluid found in the ventricles of the brain, in the central canal of the spinal cord, and in association with the meninges. 293

cerebrum (sair-uh-brum, suh-ree-brum) Main part of the brain consisting of two large masses, or cerebral hemispheres; the largest part of the brain in mammals. 295

cervix (sur-viks) Narrow end of the uterus, which projects into the vagina. 370

cesarean section Birth by surgical incision of the abdomen and uterus. 405

chancre Sore that appears on the skin; first sign of syphilis. 387

checkpoint A regulatory location in the cell cycle at which the cell assesses whether to proceed with cell division. 422, 450

chemical evolution Increase in the complexity of chemicals over time that could have led to the first cells. 518

chemical signal Molecule that brings about a change in a cell, tissue, organ, or individual when it binds to a specific receptor. 342

chemoreceptor (kee-moh-rih-sep-tur) Sensory receptor sensitive to chemical stimuli—for example, receptors for taste and smell. 205, 316

chlamydia (kluh-mid-ee-uh) Sexually transmitted disease caused by the bacterium *Chlamydia trachomatis;* can lead to pelvic inflammatory disease. 386

chlorofluorocarbons (CFCs) (klor-oh-floor-oh-kar-buns) Organic compounds containing carbon, chlorine, and fluorine atoms. CFCs, such as Freon, can deplete the ozone shield by releasing chlorine atoms in the upper atmosphere. 562, 579

cholesterol Form of lipid; structural component of plasma membrane and precursor for steroid hormones. 178

chondrocyte Type of cell found in the lacunae of cartilage. 242

chordae tendineae (kor-dee ten-din-ee-ee) Tough bands of connective tissue that attach the papillary muscles to the atrioventricular valves within the heart. 94

chorion (kor-ee-ahn) Extraembryonic membrane that contributes to placenta formation. 397

chorionic villi (kor-ee-ahn-ik vil-eye) Treelike extensions of the chorion that project into the maternal tissues at the placenta. 400

choroid (kor-oyd) Vascular, pigmented middle layer of the eyeball. 322

chromatin (kroh-muh-tin) Network of fine threads in the nucleus composed of DNA and proteins. 53

chromosome (kroh-muh-som) Chromatin condensed into a compact structure. 53

chronic bronchitis Obstructive pulmonary disorder that tends to recur; marked by inflamed airways filled with mucus and degenerative changes in the bronchi, including loss of cilia. 211

chyme (kym) Thick, semiliquid food material that passes from the stomach to the small intestine. 176

ciliary body (sil-ee-air-ee) Structure associated with the choroid layer that contains ciliary muscle and controls the shape of the lens of the eye. 322

cilium (pl., cilia) (sil-ee-um) Short, hairlike projection from the plasma membrane, occurring usually in large numbers. 55

circadian rhythm (sur-kay-dee-un) Biological rhythm with a 24-hour cycle. 359

circumcision Removal of the prepuce (foreskin) of the penis. 368

cirrhosis (sih-roh-sis) Chronic, irreversible injury to liver tissue; commonly caused by frequent alcohol consumption. 179

citric acid cycle Cycle of reactions in mitochondria that begins with citric acid; it breaks down an acetyl group as CO_2, ATP, NADH, and $FADH_2$ are given off; also called the Krebs cycle. 58

cleavage Cell division without cytoplasmic addition or enlargement; occurs during the first stage of animal development. 396

cleavage furrow Indentation that begins the process of cleavage, by which human cells undergo cytokinesis. 427

clonal selection model Concept that an antigen selects which lymphocyte will undergo clonal expansion and produce more lymphocytes bearing the same type of antigen receptor. 143

cloning Production of identical copies; can be either the production of identical individuals or, in genetic engineering, the production of identical copies of a gene. 506

clotting Process of blood coagulation, usually when injury occurs. 123

coccus Spherical bacterial cell. 134

cochlea (kohk-lee-uh, koh-klee-uh) Portion of the inner ear that resembles a snail's shell and contains the spiral organ, the sense organ for hearing. 328

cochlear nerve Either of two cranial nerves that carry nerve impulses from the spiral organ to the brain; also called the auditory nerve. 330

codominance Inheritance pattern in which both alleles of a gene are equally expressed. 481

codon Three-base sequence in mRNA that causes the insertion of a particular amino acid into a protein or termination of translation. 497

coenzyme (koh-en-zym) Nonprotein organic molecule that aids the action of the enzyme to which it is loosely bound. 58

collagen fiber (kahl-uh-jun) White fiber in the matrix of connective tissue; gives flexibility and strength. 66

collecting duct Duct within the kidney that receives fluid from several nephrons; the reabsorption of water occurs here. 224

colon (koh-lun) The major portion of the large intestine, consisting of the ascending colon, the transverse colon, and the descending colon. 180

colony-stimulating factor (CSF) Protein that stimulates differentiation and maturation of white blood cells. 121

color blindness Deficiency in one or more of the three types of cone cells responsible for color vision. 485

color vision Ability to detect the color of an object; dependent on three types of cone cells. 324

columnar epithelium (kuh-lum-nur ep-uh-thee-lee-um) Type of epithelial tissue with cylindrical cells. 74

community Assemblage of populations interacting with one another within the same environment. 2

compact bone Type of bone that contains osteons consisting of concentric layers of matrix and osteocytes in lacunae. 68, 240

complementary DNA (cDNA) DNA that has been synthesized from mRNA by the action of reverse transcriptase. 506

complementary paired bases Hydrogen bonding between particular bases; in DNA, thymine (T) pairs with adenine (A), and guanine (G) pairs with cytosine (C); in RNA, uracil (U) pairs with A, and G pairs with C. 38, 492

complement system Series of proteins in plasma that form a nonspecific defense mechanism against a microbe invasion; it complements the antigen–antibody reaction. 142

compound Substance having two or more different elements united chemically in a fixed ratio. 22

conclusion Statement made following an experiment as to whether the results support the hypothesis. 10

cone cell Photoreceptor in retina of eye that responds to bright light; detects color and provides visual acuity. 324

congenital hypothyroidism Condition resulting from improper development of the thyroid in an infant; characterized by stunted growth and mental retardation. 349

connective tissue Type of tissue that binds structures together, provides support and protection, fills spaces, stores fat, and forms blood cells; adipose tissue, cartilage, bone, and blood are types of connective tissue. 66

constipation (kahn-stuh-pay-shun) Delayed and difficult defecation caused by insufficient water in the feces. 181

consumer Organism that feeds on another organism in a food chain; primary consumers eat plants, and secondary consumers eat animals. 548

contraceptive (kahn-truh-sep-tiv) Medication or device used to reduce the chance of pregnancy. 379

contraceptive implant Birth control method using synthetic progesterone; prevents ovulation by disrupting the ovarian cycle. 379

contraceptive injection Birth control method using progesterone or estrogen and progesterone together; prevents ovulation by disrupting the ovarian cycle. 379

contraceptive vaccine Under development, this birth control method immunizes against the hormone HCG, crucial to maintaining implantation of the embryo. 379

control group Sample that goes through all the steps of an experiment but lacks the factor or is not exposed to the factor being tested; a standard against which results of an experiment are checked. 11

convulsion Sudden attack characterized by a loss of consciousness and severe, sustained, rhythmic contractions of some or all voluntary muscles. 277

cornea (kor-nee-uh) Transparent, anterior portion of the outer layer of the eyeball. 322

coronary artery (kor-uh-nair-ee) Artery that supplies blood to the wall of the heart. 95

coronary bypass operation Therapy for blocked coronary arteries in which part of a blood vessel from another part of the body is grafted around the obstructed artery. 108

corpus callosum Bridge of nerve tracts that connects the two cerebral hemispheres. 295

corpus luteum (kor-pus loot-ee-um) Yellow body that forms in the ovary from a follicle that has discharged its secondary oocyte; it secretes progesterone and some estrogen. 375

cortisol (kor-tuh-sawl) Glucocorticoid secreted by the adrenal cortex that responds to stress on a long-term basis; reduces inflammation and promotes protein and fat metabolism. 352

covalent bond (coh-vay-lent) Chemical bond in which atoms share one pair of electrons. 22

cramp Muscle contraction that causes pain. 277

cranial nerve Nerve that arises from the brain. 302

creatinine (kree-ah-tuhn-een) Nitrogenous waste; the end product of creatine phosphate metabolism. 220

Cro-Magnon (kroh-mag-nun) Common name for first fossils to be designated *Homo sapiens*. 537

crossing-over Exchange of segments between nonsister chromatids of a tetrad during meiosis. 429

cuboidal epithelium (kyoo-boyd-ul) Type of epithelial tissue with cube-shaped cells. 74

cultural eutrophication Enrichment of water by inorganic nutrients used by phytoplankton. Often, overenrichment caused by human activities leads to excessive bacterial growth and oxygen depletion. 561

culture Total pattern of human behavior; includes technology and the arts, and depends upon the capacity to speak and transmit knowledge. 7, 535

Cushing syndrome (koosh-ing) Condition resulting from hypersecretion of glucocorticoids; characterized by thin arms and legs and a "moon face," and accompanied by high blood glucose and sodium levels. 353

cutaneous receptor Sensory receptors for pressure and touch found in the dermis of the skin. 319

cyclic adenosine monophosphate (cAMP) (sy-klik, sih-klik) ATP-related compound that acts as the second messenger in peptide hormone transduction; it initiates activity of the metabolic machinery. 343

cyclin Protein that regularly increases and decreases in concentration during the cell cycle. 450

cystic fibrosis (CF) A generalized, autosomal recessive disorder of infants and children in which there is widespread dysfunction of the exocrine glands. 477

cystitis Inflammation of the urinary bladder. 233

cytokine (sy-tuh-kyn) Type of protein secreted by a T cell that stimulates cells of the immune system to perform their various functions. 141

cytokinesis (sy-tuh-kyn-ee-sus) Division of the cytoplasm following mitosis and meiosis. 422

cytoplasm (sy-tuh-plaz-um) Contents of a cell between the nucleus and the plasma membrane that contains the organelles. 46

cytosine (C) (sy-tuh-seen) One of four nitrogen bases in nucleotides composing the structure of DNA and RNA. 38

cytoskeleton Internal framework of the cell, consisting of microtubules, actin filaments, and intermediate filaments. 55

cytotoxic T cell (sy-tuh-tahk-sik) T cell that attacks and kills antigen-bearing cells. 145

D

data Facts or pieces of information collected through observation and/or experimentation. 9

daughter cell Cell that arises from a parent cell by mitosis or meiosis. 424

dead air space Volume of inspired air that cannot be exchanged with blood. 204

defecation (def-ih-kay-shun) Discharge of feces from the rectum through the anus. 18

deforestation (dee-for-eh-stay-shun) Removal of trees from a forest in a way that continuously reduces the size of the forest. 571

dehydration reaction Chemical reaction resulting in a covalent bond with the accompanying loss of a water molecule. 28

delayed allergic response Allergic response initiated at the site of the allergen by sensitized T cells, involving macrophages and regulated by cytokines. 150

deletion Change in chromosome structure in which the end of a chromosome breaks off or two simultaneous breaks lead to the loss of an internal segment; often causes abnormalities (e.g., cri du chat syndrome). 441

denaturation (dee-nay-chuh-ray-shun) Loss of normal shape by an enzyme so that it no longer functions; caused by a less-than-optimal pH or temperature. 36

dendrite (den-dryt) Branched ending of a neuron that conducts signals toward the cell body. 287

denitrification Conversion of nitrate or nitrite to nitrogen gas by bacteria in soil. 559

dense fibrous connective tissue Type of connective tissue containing many collagen fibers packed together; found in tendons and ligaments, for example. 67

dental caries (kar-eez) Tooth decay that occurs when bacteria within the mouth metabolize sugar and give off acids that erode teeth; a cavity. 173

deoxyhemoglobin Hemoglobin not carrying oxygen. 118

depolarization When the charge inside the axon changes from positive to negative. 289

dermis (dur-mus) Region of skin that lies beneath the epidermis. 77

desertification Denuding and degrading a once-fertile land, initiating a desert-producing cycle that feeds on itself and causes long-term changes in the soil, climate, and biota of an area. 571

desmosome Includes arrangement of protein fibers that tightly hold the membranes of adjacent cells together and prevent overstretching. 94

detrital food chain (dih-tryt-ul) Straight-line linking of organisms according to who eats whom, beginning with detritus. 550

detrital food web (dih-tryt-ul) Complex pattern of interlocking and crisscrossing food chains, beginning with detritus. 550

detritus feeder Any organism that obtains most of its nutrients from the detritus in an ecosystem. 548

development Group of stages by which a zygote becomes an organism or by which an organism changes during its life span; includes puberty and aging, for example. 5

diabetes mellitus (dy-uh-bee-teez mel-ih-tus, muh-ly-tus) Condition characterized by a high blood glucose level and the appearance of glucose in the urine, due to a deficiency of insulin production and failure of cells to take up glucose. 227, 354

dialysate Material that passes through the membrane in dialysis. 234

diaphragm (dy-uh-fram) **1.** Dome-shaped horizontal sheet of muscle and connective tissue that divides the thoracic cavity from the abdominal cavity. **2.** A birth control device consisting of a soft rubber or latex cup that fits over the cervix. (1) 81, 175; (2) 379

diarrhea (dy-uh-ree-uh) Excessively frequent bowel movements. 181

diastole (dy-as-tuh-lee) Relaxation period of a heart chamber during the cardiac cycle. 97

diastolic pressure (dy-uh-stahl-ik) Arterial blood pressure during the diastolic phase of the cardiac cycle. 100

diencephalon (dy-en-sef-uh-lahn) Portion of the brain in the region of the third ventricle that includes the thalamus and hypothalamus. 298

differentiation Development of an unspecialized cell into one with a more specialized structure and function. 396, 448

diffusion (dih-fyoo-zhun) Movement of molecules or ions from a region of higher to lower concentration; it requires no energy and stops when the distribution is equal. 50

digestion Breaking down of large nutrient molecules into smaller molecules that can be absorbed. 170

digestive system Organ system including the mouth, esophagus, stomach, small intestine, and large intestine (colon) that receives food and digests it into nutrient molecules. Also has associated organs: teeth, tongue, salivary glands, liver, gallbladder, and pancreas. 80

dihybrid Individual that is heterozygous for two traits; shows the phenotype governed by the dominant alleles but carries the recessive alleles. 473

diploid (2n) Cell condition in which two of each type of chromosome are present in the nucleus. 424

disaccharide (dy-sak-uh-ryd) Sugar that contains two units of a monosaccharide (e.g., maltose). 29

distal convoluted tubule Final portion of a nephron that joins with a collecting duct; associated with tubular secretion. 224

diuretic (dy-uh-ret-ik) Drug used to counteract hypertension by causing the excretion of water. 231

diverticulosis A condition in which portions of the digestive tract mucosa have pushed through other layers of the tract forming pouches where food may collect. 171

DNA (deoxyribonucleic acid) Nucleic acid polymer produced from covalent bonding of nucleotide monomers that contain the sugar deoxyribose; the genetic material of nearly all organisms. 37, 492

DNA ligase (ly-gays) Enzyme that links DNA fragments; used during production of rDNA to join foreign DNA to vector DNA. 506

DNA replication Synthesis of a new DNA double helix prior to mitosis and meiosis in eukaryotic cells and during prokaryotic fission in prokaryotic cells. 493

domain The primary taxonomic group above the kingdom level; all living organisms may be placed in one of three domains. 6

dominant allele (uh-leel) Allele that exerts its phenotypic effect in the heterozygote; it masks the expression of the recessive allele. 468

dopamine Neurotransmitter in the central nervous system. 291

dorsal-root ganglion (gang-glee-un) Mass of sensory neuron cell bodies located in the dorsal root of a spinal nerve. 302

double helix Double spiral; describes the three-dimensional shape of DNA. 492

drug abuse Dependence on a drug, which assumes an "essential" biochemical role in the body following habituation and tolerance. 307

Duchenne muscular dystrophy Chronic progressive disease affecting the shoulder and pelvic girdles, commencing in early childhood. Characterized by increasing weakness of the muscles, followed by atrophy and a peculiar swaying gait with the legs kept wide apart. Transmitted as an X-linked trait; affected individuals, predominantly males, rarely survive to maturity. Death is usually due to respiratory weakness or heart failure. 278, 485

duodenum (doo-uh-dee-num) First part of the small intestine where chyme enters from the stomach. 176

duplication Change in chromosome structure in which a particular segment is present more than once in the same chromosome. 441

E

ecological footprint A measure of an individual's demand on the Earth's resources and ecosystems. 570

ecological pyramid Pictorial graph based on the biomass, number of organisms, or energy content of various trophic levels in a food web—from the producer to the final consumer populations. 552

ecosystem (ek-oh-sis-tum, ee-koh-) Biological community together with the associated abiotic environment; characterized by energy flow and chemical cycling. 2, 547

ectopic pregnancy Implantation of the embryo in a location other than the uterus, most often in an oviduct. 398

effacement During the first stage of labor, the uterine contractions of labor occur in such a way that the cervical canal slowly disappears as the lower part of the uterus is pulled upward toward the baby's head. 409

effector Muscle or gland that responds to stimulation. 287

egg Female gamete having the haploid number of chromosomes fertilized by a sperm, the male gamete. 370

elastic cartilage Type of cartilage composed of elastic fibers, allowing greater flexibility. 68

elastic fiber Yellow fiber in the matrix of connective tissue, providing flexibility. 66

electrocardiogram (ECG) (ih-lek-troh-kar-dee-uh-gram) Recording of the electrical activity associated with the heartbeat. 99

electron Negative subatomic particle, moving about in an energy level around the nucleus of an atom. 20

electron transport chain Passage of electrons along a series of membrane-bound carrier molecules from a higher to lower energy level; the energy released is used for the synthesis of ATP. 59

element Substance that cannot be broken down into substances with different properties; composed of only one type of atom. 20

elimination Process of expelling substances from the body. 171

embolus (em-buh-lus) Moving blood clot that is carried through the bloodstream. 106

embryo (em-bree-oh) Immature developmental stage not recognizable as a human being. 398

embryonic development Period of development from the second through eighth weeks. 396

embryonic disk Stage of embryonic development following the blastocyst stage that has two layers; one layer will be endoderm, and the other will be ectoderm. 398

emerging diseases Diseases caused by pathogens that are newly recognized in the last 20 years. 166

emphysema (em-fih-see-muh) Degenerative lung disorder in which the bursting of alveolar walls reduces the total surface area for gas exchange. 211

emulsification (ih-mul-suh-fuh-kay-shun) Breaking up of fat globules into smaller droplets by the action of bile salts or any other emulsifier. 31

endochondral ossification Ossification that begins as hyaline cartilage and subsequently replaced by bone tissue. 243

endocrine gland (en-duh-krin) Ductless organ that secretes a hormone or hormones into the bloodstream. 74, 340

endocrine system Organ system involved in the coordination of body activities; uses hormones as chemical signals secreted into the bloodstream. 340

endomembrane system A collection of membranous structures involved in transport within the cell. 54

endometrium Mucous membrane lining the interior surface of the uterus. 372

endoplasmic reticulum (ER) (en-duh-plaz-mik reh-tik-yuh-lum) System of membranous saccules and channels in the cytoplasm, often with attached ribosomes. 53

eosinophil (ee-oh-sin-oh-fill) White blood cell containing cytoplasmic granules that stain with acidic dye. 122

epidemic More cases of a disease than expected in a certain area for a certain period of time. 156

epidemiology The study of diseases in populations; includes the causes, distribution, and control of these diseases. 155

epidermis (ep-uh-dur-mus) Region of skin that lies above the dermis. 76

epididymis (ep-uh-did-uh-mus) Coiled tubule next to the testes where sperm mature and may be stored for a short time. 368

epiglottis (ep-uh-glaht-us) Structure that covers the glottis during the process of swallowing. 173, 198

epinephrine (ep-uh-nef-rin) Hormone secreted by the adrenal medulla in times of stress; adrenaline. 352

episiotomy (ih-pee-zee-aht-uh-mee) Surgical procedure performed during childbirth in which the opening of the vagina is enlarged to avoid tearing. 409

episodic memory Capacity of brain to store and retrieve information with regard to persons and events. 300

epithelial tissue (ep-uh-thee-lee-ul) Type of tissue that lines hollow organs and covers surfaces; also called epithelium. 72

erectile dysfunction Failure of the penis to achieve or maintain erection. 368

erythropoietin (EPO) (ih-rith-roh-poy-ee-tin) Hormone, produced by the kidneys, that speeds red blood cell formation. 220

esophagus (ih-sahf-uh-gus) Muscular tube for moving swallowed food from the pharynx to the stomach. 173

essential amino acids Amino acids required in the human diet because the body cannot make them. 175

essential fatty acid Fatty acid required in the human diet because the body cannot make it. 186

estrogen (es-truh-jun) Female sex hormone that helps maintain sex organs and secondary sex characteristics. 358, 376

eukaryotic cell Type of cell that has a membrane-bound nucleus and membranous organelles. 46, 520

evaporation Conversion of a liquid or a solid into a gas. 554

evolution Descent of organisms from common ancestors with the development of genetic and phenotypic changes over time that make them more suited to the environment. 5

evolutionary tree Diagram that describes the evolutionary relationship of groups of organisms; a common ancestor is presumed to have been present at points of divergence. 530

excretion Removal of metabolic wastes from the body. 218

exocrine gland Gland that secretes its product to an epithelial surface directly or through ducts. 340

exophthalmic goiter (ek-sahf-thal-mik) Enlargement of the thyroid gland accompanied by an abnormal protrusion of the eyes. 349

experiment Artificial situation devised to test a hypothesis. 11

experimental variable Value expected to change as a result of an experiment; represents the factor being tested by the experiment. 11

expiration (ek-spuh-ray-shun) Act of expelling air from the lungs; also called exhalation. 197, 201

expiratory reserve volume (ik-spy-ruh-tor-ee) Volume of air that can be forcibly exhaled after normal exhalation. 204

exponential growth Growth at a constant rate of increase per unit of time; can be expressed as a constant fraction or exponent. 568

external respiration Exchange of oxygen and carbon dioxide between alveoli and blood. 206

exteroceptor Sensory receptor that detects stimuli from outside the body (e.g., taste, smell, vision, hearing, and equilibrium). 316

extinction Total disappearance of a species or higher group. 8

extraembryonic membrane (ek-struh-em-bree-ahn-ik) Membrane that is not a part of the embryo but is necessary to the continued existence and health of the embryo. 397

F

facial tic Involuntary muscle movement of the face. 277

facilitated transport Use of a plasma membrane carrier to move a substance into or out of a cell from higher to lower concentration; no energy required. 51

familial hypercholesterolemia (FH) Inability to remove cholesterol from the bloodstream; predisposes individual to heart attack. 481

farsighted Vision abnormality due to a shortened eyeball from front to back; light rays focus in back of retina when viewing close objects. 326

fat Organic molecule that contains glycerol and fatty acids; found in adipose tissue. 31

fatty acid Molecule that contains a hydrocarbon chain and ends with an acid group. 32

female condom Large polyurethane tube with a flexible ring that fits onto the cervix. Functions as a contraceptive and helps minimize the risk of transmitting infection. 379

fermentation Anaerobic breakdown of glucose that results in a gain of two ATP and end products, such as alcohol and lactate. 60

fertilization Union of a sperm nucleus and an egg nucleus, which creates a zygote. 394

fetal development Period of development from the ninth week through birth. 403

fiber Structure resembling a thread; also, plant material that is nondigestible. 180

fibrin (fy-brun) Insoluble protein threads formed from fibrinogen during blood clotting. 123

fibrinogen (fy-brin-uh-jun) Plasma protein that is converted into fibrin threads during blood clotting. 118

fibroblast (fy-bruh-blast) Cell in connective tissues that produces fibers and other substances. 67

fibrocartilage (fy-broh-kar-tul-ij, -kart-lij) Cartilage with a matrix of strong collagenous fibers. 68

fibromyalgia Chronic, widespread pain in muscles and soft tissues surrounding joints. 277

fibrous connective tissue Tissue composed mainly of closely packed collagenous fibers and found in tendons and ligaments. 242

fimbria (pl., fimbriae) Small, bristlelike fiber on the surface of a bacterial cell, which attaches bacteria to a surface. 134, 370

first messenger Chemical signal, such as a peptide hormone, that binds to a plasma membrane receptor protein and alters the metabolism of a cell because a second messenger is activated. 344

flagellum (pl., flagella) (fluh-jel-um) Slender, long extension that propels a cell through a fluid medium. 55, 134

fluid-mosaic model Model for the plasma membrance based on the changing location and pattern of protein molecules in a fluid phospholipid bilayer. 49

focus Bending of light rays by the cornea, lens, and humors so that they converge and create an image on the retina. 323

follicle (fahl-ih-kul) Structure in the ovary that produces a secondary oocyte and the hormones estrogen and progesterone. 374

follicle-stimulating hormone (FSH) Hormone secreted by the anterior pituitary gland that stimulates the development of an ovarian follicle in a female or the production of sperm in a male. 346, 358

fontanel (fahn-tun-el) Membranous region located between certain cranial bones in the skull of a fetus or infant. 248, 403

food chain Order in which one population feeds on another in an ecosystem, from detritus (detrital food chain) or producer (grazing food chain) to final consumer. 550

food web In ecosystems, complex pattern of interlocking and crisscrossing food chains. 550

foramen magnum (fuh-ray-mun mag-num) Opening in the occipital bone of the vertebrate skull through which the spinal cord passes. 248

formed element Constituent of blood that is either cellular (red blood cells and white blood cells) or at least cellular in origin (platelets). 116

fossil Any past evidence of an organism that has been preserved in the Earth's crust. 520

fossil fuel Fuel, such as oil, coal, or natural gas, that is the result of partial decomposition of plants and animals coupled with exposure to heat and pressure for millions of years. 555, 576

fossil record History of life recorded from remains from the past. 520
fovea centralis Region of the retina consisting of densely packed cones; responsible for the greatest visual acuity. 323
fragile X syndrome Most common inherited form of mental retardation; results from mutation to a single gene and results in deficiency of a protein critical to brain development. 481
functional genomics Study of all the nucleotide sequences, including structural genes, regulatory sequences, and noncoding DNA segments, in the chromosomes of an organism. 511

G

GABA (gamma aminobutyric acid) Major inhibitory neurotransmitter in the CNS. 291
gallbladder Organ attached to the liver that serves to store and concentrate bile. 178
gallstone Crystalline bodies formed by concentration of normal and abnormal bile components within the gallbladder. 178
gamete (ga-meet, guh-meet) Haploid sex cell; the egg or sperm, which join in fertilization to form a zygote. 384, 429
gamma globulin Large proteins found in the blood plasma and on the surface of immune cells, functioning as antibodies (IgG). 148
ganglia (sing., ganglion) Collections of nerve cell bodies found outside the central nervous system in the peripheral nervous system. 302
gap junction Junction between cells formed by the joining of two adjacent plasma membranes; it lends strength and allows ions, sugars, and small molecules to pass between cells. 94
gastric gland Gland within the stomach wall that secretes gastric juice. 176
gastrulation Stage of animal development during which germ layers form, at least in part, by invagination. 398
gene Unit of heredity existing as alleles on the chromosomes; in diploid organisms, typically two alleles are inherited—one from each parent. 5
gene cloning Production of one or more copies of the same gene. 506
genetic engineering Alteration of DNA for medical or industrial purposes. 507
genotype (jee-nuh-typ) Genes of an individual for a particular trait or traits; often designated by letters, for example, *BB* or *Aa*. 468
gerontology (jer-un-tahl-uh-jee) Study of aging. 410
gland Epithelial cell or group of epithelial cells specialized to secrete a substance. 74
glaucoma (glow-koh-muh, glaw-koh-muh) Increasing loss of field of vision; caused by blockage of the ducts that drain the aqueous humor, creating pressure buildup and nerve damage. 322
global warming Predicted increase in the Earth's temperature, due to human activities that promote the greenhouse effect. 557
globulin Type of protein in blood plasma. There are alpha, beta, and gamma globulins. 118
glomerular capsule (gluh-mair-yuh-lur) Double-walled cup that surrounds the glomerulus at the beginning of the nephron. 223
glomerular filtrate Filtered portion of blood contained within the glomerular capsule. 227
glomerular filtration Movement of small molecules from the glomerulus into the glomerular capsule due to the action of blood pressure. 227
glomerulus (gluh-mair-uh-lus, gloh-mair-yuh-lus) Cluster; for example, the cluster of capillaries surrounded by the glomerular capsule in a nephron, where glomerular filtration takes place. 223
glottis (glaht-us) Opening for airflow in the larynx. 173, 199
glucagon (gloo-kuh-gahn) Hormone secreted by the pancreas that causes the liver to break down glycogen and raises the blood glucose level. 354
glucocorticoid (gloo-koh-kor-tih-koyd) Type of hormone secreted by the adrenal cortex that influences carbohydrate, fat, and protein metabolism; see cortisol. 352
glucose (gloo-kohs) Six-carbon sugar that organisms degrade as a source of energy during cellular respiration. 29
glutamate Major excitatory CNS neurotransmitter. 291
glycemic index (GI) Blood glucose response of a given food. 185
glycogen (gly-koh-jun) Storage polysaccharide composed of glucose molecules joined in a linear fashion but having numerous branches. 29
glycolysis Anaerobic breakdown of glucose that results in a gain of two ATP molecules. 58
Golgi apparatus (gohl-jee) Organelle, consisting of saccules and vesicles, that processes, packages, and distributes molecules about or from the cell. 54
gonad (goh-nad) Organ that produces gametes; the ovary produces eggs, and the testis produces sperm. 358
gonadotropic hormone (goh-nad-uh-trahp-ic, -troh-pic) Chemical signal secreted by the anterior pituitary that regulates the activity of the ovaries and testes; principally, follicle-stimulating hormone (FSH) and luteinizing hormone (LH). 34
gonadotropin-releasing hormone (GnRH) Hormone secreted by the hypothalamus that stimulates the anterior pituitary to secrete follicle-stimulating hormone (FSH) and luteinizing hormone (LH). 369
gout Joint inflammation caused by accumulation of uric acid. 220
granular leukocyte (gran-yuh-lur loo-kuh-syt) White blood cell with prominent granules in the cytoplasm. 122
gravitational equilibrium Maintenance of balance when the head and body are motionless. 334
gray matter Nonmyelinated axons and cell bodies in the central nervous system. 293
grazing food chain Straight-line linking of organisms according to who eats whom, beginning with a producer. 550
grazing food web Complex pattern of interlocking and crisscrossing food chains that begins with populations of autotrophs serving as producers. 550
greenhouse effect Reradiation of solar heat toward the Earth because gases, such as carbon dioxide, methane, nitrous oxide, and water vapor, allow solar energy to pass through toward the Earth but block the escape of heat back into space. 557
greenhouse gases Gases involved in the greenhouse effect. 557, 576
growth Increase in the number of cells and/or the size of these cells. 5
growth factor Chemical signal that regulates mitosis and differentiation of cells that have receptors for it; important in such processes as fetal development, tissue maintenance and repair, and hematopoiesis; sometimes a contributing factor in cancer. 450
growth hormone (GH) Substance secreted by the anterior pituitary; controls size of individual by promoting cell division, protein synthesis, and bone growth. 347
growth plate Cartilaginous layer within an epiphysis of a long bone that permits growth of bone to occur. 243
growth rate A percentage that reflects the difference between the number of persons in a population who are born and the number who die each year. 568
guanine (G) (gwah-neen) One of four nitrogen-containing bases in nucleotides composing the structure of DNA and RNA; pairs with cytosine. 38

H

hair cell Cell with stereocilia (long microvilli) that is sensitive to mechanical stimulation; mechanoreceptor for hearing and equilibrium in the inner ear. 328
hair follicle Tubelike depression in the skin in which a hair develops. 77
haploid (n) (hap-loyd) The n number of chromosomes—half the diploid number; the number characteristic of gametes that contain only one set of chromosomes. 428
hard palate (pal-it) Bony, anterior portion of the roof of the mouth. 172
hay fever Seasonal variety of allergic reaction to a specific allergen. Characterized by sudden attacks of sneezing, swelling of nasal mucosa, and often asthmatic symptoms. 150
heart Muscular organ located in the thoracic cavity whose rhythmic contractions maintain blood circulation. 94

heart attack Damage to the myocardium due to blocked circulation in the coronary arteries; also called a myocardial infarction (MI). 106

heartburn Burning pain in the chest that occurs when part of the stomach contents escape into the esophagus. 174

heart failure Syndrome characterized by distinctive symptoms and signs resulting from disturbances in cardiac output or from increased pressure in the veins. 108

helper T cell T cell that secretes cytokines that stimulate all types of immune system cells. 146

hemodialysis (he-moh-dy-al-uh-sus) Cleansing of blood by using an artificial membrane that causes substances to diffuse from blood into a dialysis fluid. 234

hemoglobin (hee-muh-gloh-bun) Iron-containing pigment in red blood cells that combines with and transports oxygen. 37, 118

hemolysis (he-mahl-uh-sus) Rupture of red blood cells accompanied by the release of hemoglobin. 120

hemophilia (he-moh-fil-ee-uh) Genetic disorder in which the affected individual is subject to uncontrollable bleeding. 123, 486

hemorrhoid (hem-uh-royd, hem-royd) Abnormally dilated blood vessels of the rectum. 181

hepatic portal vein Vein leading to the liver and formed by the merging blood vessels leaving the small intestine. 104

hepatic vein Vein that runs between the liver and the inferior vena cava. 104

hepatitis (hep-uh-ty-tis) Inflammation of the liver. Viral hepatitis occurs in several forms. 179

herbivore (hur-buh-vor) Primary consumer in a grazing food chain; a plant eater. 548

heterotroph Organism that cannot synthesize organic molecules from inorganic nutrients and therefore must take in organic nutrients (food). 517, 548

heterozygous Possessing unlike alleles for a particular trait. 468

hexose Six-carbon sugar. 29

hippocampus (hip-uh-kam-pus) Portion of the limbic system where memories are stored. 300

histamine (his-tuh-meen, -mun) Substance, produced by basophils in blood and mast cells in connective tissue, that causes capillaries to dilate. 141

homeostasis (hoh-mee-oh-stay-sis) Maintenance of normal internal conditions in a cell or an organism by means of self-regulating mechanisms. 4, 84

hominid (hahm-uh-nid) Member of the family Hominidae, which contains australopithecines and humans. 530

hominin An extinct or modern species of humans. 530

Homo erectus (hoh-moh ih-rek-tus) Hominid who used fire and migrated out of Africa to Europe and Asia. 535

Homo habilis (hoh-moh hab-uh-lus) Hominid of 2 MYA who is believed to have been the first tool user. 535

homologous chromosome (hoh-mahl-uh-gus, huh-mahl-uh-gus) Member of a pair of chromosomes that are alike and come together in synapsis during prophase of the first meiotic division. 428

homologous structure Structure similar in two or more species because of common ancestry. 523

homologue Member of a homologous pair of chromosomes. 428

Homo sapiens (hoh-moh say-pe-nz) Modern humans. 536

homozygous dominant Possessing two identical alleles, such as *AA*, for a particular trait. 468

homozygous recessive Possessing two identical recessive alleles, such as *aa*, for a particular trait. 468

hormone (hor-mohn) A protein or steroid produced by a cell that affects a different cell, the so-called target cell 178, 340

human chorionic gonadotropin (hCG) (kor- ee-ahn-ik, goh-nad-uh-trahp-in, -troh-pin) Hormone produced by the chorion that functions to maintain the uterine lining. 378, 398

human immunodeficiency virus (HIV) Virus responsible for AIDS. 156

human leukocyte antigen (HLA) Protein in a plasma membrane that identifies the cell as belonging to a particular individual and acts as an antigen in other organisms. 145

Huntington disease Genetic disease marked by progressive deterioration of the nervous system due to deficiency of a neurotransmitter. 478

hyaline cartilage (hy-uh-lin) Cartilage whose cells lie in lacunae separated by a white, translucent matrix containing very fine collagen fibers. 67

hydrogen bond Weak bond that arises between a slightly positive hydrogen atom of one molecule and a slightly negative atom of another, or between parts of the same molecule. 24

hydrolysis reaction (hy-drahl-ih-sis re-ak-shun) Splitting of a compound by the addition of water, with the H^+ being incorporated in one fragment and the OH^- in the other. 28

hydrolyze To break a chemical bond between molecules by insertion of a water molecule. 170

hydrophilic (hy-druh-fil-ik) Type of molecule that interacts with water by dissolving in water and/or forming hydrogen bonds with water molecules. 26

hydrophobic (hy-druh-foh-bik) Type of molecule that does not interact with water because it is nonpolar. 26

hypertension Elevated blood pressure, particularly the diastolic pressure. 106

hypothalamic-inhibiting hormone (hy-poh-thuh-lah-mik) One of many hormones produced by the hypothalamus that inhibits the secretion of an anterior pituitary hormone. 345

hypothalamic-releasing hormone One of many hormones produced by the hypothalamus that stimulates the secretion of an anterior pituitary hormone. 345

hypothalamus (hy-poh-thal-uh-mus) Part of the brain located below the thalamus that helps regulate the internal environment of the body and produces releasing factors that control the anterior pituitary. 298, 345

hypothesis (hy-pahth-ih-sis) Supposition that is formulated after making an observation; it can be tested by obtaining more data, often by experimentation. 9

I

immediate allergic response Allergic response that occurs within seconds of contact with an allergen, caused by the attachment of the allergen to IgE antibodies. 150

immune system White blood cells and lymphatic organs that protect the body against foreign organisms and substances and also cancerous cells. 80, 121

immunity Ability of the body to protect itself from foreign substances and cells, including disease-causing agents. 140

immunization (im-yuh-nuh-zay-shun) Use of a vaccine to protect the body against specific disease-causing agents. 148

immunosuppressive Inactivating the immune system to prevent organ rejection, usually via a drug. 150

implantation Attachment and penetration of the embryo into the lining of the uterus (endometrium). 370, 398

incomplete dominance Inheritance pattern in which the offspring has an intermediate phenotype, as when a red-flowered plant and a white-flowered plant produce pink-flowered offspring. 481

incontinence The inability of the body to control urination or defecation. 407

incus (ing-kus) The middle of three ossicles of the ear that serve to conduct vibrations from the tympanic membrane to the oval window of the inner ear. 328

infant respiratory distress syndrome Condition in newborns, especially premature ones, in which the lungs collapse because of a lack of surfactant lining the alveoli. 201

infectious diseases Diseases caused by pathogens such as bacteria, viruses, fungi, parasites, protozoans, and prions. 156

infectious mononucleosis Acute, self-limited infectious disease of the lymphatic system caused by the Epstein–Barr virus and characterized by fever, sore throat, swelling of lymph nodes and spleen, and the proliferation of monocytes and abnormal lymphocytes. 122

inferior vena cava (vee-nuh kay-vuh) Large vein that enters the right atrium from below and carries blood from the trunk and lower extremities. 143

infertility Inability to have as many children as desired. 382
inflammatory response Tissue response to injury that is characterized by redness, swelling, pain, and heat. 140
ingestion The taking of food or liquid into the body by way of the mouth. 170
initiation Mutation of a single cell that may lead to cancer development. 449
inner cell mass An aggregation of cells at one pole of the blastocyst, destined to form the embryo proper. 397
inner ear Portion of the ear consisting of a vestibule, semicircular canals, and the cochlea, where equilibrium is maintained and sound is transmitted. 328
insertion End of a muscle attached to a movable bone. 264
inspiration (in-spuh-ray-shun) Act of taking air into the lungs; also called inhalation. 197, 201
inspiratory reserve volume (in-spy-ruh-tohr-ee) Volume of air that can be forcibly inhaled after normal inhalation. 203
insulin (in-suh-lin) Hormone secreted by the pancreas that lowers the blood glucose level by promoting the uptake of glucose by cells, and the conversion of glucose to glycogen by the liver and skeletal muscles. 354
integrase Viral enzyme that enables the integration of viral genetic material into a host cell's DNA. 160
integration Summing up of excitatory and inhibitory signals by a neuron or by some part of the brain. 292, 317
integumentary system (in-teg-yoo-men-tuh-ree, -men-tree) Organ system consisting of skin and various organs, such as hair, found in skin. 75
intercalated disk (in-tur-kuh-lay-tud) Region that holds adjacent cardiac muscle cells together; disks appear as dense bands at right angles to the muscle striations. 69, 263
interferon (in-tur-feer-ahn) Antiviral agent produced by an infected cell that blocks the infection of another cell. 142
interkinesis Period between meiosis I and meiosis II, during which no DNA replication takes place. 428
interleukin (in-tur-loo-kun) Cytokine produced by macrophages and T cells that functions as a metabolic regulator of the immune response. 149
intermediate filament Ropelike assemblies of fibrous polypeptides in the cytoskeleton that provide support and strength to cells; so called because they are intermediate in size between actin filaments and microtubules. 55
internal respiration Exchange of oxygen and carbon dioxide between blood and tissue fluid. 208
interneuron Neuron located within the central nervous system that conveys messages between parts of the central nervous system. 287
interoceptor Sensory receptor that detects stimuli from inside the body (e.g., pressoreceptors, osmoreceptors, and chemoreceptors). 316
interphase Cell cycle stage during which growth and DNA synthesis occur when the nucleus is not actively dividing. 422
interstitial cell (in-tur-stish-ul) Hormone-secreting cell located between the seminiferous tubules of the testes. 369
intervertebral disk (in-tur-vur-tuh-brul) Layer of cartilage located between adjacent vertebrae. 251
intramembranous ossification Ossification that forms from membranelike layers of primitive connective tissue. 242
intrauterine device (IUD) (in-truh-yoo-tur-in) Birth control device consisting of a small piece of molded plastic inserted into the uterus; believed to alter the uterine environment so that fertilization does not occur. 379
invasive Describes cells, such as tumor cells, that invade normal cells. 580
inversion Change in chromosome structure in which a segment of a chromosome is turned around 180°; this reversed sequence of genes can lead to altered gene activity and abnormalities. 441
invertebrate Animal without a vertebral column or backbone. 77
ion (eye-un, -ahn) Charged particle that carries a negative or positive charge. 22
ionic bond (eye-ahn-ik) Chemical bond in which ions are attracted to one another by opposite charges. 22
iris (eye-ris) Muscular ring that surrounds the pupil and regulates the passage of light through this opening. 322
isotope (eye-suh-tohp) One of two or more atoms with the same atomic number but a different atomic mass due to the number of neutrons. 21

J

jaundice (jawn-dis) Yellowish tint to the skin caused by an abnormal amount of bilirubin (bile pigment) in the blood, indicating liver malfunction. 179
joint Articulation between two bones of a skeleton. 240
juxtaglomerular apparatus (juk-stuh-gluh-mer-yuh-lur) Structure located in the walls of arterioles near the glomerulus; regulates renal blood flow. 230

K

kidney Organ in the urinary system that produces and excretes urine. 218
kingdom One of the categories used to classify organisms; the category above phylum. 7
kinocilium The largest stereocilium. 334

L

lacteal (lak-tee-ul) Lymphatic vessel in an intestinal villus; it aids in the absorption of lipids. 177
lactose intolerance (lak-tohs) Inability to digest lactose because of an enzyme deficiency. 178
lacuna (pl., lacunae) (luh-koo-nuh, -kyoo-nuh) Small pit or hollow cavity, as in bone or cartilage, where a cell or cells are located. 67
Langerhans cell Specialized epidermal cell that assists the immune system. 76
lanugo (luh-noo-goh) Short, fine hair that is present during the later portion of fetal development. 403
large intestine Last major portion of the digestive tract, extending from the small intestine to the anus and consisting of the cecum, the colon, the rectum, and the anal canal. 180
laryngitis (lar-un-jy-tis) Infection of the larynx with accompanying hoarseness. 209
larynx (lar-ingks) Cartilaginous organ located between the pharynx and the trachea that contains the vocal cords; also called the voice box. 199
learning Relatively permanent change in behavior that results from practice and experience. 300
lens Clear, membranelike structure found in the eye behind the iris; brings objects into focus. 322
leptin Hormone produced by adipose tissue; acts on the hypothalamus to signal satiety. 359
leukemia (loo-kee-mee-uh) Cancer of the blood-forming tissues leading to the overproduction of abnormal white blood cells. 122, 451
ligament (lig-uh-munt) Tough cord or band of dense fibrous connective tissue that joins bone to bone at a joint. 67, 240
limbic system Associates various brain centers, including the amygdala and hippocampus; governs learning, memory, and various emotions such as pleasure, fear, and happiness. 299
lineage Evolutionary line of descent. 530
lipase (ly-pays, ly-payz) Fat-digesting enzyme secreted by the pancreas. 176
lipid (lip-id, ly-pid) Class of organic compounds that tends to be soluble only in nonpolar solvents such as alcohol; includes fats and oils. 31
liver Large, dark red internal organ that produces urea and bile, detoxifies the blood, stores glycogen, and produces the plasma proteins, among other functions. 178
locus (pl., loci) Particular site where a gene is found on a chromosome. Homologous chromosomes have corresponding gene loci. 468
long-term memory Retention of information that lasts longer than a few minutes. 300
long-term potentiation (LTP) (puh-ten-shee-ay-shun) Enhanced response at synapses within the hippocampus; likely essential to memory storage. 300

loop of the nephron (nef-rahn) Portion of the nephron lying between the proximal convoluted tubule and the distal convoluted tubule that functions in water reabsorption. 224

loose fibrous connective tissue Tissue composed mainly of fibroblasts widely separated by a matrix containing collagen and elastic fibers. 67

lumen (loo-mun) Cavity inside any tubular structure, such as the lumen of the digestive tract. 171

lung cancer Malignant growth that often begins in the bronchi. 211

lungs Paired, cone-shaped organs within the thoracic cavity; function in internal respiration and contain moist surfaces for gas exchange. 200

luteinizing hormone (LH) Hormone that controls the production of testosterone by interstitial cells in males and promotes the development of the corpus luteum in females. 346, 358, 369

lymph (limf) Fluid, derived from tissue fluid, that is carried in lymphatic vessels. 68, 105, 138

lymphatic nodule Mass of lymphatic tissue not surrounded by a capsule. 140

lymphatic organ Organ other than a lymphatic vessel that is part of the lymphatic system; includes lymph nodes, tonsils, spleen, thymus gland, and bone marrow. 138

lymphatic system (lim-fat-ik) Organ system consisting of lymphatic vessels and lymphatic organs that transport lymph and lipids; aids the immune system. 80, 92

lymph node Mass of lymphatic tissue located along the course of a lymphatic vessel. 139

lymphocyte (lim-fuh-syt) Specialized white blood cell that functions in specific defense; occurs in two forms—T cell and B cell. 123

lymphoma Cancer of lymphatic tissue (reticular connective tissue). 451

lysosome (ly-suh-sohm) Membrane-bound vesicle that contains hydrolytic enzymes for digesting macromolecules. 54

lysozyme Enzyme found in tears, milk, saliva, mucus, and other body fluids that destroys bacteria by digesting their cell walls. 140

M

macromolecule Extremely large biological molecule; refers specifically to proteins, nucleic acids, polysaccharides, lipids, and complexes of these. 28

macrophage (mak-ruh-fayj) Large phagocytic cell derived from a monocyte that ingests microbes and debris. 142

mad cow disease Slowly progressive fatal disease affecting the central nervous system of cattle; transmissible to humans. 134

major histocompatibility complex (MHC) Cluster of genes on chromosome 6 concerned with self-antigen production; matching these is critical to success of organ transplants. The MHC includes the human leukocyte antigen (HLA) genes. 145

male condom Sheath used to cover the penis during sexual intercourse; used as a contraceptive and, if latex, to minimize the risk of transmitting infection. 379

malleus (mal-ee-us) The first of three ossicles of the ear that serve to conduct vibrations from the tympanic membrane to the oval window of the inner ear. 328

Marfan syndrome Congenital disorder of connective tissue characterized by abnormal length of the extremities. 478

mass An atom's quantity of matter. 21

mass number The sum of the number of protons and neutrons in the nucleus of an atom. 21

mast cell Cell to which antibodies, formed in response to allergens, attach, causing it to release histamine, thus producing allergic symptoms. 122, 141

mastoiditis (mas-toyd-eye-tis) Inflammation of the mastoid sinuses of the skull. 248

matrix (may-triks) Unstructured semifluid substance that fills the space between cells in connective tissues or inside organelles. 67

matter Anything that takes up space and has mass. 20

mechanoreceptor (mek-uh-noh-rih-sep-tur) Sensory receptor that responds to mechanical stimuli, such as that from pressure, sound waves, and gravity. 316

medulla oblongata (muh-dul-uh ahb-lawng-gah-tuh) Part of the brain stem that is continuous with the spinal cord; controls heartbeat, blood pressure, breathing, and other vital functions. 298

medullary cavity (muh-dul-uh-ree) Cavity within the diaphysis of a long bone containing marrow. 240

megakaryocyte (meg-uh-kar-ee-oh-syt,-uh-syt) Large cell that gives rise to blood platelets. 123

meiosis (my-oh-sis) Type of nuclear division that occurs as part of sexual reproduction in which the daughter cells receive the haploid number of chromosomes in varied combinations. 428

melanocyte Melanin-producing cell found in skin. 76

melanocyte-stimulating hormone (MSH) Substance that causes melanocytes to secrete melanin in lower vertebrates. 347

melatonin (mel-uh-toh-nun) Hormone secreted by the pineal gland that is involved in biorhythms. 359

membrane attack complex Group of complement proteins that form channels in a microbe's surface, thereby destroying it. 142

memory Capacity of the brain to store and retrieve information about past sensations and perceptions; essential to learning. 300

memory T cell T cell that differentiates during an initial infection and responds rapidly during subsequent exposure to the same anitgen. 146

meninges (sing., meninx) (muh-nin-jeez) Protective membranous coverings about the central nervous system. 82, 293

meningitis Condition that refers to inflammation of the brain or spinal cord meninges (membranes). 82

menopause (men-uh-pawz) Termination of the ovarian and uterine cycles in older women. 376

menstruation (men-stroo-ay-shun) Loss of blood and tissue from the uterus at the end of a uterine cycle. 377

messenger RNA (mRNA) Type of RNA formed from a DNA template that bears coded information for the amino acid sequence of a polypeptide. 495

metabolism All of the chemical reactions that occur in a cell. 4, 57

metaphase Mitotic phase during which chromosomes are aligned at the equator of the mitotic spindle. 426

metastasis (muh-tas-tuh-sis) Spread of cancer from the place of origin throughout the body; caused by the ability of cancer cells to migrate and invade tissues. 450

microtubule (my-kro-too-byool) Small cylindrical structure that contains 13 rows of the protein tubulin around an empty central core; present in the cytoplasm, centrioles, cilia, and flagella. 55

micturition Emptying of the bladder; urination. 219

midbrain Part of the brain located below the thalamus and above the pons; contains reflex centers and tracts. 298

middle ear Portion of the ear consisting of the tympanic membrane, the oval and round windows, and the ossicles; where sound is amplified. 328

mineral Naturally occurring inorganic substance containing two or more elements; certain minerals are needed in the diet. 187, 578

mineralocorticoid (min-ur-uh-loh-kor-tih-koyd) Type of hormone secreted by the adrenal cortex that regulates water–salt balance, leading to increases in blood volume and blood pressure. 352

mitochondrion (my-tuh-kahn-dree-un) Membrane-bound organelle in which ATP molecules are produced during the process of cellular respiration. 57

mitosis (my-toh-sis) Type of cell division in which daughter cells receive the exact chromosomal and genetic makeup of the parent cell; occurs during growth and repair. 421, 422, 424

mitotic spindle Microtubule structure that brings about chromosomal movement during nuclear division. 425

mole A unit of scientific measurement for atoms, ions, and molecules. 26

molecular clock Mutational changes that accumulate at a presumed constant rate in regions of DNA not involved in adaptation to the environment. 530

molecule Union of two or more atoms of the same element; also, the smallest part of a compound that retains the properties of the compound. 2, 22

monoclonal antibody One of many antibodies produced by a clone of hybridoma cells that all bind to the same antigen. 148

monocyte (mahn-uh-syt) Type of agranular white blood cell that functions as a phagocyte and an antigen-presenting cell. 122

monohybrid Individual that is heterozygous for one trait; shows the phenotype of the dominant allele but carries the recessive allele. 470

monosaccharide (mahn-uh-sak-uh-ryd) Simple sugar; a carbohydrate that cannot be decomposed by hydrolysis (e.g., glucose). 29

monosomy One less chromosome than usual. 437

morphogenesis Emergence of shape in tissues, organs, or entire embryo during development. 396

morula Spherical mass of cells resulting from cleavage during animal development, prior to the blastula stage. 397

mosaic evolution Concept that human characteristics did not evolve at the same rate; for example, some body parts are more humanlike than others in early hominids. 532

motor neuron Nerve cell that conducts nerve impulses away from the central nervous system and innervates effectors (muscles and glands). 287

motor unit Motor neuron and all the muscle fibers it innervates. 271

movement Motion. 171

MRSA Methicillin-resistant *Staphylococcus aureus*, a type of bacterium that causes "Staph" infections that is no longer susceptible to certain antibiotics including methicillin. 168

mucosa Membrane that lines tubes and body cavities that open to the outside of the body; mucous membrane. 171

mucous membrane (myoo-kus) Membrane lining a cavity or tube that opens to the outside of the body; also called mucosa. 82

multicellular Referring to organism composed of many cells; usually has organized tissues, organs, and organ systems. 2

multifactorial trait Controlled by several allelic pairs; each dominant allele contributes to the phenotype in an additive and like manner. 480

multiple allele (uh-leelz) Inheritance pattern in which there are more than two alleles for a particular trait; each individual has only two of all possible alleles. 481

multiple sclerosis (MS) Disease in which the outer myelin layer of nerve fiber insulation becomes scarred, interfering with normal conduction of nerve impulses. 151

multiregional continuity hypothesis Proposal that modern humans evolved independently in at least three different places: Asia, Africa, and Europe. 536

muscle dysmorphia Mental state in which a person considers his or her body to be underdeveloped and becomes preoccupied with bodybuilding and diet; affects more men than women. 192

muscle fiber Muscle cell. 263

muscle tone Continuous, partial contraction of muscle. 272

muscle twitch Contraction of a whole muscle in response to a single stimulus. 272

muscular dystrophy (mus-ku-lar dis-tro-fee) Progressive muscle weakness and atrophy caused by deficient dystrophin protein. 277

muscularis Two layers of muscle in the gastrointestinal tract. 171

muscular system System of muscles that produces movement, both within the body and of its limbs; principal components are skeletal, smooth, and cardiac muscle. 80

muscular (contractile) tissue Type of tissue composed of fibers that can shorten and thicken. 69

mutagen (myoo-tuh-jun) Agent, such as radiation or a chemical, that brings about a mutation. 453

mutation Alteration in chromosome structure or number and also an alteration in a gene due to a change in DNA composition. 5, 494

myalgia Muscular pain. 277

myasthenia gravis (mi-as-thee-ne-ah grav-is) Chronic disease characterized by muscles that are weak and easily fatigued. It results from the immune system's attack on neuromuscular junctions so that stimuli are not transmitted from motor neurons to muscle fibers. 151, 278

myelin sheath (my-uh-lin) White, fatty material, derived from the membrane of Schwann cells; forms a covering for nerve fibers. 288

myocardium (my-oh-kar-dee-um) The middle of the three layers of the wall of the heart; composed of cardiac muscle. 94

myofibril (my-uh-fy-brul) Contractile portion of muscle cells that contains a linear arrangement of sarcomeres and shortens to produce muscle contraction. 267

myoglobin Pigmented molecule in muscle tissue that stores oxygen. 274

myosin (my-uh-sin) One of two major proteins of muscle; makes up thick filaments in myofibrils of muscle fibers. See actin. 267

myxedema (mik-sih-dee-muh) Condition resulting from a deficiency of thyroid hormone in an adult. 349

N

NAD^+ (nicotinamide adenine dinucleotide) Coenzyme that functions as a carrier of electrons and hydrogen ions, especially in cellular respiration. 58

nail Protective covering of the distal part of fingers and toes. 77

nasal cavity One of two canals in the nose, separated by a septum. 198

natural selection Mechanism resulting in adaptation to the environment. 525

Neandertal (nee-an-dur-thahl) Hominid with a sturdy build who lived during the last Ice Age in Europe and the Middle East; hunted large game and left evidence of being culturally advanced. 537

nearsighted Vision abnormality due to an elongated eyeball from front to back; light rays focus in front of retina when viewing distant objects. 326

negative feedback Mechanism of homeostatic response in which a stimulus initiates reactions that reduce the stimulus. 86

nephron (nef-rahn) Microscopic kidney unit that regulates blood composition by glomerular filtration, tubular reabsorption, and tubular secretion. 220

nerve Bundle of long axons outside the central nervous system. 70

nerve signal Action potential (electrochemical change) traveling along a neuron. 288

nervous system Organ system consisting of the brain, spinal cord, and associated nerves that coordinates the other organ systems of the body. 81

nervous tissue Tissue that contains nerve cells (neurons), which conduct impulses, and neuroglia, which support, protect, and provide nutrients to neurons. 70

neuroglia (noo-rahg-lee-uh, noo-rohg-lee-uh) Nonconducting nerve cells that are intimately associated with neurons and function in a supportive capacity. 70, 287

neuromuscular junction Region where an axon terminal approaches a muscle fiber; the synaptic cleft separates the axon terminal from the sarcolemma of a muscle fiber. 270

neuron (noor-ahn, nyoor-) Nerve cell that characteristically has three parts: dendrites, cell body, and axon. 70, 287

neurotransmitter Chemical stored at the ends of axons that is responsible for transmission across a synapse. 291

neutron (noo-trahn) Neutral subatomic particle, located in the nucleus and having a weight of approximately one atomic mass unit. 20

neutrophil (noo-truh-fil) Granular leukocyte that is the most abundant of the white blood cells; first to respond to infection. 122

niche (nich) Role an organism plays in its community, including its habitat and its interactions with other organisms. 549

nitrification Process by which nitrogen in ammonia and organic molecules is oxidized to nitrites and nitrates by soil bacteria. 559

nitrogen fixation Process whereby free atmospheric nitrogen is converted into compounds, such as ammonium and nitrates, usually by bacteria. 558

node of Ranvier (rahn-vee-ay) Gap in the myelin sheath around a nerve fiber. 288

nondisjunction Failure of homologous chromosomes or daughter chromosomes to separate during meiosis I and meiosis II, respectively. 437

nonrenewable resource Minerals, fossil fuels, and other materials present in essentially fixed amounts (within human timescales) in our environment. 570

norepinephrine (NE) (nor-ep-uh-nef-rin) Neurotransmitter of the postganglionic fibers in the sympathetic division of the autonomic system; also, a hormone produced by the adrenal medulla. 291, 352

nuclear envelope Double membrane that surrounds the nucleus and is connected to the endoplasmic reticulum; has pores that allow substances to pass between the nucleus and the cytoplasm. 53

nuclear pore Opening in the nuclear envelope that permits the passage of proteins into the nucleus and ribosomal subunits out of the nucleus. 53

nucleolus (noo-klee-uh-lus, nyoo-) Dark-staining, spherical body in the cell nucleus that produces ribosomal subunits. 53

nucleoplasm (noo-klee-uh-plaz-um) Semifluid medium of the nucleus, containing chromatin. 53

nucleotide Monomer of DNA and RNA consisting of a 5-carbon sugar bonded to a nitrogen-containing base and a phosphate group. 38

nucleus (noo-klee-us, nyoo-) Membrane-bounded organelle that contains chromosomes and controls the structure and function of the cell. 20, 53, 298

nutrient Chemical substance in foods that is essential to the diet and contributes to good health. 185

O

obesity (oh-bee-sih-tee) Excess adipose tissue; exceeding ideal weight by more than 20%. 183

oblique At an angle between horizontal and vertical. 176

oil Substance, usually of plant origin and liquid at room temperature, formed when a glycerol molecule reacts with three fatty acid molecules. 31

oil gland Gland of the skin associated with hair follicle; secretes sebum; also called sebaceous gland. 79

olfactory cell (ahl-fak-tuh-ree, -tree, ohl-) Modified neuron that is a sensory receptor for the sense of smell. 321

omnivore (ahm-nuh-vor) Organism in a food chain that feeds on both plants and animals. 548

oncogene (ahng-koh-jeen) Cancer-causing gene. 450

oncologist Physician who specializes in one or more types of cancer. 451

oncology The study of cancer. 451

oogenesis (oh-uh-jen-uh-sis) Production of an egg in females by the process of meiosis and maturation. 374, 436

opportunistic infection Infection that has an opportunity to occur because the immune system has been weakened. 156

optic chiasma X-shaped structure on the underside of the brain formed by a partial crossing-over of optic nerve fibers. 326

optic nerve Either of two cranial nerves that carry nerve impulses from the retina of the eye to the brain, thereby contributing to the sense of sight. 323

optic tract Groups of neurons from the optic nerve that sweep around the hypothalamus. Most fibers synapse with neurons in nuclei within the thalamus. 326

orbitals Pathways in which electrons travel around the nucleus of an atom. 20

organ Combination of two or more different tissues performing a common function. 2, 75

organelle (or-guh-nel) Small membranous structure in the cytoplasm having a specific structure and function. 46

organic Refers to a molecule that always contains carbon and hydrogen, and often contains ozygen as well; organic molecules are associated with living things. 28

organic molecule Type of molecule that contains carbon and hydrogen—and often contains oxygen also. 28

organism Individual living thing. 2

organ system Group of related organs working together. 2, 75

origin End of a muscle attached to a relatively immovable bone. 264

osmosis (ahz-moh-sis, ahs-) Diffusion of water through a selectively permeable membrane. 50

osmotic pressure Measure of the tendency of water to move across a selectively permeable membrane; visible as an increase in liquid on the side of the membrane with higher solute concentration. 50, 118

ossicle (ahs-ih-kul) One of the small bones of the middle ear—malleus, incus, and stapes. 328

ossification Formation of bone tissue. 242

osteoblast (ahs-tee-uh-blast) Bone-forming cell. 242

osteoclast (ahs-tee-uh-klast) Cell that causes erosion of bone. 242

osteocyte (ahs-tee-uh-syt) Mature bone cell located within the lacunae of bone. 240, 242

osteoporosis Condition in which bones break easily because calcium is removed from them faster than it is replaced. 188

otitis media (oh-ty-tis mee-dee-uh) Infection of the middle ear, characterized by pain and possibly by a sense of fullness, hearing loss, vertigo, and fever. 209

otolith (oh-tuh-lith) Calcium carbonate granule associated with ciliated cells in the utricle and the saccule. 334

outbreak A disease epidemic that is confined to a local area. 156

outer ear Portion of ear consisting of the pinna and auditory canal. 328

out-of-Africa hypothesis Proposal that modern humans originated only in Africa; then migrated out of Africa and supplanted populations of early *Homo* in Asia and Europe about 100,000 years ago. 536

oval window Membrane-covered opening between the stapes and the inner ear. 328

ovarian cycle (oh-vair-ee-un) Monthly follicle changes occurring in the ovary that control the level of sex hormones in the blood and the uterine cycle. 374

ovary Female gonad that produces eggs and the female sex hormones. 358, 370

oviduct (oh-vuh-dukt) Tube that transports eggs to the uterus; also called uterine tube. 371

ovulation (ahv-yuh-lay-shun, ohv-) Release of a secondary oocyte from the ovary; if fertilization occurs, the secondary oocyte becomes an egg. 375

oxygen debt Amount of oxygen needed to metabolize lactate, a compound that accumulates during vigorous exercise. 273

oxyhemoglobin (ahk-see-hee-muh-gloh-bin) Compound formed when oxygen combines with hemoglobin. 118, 206

oxytocin (ahk-sih-toh-sin) Hormone released by the posterior pituitary that causes contraction of uterus and milk letdown. 345

ozone hole Seasonal thinning of the ozone shield in the lower stratosphere at the North and South poles. 562

ozone shield Accumulation of O_3, formed from oxygen in the upper atmosphere; a filtering layer that protects the Earth from ultraviolet radiation. 562

P

pacemaker See SA (sinoatrial) node. 98

pain receptor Sensory receptor that is sensitive to chemicals released by damaged tissues or excess stimuli of heat or pressure. 316

pancreas (pang-kree-us, pan-) Internal organ that produces digestive enzymes and the hormones insulin and glucagon. 178, 354

pancreatic amylase (pang-kree-at-ik am-uh-lays, -layz) Enzyme in the pancreas that digests starch to maltose. 178

pancreatic islets (islets of Langerhans) Masses of cells that constitute the endocrine portion of the pancreas. 354

pandemic An increase in the occurrence of a disease within a large and geographically widespread population (often refers to a worldwide epidemic). 156

Pap test Analysis done on cervical cells for detection of cancer. 370

parasympathetic division That part of the autonomic system that is active under normal conditions; uses acetylcholine as a neurotransmitter. 298

parathyroid gland (par-uh-thy-royd) Gland embedded in the posterior surface of the thyroid gland; it produces parathyroid hormone. 350

parathyroid hormone (PTH) Hormone secreted by the four parathyroid glands that increases the blood calcium level and decreases the blood phosphate level. 350

parent cell Cell that divides so as to form daughter cells. 424

Parkinson disease Progressive deterioration of the central nervous system due to a deficiency in the neurotransmitter dopamine. 298

parturition (par-tyoo-rish-un, par-chuh-) Processes that lead to and include birth and the expulsion of the afterbirth. 408

passive immunity Protection against infection acquired by transfer of antibodies to a susceptible individual. 147

pathogen (path-uh-jun) Disease-causing agent. 134

pectoral girdle (pek-tur-ul) Portion of the skeleton that provides support and attachment for an arm; consists of a scapula and a clavicle. 253

pelvic girdle Portion of the skeleton to which the legs are attached; consists of the coxal bones. 254

pelvis Bony ring formed by the sacrum and coxae. 254

penis External organ in males through which the urethra passes; also serves as the organ of sexual intercourse. 368

pentose (pen-tohs, -tohz) Five-carbon sugar. Deoxyribose is the pentose sugar found in DNA; ribose is a pentose sugar found in RNA. 29

pepsin (pep-sin) Enzyme secreted by gastric glands that digests proteins to peptides. 176

peptide bond Type of covalent bond that joins two amino acids. 35

peptide hormone Type of hormone that is a protein, a peptide, or derived from an amino acid. 343

pericardium (pair-ih-kar-dee-um) Protective serous membrane that surrounds the heart. 94

periodontitis Inflammation of the periodontal membrane that lines tooth sockets, causing loss of bone and loosening of teeth. 173

periosteum (pair-ee-ahs-tee-um) Fibrous connective tissue covering the surface of bone. 240

peripheral nervous system (PNS) (puh-rif-ur-ul) Nerves and ganglia that lie outside the central nervous system. 286

peristalsis (pair-ih-stawl-sis) Wavelike contractions that propel substances along a tubular structure such as the esophagus. 170

peritonitis (pair-ih-tuh-ny-tis) Generalized infection of the lining of the abdominal cavity. 82, 171

peritubular capillary network (pair-ih-too-byuh-lur) Capillary network that surrounds a nephron and functions in reabsorption during urine formation. 223

Peyer's patches Lymphatic organs located in the small intestine. 140

phagocytosis (fag-uh-sy-toh-sis) Process by which amoeboid-type cells engulf large substances, forming an intracellular vacuole. 52

pharynx (far-ingks) Portion of the digestive tract between the mouth and the esophagus that serves as a passageway for food and also for air on its way to the trachea. 173, 198

phenotype (fee-nuh-typ) Visible expression of a genotype—for example, brown eyes or attached earlobes. 468

pheromone Chemical signal released by an organism that affects the metabolism or influences the behavior of another individual of the same species. 343

phospholipid (fahs-foh-lip-id) Molecule that forms the bilayer of the cell's membranes; has a polar, hydrophilic head bonded to two nonpolar, hydrophobic tails. 33

photoreceptor Sensory receptor in retina that responds to light stimuli. 316

photosynthesis A process by which plants, algae, and some bacteria harvest the energy of the sun and convert it to chemical energy. 4

photovoltaic (solar) cell An energy-conversion device that captures solar energy and directly converts it to electrical current. 576

pH scale Measurement scale for hydrogen ion concentration. 27

pilus Elongated, hollow appendage on bacteria used to transfer DNA from one cell to another. 134

pineal gland (pin-ee-ul, py-nee-ul) Endocrine gland located in the third ventricle of the brain; produces melatonin. 359

pinna Part of the ear that projects on the outside of the head. 328

pituitary dwarfism (pih-too-ih-tair-ee, -tyoo-) Condition in which an affected individual has normal proportions but small stature; caused by inadequate growth hormone. 347

pituitary gland Endocrine gland that lies just inferior to the hypothalamus; consists of the anterior pituitary and posterior pituitary. 345

placebo Treatment that is an inactive substance (pill, liquid, etc.) administered as if it were a therapy in an experiment but that has no therapeutic value. 11

placenta (pluh-sen-tuh) Structure that forms from the chorion and the uterine wall and allows the embryo and then the fetus to acquire nutrients and rid itself of wastes. 378, 397

plaque (plak) Accumulation of soft masses of fatty material, particularly cholesterol, beneath the inner linings of the arteries. 106

plasma (plaz-muh) Liquid portion of blood; contains nutrients, wastes, salts, and proteins. 116

plasma cell Cell derived from a B lymphocyte specialized to mass-produce antibodies. 144

plasma membrane Membrane surrounding the cytoplasm that consists of a phospholipid bilayer with embedded proteins; functions to regulate the entrance and exit of molecules from the cell. 46

plasma protein Protein dissolved in blood plasma. 118

plasmid (plaz-mid) Self-replicating ring of accessory DNA in the cytoplasm of bacteria. 134, 506

platelet (thrombocyte) (playt-lit) Component of blood necessary to blood clotting; also called a thrombocyte. 68, 123

pleura Serous membrane that encloses the lungs. 82, 200

pneumonectomy (noo-muh-nek-tuh-mee, nyoo-) Surgical removal of all or part of a lung. 213

pneumonia (noo-mohn-yuh, nyoo-) Infection of the lungs that causes alveoli to fill with mucus and pus. 210

polar Combination of atoms in which the electrical charge is not distributed symmetrically. 24

polar body In oogenesis, a nonfunctional product; two to three meiotic products are of this type. 374, 436

pollution Any environmental change that adversely affects the lives and health of living things. 570

polygenic trait Trait is controlled by several allelic pairs; each dominant allele contributes to the phenotype in an additive and like manner. 479

polymerase chain reaction (PCR) (pahl-uh-muh-rays, -rayz) Technique that uses the enzyme DNA polymerase to produce millions of copies of a particular piece of DNA. 504

polyp (pahl-ip) Small, abnormal growth that arises from the epithelial lining. 82

polypeptide Polymer of many amino acids linked by peptide bonds. 35

polyribosome (pahl-ih-ry-buh-sohm) String of ribosomes simultaneously translating regions of the same mRNA strand during protein synthesis. 54, 500

polysaccharide (pahl-ee-sak-uh-ryd) Polymer made from sugar monomers; the polysaccharides starch and glycogen are polymers of glucose monomers. 29

pons (pahnz) Portion of the brain stem above the medulla oblongata and below the midbrain; assists the medulla oblongata in regulating the breathing rate. 298

population Organisms of the same species occupying a certain area. 2

positive feedback Mechanism in which the stimulus initiates reactions that lead to an increase in the stimulus. 87, 345

posterior pituitary Portion of the pituitary gland that stores and secretes oxytocin and antidiuretic hormone, which are produced by the hypothalamus. 345

precapillary sphincter Smooth muscle ring that controls blood flow through a capillary bed. 94

precipitation Water deposited on the Earth in the form of rain, snow, sleet, hail, or fog. 554

pre-embryonic development Development of the zygote in the first week, including fertilization, the beginning of cell division, and the appearance of the chorion. 396

prefrontal area Association area in the frontal lobe that receives information from other association areas and uses it to reason and plan actions. 297

primary germ layer One of the three layers (ectoderm, mesoderm, and endoderm) of embryonic cells that develop into specific tissues and organs. 38

primary motor area Area in the frontal lobe where voluntary commands begin; each section controls a part of the body. 296

primary somatosensory area (soh-mat-uh-sens-ree, -suh-ree) Area dorsal to the central sulcus where sensory information arrives from skin and skeletal muscles. 296

primate (pry-mayt) Animal that belongs to the order Primates; includes prosimians, monkeys, apes, and humans, all of whom have adaptations for living in trees. 527

principle Theory generally accepted by an overwhelming number of scientists; a law. 9

prion An infectious particle that is the cause of diseases, such as scrapie in sheep, mad cow disease, and Creutzfeldt–Jakob disease in humans; it has a protein component, but no nucleic acid has been detected. 137

producer Photosynthetic organism at the start of a grazing food chain that makes its own food (e.g., green plants on land and algae in water). 548

product Substance that forms as a result of a reaction. 57

progesterone (proh-jes-tuh-rohn) Female sex hormone that helps maintain sex organs and secondary sex characteristics. 358, 376

progression In cancer, a second mutation that allows cells to invade surrounding tissues. 450

prokaryotic cell Type of cell that lacks a membrane-bounded nucleus and organelles. 46, 520

prolactin (PRL) (proh-lak-tin) Hormone secreted by the anterior pituitary that stimulates the production of milk from the mammary glands. 347

promotion In cancer, development of a group of cells from a single mutated cell. 450

prophase (proh-fayz) Mitotic phase during which chromatin condenses so that chromosomes appear; chromosomes are scattered. 425

proprioceptor (proh-pree-oh-sep-tur) Sensory receptor in skeletal muscles and joints that assists the brain in knowing the position of the limbs. 318

prosimian Member of a group of primates that includes lemur and tarsiers and may resemble the first primates to have evolved. 527

prostaglandin (prahs-tuh-glan-din) Hormone that has various and powerful local effects. 359

prostate gland (prahs-tayt) Gland located around the male urethra below the urinary bladder; adds secretions to semen. 368

protease Enzyme capable of breaking peptide bonds in a protein. 160

protein Molecule consisting of one or more polypeptides. 35

protein-first hypothesis In chemical evolution, the proposal that protein originated before other macromolecules and allowed the formation of protocells. 519

proteomics The study of all proteins in an organism. 512

prothrombin (proh-thrahm-bin) Plasma protein converted to thrombin during the steps of blood clotting. 123

prothrombin activator Enzyme that catalyzes the transformation of the precursor prothrombin to the active enzyme thrombin. 123

protocell In biological evolution, a possible cell forerunner that became a cell once it could reproduce. 519

proton Positive subatomic particle, located in the nucleus and having a weight of approximately one atomic mass unit. 20

proto-oncogene (proh-toh-ahng-koh-jeen) Normal gene that can become an oncogene through mutation. 450

provirus Latent form of a virus in which the viral DNA is incorporated into the chromosome of the host. 160

proximal convoluted tubule Highly coiled region of a nephron near the glomerular capsule, where tubular reabsorption takes place. 224

pseudostratified columnar epithelium Appearance of layering in some epithelial cells when, actually, each cell touches a baseline and true layers do not exist. 74

pulmonary artery (pool-muh-nair-ee, puul-) Blood vessel that takes blood away from the heart to the lungs. 96

pulmonary circuit Circulatory pathway that consists of the pulmonary trunk, the pulmonary arteries, and the pulmonary veins; takes oxygen-poor blood from the heart to the lungs and oxygen-rich blood from the lungs to the heart. 103

pulmonary fibrosis (fy-broh-sis) Accumulation of fibrous connective tissue in the lungs; caused by inhaling irritating particles, such as silica, coal dust, or asbestos. 211

pulmonary tuberculosis Tuberculosis of the lungs, caused by the bacillus *Mycobacterium tuberculosis.* 210

pulmonary vein Blood vessel that takes blood from the lungs to the heart. 96

pulse Vibration felt in arterial walls due to expansion of the aorta following ventricle contraction. 100

Punnett square (pun-ut) Gridlike device used to calculate the expected results of simple genetic crosses. 470

pupil (pyoo-pul) Opening in the center of the iris of the eye. 322

Purkinje fibers (pur-kin-jee) Specialized muscle fibers that conduct the cardiac impulse from the AV bundle into the ventricles. 98

pus Thick, yellowish fluid composed of dead phagocytes, dead tissue, and bacteria. 141

pyelonephritis Inflammation of the kidney due to bacterial infection. 233

R

radioisotope Unstable form of an atom that spontaneously emits radiation in the form of radioactive particles or radiant energy. 21

reactant (re-ak-tunt) Substance that participates in a reaction. 57

recessive allele (uh-leel) Allele that exerts its phenotypic effect only in the homozygote; its expression is masked by a dominant allele. 468

recombinant DNA DNA that contains genes from more than one source. 506

rectum (rek-tum) Terminal end of the digestive tube between the sigmoid colon and the anus. 180

red blood cell (erythrocyte) Formed element that contains hemoglobin and carries oxygen from the lungs to the tissues; also called erythrocyte. 68, 118

red bone marrow Blood-cell-forming tissue located in the spaces within spongy bone. 139, 240

reduced hemoglobin (hee-muh-gloh-bun) Hemoglobin carrying hydrogen ions. 208

referred pain Pain perceived as having come from a site other than that of its actual origin. 360

reflex Automatic, involuntary response of an organism to a stimulus. 302

refractory period (rih-frak-tuh-ree) Time following an action potential when a neuron is unable to conduct another nerve impulse. 290

renal artery (ree-nul) Vessel that originates from the aorta and delivers blood to the kidney. 218

renal cortex (ree-nul kor-teks) Outer portion of the kidney that appears granular. 220

renal medulla (ree-nul muh-dul-uh) Inner portion of the kidney that consists of renal pyramids. 220

renal pelvis Hollow chamber in the kidney that lies inside the renal medulla and receives freshly prepared urine from the collecting ducts. 220

renal vein (ree-nul) Vessel that takes blood from the kidney to the inferior vena cava. 218

renewable resource Resources normally replaced or replenished by natural processes; resources not depleted by moderate use. Examples include solar energy, biological resources such as forests and fisheries, biological organisms, and some biogeochemical cycles. 570

renin (ren-in) Enzyme released by kidneys that leads to the secretion of aldosterone and a rise in blood pressure. 230, 352

replacement reproduction Population in which each person is replaced by only one child. 569

repolarization When the charge inside the axon resumes a negative charge. 289

reproduce To produce a new individual of the same type. 5

reproductive cloning Genetically identical to the original individual. 404

reproductive system Organ system that contains male or female organs and specializes in the production of offspring. 81

residual volume Amount of air remaining in the lungs after a forceful expiration. 204

respiratory control center Group of nerve cells in the medulla oblongata that sends out nerve impulses on a rhythmic basis, resulting in involuntary inspiration on an ongoing basis. 205

respiratory pump Mechanism whereby reductions in thoracic pressure during the breathing cycle tend to aid the return of blood to the heart from peripheral veins. 101

respiratory system Organ system consisting of the lungs and tubes that bring oxygen into the lungs and take carbon dioxide out. 80

resting potential Polarity across the plasma membrane of a resting neuron due to an unequal distribution of ions. 288

restriction enzyme Bacterial enzyme that stops viral reproduction by cleaving viral DNA; used to cut DNA at specific points during production of recombinant DNA. 506

reticular fiber (rih-tik-yuh-lur) Very thin collagen fibers in the matrix of connective tissue, highly branched and forming delicate supporting networks. 66

reticular formation (rih-tik-yuh-lur) Complex network of nerve fibers within the central nervous system that arouses the cerebrum. 298

retina (ret-n-uh, ret-nuh) Innermost layer of the eyeball that contains the rod cells and the cone cells. 323

retinal (ret-n-al, -awl) Light-absorbing molecule that is a derivative of vitamin A and a component of rhodopsin. 324

retrovirus RNA virus containing the enzyme reverse transcriptase that carries out RNA to DNA transcription. 160

reverse transcriptase Enzyme that speeds the conversion of viral RNA to viral DNA. 160

rheumatic fever Disease caused by bacterial infection, characterized by fever, swelling and pain in the joints, sore throat, and cardiac involvement. 151

rheumatoid arthritis Persistent inflammation of synovial joints, often causing cartilage destruction, bone erosion, and joint deformities. 151

rhodopsin (roh-dahp-sun) Light-absorbing molecule in rod cells and cone cells that contains a pigment and the protein opsin. 324

ribosomal RNA (rRNA) (ry-buh-soh-mul) Type of RNA found in ribosomes where protein synthesis occurs. 495

ribosome (ry-buh-sohm) RNA and protein in two subunits; site of protein synthesis in the cytoplasm. 53

rigor mortis Contraction of muscles at death due to lack of ATP. 276

RNA (ribonucleic acid) (ry-boh-noo-klee-ik) Nucleic acid produced from covalent bonding of nucleotide monomers that contain the sugar ribose; occurs in three forms: messenger RNA, ribosomal RNA, and transfer RNA. 37, 494

RNA-first hypothesis In chemical evolution, the proposal that RNA originated before other macromolecules and allowed the formation of the first cell or cells. 519

RNA polymerase (pahl-uh-muh-rays) During transcription, an enzyme that joins nucleotides complementary to a DNA template. 498

rod cell Photoreceptor in retina of eyes that responds to dim light. 324

rotational equilibrium Maintenance of balance when the head and body are suddenly moved or rotated. 333

rotator cuff Tendons that encircle and help form a socket for the humerus and also help reinforce the shoulder joint. 253

round window Membrane-covered opening between the inner ear and the middle ear. 328

rugae Deep folds, as in the wall of the stomach. 176

runoff Water—from rain, snowmelt, or other sources—that flows over the land surface, adding to the water cycle. 554

S

SA (sinoatrial) node (sy-noh-ay-tree-ul) Small region of neuromuscular tissue that initiates the heartbeat; also called the pacemaker. 98

saccule (sak-yool) Saclike cavity in the vestibule of the inner ear; contains sensory receptors for gravitational equilibrium. 334

salivary amylase (sal-uh-vair-ee am-uh-lays, -layz) Secreted from the salivary glands; the first enzyme to act on starch. 172

salivary gland Gland associated with the mouth that secretes saliva. 172

saltatory conduction Movement of nerve impulses from one neurofibral node to another along a myelinated axon. 290

saltwater intrusion Movement of salt water into freshwater aquifers in coastal areas where groundwater is withdrawn faster than it is replenished. 573

sarcolemma (sar-kuh-lem-uh) Plasma membrane of a muscle fiber; also forms the tubules of the T system involved in muscular contraction. 267

sarcoma Cancer that arises in muscles and connective tissues. 451

sarcomere (sar-kuh-mir) One of many units arranged linearly within a myofibril whose contraction produces muscle contraction. 267

sarcoplasmic reticulum (sar-kuh-plaz-mik rih-tik-yuh-lum) Smooth endoplasmic reticulum of skeletal muscle cells; surrounds the myofibrils and stores calcium ions. 267

saturated fatty acid Fatty-acid molecule that lacks double bonds between the atoms of its carbon chain. 32

Schwann cell Cell that surrounds a fiber of a peripheral nerve and forms the myelin sheath. 288

science Development of concepts about the natural world, often by using the scientific method. 9

scientific method Process of attaining knowledge by making observations, testing hypotheses, and coming to conclusions. 9

scientific theory Concept supported by a broad range of observations, experiments, and conclusions. 9

sclera (skleer-uh) White, fibrous, outer layer of the eyeball. 322

scoliosis Abnormal, laterial (side-to-side) curvature of the vertebral column. 250

scrotum (skroh-tum) Pouch of skin that encloses the testes. 367

secondary oocyte In oogenesis, the functional product of meiosis I; becomes the egg. 436

second messenger Chemical signal such as cyclic AMP that causes the cell to respond to the first messenger—a hormone bound to a receptor protein in the plasma membrane. 344

selectively permeable Having degrees of permeability; the cell is impermeable to some substances and allows others to pass through at varying rates. 46, 50

semantic memory Capacity of the brain to store and retrieve information with regard to words or numbers. 300

semen (see-mun) Thick, whitish fluid consisting of sperm and secretions from several glands of the male reproductive tract. 368

semicircular canal (sem-ih-sur-kyuh-lur) One of three tubular structures within the inner ear that contain sensory receptors responsible for the sense of rotational equilibrium. 328, 333

semilunar valve (sem-ee-loo-nur) Valve resembling a half moon located between the ventricles and their attached vessels. 95

seminal vesicle (sem-uh-nul) Convoluted structure attached to the vas deferens near the base of the urinary bladder in males; adds secretions to semen. 368

seminiferous tubule (sem-uh-nif-ur-us) Long, coiled structure contained within chambers of the testis; where sperm are produced. 369

sensation Conscious awareness of a stimulus due to nerve impulses sent to the brain from a sensory receptor by way of sensory neurons. 316

sensory adaptation Phenomenon of a sensation becoming less noticeable once it has been recognized by constant repeated stimulation. 317

sensory neuron Nerve cell that transmits nerve impulses to the central nervous system after a sensory receptor has been stimulated. 287

sensory receptor Structure that receives either external or internal environmental stimuli and is a part of a sensory neuron or transmits signals to a sensory neuron. 287, 316

septum Wall between two cavities; in the human heart, a septum separates the right side from the left side. 94

serosa Membrane that covers internal organs and lines cavities without an opening to the outside of the body. 171

serotonin A neurotransmitter. 291

serous membrane (seer-us) Membrane that covers internal organs and lines cavities without an opening to the outside of the body; also called serosa. 82

Sertoli cell Cell associated with developing germ cells in seminiferous tubule; secretes fluid into seminiferous tubule and mediates hormonal effects on tubule. 369

serum (seer-um) Light yellow liquid left after clotting of blood. 123

severe combined immunodeficiency disease (SCID) Congenital illness in which both antibody- and cell-mediated immunity are lacking or inadequate. 122, 150

sex chromosome Chromosome that determines the sex of an individual; in humans, females have two X chromosomes and males have both an X and Y chromosome. 484

sex-linked Refers to allele that occurs on the sex chromosomes but may control a trait that has nothing to do with the sex characteristics of an individual. 484

short-term memory Retention of information for only a few minutes, such as remembering a telephone number. 300

sickle-cell disease Genetic disorder in which the affected individual has sickle-shaped red blood cells subject to hemolysis. 120, 478

simple goiter (goy-tur) Condition in which an enlarged thyroid produces low levels of thyroxine. 349

sinkhole Large surface crater caused by the collapse of an underground channel or cavern; often triggered by groundwater withdrawal. 573

sinus (sy-nus) Cavity or hollow space in an organ such as the skull. 248

sinusitis (sy-nuh-sy-tis) Infection of the sinuses, caused by blockage of the openings to the sinuses and characterized by postnasal discharge and facial pain. 208

sister chromatids One of two genetically identical chromosomal units that are the result of DNA replication and are attached to each other at the centromere. 421

skeletal muscle Striated, voluntary muscle tissue found in muscles that move the bones. 69, 264

skeletal muscle pump Pumping effect of contracting skeletal muscles on blood flow through underlying vessels. 101

skeletal system System of bones, cartilage, and ligaments that works with the muscular system to protect the body and provide support for locomotion and movement. 80

skill memory Capacity of the brain to store and retrieve information necessary to perform motor activities, such as riding a bike. 300

skin Outer covering of the body; can be called the integumentary system because it contains organs such as sense organs. 75

skull Bony framework of the head, composed of cranial bones and the bones of the face. 248

sliding filament model An explanation for muscle contraction based on the movement of actin filaments in relation to myosin filaments. 269

small intestine Long, tubelike chamber of the digestive tract between the stomach and large intestine. 176

small RNA Short RNA molecules that help to regulate gene expression. 495

smooth (visceral) muscle Nonstriated, involuntary muscle tissue found in the walls of internal organs. 69, 263

sodium–potassium pump Carrier protein in the plasma membrane that moves sodium ions out of and potassium ions into cells; important in nerve and muscle cells. 289

soft palate (pal-it) Entirely muscular posterior portion of the roof of the mouth. 172

somatic system That portion of the peripheral nervous system containing motor neurons that control skeletal muscles. 302

spasm Sudden, involuntary contraction of one or more muscles. 277

species Group of similarly constructed organisms capable of interbreeding and producing fertile offspring; organisms that share a common gene pool. 2

sperm Male gamete having a haploid number of chromosomes and the ability to fertilize an egg, the female gamete. 369

spermatogenesis (spur-mat-uh-jen-ih-sis) Production of sperm in males by the process of meiosis and maturation. 369, 436

sphincter (sfingk-tur) Muscle that surrounds a tube and closes or opens the tube by contracting and relaxing. 174

spinal cord Part of the central nervous system; the nerve cord that is continuous with the base of the brain plus the vertebral column that protects the nerve cord. 293

spinal nerve Nerve that arises from the spinal cord. 302

spiral organ Organ in the cochlear duct of the inner ear responsible for hearing; also called the organ of Corti. 330

spirillum A group of bacteria that exhibit a variety of spiral or undulating shapes or are comma-shaped. 134

spleen Large, glandular organ located in the upper left region of the abdomen; stores and purifies blood. 139

spongy bone Porous bone found at the ends of long bones where red bone marrow is sometimes located. 68, 240

sprain Injury to a ligament caused by abnormal force applied to a joint. 277

squamous epithelium (skway-mus, skwah-) Type of epithelial tissue that contains flat cells. 73

standard error Number used in evaluating statistical data to show the range of error in the data. 14

stapes (stay-peez) The last of three ossicles of the ear that serve to conduct vibrations from the tympanic membrane to the oval window of the inner ear. 328

starch Storage polysaccharide found in plants; composed of glucose molecules joined in a linear fashion with few side chains. 29

stereocilia (sing., stereocilium) Long, flexible microvilli that superficially resemble cilia. Within the inner ear, these signal changes in body position and help to maintain balance and equilibrium. 328

steroid (steer-oyd) Type of lipid molecule having a complex of four carbon rings; examples are cholesterol, progesterone, and testosterone. 33

steroid hormone One of a group of hormones derived from cholesterol. 343

stimulus Change in the internal or external environment that a sensory receptor can detect, leading to nerve impulses in sensory neurons. 289, 316

stomach Muscular sac that mixes food with gastric juices to form chyme, which enters the small intestine. 175

strain Injury to a muscle resulting from overuse or improper use. 277

striae gravidarum Linear, depressed, scarlike lesions occurring on the abdomen, breasts, buttocks, and thighs due to the weakening of the elastic tissues during pregnancy. 408

striated (stry-ayt-ud) Having bands; in cardiac and skeletal muscle, alternating light and dark crossbands produced by the distribution of contractile proteins. 69

stroke Condition resulting when an arteriole in the brain bursts or becomes blocked by an embolism; also called cerebrovascular accident. 106

subcutaneous layer (sub-kyoo-tay-nee-us) Tissue layer that lies just beneath the skin and contains adipose tissue. 75

submucosa Layer of connective tissue underneath a mucous membrane. 171

subsidence Occurs when a portion of the Earth's surface gradually settles downward. 573

substrate Reactant in a reaction controlled by an enzyme. 57

sudden infant death syndrome (SIDS) Any sudden and unexplained death of an apparently healthy infant aged one month to one year. 205

superior vena cava (vee-nuh kay-vuh) Large vein that enters the right atrium from above and carries blood from the head, thorax, and upper limbs to the heart. 103

surfactant Agent that reduces the surface tension of water; in the lungs, a surfactant prevents the alveoli from collapsing. 201

sustainable Ability of a society or ecosystem to maintain itself while also providing services to human beings. 586

suture (soo-chur) Type of immovable joint articulation found between bones of the skull. 255

sweat gland Skin gland that secretes a fluid substance for evaporative cooling; also called sudoriferous gland. 79

sympathetic division The part of the autonomic system that usually promotes activities associated with emergency (fight-or-flight) situations; uses norepinephrine as a neurotransmitter. 304

synapse (sin-aps, si-naps) Junction between neurons consisting of the presynaptic (axon) membrane, the synaptic cleft, and the postsynaptic (usually dendrite) membrane. 290

synapsis (sih-nap-sis) Pairing of homologous chromosomes during prophase I of meiosis I. 428

synaptic cleft (sih-nap-tik) Small gap between presynaptic and postsynaptic membranes of a synapse. 290

syndrome Group of symptoms that appear together and tend to indicate the presence of a particular disorder. 421

synovial joint Freely movable joint having a cavity filled with synovial fluid. 82, 255

synovial membrane Membrane that forms the inner lining of the capsule of a freely movable joint. 82

systemic circuit Blood vessels that transport blood from the left ventricle and back to the right atrium of the heart. 103

systemic lupus erythematosus (SLE) Syndrome involving the connective tissues and various organs, including kidney. 151

systole (sis-tuh-lee) Contraction period of the heart during the cardiac cycle. 97

systolic pressure (sis-tahl-ik) Arterial blood pressure during the systolic phase of the cardiac cycle. 100

T

taste bud Sense organ containing the receptors associated with the sense of taste. 320

Tay–Sachs disease Lethal genetic disease in which the newborn has a faulty lysosomal digestive enzyme. 477

T cell (T lymphocyte) Lymphocyte that matures in the thymus. Cytotoxic T cells kill antigen-bearing cells outright; helper T cells release cytokines that stimulate other immune system cells. 139

T-cell receptor (TCR) Molecule on the surface of a T lymphocyte to which an antigen binds. 145

technology The science or study of the practical or industrial arts. 14

tectorial membrane (tek-tor-ee-ul) Membrane that lies above and makes contact with the hair cells in the spiral organ. 330

telomere (tel-uh-meer) Tip of the end of a chromosome. 448

telophase (tel-uh-fayz) Mitotic phase during which daughter chromosomes are located at each pole. 427

template (tem-plit) Pattern or guide used to make copies; parental strand of DNA serves as a guide for the production of daughter DNA strands, and DNA also serves as a guide for the production of messenger RNA. 493

tendinitis (ten-din-eye-tis) An inflammation of muscle tendons and their attachments. 277

tendon (ten-dun) Strap of fibrous connective tissue that connects skeletal muscle to bone. 67, 264

testes (sing., testis) (tes-teez, tes-tus) Male gonads that produce sperm and the male sex hormones. 358, 367

test group Group exposed to the experimental variable in an experiment; compare to control group. 11

testosterone (tes-tahs-tuh-rohn) Male sex hormone that helps maintain sexual organs and secondary sex characteristics. 358, 370, 406

tetanus (tet-n-us) Sustained muscle contraction without relaxation. 272

tetany (tet-n-ee) Severe twitching caused by involuntary contraction of the skeletal muscles due to a calcium imbalance. 350

thalamus (thal-uh-mus) Part of the brain located in the lateral walls of the third ventricle that serves as the integrating center for sensory input; it plays a role in arousing the cerebral cortex. 298

therapeutic cloning Used to create mature cells of various cell types, to learn about specialization of cells, and to provide cells and tissue to treat human illnesses. 404

thermoreceptor Sensory receptor that is sensitive to changes in temperature. 316

threshold Electrical potential level (voltage) at which an action potential or nerve impulse is produced. 289

thrombin (thrahm-bin) Enzyme that converts fibrinogen to fibrin threads during blood clotting. 123

thrombocytopenia Insufficient number of platelets in the blood. 123

thromboembolism (thrahm-boh-em-buh-liz-um) Obstruction of a blood vessel by a thrombus that has dislodged from the site of its formation. 106

thrombus (thrahm-bus) Blood clot that remains in the blood vessel where it formed. 106

thymine (T) (thy-meen) One of four nitrogen-containing bases in nucleotides composing the structure of DNA; pairs with adenine. 38

thymosin Peptide secreted by the thymus gland that increases production of certain types of white blood cells. 358

thymus gland Lymphatic organ, located along the trachea behind the sternum, involved in the maturation of T lymphocytes in the thymus gland. Secretes hormones called thymosins, which aid the maturation of T cells and perhaps stimulate immune cells in general. 139, 349

thyroid gland Endocrine gland in the neck that produces several important hormones, including thyroxine, triiodothyronine, and calcitonin. 349

thyroid-stimulating hormone (TSH) Substance produced by the anterior pituitary that causes the thyroid to secrete thyroxine and triiodothyronine. 346

thyroxine (T_4) (thy-rahk-sin) Hormone secreted from the thyroid gland that promotes growth and development; in general, it increases the metabolic rate in cells. 349

tidal volume Amount of air normally moved in the human body during an inspiration or expiration. 203

tight juntion Junction between cells when adjacent plasma membrane proteins join to form an impermeable barrier. 56

tissue Group of similar cells that perform a common function. 66

tissue fluid Fluid that surrounds the body's cells; consists of dissolved substances that leave the blood capillaries by filtration and diffusion. 68, 92

tonicity (toh-nis-ih-tee) Osmolarity of a solution compared with that of a cell. If the solution is isotonic to the cell, there is no net movement of water; if the solution is hypotonic, the cell gains water; and if the solution is hypertonic, the cell loses water. 50

tonsillectomy (tahn-suh-lek-tuh-mee) Surgical removal of the tonsils. 209

tonsillitis Infection of the tonsils that causes inflammation and can spread to the middle ears. 209

tonsils Partially encapsulated lymph nodules located in the pharynx. 140, 198

total artificial heart (TAH) A mechanical replacement for the heart, as opposed to a partial replacement. 109

toxin Poisonous substance produced by living cells or organisms. Toxins are nearly always proteins that are capable of causing disease on contact or absorption with body tissues. 92, 135

tracer Substance having an attached radioisotope that allows a researcher to track its whereabouts in a biological system. 21

trachea (tray-kee-uh) Passageway that conveys air from the larynx to the bronchi; also called the windpipe. 200

tracheostomy (tray-kee-ahs-tuh-mee) Creation of an artificial airway by incision of the trachea and insertion of a tube. 200

tract Bundle of myelinated axons in the central nervous system. 293

transcription Process whereby a DNA strand serves as a template for the formation of mRNA. 496

transcription factor In eukaryotes, protein required for the initiation of transcription by RNA polymerase. 503

trans fat Fats, which occur naturally in meat and dairy products of ruminants, that are also industrially created through partial hydrogenation of plant oils and animal fats. 32

transfer RNA (tRNA) Type of RNA that transfers a particular amino acid to a ribosome during protein synthesis; at one end, it binds to the amino acid, and at the other end it has an anticodon that binds to an mRNA codon. 495

transgenic organism Free-living organism in the environment that has a foreign gene in its cells. 506

translation Process whereby ribosomes use the sequence of codons in mRNA to produce a polypeptide with a particular sequence of amino acids. 496

translocation Movement of a chromosomal segment from one chromosome to another nonhomologous chromosome, leading to abnormalities (e.g., Down syndrome). 441

triglyceride (trih-glis-uh-ryd) Neutral fat composed of glycerol and three fatty acids. 31

triiodothyronine (T_3) Hormone produced by the thyroid gland that contains three iodine atoms; the metabolically active form of the thyroid hormones. 349

trisomy One more chromosome than usual. 437

trophic level Feeding level of one or more populations in a food web. 550

trophic relationship In ecosystems, feeding relationships such as grazing food webs or detrital food webs. 550

tropomyosin (trahp-uh-my-uh-sin, trohp-) Protein that functions with troponin to block muscle contraction until calcium ions are present. 270

troponin (troh-puh-nin) Protein that functions with tropomyosin to block muscle contraction until calcium ions are present. 270

trypsin (trip-sin) Protein-digesting enzyme secreted by the pancreas. 178

T (transverse) tubule Membranous channel that extends inward. 267

tubal ligation Method for preventing pregnancy in which the uterine tubes are cut and sealed. 382

tubular reabsorption Movement of primarily nutrient molecules and water from the contents of the nephron into blood at the proximal convoluted tubule. 227

tubular secretion Movement of certain molecules from blood into the distal convoluted tubule of a nephron so that they are added to urine. 228

tumor (too-mur) Cells derived from a single mutated cell that has repeatedly undergone cell division; benign tumors remain at the site of origin, and malignant tumors metastasize. 449

tumor marker test Blood test for a substance, such as a tumor antigen, that indicates a patient has cancer. 458

tumor suppressor gene Gene that codes for a protein that ordinarily suppresses cell division; inactivity can lead to a tumor. 450

tympanic membrane (tim-pan-ik) Located between the outer and middle ear where it receives sound waves; also called the eardrum. 328

U

umbilical cord Cord connecting the fetus to the placenta through which blood vessels pass. 400

unsaturated fatty acid Fatty acid molecule that has one or more double bonds between the atoms of its carbon chain. 31, 32

uracil (U) (yoor-uh-sil) The base in RNA that replaces thymine found in DNA; pairs with adenine. 38, 494

urea (yoo-ree-uh) Primary nitrogenous waste of humans derived from amino acid breakdown. 178, 219

uremia High level of urea nitrogen in the blood. 233

ureter (yoor-uh-tur) One of two tubes that take urine from the kidneys to the urinary bladder. 218

urethra (yoo-ree-thruh) Tubular structure that receives urine from the bladder and carries it to the outside of the body. 219, 368

urethritis Inflammation of the urethra. 233

uric acid (yoor-ik) Waste product of nucleotide metabolism. 220

urinary bladder Organ where urine is stored before being discharged by way of the urethra. 219

urinary system Organ system consisting of the kidneys and urinary bladder; rids the body of nitrogenous wastes and helps regulate the water–salt balance of the blood. 80

uterine cycle (yoo-tur-in, -tuh-ryn) Monthly occurring changes in the characteristics of the uterine lining (endometrium). 376

uterus (yoo-tur-us) Organ located in the female pelvis where the fetus develops; also called the womb. 370

utricle (yoo-trih-kul) Saclike cavity in the vestibule of the inner ear that contains sensory receptors for gravitational equilibrium. 334

V

vaccine Antigens prepared in such a way that they can promote active immunity without causing disease. 148

vagina Organ that leads from the uterus to the vestibule and serves as the birth canal and organ of sexual intercourse in females. 372

valve Membranous extension of a vessel of the heart wall that opens and closes, ensuring one-way flow. 94

vas deferens (vas def-ur-unz, -uh-renz) Tube that leads from the epididymis to the urethra in males. 368

vasectomy Method for preventing pregnancy in which the vasa deferentia are cut and sealed. 382

vector (vek-tur) In genetic engineering, a means to transfer foreign genetic material into a cell (e.g., a plasmid). 506

ventilation Process of moving air into and out of the lungs; also called breathing. 197

ventricle (ven-trih-kul) Cavity in an organ, such as a lower chamber of the heart or the ventricles of the brain. 293

venule (ven-yool, veen-) Vessel that takes blood from capillaries to a vein. 94

vermiform appendix Small, tubular appendage that extends outward from the cecum of the large intestine. 180

vernix caseosa (vur-niks kay-see-oh-suh) Cheeselike substance covering the skin of the fetus. 403

vertebral column (vur-tuh-brul) Series of joined vertebrae that extends from the skull to the pelvis. 250

vertebrate (vur-tuh-brit, -brayt) An animal with a vertebral column. 7

vesicle Small, membrance-bounded sac that stores substances within a cell. 54

vestibule (ves-tuh-byool) Space or cavity at the entrance of a canal, such as the cavity that lies between the semicircular canals and the cochlea. 329

vestigial structure Remains of a structure that was functional in some ancestor but is no longer functional in the organism in question. 523

villus (pl., villi) (vil-us) Small, fingerlike projection of the inner small intestinal wall. 177

virus Noncellular, parasitic agent consisting of an outer capsid and an inner core of nucleic acid. 135

visual accommodation Ability of the eye to focus at different distances by changing the curvature of the lens. 323

vital capacity Maximum amount of air moved into or out of the human body with each breathing cycle. 201

vitamin Essential requirement in the diet, needed in small amounts. Vitamins are often part of coenzymes. 189

vitamin D Required for proper bone growth. 76

vitreous humor (vit-ree-us) Clear, gelatinous material between the lens of the eye and the retina. 323

vocal cord Fold of tissue within the larynx; creates vocal sounds when it vibrates. 199

vulva External genitals of the female that surround the opening of the vagina. 372

W

water (hydrologic) cycle Interdependent and continuous circulation of water from the ocean to the atmosphere, to the land, and back to the ocean. 554

Wernicke's area Brain area involved in language comprehension. 297

white blood cell (leukocyte) Type of blood cell that is transparent without staining and protects the body from invasion by foreign substances and organisms; also called a leukocyte. 68, 121

white matter Myelinated axons in the central nervous system. 293

XDR TB Extensively drug-resistant tuberculosis, a type of bacterium that causes tuberculosis (TB) that is no longer susceptible to almost all of the drugs normally used to treat TB. 168

xenotransplantation Use of animal organs, instead of human organs, in human transplant patients. 150, 509

X-linked Refers to allele located on an X chromosome, but may control a trait that has nothing to do with the sex characteristics of an individual. 484

yolk sac Extraembryonic membrane that encloses the yolk of birds; in humans, it is the first site of blood cell formation. 397

Z

zygote (zy-goht) Diploid cell formed by the union of sperm and egg; the product of fertilization. 370, 394

Credits

Photographs

Chapter 1

Opener: Courtesy NASA/JPL/University of Arizona; 1.1(leech): © St. Bartholomews Hospital/Photo Researchers, Inc.; (mushrooms): © IT Stock/age fotostock RF; (bacteria): © Dr. Dennis Kunkel/Phototake; (meerkat): © Jami Tarris/Getty Images; (cotton): © Courtesy USDA, David Nance, photographer; (Giardia): © Dr. Fred Hossler/Visuals Unlimited; 1.3a: © Vol. 124/Corbis RF; 1.3b: © John Cancalosi/Peter Arnold/Photolibrary; 1.3c: © Ingram Publishing/SuperStock RF; 1.4a(seedling): © Herman Eisenbeiss/Photo Researchers, Inc.; (tree): Courtesy Paul Wray, Iowa State University; 1.4b(fertilization): © Dr. David Phillips/Visuals Unlimited; (fetus): © Derek Bromhall/OSF/Animals Animals; 1.6(protist): © Michael Abby/Visuals Unlimited; (fungi): © Rob Planck/Tom Stack; (plant): © Pat Pendarvis; (animal): © Corbis RF; (Archaea): © Ralph Robinson/Visuals Unlimited; (Bacteria): © A.B. Dowsett/SPL/Photo Researchers, Inc.; 1.7a: © Don and Pat Valenti/DRK Photo; 1.7b: © K. Tumanowicz/Photo Researchers, Inc.; 1.8: © Getty RF; 1.9(main): © Tony McDonough/epa/Corbis; (inset): © Eye of Science/Photo Researchers, Inc.; 1.10a–b: © blickwinkel/Alamy; 1.10c: © Phanie/Photo Researchers, Inc.; 1.12a: © Getty Images/Digital Vision RF; 1.12b: © Luiz C. Marigo/Peter Arnold/Photolibrary; 1A: © Corbis/Sygma.

Chapter 2

Opener: © liquidlibrary/PictureQuest RF; 2.3a: © Biomed Commun./Custom Medical Stock Photo; 2.3b(patient): Courtesy National Institutes of Health; (brain scan): © Mazzlota et al./Photo Researchers, Inc; 2.4a(peaches): © Tony Freeman/PhotoEdit; 2.4b: © Geoff Tompkinson/SPL/Photo Researchers, Inc.; 2.5b(crystals, shaker): © Evelyn Jo Johnson; 2.8a: © The McGraw-Hill Companies, Inc./Jill Braaten, photographer; 2.8b: © Amanda Langford/SuperStock RF; 2.10a: © Ray Pfortner/Peter Arnold/Photolibrary; 2.10b: © Frederica Georgia/Photo Researchers, Inc.; 2.13: © Jeremy Burgess/SPL/Photo Researchers, Inc.; 2.14: © Don W. Fawcett/Photo Researchers, Inc.; 2.15: © Science Source/J.D. Litvay/Visuals Unlimited; 2.20b: © Warren Toda/epa/Corbis; 2.20c: © Tony Marsh/Reuters/Corbis; p. 34(woman, curly hair): © Comstock/PunchStock RF; p. 35(hemoglobin): © P. Motta & S. Correr/Photo Researchers, Inc.; (man, muscle): © Getty RF.

Chapter 3

Opener: © Bruce Dale/National Geographic/Getty Images; 3.1(red blood cells): © Prof. P. Motta, Dept. of Anatomy, Univ. LaSapienza Rome/SPL/Photo Researchers, Inc.; (nerve cells): © Dr. Dennis Kunkel/Visuals Unlimited; (cartilage cells): © Ed Reschke; 3.3a: © David M. Phillips/Visuals Unlimited; 3.3b: © Alfred Pasieka/Photo Researchers, Inc.; 3.3c: © Warren Rosenberg/Biological Photo Service; 3Aa: Courtesy Sierra Blakely via Creative Commons/Wikipedia; 3Ab: © M. Schliwa/Visuals Unlimited; 3.4a: © Alfred Pasieka/Photo Researchers, Inc.; 3.9a(all): © Dennis Kunkel/Phototake; 3.13(nuclear pores): Courtesy E.G. Pollock; 3.13(ER): © R. Bolender & D. Fawcett/Visuals Unlimited; 3.15b: © Y. Nikas/Photo Researchers, Inc.; 3.15c: © David M. Phillips/Photo Researchers, Inc.; 3.16: © Dr. Don W. Fawcett/Visuals Unlimited; 3B: © Biophoto Associates/Photo Researchers, Inc.

Chapter 4

Opener: © PHANIE/Photo Researchers, Inc.; 4.2(all), 4.5a: © Ed Reschke; 4.5b: © The McGraw-Hill Companies, Inc./Dennis Strete, photographer; 4.5c–4.7: © Ed Reschke; 4.9: © John D. Cunningham/Visuals Unlimited; 4.10a: © Ken Greer/Visuals Unlimited; 4.10b: © James Stevenson/SPL/Photo Researchers, Inc.; 4B(top, both): © AFP/Getty Images; (bottom, both): © Paul Cooper Photography, Paris; 4C: © Stockbyte/Getty RF.

Chapter 5

Opener: © Johannes Kroemer/Corbis; 5.2(left): © Ed Reschke; (right): © Biophoto Associates/Photo Researchers, Inc.; 5.3b: © SIU/Visuals Unlimited; 5.4b: © Dr. Don W. Fawcett/Visuals Unlimited; 5.5d: © Biophoto Associates/Photo Researchers, Inc.; 5.6b-c: © Ed Reschke; 5.6d: © Lester Lefkowitz/Corbis; 5.7: © Comstock Images/PictureQuest; 5.14(normal): © Ed Reschke; (plaque): © Dr. Gladden Willis/Visuals Unlimited; p. 107(Thomas): Courtesy of The Alan Mason Chesney Medical Archives of The Johns Hopkins Medical Institutions; 5.15a: © Pascal Goethgheluck/SPL/Photo Researchers, Inc.; 5.16(right): Courtesy SynCardia Systems, Inc.; 5A: © Getty RF.

Chapter 6

Openers: (main): © SPL/Photo Researchers, Inc.; (inset): © Andrew Syred/Photo Researchers, Inc.; 6.1: © Getty RF; 6.3a: © Andrew Syred/Photo Researchers, Inc.; 6.3c: © Lennart Nilsson, Behold Man, Little Brown and Company, Boston; p. 120(sickled cell): © Phototake, Inc./Alamy; 6.7b: © Eye of Science/Photo Researchers, Inc.; 6A: © /Getty RF.

Chapter 7

Opener: © Dr. Ken Greer/Visuals Unlimited; 7.1b: © Dr. David M. Phillips/Visuals Unlimited; 7.1c: © Dr. Dennis Kunkel/Visuals Unlimited; 7.1d: © Dr. Gary D. Gaugler/Phototake; 7.4: Courtesy Centers for Disease Control and Prevention (CDC); 7A: © Bettman/Corbis; 7.7(marrow): © R. Valentine/Visuals Unlimited; (thymus, spleen): © Ed Reschke/Peter Arnold/Photolibrary; (lymph): © Fred E. Hossler/Visuals Unlimited; 7.11b: Courtesy Dr. Arthur J. Olson, Scripps Institute; 7.13: © Steve Gschmeissner/Photo Researchers, Inc.; 7.14a: © Michael Newman/PhotoEdit; 7.15a: © John Lund/Drew Kelly/Blend Images/Corbis RF; 7.15b: © Digital Vision/Getty RF; 7.15c: © PhotoDisc Collection/Getty RF; 7.17: © Southern Illinois University/Photo Researchers, Inc..

Supplement Chapter

Opener: © Bettmann/Corbis; S.6: © Elmer Koneman/Visuals Unlimited; SA: © 2000–2006 Custom Medical Stock Photo; S.7: © ISM/Phototake; S.9: © Shehzad Noorani/Peter Arnold/Photolibrary; S.10: © China Photo / Reuters/Corbis; S.12: © Cordelia Molloy/Photo Researchers, Inc.

Chapter 8

Opener: © Bubbles Photolibrary/Alamy; 8A(bike): © Corbis RF; 8A(man): © Stockbyte/Getty RF; 8.5c: © Dr. Fred Hossler/Visuals Unlimited; 8.6(villi): © Manfred Kage/Peter Arnold/Photolibrary; (microvilli):Reprinted from Medical Cell Biology, Charles Flickinger, copyright 1979, with permission from Elsevier.; 8B(all): Courtesy Given Imaging, Ltd.; p. 184(couple running): © BananaStock/age fotostock RF; p. 184(scale, foods): © Photodisc/Getty RF; 8.12: © Cole Group/Getty RF; 8.13: © Volume 20/Photodisc/Getty RF; 8C: © Evelyn Jo Johnson; 8.16a: © Ted Foxx/Alamy; 8.16b: © Donna Day/Stone/Getty Images; 8.16c: © Corbis RF.

Chapter 9

Opener: © Sean O'Brien/Custom Medical Stock Photo; 9.4(left): © CNRI/Phototake; 9.5: © Dr. Kessel & Dr. Kardon/Tissues & Organs/Visuals Unlimited; 9.9: © Burger/Photo Researchers, Inc.; 9.13a: © Matt Meadows/Peter Arnold/Photolibrary; 9.13b: © Biophoto Associates/Photo Researchers, Inc.

Chapter 10

Openers: (normal kidney, left): © SIU/Visuals Unlimited/Getty Images; (polycystic kidney, right): © Biophoto Associates/Photo Researchers, Inc.; 10.3b: © Ralph T. Hutchings/Visuals Unlimited; 10.4: © James Cavallini/Photo Researchers, Inc.; 10.5(both): © Science Photo Library/Getty RF; 10.6a: © Joseph F. Gennaro Jr./Photo Researchers, Inc.; 10B: © Ian Hooton/Photo Researchers, Inc.; 10.12: © AJPhoto/Photo Researchers, Inc.; 10C: © AP Photo/Brian Walker.

Chapter 11

Opener: © Southern Stock Corp/Corbis; 11.1(hyaline, compact bone): © Ed Reschke; (osteocyte): © Biophoto Associates/Photo Researchers, Inc.; 11A(young woman walking): © Corbis RF; 11Aa-b: © Michael Klein/Peter Arnold/Photolibrary; 11Ac: © Bill Aaron/PhotoEdit; 11.5b: © Tony Freeman/PhotoEdit; 11.8b: © Corbis RF; 11.13a: © Gerard Vandystadt/Photo Researchers, Inc.; 11B(both): © Scott Camazine/Photo Researchers, Inc.

Chapter 12

Opener: © Erproductions Ltd/Blend Images/Corbis; 12.1(all): © Ed Reschke; 12.4(both): © The McGraw-Hill Companies, Inc./J.W. Ramsey, photographer; 12.6(gymnast): © Corbis RF; (myofibril): © Biology Media/Photo Researchers, Inc.; 12.7(top right): © Victor B. Eichler; 12.12(man): © Lawrence Manning/Corbis; (muscle fibers): © G.W. Willis/Visuals Unlimited; (woman): © Corbis RF; 12A(left): © ImageState/PunchStock RF; (center): © Nancy Ney/Photodisc/Getty RF; (right): © Getty Images/Stockbyte RF; 12A(both): © Bettman/Corbis; 12B: © Focus on Sport/Getty Images; 12.14: © Getty RF.

Chapter 13

Opener: © Peter Duddek/Visum/The Image Works; 13.2a(myelin): © M.B. Bunge/Biological Photo Service; 13.2c(cell body): © Manfred Kage/Peter Arnold/Photolibrary; 13.5: Courtesy Dr. E.R. Lewis, University of California Berkeley; 13Aa: Drawing is from "Histologie due Systeme Nerveux de l'Homme et des Vertebres. Courtesy of Instituto Cajal, Madrid, Spain; 13Ab: © David Becker/Photo Researchers, Inc.; 13.7a: © Karl E. Deckart/Phototake; 13.7d: © The McGraw-Hill Companies, Inc./Rebecca Gray, photographer and Don Kincaid, dissections; 13.8b: © Colin Chumbley/Science Source/Photo Researchers, Inc.; 13.13(all): © Marcus Raichle; 13.14: © Dr. Richard Kessel & Dr. Randy Kardon/Visuals Unlimited; 13.18: © Betts Anderson Loman/PhotoEdit; 13.19: © Vol. 94 PhotoDisc/Getty RF.

Chapter 14

Openers: (both): © AP Images/Gene J. Puskar; 14.4(all tastebuds): © Omikron/SPL/Photo Researchers, Inc.; 14.8a: © Lennart Nilsson, from The Incredible Machine/Scanpix; 14.9b: © Biophoto Associates/Photo Researchers, Inc.; 14A: © Pascal Goetgheluck/Photo Researchers, Inc.; 14.13(stereocilia): © P. Motta/SPL/Photo Researchers, Inc.; 14Ba(both): Courtesy Robert S. Preston and Joseph E. Hawkins, Kresge Hearing Research Institute, University of Michigan.

Chapter 15

Opener: © Russ Curtis/Photo Researchers, Inc.; 15.8a: © AP/Wide World Photos; 15.8b: © General Photographic Agency/Getty Images; 15.9(all): From Clinical Pathological Conference, "Acromegaly, Diabetes, Hypermetabolism, Proteinura and Heart Failure," American Journal of Medicine 20 (1956) 133, with permission from Excerpta Medica, Inc.; 15.10a: © Bruce Coleman, Inc./Alamy; 15.10b: © Medical-on-Line/Alamy; 15.10c: © Dr. P. Marazzi/Photo Researchers, Inc.; 15.14a: © Custom Medical Stock Photos; 15.14b: © NMSB/Custom Medical

Stock Photos; 15.15(both): Atlas of Pediatric Physical Diagnosis, Second Edition by Zitelli & Davis, 1992. Mosby-Wolfe Europe Limited, London, UK; 15.16: © Peter Arnold, Inc./Alamy; 15A: © Bettmann/Corbis; 15.20: © Evelyn Jo Johnson.

Chapter 16

Opener: © UpperCut Images/Getty RF; 16.3b: © Anatomical Travelogue/Photo Researchers, Inc.; 16.4b: © Ed Reschke; 16.4d: © Dr. Dennis Kunkel/Visuals Unlimited; 16A: © Getty Images; 16.8a: © Ed Reschke/Peter Arnold/Photolibrary; 16.14a: © The McGraw-Hill Companies, Inc./Christopher Kerrigan, photographer; 16.14b: © Michael Keller/Corbis; 16.14d: © SIU/Visuals Unlimited; 16.14e: © Keith Brofsky/Getty Images; 16.14g: © Phanie/Photo Researchers, Inc.; 16.14h: © LADA/Photo Researchers, Inc.; 16B: © Getty RF; 16.16: © CC Studio/SPL/Photo Researchers, Inc.; 16.17: © Dr. Olivier Schwartz/Institut Pasteur/Photo Researchers, Inc.; 16.18(both): Courtesy Centers for Disease Control, Atlanta, GA. (Crooks, R. and Baur, K. Our Sexuality, 8/e, Wadsworth 2000 17.4, pg. 489; 16.19a: © Science VU/CDC/Visuals Unlimited; 16.19b: © Courtesy Centers for Disease Control and Prevention, Dr. N.J. Fiumara, Dr. Gavin Hart; 16.19c: © Centers for Disease Control and Prevention; 16.19d: © Science VU/Visuals Unlimited; 16C(left): © Vol. 161/Corbis RF; (right): © David Raymer/Corbis; 16D: © Argus Fotoarchiv/Peter Arnold/Photolibrary.

Chapter 17

Opener: © John Slater/Getty RF; 17.1: © David M. Phillips/Visuals Unlimited; 17.6: © Lennart Nilsson, A Child is Born, Dell Publishing/Scanpix; 17.8: © Neil Bromhall/Getty Images; 17A: © Associated Press; 17.10: Courtesy Dr. Howard Jones, Eastern VA Medical School; 17.11c: © Rune Hellestad/Corbis; p. 409(Semmelweis): © Getty Images; 17.12: © Image Source/Getty RF; 17.13: © Ronnie Kaufman/Corbis.

Chapter 18

Opener: © Kevin Mazur/Contributor/WireImage/Getty Images; 18.1: © CNRI/SPL/Photo Researchers, Inc.; 18.6(top): © Scott Camazine/Photo Researchers, Inc.; (bottom): © Edward Kinsman/Photo Researchers, Inc.; 18.9(early prophase, prophase, metaphase, anaphase, telophase): © Ed Reschke; (early metaphase): © Michael Abbey/Photo Researchers, Inc.; 18.17a: © Scott Wintrow/Getty Images; 18.17b: © CNRI/SPL/Photo Researchers, Inc.; 18.20b: Courtesy The Williams Syndrome Association; 18.21b(both): From N.B. Spinner et al., American Journal of Human Genetics 55 (1994):p. 239. The University of Chicago Press; 18A: © Corbis/Sygma.

Chapter 19

Opener: © Strauss/Curtis/Corbis; 19.1a: © Biology Media/Photo Researchers, Inc.; 19.1b: © SPL/Photo Researchers, Inc.; 19.1c: © Martin M. Rotker/Photo Researchers, Inc.; 19.8A: © Custom Medical Stock Photo, Inc.; 19.8B: © ISM/Phototake; 19.8C: © Dr. P. Marazzi/Photo Researchers, Inc.; 19.8D: © Custom Medical Stock Photo, Inc.; 19.8E: © Dr. Ken Greer/Visuals Unlimited; 19.9: © UHB Trust/Getty Images; 19.10: © BSIP/Phototake; 19.11: © Simon Fraser/Royal Victoria Infirmary/Photo Researchers, Inc.; 19D: © Obstetrics & Gynecology/Science Photo Library; 19E: © Visuals Unlimited/Corbis.

Chapter 20

Opener: © Getty RF; 20.2a: © HFPA, 63rd Golden Globe Awards; 20.2b: © Dynamic Graphics/PictureQuest RF; 20.2c-f: © The McGraw-Hill Companies, Inc./Bob Coyle, photographer; 20.2g: © Jupiterimages/Getty RF; 20.2h: © Creatas/PunchStock RF; 20Aa-b: © Brand X/SuperStock RF; 20.11(right): © Pat Pendarvis; 20.12(both brain tissue slides): Courtesy Dr. Hemachandra Reddy, The Neurological Science Institute, Oregon Health & Science University; (woman with Huntington): © Steve Uzzell; 20.13: Courtesy University of Connecticut/Peter Morenus, photographer; 20.14: © H. Reinhard/Arco Images/Peter Arnold/Photolibrary; 20.15: © Mediscan/Medical-On-Line; 20B(queen): © Stapleton Collection/Corbis; (prince): © Huton Archive/Getty Images; 20.19(left, right): Courtesy Dr. Rabi Tawil, Director, Neuromuscular Pathology Laboratory, University of Rochester Medical Center; (center): Courtesy Muscular Dystrophy Association.

Chapter 21

Opener: © Blend Images/Alamy RF; 21A: © Science Source/Photo Researchers, Inc.; 21.11: Courtesy Alexander Rich; 21.14(sequencer): © Lawrence Berkeley Nat'l Lab/Roy Kaltschmidt, photographer; 21.18a: © Maximilian Stock Ltd./Photo Researchers, Inc.; 21.18b: Courtesy Robert H. Devlin, Fisheries and Oceans Canada; 21.18c: © Richard Shade; 21.18d: © Dr. Brad Mogan/Visuals Unlimited, Inc.; 21.18e: © Jerry Mason/Photo Researchers, Inc.; 21.19(all): Courtesy Monsanto; 21.20b(both): Courtesy Eduardo Blumwald; 21Ba: Courtesy USDA/Doug Wilson, photographer; 21Bb: © James Shaffer/PhotoEdit; 21Bc: Courtesy Carolyn Merchant/UC Berkeley; 21.23a(top) and 21.23b (left): © Digital Vision/Getty RF; 21.23a(bottom): © Getty RF; 21.23b(right): © The McGraw-Hill Companies, Inc./Bob Coyle, photographer; 21.23c(left): © Digital Vision/Getty RF; (right): © Getty RF.

Chapter 22

Opener: © Tom McHugh/Photo Researchers, Inc.; 22.3a: © Doug Sherman; 22.3b: Reprinted by permission from Macmillan Publishers Ltd: Nature (A Cretaceous Terrestrial Snake With Robust Hindlimbs and a Sacrum), copyright 2006; 22.4(fossil, left): © Jean-Claude Carton/Bruce Coleman/PhotoShot; (drawing, right): © Joe Tucciarone; 22.6a: © Juan & Carmecita Munoz/Photo Researchers, Inc.; 22.6b: © Michael Maconachie/Papilio/Corbis; 22.8(both): © Carolina Biological Supply/Phototake; 22.12(gibbon): © Hans & Judy Beste/Animals Animals; (orangutan, chimps, gorillas): © Creatas/PunchStock RF; 22.15a: © Dan Dreyfus and Associates; 22.15b: © John Reader/Photo Researchers, Inc.; 22Aa: © Peter Schouten/National Geographic Society/Reuters/Corbis; 22Ab(right): © AP Photo/Richard Lewis; 22.17: © National Museum of Kenya; 22.19: © The Field Museum #A102513c; 22.20: Transp. #608 Courtesy Department of Library Services, American Museum of Natural History; 22B: © Peter Bowater/Photo Researchers, Inc.; 22.21a: © PhotoDisc/Getty RF; 22.21b: © Sylvia S. Mader; 22.21c: © Adam Crowley/Getty Images.

Chapter 23

Opener: © Image Source RF; 23.1(temperate, desert, rain forest, prairie): © Corbis RF; (savanna): © Gregory Dimijan/Photo Researchers, Inc.; (taiga): © Brand X Pictures/PunchStock RF; (autumn tundra): © Joseph Van Os/Getty Images; 23.1(winter tundra): © Oxford Scientific/Getty Images; 23.2a, d: © Corbis RF; 23.2b: © The McGraw-Hill Companies, Inc./Carlyn Iverson, photographer; 23.2c: © David Hall/Photo Researchers, Inc.; 23.3a(left): © Ed Reschke/Peter Arnold/Photolibrary; (right): © Authors Image/PunchStock RF; 23.3b(left): © Corbis RF; (right): © Gerald C. Kelley/Photo Researchers, Inc.; 23.3c(left): © Bill Beatty/Visuals Unlimited; (right): © Joe McDonald/Visuals Unlimited; 23.3d(left): © SciMAT/Photo Researchers, Inc.; (right): © IT Stock/age fotostock RF; 23.5: © George D. Lepp/Photo Researchers, Inc.; 23A(left): © Brand X Pictures/PunchStock RF; (right): © Barry Lewis/Corbis; 23.13a: © Thomas Kitchin/Tom Stack & Assoc.; 23.13b: © John Shaw/Tom Stack & Assoc.; 23.13c: © Sue Baker/Photo Researchers, Inc.; 23B: Courtesy NASA.

Chapter 24

Opener: © Author's Image/Punchstock RF; 24.2c: © Still Pictures/Peter Arnold/Photolibrary; 24.3(land): © Vol. 39/PhotoDisc/Getty RF; (water): © Evelyn Jo Johnson; (food): © The McGraw-Hill Companies, Inc./John Thoeming, photographer; (energy): Photo Link/Getty RF; (minerals): © PhotoLink/Getty RF; 24.4b: © Melvin Zucker/Visuals Unlimited; 24.5b: © Carlos Dominguez/Photo Researchers, Inc.; 24.6b: © PhotoLink/Getty RF; 24.7a: © Comstock Images/Alamy RF; 24.7b: © Jeremy Samuelson/FoodPix/Getty Images; 24.7c: © Stockbyte/PunchStock RF; 24.8: © AP Photo/Mark Gormus; 24.9a: © Jodi Jacobson/Peter Arnold/Photolibrary; 24.9b: © Inga Spence/Visuals Unlimited; 24.9c: © View Stock/Alamy; 24.10a: © Laish Bristol/Visuals Unlimited; 24.10b: © Inga Spence/Visuals Unlimited; 24.10c: Courtesy V. Jane Windsor, Division of Plant Industry, Florida Department of Agriculture & Consumer Services; 24.11: © Jim Richardson/Corbis; 24.12a: © Corbis RF; 24.12b: © Getty RF; 24.12c: © Argus Fotoarchiv/Peter Arnold/Photolibrary; 24.12d: © Photo Link/Getty RF; 24.13a: Courtesy DaimlerChrysler; 24.13b: © Warren Gretz/DOE/NREL; 24.15b: © Gunter Ziesler/Peter Arnold/Photolibrary; 24.16a: © Shane Moore/Animals Animals; 24.16b(both): © Peter Auster/University of Connecticut; 24.17(periwinkle): © Steven P. Lynch; (armadillo): © PhotoDisc/Getty RF; (fishing): © Herve Donnezan/Photo Researchers, Inc.; (bat): © Merlin D. Tuttle/Bat Conservation International; (lady bug): © Anthony Mercieca/Photo Researchers, Inc.; (rubber): © Bryn Campbell/Stone/Getty Images; 24A: Courtesy USDA/Lila De Guzman, photographer; 24.18(deforestation): © Carlos Dominguez/Photo Researchers, Inc.; (overfishing): © Shane Moore/Animals Animals; (water pollution): © Michael Gadomski/Animals Animals; (air pollution): © Thomas Kitchin/Tom Stack & Assoc.; (habitat loss): © Bill Zimmer/Associated Press; (erosion): © USDA/Nature Source/Photo Researchers, Inc.; 24.19(farming): © Inga Spence/Visuals Unlimited; (delta preservation): © Peter DeJong/Associated Press; (recycling): © Jeffrey Greenberg/Photo Researchers, Inc.; (mass transit): Courtesy DaimlerChrysler; (temperature control): © Reuters/Corbis; (integrated pest management): Courtesy V. Jane Windsor, Division of Plant Industry, Florida Department of Agriculture & Consumer Services; 24.20: © Dan McCoy/Rainbow; p. 590(land): © Vol. 39/PhotoDisc/Getty RF; (water): © Evelyn Jo Johnson; (food): © The McGraw-Hill Companies, Inc./John Thoeming, photographer; (energy):Photo Link/Getty RF; (minerals): © PhotoLink/Getty RF.

Line Art Credits

Chapter 5

5.15c: Source: American Heart Association; 5.16: Figure from www.texasheart.org AB/OMED Replacement Heart. Reprinted with permission from ABIOMED, Inc.

Index

Note: page numbers followed by *t* and *f* refer to tables and figures respectively.

A

Abdominal aorta, 103*f*
Abdominal cavity, 81, 82*f*
Abduction, 256*f*
Abiotic components of ecosystem, 548
ABO blood types, 125–126, 125*f*, 126*f*, 128, 481, 483*f*
Abortion, spontaneous, 429, 442
Absorption in digestive system, 171, 175
Abstinence, sexual, 379, 380*t*
Accommodation, visual, 323–324, 323*f*
ACE (angiotensin converting enzyme) inhibitors, 106
Acetabulum, 254, 254*f*
Acetylcholine (ACh)
  in Alzheimer disease, 412
  disorders of, 278
  in muscular system, 269*f*, 270, 272
  in nervous system, 291, 304, 304*t*, 306, 308
Acetylcholinesterase (AChE), 291, 412
Acetylsalicylic acid, 306
ACh. *See* Acetylcholine
AChE (acetylcholinesterase), 291, 412
Achilles tendon, 266*f*
Acid(s)
  amino. *See* Amino acids
  carbonic, 232
  essential fatty, 186
  fatty, 32, 32*f*, 185, 186, 188, 272, 273*f*, 274
  folic, 120, 190*t*
  lactic, 232
  nicotinic, 190*t*
  pantothenic, 190*t*
  trans fatty, 32, 32*f*, 186, 188
  uric, 220, 225*f*
  water and, 26–28
Acid-base balance, 23, 84, 116, 220, 232, 232*f*
Acid-base buffer systems, 232, 232*f*
Acid deposition, 27, 27*f*, 559–561, 559*f*, 576, 580
Acidic solutions, 26
Acidosis, 232
Acid rain, 27, 27*f*, 559–561, 559*f*, 576, 580
Acid reflux, 174
Acne, 79, 83
Acquired defenses, 143–147, 143*f*, 143*t*, 144*f*, 145*t*, 146*f*
Acquired immunity, 147–149, 147*f*, 149*f*
Acquired immunodeficiency syndrome (AIDS). *See also* HIV
  in HIV progression, 158–159, 385
  immune system and, 150–151
  vaccine for, 162, 163
Acromegaly, 244, 347, 348*f*
Acromion process, 253, 253*f*
Acrosome, of sperm, 369, 370*f*, 394, 394*f*
ACTH (adrenocorticotropic hormone)
  adrenal gland and, 351, 351*f*, 353
  in endocrine system, 342*t*, 346, 346*f*
Actin, 35, 69, 267, 268*f*
Actin filament, 55, 270*f*, 427
Action potential, 288*f*–289*f*, 289
Action potential propagation, 290
Active immunity, 147–148
Active site, 57
Active transport, 51, 51*f*
Acute bronchitis, 210
Acute lymphocytic leukemia (ALL), 115
AD (Alzheimer disease), 412, 412*f*
  drugs for, 306
  hippocampus and, 300
  proteins function and, 36
  treatment of, 412
ADA (adenosine deaminase), 122, 513
Adam's apple, 173, 199, 370
Adaptation
  biocultural evolution and, 538
  in evolution, 6, 520, 539
  in natural selection, 526
Addiction, nicotine, 212
Addison disease, 353, 353*f*
Adduction, 256*f*
Adductor longus, 265, 266*f*
Adenine (A), 38, 38*t*, 492, 492*f*, 494, 495*f*
Adenocarcinomas, 451
Adenoids, 209
Adenosine deaminase (ADA), 122, 513
Adenosine diphosphate (ADP), 38–39, 39*f*, 59, 271, 273
Adenosine triphosphate. *See* ATP
Adenovirus, 135*f*, 513
ADH (antidiuretic hormone), 230, 231, 342*t*, 345, 346*f*, 360
Adhesion junctions, 56, 56*f*
Adipose tissue, 66*f*, 67, 67*f*, 75*f*, 77, 218, 359
Adolescence, 244, 410
ADP (adenosine diphosphate), 38–39, 39*f*, 59, 271, 273
Adrenal cortex, 230*f*, 341*f*, 342*t*, 351, 351*f*, 352–353, 352*f*
Adrenal cortex malfunction, 353, 353*f*
Adrenal glands, 218*f*, 341*f*, 344, 351–354, 351*f*
Adrenal medulla, 99, 342*t*, 351, 351*f*, 352
Adrenocorticotropic hormone (ACTH)
  adrenal gland and, 351, 351*f*, 353
  in endocrine system, 342*t*, 346, 346*f*
Adult stem cells, bioethics and, 61. *See also* Stem cells
Aequorin, 46
Aerobic exercise, 60, 273, 274
Afferent arteriole, 223, 223*f*, 230, 230*f*
Afferent nerves, 286*f*
AFP (alpha-fetoprotein) test, 415, 459
Africa
  cancer in, 455
  early hominins in, 532
  female circumcision in, 373
  HIV and, 156, 156*f*, 157
Afterbirth, 408*f*, 410
Age structure, in more- *vs.* less-developed countries, 569–570, 569*f*
Agglutination, 125, 126, 126*f*, 128
Aging, 410
  and body systems, 413–414
  cardiovascular system in, 413, 414
  definition of, 410, 410*f*
  extrinsic factors, 411
  as genetic, 411
  hypotheses of, 411
  reproductive system in, 414
  skin in, 413
  as whole-body process, 411
Agranular leukocytes, 121*f*, 122
Agricultural runoff, 573, 574
Agriculture
  environmental impact, 559
  and food production, 574–575, 574*f*, 575*f*
  genetic engineering in, 507–508, 507*f*, 508*f*, 575
  global warming and, 557
  green revolutions, 575
  history of, 538
  livestock, 575, 575*f*
  pest control, biological, 574, 574*f*
  salt-tolerant crops, 508, 573
  and sustainability, 586, 587–588
  value of biodiversity to, 583*f*, 584
  water use in, 553, 572, 572*f*, 573, 573*f*, 574
AID (artificial insemination by donor), 384
AIDS (acquired immunodeficiency syndrome). *See also* HIV
  in HIV progression, 158–159, 385
  immune system and, 150–151
  vaccine for, 162, 163
ALA (linolenic acid), 33
Alagille syndrome, 442, 442*f*
Albumins, 38, 118, 179*t*, 233
Alcohol
  abuse of, 307, 307*t*
  and ADH secretion, 360
  and bone strength, 246
  cancer and, 454
  as chemical group, 31
  dehydration from, 231
  fermentation and, 60
  and health, 192
  infertility and, 382
  and nutrition, 188
  and osteoporosis, 411
  in pregnancy, 401
  stomach's absorption of, 175
Aldosterone, 220, 230, 230*f*, 231, 342*t*, 344, 352, 353, 360
Algal blooms, 559
Alien species, 580, 580*f*
Alkalosis, 232
Alkylating agents, in chemotherapy, 460
ALL (acute lymphocytic leukemia), 115
Allantois, 397, 397*f*, 398*f*, 400
Alleles, 468, 468*f*
Allen's rule, 539
Allergens, 150
Allergies, 150
Alli, 184
Allografting, 65, 76
All-or-none law, 271
Alpha-fetoprotein (AFP) test, 415, 459
ALS (amyotrophic lateral sclerosis), 277
Alveolar pressure, 202
Alveolus, 200, 201, 201*f*, 202
Alzheimer disease (AD), 412, 412*f*
  drugs for, 306
  hippocampus and, 300
  proteins function and, 36
  treatment of, 412
Ambiguous sex determination, 406–407, 406*f*
Ambulocetus natans, 522, 522*f*
Ames test, 453
Amino acids
  essential, 185–186, 185*f*
  gene expression and, 496, 496*t*
  in proteins, 35, 35*f*, 496, 496*t*
  shared, as evidence for evolution, 524, 524*f*
  urinary system and, 219
Amino group, 35, 35*f*, 36*f*
Ammonia, 232, 232*f*
Ammonium, in nitrogen cycle, 558, 558*f*
Amnesia, 300
Amniocentesis, 421, 421*f*, 443
Amnion, 397, 397*f*, 398*f*
Amniotic cavity, 397*f*, 398*f*, 399*f*
Amniotic fluid, 397, 398
Amniotic membrane, 409
Amphetamine psychosis, 308
Ampulla, 333, 333*f*
Amputation, phantom sensation and, 319
Amygdala, 299*f*, 300
Amyloid plaques, 412, 412*f*
Amyotrophic lateral sclerosis (ALS), 277
Anabolic steroids, 34, 280, 358
Anaerobic, 58, 273, 273*f*
Anal canal, 181*f*
Anal intercourse, and STDs, 388
Analogous structures, 523
Anal sphincter, 181, 181*f*
Anandamide, marijuana and, 309
Anaphase, 426, 427*f*, 434*f*, 434*t*, 435*t*
Anaphase I, 430*f*, 434*f*, 434*t*
Anaphase II, 430*f*, 435*t*
Anaphylactic shock, 150
Anatomical evidence, of evolution, 523–524, 523*f*, 524*f*
Androgen insensitivity, 342
Androgen insensitivity syndrome, 406–407, 406*f*
Androgens, 342*t*, 358
Andropause, 411, 414
Anecdotal data, 13
Anemia, 120
  Fanconi's, 443
  iron-deficiency, 20
Aneurysm, 105, 106
Angina pectoris, 106, 107
Angiogenesis, 450
Angiogram, 220
Angioplasty, 108
Angiotensin, 351
Angiotensin-converting enzyme (ACE) inhibitors, 106
ANH (atrial natriuretic hormone), 230, 352*f*, 353
Animalia, 527*t*
Animal kingdom, 6*f*, 7, 7*f*
Animals
  humans as related to, 6–7, 6*f*
  transgenic, 507*f*, 508–509, 509*f*
Annelids, 496
Anorexia nervosa, 191, 191*f*
Antagonist muscle, 264
Anterior compartment, 322, 323*f*
Anterior pituitary, 341*f*, 342*t*, 345–347, 346*f*
  and female reproductive system, 375–376, 376*f*, 378
  and male reproductive system, 369–370, 371*f*
Anterograde amnesia, 300
Anthrax, 10
Anthropoids, 527–528, 530
Antibacterial soap, 137
Antibiotics
  antitumor, 460
  development of, 155
  overuse of, 209
  in pregnancy, 401
  resistance to, 134, 166, 167–168, 167*f*, 168*f*, 387
  for tuberculosis, 164
Antibodies, 116, 121
  binding site of, 126
  blood type and, 125–127, 125*f*–127*f*
  classes of, 144–145, 145*t*
  IgA, 145, 145*t*
  IgD, 145, 145*t*
  IgE, 145, 145*t*, 150
  IgG, 144–145, 145*t*, 148, 148*f*, 150
  IgM, 145, 145*t*
  monoclonal, 148–149, 149*f*
  to nicotine, 213
  as proteins, 35
  structure of, 144, 144*f*
Antibody-mediated immunity, 143–145, 143*f*, 143*t*, 144*f*, 145*t*
Antibody titer, 148
Anticipation, in trinucleotide repeat disorders, 486

Anticoagulants, 128
Anticodon, 499, 499*f*, 502
Antidepressants, 306, 401
Antidiuretic hormone (ADH), 230, 231, 342*t*, 345, 346*f*, 360
Antigen-binding site, 144, 144*f*
Antigen-presenting cell (APC), 145, 146*f*, 461–463, 461*f*
Antigens, 121
  acquired immunity and, 143
  blood types and, 125–127, 125*f*–127*f*
Antihistamines, 141
Anti-inflammatory medications, 142
Antimalarial drugs, 165
Antimetabolites, in chemotherapy, 460
Anti-Müllerian hormone, 406
Antioxidants, 189–190
Antipsychotics, 306
Antiretroviral drugs (ARVs), 159, 385
Antitumor antibiotics, in chemotherapy, 460
Anus, 170*f*, 180, 181, 181*f*
Anvil (incus), 328–329, 329*f*
Aorta, 95, 95*f*, 102*f*, 103
Aortic arch, 402*f*
Aortic bodies, 205, 320
APC (antigen-presenting cell), 145, 146*f*, 461–463, 461*f*
Apes
  African *vs.* Asian, 528*f*
  humans and, 7, 527, 528
Apex, of heart, 95*f*
Apoptosis, 144, 145, 146*f*, 422, 423*f*, 448, 450
Appendicitis, 171
Appendicular skeleton, 248*f*, 253–255, 253*f*, 254*f*
Appendix, 171
Applegate, Christina, 419, 444
Aquaculture, 583
Aquaporins, 50, 50*f*, 228, 230
Aquatic ecosystems, 547, 547*f*
Aqueous humor, 322, 323*f*
Aquifers, 554, 573
Arachidonate, prostaglandins and, 359
Arbor vitae, 298
Archaea, 6–7, 6*f*, 7*f*, 48, 527*f*
Archaeopteryx, 521–522, 521*f*
Ardipithecus kadabba, 530
Ardipithecus ramidus, 530, 533*f*
Areolar tissue. *See* Loose fibrous connective tissue
Arm
  bones of, 253, 253*f*
  motor area for, 296*f*
Armadillo, nine-banded, 582*f*, 583
Arrector pili, 75*f*, 79, 281
Arsenic, 187, 578
ART (assisted reproductive technologies), 383–384
Arterioles, 93–94, 93*f*, 101*f*
Arteriovenous shunt, 93*f*, 94
Artery(ies), 93–94, 93*f*, 103*f*
  blood flow in, 101*f*
  blood pressure and, 100
  brachiocephalic, 95*f*, 96*f*
  carotid, 95*f*, 96*f*, 100, 102*f*, 103*f*
  clogged, 106–108, 106*f*, 108*f*
  coronary, 95–96, 95*f*
  dorsal, of penis, 368*f*
  femoral, 103*f*
  iliac, 102*f*, 103*f*, 402, 402*f*
  intestinal, 102*f*
  mesenteric, 103*f*
  pulmonary, 95*f*, 96, 96*f*, 102*f*, 103, 103*f*, 201*f*, 207*f*, 402*f*
  pulse in, 100
  radial, 100
  renal, 102*f*, 103*f*, 218, 218*f*, 221*f*
  in skin, 75*f*
  subclavian, 95*f*, 96*f*, 103*f*
  umbilical, 402, 402*f*
Arthritis
  osteoarthritis, 257, 257*f*
  rheumatoid, 151, 151*f*, 257
Articular cartilage, 240, 241*f*, 244*f*, 255*f*, 257
Articular facets, 251*f*
Articulations, 255–256, 255*f*, 256*f*
Artificial bladder, 235, 235*f*
Artificial heart, 109, 109*f*
Artificial hips, 257–258
Artificial insemination, 383
Artificial insemination by donor (AID), 384
Artificial knee, 239, 258
Artificial selection, 526
Artificial skin, 65, 76, 88
ARVs (antiretroviral drugs), 159, 385
Asbestos, 210*f*, 211, 453
Ascending colon, 180, 181*f*
Asexual reproduction, 432
Asian apes, 528*f*
A site, of ribosome, 499, 499*f*
Aspartame, 478
Aspartic acid, 35*f*
Aspirin, 107, 142, 192, 306, 319
Assembly, in HIV life cycle, 160, 161*f*
Assimilation, in plants, 559
Assisted reproductive technologies (ART), 383–384
Association areas, 297
Aster, 425, 425*f*, 426*f*
Asthma, 200, 210*f*, 211
Astigmatism, 326*f*, 328
Astrocytes, 71, 71*f*, 72
Atherosclerosis, 32, 34, 91, 106, 106*f*
Atherosclerotic plaques, 32
Athletes
  doping by, 120, 280, 280*f*
  lactate buildup in, 60
Atkin's diet, 184
Atlas (bone), 250
Atmosphere
  early, 518–519, 518*f*, 519*f*
  greenhouse gases in, 557, 557*f*, 576
Atom(s), 2, 3*f*, 20–21, 21*f*
Atomic mass, 20, 20*f*
Atomic mass units (amu), 21
Atomic number, 20, 20*f*, 21
ATP (adenosine triphosphate)
  and active transport, 51
  body temperature and, 279
  and cellular respiration, 57, 58–60, 59*f*
  cycle of, 60, 60*f*
  and muscular action, 269, 270–271, 270*f*, 273–274, 273*f*
  production of, 57, 58–60, 59*f*, 343, 343*f*
  rigor mortis and, 271
  sources of, 273–274, 273*f*
  structure and function of, 38–39, 39*f*
  as universal, 524
Atrial natriuretic hormone (ANH), 230, 352*f*, 353
Atrioventricular bundle, 98, 98*f*
Atrioventricular (AV) node, 98, 98*f*
Atrioventricular valves, 94–97, 96*f*, 97*f*
Atrium, 94, 95*f*, 96, 97, 97*f*, 98, 98*f*
Attachment, in HIV life cycle, 160, 161*f*
Auditory association area, 296*f*, 297
Auditory canal, 249*f*, 328, 329, 329*f*
Auditory tube, 329, 329*f*
Auricle, 94
Australopithecines, evolution of, 532, 532*f*
Australopithecus afarensis, 532, 532*f*, 533*f*
Australopithecus africanus, 532, 533*f*
Australopithecus anamensis, 533*f*
Australopithecus boisei, 532
Australopithecus garhi, 533*f*
Australopithecus robustus, 532
Autism, 486
Autodigestion, 55
Autografting, 65, 76
Autoimmune disorders, 151, 151*f*
Automatic external defibrillators (AEDs), 99
Autonomic motor nerves, 286*f*
Autonomic system, 304, 304*t*, 305*f*
Autosomal dominant disorders, 476, 477*f*, 478
Autosomal recessive disorders, 476–477, 476*f*
Autosomes, 420, 420*f*
Autotransplantation, of bone marrow, 461
Autotrophs (producers), 548, 548*f*, 549, 549*f*, 550*f*, 552*f*
Avian flu, 166, 167
Axial skeleton, 248–252, 248*f*–251*f*
Axillary lymph nodes, 138*f*
Axis (bone), 250
Axon, 70, 71*f*, 287, 287*f*
Axon branch, 269*f*
Axon terminal, 269–270, 269*f*, 287*f*, 290, 290*f*, 291*f*
AZT (zidovudine), 162

B

Bacillus, 134, 134*f*
Bacteria, 2*f*
  coliform, 181
  denitrifying, 559
  in HGH production, 348
  in insulin production, 357
  medicines from, 582–583
  nitrogen-fixing, 557
  oil-eating, 507, 507*f*
  as pathogen, 134–135, 134*f*, 140, 142*f*, 210
  as prokaryotic cell, 48
  refrigeration and, 136
  reproduction of, 135, 135*f*
  resident, 141
  STDs caused by, 386–387, 387*f*
  transgenic, 507, 507*f*
  as vector, 506, 506*f*
Bacteria domain, 6–7, 6*f*, 7*f*, 48, 527*f*
Bacterial meningitis, 82
Bacterial vaginosis (BV), 389
Balance, 318, 332–334, 333*f*
Baldness, 358, 370
Ball-and-socket joints, 255*f*, 256
Banting, Frederick, 357, 357*f*
Baroreceptors, 316
Barr body, 437–438
Barriers to entry, 140–141
Basal cell carcinoma, 76, 76*f*
Basal metabolism, 190
Basal nuclei, 298
Basement membrane, 72–73, 72*f*–73*f*
Base pairing, 37*f*, 38, 492, 492*f*, 494–495, 499, 501*f*
Bases, water and, 26–28
Basic solutions, 26–27
Basilar membrane, 330, 330*f*
Basophils, 117*f*, 121*f*, 122
Bats, 523*f*, 583*f*
B-cell receptor (BCR), 143–144, 143*f*
B cells, 122, 139, 143–145, 143*f*, 143*t*
B-complex vitamins, 180
Beaches, human habitation of, 570–571, 571*f*
Beagle, HMS, 520
Becker muscular dystrophy, 281
Bees, 583, 584
Behavioral inheritance, 480–481
Belly button, 409
Benign neuroma, 449
Benign prostatic hyperplasia (BPH), 222, 222*f*
Benign tumor, 449
Benzene, 453
Benzidine, 453
Bergmann's rule, 539
Best, Charles, 357, 357*f*
Beta blockers, 106
Beta-carotene, 454
Beta-hexosaminidase A, 62
bGH (bovine growth hormone), 508
Bicarbonate ions, 28, 119, 206, 207*f*, 208, 220, 232, 232*f*, 234, 554–555, 555*f*
Biceps brachii, 264, 265, 265*f*, 266*f*
Biceps femoris, 266*f*
Bicuspid valve, 95, 96*f*
Bifocals, 327
Bilayer. *See* Phospholipid bilayer
Bile, 176, 178
Bile canals, 179*f*
Bile duct, 179*f*
Bilirubin, 178, 179*t*, 180, 226
Binary fission, 135, 135*f*, 432
Binding site, of antibody, 126
Binge-eating disorder, 192
Binocular vision, 528
Binomial name, 526
Biochemical evidence for evolution, 524, 524*f*
Biodiversity, 579–585
  agricultural value in, 583, 583*f*
  alien species and, 580, 580*f*
  biogeochemical cycles and, 585
  climate and, 585
  consumptive use value in, 582*f*, 583–584
  definition of, 579
  direct value of, 581–584, 582*f*, 583*f*
  disease and, 580*f*, 582
  ecotourism and, 585
  habitat loss and, 580, 580*f*
  humans as threat to, 8, 15
  indirect value of, 584–585
  loss of, 580–582, 580*f*
  medicinal value in, 582–583, 582*f*
  overexploitation and, 580–582, 580*f*
  pollution and, 580, 580*f*
  soil erosion and, 585
  waste disposal and, 585
Bioethics
  biodiversity preservation, 8
  biotechnology, 14–15
  circumcision, 373
  cloning, 404
  DNA fingerprinting, 505
  drinking water, 556
  face transplants, 78
  female circumcision, 373
  fertility treatments, 383
  growth hormone treatments, 348
  humans as experimental subjects, 10
  informed consent, 11
  issues in, 14–16
  marijuana, medical, 309
  noise pollution, 331–332
  population growth, 538
  prenatal selection of children, 443
  smoking bans, 209
  stem cell research, 61, 404
Biogeochemical cycles, 552–563, 585
Biogeographical evidence for evolution, 522, 523*f*
Biogeography, 522
Bioinformatics, 512
Biological magnification, 563, 578, 579, 579*f*
Biological pest control, 574, 574*f*
Biomass, 552, 552*f*
Biomes, 546*f*7, 547
Bioreactors, 507, 507*f*
Biosphere, 2, 3*f*, 7, 8*f*, 547
Biosynthesis, in HIV life cycle, 160, 161*f*
Biotechnology, 14–15, 37, 506–509, 507*f*–509*f*
Biotic communities, in biogeochemical cycle, 553, 553*f*
Biotic components of ecosystem, 548–549, 548*f*
Biotic potential, 568
Biotin, 190*t*
Bipedal posture, evolution of, 530
Bipolar cell layer, 325, 325*f*
Bird flu, 166
Bird forelimb, 523*f*
Birth, 87, 345, 408–410, 408*f*
  cesarean section in, 406
  circulatory changes in, 402–403, 402*f*
  premature, 405
  STDs in, 387
Birth control, 379–382, 380*t*, 381*f*, 569*f*

Birth control pills, 378–379, 378*f*, 379, 380*t*, 381*f*, 389, 401
Birth defects, 401
Birthrate, 568, 569
Bisphosphonates, 246
Bitter taste, 321
Black Death, 155
Blackheads, 79
Bladder, urinary, 218*f*, 219, 235, 235*f*, 367*f*, 372*f*
Bladder cancer, 451*f*, 453*f*, 459
Blalock, Alfred, 107
Blalock-Taussig surgery, 107
Blastocyst, in pre-embryonic development, 396*f*, 397
Blastoma, 451*f*
Blind spot, 325, 325*f*
Blood
  calcium levels, 220, 233, 244, 245, 278–279, 350, 350*f*
  cancer of, 66, 122
  in capillaries, 101, 101*f*
  cholesterol levels, 19, 34, 39, 111, 178
  circulation of, 79
  clotting of, 123–125, 123*f*, 141*f*, 350
  compatibility of, 126, 126*f*
  composition of, 116–118, 117*f*
  as connective tissue, 68, 68*f*
  donation of, 124, 124*f*
  doping, 120, 280, 280*f*
  exchange function of, 92, 92*f*, 116
  flow of, fetal, 402–403, 402*f*
  flow regulation, 100–102, 101*f*
  formed elements of, 116, 117*f*
  functions of, 116
  in glomerular filtration, 225–227, 225*f*
  glucose level, 30, 84, 178, 354, 355*f*
  in heart, passage of, 96–97
  homeostasis, 79, 116
  in immune system, 116
  overview of, 116–118, 117*f*
  oxygenation of, 92
  pH of, 28, 116, 118, 205–206, 220, 320
  sodium in, 23, 352–353, 353*f*
  in systemic circuit, 103–104, 103*f*, 104*f*
  transfusion of, 125–127, 128
  as transport medium, 116
  type inheritance, 481–483, 483*f*
  typing, 125–127, 125*f*–127*f*, 128, 481, 483*f*
  in veins, 101–102, 101*f*
  volume, in pregnancy, 407
Blood cells. *See* Red blood cells; White blood cells
Blood clots, 106–107
Blood glucose test, 339
Blood pressure, 100
  aging and, 413
  capillary exchange and, 104–105, 104*f*
  high (hypertension), 100, 101*t*, 106, 230
  homeostasis, 69, 86, 93–94, 220, 230, 295, 352–353, 352*f*, 360
  low (hypotension), 100, 101*t*
  measurement of, 100, 100*f*
  normal values, 101*t*
  reflex arcs and, 295
  sodium and, 23, 352–353, 353*f*
Blood tests, 34
Blood-type crossmatching, 126
Blood vessels
  blood flow rates in, 101*f*
  in coronary bypass surgery, 107–108, 108*f*
  disorders of, 105–108, 106*f*, 108*f*
  in homeostasis, 86, 87, 87*f*
  in microscopy, 45*f*
  types of, 93–94
Blushing, 77
B lymphocytes, 122, 139, 143–145, 143*f*, 143*t*
BMI (body mass index), 183, 183*f*
Body (of sternum), 251*f*, 252
Body cavities, 81, 82*f*
Body mass index (BMI), 183, 183*f*
Body membranes, 82, 82*f*
Body shape, environmental conditions and, 539
Body stalk, 398*f*
Body temperature
  as homeostatic, 4
  inflammatory response and, 141
  muscles and, 279–280
  and protein function, 36
  regulation of, 77, 86–87, 87*f*
  scrotum and, 369
  water vaporization and, 25*f*, 26
Body weight
  in female infertility, 382
  leptin and, 359
  nutrition and, 183–184
  in pregnancy, 407
Bolus, 173, 173*f*
Bombay syndrome, 127
Bonds
  covalent, 23–24, 23*f*
  disulfide, 36, 36*f*
  double, 24
  hydrogen, 24, 25*f*, 26, 36, 36*f*
  ionic, 22–23, 23*f*
  peptide, 35, 36, 36*f*
  triple, 24
Bonds, Barry, 280
Bone(s). *See also* Skeletal system
  anatomy of, 240, 241*f*
  of appendicular skeleton, 248*f*, 253–255, 253*f*, 254*f*
  of arm, 253, 253*f*
  articulations of (joints), 255–256, 255*f*, 256*f*
  in axial skeleton, 248–252, 248*f*–251*f*
  blood cell production, 279
  calcium blood levels and, 220, 244, 245, 278–279, 350, 350*f*
  compact, 67*f*, 68, 240, 241*f*
  as connective tissue, 68, 69*f*
  coxal, 254, 254*f*
  ethmoid, 249, 249*f*
  facial, 249, 249*f*, 250*f*
  fetal development of, 403
  fracture of, 247, 247*f*
  frontal, 248, 248*f*, 249, 249*f*, 250*f*
  growth of, 242–244, 243*f*, 244*f*
  hormones and, 244
  hyoid, 250, 250*f*
  lacrimal, 249*f*
  of leg, 254, 254*f*
  muscle connection to, 264, 265*f*
  nasal, 249, 249*f*, 250*f*
  number of, 240
  occipital, 248, 248*f*, 249*f*
  osteoporosis, 188, 245, 246, 275, 411
  palatine, 249*f*
  parietal, 248, 248*f*
  repair of, 247, 247*f*
  sphenoid, 249, 249*f*
  spongy, 68, 240, 241*f*, 243, 243*f*
  temporal, 248, 248*f*, 249, 249*f*, 250*f*, 329*f*
  vomer, 249, 249*f*
  Wormian, 240
  zygomatic, 248*f*, 249, 249*f*
Bone collar, 243, 243*f*
Bone marrow, 116, 120*f*, 240, 241*f*, 243, 279. *See also* Red bone marrow; Yellow bone marrow
Bone marrow transplant, 461
Bone remodeling, 245, 245*f*, 247, 247*f*
Bony callus, 247, 247*f*
Botox, 83, 271
Bovine growth hormone (bGH), 508
Bovine spongiform encephalopathy (BSE), 137
Bowman's (glomerular) capsule, 223–224, 223*f*, 228, 230*f*
BPH (benign prostatic hyperplasia), 222, 222*f*
Brachiocephalic artery, 95*f*, 96*f*
Brachioradialis, 265
Brain, 286, 286*f*, 294*f*, 295–299
  auditory pathway to, 329–330
  cancer of, 451*f*, 452
  development of, 400*f*
  lobes of, 296, 296*f*
  in primates, 528
  respiration and, 205, 205*f*, 232
  smell information to, 321–322, 321*f*
  stem, 295*f*, 298–299
  taste information to, 320*f*, 321
  ventricles of, 293, 293*f*, 295
  visual pathway to, 324–326, 324*f*–326*f*
Branchiole, 201*f*
Braxton Hicks contractions, 408
BRCA1 gene, 419, 438, 444, 452, 459
BRCA2 gene, 452
Breaking water, in pregnancy, 409
Breast cancer, 149, 419, 438, 444, 451*f*, 452, 455, 457, 458, 458*f*, 458*t*, 461
Breast exam, 457
Breast feeding, 148, 148*f*, 161, 579
Breath holding, 205
Breathing. *See* Respiration
Breech birth, 405
Broca's area, 296*f*, 297, 301, 301*f*
Bronchi, 197*f*
Bronchial cancer, 451*f*
Bronchial tree, 200
Bronchiole, 197*f*, 200
Bronchitis, 74, 210, 210*f*, 211
Brush border, 177
BSE (Bovine spongiform encephalopathy), 137
Bubonic plague, 155
Budding, in HIV life cycle, 160, 161*f*
Buffers (chemical), 27–28, 232, 234
Bulbourethral glands, 222, 367*f*, 368, 368*t*
Bulimia nervosa, 191, 192*f*
Burke, Chris, 439, 439*f*
Burkitt lymphoma, 442, 455
Burns, 65
Bursa, 255*f*, 256, 264
Bursitis, 277
BV (bacterial vaginosis), 389
Bypass, coronary, 107–108, 108*f*

C

CA-125 test, 459
Cabbage family, 454
Cactuses, 522
Cadmium, 453, 578
Caffeine, 231, 246
Calcaneus, 254
Calcification, of bone, 242–244
Calcitonin, 245, 246, 278, 342*t*, 350, 350*f*
Calcitriol (activated Vitamin D), 350, 350*f*
Calcium
  blood levels, 220, 233, 244, 245, 278–279, 350, 350*f*
  and bone health, 411
  calcitonin and, 350, 350*f*
  as mineral, 187*t*, 188
  in muscle contraction, 267, 268, 270, 270*f*
  in neurotransmission, 291
  osteoporosis and, 246
Calcium carbonate granules, 334
Calcium chloride, 23
Calico cats, 438
California's Geysers project, 576
Calories, 25, 184
cAMP (cyclic adenosine monophosphate), 343, 343*f*
Campylobacter jejuni, 134*f*
Canaliculi, 241*f*, 242
Cancer. *See also* specific types
  aging and, 411
  alcohol and, 454
  angiogenesis and, 450
  apoptosis and, 448
  asbestos and, 211
  bladder, 451*f*, 453*f*, 459
  blood, 66, 122
  brain, 451*f*, 452
  breast, 149, 419, 438, 444, 451*f*, 452, 455, 457, 458, 458*f*, 458*t*, 461
  bronchial, 451*f*
  cases, by site and sex, 451*f*
  causes and prevention, 452–455
  cells in, 448–452, 448*f*–451*f*
  cervical, 13, 159, 365, 371–372, 385, 389, 453*f*, 455, 456, 458*t*
  chemotherapy for, 22, 22*f*, 460–461, 460*f*
  classification by tissue type, 66
  colon, 182, 451, 451*f*, 452, 456, 458*t*, 461
  colorectal, 451, 451*f*, 458*t*
  common, 451–452
  cytokines and, 149, 461*f*, 463
  deaths, by site and sex, 451*f*
  diagnosis of, 455–459
  diet and, 189, 454, 455
  differentiation and, 448
  early detection of, 458*t*
  endometrial, 458*t*
  environmental carcinogens, 453–455
  esophageal, 451*f*, 453*f*
  exercise and, 275
  formation of, 449–450, 449*f*
  genetic mutation and, 450–451, 450*f*, 451*f*, 453
  genetic testing for, 459
  growth factors and, 449
  heredity and, 452, 452*f*
  HIV and, 455
  Hodgkin lymphoma, 452, 461, 582
  imaging of, 456–458, 458*f*
  immunotherapy for, 461–463, 461*f*
  interleukins and, 149
  kidney, 451*f*, 453*f*
  larynx, 453*f*
  leukemia, 66, 115, 122, 451, 451*f*, 452, 461, 582
  liver, 451*f*, 455
  lung, 211–213, 213*f*, 451, 451*f*
  lymphoma, 66, 451, 455, 461
  melanoma, 76, 76*f*, 456, 456*f*, 459, 461
  metastasis of, 450
  monoclonal antibodies in treating, 463
  nasopharyngeal, 455
  Non-Hodgkin lymphoma, 451*f*, 452, 455
  obesity and, 454
  oral, 451, 451*f*
  organic chemicals and, 453–455
  oropharyngeal, 453*f*
  ovarian, 451*f*, 452, 461
  pancreatic, 451*f*, 453*f*
  pharyngeal, 451*f*
  plasma cell tumors, 452
  pollution and, 453–455
  prevention of, 454
  prognosis, factors in, 451
  prostate, 222, 451*f*, 455, 458*t*
  radiation and, 21, 453, 454
  radiation therapy for, 459–460, 460*f*
  radon gas and, 453
  rectal, 451*f*, 452, 458*t*
  replication of, 448
  retinal, 452, 452*f*
  screening for, 454, 456–458, 458*t*
  self-examination, 456, 457
  in situ, 449
  skin, 76–77, 76*f*, 83, 451*f*, 453, 456, 456*f*
  smoking and, 389, 451, 453, 453*f*, 454
  spinal, 452
  stomach, 453*f*
  surgery for, 459
  T-cell, 145
  telomerase and, 448, 459
  testicular, 457, 458*t*
  thyroid, 451*f*, 452, 459
  translocation of chromosomes and, 442
  treatment of, 459–463
  tumor formation in, 448–449, 449*f*
  tumor marker tests for, 458–459
  types of, 451–452, 451*f*
  uterine, 451*f*
  viruses and, 455
  warning signs of, 209, 455–456
Candida albicans, 389
Canine teeth, 172*f*
Cannabinoids, 309–310
Cannabis abuse, 307*t*, 309–310

Cannabis psychosis, 310
Canseco, Jose, 280, 280*f*
CAPD (continuous ambulatory peritoneal dialysis), 234
Capillaries, 93*f*, 94, 119*f*
blood flow in, 101, 101*f*
gravity and, 92
peritubular network of, 223, 223*f*, 225*f*, 227
in skin, 75*f*
Capillary exchange, 104–105, 104*f*
Capitulum, 253, 253*f*
Capsid, 135, 135*f*, 159, 159*f*
Capsule, of bacteria, 134, 134*f*
Carbaminohemoglobin, 119, 208
Carbohydrates, 29–31
complex, 29–30, 30*f*
in diet, 30–31, 185, 185*f*, 185*t*
digestion of, 176, 177, 177*f*, 177*t*
as nutrients, 185, 185*f*, 185*t*
on plasma membrane, 49, 49*f*
simple, 29
Carbon, isotopes of, 21
Carbon atom, 21, 21*f*
Carbon cycle, 554–558, 555*f*
Carbon dioxide
acid-base balance and, 232
breathing control and, 205–206
in carbon cycle, 554–558, 555*f*
formulas for, 24
in gas exchange, 206–208, 207*f*
as greenhouse gas, 557, 557*f*
in hyperventilation, 206
in pregnancy, blood levels of, 407
red blood cells and, 118–119
Carbonic acid, 28, 119, 232
Carbonic anhydrase, 119, 206, 208
Carbon monoxide, 116, 118
Carboxyl group, 35, 35*f*, 36*f*
Carcinoembryonic antigen (CEA), 459
Carcinogenesis, 449–450, 449*f*
Carcinogens, environmental, 453–455
Carcinoma, 66, 76, 76*f*, 451
Cardiac cycle, 97, 97*f*. *See also* Heartbeat
Cardiac muscle, 69, 70*f*, 96*f*, 263, 263*f*
Cardiac septum, 96*f*
Cardiac veins, 95, 95*f*
Cardiopulmonary resuscitation (CPR), 99
Cardiovascular disease, 52, 105–110, 106*f*
aging and, 411
cost of, 110
diet and, 186
gum disease and, 173
prevention of, 110
saturated fats and, 32
treatment of, 106–109, 108*f*, 109*f*
Cardiovascular pathways, 102–104, 102*f*, 103*f*
Cardiovascular system, 79, 80*f*. *See also* Artery(ies); Blood; Heart; Vein(s)
aging of, 413, 414
circulation in, 92, 92*f*
endocrine system and, 361*f*
features of, 100–102, 101*f*, 102*f*
fetal, 402–403, 402*f*
functions of, 92
in homeostasis, 84, 85*f*, 92*f*, 127–129, 129*f*
lymphatic system in, 92
muscular system and, 279*f*
overview, 92–93, 92*f*
urinary system and, 234*f*
Cardioverter-defibrillator, implantable (ICD), 109
Caries, dental, 173
Carnivores, 548, 548*f*, 550*f*, 552*f*
Carotene, 76
Carotid artery, 95*f*, 96*f*, 100, 102*f*, 103*f*
Carotid bodies, 205, 320
Carpals, 248*f*, 253, 253*f*
Carrier proteins, 35, 49, 51, 288–289
Carrying capacity, 568
Carson, Rachel, 579
Cartilage, 44*f*, 67–68, 67*f*, 69*f*, 240, 242
Cartilage model, 243, 243*f*
Cartilaginous joints, 255
Cassini-Huygens space probe, 1
Cataracts, 77, 327
Cats, 438, 480, 523*f*
Caveman diet, 184
Cavities, body, 81, 82*f*
CCD (colony collapse disorder), 584
CCK (cholecystokinin), 180–181, 180*f*
CD4 T lymphocytes, 157–158, 160
cDNA (complementary DNA), 506
CEA (carcinoembryonic antigen), 459
Cecum, 180, 181*f*
Cell body, of neuron, 70, 71*f*
Cell cycle, 422–424, 422*f*, 423*f*, 450
Cell division. *See* Cell cycle
Cell junctions, 56, 56*f*
Cell-mediated immunity, 145–147, 146*f*
Cells. *See also* Nucleus, cellular; *specific cells*
as basis of life, 44
cancerous, 448–452, 448*f*–451*f*
evolution of, 518–519, 518*f*, 519*f*
internal structure of, 46, 47*f*
metabolism of, 57–60, 57*f*, 59*f*
in microscopy, 44–45, 45*f*, 45*t*
movement of, 55
number of, in human body, 4
in organization of life, 2, 3*f*
sizes of, 44, 45*f*
surface area-to-volume ratio, 44, 45*f*
variety of, 44, 44*f*
Cell theory, 9, 44, 518
Cellular membrane, 49–52, 49*f*–52*f*
bacterial, 134*f*
in evolution of eukaryotic cell, 48, 48*f*
phospholipid bilayer, 33, 33*f*, 46, 49, 49*f*
Cellular respiration, 57, 58–60, 59*f*, 273, 273*f*, 274
Cellulose, 30, 30*f*
Cell wall, 134*f*
Cementum, 172*f*
Central canal
in bone, 240, 241*f*
in spinal cord, 294*f*, 303*f*
Central nervous system (CNS), 72–73, 286, 286*f*, 291, 293–299, 317, 317*f*. *See also* Brain; Spinal cord
Centriole, 47*f*, 425, 425*f*
Centromere, 420*f*, 421, 424, 425, 425*f*, 427, 427*f*
Centrosome, 47*f*, 55, 425, 425*f*, 426*f*
Cerebellum, 295*f*, 297, 298, 334
Cerebral cortex, 296
Cerebral hemispheres, 295–296, 295*f*, 298, 301
Cerebral palsy, 297
Cerebrospinal fluid, 293
Cerebrovascular accident (CVA), 106
Cerebrum, 295–298, 295*f*
Cerumen, 32
Cervical cancer, 13, 159, 365, 371–372, 385, 389, 453*f*, 455, 456, 458*t*
Cervical cap, 380*t*
Cervical nerves, 305*f*
Cervical shield, 380*t*
Cervical vertebrae, 250, 251*f*
Cervix, 371, 371*t*, 372*f*
Cesarean section, 405
CFCs (chlorofluorocarbons), 562, 579
cGMP (cyclic guanosine monophosphate), 368–369
Chalfie, Martin, 46
Chancre, from syphilis, 387, 387*f*
Channel proteins, 35, 49, 477, 477*f*
Chantix, 213
Charges, in atoms, 21
Charnley, John, 257–258
Checkpoints, in cell cycle, 422–423, 423*f*, 450
Chemical barriers, in immune system, 140
Chemical cycling, 549–550, 549*f*
Chemical digestion, 171
Chemical evolution, 518
Chemicals, cancer caused by, 453–455
Chemical signals, hormones as, 342–343, 343*f*, 344lf
Chemoautotrophs, 548
Chemoreceptors, 205–206, 316, 316*t*, 317, 320
Chemotherapy, 22, 22*f*, 460–461, 460*f*
Chernobyl Power Station, 453, 576
Chewing, 296*f*
Chewing tobacco, 212
Chick embryo, 524*f*
Chicken pox, 160
Chicken pox vaccine, 147*f*
Childbed fever, 409
Childbirth, 345
Chimpanzee, 528*f*
in evolutionary tree, 531*f*
genome of, 511, 511*f*
skeleton, *vs.* human skeleton, 529, 529*f*
China, cancer in, 455
Chlamydia, 386–387
Chlamydia trachomatis, 386
Chloride ions, 187*t*, 227, 228–230, 477, 477*f*, 513
Chlorine atoms, ozone shield and, 563, 563*f*
Chlorofluorocarbons (CFCs), 562, 579
Chloroplast, evolution of, 48, 48*f*
Choking, 84, 198, 199*f*
Cholecystokinin (CCK), 180–181, 180*f*
Cholesterol. *See also* HDL (high-density lipoprotein); LDL (low-density lipoprotein)
blood levels, 19, 34, 39, 111, 178
in diet, 34, 106
exercise and, 275
and liver, 178
in plasma membrane, 49, 49*f*
Cholinesterase inhibitors, 412
Chondroblasts, 67
Chondrocytes, 67, 241*f*, 242
Chordae tendineae, 94, 96*f*
Chordata, 527*t*
Chorion, 396*f*, 397, 397*f*, 398, 398*f*
Chorionic villi, 395*t*, 397, 398*f*, 400, 401–407
Chorionic villi sampling (CVS), 415, 421, 421*f*
Choroid, 322, 322*t*, 323*f*, 325*f*
Christmas disease, 125
Chromatids, sister, 420*f*, 421–422, 424–426, 425*f*, 427*f*, 429, 430*f*–431*f*
Chromatin, 47*f*, 53, 53*f*, 422, 426*f*
Chromium, 187, 578
Chromosome number, in various organisms, 420
Chromosomes, 420–421, 492
abnormalities in, 429
of cancer cells, 448
cell cycle and, 422–423
changes in number of, 437–439, 437*f*
deletion of, 440*f*, 441, 441*f*
duplication of, 440*f*, 441
fetal, obtaining, 421, 421*f*
in gamete formation, 469
homologous, 428, 428*f*, 430*f*–431*f*
inheritance of, 437–442
inversion of, 440*f*, 441, 441*f*
in karyotype, 420–421, 420*f*
in meiosis, 366–367, 367*f*, 428, 428*f*, 429, 430*f*–431*f*
in mitosis, 366, 425–426, 425*f*, 426, 426*f*
in monosomy, 437
in nucleus, 53, 53*f*
in oogenesis, 436, 436*f*
in spermatogenesis, 436, 436*f*
structural changes in, 440*f*–442*f*, 441–442
translocation of, 440*f*, 441
in trisomy, 437, 438–439, 439*f*, 441
Chronic bronchitis, 211
Chronic myelogenous leukemia, 442
Chronic obstructive pulmonary disease (COPD), 211
Chylomicrons, 177, 177*f*
Chyme, 176
Cigarettes. *See* Smoking
Cilia, 55, 55*f*, 200, 200*f*
Ciliary body, 322, 322*t*, 323*f*
Ciliary muscle, 323–324, 323*f*
Ciliated columnar epithelium, 73*f*
Circadian rhythms, 359, 361*f*
Circumcision
female, 373
male, 368, 373
Circumduction, 256*f*
Cirrhosis, 179
Cities, sustainability in, 588
Citric acid cycle, 58, 59, 59*f*
Civet cats, 166, 166*f*
Civilization, origin of, 538
Clavicle, 248*f*, 253, 253*f*
Cleavage, in HIV life cycle, 160, 161*f*
Cleavage, in pre-embryonic development, 396, 396*f*
Cleavage furrow, 427, 427*f*
Clemens, Roger, 280
Climate, biodiversity and, 585
Climate change, 557, 557*f*, 576, 580
Clinton, Bill, 16
Clitoris, 372, 372*f*, 374
Clogged arteries, 106–108, 106*f*, 108*f*
Clonal expansion, 143*f*, 144, 145, 146*f*
Clonal selection theory, 143–144, 143*f*, 145, 146*f*
Cloning, 15, 404
Cloning, gene, 506, 506*f*
Clostridium botulinum, 83
Clostridium tetani, 135, 272
Clotting, 106, 116, 123–125, 123*f*, 141*f*, 350
Clotting factor IX, 125, 486
Clotting factor VIII, 123–125, 486
CNS (central nervous system), 72–73, 286, 286*f*, 291, 293–299, 317, 317*f*. *See also* Brain; Spinal cord
Coal, as fuel, 576, 578
Coastal wetlands, 570–571, 576, 580
Cocaine, 307*t*, 308
Cocaine psychosis, 308
Coccus, 134, 134*f*
Coccyx, 248*f*, 250, 251*f*
Cochlea, 329, 329*f*, 330, 330*f*
Cochlear canal, 330*f*
Cochlear implants, 315, 334
Cochlear nerve, 329*f*, 330*f*
Codeine, 306
Codominance, 481
Codons, 497, 497*f*, 501, 502, 503
Coelom. *See* Ventral cavity
Coenzymes, 58, 189
Cohesion, in water, 26
Cold sores, 386
Coler, Pascal, 78, 78*f*
Coliform bacteria, 181
Colitis, 181
Collagen, 34, 66, 68, 77, 411
Collagen fibers, 66, 66*f*
Collecting duct, 221*f*, 224, 230*f*, 231, 231*f*
Colon, 180, 181–182, 181*f*
Colon cancer, 182, 451, 451*f*, 452, 456, 458*t*, 461
Colon cancer screening, 456
Colonoscopy, 182, 456
Colony collapse disorder (CCD), 584
Colony-stimulating factor (CSF), 121
Color, of skin, 479–480, 539
Colorado River, 572
Color blindness, 484–485, 484*f*
Colorectal cancer, 451, 451*f*, 458*t*
Color vision, 324
Columnar epithelium, 73*f*, 74
Combined hormone pill, 380*t*
Combined hormone vaginal ring, 380*t*
Common carotid artery, 95*f*, 96*f*
Common descent, 520–524
Common hepatic duct, 179*f*
Community, 2, 3*f*
Compact bone, 67*f*, 68, 240, 241*f*
Comparative anatomy, 523–524, 523*f*, 524*f*
Comparative genomics, 511–512, 511*f*

Competition for resources, in natural selection, 525*f*, 526
Complementary base pairing, 37*f*, 38, 492, 492*f*, 494–495, 499, 501*f*
Complementary DNA (cDNA), 506
Complement system, 142, 142*f*
Complex carbohydrates, 29–30, 30*f*
Compound light microscope, 44–45, 45*f*, 45*t*
Compounds, 22–24
Compression, of eyes, 317
Computed tomography (CT), 458
Conclusion, 9–10, 9*f*, 10–11, 12*f*, 13
Condom
female, 379, 380*t*, 381*f*
male, 379, 380*t*, 381*f*
STDs and, 388
Cone cells, 316, 316*t*, 322*t*, 323, 323*f*, 324–325, 324*f*, 325*f*, 528
Congenital defects, 401
Congenital hypothyroidism, 349, 349*f*
Congenital syphilis, 387
Conjunctivitis, 328
Connective tissue, 66–69, 66*f*–69*f*, 242, 368*f*
Consent, informed, 11
Conservation, of water, 573–574, 573*f*, 588
Constipation, 181, 407
Consumers (heterotrophs), 519, 548–549, 548*f*, 549*f*, 551*f*
Consumption. *See* Tuberculosis
Contact dermatitis, 150
Contact inhibition, 448
Continuous ambulatory peritoneal dialysis (CAPD), 234
Continuous positive airway pressure (CPAP) device, 214
Contour farming, 574, 574*f*
Contraceptive implants, 379, 381*f*
Contraceptive injections, 379, 380*t*, 381*f*
Contraceptives, 379–382, 380*t*, 381*f*
Contraceptive sponges, 380*t*
Contraceptive vaccines, 379
Contractile ring, 427
Contraction, of muscles, 267–274, 268*f*–270*f*
Contractions, in birth, 408–409
Control group, 11, 12*f*
Controlled study, design of, 11, 12*f*
Convulsion, 276–277
COPD (chronic obstructive pulmonary disease), 211
Copper, 187, 187*t*, 578
Copper intrauterine device, 380*t*
Coracoid process, 253, 253*f*
Coral reefs, 547, 547*f*, 576, 580
Corn, genetically engineered, 507–508
Cornea, 322, 322*t*, 323*f*
Corona radiata, 394, 394*f*, 396*f*
Coronary arteries, 95–96, 95*f*
Coronary artery disease, 95
Coronary bypass, 107–108, 108*f*
Corpus callosum, 295, 295*f*, 298, 299*f*
Corpus callosum surgery, 301
Corpus luteum, 375, 375*f*, 376, 376*f*, 377*t*, 378, 378*f*, 398
Cortical granules, 394*f*, 395
Corticotropin releasing hormone (CRH), 351
Cortisol, 342*t*, 346, 352
Costal cartilage, 248*f*
Cotton, 2*f*, 15
Covalent bond, 23–24, 23*f*
Cover crops, 588
COX (cyclooxygenase), 306
Coxal bone, 248*f*, 254, 254*f*
CPAP (continuous positive airway pressure) device, 214
CPR (cardiopulmonary resuscitation), 99
Crack abuse, 308
Cramps, 277
Cranberry juice, 233
Cranial cavity, 81, 82*f*
Cranial nerves, 286*f*, 302, 302*f*, 305*f*
Craniosacral portion. *See* Parasympathetic division
Cranium, 248–249
Creatine, 273, 273*f*
Creatine phosphate, 220, 273, 273*f*
Creatine phosphate (CP) pathway, 273, 273*f*
Creatinine, 220, 225*f*, 228
Crenation, 50, 51*f*
Creutzfeldt-Jakob disease, 36, 137, 348
CRH (corticotropin releasing hormone), 351
Crick, Francis, 493
Cri du chat syndrome, 442
Criminals, DNA fingerprinting and, 504–505, 504*f*
Cristae, 57, 57*f*
Crohn's disease, 181
Cro-Magnons, 537–539, 537*f*
Cross bridge, 268*f*, 270*f*, 271
Crossing-over, 429, 432, 432*f*
Crown, of tooth, 172*f*, 173
Crystal jellyfish (*Aequorea victoria*), 46
Crystal meth, 308
CSF (colony-stimulating factor), 121
CT (computed tomography), 458
CTFR gene, 52
Cuboidal epithelium, 72*f*, 74
Cultivars, 588
Cultural eutrophication, 559, 561, 579
Cultural heritage, 7–8
Culture, 535
of Cro-Magnon, 538, 539
definition of, 7
of early *Homo* species, 534, 535, 536, 538
of Neandertals, 537
Cunnilingus, and STDs, 388
Cupula, 333, 333*f*, 334
Cured foods, 454
Cushing syndrome, 353, 353*f*
Cutaneous receptors, 319, 319*f*
Cuticle, 77, 77*f*
CVA (cerebrovascular accident), 106
CVS (chorionic villi sampling), 415, 421, 421*f*
Cyclic adenosine monophosphate (cAMP), 343, 343*f*
Cyclic guanosine monophosphate (cGMP), 368–369
Cyclin, 450
Cyclin D, 450
cyclooxygenase (COX), 306
Cymbalta, 306
Cysteine, 35, 35*f*
Cystic fibrosis, 52, 211, 477–478, 477*f*, 509, 513
Cystitis, 233
Cytochrome *c*, 524, 524*f*
Cytokines, 141, 143*f*, 145, 146, 146*f*, 149, 416*f*, 461*f*, 463
Cytokinesis, 422, 422*f*, 427
Cytoplasm, 46, 47*f*
Cytosine (C), 38, 38*t*, 492, 492*f*, 494, 495*f*
Cytoskeleton, 47*f*, 55
Cytotoxic T cells, 122, 143*t*, 145–146, 146*f*, 150, 151, 461*f*, 463

D

Dams, 572–573, 576
Dandelions, 580
Dandruff, 76
Dart, Raymond, 531*f*
Darwin, Charles, 520, 521, 522, 525–526, 530
Data, 9–10, 12*f*, 13, 13*f*, 14
Daughter cells, 61, 422, 424, 428, 430*f*, 435*f*
Daughter chromosomes, 421, 427*f*
DDT, 579, 579*f*
Dead air space, 204
Deafness, 315, 334
Death, 276
Death rate, 568, 569
Decibel levels, 331*t*
Decomposers, 134, 548*f*, 549, 549*f*, 550, 551*f*, 585
Defecation, 180–181
Defibrillation, 99
Deforestation, 571, 572*f*
Degenerating zone, 244, 244*f*
Dehydration, from alcohol, 231
Dehydration reaction, 28, 28*f*, 29, 29*f*
Delayed allergic response, 150
Delayed onset muscle soreness (DOMS), 276
Deletion, 440*f*, 441, 441*f*
Deletion syndromes, 441, 441*f*
Deltoid, 265, 266*f*
Deltoid tuberosity, 253*f*
Demylinating disorders, 314
Denaturation, 36, 87
Dendrite, 70, 71*f*, 287, 287*f*, 291*f*
Denitrification, in nitrogen cycle, 558*f*, 559
Dense fibrous connective tissue, 67, 67*f*, 69*f*
Dental caries, 173
Dental hygiene, cardiovascular disease and, 110
Dentin, 172*f*, 173
Deoxyhemoglobin, 118, 206, 208
Deoxyribonucleic acid. *See* DNA
Deoxyribose, 38, 38*t*, 492, 492*f*
Depletion, of ozone shield, 562, 562*f*
Depolarization, 288*f*–289*f*, 289
Depo-Provera, 380*t*
Depression (psychiatric condition), 275, 306
Dermatitis, 150
Dermatology, 77
Dermis, 75*f*, 76*f*, 77
Descending colon, 180, 181*f*
Descent, common, 520–524
Desert ecosystem, 546*f*, 547
Desertification, 8, 8*f*, 571, 571*f*
Desmosomes, 94, 96*f*
Detergents, 31
Detrital food chains, 550
Detrital food web, 550, 551*f*
Detritus feeders, 548–549
Detrol LA, 219
Devauchelle, Bernard, 78
Development, as characteristic of life, 5
Development, human
after birth, 410–414
birth defects and, 401
differentiation in, 396
embryonic, 395*t*, 398–400, 398*f*, 399*f*
fertilization in, 394–395, 394*f*
fetal, 395*t*, 401–407
fifth to seventh month, 403, 403*f*
fifth week of, 400, 400*f*
fourth month of, 403
fourth week of, 400
gastrulation in, 398
gender differentiation in, 403
homology with vertebrate development, 524, 524*f*
morphogenesis in, 396
overview of, 5, 395*t*
pre-embryonic, 395*t*, 396–397, 396*f*
processes of, 396
second week of, 398–399
of sex organs, 405–407, 405*f*
sixth to eighth week of, 400
stages of, 396–400
third month of, 403
third week of, 399–400
DEXA (dual energy X-ray absorptiometry), 246
DHT (dihydrotestosterone), 406
Diabetes insipidus, 345
Diabetes mellitus, 84, 226, 227, 339, 354–356, 362
aging and, 411
gestational, 356, 408, 415
growth hormone and, 347
type 1, 91, 111, 178, 491
type 2, 51, 178
Dialysate, 234
Dialysis, 234, 234*f*
Diaphragm (contraceptive), 379, 380*t*, 381*f*
Diaphragm (muscle), 81, 82*f*, 175, 197*f*, 202, 202*f*, 203
Diaphysis, 240, 243*f*, 244*f*
Diarrhea, 181
Diastole, 97, 97*f*
Diastolic pressure, 100, 100*f*
Diencephalon, 295*f*, 298
Diet. *See also* Eating disorders; Food; Nutrition
acid reflux and, 174
and aging, 411
and atherosclerosis risk, 106
calcium in, 246
cancer and, 189, 454, 455
carbohydrates in, 30–31, 185, 185*f*, 185*t*
and cardiovascular disease, 110
cholesterol in, 34
eating disorders, 191–192, 191*f*
fats in, 32–33, 33*f*
lipids in, 186, 186*f*, 187*t*
meal planning, 190–191
minerals in, 187–188, 187*t*
in pregnancy, 401
protein in, 185–186, 185*f*
service size, 192
supplements, 190
vitamins in, 189–190, 189*t*, 190*t*
Dietary fiber, 30–31, 30*f*, 180, 185, 454
Dieting, 68, 183, 184
Diet soda, 478
Differentiation
cancer cells and, 448
in embryonic development, 396
Diffusion, 50, 50*f*, 105, 206–208
Digestion, 170
Digestive enzymes, 176, 177*t*
Digestive system, 80, 80*f*
accessory organs, 178–179, 179*f*
colorectal disorders, 181–182
development of, 398*f*, 400*f*
digestive tract wall, 171, 171*f*
endocrine system and, 361*f*
esophagus in, 173, 173*f*, 174
gallbladder in, 178, 179*f*
in homeostasis, 84, 85*f*, 129*f*, 180, 181
large intestine in, 170*f*, 180–182, 181*f*
liver in, 170*f*, 178, 179*f*
mouth in, 172–173, 172*f*
muscular system and, 279*f*
overview of, 170–172, 170*f*, 171*f*
pancreas in, 170*f*, 176, 178, 179–180, 179*f*, 180*f*
pharynx in, 173
secretions, regulation of, 179–180, 180*f*
small intestine in, 176–177, 176*f*, 177*f*
stomach in, 175–176, 175*f*
teeth in, 172*f*, 173
tongue in, 173
urinary system and, 234*f*
Digits, 77, 253
Dihybrid cross, 473, 473*f*
Dihydrotestosterone (DHT), 406
Dimer, 144
Dinoire, Isabelle, 78, 78*f*
Diphtheria, tetanus, pertussis (DTP) vaccine, 147*f*
Diploid (2n) number, 366, 424
Dirie, Waris, 373, 373*f*
Disaccharide, 29, 29*f*
Disease. *See also* Cancer; Infectious diseases; *specific diseases*
and biodiversity loss, 580*f*, 582
muscular, 277–278
sexually transmitted, 385–389
Distal convoluted tubule, 223*f*, 224, 225*f*, 228–230, 230*f*
Distance vision, 326–328, 326*f*
Disulfide bond, 36, 36*f*
Ditropan XL, 219
Diuretics, 106, 231–232
Diverticulitis, 171

Diverticulosis, 171, 181
Diving, 205
Division, cell. *See* Cell cycle
Dizziness, 334
DNA (deoxyribonucleic acid)
  in antibody production, 144
  bacterial, 134, 134*f*
  base pairing in, 37*f*, 38, 492, 492*f*, 494–495, 499, 501*f*
  in cancer cells, 448
  as characteristic of life, 5
  complementary, 506
  data and evolution, 527
  double helix shape of, 37*f*, 38, 492, 492*f*
  in Fanconi's anemia, 443
  free radicals and, 189
  function of, 37–38
  in HIV life cycle, 160
  in interphase, 422, 423*f*, 425*f*, 433
  microsatellites in, 459
  mitochondrial, 504, 527, 537, 540
  mutation in, 5, 450–451, 450*f*, 451*f*, 453, 494, 497
  origin of, 519
  polymerase chain reaction and, 504–505, 504*f*
  radiation damage to, 21
  repair of, in cell cycle, 423
  replication of, 38, 493–494, 494*f*
  steroid hormones and, 344, 344*f*
  structure of, 37*f*, 38, 492–493, 492*f*, 495*t*
  in study of evolution, 527
  technology using, 503–510
  transcription and, 496, 498–499, 498*f*, 501, 501*f*
  as universal, 503, 524
  viral, 135, 135*f*
DNA fingerprinting, 481, 504–505, 504*f*
DNA helicase, 494
DNA ligase, 506, 506*f*
DNA polymerase, 494, 494*f*, 504
DNA sequencing, 503–504, 503*f*
Docosahexaenoic acid (DHA), 33
Dogs, breeding of, 526
Domains, 6–7, 6*f*, 7*f*, 526, 527*f*
Domestic livestock, 575, 575*f*
Dominant allele, 468, 468*f*, 470
DOMS (delayed onset muscle soreness), 276
Dopamine, 291, 306, 307, 308
Doping, of blood, 120, 280, 280*f*
Doppler instrument, 109
Dorsal artery, of penis, 368*f*
Dorsal cavity, 81, 82*f*
Dorsal nerve, of penis, 368*f*
Dorsal root, 293, 294*f*
Dorsal root ganglion, 294*f*, 302, 303*f*
Dorsal vein, of penis, 368*f*
Double-blind study, 11
Double bond, 24
Double helix DNA structure, 37*f*, 38, 492, 492*f*
Douching, 380*t*, 389
Down syndrome, 437, 438–439, 439*f*, 441
Drinking water, 556, 556*f*, 571–572, 585
Drought-tolerant crops, 573
Drug abuse, 110, 226, 306–310, 307*f*, 308*f*, 388, 401
Drug dependence, 307
Drug resistance
  to antibiotics, 134, 166, 167–168, 167*f*, 168*f*, 387
  multidrug resistance, 461
  in viruses, 162
Drugs, medical. *See* Medications
Drug therapy, 306
DTP (diphtheria, tetanus, pertussis) vaccine, 147*f*
Dual energy X-ray absorptiometry (DEXA), 246
Dubernard, Jean Michel, 78
Dubois, Eugene, 535
Duchenne muscular dystrophy, 278, 485
Ductus arteriosus, 402, 402*f*
Ductus deferens, 368
Ductus venosus, 402, 402*f*
Due date, of baby, 397
Duodenum, 176, 179*f*
Duplication, 440*f*, 441
Duplication division, 366, 422, 424
Dutasteride, 222
Dwarfing of mammals, on islands, 534
Dwarfism, pituitary, 244, 347, 347*f*, 348
Dynamic polarization, 292
Dystrophin, 262, 278, 281, 485

E

E. Coli, 7*f*, 180
Ear
  air pressure equalization in, 329
  anatomy of, 328–330
  in hearing, 329–330
  inner, 316*t*, 329, 329*f*, 333*f*
  middle, 328–329, 329*f*
  noise pollution and, 331–332, 331*f*, 331*t*
  outer, 328, 329*f*
  swimmer's, 32
Eardrum (tympanic membrane), 328–329, 329*f*
Early atmosphere, 518–519, 518*f*, 519*f*
Earth, primitive, 518–519, 518*f*, 519*f*
Earwax, 32, 328
Earworms, 328
Eating disorders, 191–192, 191*f*. *See also* Diet; Food
EBV (Epstein-Barr virus), 122, 455
ECG (electrocardiogram), 98*f*, 99
Ecological pyramids, 550–552, 552*f*
Economic well-being, 589
*Eco*RI enzyme, 506
Ecosystems
  aquatic, 547, 547*f*
  biogeochemical cycles in, 552–563
  biotic components of, 548–549, 548*f*
  carbon cycle in, 554–558, 555*f*
  definition of, 547
  ecological pyramids in, 550–552, 552*f*
  energy flow in, 549–552, 549*f*
  failure, human impact of, 545
  food webs, 550, 551*f*
  human impact on, 547, 554, 555–558, 559–561, 559*f*
  nature of, 547–550
  nitrogen cycle in, 558–561, 558*f*
  in organization of life, 2, 3*f*
  phosphorus cycle in, 560*f*, 561–563
  theory of, 9
  trophic levels in, 550, 551*f*
  types of, 546*f*, 547, 547*f*
  water cycle in, 554, 554*f*
Ecotourism, 585
Ecstasy, 308
Ectoderm, 398, 399*f*
Ectopic pregnancy, 398
Effacement, in birth, 409
Effectors, 287, 287*f*, 303, 303*f*
Efferent arteriole, 223, 223*f*, 225*f*, 227, 230*f*
Efferent nerves, 286*f*
EGD (esophagogastroduodenoscopy), 158
Egg (ovum), 371, 374–375, 375*f*
  in fertilization, 394–395, 394*f*, 396*f*
  genetic testing of, 475, 475*f*
  vs. sperm, 367
Eicosapentaenoic acid (EPA), 33
Eighth week, of development, 400
Ejaculation, 368
Ejaculatory duct, 367*f*, 368
Elastic cartilage, 68, 242
Elastic fiber, 66, 66*f*, 67, 67*f*
Elastin, 66, 77, 441
Electrocardiogram (ECG), 98*f*, 99
Electromyography (EMG), 262
Electron(s), 20–21, 21*f*
Electron microscopes, 44–45, 45*f*, 45*t*
Electron transport chain, 59, 59*f*
Electropheragram, 503, 503*f*
Electrophoresis, gel, 504, 504*f*
Elements, 20, 20*f*
Elephants, 582
Elimination, 171
Elongation, in polypeptide synthesis, 500, 500*f*
Embolus, 106, 123
Embryo
  chick, 524*f*
  frozen, 383
  human, 242, 371, 372, 378, 398, 398*f*
  pig, 524*f*
Embryonic development, 395*t*, 398–400, 398*f*, 399*f*
Embryonic disk, 398, 398*f*
Embryonic Stem Cell Research Oversight (ESCRO) committees, 404
Embryonic stem cells, 61, 72, 117, 404
Embryonic testing, 443
Emergency contraception, 380*t*, 382
Emerging diseases, 166–167, 582
EMG (electromyography), 262
Emotions, 299
Emphysema, 210*f*, 211
Emulsification, 31, 176, 178
Enamel, tooth, 172*f*, 173
End cells, 61
Endochondral ossification, 243–244, 243*f*
Endocrine glands, 74, 340–344, 341*f*, 342*t*
Endocrine system, 81, 81*f*. *See also* Hormone(s)
  adrenal glands in, 218*f*, 341*f*, 344, 351–354, 351*f*
  cardiovascular system and, 361*f*
  digestive system and, 361*f*
  disorders, diagnosis of, 364
  feedback in, 341
  glands in, 340–344, 341*f*, 342*t*
  gonads in, 341*f*
  in homeostasis, 84, 85*f*, 129, 129*f*, 279*f*, 360, 361*f*
  hypothalamus in, 341*f*, 345, 346*f*
  integumentary system and, 361*f*
  muscular system and, 279*f*, 361*f*
  negative feedback in, 345, 347*f*
  nervous system and, 360, 361*f*
  ovaries in, 341*f*, 358, 358*f*
  pancreas in, 341*f*, 354–356, 354*f*
  parathyroid glands in, 341*f*, 342*t*, 350, 350*f*
  pineal gland in, 294*f*, 298, 342*t*, 359
  pituitary gland in, 341*f*, 345–347, 346*f*
  reproductive system and, 360, 361*f*
  respiratory system and, 361*f*
  skeletal system and, 361*f*
  testes in, 341*f*, 358, 358*f*
  thymus gland in, 138*f*, 139, 139*f*, 341*f*, 358, 511
  thyroid gland in, 21, 22*f*, 341*f*, 342*t*, 349–350, 349*f*, 350*f*
  urinary system and, 234*f*, 361*f*
Endocytosis, 52, 52*f*
Endoderm, 398, 399*f*
Endolymph, 333*f*
Endomembrane system, 48*f*, 54–55, 54*f*
Endometrial cancer, 458*t*
Endometriosis, 382
Endometrium, 372, 376–377, 377*f*, 377*t*, 378*f*, 397*f*, 398, 401
Endoplasmic reticulum (ER), 47*f*, 53, 53*f*
  in endomembrane system, 54, 54*f*
  rough, 47*f*, 53*f*, 54, 54*f*
  smooth, 47*f*, 53*f*, 54, 54*f*
Endorphins, 275, 294, 306, 307, 460
Endoscopy, 11, 12*f*, 182
Endosymbiosis, 48
Endothelium, 92, 93*f*
Endurance, muscular, 275
Energy
  acquisition of, as characteristic of life, 4, 4*f*
  ATP as carrier of, 38–39, 39*f*
  flow in ecosystems, 549–552, 549*f*
  in lipids, 31
  for muscle contraction, 272–274, 273*f*
  in primitive Earth, 518–519, 518*f*
  as resource, 575–578, 577*f*, 578*f*
  resting potential, in neurons, 288, 288*f*–289*f*
Enlarged prostate, 222
Entry, in HIV life cycle, 160, 161*f*
Entry barriers, 140
Entry inhibitors, 162
Environment, internal, 84
Environmental carcinogens, 453–455
Environmental conditions, and body shape, 539
Enzymes
  active site of, 57
  cascades of, 343, 343*f*
  digestive, 176, 177*t*
  gene expression and, 496
  hydrolytic, 54
  in metabolism, 57–58, 58*f*, 62
  as protein type, 35
  repair, for DNA, 494
  restriction, 506
Eosinophils, 117*f*, 121*f*, 122
EPA (eicosapentaenoic acid), 33
Epidemic, definition of, 156
Epidemiology, 155
Epidermis, 75*f*, 76–77, 76*f*
Epididymis, 367*f*, 368, 368*t*, 370*f*
Epiglottis, 173, 173*f*
Epilepsy, 301
Epinephrine, 99, 149, 342*t*, 343–344, 351*f*, 352
Epiphyseal growth plate, 243–244, 243*f*, 244*f*
Epiphysis, 240, 243*f*, 244*f*
Episiotomy, 409
Episodic memory, 300
Epithelial tissue, 66, 72–74, 72*f*–74*f*
EPO (erythropoietin), 119–120, 120*f*, 220, 233, 359, 460
Epstein-Barr virus (EBV), 122, 455
Equilibrium, 318, 332–334, 333*f*
ER. *See* Endoplasmic reticulum
Erectile dysfunction, 368–369
Erectile tissue, 368, 368*f*
Erection, 305*f*
Erosion, of beaches, 570, 571*f*
Erythroblasts, 117*f*
Erythrocytes. *See* Red blood cells
Erythropoietin (EPO), 119–120, 120*f*, 220, 233, 359, 460
Escherichia coli (E. coli), 7*f*, 180
ESCRO (Embryonic Stem Cell Research Oversight) committees, 404
Esophageal cancer, 451*f*, 453*f*
Esophagogastroduodenoscopy (EGD), 158
Esophagus, 170*f*, 173, 173*f*, 174, 175*f*
Essential amino acids, 185–186, 185*f*
Essential fatty acids, 186
Estrogen, 378, 378*f*
  aging and, 411
  in birth-control pills, 379
  and bone strength, 246
  in endocrine system, 358, 358*f*
  and fetal development, 406
  functions of, 342*t*
  in men, 374
  osteoporosis and, 245
  placenta and, 401
  in pregnancy, 407, 407*t*
  in reproductive system, 375, 376–377, 376*f*, 377*f*, 377*t*
  as steroid, 34
  vs. testosterone, 34, 34*f*
Ethambutol, 164
Ethmoid bone, 249, 249*f*
Ethnicity, 539, 539*f*
Eukarya domain, 6–7, 6*f*, 7*f*, 527*f*, 527*t*
Eukaryotic cells. *See also* Cells
  evolution of, 48, 48*f*, 520, 521
  internal structure of, 46, 47*f*
  vs. prokaryotic cells, 46

Euphorbia, 522
Eustachian tube, 329, 329f
Eutrophication, cultural, 559, 561, 579
Evaporation
  and salination, 573
  in water cycle, 554, 554f
Eversion, 256f
Evolution, 520–526
  anatomical evidence for, 523–524, 523f, 524f
  biochemical evidence for, 524, 524f
  biogeographical evidence for, 522, 523f
  as characteristic of life, 5–6
  chemical, 518
  common ancestors in, 530, 531f, 539
  common descent and, 520–524
  definition of, 5
  DNA evidence of, 504, 527
  of domains of life, 6f
  of eukaryotic cells, 48, 48f
  fossil evidence for, 520–522, 521f
  of hominins, 530–533, 531f, 532f
  of humans, 533–540
  vs. intelligent design, 525
  mosaic, 532
  natural selection in, 525–526, 525f
  population growth and, 538
  as theory or principle, 9
  as unifying concept, 9
Evolutionary tree, 530, 531f
Exchange pool, in biogeochemical cycle, 553, 553f
Excitatory signal, 291f
Excitatory synapse, 291f
Excitotoxicity, 412
Excretion, 218, 219–220. *See also* Urinary system
Exercise
  acid reflux and, 174
  aerobic, 60, 273, 274
  and body fat reduction, 272
  for bone strength, 245
  cancer and, 275
  cholesterol and, 275
  depression and, 275
  fuel sources for, 272, 273f
  importance of, 275
  lactate buildup in, 60
  and muscle tone, 272
  oxygen deficit in, 60
  regimen, 275t
  and water-salt balance, 231
  and weight control, 184, 191
Exocrine glands, 74, 340
Exocytosis, 52, 52f
Exons, 499
Exophthalmic goiter, 349–350, 349f
Exotic species, 580, 580f
Experimental variables, 11
Experiments, 9–11, 9f
Expiration, 197, 201, 202f, 203
Expiratory reserve volume, 204, 204f
Exponential population growth, 568
Extension, 256f
Extensively drug-resistant tuberculosis (XDR-TB), 168
Extensor carpi, 266f
Extensor digitorum, 265, 266f
Extensor digitorum longus, 266f
External auditory canal, 249f
External obliques, 265, 266f
External respiration, 206, 207f
Exteroceptors, 316, 316t
Extinction, 8
Extraembryonic cavity, 397f
Extraembryonic membranes, 397, 397f, 398, 398f
Extrinsic factors, in aging, 411
Ex vivo gene therapy, 512–513, 512f
Eye
  abnormalities of, 326–328, 326f
  anatomy of, 322–326, 322t, 323f–325f
  blind spot in, 325, 325f
  cancer of, 452, 452f
  color, 322
  compression of, 317
  physiology of, 322–326, 323f–325f
  in vision, 323–328

F

Face
  bones of, 249, 249f, 250f
  hominin, 530
  muscles of, 265, 265f
  somatosensory area for, 296f
  transplantation of, 78, 78f
Facial expressions, 296f
Facial tics, 277
Facilitated transport, 51, 51f
FAD (flavinadenine dinucleotide), 189
Fallopian tubes. *See* Oviducts
Familial hypercholesterolemia (FH), 481, 481f
Family pedigrees, 476, 476f, 482f, 485, 485f
Fanconi's anemia, 443
Farming. *See* Agriculture
Farsightedness, 326–328, 326f
Fascicle, 264, 264f
Fasting plasma glucose (FPG) test, 491
Fast-twitch muscle fibers, 274, 274f
Fat(s), 31–32, 32f
  in citric acid cycle, 59
  in diet, 32–33, 33f
  digestion of, 176, 177f, 177t, 178–179
  storage in bone, 239
Fatigue, muscle, 272, 272f, 274
Fatty acids, 32, 32f, 185, 186, 188, 272, 273f, 274
Fecal occult blood test, 456
Feces, 180
Feedback inhibition, 57
Feedback mechanisms
  in endocrine system, 341
  negative, 86–87, 86f, 87f, 341, 345, 347f
  positive, 87, 345, 557. 557f, 580
Feet
  development of, 400
  human *vs.* chimpanzee, 529, 529f
  motor area for, 296f
  somatosensory area for, 296f
  swelling of, 92
Fellatio, and STDs, 388
Female circumcision, 373
Female condom, 379, 380t, 381f
Female genitalia, 371–374, 372f
  cutting/mutilation of, 373
  development of, 405–407, 405f
Female hormone levels, 375–378, 376f–378f, 377t
Female infertility, 382
Female orgasm, 374
Female reproductive system, 371–374, 371t, 372f
Female sex drive, 374
Femoral artery, 103f
Femoral vein, 103f
Femur, 248f, 254, 254f
Fen-phen, 184
Fermentation, 50, 60, 273, 273f
Fertilization, 371, 394–395, 394f, 396f, 436, 436f
  assisted reproductive technologies, 383–384
  reproductive system and, 378–379, 378f
Fertilizers, 559, 574
Fetal development, 395t, 401–407
Fetal heart tones (FHTs), 109
Fetus, 372
  blood flow to, 402–403, 402f
  bone development in, 403
  chromosomal testing in, 421, 421f
  drug abuse and, 307, 307f
  hearing by, 403
  heart development in, 109
Fever, 87
Fever blisters, 386
FH (familial hypercholesterolemia), 481, 481f
Fiber, dietary, 30–31, 30f, 180, 185, 454
Fibrillation, ventricular, 98f, 99
Fibrillin, 66–67, 478
Fibrin, 125, 125f
Fibrinogen, 116, 118, 123, 123f, 179t
Fibrin threads, 123, 123f
Fibroblast, 67, 67f, 242
Fibrocartilage, 68, 242, 255
Fibrocartilaginous callus, 247, 247f
Fibromyalgia, 277
Fibrous connective tissue, 67, 67f, 69f, 242
Fibrous joints, 255
Fibula, 248f, 254, 254f
Fifth month, of development, 403, 403f
Fifth week, of development, 400, 400f
Fight or flight response, 300, 304, 352
Fimbria, 134, 134f, 371, 372f, 396f
Finasteride, 222
Fingerprinting, DNA, 504–505, 504f
Fingerprints, 76
Fingers
  motor area for, 296f
  somatosensory area for, 296f
Fire, first human use of, 536
First messenger, 343f, 344
Fish
  choice of, 581
  mercury in, 578, 579, 581
Fishing, 582f, 583–584
  fisheries decline, 563
  overfishing, 567, 581, 581f
Flagella, 55, 55f, 134, 134f
  of sperm, 369, 370f, 394, 394f
Flavinadenine dinucleotide (FAD), 189
Flexibility, 274
Flexion, 256f
Flexor carpi, 266f
Floating kidney, 218
Fluid, tissue, 68, 84, 92, 103, 138, 138f
Fluid connective tissues, 68, 68f, 69f
Fluid-mosaic model, 49
Fluoride, for teeth, 173
FMRP (Fragile X mental retardation protein), 486
Foam, spermicidal, 380t
Focus, in vision, 323–324, 323f
Folacin, 190t
Folate (folic acid), 120, 190t, 401
Folic-acid-deficiency anemia, 120
Follicles
  in ovarian cycle, 374, 375f
  in uterine cycle, 377, 377f
Follicle-stimulating hormone (FSH), 377f, 377t, 379
  in females, 375–376, 376f
  in males, 369–370, 371f
  obesity and, 382
  pituitary gland and, 358, 358f
Follicular phase, of ovarian cycle, 375–376, 377t, 378f
Fontanels, 248, 403
Food. *See also* Diet; Digestive system; Eating disorders
  allergies to, 149
  carbohydrates in, 30–31, 185, 185f, 185t
  fat tax, 110
  fish, 578, 579, 581, 582f, 583–584
  genetically modified, 14–15, 507–508, 510
  lipids in, 186, 186f, 187t
  meal planning, 190–191
  minerals in, 187–188, 187t
  protein in, 185–186, 185f
  as resource, 574–575, 574f, 575f
  sterilization of, with radiation, 22, 22f
  vitamins in, 189–190, 189t, 190t
  water in production of, 553
Food chains, 550, 578
Food Guide Pyramid, 189–190, 189f, 191
Food labels, 32, 33f, 188
Food poisoning, 181
Food web, 550, 551f
Footprints, 76
Foramen magnum, 248
Foramen ovale, 402, 402f, 403
Forced expiration, 203
Forelimbs
  in comparative anatomy, 523f
  in primates, 528
Forensic identification of remains, 252
Forensic urinalysis, 226
Foreskin
  female, 372
  male, 367f, 368, 368f
Forest, 546f, 547, 555–557, 571, 572f, 588
Formed elements, of blood, 116, 117f
Fossa ovalis, 401
Fossil(s)
  as evidence of evolution, 520–522, 521f
  hominin, 530
Fossil fuels, 27, 555, 559, 561, 574, 575, 576
Fossil record, 521
Fourth month, of development, 403
Fourth ventricle, 293f, 295, 295f
Fourth week, of development, 400
Fovea centralis, 322t, 323, 323f, 324
Fox, 6f
Fox, Sidney, 519
FOX03A gene, 411
FPG (fasting plasma glucose) test, 491
Fracture, bone, 247, 247f
Fragile X mental retardation protein (FMRP), 486
Fragile X syndrome, 485–486
Franklin, Rosalind, 493, 493f
Freckles, 76
Free nerve endings, cutaneous, 319, 319f
Free radicals, 189, 411
Freshwater, biodiversity and, 585
Freshwater ecosystems, 547, 547f
Frontal bone, 248, 248f, 249, 249f, 250f
Frontalis, 265
Frontal lobe, 296, 296f, 321f
Frozen embryos, 383
Fructose, 29
FSH (follicle-stimulating hormone), 377f, 377t, 379
  in females, 375–376, 376f
  in males, 369–370, 371f
  obesity and, 382
  pituitary gland and, 358, 358f
Fuel cells, 577–578, 578f
Functional genomics, 511–512, 511f
Functional groups, 33
Fungi, as pathogens, 210
Fungi kingdom, 6f, 7, 7f
Fusion, in HIV life cycle, 160, 161f

G

G_1 checkpoint, 422, 423f
G_1 stage, 422, 422f
G_2 checkpoint, 423, 423f
G_2 stage, 422, 422f
GABA neurotransmitter, 291, 306, 307
Gain-of-function mutations, 450
Galactose, 29
Galápagos Islands, 522, 525
Gallbladder, 31, 170f, 178, 179f, 305f
Gallstones, 178
Gamete, 429
Gamete formation, 469–470, 472f
Gamete intrafallopian transfer (GIFT), 384
Gametogenesis, 432
Gamma aminobutyric acid (GABA), 291, 306, 307
Gamma globulin, 148
Gamma hydroxybutyrate (GHB), 226
Ganges River, 573
Ganglia, 302
Ganglion cells, in retina, 325, 325f
Gangliosides, 62
Gap junction, 56, 56f, 94, 96f
Gardnerella vaginosis, 389
Gart gene, 439f
Gaseous cycle, 553

Gas exchange, 206–208, 207*f*
  alveoli and, 201, 201*f*
  in pulmonary circuit, 103, 103*f*
  respiratory system and, 207*f*
Gastric glands, 175*f*, 176
Gastric juice, 175*f*, 176
Gastric pit, 175*f*, 176
Gastrin, 180
Gastrocnemius, 266*f*
Gastroesophageal reflux disease (GERD), 174, 192
Gastroesophageal sphincter, 175*f*
Gastrointestinal tract, 170, 170*f*, 400*f*. *See also* Digestive system
Gastrointestinal tract wall, 171, 171*f*
Gastrulation, 398
Gate control theory of pain, 294
GDNF (glial-derived neutrophic factor), 71
Gehrig, Lou, 277
Gel electrophoresis, 504, 504*f*
Gene(s). *See also* Chromosomes; DNA
  activation of, in cell cycle, 423, 423*f*
  aging and, 411
  as characteristic of life, 5
  on DNA, 37
  human, number of, 511
Gene cloning, 506, 506*f*
Gene expression, 37, 496–503
  enzymes and, 496
  genetic code, 497, 497*f*
  overview of, 496–497, 497*f*, 501–502, 501*f*
  participants in, 501*t*
  posttranscriptional control in, 502
  posttranslational control in, 496
  pretranscriptional control in, 496
  proteins and, 496, 496*t*
  regulation of, 502–503, 503*f*
  transcription, 496, 497*f*, 498–499, 498*f*
  transcriptional control in, 502
  translation, 496–497, 497*f*, 499–501, 499*f*, 500*f*
Gene guns, 507*f*
Gene pharming, 508, 509*f*
Gene therapy, 15, 62, 108, 281, 512–513, 512*f*
Genetically modified (GM) animals, 507*f*, 508–509, 509*f*
Genetically modified (GM) bacteria, 507, 507*f*
Genetically modified (GM) foods, 14–15, 507–508, 510
Genetically modified (GM) organisms, 14–15
Genetically modified (GM) plants, 14–15, 507–508, 507*f*, 508*f*
Genetic change, as molecular clock, 530
Genetic code, 497, 497*f*, 503, 524
Genetic counselors, 487
Genetic disorders
  autosomal dominant, 476, 477*f*, 478
  autosomal recessive, 476–477, 476*f*
  nonsyndromic deafness as, 315
  X-linked dominant, 485
  X-linked recessive, 485–486, 486*f*
Genetic engineering, 506–509, 507*f*–509*f*, 575
Genetic mosaics, 438
Genetic mutation, 5, 450–451, 450*f*, 451*f*, 453, 494, 497
Genetic recombination, 432, 432*f*
Genetic testing, 443, 459, 475, 475*f*
Genetic theory, 9
Genital herpes, 386, 386*f*
Genitals
  abnormal development of, 406–407, 406*f*
  development of, 405–407, 405*f*
  female, 371–374, 371*t*, 372*f*
  male, 367–371, 367*f*, 368*f*, 368*t*
Genital warts, 385–386
Genome
  bioinformatics, 512
  human, 511
  modification of, 512–513, 512*f*
Genomics, 511–513
  comparative, 511–512, 511*f*
  functional, 511–512, 511*f*
Genotype, 468, 469
Genotypic ratio, 470
Genuine Progress Indicator (GPI), 589
Geothermal energy, 576
GERD (gastroesophageal reflux disease), 174, 192
Germ layers, 398–399, 399*f*
Gerontology, 410
Gestational diabetes, 356, 408, 415
Gey, George, 462
Gey, Margaret, 462
GH (growth hormone), 342*t*, 343, 346*f*, 347, 347*f*, 364
GHB (gamma hydroxybutyrate), 226
GI (glycemic index), 185, 185*t*
Giardia, 2*f*
Gibbon, 527*f*, 531*f*
GIFT (gamete intrafallopian transfer), 384
Gigantism, 244, 347, 347*f*
Gingivitis, 173
Giraffes, evolution of, 525–526, 525*f*
Glands
  adrenal, 218*f*, 341*f*, 344, 351–354, 351*f*
  bulbourethral, 222, 367*f*, 368, 368*t*
  Cowper. *See* bulbourethral, *above*
  definition of, 74
  endocrine, 74, 340–344, 341*f*, 342*t*
  exocrine, 74, 340
  gastric, 175*f*, 176
  in immune system, 138*f*, 139, 139*f*
  oil, 75*f*, 79
  parathyroid, 341*f*, 342*t*, 350, 350*f*
  pineal, 294*f*, 298, 342*t*, 359
  pituitary, 87, 294*f*, 341*f*, 342*t*, 345–347, 346*f*
  prostate, 219, 222, 222*f*, 243, 367*f*, 368, 368*t*
  salivary, 170*f*, 172, 340
  sweat, 74*f*, 75*f*, 79
  thymus, 138*f*, 139, 139*f*, 341*f*, 358, 411
  thyroid, 21, 22*f*, 341*f*, 342*t*, 349–350, 349*f*, 350*f*
Glandular epithelium, 74
Glans clitoris, 372, 372*f*, 374
Glans penis, 367*f*, 368, 368*f*
Glasses (corrective lenses), 321, 326–328, 326*f*
Glaucoma, 322–323, 327
Glenoid cavity, 253, 253*f*
Glial cells, 72
Glial-derived neutrophic factor (GDNF), 71
Global biogeochemical cycles, 552–563
Global warming, 557, 557*f*, 576, 580
Globin, 68, 119
Globulins, 118
Glomerular (Bowman's) capsule, 223–224, 223*f*, 228, 230*f*
Glomerular filtrate, 227
Glomerular filtration, 225–227, 225*f*
Glomerulus, 223, 223*f*, 230*f*
Glottis, 173, 173*f*, 197*f*, 199, 199*f*
Glucagon, 342*t*, 354, 354*f*
Glucocorticoids, 342*t*, 351*f*, 352, 353
Glucose
  ATP and, 39
  blood levels, 30, 85, 178, 354, 355*f*
  in diabetes mellitus, 354–356
  in glycolysis, 58
  insulin and, 354
  liver and, 178, 179*t*
  as muscle energy source, 272, 273*f*, 274
  reabsorption of, 227*t*
  as simple carbohydrate, 29
  structure of, 29
  in urine, 226, 227
Glucose tolerance test, 355, 355*f*
Glutamate, 291
Glutamic acid, 35*f*
Glutamine, 38, 478
Gluteus maximus, 263, 265, 266*f*
Gluteus minimus, 265
Glycemic index (GI), 185, 185*t*
Glycerol, 31, 32*f*
Glycogen, 29–30, 30*f*, 78, 179*t*, 267*t*, 272, 273*f*, 274
Glycolipid, 49, 54
Glycolysis, 49*f*, 58
Glycoprotein, 49, 54
GM. *See entries under* Genetically modified
GNP (gross national product), 589
GnRH (gonadotropin-releasing hormone), 358, 369–370, 371*f*, 374–375, 375*f*, 382
Goblet cell, 176*f*, 200, 200*f*
Goiters, 349, 349*f*
Golgi, Camillo, 54
Golgi apparatus, 47*f*, 54, 54*f*, 62
Gonadotropic hormones, 342*t*, 346, 346*f*, 369–370, 371*f*
Gonadotropin-releasing hormone (GnRH), 358, 369–370, 371*f*, 374–375, 375*f*, 382
Gonads, 341*f*, 342*t*. *See also* Ovary(ies); Testes
  abnormal development of, 406–407, 406*f*
  development of, 405–407, 405*f*
  in endocrine system, 358, 358*f*
  female, 371
  male, 369, 370*f*
Gonorrhea, 387
Gonorrhea proctitis, 387
Goose bumps, 281
Gorilla, 528*f*, 531*f*
Gout, 220
gp120s, 159*f*, 160
GPI (Genuine Progress Indicator), 589
Graafian follicle, 374–375, 375*f*
Gracile *Australopithecines*, 532
Grafting, skin, 76
Grain, refined, 185
Grameen Bank, 589
Granular leukocytes, 121*f*, 122
Granzymes, 146, 146*f*
Graphs, 13–14, 13*f*
Grasslands, 546*f*, 547
Gravitational equilibrium, 334
Gravitational equilibrium pathway, 333*f*, 334
Gray matter, 288, 293, 294*f*, 303*f*
Grazing food chains, 550
Grazing food web, 550, 551*f*
Greater trochanter, 254, 254*f*
Greater tubercle, 253*f*
Greenbelts, 588
Green fluorescent protein (GFP), 46
Greenhouse effect, 557, 557*f*, 576
Greenhouse gases, 557, 557*f*, 576
Green revolutions, 575
Green roofs, 588, 589*f*
Gross national product (GNP), 589
Ground substance, of connective tissue, 66, 66*f*
Groundwater depletion, 573, 573*f*
Groundwater mining, 554
Groundwater table, 554
Growth. *See also* Development, human
  as characteristic of life, 4
  in childhood, 410
  in embryonic development, 396
Growth factors, 449, 450, 451*f*
Growth hormone (GH), 244, 342*t*, 343, 346*f*, 347, 347*f*, 364
Growth plate, 241*f*, 243–244, 243*f*, 244*f*
Growth rate, population, 568
Guanine (G), 38, 38*t*, 492, 492*f*, 494, 495*f*
Gum disease, 106, 173
Gums, 172*f*
Gunston, Frank, 258

HAART (highly active antiretroviral therapy), 162, 385
Habitat loss, 580, 580*f*
Haemophilus influenzae, 82
Hair
  color of, 79
  nasal, 198
Hair cells
  in semicircular canals, 316*t*, 330, 330*f*, 331*f*, 333, 333*f*, 334
  in spiral organ, 316*t*
  in vestibule, 316*t*
Hair follicle, 75*f*, 77–79
Hair root, 75*f*
Hair shaft, 75*f*
Hairy leukoplakia, 158
Halogenated hydrocarbons, 578
Hammer (malleus), 328–329, 329*f*
Hand
  bones of, 253, 253*f*
  development of, 400
  motor area for, 296*f*
  somatosensory area for, 296*f*
Hangover, 231
Haploid (n) number, 366–367, 374, 428, 429
Hard palate, 172, 172*f*, 173*f*
Hare, Patagonian, 522, 523*f*
Hay fever, 149
Hazardous waste, 578–579
hCG (human chorionic gonadotropin), 149, 378, 379, 393, 398, 418
HDL (high-density lipoprotein), 19, 34, 39, 110, 116, 186, 275. *See also* Cholesterol
HDN (hemolytic disease of newborn), 126, 127*f*
Health span, efforts to increase, 410, 414*f*
Hearing, 328–332, 329*f*, 403
Hearing loss, 331–332, 331*f*, 331*t*
Heart, 94–99, 95*f*, 96*f*, 97*f*, 98*f*. *See also* Cardiovascular system
  artificial, 109, 109*f*
  in autonomic nervous system, 305*f*
  blood passage through, 96–97
  blood supply for, 95
  circulation in, 95–96
  development of, 399, 400*f*
  disorders of, 108–109
  muscle in, 69, 70*f*, 96*f*, 263, 263*f*
  rate, 100
  structure of, 94–95, 95*f*, 96*f*
  transplants, 109
Heart attack, 106
Heartbeat
  cardiac cycle, 97, 97*f*
  control of, 98–99, 98*f*
Heartburn, 174–175, 407
Heart disease. *See* Cardiovascular disease
Heart failure, 108–109
Heart murmur, 97
Heat, water and, 25
Heavy metals, 578, 585
Height. *See also* Dwarfism, pituitary
  growth hormone and, 347, 347*f*
  as polygenic trait, 480*f*
Heimlich maneuver, 198, 199*f*
HeLa cells, 462, 462*f*
Helicobacter pylori, 10–11, 11*f*, 13, 106, 140, 166
Helper T cells, 143*f*, 143*t*, 145–146
Hematoma, 247, 247*f*
Heme, 68, 118, 119, 119*f*
Hemispheres, cerebral, 295–296, 295*f*, 298, 301
Hemodialysis, 234, 234*f*
Hemoglobin, 20, 35, 35*f*, 36, 37, 68, 118, 119, 119*f*, 206, 207*f*, 208, 478
Hemolysis, 50, 120
Hemolytic anemia, 119
Hemolytic disease of newborn (HDN), 126, 127*f*
Hemophilia, 123–125, 382*f*, 455, 482–483, 486
Hemorrhoids, 102, 181
Hepatic duct, common, 179*f*
Hepatic portal system, 104
Hepatic portal vein, 102*f*, 104, 178, 179*f*

Hepatic vein, 102*f*, 104
Hepatitis, 147*f*, 179, 386
Hepatitis A, 147*f*, 179, 386, 454
Hepatitis B, 147*f*, 148, 179, 386, 454, 455
Hepatitis C, 142, 179, 386, 455
Hepatitis D, 386
Hepatitis E, 386
Hepatitis G, 386
Herbicide resistance, 507–509
Herbicides, 455, 574, 575
Herbivores, 548, 548*f*, 550*f*, 552*f*
Herceptin, 149
Herculex corn, 508
Heritage, cultural, 7–8
Heroin, 306, 307*t*, 308–309
Herpes simplex virus type 1 (HSV-1), 386
Herpes simplex virus type 2 (HSV-2), 386, 386*f*
Herpes viruses, 122
Heterotrophs (consumers), 519, 548–549, 548*f*, 549*f*, 551*f*
Hex A enzyme, 477
Hexose, 29
HGH (human growth hormone), synthetic, 347, 348
High-density lipoprotein (HDL), 19, 34, 39, 110, 116, 186, 275. *See also* Cholesterol
Highly active antiretroviral therapy (HAART), 162, 385
High-performance liquid chromatography (HPLC) test, 467
Hindlimbs, in primates, 528
Hinge joint, 255*f*, 256
Hippocampus, 299*f*, 300
Hip replacement, 257–258
Histamine, 141, 141*f*, 142, 150
Histocompatibility complex, 145
HIV (human immunodeficiency virus), 146, 385, 385*f*. *See also* AIDS
  acute phase of, 156–162
  cancer and, 455
  circumcision and, 373
  deaths from, 157*t*
  immune deficiency in, 122
  life cycle of, 160, 161*f*
  mutation of, 163
  opportunistic infections in, 156
  origin of, 156, 157
  phases of infection, 157–159, 159*f*
  prevalence of, 157, 157*t*, 158*f*
  prevention of, 161
  reverse transcription by, 159*f*, 160, 161*f*
  RNA in, 135–136, 159–160, 159*f*
  structure of, 159–160, 159*f*
  testing for, 158, 161
  transmission of, 160–161
  treatment of, 161–162, 385
  types of, 157
  vaccine for, 162, 163
HIV-1, 156
HIV-2, 156
HLA (human leukocyte antigens), 145
Hoarseness, 209
Hodgkin lymphoma, 452, 461, 582
Holding, of breath, 205
Homeostasis. *See also* Acid-base balance
  acid-base balance, 23, 84, 116, 220, 232, 232*f*
  aging and, 412
  blood, 79, 116
  blood glucose levels, 30, 85, 178, 354, 355*f*
  blood pH, 28, 116, 118, 205–206, 220, 320
  blood pressure, 69, 86, 93–94, 220, 230, 295, 352–353, 352*f*, 360
  body systems in, 84, 85*f*
  calcium, 220, 233, 244, 245, 278–279, 350, 350*f*
  cardiovascular system and, 84, 85*f*, 92*f*, 127–129, 129*f*
  as characteristic of life, 4
  cooperation in, 129*f*
  definition of, 84
  digestive system and, 84, 85*f*, 129*f*, 180, 181
  endocrine system in, 84, 85*f*, 129, 129*f*, 279*f*, 360, 361*f*
  gas exchange and, 206
  hypothalamus and, 298, 360
  interoceptors and, 316
  kidneys and, 228–233, 229*f*, 230*f*, 234*f*
  lymphatic system in, 85*f*, 128, 129*f*, 137
  muscular system and, 85*f*, 128, 129*f*, 264, 278–281, 279*f*
  negative feedback in, 86–87, 86*f*, 87*f*
  nervous system in, 84, 85*f*, 129*f*
  plasma proteins and, 117*f*, 118
  reflexes and, 304
  respiratory system in, 84, 85*f*, 129*f*
  as scientific theory, 9
  skeletal system in, 129, 129*f*, 278–281, 279*f*
  skin and, 75
  stimuli response and, 5
  urinary system in, 84, 85*f*, 128, 129*f*, 218, 219, 228–233, 229*f*, 230*f*, 234*f*
  water-salt balance, 188, 220, 228–231, 230*f*, 352–353, 352*f*, 360, 413
Hominids, 527*t*, 530, 531*f*
Hominines, 530, 531*f*
Hominins, evolution of, 530–533, 531*f*, 532*f*
Hominoids, 531*f*
Homo (genus), evolution of, 527*t*, 533–540, 533*f*
Homo erectus, 533*f*, 535–536, 538
Homo ergaster, 533*f*, 535–536, 535*f*, 536*f*
Homo floresiensis, 534, 534*f*
Homo habilis, 533*f*, 535, 538
Homo heidelbergensis, 533*f*
Homologous chromosomes, 428, 428*f*, 430*f*–431*f*
Homologous structures, 523, 523*f*, 524*f*
Homologues, 428
Homo neandertalensis (Neandertals), 517, 533*f*, 537–539, 537*f*
Homo rudolfensis, 533*f*
Homo sapiens, 527*t*, 533*f*, 536–537, 536*f*
Homozygous dominant genotypes, 468
Homozygous recessive genotypes, 468, 470
Honeybees, 584, 584*f*
Hormone(s). *See also* Endocrine system; Epinephrine; Norepinephrine
  in abnormal genital development, 406–407
  actions of, 340, 340*f*, 343–344, 343*f*, 344*f*
  adrenocorticotropic, 342*t*, 346, 346*f*, 351, 351*f*, 353
  aging and, 411
  antidiuretic, 230, 231, 342*t*, 345, 346*f*, 360
  anti-Müllerian, 406
  atrial natriuretic, 230, 352*f*, 353
  blood transport of, 116
  bone growth and, 244
  and cell cycle, 423
  as chemical signals, 342–344, 343*f*, 344*f*
  definition of, 178, 244
  in endocrine system, 340–341
  environmental pollutants and, 580
  female levels, 375–378, 376*f*–378*f*, 377*t*
  follicle-stimulating, 377*f*, 377*t*, 379
    in females, 375–376, 376*f*
    in males, 369–370, 371*f*
    obesity and, 382
    pituitary gland and, 358, 358*f*
  glucocorticoids, 342*t*, 351*f*, 352, 353
  gonadotropic, 342*t*, 346, 346*f*, 369–370, 371*f*
  gonadotropin-releasing, 358, 369–370, 371*f*, 374–375, 375*f*, 382
  growth, 244, 342*t*, 343, 346*f*, 347, 347*f*, 364
  in homeostasis, 84
  human chorionic gonadotropin, 149, 378, 379, 393, 398, 418
  hypothalamic, 342, 342*t*, 346*f*
  hypothalamic-inhibiting, 345, 346*f*
  hypothalamic-releasing, 342*t*, 345, 346*f*, 351*f*
  luteinizing, 377*f*, 377*t*, 379
    in endocrine system, 346, 346*f*
    in females, 375–376, 376*f*
    in males, 369–370, 371*f*
    pituitary gland and, 358, 358*f*
    urinary problems and, 222
  male, regulation of, 369–370, 371*f*
  melanocyte-stimulating, 342*t*, 347
  mineralocorticoids, 342*t*, 351*f*, 352–353, 352*f*
  Müllerian, 406
  vs. neurotransmitters, 340*f*
  parathyroid, 245, 278, 342*t*, 350, 350*f*
  peptide, 343–344, 343*f*, 407*t*, 408
  placental, 401, 407–408, 407*t*
  as proteins, 35
  in puberty, 410
  sex, 244, 246, 342*t*, 358, 358*f*, 401, 414
  steroid, 344, 344*f*
  target cells for, 340
  thymic, 139
  thyroid, 244, 342*t*, 349–350, 349*f*
  thyroid-inhibiting, 345
  thyroid-releasing, 345, 347*t*
  thyroid-stimulating, 342*t*, 345, 346, 346*f*, 347*f*, 364
  urinary system and, 220
Hormone-receptor complex, 344, 344*f*
Hormone replacement therapy, 376
Hormone skin patch, for birth control, 381*f*
Horse forelimb, 523*f*
Horseshoe crab, 583
Hox gene, 524
HPLC (high-performance liquid chromatography) test, 467
HPVs (human papillomaviruses), 13, 385, 389, 454, 455
HSV-1 (herpes simplex virus type 1), 386
HSV-2 (herpes simplex virus type 2), 386, 386*f*
HTLV-1 (human T-cell lymphotropic virus), 455
Humalog, 513
Human beings
  apes and, 7, 527, 528
  and biosphere, 8, 8*f*
  carbon cycle and, 555–558
  classification of, 526–529, 527*f*, 527*t*
  cultural heritage of, 7
  evolution of, 533–540
  land use by, 570–571, 571*f*
  nitrogen cycle and, 559–561, 559*f*
  population, 8
  population growth of, 538, 568–570, 568*f*
  as primates, 527–529, 529*f*
  resource use by, 570–579
  skeleton of, *vs.* chimpanzee skeleton, 529, 529*f*
  variation in, 539, 539*f*
  water cycle and, 554
  water use by, 571–574, 572*f*, 573*f*
Human chorionic gonadotropin (HCG), 149, 378, 379, 393, 398, 418
Human cloning, 404
Human Genome Project, 503, 511
Human growth hormone (HGH), synthetic, 347, 348
Human immunodeficiency virus. *See* HIV
Human leukocyte antigens (HLA), 145
Human life cycle, 366–367, 366*f*
Human papillomaviruses (HPVs), 13, 385, 389, 454, 455
Human T-cell lymphotropic virus (HTLV-1), 455
Humeral immunity, 144
Humerus, 248*f*, 253, 253*f*, 255*f*
Humulin N, 513
Hunchback (kyphosis), 250
Huntingtin, 478
Huntington disease, 478, 479*f*
Hyaline cartilage, 67, 67*f*, 240, 241*f*, 242, 243, 255, 256
Hybridoma, 149*f*
Hybridoma cells, 148, 149*f*
Hydrocarbons, 561
Hydrocephalus, 293
Hydrochloric acid, 26, 171, 176
Hydrofluorocarbons, as greenhouse gas, 557
Hydrogen, as fuel, 577–578, 578*f*
Hydrogenation, 32, 187
Hydrogen atom, 21, 21*f*
Hydrogen bonds, 24, 25*f*, 26, 36, 36*f*
Hydrogen ions, 27, 119, 205, 206, 228, 232, 232*f*
Hydrolization, 170
Hydrologic (water) cycle, 554, 554*f*
Hydrolysis reaction, 28, 28*f*
Hydrolytic enzymes, 54
Hydrophilic head, of phospholipid, 33
Hydrophilic molecules, 26
Hydrophobic molecules, 26
Hydrophobic tail of phospholipid, 33
Hydropower, 576, 577*f*
Hydrothermal vents, 548
Hydroxide, 189
Hydroxide ions, 27
Hydroxyurea, 478
Hymen, 374
Hyoid bone, 250, 250*f*
Hypercholesterolemia, familial, 481, 481*f*
Hyperglycemia, 354–355
Hyperparathyroidism, 350
Hypersensitivity reactions, 150–151, 151*f*
Hypertension, 100, 101*t*, 106, 230
Hyperthyroidism, 349–350, 349*f*
Hypertonic, 50, 51*f*
Hyperventilation, 206
Hypoglycemia, 354–355
Hypoparathyroidism, 350
Hypophysis. *See* Pituitary gland
Hypotension, 100, 101*t*
Hypothalamic-inhibiting hormones, 345, 346*f*
Hypothalamic-releasing hormones, 342*t*, 345, 346*f*, 351*f*
Hypothalamus, 86, 87*f*, 298, 299*f*
  adrenal system and, 351–352, 351*f*
  in brain anatomy, 295*f*
  in endocrine system, 341*f*, 345, 346*f*
  and female reproductive system, 375–376, 376*f*, 378
  in homeostasis, 298, 360
  hormones, 342, 342*t*, 346*f*
  and male reproductive system, 369–370, 371*f*
  urinary system and, 230
Hypothesis, 9, 9*f*, 11
Hypothyroidism, 349, 349*f*
Hypotonicity, 50, 51*f*
Hysterectomy, 372
H zone, 268*f*, 269

I band, 269
IBD (inflammatory bowel disease), 171, 181
IBS (irritable bowel syndrome), 171, 181
Ibuprofen, 192, 319
ICD (implantable cardioverter-defibrillator), 109
Ice, 25*f*, 26
Ice (crystal meth), 308
ICSI (intracytoplasmic sperm injection), 384
Identification, of skeletal remains, 252
IgA antibodies, 145, 145*t*
IgD antibodies, 145
IgE antibodies, 145, 145*t*, 150
IGF-1 (insulin-like growth factor 1), 347, 364
IgG antibodies, 144–145, 145*t*, 148, 148*f*, 150
IgM antibodies, 145, 145*t*
Iliac artery, 102*f*, 103*f*, 402, 402*f*

Iliac vein, 102*f*, 103*f*
Ilium, 254, 254*f*
Imaging
 in cancer diagnosis, 456–458, 458*f*
 use of radiation in, 21, 22*f*
Immediate allergic response, 150
Immune complex, 144
Immune system, 80, 80*f*
 acquired defenses in, 143–147, 143*f*, 143*t*, 144*f*, 145*t*, 146*f*
 acquired immunity in, 147–149, 147*f*, 149*f*
 aging and, 411
 allergies and, 150
 antibody-mediated, 143–145, 143*f*, 143*t*, 144*f*, 145*t*
 barriers in, 140–141
 blood in, 116
 cell mediated, 145–147, 146*f*
 complement system in, 142, 142*f*
 HIV infection and, 385
 hypersensitivity reactions, 150–151, 151*f*
 inflammatory response in, 141–142, 141*f*
 innate, 140–142, 141*f*
 lymphatic system and, 137–140, 138*f*, 139*f*
 pathogens and, 134–137, 134*f*, 135*f*, 137*f*
 proteins in, 35
 tonsils in, 209
 white blood cells in, 121–122
Immunization, 147*f*, 148. *See also* Vaccines
Immunosuppressive drugs, 150, 285
Immunotherapy, 149, 461–463, 461*f*
Implantable cardioverter-defibrillator (ICD), 109
Implantation, 371, 396*f*, 398
Impotency, 368
Incisors, 172*f*
Incomplete dominance, 481, 481*f*
Incontinence, in pregnancy, 407–408
Incus (anvil), 328–329, 329*f*
Independent alignment of chromosomes, 432, 432*f*
Index of Sustainable Economic Welfare (ISW), 589
Induced pluripotent stem cells (*iPS* cells), 61
Induction, of labor, 345
Industrial chemicals, pregnancy and, 401
Industrial revolution, 538
Industry, water use in, 572, 572*f*
Infancy, 410
Infection
 identification of, 148–149
 opportunistic, 156, 158–159
Infectious diseases. *See also specific diseases*
 deaths from, 155
 definition of, 155
 emerging, 166–167, 582
 re-emerging, 167
Infectious mononucleosis, 122
Inferior vena cava, 95*f*, 96, 102*f*, 103, 103*f*, 402, 402*f*
Infertility, 382–384
Inflammatory bowel disease (IBD), 171, 181
Inflammatory response, 141–142, 141*f*, 352
Influenza viruses, 135*f*
Informed consent, 11
Ingestion, 170
Inguinal canal, 369
Inguinal lymph nodes, 138*f*, 139
INH (isoniazid), 164
Inheritance. *See also* Genetic disorders
 behavioral, 480–481
 of blood types, 481–483, 483*f*
 of cancer risk, 452, 452*f*
 of chromosomes, 437–442
 codominance, 481
 of color blindness, 484–485, 484*f*
 dihybrid cross and, 473, 473*f*
 gamete formation and, 469–470, 472*f*
 genotypic ratio, 470
 incomplete dominance, 481, 481*f*
 multifactorial, 480–481
 multiple allele, 481–483, 483*f*
 one-trait crosses, 470–472, 471*f*
 pedigrees for, 476, 476*f*, 482*f*, 485, 485*f*
 phenotypic ratio, 470
 polygenic, 479–481, 480*f*
 probability in, 471–472
 sex-linked, 484–487, 484*f*–486*f*
 of skin color, 479–480
 two-trait crosses, 472–474, 472*f*, 473*f*, 474*f*
 of X-linked alleles, 484–487, 484*f*–486*f*
 of X-linked disorders, 485–486
Inhibin, 370
Inhibitory signal, 291*f*
Inhibitory synapse, 291*f*
Initiation, in polypeptide synthesis, 500, 500*f*
Injections, intramuscular, 266
Innate immunity, 140–142, 141*f*
Inner cell mass, in pre-embryonic development, 396*f*, 397
Inner ear, 316*t*, 329, 329*f*, 333*f*
Insemination
 artificial, 383
 intrauterine, 383, 384
Insertion, of muscle, 264, 265*f*
In situ cancer, 449
Insoluble fiber, 30–31, 30*f*
Insomnia, 298
Inspiration, 197, 201, 202, 202*f*
Inspiratory reserve volume, 203, 204*f*
Insulin, 84, 178, 341, 342*t*
 aging and, 411
 discovery of, 357
 functions of, 30
 vs. neurotransmitters, 340*f*
 pancreas and, 354, 354*f*
 synthetic, 14, 357, 362, 506, 513
Insulin-like growth factor 1 (IGF-1), 347, 364
Insulin pump, 356, 362, 513
Integrase, 159*f*, 160
Integrase inhibitors, 162
Integrated pest management, 588
Integration (sensory), 317
Integration (synaptic), 291*f*, 292
Integration (viral), in HIV life cycle, 160, 161*f*
Integumentary system, 75–79, 75*f*, 79, 80*f*
 endocrine system and, 361*f*
 in homeostasis, 85*f*
 urinary system and, 234*f*
Intelligent-design theory, 525
Intercalated disks, 70, 70*f*, 94, 96, 96*f*, 263
Intercostal muscles, 202, 202*f*, 203
Intercourse, 372, 374, 388
Interferons, 142, 149
Interkinesis, 428, 431*f*
Interleukins, 149
Intermediate filaments, 47*f*, 55
Internal environment, 84
Internal obliques, 265
Internal respiration, 207*f*, 208
Internet, scientific information on, 13
Interneuron, 287, 287*f*, 294–295, 303*f*
Interoceptors, 316
Interphase, 422, 422*f*
Interstitial cells, 369, 370*f*
Intervertebral disks, 251, 251*f*, 293
Intervertebral foramina, 250, 251*f*, 293
Intestinal arteries, 102*f*
Intestine
 large, 170*f*, 180–182, 181*f*
 small, 176–177, 176*f*, 177*f*
Intracytoplasmic sperm injection (ICSI), 384
Intramembranous bones, 242
Intramembranous ossification, 242–243
Intramuscular injections, 266
Intrauterine device (IUD), 379, 380*t*, 381*f*
Intrauterine insemination (IUI), 383, 384
Introns, 499, 502
Invagination, 48
Invasive species, 580
Inversion, 256*f*, 440*f*, 441, 441*f*
Invertebrates, in kingdom Animalia, 7
In vitro fertilization (IVF), 383, 384, 475, 475*f*
In vivo gene therapy, 513
Iodine, 187, 187*t*, 349
Ion(s), 23
Ion exchange, 224
Ionic bond, 22–23, 23*f*
Ionizing radiation, and cancer risk, 453
iPS (induced pluripotent stem) cells, 61
Iris, 322, 322*t*, 323*f*
Irish potato famine, 508
Iron, 119, 187, 187*t*, 188
Iron-deficiency anemia, 120
Irrigation, 573, 573*f*, 574
Irritable bowel syndrome (IBS), 171, 181
Ischium, 254, 254*f*
Islets of Langerhans (Pancreatic islets), 354, 354*f*
Isoniazid (INH), 164
Isotonicity, 50, 51*f*
Isotopes, 21–22
ISW (Index of Sustainable Economic Welfare), 589
IUD (intrauterine device), 379, 380*t*, 381*f*
IVF (in vitro fertilization), 383, 384, 475, 475*f*

J

Jacobs syndrome, 438
Jarvik 2000, 109
Jaundice, 119, 179
Jawbone, 172*f*
Jelly, spermicidal, 379, 380*t*, 381*f*
Jellyfish sting, neutralizing, 233
Jet lag, 298
Johanson, Donald, 532
Joint replacement surgery, 257–258
Joints, 255–256, 255*f*, 256*f*, 274
Jones, Marion, 280
Journals, scientific, 12
Jugular vein, 102*f*, 103*f*
Junctions, cell, 56, 56*f*
Jupiter, moons of, 1
Juxtaglomerate apparatus, 230, 230*f*

K

Kaposi's sarcoma, 159
Kaposi's sarcoma-associated herpesvirus (KSHV), 455
Karyotype, 420–421, 420*f*, 438–439, 439*f*
Keratin, 34, 74, 76
Keratinization, 76, 76*f*
Keratinocytes, 76
Ketones, 226
Kidneys. *See also* Urinary system
 acid-base balance and, 232, 232*f*
 in aging, 413
 in blood pressure regulation, 352–353, 352*f*
 cancer of, 451*f*, 453*f*
 dialysis, 234, 234*f*
 disorders of, 233–234
 floating, 218
 functions of, 92, 92*f*, 120*f*
 and homeostasis, 228–233, 229*f*, 230*f*, 234*f*
 hormones secreted by, 359
 structure of, 221*f*–224*f*, 222–224
 transplant, 234, 235
 in urinary system, 218, 218*f*, 219–220
Kidney stones, 233
Kingdom, 6*f*, 7, 7*f*
Kinocilium, 333*f*, 334
Klinefelter syndrome, 440
Knee-jerk reflex, 71, 318
Knee replacement, 239, 257, 258
Koch, Robert, 10, 162
Koch's postulates, 10
Krause end bulbs, 319, 319*f*
Kreb's cycle. *See* Citric acid cycle
KSHV (Kaposi's sarcoma-associated herpesvirus), 455
Kudzu, 580
Kwashiorkor, 575
Kyphosis, 250

L

Labels, food, 32, 33*f*, 188
Lab-grown bladder, 235, 235*f*
Labia majora, 372, 372*f*, 374, 405*f*, 406
Labia minora, 372, 372*f*, 374, 405*f*, 406
Labor, 408–410
 false, 408
 induction of, 345
Lacks, Henrietta, 462, 462*f*
Lacrimal bone, 249*f*
Lactase, 177*t*, 178
Lactate, 60, 273, 273*f*, 274
Lactation, hormones and, 347, 376
Lacteal, 176*f*, 177
Lactic acid, 232
Lactose intolerance, 178
Lacunae, 67, 67*f*, 241*f*
Lakes, 547
Lamarck, Jean-Baptiste, 525–526, 525*f*
Land, as resource, 570–571, 571*f*
Land mammals, evolution of, 522
Landsteiner, Karl, 128
Langerhans cells, 76
Language. *See also* Speech
 as basis of culture, 535, 538
 brain in, 301, 301*f*
 development of, 410, 511, 539
Lanugo, 403
Large intestine, 170*f*, 180–182, 181*f*
Laryngitis, 209
Larynx, 173*f*, 197*f*, 199, 209, 250, 250*f*, 453*f*
Laser treatments, for skin, 83*f*
LASIK (laser-assisted in situ keratomileusis) eye surgery, 327, 327*f*, 328
Latent tuberculosis, 164
Lateral epicondyle, 254*f*
Lateral malleolus, 254, 254*f*
Lateral sulcus, 296*f*
Lateral ventricles, 293*f*, 294*f*, 295
Latin America, HIV in, 156*t*, 157
Latissimus dorsi, 266*f*
Laxatives, 181
LDCs (less-developed countries), 568–570, 568*f*, 569*f*, 575, 579, 586
LDL (low-density lipoprotein), 19, 34, 39, 110, 116, 186
L-dopa, 61, 61*f*
Lead, 578
Learning, limbic system and, 300
Leech, medicinal, 2*f*
Left ventricular assist device (LVAD), 109
Leg
 bones of, 254, 254*f*
 motor area for, 296*f*
 somatosensory area for, 296*f*
Legionnaires' disease, 166–167
Legislation
 noise pollution, 332
 smoking bans, 209
Lemurs, in evolutionary tree, 531*f*
Lens, of eye, 322, 322*t*, 323–324, 323*f*
Lenses, corrective, 321, 326–328, 326*f*
Leon, Arthur, 275
Leprosy, 583
Leptin, 67, 359, 382
Less-developed countries (LDCs), 568–570, 568*f*, 569*f*, 575, 579, 586
Lesser trochanter, 254, 254*f*
Leukemia, 66, 115, 122, 451, 451*f*, 452, 461, 582
Leukocytes. *See* White blood cells
Lewontin, Richard, 539, 540

Leydig cells, 406
LH. *See* Luteinizing hormone
Life
as cellular, 44
characteristics of, 2–6
classification of (domains), 6–7, 6*f*, 7*f*
domains of, 527*f*
extraterrestrial, 1, 16
molecules of, 28–29, 28*f*
organization of, 2, 3*f*
origin of, 518–520, 518*f*, 519*f*
water and, 24–29
Life cycle
of HIV, 160, 161*f*
human, 366–367, 366*f*
of *Plasmodium*, 165
Ligaments, 67, 255–256, 255*f*
Ligamentum arteriosum, 401
Ligamentum teres, 402*f*
Ligase, 494
Light microscope, compound, 44–45, 45*f*, 45*t*
Limb(s), phantom, 319
Limb buds, 400, 400*f*
Limbic system, 299–300, 299*f*, 306
Limulus amoebocyte lysate, 583
Lineage, 530
Linolenic acid (ALA), 33
Lipase, 176, 177*f*, 177*t*
Lipids, 31–33, 32*f*, 106, 186, 186*f*, 187*t*
Lipitor, 19
Lipoproteins, 116
Liposuction, 68
Lips, somatosensory area for, 296*f*
Liquid, water as, 24–25
Liver
aging and, 413
alcohol and, 307
in autonomic nervous system, 305*f*
cancer of, 451*f*, 455
development of, 400*f*
in digestive system, 170*f*, 178, 179*f*
disorders of, 179
functions of, 31, 33, 84, 92, 104, 118, 120, 123, 178, 179*t*
glucose storage in, 30
proteins and, 59, 186
transplantation, 179
Liver cells, as polyploid, 366
Livestock, 575, 575*f*
Lobes, of brain, 296, 296*f*
Lobule, 369, 370*f*
Locus, 468
Long bone, 240, 241*f*
Longitudinal fissure, 295, 296*f*
Long-term memory, 300
Long-term potentiation (LTP), 300
Loop of nephron, 223*f*, 224, 225*f*, 229–230, 231*f*
Loose fibrous connective tissue, 67, 67*f*, 69*f*
Lordosis (swayback), 250
Loss-of-function mutations, 450, 450*f*
Lou Gehrig's disease, 277
Low-density lipoprotein (LDL), 19, 34, 39, 110, 116, 186
Lower gastroesophageal sphincter, 174
Lower limbs, bones of, 254, 254*f*
Lower respiratory tract, 197*f*, 200–201, 201*f*
Lower respiratory tract disorders, 210–213
LTP (long-term potentiation), 300
Lucy (australopithecine), 532, 532*f*
Lumbar nerves, 305*f*
Lumbar vertebra, 250, 251*f*
Lumen, 171, 171*f*
Lumphatic capillaries, 105, 105*f*
Lunelle, 380*t*
Lung(s), 197*f*. *See also* Respiratory system
aging and, 413
in autonomic nervous system, 305*f*
cancer of, 211–213, 213*f*
cohesion of water in, 26
in pulmonary circuit, 102*f*, 103
in respiratory system, 200–201, 201*f*, 202*f*
of smoker, 213*f*
Lung cancer, 211–213, 213*f*, 451, 451*f*
Lung packing, 205
Lung transplantation, 211
Lung volume reduction surgery (LVRS), 211
Lunula, 77, 77*f*
Lupus, 133, 151
Luteal phase, of ovarian cycle, 376, 377*t*, 378*f*
Luteinizing hormone (LH), 377*f*, 377*t*, 379
in endocrine system, 346, 346*f*
in females, 375–376, 376*f*
in males, 369–370, 371*f*
pituitary gland and, 358, 358*f*
urinary problems and, 222
LVAD (left ventricular assist device), 109
LVRS (lung volume reduction surgery), 211
Lymph, 68, 92, 105, 138, 177
Lymphatic capillaries, 137, 138
Lymphatic duct, right, 138, 138*f*
Lymphatic nodules, 140
Lymphatic organs, 138–140, 138*f*
Lymphatic system, 80, 80*f*, 92
in homeostasis, 85*f*, 128, 129*f*, 137
in immune system, 137–140, 138*f*, 139*f*
Lymphatic vessels, 68, 138, 138*f*
Lymph nodes, 68, 138*f*, 139–140, 139*f*
Lymphoblasts, 117*f*
Lymphocytes, 80, 117*f*, 118, 121*f*, 122, 137, 139, 142, 143. *See also* B cells; T cells
Lymphoid tissue, cancers of, 66
Lymphoma, 66, 451, 455, 461
Lysine, 35*f*
Lysis, 50, 51*f*
Lysosome, 43, 47*f*, 54–55, 54*f*
Lysozyme, 140

M

MacLeod, John, 357
Macromolecules, 28, 28*f*, 519
Macrophages, 141–142, 141*f*, 145
Macrosomia, 356
Mad cow disease, 137
Magnesium, 187*t*, 188
Magnetic resonance imaging (MRI), 458
Magnification
biological, 563, 578, 579, 579*f*
in microscopes, 45, 45*f*, 45*t*
Major histocompatibility complex (MHC), 145, 146*f*
Malaria, 148, 155, 164–167, 165*f*, 166*f*
Male condom, 379, 380*t*, 381*f*
Male hormonal contraception (MHC), 392
Male hormone regulation, 369–370, 371*f*
Male infertility, 382
Male orgasm, 369
Male reproductive system
development of, 405–407, 405*f*
structure and function, 367–371, 367*f*, 368*f*, 368*t*
Malignant tumor, 449, 449*f*, 450. *See also* Cancer
Malleolus, lateral, 254, 254*f*
Malleolus, medial, 254, 254*f*
Malleus (hammer), 328–329, 329*f*
Mallon, Mary, 136
Maltase, 177*f*, 177*t*
Maltose, 29, 29*f*
Mammalia, 527*t*
Mammals, land, evolution of, 522
Mammary glands, 74*f*
Mammography, 456–458, 458*f*
Mandible, 248*f*, 249, 249*f*, 250*f*
Manganese, 187, 187*t*
Manubrium, 251*f*, 252
Marfan syndrome, 66–67, 478
Marijuana abuse, 307*t*, 309–310
Marinol, 460
Marshall, Barry James, 10–11, 11*f*, 13
Marsupials, evolution of, 522
Mass, atomic, 21
Masseter, 266*f*
Mass number, 21
Mass transit, 587*f*, 588
Mast cell, 66*f*, 122, 141, 141*f*, 142
Mastication, 296*f*
Mastoiditis, 248
Mastoid process, 265
Mastoid sinuses, 248
Materials acquisition, as characteristic of life, 4, 4*f*
Matrix
of bone, 241*f*, 242–243, 243*f*
of connective tissue, 67
of HIV virus, 159–160, 159*f*
of mitochondrion, 57, 57*f*
Matter, 20
Mature mRHA, 498
Maxilla, 248*f*, 249, 249*f*, 250*f*
Maximum inspiratory effort, 203, 204*f*
McGwire, Mark, 277, 280
M checkpoint, 423, 423*f*
MDCs (more-developed countries), 568–570, 568*f*, 569*f*, 575, 586
MDMA (methylene dioxymethamphetamine), 308
Meal planning, 190–191
Measles, mumps, and rubella (MMF) vaccination, 172
Measles vaccine, 147*f*
Mechanical digestion, 170, 173
Mechanoreceptors, 316, 316*t*, 332–333, 333*f*
Medial condyle, 254, 254*f*
Medial malleolus, 254, 254*f*
Medial umbilical ligaments, 402*f*
Medical marijuana, 309
Medications
environmental diversity and, 582–583, 582*f*
gene pharming and, 508, 509*f*
immunosuppressive drugs, 150, 285
and pregnancy, 401
Medulla oblongata, 302
in brain anatomy, 295*f*
functions of, 298
and heartbeat, 99
in homeostasis, 360
and respiration, 232
in respiration control, 205
Medullary cavity, 240, 241*f*, 243, 243*f*
Meerkat, 2*f*
Megakaryoblasts, 117*f*
Megakaryocytes, 123
Meiosis, 366–367, 367*f*, 428–429, 430*f*–431*f*
in females, 374, 375*f*
and genetic inheritance, 469, 472, 472*f*
I, 428, 428*f*, 429, 430*f*–431*f*
II, 428*f*, 429, 430*f*–432*f*
mistakes during, 429
vs. mitosis, 433, 434*f*–435*f*, 434*t*, 435*t*
nondisjunction in, 437, 437*f*
in oogenesis, 436, 436*f*
overview of, 428–429, 428*f*
significance of, 432
in spermatogenesis, 369, 370*f*, 436, 436*f*
stages of, 429, 430*f*–431*f*
Meissner corpuscles, 319, 319*f*
Melacine, 461
Melanin, 76, 79, 83, 480
Melanocytes, 76, 76*f*, 79, 408
Melanocyte-stimulating hormone (MSH), 342*t*, 347
Melanoma, 76, 76*f*, 456, 456*f*, 459, 461
Melatonin, 298, 342*t*, 359, 361*f*
Memantine, 412
Membrane
amniotic, 409
bacterial, 134*f*
basement, 72–73, 72*f*–73*f*
basilar, 330, 330*f*
body, 82, 82*f*
extraembryonic, 397, 397*f*
mucous, 82, 140, 198
otolithic, 333*f*, 334
periodontal, 172*f*
plasma. *See* Plasma membrane
serous, 82
synovial, 82
tectorial, 330, 330*f*
tympanic, 328–329, 329*f*
Membrane attack complex, 142, 142*f*
Memory, 300
aging and, 413
episodic, 300
limbic system and, 300
long term, 300
semantic, 300
short term, 300
skill, 300
smell and, 322
Memory B cells, 143*f*, 143*t*, 144, 148
Memory T cells, 143*t*, 146, 146*f*, 148
Mengele, Joseph, 16
Meninges, 82, 293, 294*f*, 295*f*
Meningitis, 82, 293
Meningococcal vaccine, 147*f*
Meniscus, 255*f*
Menopause, 376, 411, 414
Menstrual cycles, synchrony in, 343
Menstruation, 359, 372, 376, 377, 377*f*, 378*f*
Mental functions, higher, 300–301, 301*f*
Mental retardation, Gart gene and, 439*f*
Mercury, 563, 578, 579, 581
Meridia, 184
Merkel disks, 319, 319*f*
Mesentary, 82
Mesenteric artery, 103*f*
Mesenteric vein, 103*f*
Mesoderm, 398, 399*f*
Messenger RNA (mRNA), 344, 344*f*, 495
formation of, 496, 498, 498*f*
processing of, 498–499, 498*f*
in transcription, 498–499, 498*f*
in translation, 496–497, 499, 499*f*, 500, 500*f*
Metabolic pathways, 57
Metabolism
basal, 190
cellular, 57–60, 57*f*, 59*f*
as characteristic of life, 4
thyroid hormones and, 349–350
urinary system and, 219–220
Metacarpals, 248*f*, 253, 253*f*
Metals, heavy, 578, 585
Metaphase, 426, 426*f*–427*f*, 434*f*, 434*t*, 435*t*
Metaphase I, 429, 430*f*, 434*f*, 434*t*
Metaphase II, 430*f*, 435*t*
Metastasis, 450
Metatarsals, 248*f*, 254, 254*f*
Methamphetamine abuse, 307*t*, 308
Methane, as greenhouse gas, 557, 557*f*
Methanosarcina mazei, 7*f*
Methicillin-resistant staph aureus (MRSA), 168
Methyl bromide, 562
Methylenedioxymethamphetamine (MDMA), 308
MHC (major histocompatibility complex), 145, 146*f*
MHC (male hormonal contraception), 392
Microbes, 134–137, 134*f*, 135*f*, 137*f*
Microglia, 71, 71*f*
Micrographs, 44, 45*f*
MicroRNA (miRNA), 495, 496, 502
Microsatellites, 459
Microscope
compound light, 44–45, 45*f*, 45*t*
electron, 44–45, 45*f*, 45*t*
Microscopy
cells in, 45–46, 46*f*
staining, 45, 46, 46*f*
Microtubule, 47*f*, 55, 55*f*, 56, 426
Microvilli, 55, 74, 177, 177*f*, 224m 224*f*
Micturition, 219, 219*f*
Midbrain, 295*f*, 298
Middle ear, 328–329, 329*f*

Miller, Stanley, 519, 519*f*
Mineralization, 520
Mineralocorticoids, 342*t*, 351*f*, 352–353, 352*f*
Minerals
  in diet, 187–188, 187*t*
  as resource, 578
  storage in bone, 239
Mining, 561*f*, 578
miRNA (MicroRNA), 495, 496, 502
Miscarriage, 429
Mites, 584
Mitochondria, 47*f*, 57–60, 57*f*, 59*f*
  cellular respiration in, 274
  evolution of, 48, 48*f*
  exercise and, 60
  in muscle fiber, 268*f*, 273
  of sperm, 394
Mitochondrial DNA, 504, 527, 537, 540
"Mitochondrial Eve," 537
Mitochondrial hypothesis of aging, 411
Mitosis, 366, 424–427
  in cell cycle, 421
  in embryonic development, 396
  in females, 375*f*
  importance of, 424, 424*f*
  vs. meiosis, 433, 434*f*–435*f*, 434*t*, 435*t*
  overview of, 422, 422*f*, 424, 425*f*
  phases of, 425–427, 426*f*–427*f*
  in spermatogenesis, 370*f*
Mitotic inhibitors, in chemotherapy, 461
Mitotic spindle, 425, 425*f*, 426
Mitral valve, 95
Molars, 172*f*
Mole (tumor type), 449
Mole (unit), 26
Molecular clock, 530
Molecular formulas, 24
Molecules, 22–24
  covalent bonding in, 23–24, 23*f*
  in diffusion, 50, 50*f*
  hydrogen bond in, 24, 25*f*
  hydrophilic, 26
  hydrophobic, 26
  ionic bonding in, 22–23, 23*f*
  of life, 28–29, 28*f*
  neurotransmitter, 291
  organic, 28–29, 28*f*
  in organization of life, 2, 3*f*
  polar, 24
Molybdenum, 187
Monkeys, in evolutionary tree, 531*f*
Monoblasts, 117*f*
Monoclonal antibodies, 148–149, 149*f*, 463
Monoculture, 574
Monocytes, 117*f*, 121*f*, 122, 141–142, 141*f*
Monohybrids, 470–472, 471*f*
Monomers, 144
Mononucleosis, infectious, 122
Monosaccharide, 29
Monosomy, 437
Monounsaturated oils, 32, 185
Mons pubis, 372, 372*f*
Montagut, Jacques, 443
Montreal Protocol, 562
Mood, neurotransmitters and, 306
More-developed countries (MDCs), 568–570, 568*f*, 569*f*, 575, 586
Morning-after pills, 382
Morphine, 306
Morphogenesis, in development before birth, 396, 398
Morula, in pre-embryonic development, 396*f*, 397
Mosaic evolution, 532
Mosaics, genetic, 438
Mosquitoes, malaria and, 164–166
Mothers, surrogate, 384
Motion sickness, 334
Motor areas, primary, 295*f*, 296, 296*f*, 297*f*, 301*f*
Motor axon, 269*f*
Motor nerves, 286*f*
Motor neurons, 269, 287, 287*f*, 294–295, 303*f*
Motor speech area, 296*f*
Motor unit, 271–272
Mouth, 170, 170*f*, 172–173, 172*f*
Movement
  of cells, 55, 55*f*
  of GI tract, 171
  of joint, 256, 256*f*
MRI (magnetic resonance imaging), 458
mRNA. *See* Messenger RNA (mRNA)
MRSA (methicillin-resistant staph aureus), 168
MS (multiple sclerosis), 151, 285, 288, 310
MSH (melanocyte-stimulating hormone), 342*t*, 347
Mucosa, 171, 171*f*, 175*f*
Mucous membranes, 82, 140, 198
Mucus, 171, 176
Müllerian ducts, 405*f*, 406
Müllerian hormone, 406
Multicellular, defined, 2
Multidrug resistance, 461
Multifactorial disorders, 480–481
Multifactorial trait, 480
Multiple allele inheritance, 481–483, 483*f*
Multiple sclerosis (MS), 151, 285, 288, 310
Multipurpose trees, 588
Multiregional continuity hypothesis, 536, 536*f*, 539
Multiuse farming, 588
Mumps vaccine, 147*f*
Murmur, heart, 97
Muscle(s)
  anabolic steroids and, 280
  body temperature and, 279–280
  cardiac, 69, 70*f*, 96*f*, 263, 263*f*
  ciliary, 323–324, 323*f*
  connection to bones, 264, 265*f*
  contraction of, 267–274, 268*f*–270*f*
  coordination of, 298
  delayed onset soreness of, 276
  disorders of, 276–278
  energy for, 272–274, 273*f*
  facial, 265, 265*f*
  fast twitch, 274, 274*f*
  fatigue of, 272, 272*f*, 274
  functions of, 264
  insertion of, 264, 265*f*
  motor units and, 271–272
  names of, 265, 266*f*
  neuromuscular junction and, 269–270, 269*f*
  number of in humans, 263
  origin of, 264, 265*f*
  pairing of, 264, 265*f*
  rigor mortis and, 271, 276
  skeletal, 69, 70*f*, 263*f*, 264
  slow twitch, 274, 274*f*
  smooth, 69, 70*f*, 263, 263*f*
  spasms in, 276
  sprains, 277
  strains, 277
  striation in, 69, 70*f*, 263, 263*f*, 267
  structure of, 264, 264*f*
  types of, 263–264, 263*f*
Muscle dysmorphia, 191*f*, 192
Muscle fibers, 69, 263, 263*f*, 264*f*, 267, 267*t*, 268*f*
Muscle spindle, 318, 318*f*
Muscle tone, 272, 318, 318*f*
Muscle twitch, 272, 272*f*
Muscular dystrophy, 262, 277–278, 281, 455, 485, 486*f*
Muscular endurance, 274
Muscularis, 171, 171*f*, 175, 175*f*
Muscular system, 80, 81*f*
  cardiovascular system and, 279*f*
  development of, 399*f*
  digestive system and, 279*f*
  disorders of, 276–278, 279*f*
  endocrine system and, 279*f*, 361*f*
  functions of, 264
  homeostasis and, 85*f*, 128, 129*f*, 264, 278–281, 279*f*
  nervous system and, 279*f*
  overview of, 263–266, 263*f*–266*f*
  reproductive system and, 279*f*
  respiratory system and, 279*f*
  urinary system and, 234*f*, 279*f*
Muscular tissue, 66, 69–70, 70*f*
Mushrooms, 2, 7*f*
Mutagen, 453
Mutation, genetic, 5, 450–451, 450*f*, 451*f*, 453, 494, 497
Myalgia, 277
Myasthenia gravis, 151, 278
Mycobacterium tuberculosis, 159, 162*f*, 164, 210
Myelin, 288
Myelin sheath, 71, 71*f*, 285, 287*f*, 288, 290, 310, 314
Myeloblasts, 117*f*
Myeloid leukemia, 453*f*
Myocardial infarction, 106
Myocardium, 94, 95
Myofibrils, 267, 267*t*, 268*f*
Myofilaments, 267, 267*t*, 268–269, 268*f*
Myoglobin, 267*t*, 274
Myosin, 35, 69, 267, 268*f*
Myosin binding sites, 270*f*
Myosin filaments, 270–271, 270*f*
Myosin head, 270*f*
Myositis, 277
Myxedema, 349

N

NAD (nicotinamide adenine dinucleotide), 58, 60, 189
Naegele's rule, 397
Nafarelin, 222
Nails, 77, 77*f*
Naproxen, 192
Nasal bone, 249, 249*f*, 250*f*
Nasal cavity, 197*f*
Nasopharynx, 173*f*, 198, 455
Natriuresis, 230
Natural family planning, 380*t*
Natural selection, 525–526, 525*f*
Nausea, 182
Neandertals (*Homo neandertalensis*), 517, 533*f*, 537–539, 537*f*
Nearsightedness, 326, 326*f*
Negative feedback, 86–87, 86*f*, 87*f*, 341
  in endocrine system, 345, 347*f*
  in female hormone regulation, 376, 376*f*
  in male hormone regulation, 369–370, 371*f*
Neisseria gonorrhoeae, 387
Neisseria meningitidis, 82
Nephroblastoma, 447, 463
Nephrons, 223–224, 223*f*, 227*t*, 229–230, 231*f*
Nephroptosis, 218
Nerve(s), 70, 71*f*
  afferent, 286*f*
  anatomy, 302, 302*f*
  autonomic motor, 286*f*
  cervical, 305*f*
  cochlear, 329*f*, 330*f*
  cranial, 286*f*, 302, 302*f*, 305*f*
  dorsal, of penis, 368*f*
  efferent, 286*f*
  lumbar, 305*f*
  motor, 286*f*
  optic, 322*t*, 323, 323*f*, 325, 325*f*, 326, 326*f*
  regeneration of, stem cells and, 72–73
  sacral, 305*f*
  sciatic, 266
  sensory, 286*f*
  in skin, 75*f*
  somatic motor, 286*f*
  somatic sensory, 286*f*
  spinal, 250, 286*f*, 293, 294*f*, 302, 302*f*
  thoracic, 305*f*
  vagus, 302, 305*f*
  vestibular, 329*f*, 332–334, 333*f*
  visceral sensory, 286*f*
Nerve fibers, 287
Nerve signals, 288–289, 288*f*–289*f*
Nervous system, 81, 81*f*
  autonomic, 304, 304*t*, 305*f*
  in breathing control, 205, 205*f*
  central, 72–73, 286, 286*f*, 291, 293–299, 317, 317*f*
  development of, 399, 399*f*
  drugs and, 306–307
  endocrine system and, 360, 361*f*
  functions of, 70, 286–287
  in homeostasis, 84, 85*f*, 129*f*
  limbic system in, 299–300, 299*f*
  muscular system and, 279*f*
  overview of, 286–292
  parasympathetic division, 179, 219*f*, 286*f*, 304, 304*t*, 305*f*
  peripheral, 302–306, 302*f*, 303*f*
    in nervous system, 286, 286*f*
    neurotransmitters in, 291
    sensation and, 317, 317*f*
  somatic, 302–304, 304*t*
  sympathetic division, 286*f*, 304 304*t*, 305*f*
  urinary system and, 219*f*, 234*f*
Nervous tissue, 2, 66, 70–71, 71*f*, 287, 293
Neural folds, 399
Neural tube, in embryonic development, 399
Neural tube defects, 401, 415
Neuroendocrine system, 360
Neurofibrillary tangles, 412, 412*f*
Neurofibromatosis, 78, 78*f*
Neuroglia, 70–71, 71*f*, 287
Neuroma, benign, 449
Neuromuscular junction, 269–270, 269*f*
Neurons, 44*f*, 70, 71*f*, 287
  aging and, 413
  structure of, 287, 287*f*, 292, 292*f*
  types of, 287, 287*f*
Neurosecretory cells, 346*f*
Neurotransmitters, 290*f*, 291, 306, 340*f*
Neutralization, of antigen, 144
Neutrons, 20–21, 21*f*
Neutrophils, 117*f*, 121*f*, 122, 141, 141*f*
Nevus, 449
Newborns, passive immunity in, 148, 148*f*
Niacin, 189
Niche, 549
Nickel, 187, 453
Nicotinamide adenine dinucleotide (NAD), 58, 60, 189
Nicotine abuse. *See* Smoking
Nicotinic acid, 190*t*
Nile River, 573
Nitrates, 226, 558*f*, 559
Nitric acid, 27
Nitrification, in nitrogen cycle, 558*f*, 559
Nitrite-cured foods, 454
Nitrogen atom, 21, 21*f*
Nitrogen cycle, 558–561, 558*f*
Nitrogen fixation, 558, 558*f*
Nitrogen oxides, 27, 558*f*, 559–561
Nitroglycerin, 106
Nitrosamines, 454
Nitrosoureas, in chemotherapy, 461
Nitrous oxide, as greenhouse gas, 557
Nociceptors. *See* Pain receptors
Nodes of Ranvier, 287*f*, 288
Noise pollution, 331–332, 331*t*
Nondisjunction, 437, 437*f*, 439, 440
Non-Hodgkin lymphoma, 451*f*, 452, 455
Nonionizing radiation, and cancer risk, 453
Nonpoint source of pollution, 561–563
Nonrenewable energy sources, 576
Nonrenewable resources, 570
Nonsyndromic deafness, 315, 334
Norepinephrine, 99, 291, 304, 304*t*, 306, 342*t*, 351*f*, 352
Nose, 198
No-till farming, 574, 574*f*
Notochord, 399*f*
Nuclear envelope, 47*f*, 53, 53*f*
Nuclear pore, 53, 53*f*
Nuclear power, 576

Nuclear radiation, cancer caused by, 453
Nuclease, 177*t*
Nucleic acids, 37–39, 177*t*. *See also* DNA; RNA
Nucleocapsid, 159, 159*f*
Nucleoid, of bacteria, 134*f*
Nucleolus, 47*f*, 53, 53*f*, 425, 426*f*
Nucleoplasm, 53
Nucleosidases, 177*t*
Nucleotides, 38, 220, 492
Nucleotide structure, 38, 38*t*
Nucleus, atomic, 20, 21*f*
Nucleus, cellular, 47*f*
  of cancer cells, 448, 448*f*
  evolution of, 48, 48*f*
  protein production and, 53–55, 53*f*, 54*f*
  structure of, 53, 53*f*
Nutrient(s)
  absorption in small intestine, 176*f*, 177, 177*f*
  classes of, 185–190
  definition of, 185
Nutrition, 183–192. *See also* Diet; Food
Nutrition labels, 32, 33*f*, 188

O

Obesity
  and acid reflux, 174
  cancer and, 454
  and cardiovascular disease, 110
  definition of, 183
  infertility and, 382
  leptin and, 359
  nutrition and, 183–184
  pregnancy and, 401
Objectivity, 9
Oblique layer, of stomach, 176
Observation, 9, 9*f*
Obstructive pulmonary disorders, 211
Occipital bone, 248, 248*f*, 249*f*
Occipital lobe, 295, 296*f*
Ocean
  as ecosystem, 547, 547*f*
  waste dumping in, 563
Odor receptors, 198
OGTT (oral glucose tolerance test), 339
Oil(s), 31–32, 32*f*
Oil, as fuel, 576
Oil (sebaceous) glands, 75*f*, 79, 140
Olecranon process, 253
Olfactory area, 295*f*
Olfactory bulb, 299*f*, 321–322, 321*f*
Olfactory cells, 316*t*, 321, 321*f*
Olfactory cortex, 322
Olfactory epithelium, 321*f*
Olfactory tract, 299*f*
Oligodendrocytes, 71, 71*f*, 72, 73*f*, 314
Omega-3 fatty acids, 33, 106, 186, 581
Omnivores, 548, 550*f*
Oncogenes, 450, 451*f*
Oncologist, 451
Oncology, 451
OncoMouse, 509
One-trait crosses, 470–472, 471*f*
On the Origin of Species (Darwin), 525
Oocyte, 375*f*, 376*f*, 436, 436*f*
Oogenesis, 374, 375*f*, 436, 436*f*
Opiates, 306
Opportunistic infections, in HIV/AIDS, 156, 158–159
Opsin, 324, 324*f*
Optic chiasma, 326, 326*f*
Optic nerve, 322*t*, 323, 323*f*, 325, 325*f*, 326, 326*f*
Optic tract, 326, 326*f*
Optic vesicle, 400*f*
Oral cancer, 451, 451*f*
Oral cavity. *See* Mouth
Oral contraceptives, 379, 380*t*, 381*f*
Oral glucose tolerance test (OGTT), 339
Orangutan, 528*f*, 531*f*
Orbicularis oculi, 265, 266*f*
Orbicularis oris, 266*f*
Orbitals, 20, 21, 21*f*
Orca (killer whale), 581–582
Organelles, 46, 48
Organic, definition of, 28
Organic chemicals, cancer caused by, 453–455
Organic molecules
  origin of, 518–519, 518*f*, 519*f*
  types and characteristics of, 28–29, 28*f*
Organisms, in organization of life, 2, 3*f*
Organization
  of cells, 46, 47*f*
  of life, 2, 3*f*
Organ of Corti (spiral organ), 316*t*, 330, 330*f*, 331, 331*f*
Organs
  engineered, 150
  in organization of life, 2, 3*f*
Organ systems, 75, 79–81, 80*f*–81*f*
  in homeostasis, 84, 85*f*
  in organization of life, 2, 3*f*
Orgasm, 305*f*
  female, 374
  male, 369
Origin, of muscle, 264, 265*f*
Origin of life, 518–520, 518*f*, 519*f*
Orlistat, 184
Oropharynx, 453*f*
Orrorin tugenensis, 530
Osmoreceptors, 316
Osmosis, 50
Osmotic pressure, 50
  of blood, 116
  in capillary exchange, 104–105, 104*f*
  plasma and, 118
Ossicles, 328–329, 329*f*
Ossification, 242–244, 243*f*
Ossification center, 243, 243*f*
Ossification zone, 243, 244*f*
Osteoarthritis, 257, 257*f*
Osteoblasts, 68, 242–245, 243*f*, 247
Osteoclasts, 68, 242, 245
Osteocytes, 67*f*, 240, 241*f*, 242
Osteon, 67*f*, 68, 240, 241*f*
Osteoporosis, 188, 245, 246, 275, 411
Otitis media, 209
Otolithic membrane, 333*f*, 334
Otoliths, 333*f*, 334
Outbreak, definition of, 156
Outer ear, 328, 329*f*
Out-of-Africa hypothesis, 536, 536*f*, 539–540
Oval window, 328–329, 329*f*, 330, 330*f*
Ovarian cancer, 451*f*, 452, 461
Ovarian cycle, 374–379, 375*f*, 377*t*
Ovariohysterectomy, 372
Ovary(ies), 396*f*
  aging and, 411
  development of, 403, 405*f*, 406
  in endocrine system, 341*f*, 358, 358*f*
  in female reproductive system, 371, 371*t*
  hormones produced by, 342*t*
Overactive bladder, 219
Overexploitation, 580–582, 580*f*
Overfishing, 567
Oviducts, 371, 371*t*, 372*f*, 396*f*, 405*f*, 406
Ovulation, 375, 375*f*, 377, 377*f*, 377*t*, 378*f*, 396*f*
Ovum. *See* Egg
Oxygen, in gas exchange, 206–208, 207*f*
Oxygen atom, 21, 21*f*
Oxygen debt/deficit, 60, 273
Oxyhemoglobin, 118, 206, 208
Oxytocin, 87, 342*t*, 345, 346*f*, 408
Ozone hole, 562, 562*f*
Ozone shield, 562, 562*f*

p16 gene, 459
P53 protein, 423
Pacemaker, 98
Pacinian corpuscles, 319, 319*f*
Paget's disease, 245
Pain
  gate control theory of, 294
  referred, 320
Pain receptors, 316, 319–320
Pairing of muscles, 264, 265*f*
Palate
  hard, 172, 172*f*, 173*f*
  soft, 172, 172*f*, 173, 173*f*
Palatine bone, 249*f*
Paleontology, 521
Pancreas
  in autonomic nervous system, 305*f*
  cancer of, 451*f*, 453*f*
  in diabetes mellitus, 355–356
  in digestive system, 170*f*, 176, 178, 179–180, 179*f*, 180*f*
  in endocrine system, 341*f*, 354–356, 354*f*
  functions of, 84
  hormones, 342*t*
Pancreatic amylase, 176, 177*f*, 177*t*, 178
Pancreatic duct, 179*f*
Pancreatic gland, 74*f*
Pancreatic islets (islets of Langerhans), 354, 354*f*
Pancreatic juice, 178, 179–180, 179*f*
Pancreatic lipase, 178
Pancreatic trypsin, 176
Pandemic, 156
Pantothenic acid, 190*t*
Papillae, 320–321, 320*f*
Papillary muscles, 96*f*
Pap test, 74, 365, 371–372, 456
Paramecium, 7*f*
Paranthropus aethiopicus, 533*f*
Paranthropus boisei, 533*f*
Paranthropus robustus, 533*f*
Paraplegia, 294
Parasympathetic division, 179, 219*f*, 286*f*, 304, 304*t*, 305*f*
Parathyroid glands, 341*f*, 342*t*, 350, 350*f*
Parathyroid hormone (PTH), 245, 278, 342*t*, 350, 350*f*
Parent cells, 424, 428
Parietal bone, 248, 248*f*
Parietal lobe, 295, 296*f*
Parkinson disease, 61, 61*f*, 71, 298, 412
Partially hydrogenated oil, 186, 188
Partial pressure, gas exchange and, 206
Parturition, 408–410, 408*f*
Passion flower, 6*f*
Passive immunity, 148–149, 148*f*
Passive immunotherapy, 463
Passive smoking. *See* Secondhand smoke
Pasteur, Louis, 10, 44
Patagonian hare, 522, 523*f*
Patella, 248*f*, 254, 254*f*
Patellar reflex, 71, 318
Paternity testing, 128, 505
Pathogens, 134–137, 134*f*, 135*f*, 137*f*, 155., 167. *See also* Bacteria; Virus(es)
Paxil, 306
PCP (*Pneumocystis jiroveci* pneumonia), 157, 159, 210
PCR (polymerase chain reaction), 504–505, 504*f*
PCT (proximal convoluted tubule), 223*f*, 224, 224*f*, 225*f*, 230*f*
PDN (polycystic kidney disease), 235
Pectoral girdle, 248*f*
Pectoralis major, 266*f*
Pedigrees, 476, 476*f*, 482*f*, 485, 485*f*
Pelvic cavity, 81, 82*f*
Pelvic girdle, 248*f*, 254, 254*f*
Pelvic inflammatory disease (PID), 382, 387
Pelvis, 254, 296*f*, 529, 529*f*
Penicillin, 134, 167, 582
Penis, 222*f*, 367*f*, 368, 368*f*, 368*t*
  circumcision of, 368, 373
  erectile dysfunction, 368–369
  urethra in, 219
Pentose, 29
Pepsidase, 177*f*
Pepsin, 176, 177*t*
Peptidases, 177*t*
Peptide bond, 35, 36, 36*f*
Peptide growth factors, 359
Peptide hormones, 343–344, 343*f*, 407*t*, 408
Peptides, 35, 36*f*, 177*f*
Perforins, 145–146, 146*f*
Pericardium, 82, 82*f*, 94, 95*f*
Periodicity, of elements, 20
Periodic table, 20, 20*f*
Periodontal membrane, 172*f*
Periodontitis, 173
Periosteum, 240, 241*f*, 243
Peripheral arterial disease (PAD), 91
Peripheral nervous system (PNS), 302–306, 302*f*, 303*f*
  in nervous system, 286, 286*f*
  neurotransmitters in, 291
  sensation and, 317, 317*f*
Peristalsis, 170, 173*f*, 174–175, 180–181
Peristaltic waves, 175*f*, 176
Peritoneum, 82, 82*f*, 171
Peritonitis, 82, 171
Peritubular capillary network, 225*f*, 227
Periwinkle, rosy, 582, 582*f*
Permafrost, 547
Permeability, selective, 50, 50*f*
Pernicious anemia, 120
Perspiration, 140
Pertussis vaccine, 147*f*
Pest control, biological, 574, 574*f*, 583
Pesticides, 574, 575, 578, 579, 583
Pest management, integrated, 588
PET (positron emission tomography), 21, 22*f*
Pet market, 580–581
Peyer's patches, 140, 171
p53 gene therapy, 463
p53 protein, 450
pH
  of acid rain, 559
  of blood, 28, 116, 118, 205–206, 220, 320
  buffer system and, 232
  homeostasis, 28, 116, 118, 205–206, 220, 320
  immune system and, 140
  kidneys and, 232, 232*f*
  and protein function, 36
  of stomach, 176
  urinary system and, 220
  of urine, 226, 226*f*
  of vagina, 372
Phagocytosis, 52, 52*f*, 116, 121, 141, 141*f*
Phalanges, 248*f*, 253, 253*f*, 254, 254*f*
Phantom pain, 319
Phantom sensation, 319
Pharmaceuticals. *See* Medications
Pharyngeal cancer, 451*f*
Pharyngeal pouch, 400*f*, 524, 524*f*
Pharynx, 170*f*, 173, 197*f*, 296*f*
Phenotype, 468, 469
Phenotypic ratio, 470, 471*f*, 473, 473*f*, 474*f*, 484*f*
Phenylalanine, 467, 478, 487
Phenylalanine hydroxylase, 467
Phenylketonuria (PKU), 467, 478, 487
Phenylpyruvate, 467
Pheromones, 343
Philadelphia chromosome, 442
Phospholipid bilayer, 33, 33*f*, 46, 49, 49*f*
Phospholipids, 31, 33, 33*f*, 54
Phosphoric acid (phosphate), 492, 492*f*, 559, 560*f*, 561, 561*f*
Phosphorus, 187*t*
Phosphorus cycle, 553, 560*f*, 561–563
Photoreceptors, 316, 316*t*, 321, 324, 324*f*
Photosynthesis, 4, 554
Photovoltaic (solar) cells, 576–577, 577*f*
pH scale, 27, 27*f*
PID (pelvic inflammatory disease), 387
Pig
  embryo, 524*f*
  xenotransplantation with, 150
Pigments, melanin, 76, 79, 83
PillCam, 182, 182*f*
Pilus, sex, 134, 134*f*

Pineal gland, 294*f*, 298, 342*t*, 359
Pinkeye, 328
Pinna, 328, 329*f*
Pinocytosis, 52, 52*f*
Pitocin, 345
Pituitary dwarfism, 244, 347, 347*f*, 348
Pituitary gland, 87, 294*f*, 341*f*, 342*t*, 345–347, 346*f*
PKD (polyclastic kidney disease), 217
PKU (phenylketonuria), 467, 478, 487
Placebo, 11
Placenta, 378
in birth, 408*f*, 410
blood flow in, 402, 402*f*
chorion and, 397, 397*f*
dangerous substances crossing, 401
hormones, 401, 407–408, 407*t*
Plant kingdom, 6*f*, 7, 7*f*
Plants, transgenic, 507–508, 507*f*, 508*f*
Plaque, atherosclerotic, 106, 106*f*, 123, 186, 274
Plasma, 68*f*, 116, 117–120, 117*f*. *See also* Blood
Plasma cells, from B cells, 143*f*, 143*t*, 144, 148, 149*f*
Plasma cell tumors, 452
Plasma membrane, 46, 49–52, 49*f*–52*f*
bacterial, 134*f*
of egg, 394–395, 394*f*
in evolution of eukaryotic cell, 48, 48*f*
phospholipid bilayer, 33, 33*f*, 46, 49, 49*f*
self proteins on, 145
Plasma proteins, 117*f*, 118
Plasmids, 134, 134*f*, 506, 506*f*
Plasmin, 123
Plasmodium, 165, 166
Platelets, 68, 68*f*, 116, 117*f*, 123, 123*f*
Pleura, 82, 82*f*, 200
Pleurisy, 200
Pluripotent stem cells, 116, 117
Pneumococcal vaccine, 147*f*
Pneumocystis jiroveci pneumonia (PCP), 157, 159, 210
Pneumonectomy, 213
Pneumonia, 210, 210*f*
PNS. *See* Peripheral nervous system
Podocyte, 223*f*, 224, 230*f*
Point source of pollution, 561
Polar body, 374, 375*f*, 436, 436*f*
Polarization, dynamic, 292
Polar molecule, 24
Polio vaccine, 147*f*
Pollination, 583, 583*f*, 584
Pollution
acid rain, 27, 27*f*, 559–561, 559*f*, 576, 580
biodiversity loss from, 580, 580*f*
biogeochemical cycle disruption and, 553
cancer and, 453–455
definition of, 570
by humans, 570–579
noise, 331–332, 331*t*
point *vs.* nonpoint sources of, 561–563
water, 561–563, 561*f*
Polyclastic kidney disease (PKD), 217
Polyculture, 574, 574*f*
Polycystic kidney disease (PDN), 235
Polygenic inheritance, 479–481, 480*f*
Polygenic trait, definition of, 479
Polymerase chain reaction (PCR), 504–505, 504*f*
Polymorphonuclear leukocytes, 122
Polynucleotide structure, 38
Polypeptide, 35, 36, 36*f*
Polyploidy, in liver cells, 366
Polyps, colon, 182
Polyribosome, 47*f*, 54, 500, 501*f*
Polysaccharides, 29–30, 30*f*
Polysomnogram, 196, 214
Polyspermy, 394
Poly-X females, 440
Ponds, 547
Pons, 295*f*, 298
Population, as organizational level of life, 2, 3*f*
Population, human
growth of, 538, 568–570, 568*f*
and resources, 8
Positive feedback, 87, 345, 557, 557*f*, 580
Positron emission tomography (PET), 21, 22*f*
Positrons, 21
Postanal tail, 524, 524*f*
Posterior compartment, 323, 323*f*
Posterior pituitary, 341*f*, 342*t*, 345, 346*f*
Posttranscriptional control, 502
Posture, 318
Potassium
in diet, 23, 187*t*
homeostasis, 220, 230
in neurotransmission, 288, 288*f*–289*f*, 289, 291
Potato blight, 508
Prairie (temperate grassland), 546*f*, 547
Preadolescent period, 410
Precapillary sphincter, 93*f*, 94, 101, 105*f*
Precipitation, in water cycle, 554, 554*f*
Precision farming, 588
Predators, 548
Pre-embryonic development, 395*t*, 396–397, 396*f*
Prefrontal area, 296*f*, 297
Pregnancy, 378–379, 378*f*, 407–408, 407*t*
alcohol in, 401
birth control pills and, 378–379, 378*f*
birth defects and, 401
blood flow in, 402–403, 402*f*
body weight in, 407
breech birth in, 405
dangerous substances in, 401
diabetes and, 356
due date and, 397
ectopic, 398
energy level in, 407
estrogen in, 401
fetal hearing and, 403
and folic acid deficiency, 120
HIV in, 161, 162
induction of labor in, 345
multifetal, 383
nutrition in, 401
obesity and, 401
passive immunity and, 148, 148*f*
in past, dangers of, 409
paternity testing and, 128
progesterone in, 401
pulmonary values in, 407
uterine relaxation in, 407
varicose veins in, 102
X-rays in, 401
Pregnancy testing, 148–149, 393, 398, 418
Prehypertension, 101*t*
Preimplantation genetic diagnosis, 475, 475*f*
Premature birth, 405
Premolars, 172*f*
Premotor area, 296*f*, 297
Prepuce, 405*f*
Presbyopia, 327
Pressoreceptors, 316
Pressure
blood. *See* Blood pressure
in capillary exchange, 104–105, 104*f*
osmotic, 50
plasma and, 118
Pressure receptors, 319
Pretranscriptional control, 502
Prevention
of cancer, 454, 455
of HIV, 161
Primary auditory area, 296*f*, 297, 301*f*
Primary follicle, 374, 375*f*
Primary germ layers, 398–399, 399*f*
Primary motor area, 295*f*, 296, 296*f*, 297*f*, 301*f*
Primary mRHA, 498
Primary olfactory area, 296*f*, 297
Primary oocyte, 436, 436*f*
Primary somatosensory area, 296, 296*f*, 297*f*
Primary spermatocytes, 436, 436*f*
Primary structure, of protein, 36, 36*f*
Primary taste areas, 296, 296*f*
Primary visual area, 296*f*, 297
Primates
common ancestor of, 530, 531*f*
evolution of, 527*t*
HIV in, 156
humans as, 527–529, 529*f*
Prime mover, 264
Primitive Earth, 518–519, 518*f*, 519*f*
Prions, 137, 137*f*
Pritikin Diet, 184
PRL (prolactin), 342*t*, 346*f*, 347
Probability, in genetics, 471–472
probability value (*p*), 14
Processing centers, 297
Producers (autotrophs), 548, 548*f*, 549, 549*f*, 550*f*, 552*f*
Product rule, in genetics, 471–472
Products, 57
Progesterone, 342*t*, 358, 358*f*, 375–378, 376*f*, 377*f*, 377*t*, 378*f*, 379, 398, 401, 407, 407*t*, 411, 423
Progestin, 380*t*, 382
Progression, in carcinogenesis, 449*f*, 450
Prokaryotic cells, 46, 48, 48*f*, 520, 521
Prolactin (PRL), 342*t*, 346*f*, 347
Proliferating zone, 244, 244*f*
Proliferative phase of uterine cycle, 377, 377*f*, 377*t*
Promotion, in carcinogenesis, 449*f*, 450
Pronucleus, sperm, 394*f*
Prophase, 425, 426*f*, 434*f*, 434*t*, 435*t*
Prophase I, 429, 430*f*, 434*f*, 434*t*
Prophase II, 430*f*, 435*t*
Proprioceptors, 318, 318*f*
Prosimians, 527–528, 530, 531*f*
Prostaglandins, 342, 359, 368
Prostate cancer, 222, 451*f*, 455, 458*t*
Prostate gland, 219, 222, 222*f*, 243, 367*f*, 368, 368*t*, 413
Prostate-specific antigen (PSA) test, 459
Protease, 159*f*, 160
Protease inhibitors, 162
Protective proteins, 142, 142*f*
Proteinase enzymes, 449
Protein-first hypothesis, 519
Proteinoids, 519
Proteins, 34–37. *See also* Enzymes
alcohol and, 307
amino acids in, 35, 35*f*, 496, 496*t*
antibodies as, 35
carrier, 35, 49, 51, 288–289
channel, 35, 49, 477, 477*f*
in citric acid cycle, 59
C-reactive, 106
denaturation of, 36, 87
digestion of, 176, 177, 177*f*, 177*t*, 178
free radicals and, 189
functions of, 34–35, 496
gene expression and, 496, 496*t*
levels of organization in, 36–37, 36*f*
liver and, 186
major histocompatibility complex, 145, 146*f*
as nutrients, 185–186
plasma, 117*f*, 118
on plasma membrane, 49, 49*f*
prions as, 137, 137*f*
protective, 142, 142*f*
proteomics, 512
receptor, 321, 341, 343*f*
self, 145
structural, 34
structure of, 36–37, 36*f*, 496
synthesis of, 37, 53–55, 53*f*, 54*f* (*See also* Gene expression)
transcription factors, 503
in urine, 226
Proteomics, 512
Prothrombin, 116, 123, 123*f*
Prothrombin activator, 123, 123*f*
Protists, 6*f*, 7, 7*f*
Protocell, development of, 519
Protonix, 192
Protons, 20–21, 21*f*
Proto-oncogenes, 423, 450, 451*f*, 452
Provirus, 160
Proximal convoluted tubule (PCT), 223*f*, 224, 224*f*, 225*f*, 230*f*
Prozac, 306
PSA (prostate-specific antigen) test, 459
Pseudomonas aeruginosa, 134*f*
Pseudostratified ciliated columnar epithelium, 73*f*, 74, 200, 200*f*
P site, or ribosome, 499, 499*f*
PTH (parathyroid hormone), 245, 278, 342*t*, 350, 350*f*
Puberty, 366, 376, 410
Pubic symphysis, 254
Pubis, 254, 254*f*
Publication, of scientific studies, 11–12
Pulmonary arteries, 95*f*, 96, 96*f*, 102*f*, 103, 103*f*, 201*f*, 207*f*, 402*f*
Pulmonary circuit, 103, 103*f*
Pulmonary fibrosis, 210*f*, 211
Pulmonary trunk, 96, 96*f*, 103, 402*f*
Pulmonary tuberculosis, 210–211, 210*f*
PulmonaSry values, in pregnancy, 407
Pulmonary veins, 95*f*, 96, 96*f*, 102*f*, 201*f*, 207*f*, 402*f*, 403
Pulp, of tooth, 172*f*, 173
Pulse, 100
Punnett square, 470, 471–472, 471*f*, 472*f*
Pupil, 322, 322*t*, 323*f*
Pupil constriction, 305*f*
Purine, 439*f*, 492, 492*f*
Purkinje fibers, 98, 98*f*
Purse seining, 581
Pus, 141
p value, 14
P wave, 99
Pyelonephritis, 233
Pyloric sphincter, 175*f*, 176
Pyrazinamide, 164
Pyrimidine, 492, 492*f*
Pyrogens, 359
Pyruvate, 58, 60

Q

Qinghao plant, 155
QRS wave, 99
Quadriceps, 318*f*
Quadriceps femoris, 265, 266*f*
Quadriplegia, 294
Quality of life, measurement of, 589
Quarternary structure, of protein, 36, 36*f*
Queen Victoria, 482–483

R

Rabbit, Himalayan, 480, 480*f*
Rabies vaccine, 155
Radial artery, 100
Radiation
and cancer, 21, 453, 454
and chromosome structure change, 441
medical uses of, 21, 22*f*
sterilization with, 22, 22*f*
tissue damage from, 21–22
Radiation therapy, 459–460, 460*f*
Radicals, free, 189, 411
Radioactive wastes, 576
Radioisotopes, 21
Radius, 248*f*, 253, 253*f*
Radon gas, 453, 454
Rain, acid, 27, 27*f*, 559–561, 559*f*, 576, 580
Rain forest, 546*f*, 547, 571, 572*f*, 580, 582
Ramon y Cajal, Santiago, 292
RAS (reticular activating system), 298–299, 299*f*, 317
Ras oncogenes, 450
RB gene, 452, 452*f*
Reactants, 57
Reading, brain areas in, 301, 301*f*
Reception, of cell cycle signal, 423, 423*f*

Receptor-mediated endocytosis, 52, 52f
Receptor protein, 321, 341, 343f
Recessive allele, 468, 468f
Recombinant DNA products, 506–509, 507f–509f
Recombinant DNA technology, 357, 362, 506, 506f
Recruitment, 272
Rectal cancer, 451f, 452, 458t
Rectum, 170f, 180, 181–182, 181f
Rectus abdominis, 265, 266f
Recycling, of bone, 245, 245f
Red blood cells, 44f, 79
  carbon dioxide transport, 118–119
  as connective tissue, 68, 68f
  in differing tonicities, 50, 50f, 51f
  formation of, 117f
  in microscopy, 45f
  oxygen transport, 118, 119f, 120f
  in pregnancy, 407
  production of, 119–120, 120f
  transport function of, 116
Red bone marrow, 116, 120f, 240, 241f, 243, 279, 359
  in immune system, 138f, 139, 139f
Red-green color blindness, 484–485, 484f
Redi, Francesco, 44
Red pulp, 139, 139f
Reduced hemoglobin, 208
Reduction division, 366–367, 428
Re-emerging diseases, 167
Referred pain, 320
Reflex actions, 294–295
Reflex arc, 303–304, 303f, 317
Reflexes, 71, 302–304, 303f, 304, 318
Reforesting, 588
Refractory period, 290, 369, 374
Refrigeration, bacteria and, 136
Remodeling, bone, 245, 245f, 247, 247f
Renal artery, 102f, 103f, 218, 218f, 220, 221f
Renal capsule, 218
Renal cortex, 220, 221f
Renal medulla, 220, 221f
Renal pelvis, 220, 221f
Renal pyramids, 220, 221f
Renal vein, 103f, 218, 218f, 221f
Renewable energy sources, 576–578, 577f, 588
Renewable resources, 570
Renin, 220, 229f, 230, 230f, 231, 352–353, 352f
Renin-angiotensin-aldosterone system, 353, 407
Repair, of bone, 247, 247f
Repair enzymes, for DNA, 494
Replacement reproduction, 569–570
Repolarization, 288f–289f, 289
Reproduction
  of bacteria, 135, 135f
  as characteristic of life, 4, 4f
  control of, 379–384
  replacement level of, 569–570
Reproductive cloning, 404
Reproductive rate, in primates, 528
Reproductive system
  abnormal development of, 406–407
  aging of, 414
  development of, 399f, 405–407, 405f, 406f
  endocrine system and, 360, 361f
  female, 371–374, 371t, 372f
  functions of, 366
  human life cycle and, 366–367, 366f
  male, 367–371, 367f, 368f, 368t
  meiosis in, 366–367, 366f, 375f
  mitosis in, 366, 366f
  muscular system and, 279f
  overview of, 81, 81f
Reservoir, in biogeochemical cycle, 553, 553f, 555
Resident bacteria, 141
Residual volume, 204, 204f
Resistance, antibiotic, 134, 166, 167–168, 167f, 168f, 387
Resources
  energy, 575–578, 577f, 578f
  food, 574–575, 574f, 575f
  land, 570–571, 571f
  minerals, 578
  nonrenewable, 570
  renewable, 570
  water, 571–574, 572f
Respiration
  cellular, 57, 58–60, 59f, 273, 273f, 274
  mechanism of, 201–204
Respiratory control center, 205, 205f, 232
Respiratory pump, 101, 102
Respiratory system
  acid-base balance and, 232
  aging and, 413
  chemoreceptors and, 205–206
  in cystic fibrosis, 477–478
  development of, 399f
  disorders of, 208–213
  endocrine system and, 361f
  gas exchange in, 206–208, 207f
  in homeostasis, 84, 85f, 129f
  lower, 200–201, 201f
  lungs in, 200–201, 201f, 202f
  muscular system and, 279f
  overview of, 80, 80f, 197, 197f
  in pregnancy, 407
  upper, 197, 199f, 208–209
  urinary system and, 234f
  ventilation control in, 205–206, 205f
Respiratory tract, 197f, 198–199, 199f
Responsibility, social, 14–16
Resting potential, 288–289, 288f–289f, 291f
Resting zone, 244, 244f
Restriction enzyme, 506
Restrictive pulmonary disorders, 211
Results, of experimentation, 11
Retardation, 439f
RET gene, 452, 459
Reticular activating system (RAS), 298–299, 299f, 317
Reticular fiber, 66, 66f
Reticular formation, 298–299, 299f
Retina, 322t, 323, 323f, 324–325, 325f
Retinal, 324, 324f
Retinoblastoma, 452, 452f
Retrieval, memory, 300
Retrograde amnesia, 300
Retroviruses, 159f, 160, 455, 513
Reverse transcriptase, 159f, 160, 161f
Reverse transcriptase inhibitors, 162
Reverse transcription, in HIV life cycle, 160, 161f
Reward circuit, 307
R group, 35
Rh blood groups, 126–127, 127f, 481–483
Rheumatic fever, 151
Rheumatoid arthritis, 151, 151f, 257
Rh factor disease, 127f
Rhodopsin, 324, 324f
RhoGAM treatment, 126–127
Rib cage, 202f, 248f, 251, 251f
Rib facets, 251f
Riboflavin, 189, 190t
Ribonucleic acid. *See* RNA
Ribose, 38, 38t
Ribosomal RNA (rRNA), 53, 495, 502, 526
Ribosomes, 47f, 53–54, 54f, 134f, 344f, 499, 499f, 501f, 502
Ribozymes, 499, 519
Ribs, 248f, 251f, 252
Rice, genetically engineered, 583
Rickets, 23, 244
Rifampin, 164
Rigor mortis, 271, 276
Rio Grande River, 572–573
Ripkin, Cal, 277
River(s)
  damming of, 572–573
  as ecosystem, 547, 547f
RNA (ribonucleic acid)
  formation of, 498, 498f, 501, 501f
  function of, 37–38, 494–496, 495f
  of HIV, 159–160, 159f
  messenger (mRNA), 344, 344f, 495
    formation of, 496, 498, 498f
    processing of, 498–499, 498f
    in transcription, 498–499, 498f
    in translation, 496–497, 499, 499f, 500, 500f
  MicroRNA (MiRNA), 495, 496, 502
  in primitive evolution, 519
  processing of, 498–499, 498f, 501
  ribosomal (rRNA), 53, 495, 502, 526
  small, 495, 499, 502
  structure of, 37f, 38, 494–496, 495f, 495t
  in transcription, 496, 497f, 498–499, 498f, 501, 501f
  transfer (tRNA), 495, 499, 499f, 500, 500f, 502
  in translation, 496–497, 497f, 499–501, 499f, 500f
  viral, 135–136, 135f
RNA-first hypothesis, 519
RNA polymerase, 498
Robust *Australopithecines*, 532
Rod cells, 316, 316t, 322t, 323, 323f, 324–325, 324f, 325f
Rohypnol ("roofies"), 226
Root, of tooth, 172f, 173
Root canal, 172f
Root hair plexus, 319, 319f
Rotation, 256f
Rotational equilibrium, 333
Rotational equilibrium pathway, 333–334, 333f
Rotator cuff, 253
Roughage. *See* Fiber, dietary
Rough endoplasmic reticulum, 47f, 53f, 54, 54f
Round window, 328, 329f, 330f
rRNA (ribosomal RNA), 53, 495, 502, 526
Rubella vaccine, 147f
Ruffini endings, 319, 319f
Rugae, 175f, 176, 218
Runoff, 558f, 560f, 561
  agricultural, 573, 574
  in water cycle, 554, 554f
Rural sustainability, 587–588

S

Saccule, 332, 333f, 334
Sacral nerves, 305f
Sacral vertebra, 250, 251f
Sacrum, 248f, 250, 251f
Safer sex, 388
Sagan, Carl, 20
Sahelanthropus tchadensis, 530, 533f
Salinization of soil, 508
Saliva, 140, 172, 340
Salivary amylase, 172, 177t
Salivary glands, 170f, 172, 340
Salivation, 296f, 305f
Salt
  in plasma, 117–118
  reabsorption of, 228–231, 230f
  table, 22, 23f
Saltatory conduction, 290
Salt-cured foods, 454
Salt marshes, 547, 547f
Salt-tolerant crops, 508, 573
Salt-water balance, 188, 220, 228–231, 230f, 352–353, 352f, 360, 413
Salt water ecosystems, 547, 547f
Saltwater intrusion, 573
Salty taste, 321
Sanatorium, 163
Sarcolemma, 267, 267t, 268f, 269–270, 269f
Sarcoma, 66, 451, 455
Sarcomeres, 267, 268–269, 268f
Sarcoplasm, 267, 267t, 268f
Sarcoplasmic reticulum, 267, 267t, 268, 268f
SARS (severe acute respiratory syndrome), 166, 167
Sartorius, 263, 266f
Satiety, 359
Saturated fatty acids, 32, 32f
Savanna (tropical grassland), 546f, 547
Saw palmetto, 222
Scanning electron microscope, 44–45, 45f
Scapula, 248f, 253, 253f, 255f
Schwann cells, 72, 73f, 287f, 288, 314
Sciatic nerve, 266
SCID (severe combined immunodeficiency disease), 122, 150, 512–513, 512f
Science
  objectivity in
  as process, 9–13
  social responsibility and, 14–16
  theories in, 9
Scientific method, 9–10, 9f
Scientific studies
  design of, 9–11, 12f
  interpretation of, 13–14
  publication of, 11–12
Scientific theories, 9
Sclera, 322, 322t, 323f
Scoliosis, 250
Scrapie, 137
Scrotal sac, 369, 369f
Scrotum, 81, 367, 367f, 405f, 406
Seafood, 582f, 583–584. *See also* Fish
Sea level, global warming and, 576
Seawater, as hypertonic, 51
Sebaceous glands. *See* Oil glands
Secondary follicle, 374, 375f
Secondary oocyte, 375f, 436, 436f. *See also* Egg
Secondary sex characteristics, 358, 370, 376, 410
Secondary spermatocytes, 436, 436f
Secondary structure, of protein, 36, 36f
Secondhand smoke, 212, 401, 453
Second messenger, 343f, 344
Second week, of development, 398–399
Secretin, 180–181, 180f
Secretions, digestive, 179–180, 180f
Secretory phase or uterine cycle, 377, 377f, 377t
Secretory vesicle, 54f
Sedentary lifestyle, 110
Sedimentary cycle, 553
Sedimentation, 520
Seizure, 276–277
Selective permeability, 46, 50, 50f
Selenium, 187, 187t
Self markers, 49, 145, 150
Semantic memory, 300
Semen, 368
Semiarid lands, human habitation in, 571, 571f
Semicircular canals, 316t, 329, 329f, 330f, 332, 333, 333f, 334
Semiconservative replication, 493, 494f
Semilunar valves, 95, 96, 96f, 97, 97f
Seminal fluid, 368, 369
Seminal vesicles, 367f, 368, 368t
Seminiferous tubule, 369, 370f
Semmelweis, Ignaz, 409
Sensation(s), 316–318, 316t, 317f
Sensation, phantom, 319
Sensory integration, 317
Sensory neurons, 287, 287f, 303f
Sensory receptors, 287, 287f, 303f, 316–318, 316t, 317f
  aging and, 413
  cutaneous receptors, 77, 319, 319f
  proprioceptors, 318, 318f
Sensory speech area, 296f
Septum
  cardiac, 94, 95f, 96f
  penile, 368f
Serosa, 171, 171f
Serotonin, 291, 306
Serous membrane, 82
Serratus anterior, 266f
Sertoli cell, 369, 370, 370f
Serum, 125
Set point, 86, 86f, 87f
Seventh month of development, 403, 403f

Seventh week, of development, 400
Severe acute respiratory syndrome (SARS), 166, 167
Severe combined immunodeficiency disease (SCID), 122, 150, 512–513, 512*f*
Sewage, 579, 585
Sex characteristics, secondary, 358, 370, 376, 410
Sex chromosomes, 420, 420*f*
  changes in number of, 438, 440
Sex hormones, 342*t*, 358, 358*f*
  aging and, 414
  and bones, 244, 246
  and pregnancy, 401
Sex-linked inheritance, 484–487, 484*f*–486*f*
Sex pilus, 134, 134*f*
Sexual arousal, 368–369, 372, 374
Sexual intercourse, 372, 374, 388
Sexually transmitted diseases (STDs), 385–389. *See also* HIV
Sexual reproduction. *See* Meiosis
Shaggy mane mushroom, 7*f*
Shape, of proteins, 36
Sheep cloning, 404, 404*f*
Shells, atomic, 21
Shimomura, Osamu, 46
Shingles, 158, 160
Shivering, 87, 87*f*
Shock, anaphylactic, 150
Short-term memory, 300
Siamese cats, 480
Sickle-cell disease, 37–38, 120, 120*f*, 478
SIDS (sudden infant death syndrome), 205
Sight, 323–328
Sigmoid colon, 180, 181*f*
Sigmoidoscopy, 456
Silent Spring (Carson), 579
Silicon, 187
Simple carbohydrates, 29
Simple columnar epithelium, 73*f*, 74
Simple cuboidal epithelium, 72*f*, 74
Simple epithelia, 72*f*–73*f*, 73–74
Simple goiter, 349, 349*f*
Simple squamous epithelial tissue, 72*f*, 73
Sinkholes, 573, 573*f*
Sinoatrial (SA) node, 98, 98*f*, 99
Sinus, 248
Sinusitis, 208
siRNA (small interfering RNA), 495, 502
Sister chromatids, 420*f*, 421–422, 424–426, 425*f*, 427*f*, 429, 430*f*–431*f*
Sixth month of development, 403, 403*f*
Sixth week, of development, 400
Skeletal muscle, 69, 70*f*, 263*f*, 264
  contraction of, 267–271, 268*f*–270*f*
  functions of, 264
  names of, 265, 266*f*
  pairing of, 264, 265*f*
  structure of, 264, 264*f*
Skeletal muscle pump, 101, 101*f*
Skeletal remains, 252
Skeletal system, 80, 81*f*. *See also* Bone(s)
  appendicular skeleton, 248*f*, 253–255, 253*f*, 254*f*
  axial skeleton, 248–252, 248*f*–251*f*
  development of, 399*f*
  endocrine system and, 361*f*
  functions of, 240
  in homeostasis, 129, 129*f*, 278–281, 279*f*
  human *vs.* chimpanzee, 529, 529*f*
  joints, 255–256, 255*f*, 256*f*, 274
  movement in, 256, 256*f*
  overview of, 240–242
Skill memory, 300
Skin. *See also* Integumentary system
  accessory organs of, 77–79, 77*f*
  aging of, 413
  artificial, 65, 76, 88
  Botox treatments, 83
  cancers of, 76–77, 76*f*, 83, 451*f*, 453, 456, 456*f*
  color of, 479–480, 539
  in development, 399*f*
  grafting, 76
  and homeostasis, 75
  in immune system, 140
  laser treatments for, 83
  as organ, 75
  regions of, 75–77, 75*f*
  sensory receptors in, 319, 319*f*
  stem cells in, 75*f*, 76
  surface area of, 75
  tanning and, 83, 83*f*
Skin grafting, 65
Skin patch, for birth control, 381*f*
Skull, 248–249, 248*f*, 249*f*
SLE (systemic lupus erythematosus), 151
Sleep apnea, 196, 214
Sliding filament model, 269
Slow-twitch muscle fibers, 274, 274*f*
Small intestine, 176–177, 176*f*, 177*f*
Smallpox, 155
Small RNAs, 495, 499, 502
Smell, 321–322, 321*f*, 511
Smog, 560*f*, 561, 562
Smoked foods, 454
Smoking, 74, 307–308, 307*t*
  and cancer, 389, 451, 453, 453*f*, 454
  government bans on, 209
  health effects of, 106, 110, 192, 200, 211, 212, 213*f*, 411
  and infertility, 382
  and nutrition, 188
  pregnancy and, 401
  quitting, 212–213
  secondhand smoke, 212, 401, 453
Smooth endoplasmic reticulum, 47*f*, 53*f*, 54, 54*f*
Smooth muscle, 69, 70*f*, 263, 263*f*
snoRNA (small nucleolar RNA), 495
snRNA (small nuclear RNAs), 495
Snuff, 212
Social responsibility, 14–16
Soda, diet, 478
Sodium
  angiotensin and, 353
  blood levels, 23, 352–353, 353*f*
  in cystic fibrosis, 477
  in diet, 188, 189*t*
  in neurotransmission, 288, 288*f*–289*f*, 289, 290*f*, 291, 292
  reabsorption of, 227, 227*t*
  in salt-tolerant plants, 508
Sodium bicarbonate, 176, 178
Sodium chloride, 22, 23*f*
Sodium hydroxide, 26
Sodium-potassium pump, 51, 51*f*, 288–289, 288*f*–289*f*
Soft palate, 172, 172*f*, 173, 173*f*
Soil
  erosion of, 574–575, 585
  replenishment of, 588
  salinization of, 508
Solar (photovoltaic) cells, 576–577, 577*f*
Solar energy, 4, 576–577, 577*f*
Solar plexus, 204
Solar power plants, 577, 577*f*
Soluble fiber, 31
Solutions
  acidic, 26
  basic, 26–27
Solvent, water as, 26
Somatic motor nerves, 286*f*
Somatic sensory nerves, 286*f*
Somatic system, 302–304, 304*t*
Somatosensory area, primary, 296, 296*f*, 297*f*
Somatosensory association area, 296*f*, 297
Somatotropic hormone, 347
Somite, 400*f*
Songs stuck in head, cause of, 328
Sosa, Sammy, 280
Sound production, 199, 199*f*
Sour taste, 321
South Beach Diet, 184
Soybeans, 508
Spasms, muscle, 276
Spastic colon, 181
Specialized connective tissue, 67–68, 69*f*
Species, definition of, 2
Specific gravity, of urine, 226, 226*f*
Speech, 199, 199*f*, 296*f*, 297, 300. *See also* Language
Speed, 308
Sperm, 368, 436, 436*f*
  vs. egg, 367
  in ejaculation, 369
  in fertilization, 394–395, 394*f*, 396*f*
  movement of, 55, 55*f*
  pronucleus, 394*f*
  structure of, 369, 370*f*
  swim speed of, 56
Spermatids, 369, 370*f*, 436, 436*f*
Spermatocyte, 436, 436*f*
Spermatogenesis, 369–370, 370*f*, 371*f*, 436, 436*f*
Spermatogonium, 369, 370*f*
Sperm count, 369, 382, 383
Spermicide, 379, 380*t*, 381*f*
Sphenoid bone, 249, 249*f*
Sphincters, 174
Sphygmomanometer, 100, 100*f*
Spinal cord, 286, 286*f*, 293–295, 293*f*, 294*f*
  cancer of, 452
  regeneration of nerves in, 72–73
Spinal nerves, 250, 286*f*, 293, 294*f*, 302, 302*f*
Spinal reflexes, 303–304, 303*f*
Spindle
  mitotic, 425, 425*f*, 426
  muscle, 318, 318*f*
Spindle fibers, 55, 425, 425*f*, 426, 426*f*, 427*f*
Spine. *See* Vertebral column
Spinous process, 251*f*
Spiral organ (Organ of Corti), 316*t*, 330, 330*f*, 331, 331*f*
Spirillum, 134, 134*f*
Spirometer, 203, 204*f*
Spleen, 138*f*, 139, 139*f*
Splicing, of mRNA, 499
Sponge, contraceptive, 380*t*
"Sponge effect," 585
Spongy bone, 68, 240, 241*f*, 243, 243*f*
Spontaneous abortion, 429, 442
Spontaneous generation, 44
Sprain, 277
Squamous epithelium, 72, 73*f*
SRY gene, 403, 406, 420, 440, 509, 509*f*
S stage, 422, 422*f*
Staining, in microscopy, 45, 46, 46*f*
Standard error, 14
Stapedius, 263
Stapes (stirrup), 328–329, 329*f*, 330, 330*f*
Staphylococcus aureus, 134*f*, 167
Starch, 29–30, 30*f*, 178
Statin drugs, 111, 412
Statistical data, 14
Statistical significance, 14
STDs (sexually transmitted diseases), 385–389. *See also* HIV
Stem cells
  bioethics and, 15, 61
  in bone marrow, 116, 117*f*, 119
  as connective tissue, 66*f*
  nerve regeneration and, 72–73
  research on, 61, 116, 404
  in skin, 75*f*, 76
  in spinal cord, 72
  uses of, 61
Stents, 108, 108*f*
Stercobilin, 180
Stereocilia, 330, 330*f*, 333–334, 333*f*
Sterilization
  with radiation, 22, 22*f*
  surgical, 382, 382*f*
Sternocleidomastoid, 265
Sternum, 248*f*, 251*f*, 252
Steroid hormones, 344, 344*f*
Steroids, 31, 33–34, 34*f*
Sterols, 34, 34*f*
Stimulus
  neuronal, 288*f*–289*f*, 289
  response to, as characteristic of life, 4–5
  of sensory receptor, 316, 317*f*
Stirrup (stapes), 328–329, 329*f*, 330, 330*f*
Stomach, 26, 140, 170*f*, 175–176, 175*f*, 305*f*
Stomach cancer, 453*f*
Stomach ulcer, 10–11, 13, 140
Stool blood test, 456
Stop codons, 500, 500*f*
Storage, memory, 300
Strain, muscle, 277
Strand, of DNA, 38
Strangulation, 250
Strata, sedimentary, 520, 521*f*
Stratified epithelia, 73*f*, 74
Stratified squamous epithelium, 73*f*, 76
Streams, 547
Strength training, 275
Strep throat, 208
Streptococcus, 166
Streptococcus pneumoniae, 82
Streptococcus pyogenes, 208
Streptomycin, 583
Stress
  adrenal glands and, 351–352, 351*f*
  and cardiovascular disease, 110
  limbic system and, 299
Stretch marks, 408
Striae gravidarum, 408
Striation, in muscles, 69, 70*f*, 263, 263*f*, 267
Strip mining, 578
Stroke, 106, 107, 298
Structural formulas, 24
Structural proteins, 34
Styloid process, 249*f*
Subatomic particles, 20–21, 21*f*
Subclavian artery, 95*f*, 96*f*, 103*f*
Subclavian vein, 102*f*
Subcutaneous layer, of skin, 75*f*, 77
Submucosa, 171, 171*f*, 198
Subsidence, 573
Substance P, 306
Substrate, 57–58, 58*f*
Sucralfate, 192
Sucrose, 29
Sudden infant death syndrome (SIDS), 205
Sudoriferous glands. *See* Sweat glands
Sugar, table, 29
Sulcus (sulci), 296
Sulfur, 187*t*
Sulfur dioxide, 27, 559
Sulfuric acid, 27
Summation, in muscle contraction, 272, 272*f*
Sum rule, in genetics, 472
Sun, as source of Earth's energy, 4, 576–577, 577*f*
Sunbathing, 454
Sunscreen, 77
Superfund, 578
Superior vena cava, 95*f*, 96, 96*f*, 102*f*, 103, 103*f*, 402*f*
Superoxide, 189
Support, as function of proteins, 34
Supporting cell, 333*f*
Supportive connective tissue, 67–68, 67*f*, 69*f*
Surface area, of cells, 44, 45*f*
Surface mining, 578
Surfactant, 201, 203
Surgery, for cancer, 459
Surrogate mothers, 384
Suspensory ligament, in eye, 323–324, 323*f*
Sustainable society, 586–589, 586*f*–588*f*
Sutures (anatomical), 255
Swallowing, 173–175, 173*f*, 296*f*
Swayback (lordosis), 250
Sweat glands, 74*f*, 75*f*, 79
Sweating, 26, 231
Sweet taste, 321
"Swimmer's ear," 32

Swine flu (H1N1), 166
Sympathetic division, 286f, 304, 304t, 305f
Synapse, 290–292, 290f, 291f
Synapsids, 522
Synapsis, 428, 429, 432, 432f
Synaptic cleft, 269f, 270, 290, 290f
Synaptic integration, 291f, 292
Synaptic vesicle, 269f
Syndrome, 421
Synergists, 264
Synovial fluid, 255f, 256
Synovial joints, 255–256, 255f
Synovial membrane, 82
Synthetic growth hormone, 347, 348
Syphilis, 16, 387, 387f
Systemic circuit, 103–104, 103f, 104f
Systemic lupus erythematosus (SLE), 151
Systole, 97, 97f
Systolic pressure, 100, 100f

T

Table sugar, 29
TAH (total artificial heart), 109, 109f
Taiga, 546f, 547
Tail, of embryo, 400f, 524, 524f
Tailbone, 250, 251f
Talus, 254, 254f
Tanning, 76–77, 83, 83f, 454
Target cells, for hormones, 340, 342t, 343f
Tarsals, 248f, 254, 254f
Tarsiers, in evolutionary tree, 531f
Taste, 316t, 320–321, 320f
Taste buds, 173, 320–321, 320f
Taste cells, 320f, 321
Taste pore, 320f, 321
Tau protein, 412, 412f
Taussig, Helen, 107
Taxoids, 461
Taxol, 461
Tay-Sachs disease, 43, 55, 62, 477, 477f
T-cell receptor (TCR), 143f, 144, 145, 146f
T cells, 122, 139, 143, 143t, 145–147, 146f, 385
TCR (T-cell receptor), 143f, 144, 145, 146f
Tears, 140, 305f
Technology
  assisted reproductive technologies, 383–384
  biotechnology, 14–15, 37, 503–510
  definition, 14
Tectorial membrane, 330, 330f
Teeth, 172f, 173, 252, 296f
Teflon, 257
Telencephalon. *See* Cerebrum
Telomerase, 448, 459
Telomeres, 448
Telophase, 427, 427f, 434t, 435f, 435t
Telophase I, 43ft, 431f, 434t
Telophase II, 431f, 435t
Temperate forest ecosystem, 546f, 547
Temperate grassland (prairie), 546f, 547
Temperature, body. *See* Body temperature
Temperature receptors, 316, 319, 319f
Template, in DNA replication, 493
Temporal bone, 248, 248f, 249, 249f, 250f, 329f
Temporal lobe, 295, 296f, 322
Tendinitis, 277
Tendons, 67, 264, 264f
Terazosin, 222
Termination, in polypeptide synthesis, 500, 500f
Tertiary structure, of protein, 36, 36f
Testes, 367, 367f, 368t, 369, 370f
  development of, 405f, 406
  in endocrine system, 341f, 358, 358f
  fetal development of, 403
  hormones produced by, 342t
  testosterone in, 34
Test group, 11, 12f
Testicular cancer, 457, 458t
Testimonial data, 13
Testing, for HIV, 161
Testis-determining factor, 406, 440
Testosterone, 34, 34f, 246, 358, 358f
  aging and, 411
  and genital development, 406
  in male reproductive system, 369–370, 371f
  prostate and, 222
  in women, 374
Tetanus (infection), 135, 147f, 272
Tetanus (muscle contraction), 272, 272f
Tetany, 350
Tetracycline, 583
Tetrahydrocannabinol (THC), 309–310
Thalamus, 295f, 298, 299f
THC (tetrahydrocannabinol), 309–310
Theory, scientific, 9
Therapeutic cloning, 404
Thermal inversion, 560f, 561
Thermoreceptors, 316, 319, 319f
Thiamine, 190t
Thigh, motor area for, 296f
Third month, of development, 403
Third ventricles, 293f, 294f, 295
Third week, of development, 399–400
Thomas, Vivien Theodore, 107
Thoracic cavity, 81, 82f, 202f
Thoracic duct, 138, 138f
Thoracic nerves, 305f
Thoracic vertebra, 251f
Threshold, in action potential, 288f–289f, 289, 291f
Thrombin, 123, 123f
Thrombocytes (platelets), 68, 68f, 116, 117f, 123, 123f
Thrombocytopenia, 123
Thromboembolism, 106, 123
Thrombus, 106, 123
Thymic hormones, 139
Thymine (T), in DNA, 38, 38t, 492, 492f
Thymosins, 139, 342t, 358
Thymus gland, 138f, 139, 139f, 341f, 358, 411
Thyroid cancer, 451f, 452, 459
Thyroid gland, 21, 22f, 341f, 342t, 349–350, 349f, 350f
Thyroid hormones, 244, 342t, 349–350, 349f
Thyroid-inhibiting hormone (TIH), 345
Thyroid-releasing hormone (TRH), 345, 347f
Thyroid-stimulating hormone (TSH), 342t, 345, 346, 346f, 347f, 364
Thyroxine (T_4), 21, 342t, 347f, 349, 364
Tibia, 248f, 254, 254f
Tibialis anterior, 266f
Tibial tuberosity, 254, 254f
Tics, facial, 277
Tidal volume, 203, 204f
Tight junctions, 56, 56f
TIH (thyroid-inhibiting hormone), 345
Tin, 578
Tissue(s)
  adipose, 66f, 67, 67f, 75f, 77, 218, 359
  connective, 66–69, 66f–69f, 242, 368f
  definition of, 66
  engineered, 150
  epithelial, 66, 72–74, 72f–74f
  erectile, 368, 368f
  muscular, 66, 69–70, 70f
  nervous, 2, 66, 70–71, 71f, 287, 293
  in organization of life, 2, 3f
  rejection of, 150
  types of, 66
Tissue fluid, 68, 84, 92, 103, 138, 138f
Tissue plasminogen activator (t-PA), 106–107
T lymphocytes, 122, 139, 143, 143t, 145–147, 146f, 385
Tobacco smoke. *See* Smoking
Toddler stage, 410
Toes, 254, 254f, 256
Tone, muscle, 272
Tongue, 173, 296f, 320f
Tonicity, 50, 51f
Tonsil(s), 138f, 140, 172f, 209
Tonsillectomy, 209
Tonsillitis, 209
Tools
  agricultural, 538
  of Cro-Magnons, 539
  of early *Homo* species, 534, 535, 536, 538
  of Neandertals, 537
  population growth and, 538
Tooth decay, 173
Total artificial heart (TAH), 109, 109f
Totipotent stem cells, 117
Touch, sensory receptors for, 319
Toxins
  bacterial release of, 135
  removal of, 92
Toxoplasmic encephalitis, 159
Trabeculae, 240, 243
Trace minerals, 187, 187t
Tracer, radioactive, 21
Trachea, 74, 84, 173, 173f, 197f, 200, 200f
Tracheostomy, 200
Tracts, in central nervous system, 293, 294, 295, 298, 299f
Transcription, 496, 497f, 498–499, 498f, 501, 501f
Transcriptional control, 502
Transcription factors, 502, 503
Transduction, 423, 423f
Trans-fatty acids, 32, 32f, 186, 188
Transfer RNA (tRNA), 495, 499, 499f, 500, 500f, 502
Transfusions, blood, 125–127, 128
Transgenic animals, 507f, 508–509, 509f
Transgenic bacteria, 507, 507f
Transgenic foods, 14–15, 507–508, 510
Transgenic organisms, 14–15
Transgenic plants, 14–15, 507–508, 507f, 508f
Transitional epithelium, 74
Translation, 496–497, 497f, 499–501, 499f, 500f, 501f
Translational control, 502
Translocation, 440f, 441
Translocation syndromes, 442, 442f
Transmission electron microscope, 44–45, 45f, 45t
Transpiration, 554
Transplantation
  bladder, lab-grown, 235, 235f
  bone marrow, 461
  face, 78, 78f
  heart, 109
  kidney, 234, 235
  liver, 179
  lung, 211
  MHC antigens and, 145
  rejection and, 150
  self proteins and, 150
Transport
  active, 51, 51f
  facilitated, 51, 51f
  as protein function, 35
Transporters. *See* Carrier proteins
Transport vesicle, 54f
Transposons, 455
Transurethral resection of prostate (TURP), 222
Transverse colon, 180, 181f
Transverse process, 251f
Transverse tubules, 267, 267t, 268, 268f, 270
Trapezius, 265, 266f
Trawling, 581, 581f
Trees, multipurpose, 588, 588f
Treponema pallidum, 387, 387f
TRH (thyroid-releasing hormone), 345, 347f
Triceps brachii, 264, 265f, 266f
Trichomonas vaginalis, 389
Trichomoniasis, 389
Triclosan, 137
Tricuspid valve, 95, 96, 96f
Triglyceride, 31, 272, 273f, 274
Triiodothyronine (T_3), 342t, 349, 364
Trinucleotide repeat disorders, 478, 486
Triple bond, 24
Trisomy, 437–438
Trisomy 21, 437, 438–439, 439f, 441
tRNA (transfer RNA), 495, 499, 499f, 500, 500f, 502
Trochanter, greater and lesser, 254, 254f
Trochlea, 253, 253f
Trophic levels, 550, 551f
Trophic relationships, 550, 551f
Tropical grassland (savanna), 546f, 547
Tropical rain forest, 546f, 547, 571, 572f, 580, 582
Tropomyosin, 268, 270, 270f
Troponin, 268, 270, 270f
Trunk, motor area for, 296f
Trypsin, 176, 177f, 177t, 178
Tryptophan, 35f, 503
TSH (thyroid-stimulating hormone), 342t, 345, 346, 346f, 347f, 364
Tsien, Roger Y., 46
T tubule, 267, 267t, 268, 268f, 270
Tubal ligation, 382, 382f
Tubercles, 164, 164f, 210
Tuberculin test, 210–211
Tuberculosis, 150, 162f, 164f
  causative agent, 162–164, 162f, 164
  discovery of, 162
  drug-resistant, 164, 167, 168
  in HIV patients, 157
  prevention, 164
  pulmonary, 210–211, 210f
  sanatoriums and, 164
  transmission, 164
  treatment, 164
Tubular reabsorption, 225f, 227, 227t
Tubular secretion, 225f, 228
Tubulin, 55
Tumor marker tests, 458–459
Tumors, 448–449, 449f
Tumor suppressor genes, 419, 423, 450, 450f, 452
Tumor suppressor proteins, 438
Tundra, 546f, 547
Turner syndrome, 440
TURP (transurethral resection of prostate), 222
Tuskegee experiments, 16
T wave, 99
Twins, 396, 480
Twitch, muscle, 272, 272f
Two-trait crosses, 472–474, 472f, 473f, 474f
Tympanic canal, 330, 330f
Tympanic membrane, 328–329, 329f
Tympanostomy tubes, 209
Typhoid fever, 136
Typing, blood, 125–127, 125f–127f, 128
Tyrosine, 467

U

Ulcer, stomach, 10–11, 13, 140
Ulcerative colitis, 181
Ulna, 248f, 253, 253f, 255f
Ultraviolet radiation, 76, 453, 562
Umami, 321
Umbilical arteries, 402, 402f
Umbilical cord, 398f, 402, 402f
  development of, 400, 400f
  extraembryonic membranes and, 397f
Umbilical vein, 402, 402f
Uncoating, 160
Unicellular organisms, evolution of, 521
Unsaturated fatty acids, 32, 32f, 33
Unsustainable features of society, 586–587, 586f
Upper limbs, bones of, 253, 253f
Upper respiratory tract, 197f, 199f, 208–209
Uracil (U), 38, 38t, 494, 495f
Uranium, 453
Urban sustainability, 588
Urea, 59, 178, 179t, 219, 225f, 227t, 228
Uremia, 233
Ureters, 218–219, 218f, 219f, 221f, 367f

Urethra, 69, 218*f*, 219
  female, 219, 372*f*, 374
  male, 219, 222, 222*f*, 367*f*, 368, 368*t*, 369
Urethral opening, male, 368*f*
Urethritis, 233
Uric acid, 220, 225*f*, 233
Urinalysis, 226, 226*f*
Urinary bladder, 218*f*, 219, 235, 235*f*, 367*f*, 372*f*
Urinary bladder cancer, 359, 451*f*, 453*f*
Urinary incontinence, in pregnancy, 407–408
Urinary system, 80, 80*f*. *See also* Kidneys
  acid-base balance and, 232, 232*f*
  bladder in, 218*f*, 219, 235, 235*f*, 367*f*, 372*f*
  cardiovascular system and, 234*f*
  development of, 399*f*
  digestive system and, 234*f*
  endocrine system and, 361*f*
  functions of, 219–220
  homeostasis and, 84, 85*f*, 128, 129*f*, 218, 219, 228–233, 229*f*, 230*f*, 234*f*
  integumentary system and, 234*f*
  kidneys in, 218, 218*f*, 219–220
  muscular system and, 234*f*, 279*f*
  nervous system and, 219*f*, 234*f*
  organs of, 218–219, 218*f*
  overview of, 218–220
  prostate and, 222, 222*f*
  respiratory system and, 234*f*
  ureters in, 218–219, 218*f*, 219*f*
  water-salt balance and, 228–231, 230*f*
Urinary tract infections, 233
Urination, 19, 219*f*, 305*f*
Urine
  in bladder, 219
  color, 226
  in diabetes mellitus, 354
  drinking of, 227
  formation of, 225–228, 225*f*, 227*t*
  in immune system, 140
  odor, 226
Urine therapy, 227
Urobilinoen, 226
Urochrome, 229*f*
Urogenital groove, 405*f*, 406
Uterine cancer, 451*f*
Uterine cycle, 372, 376–377, 377*f*, 377*t*
Uterine tubes. *See* Oviducts
Uterus, 371, 371*t*, 372*f*
  in childbirth, 408–409, 408*f*
  contraction of, 345
  development of, 405*f*, 406
  in pregnancy, 407
Utricle, 332, 333*f*, 334
Uvula, 172, 172*f*

V

Vaccines, 147*f*, 148
  for Alzheimer Disease, 412
  anti-sperm, 379
  for bacterial meningitis, 82
  contraceptive, 379
  for hepatitis, 386, 454, 455
  for HIV, 162, 163
  for HPV, 385, 389, 454
  infectious disease and, 155
  measles, mumps, and rubella (MMF), 172
  rabies, 155
  smallpox, 155
Vagina, 140, 371*t*, 372, 372*f*, 374, 405*f*, 406
Vaginal infections, 389
Vaginal orifice, 372*f*
Vaginal ring contraceptive, 380*t*
Vaginitis, 389
Vaginosis, 389
Vagus nerve, 302, 305*f*
Valine, 35*f*
Valium, 306
Valves
  atrioventricular, 94–97, 96*f*, 97*f*
  semilunar, 95, 96, 96*f*, 97, 97*f*
  in veins, 94, 101, 101*f*
Vanadium, 187
Vaporization, of water, 25–26, 25*f*
Variable, experimental, 11
Variation
  in humans, 539, 539*f*
  in natural selection, 526
Varicella vaccine, 147*f*
Varicella zoster, 160
Varicose veins, 102
Vascular endothelial growth factor (VEGF), 108
Vas deferens, 367*f*, 368, 368*t*, 369, 370*f*, 382, 382*f*
Vasectomy, 382, 382*f*
Vector, 164, 506, 506*f*, 513
VEGF (vesicular endothelial growth factor), 108
Vein(s), 93*f*, 94, 103*f*
  blood flow in, 101–102, 101*f*
  cardiac, 95, 95*f*
  dorsal, of penis, 368*f*
  femoral, 103*f*
  hepatic, 102*f*, 104
  hepatic portal, 102*f*, 104, 178, 179*f*
  iliac, 102*f*, 103*f*
  jugular, 102*f*, 103*f*
  mesenteric, 103*f*
  pulmonary, 95*f*, 96, 96*f*, 102*f*, 201*f*, 207*f*, 402*f*, 403
  renal, 103*f*, 218, 218*f*, 221*f*
  in skin, 75*f*
  subclavian, 102*f*, 103*f*
  umbilical, 402, 402*f*
  valves in, 94, 101, 101*f*
  varicose, 102
Vena cava
  inferior, 95*f*, 96, 102*f*, 103, 103*f*, 402, 402*f*
  superior, 95*f*, 96, 96*f*, 102*f*, 103, 103*f*, 402*f*
Ventilation, 197, 203–204
Ventilation control, 205–206, 205*f*
Ventral cavity, 81, 82*f*
Ventral root, 293, 294*f*
Ventricles, 97, 97*f*, 98, 98*f*
  brain, 293, 293*f*, 295
  cardiac, 94, 95*f*, 96, 96*f*, 97, 402*f*
Ventricular fibrillation, 98*f*, 99
Venules, 93*f*, 94, 101*f*, 103, 104*f*, 176*f*
Vermiform appendix, 180, 181*f*
Vernix caesosa, 403
Vertebrae, 250, 251*f*
Vertebral canal, 25o, 81, 82*f*, 251*f*
Vertebral column, 248*f*, 250, 251*f*, 529, 529*f*
Vertebrate(s)
  classification of, 7
  genomes of, 511–512
Vesicle, 47*f*, 54, 54*f*
Vesicular (Graafian) follicle, 374–375, 375*f*
Vessels. *See* Blood vessels
Vestibular canal, 330, 330*f*
Vestibular nerve, 329*f*, 332–334, 333*f*
Vestibule, 316*t*, 329, 329*f*
Vestigial structures, 523–524
Victoria, Queen of England, 482–483
Villi, 176*f*, 177, 177*f*
Vinyl chloride, 453
Viral hepatitis, 179
Viral infection, and atherosclerosis risk, 106
Viral load, in HIV, 158
Viral meningitis, 82
Virus(es)
  cancer caused by, 455
  and chromosome structure change, 441
  drug resistance in, 162
  in gene therapy, 513
  as pathogen, 135–136, 135*f*, 210
  sexually transmitted diseases caused by, 385–386, 385*f*, 386*f*
Visceral muscle. *See* Smooth muscle
Visceral sensory nerves, 286*f*
Vision, 322–328
  abnormalities in, 326–328, 326*f*
  binocular, 528
Vision correction, 321, 326–328, 326*f*
Visual accommodation, 323–324, 323*f*
Visual association area, 296*f*, 297
Visual cortex, 301*f*, 325, 326, 326*f*
Vital capacity, 203
Vitamin(s), 57, 180, 189–190, 189*t*, 190*t*
Vitamin A, 178, 179*t*, 189, 189*t*, 324, 401, 454
Vitamin B_1, 190*t*
Vitamin B_2, 190*t*
Vitamin B_{12}, 121, 178, 190*t*
Vitamin C, 189, 190*t*, 454
Vitamin D
  functions and sources of, 188, 189*t*, 244, 246, 278–279, 350, 350*f*
  kidneys and, 220, 233
  in liver, 178, 179*t*
  metabolism of, 350
  overview of, 190
  ultraviolet radiation and, 76
Vitamin E, 178, 179*t*, 189, 189*t*
Vitamin K, 123, 178, 179*t*, 180, 189*t*
Vitreous humor, 323, 323*f*
Vocal cords, 199, 199*f*
Vocalization, 296*f*
Voice, 199, 199*f*
Voluntary muscle. *See* Skeletal muscle
Vomer bone, 249, 249*f*
Vomiting, 174
Von Willebrand disease, 125
Vulva, 372, 372*f*

W

Warren, J. Robin, 10–11
Warts, genital, 385–386
Waste disposal, 563, 585. *See also* Decomposers
Wastewater, treatment of, 573*f*
Wasting diseases, 137
Water
  absorption in large intestine, 180
  acids and, 26–28
  bases and, 26–28
  cohesion in, 26
  conservation of, 573–574, 573*f*, 588
  dissociation of, 26
  drinking, 556, 556*f*, 571–572, 585
  heat and, 25
  human use of, 553, 571–574, 572*f*
  as ice, 25*f*, 26
  life and, 24–29
  as liquid, 24–25
  molecule, 23*f*, 24, 25*f*, 26
  in osmosis, 50
  in plasma, 117, 117*f*
  properties of, 24–26, 25*f*
  reabsorption of, 227*t*, 228–231, 230*f*
  safe coliform levels, 181
  sea, as hypertonic, 51
  as solvent, 26
  supplies, increasing of, 572–473
  vaporization of, 25–26, 25*f*
Water (hydrologic) cycle, 554, 554*f*
Water on the brain (hydrocephalus), 293
Water pollution, 561–563, 561*f*, 575
Water-salt balance, 188, 220, 228–231, 230*f*, 352–353, 352*f*, 360, 413
Watson, James, 493
Waxes, 31–32
Weight, body
  in female infertility, 382
  leptin and, 359
  nutrition and, 183–184
  in pregnancy, 407
Weight control, 183–192
Weight loss, 68, 183, 184
Weight training, 275
Wernicke's area, 296*f*, 297, 301, 301*f*
Wetlands, 570–571, 588
Whale(s), 522, 522*f*, 523*f*
White blood cells, 79, 116, 119, 121–122, 121*f*, 141, 141*f*
  as chromosome source, 420
  as connective tissue, 66*f*, 68, 68*f*
  disorders of, 122
  types of, 121, 121*f*, 122
  in urinalysis, 226, 227
White-handed gibbon, 528*f*
White matter, 288, 293, 294*f*, 297–298, 303*f*
White pulp, 139, 139*f*
Whole-body theory of aging, 411
Wilkins, Maurice, 493
Williams syndrome, 441, 441*f*
Wilms disease, 447, 463
Wind knocked out, 204
Windpipe. *See* Trachea
Wind power, 576, 577*f*
Wisdom teeth, 173
Withdrawal method, 379, 380*t*
Woese, Carl, 527
Wolffian ducts, 405*f*, 406, 407
Womb, 372
Wormian bones, 240
Wrinkles, skin, 83
WTI gene, 463

X

Xanax, 306
XDR-TB (extensively drug-resistant tuberculosis), 168
Xenical, 184
Xenotransplantation, 150, 509
Xiphoid process, 251*f*, 252
X-linked alleles, 484–487, 484*f*–486*f*
X-linked dominant disorders, 485
X-linked recessive disorders, 485–486, 486*f*
X-rays, 401, 453
XX male syndrome, 406
XY female syndrome, 40

Y

Y chromosome, and genital development, 405*f*
Yeast fermentation, 60
Yeast infection, 389
Yellow bone marrow, 240, 241*f*, 279
Yellow River, 572
Yolk, 397
Yolk sac, 397, 397*f*, 398, 398*f*, 399*f*

Z

Zebra mussels, 580
Zidovudine (AZT), 162
Zinc, 187, 187*t*, 188, 578
Z lines, 267, 268*f*, 269
Zona pellucida, of egg, 394–395, 394*f*, 396*f*
Zone Diet, 184
Zygomatic bone, 248*f*, 249, 249*f*
Zygote, 366, 371, 378, 396, 396*f*, 429